Sriwijaya International Conference on Engineering, Science and Technology (SICEST 2016)

MATEC Web of Conferences Volume 101 (2017)

Bangka Island, Indonesia
9 - 10 November 2016

Editors:

I. Iskandar
S. Ismadji
T.E. Agustina
I. Yani
L.N. Komariah
S. Hasyim

ISBN: 978-1-5108-3749-2

Printed from e-media with permission by:

Curran Associates, Inc.
57 Morehouse Lane
Red Hook, NY 12571

Some format issues inherent in the e-media version may also appear in this print version.

Printed by Curran Associates, Inc. (2017)

For additional information, please contact EDP Sciences – Web of Conferences
at the address below.

EDP Sciences – Web of Conferences
17, Avenue du Hoggar
Parc d'Activité de Courtabœuf
BP 112
F-91944 Les Ulis Cedex A
France

Phone: +33 (0) 1 69 18 75 75
Fax: +33 (0) 1 69 28 84 91

contact-edps@webofconferences.org

Additional copies of this publication are available from:

Curran Associates, Inc.
57 Morehouse Lane
Red Hook, NY 12571 USA
Phone: 845-758-0400
Fax: 845-758-2633
Email: curran@proceedings.com
Web: www.proceedings.com

TABLE OF CONTENTS

ISSN (electronic edition): 2261-236X

Preface

Sriwijaya International Conference on Engineering, Science and Technology 2016 (SICEST 2016), is the first international conference organized by the Faculty of Engineering Universitas Sriwijaya. Theme of the event was "Building a better future through innovation in engineering, science and technology".

Through SICEST conference, the organizing committee has successfully recognized audience to the 300 presented articles from 650 submissions. The forum was arranged in 7 symposia and attended by more than 450 participants. They who contributed were scientists, researchers, engineers, students, professional stakeholders, consultants, government officials, marketers and professional users. All the symposia attended by various participants from 20 countries and from 137 reputable affiliations.

In order to make SICEST conference proceedings widely usable by the scientific community, we are proud having a partnership with MATEC as one of reputable partner with highly flexible publishing solutions. *MATEC Web of Conferences* are believed could increase the visibility and enhance the profile of SICEST conference, as well as to highlight the work of the paper contributors. It also provides a long-term open-access record of the conferences.

After a thorough consideration based on the quality of paper submitted, SICEST Scientific Committee has selected 115 articles for publication via Matec Web of Conferences. We have conducted peer-review in accordance with the policy on publishing integrity of the journal, in order to safeguard good scientific practice in publishing.

All submitted articles to SICEST 2016 were gone through a double blind peer review by competent reviewers. After that stage, many papers were invited to revise and resubmit in maximum of 3 weeks. The final decision was made on the basis of the revised paper only. Reviews were made explicit how the reviewers think the paper could be improved.

We express our deepest gratitude to all members of the international scientific committee, keynote and invited speakers, reviewers, sponsors, all paper contributors, participants in the conference and members of the SICEST organizing committee for their valuable contribution to the successful organization of SICEST2016.

Dr. Muhammad Faizal
Executive Chairman of SICEST 2016
Sriwijaya International Conference on Engineering, Science and Technology

Palembang-Indonesia, February 2017

edp sciences

MATEC Web of Conferences

Statement of Peer review

In submitting conference proceedings to MATEC Web of Conferences, I certify to the Publisher that I adhere to the **Policy on Publishing Integrity** of the journal in order to safeguard good scientific practice in publishing.

1. All articles have been subjected to peer review administered by the proceedings editors.
2. Reviews have been conducted by expert referees, who have been requested to provide unbiased and constructive comments aimed, whenever possible, at improving the work.
3. Proceedings editors have taken all reasonable steps to ensure the quality of the materials they publish and their decision to accept or reject a paper for publication has been based only on the merits of the work and the relevance to the journal.

Title, date and place of the conference

Sriwijaya International Conference on Engineering, Science and Technology (**SICEST 2016**)

November 9-10, 2016
Santika Hotel Bangka Island, Indonesia

Proceedings editor(s):

Prof. Iskhaq Iskandar
Suryadi Ismadji, PhD
Tuty Emilia Agustina, PhD
Irsyadi Yani, PhD
Dr. Leily Nurul Komariah
Dr. Saloma Hasyim

17, avenue du Hoggar - PA de Courtabœuf – BP 112 - 91944 Les Ulis Cedex A (France)
Tél. : 33 (0)1 69 18 75 75 - Fax : 33(0)1 69 07 45 17 – www.edpsciences.org

DOI: 10.1051/matecconf/201710101001

Metal supported on natural zeolite as catalysts for conversion of ethanol to gasoline

Anis Kristiani[1,*], *Sudiyarmanto* Sudiyarmanto[1], *Fauzan* Aulia[1], *Luthfiana* Nurul Hidayati[1] and *Haznan* Abimanyu[1]

[1]Research Center for Chemistry, Indonesian Institute of Sciences, Kawasan PUSPIPTEK Serpong, South Tangerang, Indonesia 15314

Abstract. A various of metal supported into natural zeolite was prepared via wet impregnation method. The transition metals impregnated are nickel, cobalt, copper and zinc. The catalytic properties both of physical and chemical properties were characterized by X-ray Diffraction (XRD), Thermo Gravimetri Analysis (TGA)-Differential Scanning Calorimetry (DSC), Surface Area Analyzer-Porositymeter and also gravimetry method for acidity measurement following by the adsorption of organic bases. The results showed that different metals impregnated into natural zeolite affected physical and chemical properties, i.e. crystalinity, surface area, pore size, pore volume and acidity. Their catalytic activity was tested for conversion ethanol to gasoline and showed high conversion up to 80-90% with the aromatics as major product.

1 Introduction

Catalysis is one of the important steps in chemical processes[1,2]. Bifunctional catalysts involved two active sides, both of metal and acid sites. The metal sites ussually come from a supported transition metal, such as Ni, Mo, W, Co, or their combinations, oxide, or sulfide of group VIII and/or group VI B[3]. While, the acid sites come from alumina-silica [4-6], various zeolites such as ZSM-5 [7], Y [8,9] and beta [10], MCM-41, and amorphous silica-alumina.

Zeolite is known to have great potential as catalysts or catalyst supports because of their properties, such as high surface area, high porosity, high adsorption capacity, and easy to separate from reactant and products. The catalytic reactions depend on catalyst's structure, acidity and reaction condition. [11,12].

Indonesia has abundant natural zeolite. One of the natural zeolite deposits in Indonesia, which has a large silica content of approximately 60% is in Gunung Kidul, Yogyakarta with the deposit of 55,000,000 m^3. These minerals are often found in the hills area of Baturagung, Gedangsari and Ngawen [13].

Bioethanol from biomass can be blended into gasoline in small percentage as oxygenate component to improve the fuel quality with low emissions of CO, CO_2 and hydrocarbon [14]. Ethanol has vapor pressure limitations where 10 % volume ethanol is blended into gasoline will give evaporative emissions and increase aldehyde emissions [15,16] and also its addition in large percentage requires the modification of the engine system. It needs for conversion of ethanol into gasoline components such as aromatics and iso-paraffins which have high octane number.

The conversion of ethanol into gasoline involved several reactions as shown generally in Figure 1. First is dehydration ethanol to ethylene. Ethylene possible will have several reactions, which are oligomerization, isomerization, cracking or aromatization to produce various kinds of hydrocarbons, both paraffin (ethene, propene, butene, etc.) and aromatics (benzene, toluene, C8-10, gasoline).

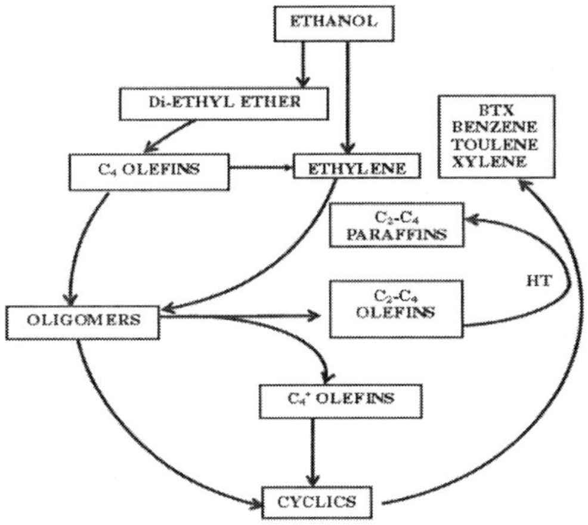

Fig.1. Reaction Pathways of Ethanol to Gasoline [16].

A recent review on the catalytic conversion of bioethanol to hydrocarbon fuels reported by Tretyakov et.al [17]. The addition of Ni into ZSM-5 improves the catalyst stability against coking [18]. Freeman et.al. [19] also investigated this reaction over Ga_2O_3/HZSM and Ga_2O_3-WO_3/alumina catalysts at 400°C and found that the addition of Ga_2O_3 to HZSM increases the selectivity to aromatics at the expense of C_2-C_4 hydrocarbons. Cr/Bentonite compared to HZSM-5 was also reported as

* Corresponding author: anis.kristiani@lipi.go.id

promising catalysts for ethanol conversion to gasoline range (C4-C12) [20].

Sudiyarmanto et al. [21] reported the conversion of ethanol into aromatics by using various transition metals, which are Cu, Co and Zn supported into commercial zeolite catalyst. The addition of transition metals resulted the yield aromatics more than 90%.

In this research, we prepared metal/natural zeolite catalysts through impregnation method and characterized their physical and chemical properties. The effects of variation of transition metals loading were studied. Their catalytic activities were tested for conversion of ethanol to gasoline.

2 Methodology

2.1. Materials

The metal precursors consist of cobalt (II) nitrate hexahydrate, copper (II) nitrate hexahydrate, nickel (II) nitrate hexahydrate, and zinc (II) nitrate hexahydrate in pure grade (Merck), while the other materials used are natural zeolite (NZ) from Gunung Kidul-Indonesia.

2.2. Catalyst preparation

The natural zeolite (NZ) from Gunung Kidul-Indonesia was activated before by using HCl [22]. A various of the transition metals (M=Ni, Cu, Co and Zn) was impregnated into natural zeolite. The procedure was conducted as follow: Aqueous mixture of 10 g of natural zeolite and 20 % of metal loading was refluxed and stirred constantly at temperature of 80 °C for 24 h. Then, catalyst samples were separated, dried overnight at 100 °C and calcined for 2 h at 500 °C as illustrated in Figure 2.

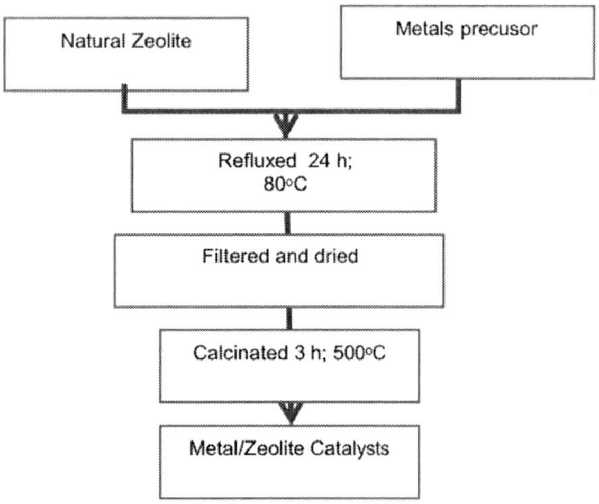

Fig.2. Diagram of preparation method.

2.3. Characterization of catalyst

For crystallographic phase identification, X-Ray Diffraction (XRD) analysis used a Rigaku diffractometer, with Cu Kα radiation at 40 kV and 30 mA, and a secondary graphite monochromator.

The calcinations temperature was determined by Thermal Gravimetric Analysis (TGA) and Differential Scanning Calorimetry (DSC) instrumentation by Linseis under heating up to 1000 °C.

The specific surface area was measured by the Brunauer-Emmet-Teller (BET) while the pore volume and pore size of the particles were measured by Barret-Joyner-Halenda (BJH) method. Both of the methods used N_2 adsorption/desorption at 77.3 K (Tristar II 3020 Micromeritics Instrument).

The total and surface acidity of samples were determined by the irreversible adsorption of organic base, which are pyridine and ammonia. While, the nature and types of groups and acid site was observed by using Fourier Transform – Infra Red (FT-IR) spectroscopy in KBr phase using Shimadzu, Prestige-21 with ATR 4000 attached to an automatic data acquisition center.

2.4. Catalytic activity test

Catalytic activity tests were tested by introducing 1% of catalyst loading in a fixed bed reactor electrically heated with adjustable temperature control. A total of 100 ml of ethanol feed was pumped into the reactor. The reaction was conducted at 350 °C for 2.5 h as described in Figure 3.

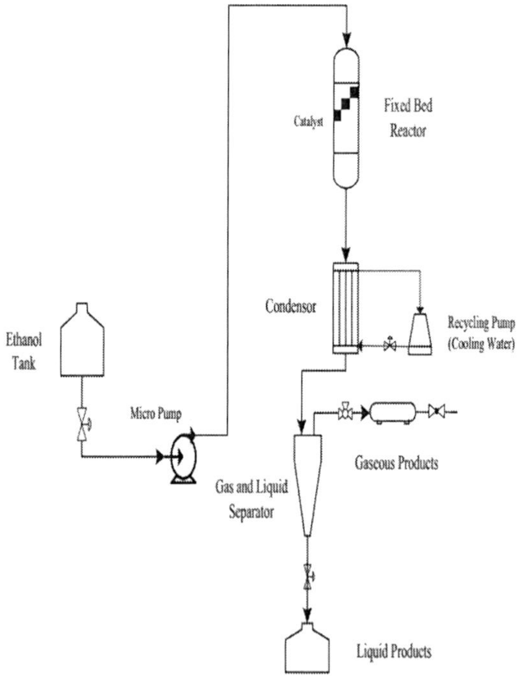

Fig.3. Schematic diagram of the reactor.

2.5. Product Analysis

The liquid product was analyzed by using Gas Chromatography Mass Spectrometry (GC-MS) Agilent 19091S-436, equipped by HP-5MS, 0.25mm x 60m x 0.25um column.

3 Result and discussion

3.1. X-Ray diffraction (XRD) analysis

X-ray diffraction pattern for the metal/natural zeolite samples are shown in Figure 4. In the natural zeolite (NZ) sample, there are five main peaks exhibited in the 2θ region 9.828°, 22.392°, 26.440°, 26.640°, and 28.016°. These peaks are the characteristic of mordenite according to JCPDS data. The loading transition metals into natural zeolite decreased the crystallinity as indicated by a decrease in the intensity of some characteristic peaks.

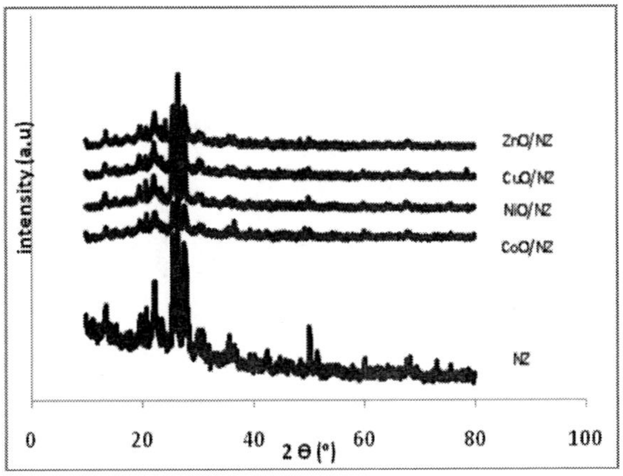

Fig.4. X-ray diffraction pattern of metal/natural zeolite.

3.2. Thermal Gravimetric Analysis (TGA) and Differential Scanning Calorimetry (DSC)

To investigate the changes in physical and chemical properties of catalysts occurred during increasing activation temperatures, the weight loss of these materials under heating process were observed. Figure 5-9 shows the effect of metal loading on the TG and DSC curves of materials previously activated under heating up to 1000 °C.

The initial weight loss observed at approximately 200 °C corresponded to the desorption of water molecular from the catalysts samples. At higher temperature at about 200 °C to 500 °C can be attributed to the decomposition of nitrate entrapped in the narrow pores of the metal/zeolite catalyst. The slow change of weight after 500 °C indicates the prolonged process of metal oxide/zeolite compound and/or gradual loss of the hydroxyl groups in the crystallyne structure. According to the TGA/DSC profile, temperature at 500 °C was selected for the calcination of the catalysts.

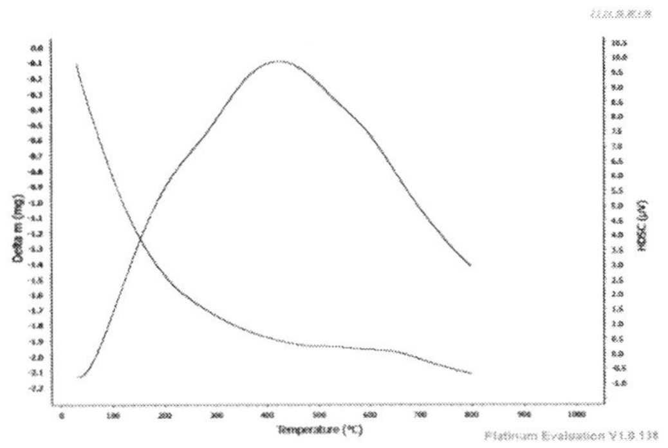

Fig.5. TGA/DSC pattern of natural zeolite

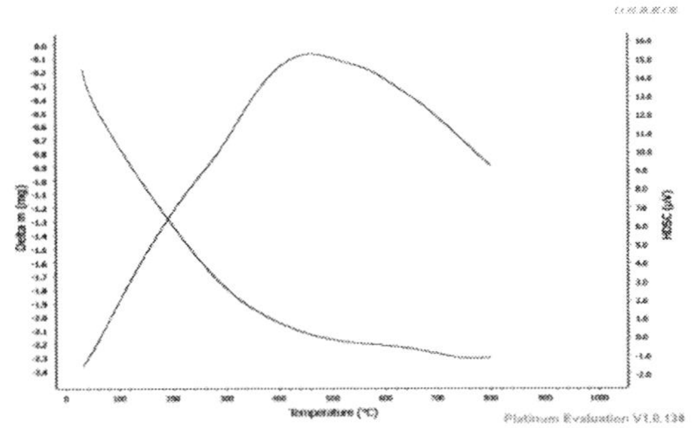

Fig.6. TGA/DSC pattern of NiO/NZ

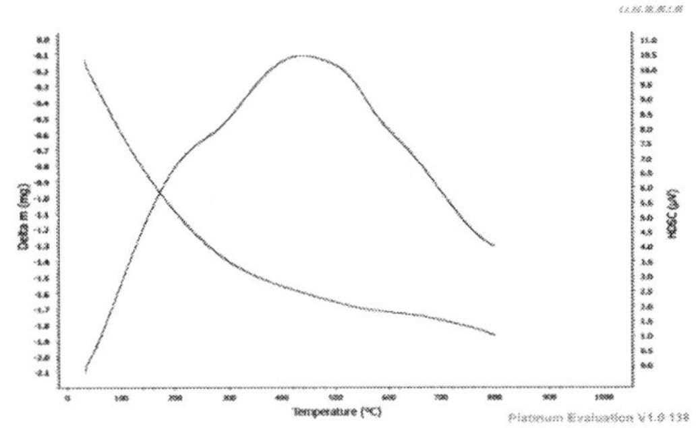

Fig.7. TGA/DSC pattern of CuO/NZ

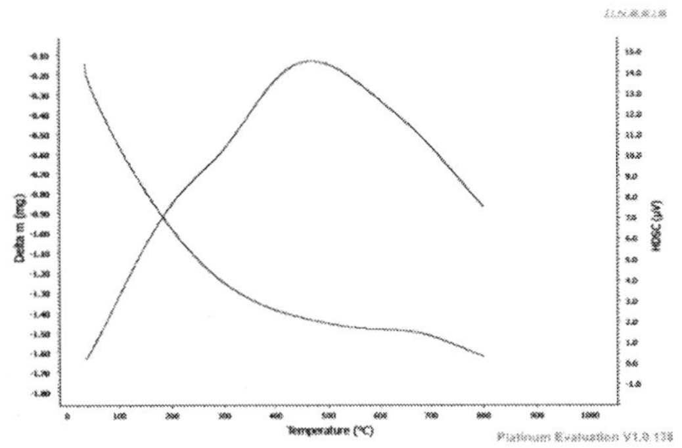

Fig.8. TGA/DSC pattern of CoO/NZ

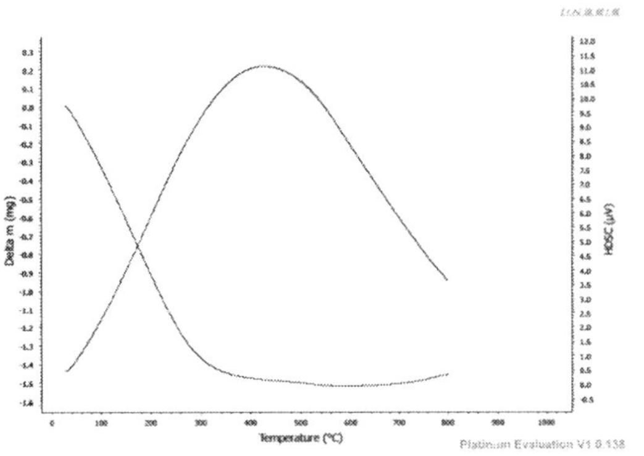

Fig.9. TGA/DSC pattern of ZnO/NZ

3.3. Specific surface area measurement

Specific surface area (S_{BET}), pore volume and pore size are shown in Table 1. It can be seen that the different transition metals supported on natural zeolite gave different specific surface area, but did not give significant changes in pore volume and pore size.

The transition metals might not enter the pore but only remained on the surface of the catalysts. Supported copper oxide on the zeolite has the highest surface area among the other transition metals. Copper metal was distributed well on the zeolite support so its spesific surface area tends to higher than others.

Table 1. Specific Surface Area, Pore Volume and Pore Size of Metal/Natural Zeolite Catalysts

Samples	Specific surface area [$m^2.g^{-1}$]	Pore volume [$cc.g^{-1}$]	Pore size [Å]
NZ	13.064	0.0033	2.099
NiO/NZ	15.471	0.0036	2.096
CuO/NZ	20.019	0.0039	2.092
CoO/NZ	18.092	0.0035	2.093
ZnO/NZ	17.234	0.0036	2.096

3.4. Acidity

Total acidity was determined using ammonia vapour while the surface acidity was determined using pyridine vapour. The nature of acid sites defined by the surface protons leading to the Bronsted sites or cationic centers due to unsaturation in coordination as the Lewis sites.

The result from the total and surface acidity of the samples are shown in Table 2. Metal loaded on zeolite tend to decrease the total acidity for all samples. While, the surface acidity for Cu and Zn supported on zeolite tend to increase slightly.

Table 2. Acidity of Metal/Natural Zeolite Catalysts

Samples	Total acidity [$mmol.g^{-1}$]	Surface acidity [$mmol.g^{-1}$]	Pores acidity [$mmol.g^{-1}$]
NZ	5.035	0.657	4.378
NiO/NZ	4.649	0.551	4.098
CuO/NZ	3.566	0.696	2.87
CoO/NZ	3.161	0.441	2.72
ZnO/NZ	3.707	1.018	2.689

Figure 10 shows NH_3 adsorbed on catalyst samples.The band at 1440 cm^{-1} is the characteristic peak of ammonium ion formed on the Bronsted acid sites and the absorption peak at 1300-1000 cm^{-1} is contributed by ammonia coordinately bonded to Lewis acid sites.

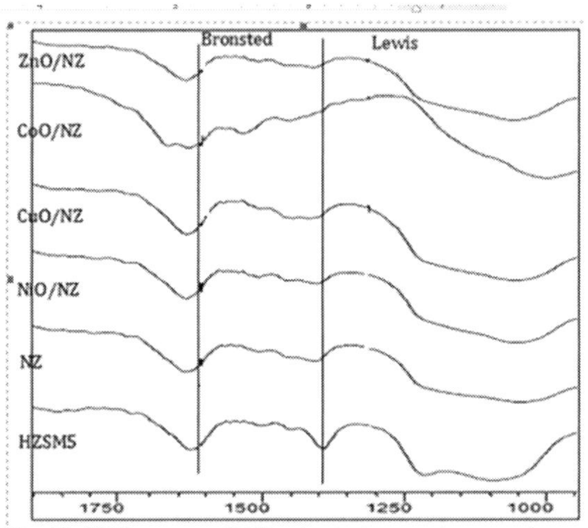

Fig. 10. FTIR spectra of NH_3 adsorbed on catalyst samples.

Figure 11 shows pyridine adsorbed on catalyst samples.The band at 1650 cm^{-1} is the characteristic peak of pyridine ion formed on the Bronsted acid sites in the surface of catalyst samples. This FTIR spectra indicated the presence of both Bronsted and Lewis acid sites on the catalyst samples.

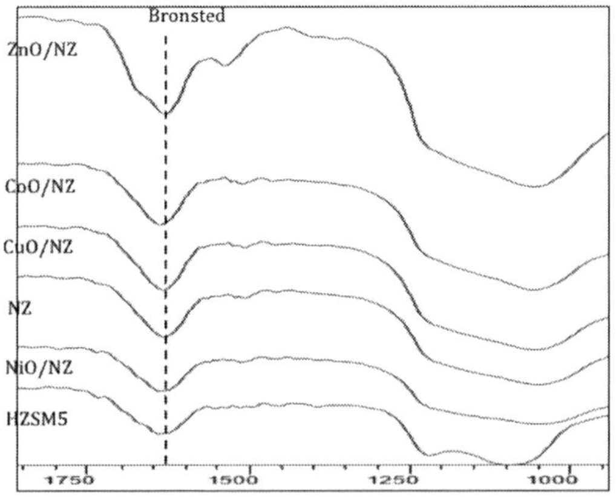

Fig. 11. FTIR spectra of pyridine adsorbed on catalyst samples.

3.5. Catalytic activity test

The catalysts were tested for ethanol to gasoline (ETG) conversion. All the prepared catalysts showed a good activity in ETG reaction. The conversion of reaction reached more than 80% at the reaction conditions studied. The liquid product contains major amount of aromatics followed by olefins, and a small fraction of parafins as shown in Figure 10.

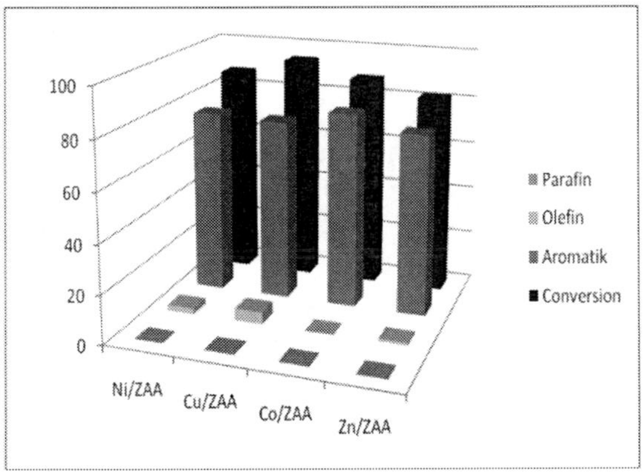

Fig. 10. Conversion and Product Distribution in Liquid Product

Among prepared catalysts, CuO/NZ catalyst revealed the highest conversion of ethanol, while ZnO/NZ performed the lowest activity. CuO/NZ catalyst shown highest selectivity of parafin and lowest of aromatic compared with other catalysts prepared. While, CoO/NZ catalyst shown highest selectivity of aromatics and lowest of parafins. Higher aromatics in gasoline fraction is less favored because it contributes knocking effect on the machine.

4 Conclusion

Different transition metals (Ni, Cu, Co and Zn) supported on zeolite by impregnation method resulted different physical and chemical properties, i.e. crystalinity, surface area, pore size, pore volume and acidity of the catalyst. The increasing of surface area and pore volume and the decreasing of pore size as the result of transition metals supported into zeolite gave better physical and chemical properties of catalyst. Copper metal was distributed well on the zeolite support so it has the highest surface area among the other transition metals. FTIR adsorption organic base indicated the nature acid site both of Bronsted and Lewis acid site on all the catalyst samples. The catalytic activity test showed the prepared catalysts gave conversion more than 80% and the major products were aromatics followed by olefins and a small fraction of parafins.

This project supported financially by Research Center of Chemistry – Indonesian Institute of Sciences. Big appreciation to PUSLABFOR MABES POLRI for operating and analyzing GC-MS for the liquid product.

References

1. J. Heveling, J. Chem. Educ., **89**, 1530–1536 (2012)
2. B. Yilmaz, and Muller, U., Top Catal, **52**, 888–895 (2009)
3. W. Sriningsih, Saerodji, M.G., Trisunaryanti, W., Triyono, Armunanto, R., and Falah, I.I., Procedia Environmental Sciences, **20**, 215 – 224 (2014)
4. F. Regali, Liotta, L.F.,Venezia, A.M., Boutonnet, M., Järås, S., Applied Catalysis A: General, **469**, 328-339 (2014)
5. Y, Yu, Gutiérrez, O.Y., Haller, G.L., Colby, R., Kabius, B., Rob van Veen, J.A., Jentys, A., Lercher, J.A., Journal of Catalysis, **304**, 135-148 (2013)
6. A.L. Santos, Reis, R.A., Rossa, V., Reis, M.M., Costa, A.L.H., Veloso, C.O., Henriques, C.A., Zotin, F.M.Z., Paredes, M.L.L., Silveira, E.B., Chiaro, S.S.X., Materials Letters, **83**, 158-160 (2012)
7. Chang N., Gu Z., Wang Z., Liu Z., Hou X., Wang J., J. Porous Mater, **18**, 589-596 (2011)
8. M.N. Uddin, Wan Daud, W.M.A., Abbas, H.F., Energy Conversion and Management, **90**, 218-229 (2015)
9. M. Dastanian and Azad F.S., Ind. Eng. Chem. Res., **49**, 11254-11259 (2010)
10. JM. Escola, Aguado, J., Serrano, D.P., Briones, L., Díaz de Tuesta J.L., Calvo, R., Fernandez, E., Energy Fuels, **26**, 3187-3195 (2012)
11. A. Galadima, and Muraza, O., Microporous and Mesoporous Materials, **213**, 169–180 (2015)
12. EF. Iliopouloua, Stefanidis, S.D., Kalogiannis, K.G., Delimitis, A., Lappas, A.A., Triantafyllidis, K.S., Applied Catalysis B: Environmental, **127**, 281–290 (2012)
13. Satriyo Krido Wahono et al., Energy Procedia, **47**, 263 – 267 (2014)
14. K. Inoue, Inaba, M., Takahara, I., and Murata, K., Catalysis Letters, **136**, 14–19 (2010)
15. TQ. Hoang, Zhu, X., Danuthai, T., Lobban, L.L., Resasco, D.E., Mallinson, R.G., Energy Fuels, **24**, 3804–3809 (2010)
16. N. Viswanadham, S.K. Saxena, J. Kumar, P. Sreenivasulu, D. Nandan, Fuel, **95**, 298–304 (2012)
17. VF. Tretyakov, Tretyakov KV, Talyshinskii RM, Catalysis in Industry, **2**, 402-420 (2010)
18. AG. Gayubo, Alonso A, Valle B, Aguayo, A.T., Olazar, M., and Bilbao, J., Fuel, **89**, 3365–3372 (2010)
19. D. Freeman, Wells RP, Hutchings GJ, J. Catal, **205**, 358-362 (2002)
20. Widjaya, R.R., Soegijono, B., and Rinaldi, N., Makara Journal of Science, **16/1**, 65-70 (2012)
21. Sudiyarmanto, Kristiani, A., Hidayati, L.N., Andreas, and Abimanyu, H., AIP Conference Proceedings, **1755**, 080007 (2016)
22. Wahono, S.K., Hernawan, Kristiani, A., Tursiloadi, S., Abimanyu, H., Energy Procedia, **47**, 263 – 267 (2014)

Preparation of molecularly imprinted polymers simazine as material potentiometric sensor

Yohandri Bow[1], *Edy* Sutriyono[2], *Subriyer* Nasir[3], and *Iskhaq* Iskandar[4]

[1]Chemical Engineering Department, Politeknik Negeri Sriwijaya, 30139 Palembang, Indonesia
[1]Candidate Doctor of Environmental Science, Universitas Sriwijaya, 30162 Inderalaya, Indonesia
[2]Geology Study Program Faculty of Engineering Faculty Universitas Sriwijaya, 30162 Inderalaya, Indonesia
[3]Chemical Engineering Department Faculty of Engineering Universitas Sriwijaya, 30162 Inderalaya, Indonesia
[4]Physics Department, Faculty of Mathematics and Natural Science Faculty Universitas Sriwijaya, 30162 Inderalaya, Indonesia

Abstract. Molecular imprinting technology is a promising technique for creating recognition elements for selected compounds and has been successfully applied for synthesis of environmental pollutants such as simazine. Simazine is a pesticide ingredient that is commonly used in agriculture, which has devastating effects on the environment if used excessively. Molecularly imprinted polymer (MIP) provides cavities to form a particular space generated by removing the template when the polymer has formed. In this study, MIP using simazine as template had been made by the cooling-heating method and used as a material potentiometric sensor for detecting simazine. A template (simazine) was incorporated into a pre-polymerization solution that contains a methacrylic acid as functional monomer, an ethylene glycol dimethacrylate as cross linker, and benzoyl peroxide as initiator. Characterization was performed by scanning electron microscope (SEM) and fourier transforms infra-red (FTIR). The FTIR spectra of the MIP showed that the peaks of amine group decrease significantly, indicating that the simazine concentration decreases drastically. Characterization by SEM images showing the broadest pore size distribution with the highest number of pores in the MIP prepared under the heating time of 150 min. The MIPs therefore could be applied as a simazine sensor.

1 Introduction

Molecularly imprinted polymers (MIPs) is a technique to produce a polymer which has a cavity as a result of the disposal of the template, where the cavity functions are to recognize the molecule with the same size, structure, and chemical-physical properties [1]. Affinity and selectivity of the template will increase by itself with increasing the value concentration [2]. Until this day, the MIPs method is still being developed because it is ease in the manufacture of polymers, low cost, and used widely in many target molecules, either as biological or chemical recognition elements, for the example in the food and medicine [3, 4].

In the synthesis of MIP, there are many parameters that must be considered because this parameter can affect the morphology, characteristics, and uses of the polymer. In the synthesis, the selection of chemical reagents becomes the important factors in producing the efficient functional MIP [5].

Selection of monomer is essential to produce very specific cavities that made for the template molecule [6]. Functional monomer type as carboxylic acids (acrylic acid, methacrylic acid, vinylbenzoic acid), sulphonic acids (2-acrylamido-2-methylpropane sulphonic acid), heteroaromatic bases (vinylpyridine, vinylimdazole). The combination of functional monomer and template determined by the equilibrium, and the functional monomer has to be made the complex where the comparison between the template and functional monomer is 1:4. The uses of methacrylic acid (MAA) excess because of its ability to act as the hydrogen binder and proton donor as well as the acceptor of hydrogen binding.

In the imprinted polymer, cross-linker fulfils three main functions. First, the cross-linker is necessary to control the morphology of matrix polymer, included as the gel, pore, or microgel powder. Second, its function is to stabilize the side of the imprinted bond. Third, give the mechanical stability of the matrix polymer [7]. From the polymerization perspective, cross bond with the high ratios normally more liking to access permanently the pores material (macropores) and to produce the material with the affordable mechanic stability. So, the amounts of the cross-linker must be high enough to maintain the stability side of the identifier. The polymer with a cross-linker ratio more than 80% is often used.

Simazine (2-chloro-4,6bis(ethylamino)-*s*-triazine) is a synthetic chemical that is used as a herbiced to control the growth of weed. Simazine is a family of s-triazine herbicide [8]. Simazine has been widely used as an herbicide to control the weed in agriculture. This herbicide is toxic and for the biodegradation product as diethylatrazine also dangerous, besides that, simazine is

* Corresponding author: yohandribow@gmail.com

also a contaminant in the water surface and groundwater [9]. Among the s-triazine herbicide, simazine is widely used at the suburb. Simazine is suspected as the chemical substance that disrupts the endocrine. Based on the water purity standards World Health (WHO), tap water that contains 10 nM (2 ppb) of simazine considered as dangerous [10]. Based on the existing literature, simazine is effective to use control the weeds with a concentration of 0,1-1 milligrams per liter [11].

In past few years, the well-established field of potentiometric sensors has undergone a quiet revolution that did not go unnoticed in the general analytical chemistry community. While it has been traditional wisdom that such sensors may reach only mediocre limits of detection (LODs) around the micromolar range, they have been improved to make an actual possible trace-level analysis at sub-nanomolar (low parts per trillion) concentration now. Obviously, this improvement asks for a new application where this technique has not been used traditionally [12].

2 Materials and methods

2.1 Preparation of NIP/MIP

By adding simazine as a template into a tube, which contains chloroform, the synthesis begins. After that, methacrylic acid (MAA) as a functional monomer, ethylene glycol dimethyl acrylate (EDMA) as a cross-linker, and benzoyl peroxide (BPO) as an initiator were added in sequence to make a solution of pre-polymer. The solution of pre-polymer then stirred for 15 min, before it was out into a refrigerator at -5°C for 60 min. Later, the cleaner solution that produced is put into little bottles. The bottle is placed into 70°C for 150 min. After it is done, the solid polymer obtained in the mix. Template disposal process is done by the resulting polymer solids finely pulverized, sieved, and did the sedimentation in the acetonitrile for 24 hours, then washed with acetic acid, aquabidest and evaporated in a methanol solution, respectively. The result is a dry powder. The resulting polymer particles further characterized using SEM and FTIR to see if the resulting polymer has been as expected.

A control sensor was prepared in every case following the same procedure but in the absence of template molecule. The control (or non-imprinted polymer-modified electrode) had at any time the same treatment as the imprinted sensor, to assure that the effects are observed only due to the features of imprinting and not to the subsequent treatments under-gone by the sensor.

2.2 Preparation of electrode

The general procedure used to prepare the potentiometric sensor based MIP is by using aluminium wire (1 mm diameter and 120 mm long). The electrodes were polished with fine alumina slurries on a polishing cloth, sonicated in distilled water and dried in air. Unsaturated resin was glued to one end of the above aluminium wire. Simazine was dipped into the aluminium wire and then a solution of chloroform was added on the sensor surface that has been coated with simazine.

3 Results and discussion

3.1. Analysis of Molecularly Imprinted Polymer (MIP) formation

Molecularly imprinted polymer (MIP) was polymer formed simultaneously with templates. MIP was synthesized by reacting monomer, cross-linker, and initiator. Monomer served as a mold or template. Cross-linker served to form bond linking the polymer chains with other polymers. While the initiator was chemical for increasing the speed of catalyst.

Polymer formation reaction between the functional monomers with template simazine occurred in situ in non-covalent interaction. This treatment was useful for hydrogen bonds formation, interaction between simazine and MAA on the cluster amines and alcohols. This stage was called pre-polymerization.

Premix form were simazine and methacrylic acid (MAA) having mole ratio of methacrylic acid (MAA) larger than simazine with the aim that when treated, it might form simazine mold surrounded by methacrylic acid, since in the same circumstances, simazine might also act as the monomer [13]. There was also cross-linker, EGDMA, involves in the reaction of mixture cross-linker called copolymerization stage. Polymerization process was addition polymerization. Addition polymerization was added by the addition of initiator, benzoyl peroxide which worked to change methacrylic acid (MAA) into radical molecules which might easily go through a stage of propagation or chain elongation. The polymerization process finished when the heated clear fluid turned into yellowish translucent solid resembling ceramic such as acrylic attached to the vial bottle.

This simazine MIP formation was influenced by the stirring, composition, and heating time. Stirring played important role in MIP formation. The stirring aimed to accelerate the polymerization process and homogenize the mixture. Unstable stirring might result bad polymer since it might fray the bond. The mixture was then in the oven for 60 min. The result was white solid. The heating time was varied with the length of time of 90 min, 120 min and 150 min. The results contained in the vial bottle were still white solid. This was possible because simazine might also act as a monomer. The composition was then converted by tripling first composition, where in the mole methacrylic acid (MAA) and EDMA were added more than simazine and more mole compared to the initial composition. The composition was altered only the amount of EDMA and MAA while the composition of benzoyl peroxide and simazine was fixed. This was because benzoyl peroxide was only acting as a catalyst in the formation of MIP. The polymerization conditions were also made permanent by using cooling-heating method. When the heating time was 60 min, the mixture was still clear liquid. Then the time was varied to 90 min, 120 min, and 150 min. When

 DOI: 10.1051/matecconf/201710101002

the oven for 150 min found to result in the vial bottle colored translucent acrylic solid form indicating that polymerization was completed and successful. The longer heating time caused bonds between simazine and MAA stronger. Based on these experiments it was found that the optimum condition of simazine MIP formation heating time of 150 min. From the experiments that had been carried out, the optimum condition for MIP formation occurred in the heating process in the oven with the temperature of 70^0C and time of 150 min.

The resulting solid then crushed to produce fine solid and the scouring aimed to produce cavities. The resulting fine solid was washed by mixture of acetic acid and methanol to remove the remnants of the reactants in the polymerization process which did not react. Washing was continued sequentially by using three different solvents i.e. acetonitrile, aquabidest, and methanol intended to extract the existing template in the MIP. At the end of washing, it was expected to produce MIP with specific simazine molecule mold to the analyte in the sample [14].

The success of MIP simazine synthesis determined by physical and chemical characteristics by Scanning Electron Microscope (SEM) and Fourier Transform Infra-Red (FT-IR).

3.2 Characterization of polymer surfaces

Scanning Electron Microscope (SEM) analysis described both the morphological characteristics of the shape and surface structure of the MIP. This was important information for the synthesis and applications of MIP [15]. Here were the results of SEM to the MIP obtained.

Fig. 1. Standard of Simazine

The SEM images show the difference between standard simazine and MIPs-based simazine as shown in Fig. 1-4.

Fig. 3 and 4 show SEM images of the polymer before removing the simazine template and its simazine MIP after the template removal, respectively. The SEM image of standard simazine given in Fig. 1 and the SEM image of non imprinted polymer (NIP) given in Fig. 2 is for comparison purposes.

Fig. 2. Non Imprinted Polymer (NIP)

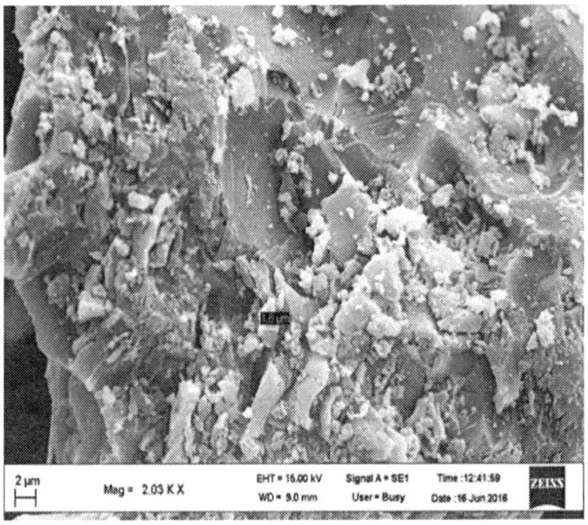

Fig. 3 Polymer before removal of simazine template

After the template removed, the MIP which contain cavities is formed. It is expected that the MIP can recognize target molecules that have the same properties as or similar with those of the template molecules.

It is seen that the simazine MIP surface after the template removal is cleaner than before template removal. This finding is indicating that the template removal process is effective [16].

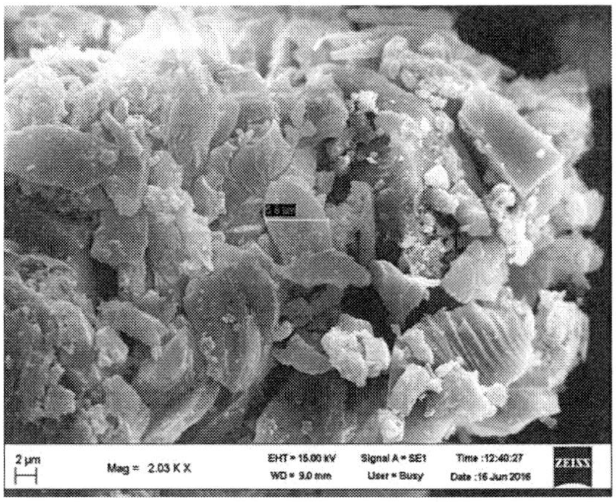

Fig. 4. The simazine MIP after template removal

Distribution of broader pore size and the number of pores that have a size in the same range with simazine particles also increases. The MIP obtained using the washing process, shows the distribution of broadest pore size with the highest number of pores. So it is expected that the capability of this MIP to capture the target molecules is highest compared to those prepared without washing. The pore size distribution by washing MIP and MIP before washing was shown in following histogram.

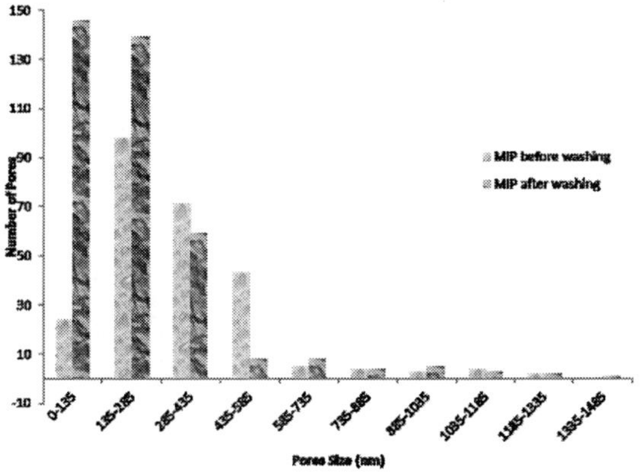

Fig. 5. Histogram of pore size distribution of MIP Simazine before washing and MIP Simazine after washing

Fig. 5 shows that the number of pores MIP simazine after washing is more than the MIP simazine before washing. The pore size of 0-285 nm is the most dominant pore size. There are significant differences between the MIP before and after being washed in the pores range. The number of pores in the range of 0-135 nm after the wash process is 146 pores and the pore after being washed are 139 pores and before the wash are 98 pores. MIP simazine after the being washed have more number of pores because the washing process releases the template.

3.3 Characterization of Molecularly Imprinted by Fourier Transform Infra-Red (FTIR)

Fourier Transform Infrared or infrared spectroscopy is a method of observing the interaction of molecules with electromagnetic radiation that is in wave numbers. The following FT-IR spectra contained in the NIP or MIP as shown in Fig. 6-8.

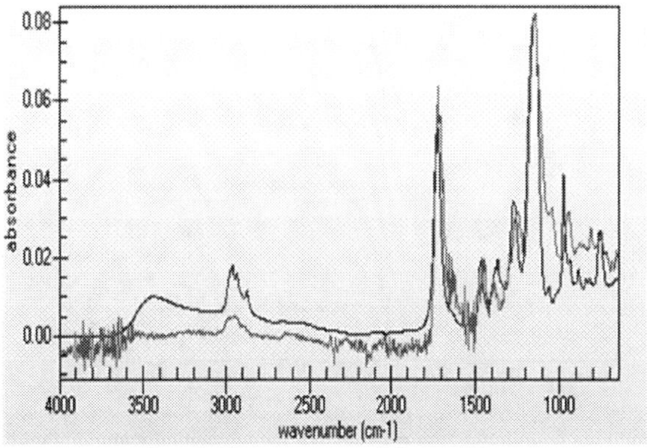

Sample ID Library Hit

Fig. 6. Spectra of Non Imprinted Polymer (NIP)

The methacrylic acid used as monomer which is the group of carboxylic acid, which has a C=O bond dominant and distinctive absorption at the wavelength of 1870 – 1540 cm^{-1} [17]. In the picture of FTIR spectra indicated that the C=O bond contained at a wavelength of 1721 cm^{-1}. Moreover, the overlap between the bonds O-H and C-H at a wavelength of 2940 cm^{1-} and there are two peaks which are a bond O-C at a wavelength of 1261 cm^{-1} and 1295 cm^{-1} which are identifying that the polymer included in the group of carboxylic acids, respectively.

Furthermore, the presence simazine as a template which is a distinctive group of amines is the N-H stretching at a wavelength of 3500 – 3220 cm^{-1}, that binds to MAA which is indicated by the bond N-H, O-H at a wavelength of 3600-3200 cm^{-1} [18]. In Fig. 8, it can be seen that there is a peak at a wavelength of 3428 cm^{-1} and 3528 cm^{-1}, that means the polymer completely contained simazine as expected.

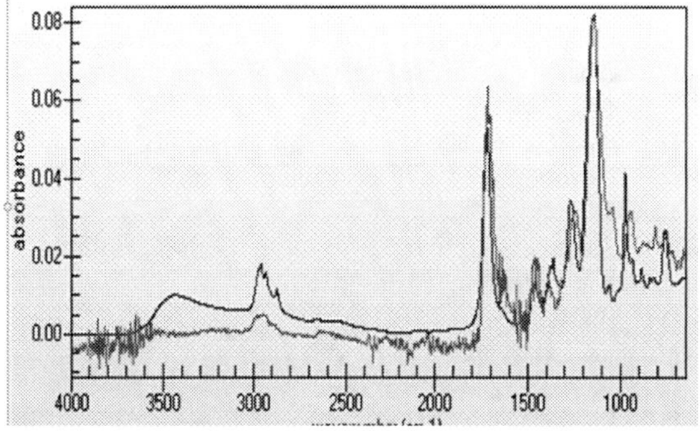

Sample ID Library Hit

Fig. 7. Spectra of MIP Simazine before washing

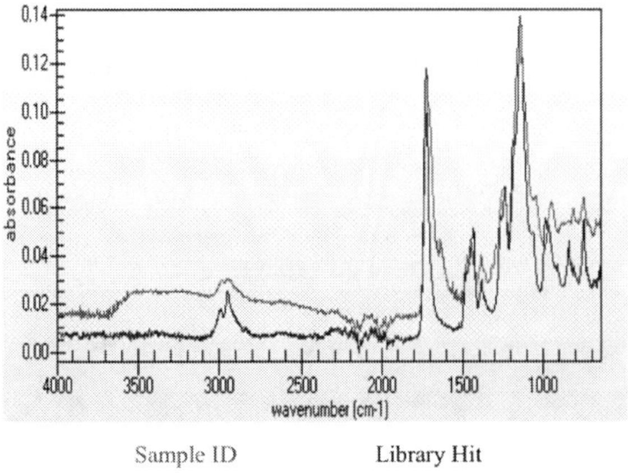

Fig. 8. Spectra of MIP Simazine after washing

4 Conclusion

The results of molecularly imprinted polymer (MIP) characterization by using Scanning Electron Microscope (SEM) indicated that the pore distribution in the MIP after washing was more than MIP before washing meaning that the template removal process was effective. The result of characterization by using Fourier Transformed Infrared (FTIR) showed that in the spectra of the MIP was peak of amine group and carboxylic group derived from simazine and methacrylic acid.

References

[1] M.C. Blanco-Lopez, M.J. Lobo-Castanon, A.J. Miranda-Orderes, P. Tun on-Blanco. Biosensor and Bioelectronics **18,** 353-362 (2003)
[2] G. Vasapollo, R. D. Sole, L. Margola, M. R. Lazzoi, A. Scardino, S. Scorrano, G. Mele. Int. J. Mol. Sci. **12**, 5908-5945 (2011)
[3] Carlos. FEMS Microbial Lett **345**, 1-12 (2013)
[4] Y. Bow, Hairul, I. Hajar. Int. J. on Adv. Sci, Eng. and Information Tech. **5:6**, 422-425 (2015)
[5] J. W. Boyd, G. P. Cobb, G. E. Southard, G. M. Murray. John Hopkin Apl Technical Digest **25:1**, 44-48 (2004)
[6] I. M. Naseer, I. Adnan, M. Adeel, A. Tajamal, Ahsan, S. E. Muhammad. Nanomaterials **3**, **615-637** (2013)
[7] Ganjali, M.R. Alizade, T. Larijani, B. Faribod, F. Norouzi. Int. J. Electrochem. Sci. **7**, 4756-4765 (2012)
[8] Saxena, Ronika, P. Garg. GERF Bulletin of Bioscience **1:1,** 1-6 (2010)
[9] Zhou
[10] Fuchiwaki
[11] Piletska
[12] E. Bakker, E. Pretsch. Trends Analyt Chem. **24:3**, 199-207 (2005)
[13] Rusdianasari, Y. Bow, A. Taqwa. Adv. Material Research **896**, 145-148 (2014)
[14] I. Kubo, R. Shoji, R. Fuchiwaki, Y. Suzuki. Electrochemistry **76** (2008) 541-544
[15] R. Liang, R. Zhang, W. Qin. Sensor and Actuaters **B. 141** (2009) 544-550
[16] A. Mujahid, F. L. Peter, L. Materials **3**, 2196-2217 (2010)
[17] R. M. Silberstein, F.X. Webster and D.J. Kiemle. *Spectrometric Identification of Organic Compound.* John Willey & Sons. Inc. (2006)
[18] B. Stuarts. *Infrared Spectroscopy: Fundamental and Application.* John Willey & Sons. Inc. (2004)

Synthesis of grafted flocculants based on several kinds of starch and its performance in water turbidity removal

Mujtahid Kaavessina [1,*], *Sperisa* Distantina[1], *Fadilah*[1]

[1]Chemical Engineering Department, Sebelas Maret University, 57126 Surakarta, Indonesia

Abstract. Taro starch, cassava starch and also their modified were prepared as a backbone of grafted flocculants. These flocculants were synthesized by combining the properties of synthetic and natural polymer via the 'grafting to' method that separated into two stages. The first stage was synthesizing of non-terminated polyacrylamide (nt-PAM) and then followed by the second stage grafting of the nt-PAM onto several kinds of starch. The intrinsic viscosity (η) and average molecular weight (Mv) of starches-g-PAM were observed and their molecular structures were analyzed by using Fourier Transform Infra-Red (FTIR). Finally, the obtained flocculants were tested to evaluate their performance for the treatment of the simulated water turbidity containing 10 g of kaolin in 1000 ml of aquadest. The results showed that all the grafted flocculants can be good comparable to some commercial organic flocculants. The percentages of the water turbidity removal for each of backbones are 93.6% (taro starch), 91.2% (cassava starch), 95.3% (modified taro starch) and 94.1% (modified cassava starch). It can be highlighted that the modified taro starch exhibited the best performance in water turbidity removal up to 95%.

1 Introduction

Wastewater treatment is a multi-stage process to restore water quality from sources of wastewater to be used in a particular need. This process is reducing or removing organic matter, solids, minerals, nutrients, disease-causing organisms or other contaminants from wastewater. Flocculation is one stage of this treatment. It agglomerates (congregates) the suspended particles into a bigger one (floc) that could be separated by sedimentation and filtration processes [1]. There are two commercial flocculants, i.e.: polymeric flocculants and inorganic coagulating agents. Nayak, et al (2002) affirmed that the polymeric flocculants perform well to form large size of flocs and strongly bound to resist high shear degradation forces in small dosage [2].

Polymer flocculants can be either synthetic or natural. Many researchers reported that synthetic polymers are more effective to bind suspended particles than natural ones. However, their shear stability and biodegradability are poor. In contrast, natural polymers are shear stable, highly biodegradable and non-toxic [3]. Based on this description, the combination properties of synthetic and natural polymer is interested to be developed.

Both blending and copolymerization is combining two or more polymers to obtain new properties that are part of the constituent. Generally, blending is to enhance physical properties of polymers (thermal and mechanical properties). Copolymerization involves chemical reaction by adding a functional group to enhance chemical properties, i,e: hydrophilicity, hydrophobicity, etc. [4].

Polyacrylamide (PAM) is a synthetic polymer which widely used as flocculants. Starch is natural polymer which abundantly available in many sources, i.e.: corn, corn, wheat, corn, potato, taro, cassava, rice, etc. Starch consists of two parts, amylose in a amorphous part (normally 20-30%) and amylopectin in a crystalline one (normally 70-80%). Both amylose and amylopectin has hydroxyl groups (-OH) that gives an opportunity to be substituted with the other groups. Therefore, polyacrylamide could be attached/grafted onto the starch as the backbone.

Many synthesized flocculants have been reported by grafting PAM onto, amylopectin [5], sodium alginate [6], Glycogen [7], sago starch [8] and corn starch [9]. It can be highlighted that graft copolymers are more effective than the linear ones due to their greater "approachability" to the suspended particles [2,10].

Recently, utilization of many starch sources as backbone was explored, developed and proven their performance as flocculating agent for industrial effluents [2, 10, 11]. Most of them have reported the utilization of native starch. Generally, the physical characteristics of native starch are clarity, odorless, and stickiness. The native starch has some drawbacks in wide range applications such as unstable viscosity to heat, acid and shear forces, weak structure and undesirable gel characteristic. These drawbacks could be tackled by starch modification. Further, this starch was called modified starch.

* Corresponding author: mkaavessina@staff.uns.ac.id

Modified starch is a starch which has been revamped its hydroxyl groups through chemically (esterification, oxidation, etc.) or physically (pre-gelatinization and hydrothermal treatment). As known, the utilization of chemicals in starch modification has good performance (high efficiency), but it has some weakness, i.e.: concerned in points of food safety, and cost of waste water treatment. In other side, physical methods shows no chemical residues that reduces cost in waste water treatment. But it performs in lower efficiency [12].

There are a few reports on the utilization of modified starch as backbone in grafted flocculant. In this investigation, we developed backbone from modified starch and compared their performance to the native one. Modification of starch was done by adding α-amylase. The utilization of α-amylase is intended to get high efficiency and safety. We report for both native and modified starch of taro and cassava.

2 Experimental details

Taro and cassava were supplied from local market in Surakarta, Indonesia. Both were made into taro and cassava starch. The enzyme α-amylase, acrylamide, $K_2S_2O_8$, TMEDA and Hydroquinone were purchased from Sigma Aldrich. Acetone and methanol were kindly supplied by Fulltime. All chemicals were used without any purification.

The preparation of starch-g-PAm was carried out with redox initiation process (synthesizing of non-terminated polyacrylamide (nt-PAM)) and then followed by grafting of the nt-PAM onto several kinds of starch. Polymerization of Acrylamide was initiated by adding $K_2S_2O_8$ and TMEDA, respectively and was done for 30 min at 30°C. This reaction obtained nt-PAM. The optimum initiator concentration ($K_2S_2O_8$ and TMEDA) was obtained by varying its concentration: 0.05, 0.1, 0.15, 0.2, and 0.25 mol/L. Backbone was prepared by gelatinized of starch at 70°C until the color changes under a N_2 atmosphere. The gelatinized starch was cooled to ambient temperature.

Grafting Nt-PAm onto gelatinized starch was carried out in a round-bottom flask equipped with a reflux condenser, a nitrogen gas inlet, and placed in a magnetic stirrer for 1.5 h. The process conditions are kept at 50°C and stirring speed at 250 rpm under a N_2 atmosphere. The product was washed by adding aquadest to make a homogeneous slurry. Finally, aseton and methanol were added sequentially to precipitate starch-g-PAM in form of white granules.

Percentage of polymer (Nt-PAm) incorporated in starch backbone was evaluated by calculating the percentage of grafting yield (%GY).

$$\%GY = \frac{W_3}{(W_1 + W_2)} \text{x}100\% \quad (1)$$

where, W_1, W_2 and W_3 are weight of starch, acrylamide, starch graft polyacrylamide, respectively (Qudsieh, 1999).

The molecular structure was determined by A Perkin-Elmer 630 IR spectrophotometer. It was used to record the IR spectra within the range of 4000-400 cm^{-1}. The IR spectra samples were prepared in solid state using a KBr pellet method.

Average molecular weight (Mv) of grafted flocculants was determined using the Mark Houwink equation (1):

$$\eta = K.Mv^a \quad (2)$$

where, K = 6,31 x 10^{-3} g/ml dan a = 0,8 for PAm in water (Qudsieh,1999).

Flocculation test was done to clarify the synthetic kaolin suspension (10 g of kaolin in 1000 mL of aquadest). Grafted flocculant (10 mg) was dissolved into 20 ml aquadest. The solution was taken 4ml as a dosage in flocculation test and poured into 500 ml of kaolin suspension. The mixture was agitated for 2 min (120 rpm) and 10 min (50 rpm), sequentially. Then, it was allowed for 1 h and its turbidity measured.

3 Results and discussion

Enzymatic modification of starch is using α-Amylase. The visual appearance of modified starch solution in water is clearer and it has lower viscosity than native starch. Meanwhile, as shown in Figure 1 it can be seen that molecular structure of modified taro starch is unchanged compared to the native one. This phenomena is also occurred for cassava starch.

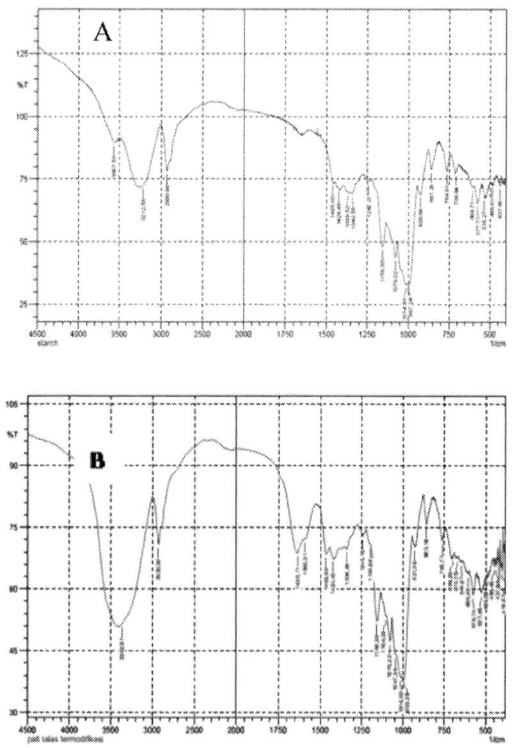

Fig. 1. IR spectrum of native taro starch (A) and modified taro starch (B).

The similar IR spectrum of samples was obtained that indicating the same of their molecular structure. The absorption band data of native and modified starch was summarized in Table 1 and 2. However, it can be highlighted at wave number 3362.07 cm^{-1}. It shows the hydroxyl group (-OH), weak intensity is occurred in native and stronger intensity can be seen in modified one. It indicated that the enzymatic modification can revamp hydroxyl groups in starch. The similar phenomenon is also for cassava starch.

The similar IR spectrum of samples obtained indicating the similarity of their molecular structures. The absorption band data of native and modified starch was summarized in Table 1 and Table 2.

Table 1. Comparison absorption band of native taro starch and modified taro starch

Wave number of taro starch		Chemical Bonding	Mode	Notification
Native	Modified			
3362.07	3269.92	–OH	Stretching	stronger intensity after modification
1579.22	1583.40	–C-O	Stretching	
1384.49, 1395.96, 1420.63, 1450.18	1370.58, 1398.55 1415.60, 1445.15	–CH–	Deforma-tion	asymmetric, symmetric
1237.21	1256.42	–C-OH	Bending	
871	900,8	–C–C–	Stretching	

It shows the weak intensity of hydroxyl group (-OH) is occurred in native and stronger intensity can be seen in modified one. It indicated that the enzymatic modification can revamp hydroxyl groups in starch. The similar phenomenon is also for cassava starch.

Table 2. Comparison absorption band of native cassava starch and modified cassava starch

Wave number of cassava starch		Chemical Bonding	Mode	Notification
Native	Modified			
3419.94	3419.92	–OH	Stretching	stronger intensity after modification
1643.42	1643.40	–C-O	Stretching	
1340.58, 1368.55, 1420.63, 1460.18	1340.58, 1368.55 1420.60, 1460.15	–CH–	Deforma-tion	asymmetric, symmetric
1242.21	1268.21	–C-OH	Bending	
890	921.1	–C–C–	Stretching	

Figure 2 shows that the greater of the initiator concentration, the percentage of grafting yield (%GY) tends to increase. It could be explained as follows: the presence of initiator acts as stimulator of the active radical formation (non-terminated polyacrylamide/nt-PAM). Increasing initiator concentration provides more opportunity of formation acrylamide chains in active radical form that will be grafted on backbone starch. Thus, %GY tends to increase. From this figure, the maximum value of %GY is obtained at 91.51% when the initiator concentration is 0.2 M. This value is maximum in the range of initiator concentration between 0.05-0.25 M.

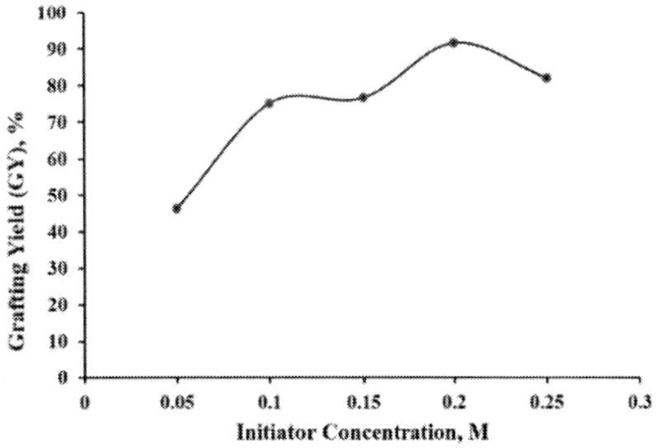

Fig. 2. The effect of initiator concentration towards percentage of grafting yield (%GY).

Above 0.2 M, the percentage of grafting yield begins to decrease. This phenomena may be caused by the formation of polyacrylamide in homopolymer form. The excess initiator during polymerization of poly acrylamide leads to the incorporation of the formed active radicals in polyacrylamide chains that configure terminated polyacrylamide (without active radicals).

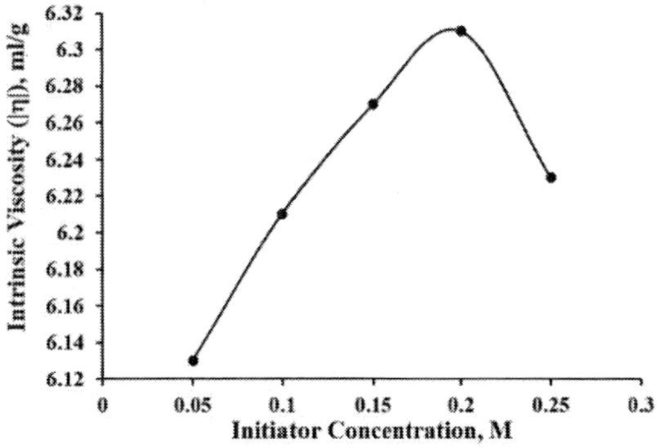

Fig. 3. The effect of initiator concentration towards intrinsic viscosity (|η|)

The effect of initiator concentration toward intrinsic viscosity was also investigated as depicted in Figure 3. It can be highlighted that similar trend with %GY was occurred. The greater of the initiator concentration, the intrinsic viscosity (|η|) shows increasing trend. The maximum value of intrinsic viscosity is 5,625x10^3 g/mol when the initiator concentration is 0.2 M. However, above 0.2 M, The depletion of intrinsic viscosity was also occurred. It can be concluded that there is a correlation between the percentage of grafting yield and intrinsic viscosity.

As the correlation with intrinsic viscosity, molecular weight of grafted flocculant has same trend. It can be seen in Figure 4. The maximum molecular weight of grafted flocculant is 5625 Da in the range of initiator concentration between 0.05-0.25 M. The maximum value was obtained at 0.2 M. Based on this initial data, the initiator concentration was set at 0.2 M for further investigation

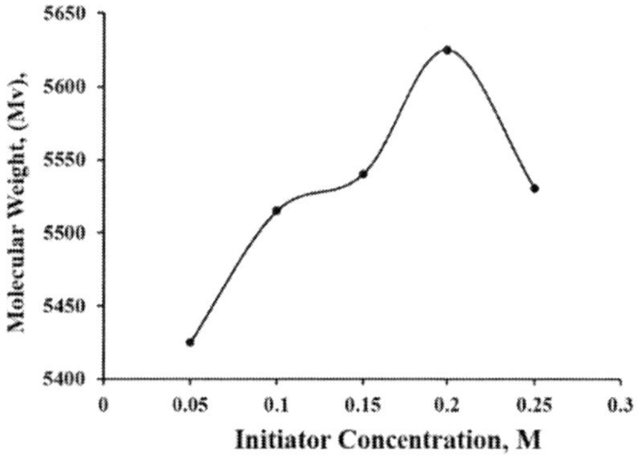

Fig. 4. The effect of initiator concentration towards Molecular weight (Mv)

Figure 5 depicts the infrared spectrum of starch-g-polyacrylamide. It can be seen that some peaks of pure starch indicating its chemical bonds: 1158.30 cm^{-1} for C-O; 3438.26 cm^{-1} for –OH; 1394.59 cm^{-1}, 2926.14 cm^{-1}, and 2853.81 cm^{-1} for C-H, and 880.34 cm^{-1} for C-C. Figure 3 also shows other peaks indicating the presence of polyacrylamide i.e.: 1193,99 cm^{-1} and 1240,28 cm^{-1} for C-N. The presence of chemical bonds for starch and polyacrylamide in sample is verifying that starch-g-polyacrylamide has successfully synthesized.

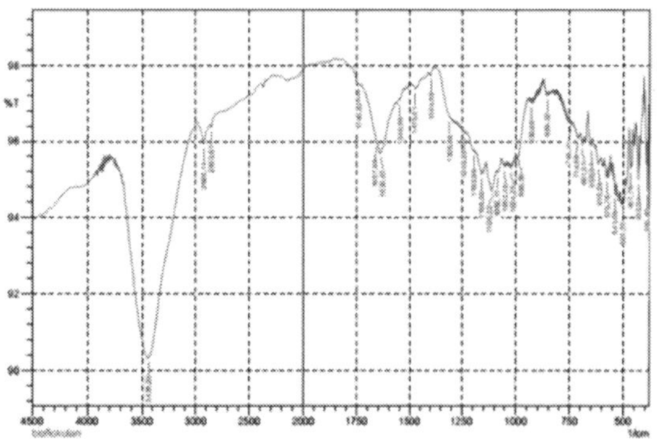

Fig. 5. IR spectrum of starch-g-polyacrylamide (modified taro starch)

Intrinsic viscosity and molecular weight of starch-g-PAm (grafted flocculant) were summarized in Table 3. It can be seen that higher molecular weight was obtained from modified taro starch as backbone. Backbone comparison between native and modified starch is monitored and it is shown that the modified one is always higher their molecular weight than the native starch. It may be explained that this phenomenon related to the alteration of hydroxyl groups.

Table 3. Summary of intrinsic viscosity |η| and average molecular weight (mv)

Graft co-polymer*	Back-bone, g	Acrylamide, M (4 ml)	|η|, ml/g	Mv.10^3
Sample A	2.5	0.1	6.9	6.288
Sample B	2.5	0.1	6.3	5.612
Sample C	2.5	0.1	7.4	6.862
Sample D	2.5	0.1	7.05	6.459

*Sample A= Taro starch-g-Pam, Sample B = Cassava starch-g-Pam, Sample C= Modified taro starch-g-Pam, Sample D= Modified cassava starch-g-Pam

Enzymatic modification using α amylase is breaking the cell wall of starch granules followed by the release of amylase or amylopectin. Thus, hydroxyl groups of both amylase and amylopectin are easier to be substituted with non-terminated polyacrylamide. This claim will be studied further for the next investigation.

Further, the graft copolymers were tested to treat the artificial turbid water containing a kaolin suspension. The results were summarized in Table 4. It shows that higher molecular weight of graft copolymer is more effective to turbidity removal in water. This phenomenon might be related to the quantity of polyacrylamide grafted in the backbone. As explained above, modified starch give the easiness of hydroxyl groups to be substituted with radical group of non-terminated polyacrylamide (nt-PAm). However, it needs further investigation to support this statement.

Table 4. Flocculation performance for each graft copolymers

Criteria	Sample *			
	A	B	C	D
pH	6.5	6.5	6.5	6.5
Temperature, °C	30	30	30	30
Dosage, mg/L	2.67	2.67	2.67	2.67
Initial Turbidity, NTU	180	180	180	180
Residual Turbidity, NTU	11.52	15.84	8.46	10.62
Removal, %	93.6	91.2	95.3	94.1

*Sample A= Taro starch-g-Pam, Sample B = Cassava starch-g-Pam, Sample C= Modified taro starch-g-Pam, Sample D= Modified cassava starch-g-Pam

4 Conclusions

Combination of synthetic and natural polymer was successfully prepared through the graft copolymerization. Non-terminated polyacrylamide can be grafted onto a backbone either in the form of native or modified starch. Average molecular weight related to the intrinsic viscosity was observed to increase after modification. The obtained starch-g-polyacrylamide performed the effectively flocculation to kaolin suspension in water. This result showed that the graft copolymer is potential to be applied for water turbidity removal or treatment process of wastewater.

This research was financially supported by Sebelas Maret University through research grant of "Hibah Peneliti Utama sebagai Rujukan Hibah MRG-UNS PNBP 2016".

References

[1] R.P. Singh, T. Tripathy, G.P. Karmakar, S.K. Rath, N.C. Karmakar, S.R. Pandey, K. Kannan, S.K. Jain, N.T. Lan, Curr. Sci. **78,** 798-803 (2000).

[2] B.R. Nayak, D.R. Biswal, N.C. Karmakar, R.P. Singh, Bull. Mater. Sci. **25,** 537-540 (2002).

[3] C.L. Swanson, R.L. Shogren, G.F. Fanta and S.H. Imam, J. Environ. Polym. Degr. **1/2,** 155 (1993)

[4] M. Kaavessina, A. Ilias, R.H. Elleithy, S.M. Al-zahrani, S.M., J. Polym. Res. **19/2,** 1-12 (2012).

[5] S.K Rath, R.P. Singh, J. Polym. Sci. **66,** 721-729 (1997).

[6] T. Tripathy, S.R. Pandey, N.C. Karkamar, R.P. Bhagat, R.P. Singh, Eur. Polym. J. **35** 2057-2072 (1999).

[7] P. Adhikary, K.N. Tiqari, R.P. Singh, J. Appl. Polym. Sci, **103,** 773-778 (2007).

[8] I.Y. Qudsieh, A. Fakhru' l-Razi, N.A. Kabbashi, M.E.S. Mirghani, K.G. Fandi, M.Z. Alam, S.A. Muyibi, M.M. Nasef, J. Appl. Polym. Sci. **109,** 3140-3147 (2008).

[9] M. Gao, Z.Q. Xu, Adv. Mat. Res. **875-877,** 44-48 (2014).

[10] C.S. Lee, J. Robinson, M.F. Chong, Process Saf. Environ. **92,** 489-508 (2014).

[11] Y.J. Sun, G.F. Guo, Y.L. Yang, Adv. Mat. Res. **815,** 432-436 (2013).

[12] R. Wongsagonsup, P. Deeyai, W. Chaiwat, S. Horrungsiwat, K. Leejariensuk, M. Suphantharika, A. Fuongfuchat, S. Dangtip, Carbohydr. Polym, **102,**790-798, (2014)

Effect of combination dope composition and evaporation time on the separation performance of cellulose acetate membrane for demak brackish water treatment

Tutuk Djoko Kusworo[1,*], Budiyono[1], *Diyono* Ikhsan[1], *Nur* Rokhati[1], *Aji* Prasetyaningrum[1], F. R. Mutiara[1], and N. R Sofiana[1]

[1]Department of Chemical Engineering, University of Diponegoro, Semarang, Indonesia

Abstract. The coastal areas in Indonesia often have a problem of clean water lack, because the water is classified as brackish water. Therefore, this research investigated the fabrication of CA membranes using phase inversion method for brackish water treatment. Investigation was conducted to study the effect of combination dope composition and evaporation time on separation performance and morphology of the memrbane. Membrane was fabricated by dry-wet phase inversion technique with variation of polymer concentration 17, 18 and 20 wt% in the total solid and evaporation time of 5, 10 and 15 seconds, respectively. The asymmetric membranes were characterized by permeability test through rejection and flux measurements using brackish water as feed. The experimental results from SEM images analysis showed that all the membranes have a thin small porous layer and thicker sub-structure of larger porous layer formed asymmetric membrane. Moreover, the greater polymer concentration is resulting smaller pore size and smaller membrane porosity. The longer evaporation time was also resulted in denser membrane active layer. The best membrane performance was observed at the composition of 20 wt% CA polymer, 1 wt % polyethylene glycol with the solvent evaporation time of 15 seconds.

1. Introduction

Membrane separation could be defined as a selective barrier between two phases. The advantages of membrane based separation are continuous process, energy efficient, modular, can be operated under mild condition, membrane properties are adjustable, wthout any required additives. With aforement benefits of membrane usage, the application of membrane become more extensive such as desalination of saline water, sea water and brackish water, the concentration and separation in industrial waste water (waste water treatment), sterilization and purification of potable water, gas separation, biomedical technology [1]. Research about membrane which focuses on conditioning of membrane production is still interesting. Many factors in fabrication process variables affect the characteristic, morphology, and separation properties of membrane. By combining several various parameters allows obtaining tailor-made membranes that are specific to a particular separation purposes.

Polymeric membrane with asymmetric structure is mostly applied in water and waste water treatment. Permeate water flux that is resulted by asymmetric membrane is higher than symmetric membrane. This is due to the dense layer of membrane is thinner than symmetric membrane. The asymmteric membrane consisting sub-layers structure with different thickness. The upper skin layer has a thickness approximately 0.1-0.5 μm and 50 to 100 μm thickness of other layer. The top layer of assymetric membrane acts as a filter medium, while the sub-layer or intermediate layer contributes as a supporting medium of the top skin layer. Hence, the asymmetric structured membranes are suitable for water treatment [2]. The membrane performance parameters were defined by the value of permeate flux and pollutant rejection efficiency. The higher permeate water flux and rejection efficiency indicates that the membrane has an excellent separation performance. However, the conditions encountered is an effort to improve membrane flux will be lowering the membrane rejection.

The membrane performance was affected by several factors such as the dope composition, solvent evap[oration time, solvent type, additive type and concentration, and shear rate [3]. In the research reported by another researcher [4-9] concluded that the dope composition was affected the performance of asymmetric membrane. They stated that the greater polymer concentration, membrane pore size and porosity will decrease. This will obtained smaller flux, while rejection increased. Evaporation time of solvent is an important factor that affected the membrane properties. The variation of evaporation time resulted the different membrane morphology and separation properties. Bhongsuwan et al. [2] perpared a nanofiltration

* Corresponding author: tdkusworo@che.undip.ac.id

 DOI: 10.1051/matecconf/201710101004

membranes with different solvent evaporation time where the membrane was evaporated in 40 seconds the rejection of tanin increased up to 21.68% and the permeate flux decreased up to 69.78%. The significant decline of permeate flux of nanofiltration membrane solvent evaporation for a while, it indicates that the evaporation time of solvent is an important factor in membrane production. Longer evaporation time will result membrane with a smaller pore size and increase the thickness of the active layer membrane. This morphology causes smaller membrane flux, while rejection will increase. These results have been confirmed with previous researchers [10-11]. Therefore, for further investigation needs to be conducted to obtain the optimum evaporation time which produces an excelent membrane separation performance.

In this study of asymmetric CA membrane fabrication, acetone is commonly used as a solvent in dope preparation [3-5]. CA was selected as polymer material because CA capable to form a thin active layer of asymmetric structure and dissolved solid can be retained in the support layer of membrane as well as chemical tolerant and resistant to foulant deposition. Therefore, the stability of membrane and separation performance are the important factors for membrane application in water treatment. Furthermore, the main purpose of this research is to study the effect of dope composition and evaporation time to enhance the performance of nanofiltration CA membrane for treating Demak brackish water.

2. Experimental Details

2.1. Materials

Cellulose acetate (CA) was used as membrane material and was purchased from MKR Chemicals Indonesia, 99.75% acetone was supplied by Mallinckrodt Chemicals, Polyethylene glycol with molecular weight 4000 g/mole was supplied by Merck, and original brackish water sample was obtained from Demak.

2.2. Fabrication of asymmetric CA

Asymmetric CA membrane was fabricated by preparation of dope polymer solution consists of 17, 18 and 20 wt-% CAs, acetone, distilled water and Polyethylene glycol as an additive with concentration of 1 wt-%. The stable and homogeneous CA solution was prepared according to the following procedure; an approrpiate amount of CA polymers were dissolved into the acetone and constantly stirred for 6 hours subsequently add an appropriate amount of polyethylene glycol. The solution was continuously mixed with a suitable stirrer for 6 hours to ensure complete dissolution of the polymer. A desired amount of distilled water was added to the homogenous solution. This dope solution was constantly agitated at high speed for at least 12 hours. After all the ingredients dissolved completely, then the dope solution allowed for 1 hour in order to removes air trapped. The membrane was casted using the technique of dry-wet phase inversion on a glass plate via a casting knife and allowed to correspond with the variation of evaporation time 5, 10 and 15 seconds. The membrane then was immersed into a coagulation bath containing distilled water in coagulation tank for 24 h at ambient temperature (30±2°C). The experimental brackish water treatment with membranes was performed using a dead-end nanofiltration system. Flux and rejection is periodically measured to obtain the profile of permselectivity performance of each membrane.

2.3. Characterization of Asymmetric CA membranes

Permeate water flux and rejection were measured via dead-end filtration. An appropriate effective area of CA membrane was mounted into membrane cell. Before performing experimental test using original samples, the membrane was compacted using distilled water for 30-45 minutes. After the membrane being compacted, the distilled water was replaced with original brackish water. The permeate volume was measured periodically every 5 minutes. Rejection was determined by performing the analysis of pollutant such as TDS, Ca^{2+}, Mg^{2+}, and turbidity of feed and permeate water. TDS was measured using TDSmeter, Ca and Mg ions were analyzed using complexometry conventional titrimetry. Schematic diagram of dead-end filtration was ilustrated in Figure 1. The flux was determined by evaluating the equation as stated in equation 1 [12]:

$$J = Vol / (P.A.\tau) \quad (1)$$

J = Flux ($L.bar^{-1}m^{-2}.h^{-1}$)

P = Pressure, bar

τ = Time (hour)

Vol = Volume of Permeate water (Liter)

A = Membrane effective Surface Area (m^2)

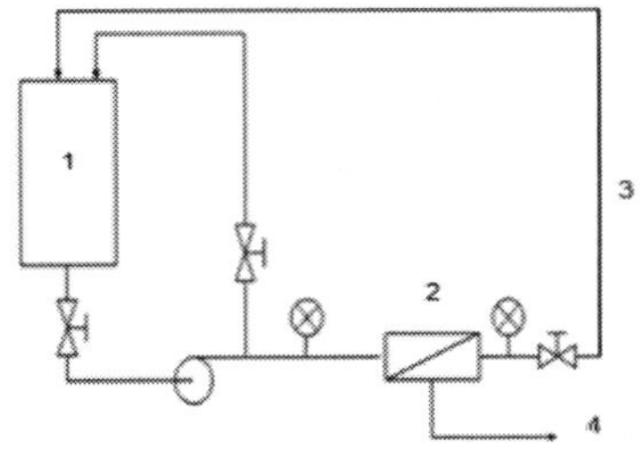

Fig. 1. Filtration cell dead-end type 1. Feed; 2. Membrane cell; 3. Retentate stream; 4 Permeate stream

The observation of asymmetric structure of membrane was performed using Scanning Electron Microscope (SEM). Membrane samples were prepared by fracturing in liquid nitrogen. The prepared membranes were mounted on an aluminium disk with a double surface tape and then the sample holder was placed and evacuated in a sputter-coater with gold. The chemical structure alteration as an effect of several variation of treatments in membrane preparation was analyzed using FTIR (Fourier Transform Infrared). The IR absorption spectra were scanned in a range from 4000 to 500 cm^{-1} with a spectral resolution of 8 cm^{-1} and averaged over 16 scans.

3. Results and discussions

3.1. Effect of Dope Composition on the Performance of CA Membrane

In this study the variation of dope composition by varying the polymer concentration in dope solution. Effect of dope composition on the performance of CA membrane at different evaporation time is shown in Figure 2. Figure 2 indicated that the higher CA concentration, the flux was decreased. This phenomenon was occured on all evaporation time. The decrease of flux was caused by lowering of membrane porosity with increasing polymer concentration. Increased of polymer concentration will be lead to the increasing number of polymer concentration at the membrane interface. This phenomenon indicated that the volume fraction of polymer was increased; hence, it has a smaller porosity. The results were in agreement with the previous study [13]. In the solid-liquid phase separation, increasing the polymer concentration will increase the density, whereas it will also reduce porosity. Polymer concentrations had a major impact on the bulk porosity and structure of the membrane. The decreasing porosity was occurred primarily in bulk porosity membrane. This is reinforced by the fact that the increase of polymer concentration was decreased the internal pore size [9].

Increasing polymer concentrations of membranes was also formed a membrane with thicker active layer. Therefore, the productivity was decreased. Increasing polymer concentration in solution will increase viscosity of dope solution. As a consequence, the interaction between the solvent and the polymer becomes more powerful, interaction between non-solvent and polymer becomes larger, thus reducing the ability of solvent to dissolve the polymer. This will accelerate the aggregation of polymer molecules on the surface to form a closer bond. The increase in viscosity will also prevent the diffusion exchange between solvent and non-solvent in the sub-layer, it makes the precipitation rate becomes slower. The rapid phase separation at the surface and slow in the sub-layer will produce asymmetric membrane with dense and thick layer [6].

Both porosity and membrane thickness decreased with increasing CA concentration. The increasing of concentration will strengthen the thermodynamic instability of dope solution, so de-mixing in this solution can be done with less amount of non-solvent. The increased viscosity of the solution due to the increased concentration will result in decreased diffusivity between the non-solvent and solvent on the system during the solidification process of dope solution. Greater use of CA makes precipitation process stops after a long time and will form a denser and thin membrane [8].

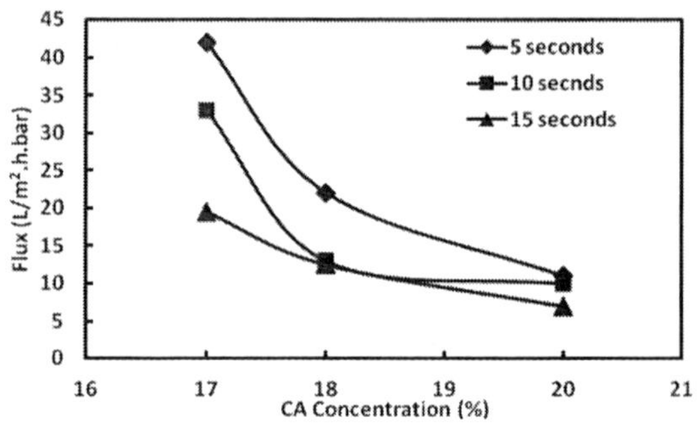

Fig. 2. The effect of polymer concentration on membrane Flux with different evaporation time

3.2. Effect Dope Composition on the rejection of CA membrane

Effect of dope composition on CA membrane rejection rate at different evaporation time was presented in Figure 3. Figure 3 showed that increasing of polymer concentration was increased the rejection. This phenomenon was occured on all evaporation time. The increase of rejection is due to higher CA concentrations inversely with decreasing flux. As discuss in section 3.1, the increase in the polymer concentration induced the formation of fine membrane pores and was also reduced the thickness of the membrane and resulted the membrane active layer will thicken. Therefore, the membrane was able passed the water, but not able to get through dissolved solids larger than the pore size. Thick active layer allows more dissolved solids are retained on the membrane section. Thus, the smaller of pore size and the thickness of active layers, gave more dissolved solids rejected, where it shows the selectivity of the membrane is great [13].

3.3. Effect of Solvent Evaporation Time on the Performance of CA membrane

Effect of solvent evaporation time on the performance of CA membrane at different polymer concentrations is shown in Figure 4. As shown in Figure 4, the increasing of evaporation time was decreased the flux. The decreasing of flux was due to longer of evaporation time caused during the evaporation process, the molecular solvent was swaped places with air molecules. Empty space left by the solvent will be occupied by air molecules. The longer of evaporation time, the closure of space by air molecule getting larger. Moreover, the membrane with smaller pore size was formed, as a result flux was decreased.

Basically, the evaporation in membrane manufacturing aims to create a more selective membrane. This is because during evaporation of solvent, the polymer concentration in the surface layer increased which will form a dense surface. Moreover, this layer will inhibit the speed of solvent exchange with non-solvent through a membrane sub-layer during coagulation processes and produced a smaller pore size

DOI: 10.1051/matecconf/201710101004

and ultimately will reduce the flux [11]. That the longer of evaporation time can also increase the active layer thickness of the membrane.

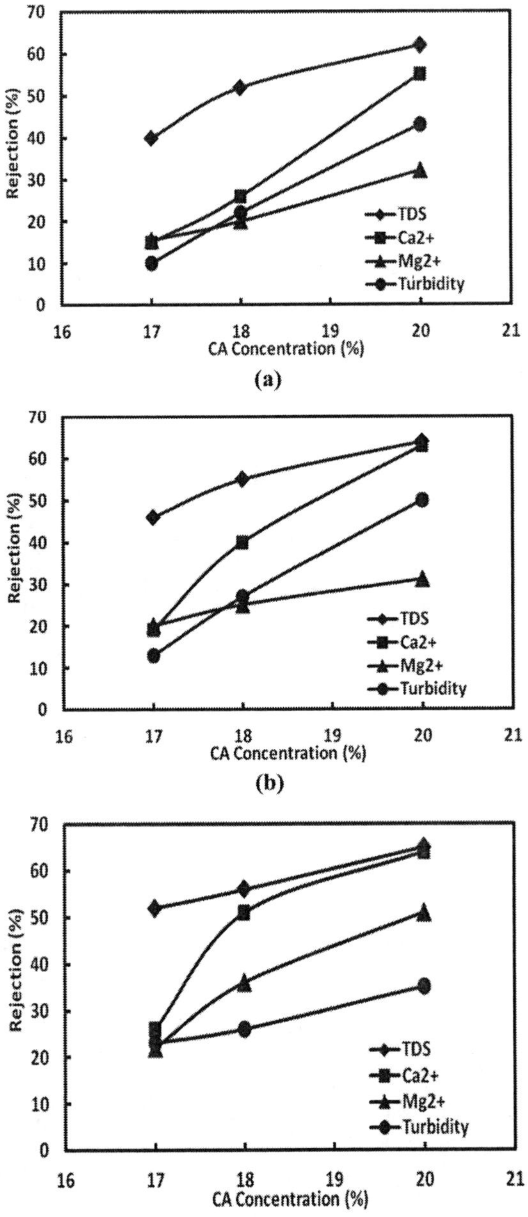

(a)

(b)

(c)

Fig. 3. Dope Composition effect on the rejection at different evaporation time: (a) 5 seconds; (b) 10 seconds; (c) 15 seconds

The outer part of the membrane will experience phase separation during evaporation process. Evaporation before immersion in bath coagulation gave a huge impact on the thickness of the active layer membrane. This process is called dry/wet phase separation that occurred during the formation of a very thin layer. Increased evaporation time make a thicker active layer membrane [10].

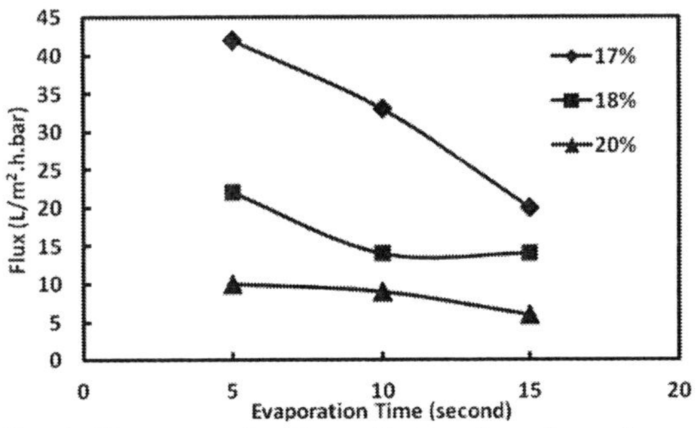

Fig. 4. The evaporation time effect on flux of membrane at different polymer concentration

3.4. Effect of evaporation time on the rejection of CA membrane

Effect of evaporation time on CA membrane rejection at different polymer concentrations is shown in Figure 5. As shown in Figure 5, increasing solvent evaporation time was resulted in the greater rejection. This phenomenon was occured at all concentrations of CA. The increase in rejection was also occured in all types of rejection. The increase in rejection due to higher evaporation time is inversely proportional to the decrease in flux. As discussed earlier, the longer evaporation time will reduce membrane pore size and thicken membrane active layer. Membrane is only able to get through water and dissolved solids that are smaller than the pore size of the membrane. Thick active layer allows more dissolved solids to be retained on the layer. Thus, the smaller of pore size and the thickness of the active layer membrane, will result in more dissolved solids are left behind that indicates great membrane selectivity[13].

3.5. Effect of dope composition on the morphology of CA membrane

Surface morphology and the appearance of cross section CA membranes can be determined through analysis using SEM (Scanning Electron Microscopy). SEM analysis results the effect of the polymer concentration of membranes is shown in Figure 6. From Figure 6, it can be seen that both membranes have an asymmetric porous structure and visible difference. At the appearance of the membrane surface with a magnification of 2500 times, the membrane b (20% CA) has a smaller pore size. While on the cross-section with a CA membrane has a 20% has smaller internal porosity than membrane with 17% CA.

This confirms a phenomenon that occurs in flux and rejection, as has been reviewed previously. Similar results were also obtained by Van de Witte et al. [20], which states the increase of polymer concentration will increase the thickness of the upper layers, lower porosity and pore inter-connectivity of membrane, resulting formation of macrovoid will be reduced.

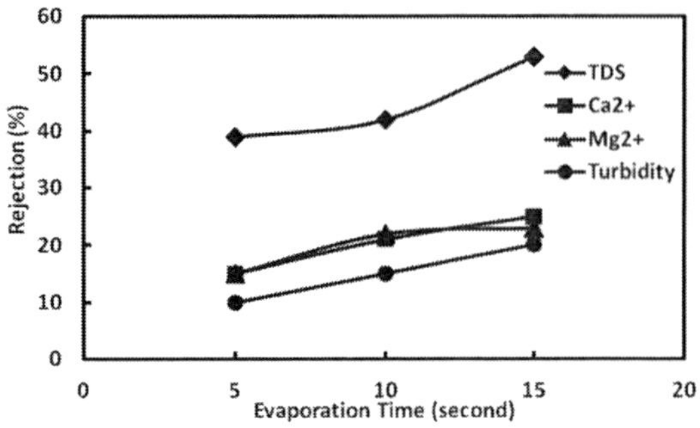

(a)

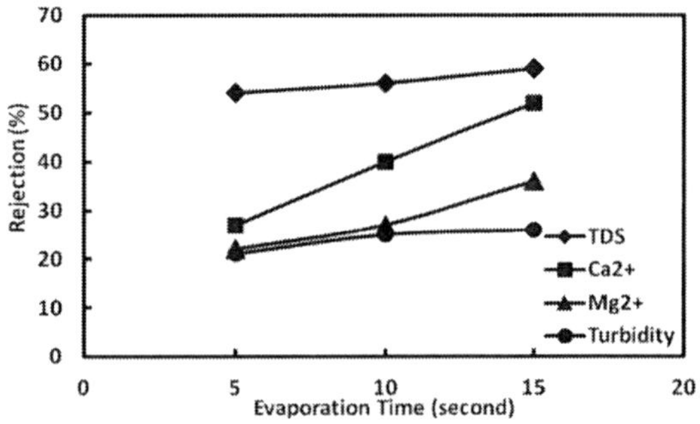

(b)

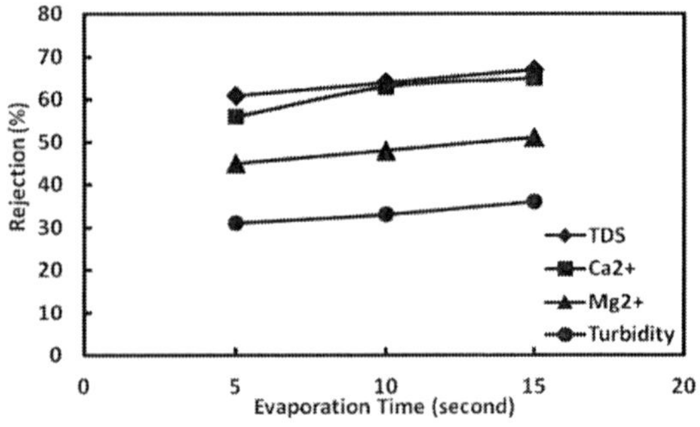

(c)

Fig. 5. Evaporation time effect on Membrane Rejection on Different Polymer Concentration: (a) 17 wt%; (b) 18 wt%; (c) 20 wt%

In Figure 6 seems to have a larger pore size. This is due to the dope solution has a lower viscosity. After the solidification of the polymer phase, low viscosity will result in the formation of pores quickly. Internal pore rapid growth led to the formation of larger pore size. With the decrease in the polymer concentration, the overall porosity increases [16].

3.6. Effect of evaporation time on the morphology of CA membrane

Effect of evaporation time at 20% polymer concentration on the membrane morphology is shown in Figure 7. From Figure 7, it can be observed that the membranes also have an asymmetric porous structure. At the appearance of the membrane surface with a magnification of 2500 times, the membrane b (evaporation time 15 seconds) has more dense surface and smaller pores. Whereas the membrane crosses section with evaporation time of 15 seconds has internal porosity of membrane smaller than the 5 seconds evaporation time. This confirms a phenomenon that occurs in flux and rejection, as has been reviewed previously. Similar results were reported by See-Toh et al. [18], which states that the shorter time evaporation will increase the size of the gap (pores) in the membrane. Moreover, Muliawati [14] states the longer of time evaporation indicates that the delayed of demixing process also longer, so that membranes produce are more dense.

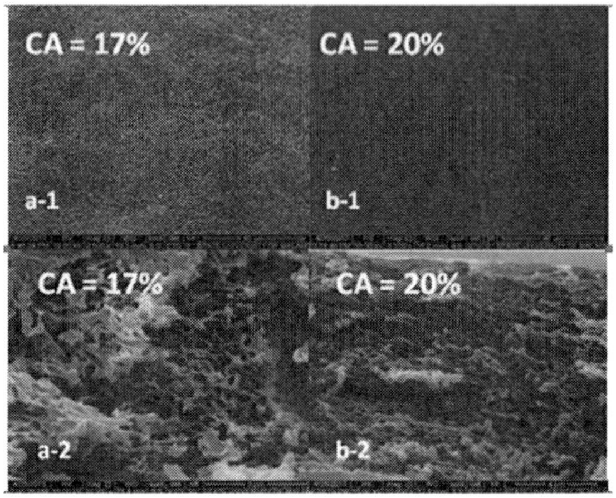

Fig. 6. Morphology of CA Membrane : a) 17% CA, 1% PEG, 1% aquadest, 15 seconds evaporation time; b) 20% CA, 1% PEG, 1% aquadest, 15 second evaporation time; 1) surface; 2) cross section

However, the longer of evaporation time (> 30 seconds) will make the membrane pore size becomes larger. It is because there are two influencing factors during membrane exposure, i.e the evaporation of solvent and water absorption in the dope solution. When the evaporation time is less than 30 seconds, the solvent evaporation has bigger influence because the short time absorption is very small.

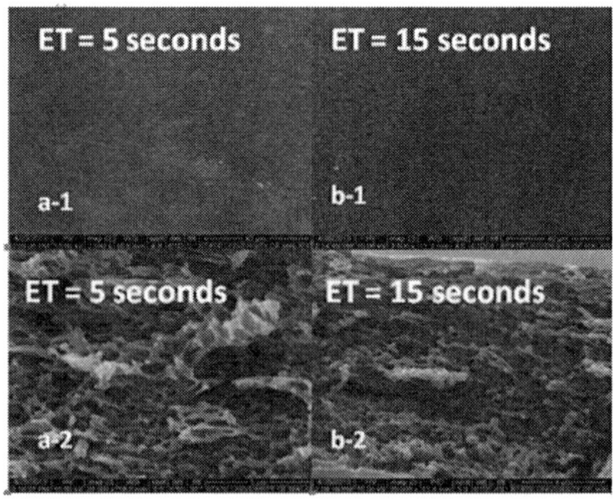

Fig. 7. Morphology of CA Membrane : (a) 20% CA, 1% PEG, 1% aquadest, 5 seconds evaporation time; (b) 20% CA, 1% PEG, 1% aquadest, 15 seconds evaporation time; 1) surface; 2) cross section

At the evaporation time of 1-2 minutes, the effect of solvent evaporation is almost balanced by the effects of water absorption, so that the pore size in the range appears at this time. However, more than 2 minutes evaporation will cause the water absorption become a major factor during the evaporation process, so it will form a membrane with a larger pore size [11].

4. Conclusion

Based on the experimental results of this research can be concluded that in this study, the concentration of polymer is very influential. The higher polymer concentration, the flux will decline, while rejection increased. This is because the increase in concentration will minimize membrane pore size. Results of SEM analysis showed that the high concentration of CA will produce asymmetric membranes with smaller pore size and porosity. Evaporation time also affects the membrane performance. The longer evaporation time, smaller flux will be obtained and rejection will increased. This is because the longer evaporation time will cause the decrease of membrane pore size. SEM analysis showed the longer evaporation time will produce increasingly dense membrane. The best membrane was achieved with a composition of 20 wt % of CA, 1 wt % of PEG, 1 wt % aquadest and the evaporation time of 15 seconds with the performance in term of flux as large as 6.08 $L.m^2.h^{-1}.bar^{-1}$, TDS rejection value of 66.99 %, rejection Ca^{2+} ions 64.71%, Mg^{2+} ion rejection 37.10%, and 53.28% turbidity rejection, respectively.

References

[1] R.W. Baker, *Membrane Technology and Applications*, *2nd ed.* (John Wiley & Sons, Ltd., 2004)
[2] D. Bhongsuwan and T.J. Bhongsuwan, Nat. Sci. **42**, 311-317. (2008)
[3] A.F. Ismail, A.R. Hasan, and B.C. Ng, J. Sci. Technol., **24**, 878-889. (2002)
[4] T.D. Kusworo, Budiyono, J. Supriyadi and D.C. Hakika, J. of Wast. Res. **4**, 1. (2014)
[5] T.D. Kusworo, Budiyono, A.I. Wibowo, G.D. Harjanto, A.D. Yudisthira and F.B. Iswanto, J. App. Sci., Eng. and Tech. (2014).
[6] M.I. Mustaffar, A.F. Ismail, and R.M. Illias, *Study on the Effect of Polymer Concentration on Hollow Fiber Ultrafiltration Membrane performance and Morphology*, University Teknologi Malaysia. (2004)
[7] A. Sofyan, and N. Pocut, *Desalination of Brakish water using nanofiltration Membrane,* (Thesis) Syiah Kuala University. (2007).
[8] E. Saljoughi, M. Sadrzadeh, and T.J. Mohammadi, Mem. Sci. **326**, 627-634. (2009)
[9] A. Akbari and R.J. Yegani, Mem. Sci. **1**, 100-107. (2012)
[10] I. Soroko, M. Markowski, F. Spill, and A.J. Livingston, Mem. Sci. **381**, 163-171. (2011)
[11] L. Gao, B. Tang, and P.J. Wu, Mem. Sci. **326**, 168-177. (2009)
[12] M.S.F. Dasilva, Disertation. *Polyamide and Polyetherimide Organic Solvent Nanofiltration Membranes,* (Disertation) University Of Nova De Lisboa. (2007)
[13] M. Mulder, *Basic Principles of Membrane Technology. 2nd edition.* (Kluwer Academic Publisher, Netherland, 1996)
[14] E.C. Muliawati, *Fabrication and characterization of Nanofiltration Membrane for Water Treatment,* (Thesis) Diponegoro University. (2012)
[15] L. Amand, and C.J. Tullin, *The Theory Behind FTIR Analysis.* (The Centre of Combustion Science and Technology, Lund University, 1999)
[16] J.C. Jansen, M. Macchione, C. Oliviero, R. Mendichi, G.A. Ranieri, and E. Drioli, Polymer **46**, 11366-11379. (2005)
[17] K. Scott, K. and R. Hughes, *Industrial Membrane Separation Technology.* (Blackie Academic and Professional, London, 1996)
[18] Y.H. See-Toh, F.C. Ferreira, and A.G.J. Livingston, Mem. Sci. **299**, 236-250. (2007)
[19] R.U. Somantri, *Pengaruh Penambahan Formamide dan Lama Penguapan Pelarut (Aseton) terhadap Membran Selulosa Asetat.* (Bogor Agricultural Institute, Bogor, 2003)
[20] P. Van de Witte, P.J. Dijkstra, J.W.A. van de Berg, and J.J. Feijen, Mem. Sci. **117**, 1-31. (1997)

MATEC Web of Conferences **101**, 01005 (2017) DOI: 10.1051/matecconf/201710101005
SICEST 2016

Characterization of fermented broccoli (*Brassica oleracea* L.) and spinach (*Amaranthus* sp.) produced using microfiltration membrane as folic acid source for smart food formula

Agustine Susilowati[1,*], *Aspiyanto*[1] and *Hakiki* Melanie[1]

[1]Research Center for Chemistry, Indonesian Institute of Sciences, 15315 Tangerang Selatan, Indonesia

Abstract. Purifying and drying both fermented biomasses of broccoli (*Brassica oleracea* L.) and spinach (*Amaranthus* sp.) by Kombucha culture has been conducted to recover concentrate and powder of folic acid. The aims of this study is to determine the differences of particles characteristics and compositions of concentrate and powder from both mentioned folic acid source through Micro Filtration (MF) membrane and without MF membrane. The best folic acid produced by MF membrane process (room temperature, stirrer rotation speed 400 rpm, pressure 40 psia and 30 minutes) and drying (30 °C, 22 cm Hg and 24 hours) were resulted in biomass of the concentrate and powder with compositions of total solids 6.29 % and 96.91 %, total polyphenol 0.25 % and 0.06 %, folic acid 58.8 µg/mL and 54.33 µg/mL, reducing sugar 105.34 mg/mL and 441.39 mg/mL, and total acids 0.57 % and 2.33 %, respectively. In optimum condition, fermented spinach concentrate contributed to particles distribution with diameter size (Ø) between 0,4 and 100 µm (75.45 %), and with Ø between 100 and 1000 µm (26.3 %), otherwise, the process without MF membrane was resulted the particles distribution respectively 74.1 % and 25.9% by each interval of Ø.

1 Introduction

Broccoli (*Brassica oleracea* L.) and spinach (*Amaranthus* sp.) fermented by Kombucha culture have potential utilization as food functional fortificant as folic acid source to enhance brain smartness and prevent anaemia [1]. Novelty aspect of this fermentation process is based on ability of Kombucha culture in degrading tea polyphenol during preparation of Kombucha tea [2, 3, 4].

Potential exploration of folic acid in spinach (*Amaranthus* sp.) will make this commodity as alternative resource of natural folic acid. Polyphenols contained in broccoli and spinach are source of folic acid produced by activity of microbes in Kombucha culture through de novo biosynthesis [5] in amino acids (adenine, hisyidine, methionine) from Kombucha culture (fungi, bacteria, yeast) and tea leaves.

This biosynthesis involves enzyme activities of dihydrofolate synthase (DHFS) and ypolyglutamate synthease (FPGS) to form dihydrofolate and polyglutamate folate, a derivative of folic acid [5], in which this process is occurred only on plant and microbe [6]. Kombucha cultures containing Acetic Acid Bacteria, such as *Acetobacter xylinum* and yeast, such as *Saccharomyces cereviseae*, *Saccharomyces ludwigii*, *Saccharomyces bisporus*, *Zygosaccharomyces* sp. and various type of khamir (*Torolupsis* sp.) [7] could possibly degrade polyphenol and amino acids in broccoli and spinach. It has been known that there are more than eight types of specific enzymes from plant, which have important roles in this synthesis, particularly esterification of ρABA with glucose [8].

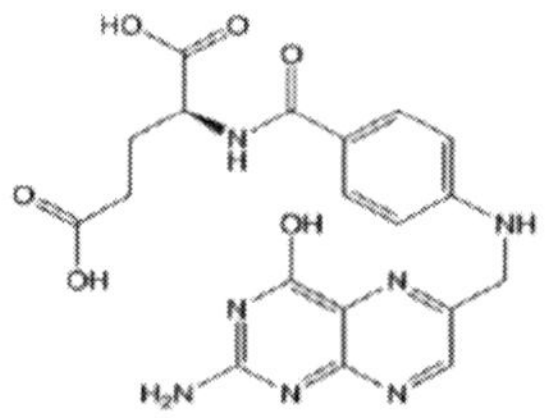

Fig. 1. Chemical Structure of Folic Acid

Process of concentrating biomass for both fermented broccoli and spinach through MF membrane (0.15 µm) equipped in dead-end stirred microfiltration cell (DESMFC) produces concentrate and permeate with different specification. MF membrane retains more folic acid in concentrate than in permeate, although particle size of folic acid is smaller than pore size in MF membrane used (0.01 µm < 0.15 µm) [9]. Its occurrence of fouling is possibly caused by operation condition factors such as pressure and time, membrane type, characteristic of biomass [10, 11]. In drying process of biomass concentrate using vacuum dryer at 30 °C for 24 hours and biomass without MF membrane process as

* Corresponding author: agustine_1408@yahoo.co.id

comparison, recovery of powder affect on composition, particles size and distribution. For smaller particles size and larger particles distribution enable better surface tension recovery on formulation so that it facilitates application process on functional property in ability of adsorption of folic acid by body.

The aims of this experimental study was to determine the differences of particles characteristic and distribution of particles size, and compositions of concentrate and powder from both fermented broccoli and spinach by Kombucha culture as folic acid source through MF membran and without MF membrane as in biomass

2 Experimental details

2.1 Materials, equipments and analysis instrument

Main materials used in this experimental work were fresh green spinach (*Amaranthus* sp.) and broccoli (*Brassica oleracea* L.) purchased from local market, sucrose, Kombucha culture (Research Centre for Chemistry – LIPI), 0.15 µm microfiltration (MF) membrane (FSM-0.15-PP, Danish Separation Systems, Denmark), distilled water, and chemicals for preparation and analysis purposes. All of the chemical reagents used were of analytical grade quality.

The main equipment used in this experimental work were glassware (cylinder glass, beaker glass, erlenmeyer flask, stoppered conical flask), high precision balance (Fujitsu, Japan), fermentation system in laboratory scale (local), laminar flow chamber (local), incubator (local), blender, sieves of 60 mesh and 80 mesh (Retsch, Germany), magnetic stirrer (HI 303 N, HANNA Instrument, Japan), pressure gauge of nitrogen (Fisher Scientific Company, England), cylindrical tank for nitrogen (local), stop watch (Hanhart Profil 2, Germany), Dead-End Stirred Ultrafiltration Cell (DESUFC) (Amicon, Model 8200, MILLIPORE, U. S. A.), vacuum dryer (Vacutherm, Heraus Type VT 6060 P, Kendro Laboratory Products, Germany), and porcelain mortar & pestle (Z247529-1EA, SIGMA-ALDRICH). Analysis instruments were UV-Vis Spectrophotometer (Model RF-550, Shimadzu, Japan) and Particle Size Analyzer (Beckman Coulter LS 100 Q, U. S. A.).

2.2 Experimental design and analysis

Experimental work performed by using biomass filtrates of fermented broccoli and spinach by Kombucha culture which were filtered via 80 mesh and further purified by MF membrane (0.15 µm) equipped in DESMFC and resulted in retentate (concentrate) and permeate [12]. Analyses were carried out on total solids (Gravimetric method), total acids (Titrimetric method), reduction sugars (Somogyi Nelson method) [13], total polyphenol (Folin-Denis method) [14], folic acid (UV-Vis Spectrophotometer) [15] and particles distribution (LS Particle Size Analyzer) (Bacman Coulter). Process and analysis were conducted in duplicate. Data were processed in descriptive analysis based on average values.

2.3 Process steps

2.3.1 Fermentation process

Fermentation process of fresh broccoli was subsequently blanched at 80 °C for 5 minutes, water was added at Brocolli to water ratio 1 part : 4 parts, pulverized and filtered via a 80 mesh sieve to produce filtrate and residue. The broccoli filtrate was then used as substrate in fermentation process. In this fermentation process, substrate was added as inoculum of broccoli (15 %, v/v vegetable filtrate) [9] , sucrose 10 % added (w/v filtrate) and stored in closed container using cloth followed by aeration in dark room at room temperature for 6 days. The whole fermentation activities were aseptically. Both similar method and concentrations of inoculum of vegetable and sucrose was applied for spinach for 15 minutes. Results of these fermentation processes were fermented broccoli and spinach. Recovery of each biomass were used as feed in purifying process of folic acid through MF membrane (0.15 µm).

2.3.2 Purifying and drying processes

A number of fermented broccoli suspension was filtered through 80 mesh sieve, and filtered via 0.15 µm MF membrane fitted in DESMFC at stirrer rotation speed 400 rpm, room temperature and pressure 40 psia for 30 minutes. Same method and process condition were done for fermented spinach suspension. The result of MF membrane process on fermented broccoli suspension were retentate of fermented broccoli or spinach (concentrate) and then dried using vacum at 30 °C for 24 hours, reduced size and sieved via 60 mesh. The flowchart of the overall process is shown in Figure 2.

2.3.3 Folic acid analysis

Analysis of folic acid was done by using spectrophotometer method according to reaction of diazotization acid p-aminobenzoylglutamic yielded after reduction reaction of folic acid and 3-aminophenol to form yellow-orange complex. One (1) mL of standard folic acid or sample was subsequently added by 1 mL of Hydrochloric (4 M), 1 mL of Sodium Nitrite 1 % (w/v), 1 mL of Sulfamic Acid 1 % (w/v) and 1 mL of 3-Aminophenol 1 % (w/v) and then vortexed to form yellow-orange complex. Further, absorbance was measured by using UV-Vis Spectrophotometer at wavelength of 460 nm [15].

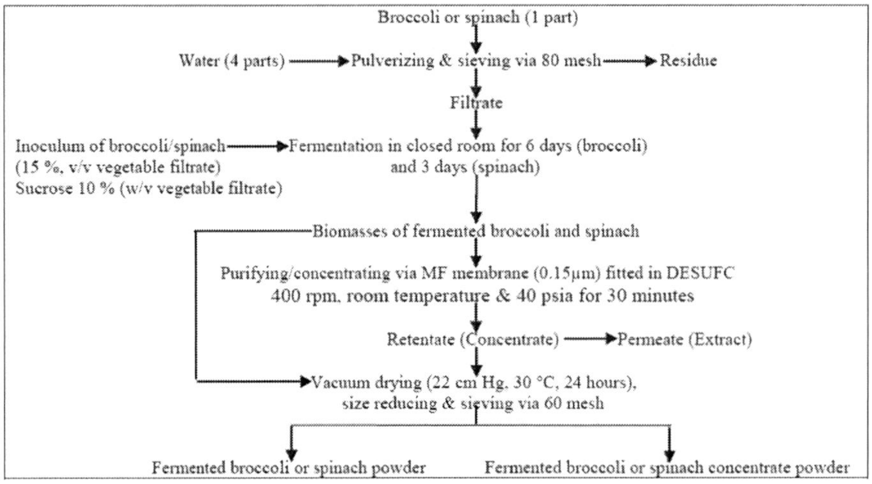

Fig.2. Flow diagram of recovery of fermented broccoli and spinach concentrate powders via microfiltration membrane process.

3 Results and discussion

3.1 Effect of purifying fermented biomass of broccoli and spinach by Dead-End Stirred Microfiltration Cell (DESMFC)

Both broccoli and spinach fermented by kombucha culture were produced through blanching, pulverizing, sieving via 100 mesh and fermentation in closed container and dark room at room temperature for 6 days and 3 days, respectively [9]. Purifying process on biomass using MF membrane (pore size 0.15 μm) fitted in dead-end stirred microfiltration cell (DESMFC) at stirrer rotation speed 400 rpm, room temperature and pressure 40 psia for 30 minutes were resulted in retentates (concentrates) of fermented broccoli and fermented spinach as thick suspension with brownish yellow and brownish turbid, as represented in Figures 3.

Purification process of fermented vegetables produced recovery of folic acid in concentrates of fermented broccoli (88.14 %) and fermented spinach (89.18 %) when compared to retentate and permeate. When compared to folic acid in fermented broccoli and fermented spinach feeds, this purification process was able to separate folic acid in concentrates, namely from 43.10 μg/mL to 28.1 μg/mL (65.2 % or 0.65 folds) and from 54.83 μg/mL to 58.90 μg/mL (107.72 % or 1.07 folds). Purification process also showed difference in compositions of component in retentate (concentrate) and permeate, in which DESMFC was able to retain more total solids, folic acid and reducing sugars in retentate (concentrate), but not able to retain total polyphenols and total acids in which it was passed freely across as permeate. Selectivity of MF membrane has significant effect on compositions of retentate (concentrate) and permeate and it is affected by operation condition, such as stirrer rotation speed, temperature, pressure, time and material type [16].

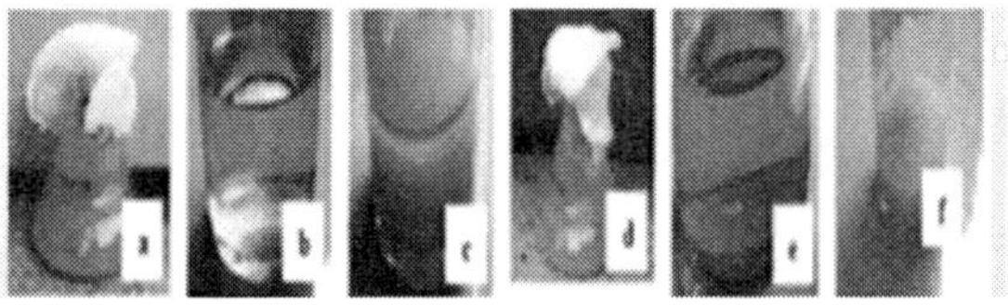

Fig. 3 (a) Spinach fermented for 3 days, (b) permeate (extract) and (c) retentate (concentrate) of fermented spinach as a result of MF (0.15μm) membraneprocess, (d) broccoli fermented for 6 days, (e) permeate (extract) and (f) retentate (concentrate) of fermented broccoli as a result of MF (0.15μm) membrane process.

Table 1. Composition of fermented broccoli and spinach as feed and as a result of purification via 0.15 μm MF membrane.

Types of Material	Composition				
	Total Solid (%)	Total Polyphenol(%)	Folic Acid (μg/mL)	Reducing Sugar (mg/mL)	Total Acid (%)
Fermented broccoli (6 days)/Feed	5,60	0,05	43,10	96,51	0,10
Fermented broccoli concentrate (**)	5,76	0,06	28,10	66,09	0,93
Fermented broccoli permeate	3,60	0,07	3,78	43,38	0,82
Fermented spinach (3 days)/Feed	7,90	0,07	54,83	86,30	0,58
Fermented spinach concentrate (**)	6,29	0.25	58,90	105,34	0,57
Fermented spinach permeate	8,25	0,08	7,14	116,43	0,47

3.2 Effect of drying on composition of fermented broccoli and spinach

Drying process at vacuum 22 cm Hg and 30 °C for 24 hours by vacuum dryer produced both fermented broccoli and spinach with different composition, as illustrated in Figure 4. Drying process increased total solids concentration on both fermented broccoli and spinach concentrates as a result of MF membrane process from 5.76 % to 94.37 % (92.83 %) and from 6.28 % to 96.43 % (93.48 %), respectively. Total solids concentration are not different compared with recovery of total solids from biomass without MF membrane process, namely 93.55 % and 96.91 %. Total solids of biomass are accumulation of the whole solids both soluble and insoluble [13].

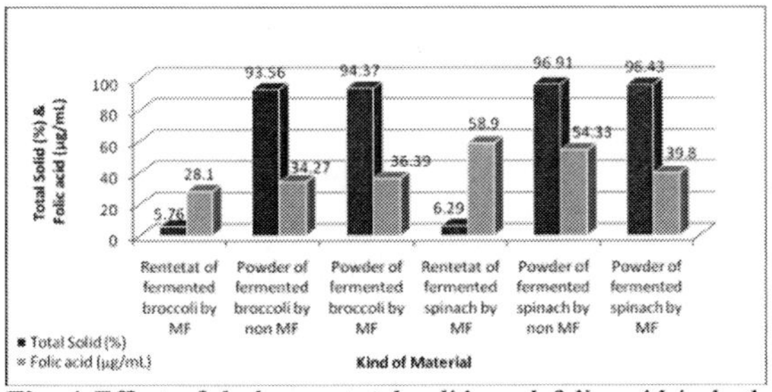

Fig. 4 Effect of drying on total solids and folic acid in both fermented broccoli and spinach with and without through MF membrane process

Different pattern observes on folic acid content, in which the concentrate of fermented broccoli produced via MF membrane pressure increase from 28.1 µg/mL to 36.4 µg/mL (29.54 %), higher than recovery of folic acid in powder from biomass without MF membrane processes (34.27 µg/mL). Whereas, in biomass of fermented spinach, folic acid decrease from 58.1 µg/mL to 39.8 µg/mL (31.5 %), and become lower than recovery of folic acid in powder from biomass without MF membrane processes (54.339 µg/mL). Fermented spinach possibly contain more than fermented broccoli in which same drying process condition gave different evaporation rate. At the same time, folic acid is integrated with other components. Folic acid was not stable, particularly when being exposed to heating, light and mechanic activation [17].

Drying process also increased organic acids as total acids, but decreased total polyphenol both fermented broccoli and fermented spinach, as represented in Figure 5. Increasing total acids concentration on fermented broccoli and fermented concentrate powders processed via MF membrane were from 0.93 % to 4.67 % (402.15 % or 4 folds) and from 0.57 % to 2.33 % (308.77 % or 3 folds). When compared to total acids from biomass without MF membrane process, these total acids concentration are similar with concentration of total acids in biomass concentrate powder without MF membrane, namely 4.67 % and 2.33 %. This is possibly caused by organic acids linked strong on solutes so that raising total solids would increase total acids concentration, as well. Total acids are all organic acids produced as metabolite by invertase enzyme activity of microbes, in which sugars are fermented to form organic acids and further to form alcohol [3.4]. Organic acids contribute fresh aroma in biomass like as fruit aroma.

Drying process increased reducing sugars for both concentrates of fermented broccoli and spinach as a result of MF membrane processes from 66.09 mg/mL to 412.18 mg/mL (523.66 % or 5.23 folds) and from 105.34 mg/mL to 545.59 mg/mL (417.93 % or 4.18 folds). Reducing sugars in fermented broccoli was lower than reducing sugar in powder from biomass without MF membrane process (537.60 mg/mL), however it was higher than in powder of biomass without MF membrane process (441.39 mg/mL).

Drying process also showed higher concentration of total polyphenol in fermented spinach powder (0.05 %) than in fermented broccoli powder (0.04 %). Based on the results of overall processes, it shown that fermented spinach produced better total polyphenol than fermented broccoli both with and without MF membrane process. Total polyphenols are all polyphenols compound and their derivatives from both vegetables type, in which their concentrations would decrease with longer fermentation time [9]. A reduction of total polyphenols were caused by heat conductive during drying process, as well.

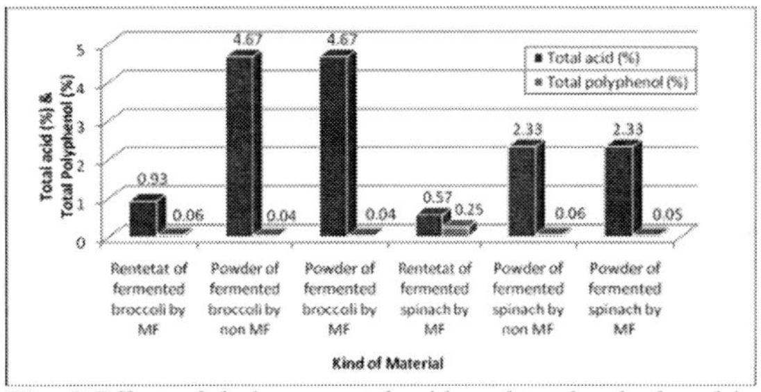

Fig. 5 Effect of drying on total acids and total polyphenol in both fermented broccoli and spinach with and without through MF membrane process.

Drying process also raise reducing sugars from both types of biomass, as demonstrated in Figure 6. Reducing sugar is all of sugars as a result of fermentation, which has reduction property as metabolite.

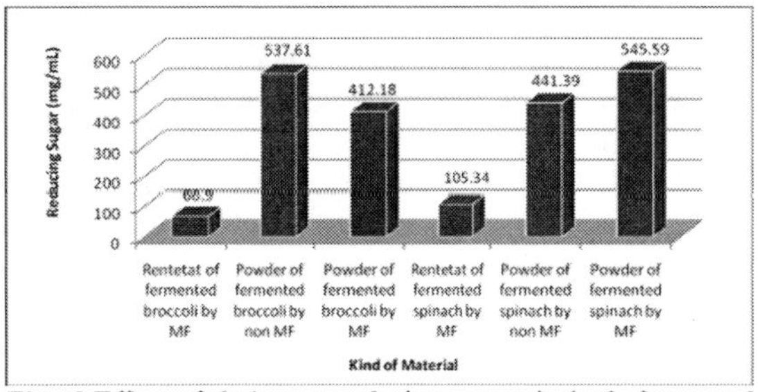

Fig. 6 Effect of drying on reducing sugars in both fermented broccoli and spinach with and without through MF membrane process.

Kombucha culture produce glucose as source of carbon to metabolite, however not of all glucose was fermented by various factors, particularly microbes internal factors and fermentation condition (temperature and time) and substrate type.

Reducing sugars in powder from both concentrates of fermented broccoli and spinach are increased as a result of MF membrane processes from 66.09 mg/mL to 412.18 mg/mL (523.66 % or 5.23 folds) and from 105.34 mg/mL to 545.59 mg/mL (417.93 % or 4.18 folds). Reducing sugars in fermented broccoli was lower than reducing sugar in powder from biomass without MF membrane process (537.60 mg/mL), however it has higher content compared with reducing sugar in powder of biomass without MF membrane process (441.39 mg/mL).

3.3 Characteristics of particles of fermented broccoli and spinach concentrates

Purification process on fermented broccoli and fermented spinach through MF membrane process produced concentrates of biomass with different particles characteristic with biomass without via MF membrane process, as shown subsequently in Figures 7 and Figure 8.

3.3.1 Fermented Broccoli

Particles distribution on biomass of fermented broccoli without MF membrane process in Figure 7a has no particles with diameter size (Ø) ranging from 0.4 µm to 1 µm or < 1 µm (0 %), but there are particles with Ø < 10 µm or between 0.4 µm and 10 µm (47.5 %), particles with Ø < 100 µm or between 0.4 µm and 100 µm (100 %) and particles with Ø < 1000 µm or between 0.4 µm and 1000 µm (100 %) from 100 % of material volume. In other words, 47.5 % of concentrate is particles with Ø between 0.4 µm and 100 µm, and particles with Ø > 100 – 1000 µm or the whole particles were dominated by particle with Ø < 100 µm. Particles distribution in biomass of fermented broccoli concentrate, as shown in Figure 7b from result of MF membrane at stirrer rotation speed 400 rpm, room temperature and pressure 40 psia for 30 minutes indicated subsequently particles with Ø between 0.4 µm and 1 µm or < 1 µm (2.15 %), particles with Ø < 10 µm or between 0.4 and 10 µm (24.1 %), particles with Ø < 100 µm or between 0.4 – 100 µm (49.2 %) and particles with Ø < 1000 µm or between 0.4 – 1000 µm (100 %) from 100 % of material volume. In other words, 75.45 % of concentrate is particles with Ø between 0.4 µm and 100 µm and 24.55 % of remain was particles with Ø > 100 – 1000 µm or the whole particles were dominated by particles with Ø < 100 µm.

Concentrate of fermented spinach produced via MF membrane process in Figure 8b indicated particles with Ø between 0.4 and 1 µm or < 1 µm (5.21 %), particles with Ø < 10 µm or between 0.4 and 10 µm (73.7 %), particles with Ø < 100 µm or between 0.4 and 100 µm (100 %) and particles with Ø < 1000 µm or between 0.4 and 1000 µm (100 %) from 100 % volume of material. In other words, 73.7 % of concentrates was particles with Ø between 0.4 µm and 100 µm, and 26.3 % of concentrate was particles > 100 – 1000 µm or the whole particles were dominated by particles with Ø < 100 µm.

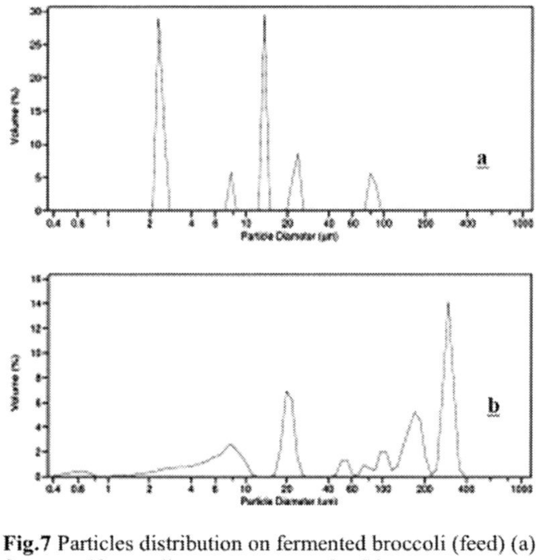

Fig.7 Particles distribution on fermented broccoli (feed) (a) and fermented broccoli concentrate powder (b).

Difference in particles size was possibility caused by contribution of particles size reduction as a consequence of MF membrane process. Its occurrence of interaction amongst particles by driving force at stirrer rotation speed 400 rpm, room temperature and pressure 40 psia for 30 minutes causes a part of particles broken to smaller size, like process of size reduction. Biomass of fermented spinach is a group of particle components and microbes (fungi, yeast, bacteria) which have various particles size. Its composition showed a domination of polyphenol derivatives with molecular weight (MW) between 191.08 Dalton (Da.) and 192.78 Da., and total acids organic as Lactic Acid derivatives with MW from 251.13 Dalton (Da.) to 252.14 Da. or particles size with Ø < 1 µm [9].

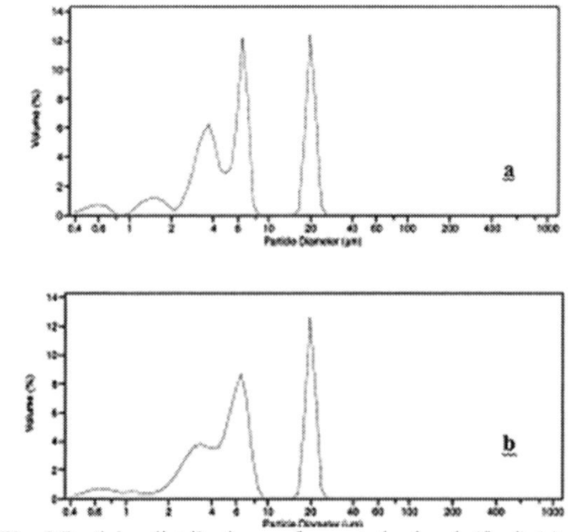

Fig. 8 Particles distribution on fermented spinach (feed) (a) and fermented spinach concentrate (b).

4 Conclusion

DESMFC mode was able to separate and recover folic acid in both fermented broccoli and spinach concentrates 0.65 folds and 1.07 folds, respectively when compared to feed and retain more total solids, folic acid, and reducing sugar in concentrate. Drying process increased total solids, folic acid, total acids and reducing sugar, but decreased total polyphenol in both fermented broccoli and spinach concentrate powders produced via MF membrane process. Based on recovery of folic acid, optimization of MF membrane and drying processes was achieved by type of fermented spinach and gave concentrate and powder with compositions of total solids 6.29 % and 96.91 %, total polyphenol 0.25 % and 0.06 %, folic acid 58.8 μg/mL and 54.33 μg/mL, reducing sugar 105.34 mg/mL and 441.39 mg/mL, and total acids 0.57 % and 2.33 %, respectively. While, fermented broccoli gave concentrate and powder with compositions of total solids 5.76 % and 94.37 %, total polyphenol 0.06 % and 0.04 %, folic acid 28.1 μg/mL and 36.39 μg/mL, reducing sugar 66.1 mg/mL and 412.18 mg/mL, and total acids 0.93 % and 4.67 %, respectively. Concentrates of fermented broccoli and spinach gave particles distribution with diameter size (Ø) between 0.4 and 100 μm of 75.45 % and 73.7 %, and particles distribution with Ø between 100 – 1000 μm of 24.55 % and 26.3 %, respectively. Whereas, biomass of fermented broccoli and spinach without MF membrane process gave particles distribution with Ø between 0.4 – 100 μm of 47.5 % and 74,1 %, and particles distribution with Ø between 100 – 1000 μm of 52.5 % and 25.9 %, respectively.

References

[1] Anonymus. 2012. Dietary Supplement Fact Sheet : Folate, Office of Dietary Supplements, National Institutes of Health. http://ods.od.nih.gov/factsheets/folate.asp, Accessed 8 December 2014.
[2] Susilowati, A., Aspiyanto, Maryati, Y., and Melanie, H. 2016 Research/Jurnal Penelitian Teh dan Kina. Vol. **19**, No 1, pp. 67 – 84, P-ISSN 1410-6507, E-ISSN 2527-2942, 544/AU2/P2MI-LIPI/06/2013.
[3] Jayabalan, R., Subathradevi, S., Marimuthu, M., Sathshkumar, and Swaminathan, K. 2008. Changes in free-radical scavenging ability of kombucha tea during fermentation, Food Chem., 2008, 109 (1) 227 – 234.
[4] Dufresne, C. and Farnworth, E. 2000. Tea, kombucha, and health : a review, Food Res. Int., 33 (6), 409 – 421.
[5] De Crécy-Lagard V., El Yacoubi B., de la Garza R.D., Noiriel A., and Hanson A. D. Comparative genomics of bacterial and plant folate synthesis and salvage : predictions and validations, BMC Genomics. 2007, 8 : 245.
[6] Hauser, P. M. and Macreadie, I. G. 2006. Isolation of the Pneumonic carinii dihydrofolate gene and functional compllementation in Saccharomyces cerevisiae, Federation of European Microbiological Societies Lett 256 (2006) 244 – 250, Published by Blackwell Publishing Ltd.
[7] Malbasa et al. 2008. Comparison of the product of kombucha fermentation on sucrose and molasses. Food Chem. 106 (3), 1039 – 1045.
[8] Quilivan, E. P., S. Roje, G. Basset, Y. Shahar-Hill, J. F. Gregory, and A. D. Hanson. 2003. The Folate precursor ρ-aminobenzoate is reversible converted to is glucose ester in the plant cytosol. J. Bio. Chem. 278 :20731 – 20737.
[9] Susilowati, A., Iskandar, Y. M., Melanie, H., Maryati, Y., Lotulung, P. D., Aryani, D. G. 2015. Pengembangan Konsentrat Sayuran Hijau dan Kacang-kacangan Terfermentasi pada jagung (Zea mays L.) pramasak sebagai sumber asam Folat untuk formula pangan pintar. Laporan Hasil Penelitian, Program Tematik, Kedeputian IPT, Tahun Anggaran 2015, Pusat Penelitian Kimia - LIPI, PUSPIPTEK, Serpong, Tangerang Selatan.
[10] Mulder, M. H. V. 1996. Basic Principles of Membrane Technology, Kluwer Academic Publishers, Dordecht, The Nederlands.
[11] Cheryan, M. 1992. Membrane Technology in Food Bioprocessing, In : R. P. Singh and M. A. Wirakartakusumah, Advances in Food Engineering, CRC Press Inc., Boca Ratan, Florida.
[12] Anonymous. 2000. Catalogue and Manual of Stirred Ultrafiltration Cells, Amicon Bioseparations, MILLIPORE, U.S.A.
[13] Association of Official Analytical Chemists. 1995. Official Methods of Analysis, Washington, D. C.
[14] Liu, Z. 2006. New techniques for tea catechins extraction, in : International Training Workshop of Tea Science, Hunan Agricultural University. China, 21 July – 10 August.
[15] Ruengsitagoon, W. and Hattanat, N. 2012. Simple spectrophotometric method for determination of folic acid, The 4th Annual Northeast Pharmacy Research Conference, Thailand.
[16] Keith Scott. 1998. Handbook of Industrial Membranes, 2nd. Edition, Elsevier Advanced Technology, Oxford, 763, 766, 767 & 768.
[17] Gilles J. C., Basset, Eoin P., Quinlivan, Jesse F. Gregory III and Andrew D. Hanson. 2005. Folate Synthesis and Metabolism in Plants and Prospects for Biofortification, CROP SCIENCE. Vol **45** : 449 – 453, Crop Science Society of America 677 S. Seoe Rd. Madison. WI 53711 USA. Accessed 18 December 2015.

Ammonium hydroxide addition and its influence on the catalytic activities of Pt-based catalysts for methane oxidation

Mardwita M[1,*], and *M. Djoni* Bustan[2]

[1]Chemical Engineering Department, Faculty of Engineering, Muhammadiyah University of Palembang, 30252 Palembang, , Indonesia
[2]Chemical Engineering Department, Faculty of Engineering Universitas Sriwijaya , 30662 Indralaya , Indonesia

Abstract. A series of binary chromium-platinum catalysts were prepared and the catalytic activity was tested in methane oxidation. The effects of aqueous ammonium hydroxide addition into binary catalyst chromium-platinum catalysts were investigated. All the catalysts were prepared by impregnation method. The reactant gases were consisted of methane-oxygen-argon. The reaction was conducted at an atmospheric pressure and the temperature was ranged from 573 to 773 K. The products were analyzed using thermal conductivity detector. Under this experiment conditions, the only products were carbon dioxide and water. The catalysts were characterized using carbon monoxide chemisorptions and x-ray photoelectron spectrometer. The catalytic activity results showed that the catalyst with the addition of aqueous ammonium hydroxide has smaller platinum metal particle size and its catalytic activity is the highest compared to the other catalysts. X-ray photoelectron spectrometer recorded higher intensities and clear peaks on high pH catalysts indicating a change of interaction between chromium and platinum on alumina. However, active state of chromium was not observed on catalyst surface.

1 Introduction

Methane oxidation or methane combustion is a method to reduce methane gas from exhaust gas engine. On the other hand, methane oxidation process may generate some energy [1, 2]. Catalysts play an important role in catalytic reaction, including in oxidation. Noble metal catalysts, such as platinum (Pt) and palladium (Pd), are well-known because of its reactivity, stability, resistance to poison and thermal [3-5]. However, these noble metal catalysts are expensive, therefore it is impossible to use only noble metal in a catalyst system [6]. An in-expensive and low loading of noble metal catalysts poses high catalytic activity is needed, hence another metal from transition metal are usually added as an additive or as a second metal in noble metal catalysts. Transition metals, such as manganese (Mn), cobalt (Co), iron (Fe), nickel (Ni), copper (Cu), and chromium (Cr), has been widely used as an additive in a catalyst system [7-9]. Binary catalyst systems, such as chromium-copper-titanium (Cr-Cu-Ti) oxide catalyst has been reported as a catalyst that has more sulphur resistance compared to platinum-palladium (Pt-Pd) catalyst in combustion reaction [10]. Another bimetallic catalyst such as $La_{3.5}Ru_{4.0}O_{13}$, Pt-Sn, Pt-Ce has been reported to show high catalytic performance [11, 12] and its effect to the catalytic performance of Pt catalyst has been investigated [8]. Among the other transition metal oxide catalyst, supported chromium metal oxide has been found as the most active catalyst in methane combustion [13, 14]. Kuznetsova et al has been studied on the chromium for hydrocarbon destruction and concluded that the Cr(IV) cations as the main active component [14] and showed the relation between support basicity and the formation of Cr(IV). Though, the mechanism of activated C-H bond on the surface of catalyst is not well-understood yet [15].

To obtain an effective and reactive catalyst in low temperature, the preparation of catalyst plays an important role. The concentration of additive metals, reducing and oxidizing conditions, the pH solutions will affect to the performance of catalyst. It must be considered that the interaction between main metal and additive occurs during the preparation is an important thing. As best of our reviews, there are no studies on catalytic activity of platinum and chromium by the addition of aqueous ammonium hydroxide (NH_4OH). Therefore, in this study we investigated the catalytic performance of platinum and chromium catalysts and also the effect of ammonium hydroxide addition during the catalyst preparation.

2 Experiment details

2.1 Catalyst preparation

A supported platinum (Pt/Al_2O_3) catalyst was prepared by impregnation method. Gamma-alumina (γ-Al_2O_3) was used as the support. Required amount of H_2PtCl_6 solution, as the source of Pt metal, was impregnated into γ-Al_2O_3, the solution was left unstirred at room

* Corresponding author: wiwitdiita@gmail.com

temperature then it was stirred slowly. The water was removed by stirring and dried-up on a hotplate at 383 K, the resulted solid was kept overnight in an oven at 333 K, finally was calcined in air at 673 K for 3 hours. The loading of Pt was 2.0 wt.% and the catalyst was denoted as $2.0Pt/Al_2O_3$.

The same procedure as described above was also applied to prepare supported chromium (Cr/Al_2O_3) catalyst in loading of 6.0 wt.%, 2.0 wt.%, and 0.5 wt.% Cr. An aqueous solution of chromium nitrate ($Cr(NO_3)_3$) was used as the source of Cr metal. Final catalysts were denoted as $6.0Cr/Al_2O_3$, $2.0Cr/Al_2O_3$, and $0.5Cr/Al_2O_3$.

Three samples of binary Cr-Pt catalysts containing 6.0 wt.% Cr, 2.0 wt.% Cr, and 0.5 wt.% Cr were prepared. The loading of Pt metal was kept constant at 2.0 wt.%. These catalysts were prepared by stepwise-impregnation method. First, binary $Cr/Pt/Al_2O_3$ catalyst was prepared by impregnating aqueous solution of $Cr(NO_3)_3$ into $2.0Pt/Al_2O_3$ catalyst. The pH solution was 4. The catalyst was prepared using the same drying and calcining process as described in preparation of $2.0Pt/Al_2O_3$ catalyst. The resulted catalysts were denoted as 6.0 Cr /2.0 Pt/ Al_2O_3, 2.0 Cr /2.0 Pt / Al_2O_3, $0.5Cr/2.0Pt/Al_2O_3$ catalyst.

Catalysts with addition of ammonium hydroxide (NH_4OH) were prepared using the same preparation method for $Cr/Pt/Al_2O_3$ catalyst described above. The aqueous solution of $Cr(NO_3)_3$ was added into $2.0Pt/Al_2O_3$ catalyst and the pH of solution was adjusted to ca. 10 by the addition of NH_4OH drop by drop during the preparation. The other catalyst was prepared by adjusting the pH to ca. 11. All the catalysts contained 6.0 wt.% loading of Cr and 2.0 wt.% loading of Pt. The catalysts were denoted as $6.0Cr/2.0Pt/Al_2O_3$ (pH 10) and $6.0Cr/2.0Pt/Al_2O_3$ (pH 11) catalyst.

2.2 Catalyst test

Before catalytic reaction, the catalyst was pretreated by calcinations in a stream of air at 773 K for 1 hour to remove impurities and to activate the metal sites, and then the catalyst was cooled into room temperature. The methane oxidation was carried out at atmospheric pressure. The micro-reactor used in this experiment was made of Pyrex glass. The reactor contained quartz wool onto which 0.25 grams catalyst was added and externally heated by a furnace. The total flow rate was closed to 26 ml/min and the fed gases consisting of methane (CH_4), oxygen (O_2), and argon (Ar) with volumetric composition 1%:5%:20% was introduced into the reactor. The reaction temperature was ranged from 573 K to 773 k and measured by a thermocouple inserted in the catalyst bed. All the products, except water were analyzed by a thermal conductivity detector (TCD). The only products were carbon dioxide (CO_2) and water (H_2O). Carbon monoxide (CO) was not detected under this experimental conditions and the conversion was calculated based on the hydrocarbon consumptions.

2.3 Catalysts characterization

The Pt particle size in catalysts was determined using CO chemisorptions apparatus on a BEL-CAT (BEL JAPAN INC). The catalyst in weight of 0.05 grams was loaded into U-tube reactor. In the apparatus, the catalyst was treated by flowing helium (He) gas at 323 K to remove impurities. Then, the temperature was increased to 573 K in flow of hydrogen (H_2) and reduced for 1 hour, finally cooled to 323 K in He flow. CO/He gas was purged into the reactor and TCD will detect the effluent gas. The decrease in peak area of CO due to its adsorption on the catalyst was used to calculate the amounts of CO adsorbed on the catalyst's surface.

X-ray photoelectron spectrometer (XPS) Shimadzu, ESCA-3400 apparatus using Mg Kα (1253.6 eV) radiation was conducted on stepwise impregnated catalyst $Cr/Pt/Al_2O_3$ to analyze the catalyst's morphology and metal state. The XPS was measured in calcined catalysts. Powder sample of catalyst was poured on the double sticky carbon on the sample's place. This sample was kept in vacuum for one day before analysis. The binding energies of catalysts were measured with accuracy of ± 0.1 eV and used C 1s at 284.3 eV as standard.

3 Results and discussion

In order to find the optimal composition of catalyst, the content of Cr loading was investigated on 0.5 wt.%, 2.0 wt.%, and 6.0 wt.%. Figure 1 shows comparison of methane oxidation over $0.5Cr/Al_2O_3$, $2.0Cr/Al_2O_3$, and $6.0Cr/Al_2O_3$ catalyst. The results showed that $0.5Cr/Al_2O_3$ catalyst has no catalytic activity under this reaction condition, meanwhile $6.0Cr/Al_2O_3$ catalyst shows the highest conversion, however a very small amount of CO was detected in products, indicates that supported Cr catalyst leads to the reaction of partial oxidation.

In binary Cr/Pt catalyst, the alumina was chosen as the support because it forms the support-metal-interaction with the Pt metal, this can give another advantages such as to prevent the Pt to form PtO metal during the catalytic reaction. The catalytic activities of Pt/Al_2O_3 and $Cr/Pt/Al_2O_3$ catalysts are shown in figure 2. It is observed that all the catalysts started the complete methane oxidation at 673 K except for $0.5Cr/Al_2O_3$ catalyst which started the catalytic activity at 723 K. Though $6.0Cr/2.0Pt/Al_2O_3$ and $2.0Cr/2.0Pt/Al_2O_3$ catalyst started at the same temperature, their conversions are far different. Besides that, we found that the addition of 2.0 wt.% Cr into supported Pt does not increase the activity so much, even the catalytic activity just little bit higher than $2.0Pt/Al_2O_3$ catalyst. The catalytic activity of $0.5Cr/2.0Pt/Al_2O_3$ and $2.0Cr/2.0Pt/Al_2O_3$ catalyst was mainly due to the Pt metal, as it has been well-known that Pt metal is an active catalyst than transition metal oxide catalyst. Complete methane oxidation which was conducted on $0.5Cr/Al_2O_3$ catalyst shows no catalytic activity, suggests that no active phase was resulted by the calcinations process on $0.5Cr/Al_2O_3$ catalyst due to its low concentration. The catalyst just existed as un-active phase, when this deposited onto $2.0Pt/Al_2O_3$ catalyst, this un-active phase covered the active site of Pt hence reduces in the catalytic performance. While in $2.0Cr/2.0Pt/Al_2O_3$ catalyst, low aggregation of Cr and Pt, exhibit some activity. Interestingly, the catalytic activity increase dramatically when 6.0 wt.% Cr was deposited onto $2.0Pt/Al_2O_3$ catalyst, it seems high aggregation and interaction between these two metals was occurred during the catalyst preparation. The activities of catalysts decrease as follows: $6.0Cr/2.0Pt/Al_2O_3$ > $2.0Cr/2.0Pt/Al_2O_3$ > $0.5Cr/2.0Pt/Al_2O_3$.

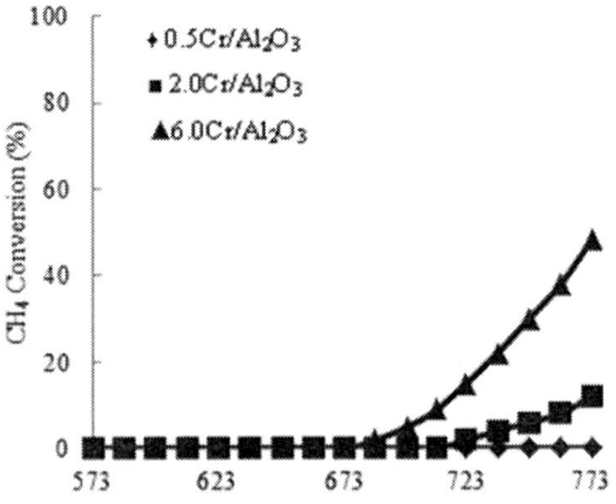

Fig. 1. Methane conversion as a function of temperature for various loading Cr/Al_2O_3 catalysts.

The catalytic activity of physical mixture catalyst ($6.0Cr/Al_2O_3$ + $2.0Pt/Al_2O_3$) showed lower catalytic activity compared to binary $6.0Cr/2.0Pt/Al_2O_3$ catalyst. This strongly suggests that on binary Cr-Pt catalyst, the Pt and Cr metal existed in interaction than existed separately. More than 50% conversion of methane was given by $6.0Cr/2.0Pt/Al_2O_3$ catalyst at 723 K, while at the same temperature the physical mixture catalyst gave 30% conversions. The results are summarized in figure 3.

The $6.0Cr/2.0Pt/Al_2O_3$ (pH 4) catalyst was prepared by stepwise impregnation method and followed the general manner. On the other hand, $6.0Cr/2.0Pt/Al_2O_3$ (pH 10) and $6.0Cr/2.0Pt/Al_2O_3$ (pH 11) catalysts were prepared by adjusting the pH of impregnated solution with addition of NH_4OH.

Figure 4 shows all the catalysts started their catalytic activities at 623 K and increase with the increases of temperatures. However, the catalytic activities of basic catalysts are relatively higher than acid catalyst. The $6.0Cr/2.0Pt/Al_2O_3$ (pH 11) showed the highest conversions among the others. It could be concluded that the catalyst prepared in basic solution is easier to oxidize the reactant gas than the acid catalyst.

Some authors have investigated the adsorption of bases on Al_2O_3 surface, they concluded that due to the addition of a base, hydroxyl groups on Al_2O_3 surface are removed and replaced by the base groups. This condition may decrease the ability of metal to adsorb on the surface of Al_2O_3 [16-17].

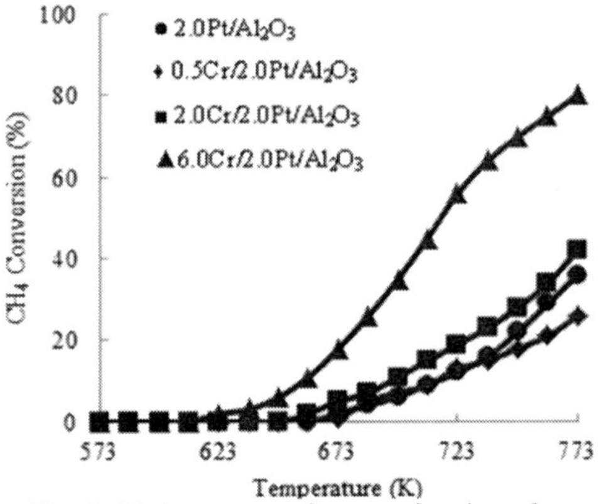

Fig. 2. Methane conversion as a function of temperature for $Cr/Pt/Al_2O_3$ catalysts.

In our experiment, it cannot be calculated the amount of Cr adsorbed on the surface of alumina or electronically adsorbed on the surface of Pt. Based on the catalytic activity results, it can be assumed that during the catalyst preparation, NH_3 cation is adsorbed on the surface of alumina and also interacted with the chromium nitrate. It is expected that the formation of chromium ionic is higher in the presence of NH_4OH, thus increase the interaction between Pt and Cr in catalyst. This interaction gave a synergistic effect that increases the catalytic activity.

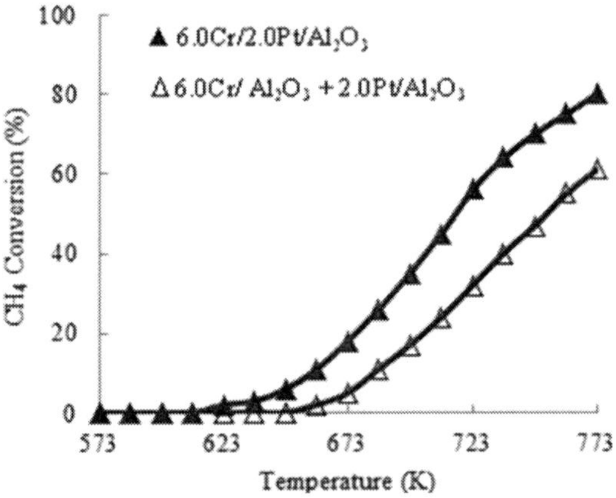

Fig. 3. Comparison of methane conversion over $6.0Cr/Al_2O_3$ + $2.0Pt/Al_2O_3$ catalyst and $6.0Cr/2.0Pt/Al_2O_3$ catalyst.

The Pt particle sizes are shown in table 1. The Pt particle sizes calculated by CO chemisorptions are 2.2, 1.8, and 1.5 nm for $6.0Cr/2.0Pt/Al_2O_3$ (pH 4), $6.0Cr/2.0Pt/Al_2O_3$ (pH 10), and $6.0Cr/2.0Pt/Al_2O_3$ (pH 11), respectively. Somehow, the Pt particle size is smaller in high pH than in low pH, suggests that the addition of NH_4OH also change the Pt dispersion on support.

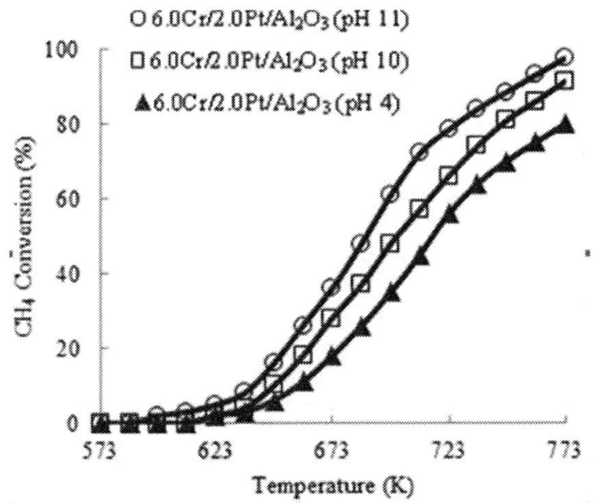

Fig. 4. Methane conversion as a function of temperature for $6.0Cr/2.0Pt/Al_2O_3$ (pH 4), $6.0Cr/2.0Pt/Al_2O_3$ (pH 10), $6.0Cr/2.0Pt/Al_2O_3$ (pH 11) catalysts.

As shown in figure 4, at 723 K the conversion of methane was 56% on $6.0Cr/2.0Pt/Al_2O_3$ catalyst, meanwhile $6.0Cr/2.0Pt/Al_2O_3$ (pH 10) and $6.0Cr/2.0Pt/Al_2O_3$ (pH 11) gave 66% and 79%, respectively. If no synergistic effect between Cr and Pt, the catalytic of these catalysts should be the same. The catalytic activities decrease as follows: $6.0Cr/2.0Pt/Al_2O_3$ (pH 11) > $6.0Cr/2.0Pt/Al_2O_3$ (pH 10) > $6.0Cr/2.0Pt/Al_2O_3$ (pH 4).

Table 1. CO chemisorptions results.

Catalyst	Pt size (nm)	T 723 K (CH4 conv, %)
$6.0Cr/2.0Pt/Al_2O_3$ (pH 4)	2.5	56
$6.0Cr/2.0Pt/Al_2O_3$ (pH 10)	1.8	66
$6.0Cr/2.0Pt/Al_2O_3$ (pH 11)	1.5	79

The binding energy (eV) was recorded on XPS using Mg Kα (1253.6 eV) radiation at 10 kV and 10 mA. The XPS was measured on calcined catalysts. The binding energies of catalysts were measured with accuracy of ± 0.1 eV and used C 1s at 284.3 eV as standard. It has been known that the interaction of Pt-alumina will results a strong metal surface interaction. Due to this condition, the peak of Pt is difficult to obtain using XPS spectra. The catalysts shown an overlaps between Pt 4f with main peak of Al 2p at 74.6 eV, on the other hand, the peak of oxygen and aluminum oxides is also appeared simultaneously. In order to observe a clear peak for the presence of Cr 2p on catalysts, the XPS was conducted on curve-fitting area of Cr in range of 555-595 eV.

The XPS spectra are showed in figure 5. For $6.0Cr/2.0Pt/Al_2O_3$ (pH 4) catalyst, somewhat a broad peak was obtained on the surface of this catalyst after 500 sweeps. On this broad peak, a low peak is located at binding energies 576.7 eV, this binding energies indicates the presence of Cr 2p3/2. Our results agree well with numbers of researchers who have determined the position of Cr 2p on supported alumina [18, 19]. The XPS spectra on curve-fitting area of Cr for $6.0Cr/2.0Pt/Al_2O_3$ (pH 10) and $6.0Cr/2.0Pt/Al_2O_3$ (pH 11) catalyst recorded higher intensities and clear peaks, the catalyst's peaks are shifted to the left compared with $6.0Cr/2.0Pt/Al_2O_3$ (pH 4) catalyst. This indicates that in high pH catalyst, more Cr-Pt metals exist as co-existence than presence separately. The highest peak's binding energy for $6.0Cr/2.0Pt/Al_2O_3$ (pH 10) catalyst is 578.3 eV, this binding energy is higher than the binding energy for Cr 2p3/2 Cr_2O_3 and closer to the binding energy for Cr 2p3/2 CrO_3. The same peak's binding energy is also recorded on $6.0Cr/2.0Pt/Al_2O_3$ (pH 11) catalyst. The synergistic effect of Cr and Pt could be enhanced by the addition of NH_4OH.

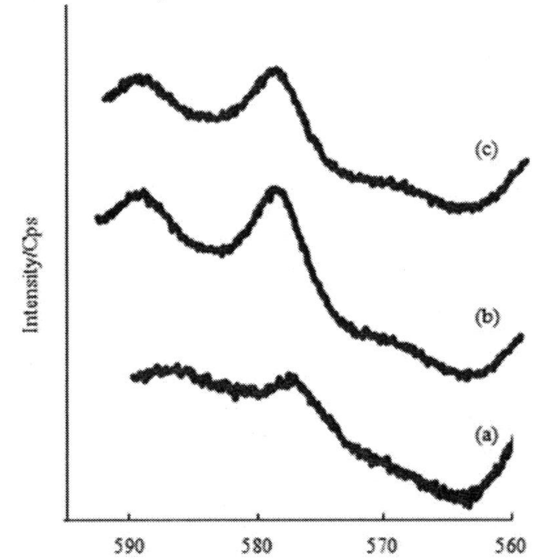

Fig. 5. Xps spectra Cr 2p for (a) $6.0Cr/2.0Pt/Al_2O_3$ (pH 4), (b) $6.0Cr/2.0Pt/Al_2O_3$ (pH 10), (c) $6.0Cr/2.0Pt/Al_2O_3$ (pH 11) catalysts.

4 Conclusions

We investigated the catalytic activity of binary chromium-platinum catalysts. The oxidation of methane was dependent on chromium loading, Pt particle size, and pH of solution. In this experiment, the pH of solution of binary catalysts was adjusted to ca. 10 and ca. 11by the addition of NH_4OH drop by drop during the catalysts preparation. The catalytic activity results showed that $6.0Cr/2.0Pt/Al_2O_3$ (pH 11) catalyst has the highest catalytic activity and CO chemisorptions results showed that this catalyst has the smallest Pt particle size compared to other catalysts. The addition of NH_4OH affected the Pt particle size. The XPS spectra showed higher intensities and clear peak of binding energy for $6.0Cr/2.0Pt/Al_2O_3$ (pH 11) catalyst, indicates more aggregation between $Cr-Pt-Al_2O_3$ occurred. However, the formation of Cr_2O_3 was not observed on the XPS results.

References

[1] P. Forzatti, G. Groppi, Catal. Today **54**, 165-180 (1999)

[2] J. H. Lee, D. L. Trimm, Fuel processing technology **42**, 339-359 (1995)

[3] T. V. Choudhary, S. Banerjee, V. R. Choudhary, Appl. Catal. A : General **234**, 1-23 (2002)

[4] M. Niwa, K. Awano, Y. Murakami, Appl. Catal. **7**, 317-325 (1983)

[5] R. B. Anderson, K. C. Stein, J. J. Feenan, L. J. E. Hofer, Ind. Eng. Chem. **53**, 809-812 (1961)

[6] Janbey, W. Clark, E. Noordally, S. Grimes, S. Tahir, Chemosphere **52**, 1041-1046 (2003)

[7] D. Roth, P. Gelin, A. Kaddouri, E. Garbowski, M. Primet, E. Tena, Catal. Today **112**, 134-138 (2006)

[8] S. H. Oh, P. J. Mitchell, R. M. Siewert, J. Catal. **132**, 287-301 (1991)

[9] S. Amone, G. Bagnasco, G. Busca, L. Lisi, G. Russo, M, Turco, Studies in Surf. Sci. and Catal. **119**, 65-70 (1998)

[10] P. Hurtado, S. Ordanez, A. Vega, F. V. Diez, Chemosphere **55**, 618-689 (2004)

[11] G. Corro, O. Vazquez-Cuchillo, F. Banuelos, A. Cruz-Lopez, J. L. G. Fierro, J. Ceramic Processing Research **9**, 615-661 (2008)

[12] M.J. Tiernan, O.E. Finlayson, Appl. Catal. B: Environmental **19**, 23-25 (1998)

[13] S. Khairulin, B. Beguin, E. Garbowski, M. Primet, J. Chem. Soc. M Faraday Trans **93**, 2217-2223 (1997)

[14] L. L.Kuznetsova, E. A. Paukshtis, C. P. Shkurina, R. A. Shkrabina, N.A. Koryabkina, D.A. Arendarskii, G.B. Barannik, Z.R. Ismagilov, Catal. Today **17**, 209-216 (1993)

[15] R. Burch, D. J. Crittle, M. J. Hayes, Catal. Today **47**, 229-234 (1999)

[16] J. Ryczkowski, W. Grzegorczyk, D. Nazimek, Appl. Catal. A : General **126**, 341-349 (1995)

[17] J. Medema, J. J. G. M. Van Bokhoven, A. E. T. Kuiper, J. Catal. **25**, 238-244 (1972)

[18] S. Engels, H. Lausch, Appl. Catal. **55**, 93-107 (1989)

[19] P. W. Park, J. S. Ledford, Langmuir **13**, 2726-2730 (1997)

New method of thermal cycling stability test of phase change material

Nandy Putra[1,*], *Muhammad* Amin, *Rizky* Achmad Luanto, *Engkos* A. Kosasih, *Nasruddin* A. Abdullah

Applied Heat Transfer Research Group Department of Mechanical Engineering Universitas Indonesia, 16424 Depok, Indonesia.

Abstract. Phase Change Material (PCM) is the most promising material as thermal energy storage nowadays. As thermal energy storage, examination on endurance of material for long-term use is necessary to be carried out. Therefore, thermal cycling test is performed to ensure thermal stability of PCM. This study have found a new method on thermal cycling test of PCM sample by using thermoelectric as heating and cooling element. RT 22 HC was used as PCM sample on this thermal cycling test. The new method had many advantages compared to some references of the same test. It just needed a small container for PCM sample. The thermoelectric could release heat to PCM sample and absorb heat from PCM sample uniformly, respectively, was called as heating and cooling process. Hence, thermoelectric had to be supported by a relay control device to change its polarity so it could heat and cool PCM sample alternately and automatically. On the other hand, the thermoelectric was cheap, easy to be found and available in markets. It can be concluded that new method of thermal cycling test by using thermoelectric as source of heating and cooling can be a new reference for performing thermal cycling test on PCM.

1 Introduction

Phase Change Materials (PCMs) are materials which are mostly tested as thermal energy storage nowadays [1]. One of the advantages of PCM which have a high capacity in storing heat/cold because the contains the latent heat [2]. Meanwhile PCMs have also disadvantage, which has a low thermal conductivity [3]. In order to increase its themal conductivity, some researcher are adding nanoparticles into PCM [4, 5]. The addition of nanoparticles can rise the rate of process in absorption and distribution of heat [6]. Unfortunatly, it can reduce the latent heat [7]. Therefore, thermal engineering is needed in process of mixing nanoparticles into PCM without reducing heat capacity of PCM. Mixing nanoparticle with PCM has been conducted by many researchers to increase latent heat [8, 9] and heat capacity [10].

If heat stored in PCM is used continuously (heating and cooling), the latent heat will decrease in certain time [11, 12]. Hence, to find out the ability of PCM in storing heat for a long period, thermal cycling test is needed to performe. The objective of this test is to know the stability and realibity of PCM for long-term use [13]. The use of PCM in long-term is expected to be proper economically.

The process of thermal cycling test requires a particular measuring instrument that is well designed. The design of thermal cycling test has no standard, it just depends on the researchers themselves [14]. The test is expected that it can be operated continuously up to 1000 cycles. One cycle is equal to one process of heating and one process of cooling.

Basically, thermal cycling test utilizes heating and cooling element to release and absorb heat so it can be run continuously. Thermoelectric module that can provide hot and cold side can be applied as a new method as thermal cycling test of PCM. Thermoelectric module works as Peltier principle of heating and cooling [15, 16]. The advantages of thermoelectric are simple, small dimension, light, and eco-friendly [17], able to heat and cool PCM in temperature ranges between 0 and 100 °C.

The study about thermal cycling test were applied in different ways by some researchers. Sharma et al used hot plate as heating element for PCM, then allowed to cool at room temperature [11, 12, 18]. Shukla A et al utilized an oven that can be set at constant temperature to heat PCM. Then, it was taken out from the oven and allowed to cool at ambient temperature [19]. Sari A conducted thermal cycling test by using a thermostatic bath. It consisted of a heater which was used to heat PCM. For cooling process, heating controller was switched off and PCM was allowed to cool at ambient temperature [20]. Sharma A and Shukla A used two chambers that were filled by water, both of them was used, respectively, as source of heating and cooling for PCM [21]. Silakhori M et al utilized stainless steel heater which was sticked on a rectangular box that made of copper and filled by PCM sample to be heated. A heat pipe heat sink with fan was also sticked on that box to help the cooling process [22].

* Corresponding author: prayudi@sttpln.ac.id

Based on some references, that thermal cycling test with thermoelectrics is not found. The objective of this study is to design a new method of thermal cycling test on PCM. This new method is predicted to have many advantages compared to recent method of some researchers.

2 Methodology

2.1 Design Concept of Thermal Cycling Test Device

In this research proposed a new thermal cycling test device that has been designed and shown in Fig 1. This new concept is that applied two thermoeletrics as source of heating and cooling. In order to have heating and cooling alternately to the PCM sample, this new design needs polarity switch to change the polarity of output voltage. A programmable DC power supply is connected to the device to trigger automatically. The others support on this device consist of 2 DC fans, heat sink, 2 clamps, 4 bolts, 4 nuts, 8 rings, DC adaptor. Fig 2 shows the overall layout of thermal cycling test device.

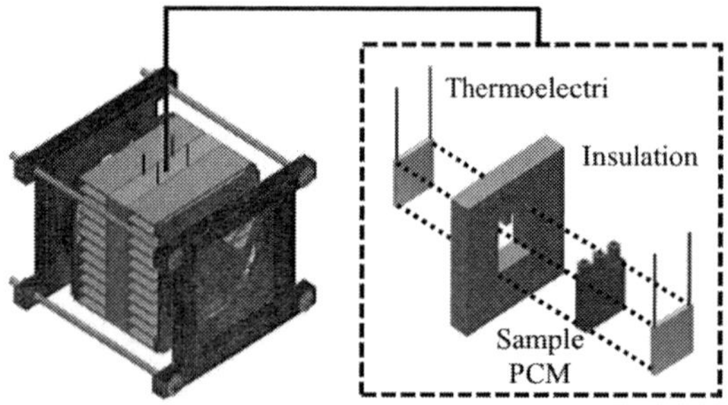

Fig. 1. The layout of thermal cycling test device.

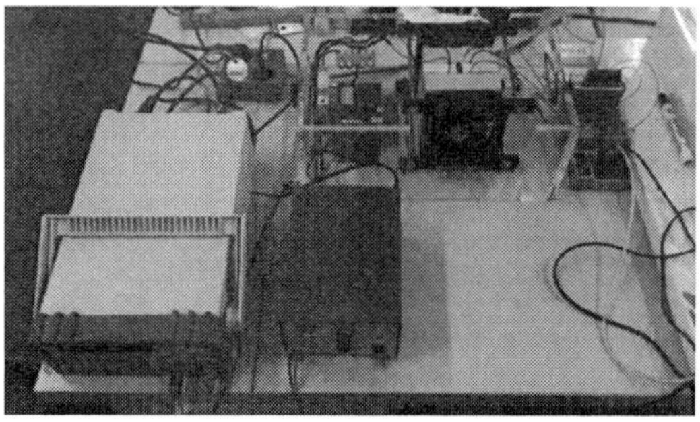

Fig. 2. Overall layout of thermal cycling test device.

In Fig 3 shows the installation process of thermal cycling test device. PCM is poured into cartridge that made of copper with dimension 40 x 40 x 14 mm. PCM cartridge is made by preparing 0.8 mm copper plate. It is formed into 6 sides, then it is connected by welding. The top side of cartridge is given with 3 nipples for the way in and out of PCM sample that will be tested on its thermal cycle. The middle nipple is also used as the access of thermocouple into PCM cartridge. The side of PCM cartridge is isolated in order to avoid heat lose to surroundings when thermal cycling test is operating for heating and cooling. Polyurethane that has low thermal conductivity is used for insulation system. It has dimension of 120 x 22 x 120 mm with hole in the center. The hole is adapted with dimension of PCM cartridge so it can enter into the insulation. Two thermoelectrics are installed on this thermal cycling test with dimension 40 x 40 mm. It have a specification of 12 V and 5 A. Heat sink is used in order to thermoelectric is not easily damaged. It is made of aluminium with dimension 120 x 35 x 120 mm. It has 12 fins with the thickness of 2.5 mm. All of those parts can be purchased cheaply and are able to be used continuously in particular to test thermal cycle of PCM.

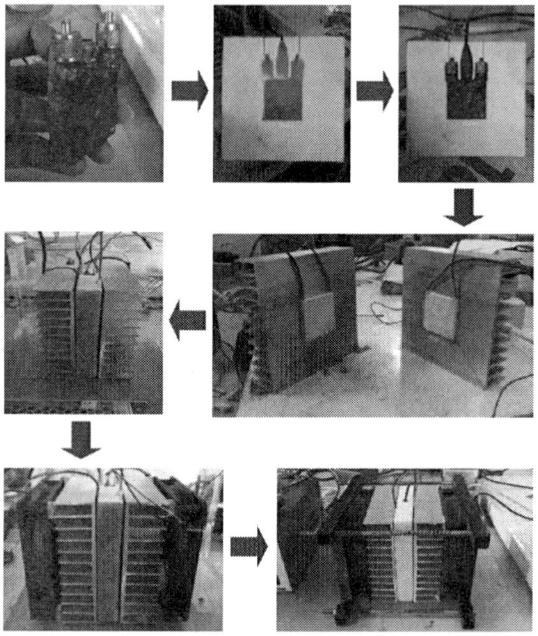

Fig. 3. Installation process of thermal cycling test device.

2.2 Thermal Cycling Test

In this study, sample of PCM RT 22 HC is used and It was purchased from Rubitherm Technologies GmbH. Thermal cycling test device is operated in this study until 1000 cycles. The amount of those cycles are similar to the research of Silakhori M et al [22]. In Fig 4 shows the schematic of thermal cycling test. The procedures of this study are fully outlined as follows: firstly the sample is poured into PCM cartridge with mass of 12 grams. The thermocouples are attached in PCM cartridge. PCM cartridge is placed into hole center of polyurethane insulation which is already formed. Installed two thermoelectric on both side of PCM cartridge. Make sure two thermocouples are fitted between PCM cartridge and thermoelectric. Stick two heat sinks on both of thermoelectric. Installed two DC fans on both of heat sink fins. Installed two clamps on both of the outside of DC fans and tightened those two clamps with bolts, nuts and rings. Voltage of polarity switch with 12 V is connected to adaptor. Trigger voltage of polarity switch is connected to programmable DC power supply. Connected thermoelectric with output voltage of polarity switch and connected DC fan with 10 Ampere DC adaptor. Connected thermocouples that already installed on their spots with NI cDAQ 9174 and connected NI cDAQ 9174 with NI 9211. Set voltage on 10 A DC adaptor with 12 V and turned on DC fan. Set voltage on programmable DC power supply with 0 and 3 V that changed alternately. Time set for 0 V is 420 seconds (7 minutes) and time set for 3 V is 180 seconds (3 minutes). The voltage of 0 V is for cooling and 3 V is for heating. Turned on thermoelectric with activating of 12 V adaptor and programmable DC power supply. Recorded the thermocouple readings through computer.

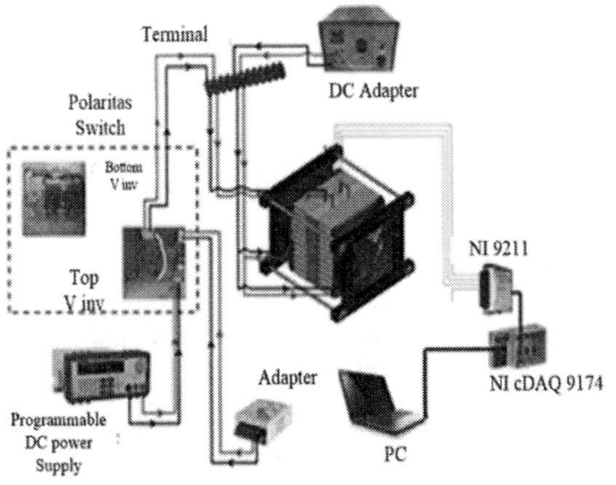

Fig. 4. Schematic of thermal cycling test.

Fig 5 shows the installation of thermocouples in PCM cartridge and in the wall of PCM cartridge (between PCM cartridge and thermoelectric). In this study, one thermocouple is installed in PCM cartridge, two thermocouples are installed between PCM cartridge and thermoelectric. Installation of thermocouple in PCM cartridge is to know the temperature of PCM samples when they are melting and freezing. Installation of thermocouple between PCM cartridge and thermoelectric is to know the temperature of thermoelectric when it is heating and cooling.

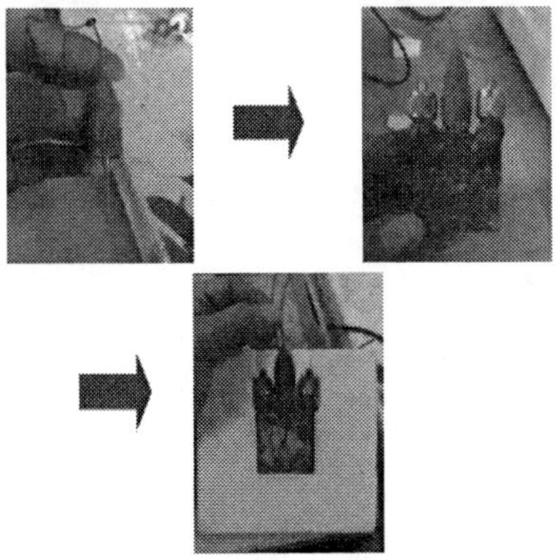

Fig. 5. Installation of thermocouples.

3 Analysis and Discussion

3.1 Results of Thermal Cycling of PCM RT 22 HC Thermoelectric Base

In Fig 6 shows the temperature distribution of thermal cycling test of PCM with sample of RT 22 HC. The thermal cycle device is operated for 300, 500, 700 and 1000 cycles. The process of heating and cooling ran for 7 minutes and for 3 minutes respectively. During the process, there are some phenomenon. When thermoelectric worked as cooling, sample starts to undergo the process of freezing. There are horizontal lines (near constant temperature) known as a process whereby the sample is changing phase. Then, when thermoelectric changes its function to heat up, temperature of sample will increase followed by the increase of thermoelectric temperature. It is caused by a lack of cooling time that is given to the sample so that phenomena appeared.

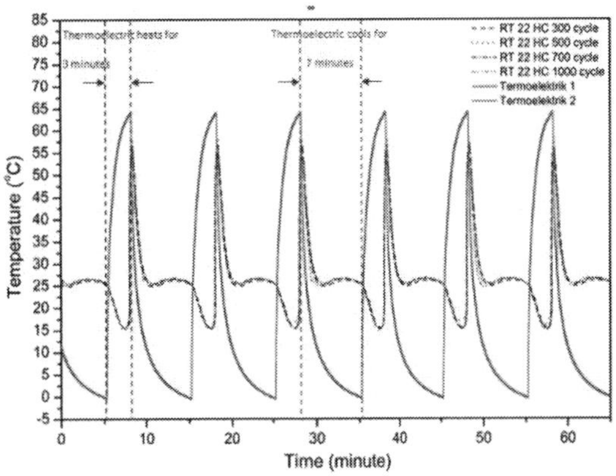

Fig. 6. The result of thermal cycling test of PCM with sample of RT 22 HC.

Table 1. Thermal properties of RT 22 HC after 0, 500 and 1000 cycles (melting).

Number of cycle (s)	Latent heat of fusion (J/g)	Melting point (°C)
0	163.31	25.37
500	196.80	25.08
1000	169.62	25.11

Table 2. Thermal properties of RT 22 HC after 0, 500 and 1000 cycles (freezing).

Number of cycle (s)	Latent heat of fusion (J/g)	Melting point (°C)
0	169.02	19.11
500	204.83	18.86
1000	177.29	18.90

The placement of thermocouple was located in the middle of sample in PCM to monitor the change of temperature, when the surface temperature of thermoelectric increases significantly, temperature of PCM still low and frozen. This is caused by the placement of thermocouple at the middle of sample,

Heat from thermoelectric needs time to reach the center of sample.

From the temperature measurement, there is no difference in measured temperature between 300, 500, 700 and 1000 cycles when sample undergo the process of heating and cooling. It proves that heat given by both thermoelectrics at each cycles are uniform, so the obtained data will be more accurate.

The latent heat of fusion and melting point of RT 22 HC after 0, 500 and 1000 cycles (melting and freezing point) were measured by Differential Scanning Calorimetry (DSC) and summarized in Table 1 and Table 2. Based on the results, the latent heat of fusion and melting point did not change significantly up to 1000 cycles. So, it is necessary retested more 1000 cycle to have the change in thermal properties of RT 22 HC.

3.2 The Comparison of Method of Thermal Cycling Test Device

Some methods of thermal cycling test device of PCM from the other researchers as comparison will be discussed.

Mehrali M et al create a thermal cycling test device of PCM as shown in Fig 7 [23]. This device have disadvantage compared to thermal cycling test device based on thermoelectric. The temperature range Mehrali's device can operate from 30 °C to 80 °C while the device based on thermoelectric can operate from 0 °C to 100 °C. This device uses fan for cooling PCM. The time for cooling PCM on this device is longer than on thermal cycling test device based on thermoelectric. The device based on thermoelectric need only 7 minutes for cooling PCM.

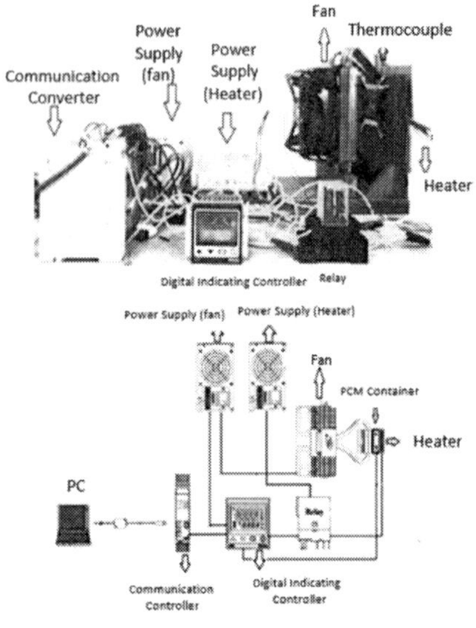

Fig. 7. Design of thermal cycling test device by Mehrali M et al [23].

Atul Sharma and A. Shukla proposed a thermal cycling test device of PCM using two water baths as shown in Fig 8 [21]. This device has disadvantage that is too complicated, needs large space and high cost. The process of cooling PCM is conducted by allowing PCM in cold water bath which its temperature has set. From both of comparison above, the purposed thermal cycling test device of PCM based on thermoelectric has advantages such as simple, easy to be moved, light, low-cost, taking more accurate data, and can be set from low to high temperature. This thermal cycling test device can be used continuously and be a new method for testing thermal stability of PCM.

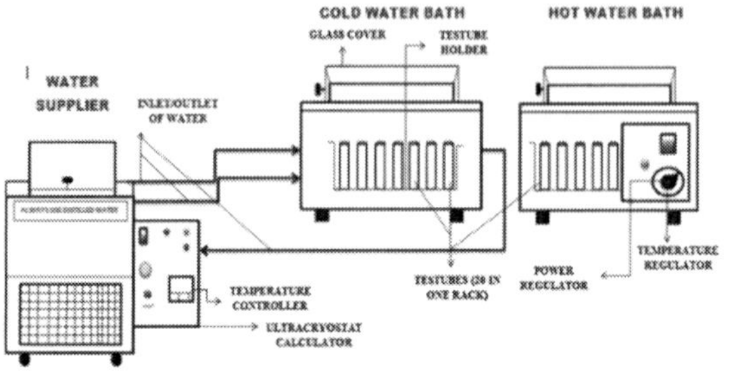

Fig. 8. Design of thermal cycling test device by A. Sharma and A. Shukla [21].

4 Conclusion

In this study, a new method has been proposed to obtained thermal stability by using thermal cycling test device based on thermoelectric. It has advantages such as simple, easy to be moved, light, low cost, taking more accurate data, can be set from low to high temperature. Heat released from thermoelectric on thermal cycling test are uniform at each cycles when it undergo the process of heating and cooling. The latent heat of fusion and melting point did not change significantly up to 1000 cycles.

The authors would like to thank KEMENRISTEK DIKTI and DRPM University of Indonesia for funding this research.

References

1. Zalba B, Marín JM, Cabeza LF, Mehling H.. *Applied thermal engineering* (2003),**23**:251-283.
2. Tatsidjodoung P, Le Pierrès N, Luo L. *Renewable and Sustainable Energy Reviews* (2013),**18**:327-349.
3. Pincemin S, Olives R, Py X, Christ M. *Solar Energy Materials and Solar Cells* (2008),**92**:603-613.
4. Prasher RS, Chang J-Y, Sauciuc I, Narasimhan S, Chau D, Chrysler G, *et al. Intel Technology Journal* (2005),**9**.
5. Saleh R, Putra N, Wibowo RE, Septiadi WN, Prakoso SP. *Experimental Thermal and Fluid Science* (2014),**52**:19-29.
6. Putra N, Prawiro E, Amin M.. *International Journal Of Technology* (2016),**7**:244-253.
7. Yuan Y, Zhang N, Li T, Cao X, Long W. Thermal

performance enhancement of palmitic-stearic acid by adding graphene nanoplatelets and expanded graphite for thermal energy storage: A comparative study. *Energy* (2016),**97**:488-497.
8. Amin M, Putra N, Kosasih EA, Prawiro E, Luanto RA, Mahlia TMI. *Applied Thermal Engineering* (2017),**112**:273-280.
9. Şahan N, Fois M, Paksoy H. *Solar Energy Materials and Solar Cells* (2015),**137**:61-67.
10. Chieruzzi M, Cerritelli GF, Miliozzi A, Kenny JM. *Nanoscale research letters* (2013),**8**:1.
11. Sharma S, Buddhi D, Sawhney R. Accelerated thermal cycle test of latent heat-storage materials. *Solar Energy* (1999),**66**:483-490.
12. Sharma A, Sharma S, Buddhi D. *Energy Conversion and Management* (2002),**43**:1923-1930.
13. Pielichowska K, Pielichowski K. Phase change materials for thermal energy storage. *Progress in Materials Science* (2014),**65**:67-123.
14. Ferrer G, Solé A, Barreneche C, Martorell I, Cabeza LF. Review on the methodology used in thermal stability characterization of phase change materials. *Renewable and Sustainable Energy Reviews* (2015),**50**:665-685.
15. He W, Zhou J, Hou J, Chen C, Ji J. Theoretical and experimental investigation on a thermoelectric cooling and heating system driven by solar. *Applied Energy* (2013),**107**:89-97.
16. Putra N, Ardiyansyah, Sukyono W, Johansen D, Iskandar FN. The characterization of a cascade thermoelectric cooler in a cryosurgery device. *Cryogenics* (2010),**50**:759-764.
17. Djafar Z, Putra N, Koestoer RA. The Utilization of Heat Pipe on Cold Surface of Thermoelectric with Low-Temperature Waste Heat. In: *Applied Mechanics and Materials*: Trans Tech Publ; (2013). pp. 410-415.
18. Raam Dheep G, Sreekumar A. *Energy Conversion and Management* (2015),**105**:13-19.
19. Shukla A, Buddhi D, Sawhney R. *Renewable Energy* (2008),**33**:2606-2614.
20. Sarı A. Thermal reliability test of some fatty acids as PCMs used for solar thermal latent heat storage applications. *Energy Conversion and Management* (2003),**44**:2277-2287.
21. Sharma A, Shukla A. *Energy and Buildings* (2015),**99**:196-203.
22. Silakhori M, Naghavi MS, Metselaar HSC, Mahlia TMI, Fauzi H, Mehrali M. *Materials* (2013),**6**:1608-1620.
23. Mehrali M, Latibari ST, Mehrali M, Metselaar HSC, Silakhori M. *Energy Conversion and Management* (2013),**67**:275-282.

Effect of ageing time 200 °C on microstructure behaviour of Al-Zn-Cu-Mg cast alloys

Diah Kusuma Pratiwi[1,*], *Nurhabibah Paramitha Eka* Utami[1]

[1]Department of Mechanical Engineering, University of Sriwijaya, 30662, Indonesia

Abstract. Al-Zn-Cu-Mg is heat treatable alloy that can be used in many hightech applications, such as aerospace and military. The main objective of this study is to investigate the influence of ageing process in microstrucure behaviour of Al-9Zn-5Cu-4Mg cast alloy by performing SEM analysis and its correlation with hardness tests of as-cast Al-9Zn-5Cu-4Mg alloy and heat treated Al-9Zn-5Cu-4Mg cast alloy. The results show the deployment of precipitation spread over the dendrite and also the presence of second phases $Mg_3Zn_3Al_2$, Cu_2FeA_{l7} , $CuAl_2$, and $CuMgAl_2$ in as-cast Al-9Zn-5Cu-4Mg alloy. The presence of all these second phases are affecting to the toughness of aluminium alloy and the presence of $MgZn_2$ leads the impairment of hardness value of heat-treated Al-9Zn-5Cu-5Mg cast alloy.

1 Introduction

Over the past fifty years, aluminum has become the metal used broadly after the use of steel. Aluminium selected as an alternative material to increase efficiency because its mechanical behaviour : light weight, easy to design, easy to recycle, good in machinability and manufacturability, good heat conductor and corrosion resistant [1]. However, Pure aluminum can not stand alone because it has low strength and unstable mechanical properties. Therefore, to improve the mechanical properties of aluminum and modify the microstructure can be done through alloying and heat treatment [2].

Aluminium 7xxx series has a higher strength when compared to any other classes of aluminum alloys. The main properties of 7xxx series alloy indispensable in some high-tech applications such as on an airplane wing is of the nature of strength, toughness and resistance properties to stress corrosion cracking and fatigue [3].

The addition of zinc will improve the homogenity of the martix aluminium. Zinc provides a solid solution strengthening effect or work hardening in aluminium. This element improve the potential of the solution in aluminium and it makes Zn usually used in in the protective layer and the sacrificial anode [4]. The addition of 9 % Zn is increasing the strength that followed by the decrease in ductility. It is happen through the mechanism of solid solution strengthening, where Zn dissolved as a substitute in the lattice of aluminum atoms. The addition of magnesium provides the improvement and strengthening in work-hardening characteristics of aluminium so it produces good corrosion resistance and weldability.

The impact energy and hardness of the matrix are enhanced when Cu addition is within 2-4 vol % [5]. The effect of addition copper in aluminium is improving the machinability with increasing matrix hardness. In other words, Copper has a greatest impact on the strength and hardness of aluminum casting alloys, in both heat treated and not heat treated condition. The addition of Cu to the ternary Al-Zn-Mg system will produce the heat-treatable compositions which contributed to the increased hardness and strength [6].

Study on evolution of eutectic structures intheir Al-Zn-Mg-Cu alloys reported that several coarse intermetallic phases can be formed below the solidus line during solidification of Aluminium 7xxx alloys as a result of solute redistribution of metals [7]. These properties are determined by the major phases formed in the alloy, the GP zone η ', η, T, and S. Thus the nature of the 7xxx alloys can be optimized by modifying the microstructure through changes in alloy composition and heat treatment variations. System Al-Zn-Mg-Cu system is quite complex, chemical composition into primary particles present is η-$MgZn_2$, T-$Al_2Mg_3Zn_3$, S-Al_2CuMg, and θ-Al_2Cu, Al_7Cu_2Fe, $Al_{13}Fe_4$ and Mg_2Si [8,9].

During the casting process, there is the hardening of atoms that remain in solution. When the temperature increase, the rate of nucleation and growth of precipitates can be accelerated. The increase in strength occurs in line with the decrease in ductility on dendritic material. T5 is an artificial aging that are commonly used in increasing reinforcement through hardening precipitates. Benefits of T5 heat treatment can be lost when done quenching after casting.

The main mechanical properties of metal in addition affected by the chemical composition but also strongly influenced by the microstructure of the metal. By heat treatment, Aluminium will show the change in

* Corresponding author: pratiwidiahkusuma@ft.unsri.ac.id

its structure. Artificial ageing in aluminum alloys above room temperature that is taking place between 100-200 °C. In this work, we were used T5 artificial ageing with heating temperature at 200 °C for 200 hours without added any solution treatment processes.

The main objective of this study is to investigate the influence of ageing process on microstrucure behaviour of Al-9Zn-5Cu-4Mg cast alloy by performing SEM analysis and its correlation with hardness tests.

2 Experimental Details

2.1 Preparation of Al-9Zn-5Cu-4Mg Alloy

Casting process of Al-9Zn-5Cu–4Mg alloy was conducted using a graphite crucible in muffle furnace with melting temperature measured stable in 750°C. Aluminium and alloy composition weight of the charge is 4.5 kg with calculation of mass balance of each alloy and the possible evaporation by 14% Zn and Mg 10% are as shown in Table 1.

Table 1. Mass Balance of Aluminum Alloys

Al	Zn (0,09)	Mg (0,04)	Cu (0,05)	Zn Assumed evaporated 14%	Mg Assumed evaporated 10 %
3615	405	180	225	461,7	198

The melting process begins with inserting aluminium into crucible for approximately 1 hour until it is completely dissolved. Then Cu, Zn, Mg inserted sequentially at Muffle furnace for 2 hours until all elements also completely dissolved.

To minimize the presence of hydrogen in liquid aluminum, can done by lowering the melting temperature so that the solubility of hydrogen in joined aluminum decreases with decreasing temperature. The degassing process must be done properly on all aluminium liquid parts. The degassing process that is useful to remove impurities dissolved performed using argon gas for approximately 2 minutes. Mold that used are permanent metal molds made from mild steel which has been coated with clay.

Coating with clay intended to casting product can be removed easily. The molten metal is cooled in a metal mold at furnace temperature until solidification occurs. After the molten metal have solidifying as a whole, the casting product then released from the mold and followed by a visual inspection of the results of the casting product's surface.

Cast products then tested for chemical compositions analysist by using emission spectrometer with type PMI– Master Pro Oxford instrument analytical GmBH the results shown in Table 2.

Table 2. Chemical Compositions of Al-9Zn-5Cu-4Mg cast alloy

Zn	Cu	Mg	Fe	Al
[%]	[%]	[%]	[%]	[%]
9.63	5.52	4.43	0.14	Rest

2.2 Cutting Samples

Casting product cutted to some parts based on samples requirement of testing to be performed. to ease process of cutting samples, small samples were cut by using hand grinder. Cutting process of samples is using a metal cutting saws 9" Type Krisbow KW1500053. Refinering process of surface was conducted by sandpapers to ensure the surface of samples completely smooth.

2.3 Heat Treatment

The experiment were prepared in two conditions, as-cast and heat-treated alloy. Artificial ageing was conducted by using Hofman K1 furnace at temperature 200 ° C for 200 hour, while as-cast aloy was not conducted any heat treatment process.

Furnace was preheated to 200°C before performing the heat treatment process. Cooling process was carried out in the heater furnace, and it takes 3 hr until the temperature in the furnace back to normal. Time required to heating the samples was 20 days with range of warm-up time approximately 10 hours each day.

2.4 Hardness Test

Artificial ageing carried out to observe the change in characteristic and the strength of Al-9Zn-5Cu-4Mg cast alloy to a long period and hight temperature heat treatment. Hardness test was performed in the two conditions of the samples, as-cast alloy and heat treated Al - 9Zn - 5Cu – 4Mg cast alloy in which each condition was conducted in 3(three) test samples to analyze the uniformity of hardness value in the alloy. Sample preparation was done by casting, sample cutting, heat treatment, surface smoothing and polishing. The load that used in this test is 10 kgf. Hardness (HV) obtained by using Equation 1 below

$$HV = \frac{2p\sin\left(\frac{\theta}{2}\right)}{L^2} = 1{,}854\ \frac{P}{L^2} \qquad (1)$$

2.5 Microstructure Observations

Microstructure observations were conducted using optical microscopy (OM), and Scanning Electron Microscope (SEM) with Energy-dispersive spectroscopy observation (EDS). All tests were carried out in two material conditions, as-cast and heat treated alloy.

Sample preparation process for these two observations are relatively similar; sample cutting,

DOI: 10.1051/matecconf/201710101008

mounting, grinding, polishing, etching and the addition of the coating process using Au element for SEM samples.

Polishing process was conducted using TiO_2 and etching process was conducted using a mixture of 12.5 ml 8 ml HNO3 + HF + 85 ml of water for optical microscope and 0.5% HF for Scanning Electron Microscope observation.

3 Results and Discussions

3.1 Mechanical Characteristics

The hardness test results for as-cast and heat-treated alloy show in Table 3 and Fig 1. It was found that as-cast alloy has the highest hardness value. In the other hand, there were the impairment value of hardness in heat-treated alloy.

Table 3. Hardness Test Results

Al-Zn-Cu-Mg Cast Alloy	Hardness (VH)
As -cast	138
As –cast	139
As –cast	142
Heat treated T5	136
Heat treated T5	136
Heat treated T5	137

As-cast alloy has higher hardness because the presence of microsegregations in its structure that embrittle the Al-Zn-Cu-Mg alloy. The previous studies reported that smaller grain boundaries will make the metal harder and stronger [10]. In the smaller grain, there is one square unit of grain boundary for each dislocation. There is a much greater chance for a dislocation to be stopped at a grain boundary in the smaller grain. Therefore, the smaller grain is stronger. In the larger grain, a dislocation can travel up to 4 units without being stopped by a grain boundary, indicating the potential for extensive plastic flow. In the small grain, no dislocation can travel more than 1 unit of distance [11].

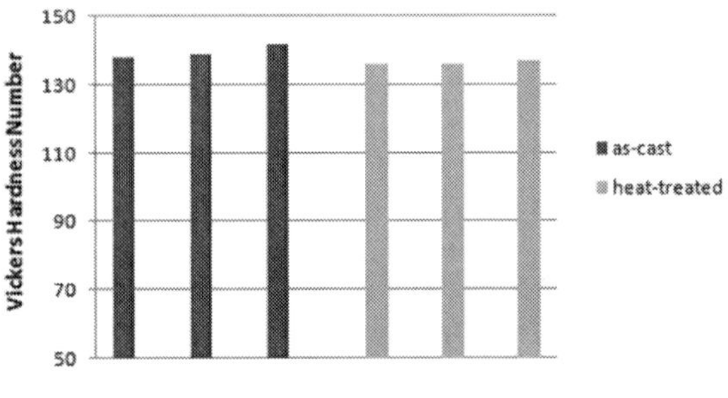

Fig. 1. Heat Treatment effects on hardness of Al-9Zn-5Cu-4Mg

The deployment of precipitation that affecting to the brittleness and the toughness of heat treated alloy, it makes the and hardness value were not exceed the value of as-cast alloy.

3.2 Microstructure Characteristic

Fig 2 and Fig 3 shows the difference of microstructure between as-cast and heat-treated Al - 9Zn - 5Cu – 4Mg cast alloy. In the as-cast alloys, the deployment of precipitation only occurs at the edge of grain boudary. Heat treatment process makes the precipitation spread to all over dendrite.

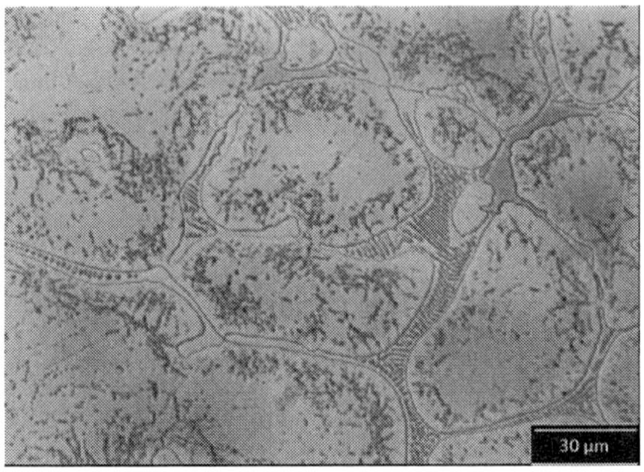

Fig. 2. Microstructure of as-cast Al-9Zn-5Cu-4Mg alloy

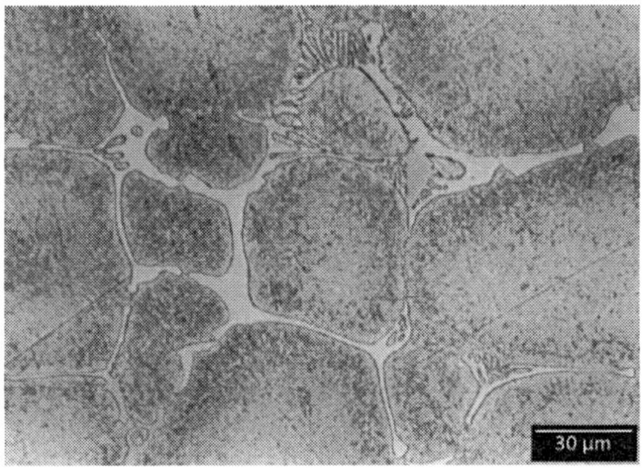

Fig. 3. Microstructure of heat-treated Al-9Zn-5Cu-4Mg cast alloy

Observation points of Scanning Electron Microscope (SEM) on dendrite of as- cast alloy and its microanalysis report by Energy - dispersive spectroscopy (EDS) observation are shown in Figure 4 (a) and 4(b). While the analysis of grain boundary of as- cast alloy and its microanalysis report from Energy - dispersive spectroscopy (EDS) observation were shown in Figure 5(a) and 5(b). The phase that may be formed in the as-cast alloy are shown in Table 4 and Table 5.

Fig. 6(a) shows the results of Scanning Electron Microscope (SEM) on dendrite of heat-treated alloy while Fig.6(b) shows and its microanalysis report by

 DOI: 10.1051/matecconf/201710101008

Energy - dispersive spectroscopy (EDS) observation. The analysis of grain boundary of heat-treated alloy is shown in Fig. 7(a), while its microanalysis report from Energy - dispersive spectroscopy (EDS) observation is shown in Fig. 7(b). The phase that may be formed in the heat-treated alloy are shown in Table 6 and Table 7.

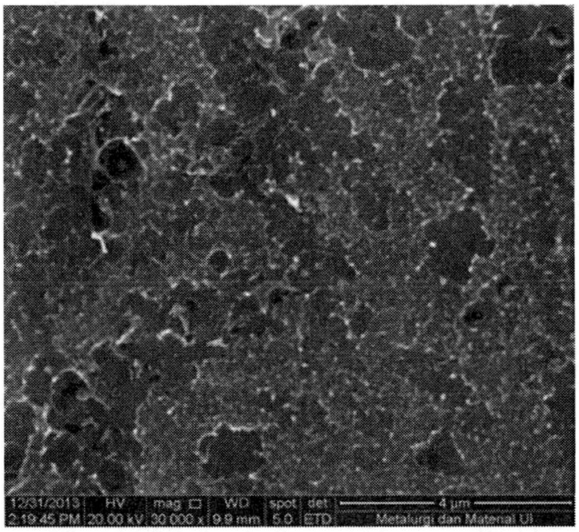

(a)

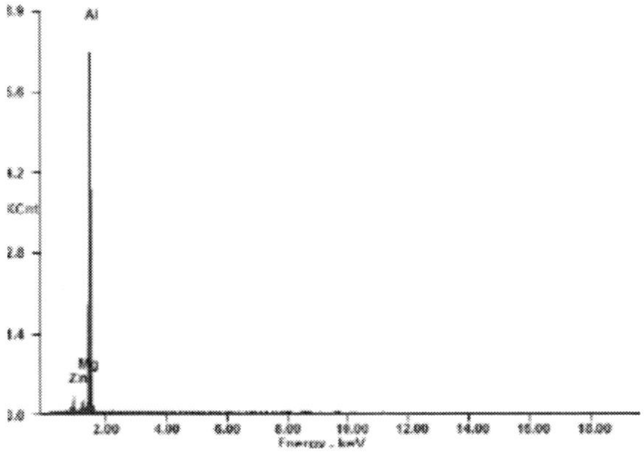

(b)

Fig 4. (a) Microstructure observation in dendrite of as-cast Al-9Zn-5Cu-4Mg Alloy (b) microanalysis report by Energy - dispersive spectroscopy (EDS)

Table 4. Summary of EDS observation on dendrite as-cast Al - 9.7Zn - 5.5Cu - 4.5Mg cast alloy based on SEM pictures

No	Average element wt.%			Phase may be formed*
	Al	Zn	Mg	
1	90.34	6.81	2.85	$Mg_3Zn_3Al_2$
2	89.40	7.30	3.30	$Mg_3Zn_3Al_2$
3	89.65	7.32	3.03	$Mg_3Zn_3Al_2$

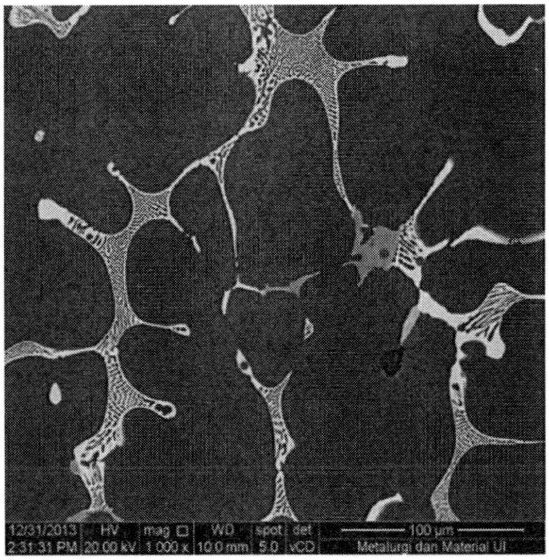

(a)

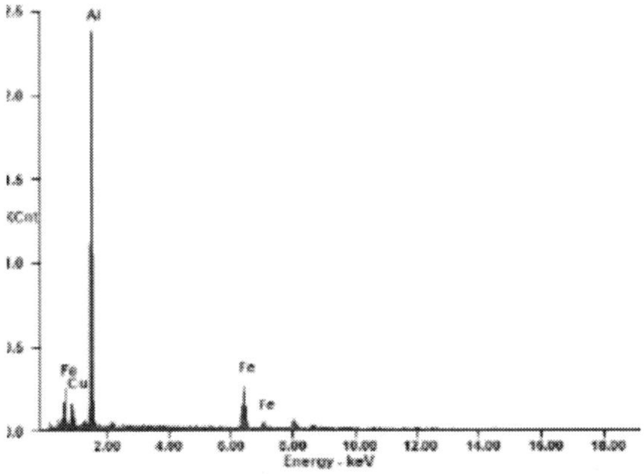

(b)

Fig 5. (a) Microstructure observation in grain boundary of as-cast Al-9Zn-5Cu-4Mg Alloy (b) microanalysis report by Energy - dispersive spectroscopy (EDS)

Table 5. Summary of EDS observation on grain boundary as-cast Al - 9.7Zn - 5.5Cu - 4.5Mg cast alloy based on SEM pictures.

No	Average element wt.%					Phase may be formed*
	Al	Zn	Mg	Cu	Fe	
1	68.59	-	-	11.31	22.09	Cu_2FeAl_7
2	24.72	27.29	23.85	24.14	-	$Mg_3Zn_3Al_2$, $CuAl_2$, $CuMgAl_2$
3	57.83	10.68	13.67	17.82	-	$Mg_3Zn_3Al_2$, $CuAl_2$, $CuMgAl_2$

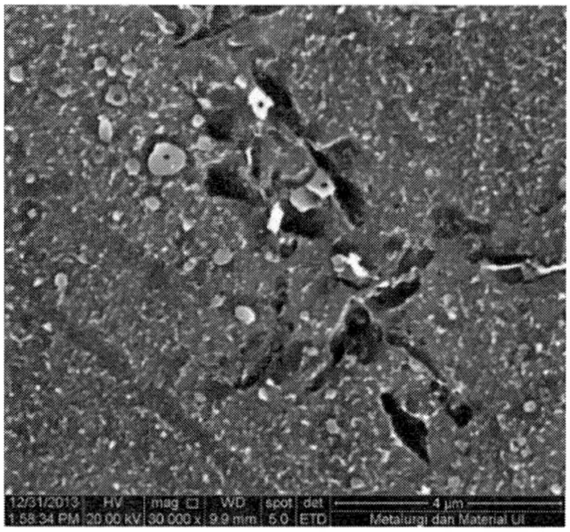

(a)

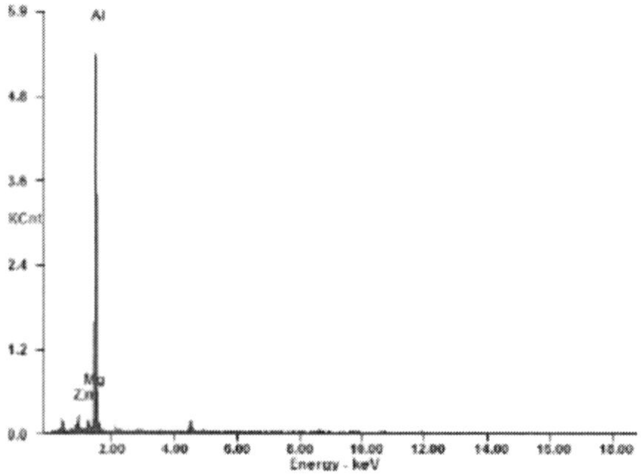

(b)

Fig 6. (a) Microstructure observation in dendrite of heat-treated Al-9Zn-5Cu-4Mg Alloy (b) microanalysis report by Energy - dispersive spectroscopy (EDS)

Table 6. Summary of EDS observation on dendrite heat-treated Al - 9.7Zn - 5.5Cu - 4.5Mg cast alloy based on SEM pictures

No	Average element wt.%			Phase may be formed*
	Al	Zn	Mg	
1	90.05	7.08	2.87	$Mg_3Zn_3Al_2$
2	89.38	3.09	7.53	$MgZn_2$
3	89.91	2.75	7.34	$MgZn_2$
4	90.52	2.75	6.73	$MgZn_2$

Based on SEM images,the precipitation in as-cast alloy is not visible clearly, $Mg_3Zn_3Al_2$ is the most dominant phase due to present at all three points of testing (see table 5), it may be possible because the comparison of Zn and Mg greater than 2.2 on as-cast alloy. While in heat-treated alloy, the precipitation is visible clearly. The holes in the surface indicate the presence of second phase precipitations which were detached from the surface due to etching and polishing processes.These precipitation will form the second phase that contained in dendrites and the grain boundaries. According to Table 7, the second phase that may be formed are $MgZn_2$ and $Mg_3Zn_3Al_2$.

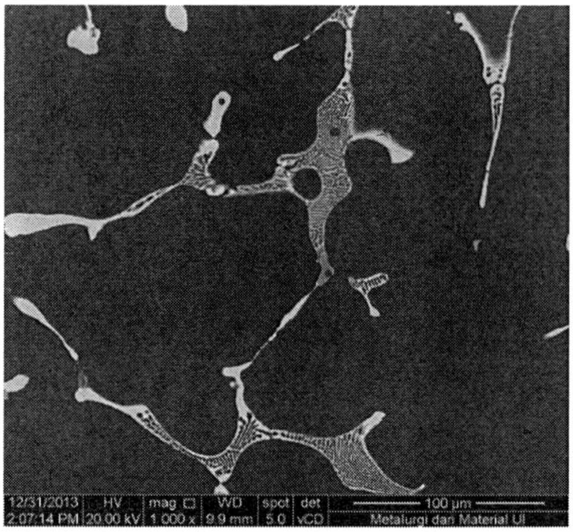

(a)

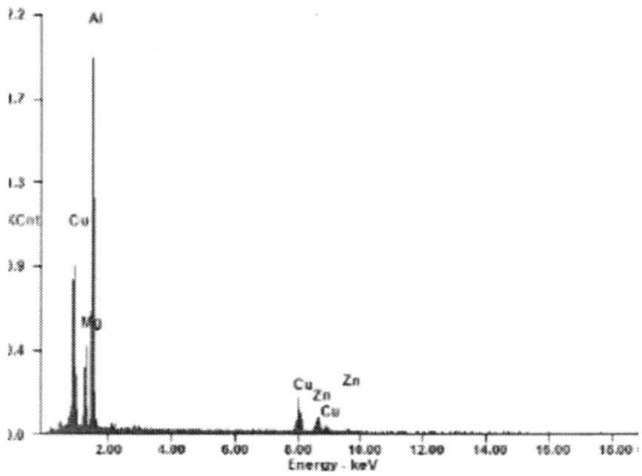

(b)

Fig 7. (a) Microstructure observation in grain boundary of heat-treated Al-9Zn-5Cu-4Mg Alloy (b) microanalysis report by Energy - dispersive spectroscopy (EDS)

Table 7. Summary of EDS observation on dendrite heat-treated Al - 9.7Zn - 5.5Cu - 4.5Mg cast alloy based on SEM pictures

No	Average element wt.%					Phase may be formed*
	Al	Zn	Mg	Cu	Fe	
1	69.51	-	-	7.70	22.79	Cu_2FeAl_7
2	25.97	25.02	25.20	23.82	-	$Mg_3Zn_3Al_2$, $CuAl_2$, $CuMgAl_2$
3	57.83	10.68	13.67	17.82	-	$Mg_3Zn_3Al_2$, $CuAl_2$, $CuMgAl_2$

*) Prediction phase based on the phase data that may appear in Al-Zn – Mg [12]

The presence of Fe element as impurities made aluminum become coarse grains, reducing the ductility of alloy, and leads the presence of Cu_2FeAl_7. SEM observations at the grain boundaries of heat-treated alloy are not really different much with as-cast alloy, the conclusions of EDS can be seen in Table 5. The presence of these second phases are very influential to the strength and brittleness.

4 Conclusions

Phases that may be formed on as-cast alloy are $Mg_3Zn_3Al_2$, $CuAl_2$, $CuMgAl_2$ and Cu_2FeAl_7. While in heat-treated alloy, the phases that may be formed are the same with the as-cast. But the long period heating process and high temperatures heat treatment generate the presence of embrittled phase $MgZn_2$ within the dendrites. It phase is very influential for toughness and the strength of Al-9Zn-5Cu-5Mg cast alloy. The Presence of $MgZn_2$ leads the impairment of hardness value of heat-treated Al-9Zn-5Cu-5Mg cast alloy.

References

1. X. Fan,D. Jiang, Q. Meng, L. Zhong, TMAH, **60** (2005)
2. M. Kaczorowaki, A. Krzynska, TSM, **8** (2008)
3. J. Linder, M. Axelsson,H. Nilsson , TIP, **28** (2006)
4. AD. Isadarea, B. Aremob, MO. Adeoyec, OJ. Olawalec, MD. Shittu. EHT, **16** (2013)
5. R.Orozco, J.Genesca, J.Juarez-Islas, EMPA, **16** (2007)
6. L. Yan, Y. Zhang, X Li, Z. Li, EZMA, **24** (2014)
7. J. Pezda. HTEMP, **12** (2012)
8. JB. Ferguson, BF. Schultz , JC Mantas, H Shokouhi, PK Rohatgi, ECHT, **4** (2014)
9. N. Nafsin, HMMA. Rashed, ECMH, **2** (2013)
10. I. Adlakha, M.A. Bhatia, K.N. Solanki, M.A. Tschopp, ASI, (2014)
11. LF Mondolfo, *Aluminium alloys : Structure and Properties* (Butterworths, London,1979)

Mechanical properties analysis of Al-9Zn-5Cu-4Mg cast alloy by T5 heat treatment

Nurhabibah Paramitha Eka Utami[1,*], *Hendri* Chandra[1]

[1]Department of Mechanical Engineering, University of Sriwijaya, 30662, Indonesia

Abstract. The Improvement of mechanical properties of Al-9Zn-5Cu-4Mg cast alloy by T5 Heat treatment 200 °C during 200 hours was studied. Al-Zn-Cu-Mg alloy can be used in wide range of aircraft industry is based on the superior characteristic of this alloy. The main objective of this study is to investigate the influence of ageing process by performing tensile and hardness tests. The results revealed thatagingprocessisaffecting to the deployment of precipitationand it is indicate the formation of second phases $MgZn_2$ by atomic reactions that lead to the change of mechanical properties of Al-9Zn-5Cu-4Mgcastalloy..

1 Introduction

The development of the material strength is one of the most important subjects because it is the first characteristic to used in industrial applications. Aluminum is an important material that widely used in the many important structures since more than 80 years ago. Aluminum also is one of the metals with high performance that is most easily produced, which led to the need for production and maintenance costs are relatively low. The development of advanced aluminum makes this material can effectively cimpete with modern composite materials [1]. To improve the mechanical properties of aluminum and modify the microstructure can be done through alloying and heat treatment [2].

Aluminium alloys are increasingly used in many important manufacturing areas, such as military, automotive, aerospace because of their well known performance, well established design methods, manufacturing and reliable inspection techniques [3]. Although the increase in the use of composites reducing the role of aluminum in a certain extent, the use of high-strength aluminum remain indispensable in the construction of the airframe. It is also supported on the grounds that aluminum is a metal that is relatively inexpensive, light weight, heat treatable, and can be given a relatively high pressure.

The demand for aluminum alloys is growing rapidly due to the unique combination of properties that make it becomes one of the most versatile engineering and construction materials. Aluminium 7xxx series has a higher strength and mechanical properties if compared with other class of aluminum alloys [4]. High-strength age-hardened 7xxx series Al–Zn–Mg–Cu alloys are widely used for aircraft structures that is high demanding operating conditions. The important properties must be considered for these applications are corrosion and damage tolerance (e.g. fracture toughness and fatigue resistance), modulus, strength, and ductility. These important properties can be produced through a variety of alloys, the exact process or a combination of both [5].

Al-Zn based alloys have excellent corrosion resistance in variety of environments. The presence of aluminum in this alloy enhances corrosion resistance well-known of zinc , which is the main constituent of the alloy. Zn usually used in in the protective layer and the sacrificial anode because it is will improve the homogenity of the matrix in Aluminium.The present of zinc in aluminum also provides a solid solution. While [6]. The addition of zinc composition more than 3 wt% in aluminum alloys could increase the strength of aluminum alloys,while the bad influence of zinc are lowered toughness and increasing brittleness. However, the addition zinc element under 3 wt% considered useless [7]. The present of 9% Zn will increase the strength but also will decrease the ductility in aluminium alloys. It is happen through the mechanism of solid solution strengthening, where Zn dissolved as a substitute in the lattice of aluminum atoms.

The addition of magnesium provides good weldability and good corrosion resistance. The present of copper in aluminium alloys is increasing matrix hardness so it generates i the improvement in the machinability. Copper take the good impact on the strength and hardness of aluminum alloys, in both heat treated and not heat treated condition [8]. The proper Mg addition amount is beneficial to enhance the mechanical properties of the graphite particles reinforced aluminum alloy matrix composites and the abrasion resistance of

*Corresponding author: nhparamitha@ft.unsri.ac.id

the materials due to uniformaly distribution of hard particles and reduction of friction coefficient [9].

Zinc combined with copper and / or magnesium will generate the heat-treatable compositions which contributed to the increased hardness and strength in the heat treatment because the presence of four major intermetallic phases that are expected in commercial Al-Zn-Cu-Mg alloys including η ($MgZn_2$), T ($Al_2Zn_3Mg_3$), S (Al_2CuMg), θ (Al_2Cu), Al_7Cu_2Fe, $Al_{13}Fe_4$, and Mg_2Si phases [4-6]. T and η phase are often present in solid solution with the presence of the four elements [10,11].

There are some artificial aging processes commonly applied to aluminium, in this experiment we were used T5 artificial aging. The difference of T5 from T4 and T6 is the heat treatment process passed without any solution treatment . The advantages of no solution treatment are significant energy savings and reduce distortion in components [12]. T5 is an artificial aging that are commonly used in increasing reinforcement through hardening precipitates. Benefits of T5 heat treatment can be lost when done quenching after casting. When the temperature increase, the rate of nucleation and growth of precipitates can be accelerated. The increase in strength occurs in line with the decrease in ductility on dendritic material.

The main objective of this study is to investigate the influence of aging process in a long period of Al-9Zn-5Cu-4Mg cast alloy by performing tensile and hardness tests. The results of a study of the heat treatments 200° C for 200 hours on the mechanical properties, it is possible to determine reactions, material behaviour for long period used and also determine conditions necessary to achieve optimum mechanical properties and performance.

2 Experimental details

2.1 Preparation of Al-9Zn-5Cu-4Mg Alloy

Al-9Zn-5Cu-4Mg cast alloy was prepared through the addition of measured amounts of Zn, Cu, and Mg to the molten aluminum in crucible grafit with melting temperature measured stable in 750° C. Aluminium and alloy composition weight of the charge is 4.5 kg with calculation of mass balance of each alloy and the possible evaporation by 14% Zn and Mg 10% are as shows in Table 1:

Table 1. Mass Balance of Aluminum Alloys

Al	Zn (0,09)	Mg (0,04)	Cu (0,05)	Zn Assumed evaporated 14%	Mg Assumed evaporated 10 %
3615	405	180	225	461,7	198

The melting process begins with preheated the furnace until reaching the specified temperatures. The next process is inserting aluminium into crucible for approximately 1 hour until it is completely dissolved. Then then followed by addition other elements Cu, Zn at muffle furnace for 2 hours or until all elements also completely dissolved.

The use of low-temperature condition can help minimize the presence of hydrogen in liquid aluminum. Degassing process is carried out to remove impurities dissolved in the molten metal using argon gas. The molten metal degassed for 2 minutes to lift impurities that contained in the molten metal and bring it to the surface.

Permanent mold made from mild steel and coated with clay to facilitate the release of cast product from the mold. After the molten metal have solidifying as a whole, the casting product then released from the mold and followed by a visual inspection on surface product.

Cast products then tested for chemical compositions analysist by using emission spectrometer and the result is shown in Table 2.

Table 2. Main Chemical Compotitions of Al-9Zn-5Cu-4Mg castalloy

Zn	Cu	Mg	Fe	Al
[%]	[%]	[%]	[%]	[%]
9.63	5.52	4.43	0.14	Rest

Analysis of chemical composition was performed by using of spectrometry method (emission spectrometer PMI – Master Pro Oxford instrument analytical GmBH type).

2.2 Cutting Samples

The cast product then cutted according to requirement of the testing to be performed. Cutting process was conducted by using a metal cutting saws 9" Type Krisbow KW1500053. Then followed with cutting process that using hand grinder to produce small samples. Refinering surface was conducted by sandpapers #120 #200 #400 #600 #800 and #1000 to ensure the surface of samples completely smooth.

2.3 Heat treatment

The experiment were prepared in two conditions, as-cast and heat-treated alloy. The as-cast samples were not conducted by the heat treatment process, whilet he heat-treated samples were heated in 200 °C for 200 hours with Heater furnace Hofmann K1. Cooling process carried out in the heater furnace for ± 3 hours until the temperature returns to normal. Time required to heating the samples was 20 days with range of warm-up time approximately 10 hours each day.

2.4 Tensile test

Tensile test carried out by using *Universal Testing Machine Tokyo Testing Machine* MI7G tipe RAT-30P. Testing was conducted by using standard JIS Z 2201. Both as-cast and heat-treated conditions were tested at room temperature in three samples.

2.5 Hardness test

The samples were tested using Vickers Hardness Tester VKH-2E which is based on an assessment of the indenter penetration pyramid. Preparation of the test specimen was conducted by cutting the sample, heat treatment, and surface smoothing with sandpaper # 80, # 120, # 240 and # 800. The change of characteristic and strength of the aluminum alloy that influenced by Artificial aging can be seen from the results of hardness tests. Hardness test was performed in the two conditions of the samples, as-cast and heat treated alloys. Hardness testing was using load 10 kgf. Hardness (HV) obtained by using Equation 1 below :

$$HV = \frac{2p\sin\left(\frac{\theta}{2}\right)}{L^2} = 1{,}854\ \frac{P}{L^2} \qquad (1)$$

2.6 Metallography observations

Microstructure observation were conducted through the process of metallography that are sample selection, cutting, mounting, grinding, polishing, etching and observation using optical microscope. Metal polishing was using TiO_2 and followed by etching using a mixture of 12.5 ml 8 ml HNO3 + HF + 85 ml of water. Metallography observations in this research are using an optical microscope, SEM and EDS.

3 Results and discussions

3.1 Mechanical characteristics

Investigation results of the mechanical characteristics of the Al-9Zn-5Cu-4Mg cast alloy are presented in Table 3. Based on data below, as-cast alloy has higher hardness value than heat-treated alloy. It is also proved by the reduction of the ultimate tensile strength value in heat treated samples.

The deployment of precipitation that caused by the heating process and the high content of Zn affecting to the brittleness and the toughness of heat treated alloy, it makes the ultimate tensile strength and hardness value were not exceed the value of as-cast alloy.

Table 3. Mechanical Properties of as-cast and heat treated Al-9Zn-5Cu-4Mg cast alloy

Al-Zn-Cu-Mg Cast Alloy	Tensile Strength (MPa)	Elongation %	Hardness, VH
As -cast	447	0.8	136
As –cast	446	1.1	136
As –cast	450	0.9	137
Heat treated T5	453	1.4	138
Heat treated T5	457	1.1	139
Heat treated T5	465	1.3	142

Artificial aging that conducted in a long period will present the second phase precipitates that affect to the mechanical properties of aluminum alloy casting. Wherein said second phase acts as a barrier that causes dislocation shift increasingly difficult when it is loaded but in the other hand it acts as embrittled phase that decrease the toughness and strenght of alloy.

The difference values in mechanical testing at any point due to unevenly distribution of atoms in aluminum cast and possibility of impurities presence that may occur due to error in casting process.

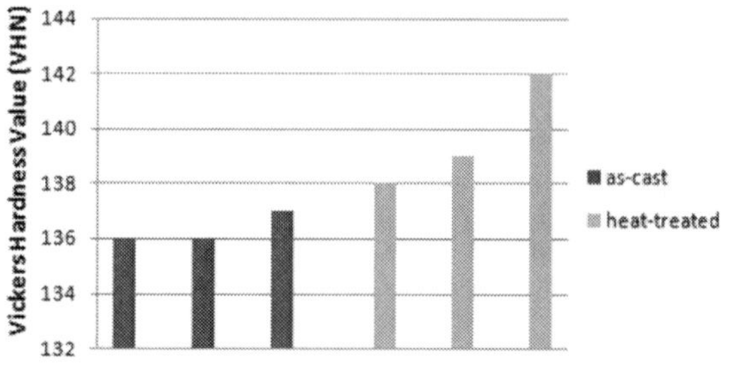

Fig. 1. Heat Treatment effects on hardness of Al-9Zn-5Cu-4Mg

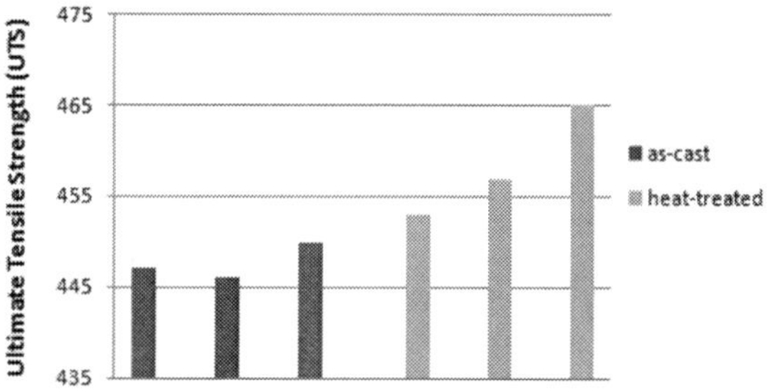

Fig. 2. Heat Treatment effects on tensile strength of Al-9Zn-5Cu-4Mg

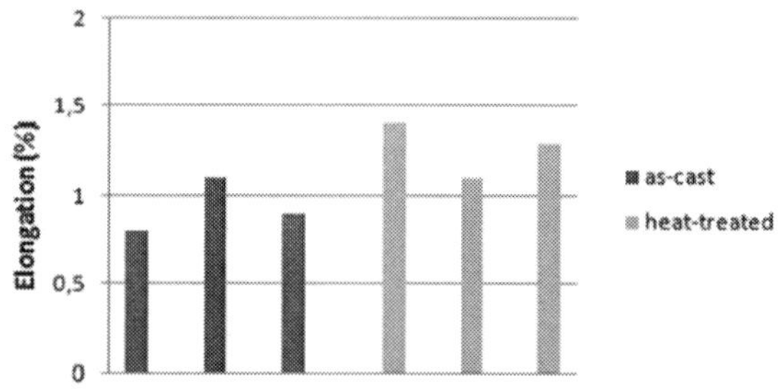

Fig. 3. Heat Treatment effects on elongation of Al-9Zn-5Cu-4Mg

3.2 Microstructure characteristic

On Fig 4 and 5, the heating process leads precipitation on the grain boundaries spread to all over dendrite. Based on EDS result of as-cast Al-9Zn-5Cu-4Mg alloy (see Table 3 and table 4) shows the presence of $Mg_3Zn_3Al_2$ precipitate dispersed finely in Aluminium, and there are also the presence of $Mg_3Zn_3Al_2$, $CuAl_2$, $CuMgAl_2$. The presence of Fe element as impurities made of aluminum become coarse grains, reducing the ductility of alloy [10] and leads the presence of Cu_2FeAl_7.The microstructure of heat-treated cast Al-9Zn-5Cu-4Mg alloy shows the presence of non-uniformly coarse grains of $MgZn_2$ (see table 5) in dendrite.

Fig 6 and Fig 7 are shown the microanalysis report by Energy - dispersive spectroscopy (EDS) of dendrite and grain boundary of as-cast alloy. While Fig 8 and Fig 9 are shown the microanalysis report by Energy - dispersive spectroscopy (EDS) of dendrite and grain boundary of heat-treated alloy.

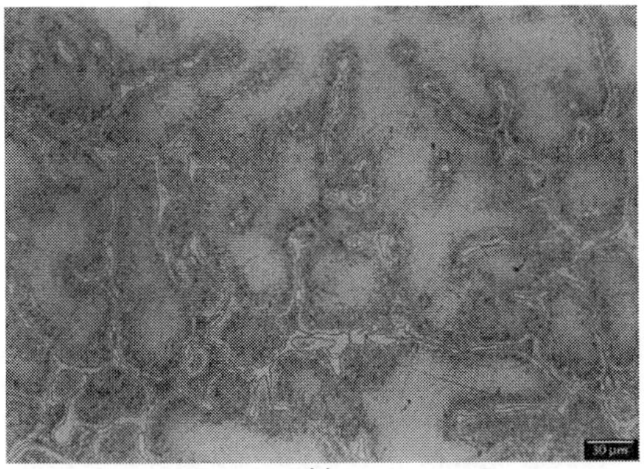

(a)

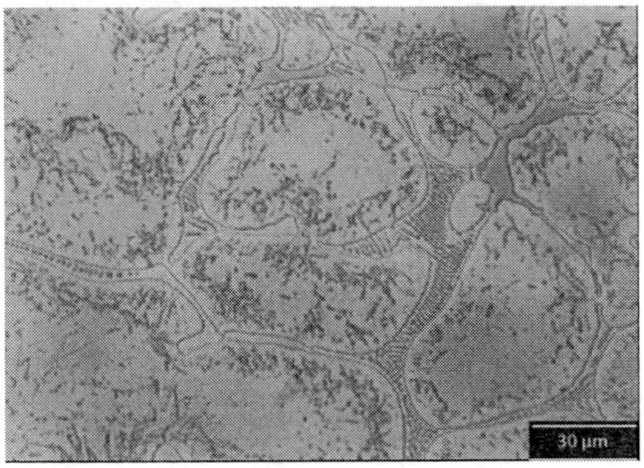

(b)

Fig. 4. Microstructure of as-cast Al-9Zn-5Cu-4Mg alloy (a) magnification 200x (b) magnification 500x

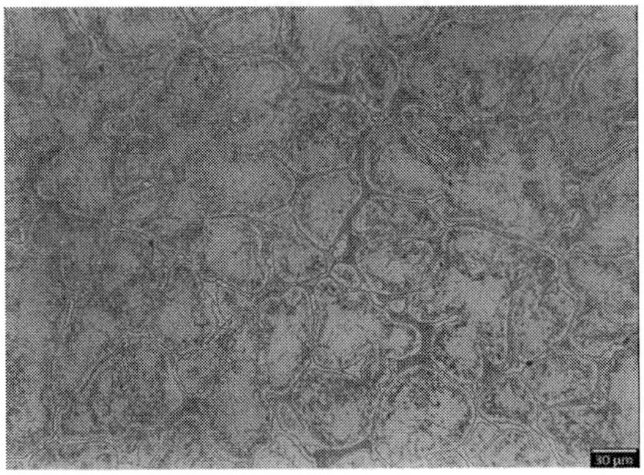

(a)

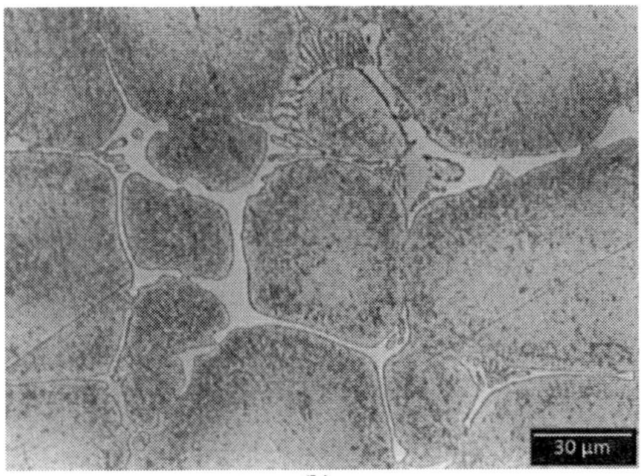

(b)

Fig 5. Microstructure of heat-treated Al-9Zn-5Cu-4Mg alloy (a) magnification 200x (b) magnification 500x

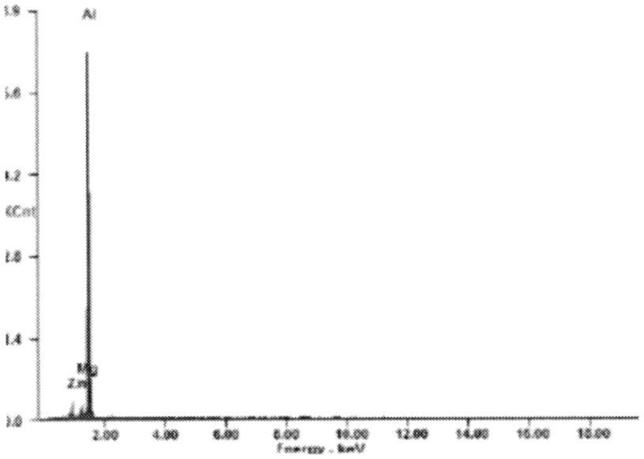

Fig 6. the microanalysis report by Energy - dispersive spectroscopy (EDS) of dendrite as-cast alloy

Table 4. Summaryof EDS observation on dendrite as-cast Al - 9.7Zn - 5.5Cu - 4.5Mg cast alloy

No	Average elementwt.%			Phase may be formed*
	Al	Zn	Mg	
1	90.34	6.81	2.85	$Mg_3Zn_3Al_2$
2	89.40	7.30	3.30	$Mg_3Zn_3Al_2$
3	89.65	7.32	3.03	$Mg_3Zn_3Al_2$

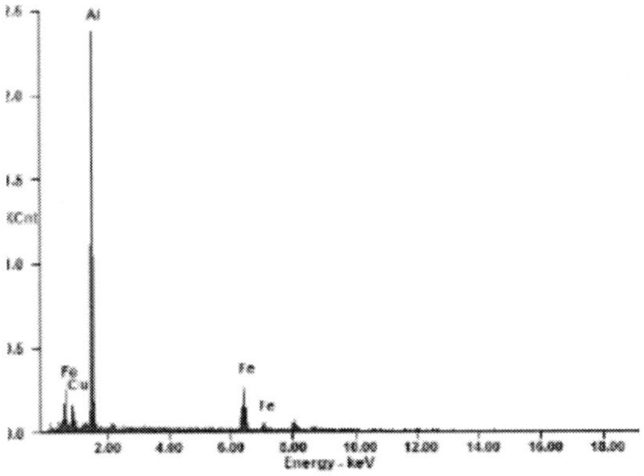

Fig 7. the microanalysis report by Energy - dispersive spectroscopy (EDS) of grain boundary as-cast alloy

Table 5. Summaryof EDS observation on grain boundary as-cast Al - 9.7Zn - 5.5Cu - 4.5Mg cast alloy

No	Average elementwt.%					Phase may be formed*
	Al	Zn	Mg	Cu	Fe	
1	68.59	-	-	11.31	22.09	Cu_2FeAl_7
2	24.72	27.29	23.85	24.14	-	$Mg_3Zn_3Al_2$, $CuAl_2$, $CuMgAl_2$
3	57.83	10.68	13.67	17.82	-	$Mg_3Zn_3Al_2$, $CuAl_2$, $CuMgAl_2$

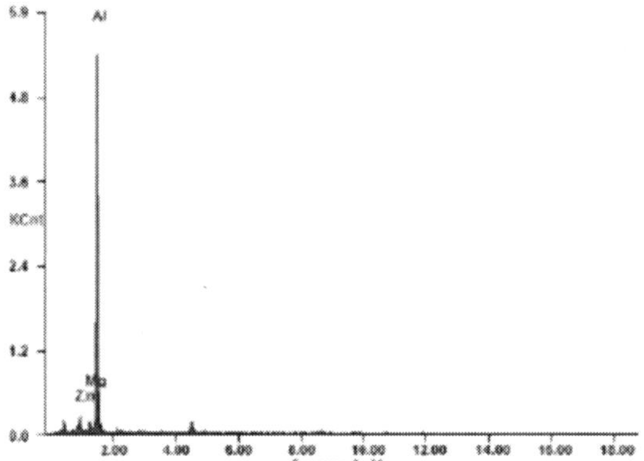

Fig 8. the microanalysis report by Energy - dispersive spectroscopy (EDS) of dendrite heat-treated alloy

Table 6. Summaryof EDS observation on dendrite heat-treated Al - 9.7Zn - 5.5Cu - 4.5Mg cast alloy

No	Average elementwt.%			Phase may be formed*
	Al	Zn	Mg	
1	90.05	7.08	2.87	$Mg_3Zn_3Al_2$
2	89.38	3.09	7.53	$MgZn_2$
3	89.91	2.75	7.34	$MgZn_2$
4	90.52	2.75	6.73	$MgZn_2$

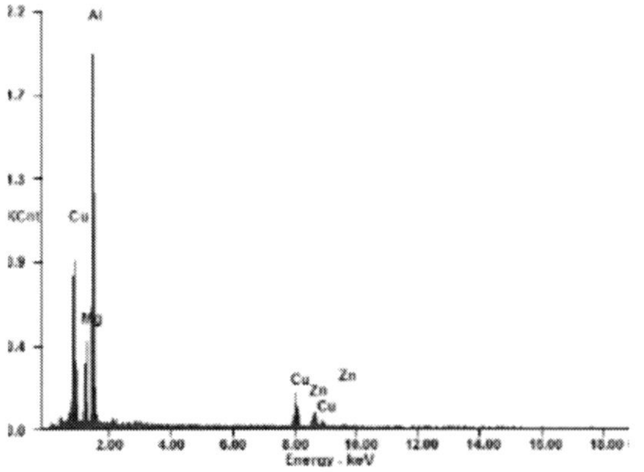

Fig 9. the microanalysis report by Energy - dispersive spectroscopy (EDS) of grain boundary heat-treated alloy

Table 7. Summaryof EDS observation on dendrite heat-treated Al - 9.7Zn - 5.5Cu - 4.5Mg cast alloy

No	Average elementwt.%					Phase may be formed*
	Al	Zn	Mg	Cu	Fe	
1	69.51	-	-	7.70	22.79	Cu_2FeAl_7
2	25.97	25.02	25.2	23.82	-	$Mg_3Zn_3Al_2$, $CuAl_2$, $CuMgAl_2$
3	57.83	10.68	13.67	17.82	-	$Mg_3Zn_3Al_2$, $CuAl_2$, $CuMgAl_2$

*) Prediction phase based on the phase data thatmayappear in Al-Zn – Mg [13]

Aluminium 7xxx series are susceptible to embrittlement because of microsegregation of magnesium zinc (MgZn2) precipitates which may lead to failure of components produced from them. This microsegregation and inherent residual stresses have serious deleterious effects on mechanical properties [14].

Heat treatment of Al-Zn-Mg-Cu alloys led to formation of $MgZn_2$ intermetallic phase in the structure. Many coarse intermetallic phases like $Al_2Mg_3Zn_3$, Mg_2Si, $MgZn_2$, AlCuMg, Al_7Cu_2Fe, Al_2Cu, and $Al_{13}Fe_4$ can be formed below the solidus line during solidification as a result of solute redistribution of metals. There was no room for solute redistribution of Mg and Zn and hence microsegregation of $MgZn_2$ was not formed.

The T5 hea-treated alloy has the highest hardness and strength because of the presence of coherent $MgZn_2$ precipitates in its structure which was developed during aging. The highest hardness value developed by age hardened samples can be attributed to precipitation of coherent and finely dispersed $MgZn_2$ phases which serves as foreign inclusion in the lattice of the host crystal in the solid solution ,it causes more lattice distortions which makes the alloy harder.

4 Conclusion

Based on the results of mechanical testings and microstructure observation,the impairment of the strength and toughness of the heat-treated Al-9Zn-5Cu-5Mg cast alloy were caused of the deployment of precipitation from the edge of dendrite to all over dendrite.It is also caused by the presence of embrittled phase $MgZn_2$ that dispersed evenly throughout the dendrite due to the heat treatment process that conducted at temperature 200° C for 200 hours. T5 heat treatment operation was found to improve hardness value, yield and ultimate tensile strengths with a corresponding decrease in ductility.

References

1. A.D. Isadare, B. Aremo, M.O. Adeoye, O.J. Olawale, M.D. Shittu, Mater. Res. J. Mater, **16** (2013)
2. M. Kaczorowaki, A. Krzynska, TSM, **8** (2008)
3. T. Dursun, C. Soutis, Mater. Des. **56** (2014)
4. J. Linder, M. Axelsson,H. Nilsson , TIP, **28** (2006)
5. P.A. Rometsch, Y. Zhang, S. Knight, Trans. Nonferrous Met. Soc. China, **24** (2014)
6. AD. Isadarea, B. Aremob, MO. Adeoyec, OJ. Olawalec, MD. Shittu. EHT, **16** (2013)
7. J.B. Ferguson, B. Schultz, J. Mantas, H. Shokouhi, P. Rohatgi, Metals (Basel), **4** (2014)
8. L. Yan, Y. Zhang, X Li, Z. Li, EZMA, **24** (2014)
9. R.S. Rana, R. Purohit, S. Das, J. Sci. Res. Publ, **2** (2012)
10. JB. Ferguson, BF. Schultz , JC Mantas, H Shokouhi, PK Rohatgi, ECHT, **4** (2014)
11. N. Nafsin, HMMA. Rashed, ECMH, **2** (2013)
12. K. Yu, W. Li, R. Wang, B. Wang, C. Li, Mater. Trans, **49** (2008)
13. LF Mondolfo, *Aluminium alloys : Structure and Properties* (Butterworths, London,1979)
14. D.A. Isadare, M.O. Adeoye, A.R. Adetunji, K.M. Oluwasegun, A.L. Rominiyi, K.J. Akinluwade, **5** (2015)

DOI: 10.1051/matecconf/201710101010

Physical and mechanical properties of membrane Polyacrylonitrile

Agung Mataram[1,*], *Syahrul* Nasution[1], *Mazari Legi* Wijaya[1], *Gurruh Dwi* Septano[1]

[1]Mechanical Engineering Department Engineering Faculty Universitas Sriwijaya, 30662 Inderalaya, Indonesia

Abstract. Tensile test is a method used to measure the strength of a material by providing the burden of the opposing forces in a straight line. This experiment used to measure a material's mechanical strength against a static force which is given slowly. This study discusses on the tensile strength polyacrylonitrile (PAN) membranes made in three variations of the specimen with a mixed composition of different PAN 15%, 17.5%, 20%, in weight). The membrane was created using Sigma Aldrich Polyacrylonitile of polymer fibers and N,N-Dimethylformamide (DMF) as solvent. The process began by mixing the two materials using a magnetic stirrer for 6 hours until the solution is assumed to be homogeneous. The solution wass stored in the bottle airtight and allowed to stand for 24 hours for the next printing. The membrane was keep in storage in tubes desiccator until its water content reduced, then the test can be done. In this study, a machine adhesion terring strength tester was used conducting Strength Test at a speed of 5 mm/min. The test results showed an increase in the value of 359 kPa on the composition of 15%, 473 kPa on the composition of 17.5%, and 653 kPa for the composition of a mixture of 20%.

1 Introduction

Microbiological contamination of some water sources that exist at the moment is very worrying to the public, according to some researchers, there are many different species of bacteria which are available in pure water and wastewater [1,2].

This application requires a membrane that can improve water quality. The membrane is a layer formed from the composition or the tangle of fine fibers used as filter or barrier for a variety of applications in the industrial world. polymer has been the material of interest to be analyzed and developed its potential as a membrane filter, application examples that are of concern at this time is a membrane filter water [3].

The main objective to be achieved in this discussion are:

- To develop a new method in the production of membrane
- To determine the mechanical properties of the membrane of polyacrylonitrile (PAN) against tensile testing

The membrane is a layer formed from the composition or the tangle of fine fibers used as filter or barrier for a variety of applications in the industrial world.

PAN is one of the versatile polymers that is commonly used for producing membranes due to its good solvent resistance property [10]. Membrane polymer produced using Polyacrylonitrile as fiber polymer and DMF as a solvent, the process of printing the membrane by the method of a flat sheet, the membrane of this type is typically used in ultrafiltration process, due to higher porosity and pore structure are interconnected, resulting in permeability is higher for water filtration [4].

Characterization of membrane was conducted to determine the properties of the membrane resulting from any research, this method will provide information on the extent to which efficiency and effectiveness can be obtained from the membrane that has been made, a technique that can be used to determine the characteristics of the membrane there are a number of ways, but a common used permeability and permeselectivity.

2 Methodology

2.1 Membrane preparation

Dominated by carbon fiber, Polyacrylonitrile fibers due to a combination of the polymer, has good mechanical properties, especially tensile strength [3].

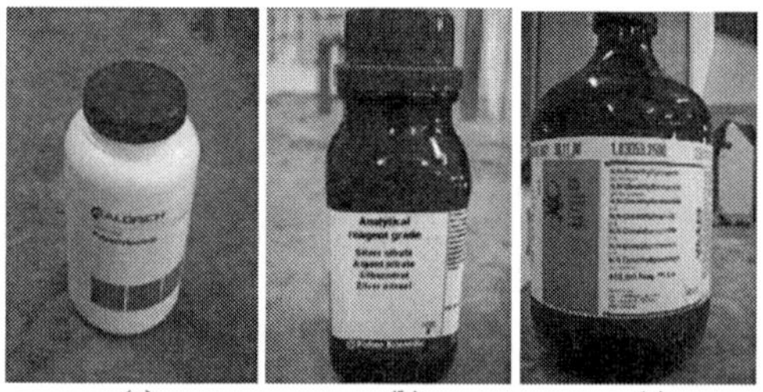

(a) (b) (c)

Fig. 1. (a) Polyacrylonitrile (PAN), (b) Silver Nitrate, (c) N,N-Dimethylformamide

* Corresponding author: amataram@unsri.ac.id

The advantage of the PAN fiber, high degree of molecular orientation and a higher melting point (PAN fiber tends to decompose below the melting point, Tm of 317-330° C), and the results were better than carbon fiber [5]. Polyacrylonitrile used (Figure 1a). Generally, carbon membrane is known by its amorphous porous structure created by the evolution of gases generated during pyrolysis of the polymeric precursor [10].

Polyacrylonitrile can form a thermally stable molecular structure of highly oriented while getting a low temperature heat treatment, which is not significantly impaired during treatment carbonization at a higher temperature, this means the fibers produced have excellent mechanical properties [6,7].

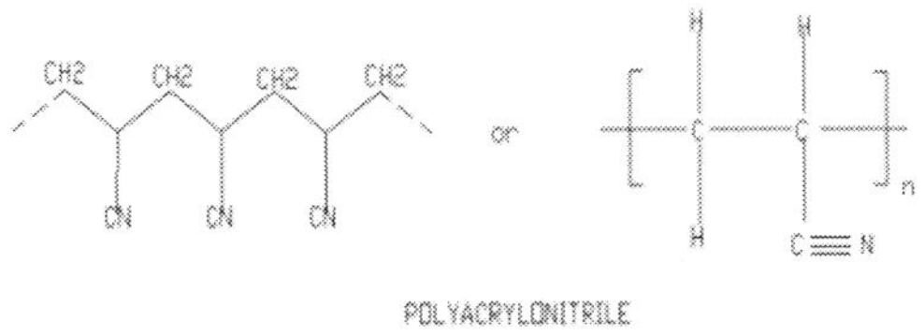

Fig 2. The molecular structure of polyacrylonitrile [10]

Silver is a chemical element in the periodic table that has the emblem of Ag and atomic number 47. arms is derived from the Latin Argentum. A transition metal is soft, white, shiny, silver has the highest electrical conductivity and heat throughout the metal and there dimineral and in free form. This metal is used in coins, jewelry, equipment tables, and photography. Silver including precious metals such as gold [8].

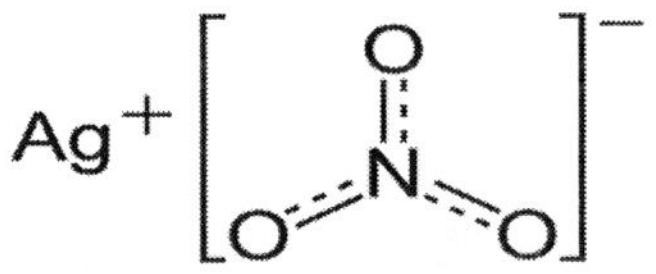

Fig. 3. Molecular Structure of Silver Nitrate

Chemical Properties Silver Nitrate (Fig 3):
- Very stable in pure air and water so it does not react with air and water at normal temperature.
- Less reactive than copper, except ozone, sulfur and hydrogen sulfide, which is rapidly discolor the silver surface.
- Not soluble in dilute acid solution, but can be dissolved in nitric acid and sulfuric acid.
- Silver and silver compounds are all soluble cyanide as the alkali metal ion Ag (CN) 2 in the free air, which is then released by adding zinc silver or aluminum as reluctant.
- Anion can precipitate as most anions forming insoluble silver salts. For example, the addition of the Ag + OH- will precipitate silver (I) oxide.

$$2Ag^{+}_{(aq)} + 2\,OH^{-}_{(aq)} \rightarrow Ag_2O_{(s)} + H_2O \qquad (1)$$

N,N-Dimethylformamide (DMF) supplied by MERCK (figure 4) and is used as a solvent without further purification, this is a strong solvent for PAN polymer, because they have desirable properties such as low volatility, non-flammable, and relatively low toxicity.

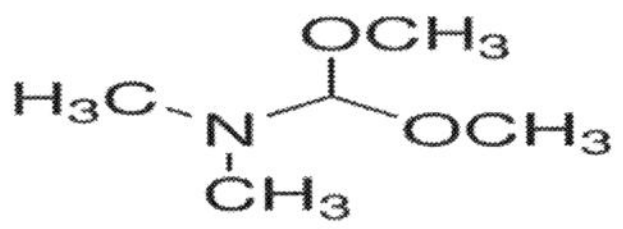

Fig. 4. N,N-Dimethylformamide

2.2 Preparation of Polymer Membranes

Membranes made in this study were conducted with two materials with different variations of the mixture on each specimen, materials used is PAN and DMF. The mixing of the two materials using a Magnetic Stirrer, this tool will mix the two ingredients with the rotation system so that the mixture becomes homogeneous [9].

Methods flat sheet is one way in the process of making the membrane author uses glass plates and duct tape as a template membrane, membrane PAN prepared by phase inversion method immersion in a tub of coagulation of water, the result of printing membranes with these methods form a flat sheet [9]

Tensile test is a method used to test the strength of a material by providing load coaxial style. This experiment used to measure the mechanical strength of a material is the force of static given slowly, the results of tensile testing is essential to engineering and product design for yield strength data material tensile testing done to complete the design information is basic mechanical strength of a material and as a data support for material specifications.

The purpose of tensile or mechanical testing is to measure the effect of force on a particular material or component, or on the adhesive or fastener that bonds two materials together.

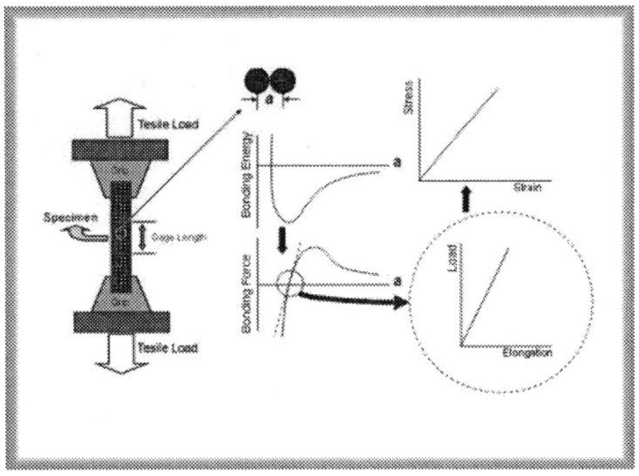

Fig. 5. Schematic of Tensile Test [souce: etomica]

As seen in Fig 5, the specimen material is pulled by tensile load at a constant rate until it fractures. During the test, the tensile load and the elongation of the gage length of the specimen are monitored and Microstructure Test tool SEM (Scanning Electron Microscope), which uses an electron microscope. SEM testing is done in stages on each membrane composition. Scanning Electron Microscopy (SEM) has been used to investigate the resultant membrane morphology [10].

SEM testing is used to view and analyze surface and texture, shape and size of objects to analyze the sample and analyze the composition of the surface of objects both quantitatively and qualitatively. At this time, through the SEM Test, the size of the pores can be performed. The voids or holes in the membrane and the bond that exists every pore in the membrane were detected as well.

Fig. 6. Adhesion Tearring Strenght Test

The test in this research conducted procedure comprising the steps the steps below:

- Preparation of the membrane material,
- Polyacrylonitrile and N, N Dimethylformamide.
- Prepare materials stirrer (magnetic stirrer).
- Considering mixture of PAN and DMF prior to the
- mixing process takes place.

recorded.

- Stir the two ingredients for 6 hours under on each specimen
- Print the membrane solution that has been homogenized using glass mold that has been provided.
- Prepare a tensile testing machine tearing adhesion strength test (Figure 6). After that perform appropriate testing procedures
- Prepare test equipment SEM, once ready to do appropriate testing procedure.

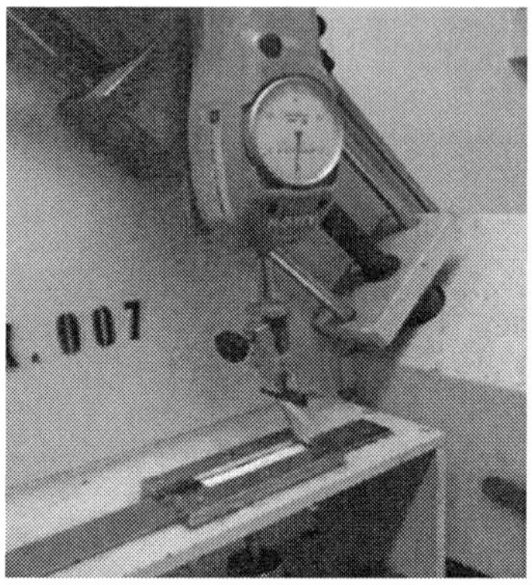

Fig. 7. Process polyacrylonitrile membrane tensile testing

3 Result and discussion

In the current work, the test results show the mechanical properties of the PAN membrane material, each specimen using a membrane tensile testing machine tearing adhesion strength test (as seen in Fig. 7).

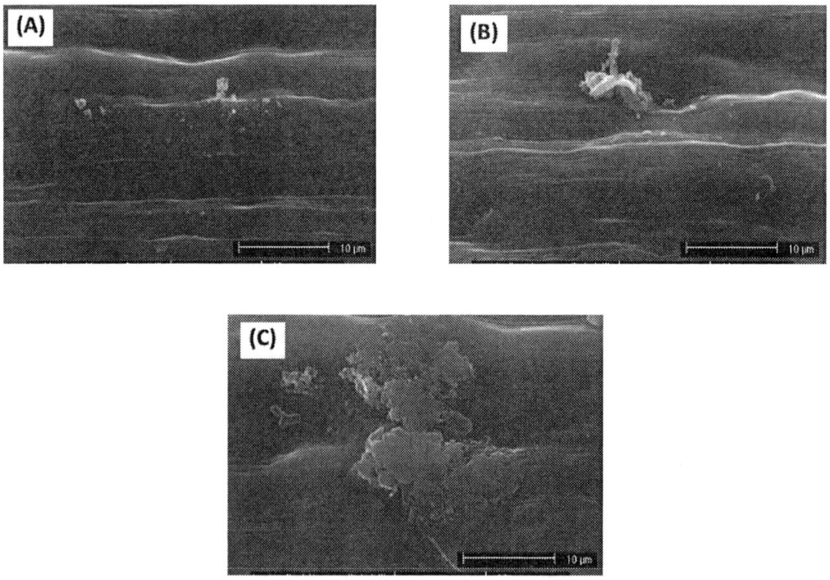

Fig. 8. SEM comparison PAN fibers with different variations (A) 20% by weight, (B) 17.5% by weight (C) 15% by weight.

This test is generally aimed to determine the mechanical properties of a material. Specifically in this study, the tensile resistance of the membrane is also evaluated. The additional results was also collected with some test using Scanning Electron Microscopy (SEM).

The results SEM micrograph of the surface of PAN fiber is shown in Fig.8, when the PAN fiber content is at 17.5% and 20%, PAN nanoparticle dispersions relatively simple homogeneous, but with the added amount (%. PAN) in a solution with a composition of 20%, occur agglomeration excessive resulting solution is more difficult to be homogeneous [10,11].

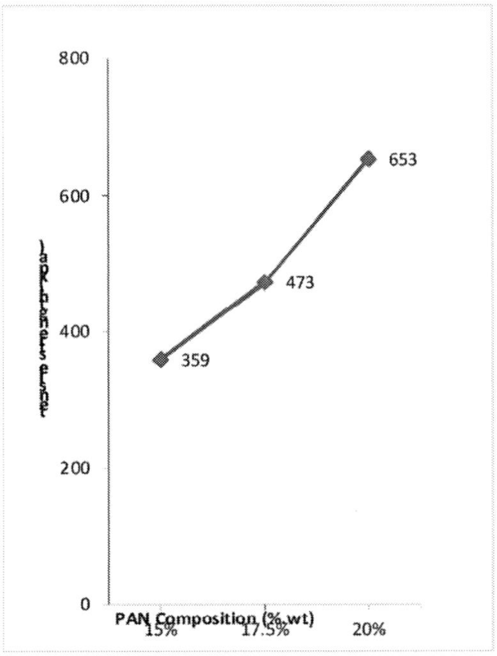

Fig 9. The tensile strength (A) of the PAN composite fiber with changes in the composition (%w of PAN) on each specimen.

Tensile test specimens were having a composition respectively 15%, 17.5%, 20% PAN. The tensile strength of the specimens showed increases in the value of 359 kPa to 600 kPa when the specimen with a composition of PAN fiber increased from 15% to 17.5%, the value of the tensile strength of the fiber membranes with composition (weight %) 20% PAN also showed an increase up 653Kpa.

The rising value of the tensile strength of the membrane of 359Kpa to 653 kPa is due to the increase amount (%w) of PAN on each specimen. It may caused by fiber braid structure of the membrane which is increasingly interacted [12].

4 Conclusion and suggestion

4.1 Conclusion

Based on the research results polyacrylonitrile membrane tensile testing, as described, it could be concluded as follows:

- The length of time and the rotation speed of the mixing process of membrane material influence on the results of the homogeneity of the membrane solution.
- The temperature at the time of stirring is also important to remove residual solvent and trapped air that can cause bubbles on the membrane
- Results of testing the best there is in the composition of 20% PAN because it can withstand loads of up to 824 kPa.

4.2 Suggestions

- To obtain a more accurate tear resistance are advised to set clamp arm so that when the testing took place the specimen is not moving, because if it moves, then the result would be different and affects the outcome of the cycle.
- Once the membrane is printed should be immediately placed in a tube desiccator until its water content is reduced, when no water should remain immersed in the coagulation so that the membrane does not become brittle thus reducing capable mechanic of the membrane itself.
- In SEM testing, to obtain better results SEM electricity entering the SEM tool must be stable and at the time of testing in order to cut into small pieces and flat membranes and flat. After that do plating / coating on non-conductive specimens in order to study the morphology of the membrane that is clearly visible.

References

1. Bonnélye, V., Gueya, L. and Del Castillo, J, *Desalination*. 222, 59–65, (2008)
2. Mulder, *M. Basic Principles of Membrane Technology. (1st ed.)* London: The Netherlands, Kluwer Academic Publisher. (1991)
3. Masuelli, M.A.. INTECH. Chap 1. (2013)
4. Thavasi, V, Singh, G. and Ramakrishna, S. *Energy and Environmental Science*. 1, 205-221, (2008)
5. Fitzer, E. and Manocha,L. M,. *Carbon Composites (Ch. 1)*. New York: Springer-Verlag. (2007)
6. Donnet, J. B. and Bansal,R. C, *Carbon Fibres. Marcel Dekker*: New York. (1984)
7. Schindler, E. and Maier,F. US Patent 4919860. Washington: U. S. Patent and Trademark Office, (1990)
8. Goldman, G, Starosvetsky, J. Armon, R, *Journal Membrane Scence*. 341, 145–152. (2009)
9. Mataram, A., Ismail, A.F., Yuliwati, Y., Matsuura, T., Zamheri, A., Rizal, S. Application of Electrospun Nanofibers. 77:1, 263-267. (2015)

10. Saufi, S. M., Ismail. A.F,. Journal of Science and Technology. 24, 843-854, (2002)
11. Sri Aprilia et al. *Report.* Jurusan Teknik Kimia, Fakultas Teknik, Universitas Syah Kuala. Banda Aceh (2003)
12. Yoneyama, H. and Nishihara. Y, US Patent 0494449. Washington: U. S. Patent and Trademark Office, (1990)

Magnetic properties of barium ferrite after milling by high energy milling (hem)

Novrita Idayanti[1], *Tony* Kristiantoro[1], *Ardita* Septiani[1] *and Ika* Kartika[2]

[1]Research Center for Electronics and Telecommunications, LIPI, Bandung 40135, Indonesia
[2]Research Center for Materials and Metallurgy, LIPI, Banten 15314, Indonesia

Abstract. Magnetic properties of barium ferrite that were mashed by High Energy Milling (HEM) has been characterized. The starting iron oxide powder (Fe_2O_3) and barium carbonate ($BaCO_3$) were prepared by powder metallurgy technique by the stages of mixing, calcining, milling, compacting, and sintering. Weight ratio of Fe_2O_3: $BaCO_3$ was 88.74 wt% : 17.52 wt%. Some additives such as calcium oxide (CaO), silicon dioxide (SiO_2), and polyvinyl alcohol (PVA) were added after calcining process. Milling by HEM was varied at 30, 60, 90, 120, and 150 minutes. The ball and container of HEM were made of stainless steel. Characterization micro structure by SEM showed that the milling time affect the grain size of the magnetic powder. The longer of milling time, the grain size was smaller and uniform. Characterization by using magnetic instrument Permagraph showed that the grain size will affect the magnetic properties of barium ferrite. Induction of remanence (Br), coercivity (HcJ), and product energy maximum (BHmax) values increased with increasing milling time. The optimal magnetic properties were obtained at the time of milling 120 minutes with value of Hc = 1.97 kG, HcJ = 2.314 kOe, and BHmax = 0.64 MGOe.

1 Introduction

Barium ferrite are known as ceramic magnet and permanent magnet materials which are based on iron oxide (Fe_2O_3) and barium carbonate ($BaCO_3$) [1]. This magnetic has a hexagonal structure, so it is also known as barium hexaferrite. The structure is magnetoplumbite/M-type with the general formula is $MFe_{12}O_{19}$ or $MO.6Fe_2O_3$, where M is Barium (Ba), strontium (Sr) or lead (Pb) [2]. Fig. 1. shows crystal structure of M-type barium hexaferrite $BaFe_{12}O_{19}$. This structure can be symbolically described as S*RSR*S* where R denotes a three-layer block containing two O4 and one BaO3 with the composition $Ba^{2+}Fe_6^{3+}O_{11}^{2-}$, while S denotes a two-layer block containing two O4 with the composition $Fe_6^{3+}O_8^{2-}$. Crystal structure of M-type barium hexaferrite $BaFe_{12}O_{19}$ [3]. The colours of spheres; red, pink, gray, blue and yellow represent iron atoms in 2a,4f1,12k,4f2, and 2b sites respectively. Small green spheres represent oxygen atoms, while larger cyan spheres represent barium atoms [4].

This magnet having high coercivity, high temperature curie, large saturation magnetization, good chemical stability, and corrosion resistive [5]. Use of ferrite magnet materials are one of the most widely in the world although they have less magnetic strength than rare earth magnets. This is because the magnets are cheaper than rear earth magnets [6].

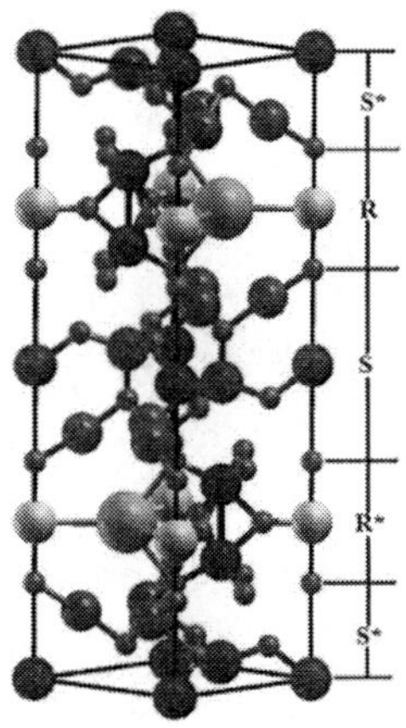

Fig.1. Cell unit of Barium Hexaferrite [4]

There are many methods have been developed for the preparation of barium ferrite to get excellent magnetic properties. YY. Meng, Li Jie, and Zhidong [7-9] have synthesize barium ferrite by sol gel method by reacting iron nitrate ($Fe(NO_3)_3$), barium nitrate ($Ba(NO_3)_2$, and citric acid were dissolved into distilled water. Co-precipitati on method has been carried out by the researchers by reacting a solution of iron in the ferric chloride with barium chloride in water was poured into an $NaOH/Na_2CO_3$ alkaline solution [10-11]. R. Nowolsieski [12] and many researchers have been using powder metallurgy for making barium ferrite magnet by

*Corresponding author: novr001@lipi.go.id

DOI: 10.1051/matecconf/201710101011

the reaction of iron oxide and barium carbonate. Among these methods, powder metallurgy process has been considered the best method and the most often use to produce barium ferrite in large scale product. The main advantage of this methodis able to produce a ceramic material that has a high melting point. Many factors in the powder metallurgy process that could affect the properties of magnetic materials such as grain size of the powder after milling. Mechanical treatment ball milling is now recognized as a powerful tool to get fine ferrite powder. In this research refining the grain size of magnetic powders do by High Energy Milling (HEM) with variations of milling time. Magnetic properties after milling observed by Permagraph, micro structure determined by using SEM and the formation of ferrite compounds by XRD.

2 Experimental details

A mixture of iron oxide (Fe_2O_3) and barium carbonate ($BaCO_3$) was made base on the composition formula of barium ferrite $BaO_{3.98}Fe_2O_3$. The calculation chemical composition based on impurities of iron oxide was obtained weight ratio of Fe_2O_3 : $BaCO_3$ was 88.74 wt% : 17.52 wt%. Both materials were mixed and milled by High Energy Milling (HEM) type ultimate gravity with various time 0, 30, 60, 90, and 120 minutes. 0 minute means that the mixture did not do milling by HEM. Alcohol was added into a mixture as much as 60% wight of mixture. The next step was calcination process in a furnace (Thermolyne) at temperature 1200°C for 3 hours. Calnined powders were characterized by SEM JEOL JCM 6000 USA to determine microstructure and grain size. Analysis of ferrite coumpound after calcination process by using XRD. Some additives such as silicon oxide (SiO_2), calcium oxide (CaO), and polyvinyl alcohol (PVA) was added approximately 0.75 – 1.5 weight % of calcined powder. The next process, the mixture was milled again by HEM for 60 minutes to obtain a homogeneous powder. After that, the mixture was dried at temperature 100°C and sieve with a shieve shaker 400 mesh and the magnet powders were ready to be pressed. The amount of the samples was 2 g, was performed in disk shape with a diameter 11.55 mm and a thickness 5 mm. The compacting pressure at 5 tons/cm^2 by hydraulic press. The compact samples were placed on an alumina substrate and were sintered at temperature 1250°C in air atmosphere with one hour holding time and heating rate was 10°C/min. Magnet samples that have been sintered tested by Permagraph Magnet Physik Germany to determine magnetic properties of the samples. Flow chart of the experiment by powder metallurgy process can be seen in Fig. 2.

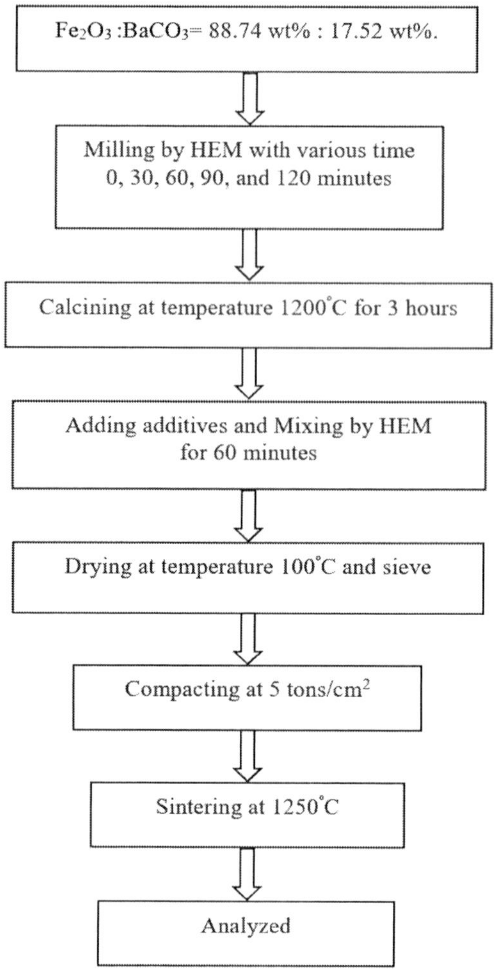

Fig.2. Flow Chart of Process

3 Results and discussion

The observation result of microstructure and grain size of samples that milled at 0 minutes until 120 minutes can be seen in Fig. 3.

SEM analysis shows that milling time can influence microstructure and grain size of the samples. Picture was observed at magnification x1000. In Fig. 3. (a) sample (0 minutes) still has large grain size and inhomogeneous. After milling process 30-120 minutes, the grain size become homogenous and smaller. Homogenous distribution of ferrite powder particles after milling is an important factor that could affect the formation of ferrite and the magnetic characteristic [13]. The grain size will affect the magnetic properties because the grain size and grain boundaries will disrupt the movement of momen magnet [14]. The increase in magnetic moment depends on the structure of the grain boundary and grain size [15]. It will be proved after the measurement of magnetic properties.

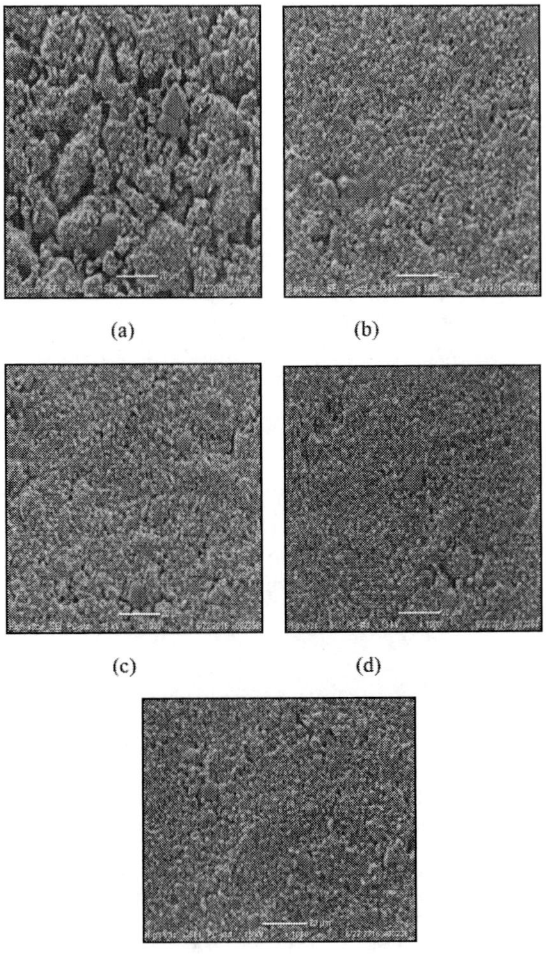

Fig.3. Microstructure and Grain Size of Samples

In Fig. 4. powder XRD patterns of calcine powder with different milling times (0 – 120 minutes) are shown. Sharp peak indicating the high crystallinity of the samples. The blunt peak showed crystalline sample is not perfect. Samples were milled at the times 0 – 60 minutes has not had sharp peaks. Sharp curve has shown after milling 90 minutes. The compound formed is barium ferrite ($BaFe_{12}O_{19}$), still there are hematite (Fe_2O_3), and magnetite (Fe_3O_4) that unfinished reacted with barium carbonate ($BaCO_3$). For the sample milled 120 minutes, the formation of barium ferrite compound is better. These effect indicate that milling process by HEM on ferrite powder will influence the formation of barium ferrite.

XRD characterization results of samples that have been sintered can be seen in Fig. 5. The curve of sintered powder are sharper than calcined powder. The coumpound is formed almost entirely is barium ferrite. Fe_3O_4 has reacted become $BaFe_{12}O_{19}$, while Fe_2O_3 compound is still there.

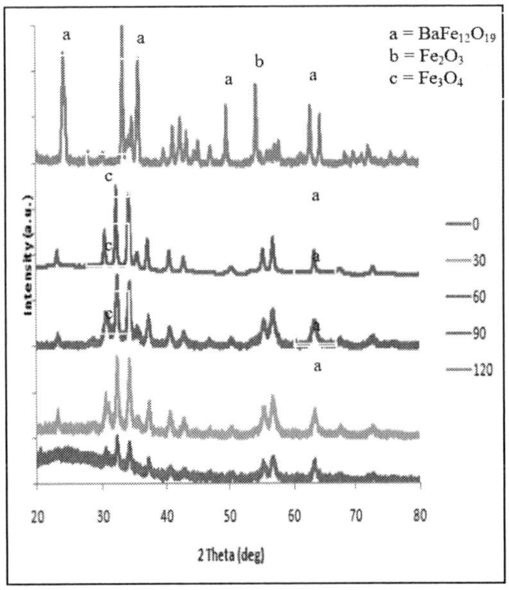

Fig.4. Diffraction Patterns of Calcined Powder

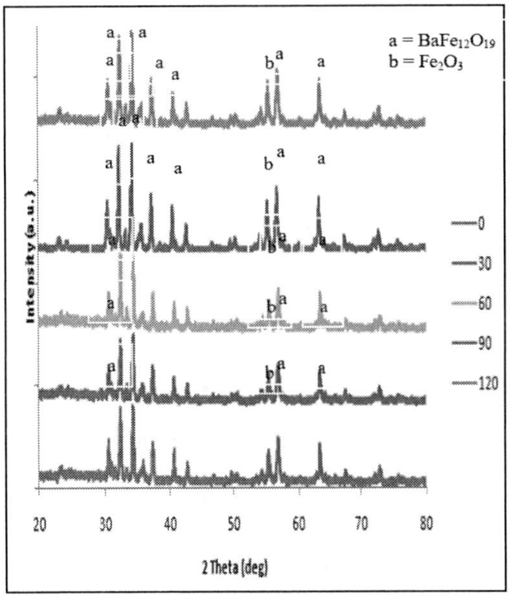

Fig.5. Diffraction Patterns of Sintered Powder

a

On Fig. 4. And Fig. 5. can be seen that after the sintering process the formation of barium ferrite magnetic material more complete and perfect. It can be shown from number of compound $BaFe_{12}O_{19}$ that appear on the XRD curve of

Fig.6 shows hysteresis loop of barium ferrite after milling 0, 30, 60, 90, and 120 minutes. The powder were observed after sintering 1250°C by using Permagraph.

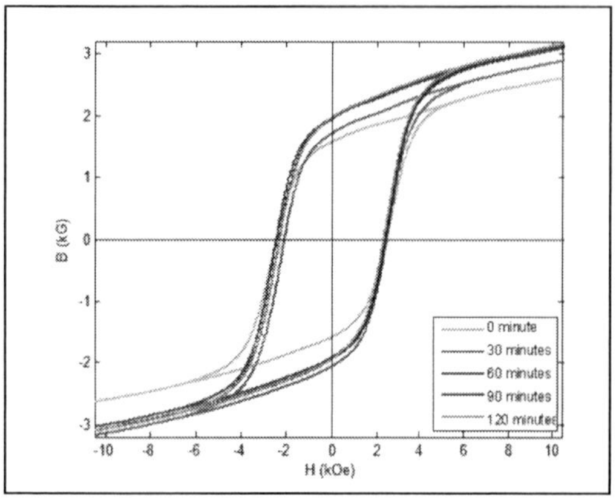

Fig.6. Hysteresis Curve of Samples

Fig. 6. Hysteresis loops for milled samples showing the effect of milling process on the shape of the demagnetization curve. At longer milling times, saturation magnetization keeps nearly constant, while that induction of remanence (Br) increase. Increasing in coercivity (HcJ) is associated to high density of internal stresses. Decreasing of product energy maximum (BHmax) as a consequence of the grain size is below of the exchange correlation length. High values of coercivity are related to particles are significantly smaller than the single-domain size. Sample which has the best magnetic characteristics is sample milled 120 minutes with the value of Br = 1.97 kG, HcJ = 2.314 kOe, BHmax = 0.64 MGOe, and density 5.01 g/cm^3. The magnetic properties of samples obtained from hysteresis curve of barium ferrite is shown in Table 1 below.

Table 1. The influence of milling time on the magnetic properties

Magnetic Properties and Density	**Variations of milling time (minutes)**				
	0	**30**	**60**	**90**	**120**
Br (kG)	1.57	1.88	1.94	1.95	1.97
HcJ (kOe)	2.453	2.244	2.419	2.339	2.314
BHmax (MGOe)	0.50	0.70	0.74	0.74	0.64
ρ (g/cm^3)	4.23	4.80	4.85	4.88	5.01

K. Sadhana [16] has done research to observe effect of grain size on the structural and magnetic properties of nanocrystalline $Al_3Fe_5O_{12}$. The grain size can influence coercive field (Hc). It was observed that Hc increases with an increase of D and further increase until Dc, Hc decreases. The boundary between single domain and multidomain particle is given by critical diameter of the particle (Dc).

4 Conclusion

Barium ferrite was sucessfully characterized after milling by HEM. The milling time influences of grain size, formation of compound, and magnetic properties of barium ferrite. The following conclusions are :

- The longer milling times will make increasingly refined powder.
- Density of ferrite powder increased.
- Increas in density can affect magnetic properties
- The formation of barium ferrite most perfect is the ferrite powder the milling time of 120 minutes.
- Magnetic properties is better because fine and homogenous powder can align magnetic moment easily.

In the next research, addition of milling time should be added to determine the influences.

The authors greatly to thank you for the financial support provided by LIPI under the Superior Research Program (Riset Unggulan) 2016.

References

1. R. Nowosielski, R. Babilas, G. Derce, L. Pajak, J. Wrona, Structure and properties of barium ferrite powder prepared by milling and annealing, *International scientific J. published monthly as the organ of the committee of materials sciences of the polish academy of sciences,* **vol. 28**, issue 12, P 736-742 (2007)
2. D. Bahadur, Influence of fuel ratios on auto combustion synthesis of barium ferrite nano particles, *J. Chemical Science*, **Vol. 118,** 1, P 15-21 (2006)
3. Z. Somogyvan, E. Syab, K. Krezhov, L.F. Kiss, D. Kaptas, I. Vincze, E. Beregi, F. Bouree, *J. of magnetism and magnetic materials,* **Vol. 304,** issue 2, P e775-e777 (2014)
4. A. Moitra, S. Kim, S.G. Kim, S.C. Erwin, Y.K. Hong, J. Park, *Computational Condensed Matter,* **Vol. 1**, P 45-50 (2014)
5. R. Nowosielski, R. Babilas, J. Wrona, Microstructure and magnetic properties of commercial barium ferrite powder, *J. of Achievements in Materials and Manufacturing Engineering*, **vol. 20,** issue 1-2 (2007)
6. M.Y. Jassem, Preparation of barium ferrite $BaFe_{12}O_{19}$ and the study of the effect of fast neutrons on its magnetic and dielectrical properties, *College of Basic Education Researchers J. University of Mosul/ College of Basic Education,* **Vol.11 No.1** (2011)
7. Y.Y. Menga , M.H. He a , Q. Zeng a , D.L. Jiao a , S. Shukla b , R.V. Ramanujan b , Z.W. Liu, Synthesis of barium ferrite ultrafine powders by a sol–gel combustion method using glycine gels, *J. of Alloys and Compounds,* **583**, P 220–22 (2014)
8. Li Jie, Zhang Huai-Wu, Li Yuan-Xun, Liu Ying-Li, and Ma Yan-Bing, The structural and magnetic properties of barium ferrite powders prepared by the sol gel method, *Chin. Phys. B ,Chinese Physical Society and IOP Publishing Ltd,* **Vol. 21, No. 1** (2012)
9. H. Yaoming Z., Ze WU, Xianyou Z., Synthesis of hexagonal barium ferrite nanoparticle by sol-gel method, *Rare Metals Elsevier,* **Vol. 25,** Issue 6, Supplement 1, P 605-608 (2006)

10. S.R.Janasi, M.Emura, F.J.G.Landgraf, D.Rodrigues, the effects of synthesis variables on the magnetic properties of coprecipitated barium ferrite powders, *J. of Magnetism and Magnetic Materials,* **238**, P 168–172 (2002)
11. Y. Atassi, I. Seyd Darwish, M. Tally, Preparation and Characterization of Nano-particle Substituted Barium Hexaferrite, *Chem. Physic Cornell Univ. library* (2014)
12. R. Nowosielski a, R. Babilas a, G. Dercz b, L. Pająk, Microstructure of polymer composite with barium ferrite powde, *J. of Achievements in Materials and Manufacturing Engineering,* **Vol. 31,** Issue 2, P 269-274 (2008)
13. W. Kaszuwara, A.Witowski, M. Leonowicz, P. Pawlik, J. Paszula, effect of milling medium on the structure and magnetic properties of mechanically alloyed barium ferrite, *Rev. Adv. Mater. Sci.* P 497-500 (2008)
14. I.R. Haris, I.P. Jones, Grain Boundarie: their character, characterization and influence on properties, Maney Publishing (2001)
15. Dr. S. Ii S, discovery of increase in magnetic moment at grain boundaries, *Scripta Materials, National Institute for Materials Science Kumamoto University* (2013)
16. K. Sadhana, S.E. Naina Vinodini, R.Sandhya and K. Praveena, effect of grain size on the structural and magnetic properties of nanocrystalline $Al_3Fe_5O_{12}$ by aqueous coprecipitation method, *Adv. Mater. Lett.* P 99-107 (2015)

Hardness improvement on low carbon steel using pack carbonitriding method with holding time variation

Poppy Puspitasari[1,3*], *Andoko* Andoko[1,3], *Heru* Suryanto[1,3], *Puput* Risdanareni[2], *Sandy* Yudha[1]

[1]Mechanical Engineering Department, Engineering Faculty, Universitas Negeri Malang, 65145 Semarang, Malang, Indonesia
[2]Civil Engineering Department, Engineering Faculty, Universitas Negeri Malang, 65145 Semarang, Malang, Indonesia
[3]Centre of Nano Research and Advanced Material, Universitas Negeri Malang, 65145 Semarang, Malang, Indonesia

Abstract. Carbonitriding is a surface hardening process of steel by heating on critical temperature, quench and followed by tempering process. In this research, pack media in carbonitriding was used as a new method. Usually, this process has been obtained using gas and liquid carbonitriding methods. The specimen used in this research is low carbon steel (St. 41) consist of 0.1517% carbon, 0.1994% of silicon, 0.5631% of manganese, 0.0224% of phosphorus, and 0.047% of sulfur. Temperature for pack carbonitriding was at 700 °C, 750 °C and 800 °C and holding time variation 60 minutes and 120 minutes respectively. The result showed that carbonitriding temperature difference affect the mechanical properties of steel St. 41. Steel hardness was increased at lower holding time (60 minutes) compared to 120 minutes. The result showed that at 700 °C and 750 °C with 60 minutes variation, the steel hardness increase from 85.7 HRB to 95.7 HRB and at 800 °C the hardness decrease to 93.1 HRB. Meanwhile, at holding time of 120 minutes, steel hardness decrease from 94.1 HRB to 92.7 HRB. This result caused by austenite phase produced at longer period of holding time.

1 Introduction

Surface hardening technique is a process to improve the properties of hardness and the performance of a component or material. Material damage usually starts from the flaw on the surface of the material due to the influence of environmental factors such as corrosion or wear due to the interaction with other components.

Various methods were used to improve the quality of the material surface, in accordance with the need of the desired product. Surface hardening technique commonly known as "surface treatment". Generally surface treatment can be obtained by two methods, first method is the addition of other chemical elements into the composition of the material (changing the chemical composition of the surface of a material), or it can also be achieved by the method of changing the phase or crystal structure contained in the material.

Some methods of surface treatment are nitriding, carburizing, carbonitriding, electric induction, and flame. Nitriding, carburizing and carbonitriding are surface treatment method by changing the surface properties with the addition of nitrogen or carbon into material properties. While the method of flame and electric induction is a surface treatment process by heating of material at a certain temperature followed by a rapid cooling process.

Carbonitriding is a steel surface hardening process on critical temperature by spraying gas. During carbonitriding process, carbon and nitrogen atom fused simultaneously into the steel by spraying gas media while the steel to the steel surface when it is heated.

Carbonitriding process is often referred as "Cyanide dry" [1]. It also known as modified form of gas carburizing, rather than a form of nitriding. The modification consists of introducing ammonia into the gas carburizing atmosphere to add nitrogen to the carburized case as it is being produced [4]. Carbonitriding process usually is used for increasing the hardness of low carbon steel surface, by heating it into carbon-nitrogen gas environment with lower temperature than carburizing temperature at 700 to 900 °C, with a depth of layer about 0.7 mm. Carbon and nitrogen-free which formed will be diffused into steel surface then react with ferrite or another alloy. Carbonitriding layer is more resistant than carburizing layer result against softening during tempering [2]. Some novel and tedious technique has been used such as plasma electrolytic carbonitriding (PEC/N) at ambient temperature and pressure and plasma electrolytic nitriding in glycerine and

* Corresponding author: poppy@um.ac.id

DOI: 10.1051/matecconf/201710101012

acetamide solution [3].

Previous research showed that bone of cow can be used as a substitute for $BaCO_3$ (Barium Carbonate) in solid carburizing process as alternative catalyst. The most rapid of carbon absorption rate occurs in a process by holding time for 25 minutes with the composition of 1kg carbon (nani charcoal) and 30% bone of cow that is 1.22 HRC/min. Increasing the value of the largest average of hardness also occurs in the holding time for 25 minutes at 23.96% [5]

Previous research showed that charcoal with high carbon composition successfully used for carburizing media. Therefore, this research propose a new method of carbonitriding using charcoal and fertilizer with rich nitrogen as solid media. Type of fertilizer for this method is urea with more than 60% of nitrogen and charcoal from wood with more than 80% of carbon source. Holding time is varied to give sufficient time for carbon and nitrogen to enter the surface of steel. The reaction of carbonitriding consists of two main part:

The transfer of carbon from charcoal

$$2CO \rightarrow C + CO_2$$

The transfer of nitrogen from ammonia

$$2NH_3 \rightarrow 2N + 3H_2$$

2 Experimental details

The experimental starts with specimen preparation using St. 41 Steel with %C 0.07% - 0.1%. The carbonitriding media consists of charcoal (coconut shell) and urea. The specimen was in cylinder shape and tied up with stainless steel wire in order to put inside the carbonitriding pack (Fig. 1). The specimen was placed inside the carbonitriding media consists of 50% charcoal and 50% urea in a pack stainless steel box. Stainless steel box has dimension of 20cm (L) x 12cm (W) x 12cm (H). Specimens was heated at 700 °C, 750 °C, 800 °C in muffle furnace and hold for 60 minutes and 120 minutes respectively. Specimens were then quenched in water for approximately 30 second and then cleaned for surface hardness testing using Rockwell Hardness Tester. The case depth analysis obtained by optical microscopy result. The measurement of case depth hardness was taken from the outer side of specimen (Fig. 2).

Fig. 1. Specimen of Carbonitrided Low Carbon Steel.

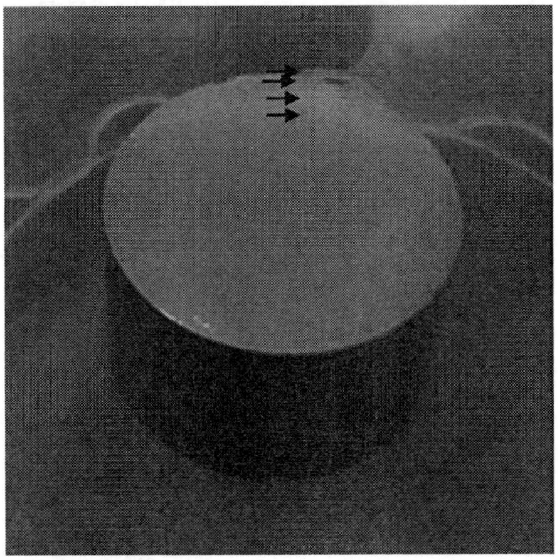

Fig. 2. Hardness Measurement from the Outer Side of Low Carbon Steel Specimen.

Chemical composition of low carbon steel was measured by X-Ray Fluorescense. The measured chemical composition of steel in this study is given in Table 1.

Table 1. Chemical composition of low carbon steel

Element	Wt %
Fe	99,0483
C	0,09073
Mn	0,48640
P	0,01246
S	0,02186
Si	0,04448
Cr	0,04218
Cu	0,14905
Ni	0,05200
W	0,00754
Mo	0,01744

3 Results and discussion

After the surface hardness data in carbonitriding process, the following step is calculating the mean value of hardness number in every carbonitriding process with temperature and time variation. Hardness investigation was taken from three different position as shown in Fig. 2. The results are summarized in Table 2.

Table 2. Hardness data of low carbon steel with carbonitriding method

No	Time Variation (minutes)	Rockwell Hardness Test (HRB)		
		700 °C	750 °C	800 °C
1	60	85.7	93.1	95.7
2	120	94.1	92.8	94.0

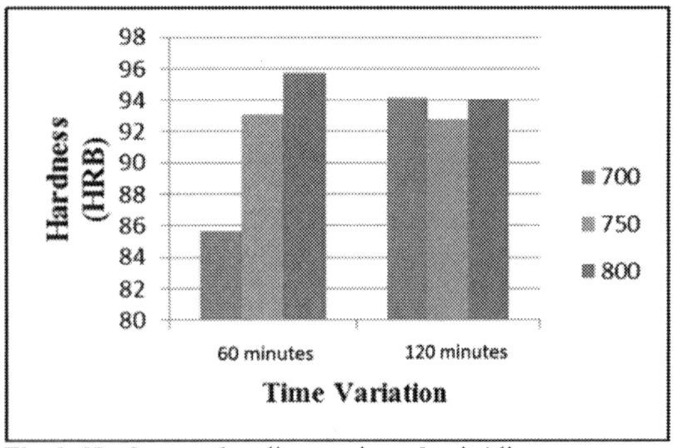

Fig. 3. Hardness value diagram in carbonitriding process time variation

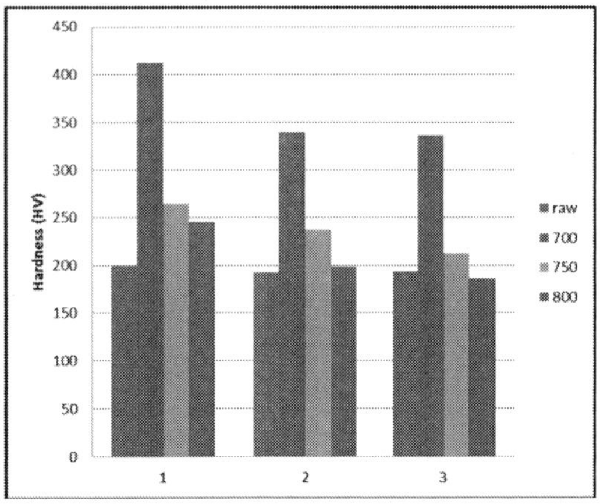

Fig. 4. Hardness value of low carbon steel in three different point

Difference hardness number after carbonitriding process as summarized in Table 1 state that, specimen I which has carbonitriding with HRB hardness difference temperature = 8.34. On specimen II which has carbonitriding at temperature of 750 °C HRB hardness difference = 1.7 and specimen III which has carbonitriding at temperature of 800 °C HRB hardness difference = 0.4. Fig. 3. and Fig. 4. Show the hardness value of specimen in the surface and three different point taken in each specimen. Hardness value decreased in second point and third point since the carbon depth did not reach the middle of specimen. It can be seen that the hardness of carbonitrided specimens are higher than raw material (non-carbonitrided specimen). This is possibly due to presence of carbides and nitrides uniformly in the fine martensitic microstructure. Carbides and nitrides were act as obstacle for dislocation motion which resulted the drop in hardness for carbonitrided specimens.

Carbonitriding process is one of carbon and nitrogen (ammonia) combination into steel surface. Addition of nitrogen on carburizing process turn it to carbonitriding process [6]. There were almost stabil hardness condition at 120 minute holding time because of the retained austenite that formed significantly under nitrogen influence. The amount of retained austenite will affect the value of hardness [7]. Hardness of specimen also affected from the formation of nitride in conjunction of nitrogen in urea. Nitride and carbide are the main phase that can increase the hardness in carbonitriding process [8-9].

Fig. 4 shows there is slightly decreas of hardness number in the depth of position from the surface. This is can be referred as transition zone which indicates the case depth of carbonitrided layer and this zone also related to microstructural stability [13].

1.3 Microstructure thickness

Carbonitriding process is one of carbon and nitrogen (ammonia) combination into steel surface. Addition of nitrogen on carburizing process turn it to carbonitriding process [6]. These are micro photo result of carbonitriding process with time holding and temperature variation.

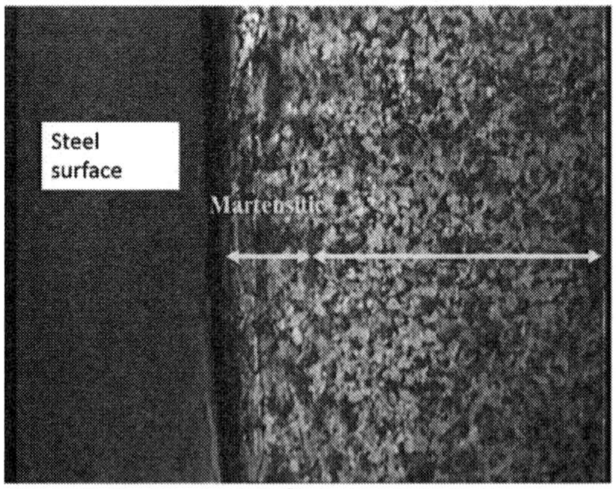

Fig. 5. Surface layer at 700 °c for 60 minutes.

DOI: 10.1051/matecconf/201710101012

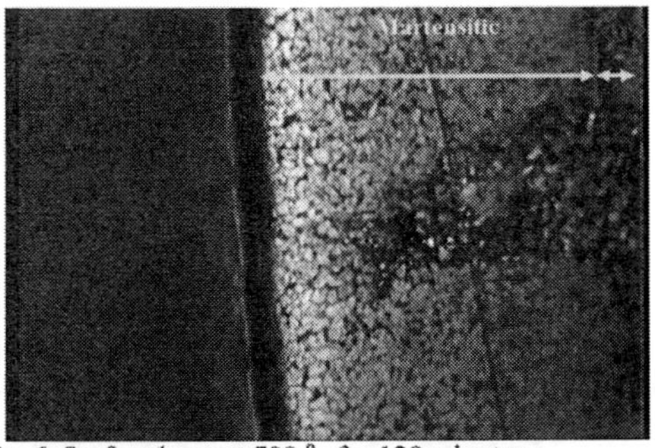

Fig .6..Surface layer at 700 °c for 120 minutes.

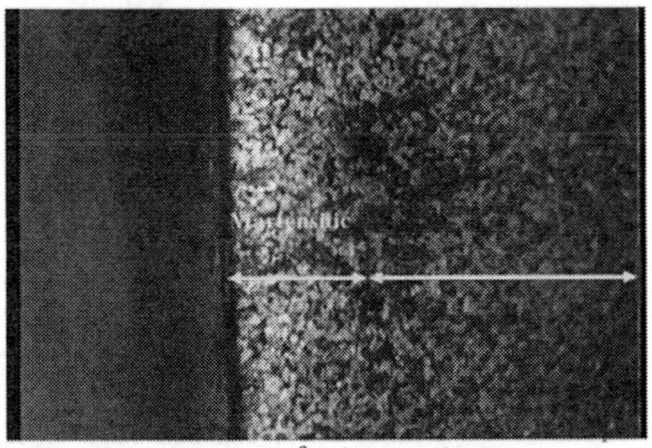

Fig. 7. Surface Layer at 750°C for 60 Minutes.

Fig. 8. Surface layer at 750 °c for 120 minutes.

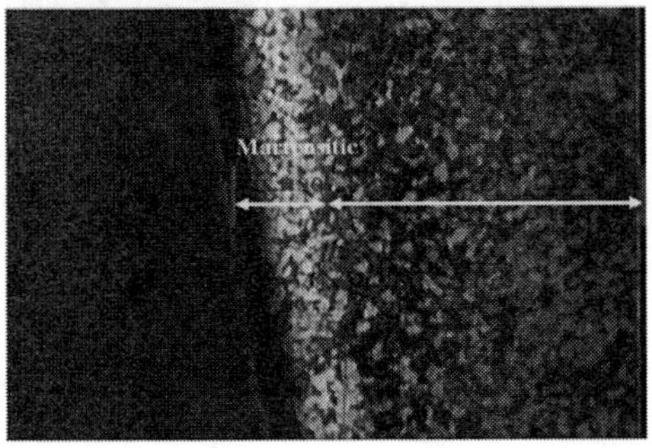

Fig. 9. Surface layer at 800 °c for 60 minutes.

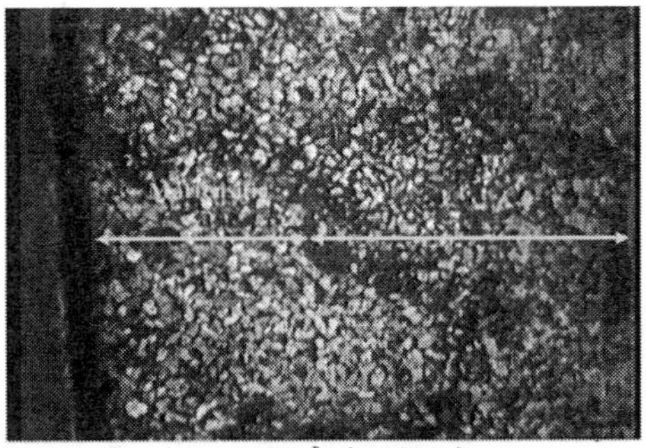

Fig 10. Surface layer at 800 °c for 120 minutes.

It is obvious that carbon atoms contents at the surface are at the largest in the whole layer (Fig. 5-Fig. 10). The thickness of the layer shows the significance increment as the carbonitriding temperature increased. The hardness number increase as the carbonitriding temperature increased, this is the reason that the higher temperature the carburizing is the main process with the presence of martensite phase. The phase transformation of the substrate at low temperature shows that nitriding process take place. The possibility of phase transformation is the presence of Fe_3C, Fe_5C_2, and Fe_2-3N [3].

Carbonitriding process find in wide use in all branches industry e.g. in power engineering, car making, food industry, agricultural machines, etc. It was found as the best among hardening technologies [10]. Microstructure shows that there is formation of carbide, ferrite, and perlite. Fig. 11 shows carbide formation at grain boundary. At this condition, the carbide formation due to the absorption of carbon in carbonitriding process.

Fig. 11. Microstructure of low carbon steel carbonitrided at 700 °c

DOI: 10.1051/matecconf/201710101012

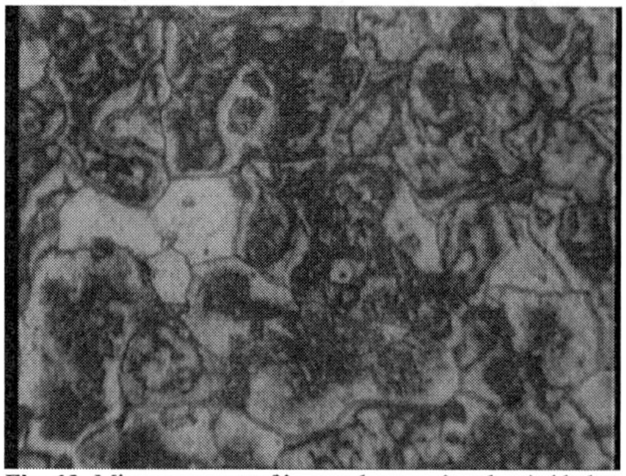

Fig. 12. Microstructure of low carbon steel carbonitrided at 750 °c

Fig. 13. Microstructure of low carbon steel carbonitrided at 800 °c

Fig. 12 and Fig. 13 show the carbide formation is not as clear as specimen at Fig 11. While the grain size become bigger and the carbide is not formed deeply into the middle of specimen. The core of specimens has ferrite-pearlite microstructure.

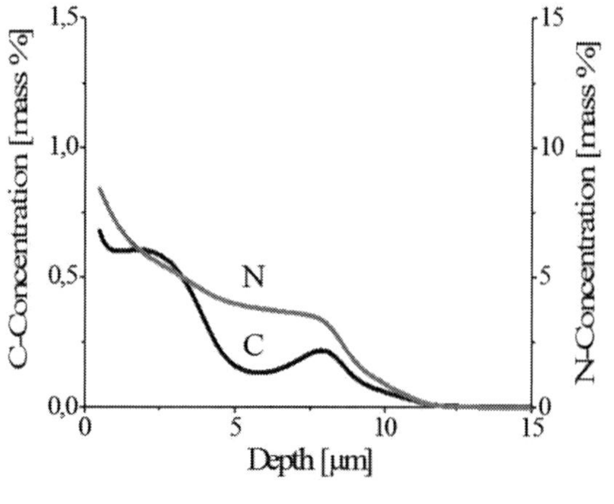

Fig. 14. Nitrogen and carbon profile analysis in depth of a layer from carbonitrided process in 550°C [11]

In carbonitriding process, the surface layer of the depth become larger in alloyed steel and this can be explained by total amount of nitrogen and carbon depth in Fig. 14. The prolongation of time in carbonitriding process will resulted the reducing carbon pressure and increasing ammonia pressure. This will increase total thickness, maximum hardness, bigger micro pores, and smallest residual compressive stress. [11]. Carbonitrided steel also has enhancement in life of component which was proved by comparison of ball bearing with and without carbonitriding process [12]. The precipitates of carbides and nitrides increases the stability of the microstructure, thereby increase the wear resistance [13].

4 Conclusion

The conclusion of this research is differences of low carbon steel surface microstructure which heated at temperature of 700 ⁰C for 60 minutes and 120 minutes of holding time. The residual of austenite trapped simultaneously with carbon and the surface layer which has thicker of austenite start to diffused resulted increment in hardness number. While at temperature of 750 ⁰C for 60 minutes of holding time, the residual of austenite trapped in lattice structure and for 120 minutes the nitrogen return into austenite phase formation stability. Therefore, the hardness decreases significantly. Specimen heated at temperature of 800 ⁰C for 60 minutes of holding time showed the forming of austenite nitrogen in stable phase and for 120 minutes holding time resulted the surface layer decompose austenite properties on nitrogen which transform with ammonia properties.

I would like to express my gratitude to Universitas Negeri Malang for the support and opportunity to join this conference.

References

[1] J. A. Sukma, 2012, Pengerasan Permukaan Baja Karbon St 40 dengan Metode Nitridasi Dalam Larutan Kalium Nitrat, 25

[2] d.scribd.com/doc/187466082/Perlakuan-Panas-Permukaan

[3] Pang Hua, LV Guo Hua, Chen Huan. Microstructure and Corrosion Performance of Carbonitriding Layer by Plasma Electrolytic Carbonitriding. China Physics Letter Vol 26, No. 8 (2009) 086805.

[4] Patiphan Juijerm. Surface Treatment Technology for Automotive Parts. Presented in Automotive Summet 2013, 20-21 June 2013, BITEC Bangkok, Thailand.

[5] Eka R.M.A. Lilipaly, Leslie Lopies, 2009, Analisis Nilai Kekerasan Baja S35C Dalam Proses Karburasi Padat

[6] J. Slycke and T. Ericsson. 1981. A Study of Reactions Occurring During the Carbonitriding Process
[7] ASM Handbook Volume 4A. 2013
[8] Yudha, Sandy. 2015. Analisis Kekerasan dan Struktur Mikro Baja Karbon Rendah terhadap Variasi Waktu Penahanan pada Metode Carbonitriding. Malang: Universitas Negri Malang
[9] Suherman. 1988. Ilmu Logam 1. Institut Teknologi Sepuluh Nopember: Surabaya
[10] S.G. Tsikh, V.I. Grishin, A.V. Supov, V.N. Lisitskii, and Yu. A. Glebova. Advancement of the Process of Carbonitriding. Metal Science and Heat Treatment Vol. 52, No. 9-10, 2010.
[11] Zumbilev, Angel. Carbonitriding of Materials in Low Temperature Plasma. Technical University of Sofia-Plovdiv Branch, Bulgaria. Article in www. Intechopen.com (accessed at 31 December 2016).
[12] J.J.C. Hoo. Rolling Contact Fatigue Testing of Bearing Steels. ASTM Special Technical Publication, Philadelphia, 169-189. 1981.
[13] Rajan, Karthikeyan, Joshi, Vinod, Ghosh, Arindan. Effect of Carbonitriding on Endurance Life of Ball Bearing Produced from SAE 52100 Bearing Steels. Journal of Surface Engineered Materials and Advanced Technology. Vol. 3, 172-177. 2013.

MATEC Web of Conferences **101**, 01013 (2017) DOI: 10.1051/matecconf/201710101013
SICEST 2016

The Effect of Rotation Stirring on Macrosegregation in Bi-Sn Alloy

Yeni Muriani Zulaida[1*], *Riyan* Afrizal[1] and *Suryana* Suryana[1]

[1] Metallurgical Department Faculty of Engineering, Sultan Ageng Tirtayasa University, 42435 Cilegon, Indonesia

Abstract. Macrosegregation is a defect that difficult to avoid in a metal alloy made by casting method. Macrosegregation can cause decreasing in mechanical properties of casting products. It will reduce their performance in industrial application. Macrosegregation is convinced occur during solidification time in liquid alloy. In the early solidified, The solids move upward/downward in liquid alloy during solidification are considered to contribute on macrosegregation formation. This movement occur due to the density differences between the solid and the surrounding liquid. This research want to observe the effect of stirring on macrosegregation formation for interfering on the movement of initial solidified solid. Stiring with rotation method is applied in this experiment at certain temperature. The temperature and the rotation speed of stirring are varied to observe the effect of rotation stirring on macrosegregation formation. The mold is covered by insulation jacket and kept the bottom part opened in order to obtain the directionally solidification. The result shows that the rotation stirring can change the macrostructure of casting but in case of composition distribution, the macrosegregation still appear. Increasing the rotation stirring will rise solid crystals up and the macrosegregation become more uniform and the treatment is better if conducting at lower temperature.

1 Introduction

Macrosegregation is a compositional inhomogeneities at large scale compared with microsegregation and can affect to the properties of the casting.. The range in scale from millimeters to centirmeters. It is seen commonly in components made by all type of casting processes and in any materials. Macrosegregation cannot be mitigated through processing of the casting after solidification complete. It may occur during solidification time. The low diffusifity of solute in solid state may be the cause. It affects to different composition at next to interface and turn to be a density differences in liquid. Beckermann summarized that the cause of macrosegregation is relative movement or flow of segregated liquid and solid during solidificatin [1]. In this case gravity force lead an important role in macrosegregation process. Typical examples are dendrites settling in the bottom of ingots or coarse primary lamellae graphite floating on the top of large cast iron castings [2]. The main driving force of fluid flow during solidification is solidification contraction. Interdendritic liquid flow can cause higher consetration of solute than average composition in the earlier solidification [3]. External force, for example centrifugal casting [4], rotating and oscillating [5] can change the solute concentration in casting.

Macrosegregation start at microsegregation level during solidification. If consentration gradient at interface are developed significantly, the flow may occur to liquid at next to interface. This can cause a segregation in the macroscopic level known as macrosegergation.

In low solubility of solute in matriks, in this case solute (Sn) dissolve into Bi, the macrosegregation in stationary casting condition can be shown in Fig.1. The the grains of primary phase that suppose to disperse evenly throughout the ingot but instead are concentrated at the bottom of ingot, it may be occur when the primary phase that is formed at the initial stage of solidification, has higher density than the surrounding eutectic liquid and due to gravity force, the primary phase settle to the bottom of ingot [2]. This cause inhomogeneities of macrostructure in ingot and furthermore can affect to properties of casting product. The observation of the grain setlling for this alloy had been conducted in previous research [6]. The primary phase grain required some times for settling so when the cooling time was accelerated, the setlling time will decrease and the pile of primary phase grain become more higher and the size smaller.

Several attempts were made to reduce the effect of the structure differences in casting product. Previous researches explained, such as, additional the electromagnetic force during centrifugal casting in pipe product[4], rotation and oscillation method in Aluminium alloy[4] or cooling rate[6] as an attempt in order to decrease macrosegregation. In this research

* Corresponding author: yeni_m_zulaida@untirta.ac.id

DOI: 10.1051/matecconf/201710101013

rotation method is used as one of attempt to decrease macrosegregation and homogenize the structure in macroscopic level. According the explanation in previous paragraph above, the alloy with low solubility of solute is used for observation on the effect of rotation.

In this work, rotation stirring method with lower rotation speed compared from previous research is applied in attempts to treat liquid during solidification in order to reduce the macrosegregation. The aim of this study was observing the effet of rotation stirring on the macrosegregation formation. Bi alloy with 30 wt pct of Sn was used for facilitating the observation since primary Bi phase could be captured by macro camera clearly[7], as seen in Fig 1.

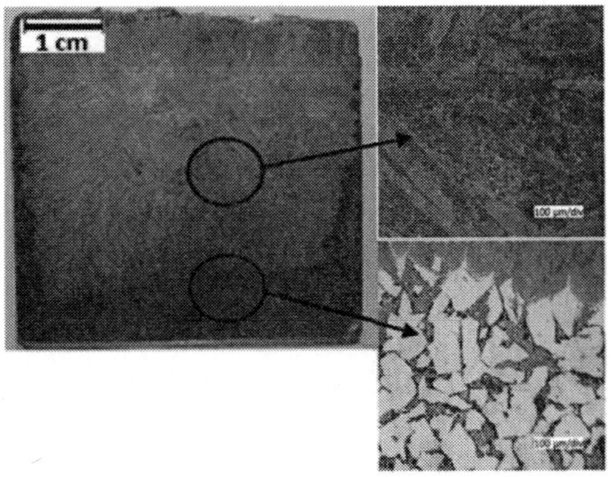

Fig.1 Primary phase (black) concentrated at the bottom ingot

2 Experimental details

This study use low melted alloy material, Sn-Bi Alloy. The alloy composition is Bi-30 wt pct Sn prepared from 99.99 pct pure Bi and Sn. This alloy reaction during solidification is considering on binary phase diagram of Sn-Bi as seen in Fig.2. The alloy weight was approximately 600 gram and then it is placed into a cylinder steel mold with 4,5 cm in diameter. The mold is covered by insulation jacket on the top and all around of the mold but opened at the bottom in order to retard heat

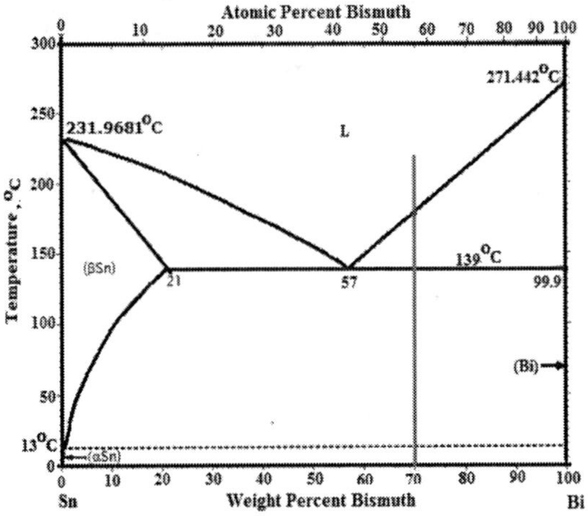

Fig.2. Sn-Bi phase diagram

transfer from the top and side direction of metal mold but start to solidify from the bottom of the ingot.

In this experiment, the alloy then is melted using electric furnace. The melting point of alloy is 180°C and overheating around 50°C. After that the molten metal filled mold is placed on rotation machine, as seen in

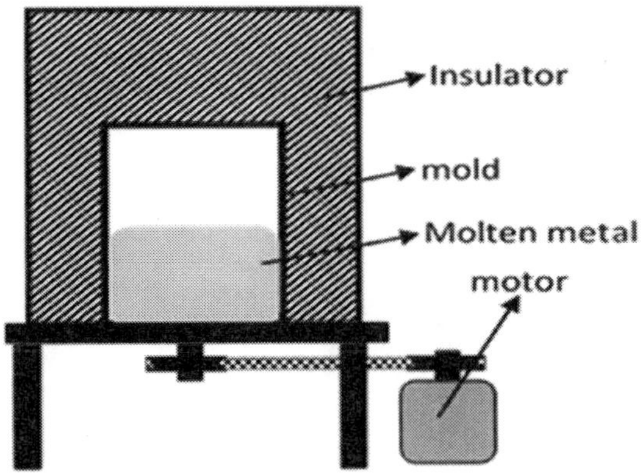

Fig.3. Schematic of rotating machine

Fig.3. And then at certain temperature, the molten is set to rotate. The speed rotations which are applied in this experiment were 50, 83 and 115 rpm. From preliminary experiment without rotation stirring, the required time of complete solidification can be obtained. The temperature changes can be measured using thermocouples that is connected with data logger. According this data, duration of rotation stirring is decided as 5 second. Rotation stirring is conducted at first in the liquid phase condition and the second in liquid-solid phase condition.

The observation of macrosegregation is held after the ingot solidify completely. The observation sampling sites located in the center of the ingot. The cylindrically Ingot is cut vertically and followed by surface preparation for macrostructure and microstructure observation. Macrostructure is observed using digital camera with macro lens and for microstructure using optical microscope. For consentration of solute distribution is analyzed using scaning electron microscope with energy disperse X-ray spectroscopy (SEM/EDS).

3 Result and Discussion

According phase diagram, when temperature reach liquidus line of Bi-30 wt pct Sn alloy, Bi primary phase start to nucleate. This nucleation should be occured at nucleation potential site, for example at mold wall. This initial phase has higher density than remain liquid and with enough time will move downward and settle in bottom of ingot. This mechanism cause the differences of distribution phase between upper side and down side of ingot. Rotation stirring process during solidification is expected can give additional force to move the Bi primary phase and spread it to all parts of casting.

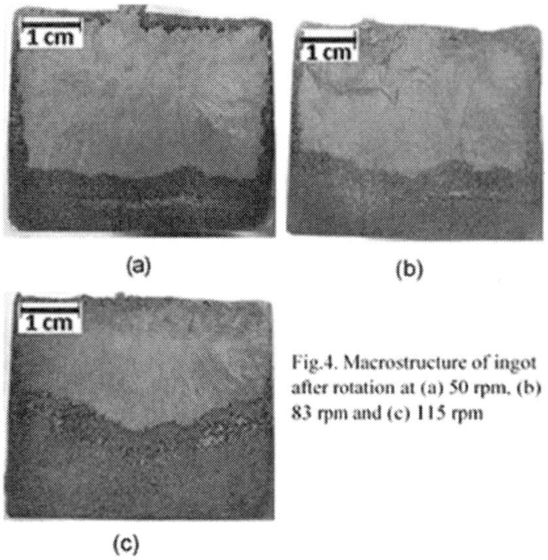

Fig.4. Macrostructure after rotation at (a) 50 rpm, (b) 83 rpm and (c) 115 rpm

From this experiment, the macrostructure after rotation become different and it can be shown in Fig.4. Dark zone in the base of the ingot indicate Bi primary phase zone and the grey zone above show columnar grains of eutectic phase. Increasing the rotation speed is resulting primary phase consentration zone higher. Stirring using rotation method can generate a force inside remain eutectic liquid. This force move the liquid and push the primary phase toward mold wall and upward. The rotation speed influence on the height of the primary phase pile. In this study, 50 rpm of rotation speed will produce the lowest force than two other rotation speed. With this condition it has not enough force to raise primary phase higher and the result the

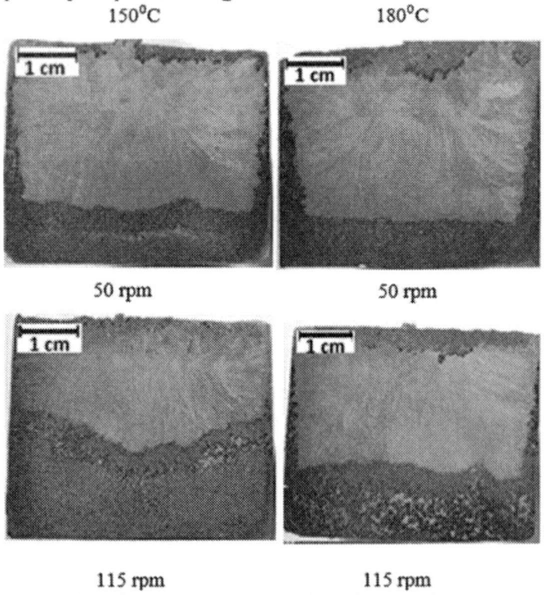

Fig.5. The effect of temperature and rotation speed on macrostucture of Bi-Sn Alloy ingot

height of primary phase pile is the lowest of all condition, eventhough The height of Bi primary phase zone is still higher compared with the ingot without rotation stirring. Above of primary phase region can be seen columnar structure.

The rotation speed can affect to the distance among one Bi primary grain to another grains. The distance is getting farther in 115 rpm than two other lower rotation speed and the region become greater as well. The large of primary phase region is about 830 mm^2 for 115 rpm of speed and it is larger than rotation speed 50 rpm and 83 rpm that only had 540 mm^2 and 675 mm^2.

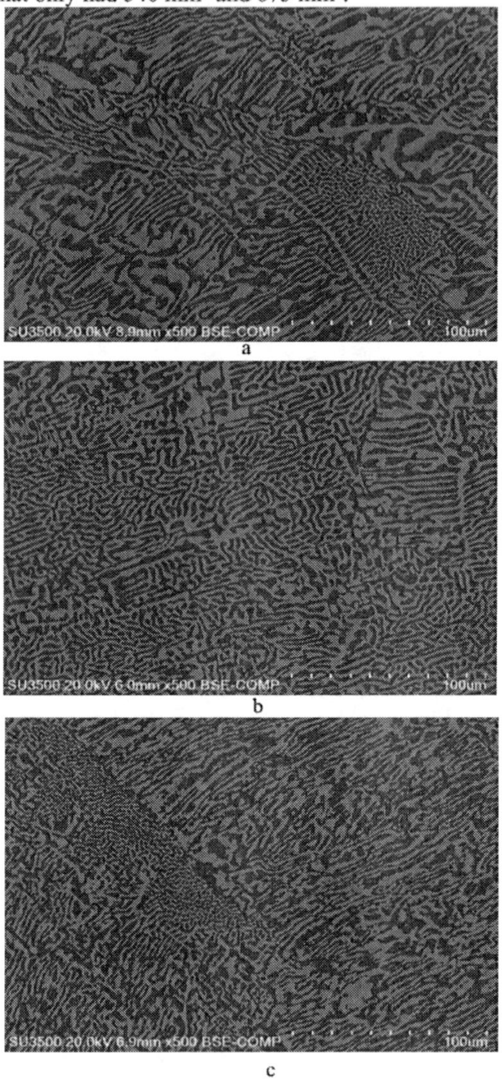

Fig.6. The microstructure of eutectic ini different rotation speed, a) 50 rpm, b) 85 rpm and c) 115 rpm

The rotation temperature when the rotation stirring process conducted influence to the uniformity of macrostructure in ingot. Base on the distribution of phases in ingot, the rotation treatment is better conducting at lower temperature or in range solidification (solid-liquid phase) condition than liquid condition. Figure 5 show that in lower temperature result, the macrostructure is more uniform at the bottom of ingot, despite it still appear there is some regions that lack of primary phase among Bi primary zone. In spite of the proper speed and temperature selection is still necessary to get a good result. The rotation stirring cannot performe at the temperature is too low, this stirring process may induce the damage in the structure and lead tearing in the ingot.

Increasing rotation speed can indeed lift the primary phase zone become higher but the problem is, when the rotation speed keep increasing will result the height of the primary phase region at the center of ingot decrease and the consequences, the region like V-shape will appear in the ingot as seen in Figure 5 at stirring speed 115 and temperatur of stirring 150°C. It is predicted that if the speed increase higher and higher. It is posibble that in the middle of the ingot will vacant with one of phase that will be form during solidification due to centrifugal force. This mechanism also can work in other alloy. So it is conclude that the treatment using rotation stirring was not effective to reduce macrosegregation if the rotation stirring speed too high.

The microstructure profile show the different result. In the Columnar structure consist eutectic phase as seen in Figure 6. From previous research by Olga Kusakova [8] summarized that cooling rate can influence on the microstructure formation. The different cooling rate could form an eutectic microstructure differently, such as, skeleton-like in Bi-Sn alloy or foil-cross section microstructure. Although in this experiment, the change of temperature during rotation treatment cannot be measured since technical matter, but from the microstructure result show, it is possible there that the cooling rate is the change during rotation stirring. It may occurs locally in ingot and cause different eutectic structure in microstructure level.

a b

Fig.7. Fragmentation of the primary phase due to rotation stirring, a) 85 rpm and b) 115 rpm

In microstructure figure, The Bi primary phase is shown as white grain. Rotation stirring can cause the primary phase break into fragments. It can be seen in Figure 7, The higher of the speed rotation, more grain/dendrite structure may be broken and distribute into different sizes. As explained in macrostructure part, that the higher rotation speed could make the Bi primary phase more distance and emerge spaces without Bi primaty phase, It is shown in Figure.7 as well. In the space appear Sn-rich phase, which is mark as black phases. The application of rotation stirring may cause solid solution phase Sn-rich phase appear.

Macrosegregation refers to spatial variations in composition that occur in casting ingot. The distribution of solute will be different in relative large scale. EDX/SEM equipment can measured the composition of Sn in Alloy. In this research the distribution of Sn as a solute in alloy can be more homogen and closer to nominal composition in rotation stirring condition 85 rpm and performed at temperature 150°C, as seen in Figure 8 and Figure 9.

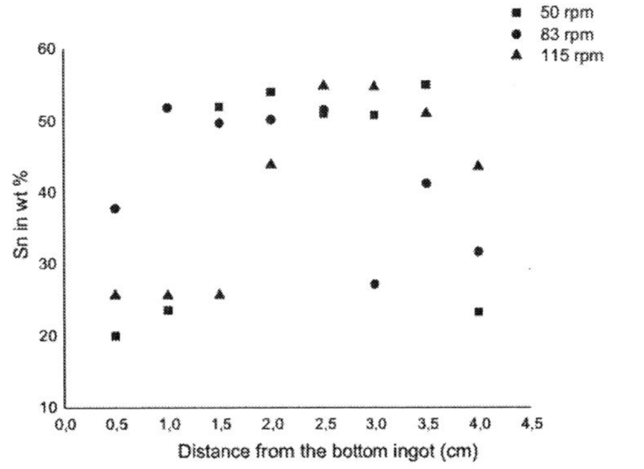

Fig.8 Distribution Sn with Rotation Stirring at temperature 180°C

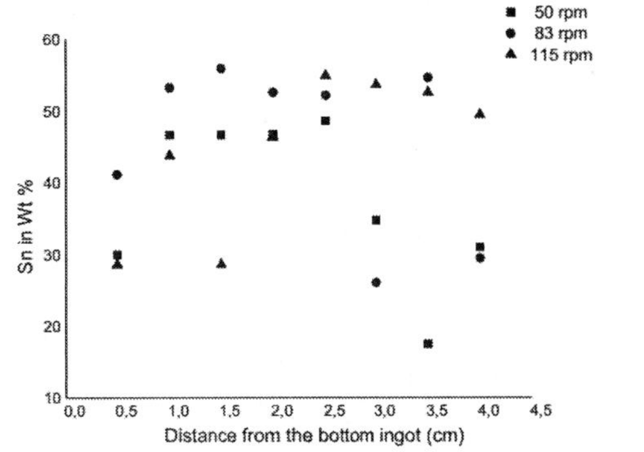

Fig.9. Distribution Sn with Rotation Stirring at temperature 150°C

4 Conclusion

In this study, rotation stirring is used as an attempt to reduce macrosegregation in ingot. Base on the result of experiment, rotation stirring can give a force to raise and spread denser primary phase in remain eutectic liquid. It is proved by the large of primary phase zone become higher from 540 mm^2 to 830 mm^2. But it is still not enough to make the structure in ingot more uniform with this condition. The appropriate selection for temperature and speed is necesarry to obtain a good result. Generally stirring using rotation method can lead more uniform macrostructure in ingot but alternative process should be considered to get better results.

This work has been supported by Ministry of Research, Technology and Higher Education through the competitive grant. Author also want to thank to PT Timah (Persero) Tbk and PT Fajar Mas Murni. Both of them have supported us for availability of material and examination.

REFERENCES

[1] C. Beckermann, "Macrosegregation," *ASM Handbook, Vol. 15 Cast.*, vol. 15, pp. 348–352, 2008.

[2] D. M. Stefanescu, *Science and Engineering of Casting Solidification*, 2nd editio. Newyork: Sprnger, 2009.

[3] R. Mehrabian, M. Keane, and M. C. Flemings, "Interdendritic Fluid Flow and Macrosegregation ; Influence of Gravity," *Metall. Trans.*, vol. 1, no. May, 1970.

[4] W. Q. Zhang, Y. S. Yang, Y. F. Zhu, and Q. M. Liu, "Structural Transition and Macrosegregation of Al-Cu Eutectic Alloy Solidified in the Electromagnetic Centrifugal Casting Process," *Metall. Mater. Trans. A*, vol. 29, no. January, pp. 1–5, 1998.

[5] M. J. Stewart, L. C. Macaulay, and F. Weinberg, "Macrosegregation in Castings Rotated and Oscillated During Solidification," *Metall. Trans.*, vol. 2, no. January, pp. 169–173, 1971.

[6] Y. M. Zulaida, "The Influence of Cooling Rate on Macrosegregation in Bi-Sn Alloy," *Adv. Mater. Res.*, vol. 1112, pp. 502–505, 2015.

[7] Y. M. Zulaida and Suryana, "Macrosegregation and Dendritic structure formation in different solute composition," in *AIP conference Proceedings*, 2016, vol. 20011.

[8] O. Gusakova, V. Shepelevich, and L. Scherbachenko, "Effect of Melt Cooling Rate on Microstructure of Sn - Bi and Sn - Pb Eutectic Alloys," *Adv. Mater. Res.*, vol. 856, pp. 236–240, 2013.

The characteristic of unsaturated polyester resin wettability toward glass fiber orientation, density and surface treatment

Asep H. Saputra[1,*] and *Dena P.* Hallatu[1]

[1]Department of Chemical Engineering, Faculty of Engineering, Universitas Indonesia, Depok 16424, Indonesia

Abstract. Wettability of composite is one of key to increase mechanical properties of composite that affected by structure of reinforcement and type of resin used. Therefore, this research focused on the effect of orientation, density and surface treatment on fiber to the characteristic of composite's wettability, which is observed by contact angle and wetting time. The fiber used in this research is fiberglass, and the method for contact angle measurement is direct observation from the camera recorder and the data record will be processed and analyzed by using image processing method. The result for those variations can be obtained from the relation of variations toward contact angle and wetting time. According to result of research, fiber with orientation 45°/45° gives lower contact angle but longer wetting time than fiber with orientation 0°/90°. For orientation 45°/45°, the differences in wetting time is 15 second longer than orientation 0°/90°. In case of fiber density, the sheet with fiber density of 900 has 7 second faster for wetting time than sheet with fiber density of 1250. The surface treatment with NaOH 5% can accelerate the wetting time until 10 second.

1 Introduction

Composite material is often preferred because it has good mechanical properties. The type of matrix, the reinforcement used, and the interaction between matrix and reinforcement can affect mechanical properties of composite material. Wettability is the ability of liquid 'wetting' the surface of a solid material, as in how composite matrix wet reinforcement particles. Optimizing the ability of a matrix to wet the reinforcement is a key for increasing the mechanical properties of composite material.

Wettability properties of a material depend on hydrophobic properties and viscosity of liquid. Viscosity affect hydrophobic properties by increasing the contact angle through liquid viscosity decrease, and the rough surface effect might not always create high hydrophobic properties [9]. Reinforcement materials have a portion in surface free energy used for composite material wettability measurement [3].

Contact angle measurement is one of the method that is used to observe and develop the wetting phenomena. This method is commonly used because the measurement characteristic does not damage experiment materials nor alter the chemical and physical properties of material [2]. Contact angle is obtained by measuring the angle between material surface and liquid in which both has different polarity. Contact angle and wettability both depend on the time of substrate drop and the time of image data taken to measure the contact angle. [4].

Several methods are commonly used in the measurement of contact angle. One of the methods is by measuring the angle between composite material surface and the liquid drop. In accordance to the development of technology in computation, tools for contact angle measurement has been developed with better accuracy and precision. The uses of camera which providing high zoom capability as a tool to capture liquid drop profile provide the measurement of profile on detail [8]. Image processing method for contact angle measurement has the potential to measure the contact angle from any type and shape of liquid drop [6]. Identification with image processing method gives several advantages, such as increasing the sensitivity of experiment and resulting in better accuracy within the number of image that can be analyzed by computer [1].
Since wettability of composite is one of key factor in composite fabrication which affected by the reinforcement and resin, the effect of fiber orientation and fiber sheet density, and fiber surface treatment to composite wettability is needed to be observed.

2 Experimental details

2.1 Materials and tools

Materials used for research are fiberglass and unsaturated polyester resin. Unsaturated polyester resin which is used in this research is EVERPOL 324 AR-2

* Corresponding author: author@e-mail.org

supplied by PT. Arindo Pacific Chemicals. Resin type is ortho with clear pale color and has 560 cps of viscosity

The type of fiberglass used has a varied orientation and sheet density. Type of fiberglass used are Biax Combi 900 (0°/90°/M300, Width 127 cm), Biax Combi 1250 (0°/90°/M450, Width 127 cm), Biax Combi 900 (+45°/-45°/M300, Width 127 cm, and Biax Combi 1250 (+45°/-45°/M450, Width 127 cm). The number of 900 and 1250 indicate the density of fiber, while 0°/90° and +45°/-45 indicate fiber orientation.

The tools used in this research are camera with tripod and dropper/pipette. Using tripod gives optimal results, because it holds camera on constant position along the research. Camera that used has capacity 60 frame per second and standard lens with 55 mm zoom.

2.2 Data collection process

The dependent variables in this research that will be observed are contact angle and wetting time. The method used for contact angle measurement is image processing method by using the tools as seen in Figure 1.

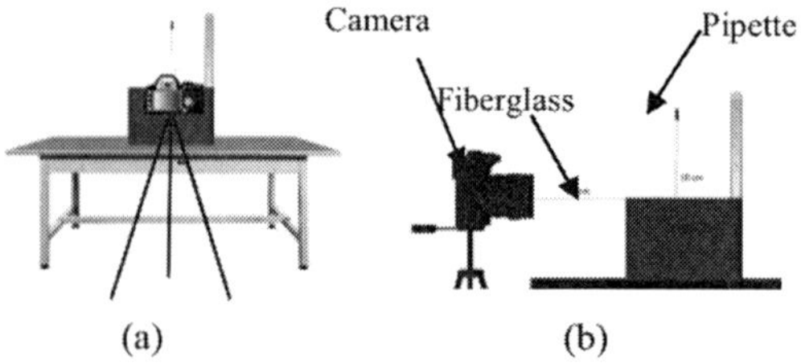

Fig. 1. Scheme of Experimental Tools; (a) front view; (b) side view

The use of sections to divide the text of the paper is optional and left as a decision for the author. Where the author wishes to divide the paper into sections the formatting shown in Table 2 should be used. The data collection process in this research is contact angle in degree (°) and wetting time in second (s). The form of liquid droplet dropped from pipette to fiber surface is assumed to be in the form of sessile drop. The data is taken with a camera to record all of the process beginning with the time of resin dropping up until the resin completely disappeared from fiber surface. To provide data with higher precision, the camera is supported with tripod. Considering that this experiment depends on tools positioning, all of the tools used such as camera, pipette and experimental table are placed in same position during the observation.

The experiment material, fiberglass, is placed in a table and pipette, which is a tool used for dropping the liquid is placed 8 cm above the table. Camera is set in line with experiment table and 25 cm from it. Resin dropped from pipette to fiberglass surface and recorded with camera. The camera is in recorder mode. The video results will be converted to image by using software converter for 60 images (frames) in 1 second. Data for contact angle is taken in frame 1 (0 s), 50 (0.83 s) and 100 (1.67 s). In that frame, contact angle will be measured using software. Meanwhile, wetting time is obtained from data recorder. Wetting time is condition when liquid droplet completely disappeared from fiberglass surface or it can be defined as condition when contact angle reaches 0 degree.

3 Results and discussion

This research is focused on the effect of fiber orientation, fiber density and surface treatment with alkali solution (NaOH 5%) which is observed through contact angle and wetting time, since wetting time is one of important thing in manufacturing process of composite materials, especially in large object with many layers.

3.1. The effect of fiber orientation in fiberglass without surface treatment

The effect of fiber orientation to the contact angle and wetting time is observed. Fiber orientation which is variation in this research are fiber with orientation 0°/90° and fiber with orientation 45°/45°. The data was obtained from the condition in which fiber has same fiber density. This data will be compared between the sheet with fiber density of 900 and the sheet with fiber density of 1250.

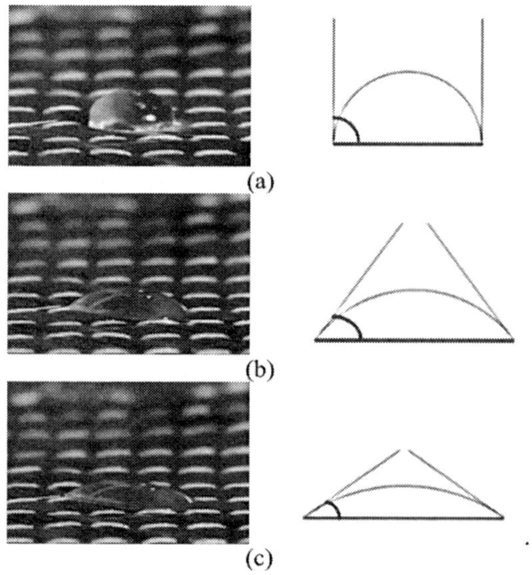

Fig. 2. Experimental Results Example for Fiber without Surface Treatment; (a) Frame 1 (0 s); (b) Frame 50 (0.83 s); (c) Frame 100 (1.67 s)

Figure 2 shows the typically form of liquid droplet in fiberglass surface. Contact angle data is get from measurement angle between fiberglass surface and liquid droplet, as seen as Figure 2 illustration. As seen in figure since the increasing of time, contact angle will be decrease and disappear from fiberglass surface.

Table 1. Contact Angle of the Sheet with Fiber Density of 900 for the Fiber without Surface Treatment.

Orientation	Contact Angle		
	Frame 1 (0 s)	**Frame 50 (0.83 s)**	**Frame 100 (1.67 s)**
0°/90°	θ = 88.752	θ = 49.223	θ = 37.168
45°/45°	θ = 88.707	θ = 49.144	θ = 36.044

Table 2. Contact Angle of the Sheet with Fiber Density of 1250 for the Fiber without Surface Treatment

Orientation	Contact Angle		
	Frame 1 (0 s)	**Frame 50 (0.83 s)**	**Frame 100 (1.67 s)**
0°/90°	θ = 91.586	θ = 61.994	θ = 45.354
45°/45°	θ = 90.355	θ = 52.277	θ = 39.073

The results for contact angle measurement for sheet with fiber density of 900 and sheet with fiber density of 1250 is given in Table 1 and Table 2. Both of sheet with fiber density of 900 and sheet with fiber density of 1250 has results with the same tendency. Contact angle result from fiber with orientation 0°/90 is higher than result from fiber with orientation 45°/45°. The difference reaches 6 degree in frame 100. Besides from the contact angle, wettability is also can be observed from the wetting time.

Table 3. Wetting Time of Fiber without Surface Treatment

Fiber Density	Fiber Orientation	Wetting Time (s)
900	0°/90°	32.00
	45°/45°	45.33
1250	0°/90°	35.00
	45°/45°	48.67

Table 3 shows that the fiber with orientation 0°/90° has faster wetting time than the fiber with orientation 45°/45°. This result happens in either sheet with fiber density of 900 or sheet with fiber density of 1250. Fiber with orientation 0°/90° has 13 second faster in wetting time than fiber with orientation 45°/45°.

The result for contact angle and wetting time measurement shows that the fiber with orientation 0°/90° has higher contact angle and faster wetting time than fiber with orientation 45°/45. It happens because in contact angle measurement, the droplet formed depends on early surface contact with the droplet while wetting time depends on fiber space areas in fiber sheet surface. The fiber with orientation 0°/90° has wider area than fiber with orientation 45°/45° that makes the wetting time faster.

3.2 The effect of fiber density in fiberglass without surface treatment

The effect of fiber density to the contact angle and wetting time is observed. Density which will be compared in research is sheet with fiber density of 900 and sheet with fiber density of 1250.

Table 4. Contact Angle of Fiber with Orientation 0°/90° in Fiber without Surface Treatment

Density	Contact Angle		
	Frame 1 (0 s)	**Frame 50 (0.83 s)**	**Frame 100 (1.67 s)**
900	θ = 88.752	θ = 49.223	θ = 37.168
1250	θ = 91.586	θ = 61.994	θ = 45.354

Table 5. Contact Angle of Fiber with Orientation 45°/45° in Fiber without Surface Treatment

Density	Contact Angle		
	Frame 1 (0 s)	**Frame 50 (0.83 s)**	**Frame 100 (1.67 s)**
900	θ = 88.707	θ = 49.144	θ = 36.044
1250	θ = 90.355	θ = 52.277	θ = 39.073

Table 4 and Table 5 shows the effect of fiber density to the contact angle. Based on the results, both of fiber with orientation 0°/90° or fiber with orientation 45°/45° shows that contact angle in the sheet with fiber density of 900 is lower than the sheet with fiber density of 1250. In addition, wettability can also be observed from the wetting time.

Table 6. Wetting Time of Fiber without Surface Treatment

Fiber Density	Fiber Orientation	Wetting Time (s)
900	0°/90°	32.00
	45°/45°	45.33
1250	0°/90°	35.00
	45°/45°	48.67

As seen also in Table 6, the results shows that sheet with fiber density of 900 has lower contact angle and faster wetting time than sheet with fiber density of 1250. Sheet with fiber density of 900 has 3 second faster than sheet with fiber density of 1250. When the density of fiber is tighter, it will be more difficult for liquid to enter the fiber surface because the space area provided in the fiber surface is too small for the liquid droplet to enter.

3.3 The effect of fiber orientation in fiberglass with surface treatment (NaOH 5%)

The surface treatment in this research is performed by submerging the fiberglass with alkali (NaOH 5%) solution. The purpose of the submersion is to discover the effect of surface treatment on fiber to the contact angle and wetting time.

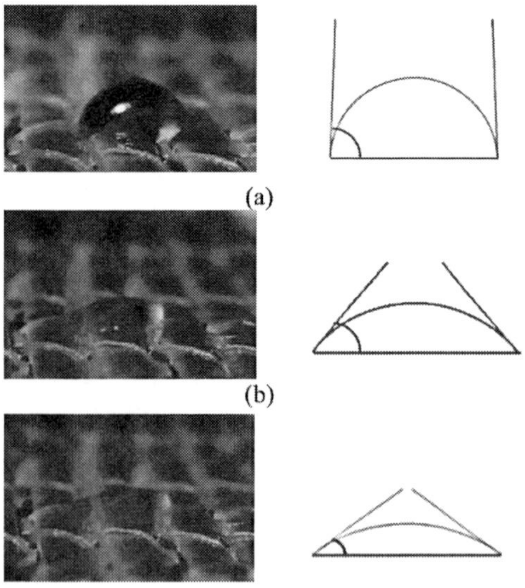

Fig. 3. Experimental Results Example for Fiber with Surface Treatment; (a) Frame 1 (0 s); (b) Frame 50 (0.83 s); (c) Frame 100 (1.67 s)

Figure 3 shows the example of image taken in research for fiberglass with surface treatment. Image data will be using in contact angle measurement with software analysis and give the exact contact angle. Contact angle data is get from measurement angle between fiberglass surface and liquid droplet, as seen as Figure 3 illustration. Since the increasing of time, contact angle will be decrease and disappear from fiberglass surface. The data was obtained from the condition in which fiber has same density. This data is then compared between the sheet with fiber density of 900 and the sheet with fiber density of 1250. The results for contact angle measurement for sheet with fiber density of 900 and sheet with fiber density of 1250 will be given in Table 7 and Table 8. Table 7 and Table 8 shows that both of the sheet has the same tendency. Contact angle resulted from fiber with orientation 0°/90 is higher than the one resulted from fiber with orientation 45°/45°. Table 9 showed that fiber with orientation 0°/90° need faster time than fiber with orientation 45°/45, it happens in either sheet with fiber density of 900 or sheet with fiber density of 1250. Furthermore, wettability can also be observed from the wetting time.

Table 7. Contact Angle of Sheet with Fiber Density of 900 in Fiber with Surface Treatment

Orientation	Contact Angle		
	Frame 1 (0 s)	**Frame 50 (0.83 s)**	**Frame 100 (1.67 s)**
0°/90°	$\theta = 87.698$	$\theta = 43.029$	$\theta = 34.397$
45°/45°	$\theta = 87.498$	$\theta = 42.000$	$\theta = 32.294$

Table 8. Contact Angle of Sheet with Fiber Density of 1250 in Fiber with Surface Treatment

Orientation	Contact Angle		
	Frame 1 (0 s)	**Frame 50 (0.83 s)**	**Frame 100 (1.67 s)**
0°/90°	$\theta = 88.727$	$\theta = 47.373$	$\theta = 40.997$
45°/45°	$\theta = 88.031$	$\theta = 46.483$	$\theta = 35.066$

Table 9. Wetting Time for Fiber with Surface Treatment

Fiber Density	Fiber Orientation	Wetting Time (s)
900	0°/90°	19.00
	45°/45°	36.33
1250	0°/90°	19.67
	45°/45°	37.00

In addition, fiber with orientation 0°/90° has higher contact angle and faster wetting time than fiber with orientation 45°/45. It happens because the contact angle measurement depends on the early surface contact with the droplet, while the wetting time depends on fiber space areas in surface of fiber sheet. The fiber with orientation 0°/90° has wider area than fiber with orientation 45°/45° which makes the wetting become faster. Both of using surface treatment or without surface treatment, the results give the same tendency, even though fiber with surface treatment gives lower contact angle and faster wetting time.

3.4 The effect of fiber orientation in fiberglass with surface treatment (NaOH 5%)

The effect of fiber density to the contact angle and wetting time is observed. The data was obtained from the condition in which fiber has same fiber orientation. This data is then compared between fiber with orientation 0°/90° and fiber with orientation 45°/45°.

Table 10. Contact Angle of Fiber with Orientation 0°/90° in Fiber with Surface Treatment

Density	Contact Angle		
	Frame 1 (0 s)	**Frame 50 (0.83 s)**	**Frame 100 (1.67 s)**
900	θ = 87.698	θ = 43.029	θ = 34.397
1250	θ = 88.727	θ = 47.373	θ = 40.997

Table 11. Contact Angle of Fiber with Orientation 0°/90° in Fiber with Surface Treatment

Density	Contact Angle		
	Frame 1 (0 s)	**Frame 50 (0.83 s)**	**Frame 100 (1.67 s)**
900	θ = 87.498	θ = 42.000	θ = 32.294
1250	θ = 88.031	θ = 46.483	θ = 35.066

Table 10 and Table 11 shows the effect of fiber density to contact angle. Based on the results, both fiber with orientation 0°/90° or fiber with orientation 45°/45° shows that contact angle in the sheet with fiber density of 900 is lower than the sheet with fiber density of 1250. In addition, wettability can also be observed from the wetting time.

Table 12. Wetting Time for Fiber with Surface Treatment

Fiber Density	Fiber Orientation	Wetting Time (s)
0°/90°	900	19.00
	1250	19.67
45°/45°	900	36.33
	1250	37.00

Table 12 shows that the sheet with fiber density of 900 has faster wetting time than the sheet with fiber density of 1250. The sheet with fiber density of 900 has 1 second faster than sheet with fiber density of 1250. When density of a fiber is tighter, it will more difficult for liquid to enter the surface of fiber because of the small space areas in the surface of fiber. Both of fiber with surface treatment or without surface treatment, it has the same tendency results.

3.5 The effect of fiber surface treatment

Surface treatment on fiber with alkali solution (NaOH 5%) decrease the contact angle and wetting time. Fiber with surface treatment has lower contact angle and also faster wetting time than fiber without surface treatment, it is happened in fiber with orientation 0°/90° and fiber with orientation 45°/45°, and also in sheet with fiber density of 900 and sheet with fiber density of 1250. The contact angle in surface treated fiber with specification 0°/90°/900 is 2 degree lower than without surface treatment and 5 degree lower in fiber specification 0°/90°/1250. Meanwhile, contact angle fiber specification 45°/45°/900 is 3 degree lower than without surface treatment and 4 degree lower in fiber specification 45°/45°/1250. The use of alkali solution gives faster wetting time up to 10 second. Surface treatment of fiber with alkali solution results the more rugged fiber with wider space area, which makes it easier for liquid droplet to enter the fiber surface.

4 Conclusion

Fiber with orientation 45°/45° gives lower contact angle but longer wetting time than fiber with orientation 0°/90°. For orientation 45°/45°, the differences wetting time is 20 second longer than orientation 0°/90°. The sheet with fiber density of 900 has lower contact angle and faster wetting time than the sheet with fiber density of 1250. The wetting time of the sheet with fiber density of 900 is 7 second faster than the sheet with fiber density of 1250. The surface treatment with NaOH 5% has the faster wetting time for 10 second.

References

1. K. Adi, S. Firdausi, R. Gernowo, B, Rahardho, I, Siena, A. Putranto. *Sistem Pencitraan Mikroskop Digital Untuk Identifikasi Bakteri Tuberkulosis*. (2012)
2. Q. Benard, M. Fois, M. Grisel. *Roughness and Fibre Reinforcement Effect Onto Wettability of Composite Surfaces*. (2006)
3. B. Bhusan. J. of Phil, 1631-1672. (2009)
4. H. Chen, B. Fei, G. Wang, H. *Cheng. Contact Angles of Single Fibers Measured in Different Temperature and Related Humidity*. (2010)
5. S. F. Chini, A. Amirfazli. Colloids and Surfaces. 177-178. (2011)
6. B. Flinn, A. C. Tracey. *Improving Adhesive Bonding of Composite Through Surface Characterization*. (2012)
7. M. Gharechahi. J. Biomat. Nanotech. 541-544. (2012)
8. D. Gustri, Ratnawulan, Gusnedi. Pillar. Phys. 09-16. (2014)
9. A.W. Hefer. *Adhesion in Bitumen-Aggregate System and Quantification of The Effects of Water on The Adhesive Bond*. (2004)
10. M. Hyers. *Stress Analysis of Fiber Reinforced Composite Materials*. S.1. (1998)
11. Y. Yuan, T. R, Lee. Surf. Sci. Tech. 3-34. (2013)

Experimental study on strength and stiffness connection of wooden truss structure

Altho Sagara[1]*, *Johannes* Adhijoso Tjondro[1], *Husain* Abdurrahman Shiddiq[1]

[1]Civil Engineering Department, Parahyangan Catholic University, 40140 Bandung, Indonesia

Abstract. Scarcity of wood is hindering the use of wood materials that have large dimension and high quality. Thus, wooden truss is expected to be a solution for this problem. However, the use of this system is not complemented with adequate research of its strength and behavior. In this study, wooden truss would be examined experimentally. The specimen is a wooden truss joist supporting the floor slab for typical residences with two connection type variations which are nail and glue. These specimens has dimension of 185 cm in length and 36 cm in height with 3 layers of a wooden board in width, and were produced by using Sengon wood with elasticity modulus of 2867 MPa, and specific gravity of 0.4. Destructive and non-destructive test were performed. From the result, we obtained a higher proportional strength for glue type connection truss which at 1123.34 kg compared to nail type connection truss which at 767.07 kg. However, unlike the nail type connection, the glue type connection has a brittle behavior. Moreover, the stiffness connection study resulted with partial fixity value of 0.07 for nail type and 0.98 for glue type connection. Furthermore, these results have also been validated using numerical model analysis

1 Introduction

These days, wood as a construction material is gaining its fame, especially with the recently high demand of green material. One of the reasons is the low amount of energy to produce those wood materials compared to other construction materials. Moreover, the wood materials are getting more demand with the latest development of fast growing wood species [1]. These eventually lead to the issues of the lack of dimension and quality of wood as structural member.

In this study, these woods with lack of dimension and quality are engineered for use in the wooden truss joist supporting a residential floor slab with connection type variations. The truss type structure is chosen because this type of structure is more likely to be the best for optimizing the use of wood with small dimension. In addition, it has more esthetically values.

To fully understand this type of structure behavior, experimental test is required and conducted in this study. This test is also conducted to obtain strength and stiffness value as part of its behavior. Furthermore, structural analysis with numerical program is also conducted and validated with the experimental results obtained previously. This is intended for practical uses in future design.

2 Experimental details

2.1 Test specimen

Wood species used in this study is Sengon (Albizia chinensis). Its mechanical properties were tested based on ASTM D2395-02 [2] and obtained the bending modulus of elasticity and specific gravity which at 2847 MPa and 400 kg/m3. This wood species is classified as a low quality of wood according to SNI 7933-2013 [3].

These specimens has dimension of 185 cm in length and 36 cm in height with 3 layers of a wooden board in width. Each board is made with 6 cm x 2 cm of dimension. This board is used in the specimen as vertical, horizontal, and diagonal bar for tension and compression. The specimen configuration and dimension can be seen in the Figure 1.

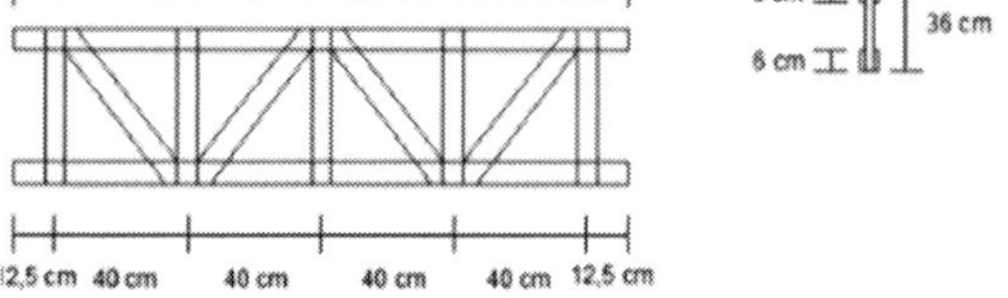

Fig. 1. Specimens' configuration and dimension

These specimens were arranged from two horizontal bars as top and bottom of the truss. For the middle part, it consists of vertical and diagonal bars. Figure 2 shows the assembly of the specimen

* Corresponding author: altho.sagara@unpar.ac.id

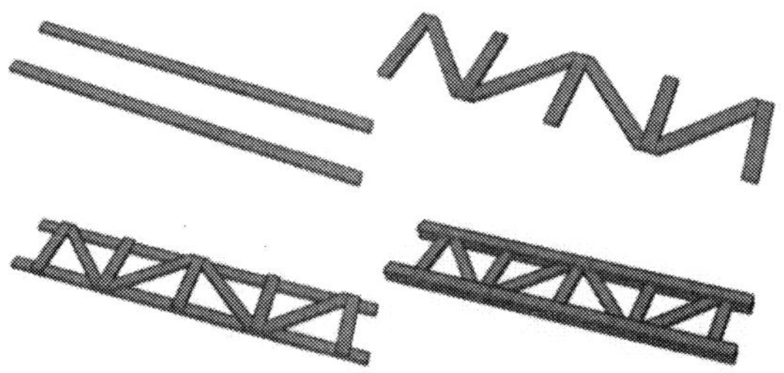

Fig. 2. The assembly of the specimen

2.2 Specimens' test specimen

Two connection types, which are nail and glue type connections, were observed in this study. These types of connection are the most common connection for practical uses in construction.

Two millimeter diameter nails with 40 mm length were used in the nail type connections. The nail has 107 kg of maximum shear strength. For the glue type, epoxy type adhesive was selected. The glue has 4.18 MPa of maximum shear stress and was applied in the specimen with 0.95 mm thickness.

Three nail type connection specimens, named A-1, A-2, and A-3, and also three glue type connection specimens, named B-1, B-2, and B-3, were produced for this study.

2.3 Destructive and non – destructive test

Destructive tests were conducted using UTM (Universal Testing Machine) in order to obtain wooden truss' strength. Three millimeter per minute speed rate was suggested by ASTM D143-94 [4]. The proportional, ultimate, and rupture condition of load and displacement can be found and then analyzed to determine the strength and behavior of the specimens. Figure 3 shows the test setup of specimens.

Fig. 3. Destructive test specimen

Non-destructive tests were also conducted in order to obtain the stiffness of specimens. The test gave the relation between load and displacement which was measured using LVDT (transducer). The load excitation used weight plates that increased gradually. Non-destructive testing from the specimen is shown in Figure 4.

Fig. 4. Non-destructive testing specimen

2.4 Experimental test result

Figure 5 – Figure 10 shows the load-displacement curve for each specimen obtained from destructive test. Three points shown in the graph represent the specimen's design, proportional, and ultimate load. Design load refers to maximum load for joist supporting residential floor slab design which equal to 266.5 kg [5]. Proportional and ultimate point is used to describe the specimen behavior.

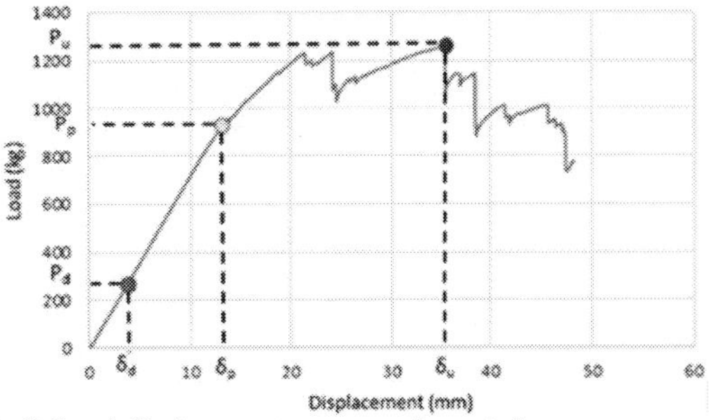

Fig.5. Load-displacement curve specimen A-1

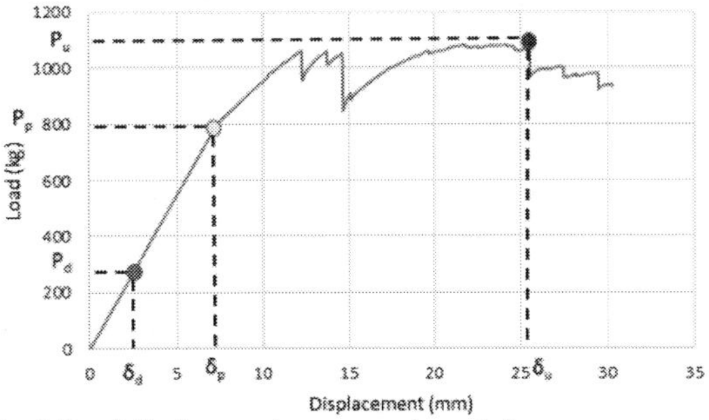

Fig.6. Load-displacement curve specimen A-2

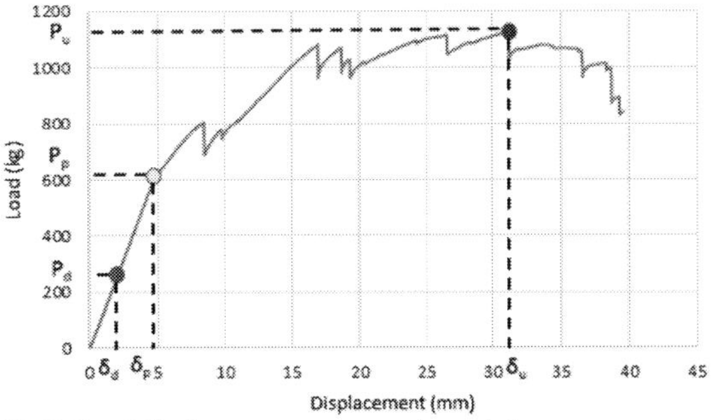

Fig.7. Load-displacement curve specimen A-3

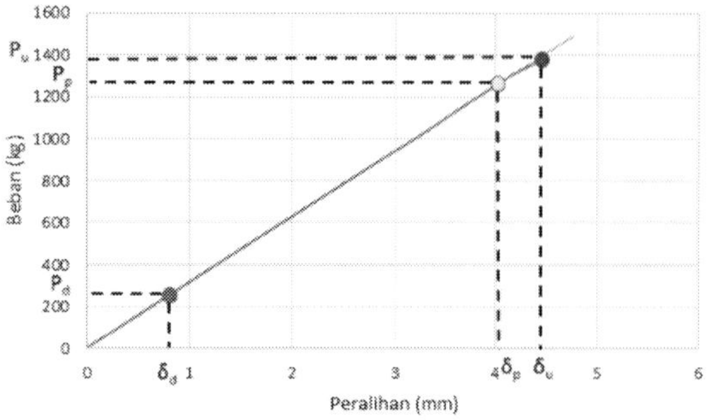

Fig.8. Load-displacement curve specimen B-1

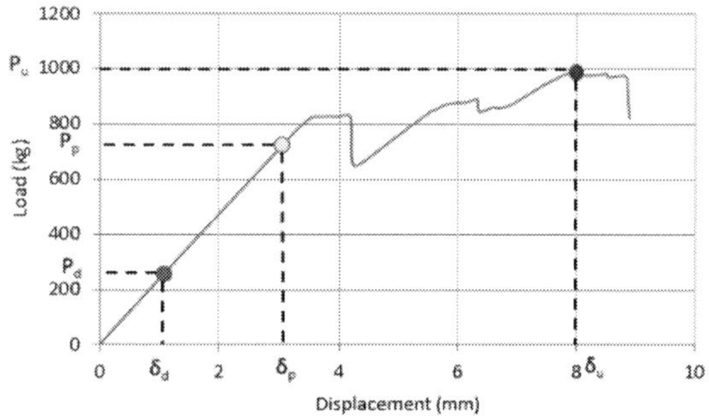

Fig.9. Load-displacement curve specimen B-2

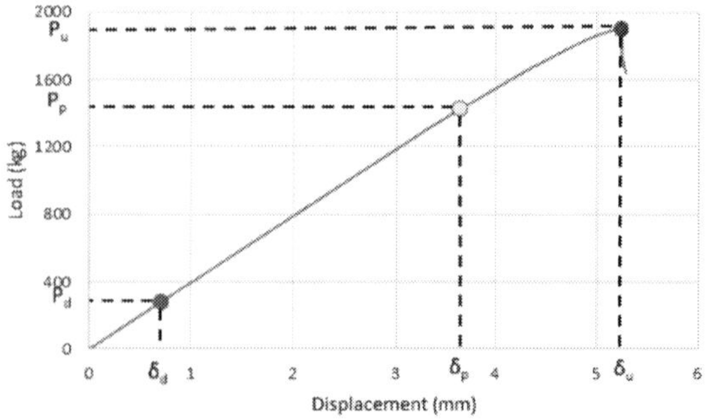

Fig.10. Load-displacement curve specimen B-3

Figure 11 shows the relation between load and displacement of the specimens obtained from non-destructive test. The allowable displacement of structural beam that refer to the latest code (L/300 which equal to 5.33 mm) [3] is also added to the curve in Figure 11

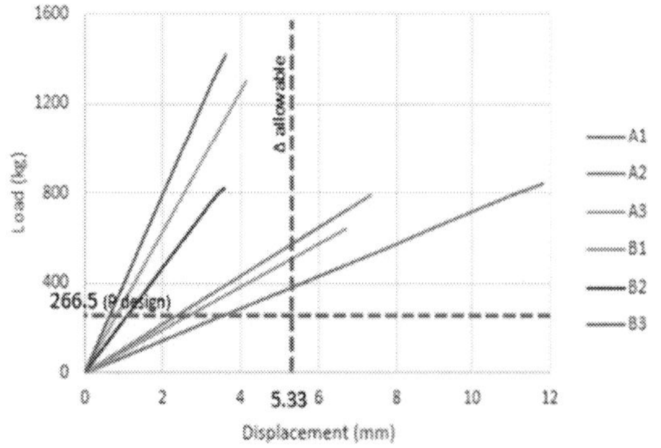

Fig.11. Non-destructive test result

3. Numerical model validation

Numerical model validation is required to obtain wooden truss connection stiffness. In addition, it can be used to validate experimental results for structural analysis design. The numerical analysis was modeled using SAP2000. All of its properties required for the analysis (except for connection stiffness) were taken from the actual experimental test. Connection stiffness was found by using trial and error value of partial fixity.

The connection stiffness was evaluated using Romstad and Subramanian equation (1) [6] below:

$$Sc = \lambda \; 4EI / L \qquad (1)$$

where Sc is connection stiffness and λ is partial fixity value of connection. The partial fixity represents the structural connection condition. Three type of partial fixity value that usually used on numerical model are rigid frame (λ = 1) that assumes the connection have sufficient rigidity to hold virtually unchanged the original angles between intersecting member, simple framing (λ = 0) that assumes insofar as gravity loading is concerned, ends of beams and girders are connected for shear only and are free to rotate under gravity load, and lastly semi-rigid framing (0 < λ < 1) that assumes the rigidity of connection in degree between rigid and simple framing condition. Figure 12 shows the modelling of partial fixity condition.

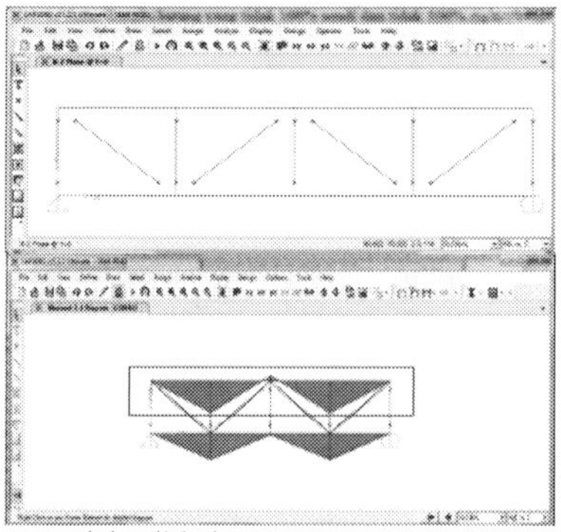
Fig.12. Test model validation

After several trial value of partial fixity, the graph in Figure 13 summarizes the relation between partial fixity and displacement. Displacement from experimental test is fitted to find partial fixity value for both type of wooden truss connection.

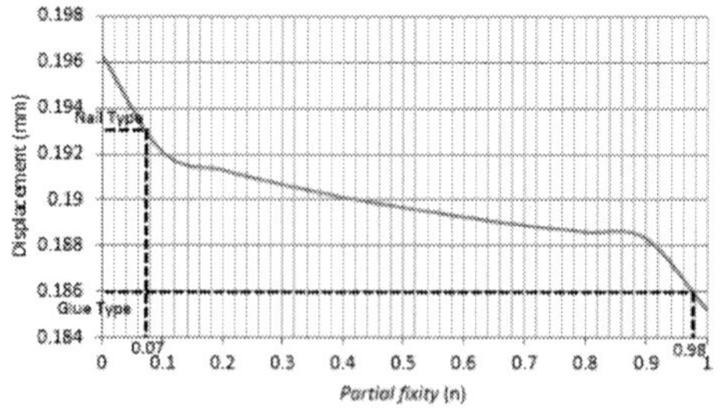

Fig.13. Relation between partial fixity and displacement

From the trial and error, it is concluded that the partial fixity is 0.07 and 0.98 for nail type and glue type connection respectively as shown in figure above.

4. Analysis and discussion

4.1 Destructive test result

The table 1 show the summary of proportional and ultimate condition for both type connections obtained from load-displacement curves.

Table 1. Destructive test result

Type	Pp (N)	δp (mm)	Pu (N)	δu (mm)
A-1	8442.8	11.82	12364.74	35.49
A-2	7843.3	7.31	10562.51	30.31
A-3	6265.9	6.70	11007.40	31.18
B-1	12714.84	4.13	13446.46	4.44
B-2	6420.43	3.58	9636.88	8.01
B-3	13861.71	3.63	18582.13	5.23

As shown in Table 1, proportional load for nail type connection ranges from 6000-8500 N which is smaller than the proportional load for glue type connection that ranges from 6000-14000 N. Likewise, this condition is also applied to specimens' ultimate load.

Table 1 also shows both of the proportional and ultimate displacement of the specimens that would be used to determine its ductility. The ductility is calculated from the ratio between ultimate and proportional displacement. It shows the ability of the specimen to resist load after yield point. Displacement ductility of both specimens has significantly large differences as shown figure in table 2.

Table 2. Average ductility

Type	Displacement Ductility	Average Ductility
A-1	3.00	
A-2	4.15	3.94
A-3	4.66	
B-1	1.08	
B-2	2.24	1.58
B-3	1.44	

From Table 2, it can be concluded that wooden truss with nail type connection is more ductile than the other type. Its ductility value ranges from 3 – 5 while the glue type connection has ductility value less than 2.5. On the other hand, specimen with glue type connection has greater strength than nail type connection.

4.2 Non – destructive test result

The glue type connections have better stiffness than nail type connections as shown in Figure 11. It is related with the ductility behavior of specimen. Based on the graph in Figure 11, it shows that allowable displacement for glue type is not in elastic condition. Therefore, the design for glue type connection is limited by proportional load unlike the nail type connection that is limited by allowable displacement

4.3 Safety factor

The ratio between ultimate and proportional load against design load give safety factor for both specimen type. Design load for this study is obtained from calculation of a 0.85 meter joist supporting a residential floor slab (=2665.6 N).

Table 3. Safety factor

Type	Pd (N)	Pp (N)	Pu (N)	SF (Pp / Pd)	SF (Pu / Pd)
A-1		7845.29	12364.7		
A-2	2665	7843.23	10562.5	2.75	4.24
A-3		6265.95	11007.4		
B-1		12714.8	13446.5		
B-2	2665	6420.43	8636.88	3.99	5.21
B-3		12735.8	18582.1		

Safety factor for nail type connection is lower than glue type connection as shown on table 3. It is 2.75 and 4.24 in proportional and ultimate condition respectively for nail type connection. As for the glue type connection, the safety factor is 3.9 and 5.21 in proportional and ultimate condition respectively. Overall, safety factor from both connection types is quite large and adequate.

4.4 Failure characteristic

From the previous test, failure characteristic from both specimens is quite different. The failure type that happened on the specimen is ductile failure on nail type connection as shown in Figure 14. The ductile failure is indicated by cracks on the element that developed gradually. This failure is totally different from brittle failure that experienced by glue type connection. In brittle failure, cracks suddenly developed as shown on figure 15

There is no connection failure in both type of specimen. It's caused by maximum shear stress of nail and glue hasn't been exceeded. It is also proven experimentally by visual checks of nail and glue condition after test.

Fig.14. Ductile failure on nail type connection specimen

DOI: 10.1051/matecconf/201710101015

Fig.15. Brittle failure on glue type connection specimen

5. Conclusion

Wooden truss with nail and glue type of connection have adequate safety factor for a residential floor joist design which range from 2.75 – 4.00.

It's also observed that the specimen with nail type connection has lower stiffness value than the glue type connection. It's also proven by model validation that resulted the partial fixity value of 0.07 and 0.98 for nail and glue type of connection respectively.

Failure characteristic of specimen is ductile for nail type of connection unlike the other specimen that experienced more brittle failure. Although the glue type connections have a brittle failure, it has 30% higher strength than the nail type connection.

The writer would like to acknowledge Bambang Suryoatmono, Dina Rubiana Widarda, Buen Sian, Julius Belen, Christian Arief and Sisi Nova from Civil Engineering Department, Parahyangan Catholic University for technical input to improve the research and their continued advisement.

References

1. Forest Products Laboratory, *Wood Handbook—Wood As An Engineering Material.* United States Department of Agriculture Forest Service. Madison, Winconsin. (2010).
2. ASTM D2395-02: *Standard Methods of Specific Gravity of Wood and Wood-Based Materials.* Annual Book of ASTM Standards volume **04.10** Baltimore, U.S.A. (2008).
3. SNI, *Spesifikasi Desain untuk Konstruksi Kayu, SNI 7973:2013.* Badan Standar Indonesia, Jakarta, Indonesia. (2013).
4. ASTM D198-99 "*Standard Test Methods of Static Tests of Lumber In Structural Sizes*" American Society for Testing and Materials, West Conschohocken. (1999).
5. A. Aghayere, *Structural Wood Design A Practice-Oriented Approach Using the ASD Method.* John Willey and Sons. Hoboken, New Jersey. (2007).
6. Romstad and Subramanian, *Non-Linier Static and Cyclic Analysis of Steel Frames With Semi-Rigid Connection.* Elsevier Ltd, Kiddlington, Oxford. (1970).
7. Stanalker, Judith J. *Structural Design in Wood.* Chapman & Hall, New York. (1997).
8. Williamson, Thomas G. APA *Engineered Wood Handbook.* McGraw-Hill, New York. (2002).
9. S. Somayaji, *Structural Wood Design.* West Publishing Company. Kellog Boulevard, St.Paul. (1990).
10. A. Buchanan, *Timber Design Guide*, New Zealand Timber Industry Federation Inc, New Zealand, p. 55. (2007).

MATEC Web of Conferences **101**, 01016 (2017) DOI: 10.1051/matecconf/201710101016
SICEST 2016

Influence of uncoated and coated plastic waste coarse aggregates to concrete compressive strength

Heru Purnomo[1*], *Gandjar* Pamudji[1], and *Madsuri* Satim[1]

[1]Civil Engineering Department, Faculty of Engineering, Universitas Indonesia, 16424 Depok, Indonesia

Abstract. The use of plastic waste as coarse aggregates in concrete is part of efforts to reduce environmental pollution. In one hand the use of plastic as aggregates can provide lighter weight of the concrete than concrete using natural aggregates, but on the other hand bond between plastic coarse aggregates and hard matrix give low concrete compressive strength. Improvement of the bond between plastic coarse aggregate and hard matrix through a sand coating to plastic coarse aggregate whole surface is studied. Sand used to coat the plastic aggregates are Merapi volcanic sand which are taken in Magelang. Three mixtures of polypropylene (PP) coarse plastic aggregates, Cimangkok river sand as fine aggregates, water and Portland Cement Composite with a water-cement ratio of 0.28, 0.3 and 0.35 are conducted. Compression test are performed on concrete cylindrical specimens with a diameter of 10 cm and a height of 20 cm. The results in general show that concrete specimens using plastic aggregates coated with sand have higher compressive strength compared to those of concrete specimens using plastic aggregates without sand coating. The bond improvement is indirectly indicated by the betterment of concrete compressive strength.

1 Introduction

Nowadays lightweight concrete has been used for building materials. As lightweight aggregate is part of lightweight concrete, lightweight aggregate can be taken from nature such as pumice, coconut shell, perlite etc, and can be made from fly ash, slag or plastic as example. Until now plastic are used daily by the modern community. This phenomenon is due its low production cost, light in weight and relatively easy to form. However as the number of people increase, the plastic waste becomes problem for most of countries around the world.

At world level, plastic production reaches almost 150 million ton per year which in turn 4.8 ton per second according to The Central Pollution Control Board [1]. In Indonesia according to Indonesia Solid Waste Association (InSWA) total plastic garbage in year 2008 has reached 5.4 million ton per year or 14% from total garbage production. The plastic waste is non bio-degradable which in turn creating environmental problem [2].

Transformation of plastic waste to a new production that benefits to the people will decrease the number of unused plastic waste. Transformation of plastic waste into aggregate for concrete making will give benefit not only for environment but also for construction world in view of reducing the own weight of the concrete [3-7]. This effort can be considered as new perspective in research activities in integrating concrete technology and environmental issue. However the weak side of using plastic as coarse or fine aggregates in concrete is in the mechanical properties of that concrete.

According to Saikia and Brito [8], this strength reduction are caused by (a) low bond strength between plastic aggregate surface and cement paste, (b) plastic characteristic which is unable to absorb water, hampering cement hydration reaction as there is limitation of water movement. Choi *et al* [9] have improved the interface characteristic between plastic aggregate and cement paste by slag (GBFS) addition in plastic aggregate material which in turn strengthens the surface of plastic aggregate as GBFS reacts with calcium hydroxide.

By improving the interface characteristic between surface of plastic aggregate and mortar through enhancement of hardness and texture of surface of plastic aggregate, performance of concrete mechanics will be better. This study covers the development of uncoated and coated plastic coarse aggregate as one of the concrete materials. The coating material chosen is Merapi volcanic sand coming from Magelang located at Java Island, Indonesia. Volcanic sand is considered having higher SiO_2 content compared to river sand which in turn it could expand interface zone between aggregates and mortar [10].

* Corresponding author: herupur@eng.ui.ac.id

DOI: 10.1051/matecconf/201710101016

2 Experimental program

2.1 Material

Portland Cement Composite is used to satisfy Indonesian Standard (SNI 15-7064-2004) and European Standard (EN 197-1:2000 (42.5 N & 42.5 R).

Sand from Cimangkok at West Java, Indonesia is used as fine aggregate. Size of the fine aggregate is less than 9.5 mm while specific gravity and fineness modulus of the fine aggregate where found to be 2.33 and 2.63 respectively. The percentage of passing is within the limits as per SNI 03-2834-2000.

The coarse aggregate is made from waste of Polypropylene (PP) plastic. Figure 1 shows the shapes of uncoated plastic aggregate. The uncoated aggregate shape is similar to the plastic aggregate developed by Pamudji, G *et al.* [7]. The aggregate has generally 10 mm of thickness and, 20 mm of length and 20 mm of width. Plastic aggregate is formed by heating the plastic up to the melting point; it is flowed then into the mould by hydraulic pressure. The process itself is conducted using manual injection plastic machine with automatic temperature control. The plastic melting temperature is in range of 1300C ± 100C. After the formation, the plastic aggregate were then cooled.

Once the uncoated coarse aggregate is formed, then it is coated with hot sand by placing the uncoated plastic aggregate into a rotating cylinder. Merapi Volcanic Sand taken at Magelang is used for coating purposes. The final shape of plastic aggregate coated with Merapi sand is shown in Figure 2. It is intended to improve bonding at the interface between coarse aggregate and mortar.

Fig.1. Shape of uncoated plastic aggregate [7]

Fig.2. Plastic aggregate coated with sand

2.2 Mix proportion of concrete

Proportion of materials is determined using mix design method that is based on absolute volume. From trial mixes, it is obtained series of material mix proportion of concrete containing plastic aggregate, denoted by M type of mix proportion. Water to cement ratio (W/C) used are 0.28, 0.3 and 0.35 respectively as presented in Table 1. Superplasticizer (SP) is used to have better workability in mixing the materials for M1 to M3 proportion.

Table 1. Concrete mix proportion

Type	W/C	Cement (kg)	Sand (kg)	Plastic Aggr. (kg)	Water (kg)	Admixture (SP) (kg)
M1	0,28	500	678	411	140	8
M2	0,30	500	738	388	150	7
M3	0,35	500	738	372	175	5

3 Results and discussion

3.1 Physical properties of material

Observation of physical properties of sand from Merapi and Cimangkok was conducted on density, water absorption, granular gradation, and fineness modulus. Test result of the physical properties of the sand is presented in Table 2.

Table 2. Physical properties of sand

Physical characteristic	Sand	
	Cimangkok	Merapi
Specific gravity (g/cm^3)		
Apparent	2.44	2.74
Bulk	2.26	2.68
SSD	2.33	2.70
Absorption (%)	3.31	2.81
Density	1.39	1.45
Fineness modulus (FM)	2.63	3.44

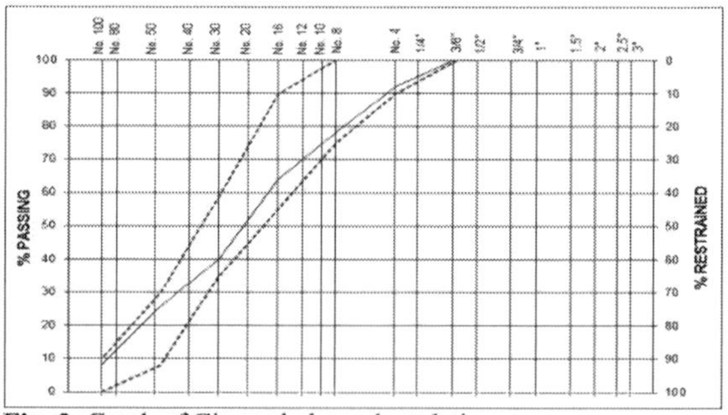

Fig. 3. Graph of Cimangkok sand gradation

While gradation of fine aggregate grains are shown in Figures 3 and 4. The gradation of both types of sand are within the envelope limits. At low percentage passing both sand behave similarly but gradually as the percentage passing grows Merapi sand tends to near the left envelope while that of Cimangkok near the right

DOI: 10.1051/matecconf/201710101016

envelope. Specific gravity of Merapi Sand is higher than that of Cimangkok Sand.

Observation of physical properties of uncoated plastic aggregate was conducted on density, bulk density and water absorption. The test result of physical properties of uncoated plastic aggregates is presented in Table 3. Its absorption is in fact very small in percentage.

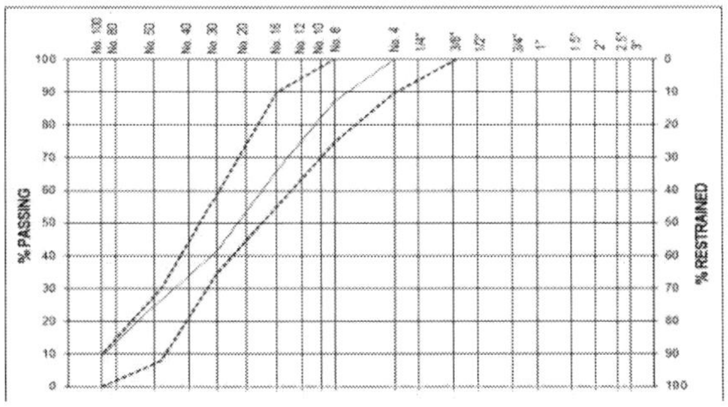

Fig. 4. Graph of Merapi sand gradation

Table 3. Properties of coated plastic aggregate

Characteristic of physic	Plastic aggregate
Specific gravity (g/cm^3)	
Apparent	1.13
Bulk	1.07
SSD	1.12
Absorption (%)	0.05
Density	0.70
Fineness Modulus (FM)	4.45

3.2 Compressive strength of plastic cube

Compressive strength of PP plastic is obtained by compression testing of plastic cube specimens. Using manual injection machine under controlled temperature, plastic cube specimens are made in cubical shape having size of 5 cm x 5 cm x 5 cm.

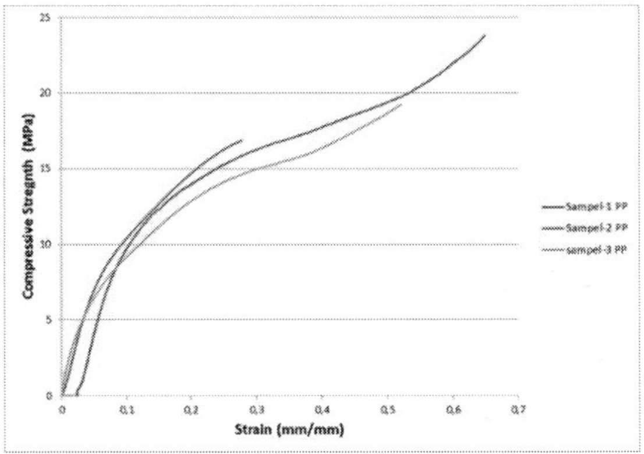

Fig. 5. Curve of - Plastic cube

Results of compressive strength versus strain of PP plastic cubes are shown in Figure 5. All the three curves show similar behavior where stress is still increasing as strain grows bigger. All the cube specimens experienced large deformation without any cracks appeared on the surface of the specimens and the depth of all specimens shortened at large strain.

As there is no concrete crushing phenomenon existed at plastic cube specimen during compression test, the compressive strength of the three PP cubes specimen is taken minimum between 7 to 8 MPa referring to 0.06 strain of the blue sample (specimen 1) linier curve limit.

3.3 Density of Concrete

Evaluation of concrete density are performed on 28 days aged concrete cylindrical specimens with a diameter of 10 cm and a height of 20 cm that serve for compression testing. Two kinds of concrete specimen are evaluated. The first type is concrete containing uncoated plastic coarse aggregate while the second type of concrete contains Merapi sand coated plastic aggregate. Evaluation result shows that density of concrete containing uncoated plastic coarse aggregate is in general between 10 to 14% lighter than that containing sand coated plastic coarse aggregate for three kind of water to cement ratios, as presented in Figure 6.

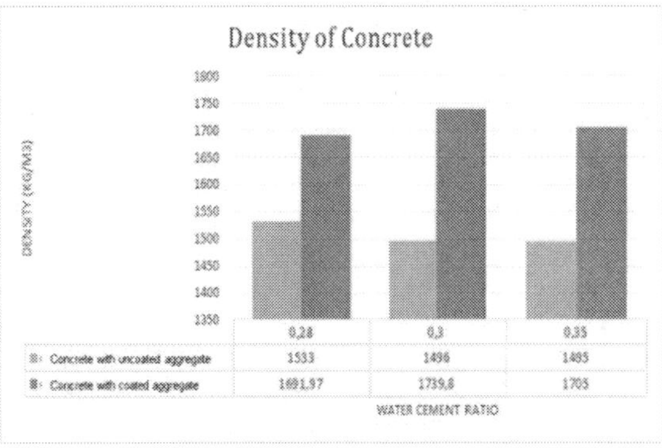

Fig. 6. Density of concrete for three kind of water to cement ratio

The density of concrete containing uncoated plastic aggregate decreases a little following an increase of water to cement ratio. However it is not the case for concrete containing sand coated plastic aggregate. The highest density for concrete containing coated plastic aggregate is obtained with water to cement ratio equal to 0.3.

3.4 Compressive strength of concrete

Compression test are performed on 28 days aged concrete cylindrical specimens with a diameter of 10 cm and a height of 20 cm. The same kind of specimens used for density evaluation, are then utilized for compression testing. Three different water to cement ratios and two coating condition are fixed for the concrete compressive strength evaluation.

Result of the compression testing shows that compressive strength of all kinds of concrete decreases

DOI: 10.1051/matecconf/201710101016

as water to cement ratio increases. Compressive strength of concrete containing sand coated plastic coarse aggregate is higher than that of concrete with uncoated plastic aggregate, as shown in Figure 7. It is valid for three kind of water to cement ratio. 7 days aged concrete compression testing results are intended for controlling the concrete strength.

For all W/C ratios, concrete specimens containing uncoated plastic aggregate show only slight increase of 7 days compressive strength to 28 days compressive strength. In this case, the 28 days highest compressive strength of concrete specimens containing uncoated aggregate which is 14.25 MPa given by the W/C ratio equals to 0.28, is still below the building minimum structural compressive strength of 17.5 MPa based on normal cylinder specimen as stated in Indonesian National Standard SNI 03-2847-2002.

For two W/C ratios, a much better increase of compressive strength for 28 days concrete specimens containing coated plastic aggregate from its 7 days compressive strength. However it is not the case for concrete specimens containing coated aggregate with W/C equals 0.28 as it might experience bad compaction. The highest compressive strength for 28 days of concrete specimens containing coated aggregate is given by the W/C equals 0.3 with a value of 18.16 MPa. It is almost near the minimum required compressive strength defined by the Indonesia National Standard SNI 03-2847-2002 as the tested specimens are smaller than cylinder stated in the Standard. As seen in Table 3, the uncoated plastic aggregate density is lower than one, resulting in disturbing the workability which in turn difficult to compact the fresh concrete as uncoated aggregate tends to separate. This situation could cause segregation of the materials. Uncoated plastic aggregate also has smooth surface and hydrophobia characteristic.

After crushing the specimens, observation to the concrete specimens containing uncoated plastic coarse aggregate indicates weak bonding at the interface zone between uncoated plastic aggregate and mortar. Examples of this situation are shown in Figure 8a, 8b and 8c. It confirms low compressive strength of concrete containing uncoated coarse aggregate.

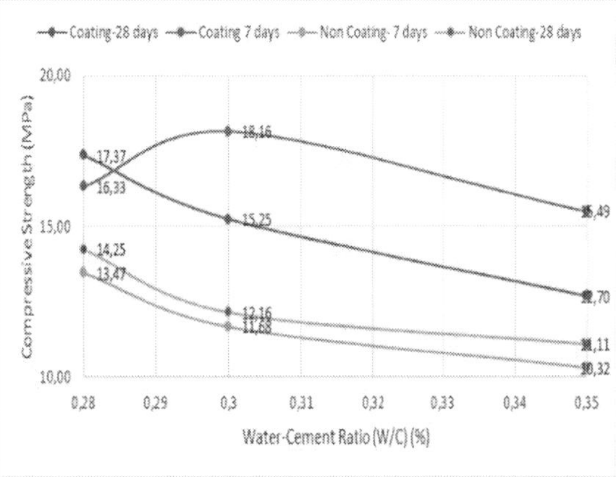

Fig. 7. Compressive Strength versus W/C and concrete age

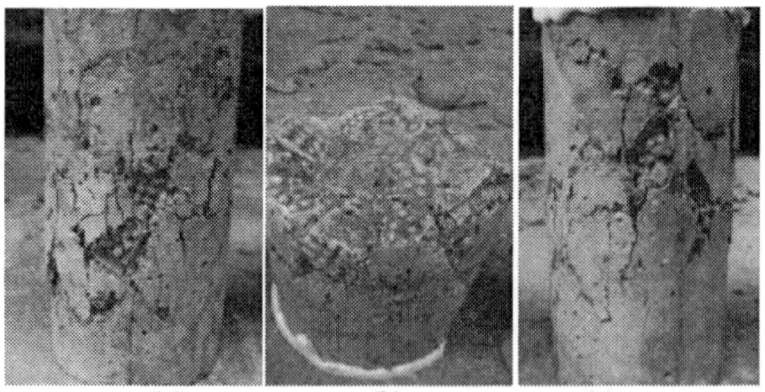

(a) (b) (c)

Fig. 8 Weak bonding between uncoated plastic aggregate and mortar

Fig. 9. Rupture of concrete cylinders containing uncoated plastic aggregate

Specimens of this kind show a relative distribution of crack on most of the specimen surface as the weak bonding are existed at interfaces of uncoated plastic aggregate and mortar in the specimens. Example of this condition is presented in Figure 9.

To overcome the problem, using the principle of Choi *et al* [9] to coat the plastic PET fine aggregates, the PP coarse aggregate is coated with sand. This method works as shown in Figure 7 and Figure 10 by the test results of compression testing of concrete containing sand coated PP coarse aggregate. Figure 10 shows a better bonding at the interface between sand coated PP coarse aggregate and mortar as less cracks appear over the whole surface of the specimens. Four concrete cylinders containing sand coated plastic aggregate show almost similar crack pattern where its base of cylinder experienced spalling.

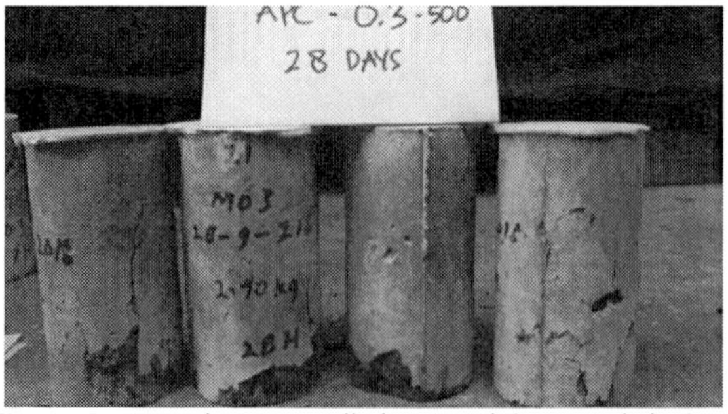

Fig. 10. Rupture of concrete cylinders containing coated plastic aggregate

4 Conclusion

Result of the compression testing shows that compressive strength of all kinds of concrete studied, decreases as water to cement ratio increases. Compressive strength of concrete containing sand coated plastic coarse aggregate is higher than that of concrete with uncoated plastic aggregate. Based on the experiments results represented by the increase of compressive strength and also lesser crack appearance on the surface of concrete cylinder specimens, it can be concluded that the sand coating of the surface of plastic coarse aggregate is able to enhance the bond between coarse aggregate and mortar.

The authors convey their gratitude to the Ministry of Research Technology and Higher Education for Fundamental Research Grant support through Fiscal Year 2016. The same gratitude is also conveyed to the Head of Laboratory of Structure and Building Materials at Department of Civil Engineering, Faculty of Engineering, Universitas Indonesia for the facilities given in the experimental works.

References

1. L. Ferreira, J. de Brito, N. Saikia, Constr. Build. Mater. **36** (2012)
2. R. Ermawati, J. Riset Ind. **5**, 3 (2011)
3. Y.W. Choi, D.J. Moon, Y.J. Kim, M. Lachemi, Constr. Build. Mater. **23** (2009)
4. M. Frigione, Waste Manage. **xxx** (2010).
5. S.C. Kou, G. Lee, C.S. Poon, W.L. Lai, Waste Manage. **29** (2009)
6. A.M. Mustafa Al Bakri, S.M. Tamizi, A.R. Rafiza, Y. Zarina, J. Asian Sci. Res. **1**, 7 (2011)
7. G. Pamudji, H. Purnomo, I. Katili, I. Imran, P. 6th Civil Eng. Conf. Asia Region Jakarta (2013)
8. N. Saikia, J. de Brito, Constr. Build. Mater. **34** (2012)
9. Y.W. Choi, D.J. Moon, J.S. Chung, S.K. Cho, Cem. Concr. Res. **35** (2005)
10. M. Tanudjaja, S. Indrawan, J. Tek. Sipil **1**, 2 (2004)

Effect of longitudinal joint on the shear-key of hollow core slab which function as an rigid diaphragm

Gambiro Soeprapto[1*], *Mukhlis* Sunarso[1], Sumarsono[2], *Ferryandy* Murdono[1], *Winda* Agustin[1], *Raynelda* Siahaan[1]

[1]PT Wijaya Karya Beton Tbk., Research and Development Departement, 17411 Bekasi, Indonesia
[2]PT Wijaya Karya Beton Tbk., Concrete Product Factory, 16969 Bogor, Indonesia

Abstract. Various types of reinforced concrete floors can be applied to building construction. The floor of the building can be made of cast in place or precast concrete. The floors should function as a rigid diaphragm, in order to distribute the lateral forces on the building frame. While the floor is made of precast concrete, especially those without overtopping, should be arranged in such a way to function as a diaphragm. In order for the precast slab can function as a diaphragm, the components must be connected to the building frame and between the component itself . Some of them are drag strut, boundary element (chord), transverse joint, collector and longitudinal joint. This paper will discuss the influence of the longitudinal joint shear key with some types. The test specimen is modeled as two HCS are mounted parallel to each other , then the both of HCS are combined with perimeter beams to obtain the same model with the actual conditions and simplify to execute. All models of the test specimen are made to represent the three types of shear key. The test specimen is loaded the shear force and observed the structure behavior and crack patterns.

1 Introduction

Various types of reinforced concrete floors can be applied to building construction. The floor of the building can be made of cast in place concrete or precast concrete. Some types of precast concrete floor including solid slab (full depth slab), half slab, double T slab dan hollow core slab (HCS).

The floors in tall buildings should function as a rigid diaphragm, in order to distribute the lateral forces on the building frame. These properties characters can be obtained, when using a cast in place concrete floor. While the floor is made of precast concrete, especially without overtopping, should be arranged in such a way to function as a diaphragm.

Hollow Core Slab (HCS) is a type of plate precast components that behave in one direction (one way slab). The width range of HCS between 80-120 cm, while the length varies with the size of the longest can up to 20 meters.

Because one way slab, so the HCS are supported by two beams at the both ends. The beams are part of the building structure. The other two sides are not connected to the parallel beam.

In order for the precast slab can function as a diaphragm, then the components must be connected to building frame associated with the building frame and between the component itself. Some of them named drag strut, boundary element (chord), transverse joint, collector and longitudinal joint.

Longitudinal joint shear key form can be filled with materials or other components. There are three types of shear key that mortar, longitudinal reinforcement and shear reinforcement are mounted transversely.

2 Theory and design

Hollow core slabs are used as floor or roof to support vertical loads. But besides that also functions as a diaphragm to restrain and distribute lateral loads Lateral load be caused lateral pressure, wind loads or seismic loads. The main problem in planning the hollow core connection diaphragm is design to distribute the load to the diaphragm, the strength and ductility of slab system to carry the loads to the lateral load-bearing element.

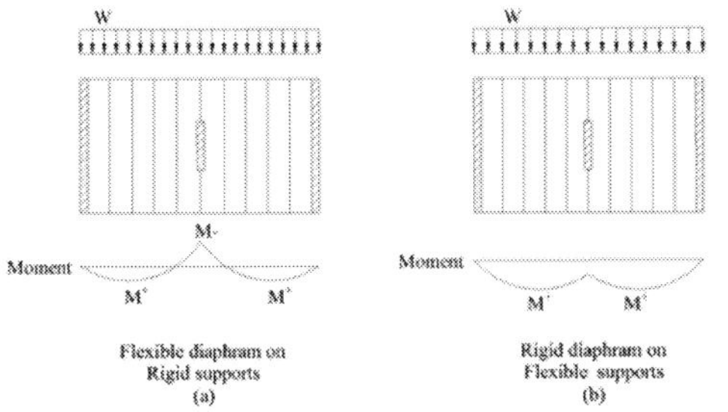

Fig. 1. Diagram bending moments.

2.1 Diaphragm element

- Boundary element : edge element around the diaphragm or perimeter of the diaphragm opening and together that bind together. Boundary element described as the pull rod / push (chord) or slider (drag strut).
- Collector : element that distribute the shear force of the diaphragm to the lateral retaining elements.
- Chord : Pull or push elements as a result of the behavior of the wing on the diaphragm to produce bending in the integrity of the diaphragm.

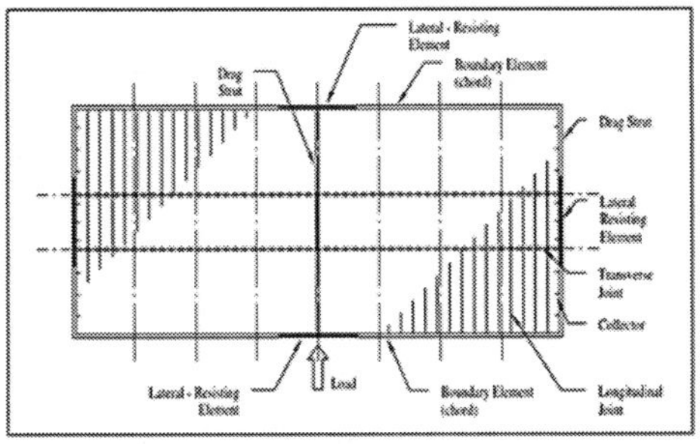

Fig. 2. Diaphragm elements.

The diaphragm must have the strength to distribute lateral loads imposed on the point to the point of lateral load resisting. Spans of the diaphragm between the lateral anchoring element behaves as a high beam and tied arch.

2.2 Longitudinal connection

Shear keyway are grouted between the slab has the capacity to distribute the longitudinal shear forces from one slab to the next slab. With shear stress 0.55 MPa, the nominal shear forces is

$$\phi V_n = \phi\,(0.55) h_n l \qquad (1)$$

l = length of connection (mm)
h_n = depth of grouted keyway. (mm)
ϕ = 0.85

If the grout strength is exceeded or ductile behavior is required, in principle, the sliding friction can be used to plan reinforcement is placed perpendicular to the longitudinal joint. This reinforcement can be placed on a transverse joint at the end of the slab rather than distributed throughout the connection.

$$A_{vf} = \frac{V_u}{\emptyset f_y \mu} \qquad (2)$$

V_u = factored ultimate shear force
μ = 1.0 for shear strength which parallel to longitudinal connection.
= 1.4 for shear strength which parallel to transversal connection.
ϕ = 0.85

3 Modelling and set up sample of testing

3.1. Modelling of sample

Shear testing the keyway Hollow Core Slab (HCS) is modeled with a 2-pieces HCS sheet is fastened with boundary elements. This boundary element serves to unite two pieces of HCS, to be given the test load. Besides this modeling represent actual conditions attached HCS.
Deformation shows the performance of the connection system to be applied

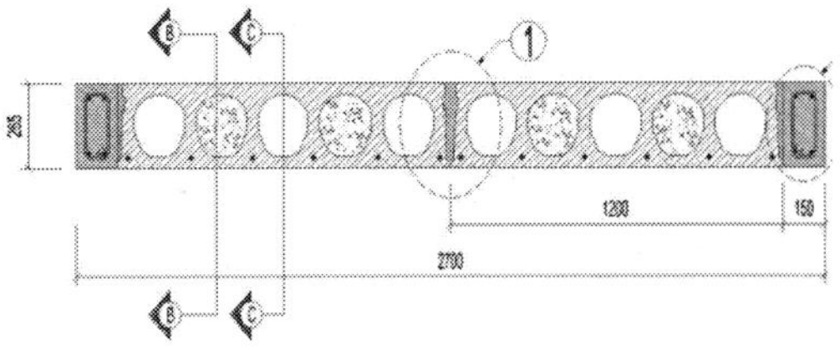

Fig. 3. Section of hollow core slab.

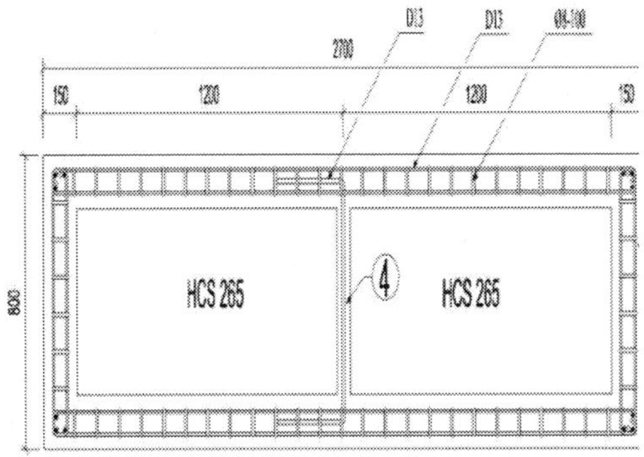

Fig. 4. Model of specimen.

 DOI: 10.1051/matecconf/201710101017

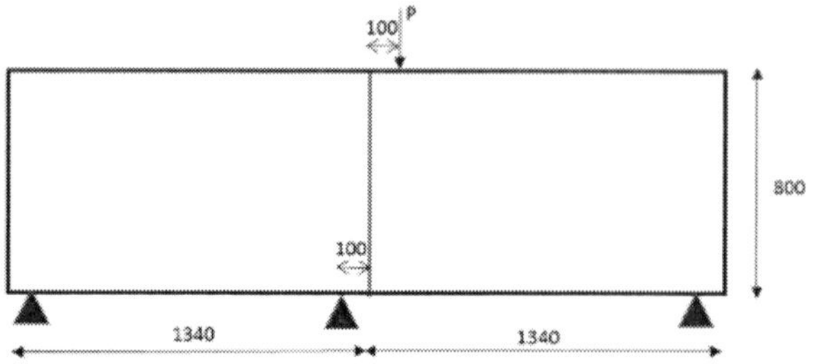

Fig. 5. Load pattern.

Fig. 6. Set up of specimen.

3.2. Sample of testing

Sample consist of 4 types :

- *Specimen type 1 (BU-1)*

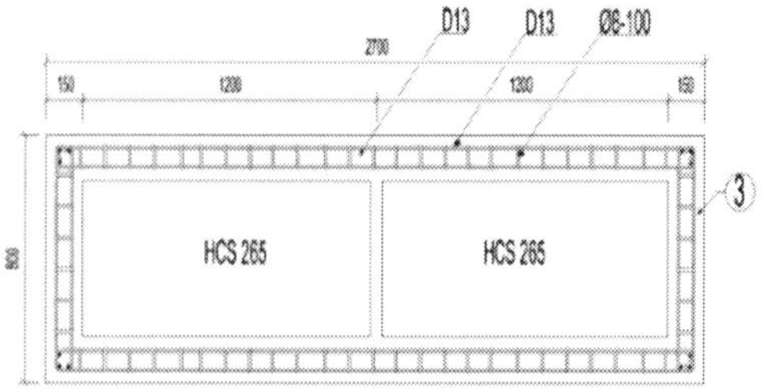

Fig. 7. Steel reinforcement for specimen type 1

No reinforment at shear keyways. Only rely on the specimen and concrete filler This specimen has function as comparison.

- *Specimen type 2 (BU-2)*

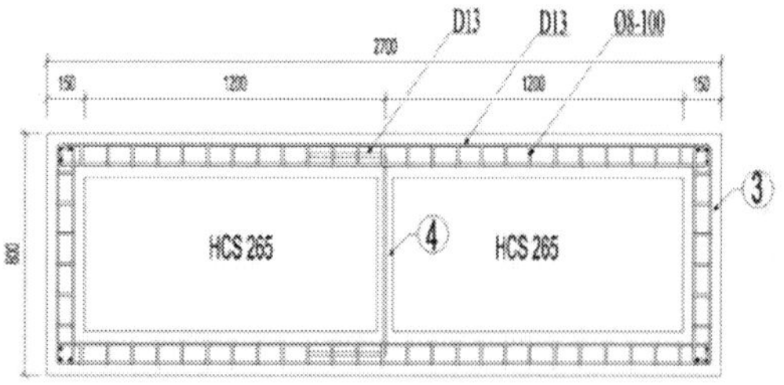

Fig. 8. Steel reinforcement for specimen type 2

Reinforcement in the shear keyway in the form of longitudinal reinforcement along the channel. It consist of 1D13 steel reinforcement.

- *Specimen type 3 (BU-3)*

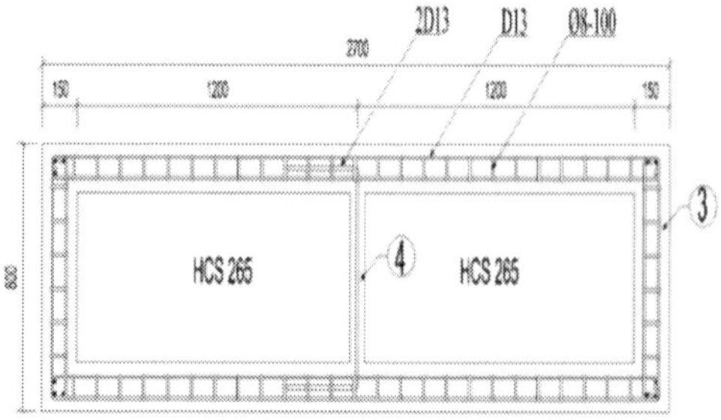

Fig. 9. Steel reinforcement for specimen type 3

Reinforcement in the shear keyway in the form of longitudinal reinforcement along the channel. It consist of 2D13 steel reinforcement.

- *Specimen type 4 (BU-4)*

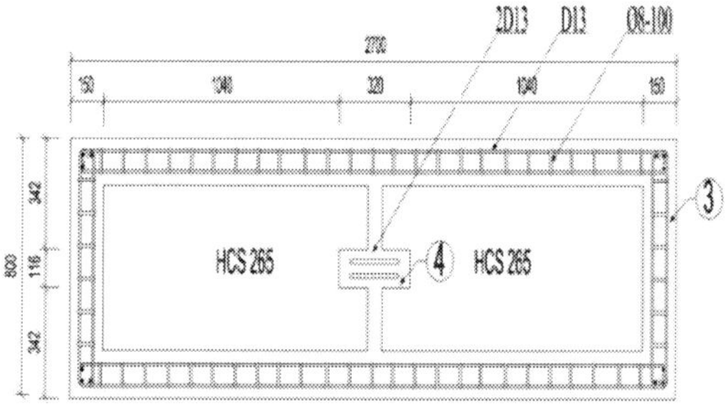

Fig. 10. Steel reinforcement for specimen type 4.

Reinforcement in the shear keyway in the form of transversal reinforcement at the middle of channel. It consist of 2D13 steel reinforcement.

- *Specimen type 5 (BU-5)*

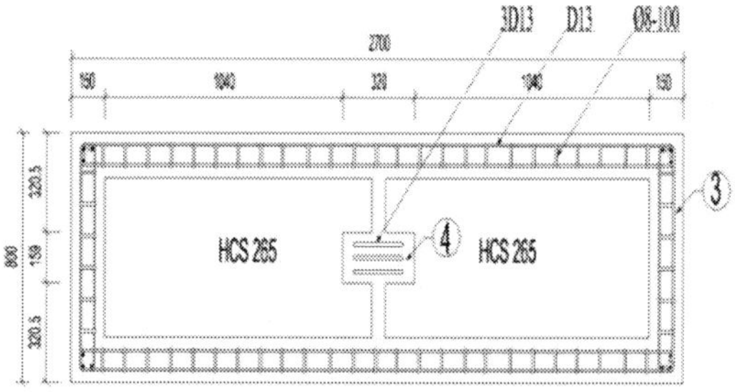

Fig. 11. Steel reinforcement for specimen type 5.

Reinforcement in the shear keyway in the form of transversal reinforcement at the middle of channel. It consist of 3D13 steel reinforcement.

DOI: 10.1051/matecconf/201710101017

4 Discussion and analysis

The test load F acts on the test object as figure 3D. But the style is observed is the force exerted on the joints between panels HCS Therefore, these forces should be determined by structural analysis.Oleh karena itu gaya-gaya tersebut harus ditentukan dengan analisa struktur. Structure of HCS is modeled as two shell elements are tied with boundary elements, so that a single unitary structure. Structural analysis is done by SAP2000.

4.1. Specimen 1

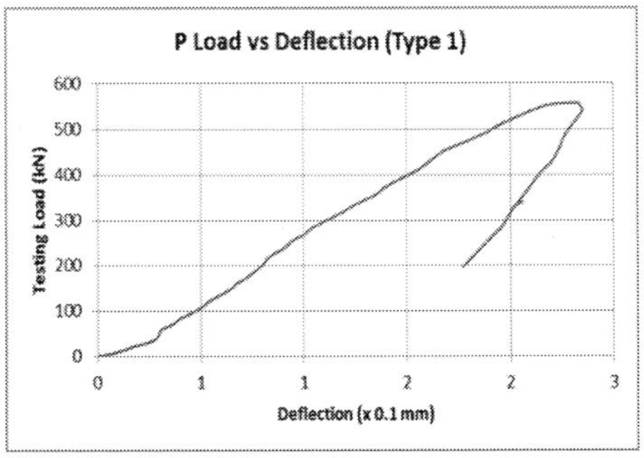

Fig. 12. Graph of P vs deflection (Type 1)

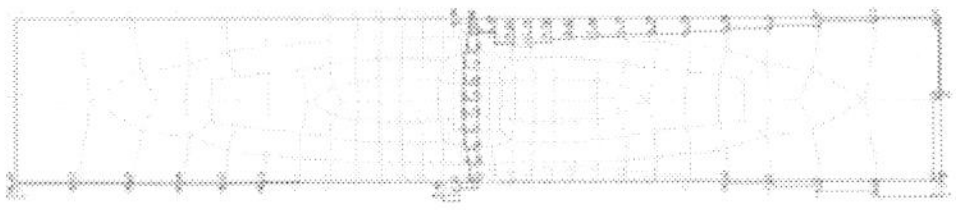

Fig. 13. Stress contour (Type 1)

BU-1 has no reinforement. Testing load is 574.87 kN. From structural analysis, the axial force is 56.41 kN at shear key.

Shear strength of longitudinal connection without reinforcement can be calculated with equation (1).

$$\phi V_n = \phi\,(0.55) h_n \ell$$
$$= 0.85 \times 0.55 \times 265 \times 300$$
$$= 37166.25\ N$$
$$= 37.166\ kN$$

The forces as testing result is 56.41 kN > 37.17 kN. These results indicate that the shear key without reinforcement and using only rely bond between shear key and concrete fillers give good results.

The magnitude of the deflection is small indicates that the structure is not deformed significantly

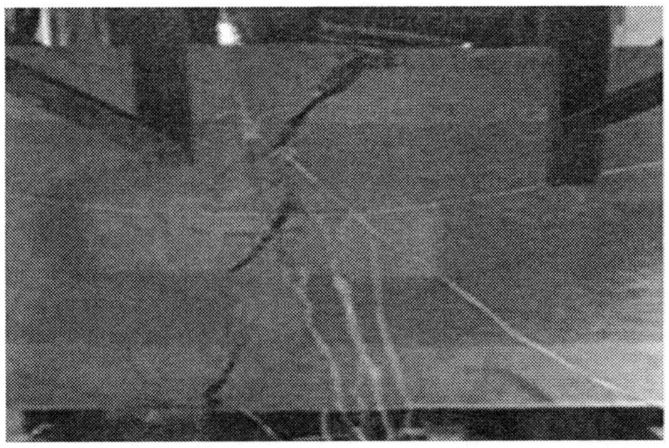

Fig. 14. Crack pattern for Specimen Type 1

Collapse pattern is almost similar to the collapse pattern as results of the finite element analysis. Stress flow that occurs as predicted, from stress due to the P load flows into the support. But the damage occurs earlier in the body slightly below the boundary elements, causing the upper boundary elements are also damaged first. As a result of the test load will stop, but the structure did not undergo plastic condition. So the ductility is small.

Deflection at maximum load is 2.04 mm (Δ_y). While the maximum deflection is 2.71 mm (Δ_u) at load 543.57 kN. So, the ductility is Δ_u/Δ_u = 2.71/2.04 = 1.33.

4.2. Specimen 2

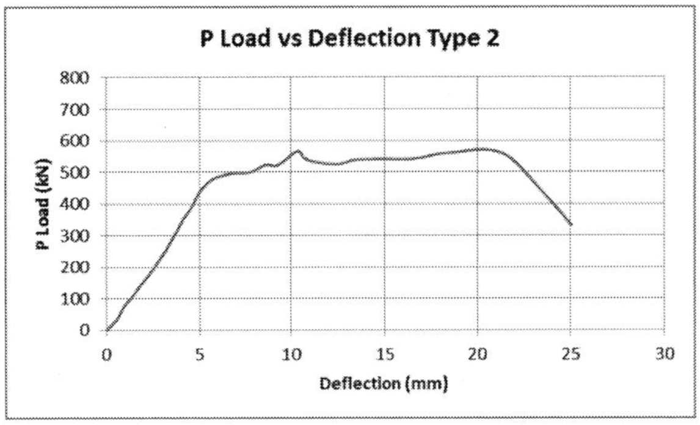

Fig. 15. Graph of P vs deflection (Type 2).

BU-2 has 1D13 reinforcement at shear key. Testing load for BU-2 is 601.45 kN. The analysis result is 57.58 kN.

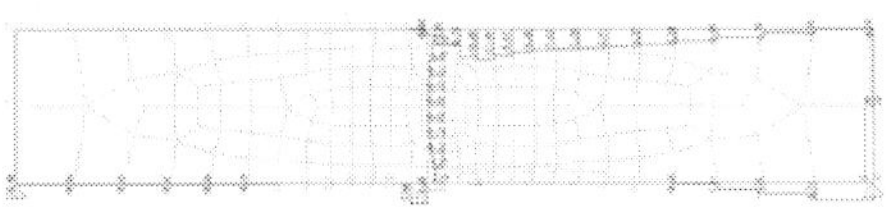

Fig. 16. Stress contour (Type 2)

Deflection at maximum load is 7.79 mm (Δ_y). While the maximum deflection is 20.99 mm (Δ_u) at load 431.74 kN. So, the ductility is Δ_u/Δ_u = 20.99/7.79 = 2.69.

Fig. 17.. Crack pattern for Specimen Type 2.

Shear strength of longitudinal connection with reinforcement can be calculated with equation (2).

$$A_{vf} = 1\ \text{D}\ 13$$
$$= 132.73\ \text{mm}^2$$

$$V_u = 132.73 \times 0.85 \times 400 \times 1.0$$
$$= 45128.98\ \text{N}$$
$$= 45.13\ \text{kN}$$

Damage from the loading point to the pivot formed a diagonal flow incomplete. When cracks diagonal touch shear connection, it cracks into line with the connection. Damage caused quite severe, because this connection is the weakest part.

4.3. Specimen 3

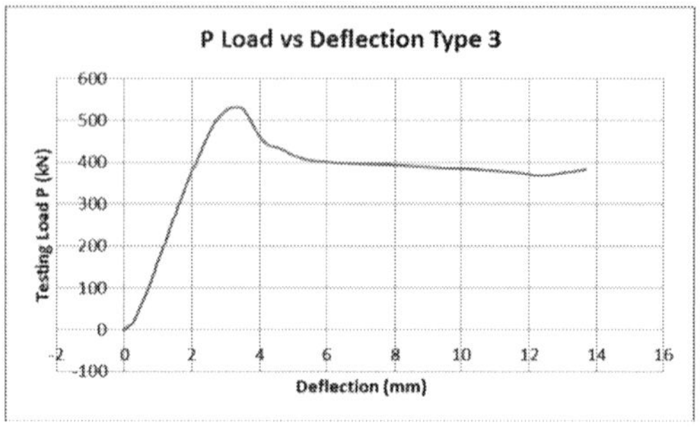

Fig. 18. Graph of P vs deflection (Type 3).

BU-3 has 1D13 reinforcement at shear key. Testing load for BU-2 is 521.6 kN. The analysis result is 49.93 kN.

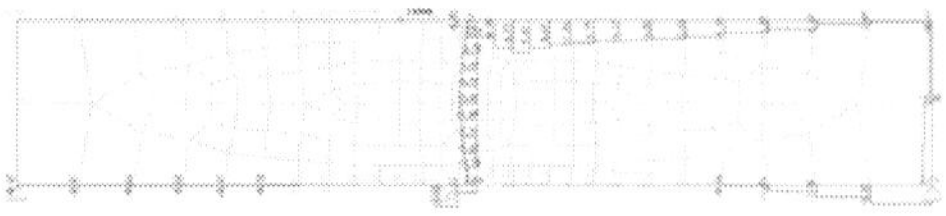

Fig. 19. Stress contour (Type 3)

Deflection at maximum load is 3.565 mm (Δ_y). While the maximum deflection is 15.14 mm (Δ_u) at load 427.23 kN. So, the ductility is Δ_u/Δ_u = 15.14/3.565 = 4.25.

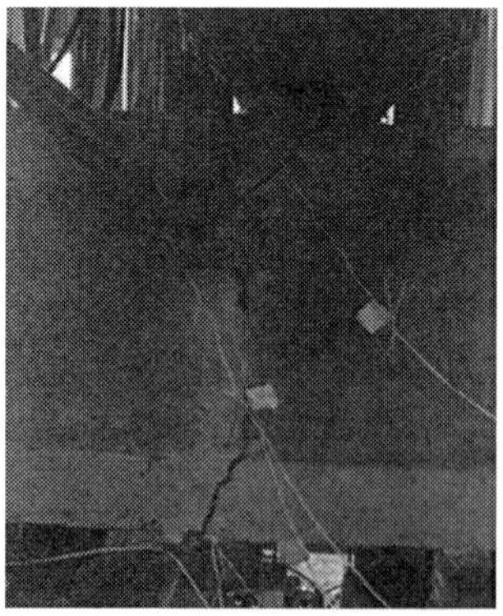

Fig. 20. . Crack pattern for Specimen Type 3.

Shear strength of longitudinal connection with reinforcement can be calculated with equation (2).

$$A_{vf} = 2\ \text{D}\ 13$$
$$= 265.46\ \text{mm}^2$$

$$V_u = 265.46 \times 0.85 \times 400 \times$$
$$= 90257.96\ \text{N}$$
$$= 90.26\ \text{kN}$$

Damage from the loading point to the pivot formed a diagonal flow incomplete. When cracks diagonal touch shear connection, the cracks being along the connection. This is due to the shear connection is the weakest part. The addition of longitudinal shear reinforcement can improve ductility.

4.4. Specimen 4

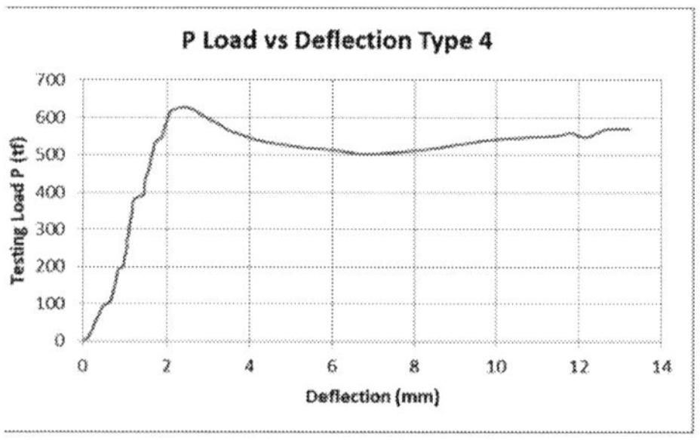

Fig. 21. Graph of P vs deflection (Type 4).

BU-4 has 2D13 reinforcement perpendicular at shear key. Testing load for BU-2 is 656.04 kN. The analysis result is 62.78 kN

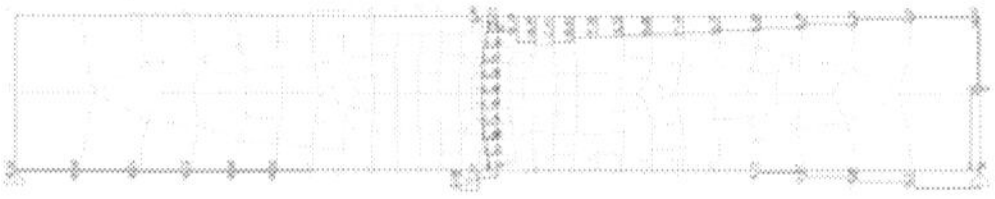

Fig. 22. Stress contour (Type 4)

Deflection at maximum load is 3.565 mm (Δ_y). While the maximum deflection is 15.14 mm (Δ_u) at load 427.23 kN. The ductility is Δ_u/Δ_u = 13.22/2.14 = 6.18.

Fig. 23. Crack pattern for Specimen Type 4

Shear strength of longitudinal connection with reinforcement can be calculated with equation (2).

$$A_{vf} = 2\ \mathrm{D}\ 13$$
$$= 265.46\ \mathrm{mm}^2$$

$$V_u = 265.46 \times 0.85 \times 400 \times 1.4$$
$$= 126361.14\ \mathrm{N}$$
$$= 126.36\ \mathrm{kN}$$

Damage from the loading point to the pivot formed a diagonal flow as prediction. The addition of longitudinal shear reinforcement can improve ductility.

4.5. Specimen 5

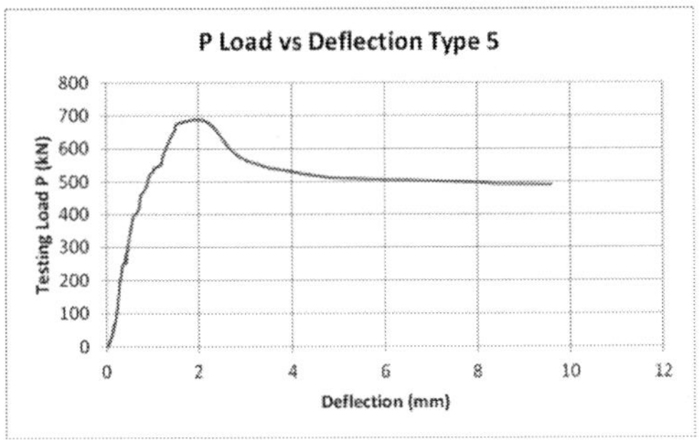

Fig. 24. Graph of P vs deflection (Type 5).

BU-5 has 3D13 reinforcement perpendicular at shear key. Testing load for BU-2 is 682.38 kN. The analysis result is 65.33 kN.

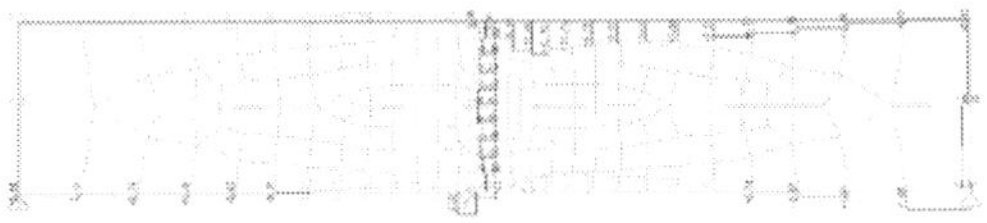

Fig. 25. Stress contour (Type 5)

Deflection at maximum load is 2.18 mm (Δ_y). While the maximum deflection is 9.6 mm (Δ_u) at load 490.30 kN. So, the ductility is $\Delta_u/\Delta_u = 9.6/2.18 = 4.40$.

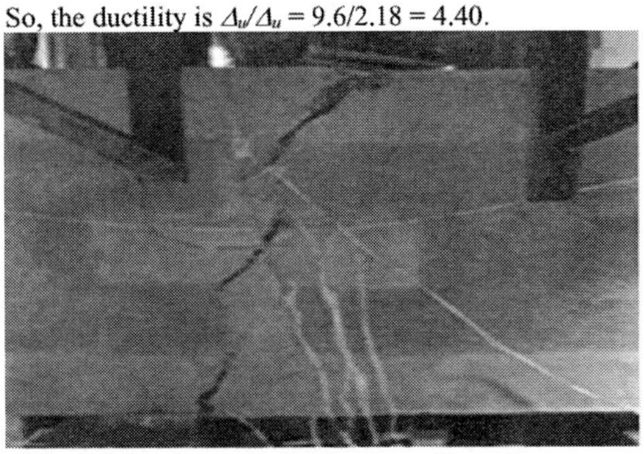

Fig. 26. Crack pattern for Specimen Type 5

Shear strength of longitudinal connection with reinforcement can be calculated with equation (2).

$$A_{vf} = 3\ \mathrm{D}\ 13$$
$$= 398.20\ \mathrm{mm}^2$$

$$V_u = 398.20 \times 0.85 \times 400 \times 1.4$$
$$= 189541.71\ \mathrm{N}$$
$$= 189.54\ \mathrm{kN}$$

The pattern of damage is similar to BU-4

Table 1. Test Result

Sample No.	$V_u(kN)$	V_{geser} (kN)	Ductility $(\Delta u/\Delta y)$
1	37.17	55.41	1.33
2	45.13	44.01	2.69
3	90.26	49.93	4.25
4	126.36	62.78	6.18
5	189.54	65.33	4.40

5. Conclusion

1. Installation and additional longitudinal reinforcement (BU-2 and BU-3) on the shear key does not give the shear strength significantly, but to improve its performance in terms of ductility.
2. Transverse reinforcement BU-4 dan BU-5 in this study does not give a significant increase compared to BU-2 and BU-3, but get a good ductility. The increase is not significant be caused the number of possibilities that only one set of reinforcement. Besides the possibility of anchoring insufficient, can affect the shear strength.
3. Longitudinal shear key between HCS with 2 or more longitudinal reinforcement likes type 3 can be used for HCS diaphragm. This reinforcement must be anchored to the supporting beam. One advantage of this system is the ease of implementation.

References

1. PCI Hollow Core Slabs Producers Committee, *PCI Manual for The Design of Hollow Core Slabs*, **Second Edition**, 4-1 – 4-9 (1998)
2. G. Soeprapto, *Laporan Penelitian Geser Hollow Core Slab*, (2016)
3. Puslitbang Permukiman Kemen. PUPERA, *Pengujian Geser Hollow Core Slab*, **1**, (2015)
4. Puslitbang Permukiman Kemen. PUPERA, *Pengujian Geser Hollow Core Slab*, **2**, (2016)
5. M. Menegotto, G. Monti, *Diaphragm Action of Precast Floors : Behavior and Modelling*, (1996)
6. MD Herlihy, R. Park, *Precast Concrete Floor Support and Diaphragm Action*,(1996)
7. R. Fenwick, D. Bull, D. Gardiner, Assesment of hollow-core floors for Seismic performance, (2010)
8. S. Dow Nakaki, G.R. Stevens, J.R. Harris, *Precast Concrete Design.*

Atomistic-continuum hybrid analysis of dislocation behavior in spinodally decomposed Fe-Cr alloys

Akiyuki Takahashi[1] and *Motoyasu* Kanazawa[1]

[1] Department of Mechanical Engineering, Faculty of Science and Technology, Tokyo University of Science, 2641 Yamazaki, Nodashi, Chiba, 278-8510, Japan

Abstract. In this study, we first present the molecular dynamics (MD) simulation of dislocation behavior in a spinodally decomposed Fe-Cr alloy. The MD simulation is used for exploring the nature of the interaction between a dislocation and the spinodal decomposition without any specific assumptions. In order to classify the interaction mechanism, dislocation dynamics (DD) simulations of the interaction between a dislocation and the spinodal decomposition are performed. In the simulations, we controlled the interaction mechanism by adding and removing the atomistic mechanism. The simulation results clearly illustrate that the atomistic mechanism can be negligible in determining the critical resolved shear stress (CRSS) of spinodally decomposed Fe-Cr alloys, and the internal stress generated by the lattice constant mismatch is a dominant mechanism. These findings are very useful for simplifying the analysis of the mechanism of material strength change due to the spinodal decomposition. Particularly in the analysis using the DD simulations, the required computational effort for simulating the dislocation behavior is greatly reduced by taking into account only the internal stress without the atomistic dislocation core influence.

1 Introduction

Duplex stainless steels consisting of ferrite and austenite phases have a high material strength, particularly the corrosion resistance, and are used as a material of primary coolant pipes in nuclear power plants. When the material is aged at temperatures in a range from 300 to 500°C, spinodal decomposition occurs in the ferrite phase, which causes an ultrafine phase separation mixing Fe-rich and Cr-rich phases. The phase separation leads to a material embrittlement and material strength change, and therefore, it is very important to understand the influence of the phase separation on the material strength for ensuring the reliability and integrity of structures. In order to investigate the material strength degeneration mechanism, an equation for the internal stress distribution arisen from the phase separation has been derived [1]. The equation can be used as a fundamental tool for investigating the influence of phase separation on the micro-scale material deformation mechanism, such as dislocation behavior. Kato conducted extensive theoretical studies on the interaction of the internal stress and dislocation and on the influence on the critical resolved shear stress (CRSS), which is a shear stress necessary for dislocations to initiate their motion in materials [2]. The information obtained by the studies is variable in clarifying the dislocation behavior in the internal stress field, and however, due to the complexity and limitation of the theoretical approach to the problem, the information is limited for dislocations with a simple shape, even though the dislocation shape must be changed a lot by the interaction with the internal stress field. In addition, the influence of the spinodal decomposition on the material strength must be controlled by not only the internal stress field, and also the other type of elements associated with atomistic chemical energies. Thus, in order to fully understand the detailed mechanism of the material strength degeneration due to the spinodal decomposition, the understanding must cover very wide range from atomistic to continuum.

Owing to a remarkable development of dislocation dynamics (DD) simulation methodology, collective behavior and complex interactions of dislocations can be simulated and calculated using computers [3-5]. Up till now, the DD method has been successfully applied to various plasticity problems of metals and alloys. Takahashi and Ghoniem have developed a dislocation dynamics-based computational method for dislocation-precipitate interaction problems, and investigated the interaction of dislocations with precipitates in terms of elasticity [6]. Furthermore, they developed a hybrid atomistic-continuum method for investigation of dislocation cores [7-8]. The method provides us with a new opportunity to study the dislocation dynamics

* Corresponding author: takahash@rs.noda.tus.ac.jp

DOI: 10.1051/matecconf/201710101018

coupled with the atomistic influence of dislocation cores. Thus, the type of simulation method could be a powerful tool for investigating the material strength degeneration mechanism due to the spinodal decomposition involving both the continuum and atomistic mechanisms.

In this study, we first present the molecular dynamics (MD) simulation of dislocation behavior in a spinodally decomposed Fe-Cr alloy. The MD simulation is used for exploring the nature of the interaction between a dislocation and the spinodal decomposition without any specific assumptions. In order to classify the interaction mechanism, DD simulations of the interaction between a dislocation and the spinodal decomposition are performed. In the simulations, we controlled the interaction mechanism by adding and removing the atomistic mechanism, say the chemical influence. In order to deal with the atomistic mechanism, we employed the discretized Peierls-Nabarro model in conjunction with the generalized stacking fault energy (GSFE). Finally, we discuss the importance of the continuum and atomistic mechanism on the material strength change due to the spinodal decomposition.

2 Computational methods

2.1 Molecular dynamics method

The interaction between a dislocation and the spinodal decomposition is first examined using the MD simulations. In the MD simulations, an embedded atom method (EAM) type of interatomic potential developed for Fe-Cr alloy is used [9]. Figure 1 shows the simulation volume for the MD simulations. The size of simulation volume is 14.71 x 11.25 x 8.52 nm, and the x, y and z axes coincide with crystal orientations of [111], [110] and [112], respectively. The average Cr atom concentration is 20%. The spinodal decomposition is made in the simulation volume by distributing the Cr atoms using the cosine function. Figure 1(b) shows the distribution of Cr atoms in Fe matrix. It can be seen that Cr atoms are well modulated using the cosine function. The magnitude of cosine function, which is equal to the magnitude of fluctuation of Cr atom concentration, is changed in a range from 0 to 20% in order to see the influence on the dislocation behavior. As a boundary condition, the atoms in the upper and lower layer in the y-direction cannot move in the y-direction, and are enforced to move by a prescribed displacement in the x-direction. The simulation volume is then relaxed using the fast inertia relaxation engine (FIRE) algorithm [10]. Using the procedure, the stress-strain behavior of the simulation volume is obtained. The critical resolved shear stress (CRSS), which is a necessary stress for dislocations to initiate their motion, is determined as the peak shear stress in the stress-strain behavior.

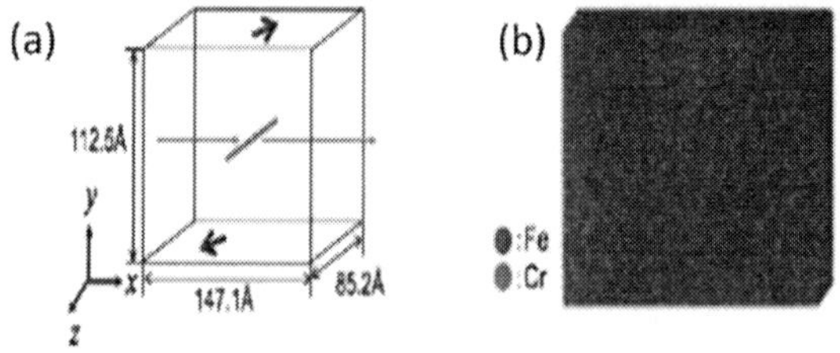

Fig. 1 MD simulation volume

2.2 Dislocation dynamics method

Although, the MD simulations can provide us with the information on the dislocation behavior and CRSS, it is impossible to classify the elementary mechanisms controlling the interaction. In order to clarify what mechanism is dominantly working in the interaction, we perform the DD simulation with the same simulation condition that we use in the MD simulations. In the DD simulation, dislocations are discretized into a number of line (dislocation) segments. The force acting on each dislocation segment is calculated based on the elasticity of dislocations. In this work, we will take into account the influence of internal stress (elastic interaction) and the atomistic mechanism (dislocation core influence). The equation of internal stress produced by the spinodally decomposition is derived by Cahn [1], and thus, the equation is utilized in the force calculation in the DD simulations. On the other hand, in order to account for the atomistic influence in the DD simulation, the discretized Peierls-Nabarro (PN) model is employed [7]. In the model, dislocation core structure is described with a collection of discrete infinitesimal dislocations. Within the dislocation core, the infinitesimal dislocations experience both the lattice restoring force and the elastic interaction between the infinitesimal dislocations. The lattice restoring force results from the stacking fault energy expanding in the dislocation core, which can be calculated using the GSFE. The GSFE can be obtained using MD calculations.

3 Simulation results

3.1 Molecular dynamics simulation results

Figures 2 and 3 show the stress-strain behavior and the snapshots of the MD simulation with the magnitude of fluctuation of 0%. In this paper, the dislocation shape is found using the common neighbor analysis (CNA) technique [11], and only the atoms extracted by the CNA are displayed in the figure. The color shows the internal stress distribution on the slip plane of dislocation calculated using the internal stress equation.

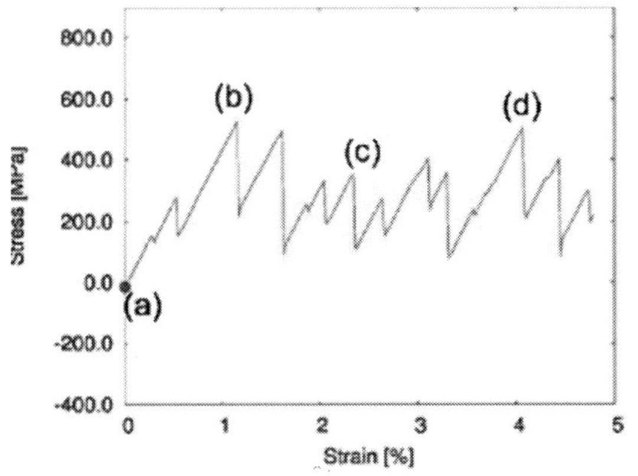

Fig. 2 Stress-strain behavior with the magnitude of Cr atom fluctuation of 0% calculated by MD simulations.The corresponding snapshots in Fig. 3 are indicated with a label of (a)-(d)

The stress component is along the direction of the Burgers vector of the dislocation on the slip plane. Based on the continuum elasticity of dislocations, the shear stress shown by the blue color gives a resistance force to the dislocation motion, while the shear stress shown by the red color gives an attractive force to the dislocation motion. It can be seen that, in this case, the stress distribution is in a striped pattern. As shown in Fig. 3, the dislocation is affected by the distribution of Cr atoms. Therefore, the shape of dislocation is not straight. In the stress-strain behavior shown in Fig 2, the stress-strain behavior has a zig-zag shape. The shape must result from the local Cr atom locations near the dislocation. Then, we found that the CRSS is about 450MPa.

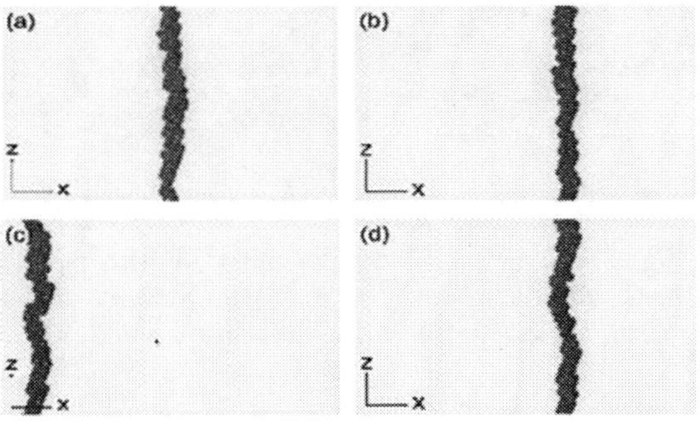

Fig. 3 Dislocation behavior in the MD simulation with a magnitude of Cr atom fluctuation of 0%. Red particles show the atoms around the dislocation core.

On the other hand, when the magnitude of fluctuation is 20%, the stress-strain behavior and the snapshots of the MD simulation are shown in Fig 4 and 5 respectively. In Fig. 5, the dislocation is first absorbed by the shear stress shown by the red color. After then, the dislocation comes to the shear stress shown by the blue color. However, in front of the shear stress, the dislocation stops, because the shear stress gives a resistance force to the dislocation. When the applied shear stress becomes large enough, the dislocation can pass the shear stress region. In the stress-strain behavior shown in Fig. 4, the stress drops down at the shear strain of 0.5%. At the shear stress, the dislocation is absorbed by the shear stress shown by red color. At the shear stress of 3.2%, the stress shows the peak value of 800MPa, which is the CRSS.

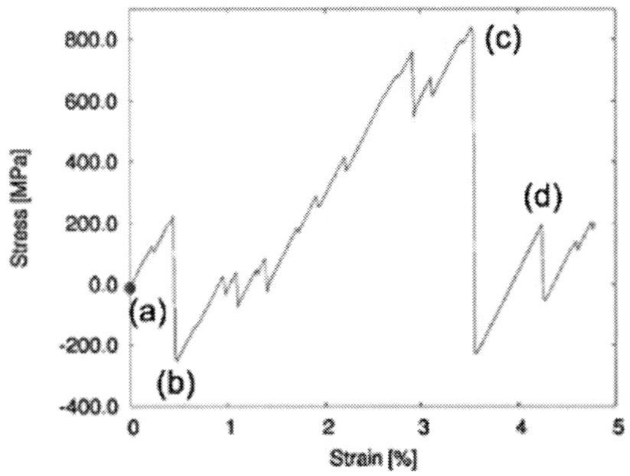

Fig. 4 Stress-strain behavior with the magnitude of Cr atom fluctuation of 20% calculated by MD simulations.

By comparing these two CRSS for the magnitudes of fluctuation of 0 and 20%, the CRSS is increased by the magnitude of fluctuation is increased. Therefore, the results suggest that the progress of the spinodal decomposition actually increase the CRSS, which must lead to a hardening of the material.The corresponding snapshots in Fig. 5 are indicated with a label of (a)-(d)

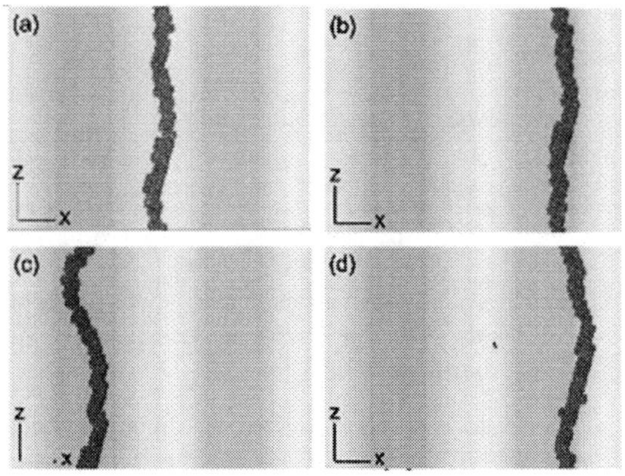

Fig. 5 Dislocation behavior in the MD simulation with a magnitude of Cr atom fluctuation of 20%. Red particles show the atoms around the dislocation core.

Because the MD simulation is based on the interatomic potential, the MD simulations can account for all the mechanism given by the spinodal decomposition, such as elastic (internal stress) interaction and atomistic dislocation core influence. However, in the past studies, only the internal stress is taken into account. Thus, it is

 DOI: 10.1051/matecconf/201710101018

necessary to clarify which mechanism dominantly works in the material strength degeneration by the spinodal decomposition.

3.2 Dislocation dynamics simulation results

In order to reveal the mechanisms working in the material strength degeneration due to the spinodal decomposition, DD simulations in conjunction with the discretized PN model are performed. The necessary information for using the discretized PN model is the GSFE of the Fe-Cr alloy. The GSFE is calculated using the MD, and is shown in Fig. 6. The peaks of the GSFE of various Cr atom concentrations do not have any clear trend, and however, the maximum slope of the GSFE has a clear Cr atom concentration dependence, where the maximum slope becomes larger as the Cr atom concentration becomes lower.

Using the GSFE, we performed DD simulation of dislocation behavior in a spinodally decomposed Fe-Cr alloy with the atomistic dislocation core influence. The snapshots of the DD simulation are shown in Fig. 7. The atomistic dislocation core influence is incorporated using the discretized PN model, and within the discretized PN model framework, the single dislocation is decomposed into 10 infinitesimal dislocations. In the figure, the color shows the assumed Cr atom concentration, where the red shows high, and the blue shows low Cr atom concentration.

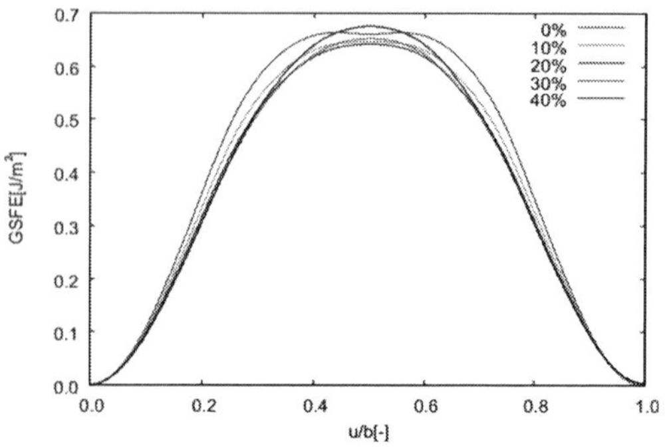

Fig. 6. Generalized stacking fault energy of Fe-Cr alloy with various Cr concentrations calculated using molecular dynamics method.

The dislocation core structure is slightly changed during the dislocation motion, particularly when the dislocation is located within the low Cr atom concentration. However, the dislocation shape keeps straight regardless the dislocation position. Along with the DD simulation with the atomistic dislocation core influence, the DD simulation without the atomistic dislocation core influence is also performed. In the simulation, the dislocation is described with a single dislocation line having the full Burgers vector. The snapshots of the simulation are shown in Fig. 8.

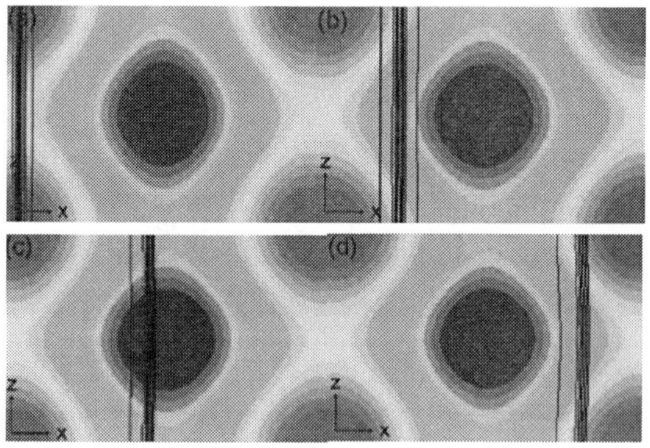

Fig.7. Dislocation behavior in spinodally decomposed Fe-Cr alloy simulated using the DD method with the atomistic dislocation core influence.

Similar to Fig. 7, the shape of dislocation is not changed, and keeps straight. The CRSS calculated by the DD simulations are shown in Fig. 9 with those calculated by the MD simulations. In the results of MD simulations, there is a wide scatter for each magnitude of Cr atom fluctuation. In the MD simulation models, the Cr atoms are distributed using random numbers so that we performed the 10 MD simulations with different simulation models created with different random seeds. In the figure, the CRSS is increased linearly as the magnitude of fluctuation is increased.

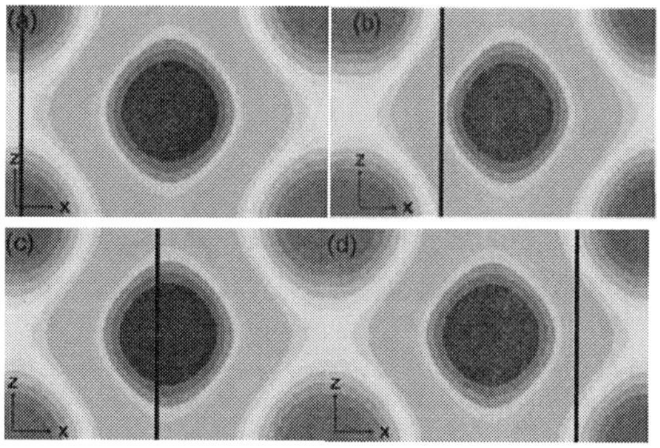

Fig.8. Dislocation behavior in spinodally decomposed Fe-Cr alloy simulated using the DD method without the atomistic dislocation core influence.

The MD simulation results are almost identical to the DD simulation results. Interestingly, the CRSS calculated by the DD simulation with the atomistic dislocation core influence is the same as that calculated without the atomistic dislocation core influence, meaning that the atomistic dislocation core influence is negligible, and the internal stress is the dominant factor controlling the material strength of spinodally decomposed Fe-Cr alloys.

 DOI: 10.1051/matecconf/201710101018

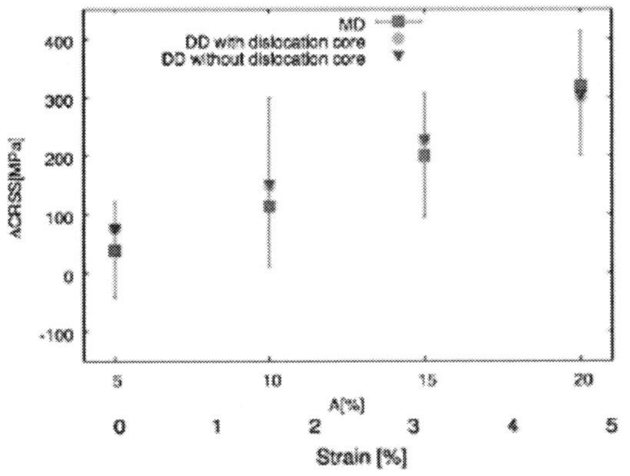

Fig. 9. Critical resolved shear stress (CRSS) for various magnitude of Cr atom fluctuation calculated by the MD and DD.

4 Conclusions

The dislocation behavior in spinodally decomposed Fe-Cr alloy was simulated using the MD method. The Cr atom distribution was assumed using a simple cosine function, and the magnitude of Cr atom fluctuation was changed in a range from 0 to 20%. The dislocation shape was extracted from the atomic volume using the CNA method. The CRSS could be calculated as the peak stress in the stress-strain behavior. In order to clarify the dominant mechanism controlling the material strength change due to the spinodal decomposition, the DD simulations with and without the atomistic dislocation core influence were performed. The atomistic dislocation core influence is incorporated into the DD simulations using the discrete PN model, while the internal stress is implemented using the classical equation for the internal stress. By comparing these simulation results of dislocation behavior and CRSS, the following summaries could be obtained.

- In the MD simulations, the dislocation motion is significantly affected by the internal stress distribution.
- The CRSS is increased linearly as the magnitude of Cr atom fluctuation is increased.
- In the DD simulation with atomistic dislocation core influence, the dislocation does not change the shape, meaning that the atomistic dislocation core influence can be negligible, and the internal stress is a dominant factor determining the material strength change due to the spinodal decomposition.
- The CRSS calculated by the MD and DD is almost identical.

These findings are very useful for simplifying the analysis of the mechanism of material strength change due to the spinodal decomposition. Particularly in the analysis using the DD simulations, the required computational effort for simulating the dislocation behavior is greatly reduced by taking into account only the internal stress without the atomistic dislocation core influence.

This study was supported by JSPS KAKENHI Grant Number JP16K05044.

References

[1] J. W. Cahn. Hardening by spinodal decomposition, ActaMetallurgica, 11 (1963) 1275-1282
[2] M. Kato. Hardening by spinodally modulated structure in b.c.c. alloys, ActaMetallurgica, 29 (1981) 79-87
[3] N.M.Ghoniem, L.Z. Sun. Fast-sum method for the elastic field of three-dimensional dislocation ensembles, Physical Review B, 60 (1999) 128-140
[4] N.M.Ghoniem, S.H. Tong, L.Z. Sun. Parametric dislocation dynamics: A thermodynamics-based approach to investigations of mesoscopic plastic deformation, Physical Review B, **61** (2000) 914-927
[5] J. El-Awady, S.B. Biner, N.M.Ghoniem. A self-consistent boundary element, parametric dislocation dynamics formulation of plastic flow in finite volumes, Journal of the Mechanics and Physics of Solids, **56** (2008) 2019-2035
[6] A. Takahashi, N.M.Ghoniem. A computational method for dislocation-precipitate interaction, Journal of the Mechanics and Physics of Solids, **56** (2008) 1534-1553
[7] S. Banerjee, N.M.Ghoniem. G. Lu, N.Kioussis. *Non-singular descriptions of dislocation cores: A hybrid ab-initio continuum approach, Philisophical Magazine*, **87** (2007) 4131-4150
[8] A. Takahashi, N.M.Ghoniem. *Structure of self-interstitial atom clusters in iron and copper, Physical Review B,* **80** (2009) 174104
[9] G. Bonny, R.Pasianot, D.Terentyev, L.Malerba. *Iron chromium potential to model high-chromium ferritic alloys, Philosophical Magazine*, **91** (2011) 1724-1746

The effect of polymer coated pumice to the stiffness and flexural strength of reinforce concrete beam

Indradi Wijatmiko[1,*], *Ari* Wibowo[1], and *Christin* Remayanti[1]

[1]Department of Civil Engineering, Faculty of Engineering, Universitas Brawijaya, Malang 65149, Indonesia

Abstract. Pumice has been proven as substitution material aggregate in lightweight aggregate concrete (LWAC). However due to its characteristic, pumice has the disadvantages of its excessive water absorption during preparation of concrete mixture which may reduce concrete strength. Therefore in order to eliminate this additional water absorption, this study investigate the effect of coated pumice in concrete mixture to the beam flexural strength and stiffness. This study performed flexural strength test on three types of sample: 1) normal reinforce concrete beam, 2) reinforce concrete beam with uncoated pumice and 3) reinforce concrete beam with coated pumice. The result showed that the lightest weight concrete occurred on the coated pumice specimen, with the reduction of water absorption was 4% compared to the uncoated pumice specimen. The stiffness of the reinforce beam with coated pumice was lower compared to the uncoated ones, this was due to the reduction of adhesion action between cement and aggregates. However, the use of coated pumice increased the flexural strength compare to the uncoated ones with 2.58%.

1. Introduction

LWAC has been known offers various advantages due to its lower density [1], higher ratio of strength/weight [2], less conductivity thermal [3,4], better durability properties [5,6], fire resistance [7] and etc. The use of lightweight concrete reduce the size of columns, beams, wall and foundation which as well reducing the dead load and minimizing the damages of structures due to earthquake [8]. The idea of LWAC is to substitute a whole or part of normal weight concrete (NWC) by a natural or synthetic lightweight materials. In Indonesia, which known also as volcanic country, pumices are enormously found in most part of country as a result of solidification of lava with the release of volcanic gases at the same time. Pumice has a porous structure which is formed due to trapped bubbles and interconnected to each other. Further, this character consequential produce lesser weight and low coefficient of thermal expansion to the concrete mixture. However, the created void causes the disintegration in concrete due to prior water absorption and the risk of flocculation during concrete pouring [9,10].

Comprehensive researches has been conducted over the last few years on the use of pumice lightweight aggregate either in structural lightweight concrete production or lightweight concrete implemented for thermal insulation and masonry blocks [11-14]. However only a few research has been conducted to overcome the disintegration concrete due to the excessive water absorption.

Polymer has long chains molecular structure which lead to the strong bonds. Polymer easily processed and commonly produce in oil-based synthetic form such as paint. One of its advantages, which is building block of basic materials, e.g. water, may becoming the solution on the prior water absorption by pumice. As the prior water absorption can be reduced, the weight of concrete becomes lighter, thus the mechanical properties of applied coated pumice as coarse aggregate yet to be discussed.

Therefore this study investigate the effect of polymer coated pumice on LWAC to its flexural strength and stiffness through series of experiments. This study compare the normal weight concrete as a control specimens with the polymer coated and uncoated pumice mixtures as coarse replacement in concrete.

2. Experimental details

To investigate the effect of polymer coated pumice to the stiffness and flexural strength of reinforce concrete beam, nine mixtures with varying volume and coating of pumice were prepared.

* Corresponding author: indradi@ub.ac.id

2.1 Material properties

Pumice aggregate was obtained from Kediri residency of Indonesia. It has been reported that part of the quality of the reserve are good, while other required to be washed for quality improvement. Coated pumice were applied by dipping a group of grain aggregate 8-16 mm into a polymer liquid. The coated pumice then dried at the room temperature 25 ± 2 °C for 24 hours.

The cement used in all mixes was ordinary Portland cement type 1, which corresponds to ASTM type 1. The fine aggregate of local river sand and tap water were used for all mixtures. The properties of aggregates can be seen in Table 1, which shows the specific gravity of coated pumice is lower than uncoated pumice, as well as its water absorption ability.

Table 1. Properties of aggregates

Aggregates	Specific gravity	Bulk density (kg/m^3) ASTM		Water absorption 24h (%)
	Ovendry	Shoveled Ovendry	Rodded Ovendry	
Coarse aggregate the				
GA	2.67	2696	2589	3.20
PA	1.85	-	-	14
CPA	1.79	-	-	10
Fine aggregate				
River sand	2.60	2530	2693	3.22

GA:Gravel Aggregate ; PA;Pumice Aggregate ; CPA : Coated Pumice Aggregate

2.2 Specimen and testing

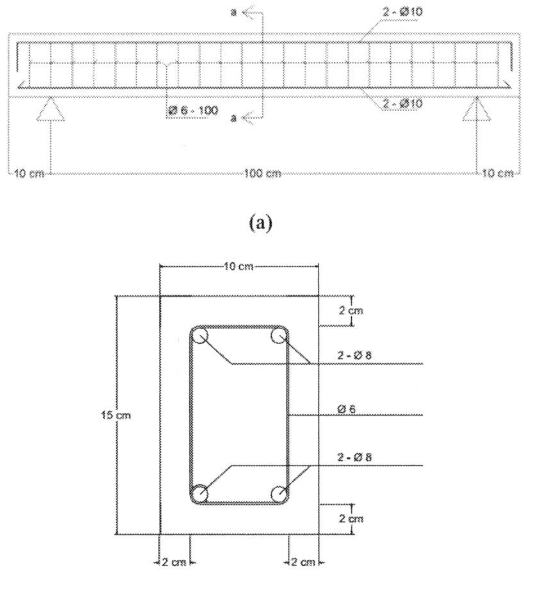

(a)

(b)

Fig. 1. (a) Specimen long section layout , (b) a-a section layout

Nine mixtures of reinforce concrete beam, consist of three normal weight concrete, three lightweight uncoated pumice aggregate concrete and three lightweight coated pumice aggregate concrete. As can be seen in figure 1a, each beam, with the dimension of 120 cm length, 10 cm width and 15 cm height, were subjected to be tested for flexural strength.

3. Result and discussion

3.1 Density and workability

As shown in Table 1, coated pumice with polymer is resulting to the decreasing of water absorption. And as consequence, the weight of a specimen also reduced. Therefore, the density of the concrete with coated pumice is the lowest due to the reduction of weight as shown in Table 2.

The workability of concrete is described to understand the effect of pumice aggregate character as coarse aggregate. Table 2 present the properties of aggregate and mix design. As a LWAC, which has lower aggregate density and produce lighter mix. Further, it is common for LWAC to have lower slump value, as confirmed from some research studies [15,16] and recommend in ACI 213R-87. LWAC coated pumice produced lowest slump value due to the lower weight as a result of water absorption reduction. Though the LWAC pumice and LWAC coated pumice has lower slump value, the workability are remain at the same stage as NWC. And in order to maintain the cohesive and avert the segregation, a higher slump value is demanded with the additional effort to preserve the ideal surface of specimens.

Table 2. Properties of mix design

Mix ID	Density (kg/cm^3)	Slump (cm)
NWC		14.5
NWC-1	2447	
NWC-2	2557	
NWC-3	2533	
LWAC : Pumice		12.2
PC-1	2169	
PC-2	2130	
PC-3	2102	
LWAC : Coated Pumice		10.8
CPC-1	2100	
CPC-2	2097	
CPC-3	2080	

NWC: Normal Weight Concrete; LWAC: Lightweight Aggregate Concrete

3.2 Compressive strength

Compressive strength test (refer to Figure 2) were conducted for 3 cylindrical specimens with 8 cm diameter and 16 cm height at the age of 28-day, as seen in Table 3.

The substitution of strong gravel aggregate by relatively weak pumice aggregate cause the decrease of compressive strength at 28-day. Further, the introduction of coated polymer on pumice resulting minor reduction in compressive strength.

DOI: 10.1051/matecconf/201710101019

Table 3. Compressive strength (28-day)

Mix ID	Density (kg/cm³)	Max. Load (kg)	f'_c (MPa)
NWC	2177	89	17.71
LWAC: Pumice	1866	64	12.74
LWAC: Coated Pumice	1990	59	11.74

Fig. 2. Compressive strength test

Reduction in compressive strength of pumice aggregate concrete can be attributed by the lack of clinker (C_3S), which resulting on slowing the hydration and decreasing the rate of heat development, especially when the polymer coat were applied. However, due to ongoing pozzolanic reaction, the differences becomes smaller in long period. It is related with the total content of silicon dioxide in pumice that can form compound with cementitious properties [17].

3.3 Flexural strength

Flexural strength test were subjected to nine reinforce beam with three categories, i.e. normal weight concrete, lightweight concrete with uncoated pumice aggregate and lightweight concrete with coated pumice aggregate. The experimental test setup in Figure 3 shows there were 2 LVST installed for subsequent readings during the applied force-controlled.

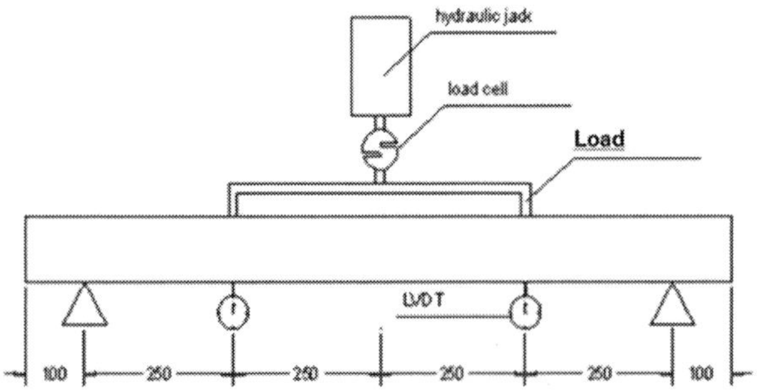

Fig. 3. Flexural strength test

The result of flexural strength test can be seen in Figure 4a, 4b, and 4c respectively for normal weight concrete, pumice aggregate concrete and coated pumice aggregate concrete.

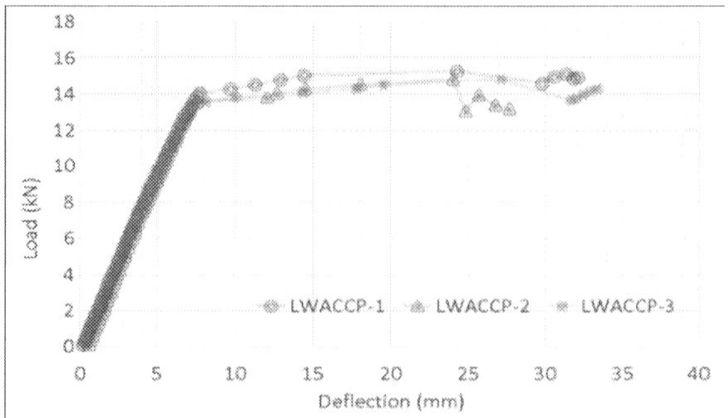

(a) Load-deflection behavior for specimens of normal weight concrete

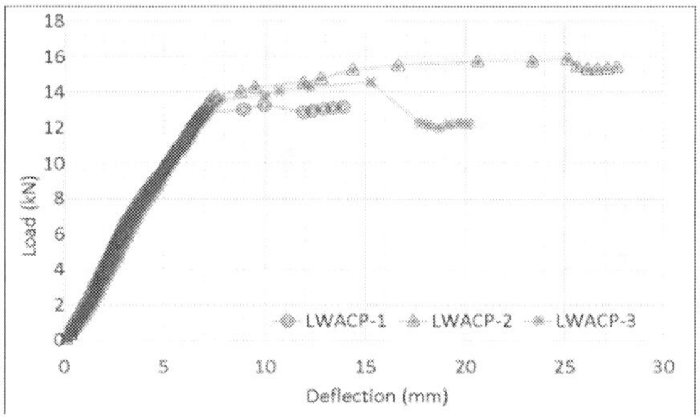

(b) Load-deflection behavior for specimens of lightweight aggregate concrete with pumice

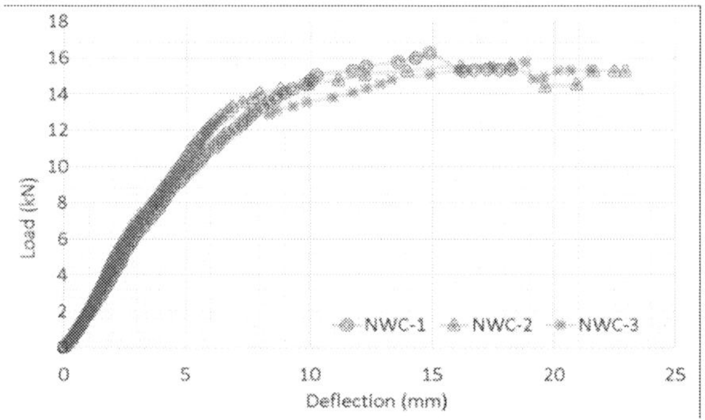

(c) Load-deflection behavior for specimens of lightweight aggregate concrete with coated pumice

Fig. 4. Detail of load-deflection behavior of specimens

Among specimens, the peak load of NWCs were higher compare with the LWACPs. The average peak load of NWC was 31.8 kN, while the LWAC with uncoated pumice had 29.2 kN and the LWAC with coated pumice had 29.93 kN. Thus with the same testing behavior and parameter, the flexural strength of NWC had similar result as the peak load. Interestingly, the application of polymer coating to the pumice increase the flexural strength up to 2.58 %, as indicated from the average flexural strength in Table 4.

Table 4. Flexural strength

Mix ID	Pu (kN)	Flexural Strength (kN.m)	Average Flexural Strength (kN.m)
NWC			
NWC-1	32.5	4.06	3.97
NWC-2	31.3	3.91	
NWC-3	31.6	3.94	
LWAC: Pumice			
LWACP-1	26.7	3.33	3.64
LWACP-2	31.8	3.97	
LWACP-3	29.1	3.64	
LWAC: Coated Pumice			
LWACCP-1	30.6	3.82	3.74
LWACCP-2	29.6	3.70	
LWACCP-3	29.6	3.70	

The idea of flexural strength is to determine the distribution of the stress and strain distribution of concrete. At the ultimate stage, the unconstrained concrete is assumed fail when it reach the compressive strain limit. The ultimate compressive has been determined by assuming a constant value i.e. 0.003 (ACI 318) or 0.0035 (BS 8110), however these value will be conservative for NWC but will be un-conservative for high strength LWAC due to the decreasing of ultimate compressive strain capacity of unconfined LWAC as a result of increasing concrete strength.

3.4 Stiffness

As seen in Figure 5, all type of specimens share similar initial stiffness. LWACCP showed the higher value of stiffness compare to the NWC and LWACP.

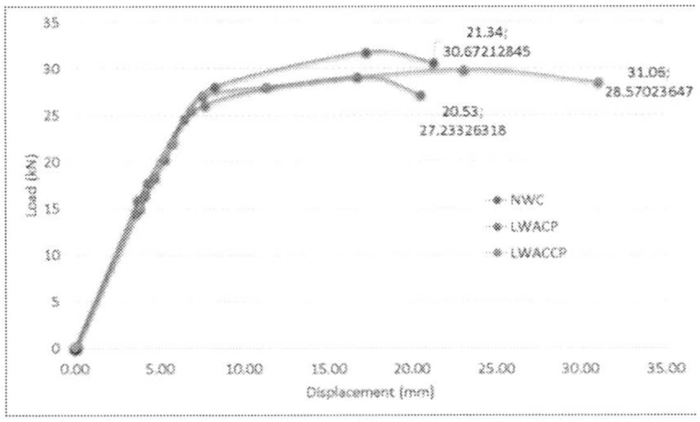

Fig. 5. Detail of load-deflection behavior of three category of specimens

Cracking reduce the beam stiffness, therefore to calculate the beam stiffness the subjected load is taken before the first crack occurred. From the observation, the first crack of NWC occurred by 15.86 kN load, while the uncoated and coated pumice had the first crack at 16.52 kN and 22.8 kN respectively.

Though the coated pumice was able to increase the flexural strength, the stiffness of coated pumice was lower than the uncoated ones. This is the due to the reduction of adhesion action between cement and aggregates

Table 5. Stiffness of specimen

Mix ID	Load (kN)	Stiffness
NWC	15.37	4.26
LWACP	15.37	4.02
LWACCP	15.37	3.89

4 Conclusions

Nine reinforce concrete beams were investigated to have a clear understanding of the effect polymer coated pumice as coarse aggregate to the flexural strength as well as its stiffness. In order to get stronger observation, three types of concrete were studied, i.e. normal weight concrete, lightweight pumice concrete and lightweight coated pumice concrete. Several outcomes can be observed as follows:

1. The Polymer coated was proven to reduce the water absorption of pumice. And as consequence, it lighter the concrete weight and reduce the density. The coated pumice has lower slump value, yet the workability still at the high stages as NWC.
2. The compressive strength of pumice mixtures on LWAC were lower than NWC, due to the lack of clinker. However, the differences becomes smaller in long period due to the content of silicon dioxide in pumice which can form compound with cementitious.
3. The flexural strength of coated pumice was slightly higher than uncoated ones. However, the coated pumice had lower stiffness characteristic due to the reduction of adhesion action between cement and aggregates

We acknowledge the support of Higher Education Excellent Research Grant (*Penelitian Unggulan Perguruan Tinggi-PUPT*) from Ministry of Research, Technology and Higher Education of the Republic of Indonesia.

References

1. N.A Libre., M.Shekarchi, M. Mahoutian, P. Soroushian, Const and Building Materials **25**, 2458-2464 (2011)
2. H. Al-Khaiat H, MN Haque, Cem Concr Res **28**, 859-866 (1998)
3. KS Al-Jabri, AW Hago, AS Al-Nuaimi, AH Al-Saidy, Cem Concr **35**, 1472-1479 (2005)
4. H. Uysal, R. Demirbofa, S. Remzi, R. Gul, Cem Concr Res **34**, 845-848 (2004)
5. CL Hwang, MF Hung, Const Build Mater **19**, 619-626 (2005)
6. J. Gao, W. Sun, K. Morino, Cem Concr Compos **19(4),** 307-313 (1997)
7. A. Bilodeu, VKR Kodur, GC Hoff, Cem Concr Compos **26(2),** 163-174 (2004)
8. IB Topcu IB, Cem Concr Res **27**, 15-21 (1997)
9. E. Yog˘urtcu, O. Cadır, K. Ramyar, S. Oner, Hazır Beton J **79** (2012)
10. S. Popovics, *Concrete materials, properties, specifications and testing*, NewJersey, US: Noyes Publications (1992)
11. R. Sahin, R. Demirboga, H. Uysal, R. Gul R, Cem Concr Res **33**, 1245–9 (2003)

12. K.M.A. Hossain, Cem Concr Res **34**, 283–91 (2004)
13. L. Gunduz, Constr Build Mater **22(5)**, 721–8 (2008)
14. L. Gunduz, Constr Build Mater **22(5)**, 747–54 (2008)
15. KMA Hossain, *IABSE Report*, Zurich, Switzerland, **81**, 145-150 (1999)
16. KMA Hossain KMA, Mag Concr Res **56(2)**, 111-120 (2002)
17. H. Ulusu, H. Y. Aruntas, O. Gencel, Cons and Building Materials **118**, 11-19 (2016)

DOI: 10.1051/matecconf/201710101020

Flexural behaviour of reinforced concrete beams with discrete steel – polypropylene fibres

Wan Amizah Wan Jusoh[1,*], *Izni* Syahrizal Ibrahim [2], and *Abdul* Rahman Mohd Sam[3]

[1]Department of Structures and Materials, Faculty of Civil Engineering and Environment, UniversitiTun Hussein Onn Malaysia, 26400 Parit Raja, Batu Pahat, Johor, Malaysia
[2]Forensic Engineering Centre, Institute for Smart Infrastructure and Innovative Construction (ISIIC), Faculty of Civil Engineering, Universiti Teknologi Malaysia, 81310 Johor Bahru, Johor, Malaysia
[3]Department of Structures and Materials, Faculty of Civil Engineering, Universiti Teknologi Malaysia, 81310 Johor Bahru, Johor, Malaysia

Abstract. This paper discusses the experimental results on the flexural test of concrete containing different proportions of steel fibre (SF) and polypropylene fibre (PPF). The flexural test was carried out under 4-point bending load and followed the relevant standards to FRC. Hooked-end deformed SF fibre with 60 mm length and fibrillated virgin PPF fibre with 19 mm length were used in this study. Meanwhile, the concrete was designed for high strength concrete of C60. The mixture included both single SF and PPF, and also the combination of both fibres; Control beam (PC), beam with 75%SF, beam with 75%SF + 25%PPF and beam with 25%PPF. The total fibre volume fraction (Vf) was fixed at 1.5%. The experimental results show that the percentage proportion of combined SF-PPF at 75-25% had the best performance for its flexural capacity. Mixture with single PPF was also found not effective in delaying the onset of tension cracks and to increase the tensile strength of the concrete. Experimental result also shows beam with 75%SF +25%PPF had their structural stiffness improved the most as compared with the others. For the compressive strength, beam with 75%SF + 25%PPF also revealed comparable performance with the control for high strength composite concrete.

1 Introduction

Plain concrete is weak in tension because it contains numerous micro cracks. These micro cracks propagate in the concrete matrix under constant applied load. Consequently, plain concrete members cannot sustain tensile stresses developed due to the applied force without the addition of reinforcing elements that are able to withstand these stresses. The addition of randomly distributed discrete fibres to the structural concrete increases its stiffness, ductility and load carrying capacity, while at the same time reduced crack development and propagation. According to the composite material theory and other findings [1], positive synergy of different fibres can complement each other to make new composite material with high performance and good economic benefits [2]–[4]. The use of two or more types of fibres in a suitable combination may potentially improve the overall properties of concrete and resulted in performance synergy [5]–[7]. In this study, steel fibre (SF) and polypropylene fibres (PPF) were combined to produce a hybrid system. Due to the lack of information on the ductile performance of hybrid fibre reinforced concrete composite (HyFRCC), an attempt was made to examine the ductility performance of HyFRCC beams. The presence of one fibre enabled more effective utilization of the potential properties of the other fibre which resulted in improved flexural rigidity, and at the same time controlled thecracking development.

2 Related previous study

A study by [8] found that concrete mixed with two different lengths of SF possessed excellent resistance to air blast loading as compared with plain concrete. In their study, the total volume fraction was fixed at 1.5%, with the mixture containing 70% long and 30% short hooked-end type steel fibre.

The investigation indicates that the steel fiber reinforced concrete panel containing of 1.5% volume fraction gave the best performance under explosive loading.In another study by researcher [9], the concrete containing 0.5% volume fraction of SF not sufficient to provide adequate resistance against blast loading and 1% of the fibres shows the best performance and significant to reduced hair line cracks on the specimen. Further investigation using three different properties of carbon and polypropylene micro fibres added to steel fibres in a concrete mixture showed that macro fibres of steel with highly deformed geometry produced better hybrid than

* Corresponding author: author@e-mail.org

those with less deformed geometry [10]. Also, lower volume fraction of fibres had better performance for hybridization than those mixed with high volume fraction.

Flexural behaviour of single and hybrid fibre concrete beams using three types of fibres was previously studied by Sekar and Ramamoorthy [11]. The fibres added in the concrete mixture were the hooked-end type SF, polyester recron fibre and coir fibre. Meanwhile, the total volume fraction was applied at 1%. They found that the compressive strength decreases when fibres were presence in the concrete. They also found that the ductility of both SF reinforced beam and hybrid fibre reinforced beam were higher compared with the control (plain). In another study, Sharmila and Thirugnanam [12] examined the significant behaviour of concrete beams added with steel, glass and recron 3s fibres under cyclic loading. It was reported that the effect of hybrid fibres influenced the behaviour of the beams by increasing the ductility characteristics by 80% and energy absorption characteristics by more than 160%. Hybrid fibres also increases energy absorption capacity and therefore an advantageous when applied for structures located at earthquake prone areas.

Besides that, other study evaluated the strengthening method using fibre reinforcement on RC beams [13]. The study found that the flexural behavior of beams strengthened with hybrid fibre reinforcement (sheet or plate) was 60% – 200% higher for its reinforcing effect than the non-reinforced beams. These results reflected excellent performance, which was attributed by the combined fibres having high reinforcing effect that leads to stable ductile failure.

Another study on hybrid fibres application revealed that the tensile strength improvement ranging between 25% and 80% compared to that of plain concrete [14]. SFs exerted a primary influence in the hybrid system when compared with PPFs. The results indicated that concrete containing PPFs can further improved residual strength in the post-peak response. The study then concluded that concrete containing combined 75% SF + 25% PPF were the most appropriate combination for improved compressive strength, flexural strength and flexural toughness [15].

Hybrid fibre reinforcement of cement composites is rapidly emerging as an innovative and promising way of improving the mechanical performance and durability ofcement-based materials. Previous researchers had carried out study on structural behaviour of glass fibre reinforced polymer (GFRP) concrete hybrid beam [15-16]. The studies found that GFRP concrete hybrid beam were presenting reasonable stiffness and also increase the strength of the beam.

3 Materials and method

3.1 Material and mix proportion

Ordinary Portland Cement (OPC) Type 1 was used in this study. 10 mm well graded size crushed granite and locally available river sand were used for the coarse and fine aggregates, respectively. The concrete strength was designed based on the DoE method to achieve compressive strength of 60 N/mm2 at 28 days, with water-to-cement ratio, w/c fixed at 0.38. Hooked-end type SF of 60 mm long and 0.75 mm diameter was used in this study (see Fig. 1), giving an aspect ratio, l/d of 80. Meanwhile, virgin fibrillated type PPF of 19 mm long was also used as shown in Fig. 2. The properties of both SF and PPF are given in Table 1 and 2, respectively. The mixture was different in proportions based on percentage of: (i) 0% (control), (ii) 75% SF + 25% PPF, (iii) 75% SF and (iv) 25% PPF. Meanwhile, the total fibre volume fraction, Vf was fixed at 1.5%. In order to improve concrete workability, superplasticizer was also added in the mixture at dosage rate of 3% of the cement content. The material proportions for 1 m3 concrete volume are summarized in Table 3.

Fig. 1. Steel fibre – hooked end

Fig. 2. Polypropylene fibre – fibrillated

3.2 Beam detail

A total of four beams with dimension of 275 mm x 380 mm x 2400 mm were used in this study. The beams were mixed with various percentage of SF and PPF. All beams were reinforced with 2 nos. of 12 mm diameter bar at the

DOI: 10.1051/matecconf/201710101020

bottom section of the beam (tension zone), while 2 nos. of 8 mm diameter bar were used as hanger bar (top section).

Table 1. Properties of steel fibre (SF)

Parameters	Capability
Type	HE 0.75/60
Shape	Hooked-End (Deformed)
Length (mm)	60
Diameter (mm)	0.75
Aspect Ratio, *l/d*	80
Tensile Strength (N/mm^2)	1100
Unit Weight (kg/m^3)	7850
Coating	None
Elastic Modulus, *E* (MPa)	205 000

Table 2. Properties of of polypropylene fibre (PPF)

Parameters	Capability
Type	Virgin PPF
Length (mm)	19
Thickness (mm)	0.05
Unit Weight (kg/m^3)	446
Tensile Strength (N/mm^2)	400
Thermal Conductivity	Low
Elastic Modulus, *E* (MPa)	3500

Table 3. Material proportions for 1 m^3 concrete volume

Beam Specimen	SF (kg/m^3)	PPF (kg/m^3)	FA (kg/m^3)	CA (kg/m^3)	Water (kg/m^3)	Cement (kg/m^3)	w/c	SP (%)
Control	-	-	794		210	522	0.38	3
75% SF	88.3		794		210	522	0.38	3
75% SF + 25%PPF	88.3	1.67	794		210	522	0.38	3
25% PPF	-	1.67	794		210	522	0.38	3

Notations:
SF – Steel Fibre
PPF –Polypropylene fibre
FA- Fine Aggregate
CA – Coarse Aggregate
w/c- Water-to-cement ratio
SP- Superplasticizer

3.3 Beam instrumentation and testing

All beams were tested under four point loading systemusing a 500 kN testing machine capacity. The beams were simply supported over a span of 1200 mm. The distance between the two point loads was kept a part at 150 mm symmetrical to the centerline of the beam. Loading was applied using a hydraulic jack at the rate of 2 kN/s until failure. Mid-span deflection was measured for every 2 kN loading increment using Linear Variable Displacement Transducer (LVDT) with 0.01 mm accuracy. The deflection of the beam was measured at three points along the beam span; mid-span and 1m from left and 1m from right supports. The beams were instrumented with embedded and external strain gauges to monitor the concrete and steel strains at different loading stages. Concrete strain gauges were installed on the beam surface, while steel strain gauges were attached at the top and bottom of the longitudinal reinforcing bars, also at mid-span. Both the electrical concrete and steel strain gauges were installed at mid-span section of the beam, where maximum bending moment was to occur. At each loading increment, deflection and strains were recorded using a data logger. At the same time, cracking development was also observed and monitored closely. The loading arrangement and instrumentation of the flexural beam test is shown in Fig. 3.

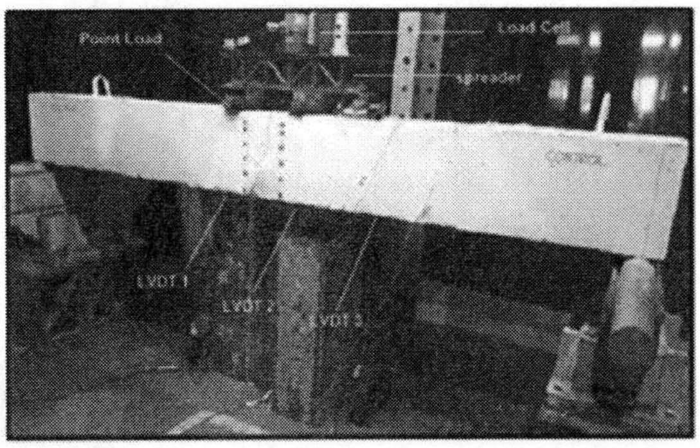

Fig. 3. Test set-up

4 Results

Flexural test was carried out on beams with varies mix proportion of SFs and PPFs as discrete reinforcement in a concrete mixture. To characterize the flexural performance and behaviour of the beams, analysis and discussion were done on its ultimate load, load-deflection behavior, load-strain behavior and cracking pattern.

4.1 Ultimate load on reinforced concrete

The result of the first crack load and ultimate load for all beams are summarized in Table 4 together with the compressive strength of the concrete. The compressive strength at 28 dayswas 60.3 N/mm^2, 68.7 N/mm^2, 66.5 N/mm^2 and 60.0 N/mm^2 for control beam, beams with 75%SF, 75%SF + 25%PPF and 25%PPF, respectively.

The table shows that the addition of fibres increased the load carrying capacity for all beams. The ultimate load for the control beam, beams with 75%SF, 75%SF + 25%PPF and 25%PPF were 77.4 kN, 97.1 kN, 97.7 kN and 70.2 kN, respectively. The finding also shows that the ultimate load carrying capacity was more significant for beam with 75%SF+25%PPF. In this study, the ultimate load for beam with 75%SF+ 25%PPF produced the highest ultimate load followed by beam with 75%SF, showing an increase of 25% and 26% from the control. Meanwhile, beam with 25%SF obtained the lowest load carrying capacity which was even lower than the control (0% fibre). The high ultimate loading capacity for beams 75%SF + 25%PPF and 75%SF was due to the sufficient bridging action of the steel fibres that prevent the crack from further expanding during the test. This is because the characteristics of SF include higher Elastic Modulus and tensile strength, and ductile in flexural mode before failure. PPF will encounter for micro cracks while SF encountered for macro cracks. The hooked end type SF also give some advantage in delaying macro cracks by gripping the particles inside concrete and bridging crack growth before failure. Meanwhile, the ultimate loading capacity of beam with 25%PPF was 9.3% lower than the control beam. In the case of PPF, the beam was attributed for arresting non-structural crack such as shrinkage crack. PPF can be used as secondary reinforcement in concrete to reduce cracking and benefit for large surface area construction such as floor slab and tunnel lining[18]–[20].

Table 4. Test results

Beam Specimen	Compressive Strength (N/mm^2)	First Crack Load (kN)	Deflection at First Crack Load (mm)	Ultimate Loading Capacity (kN)	Deflection at Ultimate Loading Capacity (mm)
Control	60.3	30	0.76	77.4	68.44
75%SF	68.7	46	1.16	97.1	16.97
75%SF + 25%PPF	66.5	50	1.32	97.7	6.93
25%PPF	60.0	34	0.66	70.2	23.1

4.2 Load - deflection response

Load-deflection relationship for all beams is shown in Fig. 4. Control beam was used to compare and discuss the result for other beams with 75%SF, 75%SF + 25%PPF and 25%PPF. As concrete weak in tension, the first tension crack occurred near the mid span region (maximum bending moment region). The number of cracks increase as the load was further increased. The load-deflection patterns for all beams at different locations are shown in Fig. 4(a), 4(b) and 4(c).

Generally, the load-deflection curves can be classified to three distinct zones; the first zone is the initial part of the curve to the cracking point, the post cracking zone, continued to the yielding point and the post yield zone, up to failure. At the initial stage the beams stiffness showed almost identical at low level of loading and up to the cracking load as this stage is controlled mainly by the concrete tensile strength The average of load and deflection curve for 75%SF and 75%SF+25%PPF proved higher stiffness compared to 25%PPF and control beam. The second zone, showed a distinct behavior in the different beams. The slope of the curve is a direct function represents the effective stiffness of the beam. Last part of the post yield zone, the beams showed the ability to withstand higher load in different rates and to gain more deformability until failure.

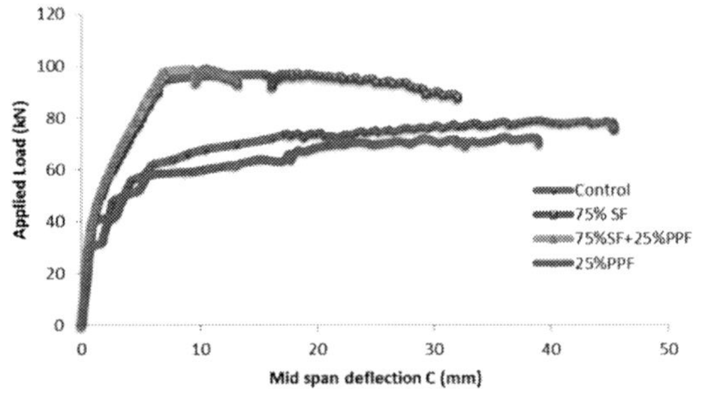

(a) Load-deflection at midspan

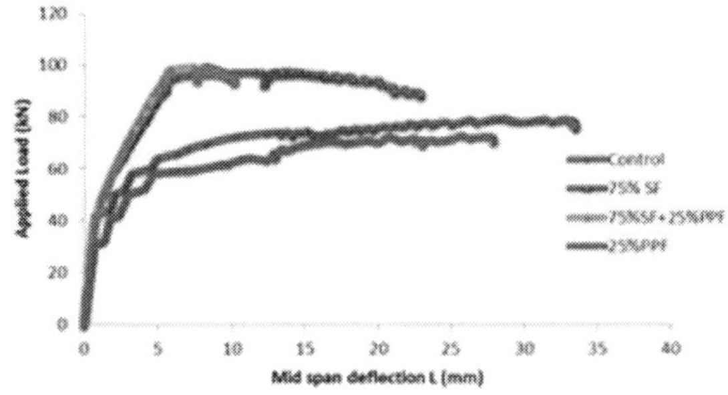

(b) Load-deflection at left of span

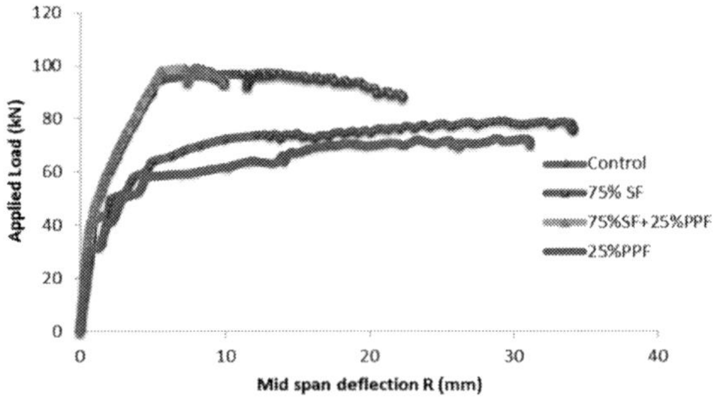

(c) Load-deflection at right of span

Fig. 4. Load-deflection relationship

4.3 Load - strain of reinforced concrete beam

Figure 5 show the load-strain behavior of the beams based on the compression zone. The behavior also indicates that the control beam behaved in brittle manner after loads of 40kN was reached while beams with 75%SF and 75%SF+25%PPF behaved in ductile manner.

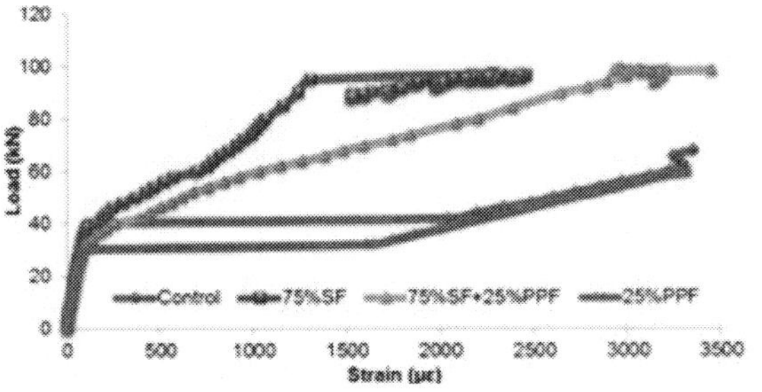

Fig. 5. Load (kN) versus strain (με) for control beam

Table 5 shows the maximum compressive strain at structural failure. Generally, the maximum compressive strain in the concrete is 0.0035, and the compressive strain for control beam in this study achieved that almost maximum compressive strain. However, the maximum compressive strain for beam with 75%SF, 75%SF lower value than that of control beam with 0.0022, 0.0029 and 25%PPF obtained almost maximum with 0.0032 respectively but the difference between them were small. The ranges of maximum compressive strain for beam 75%SF, 75%SF lower observed from this study were from 0.0020 to 0.0029 and beam with 25%PPF observed 0.0029-0.0032.

Table 5. Maximum compressive strain for control and RC beams with fibres

Type of Beam	Maximum compressive strain, ε
Control (PC)	0.0033
75% SF	0.0022
75% SF+ 25% PPF	0.0029
25% PPF	0.0031

4.4 Cracking and failure mode

All beams failed in flexure with cracking developed at the tension zone as shown in Fig. 6. It was observed that all beams developed fine cracks from the end at the tension zone under a relatively small load of about 30 – 40% of their ultimate load. The first noticeable crack was formed between the locations of the midspan followed by two point loads in the region of maximum bending moment. From Fig. 5, as the beam's ductility increases, more hairline cracks appeared in the flexural zone. Beam with 75%SF+25%PPF had a total of 10 cracking line compared to only 8 lines for beam with 75%SF. Meanwhile, for beam with 25%PPF, 6 crack lines were found as compared with only 4 lines for the control beam. The number given along each crack represents the load at which the crack was extended. This shows that by combining fibres with different properties, they are more capable to resist higher load and reduced crack opening at ultimate load. This is because the bridging effect of both SF and PPF to arrest crack development and propagation during the early stage of concrete casting (wet) and hardening (dry) process enhanced the structural performance of the beam. For control beam and beam with 25%PPF, it can resist lower ultimate loading capacity with lesser cracks than beams with 75%SF and 75%SF + 25%PPF. The crack width at ultimate load was recorded at 2.48 mm, 0.40 mm, 0.60 mm and 1.80 mm for control beam, beams with 75%SF + 25%PPF, 75%SF and 25%PPF, respectively. Table 6 summarized the number of cracks, crack spacing and depth for all beams at ultimate load.

(a) Control (PC)

(b) RC + 75% SF

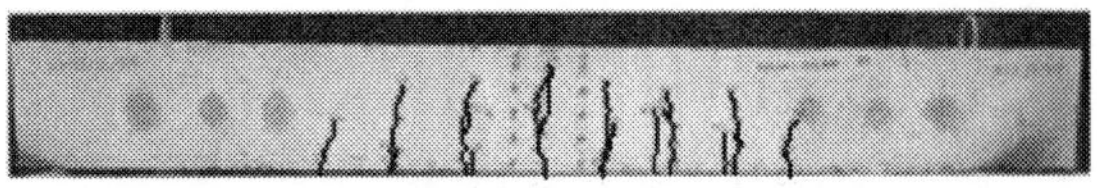

(c) RC + 75% SF+ 25% PPF

(d) RC + 25% PPF

Fig. 6. Failure mode of OPC concrete beam

Table 6. Structural failure mode

Beams	Number of Cracks	Average Crack Spacing (mm)	Crack Depth (cm)
Control (PC)	4	19.8	34.3
75% SF	8	11.2	20.6
75% SF+ 25% PPF	10	12.2	19.2
25% PPF	6	13.6	20.4

5 Conclusions

1. The crack pattern of beams for all beams shows it's failed in flexural zone. The number of cracks increased and they show better distributed especially in beams 75%SF+25%PPF indicating higher ductility of the beam.
2. Beam with 75%SF + 25%PPF had the highest

structural stiffness of 32% compared with the control beam. Meanwhile, beams with 75%SF and 25%PPF increased by 29% and 7%, respectively for its flexural strength.

3. The ultimate load capacity and stiffness of the beams increased when SF and PPF were added together in a single mixture. This indicates that both properties were a good combination for structural applications. However, adding PPF only was not suitable for structural purposes.
4. Vertical cracks appeared during the initial load until the beams failed. The crack opening and propagation increases when the applied load was further increased.

This research was funded by the Research University Grant (RUG) No. 09H36. Invaluable appreciation goes to all staffs and Assistant Engineers in the Structures and Materials Laboratory, Faculty of Civil Engineering and Faculty of Chemical, UniversitiTeknologi Malaysia (UTM). Special acknowledgment also goes to the Ministry of Higher Education Malaysia (MoHE) and UniversitiTun Hussein Onn Malaysia (UTHM) for the scholarship sponsored throughout the study.

References

1. N. Banthia, R. Gupta, Mater. Struct., **37,** 707–716 (2004)
2. C. X. Qian, P. Stroeven, *Cem. Concr. Res.*, **30,** 63–69 (2000)
3. S. P. Singh, A. P. Singh, and V. Bajaj, Asian J. Civ. Eng., **11,** 495–507 (2010)
4. P. Wang, Z. Huang, J. Jiang, Y. Wu, Am. Soc. Civ. Eng., **11,** 677–683 (2012)
5. M. Hsie, C. Tu, P. S. Song, Mater. Sci. Eng. A, **494,** 153–157 (2008)
6. J. S. Lawler, D. Zampini, S. P. Shah, J. Mater. Civ. Eng., **17,** 595–604 (2005)
7. L. Sorelli, N. Banthia, G. A. Plizzari, J. Am. Concr. Inst., **224,** 161–178 (2004)
8. M. A. Yusof, N. M. Nor, A. Ismail, N. C. Peng, R. M. Sohaimi, M. A. Yahya, Adv. Mater. Sci. Eng., **13,** 1-7 (2013)
9. M. A. Yusof, N. M. Nor, M. Fauzi, M. Zain, N. C. Peng, A. Ismail, R. M. Sohaimi, A. Mujahid, A. Zaidi, Concrete, **7,** 159–166 (2011)
10. N. Banthia, M. Sappakittipakorn, *Cem. Concr. Res.*, **37,** 1366–1372 (2007)
11. C. Sekar, N. Ramamoorthy, Int. J. Eng. Res. Technol., **2,** 2148–2157 (2013)
12. S. Sharmila, G. Thirugnanam, Int. J. Sci. Environ. Technol., **2,** 725–734 (2013)
13. S.T. Yi, S. G. Cho, *Proc. ICE - Struct. Build.*, **166,** 537–546 (2013)
14. L. Xu, L. Huang, Y. Chi, G. Mei, ACI Struct. J., **113,** 219–230 (2016)
15. S. P. Singh, A. P. Singh, V. Bajaj, Asian J. Civ. Eng., **11,** 495–507 (2010)
16. M. D. Krall, Tests on Concrete Beams with GFRP Flexural and Shear Reinforcements and Analysis Method for Indeterminate Strut-and-Tie Models with Brittle Reinforcements, Master Thesis , University of Waterloo, Canada, 2014
17. C. E. Bakis, L. C. Bank, V. L. Brown, E. Cosenza, J. F. Davalos, J. J. Lesko, A. Machida, S. H. Rizkalla, T. C. Triantafillou, J. Compos. Constr., **62,** 73–87 (2002).
18. P.K.R. Dilip, J. Softw. Hardw. Res. Eng., **23,** 47–51 (2014)
19. S. K. Verma, M. Dhakla, A. Garg, Int. J. Eng. Innov. Technol., 4, **10,** 90–94 (2015)
20. A. E. Yurtseven, Determination of mechanical properties of hybrid fiber einforced concrete, Master Thesis, Middle East Technical University, 2004

Mechanical and physical properties of metakaolin based geopolymer paste

Puput Risdanareni[1], *Poppy* Puspitasari[2], *Edi* Santoso[1], *Edo* Prasetya Adi[1]

[1] Civil Engineering Department, Engineering Faculty, Universitas Negeri Malang, Malang, Indonesia
[2] Mechanical Engineering Department, Engineering Faculty, Universitas Negeri Malang, Malang, Indonesia

Abstract. This paper presents the result of studies on utilizing metakaolin obtained from Belitung as fly ash substitute material to produce geopolymer paste. Mechanical properties were assessed by compressive strength while physical properties were assessed by setting time, porosity, microstructure test and density test. The experiment was performed on geopolymer paste with 0, 25%, 50%, 75% and 100% fly ash replacement with metakaolin. Sodium Hidroxide (NaOH) with concentration of 10 Molars and Sodium Silicate (Na2SiO3) were used as alkaline activator. In addition, activator ratio between sodium silicate to sodium hidroxide of 2 was applied. The result obtains that increasing metakaolin contain in mixture deliver longer setting time, higher open pore number, lower density and lower compressive strength of geopolymer paste. However, adding metakaolin up to 25 % into geopolymer paste mixture could increase mechanical and physical properties of geopolymer paste.

1 Introduction

Recently, research on investigating alternative material as cement substitute becomes a massive issue in construction industry. Previous research proved that utilizing cement contribute bad environment effect due to high CO_2 emitted during its production [1, 2]. Hence, reducing cement use by replacing or substituting it with other eco green material will be an enable solution for a better future.

Geopolymer material, a new eco green material synthesized from precursor and alkaline activator revealed as a promising construction material to reduce utilization of Portland cement [3]. Precursor which has high silica and alumina content is required to produce geopolymer binder. Fly ash, a waste material derived from coal combustion process which rich in silica and alumina become a popular precursor in producing geopolymer binder [4]. Hence, sodium hydroxide (NaOH) and sodium silicate (Na_2SiO_3) are used as alkaline activator.

As a waste material, controlling and predicting quality of geopolymer binder produced from fly ash became a though jobs [5]. Therefore, utilizing alternative precursor materials besides fly ash that has stable properties are required in producing geopolymer binder.

Metakaolin, a material produced from kaolin calcinations which has similar content to fly ash was considering as alternative precursor material besides fly ash. Based on statistic data from ministry of energy and mining resources of Indonesia in 2010, availability of Kaolin in Indonesia is abundant and less utilization. Hence, in this study metakaolin (MK) was used as precursor replacing fly ash (FA) to produce geopolymer paste. In order to find optimum doses of metakaolin as fly ash replacing, five metakaolin doses variation of 0%, 25%, 50%, 75% and 100% was performed.

Previous research stated that concentration of NaOH and ratio between NaOH and Na_2SiO_3 played important role in developing mechanical and physical properties of geopolymer paste [4]. Based on previous study by Ekaputri, NaOH concentration of 10 molars and alkaline activator ratio of 2 was used in this research due to its good performance previously [6].

In general, investigating mechanical properties and physical properties of geopolymer paste which produced from metakaolin and fly ash become the first aim of this research. Giving recommendation of optimum composition of metakaolin as fly ash replacement is the second aim of this study.

2. Material and method

2.1 Materials

Fly ash class C obtained from PT. Petrokimia gresik and kaolin class N from Belitung Island with density of 2.47 gr/cm^3 and 2.36 gr/cm^3 was used as raw material. Kaolin was calcined in temperature of 700° Celcius in order to perform metakolin. Both metakaolin (MK) and fly ash (FA) was sieved and passes sieve number 200. Chemical composition of fly ash and metakaolin obtained from XRF test was presented at Table 1.

A clear gel of Na_2SiO_3 and a flake shape of NaOH were used as alkaline activator. NaOH solution with concentration of 10 molars and activator ratio between

* Corresponding author: puput.risdanareni.ft@um.ac.id

Na_2SiO_3 and NaOH of 2 was performed in this research. NaOH solution was prepared by dissolving NaOH flake into distilled water one day prior before it use.

Table 1. Chemical composition of fly ash and metakaolin

Particles	FA % by mass	MK % by mass
SiO_2	35.91	31
Al_2O_3	16.02	53.5
CaO	14.43	1.1
Fe_2O_3	12.34	6.58
K_2O	1.28	5.79
MgO	9.09	0.12
Na_2O	5.87	0.04
TiO_2	0.66	0.919

2.3 Method

Flow chart diagram for this research was ilustrated at figure 1. Preparing material such as fly ash, metakaolin, and alkaline activator was conducted at first stage of this research. Next step was perform preliminary material test such as XRF, XRD and SEM. Consistency normal and setting time was conducted in order to investigate physical propertis of fresh geopolymer paste. After manufacturing paste and cure paste at room temperature, paste speciment was tested for its density, compressive strength and pororosity. Last stage of this research is performed data analysis for all testing.

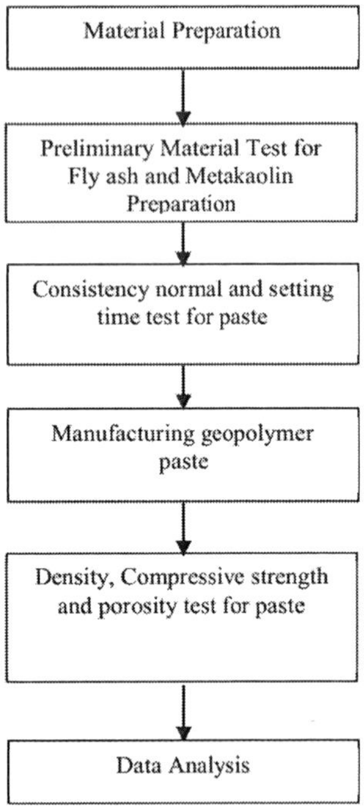

Fig. 1. Experimental Flow Chart Diagram

2.4 Specimens and mixture proportions

Specimens of geopolymer paste were cylinder with the diameter of 2 cm and the height of 4 cm. Five mixture proportions of 0%, 25%, 50%, 75% and 100% based on metakaolin content was produced in this study. Ratio between alkaline activator and precursor used in this research was 0.6. Table 2 shows mixture proportion details of geopolymer paste.

Table 2. Mixture proportion details

Code	MK % by mass	Num (pcs)	Precursor		Alkaline activator	
			MK (gr)	FA (gr)	NaOH (gr)	Na_2SiO_3 (gr)
MK 1	0%	6	0	109	24	48
MK 2	25%	6	27	81	24	48
MK 3	50%	6	54	54	24	48
MK 4	75%	6	81	27	24	48
MK 5	100 %	6	109	0	24	48

2.5 Laboratory test

Compressive strength test was performed according to ASTM C39-03 guidelines. Test was experienced on geopolymer paste with the age of 7, 14, 21 and 28 days using Universal Testing Machines (UTM) at Laboratory Structure of ITS Surabaya. The testing result was evaluated by taking average value from 6 specimens.

Setting time test was referred to ASTM C 191-92. Hence, porosity test was conducted in geopolymer paste with the age of 28 days as referred to AFNOR NF B 49104. Documentation of setting time and compressive strength test were presented at figure 2.

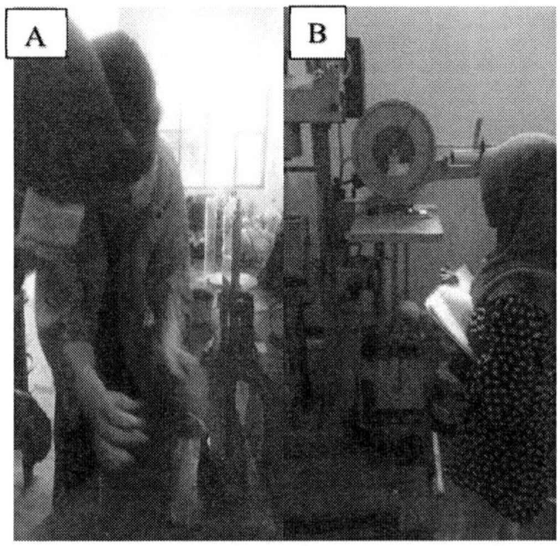

Fig. 2.(a) Setting Time Test , (b) Compressive Strength Test

2.5 Curing

Immediately after moulding, specimens were covered with plastic cover and left in room temperature until testing date to reduce evaporation. Picture of several paste specimens was presented at figure 3.

Fig. 3. Geopolymer paste specimens

3. Result and discussion

3.1 Density

The density test results of geopolymer paste are presented on table 3. Based on data presented at table 3, it can be seen that all samples have ranged of density between 1.8 to 2.1 gr / cm^3. Hence, it is proven that metakaolin levels affect the density of geopolymer paste. The higher MK content in the mixture, the lower density of geopolymer paste produced.

The low density of paste with low MK content probably caused by the high amount of open pore occurred [7]. In addition, metakaolin which has similar character with clay experience easy expand and shrinkage when reacts with water. So that, open pore occurred in sample with high content of MK become higher than others [8]. Moreover, this condition will be discussed further in this paper.

Table 3. Density test result

Sample code	Density (gr/cm^3)
MK 1	2.10
MK 2	2.07
MK 3	2.07
MK 4	1.9
MK 5	1.8

3.2 Compressive strength

The compressive strength test result of geopolymer paste was presented in Table 4. All specimens experienced optimum compressive strength at the age of 28 days. Compressive strength of all the samples at 28 days was in the range of 4.73 to 38.32 MPa. Sample MK 2 which has metakaolin content of 25 % owned highest compressive strength among the others. It can be concluded that the addition of 25% of MK to the mix provides optimum compressive strength results. The compressive strength experienced to decrease in paste with MK content of more than 25%. These results has a good agreement with previous studies stated that the level of class N pozzolan optimum in geopolymer paste mixture is 25 to 50 % [9].

Table 4. Compressive strength test result

Sample code	Compressive strength of each age (MPa)			
	7 days	14 days	21 days	28 days
MK 1	5.97	24.23	30.64	38.32
MK 2	2.09	25.11	30.72	40.03
MK 3	11.22	20.21	24.23	27.65
MK 4	3.14	8.60	9.99	12.57
MK 5	0.68	1.51	2.23	4.73

The relationship between the compressive strength of geopolymer paste at each age is presented in Figure 4. According to the graph presented in Figure 4, it can be seen that the samples which contained more than 25% MK was experienced to slow react, which is indicated by the slow increasing of compressive strength at each testing age. Paste with a level MK content of 25% and paste without MK content has significant increasing of compressive strength at early age and has not showed stability compressive strength at the age of 28 days. Reactivity of Mk was presumable as the causes of slow reacting of sample contained MK levels over 25%. The previous study by Van Jaarveld mentions that the reactivity of the precursor has a great influence in the mechanical properties of geopolymer paste [10]. The less reactive of MK possible occur because the imperfections combustion process of kaolin into metakaolin. Previous research by Triani states that the temperature and the combustion quality are very essential in the activation process of kaolin [11].

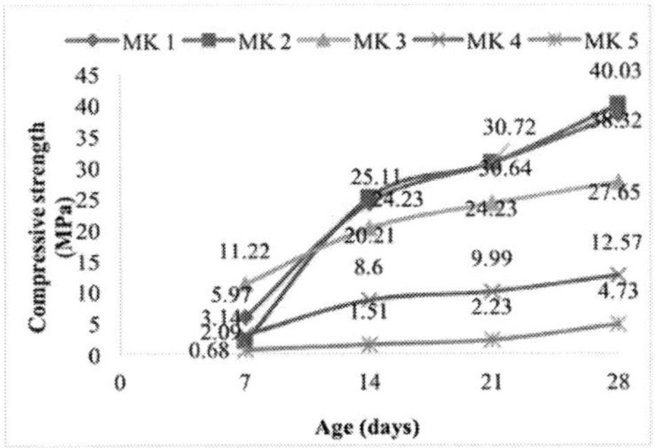

Fig. 4. Compressive strength of geopolymer paste at different age

3.3 Setting time

Setting time test result was presented at table 5. Initial and final setting time result of all variants has range value from 165 minutes to 2 days. The shortest setting time was owned by sample MK1 which contain 0% MK, while the longest setting time was owned by sample MK 5 which contain 100% MK. Hence, it can be concluded that MK contained in the mixture has directly proportional relationship with setting time of its geopolimer fresh paste.

Table 5. Setting time test result

Sample Code	Initial setting time (Minutes)	Final setting time (Minutes)
MK 1	165	240
MK 2	240	345
MK 3	720	960
MK 4	1080	1500
MK 5	more than 2 days	

The higher the levels of metakaolin in the mixture, the longer setting time of fresh geopolymer paste occured. This is consistent with previous research by Risdanareni which states that the addition of pozzalan class N to the geopolymer paste mixture can prolong setting time [12]. Additionally CaO content in the mixture also affect the value of setting time. Fly ash has higher a CaO content compared to metakaolin so that a mixture that containing a lot of metakaolin will have fewer CaO content that deliver long setting time [13].

3.4 Porosity

The amount of open and close pore of geopolymer paste was illustrated in figure 2. Open pore is a pore that is permeable so that it can be penetrated by water and air, while the close pore is impermeable so that water and air cannot pass [14].

Referring from data displayed in figure 5, it can be seen than sample MK 2 owned lowest open pore number and highest close pore number. Sample with MK content more that 25 % experienced high open pore and low close pore. The higher MK content in the mixture the higher open pore occurred. The characteristic of metakaolin presumed cause this condition.

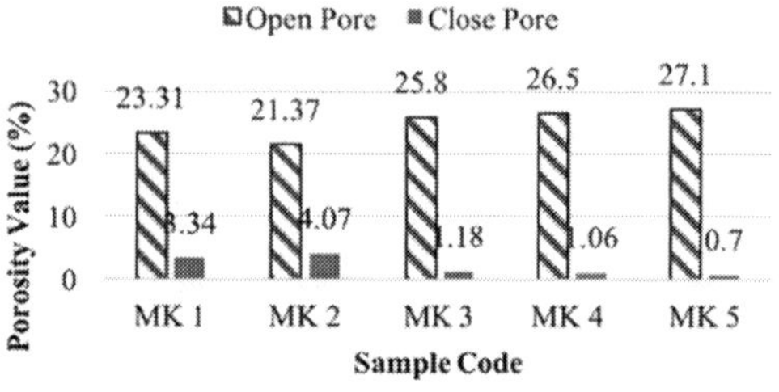

Fig 5. Porosity test result

The result of open pore test is closely associated with the compressive strength and density test result. The test results of open pore, compressive strength and density of geopolymer paste was mentioned in table 6. Based on data presented at table 6, it can be seen that sample which has high number of open pore would have a low compressive strength and density value. So it can be concluded that the number of open pore is inversely proportional to the compressive strength and density of geopolymer paste. This is consistent with previous studies which stated that increasing the number of open pore will decrease the compressive strength and density of geopolymer paste [14].

Table 6. Relationship between mechanical and physical properties test

Sample Code	Open pore (%)	Compressive strength (MPa)	Density (gr/cm^3)
MK 1	23.31	38.32	2,109
MK2	21.37	40.03	2,069
MK3	25.8	27.65	2,069
MK4	26.5	12.57	1,870
MK5	27.1	4.73	1,830

3.5 Microstructure of Precursor

SEM (Scanning Electron Microscopy) test was conducted in order to investigated morphology of fly ash and metakaolin. Microstructure test result of fly ash and metakolin was presented at figure 6. Previous research by Tajunnisa stated that microstructure of precursor influences mechanical properties of geopolymer paste. SEM test result can be used to predict mechanical and physical characteristic of geopolymer paste. [15] Fernadez was also stated that particle size and particle distance play an important role in developing properties of geopolymer binder. [7] Particle distance play an important role in developing mechanical properties of geopolimer paste because the geopolyerization only occur at surface area of precursor. [16] So it can be concluded that, the larger particle distance of precursor, the lower mechanical properties of geopolymer paste occur.

Fig 6. (a) microstructure of metakaolin, (b) microstructure of fly ash

Refer to figure displayed at figure 6 it can be seen that fly ash has spherical shape while metakaolin has irregular shape (bulky). Due to its irregular shape, metakaolin has large particle distance compared to fly ash which has homogenous spherical shape. This result has a good agreement with compressive strength result of geopolymer pasta. The optimum metakaolin content in the mixture is 25% and compressive strength is experienced to decrease with increasing of metakaolin content. Hence it can be concluded that metakolin which has bulky shape and has large particle distance can decrease the mechanical properties of geopolimer paste.

4 Conclusion

The result evidenced that metakaolin content as fly ash replacement has a great influence in developing mechanical and physical properties of geopolymer paste such as density, compressive strength, setting time and porosity. The high level of metakaolin content will decrease the compressive strength and density of geopolymer paste. However, increasing metakaolin content will deliver longer setting time and higher open pore. Microstructure of precursor was also influence the physical and mechanical properties of geopolymer paste. Precursor which has short distance particle and homogenous shape will deliver high compressive strength of pasta produced. Based on several testing on investigating mechanical and physical properties of geopolymer paste in this research, it can be concluded that replacing 25 % of fly ash with metakaolin proved to give better mechanical and physical properties.

The authors gratefully acknowledge The Directorate General for Higher Education, Ministry of Research, Technology and Higher Education, Indonesia, who provided the research grant through the 'Penelitian Kerjasama Antar Perguruan Tinggi (PEKERTI)' scheme in 2016.

References

1. C.A Hendriks, *Proceedings of the 4th International Conference on Greenhouse Gas Control Technologies, Interlaken, Austria, Aug. 30–Sept.* **2**, IEA GHG R&D Programme, UK(1998)
2. E. Gartner, Cement and Concrete Research, **34**, 1489–1498 (2004)
3. Davidovits, Journal of Thermal Analysis Vol. **37**, (1991)
4. D. Hardjito, *Studies on Fly Ash-Based Geopolymer Concrete*, Curtin Univ of Technology, (2005)
5. R. Nurwidayati, M.B. Ulum, J.J. Ekaputri, Triwulan, P. Suprobo, Materials Science Forum, Vol. **841**, pp 118-125, (2016)
6. J. Ekaputri, Triwulan, Jurnal Teknik Sipil ITB Vol **20** No.1 (2013)
7. Ferna´ndez-Jime´nez, Palomo, *Characterisation of Fly ashes Potential Reactivity as Alkaline Cement*, Fuel Vol. **82** pp 2259–226, (2003)
8. F. Slaty, H. Khoury, J. Wastriels, H. Rahier, *Characterization of Alkali Activated Kaolinitic Clay*, Applied Clay Science, (2003)
9. P. Risdanareni, A. Karjanto, J.J. Ekaputri, P. Puspitasari, F. Khakim, Material science Forum Vol. **857**, (2016)
10. V. Jaarsveld, J. Van Deventer and G. Lukey, Materials Letters, vol. **57**, p. 1272–1280, (2003)
11. D.N.D Triani, J.J Ekaputri, Triwulan, S. Hardono, T.E Susanto, Material Science Forum Vol. **841** pp 111-117, (2016)
12. P. Risdanareni, A. Karjanto, F. Khakim, Material science Forum Vol. **841**, (2015)
13. Antoni, S.W. Wijaya, D. Hardjito, Material Science Forum Vol **841**, (2015)
14. P. Risdanareni, J.J Ekaputri, M.M.A.B Abdullah, Applied Mechanics and Materials Vol. **754-755**, (2015)
15. Y. Tajunnisa, M. Sugimoto, T. Sato, J.J. Ekaputri, M.Shigeishi, International Conference and Exhibition of Structural Faults and Repair 2016, Edinburg, 17th-19th May, (2016)
16. J. Duchesne , L. Duong , T. Bostrom, R. Frost, Waste Biomass Valor, **1:367–377**,(2010)

Experimental determination of monoethanolamine protonation constant and its temperature dependency

Sholeh Ma'mun*, Kamariah, Sukirman, *Desi* Kurniawan, *Eleonora* Amelia, *Vitro* Rahmat, and *Deasy R.* Alwani

Department of Chemical Engineering Faculty of Industrial Technology Islamic University of Indonesia, 55501 Yogyakarta, Indonesia

Abstract. Carbon dioxide as one of the major contributors to the global warming problem is produced in large quantities by many important industries and its emission seems to rise from year to year. Amine-based absorption is one of the methods to capture CO_2 from its sources. As a reactive system, mass transfer and chemical reaction take place simultaneously. In a vapor-liquid equilibrium model for the CO_2-amine-water system, some parameters such as mass transfer coefficients and chemical equilibrium constants need to be known. However, some parameters could be determined experimentally and the rests could be regressed from the model. The protonation constant (pK_a), as one of the model parameters, could then be measured experimentally. The purpose of this study is to measure the pK_a of monoethanolamine (MEA) at a range of temperatures from 303 to 330K by a potentiometric titration method. The experimental data obtained were in a good agreement with the literature data. The pK_a data from this work together with those from the literature were then correlated in an empirical correlation to be used for future research.

1 Introduction

Carbon dioxide (CO_2) is known as one of the major contributors to the global warming problem. Carbon dioxide is produced in large quantities from coal-fired power plants, steel production, petrochemical manufacturing, cement production, and natural gas purification as well as the exhaust gas from the transportation sectors. In the past decades, CO_2 recovery from flue gas streams was conducted mainly for Enhanced Oil Recovery (EOR) operations. Moreover, CO_2 was also produced for other industrial applications such as carbonation of brine, welding as an inert gas, food and beverage carbonation, dry ice, urea production, and soda ash industry [1, 2]. However, environmental concerns such as the global climate change are now focused as one of the most important and challenging environmental issues facing the world community and drive an intensive research on CO_2 capture and sequestration [3].The CO_2 emission seems to rise from year to year. In 2005, 26.3 Gt of CO_2 was emitted globally and reached 32 Gt in 2010 [4]. Necessary efforts to reduce the CO_2 emission are, therefore, required. Absorption is one of the methods to capture CO_2 from its emission sources.

A commercial absorbent for CO_2 removal requires both a high net cyclic capacity and high reaction rate for CO_2 as well as high chemical stability. In addition, other criteria that need to be considered are, according to [5], base strength (pK_a), molecular weight per NH functionality, slope of equilibrium curve, heat of absorption, viscosity, nature of degradation products, nitrosamine formation potential, cost and availability, and HSE (Health, Safety, and Environment). Research on solvent development has been conducted to meet those criteria by testing alkanolamines and their blends, sodium carbonate solutions, chilled ammonia, and amino acid salts [3, 6-13].

Aqueous solutions of alkanolamines are the most commonly used as chemical absorbents for the CO_2, such as CO_2 removal from natural gas, from refinery and synthesis gas streams. Among them, aqueous monoethanolamine ($H_2NCH_2CH_2OH$, MEA) as a primary amine has been used extensively for this purpose. Fig. 1 shows the MEA structure. It has several advantages over other commercial alkanolamines, such as high reactivity, low solvent cost, low molecular weight and, thus, high absorbing capacity on a mass basis and reasonable thermal stability and thermal degradation rate. However, MEA has some drawbacks such as high enthalpy of reaction with CO_2 leading to higher regeneration energy consumption in desorber, formation of a stable carbamate and also formation of degradation products with COS or oxygen bearing-gases, inability to remove mercaptans, vaporization losses because of relatively high vapor pressure [14, 15], more corrosive than many other alkanolamines [15, 16] and non-selective for the removal of H_2S in the presence of CO_2 [14].

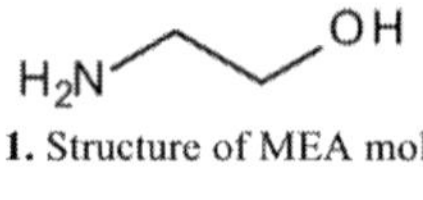

Fig 1. Structure of MEA molecule

The objective of this study is to evaluate the protonation constant (pK_a) of MEA at a range of temperatures from 303 to 333K using a potentiometric titration method.

2 Theory

Carbon dioxide capture using amine-based absorption is categorized as a chemical absorption process where mass transfer and chemical reaction occur simultaneously. A series of physical and chemical reaction equilibria for the CO_2-amine-H_2O system can, according to [17], be written as follows:

$$CO_{2\,(g)} \leftrightarrow CO_{2\,(l)} \quad (1)$$

$$2H_2O \leftrightarrow H_3O^+ + OH^- \quad (2)$$

$$CO_{2\,(l)} + 2H_2O \leftrightarrow HCO_3^- + H_3O^+ \quad (3)$$

$$HCO_3^- + H_2O \leftrightarrow CO_3^= + H_3O^+ \quad (4)$$

$$AmH^+ + H_2O \leftrightarrow Am + H_3O^+ \quad (5)$$

$$Am + CO_{2\,(l)} + H_2O \leftrightarrow AmCOO^- + H_3O^+ \quad (6)$$

$$^+HAmCOO^- + H_2O \leftrightarrow AmCOO^- + H_3O^+ \quad (7)$$

where g, l, and Am refer to gas, liquid, and amine molecule, respectively.

From (1) to (7), the chemical equilibrium constants as a function of the activity coefficients γ_i and mole fraction x_i can be formulated as follows

$$K_j = \prod_i a_i^{v_{ij}} = \prod_i (\gamma_i x_i)^{v_{ij}} \quad j = 1, 2, \ldots, R \quad (8)$$

where v_{ij} is the stoichiometric coefficient for component i in reaction j and R is the number of chemical reactions.

The equilibrium constant in (5) is the protonation constant (K_a) of amine in which its value can be experimentally measured.

The protonation constant, also known as the acid dissociation constant, can be calculated from the reaction of a weak acid and a strong base where K_a value is equal to the concentration of H^+ ions in the solution. This occurs when the total volume of titrant added is equal to half of the equivalence volume (V_e) required.

Let a weak acid HA react with a strong base BOH in solution

$$HA + BOH \leftrightarrow BA + H_2O \quad (9)$$

in which the remaining weak acid HA and the salt formed BA dissociate in the solution. Since HA is a weak acid with small K_a value, the amount of HA in the solution is, therefore, assumed to be constant. The number of A- ions in the solution are mostly from BA dissociation, i.e. A- ions from HA dissociation is negligible. In this condition, the weak acid HA concentration in the solution is equal to the concentration of A- ions, so that the protonation constant (K_a) can be expressed as follows

$$K_a = \frac{[H^+][A^-]_{total}}{[HA]_{left}} \approx [H^+] \quad (10)$$

Equation (10) indicates that the K_a value of the weak acid HA approximately is equal to the concentration of H^+ ions or, in other words, the pK_a equals the pH of the solution.

The temperature dependency of the protonation constant (K_a) can, according to [18], be expressed in the form

$$\ln K_a = A + \frac{B}{T} + C \ln T + DT + ET^2 \quad (11)$$

where T is the temperature in K, while A, B, C, D and E are the coefficients that can be determined by use of the regression method which gives the Sum of the Squared Errors (SSE) minimum. The minimization of errors will produce linear systems expressed mathematically as

$$\sum \ln K_a = nA + B\sum\frac{1}{T} + C\sum \ln T + D\sum T + E\sum T^2 \quad (12)$$

$$\sum\frac{\ln K_a}{T} = A\sum\frac{1}{T} + B\sum\frac{1}{T^2} + C\sum\frac{\ln T}{T} + D + E\sum T \quad (13)$$

$$\sum(\ln T \ln K_a) = A\sum \ln T + B\frac{\ln T}{T} + C\sum(\ln T)^2 + D\sum(T \ln T) + E\sum(T^2 \ln T) \quad (14)$$

$$\sum(T \ln K_a) = A\sum T + B + C\sum(T \ln T) + D\sum T^2 + E\sum T^3 \quad (15)$$

$$\sum(T^2 \ln K_a) = A\sum T^2 + B\sum T + C\sum(T^2 \ln T) + D\sum T^3 + E\sum T^4 \quad (16)$$

where n is the number of data. From (12) to (16), it is obtained a set of linear equations. Some methods can be applied to solve the linear equation systems, e.g. the Gauss elimination method [19]. This solution method will then be used in this work.

3 Materials and method

3.1 Materials

The materials used in this work consist of monoethanolamine (MEA) obtained from Sigma-Aldrich with purity of min. 98%, sulfuric acid (H_2SO_4) from Merck with purity of 96.0%, and distilled water. Moreover, MEA received from the manufacturer was used directly without further purification.

3.2 Apparatus and procedure

Figure 2 shows the experimental setup consisting of a 300-mL jacketed reactor, a pH meter (Lutron PH-201 with pH electrode PE-03 and accuracy up to 0.01), a TFA® digital thermometer with accuracy up to 0.1°C, a memmert water bath, an OHAUS® digital balance with readability up to 0.0001 g, a magnetic stirrer, and a 5-mL syringe.

A certain amount of MEA min. 98%, ~0.5 g, was added into a beaker glass containing ~50 mL of distilled water. The solution was then stirred some time to make it homogeneous. A certain amount of titrant, i.e. 0.1 M H_2SO_4 solution, in a syringe was added into the beaker glass. The reaction taken place was exothermic, so that the solution temperature was slightly increased. It took some time until the temperature and pH of the solution stable. The addition of 0.1 M H_2SO_4 from the syringe was repeated until the equivalence point is reached. When the titration was about the equivalence point, the titrant must be added with a very small amount, i.e. ~0.02 mL, to ensure that the pH did not change much, so that the equivalence point was not exceeded.

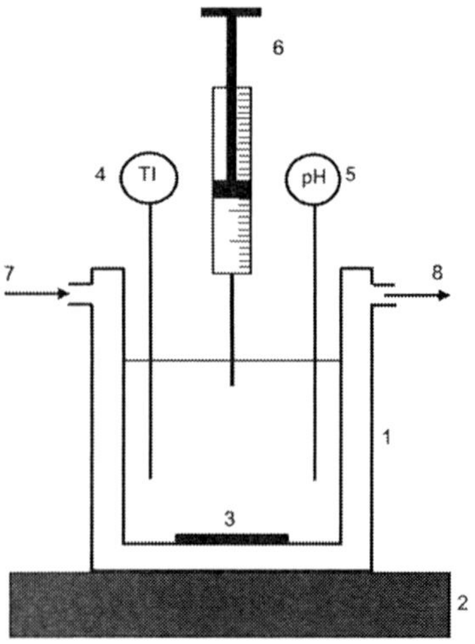

Fig 2. Experimental setup for pK_a measurements (1: jacketed reactor, 2: magnetic stirrer, 3: stirrer, 4: thermometer, 5: pH meter, 6: syringe, 7: heating medium inlet, 8: heating medium outlet)

4 Results and discussion

The pK_a measurements of MEA were conducted at a range of temperatures from 303 to 333K using the potentiometric titration method with 0.1 M H_2SO_4 solution as the titrant.

As seen in Fig. 2, the titrant was discretely added. The amount of titrant added was calculated from the syringe weight difference. The titration was terminated after the solution reached pH ~2.0, i.e. excess of H_2SO_4. This is to ensure that all MEA molecules have reacted with H_2SO_4.

From the titration curve obtained, the pK_a value can then be determined graphically. Figure 3 shows the titration curve for Run 1 in which the equivalence point was reached when the titrant volume (V_e) added was ~45 mL. It was quite challenging to determine the precise equivalence point directly from the curve. Another step is therefore required, i.e. determination of pH slope (ΔpH/ΔV) for every single point, as seen in Fig. 4. The equivalence point is reached when the slope (ΔpH/ΔV) gives minimum value. However, the precise equivalence point cannot still be read from the figure; therefore enlarging the figure scale around the equivalence point is required. It can be seen from Fig. 5 that the equivalence volume (V_e) for Run 1 is 45.007 mL.

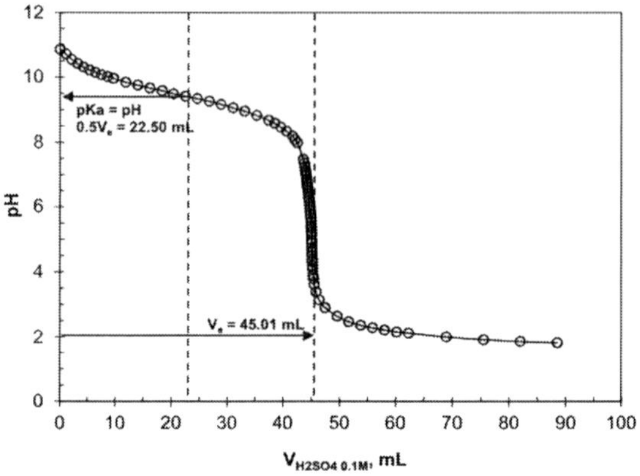

Fig 3. Titration curve of MEA at 303K (V_e = the volume of titrant at the equivalence point)

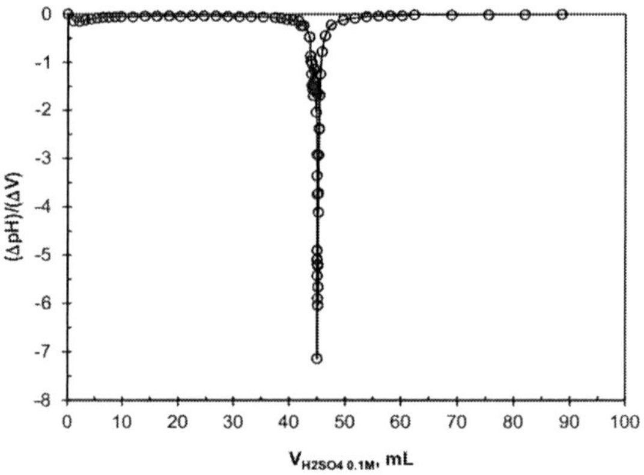

Fig 4. Titration slope (ΔpH/ΔV) of MEA at 303K

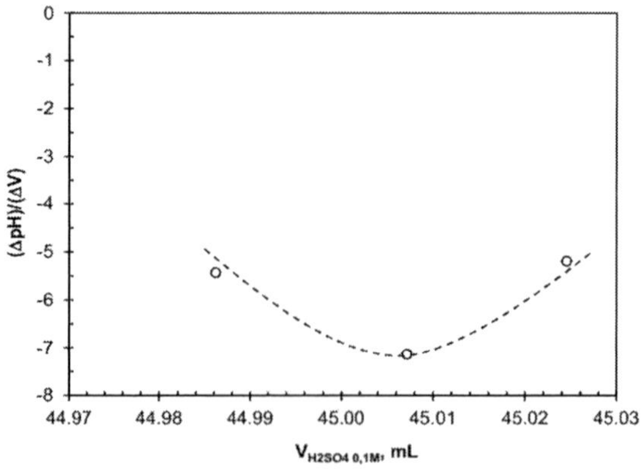

Fig 5. Titration slope (ΔpH/ΔV) around the equivalence point at 303K

Furthermore, the pK_a value occurs when the titrant volume equals half of the equivalence volume. At this point, pK_a is equal to pH. It can be seen from Fig. 3 that pK_a is 9.41. Similar procedure was applied to the other data and the results are then summarized in Table 1.

Table 1. pKa values at various temperature

Run	T, K	Number of data	pK_a	$pK_{a\text{-average}}$
1	303	79	9.41	9.43
2	303	85	9.44	
3	313	76	9.18	9.14
4	313	86	9.10	
5	323	89	8.88	8.90
6	323	82	8.91	
7	333	87	8.68	8.67
8	333	85	8.66	

Table 2. Comparison of experimental data with those obtained from literature

T, K	$-\ln K_a$						
	[20]	[21]	[22]	[23]	[24]	[25]	This work
273	23.73						
278	23.33						
283	22.95						
288	22.57	22.57				22.25	
293	22.21			22.08			
298	21.87	21.88	21.90	21.74	21.85	21.72	
303	21.53			21.42			21.71
308	21.20	21.20		21.10		21.20	
313	20.88			20.80	20.84		21.05
318	20.59	20.56		20.50		20.64	
323	20.29			20.21	20.25		21.49
333					19.71		19.96
343					19.08		
353					18.55		
363					18.18		

The pK_a values of MEA obtained from this work are then compared to those obtained from the literature as given di Table 2 and Fig. 6. It can be seen from Fig. 6 that the results, in general, agree well with the literature data, except at 333K, where the ln K_a from this study is a bit lower compared to that from [24]. Even though the vapor pressure of MEA at 333K is relatively low (P^o = 0.66 kPa [26]), the evaporation loss during titration might slightly reduce the amount of MEA in the solution, thereby slightly affecting the measurement result.

There are 38 data points of K_a obtained from this work together with the data from the literature as seen in Table 2. By use of the linear regression method, the protonation constant K_a can then be correlated as a function of temperature in an empirical equation form as follows:

$$\ln K_a = -164.40 + \frac{4.36\times10^{-4}}{T} + 25.50\ln T + 1.37\times10^{-8}T + 3.11\times10^{-5}T^2 \qquad (17)$$

Equation (17) could be useful to determine the protonation constant with its temperature dependency for future work.

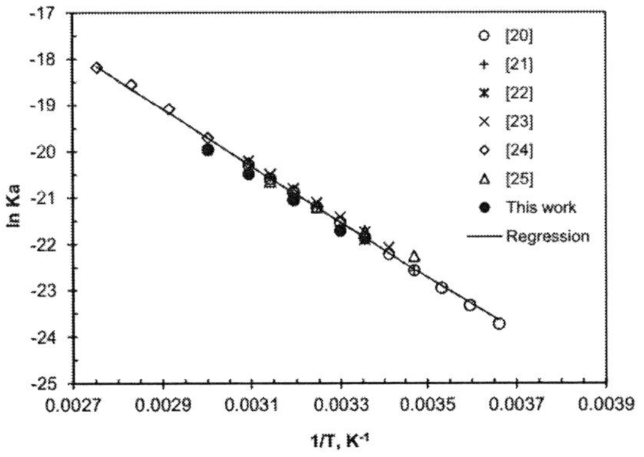

Fig 6. Comparison of the measured K_a of MEA with those obtained from literature [20-25]

5 Conclusion

The protonation constant (pK_a) measurements of MEA were conducted at a range of temperatures from 303 to 330K by a potentiometric titration method to complement the existing data in the literature. The experimental pK_a data obtained in this work, in general, agree well with the results published in the literature. The pK_a data from this work together with those from the literature were then correlated in an empirical form to easily be used for future research.

The authors acknowledge the Islamic University of Indonesia for the financial support through the Department of Chemical Engineering Research Grant 2016.

References

1. A.B. Rao, E.S. Rubin. A, Environ. Sci. Technol. **36,** 4467 (2002)
2. T. Suda, M. Iijima, H. Tanaka, S. Mitsuoka, T. Iwaki, Environ. Prog. **16,** 200 (1997)
3. S. Ma'mun, H.F. Svendsen, K.A. Hoff, O. Juliussen, Energy Convers. Manage. **48,** 251 (2007)
4. Van der Hoeven, *World Energy Outlook 2012* (International Energy Agency, Paris, France, 2012)
5. K.A. Hoff, E.F. da Silva, I. Kim, A. Grimstvedt, S. Ma'mun, Energy Procedia **37,** 292 (2013)
6. S. Ma'mun, R. Nilsen, H.F. Svendsen, J. Chem. Eng. Data **50,** 630 (2005)
7. S. Ma'mun, V.Y. Dindore, H.F. Svendsen, Ind. Eng. Chem. Res. **46,** 385 (2007)

8. R. Idem, M. Wilson, P. Tontiwachwuthikul, A. Chakma, A. Veawab, A. Aroonwilas, D. Gelowitz, Ind. Eng. Chem. Res. **45,** 2414 (2006)
9. Y. Lu, X. Ye, Z. Zhang, A. Khodayari, T. Djukadi, Energy Procedia **4,** 1286 (2011)
10. Alstom, *Chilled ammonia-based wet scrubbing for post-combustion CO_2 capture* (DOE/NETL report No. 401/021507, 2007)
11. H. Knuutila, U.E. Aronu, H.M. Kvamsdal, A. Chikukwa, Energy Procedia **4,** 1550 (2011)
12. S. Ma'mun, I. Kim, Energy Procedia **37,** 331 (2013)
13. S. Ma'mun, Energy Procedia **51,** 191 (2014)
14. F.-Y. Jou, A.E. Mather, F.D. Otto, Can. J. Chem. Eng. **73,** 140 (1995)
15. Y. Liu, L. Zhang, S. Watanasiri, Ind. Eng. Chem. Res. **38,** 2080 (1999)
16. R.W. Bucklin, Oil Gas J. 204 (1982)
17. S. Ma'mun, J.P. Jakobsen, H.F. Svendsen, O. Juliussen, Ind. Eng. Chem. Res. **45,** 2505 (2006)
18. R.H. Weiland, T. Chakravarty, A.E. Mather, Ind. Eng. Chem. Res. **32,** 1419 (1993)
19. E. Kreyszig, *Advanced engineering mathematics, 10th ed.* (John Wiley & Sons, Inc., USA, 2011)
20. R.G. Bates, G.D. Pinching, J. Res. Natl. Bur. Stand. **46,** 349 (1951)
21. S.P. Datta, A.K. Grzybowski, J. Chem. Soc. 3068 (1962)
22. J.-H. Kim, C. Dobrogowska, L.G. Hepler, Can. J. Chem. **65,** 1726 (1987)
23. E.S. Hamborg, G.F. Versteeg, J. Chem. Eng. Data **54,** 1318 (2009)
24. Kim, C.M. Jens, A. Grimstvedt, H.F. Svendsen, J. Chem. Thermodynamics **43,** 1754 (2011)
25. D. Fernandes, W. Conway, X. Wang, R. Burns, G. Lawrance, M. Maeder, G. Puxty, J. Chem. Thermodynamics **51,** 97 (2012)
26. R.K. Sinnot, *Coulson & Richardson's Chemical Engineering Volume 6 - Chemical Engineering Design, 4th ed.* (Elsevier Butterworth-Heinemann, Oxford, UK, 2005)

Lactic acid production from date juice using *lactobacillus casei ATCC 393* in batch fermentation

Mujtahid Kaavessina[1,*], *Fitriani* Khanifatun[2], and *Saeed M.* Alzahrani[3]

[1]Chemical Engineering Department, Sebelas Maret University, 57126 Surakarta, Indonesia
[2]Educational Staff of Biology, Department of Education, 63202 Ngawi, Indonesia
[3]Chemical Engineering Department, King Saud University, 11421 Riyadh, Saudi Arabia

Abstract. Lactobacillus casei ATCC 393 was employed as a fermentative organism to convert sugars from date juice into lactic acid. Both glucose and fructose in date juice were fermented directly without any pre-treatment. The influences of supplementation of yeast extract and date juice concentration on some fermentation parameters, such as: cell growth rate, sugar conversion, productivity and yield, were investigated using this bacterium in batch fermentation. The results showed that by adding yeast extract about 20g/l in a date juice medium, the maximum specific growth rate of bacteria (μ_m) enhanced from 0.1229 to 0.1819 g/l. Meanwhile, increasing date juice concentration from 86.6942 to 158.9181 and 229.5367 g/l enhanced the μ_m from 0.1819 to 0.2107 and 0.1916 g/l, respectively. It indicated that the optimum value for μ_m is 0.2107 g/l in this concentration range. In the date juice concentration of 158.9181 g/l, the optimum lactic acid can be produced is 117.8301 g/l with yield of 92.685% for 48 h.

1 Introduction

Lactic acid, one of the most important organic acids, and its derivatives has been utilized in many applications such as in the food, textile, pharmaceutical, cosmetic and chemical industries [1]. Even, it became a prime candidate to be developed as a biodegradable polymer. Polymerization of lactic acid obtained poly (lactic acid) which has comparable mechanical properties, transparency, and UV light barrier to many conventional polymer (polystyrene, polyethylene, etc.) [2].

Recently, the global poly (lactic acid) market was expanding rapidly followed by increasing of lactic acid demand. Several factors stimulated this growth such as: sustainability of raw materials and government policy for bio-based and biodegradable product to tackle the waste problem. The global market of lactic acid is predicted to reach 1076.9 thousand tonnes in 2016 [3]. However, the global production of lactic acid is only 120 thousand tonnes in 2006 [4], thus the minimal production growth of lactic acid is 25% per years until 2016 to balance the gap between production and demand.

Lactic acid can be produced through chemical synthesis and microbial fermentation. The fermentation is an effective and attractive method due to produce lactic acid in high purity of one stereoisomer. The high purity of L(+) or D(-) lactic acid can be produced depending on a microbial strain and source of carbon (substrate) [4,5]. The economics of lactic acid fermentation is affected by many factors: raw material, purification, etc. The cost of the raw materials spends approximately 60-80% of the total production cost [5]. Thus, it is important to explore some potential of agriculture product to get cheap and abundantly existing material. It can be summarized that there are three big groups of substrate: sugar, starchy material and lignocellulose.

As well known, sugar was reported as the preferred carbon sources. However, it is very expensive to use as the feedstock for lactic acid fermentation. Date is one of the promising biomass for lactic acid production without complicated pretreatment. Date contains between 70-80 wt% of fermentable sugars, mainly glucose and fructose in a balance ratio which can be consumed directly by lactic acid bacteria [6] Besides that, as reported by Al-Hooti et al. [7] and Al-Farsi et al. [8], date contains some minerals and low range of vitamins. In Arabic countries, a lot of dates are being wasted due to overproduction and poor handling low quality dates. Thus, production of lactic acid from dates is very attractive.

Lactobacillus casei, a genus of facultative anaerobic bacteria, is one of the bacteria that able to convert some sugars to lactic acid. During its growth, it consumes sugars as energy sources and converts to lactic acid. In this work, *Lactobacillus casei* ATCC 393 was employed to produce lactic acid from date juice. We investigated the effect of yeast extract as nitrogen source and initial sugar concentration on lactic acid production in batch fermentation.

* Corresponding author: mkaavessina@staff.uns.ac.id

2 Experimental part

2.1 Materials

Some chemicals such as $MgSO_4$, $MnSO_4$, K_2HPO_4, KH_2PO_4 and $FeSO_4$ were kindly supplied by LOBA Chemie (India). Tween 80 and yeast extract as a fatty acid and a nitrogen source of bacteria, respectively, were purchased from Sigma-Aldrich. Further, all chemicals were used in this investigation without any purification.

2.2 Microorganism

Lactobacillus casei ATCC 393 is a homofermentative bacterium to produce lactic acid. This strain was kindly purchased from ATCC and stored as a freeze-dried inoculum. To revive it, 0.5 ml of a liquid medium was added to the culture with a sterile Pasteur pipette and mixed well. The mixture was then transferred to a test tube containing 5 ml of broth medium and incubated for 24 h on a shaking water bath maintained at 38°C. Further, this mixture was as a parent culture and subsequently kept it in the refrigerator at 10°C.

For inoculation purpose, 2 ml of culture was transferred to 100 ml broth (in 500 ml erlenmeyer) and incubated for 48 h at 37°C. The composition of inoculum broth is: yeast extract (20 g/l), peptone (5 g/l), dextrose (10 g/l), monopotassium phosphate (2 g/l) and polysorbate 80/tween 80 (0.1 g/l). Prior to usage, the pH of broth was adjusted at 6.5 [9].

2.3 Preparation of date juice as a substrate

Date juice was purchased from a local market in Riyadh, Saudi Arabia. The main composition of date juice concentrate is mainly fructose (1000g/l) and glucose (1000g/l). The date juice was then diluted with tap water to desired concentration of glucose and centrifuged at 6000 rpm for 3 mins to separate the cellulosic debris and fibres from the supernatant.

2.4 Fermentation

The batch fermentation was carried out in 100 ml working volume using 250 ml·erlenmeyer without pH control. In this investigation, the effect of nutrient was studied by carried out fermentation in two different media: salted date juice with and without supplementation of yeast extract. Investigation of initial sugar concentration effect was conducted using salted date juice with supplement at various concentration as tabulated in Table 1.

Table 1 List of the date juice samples prepared for this study

Sample name	Concentration of sugar, g/l	Yeast extract, g/l	Salt
DJ_a	86.6942	0	Added
DJ_b	86.6942	20	Added
DJ_c	158.9181	20	Added
DJ_d	229.5367	20	Added

The salt was containing 0.2 g/l $MgSO_4$, 0.03 g/l $MnSO_4$, 0.3 g/l K_2HPO_4, 0.3 g/l KH_2PO_4, 0.02 g/l $FeSO_4$, 1 ml/l Tween 80. Yeast extracts as a nitrogen source was supplemented about 20 g/l. Then, all media were sterilized at 121°C for 15 mins. The fermentation was done by adding culture of *lactobacillus casei* ATCC 393 with inoculum size of 10% (v/v of solution) into each media. For certain time, each sample was analyzed their dry cell weight (DCW) and composition of substrate and product. Table 1 shows the list of date juice samples prepared for this study.

2.5 Characterization

The bio-cell concentration was calculated by measuring the weight of dried cells in 1.5 ml of tested sample at certain time and then converted to the weight of dried cells in 1 liter. The concentrations of glucose, fructose and lactic acid were determined using an Agilent 1260 infinity HPLC, USA. The mixture of acetonitrile in water (3:1) was used as mobile phase during the analysis.

Cell growth of bacteria can be expressed in an equation as follows:

$$\frac{dX}{dt} = \mu X \quad (1)$$

Where, μ is a specific growth rate of bacteria. The integration of Equation 2can be written:

$$\mu = \frac{\ln(X_t) - \ln(X_o)}{t - t_o} \quad (2)$$

By analyzing in growth period (log phase), maximum specific growth rate (μ_m) can be obtained as a slope of plotting between ln(X) versus time. The μ_m is a constant value. The yield parameters in fermentation are expressed by $Y_{P/S}$, $Y_{P/X}$ and $Y_{X/S}$, which can be determined by those equations [9]:

$$Y_{P/S} = \frac{\text{Mass of lactic acid formed}}{\text{Mass of substrate consumed}} = \frac{\Delta P}{\Delta S} \quad (3)$$

$$Y_{P/X} = \frac{\text{Mass of lactic acid formed}}{\text{Mass of new cells formed}} = \frac{\Delta P}{\Delta X} \quad (4)$$

$$Y_{X/S} = \frac{\text{Mass of new cells formed}}{\text{Mass of substrate consumed}} = \frac{\Delta P}{\Delta S} \quad (5)$$

3 Results and discussion

The date juice was utilized as a substrate with different supplement and concentration. The Lactobacillus casei ATCC 393 was employed as a fermentative organism for converting sugars to lactic acid. The bio-cells growth and lactic acid production was presented and discussed here.

3.1 Bio-cells growth

Figure 1 shows the effect of yeast extract and initial sugar concentration of date juice on cell growth in batch fermentation. The growth of Lactobacillus casei showed similar behaviour that can be divided into two regions. First region (0-12 hrs) refers to the growth period and the other refers to the stationary period (12-48 hrs). The lag period was not monitored in this investigation. Probably due to the adaption time was very fast (less than 4 hrs).

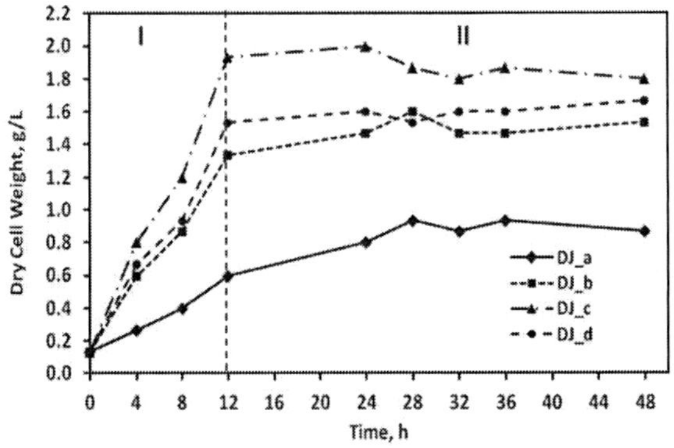

Fig. 1. Effect of yeast extract and initial sugar concentration of date juice on cell growth in batch fermentation.

The effect of yeast extract on bio-cells growth can be seen in comparison between samples of DJ_a and DJ_b (Fig. 1). The presence of yeast extract enhanced cell mass significant. For example at 24 hrs, the cell concentration is 0.87 g/l when grown in medium without supplementation of yeast extract while the cell growth becomes 1.47 g/l when the medium was added with yeast extract. This result was also in agreement with others [11, 12]. The slow cell growth on the date juice may be due to the deficiency of nitrogen in the medium. Some researchers investigated some nitrogen sources (yeast extract, peptone, urea, ammonium sulphate and corn step liquor) and concluded that yeast extract is the best as a nitrogen source than others [11, 13]. Beside as a nitrogen source, yeast extract has also rich in B vitamins.

The effect of sugar concentration in the date juice medium is monitored in samples of DJ_b, DJ_c and DJ_d. It is founf that the cell growth was increasing significantly. For example at 24 hrs, the cell concentration was enhancing from 1.47 g/l to 2.00 g/l when the sugar increased from 86.6942 g/l to 158.9181 g/l. However, the negative effect is monitored when the sugar concentration is at 229.5367 g/l at which the cell concentration tends to decrease to 1.6 g/l. It is possibly due to the high concentration of sugars will restrict the diffusion of sugar and nutrition into the cell. Thus, the cell growth was inhibited. It can be highlighted that the optimum initial sugar concentration is 158.9181 g/l and containing glucose and fructose in a balance ratio to grow the bio-cells.

3.2 Lactic acid production

Figure 2 shows that the lactic acid production is affected by adding yeast extract in the date juice medium. Due to the fermentation entangles the life cycle of *Lactobacillus casei*, the lactic acid production is in accordance with its growth. Figure 2(a) and 2(b) show that the supplemented date juice is more consumable with *L. casei* than that of date juice without yeast extract. Both glucose and fructose were simultaneously consumed. Fig. 2(a) shows that lactic acid production increase slightly, probably due to the bacteria growth was inhibited in date juice with lack of nitrogen. The lactic acid concentration and productivity after 48 hrs were only 17.2578 g/l and 0.3595 g/(l h), respectively. The addition of yeast extract can tackle the deficiency of nitrogen in a medium. Fig. 2(b) shows the lactic acid production can be enhanced significantly become 67.9121 g/l (48 hrs).

The influence of diluted and concentrated date juice on lactic acid fermentation was investigated. The various initial concentrations of sugar are 86.6942, 158.9181 and 229.5367 g/l. As shown in Fig. 3, the final lactic acid concentration improved and also the concentration of sugars was more consumable with an increase initial of sugar. Further, the quantification of date juice fermentation by *Lactobacillus casei* ATCC 393 is summarized in Table 2.

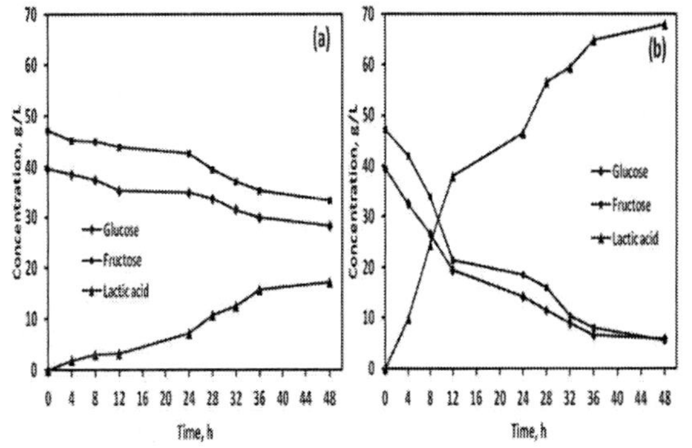

Fig. 2. Effect of yeast extract on lactic acid production in batch fermentation: (a) without yeast extract and (b) with yeast extract (20g/l)

The maximum specific growth rate (μ_m) and maximum quantity of *L. casei* is in DJ_c medium: 0.2107 1/h and 2 g/l, respectively. It could be explained that in this medium, the cells could be growth without any interferences. Nitrogen from yeast extract and some nutrient suited to their growth. Compared with DJ_b and DJ_d medium, the μ_m and maximum quantity of cells were lower than that of DJ_c, although the yeast extract and some nutrient are added in same quantity. It exhibited that the DJ_b medium is not matching with the carbon demand of cells due to the lower initial concentration of sugars. For DJ_d, the higher concentration sugars increased the viscosity of substrate, thus the diffusion of sugars into the cell was inhibited.

DOI: 10.1051/matecconf/201710102002

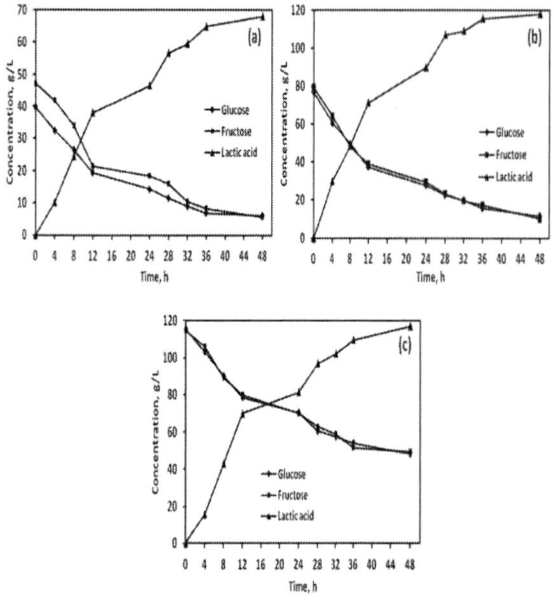

Fig. 3. Effect of initial concentrations of sugar on lactic acid production in batch fermentation: (a) 86.6942, (b) 158.9181 and (c) 229.5367 g/l

The highest total lactic acid production was obtained in the DJ_c medium, 117.8301 g/l for 48 h. Both glucose and fructose in the date juice medium were utilized about 85.0137 and 87.3404 %, respectively. Compared to the DJ_d medium, the total lactic acid produced is 116.8910 g/l and total sugars utilized were about 57 %.

Table 2 The fermentation parameters of lactic acid production from date juice using *Lactobacillus casei* ATCC 393

Parameter	Date juice			
	(DJ_a)	(DJ_b)	(DJ_c)	(DJ_d)
Maximum cells, g/l	0.930	1.600	2.000	1.670
Maximum specific growth rate (μ_m), 1/h	0.123	0.182	0.211	0.192
Initial total sugars [a], g/l	86.694	86.694	158.918	229.537
Residual total sugars (48 hrs), g/l	61.579	11.462	21.523	97.910
Total lactic acid (48 hrs), g/l	17.258	67.912	117.830	116.891
Percentage of utilized glucose (48 hrs), %	28.619	85.023	85.014	57.981
Percentage of utilized fructose (48 hrs), %	29.265	88.251	87.340	56.705
Yield , $Y_{P/S}$ [b]	0.724	0.906	0.927	0.903
pH range	6.5-6	6.5-5.5	6.5-5.3	6.5-5.6
Productivity (48 hrs), g/(l h)	0.359	1.415	2.455	2.435

[a]) Total sugars refer to the summation of glucose and fructose
[b]) $Y_{P/S}$ is the ratio between lactic acid formed and substrate consumed, g/g

Considering the maximum specific growth rate, the bio-cells in DJ_d was lower than that in DJ_c. And also as reported by Yu et al. [14] the presence of lactic acid and concentrated sugars in a medium was also inhibiting the cell growth. It could be predicted that it will take a long time to increasing the conversion of DJ_d till about 80%.

Again, this result exhibited that the optimum concentration of date juice is 158.9181 g/l.

The yield of lactic acid was also increased from 72.38% to 90.55% when yeast extract was supplemented in a medium. It could be attributed to be more appropriate of nutritive value was found in supplemented date juice with yeast extract [15] Increasing the concentration of sugars was not significant affected in the yield. It was varying from 90.50% to 92.68 and 90.30% when initial concentrations of sugar were 86.6942, 158.9181 and 229.5367 g/l, respectively.

Table 3 Comparison of lactic acid fermentation in batch from various carbon sources and microorganisms.

C and N sources (g l^{-1})	Titer (g l^{-1})	Yield (%)	Productivity (gl^{-1}h^{-1})	Strain	Ref.
Date juice w/o YE[a]	86.7	0.72	0.36 (48 h)	*Lactobacillus casei*	This study
Date juice (YE:20g /l)	86.7	0.91	1.42 (48 h)	*Lactobacillus casei*	This study
Date juice (YE:20g /l)	158.9	0.93	2.45 (48 h)	*Lactobacillus casei*	This study
Date juice (YE:20g /l)	229.5	0.9	2.43 (48 h)	*Lactobacillus casei*	This study
Glucose w/o YE	60	0.78	0.67 (48 h)	*Lactococcus lactis*	[19]
Glucose (YE:20g /l)	100	-	2.1 (48 h)	*Lactobacillus casei*	[1]
Wheat hydrolisate (CSL[b]:25g/l)	200	0.92	4.14 (24 h)	*Enterococcus faecalis*	[16]
Glucose w/o YE	112	0.94	5	*Rhyzopus orizae*	[17]
Sugarcane w/o YE	133	0.95	1.66 (72 h)	*Lactobacillus delbrueckii*	[15]
Date juice ($(NH_4)_2SO_4$: 10g/l)	80	-	0.38 (20 h)	*Lactobacillus rhamnosus*	[13]

[a] YE: Yeast Extract
[b] CSL: Corn Steep Liquor

Table 3 summarized the results of lactic acid fermentation conducted in this work and the other reports. As a result of these kinds of substrate, sugars from date juice supplemented by yeast extract can be utilized with higher productivity. Beside yeast extract, the deficiency of nitrogen in medium can be fulfilled with corn steep liquor, ammonium sulphate, urea, etc. [13, 16]. *Rhyzopus orizae*, one species of fungi, showed its capability to produce lactic acid [17, 18]. From table 3, the capability of microorganisms is not same and they are not always having similar optimum conditions. These conditions are pH, nutrients, temperature, oxygen demand, etc. Therefore, study of the optimum condition during the lactic acid fermentation with different microorganisms is still a major challenge.

4 Conclusions

This investigation result showed that the date juice has a potential to be utilized as a substrate in lactic acid fermentation using *Lactobacillus casei* ATCC 393. Yeast extract is supplemented in a date juice medium to supply the nitrogen. The optimum production of lactic acid is about 117.8301 g/l for 48 hrs with initial sugar concentration of 158.9181 g/l. Moreover, the productivity and yield of lactic acid from date juice were not only high but also cost effective due to the date juice can be utilized directly.

This research was financially supported by Sebelas Maret University through research grant of "Hibah Research Mandatory (RM-UNS) PNBP 2016".

References

1. M. Hujanen, Y.Y. Linko. Appl. Microbiol Biotechnol., **45**, 307-313 (1996).
2. R. Auras, B. Harte, S. Selke, Macromol. Biosci. **4**, 835-864 (2004).
3. Anonym, 2011, Global lactic acid & poly lactic acid (PLA) market by raw materials, types, applications and potential opportunities (forecast to 2016), in http://www.marketsandmarkets.com/ Market-Reports/polylacticacid-387.html.
4. Datta, M. Henry, J. Chem. Technol. Biotechnol, **81**, 1119-1129 (2006).
5. A.K. Yadav, N.K. Bipinraj, A.B. Chaudhari, R.M. Kothari, Starch – Stärke, **63**, 632-636 (2011)
6. N. Nancib, A. Nancib, J. Boudrant (1997)., Bioresour. Technol., **60**, 67-71 (1997).
7. S.N. Al-Hooti, J.S. Sidhu, J.M. Al-Saqer, A. Al-Othman, Food Chem., **79**, 215-220 (2002).
8. M. Al-Farsi, C. Alasalvar, M. Al-Abid, K. Al-Shoaily, M. Al-Amry, F. Al-Rawahy, Food Chem., **104**, 943-947 (2007).
9. R.M. Atlas, *Handbook of Microbiological Media, 4 ed.*, ASM Press, CRC Press, Taylor and Francis Group, 2010.
10. H.S. Fogler, Elements of chemical reaction engineering, 4 ed., Prentice hall international series, Massachusetts, 2008.
11. M. Altaf, B. Naveena, G. Reddy, Food Technol. Biotech., **43**, 235-239 (2005).
12. D. Roy, J. Goulet, A. LeDuy, Appl. Microbiol. Biotechnol., **24**, 206-213 (1986).
13. N. Nancib, A. Nancib, A. Boudjelal, C. Benslimane, F. Blanchard, J. Boudrant, Bioresour. Technol., **78**, 149-153 (2001).
14. L. Yu, X. Pei, T. Lei, Y. Wang, Y. Feng, J. Biotechnol. **134**, 154-159 (2008).
15. B. Calabia, Y. Tokiwa, Biotechnol. Lett., **29**, 1329-1332 (2007).
16. H. Oh, Y.-J. Wee, J.-S. Yun, S. Ho Han, S. Jung, H.-W. Ryu. Bioresour. Technol. **96**, 1492-1498 (2005).
17. A. Tay, S.-T. Yang, Biotechnol. Bioeng, **80**, 1-12. (2002)
18. Z.Y. Zhang, B. Jin, J.M. Kelly, Biochem. Eng. J., **35**, 251-263, (2007)
19. L.S. Cock, A.R. Stouvenel, *Electron. J. Biotechnol.*, **9**, 40-45. (2006)

DOI: 10.1051/matecconf/201710102003

Laboratory and pilot plant scale study on water dechlorination by medium pressure ultraviolet (UV) radiation

Maryani Paramita Astuti[1], *Rongjing* Xie[2], and *Nicky Satyadharma* Aziz[3]

[1]Environmental Engineering Study Program Faculty of Engineering, President University, 17550 Cikarang Indonesia
[2]Water Supply (Plants) Department, PUB, 228231 Scotts Road, Singapore
[3]Infrastructure Civil Department, Surbana Jurong Consultants Pte. Ltd, 150168 Bukit Merah, Singapore

Abstract. Ultra violet (UV) dechlorination was performed to eliminate residual chlorine as the byproducts of chlorination process. UV dechlorination utilizes photon energy generated by medium pressure (MP) UV lamp to produce powerful hydroxyls which in return break down chemical bond of the residual chlorine. This study was undertaken to investigate the removal of residual chlorine under a medium pressure UV radiation and evaluate the influence of UV fluence on chlorine removal efficiency based on both laboratory and plant scale experiments. In laboratory experiments, water samples were exposed to a UV collimated beam apparatus equipped with a medium pressure (MP) UV lamp over a specified exposure time. Chlorine concentrations were measured before and after UV exposure to calculate its chlorine removal efficiency. Results showed that chlorine residual decreased over time and the removal efficiency increased as the UV fluence increased. The maximum UV fluence applied in the laboratory experiments (513 mJ/cm2) resulted in less than 25% of total chlorine reduction. The field experiments were conducted at a waterworks in Singapore with total capacity of 22 l/s using an existing UV system. The plant scale studies supported laboratory findings with about 9% of chlorine removal efficiency. The chlorine decay rates (fluence-based first order constant) were also calculated.

1 Introduction

Chlorination remains as the most widely used method for disinfection of water and wastewater in the United State [1] and probably worldwide. Some advantages of this process include: (1) relatively low cost, (2) high efficiency, and (3) ease of use [2]. In this process, chlorine is commonly introduced in the forms of chlorine gas (Cl_2), sodium hypochlorite (NaOCl), or calcium hypochlorite ($Ca(OCl)_2$). When each of these chemical is added to water, rapid hydrolysis occurs to form hypochlorous acid, HOCl, which is further hydrolyzed to yield hypochlorite ion, OCl^-, as shown in the reaction (1) through (4). Reaction (4) is strongly dependent on pH of the solution with equilibrium constant pKa = 7.5 (at 25°C) [3]. Both hypochlorous acid and hypochlorite ions are commonly referred to as "free chlorine residual" [2].

$$Cl_2 + H_2O \rightarrow HOCl + H^+ + Cl^- \quad (1)$$

$$NaOCl + H_2O \rightarrow Na^+ + HOCl + OH^- \quad (2)$$

$$Ca(OCl)_2 + 2H_2O \rightarrow Ca^+ + 2HOCl + 2OH^- \quad (3)$$

$$HOCl \leftrightarrow H+ + OCl^- \quad (4)$$

In the presence of ammonia, hypochlorous acid and hypochlorite ion react with the ammonia to form monochloramine (NH_2Cl), dichloramine ($NHCl_2$), and trichloramine (NCl_3), which referred to as "combined chlorine residual". Chlorination that converts all ammonia to either trichloramine or nitrogen gas is known as "break point chlorination". After the break point, all ammonia is converted and thus the addition of chlorine resulted in formation of free chlorine residual.

Despite the extensive use of chlorination, chlorine residual is of concern to drinking water and wastewater treatment. Chlorine generates unpleasant odor and affects taste of drinks and liquids [4]. Recent studies suggest that residual chlorines may react further with dissolved organic matter by oxidation, addition and substitution reactions to form a range of DBPs (disinfection by products) such as the trihalomethanes (THMs), haloacetic acids (HAAs), and chlorite [5] that may pose unintended health hazards. Whilst for industrial water, chlorine residual could damage delicate process equipment such as reverse osmosis (RO) and deionization (DI) resin units and could accelerate corrosion of vessels, valves and piping.

In the case of wastewater treatment, toxicity of chlorine to aquatic life is another concern which had been studied extensively during the late 1970s and early 1980s. Further research on chlorine toxicity on early life stages of freshwater mussels showed that the impact of long-term exposure to low doses of chlorine may result in sub-lethal impairment [6]. Review of literature [7] concludes that DBPs formed in chlorinated wastewaters can be toxic and may have a deleterious impact on aquatic organisms. Therefore, water dechlorination process is needed to remove residual chlorine to meet specified standard of the treated water.

Ultraviolet (UV) dechlorination utilizes photon energy generated by low pressure (LP) and medium pressure (MP) UV lamps that contain an inert gas, e.g., argon, and a small amount of liquid mercury [8]. When a potential energy is applied to the lamp, some of the liquid mercury vaporizes and collides with free electrons and ions, and thus increases the energy state of mercury vapor which tends to return to its normal energy state by discharging energy in the form of UV light [9]. This discharged energy is then used to break down chemical bond of the residual chlorine. The LP lamps emit essentially monochromatic light at 254 nm wavelength, whilst the MP lamps radiate polychromatic lights ranging from about 205 nm to above 500 nm wavelength [8].

In the photodecomposition reactions of free chlorine in water, various primary reactive intermediates are formed which are rapidly converted into hydroxyl radical (•OH) and chlorine radical (•Cl). These reactions are presented in reaction (5) through reaction (8). The rate of free chlorine destruction under UV irradiation is best estimated by a first-order kinetics [10].

$$HOCl + uv\ photons \rightarrow \cdot OH + \cdot Cl \quad (5)$$

$$OCl^- + uv\ photons \rightarrow \cdot OH + \cdot \mathrm{Cl} \quad (6)$$

$$\cdot O^- + H_2O \rightarrow \cdot OH + \mathrm{OH}^- \quad (7)$$

UV dechlorination is believed to outperform other dechlorination technologies since there is no mixing or contact tanks required, no chemical addition required, and only minimum operation and maintenance involved. It leaves no impact on the taste, odor, color, and pH of water as well. However, chlorine removal efficiency needs to be considered prior to applying this method. Therefore, this study was undertaken to investigate the effectiveness of a medium pressure UV technology in water dechlorination and evaluate the influence of UV fluence dose on chlorine removal efficiency based on laboratory experiments and pilot plant scale experiments. In addition, based on the experimental results obtained in this study, fluence-based chlorine decay rates under UV exposure were determined.

2 Experimental details

Water samples containing an equal total chlorine concentration were exposed to polychromatic UV irradiation produced by medium pressure (MP) UV lamps. MP lamps were operated at temperatures range of 600-900°C and at mercury vapour pressures of 2 – 200 psi [9]. At the beginning and end of the UV exposure, the total and free chlorine concentration in each sample was measured in order to calculate the amount of chlorine removed due to UV exposure.

The experiments were carried out in laboratory scale and in pilot plant scale. Water samples used in the experiments were the drinking water produced by a waterworks in Singapore. All experiments were conducted at ambient temperature (23 ±2°C).

The initial total chlorine concentration was designed at 2 mg/L. Since chlorine is a highly oxidative compound, an equal chlorine concentration was difficult to achieve. Therefore, the concentrations were set of to be within the range of 1.90 mg/L to 2.2 mg/L.

2.1 Laboratory scale experiments

The laboratory experiments were conducted at CAWT (Centre for Advanced Water Technology) Laboratory in Singapore using a collimated beam UV apparatus (Atlantium Technologies, Ltd) equipped with an MP UV lamp. Each sample was placed in a 6-cm petri dish and was exposed to UV light over specified exposure time. A controlled experiment was designed, i.e., a sample which was placed in the dark on a bench far-off from the UV apparatus over the duration of the experiment.

Schematic of the laboratory experiments set up is shown in Figure 1. The surface of the samples to be exposed was about 2 cm from the UV apparatus. Appropriate amounts from a sodium hypochlorite dosing solution made from a sodium hypochlorite solution (NaOCl 12%) were added to the water samples to obtain the desired initial chlorine concentration. The chlorine concentrations were measured before and after UV exposure by a DPD colorimeter (HACH DR/890).

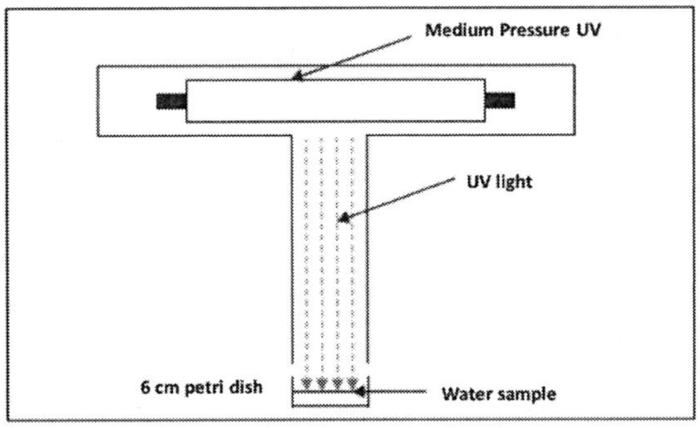

Fig. 1. Schematic of laboratory experiments

2.2 Pilot plant scale experiments

The pilot plant scale (with a capacity of 22 L/s) experiments were performed at a waterworks in Singapore. The raw water of this waterworks was extracted from reservoirs and then treated which would then be treated via coagulation, sedimentation and sand filtration, followed by chlorination as disinfection unit.

The UV system used in the experiments was Atlantium Hydro-Optic R-200. The sodium hypochlorite dosing solution was made from a commercial solution of

DOI: 10.1051/matecconf/201710102003

sodium hypochlorite (NaOCl 12%). This dosing solution was injected at the inlet of the UV system. The chlorine concentrations were then measured at the inlet after the inline static mixer and at the outlet of the UV system by a DPD colorimeter (HACH DR/890).

2.3 Chlorine residual analysis

In the experiments, concentration of total and free chlorine were measured by N,N-diethyl-p-phenylenediamine (DPD) colorimetric method using Hach pocket chlorine colorimeters (DR/890) with two types of reagents—DPD Free Chlorine Reagent and DPD Total Chlorine Reagent—for 10-ml of sample reagent.

The DPD amine (reagent) was added to the water sample and was oxidized by chlorine contained in the sample to two oxidation products, Würster dye and imine. Würster dye was relatively stable and would form a magenta color at neutral pH, whilst imine was relatively unstable and colorless that would be formed at higher oxidant levels, i.e., higher chlorine concentration. The intensity of the magenta color was then measured photo-metrically which represented concentration of chlorine in the samples. The DPD-chlorine reactions are illustrated in Figure 2.

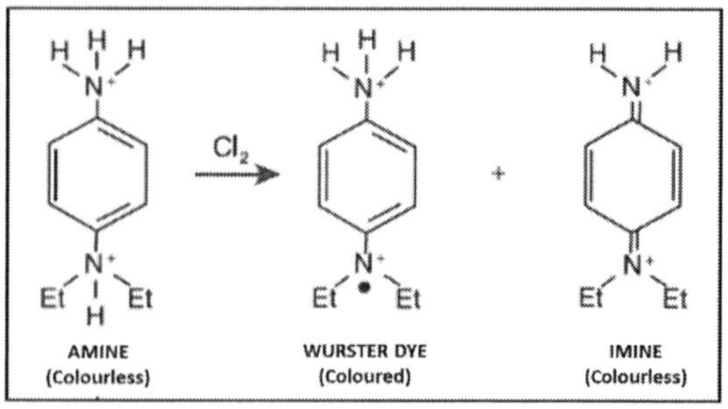

Fig. 2. DPD-chlorine reaction [12]

2.4 UV fluence variations

UV fluence refers to multiplication of UV intensity by the exposure time. The UV intensity of the UV collimated beam used in the laboratory experiments was set at a fixed value, UV fluence variations were achieved by varying exposure times of the samples to UV beams. The time variations were 10, 20, and 30 minutes. The fluence was measured by a radiometer Model SEL240 UV Curing Radiometer, Internal Light Inc. This device consists of a highly sensitive electronic ammeter connected to a UV sensor head, which generates a current proportional to the incident fluence rate. The device was used to measure UV intensity of a collimated beam apparatus by setting its radiometer detector head at the same height as the surface of samples in petri dish. Results of the UV intensity of the UV beam are presented in Figure 3.

In the experiments, 2 cm distance of the sample surface and the UV apparatus was selected, which per Figure 3 was equal to the UV intensity of 0.285 mW/cm^2. The UV fluence was calculated by multiplying UV intensity by the exposure time, which were 171, 342, and 513 mJ/cm^2.

In the pilot plant scale experiments, UV fluence were 80, 130, and 180 mJ/cm^2. For each of the trials, the UV system—Atlantium Hydro-Optic R200—was adjusted automatically to the specified UV fluence by adjusting the UV intensity value.

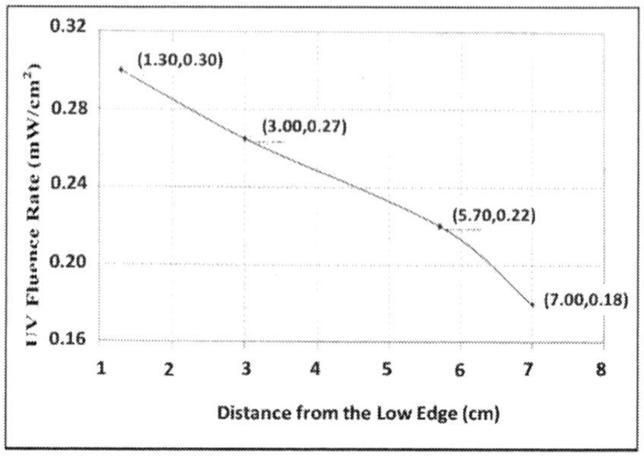

Fig. 3. UV intensity of collimated beam apparatus at varied exposure distance

2.5 Chlorine decay rates (fluence-based first order rate constant) analysis

Results from the experiments were used to calculate chlorine decay rates assuming a first order kinetics. Reference [10] states the rate of chlorine destruction under UV exposure was best estimated by a first-order kinetics which is usually expressed as:

$$ln \frac{[C]_o}{[C]_t} = k_1{}'t \quad (8)$$

Where $[C]_o$ and $[C]_t$ are the chlorine concentration before and after exposure to UV, respectively for a time of t min; k is the first order constant in unit of min^{-1}.

Since the value of k is affected by the experimental conditions (*e.g.*, irradiance, absorbance, path length, and *etc*) which are difficult to reproduce, this expression can be modified into:

$$ln \frac{[C]_o}{[C]_F} = k_1'F \quad (9)$$

$$\frac{[C]_F}{[C]_o} = e^{-k_1'F} \quad (10)$$

where $[C]_o$ and $[C]_F$ are the chlorine concentration before and after exposure to the UV fluence of *F*; k_1' is the fluence-based first order rate constant [10].

3 Results and discussions

3.1 Chlorine removal efficiencies

Results obtained from both laboratory and plant scale experiments indicate that chlorine concentration decreases upon exposure of UV light. The total chlorine removal efficiencies illustrated in Figure 4 suggest that the removal efficiency increases as UV fluence increases. It is because higher fluence transfers greater amounts of energy to the samples which then utilized to break the chemical bonds of residual chlorine in the samples. These results are consistent with those reported [10, 13, 14], suggesting that higher UV fluence is required to achieve higher removal efficiencies.

The control experiments (dark sample) showed small decreases in chlorine concentration (<4.8%). These amounts of chlorine are considered as chlorine demand of the water sample, i.e., the amount of chlorine required to remove impurities in the water sample. This suggests that the removal efficiencies achieved in this study are inclusive of chlorine demand of the water samples.

As shown in Figure 4, the highest UV fluence utilized in the laboratory experiments, 513 mJ/cm^2, results in about 25% of total chlorine removal efficiency. The pilot plant scale experiments with typical UV fluence dose resulted in only about 9% of maximum removal efficiency at the highest UV fluence of 180 mJ/cm^2. The graph also illustrates that the removal efficiencies yield in laboratory experiments were generally higher than those achieved in pilot plant experiments.

At about the same UV fluence, i.e., 171 mJ/cm^2 in laboratory scale and 180 mJ/cm2 in pilot plant scale, the removal efficiencies achieved in laboratory scale are about 17% whilst the pilot plant scale only resulted in about 10%. This likely occurred because the water sample was in contact with the open air in the laboratory scale, whereas in pilot plant scale set up the water sample was inside the inlet pipe and UV reactor at all times. Therefore, some loss of chlorine in the laboratory scale experiment could be attributed to the relative volatility and instability of chlorine compounds in water. Further studies are needed to confirm this hypothesis.

Watts and Linden [14] showed that nitrate, NO_3, is among the products formed by UV reactions of water containing monochloramine and under some conditions, photolysis of nitrate led to formation of nitrite, NO_2. Such products are considered as drinking water, surface water, and groundwater contaminants, which pose threat to human health upon consumption. In fact, The National Drinking Water Quality Standard released by USEPA sets the MCL (maximum contaminant level) of nitrate at 10 mg/L and nitrite at 1 mg/L.

The formation of these compounds in deionized (DI) water during photolysis of aqueous monochloramine as shown in Figure 5 indicates that concentration of nitrite and nitrate formed by the photolysis reaction increases as the UV fluence increases.

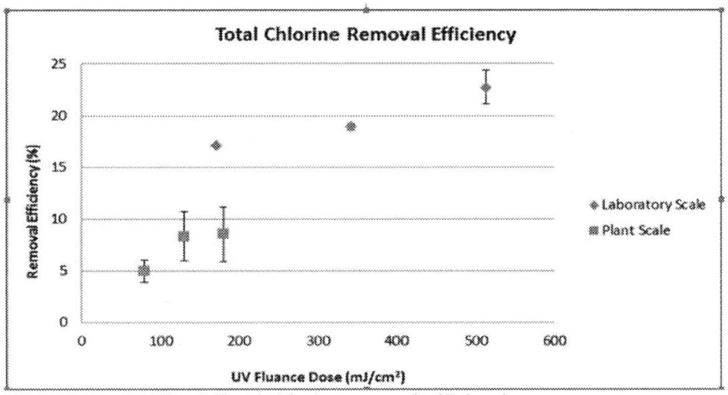

Fig. 4. Total chlorine removal efficiencies

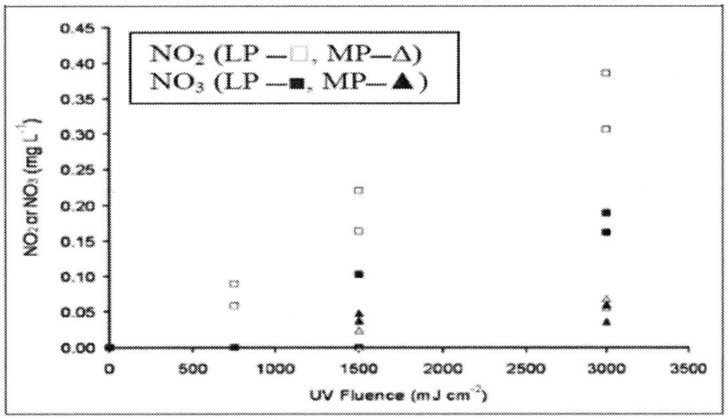

Fig. 5. Nitrogenous photoproduct formation in chloraminated (4.5mgCl/L) DI water after UV exposure [14]

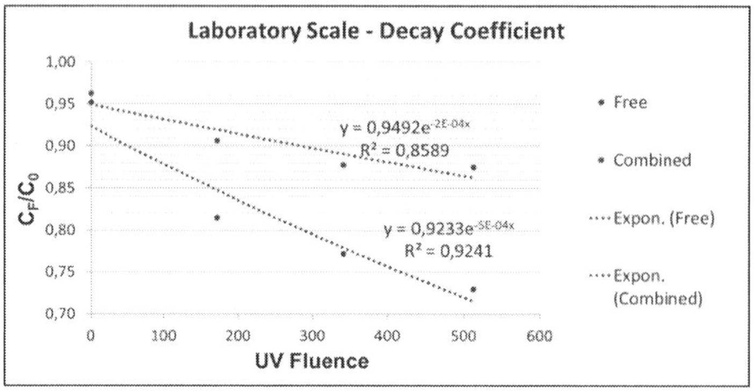

Fig.6. Chlorine decay constant – laboratory scale

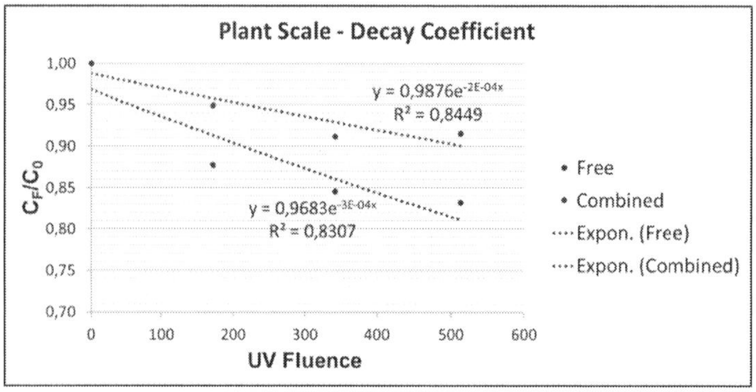

Fig.7. Chlorine decay constant – pilot plant scale

Therefore, further study conducted to investigate the required UV fluence to achieve better removal efficiencies should consider the formation of these compounds. Figure 5 also shows that there was little nitrate and nitrite formation in the range of UV fluence applied in this study.

3.2 Chlorine Decay Rates (fluence-based first order rate constant)

Results of chlorine decay constant from laboratory scale (Figure 6) and pilot plant scale (Figure 7) showed that the decay rates of free chlorines were higher than those of combined chlorines. This finding suggests that polychromatic UV light used in the experiments is more effective in decaying free chlorine than combined chlorine.

This is likely due to the maximum photon absorption of free chlorine at 290 nm. Such wavelength falls in the range of polychromatic UV light spectrum by the MP UV lamp. However, monochloramine's maximum photon absorbance is approximately at 246 nm which is close to the wavelength of monochromatic spectrum produced by LP UV Lamp [13]. The R^2 values of correlations from the experiments data are greater than 0.83, revealing near significant relationship although more data points were preferred to obtain better R^2 values and to accurately determine the kinetic reactions of free and combined chlorine under UV light exposure.

4 Conclusions

In this study, the effectiveness of a medium pressure UV technology in water dechlorination was investigated in laboratory scale and plant scale experiments. The results proved that UV light radiation reduces chlorine concentration in water samples. Higher UV fluence dose generated higher total chlorine removal efficiencies. The highest UV fluence dose applied in the laboratory scale experiments, i.e., 513 mJ/cm^2, resulted in about 25% of chlorine removal, whilst the maximum fluence dose utilized in the plant scale experiments, i.e., 30mJ/cm^2, resulted in about 9% of removal efficiency. The fluence-based free and combined chlorine decay rates were found and the results show that under exposure of polychromatic UV light generated by MP UV lamp, free chlorine was degraded faster than combined chlorine.

The authors would like to thank Nanyang Technological University (NTU) and Center for Advanced Water Technology (CAWT) for research facilities used in this project. Dr. Benoit Jean Guieysse, Professor at Massey

University of New Zealand (previously Assistant Professor at NTU) for his valuable guidance and support, Mylene Gomez, Fu Yong, and other researchers researchers at CAWT laboratory for their timely assistance and technical support.

References

1. J.K. Howe, D.W. Hand, J.C. Crittenden, R.R. Trussell, G. Tchobanoglous,: John Wiley & Sons, (2012)
2. D.C. Gang, T.E. Clevenger, S.K. Banerji, J. Env. Informatics, **1,** 21 (2003)
3. G. Tchobanoglous, F.L. Burton, H.D. Stensel, Metcalf and Eddy. McGraw-Hill Education, (2003)
4. Spellman, R. Frank, *Handbook of Water and Wastewater Treatment Operations*, Boca Raton: CRC Press, (2013)

A. Herbert, D. Forestier, D. Lenes, D. Benanou, S. Jacob, C. Arfi, L. Lambolez, Y. Levi, J. Water Rsch, **44**, 3147 (2010)

5. T.W. Valenti, D.S. Cherry, R.J. Currie, R.J. Neves, J.W. Jones, , J. Env. Toxicology and Chemistry. **25**, 2512 (2006)
6. K. Watson, G. Shaw, F.D.L. Leusch, N.L. Knight, J. Water Rsch. **46**, 6069 (2012)
7. K.G. Linden, C.M. Sharpless, J. Env. Sci. Tech, **37**, 1933 (2003)
8. S.H. Clarke, U.S. Army Center for Health Promotion and Preventive Medicine, (2006)
9. Y. Feng, D.W. Smith, J.R. Bolton, J. Env. Eng. Sci, **6**, 277 (2007)
10. J.R. Bolton, *Ultraviolet Applications Handbook*, Edmonton: ICC Lifelong Learn Inc, (2010)
11. D.L. Harp, Technical Information Series--Booklet No17, (2002)

B. Örmeci, G.A. Ishida, K.G. Linden, J. Water SRT - Aqua, **54** (2005)

12. M.J. Watts, K.G. Linden, J. Water Rsch, **41**, 2871 (2007)
13. R.J. Xie, M.J. Gomez, Y.J. Xing, J. Env. Sci. Hlth A Toxicity Hazard Substances, **42**, 2047(2007)

Combination of $CaCO_3$ and $Ca(OH)_2$ as agents for treatment acid mine drainage

Poedji Loekitowati Hariani[1,*], *Salni* Salni[2], and *Fahma* Riyanti[1]

[1]Department of Chemistry Faculty of Mathematics and Natural Sciences Sriwijaya University, 30139 Palembang, Indonesia
[2]Department of Biology Faculty of Mathematics and Natural Sciences Sriwijaya University, 30139 Palembang, Indonesia

Abstract. Acid mine drainage (AMD) has characteristic very low pH solution and containing metal ions in high concentration. This paper presents the use of $CaCO_3$, $Ca(OH)_2$ and the combination of both to increase the pH and decreased the concentration of Fe and Mn ions for acid mine drainage. The research variables are the effect of reactant dosage, contact time and temperature by batch studies. The AMD before treatment has pH solution of 3.38, Fe and Mn ions concentration of 44.6 and 7.19 mg/L, respectively. The dosage of $CaCO_3$ to increased pH solution about 7.0 was found 2400 mg/L at contact time 60 minutes and temperature 40 ^{0}C. The amount of $Ca(OH)_2$ for the neutralization of AMD solution smaller than $CaCO_3$ is 210 mg/L at contact time 45 minutes and temperature 40 ^{0}C. The combination dosage of $CaCO_3$ 1000 mg/L and $Ca(OH)_2$ 90 mg/L can increased the pH of AMD solution to 7.10 and reduction concentration of Fe to 3.53 and Mn to 4.51 mg/L. Therefore, the integrated of $CaCO_3$ and $Ca(OH)_2$ has the potential to be applied to treatment acid mine drainage.

1 Introduction

Acid mine drainage (AMD) is an important problem in the environment and a serious concern in many countries. The AMD contains of heavy metal from the oxidation of sulphidic minerals and producing low of pH solution. The treatment of AMD must be done on the exhaust into the environment because this substance is harmful to aquatic life at low concentration, not biodegradable and media surroundings. The AMD contains a lot of heavy metal ions such as Cu^{2+}, Fe^{3+}, Mn^{2+}, Zn^{2+}, Cd^{2+} and Pb^{2+}. The heavy metal ions are not biodegrable and tends to accumulate in the bodies in living organis [1]..

The AMD solution usually is orange colour from precipitation of iron oxide and hydroxide [2]. Characteristic of AMD of each region is different depending on typical mine water for the individual deposits [3]. The characteristic of AMD solution is influenced by several factors such as bacteria, temperature, starting pH and alternative oxidants like iron or manganese [4].

The treatment of AMD must be efficient and continual [5]. The method is often used for AMD treatment is oxidation, coagulation/flocculation, neutralization and precipitation of metal ions. The oxidants for the AMD such as $Ca(OCl)_2$, NaClO, CaO_2, H_2O_2, whereas coagulands often to treatment of AMD are $Al_2(SO_4)_3$, $FeSO_4$, $Fe_2(SO_4)_3$, $NaAlO_2$ [2].

Some materials can be used for the neutralization process such as $CaCO_3$ [3, 6], $Mg(OH)_2$ [7], fly ash [8], NaOH [9]. The use of $CaCO_3$ for the neutralization process of AMD was patented by the US Geological Survey Leetown Science Center [10]. It has been reported that the pH of AMD treatment using $CaCO_3$ (limestone) more economical but slow rate of dissolution with effectiveness ± 30%) [11]. The other research, effect neutralization of AMD with limestone in the reactor during 48 hours produced an armor coating in the bottom [6]. It causes the decline in the effectiveness of the use of $CaCO_3$. Two step reaction process of limestone with sulfuric acid in the AMD as follows:

$$CaCO_3 + H_2SO_4 \rightarrow CaHSO_4^+ + HCO_3^- \quad (1)$$

$$HCO_3^- + CaHSO_4^+ \rightarrow CO_2 + H_2O + CaSO_4 \quad (2)$$

At the end of the reaction occurs of precipitated calcium sulfate. The other studies have shown that calcium carbonate to raise the pH by consuming hydrogen ions and adding alkanity to form bicarbonate ions, according the following reaction: [12]

$$CaCO_3 + 2H^+ \rightarrow Ca^{2+} + H_2O + CO_2 \quad (3)$$

$$CaCO_3 + H_2CO_3 \rightarrow Ca^{2+} + 2HCO_3^- \quad (4)$$

The metal ions can be precipitate to form hydroxides or oxyhydrosides.

* Corresponding author: pujilukitowati@yahoo.com

Several methods have been developed for treatment of AMD. In this paper, integrated of limestone ($CaCO_3$) and hydrated lime ($Ca(OH)_2$) was used for neutralization of AMD solution. Effectiveness $Ca(OH)_2$ for AMD treatment three times greater than $CaCO_3$, but the cost of processing using these materials more expensive [11]. Hydrated lime has the ability to raise the pH quickly and precipitate the metal ions as hydroxide such as manganese at pH 9 to 9.5. Reaction of $Ca(OH)_2$ with hydrogen ions or metal ios as follow: [13]

$$Ca(OH)_2 + 2H^+ \rightarrow Ca^{2+} + 2H_2O \quad (5)$$

$$Ca(OH)_2 + Me^{2+}/Me^{3+} \rightarrow Me(OH)_2/Me(OH)_3 + Ca^{2+} \quad (6)$$

The other research, the use of $CaCO_3$ on AMD treatment can increase the pH solution from 2.9 to over 7 at the contact time of 48 hours [6]. Beside that, the addition of $Ca(OH)_2$ do not only neutralized, but also the OH^- anionic ions increased the rate of speciation with reaction to metal ions and increased the pH of AMD. $CaCO_3$ and $Ca(OH)_2$ can also destabilize the hydrolysis of acid mine drainage so that the value of TSS decrease [14].

At present, the combination of $CaCO_3$ and $Ca(OH)_2$ to treatment of AMD $_2$ is expected to increase the pH and reduction the heavy metal ions at the faster contact time. $Ca(OH)_2$ has a greater solubility than $CaCO_3$ are 1850 and 14 mg/L, respectively [7]. The parameters of studied are the effect of reactant dosage, contact time and temperature by batch studies.

2 Experimental details

2.1.1 Materials

The reagents such as $CaCO_3$, $Ca(OH)_2$ were analytical grade by Merck and all solution were prepare with deionized water. Acid mine drainage collected from a mine in Tanjung Enim, South Sumatera, Indonesia.

2.1.2 Effect of dosage, contact time and temperature

The effect of dosage was added of $CaCO_3$ and $Ca(OH)_2$ to 1 L of AMD solution at constant stiring (120 rpm) at room temperature for 60 minutes. The amount of $CaCO_3$ was added from 100-2500 mg/L (interval 100 mg) and $Ca(OH)_2$ 10-240 mg/L (interval 10 mg). The data of effect contact time obtained from 0-100 minutes with interval 5 minutes. The effect of the temperature was performed using thermostatic water bath at temperature of 30-90 ^{0}C. Data is the observed pH of the solution. The pH solution is obtained using pH meter Orion star A2111.

2.1.3 Neutralization of AMD using combination $CaCO_3$ and $Ca(OH)_2$

For this purpose $CaCO_3$ and $Ca(OH)_2$ were added to 1 L of AMD. The amount of $CaCO_3$ 1000 mg while the $Ca(OH)_2$ in the range 10-120 mg (interval 10 mg) with contact time and temperature using the latest result of previous studies. Determination of metal ions (Fe and Mn) in AMD before and after treatment were determined using Atomic Absorption Spectroscopy Shimadzu AA 7000 with atomization by N_2O-acetilene flame. The wave number for obtaining Fe and Mn ions at 248.3 and 279.8 nm, respectively.

3 Result and discussion

There are two technologies for treatment of acid mine drainage (AMD), these are active treatment and passive treatment processes. The active treatment method is the addition of chemicals to raise the pH and precipitate the metal ions. The active treatment using chemical reagents. This method is very effective for the treatment of AMD. The passive treatment includes aerobic and anaerobic wetland. This method is natural process.

In traditional treatment of AMD, $CaCO_3$ has been used for neutralization of AMD in many countries. There used in anoxic limestone drain and open limestone channel [9]. In this study, the combination of $CaCO_3$ and $Ca(OH)_2$ to treatment of AMD was evaluated. Before integrating of both, investigated the optimum conditions of treatment in each material includes the effect of dosage of $CaCO_3$ and $Ca(OH)_2$, contact time and temperature.

Neutralization of AMD using $CaCO_3$ and $Ca(OH)_2$ are presented in Fig 1. and Fig 2. The AMD solution has initial pH of 3.38. The AMD is classified as types of 1 that has pH solution very lower. In the AMD, pyrite is oxidized to soluble iron and sulphuric acid as follows : [15]

$$2FeS_2 + 7O_2 + 2H_2O \rightarrow 2Fe^{2+} + 4SO_4^{2-} + 4H^+ \quad (7)$$

Fe^{2+} is oxidaxed to Fe^{3+} by oxygen and hydrogen and oxidation of sulfure by Thiobacillus and Ferroplasma bacteria.

The oxidation process is influenced by pH solution, this reaction occurs fast at low pH (<4). Ferrous hydroxide formed is namely *yellow boy*. [5,16]

$$2Fe^{2+} + 7O_2 + 2H_2O \rightarrow 4Fe^{3+} + 4H_2O \quad (8)$$

$$4Fe^{3+} + 12\ H_2O \rightarrow 4Fe(OH)_3 + 12H^+ \quad (9)$$

The result obtained that the increased dosage of $CaCO_3$ and $Ca(OH)_2$ increased of pH and reached at pH 7 on the addition $CaCO_3$ of 2400 mg while $Ca(OH)_2$ of

210 mg. Ca (OH)$_2$ has the ability to increase the pH solution of AMD better than $CaCO_3$. The positive correlation that fact with increasing dosage of neutralization agents, more hydrogen ions can be neutralized using $CaCO_3$ and $Ca(OH)_2$.

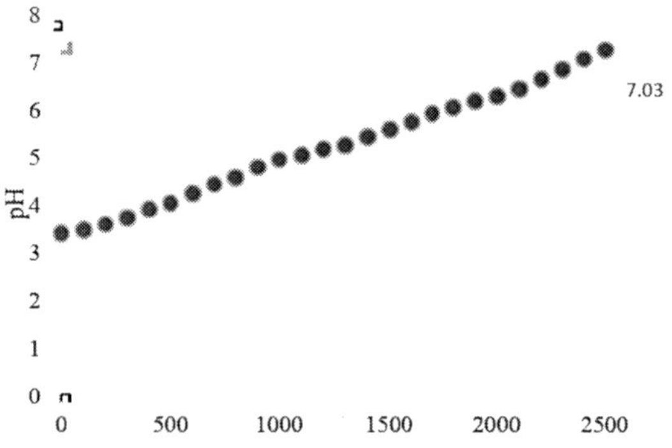

Fig. 1. Effect of dosage $CaCO_3$ for pH solution

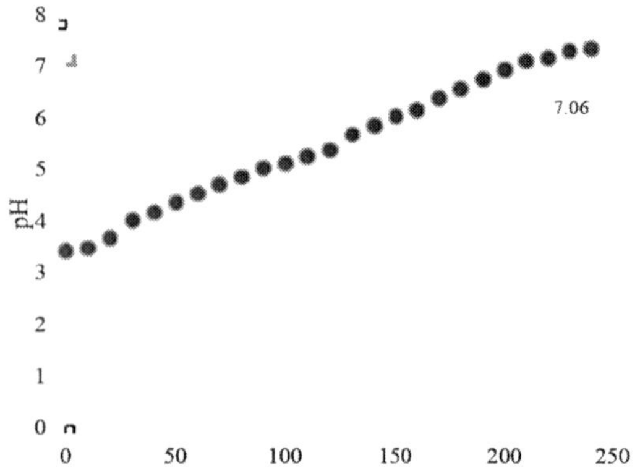

Fig. 2. Effect of dosage $Ca(OH)_2$ for pH solution

The both of agents have the same acid equivalent is 2 but $Ca(OH)_2$ has an efficiency factor of 0.8 higher than $CaCO_3$ of 0.4 [17]. Efficiency factor is an empirical estimate for neutralizing acidity. Besides than, $Ca(OH)_2$ is a strong alkaline has a neutralization efficiency of 90 % greater than $CaCO_3$ only 30 % [11]. The alkaline necessary to raise the pH solution and produce hydroxide to precipitated with metal ions.

Table 1 shows some of the materials used to raise the pH solution of AMD. Compared to the data in table, the results of this research requires fewer doses to increase pH of AMD. The type and dosage of materials have effect to the increase of pH solution.

Figure 3 showed effect contact time of $CaCO_3$ and $Ca(OH)_2$ on the neutralization of AMD. The data obtained that the equilibrium time required for the neutralization was almost 60 minutes for $CaCO_3$ and 45 minutes for $Ca(OH)_2$.

Table 1. The materials to treatment of AMD

Materials	Dosage	pH of Acid mine drainage	
		Initial pH	Final pH
$Mg(OH)_2$ [7]	0.25 g/L	3.4	8.3
$Ba(OH)_2.8H_2O$ [7]	7.35 g/L	3.4	12
Fly ash [8]	AMD:Fly ash (2:1)	2.7	11.5
Fly ash [8]	AMD:Fly ash (1:1)	2.7	11.5
Fly ash [8]	AMD:Fly ash (3:1)	2.7	9.5
Limestone [6]	-	2.9	6.5
Bentonite [18]	1 g/L	2.7	7.5

The contact time affect the number of hydrogen ions that can be neutralized by chemical agents. The longer the contact time so the more hydrogen ions that can be neutralized. From the result, the neutralization using $Ca(OH)_2$ faster than $CaCO_3$. In this work indicated that an increase in contact time resulted in increased of pH solution. After the equilibrium time, the pH relatively constant at pH 7. It is clear that the pH value dependence of contact time. The longer of the contact time, the more neutralization process. Furthermore, pH relatively constant at about pH 7 for a limited amount of reagents.

To compare the other research, the optimal dosage to reach at pH 7 on acid mine water in the locality of Jiří Mine in the Sokolov Region is 150 mg/L of $Ca(OH)_2$ with contact time 30 minutes [3]. The reduction of sulfate from AMD also dependent dosage of $CaCO_3$ and $Ca(OH)_2$. The research indicated that the relationship between sulfate solubilization and pH was direct and linear [19].

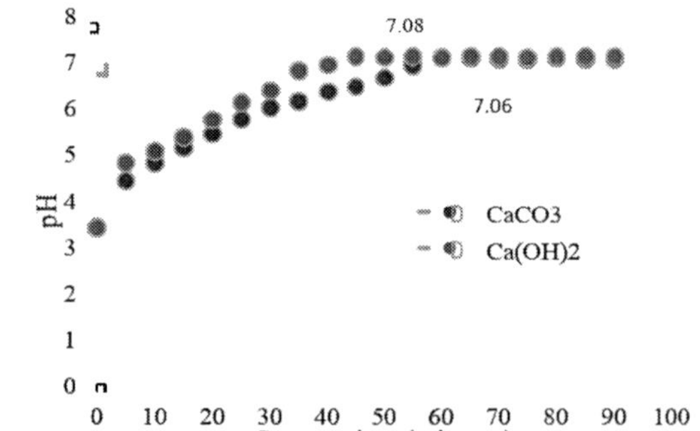

Fig. 3. Effect of contact time for Ph solution

The effect of temperature for AMD neutralization is presented in Fig 4. On evaluating the result, it was clear that an increase of pH at the temperature 20-40 ^0C, and then the pH decreased at higher temperature. The solubility indicated the maximum concentration of substance that can be dissolved at solution (ussulaly at room temperature). Beside pH solution, the solubility of compound is also dependent by temperature. The increase of temperature so the greater solubility of compounds.

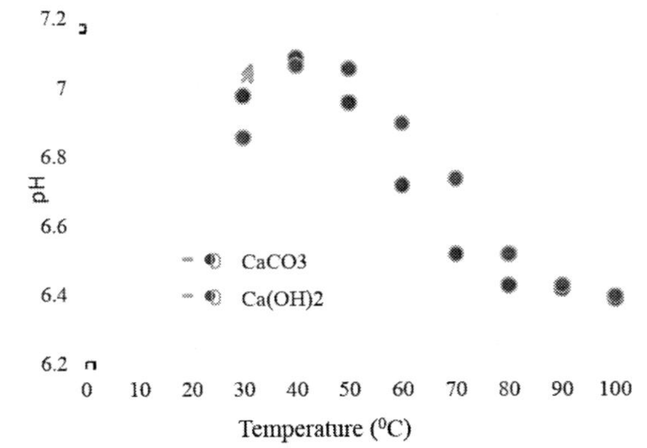

Fig.4. Effect of temperature for pH solution

The chemical reagents of $CaCO_3$ and $Ca(OH)_2$ have the same pattern of obtained pH 7 at temperature 40 ^0C. If the temperature is increased, the average kinetics also increases, it destabilized the solid state and thus the dissolve of the precipitate. This causes solubility of hydrogen ions and then the pH solution is decreased. The same result observed the influence of temperature for neutralization of AMD solution using NaOH. The optimum temperature at 20 ^0C, there was no change in pH at temperature 40 and 60 ^0C. The pH dropped to 2.2 at temperature 90 ^0C [9].

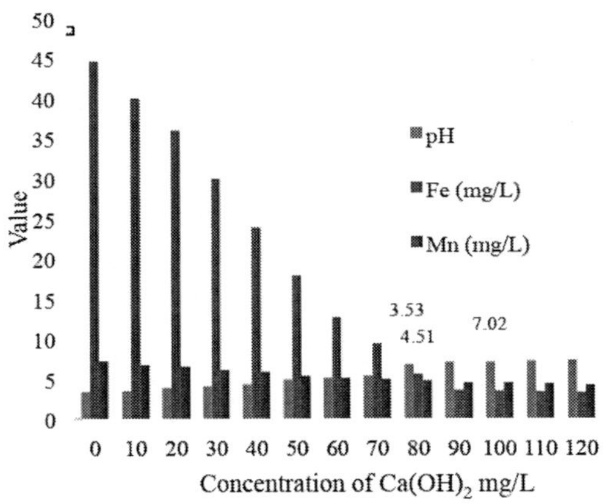

Fig. 5. Combination of $CaCO_3$ with $Ca(OH)_2$ to treatment of AMD solution

The combined of $CaCO_3$ and $Ca(OH)_2$ is expected to reduce the dosage of $CaCO_3$ and improve efficiency process and economical. The concentration of Fe and Mn ions in AMD solution during the addition of 1000 mg/L $CaCO_3$ and variation dosage of $Ca(OH)_2$ from 10-120 mg/L showed in Fig 5.

The process at contact time 45 minutes and temperature 40 ^0C. The AMD have contain of Fe is 44.6 mg/L while the Mn is 7.19 mg/L. Characteristic of AMD solution exceeds of acid mine drainage quality standard. The maximum of Fe is 7 mg/L, Mn is 4 mg/L and pH solution in the range 6-9 [20]. We can see that after neutralization using $CaCO_3$ and $Ca(OH)_2$ was reached Fe and Mn ions of 3.53 and 4.51 mg/L, respectively.

The pH solution increased from 3.38 to 7.02. The effectively decrease of Fe and Mn ions were 92.42 and 37.27 %. The effectivity to removal Fe ions from AMD greater than Mn ions. The oxidixed and precipitation of Fe ions about at pH 7.0 while Mn ions at higher pH is 8. The other studies shows that $CaCO_3$ and $Ca(OH)_2$ effectif to reduced TSS by mechanism coagulation and floculation [21]. The study also investigated interaction double layer metal ions on both reagents of the aqua-colloids.

The use of bottom ash, bentonite and fly ash to reduce the concentration of iron ions obtained at the optimum weight of 3, 4, 4 g in 100 mL of AMD solution, respectively [18]. In this study, has a smaller dosage than the result. Another study shows that adsorption capacity for adsorption of Mn ions in AMD is 6.03 mg/g using bone char [22].

The AMD solution is very complex, there is competition between the metal ions to form precipitate. The metal ions form precipitates depending value of Solubility Product Constant (KSP). The Ksp is the equilibrium constant, indicated that saturated solutions of ionic dissolving in an aqueous. The metal ions have small solubility product constants will be precipitated first. Example, Ksp $Fe(OH)_3 = 6.10^{-38}$ smaller than Ksp $Mn(OH)_2 = 2.10^{-13}$, so Fe ions to form precipitate earlier than Mn ions [23].

4 Conclusions

The combination of $CaCO_3$ and $Ca(OH)_2$ can be used for neutralization and reduce metal ions on acid mine drainage (AMD) solution. Integrated of both with a dosage of $CaCO_3$ 1000 mg/L and $Ca(OH)_2$ 90 mg/L can raise the pH solution from 3.38 to 7.02 and decrease of Fe and Mn ions with efficiency 92.42 and 37.27 %, respectively. In this study show that combination of $CaCO_3$ and $Ca(OH)_2$ an effectif to treatment of AMD.

This research was supported by The Directorate General of Technology Research and Higher Education (Ristekdikti), Republic Indonesia with "Unggulan Perguruan Tinggi Programme in 2016, Sriwijaya University".

References

1. N. Moreno, X. Querol, X., C. Ayora. Environm. Sci. and Tech., **35**, 3526-3534 (2001)
2. T. Motsi. Thesis. University of Birmingham, United Kingdom (2010)
3. D. Kirby. Dissertation. Marshall University, Virginia (2014)
4. S. Heviankova, I. Bestova, M. Kyncl, L. Simkova, M. Zechner. J. of the Polish Mineral Eng. Soc., 159-166 (2013)
5. C. Zipper, J. Skousen, C. Jage. Virginia Cooperative Extension. Powell River Project, 1-14 (2011)
6. J.M. Hammarstrom, P.L. Sibrell, H.E. Belkin, Appl. Geochem., **18**, 1705-1721 (2003)
7. V. Bologo, J.P. Maree, C.M. Zwinovanda. Proceeding of the International Mine water Conference, Pretoria, South Africa, 372-380 (2009)
8. V.R.K. Vadapalli, M.J. Klink, O. Etchebers, L.F. Petrik, W. Gitari, R.A. White, D. Key, E. Iwuoha, S AFR J SCL, **104**, 317-322 (2008)
9. E. Michalkova, M. Schwarz, P. Pulisova, B. Masa, P. Sudovsky, Pol. J. Environt Stud., **22(4)**, 1111-1118 (2013)
10. B.J. Watten. US patent No. 5.914.046. US Department of Commerce, Washington DC, (1999)
11. J. Skousen. T. Hilton, B. Faulkner, American Society of Agronomy and the American Society for Surface Mining and Reclamation, (1997)
12. Y. Paul, S.A. Banwart, Hedin, S. Robert, Kluwer Academic Press, Netherlands, (2002)
13. B. Lottermoser, Springer Berlin Heilderberg, London-New York, (2010)
14. I.O. Ntwampe, F.B. Waanders, J.R. Bunt, J. Chem. Eng. Mater. Sci., **6(3)**, **15-33** (2015)
15. C.N. Sawyer, Mc Catty, P.L., G.F. Parkin. Mc Graww Hill International Editions, (1994)
16. H.L. Yadav, A. Jamal, Intern. J. New Tech. Sci. Eng., **2(3)**, 77-84 (2015)
17. C.A. Cravotta, D.L. Parkhurst, B. Means, R. Mc Kenzie, W. Arthur. Proceeding of National Meeting of The American Society of Mining and Reclamation, Pittsburgh, PA Bridging Reclamation, Scince and the Community, 1413-1436 (2010)
18. E.O. Orakwue, V. Asokbunyarat, E.R. Rene, P.N.L. Lens, A. Annachhatre, Water Air Soil Pollut, **227(74)**, 1-12 (2016)
19. S. Rose, W. C. Elliot. Appl. Geochem., **15**, 27-34 (2000)
20. Governor Regulation in South Sumatera No 8 (2012).
21. I.O. Ntwampe, F.B. Waanders, J.R. Bunt, J. Chem. Eng. Mater. Sci., **6(3)**, 34-51 (2015)
22. D.C. Sicupira, T.T. Silva, A.C.Q. Ladeira, M.B. Mansur, Brazilian J. Chem. Eng., **32(02)**, 577-584 (2015)
23. W.B. Euler, L.J. Kirschenbaum, B. Ruekberg, Jour. Chem. Ed., **77(8)**, 1039-1040

Partial oxidation of methane to methanol over catalyst ZSM-5 from coal fly ash and rice husk ash

Fusia Mirda Yanti[1,*], *S.D. Sumbogo* Murti[1], and *Yuni* K. Krisnandi[2], *Adiarso*[1]

[1]Agency for Assessment and Application of Technology, 10340 Jakarta, Indonesia
[2]Department of Chemistry Faculty of Mathematics and Natural Science Universitas Indonesia, 16424 Depok, Indonesia

Abstract. Methane is one of the greenhouse gases that can be converted into liquid fuels such as methanol to retain most of the energy of methane and produce a cleaner environment. The conversion of methane to methanol using ZMS-5 represents a breakthrough in the utilization of methane. However, material sources for zeolite synthesis as catalyst usually are pro-analysis grade materials, which are expensive. Therefore, in this research, coal fly ash and rice husk ash were used as raw materials for mesoporous ZSM-5 zeolite synthesis. First, coal fly ash and rice husk were subjected to pre-treatment to extract silicate (SiO_4^{4-}) and aluminate (AlO_4^{5-}) and impurities separation. The ZSM-5 zeolite was synthesized through hydrothermal treatment using two types of templates. After ZSM-5 was synthesized, it was modified with Cobalt through impregnation method. The catalytic activity of both ZSM-5 and Co/ZSM-5 zeolites as heterogeneous catalysts in partial oxidation of methane were preliminary tested and compared with that commercial one. The result showed that the zeolite catalyst ZSM-5 from fly ash coal and rice husk ash has the potential to be used as catalysts in the partial oxidation of methane to methanol.

1 Introduction

Methane (CH_4) and carbon dioxide (CO_2) are major components of biogas which are generated in landfills as the result of waste decomposition and in the treatment of wastewater [1]. Methane, the most plentiful component of natural gas, is a useful material for synthesis of more valuable product such as ethylene, methanol or formaldehyde. Methane is also a part of greenhouse gases, so it is desirable to convert methane into liquid transportable fuels such as methanol because it retains most of methane energy and produce clean environment [2]. Recently, in order to utilize methane, partial oxidation reaction becomes one of main research interests. Partial oxidation is a technically mature process in which gas or a heavy hydrocarbon fuel (heating oil) is mixed with a limited amount of oxygen in an exothermic process. Research in methane conversion to methanol and formaldehyde through partial oxidation using ZSM-5 zeolite as heterogeneous catalyst at 423 K has been reported [3].

Methanol is one bulk chemical being widely used in chemical industry. Methanol is used as solvent, a gasoline additive and a chemical feedstock for production of biodiesel and hundreds of other chemical [4]. Besides that, methanol also is widely regarded as promising alternative for automobile fuel. Introduction of automobiles and other transportation vehicles that are able to bum pure or blended methanol would be highly benefical as per enviromental energy considerations [5]. Some of positives effects of developing a world-wide market for methanol as a fuel are: (a) decrease in dependency on foreign energy sources, (b) conservation of existing petroleum based energy reserves, (c) increased competition among energy providers resulting in lower cost and (d) enviromental benefits from improved emissions inherent in the combustion of methanol as compared to petroleum-base fuels. Adoption of methanol as a primary automobile or transportation fuel would necessitate a large increase in the world's methanol production capacity. This would make methane a much more attractive and valuable energy sources which is easily transportable as compared to its origin [6].

Beznis *et al..*. [7] has been investigated the oxidation of methane to methanol and formaldehyde over Co-ZSM-5 molecular sieves. Eugeny *et al.* [8] also has been investigated about oxidation methane to methanol on the surface of FeZSM-5 zeolite. Krisnandi *et al.* [1] also has been researched partial oxidation of methane over heterogeneous catalyst Co/ZSM-5 with two type structures: microporous and hierarchical. The result showed that the hierarchical Co/ZSM-5 catalyst give the highest conversion of methane to methanol. However, material sources for zeolite synthesis as catalyst usually are pro-analysis grade materials, which are expensive. Therefore, in this research, coal fly ash and rice husk ash were used as raw materials for mesoporous ZSM-5 zeolite synthesis.

* Corresponding author: fusia.mirda@bppt.go.id

2 Experimental

2.1 Materials

The materials used were: Coal Fly Ash from Paiton, Indonesia, rice husk from Bogor, Indonesia, sodium hydroxide (NaOH) pellet, Chloride acid (HCl) 1M, Tetrapropylammonium hydroxide, TPAOH 1M (Sigma Aldrich), Poly Diallyl Dimethyl Ammonium Chloride (PDDA 35 %) (Aldrich), Cobalt (II) Nitrate ($Co(NO_3)_2.6H_2O$) (Sigma Aldrich), methane gas (99.99 % BOC), N_2 gas (99.99 % UHP) and deionized water.

2.2 Pretreatment of Coal Fly Ash

Coal fly ash was dried and sieved using a 200 mesh sieve, and then activated in a furnace at a temperature of 800 °C for 2 hours, followed by refluxing with 1M HCl for 3 hours at 90 °C. After this, the mixture was washed with distilled water to pH neutral. The products were filtered with filter paper and then dried in an oven with a temperature of 105 °C for 6 hours. Characterization of product was carried out by AAS, FTIR, and XRD

2.3 Pretreatment of Rice Husk Ash

Risk husk samples (500 grams) was washed and then dried in an oven at 105 ^{0}C for 1 hour. After that, the sample was crushed and sieved into 60 meshes, then burned with a furnace at 700 °C for 4 hours. Sample was characterized by AAS, FTIR and XRD

2.4 Synthesis of ZSM-5 Catalyst

Synthesis of ZSM-5 catalyst was prepared with a molar ratio: 12 NaOH: 30 SiO_2: 0.75 Al_2O_3: 6 TPAOH: 1800 H_2O [9]. In this experiment, 2.0286 g of coal fly ash was mixed with 30 ml of 0.4 M NaOH solution (mixture A) at 100 °C. Then 5.7187 g rice husk ash was added with 70 ml of 0.4 M NaOH and then stirred at 100 °C (mixture B). Furthermore, mixture's B was slowly added to the mixture A, then 4.0552 g TPAOH was added to the mixture, and glacial acetic acid was added to adjust pH up to 11. Stirring was continued for 3 hours at a temperature of 100 ˚C. After that, added 1 g of PDDA (10 wt%) to the mixture and stirred again for 15 hours at room temperature. Then the reaction mixture was transferred into the autoclave and heated for 144 hours at a temperature of 150 ˚C in the oven. The reaction product was be cooled, washed with aquademin, filtered, and dried at room temperature for one night. After that, the product was calcined for 8 hours at a temperature of 500 ˚C. The synthesized crystals of mesoporous ZSM-5 was characterized by XRD instruments, FTIR, SEM, and pore size analyzer (BET).

Post-synthesis modification was carried out to obtain Co/ZSM5 by following the procedure as reported previously [1], i.e. by mixing 0.2495 M cobalt solution (from $Co(NO_3)_2$) with the zeolite. After the impregnation process, the mixture was calcined at 500 °C to obtain Co/ZSM-5

2.5 Application for Partial Oxidation of Methane to Methanol

Testing of catalytic partial oxidation of methane to methanol in a stainless steel vessel as a batch reactor was carried out. The testing procedure has been reported previously [1]. The catalyst used for each reaction is 0.5 g, which has been activated at 500 ^{0}C. The reaction was performed at 150 ^{0}C with methane pressure of 0.75 bar and a nitrogen pressure of 2 bar with a variety of reaction time of 30 minutes and 60 minutes then extracted with ethanol, and characterized using Gas Chromatography-FID.

3 Results and Discussion

3.1. Synthesis of ZSM-5 catalyst

The mesoporous ZSM-5 zeolite from fly ash and rice husk ash were synthesized through hydrothermal treatment using two types of templates (TPAOH and PDDA). First, coal fly ash and rice husk were subjected to pre-treatment in order to extract silicate (SiO_4^{4-}) and aluminate (AlO_4^{5-}) and separated from the impurities. The ZSM-5 zeolites were synthesized using TPAOH as the micropore template and structure-directing agent for MFI-type zeolite, whereas in order to create mesoporosity within the structure, PDDA polycation was added. Soft templates such as PDDA have two possible positions in the zeolite, during the crystallization process, in which it could coat on the inside and outside of the zeolite particles

Figure 1 showed diffraction pattern of coal fly ash, rice husk, and catalyst ZSM-5. Figure 1 showed the appearance of certain peak of rice husk in the position 2 theta at 21 ^{0,} indicative of amorphous silica, and the coal fly ash contained silica quartz peaks with low intensity at position 2 theta 26.5 ^{0}. In addition there is a silica quartz, the coal fly ash also contained the peak of hematite (Fe_2O_3) at position 2 theta 24 ^{0}; 33 ^{0}.49 ^{0}; 54 ^{0}, 62 ^{0} and magnetite (Fe_3O_4) at position 2 theta 31 $^{0;}$ 36 ^{0}; 43 ^{0}; 57 ^{0}; 63^0[10]. Diffraction patterns of ZM-5 from fly ash dan rice husk ash showed the appearance of certain peaks in the position 2 theta between 7 ° to 9 ° and 22° to 25° (Figure 1), indicative of successfully synthesized ZSM-5 structure [11].

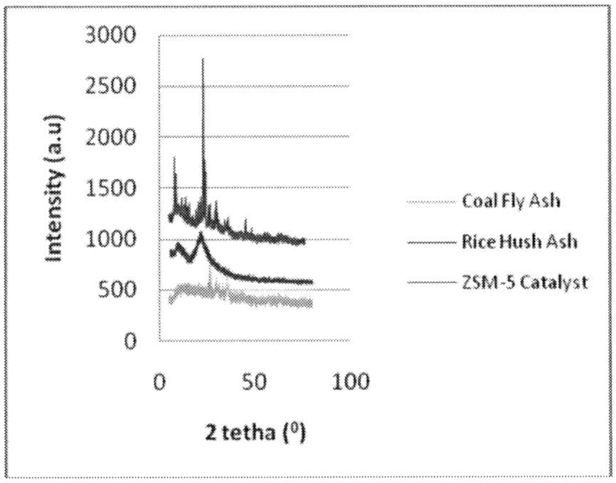

Fig 1. XRD Pattern of Coal Fly Ash, Rice Husk Ash and ZSM-5 Catalyst

Figure 2 shows the FTIR spectra of ZSM-5. The characteristic absorption bands of ZSM-5 were found at around 1250-950 cm^{-1} ($v_{asimetri}$ T-O), 1100-700 cm^{-1} ($v_{simetri}$ T-O), and also were found at round 650- 500 cm^{-1} indicated the presence of double 5- ring in ZSM-5.[12].

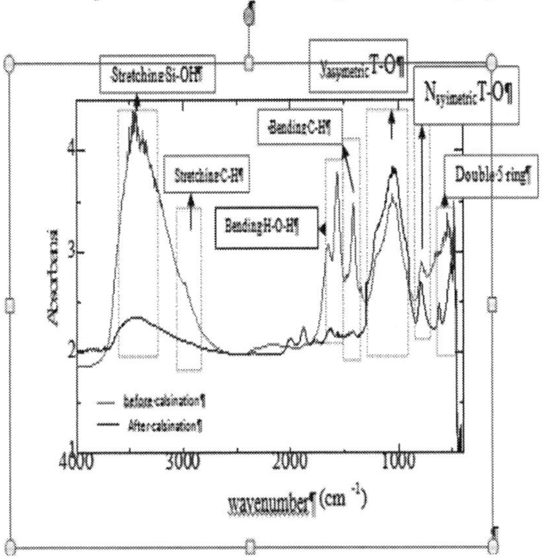

Fig 2. Spectra FTIR of ZSM- 5 zeolite catalyst

Table 1 shows the chemical content of fly ash, rice husk ash, ZSM-5 zeolite and the Si/Al ratio of synthesized ZSM-5. Rice husk ash contains the main components such as silica, so it is potentially used as a source of silica for synthesis of ZSM-5 catalyst. Coal fly ash main content are SiO_2, Al_2O_3, Fe_2O_3, and CaO. Ratio of Si/Al ratio to ZSM-5 from fly ash is 16.73. Catalyst ZSM-5 synthesized from fly ash coal and rice husk ash has other components beside Si and Al like Fe, Ca, Mg and Mn.

Table 1. Chemical content (%wt) of rice husk ash, fly ash,and ZSM-5 catalyst.

Component	Rice husk ash	Fly ash	ZSM-5 Catalyst
SiO_2	93.73	30.19	66.34
Al_2O_3	0.63	14.28	2.91
Fe_2O_3	0.00	20.20	5.67
CaO	0.00	19.93	12.79
MgO	0.26		2.74
Na_2O	0.16	1.65	1.25
K_2O	2.14	1.19	1.36
TiO_2	-	0.79	-
PO_5	-	0.42	-
Mn_3O_4	0.11	0.22	0.12
SO_3	-	1.58	-

The morphology of crystals ZSM-5 is hexagonal form with unsmooth or rough surface indicating the formation of mesoporous feature on the surface of the crystal (Figure.3).

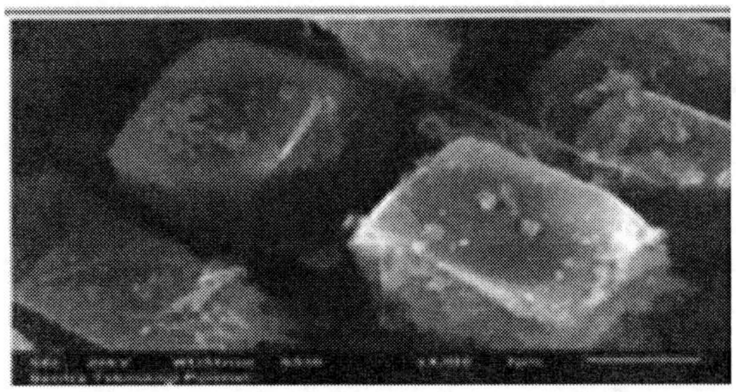

Fig.3 . SEM image of ZSM-5 zeolite

The results of surface area analyses of ZSM-5 are summarized in Table 2.

Table 2. Surface area analyses of ZSM-5 calculated from BET

Sample	S BET (m2/g)	S ext (m2/g)	V tot (cm3/g)	V meso (cm3/g)	V micro (cm3/g)	Average diameter pore (nm)
ZSM-5	43.759	43.759	0.287	0.287	-	26.2618

Table 2 shows that the surface area of ZSM-5 catalyst from fly ash and rice husk ash is low, due to ZSM-5 from fly ash and rice husk ash still has many impurities especially Fe which could as inhibitor in zeolititation process. From Table 2 also shows the pore diameter of ZSM-5 catalyst of 26.26 nm, indicating formed mesoporous (IUPAC: mesoporous 2-50 nm). The adsorption isotherm curve for ZSM-5 is shown in Figure 4.

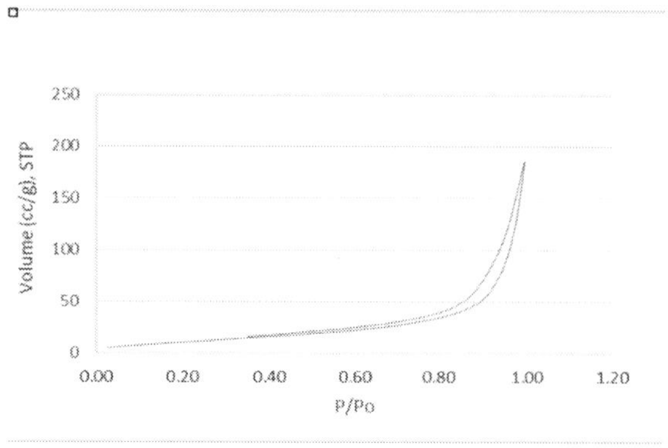

Fig 4. Isotherm adsorption-desorption plots of ZSM-5 catalyst

3.2 Catalytic test: partial oxidation of methane to methanol

Catalytic partial oxidation of methane to methanol was conducted in a vessel of stainless steel as a batch reactor by following the procedure have been reported previously [1]. Prior to the catalytic reaction, the ZSM-5 catalyst must be first activated at a temperature of 500 °C with the aim to form cobalt oxide species which constitute the active site of the catalyst Co/ZSM-5. Furthermore, flushing was carried out using nitrogen gas to remove organic impurities, water vapour and excess oxygen in the reactor. The condition of the reaction was adapted from the work reported by Krisnandi *et al.*. [1] and Bezniz et.al [7]. The first set of experiments was carried out at 150 °C with reaction time of 30 minutes and 60 minutes, and CH_4 pressure (Pmethane) of 0.75 Bar in 200 cm^3 stainless steel vessel.

Fig 5. Batch reactor for partial oxidation of methane to methanol

The results of this experiment are summarized in Figure 6

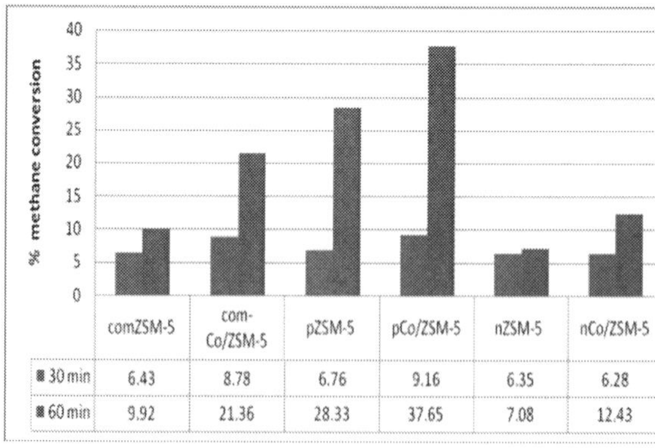

Fig 6. Percentage of methane conversion to methanol in reaction carried out at 150 °C, t= 30 and 60 min and methane partial pressure of 0.75 Bar

Figure 6 showed that the catalytic activity of ZSM-5 and Co/ZSM-5 from fly ash and rice husk was still inferior (7.08% and 12.43% conversion of methane) compared to the pro-analysis materials (28.33% and 37.65%) and commercial one (9.92% and 21.36%). However, they were potential to be used as catalyst in the partial oxidation of methane to methanol. The activity of ZSM-5 and Co/ZSM-5 from fly ash and rice husk ash inferior compared to the pro-analysis materials because ZSM-5 from fly ash and rice husk has lower surface area than ZSM-5 pro-analysis materials. The cobalt-loaded ZSM-5 catalysts gave higher conversion than the ZSM-5 without cobalt-loaded because impregnated Co/ZSM-5, mainly containing cobalt oxide species at the zeolite outer surface, were more selective towards methanol [7]

The mechanisms (modified from[13]) that occurred during the partial oxidation of methane to methanol are as follows:

Step 1, C-H activation
OM-ZSM-5 + CH_4 - ► (•HO)-M-ZSM-5 + •CH_3
Step 2, formation of hidroxy
(•HO)-M-ZSM-5 + •CH_3 - ►(CH_3)-(HO)-M-ZSM-5
Step 3, formation of methanol
(CH_3)-(HO)-M-ZSM-5 ► (CH_3OH)-M-ZSM-5
Step 4, desorption of metanol
(CH_3OH)-M-ZSM-5 - ►M-ZSM-5 + CH_3OH

4 Conclusions

The results showed that the zeolite catalyst ZSM-5 from fly ash coal and rice husk ash has the potential to be used as catalysts in the partial oxidation of methane to methanol although the catalytic activity of ZSM-5 and Co / ZSM-5 from fly ash coal and rice husk still inferior (7:08% and methane conversion of 12.43%) compared with the catalyst of pro-analysis materials (28.33% and 37.65%) and the commercial catalyst (9.92% and 21.36%).

This research was funded by : Hibah PUPT BOPTN No. 0545/UN2.R12/HKP/05.00/2015 and BPPT SK No. No.211a/2014

References

1. Z. Zakaria, S.K. Kamarudin. *Renewable and Sustainable Energy Reviews*. 65. 250-261 (2016(
2. Bill, A. Wokam, B. Eliasson, E. Killer, U. Kogelschatz, Greenhousegas Chemistry. *Energy Conversion and Management*, **38** : S415-S422 (1997)
3. Y. Krisnandi, A.P. Putra Bimo, M. Bahtiar, Zahara, M. Abdullah, F. Howe Russell. *Procedia Chemistry*, **14**, 508-515 (2015)
4. Qijian Zhars, Dehua, Qiming Zhu. *Journal of Natural Gas Chemistry*, **12**, 81-89 (2013)
5. Morleza Sohrabi and Leila Vafajoo. *World Applied Science Journal*, **6**, 339-346 (2009)
6. Priyank Khirsariya, Raju K. Mewada. *Procedia Engineering*, **51**, 409-415 (2013)
7. Beznis, Nadzeya V., Bert M. Weckhuysen, Johannes H. Bitter. Catalysis Letters, **136**, 52-56 (2010)
8. V. Eugeny, Starokon, V. Mikhail, Perfenov, S. Sergey, I. Arzumanov, Panov. *Journal of Catalyst*. **300.** 47-54 (2013)
9. K. Kordatos, S Gavela, A. Ntiziouni, K.N. Pistiolas, A. Kyritsi, V. Kasselouri-Rigopaulou. *Microporous and Mesoporous Material*. 189-196 (2008)
10. J. Izidoro J, D.A. Fungaro, F. Santos, S. Wang. Fuel Processing Technology, **97**, 38-44 (2012)
11. M.M.J. Treacy and J.B. Higgins. *4th revised ed. Amsterdam, Elsevier*, (2001)
12. S. Kulprathipanja. Weinhem, *Wiley-VCH Verlag Gmbh and Co. KgaA. (2010)*
13. M.F. Fellah, Inal, Isik. *J. Phys. Chem*, **114**, 3042-3051 (2010)

Adsorption of Jumputan liquid waste by betel nuts activated carbon in a continuous fixed-bed adsorber

Lia Cundari*, *Aris* Setiawan Kemit, and *Baharuddin* Rasyid Usman

Chemical Engineering Department, Faculty of Engineering, Universitas Sriwijaya, 30139 Inderalaya, Indonesia

Abstract. *Jumputan* is one of South Sumatera art crafts that increasingly produced by textile industries in the region. Dyeing and washing process are applied in *jumputan* production that will contribute to the increase of chemical content in waste water. This research will address the solution to this problem by investigating the waste water treatment using continuous adsorption process with activated carbon made from betel nuts as an adsorbent. Activated carbon was produced at carbonization temperature 500°C and activated by 0.5 M HCl and sized 18 mesh sieve. Adsorption was performed using Fixed-bed adsorber made from fiberglass with 100 cm height and 10 cm diameter. *Jumputan* liquid waste was fed from the top of the column at 58 mL/min. The height of bed was varied at 5, 15, 25 cm and sampling time was performed in 30, 60, 90, 120, 150 min to determine COD reduction. Optimum condition of this process is 25 cm of bed height and 150 min of sampling time that reduces COD value of waste water to 99.69%.

1 Introduction

South Sumatera province especially Palembang, has popular art crafts such as *songket*, *jumputan*, and wood cabinets craved. One of craft that intensively industrialized is *jumputan*, a well-known fabric with traditional painting. People used to wear *jumputan* not only in a ceremonial occasion but also in daily life. Fabrication of *jumputan* involves dyeing and washing process. Dyeing process utilizes synthetic dyes, such as Napthol, Indigosol, Rapide, Ergan Soga, Kopel Soga, Chroom Soga, and Procion [1]. Synthetic dyes were preferable than natural due to its low cost, durability and diverse of colours.

Dyeing and washing processes bring out liquid waste that contains hazardous chemicals. If that liquid waste directly discharged into the river, it will cause environmental pollution. According to Minister of Environment pursuant regulation No. 5 of 2014 on the quality of raw sewage, the water discharged into water bodies must be below the threshold [2]. In the fabrication of *jumputan* lots of liquid waste has resulted because some of the synthetic dyes can not be absorbed by *jumputan textile*. The waste is dumped into Musi river which can cause damage to ecosystems that exist in the river. These synthetic dyes will reduce oxygen demand that water organism needed.

Musi River turbidity has doubled exceeding the maximum allowable turbidity [3]. According to [4], maximum allowable turbidity is 600 NTU. Turbidity is caused by the disposal of wastewater from domestic and industrial, as well as sedimentation of solid particles that lead to settling of Musi river.

Besides turbidity, the degree of water quality also can be reflected from the content of chemical oxygen demand (COD), biological oxygen demand (BOD), total suspended solid (TSS), and the content of chrome, phenols, ammonia, sulfides and oil/fats [2]. COD is the amount of oxygen consumed by organic compounds and inorganic matter which were oxidized in water [5]. Lots of organic and inorganic matter indicates higher COD that means a bad water quality.

Various methods have been developed to treat liquid waste like coagulation-flocculation [6], advanced oxidation process [6], mechanical filtration, adsorption, ion exchange and membrane systems. One of wastewater treatment technology that was cheap, easy to obtain, easy to operate, and effective is adsorption. Adsorption is a process to absorb contaminant in the feed with solid material as an adsorbent. The contaminant accumulates on the surface of the adsorbent. There are a lot of factors that influence adsorption processes, such as surface area contact of adsorbent, type of the feed, molecules structure and concentration of the feed, temperature operation, pH, agitation speed (for a batch system), and contact time between feed and adsorbent. Widely adsorbent that uses in many adsorption process is activated carbon.

Activated carbon based on organic materials that used in large and small industries either as a catalyst, deodorizing, absorption color, substance purification and so on. Activated carbon as adsorbent has been widely studied, either in a pure state [7], [8], [9], [10], [11], [12], [13], [14], [15], [16]; or combined with other ingredients [17].

* Corresponding author: icun_hyang02@yahoo.com

Yuliasari et al (2011) removed COD on *Jumputan* wastewater by using *lumut hati* powder. COD value analyzed with spectrophotometry method. 100 mesh *lumut hati* powder contacted with 50 mL waste media. Variation of mass *lumut hati* (5, 15, 23, 35, and 50 mg) was made. The mixture was shaken at 150 rpm for 60 minutes and after that settled for 10 minutes. The filtrate was taken and analyzed. Optimum condition was 60 minutes operation and 25 mg lumut hati powder, COD reduced from 193.5 mg/L to 48.69 mg/L or 74.84% COD reduction [10].

Batch investigation on *jumputan* wastewater treatment by using activated carbon, zeolite, and fenton reagent, within 30 minutes operation, COD reduction was 38% and color degradation was 10% [9]. The effect of temperature and time on procion synthetic wastewater treatment with fenton reagent was studied. This process called advanced oxidation processes (AOPs) method. Red and blue procion synthetic dyes were used with concentration 150-250 mg/L, agitation 200 rpm, Fenton reagent concentration H_2O_2 80 mM and $FeSO_4.7H_2O$ 4 mM, pH 3, agitation time 0 – 60 minutes, and temperature 25 – 55 °C. This article reported 100% dye degradation could be obtained in temperature 55°C and 60 minutes agitation time. That condition was applicated to *jumputan* wastewater. For 100% dye degradation, optimum condition was at temperature 55°C and agitation time 120 minutes with reduced COD 66% [1].

Cundari et al (2015) conducted preliminary research by using betel nuts activated carbon to adsorb *jumputan* wastewater in a batch system. Researchers analyzed the best betel nuts activated carbon that could adsorb *jumputan* wastewater. The temperature carbonization and HCl concentration were varied. The carbonization temperature was 400°C, 500°C, and 600°C with HCl concentration 0.1 M, 0.3 M, and 0.5 M. This research reported the optimum carbonization temperature was 500°C with 0.5 M HCl concentration. The results showed 98.61% reduction in COD; 98.50% reduction in BOD; 87.50% reduction in TSS; and 96.30% reduction in Chrome.

Most research on activated carbon is done in the batch system, only a few are using continuous flow fixed bed adsorber. The data obtained by a batch system is inapplicable to get the equilibrium data due to insufficient time (Uddin, et.al. in [11]). The equilibrium data of absorption the basic red 46 (BR46) using granular activated carbon (12-40 mesh (0.425 to 1.70 mm)) in a fixed-bed column reported [11] indicates that Freudlich isotherm model is more suitable than Langmuir. They use a column with the following dimensions: inside diameter = 12 mm, height of column 210 mm, the height of bed 100 mm, and flow rate 50 mL/min.

Gimenes (2013) also reported on a continuous adsorption. The copper absorption by using bentonite (0.855 mm) in regenerable Calcined Clay Column. Acrylic column dimensions are: inside diameter = 1.4 cm, the height of column 24 cm, height of bed 14 cm, and flow rate 4 mL/min. The test resulted that the adsorption process using a fixed-bed column was promising but still requires further investigation to determine its technical and economical feasibility.

The type of adsorbent used in this study is activated carbon made from betel nuts. The preliminary assessment of the *jumputan* wastewater treatment using betel nuts activated carbon had been conducted in a batch process [7]. Wastewater treatment required a comprehensive system and operated continuously. Compared to batch process, continuous treatment is more applicable due to time efficiency, low operating cost, and quantity of wastewater that can be treated [18].

This research will examine chemical oxygen demand (COD) degradation in *jumputan* wastewater treatment by using continuous fixed-bed adsorber with the various height of bed and sampling time. Aims of this research are to improve the quality Musi river, due to continuously increased turbidity, up to double maximum turbidity numbers [3], and to get appropriate technology for processing *jumputan* liquid waste that can be directly installed and used by craftsmen.

2 Experimental Details

Betel nuts activated carbon was produced at 500 °C carbonization temperature and activated using 0.5 M HCl concentration based on the method by [7]. *Jumputan* liquid waste is taken from *jumputan* crafting area in Kertapati, Palembang. Figure 1 illustrates schematic of apparatus for adsorption process.

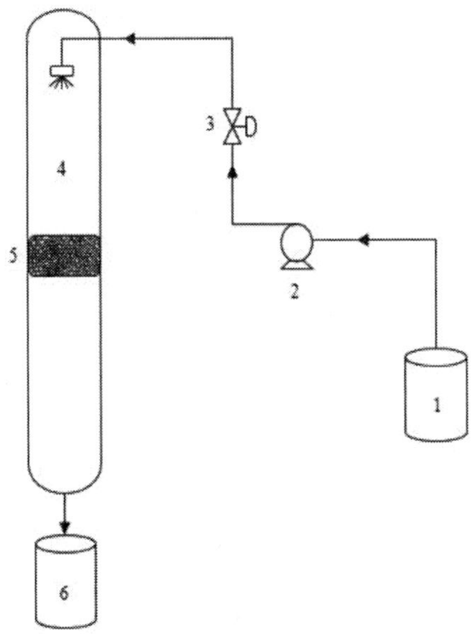

Fig.1. Schematic Apparatus of Adsorption Process

Description:
1) *Jumputan* Liquid Waste Tank
2) Pump
3) Valve
4) Continuous Fixed Bed Adsorber
5) Bed
6) Outlet Tank

Betel nuts activated carbon was sieved to 18 mesh and was placed inside the continuous fixed-bed adsorber. The dimension of the column was 10 cm diameter and 100 cm height. The height of bed varies from 15, 25, and 35 cm. Another variable is sampling time that varied in 30, 60, 90, 120, and 150 minutes. Volumetry method is used to analyze COD of liquid in the outlet tank.

Jumputan liquid waste in the tank (1) is filtered and then pumped to the top of the adsorber. The inlet will be distributed into the column by using a sprayer. Adsorber (4) which is used in this study is a fixed bed type, wherein the adsorbent is placed permanently in the bed. This bed (5) is placed in the middle column. The adsorption process is carried out continuously. Because of that, this apparatus calls continuous fixed-bed adsorber.

After the mass transfer between *jumputan* liquid waste and adsorbent, the content of dye, biological, chemical, and heavy metals will be absorbed by the adsorbent. Outlet liquid will be removed from the bottom of the column to the storage tanks (6). The samples are taken from the outlet with variation sampling time and analyzed. This study examines chemical oxygen demand in the samples.

3 Results and Discussion

The quality of activated carbon is determined by moisture content, iodine adsorption, methylene blue adsorption, color, surface area, and volume of pores. In this research, two of quality analyses are done. That is moisture content and iodine adsorption. The presence of water in the activated carbon relating to hygroscopic nature, wherein activated carbon has a great affinity to water. This calls for moisture content. Because of this hygroscopic properties, betel nuts activated carbon is used as the adsorbent. The moisture content of 18 mesh betel nuts activated carbon results 2.98%. This point is compliance with quality standards based on SNI 06-3730-1995, where the maximum moisture content of granular activated carbon is 4.5%.

The purpose of moisture content test is to determine the amount of water remains in the activated carbon after going through the activation process. Water molecules contained in the activated carbon will affect the absorption rate because the water molecules that bound to the activated carbon will cover the pores. Large pores will increase surface area which means increase its ability to adsorb.

Adsorption testing of iodine on activated carbon occurs to determine the activated carbon ability to adsorb molecules that have a small diameter and determine the ability to adsorb the color solution. Gimenes (2013) described adsorption of iodine indicates the ability of carbon to adsorb low molecular weight component. Hydrocarbon compound that remains on the surface of activated carbon will reduce adsorption ability. Laboratory analyses of adsorption of iodine obtain 485.85 mg/g. This number is below the quality standards according to SNI 06-3730-1995 which at least 750 mg/g. The iodine adsorption is changing rapidly. So, it could be an error in analyzation. This iodine absorption number must be analyzed at a laboratory with ISO certified to get the right measurements.

One of the indicators that can be used to determine water quality is COD value. The content of chemical oxygen demand in *jumputan* wastewater is analyzed with volumetry method appropriate with SNI 6989.2-2009. The content of COD is shown in Table 1 as follows:

Table 1. Chemical Oxygen Demand (mg/L)

Height of Bed (cm)	Sampling Time (minutes)					
	0	30	60	90	120	150
5	640	119.5	35	2	16	28
15	640	70	15	30	26	2
25	640	80	37	75	88.5	2

According to [2], maximum COD threshold allowed is 150 mg/L. Based on the preliminary test, COD level of *Jumputan* wastewater before treatment is 640 mg / L (time sampling 0 minutes). This value is four times greater than the threshold of COD in textile waste allowed. That value indicates there are a lot of organic and inorganic matters in the water. So, appropriate treatment to reduce the COD content is needed.

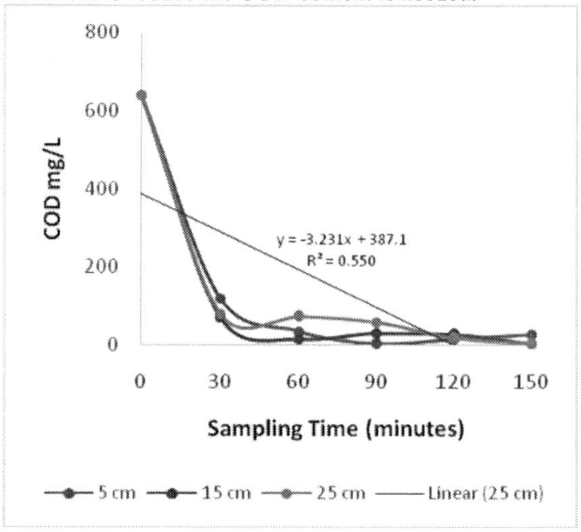

Fig. 2. COD remain in *Jumputan* liquid waste

One of the treatments is adsorption process by using continuous fixed-bed adsorber. Betel nuts activated carbon is placed in bed inside the column as an adsorbent. In this research, there is three factors influence adsorption, that is the amount of adsorbent, the surface area of adsorbent, and contact time. The amount of activated carbon is comparable to the bed height. The higher bed contains more activated carbon. Therefore bed height is varied (5, 15, and 25 cm). The size particle of betel nuts activated carbon is uniform (18 mesh sieved). Variation of contact time between jumputan liquid waste and betel nuts adsorbent is applied. Those are 30, 60, 90, 120, and 150 minutes. The relationship between the height of the bed and sampling time with COD remain in the waste is shown in Figure 2.

The relationship between the height of the bed and sampling time with COD remain in the waste is shown in Figure 2. The relationship between the height of the bed and sampling time with remaining COD in wastewater is fluctuating. By using 5 cm bed height, optimum COD remaining occurs at sampling time 90 minutes with COD measurement 2 mg/L. At 15 and 25 cm bed height, optimum COD remain in *jumputan* waste is 2 mg/L with the sampling time of 150 minutes. Based on table 1, all of the COD remaining after adsorption process in the *jumputan* liquid waste have met the standard wastewater characteristics as mentioned in Minister of Environment pursuant regulation No. 5 of 2014 that constrain maximum eligible COD discharged is 150 mg/L.

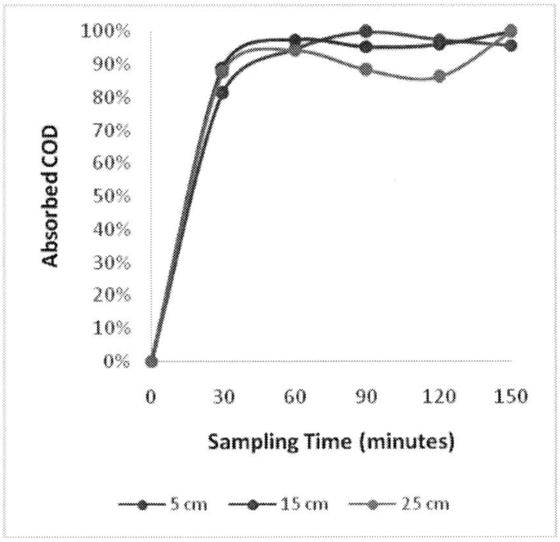

Fig. 3. Percentage of Adsorbed COD

Figure 3 shows the percentage of adsorbed COD. From the graph can be seen that the highest COD adsorbed is the bed with 5 cm high and 90 minutes sampling time, with amount 99.69%. In 15 and 25 cm bed height, the percentage of COD adsorbed at 150 minutes sampling time generated optimum condition with an amount equal to 99.69%. Compared with a previous study conducted by [7], this value is better. They obtained COD absorption percentage of 98.6111%. From the two differences of absorbed COD percentage even with batch or continuous process surely that betel nuts activated carbon has ability to adsorb organic and inorganic components in *jumputan* liquid waste.

Factor that can reduce the content of COD in industrial wastewater is the quality and quantity of activated carbon. The quality of activated carbon is showed by water content and iodine adsorption. The water content of betel nuts activated carbon is appropriate with a standard. In another word, the activated carbon is good adsorbent. Higher percentages of adsorbed COD also indicate a higher quality of activated carbon. Compared with [1], [7], [9], and [10], removed COD on *Jumputan* wastewater by betel nuts activated carbon is the highest.

The quantity of activated carbon is directly proportional to the height of the bed and contact time between the adsorbent and *jumputan* wastewater. The highest bed will be placed by the most of the activated carbon, which means the best height of the bed is 25 cm due to its ability to adsorb COD component. Adsorption process in packed bed column is largely dependent on bed height [20]. This research results suitable with that statement. From linearity in figure 2., the best height of the bed is 25 cm with linear equation $y = -3.231x + 387.1$ with a correlation coefficient (R) is 0.550. R value for 5 and 15 cm height bed is smaller than 0.550.

Contact time between the adsorbent and *jumputan* liquid waste can be seen from sampling time. In this investigation, the time limited to 150 minutes. Activated carbon efficiency decline with time addition [21]. Degradation ability of activated carbon to adsorb organic and inorganic components due to the pores on the surface of carbon has saturated. At 150 minutes, with 25 cm bed height, COD degrades from 640 mg/L to 2 mg/L. From figure 2., COD continuously declines from its original. To get the whole description of COD reduction, time contact must be added. That means the activated carbon used has not reached saturation yet. The 150 minutes sampling time indicates that betel nuts activated carbon has a long lifetime.

4 Conclusion

Betel nuts activated carbon has ability to adsorb organic and inorganic components in *jumputan* liquid waste. Optimum variable to reduce COD in *Jumputan* waste liquid is 25 cm bed height and 150 minutes sampling time. Depreciation of COD value is 99.69% from its original.

The financial support given by Engineering Faculty of Universitas Sriwijaya, Palembang, Indonesia for this research work is gratefully acknowledged

References

1. T. E., Agustina, & M., Amir, J. Teknik Kimia Universitas Sriwijaya **18**, 3, 54-61 (2012)
2. B., Kambuaya, Kementrian Lingkungan Hidup, Indonesia (2014)
3. Kamsari, Center of Marine and Coastal Resources Institut Pertanian Bogor, Retrieved februari 12, 2016, from Institut Pertanian Bogor: www.pkspl.ipb.ac.id (2015, april 25)
4. S., Novita, Universitas Sumatera Utara, Retrieved februari 16, 2016, from Universitas Sumatera Utara: www.repository.usu.ac.id (2013)
5. Q., Yang, Z., Liu & J., Yang, J. Water and Protection , 286-289 (2009)
6. M., Rahbar, E., Alipour & R., Sedighi,

International J. of Environment Science and Technology, **3**, 1, 79-88 (2006)
7. L., Cundari & dkk., *Added Value of Energy Resources,* Palembang: Universitas Sriwijaya (2015)
8. A., Nurfitriyani & dkk., J. Online Institute Teknologi Nasional, **1**, 2 (2013)
9. T. E., Agustina & dkk., *Added Value of Energy Resources (AVoER)*, 260-266, Palembang: Universitas Sriwijaya (2011)
10. N., Yuliasari, Miksusanti, & E., Setiyowati, J. Penelitian Sains , **14** (1(C)), 29-32 (2011)
11. Riyanto, J. Kimia, 1-9 (2010)
12. Yuliusman & Adelina, *Seminar Rekayasa Kimia dan Proses,* Depok: Universitas Indonesia (2010)
13. A., Surest & dkk., J. Teknik kimia, 17-22 (2008)
14. F., Ramdja & dkk., J. Teknik Kimia , **14** (4), 1-7 (2008)
15. A., Astuti, Semarang: Universitas Dipenegoro (2007)
16. Y., Danarto & E. K., Artati, EKUILIBRIUM, **4** (2), 86-91 (2005)
17. M. A., Fauziah & dkk., Semarang: Universitas Diponegoro (2014)
18. H., Halim & K., Mee, *International Conference on Enviromental and Industrial Innovation, IPCBEE ,* **12**, 263-267 (2011)
19. M. L., Gimenes, AIDIC, **32**, 2023-2028 (2013)
20. Ikawati dan Melati, Jurusan Teknik Kimia, Fakultas Teknik, Universitas Diponegoro (2010)
21. S., Sughashini & K., Begum, J. of Chemistry, 1-11 (2012)
22. R. D., Siregar, T. A., Zaharah & N., Wahyuni, JKK, 62-66 (2015)

Recovery of H_2SO_4 from spent acid waste using bentonite adsorbent

Marwan Asof[1,*], *Susila Arita* Rachman[2], *Winny Andalia* Nurmawi[3], *Cindy* Ramayanti [2]

[1]Department of Mining Engineering Universitas Sriwijaya, 30662 Palembang, Indonesia
[2]Department of Chemical Engineering Universitas Sriwijaya, 30662 Palembang, Indonesia
[3]Department of Industrial Engineering, Universitas Tridinanti, 30145 Palembang, Indonesia

Abstract. Spent acid waste is produced in the petroleum processing industry at the alkylation process unit that requires sulfuric acid as a catalyst to produce iso-octane. Therefore, spent acid contains sulfuric acid attached to a long chain carbon and other compounds. This compound is corrosive and has a pH below 1.0, so that it cannot be directly discharged into the environment. The processing and utilization of spent acid to meet environmental quality standards need to be conducted. This study aimed to conduct waste treatment of spent acid and to recover sulfuric acid from the spent acid solution because the level of H_2SO_4 is high at about 82%, then the sulfuric acid can be reused in the alkylation process or other processes. Adsorption method was selected to process the waste using bentonite as adsorbent in 3 stages adsorption columns. The results showed that the recovery of H_2SO_4 was successful in which the purity of H_2SO_4 generated after distillation reached 94.94%.

1 Introduction

The production process of petroleum processing industry, especially in the alkylation unit produces spent acid waste. Spent acid waste is classified as hazardous and toxic material and belongs to B3 waste. According to the regulation of Ministry of Environment Republic of Indonesia No 18/2009, industries must manage or perform waste treatment of B3 wastes in order to meet the Indonesian National Standard (SNI) or international standards, or other standards that apply nationally and internationally. Spent acid is one of the B3 waste that requires special treatment to be able to be reused in the process or transformed into other useful products.

The spent acid used in this study consists of compounds of waste from the alkylation unit in the petroleum processing industry. The main component of the spent acid is sulfuric acid, a major catalyst in the reaction between iso-butane and iso butylene that will react to form isooctane. Some of the constituent formed from alkylation unit other than sulfuric acid are distillate, alkylate, diethyl sulfate and dimethyl sulfate. The physical properties of spent acid are dark, smelled very sharp with a pH <1, wherein the initial concentration of sulfuric acid in the spent acid treated solution is 82%.

Problems in managing acid waste from oil and gas industry process are very important things to be resolved, so that the environment can be protected from contamination. Management of acid waste in this study is focused on recovery of H_2SO_4 contained in the waste solution using bentonite adsorption methods. In this case, bentonite is activated by physical adsorption in a stratified adsorbent column.

Researches on the treatment of spent acid wastewater have been performed by many researchers, among others, [1] conducted a spent acid recycle process to recover molybdenum contained in the mixture of waste in a way to neutralize the ammonia gas in the spent acid containing molybdenum, and then followed by crystallization and filtration. Hydrogen reduction of molybdenum oxide produces commercial products such as molybdenum metal powder. Fluid that contains a lot of nitrogen from this process can be used as fertilizer. [2] was successfully recovered the acid from the spent acid of steel industry (electroplating) which contains heavy metals included in B_3 category with a high enough concentration.

[3][4] succeeded in recovering HCl, Fe and $ZnCl_2$ from iron material production waste (galvanizing plants) using merger technology between ion exchange with membrane electrowinning, while [5] and [6] tried to minimize the spent acid from galvanic plant using hydrochloric acid as an impurities cleaning compound from metals by conducting hydrochloric acid regeneration using Kleingarn acid management system. It showed a positive result because it can reduce the spent acid wastes discharged into surface water.

*Corresponding author: marwan_asof@yahoo.com

Bentonite is a potential mineral and is scattered in almost every island in Indonesia. On the Sumatra island, bentonite is spread almost throughout the region, especially in the Jambi area. [7] showed that bentonite in Jambi is generally composed of four types of minerals; kaolinite, monmorrilonit, quartz and cristobalite. Bentonite of clay type having a major content of smectite minerals (montmorillonite) with levels of 85 -95%, is soft with a hardness 1 on the Mohs scale, the density between 1.7 to 2.7, breakable, plastic, and high colloidal.

Based on the physical properties, bentonite is differentiated as Na-bentonite and Ca-Bentonite. Na bentonite or swelling bentonite can swelling up to eight times when immersed in water and will be suspended when dispersed in water. In the dry state the colour is white or cream. In a wet state and when exposed to sunlight, it will turn shiny. High ratio of soda and high lime will produce colloidal suspension having a pH of 8.5-9.8 and cannot be activated. The exchange positions are occupied by sodium ions (Na^+) that have a large content of Na^+ between the layers and have swelling properties.

In the Ca-Bentonite-non swelling, bentonite content of Ca^{2+} and Mg^{2+} is relatively higher when compared with the content of Na^+. This kind of bentonite is less swell when immersed in water, remain dispersed in the water and will quickly settle or not form a suspension, but naturally or after being activated it has a good adsorption property [8]. Comparison of Na and Ca content is low and colloidal suspensions have a pH: 4-7. Ion exchange position is more occupied by ions of calcium and magnesium. In the dry state, it is rapid slaking, gray, blue, yellow, red and brown. This bentonite can be used as a catalyst support, whereas the modified bentonite can be used as a catalyst.

Indonesia needs bentonite as the filler or fillers in the paint industry, metal casting, oil drilling and as one type of adsorbent frequently used as a purification material of palm oil or coconut oil (bleaching agent) to absorb the stain and impurities in oil continuous increase. Absorption capability of bentonite can be enhanced by activation both physically and chemically that are useful to increase the surface area and modify the structure of bentonite. [9] Liisa Carlson, 2004 stated the nature and characteristics of bentonite are highly dependent on the geographic location of the region and the largest content is Na-montmorillonite, other parts are hematite, smectite, iron, pyrite and others. Smectite mineral type composed of aluminosilicate framework and form a layered structure has a negative charge evenly across its surface and a good cation exchanger.

An adsorption is an event of adsorption of molecules at the surface or between phases, where the molecules of the material are collected on the adsorbent. The materials to be adsorbed and adsorbent materials are two distinct phases. Then in the event of adsorption, the adsorbed material will be collected at the interface of the two phases. Adsorption is very widely used in industry, especially in wastewater treatment at industrial level. Adsorption known is chemical adsorption and physical adsorption. Chemical adsorption occurs because of the chemical bonding between the solute molecules (solute) and adsorbent molecules usually progresses slowly but irreversibly, while physical adsorption occurs because of the weak attraction of molecules between the two objects by Van der Waals forces on the surface of the adsorbent, so that it will form a thin layer of fine particles on the surface. The numbers of substances absorbed depends on the partial pressure of the molecule (PA). Some molecules diffuse laterally or may react on the surface (chemisorption). The speed of adsorption and desorption determines the equilibrium on the surface. The success rate of adsorption process on the surface of solids is highly dependent on the characteristics and concentrations of the compound to be absorbed, the molecular size of the adsorbate (the material to be absorbed), the surface area of solids (adsorbent), the degree of acidity of the solution [10], temperature, mass and particle size, contact time and others. To recover H_2SO_4 that the levels are still high, this research is used adsorption using bentonite adsorbent that has been activated physically.

2 Materials and methods

2.1. Materials and equipment

The raw material used was a solution of spent acid from alkylation unit of the oil and gas industry. The adsorbent used was bentonite originating from Jambi with a particle size of 100μm.Research equipment used was adsorber column operated in 3 stages as shown in Figure 1. Columns made of pyrex with a column diameter of 5 cm with a height of 70 cm. The top of columns were equipped with liquid distributor made of perforated plate, and the below were given sponge and installed vacuum pumps to accelerate the flow down.

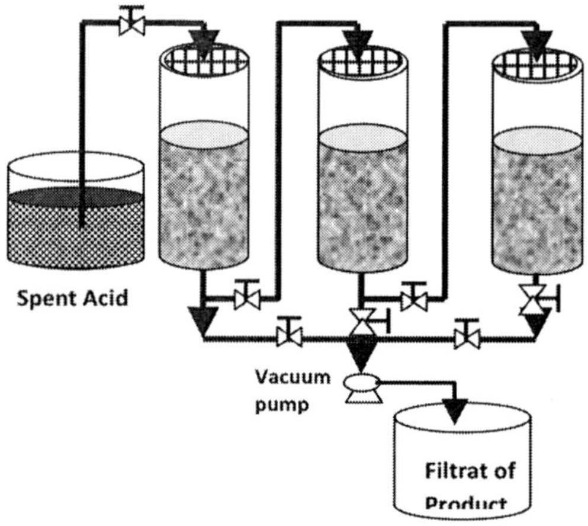

Fig.1. The schematic representation of the Series Adsorber Column

1.2 Research procedure

Physical activation of bentonite was conducted by heating at a temperature of 850 °C for 3 hours, then the adsorbent was washed with distilled water and dried in an oven at 110°C for 3 hours. The adsorption capacity of the adsorbent was investigated by conducting a variation of the adsorbent weight of 100, 150 and 200 grams that were inserted into the adsorption column.

Adsorption column was filled with 100 grams of bentonite adsorbent with a size of 100 μm. Above the column was mounted a liquid distributor so that the flow entered evenly through all the volume of the adsorbent. Under the column was mounted vacuum pump in order to make the spent acid penetrate more quickly into the adsorbent and was collected as filtrate under the adsorbent column. Slowly the spent acid from column 1 was filtrated and to be input in the 2nd column and the 2nd column output became the input to the 3rd column. Filtrate from at each output column was measured its volume and analysed its H_2SO_4 concentration. Then repeat the same procedure for the adsorbent weight of 150 grams and 200 grams.

The analysis of H_2SO_4 concentration was done by titration method. This method aimed to determine the changes in the sulfuric acid concentration of spent acid solution after passing through several stages of the adsorption process. SEM-EDS was performed to observe the structure of bentonite to determine bentonite morphology and the composition of the constituent elements in the bentonite.

3 Results and discussion

3.1. The effect of adsorbent mass on the filtrate volume and H_2SO_4 purity

Research results in Figure 2 show that the filtrate volume that can be recovered using adsorbent of bentonite is quite large. The highest and the lowest percentages of filtrate volume on the first output column are 72% and 40%, respectively. In the 2nd column, the highest and the lowest percentages are 53.4% and 29.4%, respectively. While on the 3rd output column, the highest percentage is 18% and the lowest is 10.8%. The heavier the adsorbent mass, the smaller the percentage of filtrate recovered, but the greater the concentration of sulfuric acid produced. It could be seen in Figure 3.

In Figure 2 and Figure 3 are shown how the mass of adsorbent affects filtrate volume and H_2SO_4 concentration produced. The heavier the adsorbent mass, the smaller the percentage of filtrate recovered. It indicates that the performance of the adsorbent is quite excellent at absorbing water and hydrocarbon impurities contained in the spent acid.

The resulting filtrate was distilled at a temperature of 110°C until the water in the solution evaporates completely so that the remaining solution in the distillation flask was analyzed. Subsequently, the result obtained was the increase in the concentration of H_2SO_4 up to 94, 94% with an adsorbent mass of 200 grams.

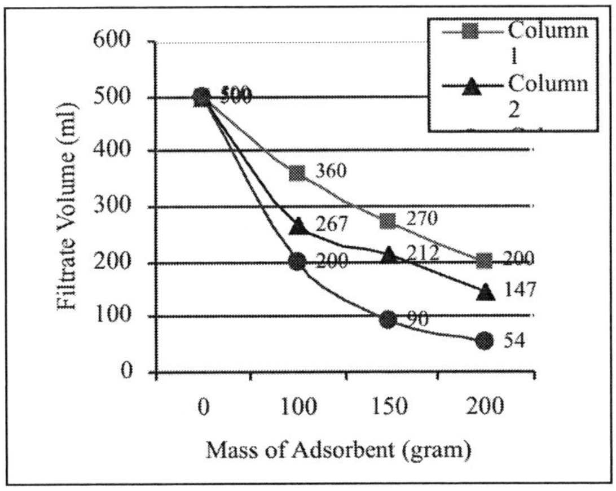

Fig.2. Results of the filtrate versus adsorbent mass

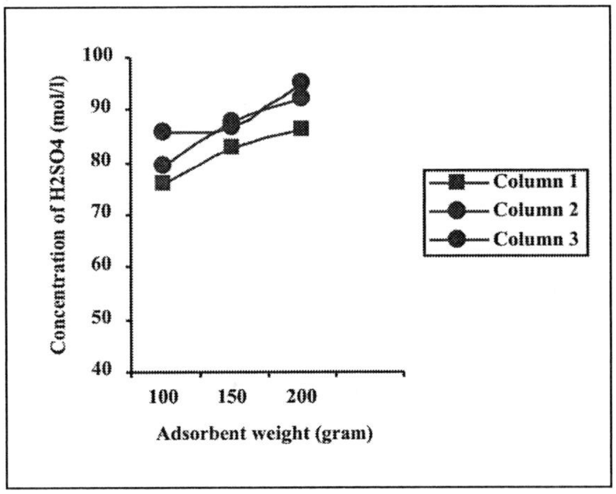

Fig. 3. Concentration of H_2SO_4 output of series adsorber column versus adsorbent mass

The purity of H_2SO_4 generated in this research was quite high. The next research would attempt to use this recovery result to activate bentonite as it has been researched by Hulya et al, 2007 and Muserref et al, 2007 who have activated bentonite with various concentrations of H_2SO_4. Their results indicate that activation by H_2SO_4 concentration > 50% can increase the ability of bentonite as a bleaching agent in soybean oil up to 70%.

3.2. Characterization of Bentonite Adsorbent

The characteristic of bentonite as adsorbent was investigated using Scanning Electron Microscopy (SEM) and Energy Disvertive Spectroscopy (EDS).

The analysis shows that the constituent elements of bentonite consist of silica, alumina, Oxygen, Nitrogen and Fe as shown in Figures 4 and Figure 5.

DOI: 10.1051/matecconf/201710102007

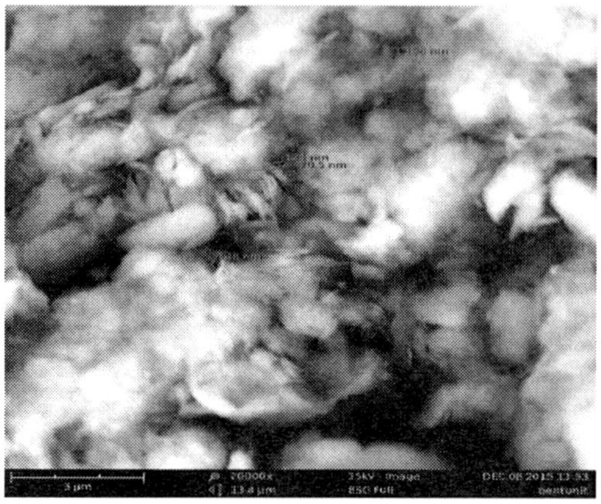

Fig. 4. Analysis of Scanning Electron Microscopy (SEM) bentonite before Adsorption Process

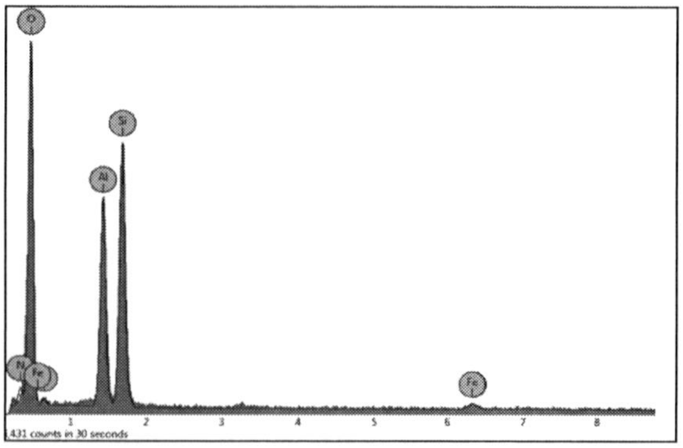

Fig. 5. SEM-EDS Spectrum of Bentonite before Adsorption Process

In the Figures 5, bentonite solids shaped layers of clay mineral mixture with a pore size of 70.5 nm to 148 nm. Meanwhile, the composition of each constituent element can be seen in the table below.

Table 1. The constituent elements of bentonite before adsorption

Element Number	Element Symbol	Element Name	Confi dence	Concen tration
8	O	Oxygen	100.0	68.2
14	Si	Silicon	100.0	14.8
13	Al	Aluminium	100.0	13.4
7	N	Nitrogen	100.0	2.7
26	Fe	Iron	Manual	1.0

In Figure 6 and Figure 7, it shown that after adsorption of spent acid solution, bentonite was contained increasing percentage of sulfur, molybdenum and potassium, while the pore sizes enlarged between 157 nm to 254 nm. The increase of bentonite pores may be due to the nature of the reactive sulfuric acid towards bentonite so as to make the bentonite lattice decays and bentonite pores enlarge. The small residual sulfur bonded to the surface of bentonite showed that most of the sulfur is adsorbed as the filtrate so that it can be recovered.

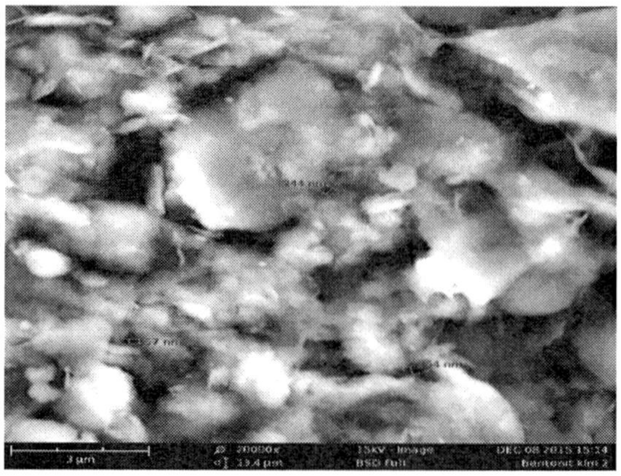

Fig.6. analysis of Scanning Electron Microscopy (SEM) bentonite after adsorption proses

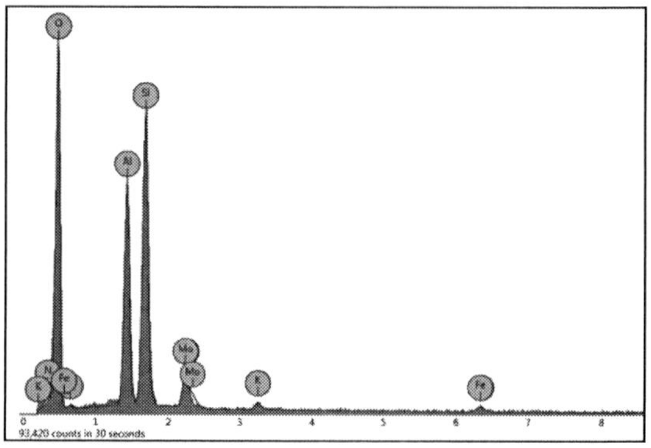

Fig.7. EDS Spectrum of Bentonite after Adsorption Process

Table 2. The constituent elements of bentonite after adsorption

Element Number	Element Symbol	Element Name	Confi-dence	Concen tration
8	O	Oxygen	100.0	66.2
14	Si	Silicon	100.0	13.8
13	Al	Aluminium	100.0	12.5
16	S	Sulfur	Manual	0.5
42	Mo	Molybdenum	100.0	3.6
7	N	Nitrogen	100.0	2.2
19	K	Potassium	Manual	0.4
26	Fe	Iron	Manual	0.8

4 Conclusion

Recovery of sulfuric acid from the spent acid of alkylation results can be achieved by adsorption method using adsorbents of bentonite that are activated physically with a temperature of 850 ° C for 3 hours. Adsorbent mass affects the filtrate volume and sulfuric acid purity produced, where adsorption processes at 3 stages are capable of raising sulfuric acid purity of 82% to 94.94%.

Authors thank to Sriwijaya University who has been financing this research through Competitive Superior Research Program, research fellows, and students for their support in the research.

References

1. J. Pak, J. Jo, C. Park, J. Kang, and D. Shin. Materials transactions **49**, no. 1, (2008).
2. A. Agrawal and K. K. Sahu. Journal of hazardous materials **171**, (2009).
3. T. I. T. G. Csicsovszki, T. Kekesi. Hydrometallurgy, J., **77** (2005).
4. M. Boucher, N. Turcotte, V. Guillemette,G. Lantagne, A. Chapotot, G. Pourcelly, R. Sandeaux, and C. Gavach. Hydrometallurgy, J., **45**, (1997).
5. C. Stocks, J. Wood, and S. Guy. Resources, conservation and recycling **44,** 2 (2005)
6. P. Rosnah and B. Mohd. Advanced Materials Research **13**, no. 6, (2013).
7. M. Naswir and S. Arita. Journal of Clean Energy Technology, **1**, no. 4 (2013).
8. D.S Bath, J.M Siregar, M.T. Lubis. J. Jurnal Teknik Kimia USU **1**, no. 1 (2012).
9. L. Carlson. Mineralogical Research of Selected Bentonites, (2004).
10. M. Mahramanlioglu, I. Kizilcikli, I.O. Bicer. Journal of Fluorine Chemistry, **115**, 41-47 (2002).

Laboratory study of salinity and surfactant concentration effects on oil recovery

Widia Yanti[1,*], *Sugiatmo* Kasmungin[1], *Rabiatul* Adawiyah[2], and *Blandina* Kolanus[2]

[1]Department of Petroleum Engineering Faculty of Earth Technology and Energy Trisakti University, 11440 Jakarta, Indonesia
[2]Student of Department of Petroleum Engineering Faculty of Earth Technology and Energy Trisakti University, 11440 Jakarta, Indonesia

Abstract. This paper perfoms a discussion of the result from two laboratory tests of surfactant injection. The first test is focused on the influence of salinity that varies between 5000 ppm- 9000 ppm with a concentration of 1% surfactant to the value of density, viscosity, interfacial tension, and Recovery Factor. The second test is performing the influence of surfactant concentration that varies between 0.1% - 1% with a value of 5000 ppm salinity to the value of residual oil saturation, density, viscosity, and interfacial tension. The result of the first test shows that along with the increase of salinity levels, the increase in viscosity and density will occured as well. However, variations in salinity does not affect to the value of interfacial tension. Meanwhile, on Recovery Factor testing, the Recovery Factor optimal value of 61.53% was obtained when the salinity levels 6,000 ppm. The result of the second test showed that the addition of surfactant concentration lead to a decrease in the residual oil saturation and interfacial tension, but it may cause an increase in density and viscosity.

1 Introduction

Based on BP Statistical Review 2013, Indonesia oil production has been declined since 1995 [1]. However, oil consumption still increase year by year due to the economic growth. This creates an unbalance condition between oil demand and supply [2]. Therefore, in order to create a stable condition, the Indonesian government together with the oil companies must work together to increase oil production from existing oil fields in Indonesia.

Chemical flooding, one of EOR methods, are now getting importance in enhanced oil recovery to recover the trapped oil after primary and secondary recovery. One of chemical flooding methods in the present study has been carried out on surfactant flooding. The chemicals with different compositions and combinations were used to recover the oil after conventional water flooding [3]. By applying primary and secondary oil recovery such as artificial lift and waterflooding, the average oil recovery still in between 5-30% [4]. However, after primary and secondary recovery, original oil in place is still large (about 50 billion barrel) [5]. It creates a prospect for applying Enhanced Oil Recovery (EOR) application (under this method, oil recovery between 5-65% from original oil in place) [4].

Enhanced oil recovery (EOR) or tertiary recovery is considered for the solution of relatively low recovery efficiencies relating to petroleum production processes [6]. From the United States Departement of Energy, it is estimated that the full use of enhanced oil recovery in United States could generate a additional 240 billion barrels of oil [7].

According to the US Department of Energy, there are three primary techniques for EOR : thermal recovery, gas injection, and chemical injection. The reservoir in this research will be injected by surfactant which is one of the chemical injection technique. The reason to select this technique is because of surfactant addition has become very attractive because of its ability to reduce surface/interfacial tension (IFT) between immiscible fluids. The interfacial tension plays a significant role in the process of enhanced oil recovery. The IFT reduces as the surfactant concentration increase [7].

Surfactants are chemical substances that adsorb on or concentrate at a surface or fluid-fluid interface and alter the interfacial properties significantly; in particular, they decrease interfacial tension (IFT) [8]. Based on the nature of the hydrophilic part the surfactant can be classified as anionic; cationic; amphoteric or Zwitterionic; and nonionic. In this research will be used AOS surfactants [7].

AOS (Alpha Olefin Sulfonate) surfactants are one of the anionic surfactants which are most widely used in EOR processes because they show relatively low adsorption on sandstone rocks whose surface charge is negative [3]. Moreover, AOS surfactants provide outstanding detergency, high compatibility with hard water, and good wetting and foaming properties with CO_2 even when the porous medium is partially saturated with oil. These properties make AOS surfactants an excellent candidate for EOR project that aim to produce more oil from underground reservoirs [9]. Beside of that,

* Corresponding author: widiayanti@gmail.com

AOS surfactants have good stability at high temperature [10].

Due to the high cost of the chemicals, the economics of the surfactant injection is very important. Thus optimum injection scheme has to be designed and considered [11].

This research is preliminary study to investigate the effects of surfactant concentration on enhanced oil recovery. The objective of this research is to determine the optimum of surfactant concentration to obtain high oil recovery.

2 Experimental details

This research was divided into two group of tests. First group discusses the influence of salinity that varies between 5000 ppm - 9000 ppm with a concentration of 1% surfactant to the value of density, viscosity, interfacial tension, and Recovery Factor (RF). The second test discusses the influence of surfactant concentration that varies between 0.1% - 1% with a value of 5000 ppm salinity to the value of residual oil saturation, density, viscosity, and interfacial tension. For more details, following an experimental procedure in this study, namely: (A) Preparation of the brine solution; (B) Preparation of the surfactant solution, (C) Measurement of physical properties of the solution, and (D) the process of surfactant injection into the reservoir model (sandpack).

2.1 Preparation of the Brine Solution

Brine was used as the aqueous phase during research. The basic ingredient of the brine solution that used in this research is NaCl with 5000 ppm concentration. There are some steps to see the influence of salinity on physical properties of the solution and Recovery Factor (RF).

First of all, by using magnetic stirrer, the 5000 ppm of NaCl mixed with 1000 ml of pure water until the NaCl dissolved completely. Then the brine solution conducted with varying salinity levels, ie 5000 ppm, 6000 ppm, 7000 ppm, 8000 ppm, and 9000 ppm.

2.2 Preparation of the Surfactant Solution

To prepare the surfactant solutions, the 5000 ppm of brine mixed with 10 ml of AOS surfactants using magnetic stirrer. Mixed them, until the AOS surfactant are dissolved completely. Repeat this steps for each salinity level (6000 ppm – 9000 ppm). In this research, AOS surfactants concentration varies : 0.1%; 0.5%; 1%; and 1.5%. It mean the AOS surfactants needed are respectively 1 ml; 5 ml; 10 ml; and 15 ml.

2.3 Measurement the Physical Properties of the Solution

The physical properties of the solution to be counted are viscosity (brine and brine+AOS surfactants), density (brine and brine+AOS surfactants), and surface tension (brine and brine+AOS surfactants). The viscosity is measured using viscometer Oswald. Then the density is measured using picnometer. And the surface tension is measured using tensionmeter Du Nouy.

2.4 The Process of Surfactant Injection into the Reservoir Model

The objective of this step is for knowing the Recovery Factor (RF) value of each solution and could determine which one has the optimum salinity concentration. First step, prepare the reservoir model using a sandpack as a media. The volume and size of the sand is respectively 54.19 ml^3 and 80 mesh. Second step, preparing a brine saturation inside the reservoir model. In this step, the pressure used is 2 psi. Third step, preparing a parafin saturation inside the reservoir model. In this step, the pressure used is 1 psi - 2 psi. Fourth step, preparing a water injection. Fifth step, preparing the AOS surfactants to be injected into reservoir model. The total of the RF value could be obtained from this step. Last step, preparing the brine to be injected into reservoir model as a preflush. Repeat this steps for each salinity level (5000 ppm – 9000 ppm) and for each AOS surfactant concentration (0.1% - 1.5%).

3 Result and discussion

3.1 Density Measurement Results

Density measurements conducted in the laboratory, using picnometer. Here, on Figure 1 below, are the result of the fifth density measurements on a brine solution that varies between 5000 ppm - 9000 ppm.

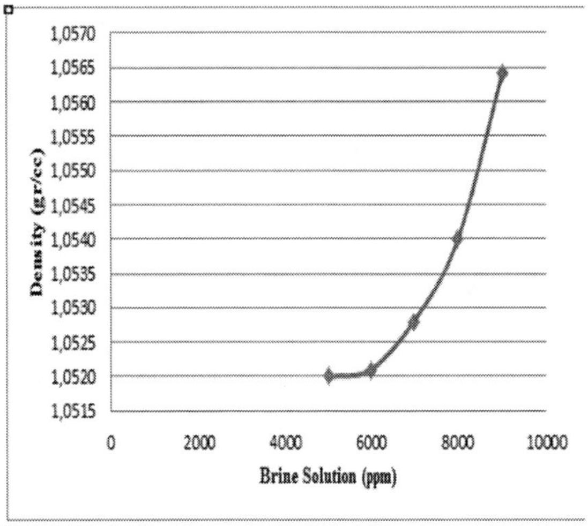

Fig. 1. The brine solution vs density

From Figure 1 above shows that the increased salinity levels will increase the value of its density. In this study, the density test conducted using a brine solution that varies between 5000 ppm - 9000 ppm.

DOI: 10.1051/matecconf/201710102008

In the second test, the AOS surfactants are added into the solution. Here are the results of measurements of the density after the adding of AOS surfactants.

Figure 2 below shows that the increased brine solution with 1% AOS surfactants, will increase the value of its density. In this study, the density test conducted using the brine solution with the adding of the 1% AOS surfactants concentration that varies between 5000 ppm - 9000 ppm.

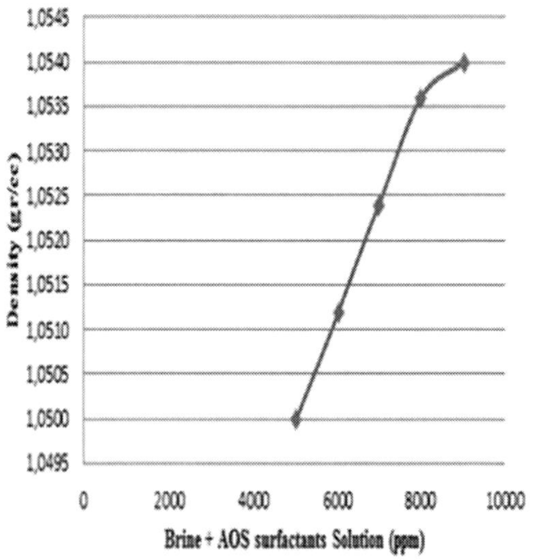

Fig. 2. The brine solution after AOS surfactant added into the solution vs density .

3.2 Viscosity Measurement Results

The viscosity measurements conducted in the laboratory using Oswald viscometer. On the measurement of the viscosity of the surfactant solution used only one concentration of the same for the fifth solution, ie 1%.

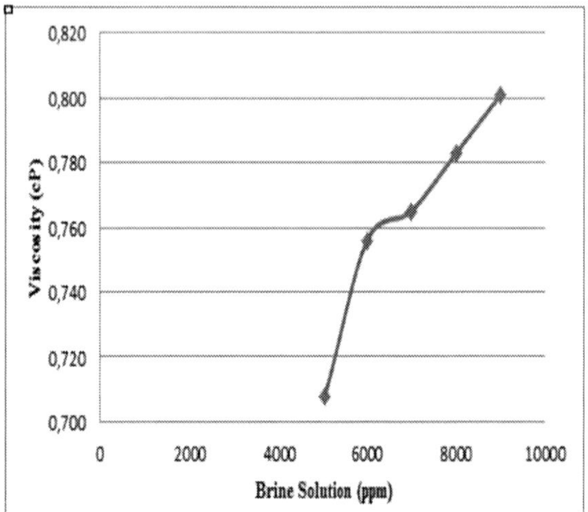

Fig. 3. The brine solution vs viscosity

In this study, it presents the results of measurements of the viscosity of the surfactant solution which varies between 5000 ppm - 9000 ppm. Figure 3 shows that the increase in salinity levels will increase the viscosity value. The results of measurements of the viscosity of the surfactant solution which varies between 0.1% - 1% were then discussed here.

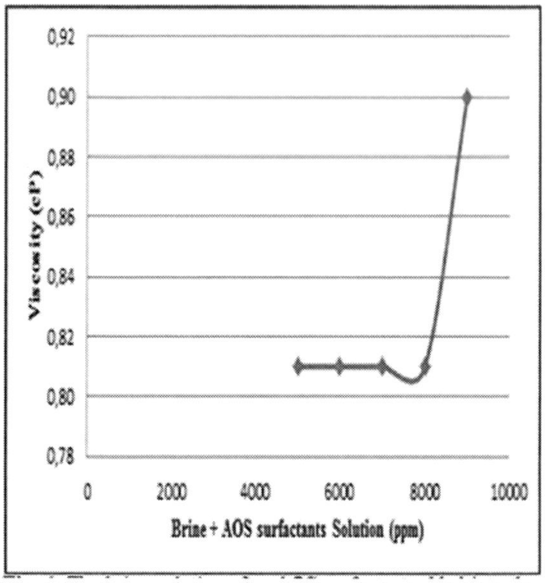

Fig. 4. The brine solution after AOS surfactant added into the solution vs viscosity

Figure 4 shows that the increase in the concentration of surfactant will increase the viscosity value. It followed by a discussion on the results of measurements of the viscosity after the adding of AOS surfactants.

3.3 Surface Tension Measurement Results

Surface tension measurements performed in the laboratory using a Du Nouy tensiometer. surface tension measurements performed twice, ie measurement of surface tension surfactant solution and measurement variation surfactant solution.

At surface tension measurement of the surfactant solution used only one concentration of the same for the fifth solution, ie 1%. The following surface tension measurement results to the surfactant solution which varies between 5000 ppm - 9000 ppm.

From figure 5, it shows that increasing salinity levels will impact the value of surface tension. Surface tension value increases with the concentration of salinity. Here are the results of surface tension measurement of the surfactant solution which varies between 0.1% - 1%.

From figure 6, it shows that the increase in the concentration of surfactant will affect the value of surface tension. Surface tension value decreased with increasing concentration of surfactant.

 DOI: 10.1051/matecconf/201710102008

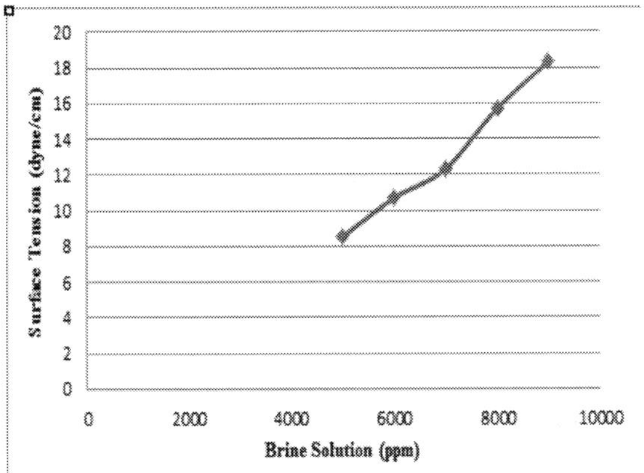

Fig. 5. The brine solution vs surface tension

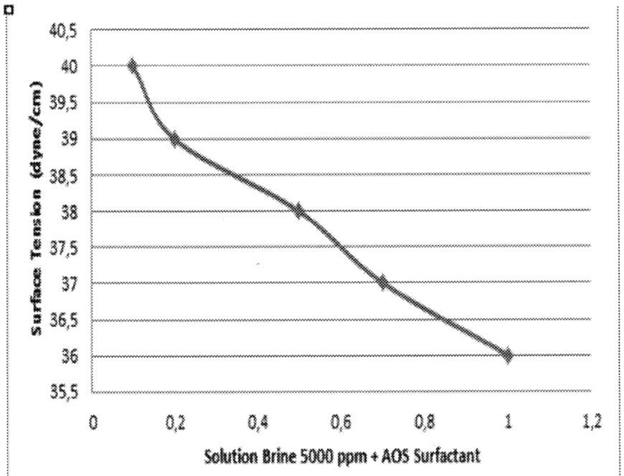

Fig. 6. The brine solution after AOS surfactant added into the solution vs surface tension

3.4 The Results of Surfactant Injection Into Reservoir Model to Determine the Optimum Salinity

The purpose of this test is to determine the optimum salitnity. Here are the results of surface injection into reservoir model.

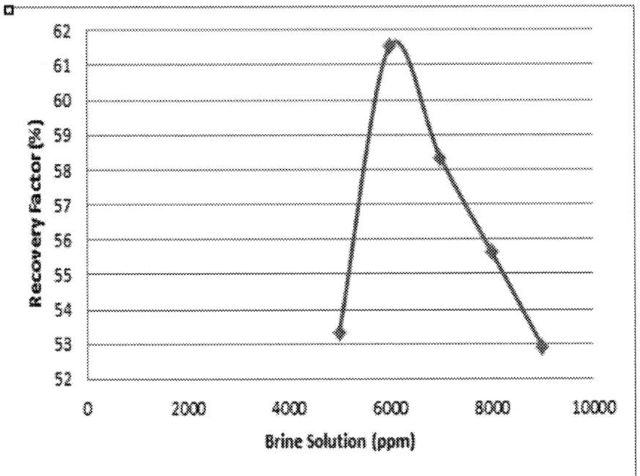

Fig. 7. The brine solution vs RF

From Figure 7, it shows that the solution to the salinity of 6,000 ppm yield the most optimum RF. In this study, each solution is conducted an injection using 1% surfactant concentration and the brine solution are varies between 5000 ppm – 9000 ppm.

3.5 The Results of Surfactant Injection Into Reservoir Model to Determine the Maximum Surfactant Concentration

The purpose of this test is to determine the optimum salitnity. Here are the results of surface injection into reservoir model.

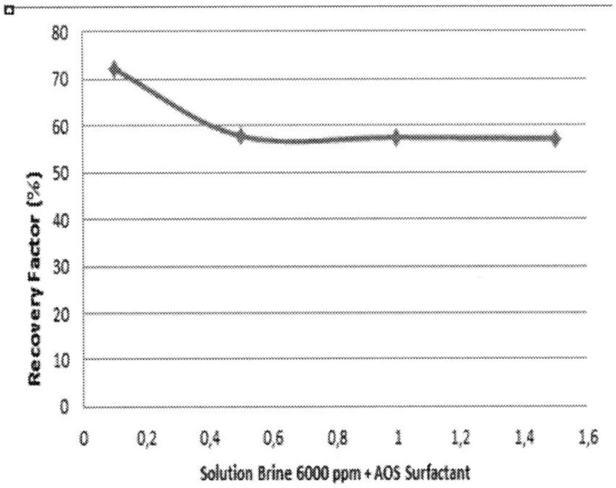

Fig. 8. The various AOS concentration vs RF

Figure 8 shows the solution to the salinity of 6,000 ppm provides the most optimal RF, then the reinjection process to determine the most optimal concentration using a salinity of 6,000 ppm. The result of the test is the surfactant concentration of 0.1% yield the most maximum RF.

4 Conclusion and recommendation

a. Conclusion

Based on the analysis and calculations have been conducted in this research, there are several conclusions.

(1) Increased levels of salinity and concentrations of surfactants will increase the value of density;
(2) Increased levels of salinity and concentrations of surfactants will increase the value of viscosity;
(3) Increased levels of salinity and concentrations of surfactants will increase the value of IFT;
(4) The solution which has the most optimum salinity is a solution with 6000 ppm salinity. This solution has the greatest RF, about 61.53%;
(5) The solution which has the most optimum surfactant concentration is a solution with 0.1% surfactant concentration. This solution has the greatest RF, about 72.22%;

(6) Applying a 6000 ppm salinity and 0.1% surfactant concentration into a solution, will caused a significant increasing of RF (is around 30%).

b. Recommendation

Recommendation for the next research is to find a critical surfactant concentration which is very important from the economic point of tertiary recovery (chemical flooding) since the use of surfactants is expensive.

Critical micelle concentration (CMC) refers to the concentration of surfactant solutions to form a large amount of micelles as one of the main parameters of each surfactant [9]. Critical micelle concentration can be determined from plot of interfacial tension (IFT) and surfactant concentration [3]. CMC values are important in all surfactant applications realted to the petroleum industry.

References

1. http://crudeoilpeak.info/peak-oil-and-fuel-subsidies-in-indonesia (accessed:August 18, 2016).
2. M. Abdurrahman, W.S. Bae, A. Novriansyah, and I. Khalid. The IRES International Conference, 9-12 (2016).
3. Samanta, A., Bera, A., Ojha, K. *et al.* J Petrol Explor Prod Technol (2012)
4. L.W. Lake, *Enhanced Oil Recovery* (1989).
5. *SKK Migas Annual Report* (2012).
6. F.D.S. Curbelo, A.I.C. Garnica, T.N.C. Dantas, E.L. B. Netto, Oil Rec. by Ion. Surf. and Swe. Eff. Stu. in Sands. Brazilian Journal of Petroleum and Gas, **V.2**, 9-16 (2008).
7. J.A. Xavier, *Effect of Varying Surfactant Concentration on Interfacial Tension in Tertiary Recovery of Crude Oil* (2011).
8. D.W. Green, and G.P. Willhite, E.O.R. Society of Petroleum Engineers Inc. (1998).
9. R. Farajzadeh, R. Krastev, and P.L.J. Zitha, Foa. Fil. Stab. w. A.O.S. Journal of ELSEVIER 35-40 (2008).
10. http://jpchemicals.in/a-o-s/ (accessed:November 01, 2016).
11. T. Babadagli, A. Al Bemani, F. Boukadi, R. Al Maamari, A Lab. Feas. Stu. of Dil. Surf. Inj. Journal of Petroleum Science and Engineering, **48**, 37-52 (2005).

* Corresponding author: widiayanti@gmail.com

Remediation of leachate by composite NZVI-activated carbon in packed column

Eka Sri Yusmartini[1*], *Muhammad* Faizal[2], *Marsi*[3]

[1]Chemical Engineering Department Faculty of Engineering Muhammadiyah University, 30263 Palembang, Indonesia
[2]Chemical Engineering Department Faculty of Engineering Sriwijaya University, 30139 Palembang, Indonesia
[3] Faculty of Agriculture Sriwijaya University, 30662 Inderalaya, Indonesia

Abstract. Improper solid waste management at final disposal site can cause in environtmental problem. The surrounding water bodies could be affected by leachate from the solid waste decomposition process because leachate contains high concentration of BOD_5, COD, ammonia, hydrocarbon suspended solid, and heavy metals. This problem has affected a final disposal site (named Sukawinatan) in Palembang, Indonesia, which did not implement leachate management system. This paper propose a method to reduce BOD_5, COD, Ammonia concentrations of the leachate in by applying composite nano zero valent iron (NZVI) – activated carbon for the remediation system of the surrounding area of Sukawinatan final disposal. The performance of the method was evaluated in this study using laboratory porous column apparatus. The results showed that the composite NZVI-activated carbon can be used to reduce concentration of pollutants from water bodies. The removal efficiency of this method on BOD_5, COD and ammonia are 94.81 %, 92.81 % dan 95.58 % respectively.

1 Introduction

One current environmental problem is related to waste management. One of the most important environmental problems issues in Indonesia is groundwater contamination caused by improper waste management. In accordance to the Regulation No. 81 Year 2012 on Household Waste Management and Waste Household Waste kind in Indonesia the waste is usually brought to Final Disposal (TPA) site. While a small portion is recycled, the remaining stays as dumping.

In general, landfill leachate is defined as any contaminated liquid effluent percolating through deposited waste and emitted within a landfill or dump site through external sources [1], of which its route of exposure and toxicity often remains unknown [2]. The waste dumped in landfills will go through chemical and biochemical decomposition process. Rainfall and surface water percolating through the solid waste, could leachate out some of the hazardous constituents [3,4]. Indicated that the leachate usually contains COD (Chemical Oxygen Demand), BOD (Biochemical Oxygen Demand), ammonia, suspended solid hydrocarbon, and heavy metals concentrations. Leachate is also rich in phenols, nitrogen, phosphorus, which if not treated before it is discharged into the environment, will become a potential source soil and groundwater contamination [5].

As pointed by numerous researchers, leachate is a serious environmental problem to disharge heavy metals continuously, if it is not handled properly [6-12]. Landfill waste management using open dumping system without proper management of the leachate before dicharge into water bodies will lead to pollution of surrounding water bodies. Groundwater is one of the water resources that could be used as a raw material for various purposes for residential/domestic, agriculture, fisheries, livestock, and industrial. To overcome this problem, groundwater remediation should be implemented in the area surrounding the landfill.

Some researchers have carried out groundwater remediation effort to reduce the effect to reduce the effect of leachate various methods. Among others were, using Permeable Reactive Barrier to reduce heavy metals in leachate, such as Mn, Zn, Cd, Al, and Cu which respectively with reduction efficiency of 95.4%, 57.3% , 64.3%, 54.0% and 30.8 %. Percent reduction of the parameters of nitrogen in the leachate reached 74.6%, while the percent reduction of COD was 80.5% [13]. The use of activated carbon to the sequencing batch reactor (SBR) was reported by [4] where the leachate treatment process provides reduction of the COD by 66.1%.

More recently, the employment of membrane filtration technologies (ultrafiltration, nanofiltration, microfiltration and reverse osmosis) has emerged to be a justifiable and viable tool in pre-treatment or in patnership with chemical treatmens for elimination of colloids and suspended matters, and fractionation in evaluation of the pre ponderant molecular mass of

* Corresponding author: eka.yusmartini@gmail.com

organic pollutants in a given leachate, achieving a COD and heavy metals rejection coefficient value of 98 and 99 %, respectively [4, 14, 15]. On the contrary, the pressure-driven processes are subjected to the fouling effect by a wide spectrum of constituents (which requires extensive pre-treatment or chemical cleaning of the membranes, resulting in short lifetime of the membranes and decreases of process productivity) and a huge generation volume of concentrate (which is unusable and need further treatments [3].

Other for the remediation of polluted water leachate is by utilizing nanoparticles technology. This eco-friendly technology can be used for the sustainable management of groundwater resources [16]. Nanometer-size nano particles have a great potential in the handling of water, waste and air pollution [17]. Research by [18] indicated that the use of organic NZVI resulted in large reduction of contaminants, not only by magnitude but also based on the rate of reduction.

This paper presents results of a study on the performance evaluation of NZVI-activated carbon composite from coconut shell in remediate waste leachate Sukawinatan Landfill in Palembang Indonesia. The surface area of activated carbon with high absorption capability was able to provide advantages in this composite system. Application of composite NZVI-activated carbon from coconut shell made using adsorption column. Parameters measured in this study were the reduction of leachate concentrations based on the reduction of COD, BOD_5, Ammonia as wel as Pb and Cr.

2 Experimental details

2.1 Materials and method

The NZVI activated carbon composite used in this study was made of NZVI-Activated carbon with ratio 1:3 (w/w) the composition [19]. Leachates used in this research were taken from the Sukawinatan Landfill, Palembang, Indonesia.

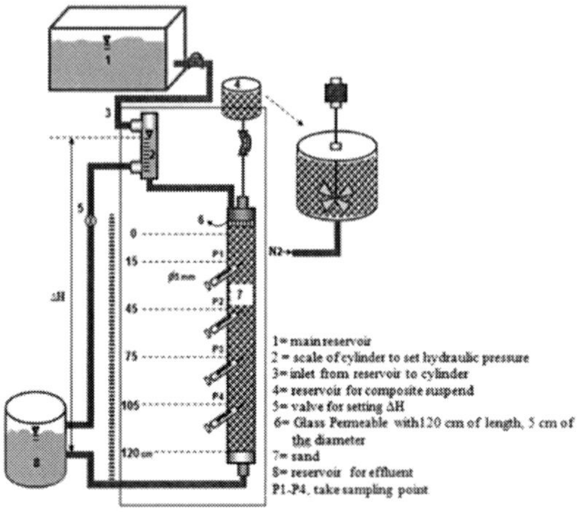

Fig 1. Packed Column Remediation

The study used laboratory porous column apparatus or packed column remediation apparatus as shown in Figure 1. This column with porous media was developed or modified based on the column used in previous study by Hosseini's column [20].

2.2. Remediation using porous column apparatus

The column done was made of 120 cm high glass cylinder with diameter of 5 cm. The column was filled with sand. The particle size of the sand ranged from 2 mm to 1 mm. The sand was compacted in the a glass cylinder. A stream of leachate was added manually to saturate the sand column from main reservoir. The concentration of the composite used in this study were 5, 8, 11 and 14 g/L. The composite was streamed into columns with fixed nitogren gas flow so that the flow remains stable. Composite flow rate is set at a speed of 4 mL/min. Samples were taken simultaneously from four (4) sampling points along the glass cylinder. The locations of the sampling points were at 15, 45, 75 and 105 cm from the surface of the column. Remediation time was varied in 60, 120, 180 and 240 minutes.. The samples were evaluated by to measure the parameters analyzed BOD_5, COD, Ammonia and concentrations of Pb and Cr.

3 Result and discussion

The initial concentration of each substance prior to treatment in the porous columns as compared to standard value are shown in Table 1.

Table 1. Concentration of hazardous elements from Sukawinatan Landfill Leachate as compare to standard range

Parameter	Unit	Value	Quality Standars*)	Comments
pH	-	5.65	6 - 9	OK
BOD_5	mg/L	924.00	50	High
COD	mg/L	1664.00	100	High
Ammonia	mg/L	47.52	1	High
Pb	mg/L	2.40	0.1	Low
Cr	mg/L	0.05	0.5	High

*) South Sumatera Governor Regulation Number 8 Year 2012

3.1 Effect of weight composites and contact time on BOD_5 reduction

Based on Table 1, the percent reduction expected to achieve the Quality Standard suggested by South Sumatera Governor Regulation Number 8 Year 2012, the parameters BOD_5, COD, Ammonia, concentrations of Pb and Cr are 94.58 %, 93.99 %, 97.89 % and 95.83 % respectively. The requirement should be met by varying contact time, the composite weight and sampling distance from the source.

Effect of weight and time remediation to the percent reduction of BOD_5 at all points of observation is shown in Figure 2. The figure indicates that the longer the contact time between the adsorbent with the adsorbate, the greater BOD_5 absorbed. The highest reduction was achieved in the composite weight of 14g and remediation time of 240 minutes. The reduction was 94.81 % with a final consentration of BOD_5 is 48 mg/L which meets the targets of Quality Standard that is 94.58 %.

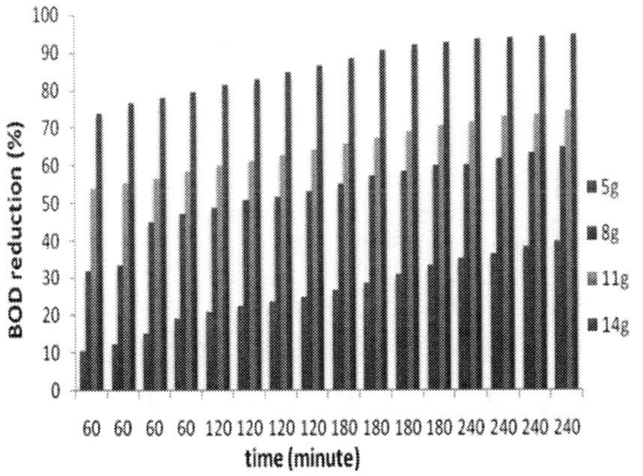

Fig.2. Effect of concentration of composite and contact time on BOD_5 reduction

3.2 Effect of weight composites and contact time to reduce COD

The effect of composite weight and contact time on the percent reduction of COD is shown in Fig.3. The highest reduction was achieved with contact time of 240 minutes weight 14 g, i.e. of 92.49 %. Compared with the target of reduction that meets the target quality i.e. 93.99 %, thus more NZVI composite and longer contact time was require to meet the reduction targets.

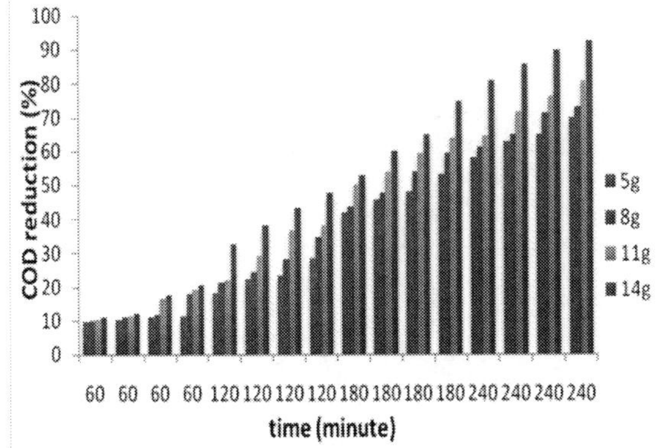

Fig.3. Effect of concentration of composite and contact time on COD reduction

COD and BOD_5 is important indicators used to access the biodegrability of organic compounds [13]. The ability of composite NZVI-activated carbon to reduces the possibility of a break in the chain of carbon from complex organic compounds, so most of the organic compounds that are non-biodegradable were converted to biodegradable. Reduction of BOD_5, COD on a variation of the weight of the composite or composite concentration shows a drastic change in the remediation of leachate pollutans proves that most of the non-biodegradable organic compounds has been converted into a biodegradable.

The analysis of the concentration of Pb and Cr resulted in leachate after remediation process was conducted using a concentration of 5 g/L, has not been detected again. This is supported by the results obtained by [24] to obtain a percent reduction of 100 % NZVI weight concentration by used of 0.4 g/L. Research by [25] get maximum % removal Cr was seen at time of 105 min. in column studies.

The analysis indicates the relative reduction in the composite pollutant leachate. The stable composite was established during reduction of pollutant in leachate. Adsorption capacity of activated carbon and reactive characteristic of NZVI inside of composite caused suspended composite be able to stable in long times [22]. The NZVI-activated carbon composites combine the physical adsorption capacity of activated carbon with the destructive reactivity of NZVI and can be used as a reactive activated carbon, which is not subject to the limitations of using activated carbon and NZVI separately [23].

3.3 Effect of weight composites and contact time to reduce ammonia

The influence of weight and contact time on the concentration of ammonia at all points of the observations is shown on Fig.4. In the figure, it shown that the longer the time of contact time between the adsorbents with adsorbate, the greater the ammonia absorbed.

The highest reduction was achieved on heavy use of composite 14g with 240 minutes of contact time with the reduction of ammonia 95.58 %. Compared with the target of reduction that meets the target quality i.e.97.89 %, thus more NZVI composite and longer contact time was require to meet the reduction targets.

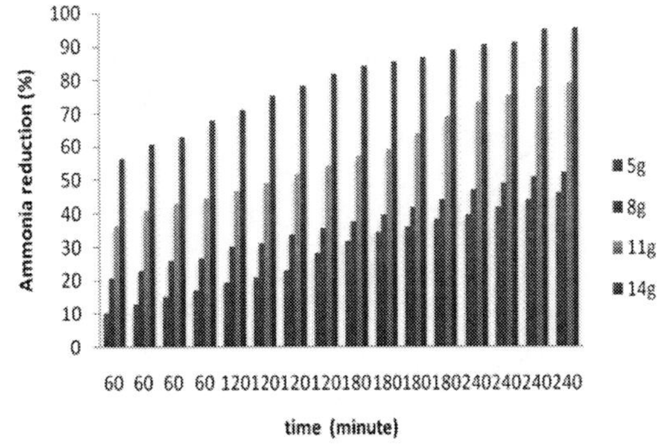

Fig.4. Effect of concentration of composite and contact time on ammonia reduction

Based on the results of a composite application against leachate on remediation, the column reached percent reduction on the parameters of BOD_5, COD and Ammonia as shown in Table 2. Reduction of BOD, COD and ammonia content in leachate were measured by the difference between initial concentration by each of pollutant's elements.

Results of analysis of Pb and Cr concentrations in leachate post remediation process performed using a concentration of 5 g/L was no longer detectable. These results are supported by the results obtained [24], get a 100% percent reduction using NZVI heavy concentration of 0.4 g / L. Reduction is the predominant mechanism for Pb and Cr removal [26].

Table 2. Reduction NZVI- activated carbon composite in leachate remediation

Parameter	Initial concen.(mg/L)	Final concen. (mg/L)	% Reduction
BOD_5	924	48	94.81
COD	1664	125	92.49
Ammonia	47.52	2.1	95.58

Results from this study, it is resulted that the stability of the composite in reducing pollutants leachate. The role of activated carbon that maintains the stability and form a barrier sorption and reactivity NZVI can maintain the stability of the suspension in a long time [22]. Percent reduction of the composite to heavy metals Pb and Cr reaches 100%.

The reduction of BOD_5 and COD reached respectively at 94.81% and 92.49% on the weight of the composite 14 g / L of contact time of 240 minutes, while [21] get a percent reduction of BOD_5 and COD respectively 35% and 47.94% by weight of 2.5 g / L., while [13] getting percent reduction of COD in leachate 66.8% using ZVI and 80.5% use zeolite media and ZVI on permeable reactive barrier system.

4 Conclusion

The results obtained in the process of remediation using NZVI-Activated Carbon prove that composites in 1:3 ratio is aneffective substance to be used for groundwater remediation of contaminated water by leachate from landfil. The substance was mixed with contaminated water in a porous column with a contact time of 240 minutes. In particular composite number, the performance of the adsorbent is relatively increased because it is supported by the stability and high absorption, besides their reactivity makes composites capabilities do not only absorb but also reduce the amount of contaminants. The removal efficiency of this method BOD_5, COD and Ammonia are 94.81 %, 92.49 %, 95.58 % respectively.

The authors would like to acknowledge all support coming from Management of Sukawinatan Landfill in Palembang and National Nuclear Energy Agency of Indoensia (BATAN) for supporting and advocationg the process of NZVI synthesis and analysis.

References

1. M.E.Taulis, Resolution Press, Christchurch, **173-190** (2005).
2. L. Koshi, E. Paris, S.Ling, T. Jones, K. Beruhe, Sci. Total Environ, **384:171-181** (2007).
3. S. Renoua, S., J.G Givaudan, S. Poulain, F. Dirassouyan and P. Moulin. J. Hazard Mater. **150:468-463** (2008)
4. K.Y. Foo, and B.H. Hameed. J. Of Hazardous Materials, **171:54-60** (2009)
5. T.H. Christensen., P. Kjeldsen and P.L. Bjerg. J. Appl.Geochem. **16:659-718** (2001)
6. H. Fan, H. Shu, HS. Yang, Wc.Chen, J. Science of Tot. Envi. **36:25-37** (2006).
7. R.S. Sletten, M.M. Benjamin, JJ.Horing, JF Ferguson. J. Water Res. **29(10):237-86** (1995)
8. T. Urase, M.Salequzzaman, T. Matsuo, K. Yamakomo, N. Suzuki, Water Scie. Technol., **36(12):349-356** (1997).
9. J.B. Christensen, DL.Jensen, C. Gron, Z. Fulip, TH. Christensen, J. Water Res. **32(1): 125-135** (1998).
10. K. Fytianos, B. Tsaniklidi, Env.Int. **24(4):477-86** (1998).
11. JC. Chen, MY. Wey, YC. Lin, J. Chemosphere 37(13): 2617-25 (1998).
12. CS. Poom, ZQ. Chen. J. Chemosphere **38(3):663-680** (1999).
13. D.Jun., Z. Yongsheng., Z. Weihong and H. Mei.. Journal Of Hazardous Materials **161: 224-230** (2009).
14. K. Ushikoshi, T. Kobayashi, K. Dematsu, A. Toji, D. Kojima, K. Matsumoto, J. Desalination **150: 121-129** (2002).
15. D. Trebounet, J.P. Schlumpf, P. Jaouen, F. Quemeneur, J. Water Res. **35:2935-2942** (2001).
16. D. Grienger., A. Fjordboge, N.B. Hartman, E.Eriksson,. Journal Of Contaminant Hyd. **118, 165-183** (2010).
17. X. Li., D. Elliot and W. Zhang.. Critical Reviews In Solid State And Materials Science. **31:111-122**(2006).
18. W. Zhang, and W.E. Daniel. Applications Of Iron Nanoparticles For Groundwater Remediation. Published Online In Wiley Interscience (2006).
19. E.S. Yusmartini., D.Setiabudidaya.,Ridwan, Marsi, Faizal. Journal Advanced Materials Research **62-65** (2015).
20. S.M. Hosseini., A.B. Ashtiani and M. Kholgi. Journal Desalination .**276:214-221** (2011).
21. Z.E. Kashitarash., S.M. Taghi, N. Kazem, A. Abbas and R. Alireza. Iranian Journal of Environmental Health Science and Engineering. **5: 9-36** (2012).
22. K. Mackenzie., Schierz, A., Georgi, A., and

23. Kopinke, F.D. Journal Global Nest, **10(1):54-61.** (2008).
24. HH.Tseng, JG Su, C. Liang, J. of Hazardous Materials, **192:500-506** (20110.
25. N. Shao-feng, I. Yong, X. Xin-Hua, J. Zhejang Univercity Science, **10:1022-1027** (2005).
26. S.H. Rahmi, G.M. Madhu, AA, Kittur and R. Surresh, International J. of Current Engineering and Technology, **37** (2013).
27. X. Li, W. Daniel, W. Elliot and W. Zhang, Critical Reviews in Solid State and Material Science, **31:111-125** (2006)

Treatment of landfill leachate by electrocoagulation using aluminum electrodes

Rusdianasari[1,*], *Ahmad* Taqwa[2], *Jaksen*[1], and *Adi* Syakdani[1]

[1]Chemical Engineering Department State Polytechnic of Sriwijaya, 30139 Palembang, Indonesia
[2]Electrical Engineering Department State Polytechnic of Sriwijaya, 30139 Palembang, Indonesia

Abstract. Leachate is wastewater decomposition of organic waste that can contaminate soil and groundwater if it is not handle properly. Contamination by leachate can be be prevented by reducing leachate level before the wastewater reaches the ground. One of the methods used is electrocoagulation. Electrocoagulation is an electrochemical water treatment method where in anoda occured the release of active coagulant as metalic ion, while in cathode occured the electrolysis reaction in a form of the release of hydrogen gas. This study is objected to investigate the effectiveness of electrocoagulation in removing Biochemical Oxygen Demand (BOD_5), Chemical Oxygen Demand (COD), NH_3-N and lead (Pb) of leachate. In this research, the processing of leachate with electrocoagulation method using aluminum electrode has been proceed with the dimension of electrode 16.5 cm x 7 cm x 0.2 cm. Optimum conditions and the effectiveness of electrocoagulation for each parameter are in the current density 30 A/m^2 with 60 minutes of time process. The electrocoagulation method is effectively reduced the value of TSS 46.80%; BOD_5 71.33%; COD 73.77%; Pb 62.5%; NH_3-N 57.92%; and for the increased of pH from 8.03 to 8.95, respectively. Electrocoagulation method are successfully reduced the levels of pollutants in accordance with environmental standard.

1 Introduction

Landfilling is the most common and easy way to dispose the solid waste. Landfill is commonly received the wastes from municipal near to a landfill. If the location of waste generated is far, the transfer station is the solution to reduce cost of waste transportation. The waste usually mixed of waste products from residential area, commercial, institutional and others.

There are three kinds of output for landfills, examples gas, liquid (leachate) and inert solids [1]. Leachate has a complex structure and high pollutant load, and its treatment is quite hard to meet the discharge standards. Leachate becomes the main pollutant wastewater since it is the most difficult to treat as it is a wastewater with complex and widely variable content generated within a landfill [2]. Leachates may contain organic contaminants in large amounts and can be measured as Biochemical Oxygen Demand (BOD_5), Chemical Oxygen Demand (COD), ammonia, and high concentration of heavy metals. It contains high concentration of pollutants which can have adverse effects on the environment [3]. Therefore, many pre-treatment and combined treatment methods have been proven to treat leachate. Until today, many methods for its treatment have been proven. Some well known treatment methods such as biological treatment methods, membrane processes, advanced oxidation techniques [4], coagulation-flocculation methods, lagoon and wetland applications [5], have been examined in the literature.

Simple, affordable, and efficient leachate treatment systems are urgently needed in developing countries because most of the conventional technologies currently in use in industrialized nations are too expensive and complex. Electrocoagulation is one of a simple method to treat wastewater efficiently [4]. This electrochemical treatment seems to be a promising treatment method due its high effectiveness, its lower maintenance cost, less need for labor and rapid achievement of results. From some research, in the past few decades, it shows that the electrocoagulation is a promising treatment method and effectively potential to treat verity type of wastewater including dyes wastewater, palm oil mill effluent, food wastewater, urban wastewater, and removing heavy metals [5].

Electrocoagulation has been used in treating wastewater containing oil and grease, suspended solids and even inorganic and organic pollutants that can be flocculated. This method has been effectively applied for the treatment of the textile dye wastewater, purification of wastewater, tannery wastewater and domestic wastewater. This method is categorized by simple equipment and easy operation. The electrocoagulation processes have lesser amount of sludge [6] and having

* Corresponding author: rusdianasari19@gmail.com

features like relatively more economic and higher treatment efficiency has been a promising method [7].

Electrocoagulation treatment methods offer an alternative to the use of chemical coagulant such as metal salts or polymers for breaking the pollutants because during the electrocoagulation process, the electrode can generate coagulated species and metal hydroxides that destabilize and aggregate the suspended particles and precipitate. The hydrogen gas released from cathode would also help to float the flocculated particles out of the water [8]. Electrocoagulation process is a consistent and coast effective method with low amount of sludge and without any sensitivity to toxic materials. In the electrocoagulation process coagulant is produced by electrolytic oxidation of an appropriate sacrificial anode by suing a direct current. The electrocoagulation mechanism for aluminum electrode could be represented as follow [9]:

$$Al_{(s)} \rightarrow Al^{3+}_{(aq)} + 3e^- \quad (1)$$

$$3H_2O + 3e^- \rightarrow 3/2H_{2(g)} + 3OH^-_{(aq)} \quad (2)$$

$$Al_3^{+}{}_{(aq)} + 3OH^-_{(aq)} \rightarrow Al(OH)_{3(s)} \quad (3)$$

The experiments were carried out in a batch mode to evaluate effect of different operational variables like pH, current density, BOD_5, COD, TSS, NH_3-N, and heavy metal removal of leachate wastewater.

2 Materials and methods

2.1 Leachate sample

In the experimental study, leachate samples were collected from Sukawinatan Landfill, which is located in Palembang city, Indonesia. The samples were collected in glass bottles transported in ice cool container to the laboratory and stored at 4^oC in order to keep the leachate characteristics unchanged.

2.2 Experimental device

The batch experimental setup is schematically shown in Fig.1.

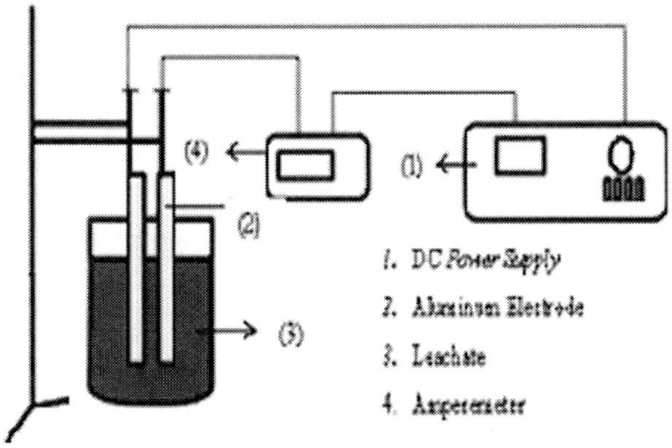

Fig. 1. Schematic diagram of experimental setup

The electrochemical unit consists of an electrocoagulation cell, a DC power supply and the aluminum electrodes. There are two monopolar electrodes that have same dimension (16.5 cm x 7 cm x 0.2 cm) as an anode and a cathode which spacing of 1 cm which were placed vertically in the floor of the cell.. In order to maintain an unchanged composition and avoid the association of the flocs in the solution, the stirrer was turned on and set at 100 rpm. All electrodes were washed with dilute HCl before every experiment conducted. Every experiment was performed at the room temperature.

2.3 Experiment procedure

The procedure started with electrocoagulation cell cleaned with distilled water and dried using dryer. The experiments were carried out in a batch mode. For each experiment, a leachate sample of 1 L was collected in the electrochemical cell with aluminum electrodes dipped into the sample. Four different of high current densities; 30 A/m^2, 50 A/m^2, 70 A/m^2 and 90 A/m^2 were applied. In each current density applied, contact times of 15, 30, 45, and 60 were used. Therefore, total experiments were carried out to determine the effect of current densities, electrocoagulation time and effectiveness of electrocoagulation. The pH of the solution was taken using pH meter. The electrodes (anode and cathode) were clamped at electrode stand. All connections in the circuit were completed by wire connection to terminal positive and negative to DC power supply, electrodes (anode and cathode), voltmeter and ampere meter.

The colour of the solution of electrolyte solution was observed before and after the process occurred. After the experiment, the treated sample was then kept undisturbed for 20 min in order to allow the flocs to settle. Subsequently, after settling the sample of supernatant was collected to perform the analysis of pH, TSS, BOD5, COD, lead metal and content of NH_3-N. All the experiments were repeated with same operational conditions using aluminium electrodes (both anode and cathode). For the treated effluent BOD_5, COD, TSS and NH_3-N were measured according to Standard Methods after each batch run.

3 Results and discussion

The electrocoagulation process was influenced by operating parameters such as current density, and reaction time on removal of BOD_5, COD, TSS, NH_3-N, and Pb concentration. Pb concentration was measured by atomic absorption spectrophotometer.

3.1 Characteristic of the landfill leachate

Several parameters such as the people living standards of the study area, landfill age, and climatic condition affected to the composition and characteristic of the landfill leachate. To this aim at different time the sampling was performed and various water treatment criteria such as pH, BOD_5, COD, TSS, NH_3-N and Pb concentration were used to assess characteristics of landfill leachate which are presented in the Table I. The conventional biological methods are ineffective to treatment of the leachate because it has high ammonia nitrogen. According to literatures to treat a wastewater with mentioned characteristic, electrocoagulation method is an effective method [10].

Table 1. Characteristics of the landfill leachate before treatment

No	Characteristics	Value	Treshold Standard*
1	pH	8.03	6-9
2	BOD_5 (mg/L)	112.6	200
3	COD (mg/L)	415	50
4	TSS (mg/L)	1304	100
5	NH_3-N (mg/L)	2.4	1
6	Pb content (mg/L)	0.08	0.1

*South Sumatra Governor Regulation No. 6 of 2012

It can be seen from the analysis that the parameters that need to improve in order to meet the disposal standards of leachate include COD, TSS, and NH_3-N.

3.2 Effect of current density to processing time for pH

This experimental work used ranges current density of 30 A/m^2, 50 A/m^2, 70 A/m^2 and 90 A/m^2. These ranges will give the data about how current density will affect the electrocoagulation efficiency in the removal of pollutants that contain in the leachate samples.

As shown in Fig. 2, we can see the increase of leachate pH. The initial pH was 8.03 categorized in base category. After treatment there was significantly increase of pH. The pH reached 8.95. The increase of pH in the electrocoagulation process was due to the process of alkalization Al^{3+} ions that were added to the water and reacted with hydroxyl ions from water hydrolysis that produced $Al(OH)_3$ and hydrogen ions.

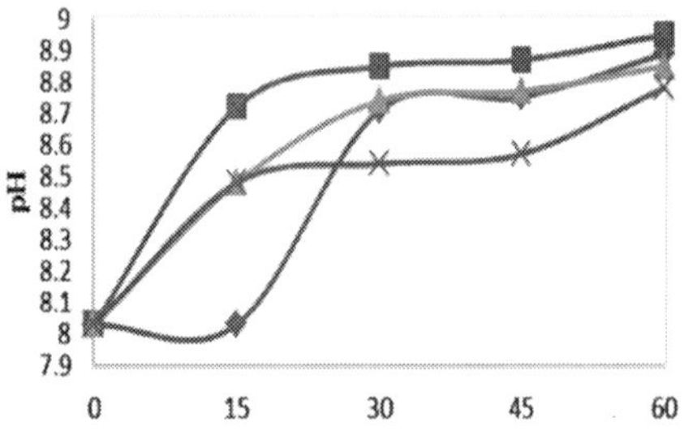

Fig. 2. The effect of current density to processing time for pH

In addition to hydrogen gas, it also generated hydroxide ions $(OH)^-$, the smaller the current density used the faster the reaction might occurred and OH^- generated also more in consequence the pH in the water would also increase. The occurrence of such reactions during the electrocoagulation process would raise the pH [11].

In Fig. 2, it can be seen that the pH tended to increase with the length of 7-8. The best results in the increase in pH of 8.95 occured in the current density of 50 A/m^2 in 60 min.

3.3 Effect of Current Density to Processing Time for TSS

TSS reduction in the electrocoagulation process occurred when the solid material of the suspended solids in the coagulant absorbed onto Al(OH)3 or absorbed into the air bubble. The results of this adsorption were floated that decrease the concentration of TSS in wastewater.

TSS sources of pollutants were chemicals both organic and inorganic that formed suspension in the waste water. Besides that the sources of TSS also came from metals that formed complex compounds either with hydroxide anions or other compound suspended in the waste solution either because of the nature of the molecular size of the compounds and polarity properties [12].

As shown in Fig. 3, we can see the results of electrocoagulation of leachate. The best result in impairment of TSS was at 60 min with current density of 30 A/m^2 in which the initial TSS value of 112.6 mg/L impaired to 59.9 mg/L. TSS reduction process were highly influential that TSS were pollutants in the form of suspension. If waste water contained high TSS, it could be concluded that the waste water was in poor quality and had the potential to damage the aquatic ecosystem in particular.

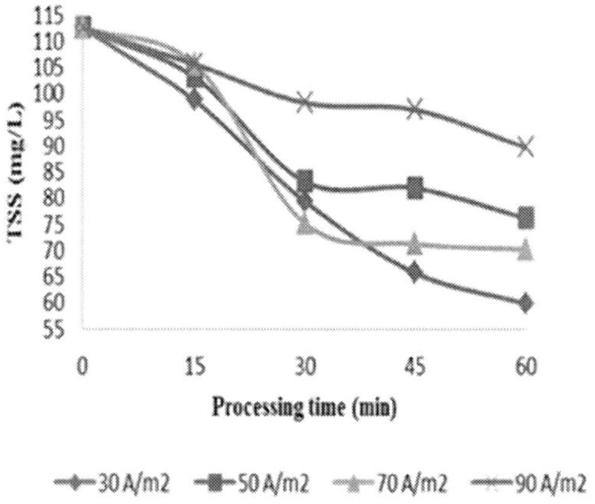

Fig. 3. The effect of current density to processing time for TSS

According to Islam *et al.* [13], the two main objectives of the coagulation are to settle the suspended colloidal particles in wastewater quickly, which settle very slowly, or maybe do not settle at all under normal conditions, later leading to residuals TSS, and to remove TSS from the leachate/wastewater and consequently to obtain clearer leachate/wastewater, which is a natural result of the earlier.

3.4 Effect of current density to processing time for BOD_5

Result of BOD_5 analysis determined the quality of water bodies, amount of oxygen needed by microorganisms to decompose organic matter contained in water in the aerobic state. BOD_5 high value played an important role in determining the ability of the water bodies in supporting the better growth of algae and aquatic organisms. The higher the number of the bacterial population, the higher the level of the water pollution would be.

In Fig. 4, the results of electrocoagulation of leachate to the value of BOD_5 decreased. The best result of BOD_5 impairment was at 60 min and current density of 30 A/m^2 with a value of BOD_5 119 mg/L. At this process time, there was significant change in the value of BOD_5 from 415 mg/L to 119 mg/L, yet the greater the current density, the greater the value of BOD_5 would slightly rise. The greater the density current and the longer the time, the electrode process would be more saturated that caused the electrocoagulation process reached its lowest point and increased in the value of BOD_5.

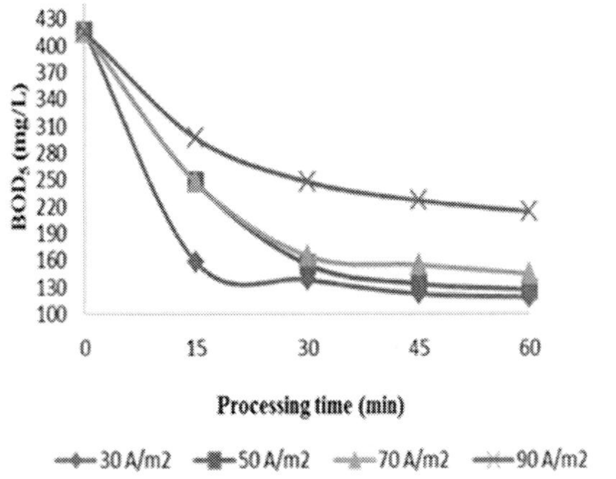

Fig. 4. The effect of current density to processing time for BOD_5

3.5 The effect of current density to processing time for COD

COD was the quantity or amount of oxidant that reacted with the sample under certain conditions. The amount of oxidant was proportional to the oxygen. Organic and inorganic compounds in the sample were oxidized subject but organic compounds were more dominant. COD was often used for measuring the quantity of pollutants in the water.

In Fig. 5, the best result of leachate waste by method of electrocoagulation in decreasing the value of COD was at 60 min with a current density of 30 A/m^2. COD value decreased drastically from 1304 mg/L to 342 mg/L.

Basically, the decline in COD values in electrocoagulation occured by destabilization. Colloidal destabilization carried out by metal cation that formed polyvalen polyhydroxide. This complex compound had high adsorption for facilitating aggregation of wide range of pollutants that formed large material easily separated by flotation technique since the density of material became smaller [14].

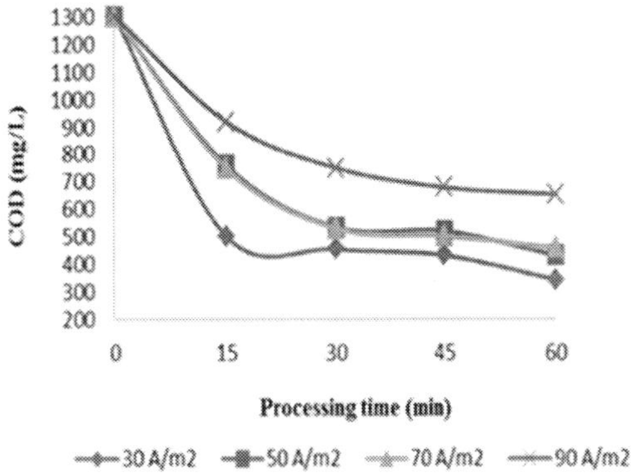

Fig. 5. The effect of current density to processing time for COD

3.6 Effect of current density to processing time for Pb metal

Based on the experiments, data showed that any change in current density and the length of time would generate different electrocoagulation efficiency. The longer the contact time, the lower the level of Pb metal in the filtrate would obtain for any variations in current density. In Fig. 6, it can be seen that Pb metal reduction of the biggest waste was achieved at runtime of 60 min and current density of 30 A/m^2. In this condition, the level of Pb metal in the filtrate was 342 mg/L.

The decreased level of Pb metal in the process of electrolysis might occur due to changes in electrical currents causing magnetic field around the electrodes. The Pb ions moved with trajectories helix shape around the electrode plate so that at that time there was tendency of Pb^{2+} ions to attach to the entire surface of the electrode plate.

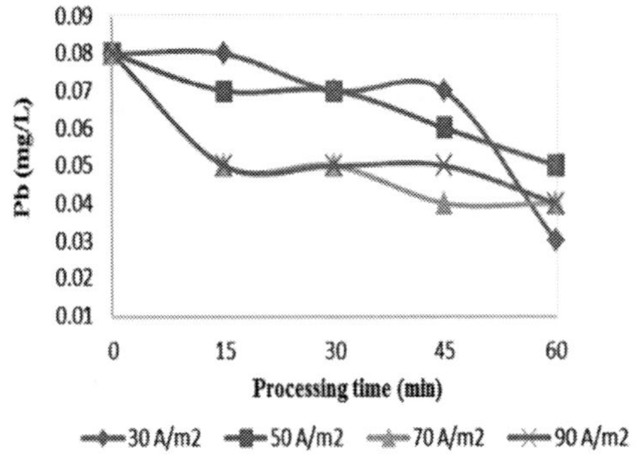

Fig. 6. The effect of current density to processing time for Pb metal

In an electrochemical process, at the same time the electric current in the anode reaction will occur in the oxidation of the anion (negative ion), an anode made of aluminum metal would undergo oxidation reactions to form Al^{3+} ion and would bind OH^- ions forming flock $Al(OH)_3$ which can bind Pb^{2+} ions and capture most of Pb that was not deposited on the cathode rod [15,16].

3.7 Effect of current density to processing time for the content of NH_3-N

The decrease content of NH_3-N in the electrocoagulation process for liquid waste from leachate was very important. This was because ammonia (NH_3) called ammonia nitrogen was produced from the decay of organic substances by bacteria. Each of ammonia released into the environment formed an equilibrium reaction with ammonium ion (NH_4^+). Ammonium then underwent nitrification to form nitrite and nitrate. Ammonia that was dissociated was more dangerous to aquatic biota instead of in the ammonium form. The value of ammonia was related to the pH value of water. The higher the pH of the water, the greater the ammonia content in the form of dissociated. High ammonia level was indication of contamination of organic material originating from domestic waste, industrial, and agricultural fertilizer runoff [17,18].

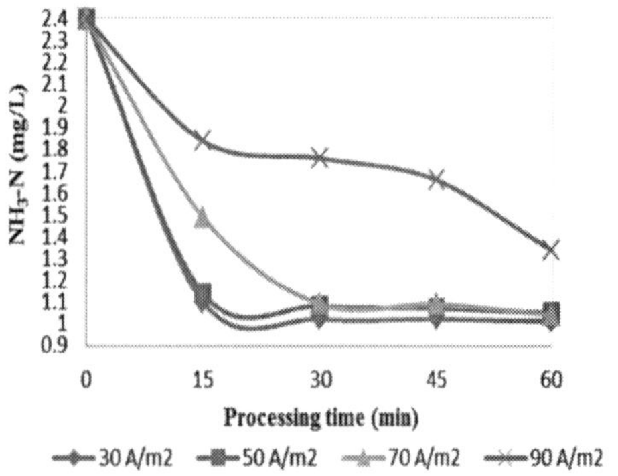

Fig. 7. The effect of current density to processing time for the content of NH_3-N

In Fig. 7, the largest content reduction of NH_3-N in leachate was achieved in 60 min at 30 A/m^2. In this condition the levels of NH_3-N in the filtrate was 1.01 mg/L from preliminary analysis of 2.4 mg/L.

3.8 The Effectiveness of electrocoagulation to the treatment leachate

The decrease of pollutants contained in the leachate might be done by electrocoagulation. After the processing pollutants contained in the effluent decreased. The effectiveness of electrocoagulation in treating leachate with the parameters of TSS, BOD_5, COD, NH_3-N, Pb levels and pH values were different.

TSS great decrease in the current density of 30 A/ m^2 with a processing time of 60 min is 59.9 mg/L. Decrease in BOD_5 and COD are good at current density of 30 A/m^2 with a processing time of 60 min were 119 mg/L and 342 mg/L. Decreased levels of Pb and NH_3-N content of the nicest on a current density of 30 A/m^2 with a processing time of 60 min were 0.03 mg/L and 1.01 mg/L. The increase in the pH value was good at current densities of 50 A/m^2 with a time of 60 min was 8.95.

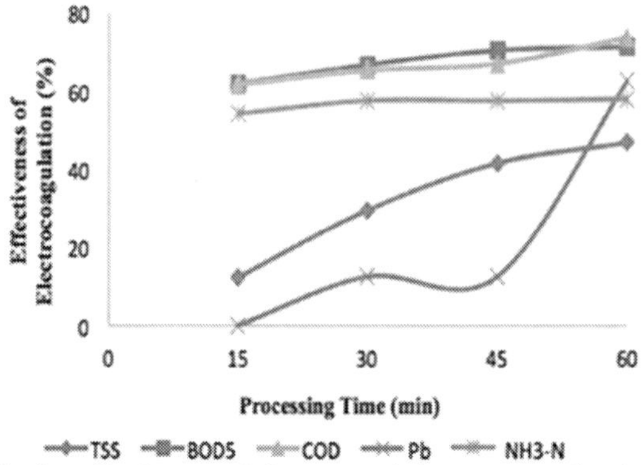

Fig. 8. Effectiveness of electrocoagulation to processing time on current density 30 A/m^2

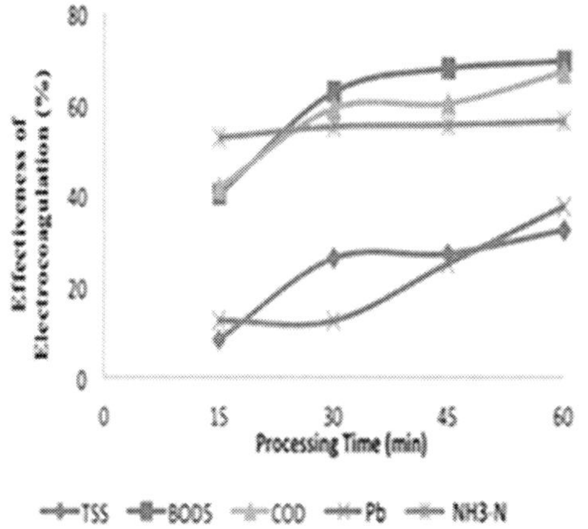

Fig. 9. Effectiveness of electrocoagulation to processing time on current density 50 A/m^2

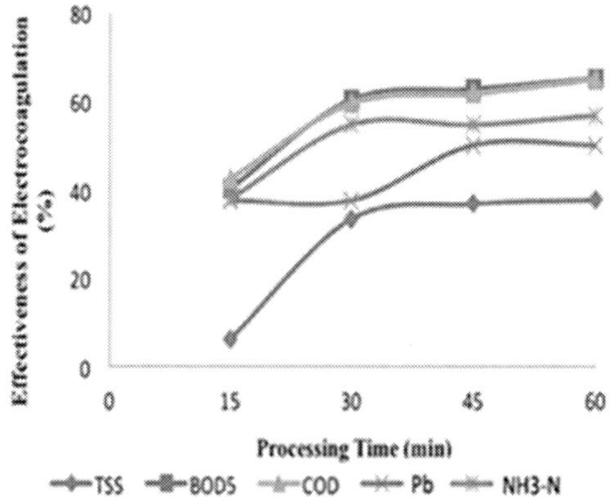

Fig. 10. Effectiveness of electrocoagulation to processing time on current density 70 A/m^2

Fig. 8-12 shown the effectiveness of the electrocoagulation versus processing time on the current density of 30 A/m^2, 50 A/m^2, 70 A/m^2, and 90 A/m^2 with a comparison between TSS, BOD_5, COD, NH_3-N, and Pb concentration and shown the effectiveness of a decrease in BOD_5, COD and NH_3-N levels.

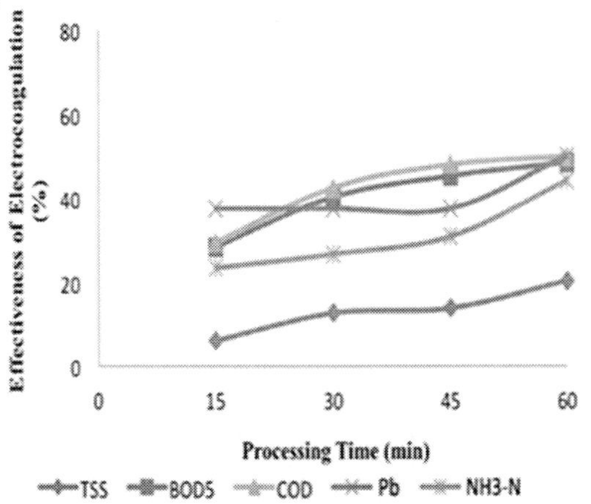

Fig. 11. Effectiveness of electrocoagulation to processing time on current density 90 A/m^2

The effectiveness for TSS was 46.80% from the initial sample of 112.6 mg/L to 59.9 mg/L. The effectiveness for BOD_5 was 71.33% from the initial sample of 415 mg/L to 119 mg/L and the effectiveness COD was 73.77% from the initial sample of 1304 mg/L to 342 mg/L. The effectiveness for Pb metal was 62.5% from the initial sample of 0.08 mg/L to 0.03 mg/L and the effectiveness for NH_3-N content was 57.92% from the initial sample of 2.4 mg/L to 1.01 mg/L.

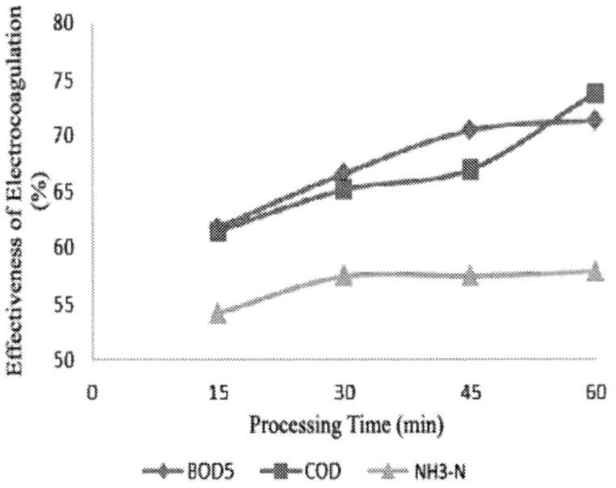

Fig. 12. The Effectiveness of Electrocoagulation to the values of BOD5, COD, and NH_3-N Content

Conclusions

The effectiveness of electrocoagulation were investigated in this study. Optimum conditions and the effectiveness of electrocoagulation for each parameter are in the current density 30 A/m^2 with 60 minutes of time process. The electrocoagulation method is effectively reduced the value of TSS 46.80%; BOD_5 71.33%; COD 73.77%; Pb 62.5%; NH_3-N 57.92%; and for the increased of pH from 8.03 to 8.95, respectively. Electrocoagulation method are succesfully reduced levels of pollutants in accordance with environmental standard.

References

1. M. K. N. Mahmad, M. R. Rozainy, I. Abustan, N. Baharun. Procedia Chem. **19**, 681-686 (2016)
2. E. S. Yusmartini, D. Setiabudidaya, Ridwan, Marsi, Faizal. J. of Phys: Conf. Series. **423** (2013)
3. B. MJK, I. MH, K. SRM. Waste Manag. **29**, 9, 2534-2541 (2009)
4. Kamaruddin, M.A, M.S. Yusoff, AzizH.A. Appl. Water Sci. **5**, 113-126 (2015)
5. M. Morozesk, M.M. Bonomo, L.D. Rocha, I.D. Duarte, E.R. Zenezi. Chemosphere **158**, 66-71 (2016)
6. C. B. Shivayogimath, C. Watawati. Int. J. of Res. in Eng. and Tech. **Nov**, 266-269 (2013)
7. F. Ilhan, U. Kurt, O. Apaydin, M.T. Gonullu. J. Hazardous Materials **154,** 381-389 (2008)
8. O. Dia, P. Drogui, G. Buelna, R. Dube, B. S. Ihsen. J. Chemosphere, 1-6 (2016
9. G. Hassani, A. Alinejad, A. Sadat, A. Esmaeili, M. Ziaei, A. A. Bazrafshan, T. Sadat. Int. J. Electrochem. Sci. **11**, 6705-6718 (2016
10. Rusdianasari, Y. Bow, A. Taqwa. Adv. Materials Res. **896**, 145-148 (2014)
11. Rusdianasari, A. Meidinariasty, I. Purnamasari. Int. J. on Adv. Sci. Eng. and Information Tech. **5**, 6, 387-391 (2015)
12. M. A. Jumaah, M. R. Othman. Int. J. of Chem. Tech. Res. **8**, 12, 604-609 (2015)
13. S.M.N. Islam, S.H. Rahman, M.M. Rahman. J. Sci. Res. **3**, 3 (2011)
14. R. Bow, S. Arita, E. Ibrahim, N. Ngudiantoro. Appl. Mech. and Mat. **391**, 29-33 (2013)
15. Y. Bow, Hairul, I. Hajar, Int. J. on Adv. Sci. Eng. and Inf. Tech. **5**, 6, 422-425 (2015)
16. M. Poveda, Q. Yuan, J. Oleszkiewicz. Int. J. of Env. Csi. Dev. 7, 4 (2016)
17. C. Phalakornkule, J. Mangmeemak, K. Intrachod, Scienceasia **36**, 1, 42-49 (2010)
18. C.B. Shivayogimath, C. Watawati. Int. J. of Ren. Energy and Env. Eng. **02**, 03 (2014)

The effect of pretreatment using sodium hydroxide and acetic acid to biogas production from rice straw waste

Budiyono[1,*], *Aryantoko* Wicaksono[1], *Arif* Rahmawan[1], *Hashfi* Hawali Abdul Matin[2], *Larasati* Gumilang Kencana Wardani[2], *Tutuk* Djoko Kusworo[1], and *Siswo* Sumardiono[1]

[1]Department of Chemical Engineering, Diponegoro University, 50275 Semarang, Indonesia
[2]Department of Environmental Engineering, Diponegoro University, 50275 Semarang, Indonesia

Abstract. Rice straw is agricultural waste containing high potency to be treated to biogas. However, the usage of rice straw is still limited due to high lignin content that will cause low biodegradability. The aim of this research was to study the effect of pretreatment using NaOH and acetic acid to biogas production from rice straw. NaOH was varied from 2%w, 4%w, and 6%w; and acetic acid was varied from 0,075 M, 0,15 M dan 0,75 M. The rice straw was cut into 1 cm size and submerged for 30 minutes in NaOH and acetic acid solution. The rice straw then filtered and neutralized before sending to anaerobic digestion process using rumen fluid bacteria. Biogas produced was measured using water displacement method. The result showed that the optimum concentration of NaOH solution was 4%w that resulted in biogas volume of 21,1 ml/gTS. Meanwhile, the optimum concentration of acetic acid pretreatment was 0,075 M that produced biogas volume of 14,5 ml/gTS. These results suggest that pretreatment using NaOH solution is more effective for decreasing the lignin content from rice straw.

1. Introduction

Indonesia is an agricultural country with a large agricultural sector, namely rice. Rice is the staple food majority of Indonesian people. Results of production are in the form of rice bran, straw and chaff. The agricultural waste which is potentially to be developed and processed into an environmentally friendly alternative energy is rice straw [19]. In some countries, rice straw has been converted and processed into other materials or products such as paper and pulp, organic fertilizers, mushroom cultivation media, building materials, handicrafts, animal feed, and in the process of burning the bricks to reduce energy consumption [22]

The amount of rice straw in Indonesia is very large because Indonesia is one of the largest rice producer in the world. Rice production was recorded in 2012 (ASEM) amounted to 69.05 million tons of dry rice husk (GKG) or an increase of 3.29 million tons (5.00 percent) compared to 2011. The increase in production occurred in Java amounted to 2.12 million ton and outside Java at 1.17 million tons. The increase in production occurs due to increasing harvested area 239.80 thousand hectares (1.82 percent) and the increase in productivity by 1.56 quintal / hectare (3.13 per cent) (CBS, 2013).

Rice straw is agricultural waste which has high cellulose content that causes their utilization is not optimal. According to Stacey (1976), there is a lot of alternative usage for rice straw i.e. to make biogas via fermentation process by converting cellulose into glucose which is the most important process in the fermentation process. Cellulose content in rice straw amounting to 33.9%, while the hemicellulose by 10.2% - 25.6% to lignin content [14].

2. Materials and methods

Biogas production process was using waste rice straw as raw material obtained from the Village Somomoro, District Plupuh Sragen. The adjuvant to speed up the process of decay and fermentation was using Rumen obtained from slaughterhouses (Slaughterhouse) in Semarang. Distilled water, NaOH (sodium Hydroxide), CH_3COOH (Acetic Acid) and H_2SO_4 (sulfuric acid) were obtained from local chemical store Indrasari in Semarang. The initial stage of the research was cutting the hay to a size of 1 cm and then added to a solution of acetic acid (0,075M, 0,15M and 0,75M) and a solution of NaOH (2% w, w 4% and 6% w). Soaking was performed for 30 minutes and then followed by filtering and washing so that the pH of the straw was neutral before it entered the next stage.

Research carried out by inserting a rice straw from initial research results with a ratio of 10 grams (weight) rumen : water : rice straw that is 2 : 1 : 1 [24]. A series laboratory test of 400 ml biodigester was operated in batch system. The main experiment apparatus consists of biodigester and biogas measurement.

* Corresponding author : budiyono@che.undip.ac.id

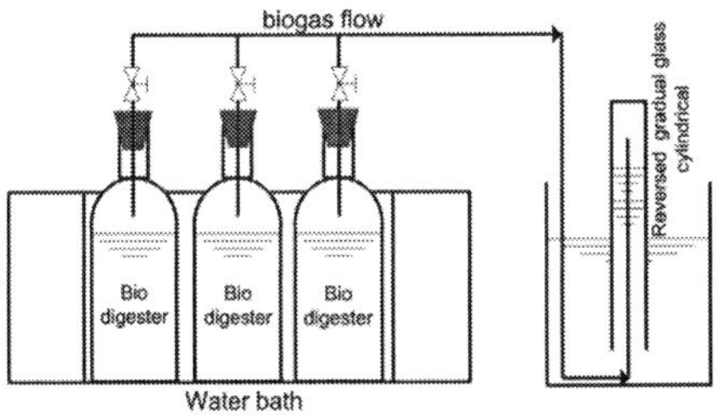

Fig. 1. Schematic diagram of series laboratory batch assessment of anaerobic digestion

Biodigester was made from polyethylene bottle plugged with tightly rubber plug and equipped with valve for biogas measurement. The temperature of biodigester was maintained at certain value using thermostatically controlled electrically heated water bath. Biogas formed was measured by "liquid displacement method" as also has been used by researchers before (3, 4-5, 21]The schematic diagram of experimental laboratory set up as shown in Figure 1.

3. Results and discussion

3.1 Effect of pretreatment on the production of biogas from straw

Pretreatment is one of the first steps to degrade lignin compounds that interfere with the process of the formation of biogas from a biomass, due to the presence of lignin compounds can inhibit the fermentation process by microorganisms. Delignification was carried out with the addition of two different chemical solutions i.e. using solution of acetic acid (CH_3COOH) [25] and using an alkaline solution (NaOH) [8]. Pretreatment is expected to degrade lignin content in rice straw and accelerates the decomposition process of rice straw to obtain optimal biogas yield [15]. Effect of early treatment on the production of biogas from rice straw can be seen in the graph below.

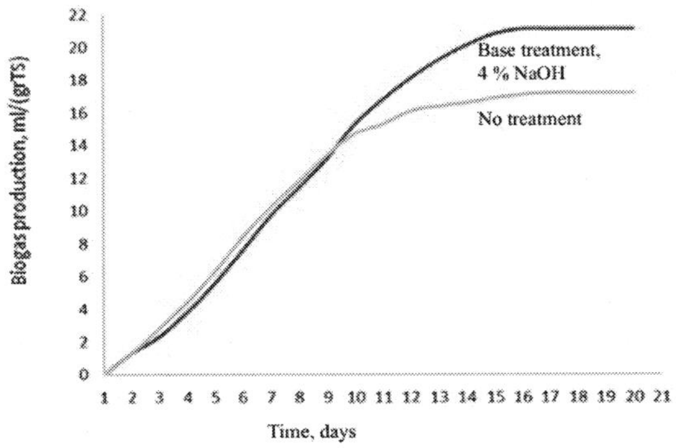

Fig. 2. Effect of Pretreatment in biogas production

In the process of biogas production from rice straw, pretreatment can increase the production of biogas. The optimum volume resulting from one of the variables pretreatment with an alkaline solution of 4% w i.e 21.1 ml/grTS whereas without pretreatment only 17.2 ml/grTS. Differences in the volume of biogas produced shows that there is an increasing volume of biogas at their initial treatment compared with rice straw without any pretreatment. This is in accordance with the research conducted by [18] which explained that early treatment will reduce lignin compounds in lignocellulose structure. Research conducted by [9] also explained that early treatment will also separate the main components namely cellulose as a source of nutrients for the biogas making bacteria. Lignin compounds would interfere with biogas forming bacteria, consequently biogas production results are not optimal. This can be seen in Figure 2 that shows the pretreatment will increase the production of biogas compared with without pretreatment.

3.2 Effect of NaOH solution on Biogas Production from Rice Straw

Pretreatment by immersion using an alkaline solution is one important step in the process of biogas production. Pretreatment on rice straw biodegradation can accelerate and raise the volume of biogas production [8]. Their initial treatment on a substrate to produce biogas is characterized by decreased levels of lignin in a substrate. In the study conducted using substrates such as rice straw pick the cellulose content is quite high at around 31%, while 28% hemicellulose, lignin with 12% in the control variable. NaOH alkaline solution is selected as the initial treatment because besides the cheap price is also effective in reducing the lignin content or degrade lignin in rice straw. Here's a graph of the results of studies using NaOH alkaline solution as a medium pretreated by soaking for 30 minutes before the main process in the manufacture of biogas.

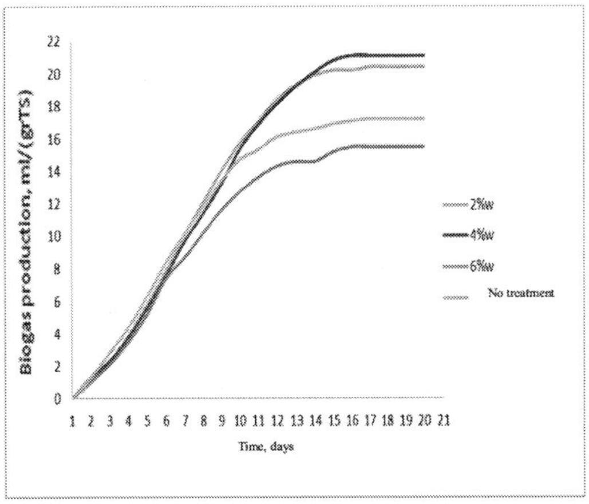

Fig. 3. Effect of Treatment NaOH to the production of biogas

In Figure 3 it can be seen that pretreatment using NaOH 4% w produced the highest volume of biogas of 21.1 ml/rTS. At NaOH 2% w only about 20.4 ml/grTS and the NaOH 6% w has 16.6 ml/grTS while biogas without pretreatment has 17.2 ml/grTS. Their initial treatment by using NaOH solution causes the volume of biogas produced is higher than the volume of biogas without pretreatment. Research conducted by Song et al. (2013) explained that with early treatment using NaOH

DOI: 10.1051/matecconf/201710102011

alkaline solution will reduce lignin compounds in rice straw. The high concentrations of NaOH that used, then increasingly reduced lignin because NaOH will change microstructure of the cell walls of rice straw to facilitate microorganisms in fermentation process for production of biogas.

However, in a study with variable pretreatment 6% w NaOH showed the opposite result with the biogas production without pretreatment. The results obtained fewer in compare without pretreatment. This is because the addition of NaOH more than 4% w during the process of pretreatment using NaOH can raise the concentration of the substrate and dissolving many carbohydrates (cellulose and hemicellulose) that are not effective in the process of pretreatment before the fermentation process [8]. At the time of cooking or soaking with NaOH causing excessive fiber (lignocellulose) becomes damaged and destroyed [23]. Therefore, the results obtained biogas is less than the biogas without pretreatment. [8] explains that methanogenic bacteria can operate optimally in the process of anaerobic digestion at 4% w NaOH is added to the substrate and if the addition of more than 4% w NaOH then methanogenic bacteria do not effective for biogas production process.

3.3 Effect of Acetic Acid Solution on Biogas Production from Rice Straw

The addition of CH_3COOH solution consists of three concentrations are at a concentration of 0.075 M, 0.15 M and at a concentration of 0.75 M. The difference in concentration on the addition of CH_3COOH solution aims to determine the optimal conditions in the delignification using an alkaline solution to produce biogas. Here is the volume of biogas produced from a wide variety of variables pretreatment using a solution of acetic acid.

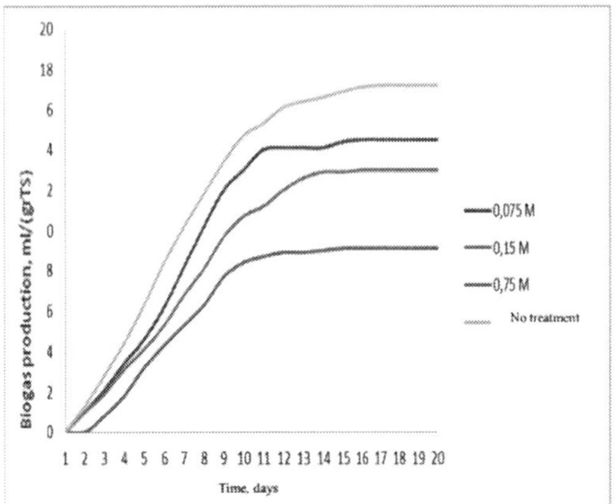

Fig. 4. Effect of early treatment of acetic acid to the production of biogas

In Figure 4 it can be seen that the volume of biogas produced in the absence of early treatment is higher compared to their initial treatment. In the control variables produce biogas volume was 17.2 ml/grTS while acetic acid treatment produces biogas lower volume is 14.5 ml/grTS at variable 0,075M, 13ml/grTS at variable 0,15M and 9,1 ml/grTS the variable 0.75 M. In the study conducted by [16] explains that the lignin compound will not be degraded in strong acids. Another study conducted by [25] explains that the treatment using a weak acid such as acetic acid at low concentrations can also reduce the lignin content during the immersion process lasts a weak acid solution. Decreased levels of lignin in the amount of 7.15% to 16.88% lignin content. During the decreased levels of lignin, cellulose is separated from the structure of lignocellulose is converted methane microorganisms that will ease the process of rice straw decomposition by microorganisms for the production of biogas (Sari et al., 2014). This is due to the increase in the rate of hydrolysis due to greater volumes of hydrogen ions that will help the process of dissolving hemicellulose and lignin.

Another study conducted [17] explained that the process of acid delignification will cause the structure of lignin and hemicellulose which binds cellulose condensed and settled so that the cellulose will be free structure that is separate from the lignin and hemicellulose then will facilitate the access of enzymes or microorganisms to transform cellulose to become monomer - monomer sugars as nutrients for fermentation biogas. The decline lignin is also accompanied by a decrease in hemicellulose and cellulose which is the main nutrient that will use microorganisms to produce biogas, so the results obtained will be slightly [25]. Treatment using a weak acid solution is less effective in degrading lignin but it is more effective to reduce the levels of hemicellulose [17]. Therefore, research using a weak acid solution such as acetic acid is less effective that the results obtained biogas will also decrease because there are some cellulose is also dissolved at the time of immersion using a weak acid solution.

3.4 Comparison of Treatment Early On Biogas Production From Rice Straw

In these discussions are expected to compare the production of biogas between rice straw is given pretreatment acidified with pretreatment under alkaline conditions. The addition of acetic acid (CH_3COOH) consists of three concentrations that are at a concentration of 0.075 M, 0.15 M and at a concentration of 0.75 M, while at the beginning of treatment using NaOH alkaline solution consists of three different variables are variables 2% w, 4% w and 6% w. Comparison of variables taken the most optimal results in each pretreatment. At the beginning of treatment in the form of the addition of acetic acid solution optimal concentration of 0.075 M, while the initial treatment using NaOH alkaline solution has a concentration of 4% optimal variable w. Differences in the two optimal concentration on each different pretreatment aims to determine the optimal conditions in the delignification chemical in producing biogas. Here is a comparison chart between the initial treatment using a solution of acetic acid and pretreatment using a solution of NaOH.

DOI: 10.1051/matecconf/201710102011

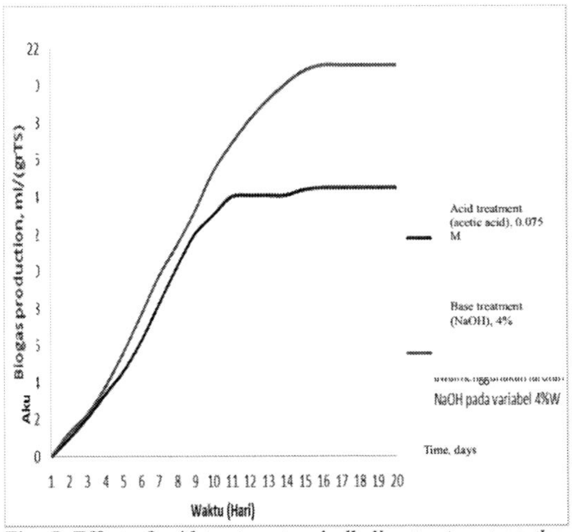

Fig. 5. Effect of acid treatment and alkaline treatment on the production of biogas

In the graph above it can be seen that pretreatment using NaOH alkaline solution is more effective in reducing the lignin compounds is compared pretreatment rice straw using acetic acid solution, so the treatment using a solution of NaOH produce biogas volume more than the treatment using acetic acid. Pretreatment with a solution of NaOH produce biogas volume of 21.1 ml/ grTS on variable 4% w, 20.4 ml/grTS on variable 2% w and 15.5 ml/grTS at 6% w variable so that at the beginning of treatment using a base NaOH most optimum is at 4% w variable with volume biogas obtained is 21.1 ml/grTS. As for the initial treatment using a solution of acetic acid to produce biogas volume of 14.5 ml/grTS at 0.075 variable M, 13 ml/grTS on variables 0.15 M and 9.1 ml/0.75 M grTS variable so that the pretreatment conditions the optimum acid is at 0,075M variable with the volume of biogas produced 14.5 ml/grTS. In the study conducted [17] explained that for decreased levels of lignin will simplify the process of rice straw decomposition by microorganisms. Cellulose is separated from the structure of lignocellulose is easier to convert into biogas microorganisms so that at the beginning of treatment using a solution of NaOH results obtained biogas is greater than the control variables or without pretreatment.

The process that occurs in the presence of the treatment using NaOH alkaline solution, namely the ester bond in the LCC (Lignin-carbohydrate complexes) were destroyed due to the reaction of hydrolysis to produce more cellulose for biogas production process. Inter bonding units and functional groups themselves lignin, cellulose and hemicellulose were also damaged. Changes in chemical composition and physical characteristics of rice straw causing it more easily degraded and increase the production of biogas [11]. At the beginning of treatment using acetic acid can also reduce lignin compounds. As the research conducted by [25] which explains that pretreatment in acidic conditions can lower the lignin content of 7.15% to 16.88%, so that microorganism easier methane compounds convert cellulose to produce biogas. However, a decrease in lignin is also accompanied by a decrease in hemicellulose and cellulose which is the main nutrient that will use microorganisms to produce methane biogas so that the results obtained will be slightly [25].

Using NaOH solution is very effective and efficient in producing biogas as effective in reducing lignin compounds that exist in fiber or straw compared with the use of acetic acid as a solvent so that the cooking process is chemically good delignification used that delignification process using NaOH alkaline solution. Results biogas obtained is also greater than the treatment using acetic acid solution ie biogas volume of 6.6 ml/grTS more use of early treatment in alkaline conditions than treatment in an acidic state.

3.5 Effect of Pretreatment to Lignocellulose Structure

Pretreatment process in lignocellulosic materials needs to be done to ease the process of hydrolysis, namely to open up the structure of lignocellulose so that the cellulose more accessible to enzymes that break down polymeric polysaccharides into the form of monomers, so as to reduce the performance of a microorganism in biogas production and reduce costs during delignification [7].

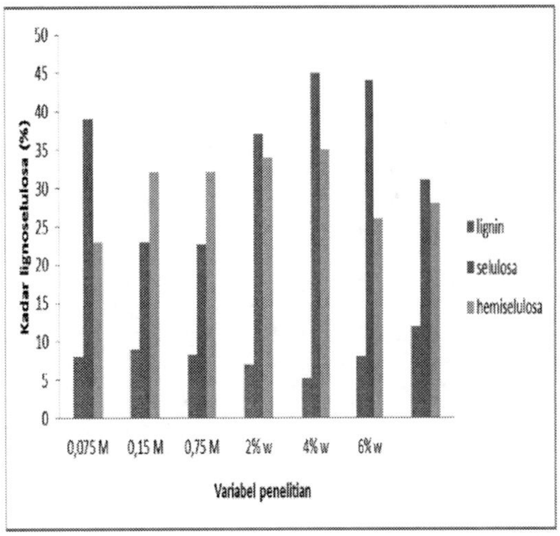

Fig. 6. Percentage change of lignin, hemicellulose and cellulose in the presence of pretreatment (acid and base) and without pretreatment

From the results of the analysis show that there have been a degradation of lignocellulosic components by six different variables after the initial treatment. Rice straw (Oryza sativa L.) were used in this study has a hemicellulose content of 28%, 31% cellulose content and lignin content of 12% which is the womb of control without the administration of pretreatment or pretreatment using either a solution of acetic acid or an alkaline solution of NaOH. Lignin in the control variable is a component that has not experienced a significant reduction in all treatments. Decrease the highest lignin component in the treatment of straw with NaOH 4% to 5.2% w lignin levels. Here is a table of analysis results in decreased levels of lignocellulose rice straw.

Table 1. Lignocellulose content analysis in acid and alkaline delignification

Variable		Lignin content (%)	Cellulose content (%)	Hemicelluloses content (%)
Acid	0,075 M	8	39	23
	0,15 M	9	23	32
	0,75 M	8,3	22,7	32
Base	2%w	7	37	34
	4%w	5,2	45	35
	6%w	8	44	26
Control		12	31	28

The table 1 above shows that overall, decreased levels of lignin in the presence of either treatment using a solution of acetic acid or an alkaline solution of NaOH. The cellulose content increased on average above 31% for the theoretical structure of lignocellulose rupture and separate cellulose from lignin and hemicellulose (Erviani et al., 2011). Characteristics of such cellulose do not dissolve in alkali, soluble in concentrated acid and insoluble in organic solvents and water (Ramdja et al., 2010). Therefore, in the variable using a solution of NaOH produces the most high cellulose content than using a solution of acetic acid as a solvent for delignification process. Cellulose is a polysaccharide component will be used microorganisms methane to be converted into simpler sugars monomer in the production of biogas. Cellulose that had high levels of naturally produced large volumes of biogas than cellulose with low levels.

In the table above shows also that the hemicellulose components decreased from 28% for the control variable to 23% to 26%. Hemicellulose has the characteristics of soluble alkali (NaOH solution is an alkaline solution) and soluble in acidic and susceptible to hydrolysis compared with cellulose [16]. Hence the hemicellulose content decreased.

3.6 The influence Lignocellulose compounds on Biogas Production From Rice Straw

In the study conducted by [13] described the structure of lignin is strong because lignin covalently bonded to the hemicellulose will result in cross link with the polysaccharide network which will add to the mechanical strength of the cell wall. The combination of lignin in the cell wall polysaccharides will prevent degradation by chemicals and pests. This causes difficulties methane microorganisms to convert polysaccharides into monomer sugar to be processed by microorganisms into biogas. Thus the initial treatment required so that the volume of biogas generated from optimal.

Components of rice straw (Oryza sativa L) has a hemicellulose content of 28%, 31% cellulose content and lignin content of 12%. Cellulose component used to produce biogas. However, the difficulties encountered in lignocellulose degradation were a heterogeneous composition of the polysaccharides found in the cell wall that lignin compounds. In the plant tissue, is difficult to degrade lignin because it has a complex and heterogeneous structure that strengthens the plant structure in resisting against microbial attack and pressure oxidation [12].

Therefore the volume of biogas produced fewer and smaller than in the presence of delignification or pretreatment. At the hay with their initial treatment, the cellulose originally bonded to the structure of lignocellulose will decompose and despite making it easier for microorganisms to convert cellulose to manufacture biogas.

4. Conclusions

Pretreatment or delignification can affect the outcome of biogas obtained because of the presence of the pretreatment compound is lignin contained in plant tissues and arranged into lignocellulose will break its structure and degraded in the presence of NaOH solution which can reduce the lignin content of 12% (variable control) to 5.2% on a 4% w variable which is the optimum variables in the study so that the results obtained biogas is greater than the biogas on the control variable. At the beginning of treatment using a solution of acetic acid can lower the lignin content of 7.15% to 16.88% but reduced levels of lignin is accompanied by decrease in hemicellulose and cellulose of network structure in rice straw. Therefore, the treatment using acid solution is less effective when used for pretreatment or delignification process.

References

1. T. Anindyawati, Cellulase potential to degrade lignocellulosic agricultural waste to organic fertilizer (2010)
2. Budiyono, I.N. Widiasa, and S. Johari, Sunarso, *World Academy of Science, Engineering and Technology*, **37**:983-988 (2010)
3. Budiyono, I. Syaichurrozi, S. Sumardiono, *World Applied Sciences Journal* **26(11)**: 1464-1472 (2013)
4. Budiyono, S. Iqbal, S. Siswo, *Research Journal of Applied Sciences, Engineering and Technology* **7(3)**:2798-2805 (2014)
5. Budiyono, I. Syaichurrozi and S. Sumardiono., *International Journal of Engineering*, **Vol. 27(2)**:177-184 (2014)
6. Budiyono, S. Sumardiono and D. Tri Mardiani, *International Journal of Engineering*, **28(2)**: 921-928 (2015)
7. Budiyono, I N. Widiasa, S. Johari, and Sunarso, *International Journal of Engineering and Technology*, **1(3)**:109-116 (2009)
8. Chandra, R., Takeuchi, H., Hasegawa, T., & R. Kumar, *Energy*, **43(1)**:273-282 (2012)
9. I.W. Gunam, N.M. Wartini, Dewi Anggreni, A. M., & P.M. Suparyana, *Jurnal Teknologi Indonesia*, 24-32 (2011)
10. Haghighi Mood, S., Hossein Golfeshan, A., Tabatabaei, M., Jouzani, G., Najafi, G., & Gholami, *Agroeconomy Research*, **27**:77-93 (2013)
11. Y. He, Y. Pang, Y. Liu, X. Li & K. Wang, *Energy & Fuels*, **22**:2775-2781 (2008)

12. A. Hendriks & G. Zeeman, *Bioresour Technology*, **100(1)**:10-18 (2009)
13. C.A. Hubbell & A.J. Ragauskas, *Bioresource Technology*, 7410-7415 (2010)
14. S. Jin & H. Chen, *Industrial Crops and Products*, 207-211 (2007)
15. Z. Lei, J. Chen, Z. Zhang & N. Sugiura, *Bioresource Technology*, 4343-4348 (2010)
16. Ramdja, R. Silalahi & N. Sihombing, *Jurnal teknik Kimia Unsri.* **17** (2010)
17. F. Sari & Budiyono, *Waste Technology*, **2(1)**:17-25 (2014)
18. I. Sari, M. Izzati & S. Haryanti, *ejournal Undip*, 45-55 (2012)
19. T. Silalertruksa & S.A. Gheewala, *Bioresource Technology*, **150**:412–419 (2013)
20. M. Stacey, Agric. Progress **51**:69-78 (1976)
21. I. Syaichurrozi, B. Budiyono & S. Sumardiono, *Bioresource Technology*, **149**:390-397 (2013)
22. Sulardjo. *Magistra* . 81-90 (2013)
23. S. Wijana, A. Mulyadi & A. Pratama, *APTA Proceeding*, 1-9 (2012)
24. A. Yulianingsih, F.D. Rahmawati, S. Amalia & Wahyudin. *Jurnal Biogas*. (2012)
25. R. Zhao, Z. Zhang, R. Zhang, M. Li, Z. Lei & M. Utsumi, *Bioresource Technology*, **101(3)**:990-994 (2010)

Storage tank materials for biodiesel blends; the analysis of fuel property changes

Leily Nurul Komariah[1,2*], *Marwani*[1], *Sucia* Aprisah[2], *Yangia* S.L Rosa[2]

[1]Department of Chemical Engineering Universitas Sriwijaya, 30139 Palembang, Indonesia
[2]Renewable Energy Laboratory Universitas Sriwijaya, 30662 Inderalaya, Indonesia

Abstract. Fuel stability is one of major problem in biodiesel application. Some of the physical properties of biodiesel are commonly changed during storage. The change in physico-chemical properties is strongly correlated to the stability of the fuel. This study is objected to observe the potential materials for biodiesel storage. The test was conducted in three kinds of tank materials, such as glass, HDPE, and stainless steel. The fuel properties are monitored in 12 weeks, while the sample was analyzed every week. Biodiesel used is palm oil based. The storage tanks were placed in a confined indoor space with range of temperature 27-34 °C. The relative humidity and sunshine duration on the location was also evaluated. The observed properties of the fuel blends were density, viscosity and water content. During 12 weeks of storage, the average density of B20 was changed very slightly in all tanks, while the viscosity was tend to increase sharply, especially in polimerics tank. Water content of B20 was increased by the increase of storage time especially in HDPE tank. In short period of storage, the biodiesel blends is found more stable in glass tank due to its versatility to prohibit oxidation, degradation, and its chemical resistance.

1 Introduction

Biodiesel has become more attractive as alternative fuel due to its environmental benefits and the fact that it is made from renewable sources. With the soaring price of petroleum-based products, biodiesel is becoming an increasingly affordable option relative to petroleum diesel. As substitute, biodiesel has been promoted to all energy users, such as transportation, industrial, commercial and institutional sectors. Biodiesel is commonly applied in a blend with petroleum diesel oil. A blend is designated by the percentage of biodiesel in the blend. For example: B20 is a fuel containing 80 percent petrodiesel and 20 percent B100.

In Indonesia, the utilization of biodiesel has been mandated as stated in Regulation of Ministry of Energy and Mineral Resources No.32/2008 and No. 23/2013 and No. 26/2015. The permit trading holder of fuel oil in Indonesia are required to blend petrodiesel with biodiesel according to minimal phasing as obliged. Currently, the minimum phasing obligation on biodiesel blends is 20%. It means B20 is already existed in the fuel market of diesel oil particularly in Indonesia.

Biodiesel fuel properties are relatively comparable to diesel engine. Therefore, behind its advantage provided, biodiesel also shows the potential shortcomings problem that need to be solved and anticipated. One of important problem is fuel storage instability. Biodiesel resulted from the production unit may be in good condition, but it can become unacceptable during distribution and storage without proper care and attention. Using biodiesel that does not meet the quality standards may caused engine failure, filter plugging, and adverse tailpipe emissions.

Biodiesel is not always compatible with the existing fuel storage and distribution infrastructure (Knothe 2005), especially in high blending ratio. Many users do not recommend storing biodiesel blends for more than six months. It is strongly related to the low stability of the fuel. So it becomes one of critical problem in biodiesel utilization.

From the various works and research, it has been identified that biodiesel oxidized and degraded faster than diesel oil [1], [2] Some physico-chemical properties are changed during storage. Acid value, density, and kinematic viscosity increase with the increase in storage time of biodiesel as reported by [1-7]. Otherwise, some authors stated that under storage duration, the acid value and flash point temperature of biodiesel are nearest to constant [8-9]. All the effects were observed more in a higher temperature of storage [7, 10]. The water content of the biodiesel showed increasing trend during storage as reported by [2, 10-11]. Biodiesel is hygroscopic it can absorb atmospheric moisture or pick up water in storage. Hence, water content of biodiesel is increased due to the methyl ester can quickly absorb moisture when storage container possibly opened or contacted with air.

From most of the research results, the fuel properties are still below all the standard values for the whole testing period in storage tanks with lower temperature [7,

* Corresponding author: leilynurul@unsri.ac.id

10,12]. The testing period is varied from 3 to 52 weeks or under one year.

Biodiesel mostly produced from unsaturated fats contributed from vegetables oil. It tends to oxidize and degrade rapidly. In addition, any previous process that removes the natural antioxidants from the oil (such as bleaching, deodorizing, or distilling) will hasten oxidation. In general, biodiesel will degrade more quickly than petro-diesel. Biodiesel can degrade due to oxidation, contact with water, and/or microbial activity.

Oxidation stability of biodiesel correlates with major fatty acid methyl ester composition i.e., different types of fatty acids provide different degree of unsaturation and long chain saturated factor [2, 13]. The oxidation of biodiesel can produce various acids or polymers, which, if in high enough concentration, can cause fuel system deposits and lead to filter clogging and fuel system malfunctions [14]. Unsaturated components in biodiesel are more chemically reactive and less stable [15]. The presence of double bond in fatty acid molecule can make a high level of reactivity with the oxygen when exposed to the air during storage [2]. From previous works, it is noted that oxygen exposure, contamination from metals and other radical initiators, water, oxygen and light exposure, and heat could all contribute to degradation of biodiesel quality [6, 16].

Most of the tests for neat biodiesel was conducted in existed diesel oil storage tank which is majorly made of metal. Some of other authors conducted some tests of biodiesel in glass storage tank, they concluded that the water content shows no significant effect on the biodiesel storage within 150 days [10]. The biodiesel are commonly stable when it stored in the dark place and without present of light [4]. There are severe effects when the fuel was exposed to daylight and air [10,16]. The stability of blends is highly influenced by storage temperature.

One of popular materials for forthcoming storage tank is High Density Poly Ethylene (HDPE). HDPE is a polyethylene thermoplastic made from petroleum. HDPE is known for its large strength-to-density ratio. It is also harder and more opaque and can withstand somewhat higher temperatures (120 °C/ 248 °F for short periods). HDPE fuel tank can possibly used for fuel storage due to its versatility. This material is known as good to reduce fuel permeation beside its great chemical resistance. A recent study conducted by Christensen & L.McCormick, 2014), they reported that biodiesel blends stored in polyethylene fuel tanks were stable for 380 days when held at 23 °C, but production of peroxides and acids was detected when the fuel was stored for 56 days at higher temperature (80 °C). However, the test performing biodiesel blends storage in HDPE and other polymeric materials are still limited.

Most tanks designed to store diesel fuel will store blends of B20 and above with no problem. Many authors reported that under 2016 hours (three months) of storage duration, the effect of degradation was happened in diesel fuel in acceptable ranges [4, 10, 12]. But in practice, it requires some additional considerations due to environmental circumstances. If the fuel stored in higher temperature and humidity it possibly triggering oxidation and further problems like fuel degradation and materials corrosion. Under normal storage conditions diesel fuel can be expected to stay in a useable condition for 12 months or longer at an ambient of 20°C, and 6-12 months at an ambient temperature higher than 30°C. Nevertheless, some of guidance stated that biodiesel should be stored at or above 40° F considering the cold flow standard. Storage for neat and blended Biodiesel fuels longer than 6 months should have antioxidants blended at point of sale.

The storage stability of biodiesel is a critical issue. For some of practical users, biodiesel *shelflife* is from four weeks to four months [**17**]. It is largely dependent of the initial storage stability of the neat form of the fuel [11]. Hence, it is suggested to avoid large storage periods from the storage stage to the dispensing stations. Oxygen exposure, water exposure, light exposure, and heat could all contribute to degradation of fuel quality [16]. As any diesel fuel ages, it can degrade via thermal and oxidative processes to form polymers and acids. This chemical change is promoted by exposure to high temperature and pressure, oxygen, acids, certain metals such as copper and zinc, or combinations of these.

It had been mentioned that the only metallic materials compatible and recommended being used with biodiesel are stainless steel and aluminum [18]. Stainless steel it far being the most compatible with biodiesel. It is compatible with high-percentage biodiesel blends due to its chemical resistance [17].

The fuel storage is majorly carried out outdoors for safety and environmental considerations. Fuel is a flammable materials and its leaks will possibly hazardous. Therefore the type, position, distance and methods of fuel storage handling is specifically regulated.

Storage tanks for a particular fuel are commonly chosen according to the characteristics of that substance. The most expected characteristics from fuel storage is lower build costs and greater chemical resistance. While steel and concrete remain one of the most popular choices for tanks, glass-reinforced plastic, thermoplastic and polyethylene tanks are increasing in popularity. They offer lower build costs and greater chemical resistance.

Table 1. Materials for Storage Tank

Type	Specific Materials
Metals	Aluminium Steel (Mild/Carbon Steel) Stainless Steel
Concrete	Concrete
Elastomer	Viton, Fluorosilicone, Hifluoer, Fluorocarbon, Chemraz, Teflon
Polymer	Polypropilene, Polethylene (HDPE), Fluoronated Materials, Nylon

Biodiesel storage methods are closely similar to conventional diesel storage methods; however, there are some considerations for optimum fuel longevity and purity. Biodiesel blends may commonly stored in existed fuel tank with special treatments or potentially placed in a new method of storage with different tank materials and conditions. Due to the nature characteristics of the fuel and its interaction with environment conditions, the new potential materials must be accounted for biodiesel storage especially in specified period (12 weeks). A series of test must be conducted to check the stability of biodiesel blends in polymeric tank storage compared to steel.

2 Experimental methods

In this current research, any changes in physico-chemical properties of the B20 was analyzed compared to B0 and B100 on 3 (three) kinds of materials of storage tank such as metal (stainless steel) and polymerics (HDPE and glass). Those tanks were place above the ground in a wide and high ceiling room, with normal temperature. No heating features attached on those tanks.

Biodiesel used in this work is palm oil based, while the petroleum diesel oil used is automotive diesel oil (ADO) or diesel oil grade#2. The fuel sample was stored in three storage tanks placed in a confined space on a rack upper ground in a big and high ceiling room. The fuel properties was monitored weekly in period of 12 weeks. The samples were analyzed periodically. The properties of fuel sample considered in this study are included density, kinematic viscosity and water content. Environmental parameters observed were temperature, sunshine hours, and reative humidity.

The palm biodiesel used in this study was supplied by PT. Sumiasih Oleochemicals Industry, West Java, Indonesia and petroleum diesel oil was taken from refinery unit III of PT. PERTAMINA, Palembang, Indonesia. The biodiesel used was palm oil based that met standards according to ASTM D613 and SNI 04-7182-2006.

Table 2. Methods for Fuel Properties Analysis

Parameters	Units	Method
Kinematic Viscosity	cSt (mm^2/s)	ASTM D445
Density	kg/m3	ASTM D941
Acid Value	mg KOH/g	ASTM D971
Water Content	mg/kg	Karl Fischer
Flash Point	°C	ASTM D93-94
Calorific Value	cal/gr	ASTM D240

The initial properties of each sample of fuel sample are summarized in Table 3.

Table 3. Initial properties of the fuel samples

Property	units	B0 (Diesel Oil)	B20	B100
Kinematic Viscosity	mm2/s	2.798	3.112	5.142
Density	g/mL	0.825	0.831	0.868
Water Content	mg/kg	83.300	261.200	1644.660

Fuel storage stability then analyzed based on the ability of a fuel to resist changes in its physicochemical characteristics brought by interaction with its environment and the materials of storage tank.

3 Results and discussion

In the current work, storage stability of biodiesel blends is represented through the changes in physical properties of the fuel during 12 weeks of storage time.

The fuel density of B20 commonly tend to increased during the storage period. It increased very slightly in all type of tanks. It is quite convincing, due to the neat biodiesel (B100) storage also showed no significant changes in fuel density in 12 weeks of storage time. It was only changed with an average of less than 1%, as shown in **Fig 1.**

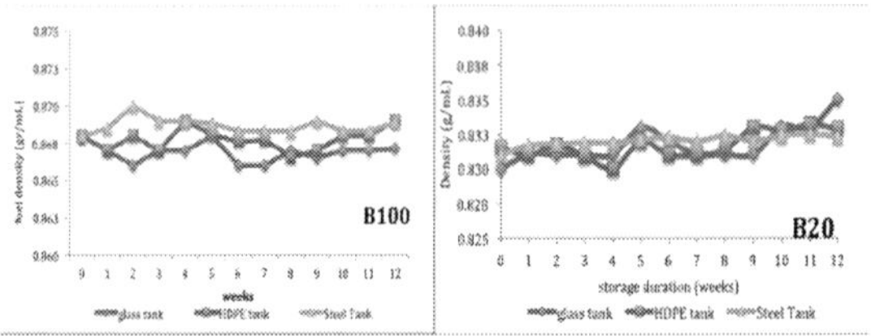

Fig 1. Profile of Density Changes in [a] Biodiesel (B100) and [b] Biodiesel blends (B20)

The increase in density was commonly caused by the formation of oxidation products, that molecularly higher in weight. The slight increase on fuel density was influenced by storage condition that relatively stable. The average ambient temperature around the place of the three storage tanks was recorded between 22-35 °C as recorded in **Fig 2a**. In that range of temperature, the effect of oxidation of B20 was not significant. It is in line with the conclusions made by Zakaria et al., (2014), they stated that the oxidation of biodiesel proceed in temperature higher than 40 °C.

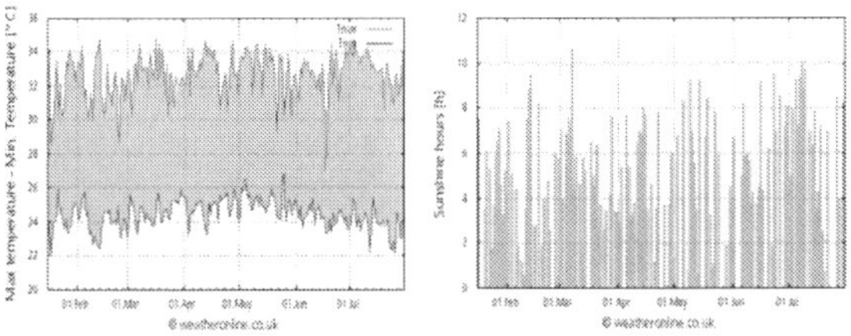

[a] Temperature [b] Sunshine hours

Fig 2. Climate Data on Storage Location – Palembang South Sumatera [a] Temperature [b]. Sunshine Hours
Source : http://www.weatheronline.co.uk, http://bmkg.go.id

As seen in **Fig 1,** the density changes of biodiesel blends in glass tank look more prominent than in HDPE and steel tank. It is confirmed that the presence of saturated fatty acid in the fuel stored in glass tank was higher in composition and purity. compared to GDPE and steel tanks. This is probably due to the conditions around the storage tanks were still allow receiving light from electricity lamps and sunlight coming in through the windows or ventilators, eventhough those tanks were placed indoor.

The increased in density is commonly linear with an increase in viscosity. In this current work, the viscosity of B20 increased more sharply with the storage time as shown in **Fig. 3**. The greatest change occurred in B20 was when it stored in the HDPE tank rather than steel tank and glass tank. It is clearly seen that the lowest changes in viscosity was occurred on steel tank, it changed from 3.112 mm2/s to 3.188 mm2/s. A storage of B20 in glass tank was more stable ini viscocity rather than HDPE tank. In 12 weeks of storage time, the viscosity of B20 in HDPE and glass tank was increased in 3.9% and 3.7%, respectively. It is also related to the average long duration of sunshine hours at the location of the storage, as presented in Fig. 2b.

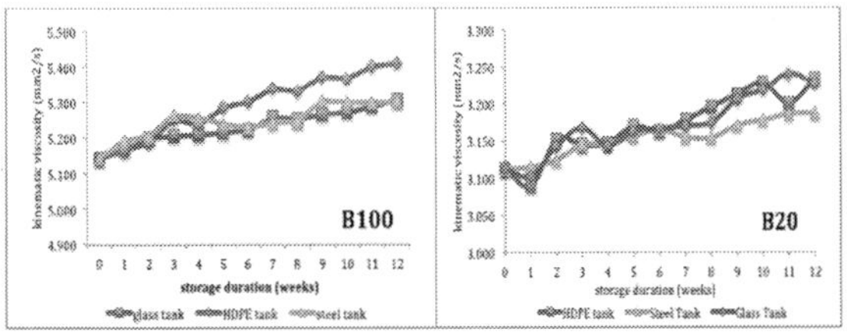

Fig. 3. Profile of Viscosity Changes in [a] Biodiesel (B100) and [b] Biodiesel blends (B20)

The linear regression equation is generated to perform a projection regardings maximum storage time in case of viscosity standard limit as stated in ASTM D975. The maximum viscosity for diesel oil grade #2 is 4.1 mm^2/s.

Table 3. Projection Maximum Time for Storing B20 under standard of kinematic viscosity

Storage Type	Regression Equation	Projection Storage Time Limit [days]
Glass Tank	y = 0.0103x + 3.1099	605
HDPE tank	y = 0.0105x + 3.1083	661
Steel Tank	y = 0.0061x + 3.1168	1128

As seen on Table 3, in such a period of 12 weeks or 84 days, the B20 storage will not caused any severe effects on the viscosity of the fuel until specified time limit. Any changes in oil viscosity is attributed as an effect of oxidation. The greater changes occurred in viscosity almost equal to the magnitude effects on oxidation caused during storage. Kinematic viscosity significantly influenced by compound structure that also sensitive to any changes in temperature and light exposure during storage. In a period of storage time, the viscosity of the methyl esters is increased by the formation of more polar, oxygen-containing molecules and also by the formation of oxidized polymeric compounds, possibly leading to the formation of insoluble products [19] . In 12 weeks of storage time, the changes in water content for B20 and B100 were examined and compared. The water content of biodiesel and its blends was significantly increased with increasing storage time as shown in **Fig. 4**.

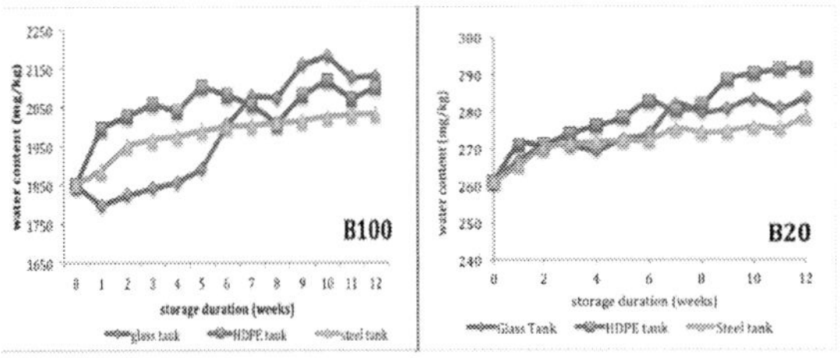

Fig 4. Profile of Water Content Changes in [a] Biodiesel (B100) and [b] Biodiesel blends (B20)

The neat biodiesel used in this study had previously been stored more than one year after production, so at the beginning of the test, it was found the water content of the sample exceeds the limit specified standard of 500 ppm.

The greater changes in water content of B100 during storage time were occurred in glass tank, however in case of B20, the highest change was taken place in HDPE tank. The changes were measured up to 11.6%. Some regression equations were also generated for evaluating the profile of water content of B20 during storage. It presented in Table 3.

Biodiesel has a polar chemical structure of carboxylate ester groups and presents the characteristics of absorbing more water than diesel. Thus, the addition of biodiesel to diesel increases the water retention capacity of the blend [20]. Meanwhile, humidity is one of important thing that affect in water content on fuel. In the period of this work, the relative humidity around location was 65-90%.

Table 4. Projection maximum time for storing b20 under standard of water content

Storage Type	Regression Equation	Projection Storage Time Limit [days]
Glass Tank	y = 1.7372x + 264.61	945
HDPE tank	y = 2.2881x + 266.1	714
Steel Tank	y = 1.0817x + 265.75	1519

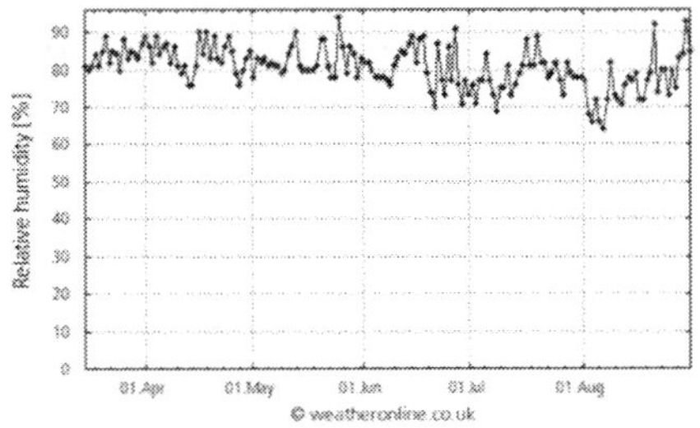

Fig 5. Relative Humidity on Storage Location (Palembang South Sumatera)
Source : http://www.weatheronline.co.uk, http://bmkg.go.id

The higher the *humidity* the faster it will soak up water from the air. The measurement of relative humidity around location of the test is presented in Fig 5.

The source of the problem in HDPE tank may came from the access points. The tank access points (e.g., man-ways, fill tube, nozzles and fittings) are used for tank access and for critical operations conducted while the tank is static or being filled or emptied. These openings allow for the movement of humid air and for ingress of contaminants resulting in product quality degradation. Humid air that enters fuel tanks during normal temperature cycling (breathing) can over time result in water condensation in the tank when the ambient temperature reaches the dew point. It can be more of a problem in higher humidity regions.

Based on physico-chemical analysis, the steel storage tank peforming the minimum changes in fuel properties. It seems more secure for storing biodiesel blends, rather than HDPE and glass tank. Steel tank is provenly solid, strong, and opaque. It can protect fuel from sunlight exposure. Otherwise, it generally has heavy weight, more complex construction and higher maintenance cost. As like other metals, steel is a good conductor, so it possible to transfer heat by temperature elevation on storage. Storage in condition with higher temperature can accelerate oxidation. At a certain condition of storage and or higher temperature, the oxidation process begins with an initial chain reaction when highly active free radicals are formed by the influence of heat, light, air, or metal ions [2]. A comparative analysis of the various tank materials indicates that steel remains a cost-effective material that meets all of the required performance criteria. In spite of that, a physical aging of biodiesel stored in steel tank may be still problematic. Polymers has become leading materials for various applications, including storage. It can be both thermal and electrical insulators. Generally, polymers are very light in weight with significant degrees of strength.

As resulted from this research, the placement of the storage tanks indoor was not extremely reducing oxidation or degradation effects. The light exposure in relatively long sunhours period per day and contact with humid air was major cause of the problem. It happened in all kind of storage tanks materials, including steel tank. The worst effect was happened in HDPE tank. The interaction of fuel with HDPE tank material was objected in several studies by many authors. The petroleum diesel was found to cause higher weight gain in the HDPE, that is, swelling, than the biodiesel [18]. Otherwise, the glass container provided the best protection to oil samples as indicated by the best sensory properties, followed by plastic and tinplate containers [21]. Metals tanks had the highest values of sedimentation after specified period of storage [22].

Glass storage tank seems not popular for fuel storage but some test led to some new findings. As like other polymeric materials, glass is inert to the corrosive environments inside and outside the tank. It acts as an insulator to retard heat transfer to the fuel. It can also be properly designed and installed to control any contact with humid air and can be produced with a resin with dark surface to protect from light exposure.

Fuel must be store in a moisture-free location away from heat and out of direct sunlight. If unable to store out of direct sunlight or warm weather; cover tanks with reflective tarp. To reduce fuel polymerization and oxidation, store in tightly sealed full tanks.

Glass tank is become attractive for those requirements. As known, oxidation is an irreversible reaction. It is a one-way path to the demise of the materials of storage. Meanwhile, physical aging of any fuel is reversible. As stated by [23], if the material that has undergone physical aging is raised to a temperature above the glass transition and then rapidly quenched, the effects of aging that have occurred previously are erased regardless of the time frame over which they have occurred. This condition will facilitate a better maintenance process that hardly handled in steel tank and HDPE tank.

4 Conclusion

Biodiesel and its blend can be stored for more than 12 weeks in metal or polymeric tanks, and it physico-chemically changed but still met the standard specifications. In a storage of B20, it is found a different patterns of changes in physical properties, such as density, viscosity and water content. The density and viscosity B20 was increased slightly during storage period. This condition was contributed by temperature around the tanks and oxigen and daylight exposure in the area of the storage. The water content of B20 was found increased significantly. The greatest changes in water content was happened in B20 stored in HDPE tank. It is due to high relative humidity that triggering oxidation and degradation in biodiesel blends.

The changes in physical properties of B20 are not only originating caused by the molecular structure, but it is very sensitive to environmental factors such as temperature, humidity, exposure to daylight or sun. Moreover, the tank configuration (roof design and access points) are very influential in the storage stability of biodiesel blends.

The selection of storage materials for biodiesel blends is necessary to consider the resistence effect on oxidation, degradation and corrosion. The material of storage tank must be able to inhibit any cause to oxidation, such as the increase of storage temperature, protect from sunlight exposure and contact with humid air. Glass tank is found better among other polimeric materials for storing biodiesel, due to its chemical resistance, and flexibility designed for controlling air access points. It can be modified in design to set in dark surface in order to protect from light exposure.

References

1. M. Shahabuddin, , M. Kalam, H. Masjuki, M. Bhuiya, M. Mofijur, *Energy* , **6**, 616-622 (2012)

2. P. Udomsap, N. Chollacoop, T Subongkoj , T. Hirotsu, *Int. J. Renewable Energy*, **4** (2), 47-59 (2009)
3. P. Mazumdar, V. B. Borugadda, V. V. Goud, (2013). *Int. J on Environmental Engineering* **3**. 13 (2013)
4. H. Yusof , N Tamaldin, M. Abdollah, A. Khalid, *ARPN J. of Engineering and Applied Sciences*, **10** (17) 7707-7711 (2015)
5. H. Meenakshi, A. Anand, R. Shyamala, R. Saratha, *Chemical Science Transactions* **2**, 99-104 (2013)
6. J. Czarnocka, M. Odziemkowska, *CHEMIK*, **69** (11) 771-776 (2015)
7. A. Khalid, N. Tamaldin, M. Jaat, M. Ali, B. Manshoor, I. Zaman, Procedia Engineering, **68**, 225-230 (2013)
8. H. Yusof, N. Tamaldin, G. Omar, M.F. Abdollah, *Int. J. of Education and Learning Systems*, **1**, 129-134 (2016)
9. C. Ezeanyanaso, V. Ajibola, E. Agbaji, E. Okonkwo, O. Okunola, Y. Alhassan, *Greener Journal of Agricultural Sciences* , **2**, 195-206 (2012).
10. Y.P. Wu, Y.F Lin, J. Y. Ye, *Biodiesel – Quality, Emissions and By-Products*. 71-88 (2011)
11. S. Kovacs , G. Polczmann , J Hancsok , *Hungarian J. Indistrial Chemistry Veszprem.* **39**.(3). 369-373 (2011)
12. H. Zakaria, A Halid, M. Farid, *Applied Mechanics and Materials* **465-466**, 316-321 (2014)
13. M. Ramos, C. Fernandez, A. Casas, L. Rodriguez, A. Perez, *Bioresource Technology*, **100**, 261-268 (2009)
14. ASTM. *Standard Specification for Biodiesel Fuel (B100) Blend Stock for Distillate Fuels*.1131-1136 (2009)
15. E. R. Mat, W.W. Yuhaidi, M. Kamaruddin, O. Hassan, *J. Advanced Research in Fluid Mechanics and Thermal Sciences*, **2** (1), 8-12 (2014)
16. Christensen, R. L. McCormick, *Fuel Processing Technology* **128**, 339-348 (2014)
17. M.A. Ortega, G.M. Alpírez, A. Eliezer, C.G. González, *FORMATEX*, 279-289 (2013)
18. E. Torsner, *Corrosion Engineering, Science and Technology*. **45:42** (2010)
19. G. Knothe, *Fuel Processing Technology* **88** 669-677 (2007)
20. P. B. L. Fregolente , Wolf Maciel , L. S. Oliveira' *Braz. J. Chem. Eng.* **32** (2015)
21. M. Rababah, H.Feng, W. Yang, K. Eriefej, M. Al-Omoush. *Int J Agriculture & Biology Engineering* **4** 66-72 (2011)
22. L. Díaz-Ballote, A. Castillo-Atoche, L. Maldonado, M.A. Ruiz-Gómez, E. Hernández, *Journal of Physics D: Applied Physics, 49* 35 (2016)
23. M. Sepe LLC, *Plastics Technology,* December (2014)

* Corresponding author: leilynurul@unsri.ac.id

MATEC Web of Conferences **101**, 02013 (2017) DOI: 10.1051/matecconf/201710102013
SICEST 2016

Bioethanol production from sodium hydroxide – dilute sulfuric acid pretreatment of rice husk via simultaneous saccharification and fermentation

Novia[1,*], *Vishnu K.* Pareek[2], *Tuty Emilia* Agustina[1]

[1]Department of Chemical Engineering Faculty of Engineering Universitas Sriwijaya, 30662 Inderalaya, Indonesia
[2]Department of Chemical Engineering School of Chemical and Petroleum Engineering Curtin University, 6102 Perth, Australia

Abstract. Degradation of lignin from lignocellulosic biomass plays a main key role in converting of lignocelluloses to bioethanol. The pretreatment of lignocelluloses is needed to enhance the fermentable sugars production from enzymatic hydrolysis. This research studied the effect of dilute sulfuric acid pretreatment on the ethanol production using alkaline-dilute acid pretreatment. This research used two stages of the pretreatment process. The first stage used NaOH solution with ratio of rice husk and NaOH solution (1:10 w/v). On the second stage, five concentrations of sulfuric acid of 1 to 5% at temperature of 121°C and various heating times (30-90 min) were investigated. The ethanol production from pretreated rice husk was conducted by the simultaneous saccharification and fermentation (SSF) for various SSF times of 72-160 hr. The experimental results indicated that an increase in concentration of sulfuric acid capable to destroy the lignocellulosic structure of rice husk. The highest value of cellulose content was achieved about 88.84 % for acid pretreated material at H_2SO_4 concentration of 2% in 90 minutes. The lignin reduction ranged from 47.46% to 79.29%. The highest ethanol concentration obtained was 13.68 g/L for the SSF time of 120 hours and the sulfuric acid concentration of 3 %.

1 Introduction

The requirement of fossil fuel has increased significantly with the increasing of human population. On the other hand, the oil production for fulfilling the fossil fuel demands have decreased regulerly. One of efforts for reducing the energy crisis is utilization of bioenergy such as bioethanol. The huge resources of biomass then the attention of bioenergy research are the lignocellulosic biomass which is highly suitable for energy uses. This is due to its great-scale accessibility, small production fee, and small impact to environment.

Utilization of plentiful lignocellulosic biomass to bioethanol as liquid fuels provides a feasible possibility to cultivate energy security and decrease greenhouse gas emissions. Regarding the source and the structure, the biomass can be categorized in several groups, with the most essential being: farmed residues (comprising agro-industrial solid wastes), hardwood, softwood, herbaceous biomass, next to cellulosic waste and municipal solid waste [1]. Softwood and hardwood comprise a higher of cellulose content, whereas the hemicellulose percentage of agricultural residues and herbaceous biomass is higher. Furthermore, wood generally contains more lignin content than the other groups. Finally, the lignin content of softwood is higher compared to hardwood.

Lignocellulosic biomass is a intractable material, comprising of cellulose, hemicellulose and lignin, structured in the form of polymers network [2]. Cellulose (C6H10O5)n is a polymer of glucose and many properties depend on the degree of polymerization, which varies between 800 and 10000 units. Cellulose is found in crystalline and non-crystalline structures. Hemicellulose represents a family of polysaccharides such as xylan, glucomannan and various other molecules, and forms highly branched structures. Lignin is also an amorphous three-dimensional polymer, with coumaryl, coniferyl and sinapyl alcohol as general building blocks. The possibility of hydrogen bond formation results in a parallel arrangement of cellulose chains. Lignocellulosic biomass is progressed to develop resistance in contradiction of enzyme hydrolysis obtained in the help of microorganism naturally. The production of lignocellulosic bioethanol comprises of various processing stages which are lignocellulosic biomass pretreatment, hydrolysis, fermentation and purification of product.

For producing of bioethanol, pretreatment is one of the most important steps to convert lignocellulosic biomass to bioethanol. It is the main key of the successful for following enzymatic hydrolysis and fermentation stages so that make the highest of the yield of the bioethanol product. The pretreatment of the lignocellulosic biomass is continued by enzymatic hydrolysis to give more fermentable sugars, the production of bioethanol from the fermentation of sugars

* Corresponding author: novia@ft.unsri.ac.id; noviasumardi@yahoo.com

and then purification of product. Both cellulose and hemicellulose can be cracked into simple sugars either by acid or enzymatically hydrolysis.

Pretreatment is a crucial for biomass-to-biofuels conversion processes. This stage is the main key for following enzymatic hydrolysis and fermentation stages for maximizing of the bioethanol product. Cellulose and hemicellulose can be cracked into simple sugars either enzymatically or by acid hydrolysis. Furthermore, the pretreatment has ideally to reduce the degradation product form. This is due to it can be the inhibitory compound on subsequent hydrolysis and fermentation processes. The pretreatment process can decrease hemicellulose content, remove cellulose crystallinity, and enhance the porosity of the materials.

The objectives of pretreatment include [3]: (1) to produce a greatly edible solids that improves sugar yields for the period of enzyme hydrolysis, (2) to minimize the sugars degradation (subject to pentose's) comprising those resulting from hemicellulose, (3) to minimize the inhibitor compounds for next fermentation stages, (4) to reduce lignin content to convert into valued co-products, and (5) to be low cost operation in reactors with medium size and minimizing heat and power supplies. The effective pretreatment must encounter the following requirements [4]: 1. Enhance the amount of sugar production or the capability to following form sugars by hydrolysis; 2. Reduce the degradation or carbohydrate loss; 3. Avoiding a by-product formation, such as inhibitory compunds to the next process.

The conversion of lignocellulosic feedstocks to advanced biofuels and other substances through a sugar-platform process includes a pretreatment stage to increase the liability of the cellulose to enzymatic hydrolysis [5].

Various pretreatments have been proposed to release the sugars locked in biomass. Among them are steam explosion [6], ammonia fibre explosion [7], organasolv [8], ozonolysis [9]. On the other hand, several of these pretreatments are not operational enough in the decreasing of the lignin content and some inhibitor compounds of yeast fermentation are formed through the hydrolysis. In several circumstances, it can be applicated as a detoxification stage of the liquid phase as well as another kind of pretreatment stage, such as an alkaline in order to reduce a higher of lignin content.

The alkaline pretreatment using NaOH, provides the swelling in biomass, the increasing of the surface area, the decreasing of the crystallinity and disrupting of the lignin structure [10]. Dilute-acid prehydrolysis is a pretreatment method that has resulted effective for different materials, and that has high potential for industrial application [11]. An acid pretreatment is one of the most effective pretreatments used for lignocellulosic biomass. On this pretreatment the hemicelluloses is detached, so it can improves the structure porosity and therefore enhances the enzymatic digestibility. The most widely applied acid is sulfuric acid (H_2SO_4), but also hydrochloric acid, phosphoric acid, nitric acid and various organic acids could be performed.

Other outhors [12] concludes that the highest enzymatic digestibility is produced while all hemicellulose is detached from the structure. Compare than the other kind of lignocellulosic materials, cellulose is extra difficult to hydrolyze on the acidic or alkaline pretreatments. This is due to its higher degree of crystallinity [13].

To enhance the biomass conversion, the other authors [14]–[16] studied the combination of acid alkaline pretreatments to produce the bioethanol from the other materials (bamboo, napiergras, sugarcane bagasse). In the other hand, previous researchers [17] investigated the two-steps of rice straw pretreated via aqueous ammonia and dilute sulfuric acid.

In this work, the stage one of the two-steps pretreatment was carried by a sodium hydroxide solution to remove lignin. After eliminating the lignin at a reasonably low temperature while provide enough cellulose and hemicellulose complete, the stage two of pretreatment was conducted using dilute sulfuric acid to solubilize hemicellulose at medium temperature (which also avoided the xylose degradation to furfural). The hydrolysate obtained through the stage two consisted xylose with the small content of lignin while the solids phase consisted generally cellulose. The objective of this study was to investigate the possibility of the sodium hydroxide-dilute sulfuric acid pretreatment and determine the effects of various either acid concentration or time heating of the pretreatment and SSF time on bioethanol production.

2 Material and Methods

2.1. Material

Rice husk samples were taken nearby (South Sumatera province of Indonesia). The moisture content was calculated 10% and they were placed in plastic bags at 5°C. The rice husk was taken from a several locations in the city of Inderalaya Ogan Ilir, South Sumatera Indonesia. The rice husk was then air-dried and stored at room temperature. The rice husks were treated using any mill and sieve machine to obtaining a particle size of 60 mesh.

2.2. Methods

2.2.1. Alkaline-Dilute Acid Pretreatment

The ratio of rice husk and NaOH solution was (1:10 w/v). The slurry resulted was continued proceed in incubator in a rotary air bath at 120 rpm and 85 °C for 1 hour. The content of lignin, cellulose and hemicellulose after alkaline pretreatment was analyzed by UV-Spectrophotometer and Chesson method. Assays were performed in duplicate by using different samples from the original raw material.

After the stage, the samples were treated by dilute acid pretreatment to promote the process of lignin content removal in the rice husk. The concentration of sulfuric acid (H_2SO_4) was varied in1%; 2%; 3%; 4% and 5% (v/v). It was occurred in the autoclave reactor at 121°C with various residence times of 30 min, 60 min and 90 min. The slurry was filtrated and separated into solid and liquid fraction. 100 ml sodium hydroxide at the concentrate of 4% (w/v) fed to the pretreated samples and inserted to the autoclave at 121°C for 30 min. The pretreated solids were splashed with water; the filtrate disclosed a neutral pH. Then they were dehydrated in the oven at 105°C. The content of lignin, cellulose and hemicellulose after alkaline pretreatment was analyzed by UV-Spectrophotometer and Chesson method. Assays were performed in duplicate by using different samples from the original raw material. The decreasing of lignin content was considered by the initial dry weight of lignin in the untreated material ($L_{UNTREATED}$) and the dry weight of lignin in the remaining treated solids t ($L_{PRETREATED}$).

2.2.2. *Simultaneous Saccharification and Fermentation*

Rice husk powders of 30 g were placed in an Erlenmeyer. Furthermore, the solution of media was added into it. The media solution consisted of yeast extract of 5 g / L, $(NH_4)_2SO_4$ of 7.5 g / L; K_2HPO_4 of 3, 5 g / L; $MgSO_47H_2O$ of 0.75 g / L and $CaCl_22H_2O$ of 1 g / L, with a ratio of 1: 10 (weight / volume). The pH of the solution was kept to ± 5. It was kept in sterile using autoclave set at the temperature of 121 °C for 60 minutes [18]. After this step, rice husk pulp was kept in cool and then added with a cellulose enzyme concentration of 3 ml (1-5 ml). The fermentation process was set between 72 to 168 hours, while the treatment of the yeast Saccharomyces cereviseae was using concentration of 5 mL. All the glass tool in this process was kept closed. It was placed on a rotary shaker at 120 rpm for each fermentation process. Then, ethanol was analyzed by 2010 Plus Series Gas Chromatography. The complete procedure of experimental work was shown in fig. 1.

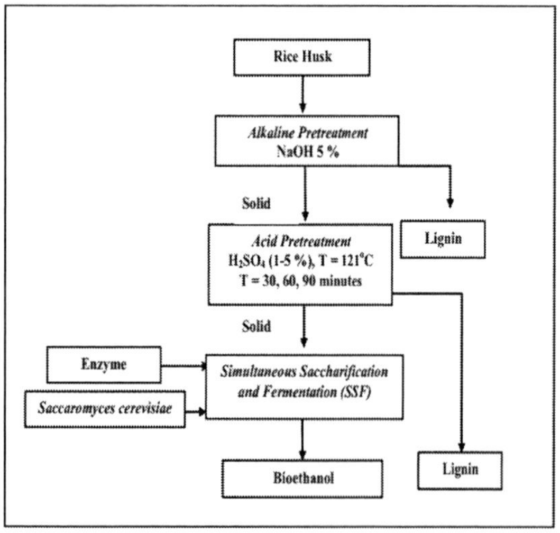

Fig. 1. Scheme of Experimental Procedure

3 Results and Discussion

3.1. Effect of sulfuric acid concentration on Cellulose, Hemicellulose and Lignin Content

The effect of sulfuric acid concentration of rice husk pretreatment on the content of cellulose, hemicellulose and lignin content, respectively, are shown in Figs. 2–4. After sulfuric acid-pretreated, the cellulose content was increased as the increase of acid concentrations. The cellulose content of the pretreated rice husks ranged from 75.02% (3%, 30 min) to 88,84% (2%, 90 min) (Fig. 2). Previous researchers [19] reported 28.72 % (1%, 30 min, 90°C) to 40.68% (2%, 60 min, 121°C of glucan preservation. During this stage, it is more favorable for the cellulose part of the biomass to be nearly unaffected.

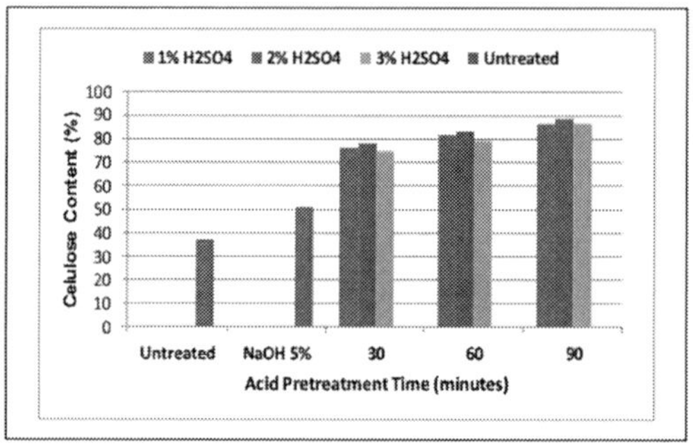

Fig. 2. Cellulose Content of Rice Husk Before and After Alkaline-Acid Pretreatment

As a result of the solubility of lignin, the cellulose content in the solid phase raised with the higher of the acid pretreatment time as shown in fig. 2. However, for similar pretreatment time, the highest cellulose content was found at the acid concentration of 2%. This is due to the solubility of some cellulose occurring over 2% of acid concentration. The highest value of cellulose content was achieved about 88.84 % for material acid pretreated at the H_2SO_4 concentration of 2% and 90 minutes acid pretreatment time. After enzymatic hydrolysis this cellulose-rich solid leads a highest concentrated of glucose solution.

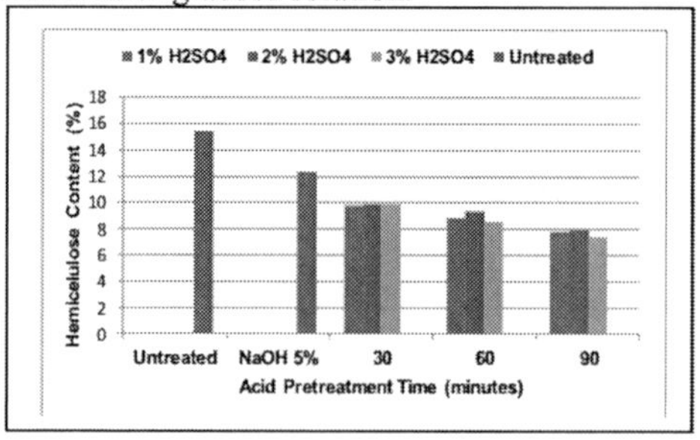

Fig. 3. Hemicellulose Content of Rice Husk Before and After Alkaline-Acid Pretreatment

Figure 3 illustrates the pretreatment caused the reducing of hemicellulose content. For similar pretreatment times, the higher temperature leads the lower value of hemicellulose content. This is due to depolymerizing of xylan to xylose then hydrated to furfural. The maximum value of hemicellulose content was achieved about 14.11 % for material alkaline pretreated at the NaOH concentration of 5%. The content of hemicellulose reduced with either the increasing concentration of H_2SO_4 or the raising of acid pretreatment time.

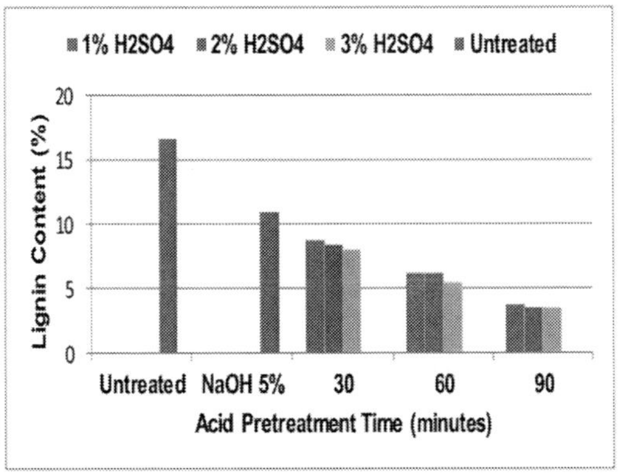

Fig. 4. Lignin Content of Rice Husk Before and After Alkaline-Acid Pretreatment

The increasing of acid concentration and pretreatment time significantly improved the delignification. The lowest of lignin content is about 3.46 % for solids pretreated at the acid concentration of 2% and 90 minutes of pretreatment time. It can be seen from fig 4 that higher concentration of acid results in a higher interruption of lignocellulosic structure, thus enabling the solubilization of lignin in the stage of pretreatment.

The result of the treatment process was an increase in the total lignin content of the material which it was not happen on the untreated samples. It is effected by the removal of carbohydrates from the structure. The lignin content of the samples varied from 3.46% (3% H_2SO_4, 90 min) to 8.78% (1% H_2SO_4, 30 min). The values was depended on the treatment conditions. On the other hand, the lignin reduction ranged from 47.46% to 79.29%. The results of this work are reasonable to Silverstein et al. (2007), who investigated 24.2% lignin declining for cotton stalks treated with 2% H_2SO_4 at 121°C for 90 min.

3.2. Effect of sulfuric acid concentration on bioethanol concentration

Rice husk was pretreated at 121°C for 60 minutes with different sulfuric acid concentrations. The simultaneous saccharification and fermentation (SSF) of pretreated samples were conducted by the commercial cellulase enzyme (Novozyme) and the *Saccharomyces Cerevisiae* for various SSF time of 72-160 hr. When 3% sulfuric acid was used, the concentration of ethanol was slightly increased comparing with the other sample.

Among the several factors that affect the enzymatic hydrolysis yield, one important thing to be noticed is the inhibition. It was occurred by the accumulation of glucose and cellobiose on the enzymes.

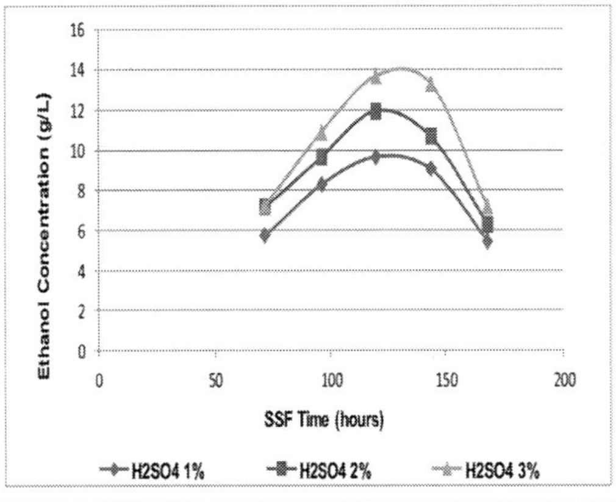

Fig. 5. Ethanol Concentration of SSF process for different sulfuric acid concentration

The simultaneous saccharification and fermentation (SSF) has been recommended as a possible solution [20]. Figure 5 shows the ethanol concentration during the SSF using the commercial cellulase enzyme (Novozyme) and the *Saccharomyces Cerevisiae*. The highest ethanol concentration obtained was 13.68 g/L for the SSF time of 120 hours and the sulfuric acid concentration of 3 %. Similar results (15.8 g/L ethanol) were obtained by previous study [16].

In order to fractionate the main components of rice husk, the acid pretreatment was successfully performed n this work. It was proceed in two stages of pretreatment using hydroxide-dilute sulfuric acid. This is reasonable result with the previous study [17].

4. Conclusion

A two-step of rice husk pretreated using sodium hydroxide continued by using dilute sulfuric acid at lower temperature was very effective in reducing lignin and hemicellulose contents, increasing the enzymatic digestibility. This pretreatment could be used to consecutively fractionate major constituents of rice husk to glucose, xylose, and lignin. The lignin content of the samples varied from 3.46% (3% H_2SO_4, 90 min) to 8.78% (1% H_2SO_4, 30 min) depending on the treatment conditions. The highest value of cellulose content was achieved about 88.84 % for material acid pretreated at the H_2SO_4 concentration of 2% and 90 minutes acid pretreatment time. The lignin reduction ranged from 47.46% to 79.29%. The highest ethanol concentration obtained was 13.68 g/L for the SSF time of 120 hours and the sulfuric acid concentration of 3 %.

The authors acknowledge Universitas Sriwijaya, Indonesia for funding the project (Hibah International Research Collaboration Grant), No. 605/UN9.3.1/LT/2015. Authors also would like acknowledge Curtin University for fasilitating the collaboration research.

References

1. L. F. Gutiérrez, Ó. J. Sánchez, and C. A. Cardona, *Bioresour. Technol.*, vol. 100, no. 3, pp. 1227–1237, 2009.
2. W. H. Chen, Y. J. Tu, and H. K. Sheen, *Appl. Energy*, vol. 88, no. 8, pp. 2726–2734, 2011.
3. G. Brodeur, E. Yau, K. Badal, J. Collier, K. B. Ramachandran, and S. Ramakrishnan, *Enzyme Res.*, vol. 2011, p. e787532, 2011.
4. Y. Sun and J. Cheng, *Bioresour. Technol.*, vol. 83, no. 1, pp. 1–11, 2002.
5. T. Persson, J. L. Ren, E. Joelsson, and A. S. Jönsson, *Bioresour. Technol.*, vol. 100, no. 17, pp. 3906–3913, 2009.
6. C. Cara, E. Ruiz, I. Ballesteros, M. J. Negro, and E. Castro, *Process Biochem.*, vol. 41, no. 2, pp. 423–429, 2006.
7. Q. Shao *et al.*, *Biotechnol. Biofuels*, vol. 3, p. 12, 2010.
8. E. Araque *et al.*, *Enzyme Microb. Technol.*, vol. 43, no. 2, pp. 214–219, 2008.
9. N. Eqra, Y. Ajabshirchi, and M. Sarshar, *Agric. Eng. Int. CIGR J.*, vol. 16, no. 1, pp. 151–156, 2014.
10. M. Balat, H. Balat, and C. Oz, *Prog. Energy Combust. Sci.*, vol. 34, no. 5, pp. 551–573, 2008.
11. N. Mosier *et al.*, *Bioresour. Technol.*, vol. 96, no. 6, pp. 673–686, 2005.
12. J. D. McMillan, *Renew. Energy*, vol. 10, no. 2–3, pp. 295–302, 1997.
13. P. Kumar, D. M. Barrett, M. J. Delwiche, and P. Stroeve, *Ind. Eng. Chem. Res.*, vol. 48, no. 8, pp. 3713–3729, 2009.
14. K. Li, J. Wan, X. Wang, J. Wang, and J. Zhang, *Ind. Crops Prod.*, pp. 1–9, 2015.
15. Z. Zhu *et al.*, *Biomass and Bioenergy*, vol. 93, pp. 269–278, 2016.
16. L. Camesasca, M. B. Ramírez, M. Guigou, M. D. Ferrari, and C. Lareo, *Biomass and Bioenergy*, vol. 74, pp. 193–201, 2015.
17. J. W. Kim *et al.*, *Bioresour. Technol.*, vol. 102, no. 19, pp. 8992–8999, 2011.
18. H. Li, N. J. Kim, M. Jiang, J. W. Kang, and H. N. Chang, *Bioresour. Technol.*, vol. 100, no. 13, pp. 3245–3251, 2009.
19. R. A. Silverstein, Y. Chen, R. R. Sharma-Shivappa, M. D. Boyette, and J. Osborne, *Bioresour. Technol.*, vol. 98, no. 16, pp. 3000–3011, 2007.
20. K. Olofsson, M. Bertilsson, and G. Lidén, *Biotechnol. Biofuels*, vol. 1, no. 1, p. 7, 2008.

* Corresponding author: novia@ft.unsri.ac.id; noviasumardi@yahoo.com

MATEC Web of Conferences **101**, 02014 (2017) DOI: 10.1051/matecconf/201710102014
SICEST 2016

Increasing biogas production from sugar cane baggase by anaerobic co-digestion with animal manure

Siswo Sumardiono[1,*], *Aldi* Budi Riyanta[1], *Hashfi* Hawali Abdul Matin[2], *Tutuk* Djoko Kusworo[1], *Bakti* Jos[1], and *Budiyono*[1]

[1]Department of Chemical Engineering, Diponegoro University, 50275 Semarang, Indonesia
[2]Department of Environmental Engineering, Diponegoro University, 50275 Semarang, Indonesia

Abstract. The result of this research is to determine the ratio of the combination of cow manure and bagasse (F/I) to the volume of biogas production, chemical pretreatment effect of raw material bagasse on biogas production volume and the effect of co-digester on biogas production. The results showed a decrease in total solid occurs when the increase of 2%, 5% to 10%, and this contributes to decreasing the production of biogas. Chemical pretreatment influences on the production of biogas through pretreatment with 2% NaOH solution with surface morphology bagasse seen with SEM image and its chemical structure changes with FTIR instrument. The highest biogas yield of 51.04 L/Kg substrate was obtained from a combination bagasse treated with 2% NaOH solution for 24 hours and as much as 20% cow's rumen.

1. Introduction

Utilization of non-renewable energy in excess can cause problems of energy crisis [50]. The use of energy sources such as fuel derived from fossil raw materials is a fuel that is not easily recyclable and requires a long process to produce the fuel. Therefore, there are the needs for new alternative energy sources that can be updated [48]. One of the energy technologies that comply with these requirements is biogas technology [16, 40].

Production of biogas from waste and organic residue with various types of anaerobic digesters method has been widely studied, both experimental and theoretical, for six decades [32].This material raw biogas can be obtained from lignocellulosic biomass and a wide range of organic wastes there are animal waste, industrial waste water and municipal solid wastes [5, 47]. Availability of lignocellulose raw materials are abundant in the world make lignocellulose into raw materials most in demand for biofuel production [46].

Biogas production has been carried out by several researchers. Biogas obtained by performing a combination of water hyacinth and sheep dung. A ratio of 4 : 12.01 : 83.90 between the hyacinth : sheep dung : water produces biogas production as much as 0.36 l / g [37, 52], in his journal did a combination of grass Pakchong with chicken manure at a ratio of 50 : 50 % produce methane 0.27 ± 0.01 L/Kg VS. Oleszek [36] comparing the production of biogas from weeds and grass varieties of walnuts. The study resulted in that canary grass produces more biogas for 406 $N.dm^3/kg$ VS than 120 varieties of wild grass $N.dm^3/Kg$ VS).

Bagasse is a lignocellulosic waste from sugar mills and agricultural processing [33]. Bagasse, like other lignocellulosic materials, has three main components: cellulose, hemicellulose and lignin [41]. Lignocellulose is composed of cellulose, hemicellulose and lignin [58]. Cellulose and hemicellulose has a long chain of sugar monomers and can be converted into bioenergy through pretreatment and hydrolysis [38]. Bagasse has a main constituent named xylose. Xylose ($C_5H_{10}O_5$) is a monosaccharide that is generated from the hydrolysis of hemicellulose in the form of xylan. Xylose also known as wood sugar as it contained in the timber -rich hemicellulose [17].

Pretreatment can increase the total yield of methane [24]. Pretreatment component initial pretreatment of biomass used as a first step to hydrolysis. This stage, complex organic molecules are broken down until get smaller components such as simple sugars, amino acids, and fatty acids. Most of the studies, conducted using alkali and acid pretreatment. Alkali used as NaOH applied to biomass pretreatment process. This makes the components of lignocellulose undergo solvation and saponification reaction components made of porous lignocellulose into the entrance of the enzyme [26]. Using acid pretreatment is also carried by [55], to break down lignocellulose component in water hyacinth.

The reform process anaerobic organic matter to biogas formation is influenced by two factors, namely, biotic and abiotic. Biotic factor such as microorganisms. Abiotic factors include: the substrate, the water content of the material/substrate, the ratio of C/N and P in the material / substrate, temperature, aeration, presence of toxic (poisonous elements), pH and stirring. So that the

*Corresponding Author: siswo.sumardiono@che.undip.ac.id

raw material (substrate) with a high C/N ratio (lignocellulose) compared with low C/N ratio (cow dung) gives the average composition corresponding input ratio desired levels [42]. Then, the condition of the anaerobic digestion process is maintained in equilibrium and dynamic conditions [39] Anaerobic co-digestion process becomes an important solution. To obtain a good digestion process needs to do a combination of the co-digestion. Co-digestion which combine with agricultural waste, animal dung to obtain the high biogas production and to achieve a C/N ratio ideal [52]. C/N ratio is 20-30 base that allows the process of digestion and biogas production is high. If the C/N ratio is high (60) biogas production will be small. Meanwhile, C/N ratio is low (10) also experienced a slow process [1].

2. Materials and methods

2.1. Raw material and inoculum

Bagasse as raw material obtained from sugar factory in Central Java Jatibarang Brebes. It dried for a week, then milled and sieved to the size of 0.85 mm (22 mesh) [2]. Inoculum used was the rumen of cows and cow dung obtained in slaughterhouses Jatibarang, Brebes in Central Java.

2.2. Batch Anaerobic Digestion

This study uses a batch digester 5 Liter with a working volume 2 Liter. Biomass raw material for laboratory-scale experiments are bagasse (AT). Characteristics % TS (total solids), % VS (volatile solids) from a variety of biomass used are shown in Table 1. with the addition of 2% NaOH solution then do the treatment for 24 hours and 48 hours besides sugarcane bagasse there is no treatment. Juntarasiri *et al* [26] reported, the addition of $NaHCO_3$ is used to get neutral pH of the raw material before process. Table 1 shows the baseline characteristics of each sample.

Table 1. Types of initial biomass feedstock

Types of Biomass	Characteristics		
	TS %	VS %	VS/TS
Bagasse	79,64	78,83	0,99
Cow dung	16,67	13,17	0,79
AT*	79,07	77,00	0,97
AT**	78,50	76,83	0,98
Rumen's cow	26,50	14,33	0,54

Raw materials and cow manure is fed to the digester with each variation of the ratio F/I 2, 5, 10% (base TS) to form slurry. Water added to the volume of work. Experiments were performed at room temperature. Digester was closed with a rubber stopper and one end of the hose was inserted inside each digester. Measuring cup was placed on the stand and clamp to measure the volume of biogas by water - displacement methods [4]. Biogas volumes were measured 2-3 days during 84 days trial of anaerobic co-digestion process.

2.3. Analytical methods

Biogas volume is measured by water displacement method. TS and VS concentration was measured using a standard method [4]. While the pH was measured using a pH stick by dissolving 5 g of sample into 50 mL of water. Cellulose and lignin content of the bagasse is determined by the Chesson method. Biogas composition were analyzed using Gas Chromatography.

3. Results and discussion

3.1. Effect of chemical pretreatment of the volume of biogas

Pretreatment before anaerobic digestion process has proven to be one simple and effective method to increase the biodegradability and biogas yield from lignocellulose materials [23].

Figure 1 shows the surface bagasse as seen in SEM image. Bagasse is undergoing treatment with NaOH 2% by immersing bagasse for 24 and 48 hours. The structure of the surface morphology bagasse more open as shown in the sample B1 and C1 which each experienced a magnification of 500x, addition of alkali can make swell lignocellulose components [30, 34]. Treatment with 2% NaOH solution on the bagasse components, such as lignin and hemicellulose will dissolve leaving a smooth surface along with the fiber [26]. This shows that most of the carbohydrate components such as lignin, hemicellulose, which is the skin surface has been dissolved by NaOH solution, so few will react with cellulose [51]. Natrium hydroxide also serves to change the structure of the composition of pure bagasse [22].

FTIR spectra (Figure 2) shows that a change in the concentration of hydroxyl functional groups of each bond is on bagasse. This indicates that the OH bond existed at the bagasse react with Na ion this case proved the value of the absorbance spectra B and C are reduced. FTIR spectra comparison indicated there was a picture where the sample A at figure 2. Bagasse without treatment and bagasse with treatment for 24 hours and 48 hours with sample B and C treated by NaOH 2%.

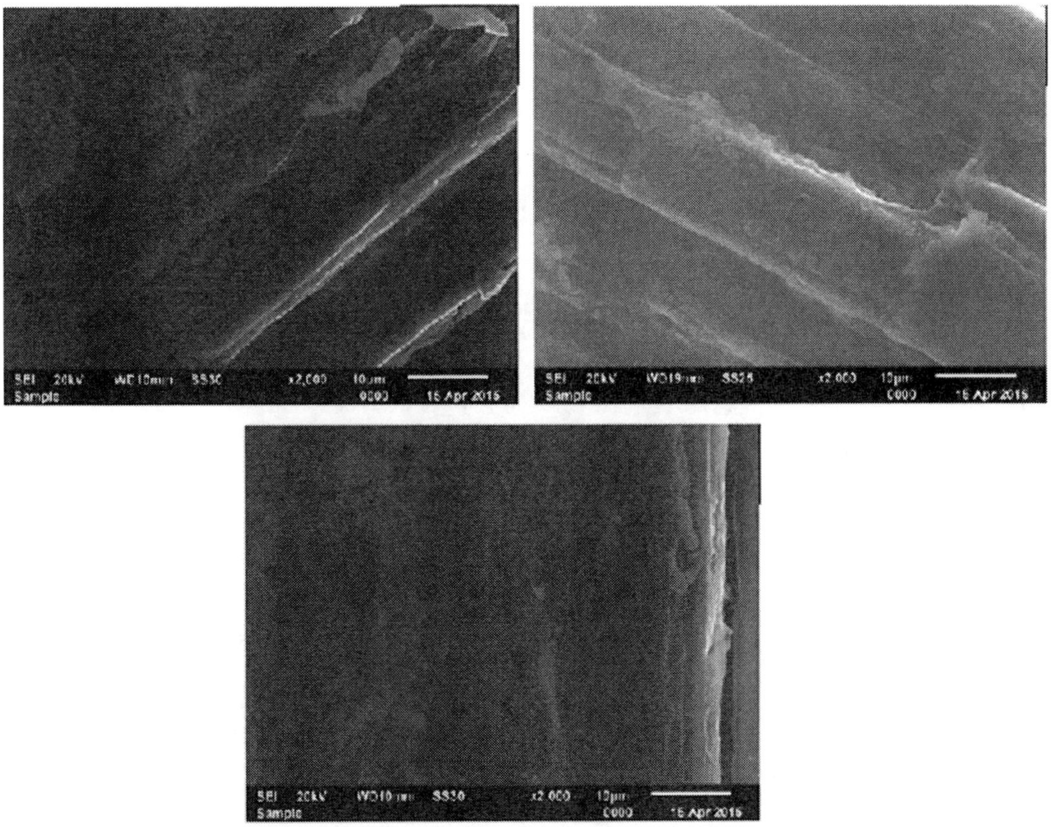

Fig 1. SEM image of bagasse of bagasse without treatment (A), bagasse undergo treatment with NaOH 2 % for 24 hours (B), and bagasse undergo treatment with NaOH 2 % for 48 hours (C).

Visible spectra are at wavelengths of more than 1200 cm-1 [49]. Structural differences have given evident with the addition of alkali (NaOH) as a treatment. The spectra are shown in the display wavelengths 3400-3500 cm-1. It shows the culmination (reduction) of the -OH functional groups. Peak around 2920 cm-1 appears there stretching of CH. Peak seen appear on down at the wavelength of 1049/cm. Bagasse spectra at 2920/cm (Figure 2. (A)) are stretching from CH_2 and down at a wavelength of 2901/cm (Figure 2. (B)) and 2851/cm (Figure 2. (C)) after a delignification process [33].

Biogas volume data is presented as a daily and cumulative biogas yield during the 84-days trial, as shown in Figure 3. Figure 3. (a) shows the volume of biogas daily on TS content of 2% and 20% cow's rumen co-digester with P1 is made of bagasse biogas without going through the treatment process, P4 bagasse with 24 hours of immersion treatment using NaOH 2% [26] and P7 is bagasse before being fed into the reactor treatment performed by soaking for 48 hours using a 2% NaOH.

Biogas started to be measured on the second day in a row and its volume was measured every other day. Measurement of the second day to day to twenty volumes of biogas continues to rise and gain first peak with P1 on day 16 amounted to 1.678 L/Kg Substrate, P4 amounted to 6.298 L / Kg Substrate on day 50, while P7 reached its peak the highest biogas volume on day 56 with a volume of 5.825 L / Kg Substrate. Figure 4. (a) it appears that no treatment effect in the acquisition of the volume of biogas.

Experiment P1 is a combination bagasse coupled with rumen as a maid in the process of digestion is seen that the highest volume at day 16 was as described González [19]. This period is the period of a single substrate. Peaks occur because the day 10 to 20 a period of single substrate. More digestible substrate will produce in advance compared to the substrate is more difficult to degrade. Rumen is a substrate that is easier to digest. Rumen bacteria contains ruminant [11]. Which is able to degrade cellulose in biomass materials. In addition to the treatment process is also a delignification process, a number of lignin will be dissolved.

This process is a process of saponification of the intermolecular ester bond that surrounds xylan, Hemicellulose and other components, like lignin and hemicellulose more. Delignification processes causing damage to the structure of lignin and release carbohydrate compound [51].

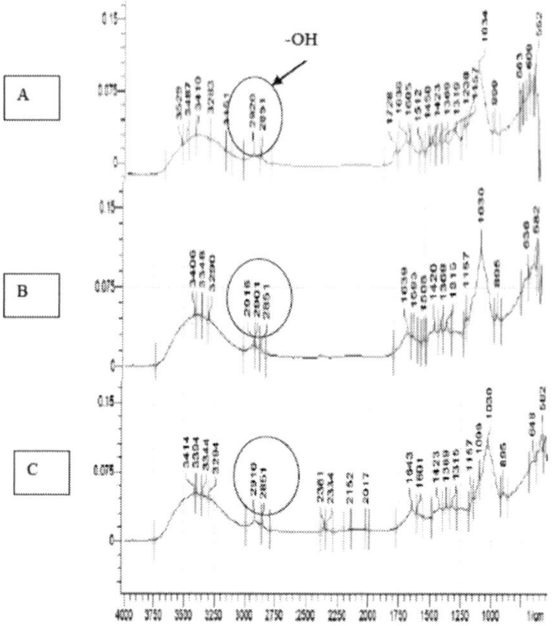

Fig 2. FTIR Spectra of bagasse without treatment (A), bagasse undergo treatment with NaOH 2 % for 24 hours (B), and bagasse undergo treatment with NaOH 2 % for 48 hours (C).

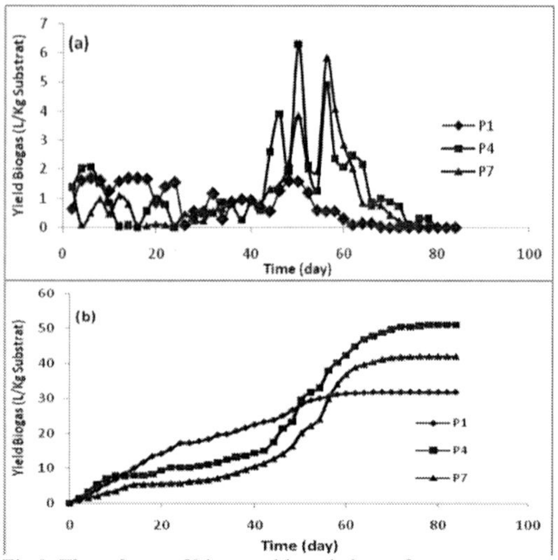

Fig 3. The volume of biogas with variations of pretreatment and total solids (TS) 2 % (a) the volume of daily biogas P1, P4 and P7 (b) Volume of cumulative biogas P1, P4 and P7

3.2. Influence of total solid content (TS)

This study will show the content of TS which affects the production of biogas. Variations of total solid used is 2%, 5% and 10%. Yield of biogas is presented in the form of biogas and volume of data accumulated by the cumulative length of time during the 84 -day trial (Figure 4.).

Experiment P1, P2 and P3 are each using TS 2%, 5%, and 10% and using a cow rumen inoculum as much as 20% of the working volume of two liters. Experiment P1, P2 and P3 biogas yield is calculated on the second day as on the first day no biogas is produced. Biogas yields continue to rise from the start to the second day up to nine reached the top first and then the volume continues to fall until day 35. Yield biogas and then rise again until day 40 to day 56 reached the top of the second and achieved a higher volume than the first peak. Experiment P1 and P3 have a similar pattern in the biogas yield accumulation. P2 has a graph which charts more gentle than the graph P1 and P3. The highest volume of biogas obtained in P2 on day 44 with a volume of 1.45 L / Kg Substrate, P1 on day 46 with a volume of 1.68 L / Kg Substrate and P3 on day 56 with a volume of 0.68 L / Kg Substrate. Accumulated biogas yield obtained at P1 reaches 31.78 L/Kg substrate, P2 as much as 18.89 L/Kg Substrate, and P3 as many 13.10 L/Kg Substrate. P1 experiments using bagasse as much as 2 % TS and use co-digester rumen viewed from the observation experiment P1 is the highest compared with P2 and P3. Figure 4.shows the production of biogas P1 , P2 to P3 tends to decrease with increase in the content of total solids (TS). With the condition that the higher the total solid of reducing the amount of substrate on anaerobic co-digestion system, so that components with high substrate will easily degradable organic components [25].

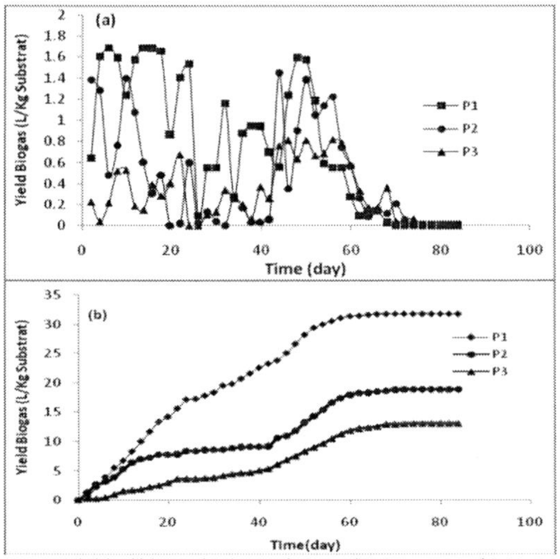

Fig 4. Yield of biogas in the variation of TS content of 2 %, 5 %, 10 % (a) = P1, P2, P3 co-digester cow rumen without treatment; (b) Yield biogas accumulation of P1, P2, P3.

Results of this research together with research conducted by [26] with the lowest cumulative biogas yield is obtained at a high TS content during the process of digestion of sewage sludge. Statistical analysis also showed the content of TS has a significant affect the biogas production [15].

The content of TS is too high causing the speed of production of organic acids in the early stages faster than the speed of digestion of acid consumption by methanogenic bacteria [25] This of course causes the activity of methanogenic bacteria is inhibited and decreases the production of biogas [31].

The content is too high TS can also occur in the inhibition process of hydrolysis process because of the low pH [31]. Microbial activity with the raw materials to be difficult because of the limited space resulted in mass transfer process is inhibited. The products of hydrolysis phase collected on the surface of the substrate and ultimately inhibits the absorption of hydrolytic enzymes. Mass transfer is limited to make the number of products available for microbial acidogenic hydrolysis slightly so will reduce product on acidogenesis processes which will be converted into biogas [43].

4. Conclusions

Biogas production on bagasse with treatment processes better than the production of biogas in the bagasse without treatment. Biogas production on bagasse best on the total solid 2% compared to the total solid higher 5% and 10%. The most optimum biogas production contained in the bagasse treated with NaOH 2% for 24 hours and the cow rumen co-digester with maximum biogas production rate of 6.3 L/kg .day and total production of biogas of 51.04 L/kg.

The authors thank to Diponegoro University via International Research and Publication (Riset Publikasi Internasional, RPI) Grant 2016 for financial support.

References

1. Achmad, K.T.B., Hidayati, Y.A., Fitriani, D., Imanudin, O. *Lucrări Ştiinţifice* - **vol. 55**, Seria Zootehnie (2015)
2. Anief, M. Gadjah Mada University Press (2010)
3. Akwaka, J.C., Kukwa, D.T. & Mwekave, S.S. *International Journal of Science and Technology*, **3(4)**, pp.222–228 (2014)
4. APHA, Standard Methods for the Examination of Water and Wastewater Part 1000. Book : *American Public Health Association, American Water Works Association, Water Environment Federation* (2005)
5. Brown, D., Shi, J. & Li, Y. *Bioresource Technology*, **124**, pp.379–386 (2012)
6. Budiyono., Syaichurrozi, I.,Sumardiono, S.. *IJE Transactions B: Applications*. **27**, No. 2, pp 177-184 (2014)
7. Budiyono, Widiasa, I.N., Johari, S and Sunarso. *Energy Research Journal* **1 (1)**: pp 6-11 (2010)
8. Budiyono, Widiasa I.N., Johari S., Sunarso. *World Academy of Science, Engineering and Technology*, **37**:983-988 (2010)
9. Budiyono, I. Syaichurrozi, S. Sumardiono. *World Applied Sciences Journal* **26 (11)**: 1464-1472 (2013)
10. Budiyono, Iqbal S., Siswo S. *Research Journal of Applied Sciences, Engineering and Technology* **7(3)**:2798-2805 (2014)
11. Budiyono, I. Syaichurrozi and S. Sumardiono. *International Journal of Engineering*, **27(2)**:177-184 (2014)
12. Budiyono, S. Sumardiono and D. Tri Mardiani. *International Journal of Engineering*, **28(2)**: 921-928 (2015)
13. Budiyono, I N. Widiasa, S. Johari, and Sunarso. *International Journal of Engineering and Technology*, **1(3)**:109-116, ISSN : 0975-4024 (2009)
14. Burak. D., Scherer P., Yenigun O., Onay T. *Environmental Science and Technology*, **40**:116–146 (2010)
15. Carneiro, Y.F., Perez, M., Romero, L. I.,. *Bioresorce Technology*, **99** pp 6994-7002 (2008)
16. Darsin, M. Design of Biogas Circulator, Seminar Nasional Kreativitas Mesin Brawijaya 2006, Universitas Barawijaya, Malang (2006)
17. Faisal, A.. Skripsi: Universitas Indonesia (2008)
18. Gaudet, M.M., Agnew, J., Fonstad, T.A.,. *In the Canadian Societyu for bioengineering / La Societe Canadiene de Genie Agroalimentaire*. Paper No. CSBE13-027 (2013)
19. González, L.M., Castro, R., Pereira, M.A., Alves, M.M., Font, X., Vicent, T.,. *Bioresource Technology* **102** pp 4734–4741 (2011)
20. Hassan, R., *International Water Technology Journal, IWTJ* (2014)

21. Hussein, I., Abdel-Shafy, A., Mona, S.M., Mansour. *Egyptian Petroleum Research Institute* pp xxx-xxx (2014)
22. Hatakka, A. Biodegradation of lignin. University of Helsinki, Viikki Biocenter, Department of Applied Chemistry and Microbiology (1994)
23. He, Y., Pang, Y., Liu, Y. P., Li, X., Wang, K., Department of Environmental Science and Engineering, and Center for Resources and Environmental Energy & Fuels, **22**, *pp* 2775–2781 (2008)
24. Hendriks, A.T.W.M. & Zeeman, G.,. *Bioresource Technology*, **100**, pp.10–18 (2009)
25. Jha, A.K., Li, J., Zhang, L., Ban, Q., Jin, Y.,. *Advances in Water Resource and Protection*, **1(2)**, pp 28-38 (2013)
26. Juntarasiri, P. S. Nijsunkij, T. Buatick , E. Jamkrajang, S. Wacharawichanant,M. Seadan, A. Wasantakorn and S. Suttiruengwong. Enhancing. *Energy Procedia* **9** pp 207 – 215 (2011)
27. Kangle, K.M. Kore, S. U. Kore. U.S. G. S, Kulkurni.,. *Universal Journal of Environmental Research and Technology*, **2(4)**, pp.210–219 (2012)
28. Kim, M., Yang, Y., Morikawa, Sakura M.S., Wang, Q., Lee M.V., Lee D.Y., *Int J Hydrogen Energy*; **37**: pp 3142-9 (2012)
29. Kim, M., Liu, C., Noh, J.W.,Yang, Y., Oh, S., Shimizu, K., Lee, D.Y., Zhang, Z., *International journal of hydrogen energy* **3 8** pp 8648-8656 (2013)
30. Kong, F., Engler, C., Soltes,. *Applied Biochemistry and Biotechnology* pp 34–35, pp 23–35 (1992)
31. *Li*, Y., Zhu, J., Wan, C., Park, S.Y.,. *American Society of Agricultur and Biological Engineers*, **54(4)**, 1415-1421 (2011)
32. *Marek*, M. Bialobrzewski, I. Zielinski, M. Debowski, M. Krzemienewski, M.,. *Renewable Energy*, **69** pp.219-225 (2014)
33. Maryana, R, Ma'rifatun, D. A. Wheni, Satriyo K.W., W. Rizal. A. *Energy Procedia*, **47** pp 250 – 254 (2014)
34. Montgomery, L.F.R., Bochmann, G.,. Book. *IEA Bioenergy*. British Library. ISBN 978-1-910154-05-2 (electronic version) (2014)
35. Mshandete, A., Bjornson, L., Kivaisi, A.K., Rubindamayugi,M.S.T., Mattiasson, B.,. *Renewable Energy*, **31**, 2385-2392 (2005)
36. Oleszek, M. Krol, A. Jerzy, T. Matyka, M. Kulik, M. *Bioresource Technology* **156** pp 303–306 (2014)
37. Patil, H. J. Raja, M. A. L. A. Raja, S. B.B. Shettyc, M. K. B. P.Kumard, P.. *Energy Procedia* **52** pp 572 – 578 (2013)
38. Petersson, A. Mette, H.T. Nielsen, H.H. Thomson, A.B. *Biomass and Bioenergy*, **31**, pp.812–819 (2007)
39. Putri, D.A. Saputro, R.R. Budiyono., Biogas Production from Cow Manure. *Int. Journal of Renewable Energy Development* **1(2)** pp 61-64 (2012)
40. Putro, Sartono, Laporan. **Vol .10**, No. 2 : 178 – 188 (2007)
41. Rabelo, S.C. H. Carrere. Maciel, F.R. Costa. A.C.. *Bioresource Technology* **102** pp. 7887-7895 (2011)
42. Sasongko, W. Tesis : Universitas Sebelas Maret (2010)
43. Sheets, J.P., Ge, X., Li, Y., *Bioresource Technology*, **144**, pp 296-303 (2015)
44. Song, Z., Yang, G., Han, X., Feng, Y. Z., Ren, G., *BioMed Research International* **Vol. 2013** Article ID 968692, pp 9 (2012)
45. Tao, Y. (He *et al.*). Z.; Guan, Y. T. *J. Cell. Sci. Technol.* 2003, **11 (1)**, 42–55 (in Chinese) (2008)
46. Teghammar, A., Evaluation. Thesis: Chalmers University of Technology (2013)
47. Tuesorn, S. Wong, W.S. Champreda, V., *Bioresource Technology*, **144**, pp.579–586 (2013)
48. Ukpai, P.A. & Nnabuchi, M.N.,. *Advances in Applied Science Research*, **3(3)**, pp.1864–1869 (2012)
49. Vivekanand, V., Olsen. E.F., Eijsink, V.G.H., Horn, S.J.,. *BioResource* **(9)1**. Pp 1311-1324 (2014)
50. Wahyuni, Sri,. Kongres Ilmu Pengetahuan Nasional (KIPNAS) ke 10 (2011)
51. Weizhang Z, Zhang Z, Qiao W, Fu P, and Liu M. *Renewable Energy*; **36**: 1875-9 (2011)
52. Wilawana, W. P. Pholchanb, and P. Aggarangsia. *Energy Procedia* **52** pp 216 – 222 (2014)
53. Yadvika, Santosh, Sreekrishnan, T.R., kohli, S., Rana, V., *Bioresource Technology*, **95**, 1-10 (2004)
54. Yavini, T.D. *Usman*, H. T. Mohammed, N and James, M. Namo. *The International Journal Of Engineering And Science*, **3(2)**, pp.1–6 (2014)
55. Yonathan, A., Prasetya, A.R. Pramudono, B. *Jurnal Teknologi Kimia dan Industri*, **Vol 2**, No 2. Pp 211-215 (2013)
56. Zhang, R., Zhang, Z., *Bioresource Technology*, **68(3)**, 235-245 (1999)
57. Zheng, Yi, Jia Zhao, Fuqing Xu, Yebo Li., *Progress in Energy and Combustion Science xxx* 1-19 (2014)
58. Zhu, J., Wan, C. & Li, Y., *Bioresource Technology*, **101(19)**, pp.7523–7 (2010)

Combustion of producer gas from gasification of south Sumatera lignite coal using CFD simulation

Fajri Vidian[1]*, *Novia*[2], *Andy* Suryatra[1]

[1]Mechanical Engineering Department, Faculty of Engineering, Universitas Sriwijaya, 30662 Inderalaya, Indonesia.
[2]Chemical Engineering Department, Faculty of Engineering, Universitas Sriwijaya, 30662 Inderalaya, Indonesia.

Abstract. The production of gasses from lignite coal gasification is one of alternative fuel for the boiler or gas turbine. The prediction of temperature distribution inside the burner is important for the application and optimization of the producer gas. This research aims to provide the information about the influence of excess air on the temperature distribution and combustion product in the non-premixed burner. The process was carried out using producer gas from lignite coal gasification of BA 59 was produced by the *updraft gasifier* which is located on Energy Conversion Laboratory Mechanical Engineering Department Universitas Sriwijaya. The excess air used in the combustion process were respectively 10%, 30% and 50%. CFD Simulations was performed in this work using two-dimensional model of the burner. The result of the simulation showed an increase of excess air, a reduction in the gas burner temperature and the composition of gas (carbon dioxide, nitric oxide and water vapor).

1 Introduction

The decreasing of crude oil as the energy sources has been followed by the increasing of the energy needs, so the needs of an alternative energy source to replace the crude oil is very crusial. Coal is one of the alternative energy to replace the crude oil. The utilization of coal in the direct combustion method will cause a problem in the pollution and a low efficiency of energy conversion. Gasification is one of the technology that can convert coal to be combustible gas with the low pollution [1-2] and high of the efficiency [2-3]. Indonesia is one of the largest coal-producing country in the world that have coal reserves about 3% from the world coal reserve [4] whereis about 57% of the reserve is the lignite coal [5]. South Sumatera is one of the largest coal suppliers in Indonesia. One of the coal product that not utilized maximally is the lignite coal (South Sumatera lignite coal called as BA 59).

The combustible gas produced from its gasification can be used for application on boiler, gas turbine, and dryer [6]. Its for the coal application, it needs an appropriate gas burner to reach the temperature distribution as expected. The method can be used to predict the temperature distribution inside the gas burner is simulation using computational fluid dynamics. Several simulations has been done by several researchers. Previous authors [7] has simulated the combustion of producer gas from the biomass gasification using computational fluid dynamics (CFD). The result of study shows the highest temperature is about 1367 ^{0}C. Other researchers [8] has simulated the combustion of producer gas from biomass gasification, the result shows the highest temperature in the combustion chamber about 1300 ^{0}K to 1700 ^{0}K. The combustion of producer gas in gas burner for applying in gas turbine is reported by [9]. The result of the work showed that the highest temperature of combustion was about 1500 ^{0}K [9].

Another work have performed by [10], they simulated the combustion of producer gas from gasification of wood chip and turkey feather.The result showed that the highest temperature can reach to 2000 ^{0}K and 1500 ^{0}K [10]. The combustion of producer gas from coconut shell gasification at swirl gas burner has simulated by [11], the result shows the application of swirl will increase the combustion temperature. The highest temperature reached about 1173^{0}K [11]. The result of the researchers showed the combustion temperature of the producer gas is effected by the fuel of the gasification process and the gas burner design with the temperature range of 1200 ^{0}K to 2000 ^{0}K. However, none of the previous studies used the application of CFD on the production of gassesfrom lignite coal gasification. The objective of this research is to obtaint he effect of excess air on the temperature distribution and the combustion products in the non-premixedburner using CFD simulation.

2 Producer gas composition

Producer gas used in the simulation is the BA 59 that produced from coal gasification from South Sumatera, Indonesia using updraft gasifier at Energy Conversion Laboratory Mechanical Engineering Department,

*Corresponding author: fajri.vidian@unsri.ac.id

Sriwijaya University. The gas composition shows in Table 1.

Table 1. Producer gas composition

Species	%(Vol)
CO	29.54
H_2	12.07
CH_4	3.4
CO_2	8.11
O_2	1.3
N_2	45.58

3 Governing equation

Combustion process of the producer gas at the stoichiometry condition define by the equation below

$$\text{producergas} + 0.26305\,(O_2 + 3.76\,N_2) \rightarrow 0.4105\,CO_2 + 0.1887\,H_2O + 1.4449\,N_2 \quad (1)$$

Combustion of producer gas in gas burner involved fluid flow and chemical reaction. Mass and momentum equation are used to drive the fluid flow.

Mass Conservation Equation

$$\frac{\partial \rho}{\partial t} + \nabla.(\rho\vec{v}) = S_m \quad (2)$$

Momentum Conservation Equation

$$\frac{\partial}{\partial t}(\rho\vec{v}) + \nabla(\rho\vec{v}\vec{v}) = -\nabla p + \nabla\bar{\bar{\tau}} + \rho\bar{g} + \bar{F} \quad (3)$$

The distribution temperature inside gas burner is solved by energy conservation equation.

$$\frac{\partial}{\partial t}(\rho E) + \nabla\big(\vec{v}(\rho E + p)\big) = \nabla\big(k_{eff}\nabla T - \textstyle\sum_j h_j \vec{J}_j + (\bar{\bar{\tau}}_{eff}\vec{v})\big) + S_h \quad (4)$$

Species movement in the gas burner is solved using the species transport equation

$$\frac{\partial}{\partial t}(\rho Y_i) + \nabla(\rho\vec{v}Y_i) = -\nabla.\vec{J}_i + R_i + S_i \quad (5)$$

The combustion process depend on the turbulence. K-Epsilon Standard equation is used to solve turbulence processd uring combustion.

$$\frac{\partial}{\partial t}(\rho k) + \frac{\partial}{\partial x_i}(\rho k u_i) = \frac{\partial}{\partial x_j}\left[\left(\mu + \frac{\mu_t}{\sigma_k}\right)\frac{\partial k}{\partial x_j}\right] + G_b + G_k - \rho\varepsilon - Y_M + S_k \quad (6)$$

The interaction between chemical reaction and turbulence in side of the gas burner are solved using the model has developed by Magnussen and Hjertager. The reaction rate will be produced by the smallest value between equation 7 and 8.

$$R_{i,r} = v'_{i,r} M_{w,i} A\rho \frac{\varepsilon}{k} min_R \left(\frac{Y_R}{v'_{R,r} M_{w,R}}\right) \quad (7)$$

$$R_{i,r} = v'_{i,r} M_{w,r} AB\rho \frac{\varepsilon}{k} \frac{\sum_P Y_P}{\sum_j^N v''_{j,r} M_{w,j}} \quad (8)$$

The mechanism of NOx production in side gas burner are modeled on thermal and prompt mechanism.

Thermal NOx mechanism

$$O + N_2 \rightleftharpoons N + NO \quad (9)$$
$$N + O_2 \rightleftharpoons O + NO \quad (10)$$
$$N + OH \rightleftharpoons H + NO \quad (11)$$

Prompt NOx mechanism

$$CH + N_2 \rightleftharpoons HCN + N \quad (12)$$
$$N + O_2 \rightleftharpoons NO + O \quad (13)$$
$$HCN + OH \rightleftharpoons CN + H_2O \quad (14)$$
$$CN + O_2 \rightleftharpoons NO + CO \quad (15)$$

Thermal and Prompt NOx transport are derived by equation below.

$$\frac{\partial}{\partial t}(\rho Y_{NO}) + \nabla.(\rho\vec{v}Y_{NO}) = \nabla.(\rho\mathcal{D}\nabla Y_{NO}) + S_{NO} \quad (16)$$

4 CFD Modeling

The combustion process was conducted in non premixed gas burner with the diameter of 45 cm and length of 180 cm [12]. The nozzle for injecting the producer gas diameter of 1 cm [12]. The schematic of of the gas is presented burner in Fig 1.

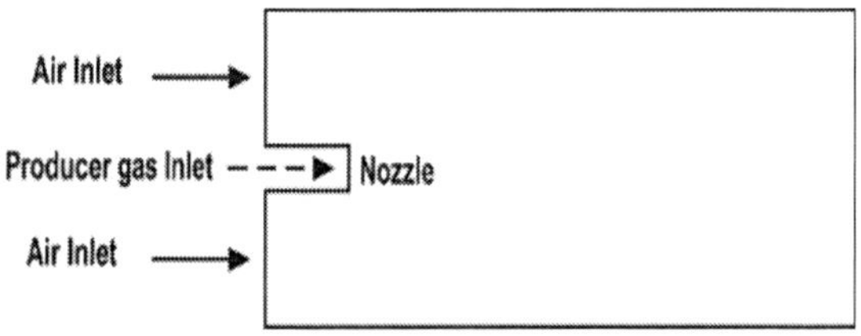

Fig.1. The scheme of gas burner

The simulation process was done in 2D of gas burner using ANSYS Fluent 16. The gas burner for domain of the calculation must be mesh to solve all the equation. The equations were solved using finite volume method. The construction of meshing are showed in Fig 2 and Fig 3.

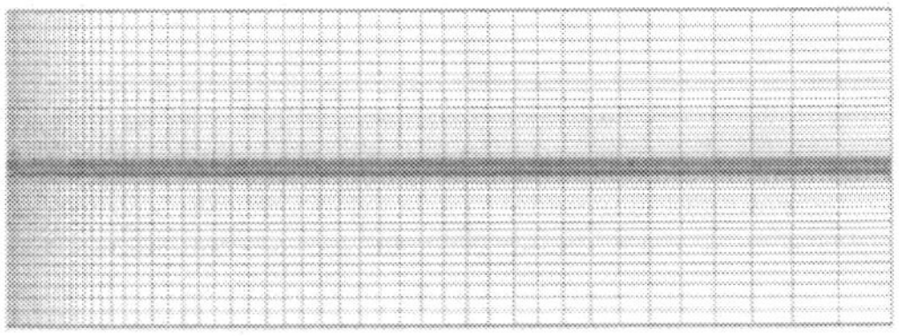

Fig.2. Mesh of the gas burner

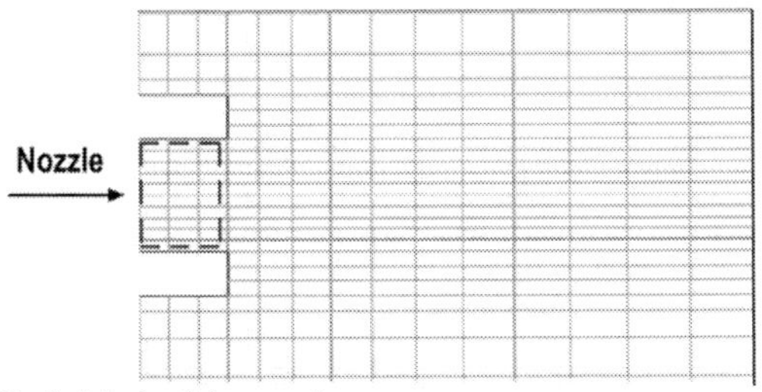

Fig.3. Mesh of the nozzle

The boundary condition for the simulation is presented in Table 2.

Table 2. Boundary Condition of The Simulation

Paremeter	Condition	Quantitity
Velocity of Producer gas	Excess air of 0%;10%;30%;50%	80 m/s
Velocity of air	Excess of air 0%	0.052 m/s
Velocity of air	Excess of air 10%	0.057 m/s
Velocity of air	Excess of air 30%	0,067 m/s
Velocity of air	Excess of air 50%	0.077 m/s
Inlettemperatur e of producergas	Excess air of 0%;10%;30%;50%	300 ^{0}K
Inlettemperatur e of air	Excess air of 0%;10%;30%;50%	300 ^{0}K

5 Result and discussion

5.1 Temperature distribution

The simulation result shows the increasing of excess air from 0 to 50% will reduce the temperature distribution in gas burner as shown in Fig 4. The increasing amount of air caused the increasing amount of nitrogen.This is due to nitrogen absorb the heat from combustion and reduce the temperature in gas burner. This result is similar with the study as reported by [13-14]. The highest temperature was about of 1910 ^{0}K. The temperature was decreased about of 100 ^{0}K for increasing of the excess air from 0% until 50 %.

In the visualization can be shown clearly the reduction of temperature distribution in the gasburner as shown in figure 5-8. The reduction zone of the yellow contour which describe the higher temperature, other wise increasing of the green zone which describe the lower temperature zone.

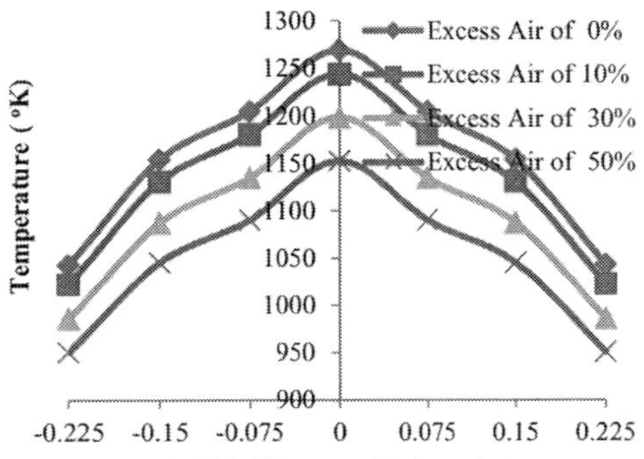

Fig. 4. Temperature distribution on middle of burne rat radial distance with difference of excess air

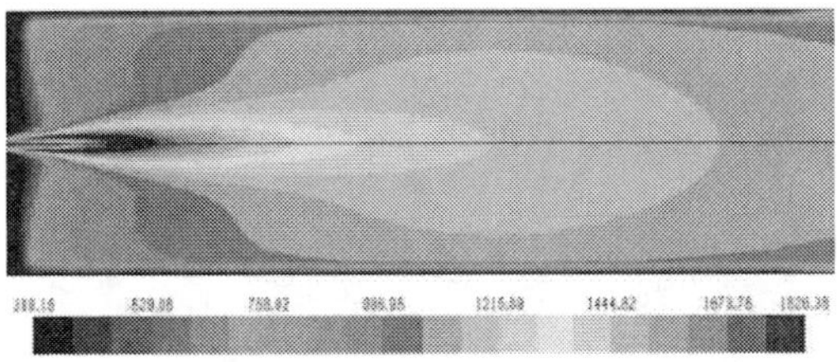

Fig.5. Temperature distribution on excess air of 0%

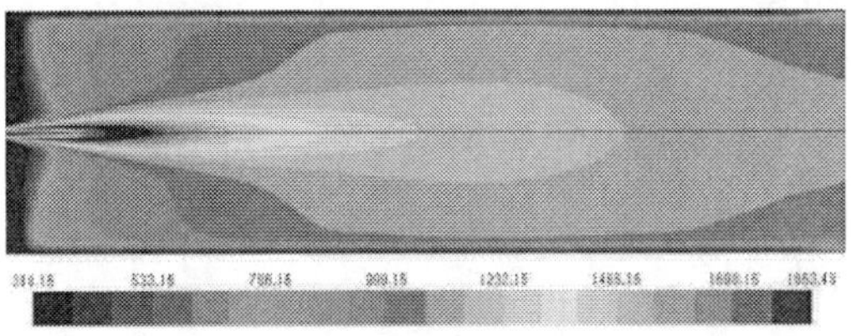

Fig.6. Temperature distribution on excess air of 10%

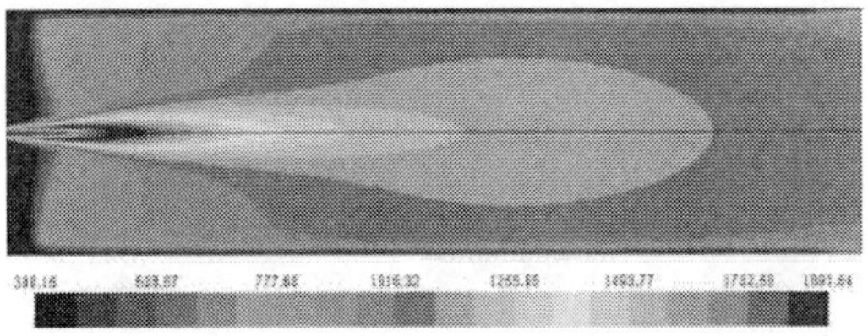

Fig.7. Temperature distribution on excess air of 30%

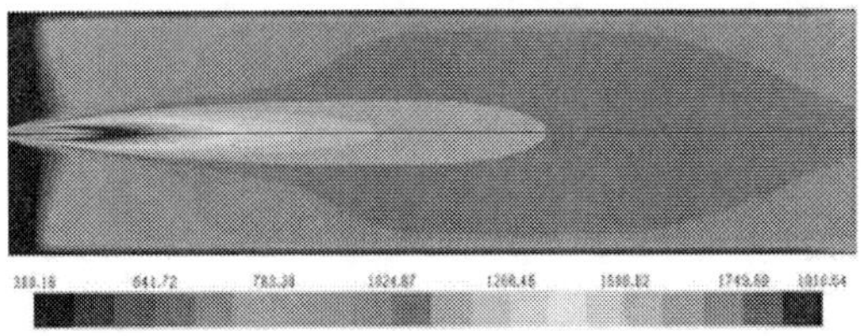

Fig.8. Temperature distribution on excess air of 50%

DOI: 10.1051/matecconf/201710102015

5.2 Carbon dioxide (CO_2) distribution

From the simulation, it showed that the excess air was increased with the decreased of the concentration distribution of CO_2 in gas burner as shown in Fig 9. This condition was occurred due to the increasing of air which is not followed by the increasing of carbon content in combustion process when the fuel flow rate is set constant. The result showed the same trend with the result has been reported by [15].

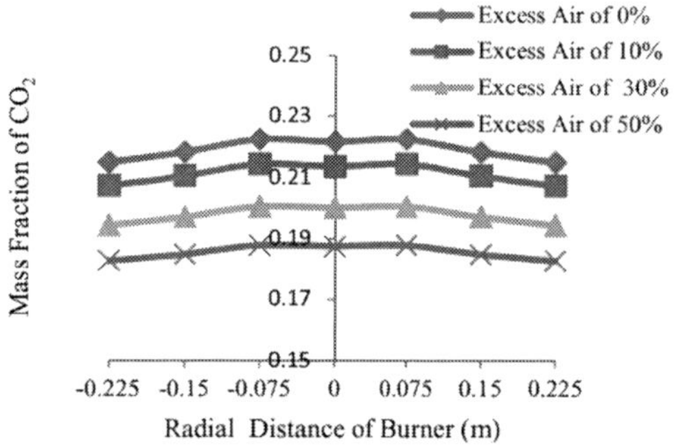

Fig. 9. Carbon dioxide distribution on middle of gas burner at radial distance with difference excess air.

The decreasing of CO_2 distribution in gas burner is shown clearly in Fig. 10-13. The area of highest temperature (brown color) shown a decrease as the increase of the excess air. The maximum of CO_2 mass fraction concentration could be reached to 28%. The increasing of excess air from 0-50% caused a decrease in mass fraction concentration about 3%.

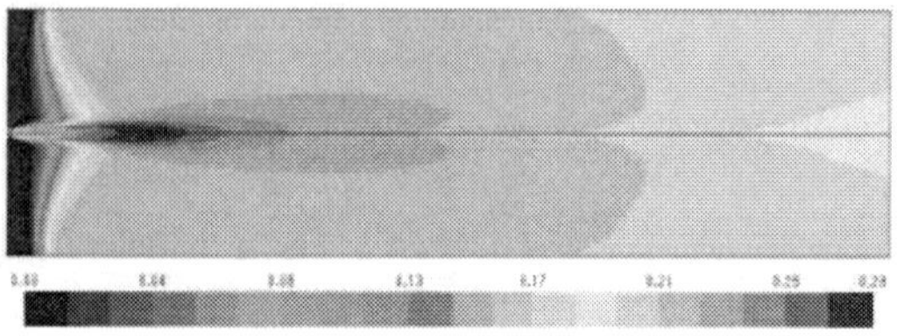

Fig.10. Carbon dioxide distribution in gas burner on excess air of 0%.

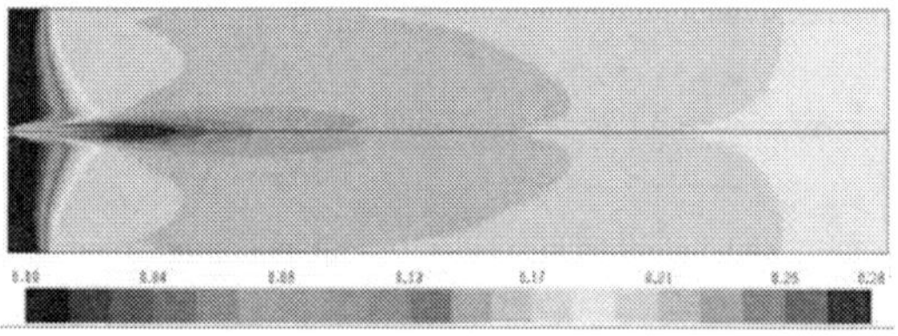

Fig.11. Carbon dioxide distribution in gas burner on excess air of 10%.

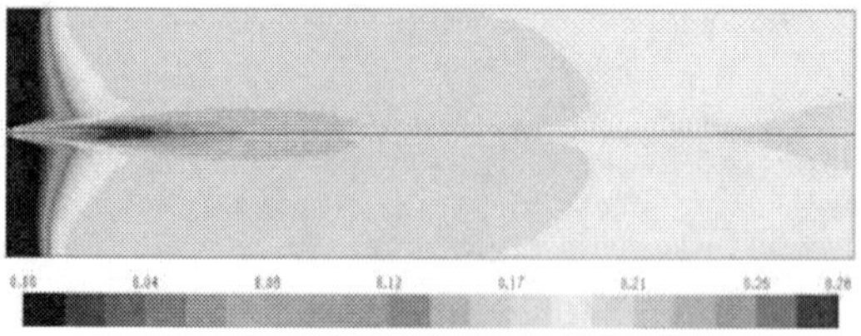

Fig.12. Carbon dioxide distribution in gas burner on excess air of 30%.

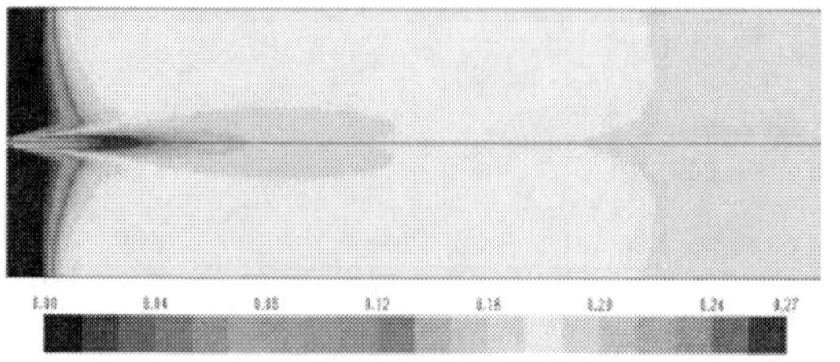

Fig.13. Carbon dioxide distribution in gas burner on excess air of 50%.

5.3 Water vapor (H_2O) distribution

The simulation result shows the increasing of excess air will decrease the distribution of H_2O in gasburner as shown in Fig 14, it is caused by the increasing of excess air was not followed by the increasing of H_2 (constant of fuel flow rate).

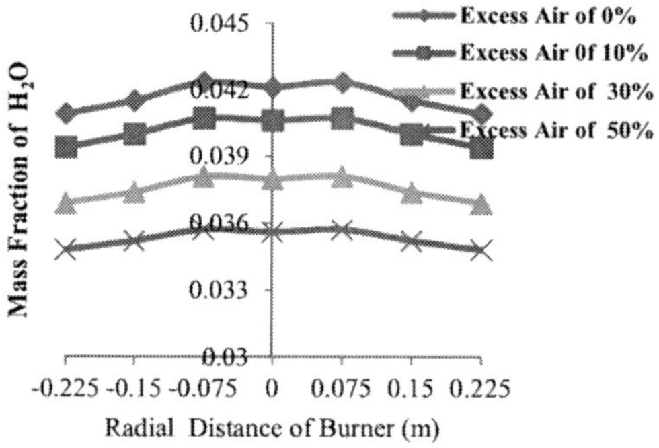

Fig. 14. Water vapor distribution on middle of gas burner at radial distance with difference excess air.

The decreasing of H_2O distribution in gas burner is shown clearly in Fig 15-18. The area of highest temperature (brown color) shown a decrease as the increase of excess air. The maximum of H_2O mass fraction concentration could be reached to 5%. The increasing of excess air from 0%-50% caused a decrease in mass fraction concentration about 0.6 %.

DOI: 10.1051/matecconf/201710102015

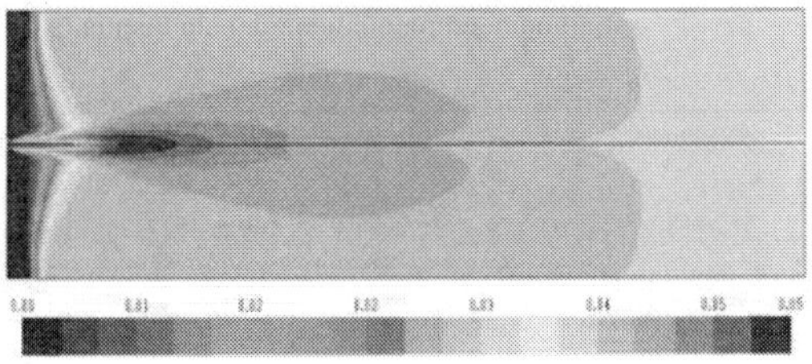

Fig. 15. Water vapor distribution in gas burner on excess air of 0%

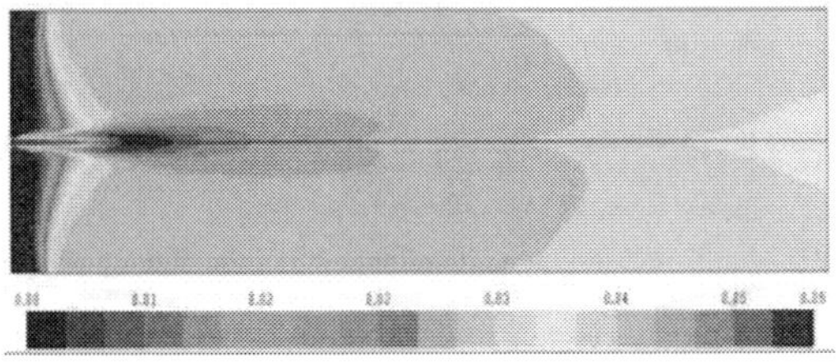

Fig. 16. Water vapor distribution in gas burner on excess air of 10%

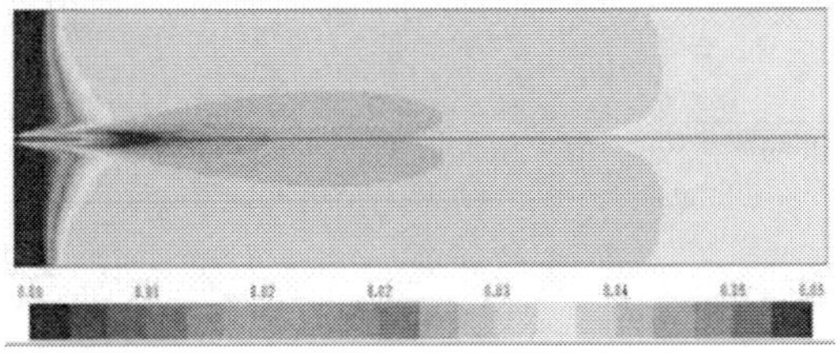

Fig. 17. Water vapor distribution in gas burner on excess air of 30%

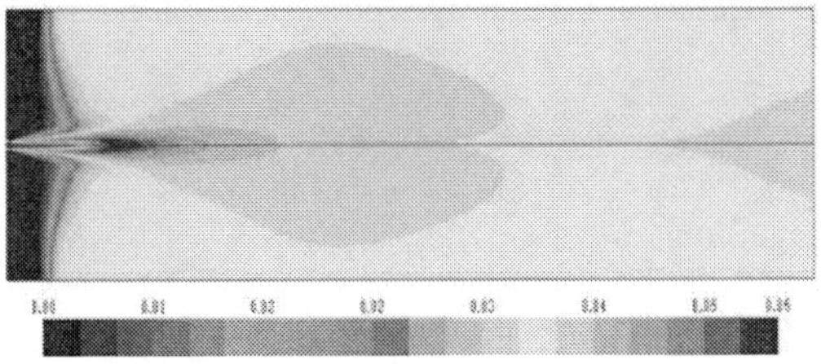

Fig. 18. Water Vapor distribution in gas burner on excess air of 50%

5.4 Nitric Oxide (NO) distribution

From the simulation process it shown an increase in excess air will caused a decrease in NO distribution in gas burner as shown in Fig 19. It is caused by the increasing of excess air which is followed by the decreasing of temperature in gas burner as shown in figure 4.The reaction 9–15 is decreased by the decreasing of temperature. This result has similarity to the simulation has reported by [16].

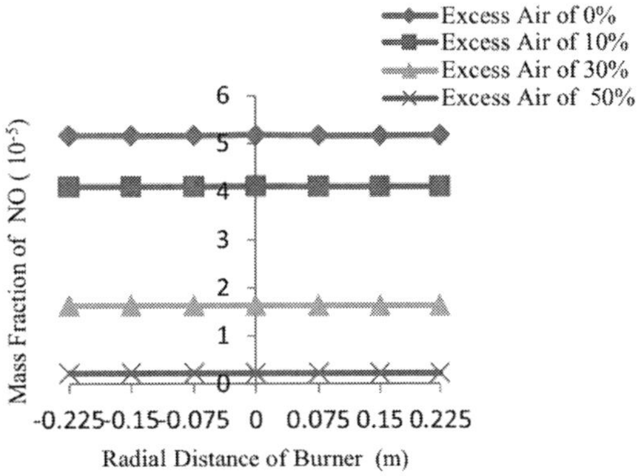

Fig. 19. Nitric Oxide distribution on middle of gas burner at radial distance with difference excess air.

The decreasing of NO distribution in gas burner shows in the visualization of contour in figure 20-23. It can be seen clearly the decreasing of the area of red contour (The highest of NO concentration) as the effect of the increasing of excess air. The maximum mass fraction of NO was about 5.37 x 10^{-3} %. The increasing of excess air from 0% until 50% will decrease mass fraction concentration about of 4.9 x 10^{-3} %.

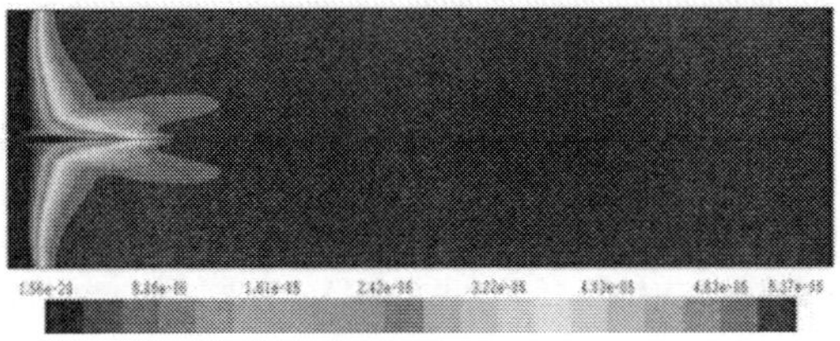

Fig. 20. Nitric oxide distribution in gas burner on excess air of 0%

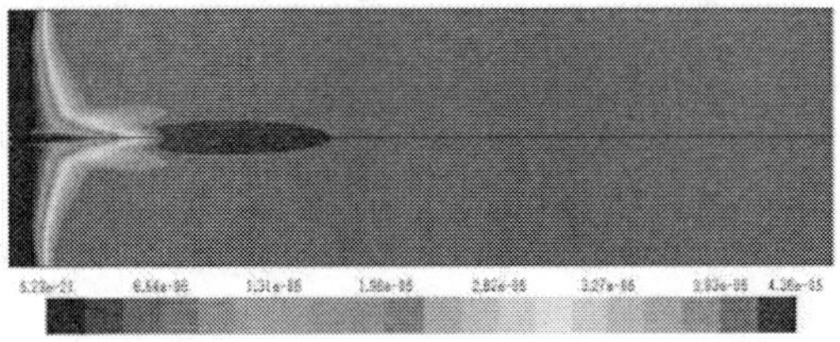

Fig. 21. Nitric oxide distribution in gas burner on excess air of 10%

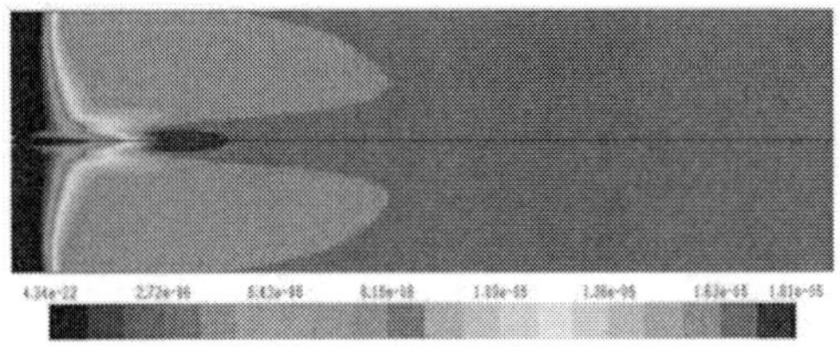

Fig. 22. Nitric oxide distribution in gas burner on excess air of 30%

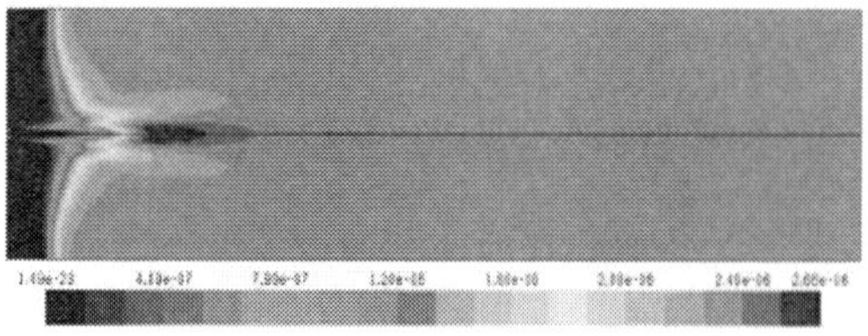

Fig. 23. Nitric oxide distribution in gas burner on excess air of 50%

6 Conclusion

From the simulation, a numberof conclusions couldbetakenas follow:

- The highest temperature can be reached by the combustion of producer gas from gasification of BA 59 coal at non premixed gas burner is 1910 ^{0}K, The highest mass fraction concentration of CO_2 about 28%, and the highest mass fraction concentration of NO about 5.37 x 10^{-3} %.
- The increasing of excess air on combustion process from 0%-50% will cause the decreasing of temperature, CO_2, H_2O, and NO distribution in side gas burner

References

1. Y. Yun, S.W. Chung, Korean Journal Chemical Engineering, **24** (4), 628-632, (2007).
2. N. Crnomarkovic, B. Repic, R. Mladenovic, O. Neskovic, M.Veljkovic, Fuel, **86,** 194-202, (2007).
3. A.G. Collot, International Journal of Coal Geology, **65,** 191-216, (2006).
4. T. Winarno, C.Drebenstedt,*Mine Planning Equipment Selection*, Springer International Publishing Switzherland, 1495-1494,(2014).
5. Asia Pacific Economic Corporation, *Final Report*, 1-67, (2013).
6. E.G. Pereira , J.N. Dasilva, J.L. Olievera, C.S. Machado, Renewable Sustainable Energy Reviewas, **16,** 4753-4762, (2015).
7. P. P. Ushekwar, S. B. Desai, M.M. Wagh, N. N. Shinde, International Research Journal of Engineering and Technology, **2**(4), 1621-1624, (2015).
8. Y. Achawangkul, N. Maruyama, M. Hirota, C. Chaichana, T. Sutabutr, IOSR Journal of Engineering, **4**(2), 17-23, (2014).
9. G.S. Sharma, M.V.S.M. Krishna, D.N. Reddy, International Journal of Mechanical and Production Engineering Research and Development, **5**(2), 21-32, (2015).
10. K. Kwiatkowski, M. Dudynski, K. Bajer, Flow Turbulence Combustion, **91,** 749–772, (2013).
11. A.Surjosatyo, Y. D. Priambodho, International Journal JurnalMekanikal, **33,** 15- 31, (2011).
12. Fluent, Incorporated : Fluent 6 UDF Manual, Lebanon (NH), Fluent Inc, 2001.
13. H.F. Elattar , R. Stanev, E. Specht, A. Fouda, " Computer and Fluids, **102,** 62-73,(2014).
14. R.K. Jain, IPASJ International Journal of Mechanical Engineering, Vol **2** (3), 20-28, (2014)
15. J.C. Lou, C.C. Huang, T.L. Yang, Journal of The Chinese Institute of Environmental Engineering, **9** (4), 219-227, (1999).
16. A.Soid, W. Kriaa, H. Mhiri, G.L. Palec, P. Bournot, *Proceeding of Word Congress on Engineering 2009,* Vol **II** WCE,July 1-3, London UK, (2009)

The effect of enzymatic pretreatment and c/n ratio to biogas production from rice husk waste during solid state anaerobic digestion (SS-AD)

Syafrudin[1,*], *Winardi* Dwi Nugraha[1], *Hashfi* Hawali Abdul Matin[1], and *Budiyono*[2]

[1]Department of Environmental Engineering, Diponegoro University, 50275 Semarang, Indonesia
[2]Department of Chemical Engineering, Diponegoro University, 50275 Semarang, Indonesia

Abstract. Rice husk is one of agricultural waste generated from the mill is potential to be converted to biogas. Otherwise, biogas production facing problem due to the high lignin content. This research is objected to study the effect of enzymatic pretreatment and C/N ratio to biogas production from rice husk by solid state anaerobic digestion (SS-AD). The laboratory scale-anaerobic digestions was used in this work. It is operated in batch system and with room temperature. Total solid (TS) was set 21%. Enzymatic pretreatment was conducted using lignase enzyme. Carbon to Nitrogen (C/N) ratio was varied from 20, 25, 30, and 35. The C/N ratio is varied by adding some quantity of technical urea to the substrate. Biogas formed was measured using water displacement method in every two days. The result showed that enzymatic pretreatment could increase biogas production from 30 to 55 %. The highest biogas production was obtained at C/N ratio 35. Specific biogas production on C/N ratio of 20, 25, 30, and 35 were respectively 12.0, 12.6, 13.5 and 18.2 ml/gr TS. The volumetric loading resulted by SS-AD was higher than liquid anaerobic digestion (L-AD). The further research need to conduct for optimization the enzyme concentration and C/N ratio.

1 Introduction

Agricultural activities are very important not only from an economic view point but also produces waste as a sustainable energy such as biogas [11] Availability of abundant lignocellulosic raw materials in the world makes lignocellulose become high demand raw materials for biofuel production [24]. Lignocellulosic material contained by crop residue is difficult to be degraded by anaerobic bacteria. Lignocellulosic often contain lignin with high concentrations.

Pretreatment on lignocellulosic raw materials also aims to facilitate the conversion of biopolymer contained in the cellulose. Pretreatment performed may increase total yield of methane [14]. Pretreatment can be performed pretreatment of physics, chemistry and biology [23]. Pretreatment of physics can be performed by size reduction to reduce the particle size and the degree of crystallinity of cellulose. The process of size reduction can increase total hydrolysis yield as much as 5-25% and reduces digestion time of approximately 23-59% [17].

The processes of anaerobic digestion are classified into two based on the content of total solids (TS) that is liquid anaerobic digestion (L-AD) and solid-state anaerobic digestion (SS-AD) [22]. Liquid anaerobic digestion (L-AD) used on the total solid content of between 0.5 to 15%, while the total solid content of>15% using a solid-state process of anaerobic digestion (SS-AD) [27]. Lignocellulosic biomass feedstock such as rice husk is suitable to be treated using the SS-AD due to the concentration of total solid on lignocellulosic biomass>15% and has a low moisture content [3, 20]. In addition, problems that occur in the L-AD-like substrate floats, fat and fiber stratification can not be found on the SS-AD [26]. [3] The volume of methane production from lignocellulosic biomass generated in the SS-AD is 2-7 times greater than with L-AD. [12] The similar result also expressed, that biogas production volume increased by 278-357% in SS-AD as compared to L-AD. Another advantage SS-AD include: reactor volume is smaller, needs less water, does not require stirring, no precipitation occurs in the bottom of the reactor, the residue of the SS-AD can be used as fertilizer because it has a low water content so it is more easily handled than waste of L-AD (16, 19].

Several studies of biogas production from lignocellulosic biomass using methods SS-AD had been conducted. [19] which conducts research biogas production from raw materials corncob yielding the largest biogas in the ratio F/I 2.43 and TS content of 22%. [21] Research with the raw materials of food waste and grass with SS-AD method. [13] Research on biogas with rice straw raw materials have also been carried out, which aims to determine the effect of type of inoculum to the production of biogas, but in these studies using meode L-AD and did not evaluate the

* Corresponding author : udin_syaf@yahoo.com

effect of the ratio F/I, TS content and physical pretreatment effect on the production of biogas.

[3] conducted a study of SS-AD (TS 18-19%) of corn cobs, straw, wheat, reeds and leaves. The raw material is dried at 40 °C for 48 hours until the moisture content of 10%, then milled 5 mm. The mixture was put into the reactor 1 L and incubated at 37 °C for 30 days. Biogas volume was measured 2-3 days. The highest biogas generated at 131.8 L/kgVS with raw materials corncob. [27] Research SS-AD (TS 22%) of corn cobs and the expired dog food. The content of dog food on the feedstock was 0%, 10%, 25%, 50% and 100%. The highest methane yield was 304.4 L/kgVS on the ratio F/I 2 and 50% of dog food and 50% of corn cobs. The ratio F/I 4 with a mixture of 25% of dog food derived methane yield 171.5 L/kg VS. The ratio F/I 6, yield of methane is 5 L/kgVS with a mixture of 0% and 10% of dog food. SS-AD study was also conducted with the raw materials of sawdust [1]. Sawdust is collected washed with water, then added 0.2% H_2SO_4 (w/w) and heated at 140 ° C for 45 minutes. The ratio of sawdust with cow dung 1:1.5. The study was conducted for 23 days. The cumulative volume of biogas generated at 1010 mL.All of previous studies have been conducted, no SS-AD method used for processing rice husks into biogas. Nitrogen needs also are important for the SS-AD because of lignocellulosic biomass has high carbon content, but low nitrogen content. Therefore, this study will focus on the effect of pretreatment and C/N ratio of the biogas production methods SS-AD.

2 Materials and methods

A series laboratory test of 600 ml biodigester was operated during this research. The main experiment apparatus consists of biodigester and biogas measurement. Biodigester were made from polyethylene bottle plugged with tightly rubber plug and was equipped with valve for biogas measurement. Biogas formed was measured by 'liquid displacement method' as also has been used byseveral researchers before [7-10]. The schematic diagram of experimental laboratory set up as shown in Figure 1. Total solid (TS) was set 21 %. Enzymatic pretreatment was conducted using lignase enzyme. Carbon to Nitrogen ratio (C/N ratio) was varied from 20, 25, 30, and 35. Implementation of the research started with the process of making organic rice husk waste (biowaste), then conducting an examination of the total composition of solids and water content. Furthermore, the pretreatment process consists of the addition of NaOH and addition of enzyme. After that, the calculation and determination of the variation of C/N ratio and the preparation of anaerobic batch reactor were conducted. Subsequently, the substrate is mixed with the inoculum and technical urea to adjust variations in C/N ratio. Samples were prepared can be inserted into the reactor, sealed in order to obtain anaerobic condition, and ready for operation. During the treatment process takes place, volume of biogas produced were observed in interval of 2 (two) days. The observation was terminated after no more biogas formed.

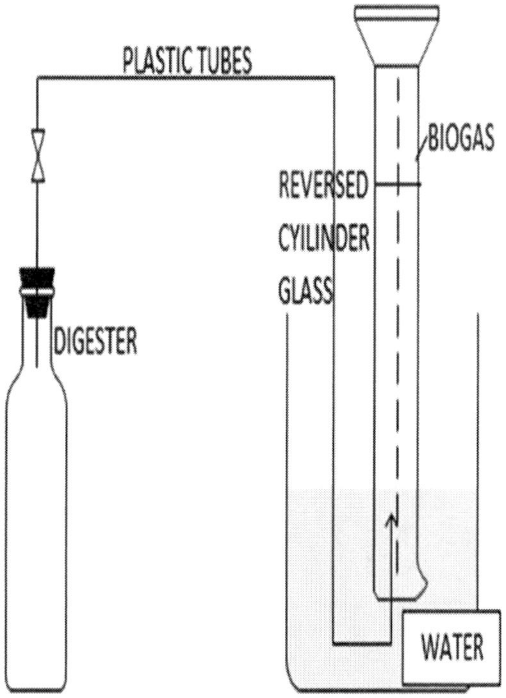

Fig. 1. Schematic diagram of series laboratory batch assessment of SS-AD

3 Results and discussion

3.1 The influence of pretreatment to biogas production using SS-AD

This phase of the research aims to determine the effect of pretreatment on production of biogas with SS-AD. Effect of Pretreatment on biogas production was observed in this study with the addition of enzyme as much as 5% volume of the solution for all the variables, the addition of NaOH as much as 3% of the volume of the solution. Biogas volume measurement results during the 60-day trial is shown as a SS-AD biogas yield is calculated from the daily biogas volume divided by the content of TS on rice husk [18]. The cumulative biogas yield per unit TS and presented in Figure 2 to 6.

Yield biogas showed that the ratio of C/N 20, pretreatment with the addition of enzymes significantly affect the production of biogas compared to pretreatment without using enzymes, can be seen in Figure 2. It can be seen that the total yield of biogas in the variable by using enzymes are higher, namely 11.9 ml/grTS and for variables without the addition of enzymes obtained biogas yield of 10.25 ml/grTS.

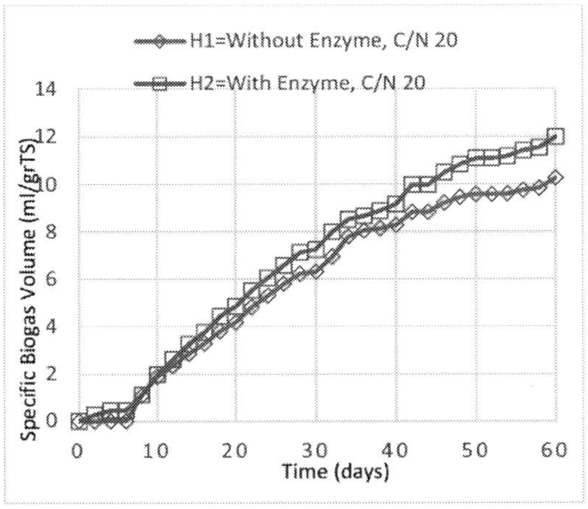

Fig. 2. The influence of enzymatic treatment to biogas yield at C/N ratio 20

Yield biogas showed that the ratio of C/N 25, pretreatment with the addition of enzymes significantly affect the production of biogas compared to pretreatment without using enzymes, can be seen in Figure 3. In the C/N ratio 25 that the data obtained for a total yield of biogas with the addition of enzymes at 12.6 ml/grTS and for variables without the addition of enzymes obtained biogas yield of 12.0 ml/grTS.

[14] Pretreatment on lignocellulosic raw materials able to facilitate the conversion of biopolymer contained in the cellulose. Pretreatment performed will increase total yield of methane. Pretreatment can be performed pretreatment of physics, chemistry and biology [23]. Pretreatment of physics can be performed by size reduction to reduce the particle size and the degree of crystallinity of cellulose.

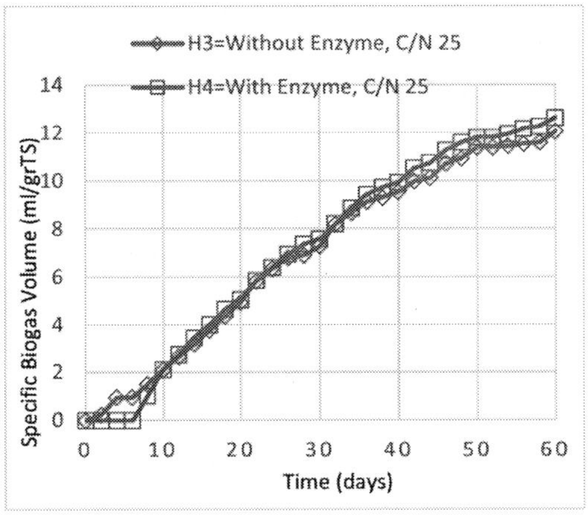

Fig. 3. The influence of enzymatic treatment to biogas yield at C/N ratio 25

In Figure 4 yield biogas showed that the ratio of C/N 30, pretreatment with the addition of enzymes have higher biogas yield than without pretreatment using enzymes but not too significant, can be seen in Figure 4. The data obtained for a total yield of biogas with the addition of an enzyme of 13.9 ml/grTS and for variables without the addition of enzymes obtained biogas yield of 13.5 ml/grTS.

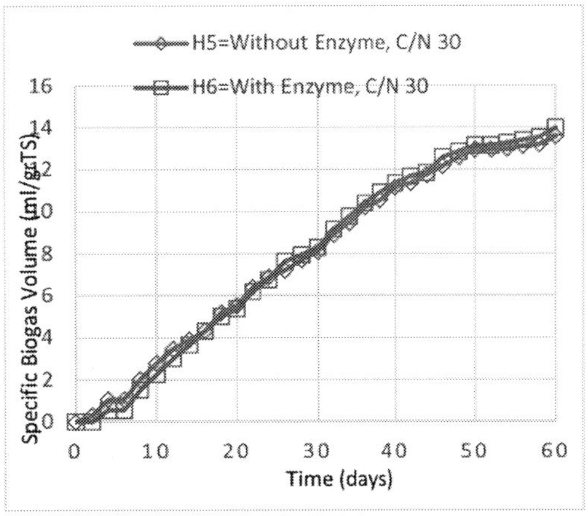

Fig. 4. The influence of enzymatic treatment to biogas yield at C/N ratio of 30

In Figure 5 biogas yield indicates that the C/N ratio of 35, with the addition of enzyme pretreatment significantly affect the production of biogas compared to pretreatment without using enzymes, can be seen in Figure 5. It can be seen that the total yield of biogas in the variable by using enzymes greater, namely 18.2 ml/grTS and for variables without the addition of enzymes obtained biogas yield of 13.5 ml/grTS.

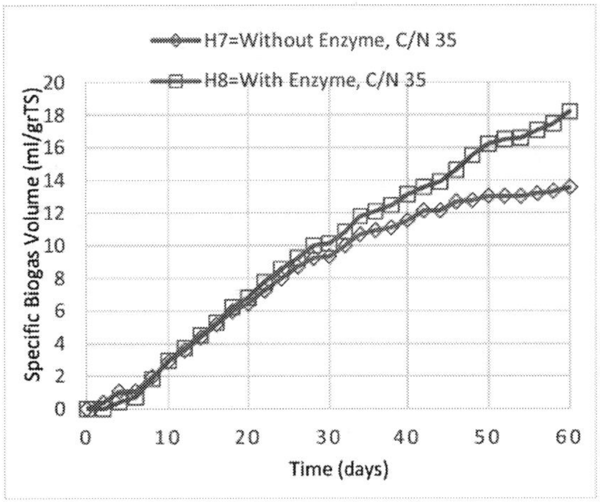

Fig. 5. The influence of enzymatic treatment to biogas yield at C/N ratio of 35

In Figure 6 under conditions of C/N ratio 25 indicated that the pretreatment with the addition of NaOH significant effect on the production of biogas compared with no use of NaOH pretreatment. It can be seen that the total yield of biogas in the variable by using NaOH greater, namely 12.6 ml/grTS and for variables without the addition of NaOH obtained biogas yield of 11.0 ml/grTS.

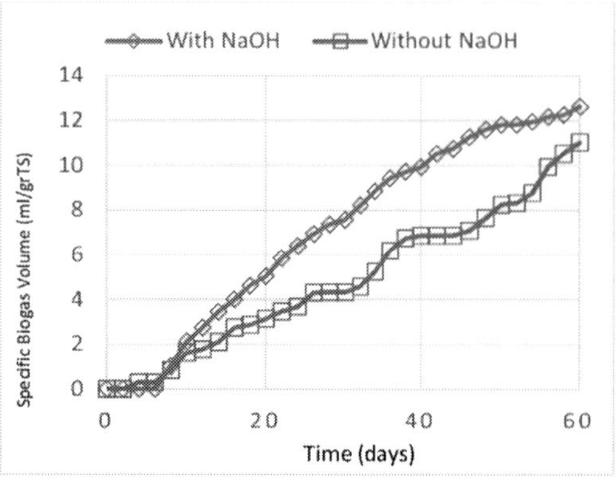

Fig. 6. The influence of NaOH pretreatment to biogas yield, C/N ratio of 25

3.2 The influence of C/N ratio to biogas production during SS-AD

In this study, the effect of C/N ratio of the biogas production was observed with a variety of C/N ratio of 20, 25, 30 and 35. Data are presented as the volume of cumulative biogas yield during the 60-day trial of SS-AD, as shown in Figure 7.

Figure 7 shows the cumulative volume of biogas per unit TS on rice husk with the addition of enzymes. In the rice husk C/N 20, the total volume of biogas that is equal to 12.0 ml/grTS. In the rice husk C/N 25 biogas cumulative volume of 12.6 ml/grTS. In the rice husk C/N 30 biogas cumulative volume of 13.5 ml/grTS, and for rice husk C/N 35 volume cumulative biogas up to the 60th day, in the amount of 18.2 ml/grTS which is the highest biogas yield, because it is based on some information obtained, so that the growth of anaerobic bacteria optimum, optimum required of C/N ratio ranges from 20:1 to 35:1 and a bit wet environmental conditions.

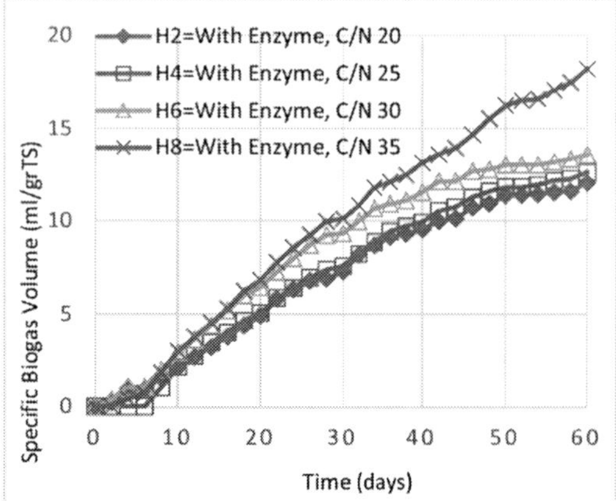

Fig. 7. The influence of C/N ratio to biogas production with enzymatic treatment

In Figure 8 can be seen the cumulative volume of biogas per unit TS on rice husk without the addition of enzymes. In the rice husk C/N 20 total biogas yield is 10.25 ml/grTS. In the rice husk C/N 25 biogas cumulative volume of 12.0 ml/gr TS. In the rice husk C/N 30 biogas cumulative volume of 13.54 ml/grTS, and the rice husks C/N 35 cumulative volume biogas up to the 60th day, which amounted to 13.56 ml/grTS which is the highest biogas yield. [2] During 104 days of observation, C/N ratio optimum is around 20-30 because it is reflected from the soluble concentration of COD with the highest value of removal efficiency in 68.21% for reactor variation 2 (C/N ratio 30) and volatile content with the highest value of removal efficiency also in 19.86% for variation 3 (C/N ratio 20). Variation of C/N ratio respectively affects the degradation process in each reactor. [15] This research results is also higher that obtained that after 52 days observation, specific biogas production was 10.78 ml/grTS with the substrate of rice husk and cow dung. Sample A (50 wt % cow dung, 50 wt % rice husk) showed a cumulative biogas production of 161.5 ml at the end of the 38th day of the experiment after which there was no further production. The production from sample B (25 wt % cow dung, 75 wt % rice husk) was not significant, while there was no production from sample C(0 wt % cow dung, 100 wt % rice husk).

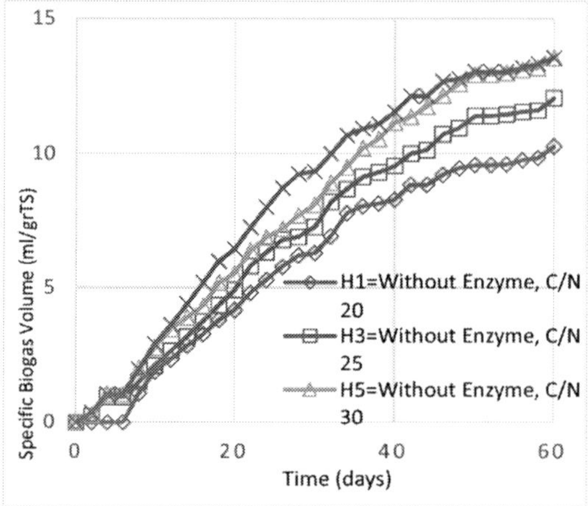

Fig. 8. The influence of C/N ratio to biogas production without enzymatic treatment

4 Conclusions

Rice husk is one of agricultural waste generated from rice mill which have a high potential to be processed into biogas. Enzymatic pretreatment could increase biogas production varied from 30 to 55 %. The highest biogas production was obtained at C/N ratio 35. Specific biogas production on C/N ratio of 20, 25, 30, and 35 were 12.0, 12.6, 13.5 and 18.2 ml/grTS, respectively. SS-AD has volumetric loading of biogas production higher than in generally liquid anaerobic digestion (L-AD). The further research need to be studied was optimization of enzyme concentration and C/N ratio.

We highly thank and appreciate to Diponegoro University through Research Professorship Program (RPP) 2016 due to the funding of this research.

References

1. Akwaka, J.C., Kukwa, D.T., Mwekave, S.S. *International Journal of Science and Technology*, **3(4)**, 222–228 (2014)
2. Arifiantari, P.N., Handajani, M., and Sembiring, T.https://jujubandung.wordpress.com/2012/06/10/pengaruh-rasio-cn-terhadap-degradasi-material-organik-dalam-sampah-pasar-secara-anaerob-2/ (2012)
3. Brown, D., Li, Y. *Bioresource Technology*, **127**, 275–280 (2013)
4. Brown, D., Shi, J., Li, Y. *Bioresource Technology*, **124**, 379–386 (2012)
5. Budiyono, Widiasa I.N., Johari S., Sunarso. *World Academy of Science, Engineering and Technology*, **37**:983-988 (2010)
6. Budiyono, I N. Widiasa, S. Johari, and Sunarso. *International Journal of Engineering and Technology*, **1(3)**:109-116, ISSN : 0975-4024 (2009)
7. Budiyono, I. Syaichurrozi, S. Sumardiono. *World Applied Sciences Journal* **26 (11)**: 1464-1472. (2013)
8. Budiyono, Iqbal S., Siswo S. *Research Journal of Applied Sciences, Engineering and Technology* **7(3)**:2798-2805. (2014)
9. Budiyono, I. Syaichurrozi and S. Sumardiono. *International Journal of Engineering*, **Vol. 27(2)**:177-184. (2014)
10. Budiyono, S. Sumardiono and D. Tri Mardiani. *International Journal of Engineering*, **Vol. 28(2)**: 921-928 (2015)
11. Chandra, R., Takeuchi, H., Hasegawa, T., & Kumar, R.. *J Appl Energy*, **94**:129-40. (2012)
12. Chen, X., Yan, W., Sheng, K., Sanati, M. *Bioresource Technology*, **154**, 215–221. (2014)
13. Gu, Y., Chen, X., Liu, Z., Zhou, X., Zhang, Y. *Bioresource Technology*, **158**, 149–155 (2014)
14. Hendriks, A.T.W.M., Zeeman, G. *Bioresource Technology*, **100**, 10–18 (2009)
15. Iyagba, E.T., Mangibo, I.A. and Mohammad, Y.S. *Scientific Research and Essay*, **4 (9)**, pp. 861-866 (2009)
16. Karthikeyan, O.P., Visvanathan, C. *Rev Environ Sci Biotechnol*, **12**, 257–284 (2013)
17. Krátký, L., Jirout, T., Nalezenec, J. Lab-scale *Acta Polytechnica*, **52(3)**, 54–59 (2012)
18. Li, Y., Park, S.Y., Zhu, J. *Renewable and Sustainable Energy Reviews*, **15(1)**, 821–826 (2011)
19. Li, Y., Zhu, J., Wan, C., Park, S.Y. *American Society of Agricultural and Biological Engineers*, **54(4)**, 1415–1421 (2011)
20. Liew, L.N., Shi, J., Li, Y. *Biomass and Bioenergy*, **46**, 125–132 (2012)
21. Liu, G., Zhang, R., El-mashad, H.M., Dong, R. *Bioresource Technology*, **100(21)**, 5103–5108 2009)
22. Mirmohamadsadeghi, S., Karimi, K., Zamani, A., Amiri, H., Horváth, I.S. *BioMed Research International*, 1–6 (2014)
23. Schimpf, U., Hanreich, A., Mähnert, P., Unmack, T., Junne, S., Renpenning, J., Lopez-ulibarri, R. *Journal of Sustainable Energy & Environment*, **4**, 53–60 (2013)
24. Teghammar, A. Thesis : *Chalmers University of Technology* (2013)
25. Xu, F., Li, Y. *Bioresource Technology*, **118**, 219–226 (2012)
26. Xu, F., Shi, J., Lv, W., Yu, Z., Li, Y. *Waste Management*, **33(1)**, 26–32 (2014)
27. Zhu, J., Zheng, Y., Xu, F., Li, Y. *Bioresource Technology*, **154**, 240–247 (2014)

MATEC Web of Conferences **101**, 02017 (2017) DOI: 10.1051/matecconf/201710102017
SICEST 2016

Fuel consumption and emission on fuel mixer low-grade bioethanol fuelled motorcycle

Setia Abikusna*, *Bambang* Sugiarto, and *Azami* Zulfan

Mechanical Engineering Department, Faculty of Engineering, Universitas Indonesia, Depok 16424, Indonesia

Abstract. Bioethanol is currently used as an alternative fuel for gasoline substitute (fossil fuel) because it can reduce the dependence on fossil fuel and also emissions produced by fossil fuel which are CO_2, HO, NOx. Bioethanol is usually used as a fuel mixed with gasoline with certain comparison. In Indonesia, the usage is still rare. Bioethanol that is commonly used is bioethanol anhydrous 99.5%. In the previous studies, bioethanol was distilled from low to high grade to produce ethanol anhydrous. But the result is only able to reach 95% or ethanol hydrous. This study is objected to design a simple mechanism in the mixing of bioethanol hydrous with the gasoline using a fuel mixer mechanism. By this mechanism, the fuel consumption and the resulting emissions from combustion engine can be analyzed. The fuel blend composition is prepared as E5, E10, and E15/E20, the result of fuel consumption and emission will be compared with pure gasoline. The using of bioethanol hydrous as a fuel mixture was tended to produce more stable bioethanol fuel consumption. However, the utilization of the mixture was found able to reduce the exhaust emissions (CO, HC, and NOx).

1 Introduction

The requirement of energy is the problem that is faced by all countries in the world, including Indonesia. Energies which are often used dominated by fossil energy, which are: petroleum, natural gas, and coal. Especially for petroleum, it is still a source that dominates energy requirement in most of the countries. Indonesia is one of the countries, which consumes crude oil as a primary fuel. In 2013, the using of fuel oil in Indonesia had achieved almost 46, 1%. By comparing the percentage of other energy need sources, the using of fuel oil is the most profoundly of all, followed by using gas fuel and coal [1].

High fossil fuel requirement relates to the using of the vehicle. In this modern era, the human is easy to obtain any means of transportation, in this case, is motor vehicles where human highly counts on them. This issue has become the main factor of the highest utilization of fossil fuel, especially in Indonesia. Jakarta is one of the cities in Indonesia, which has the highest motor vehicles growth number. We are able to see the condition from the increasing number of motor vehicles in Jakarta. In 2014, the utilization of motor vehicles reached at 17.423.967 in Jakarta. Its growth is 9, 93% per year [2]. It signifies the dependence on fossil fuel getting higher. The lack of fossil energy source, namely petroleum, and the requirement of fuel vehicles keeps increasing; consequently, it causes renewable energy development which is non-fossil energy to be severely required.

One of the alternative solutions is biomass energy whether it is in pure form or mixed with fossil fuel. It is one attempt to reduce the very high fossil fuel consumption.

One of biomass utilization technology results which can be used as an alternative fuel or as a mixture of gasoline on motor vehicles is anhydrous bioethanol, it's preferred because of its less water content, moreover, it is able to be mixed homogeneously with gasoline. Either bioethanol anhydrous or bioethanol hydrous can be mixed. The problem is the low percentage of bioethanol hydrous and less water content, however, it is not able to be directly mixed with gasoline because of its water content causes the non-homogeneous mixture. Thus, bioethanol hydrous usually has water content about 4, 9% - 5%, where gasoline water content in maximum is about 7, 4%.

Based on previous research, it discussed about a mixing mechanism between bioethanol anhydrous and gasoline by using mechanism fuel mixer into the combustion chamber through the opening valve. The problem of using bioethanol hydrous as fuel is its low exploitation. Besides, its influence on the machine performance has not been shown widely. Hence, the researcher of this advanced research is focussing the investigation on finding out the performance of fuel consumption in the dynamic loaded motor by using mixing gasoline mechanism with 96% bioethanol hydrous using fuel mixer.

This research has been conducted for finding out the machine performance of fuel consumption in a dynamic condition from a loaded motor vehicle and mixed fuel between gasoline and bioethanol by controlled ratio.

* Corresponding author: setia.abikusna@gmail.com

2 Research methodology

2.1 Valve opening controller

Taking calibration data, firstly, it is determined in the preliminary data that it is used in the calculation of calibration result. Data are:

- Source speed data from hair dryer or air (V_{in}): 16.68 m/s (air speed flows into carburetor as effect of combustion chamber inhaling)
- Large of A outlet (pipe diameter, d = 0.01905 m) : 0.00028 m^2

The pipe used is a 43.7 cm long, the measuring was conducted when the current has been fully developed. It concerns that flow rate profile is easier to be drawn and more accurate when it is measured after it is fully developed. Conducting measurement with the mechanism will also be minimized any error that occurs at real speed value [3]. To detect whether flow exhaustion is fully developed is by its appropriation with Reynold number of the fluid that flows. Its fluid is air which has a density equal to 1.2 kg/m^3 and viscosity equal to 0.000018kg/m.s. Reynold number equal to $\rho VD/\mu$ = 20266. The flow is turbulence so to determine the length of it has to be in fully developed condition by the formula [$le= Dx4.4(Re)^{1/6}$)] and the value equals to 0.437 m = 43.7 cm. The distance of pipe will be able to used to measure the flow rate.

Then preliminary assumption for calibration data that will be used for calculation is as follow:

- Assumption V_{in} = V (to gasoline carburetor venture) = V_b (gasoline vapor)
- Assumption gasoline vapor debit rate $Q = A \times V_{in} = 0.00467\ m^3/s$

Then, to get a comparison between bioethanol volume equal to 5%, 10% and 15% from the total volume mixture of bioethanol and gasoline, needed precise bioethanol vapor flow rate data.

Taking calibration data is conducted by flowing gas source from hairdryer then attends to speed value and pressure drop in every valve-opening 0° until 450^0. We can refer to that appropriate speed to get a comparison between bioethanol hydrous E5, E10, and E15 is in valve-opening 180°, 240°, and 310°. Here, this valve-opening becomes the controller when taking the real testing data.

2.2 Fuel mixer installation

In this study, the installation of fuel mixer and check on the condition of the machine were conducted; whether the machine runs properly or had trouble, in order to avoid the disorderly when data retrieval took place. Fuel mixer is able to be seen as the figure below.

Fuel mixer was mounted with two carburetors on the right and the left side. Both of them had the function to deliver the fuel to the combustion chamber. The fuel in both of carburetors could be delivered and atomized into the combustion chamber. Fuel was separated in each valve of gasoline and bioethanol. An instrument on the valve that is functioned to regulate the amount of bioethanol insertion whether it is an appropriate amount or not is needed.

Fig. 1. Fuel mixer

2.3 Variable testing

The variables were used in this research, namely:
Fixed Variables

- Motorcycle test
- Trajectory and distance test.
- Motor speed

Independent Variable

- Utilized Fuel mixture composition of bioethanol and gasoline

Dependent Variables

- Fuel consumption volume.
- Process data of fuel consumption.

2.4 Limited problems

Limitations which were used in research namely:

- Using dynamic motor fuel namely Suzuki Thunder 4 stroke 125 cc motorcycle, produced in 2007 and the condition was good about 70%.
- All capacities of each carburetor were considered equal,
- The phase for both of fuels which delivered into the combustion chamber was considered one phase, that is gas.
- Setting speed from bioethanol carburetor and its capacity were controlled by a gate valve.
- Bioethanol which was used, namely 96% bioethanol hydrous.
- Testing was conducted on dynamic condition (had running load) based on SNI 09-4405-1997(the durability test) and SNI 06-3763-1995 (the stability test).
- Dynamic testing was conducted at a constant velocity, 30km/hour by tolerance about ±5km/hour.
- The analysis was only focused on the machine performance result, that is fuel consumption to an addition of bioethanol hydrous as fuel.

DOI: 10.1051/matecconf/201710102017

2.5 Procedure of taking data

The following will explain the procedure of the fuel consumption data based on SNI 09-4405-1997 and SNI 06-3763-1995:

1. To prepare bioethanol and gasoline is by measuring the volume with measuring cup, then bring it to the bottle (tank) which had been connected to the carburator.
2. To ignite motorcycle test.
3. To prepare *stopwatch* for measuring traveling time.
4. To start taking data of motor vehicle dynamic test for the distance about 100 km to SNI 09-4405-1997, 300 m and 500 m to SNI 06-3763-1995 with constant speed about 30 km/hour (tolerance ±5 km/hour).
5. To fill the fuel addition which its volume had been measured (if needed).
6. To measure the residue in order to get consumption volume.
7. To repeat the data collection by opening valve about 180°, 240°, and 310° for obtaining ratios for E5, E10, and E15.

3 Result and discussion

3.1 Fuel consumption

On the durability test was obtained the result of fuel consumption using fuel mixture mechanism to mix fuel between bioethanol and gasoline. This test was based on SNI 09-4401-1997 with constant speed 30 km/hour (tolerance 5 km/hour) on 4 fuel conditions, 5% bioethanol hydrous, 10% bioethanol hydrous, and 15% bioethanol hydrous, and using pure gasoline. Testing result is shown in the table below:

Table 1. Testing result

Type of fuel	Timing test (hour)	Fuel consumption volume (L)
E0	3,74	4,34
E5	3,69	4,05
E10	3,71	3,90
E15	3,72	3,55

As shown in table 1. that data would be processed in order to get fuel consumption value for each type of fuel. The formula to measure fuel consumption is:

$$FC = \frac{Vf \times 3600}{t \times 1000} \quad (1)$$

where:

FC = fuel consumption (l/h)

Vf = fuel consumption at 1 sec (ml)

t = time (s)

Based on the result of calculation from above formula accordingly, the result of fuel consumption, green line for pure gasoline, which is shown in the graph below:

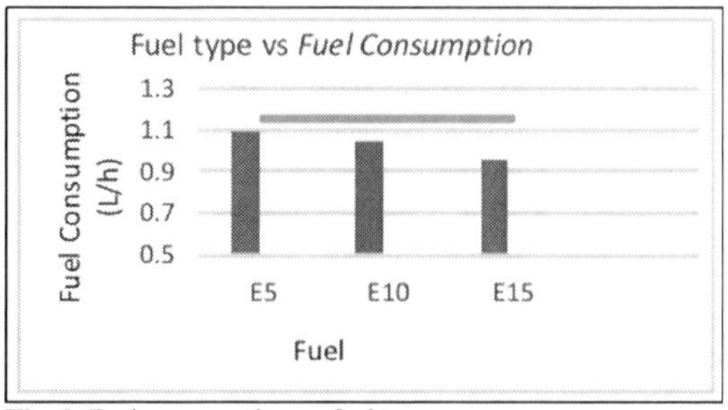

Fig. 2. Fuel consumption vs fuel type

As seen on Table 1, the fuel consumption value for each sort of test fuel was presented. Based on the graph shown in Fig 2., the fuel consumption value for fuel E15 is definitely lower than another fuel. The fuel consumption of E15 equals to 0,95 L/hour. From that graph, the consumption of pure fuel was found higher than fuel consumption value of mixed fuel between bioethanol and gasoline. It can be explained that requirement of bioethanol in mixture fuel causes lower fuel consumption or more economical than using pure gasoline. It is occurred because bioethanol has lower heating value than gasoline; therefore, it may increased thermal efficiency. The increase in thermal efficiency is commonly due to higher flame speed of the fuel than bioethanol. Consequently, the higher content of bioethanol in the mixture the more efficient in engine operation will be occurred. The increasing of efficiency was in contrast to fuel consumption value; furthermore, it could be said more economical.

The stability test results in the form of fuel consumption using fuel mixer mechanism to mixture fuel between bioethanol and gasoline. This test was conducted based on SNI 06-3763-1995 with the distance of 300 m and 500 m at a constant speed of 30 km/hour (tolerance 5 km/hour) at 4 fuel conditions, which are 5% bioethanol hydrous, 10% bioethanol hydrous, 15% bioethanol hydrous, and pure gasoline. Testing results are mentioned in figures below:

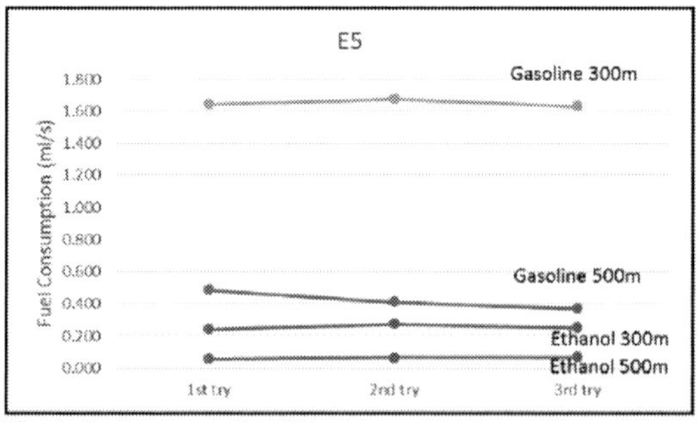

Fig. 3. Fuel consumption E5

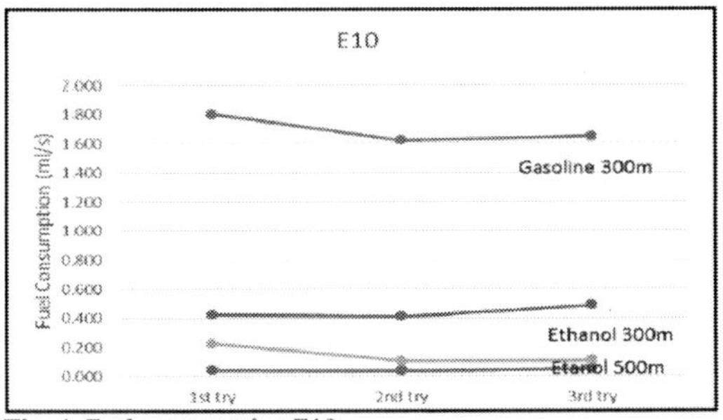

Fig. 4. Fuel consumption E10

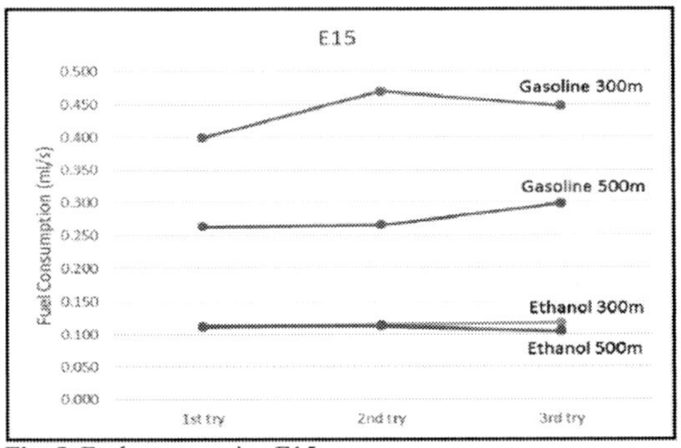

Fig. 5. Fuel consumption E15

From the results of the test, fuel consumption values became greater by using bioethanol hydrous as a fuel mixture than pure fuel gasoline. The higher fuel consumption is effected by the water content in bioethanol hydrous. Therefore, more fuel are needed to complete combustion. In E15, it is noted that bioethanol content is more than enough to assist water combustion in the combustion chamber and also because flame speed is higher by high bioethanol, thus, unburden fuel become less than other fuel mixture variations.

From the three of fuel mixtures, E15 is a fuel mixture with more stable fuel consumption. It is precisely shown from the ratio of fuel consumption in each test and each interval variation for bioethanol fuel blends.

3.2 Emission

The higher content of CO in the exhaust gasses of combustion shows an incomplete combustion. In Figure 8, it is shown that there are large contents of CO (between 2.5% - 3%) when the system used the gasoline fuel without bioethanol mixture. It shows that the air and fuel mixture is lower compared to fuel that has been mixed with bioethanol. It can be seen that the more addition volume of bioethanol, the lower content of CO generated. The lowest CO content is in the addition of bioethanol about 20% that could lower CO gas emission until 0.1% at 5.000 rpm. This is because the ethanol has –OH molecule that helps to boost the occurrence of the perfect combustion reactions. The more fuel that binds oxygen could trigger the process of oxidation so that CO can be turned into CO_2.

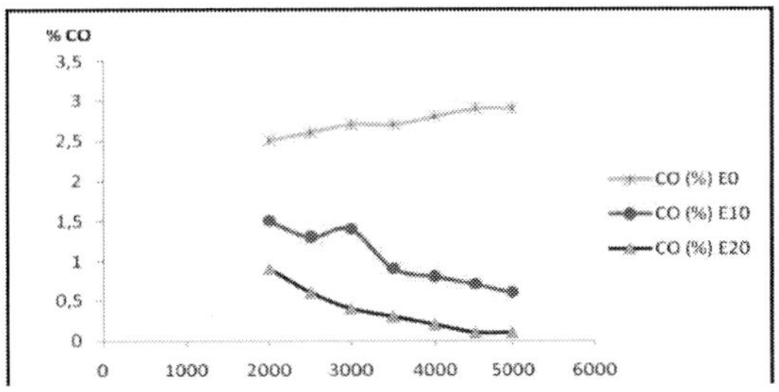

Fig. 6. CO vs engine speed

The levels of CO and CO_2 is found very close. A declined levels of CO in exhaust emissions will increase the level of CO_2. CO_2 is the combustion reaction product components. In other words, the CO_2 is a gas that is more desired from any combustion reaction product rather than CO. The higher level of CO_2 in the system is an indicator of a perfect combustion. It is due to the oxygen molecule –OH in ethanol will join in the exhaust of CO so that it will generate more CO_2 emissions, as shown in Fig. 7.

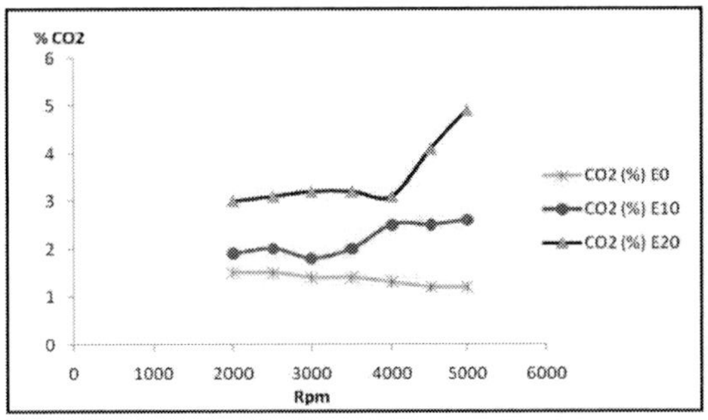

Fig. 7. CO_2 vs engine speed

More ethanol in fuel mixture was clearly performing an increased in the CO_2 products in exhaust gas as the increased in engine speed as well.

When gasoline vapor is heated at high temperature, oxidation will promoted to occured, the consequence is incomplete combustion. This unburned gasoline is produced from combustion chamber as unburnt hydrocarbon (HC) gasses.

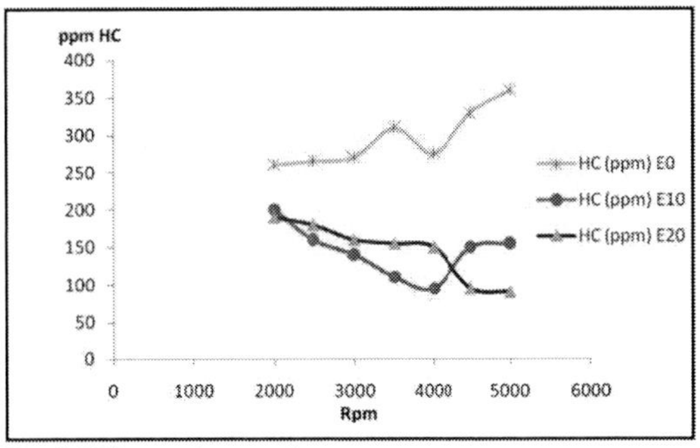

Fig. 8. HC vs engine speed

In Fig. 8, it is shown that the addition of bioethanol give insignificant changes in the HC exhaust emissions. A variations in engine speed fuelled with E10 and E20 performing the uncertain trend in HC emissions. While the usage of E0 caused an increase in HC emissions with the increase of engine speed.

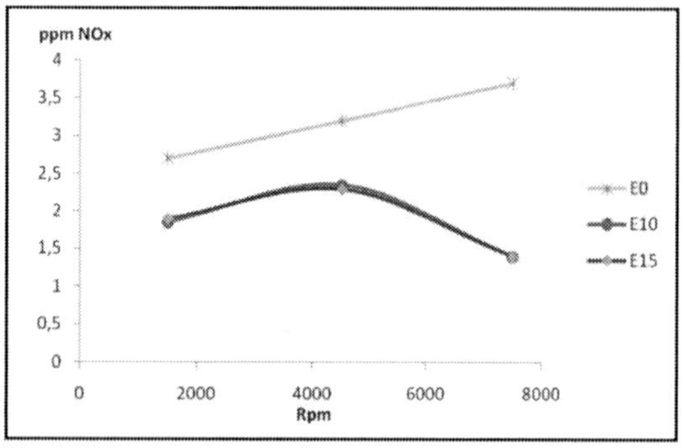

Fig. 9. NOx vs engine speed

The effect of engine speed variations on NOx emission was shown in Fig 9. In general, the addition of ethanol in fuel was extremely performing a lower NOx emissions. It was found reduced to 42% compared to the effect in using of gasoline as fuel. This is due to ethanol has a higher heating value than gasoline. That energy is used to evaporate the ethanol for a better combustion process, As the consequences, the exhaust gas temperature will be lower and so the formed NOx in the exhaust gas will be reduced.

4 Conclusion

Based on the data from the testing result, the higher bioethanol contents in fuel mixture with gasoline accordingly, the lower fuel consumption value of fuel mixture. The using of bioethanol hydrous as a fuel mixture produces more stable fuel consumption compared to pure gasoline. The addition of bioethanol, in general, is also able to decrease the exhaust emissions (CO, HC, and NOx). From the test in this study, the increase in CO_2 content in the exhaust gas is confirmed as the occurrence of a complete combustion. The CO_2 levels were relatively increased by the increase of engine speed in each of fuel mixtures.

References

1. Pusdatin Kementrian Energi dan Sumber Daya Mineral, Handbook, (2014)
2. Badan Pusat Statistik Indonesia (2015)
3. M. Ghazikhani, M. Hatami, B. Safari, D. Domiri Ganzi, *Case Studies in Thermal Engineering*, **2**, 82-90 (2014)
4. B. Sugiarto, Y. Whulanza, *Report*, *Universitas Indonesia* (2014)
5. A.S. Kyriakides, L. R. Gracia, S. Voutetakis, D. Ipsakis, P. Seferlis, S. Papadopoulou, *Fuel* **108** 208-215 (2013)
6. R. Munsin, Y. Laoonual, S. Jugjai, Y. Imai, *Fuel Fuel* **106** 586–592 (2013)
7. C. Tadeu, C. de Melo, G. B. Machado, C.R.P. Belchior, M..J. Colaço, J.E.M. Barros, E. J. de Oliveira, D. G. de Oliveira, *Fuel* **97** 796-804 (2012)
8. R. C. Costa, J. R. Sodre, Jose, *Fuel* **89** 287-293 (2010)
9. D. Turner, H. Xu, R.F. Cracknell, V. Natarjan, X. Chen, *Fuel* **90** 1999-2006 (2010)
10. Y.C. Yao, J.H. Tsai, I.T. Wang, *International Journal of Applied Energy*, *Applied Energy*, **102** 93-100 (2013)
11. M. Koç, Y. Sekmen, T. Topgül, H. S. Yücesu, *Renewable Energy* **34** 2101-2106 (2009)
12. I. Schifter, L. Diaz, R. Rodriguez, J.P. Gómez, U. Gonzalez, *Fuel* **90** 3589-3592 (2011)

Challenges and opportunities of microbial fuel cells (MFCs) technology development in Indonesia

Bimastyaji Surya Ramadan[1,*] and *Purwono*[2]

[1]Environmental Engineering Department Faculty of Civil and Environmental Engineering Bandung Institute of Technology, 40132 Bandung, Indonesia
[2]Environmental Engineering Department Faculty of Engineering Diponegoro University, 50275 Semarang, Indonesia

Abstract. Indonesian government has committed to realize the goals of sustainable development in the field of energy as stipulated in Government Regulation Number 79/2014 on national energy policy. A feasibility study of the utilization of alternative energy is important for developing countries like Indonesia. It is expected to reduce dependence on fossil fuel use and meet the energy needs on rural areas in Indonesia. Microbial fuel cells (MFCs) is a potential source of electrical energy from waste that is rich in organic matter. Trends in research and development of Microbial Fuel Cells (MFCs) technology are increasing every year due to great opportunity to address a wide range of issues related to renewable energy needs, restoration of contaminated environment, water treatment electricity generators in remote areas and many more. MFCs can be used to treat domestic waste, biomass, algae, landfill leachate, agricultural runoff, and industrial waste. MFCs technology is a technology solution for cheap, fast, simple. MFCs use of technical challenges including low electricity production, current instability, and high internal resistance. Many challenges must be address, including a more detailed analysis in energy production, consumption, and application, understanding the relationship between the amount of electricity and contaminant removal, promoting the elimination of nutrients and optimizing system configuration and operations.

1 Introduction

Microbial Fuel Cells (MFCs) is a bio-electrochemical devices that capable to produce electrical current directly from organic substances due to metabolic activity of microorganism / bacteria (release of electrons outside the cells) from waste or spesific substrate [1]. Some microorganisms (exoelectrogens) can release electrons through direct contact with electrode, nanowires, or mediator. Electrodes are used as acceptor (anode) and donor electrons (cathode). MFCs can independently treat wastewater or integrated with other treatment units [2, 3]. This interesting technology can be more competitive due to its sustainability in generating green energy.

MFCs is becoming more interesting and attracting many researcher attention. It can be seen from the number of researchers which doing project on this field and the number of papers published and cited [4]. MFCs technology can be used to generate electricity, water, hydrogen and other chemicals compound [5, 6]. A wide range of research has been made either by engineering the constituent materials (external circuits, cathode, anode, or membrane), architecture and design, as well as microorganisms that work in it to improve efficiency [2]. MFCs are also applied on field scale to monitor remote areas and combined with other treatment unit to treat wastewater. Although electricity production is still in a low level (maximum power production efficiency is ± 60%), but this issue is regarded as a challenge for researchers to improve MFCs performance [7].

One achievement of sustainable development goals (SDGs) is access to clean, affordable and sustainable energy. Energy is a major contributor to climate change, which is producing greenhouse gases about 60% from total emission. Renewable energy becoming more preferable because of sustainability and more carbon-neutral when compared with fossil fuels. Renewable energy development is expected to reduce dependence on fossil fuels and increase energy independence in each country [8]. Microbial fuel cells is one of wastewater processing technology with high energy efficiency. MFCs can be one of alternative energy source to achieve energy independence.

Indonesia Government Regulation number 79/2014 about National Energy Policy is a clear form of government efforts to realize sustainable development goals in the energy field. The policy is aimed for reducing fossil fuels uses to less than 25% and increasing renewable energy uses more than 23%. Some region in Indonesia are not entirely energy self-sufficient so that it has not covered by electricity (±.44% in rural areas and 10% in urban areas). Renewable energy has given a great benefit especially for areas that are not reached by electricity including increase the economic, social, and public health value.

* Corresponding author: bimastyaji@s.itb.ac.id

In Indonesia, those kinds of energy are still minority used. There are some examples of renewable energy sources: geothermal, biofuels, liquefied coal, biomass and waste, solar, wind energy, and micro / mini hydro. Biomass has a great chance to fulfill the need of energy and also promote sustainability of energy sources. MFCs is a kind of energy derived from biomass which is directly producing electricity in order to supply energy to transportation sector and electricity needs [7, 9-10]. This article explains how MFCs technology is potential to develop, both in terms of research and its application in Indonesia.

2 Microbial fuel cells (MFCs)

Microorganisms in MFCs convert biodegradable organic matter in waste into electricity through a quite complicated biochemical process in anaerobic or anoxic conditions. MFCs typically consist of two chambers, anode and cathode which is separated by a separator such as membrane and/or salt bridges. Each chamber has electrode as an electrons conductor and microorganism life support. Chemicals / organic substrate are filled into anode chamber as an electron donor. Micro-organism break down the chemical substances and resulting free electrons which is transferred into electrode via nanowire, mediator, or direct contact with electrode. In the bio-cathode MFCs system, oxygen which are typically injected into system by aerator, can be replaced by plant or algae. Electron that is flowing from anode to cathode through external wire and resistor results an electrical current [2].

MFCs are able to convert chemical into electrical current directly and perform biochemical reactions in the absence of additional energy in the cathode chamber. However, this technology is still difficult to apply due to highly cost of MFC material fabrication (electrode, proton exchange membrane and mediator) and low power generation [11]. Various kinds of those limitations have been studied and solved by combining two or more technologies such as constructed wetland MFC [12], activated sludge MFC, and photosynthetic MFC [11]; optimizing the design such as single chamber MFC (membrane-less MFC), cascade MFC, tubular MFC, stacked MFC, bio-cathode, and air-cathode MFC [6, 11]; and applying MFC without mediator / mediator-less MFC [13].

3 Current research in Indonesia

MFCs research in Indonesia has been carried out on a laboratory scale. Dani Permana [14] conducted a preliminary investigation research on electricity production using yeast microbial fuel cell (S. cereviceae) with methylene blue as mediator. The results showed that methylene blue may indicated increasing fuel cell voltage, but there is a need for further research to confirm it. Studies using yeast is also done by Samudro [15] in which the yeast is able to decrease COD concentration on domestic waste and increase MFC's electricity production. Engineered microorganisms also done by Purwati [16] using E. coli bacteria as degrading microorganisms. In addition to a variety of microorganisms, there is also a wide research in different substrate that microorganism in MFCs can degrade. Hermawan [17] and Purwono [18] reported that industrial wastewater has been used as substrate and microorganism sources for MFCs process. Other substrates such as wastewater fisheries [19], vegetable bins [20], cow rumen [21] and wastewater from tempeh industry [22] may also be used as MFCs substrate.

4 Challenges and opportunity

4.1 Energy status

Indonesia is an archipelago geographically country (± 17,500 islands). It has fourth largest population in the world after China, India and United States which is ± 255 millions or 3.5% of the total world population. According to energy resources perspective, Indonesia has enormous energy resources consist of renewable and non-renewable resources. Population growth and abundance of natural resources make energy consumption levels of Indonesian people increase every year. More than 75% of energy used comes from non-renewable energy resources such as oil, natural gas and coal. On the other hand, the use of renewable energy sources is still very low when compared with other countries. Exploration and exploitation is causing environmental pollution and limited amount of non-renewable energy resources lead to attract more attention on renewable energy development. In 2009, Indonesia's energy needs are supplied by six major sources include crude oil, biomass, natural gas, coal, hydropower, and geothermal. The amount and share of those energy supply can be seen in Figure 1 [10].

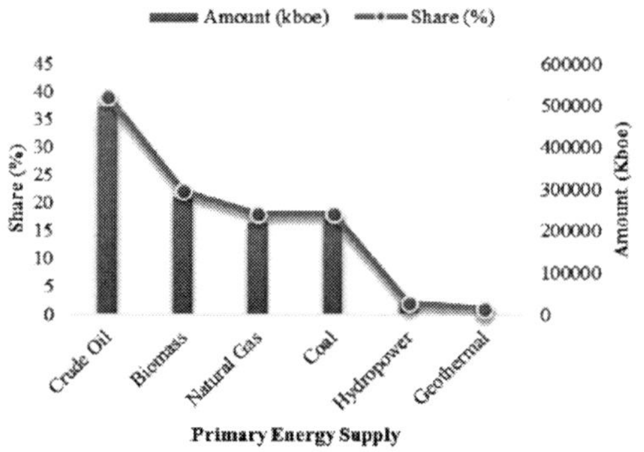

Fig. 1 Energy supply amount and share in Indonesia

Indonesia has a conducive geographical and geological conditions to develop renewable energy. Even so, these resources have not been utilized properly. Renewable energy that has been developed in Indonesia include hydropower, geothermal, solar, wind, and biomass. Among those five sources of energy, biomass is capable of producing the greatest amount of energy. It is a natural energy source that comes from plants, animal, and waste produced by living things. Biomass can produce three kinds of fuel products, namely liquid, solid and gas. Rice residue is the largest contributor of biomass fuel source followed by energy potential of rubber wood, sugar residues, and palm oil residues. Biodiesel and biogas are alternative fuels from biomass that are quite popular today. Biodiesel can replace diesel oil, because it is more environmentally friendly, capable to degrade by microorganisms, non-toxic, contains no sulphur and not produce black exhaust gas. Biogas can also be used as a source of potential energy primarily for location around livestock and landfill [10].

4.2 Water resources

High population density in Indonesia pose a serious threat to the environment. Economic pressures and rapid industrial growth led to the emergence of critical areas that are highly vulnerable to disasters such as floods, soil erosion and sludge production, lack of adequate water storage, watershed degradation, droughts, and water quality degradation. As a result, clean water supply is an issue that needs to be addressed. Low water service caused Indonesia has not been able to achieve Millennium Development Goals (MDGs) by 2013 because less than 50 % of population does not get access to clean water (from target 68.8 %). Water pollution causes clean water price more expensive and worst. It is coming from human activities that are not environmentally responsible to water resources.

Regulation No. 7/2004 about water resources management in Indonesia is the first step for reform political administration to focus on water resources management issues. This law make stakeholders promise a commitment to manage water resources properly. Indonesian government together with stakeholders undertake water resource management that includes planning, monitoring and evaluation. Integrated water resources management is expected for effectively implemented in Indonesia. Nevertheless challenges faced quite heavy. It is form an institutional frameworks, facilities management, and environmental management. Recommended actions to address these challenges are integrated planning, water payment principles, a good management approach and coordinating model with all of stakeholders [23].

4.3 MFCs Development Challenges and Opportunity in Indonesia

MFCs Technology has many challenges to be faced as an alternative renewable energy source. MFCs can meet electricity needs in rural areas. Lee [11] states that algae can be used in the cathode chamber of MFCs system (bio-cathode) and anode filled with wastewater. Algae are capable of producing electrons acceptor in cathode chamber needed for the fuel cell takes place. MFC integrated with algae is divided into two reactors configuration: algae as bio-anode and algae as bio-cathode. MFCs with anode algae reactor is commonly called as microbial solar cells or photosynthetic microbial fuel cells. In this case, the algae serve as a producer of renewable energy. Meanwhile, cathode algae MFC is a reactor in which algae has a synergistic effect between algae growth and cathode reaction. MFC-based algae bioreactor is known to effectively remove COD and ammonium up to 90 % as well as the production of electricity and biomass [24]. As an archipelagic nation with a fairly extensive sea and adequate sunlight (located on equator line area), Indonesia can take algae potential and combine it with MFCs.

MFCs are widely applied as wastewater treatment unit. Microorganisms in MFCs act as biocatalysts that could accelerate anaerobic treatment process on removing organic matter and chemical oxygen demand (COD) while the concentration is high. Anaerobic process that occur in MFCs is less energy consume make MFCs classified as an effective and efficient technology. Some researchers report that MFCs can be combined with trickling filters, activated sludge and anaerobic digester to improve treatment performance. A recent study explains that MFCs can produce electricity and hydrogen gas. It can remove more than 90 % of COD and reach more than 80% of *coloumbic efficiency*. MFCs is also known to be used in wastewater with high salinity. Since MFCs are part of anaerobic process, MFCs can also be used to eliminate strong odour pollution that is often found in aerobic wastewater treatment such as activated sludge. MFCs can also be applied to wastewater with high concentrations of metals such as landfill leachate. Combination of membrane bioreactor and MFC has a good performance as capable of removing biomass and solids with high concentrations [25].

MFCs can also be used to restore land which contaminated with organic and inorganic chemicals. Sediment MFC (SMFCs) has shown a better recovery process of contaminated sediment by heavy metals pollution. From some studies, it has suggested that MFCs success to restore contaminated land by phenols and petroleum. MFCs are known to eliminate nitrate contained in groundwater [24]. In addition, SMFC also be used as biosensors that able to monitor environmental quality parameters such as groundwater and surface water in real time. When compared with conventional BOD monitoring, MFC biosensor is able to save costs due to the electricity needed for monitoring that are generated automatically by means of SMFCs [13]. The first step to implement biosensor is by adjusting the architectural design in remote area. BOD sensor based MFCs are known to operate and able for continuous monitoring for 5 years [25].

Other kind of MFCs development are microbial desalination cells (MDCs). Seawater desalination potency in Indonesia are open widely due to Indonesia consider as a country surrounded by vast oceans. MDCs are the answer for water crisis in urban and remote areas located close to the sea [5]. Reverse Osmosis (RO) system need large capital and energy consumption. MDCs significantly reduced energy requirements up to 0.5 kWh/m^3 from 3-4 kWh/m^3 in the common desalination process. More than 60 % of energy can be saved through the MDCs. MDCs can also ease high concentrations salinity, which is more than 90 % of salt can be removed and produce 31 W/m^3 power density [24].

Today, MFCs hybrid technology has been widely explored. MFCs are known to increase treatment process efficiency. When MFCs combined with aerobic nitrification, nitrate allowance increase and followed by electricity production. MFC-based electron *fenton* process was also developed to degrade *recalcitrant* organic compounds where MFCs role as the energy supplier for electron Fenton process. Various kinds of MFCs development continue to be done in order to compete in future market [24].

In developing countries such Indonesia, MFCs are promising to be applied because it is cheap, simple, fast and able to measure water quality parameters at the monitoring location. Many studies have been done show that indigenous bacteria contained in the Indonesian environment can support MFCs system. Although MFCs has many benefits and opportunities, this technology still needs to be studied more deeply to be applied in Indonesia. Low electricity production, high internal resistance and high material cost are the major obstacle of MFCs technology implementation [4].

Even though many studies reported that MFCs power generation and process effectiveness is quite good enough to implement but it was all achieved in a lab-scale. Stability, long-term performance, efficiency and scaling up process from lab scale to full scale are future challenges to be faced in by many researchers. Currently, it has developed open air bio-cathodes which is an example of MFCs engineered design to increase power output. Certain potential substrates is also being studied for possible implementation of MFCs in various types of wastewater. To improve the consistency and stability process, it is also conducted some research in a range of concentration, substrate, material and certain microorganisms [24, 26]. By improving cooperation among disciplines, MFCs may be a potential technology for solving water and renewable energy problem in Indonesia.

5 Conclusion and Perspective

MFCs is a technology that is experiencing rapid growth in terms of both research and application. MFCs capable to produce clean energy from the activity of microorganisms that could potentially be part of the renewable energy sources in Indonesia. MFCs technology are cheap, simple, fast and able to measure water quality parameters, also meet electricity needs in rural areas. Challenges of MFCs technology development in indonesia are low electricity, stability, long-term performance, and scaling up process from lab scale to full scale. Indonesia can take algae potential, wastewater from local industry as substrate, material, seawater desalination and indigenous microorganisms for combining it with MFCs.

References

1. Min Sun, Zha Lin-Feng, Li Wen-Wei and Yu Han-Qing, *The Royal Society of Chemistry,* **45**, 2847-2870 (2016).
2. BE Logan, Microbial Fuel Cells, Hoboken, New Jersey: John & Wiley Corp (2008).
3. Huang De-Yin, Zhou Shun-Gui, Qing Chen, Bo Zhao, Yong Yuan, Li Zhuang, *Chemical Engineering Journal* **172**, 647– 653 (2011).
4. Deepak Pant, Gilbert Van Bogaert, Ludo Diels, Karolien Vanbroekhoven, *Bioresource Technology* **101**,1533-1543 (2010).
5. Li Xiaojin Li, Ibrahim M. Abu-Reesh and Zhen He, *Agriculture* **5**, 367-388 (2015).
6. T Oh Sung., Jung Rae Ki, Giuliano C. Premier, Tae Ho Lee, Changwon Kim, William T. Sloan, *Biotechnology Advances* **28**, 871–881 (2010).
7. Ieropoulos, I., Greenman, J. and Melhuish, C., *Physical Chemistry Chemical Physics* **14**, 94-98 (2011).
8. Kaygusuz Kamil, *Renewable and Sustainable Energy Reviews* **16**, 1116 – 1126 (2012).
9. Anonymous, *Peraturan Pemerintah Republik Indonesia Nomor 7 Tahun,* Jakarta (2014).
10. M.H. Hasan, T.M.I. Mahlia, Hadi Nur, *Renewable and Sustainable Energy Reviews* **16**, 2316– 2328 (2012).
11. Lee Duu-Jong, Chang Jo-Shu, Lai Juin-Yih Lai, *Bioresource Technology,* **198**, 891-895 (2015).
12. Doherty Liam, Zhao Yaqian, Zhao Xiaohong, Hu Yuansheng, Hao Xiaodi, Lei Xu, Liu Ranbin, *Water Research* **85**, 38-45 (2015).
13. N Zeng, G Li, C Dorea, W Sloan, in *8th World Wide Workshop for Young Environmental Scientists www.yes-2009-Fr (19): urban waters : resource or risks? Jun 2009*, Arcueil, France (2011).
14. Dani Permana, Desi Rosdianti, Safri Ishmayana, Saadah D. Rachman, Herlian Eriska Putra, Diana Rahayuningwulan, Hari Rom Hariyadi, *Procedia Chemistry,* **17**, 36-43 (2015).
15. Ganjar Samudro, Sri Sumiyati, Bimastyaji S. Ramadan, Lintang Iradati, *Prosiding Seminar Nasional Sains dan Teknologi Fakultas Teknik* **1**, 1 (2015).
16. Endang Purwati, Melda Latif, Hendri Purwanto dan Oki Andrean, *Inovasi Sumber Energi,* **40**, 132-141 (2014).
17. Khanigia Vanessa Hermawan, Djaenudin, M. Rangga Sururi, *Reka Lingkungan, Jurnal Online Institut Teknologi Nasional* **2** (2014).
18. Purwono, Hermawan, and Hadiyanto, *Jurnal Presipitasi* **12** (2015).
19. Bustami Ibrahim, Pipih Suptijah dan Syeila Rosmalawati, *JPHPI* **17** (2014).

20. Muhamad Imaduddin, Hermawan, dan Hadiyanto, *Jurnal Sains Dasar* **3**, 196-204 (2014).
21. Rizki Januarita, Azka Azizah, Anis Ulfa W A, Hilma Syahidah, Ganjar Samudro, *ejournal-s1 Undip*, Semarang (2015).
22. Tania Surya Utami, Rita Arbianti, Deni Novitasari, Ester Kristin and Astry Eka Citrasari, in *Proceedings of The 5th Sriwijaya International Seminar on Energy and Environmental Science & Technology*, Palembang (2014).
23. Mohamad Ali Fulazzaky, *Water* **6**, 2000-2020 (2014).
24. Lei Xu, Zhao Yaqian, Doherty Liam, Hu Yuansheng and Hao Xiaodi, *Critical Reviews in Environmental Sciences and Technology* **46**, 60-91 (2015).
25. Mostafa Rahimnejad, Arash Adhami, Soheil Darvari, Alireza Zirepour, Sang-Eun Oh, *Alexandria Engineering Journal* **54**, 745-756 (2015).
26. Zhang Quanguo, Hu Jianjun, Lee Duu-Jong, *Bioresource Technology* **217**, 121-128 (2016).

* Corresponding author: bimastyaji@s.itb.ac.id

Bio-coal briquettes made from South Sumatera low rank coal and palm shell charcoal for using in small industries

Riman Sipahutar[1,*], *Irwin* Bizzy[1], *Muhammad* Faizal[2], *Olistiyo* Maussa[1]

[1]Mechanical Engineering Department, Engineering Faculty, Sriwijaya University, 30139 Palembang, Indonesia
[2]Chemical Engineering Department, Engineering Faculty, Sriwijaya University, 30139 Palembang, Indonesia

Abstract: The objective of this study is to blend the South Sumatera low rank coal and palm shell charcoal for producing bio-coal briquettes which have better fuel properties. The experimental study for making bio-coal briquettes was carried out to examine the effect of raw material composition and binder type on the quality of the briquettes produced. A screw conveyor machine equipped with a three blade stirred and designed with the length of 40 cm, mixing process diameter of 10 cm and the capacity of 2 kg bio-coal briquettes per hour was used to produce bio-coal briquettes ready to use in small industries. Proxymate analyses of the South Sumatera low rank coal, palm shell charcoals and bio-coal briquettes were conducted in accordance with the American Society of Testings and Materials (ASTM) standards and the calorific value was determined by using a Bomb calorimeter. The experimental results showed that the calorific value of bio-coal briquette was greatly influenced by the raw material composition and the binder type. The highest calorific value was 6438 (cal/g) at the sampel of SSC65-PSC20-B15(2).

1 Introduction

The government of Indonesia has already proclaimed the increased use of coal for domestic interest and the decreased export of coal through a National Energy Policy. Coals of different varieties and their characterization are very vital in deciding their suitable applications [1]. The use of energy derived from coal will be incrased approximately 33 % of Indonesia's total energy by 2025. South Sumatera Province as one of the producers of coal in Indonesia has the potential to support that national policy. Most coal reserves owned by the provincial government of South Sumatera is low-quality brown (lignite) or sub-bituminous coals having calorific value of 5.000-6.500 kcal/kg but low sulphur and ash contents. Based on the Indonesian National Standard (SNI 13-6011-1999), South Sumatera low rank coals can be categorized as brown coals which are soft, easily crushed, and having high moisture content (10-70 %). It has been proved that blending coal with biomass give rise to briquettes which have better fuel properties and less harmful emissions compared to raw coal briquettes [2]. In addition, lignite tends to ignite spontaneously due to it's high moisture content and porous structure [3].

Currently, the prices of fuel oil and LPG for household and home industry are still quite high. The use of bio-coal briquettes as an alternative fuel has been proclaimed by the Indonesian government but still faces many problems such as poor quality of briquettes and combustion equipments as well as high levels of pollution produced.

Low rank coals usually contain a high amount of moisture, rendering low energy output and low fuel efficiency compared to high-rank coals [4,5]. The use of low rank coal as the primary raw material in the manufacture of bio-coal briquettes is not ready-made, especially with regard to the composition and type of binder required, the size of coal particles and the necessary pressure on the briquetting process. It is necassary to blend the coal for maximazing the efficiency and avoiding environmental hazards especially when coals are highly used for energy purposes [6]. The raw materials in the manufacture of bio-coal briquettes were South Sumatera low rank coals and palm shell charcoals existing in numerous amounts in the province of South Sumatera. Binder in briquetting process was a solution consisting of starch powder, water and caustic soda.

2 Experimental details

2.1 Sources and preparation of materials

South Sumatera low rank coal samples were collected from coal mine site of PT Bukit Asam (Persero), Tbk located in Tanjung Enim, South Sumatera – Indonesia.

*Corresponding author: riman_sipahutar@yahoo.com

The coal samples are labelled as SSC. The palm shell charcoals were collected from PT Global Deorub Industry located at Jl. Pangeran Sido Ing Kenayan No. 88, Kelurahan Karang Anyar, Palembang, South Sumatera – Indonesia. The materials of binder for briquetting were starch powder and caustic soda purchased in open market. The coals and palm shell charcoals were dried and pulverized before sieving to 100 mesh size.

2.2 Proximate analyses and calorific value of samples

Proximate analyses of the South Sumatera low rank coals, palm shell charcoals and bio-coal briquettes were conducted in accordance with American Society of Testing and Materials (ASTM) and the calorific value was determined by using a Bomb Calorimeter.

2.2.1 Moisture content

The moisture content of coals, charcoals and biocoal briquettes was determined based on American Society for Testing and Materials Standards [7]. Each sample (1 g) was weighed in a crucible that has been weighed previously. Next, the crucible plus sample was placed in a cold muffle furnace and heated to 104°C for one hour to evaporate the moisture. The heated crucible was then cooled in a desiccator. The cold crucible was reweighed to get the percentage of weight loss as the sample percentage of moisture content.

2.2.2 Volatile Matter

The volatile matter of coals, charcoals and biocoal briquettes was determined based on ISO-562 :2019 (e). Each moisture free sample (1 g) was weighed in a crucible covered with lid. Next, the lid covered crucible and its content was placed in a muffle furnace and heated to 925°C. It was removed after staying exactly 7 minutes in the hot zone of the furnace just before attaining the ignition temperature and was then cooled in a desiccator. The cold crucible with its content was reweighed to get the percentage of weight loss as the sample percentage of volatile matter.

2.2.3 Ash

The ash content of coals, charcoals and biocoal briquettes was determined based on American Society for Testings and Materials Standards [8]. Each moisture free sample (1 g) was weighed in a clean crucible. Next, the uncovered crucible with its content was placed in a cold muffle furnace and heated gradually so that the temperature increased to 500°C in one our and 750°C in two hours. The hot crucible was then cooled and covered in a desiccator. The cold crucible with its content was reweighed to get the percentage of remaining weight as the sample percentage of ash content.

2.2.4 Fixed Carbon

The fixed carbon of coals, charcoals and biocoal briquettes was determined based on American Society for Testings and Materials Standards [9] The fixed carbon content of coal was determined .by substracting the percentages of moisture, volatile matter and ash from the original mass of the coal sample.. It can be written as follows:

$$FC = 100 - (MOS + VM + A) \quad (1)$$

where FC is the percentage of fixed carbon, MOS is the percentage of moisture content, VM is the percentage of volatile matter and A is the percentage of ash in the sample analized.

2.2.5 Determination of calorific value of samples

The calorific value of the samples was determined based on American Society for Testing and Materials standards [10]. The Ballistic Bomb Calorimeter used was fist calibrated using a standard sample of benzoic acid whose known calorific value is 6.32 kcal/g. A small mass quantity (0.5 g) of different samples were placed in a crucible. The body of bomb calorimeter was screwed in its position and the thermocouple wire was plugged into it. The pressure release valve was closed and oxygen was allowed to enter the bomb until the pressure rose to 25 bars. Then, the firing knob was pressed and released to fire the bomb. Heat was released and the maximum deflection of galvanometer scale was recorded. The maximum deflection of galvanometer scale was then converted to energy value of the sample material by comparing the rise in galvanometer deflection with that obtained when a sample of known calorific value of benzoic acid is combusted as shown in equation below:
Heat released from sample,

$$Q = \frac{\text{Galvanometer Deflection x Calibration constant}}{\text{Mass of sample}} \quad (2)$$

The above equation is equal to:

$$Q = \frac{(\theta_3 - \theta_1) \times \gamma}{Z} \text{ kcal/kg} \quad (3)$$

where θ_3 is the galvanometer deflection with sample, θ_1 is the galvanometer deflection without sample, Z is the mass of sample = z g, and γ is the calibration constant.

The complete experiments were repeated for all different samples of South Sumatera low rank coals, palm shell charcoals and bio-coal briquettes.

2.3 Preparation of the biocoal briquette samples

The South Sumatera low rank coal and palm shell charcoal samples were ground before sieving to 100 mesh size. Then, the samples were labelled as SSC and PSC, respectively. The starch powder, pure water and caustic soda liquid were also prepared as binder materials and labelled as SP, WTR and CSL, respectively. The materials of bio-coal briquettes consist of the South Sumatera low rank coals, the palm shell charcoals, and the binders. There are two types of binders used in this research, i.e. the combination of starch powder and water with the mass ratio of 1 : 2, and the combination of starch powder, water and caustic soda liquid with the mass ratio of 1 : 1 : 1. The South Sumatera low rank coals (labelled as SSC), the palm shell charcoals (labelled as PSC) and the binders (labelled as B) were blended at various mixing proportions of 45 : 40 : 15(1), 55 : 30 : 15(1), 65 : 20 : 15(1), 45 : 40 : 15(2), 55 : 30 : 15(2), and 65 : 20 : 15(2). The proportion of 15(1) means that the mass percentage of the binder is 15% with the composition of SP and WTR at the mass ratio of 1 : 2 (labelled as B15(1)). The proportion of 15(2) means that the mass percentage of the binder is 15% with the composition of SP, WTR and CSL at the mass ratio of 1 : 1 : 1 (labelled as B15(2)). The samples were weighed using a digital weighing balance with maximum load of 500 g and accuracy of 0.01 g. The various concentrations were loaded into the mixing compartment of a screw conveyor machine.

2.4 Screw conveyor machine

The screw conveyor machine designed and manufactured in this research can be seen in the Fig. 1. The machine was equipped with a stirrer for mixing the briquitting materials. The compaction process to produce biocoal briquettes occurs due to the movement of conveyor screw pressing briquetting materials through a channel with a smaller cross sectional area. The machine was designed with the length of 40 cm, the diameter of 10 cm, and the capacity of 2 kg briquettes per hour. Before running the machine, the components of briquetting process should be prepared in accordance with the composition expected. The raw materials of South Sumatera low rank coal and palm shell charcoal should be ground to the size of 100 mesh and the bonding solution should also be prepared as much as 15% (by weight) of the briquetting mixture.

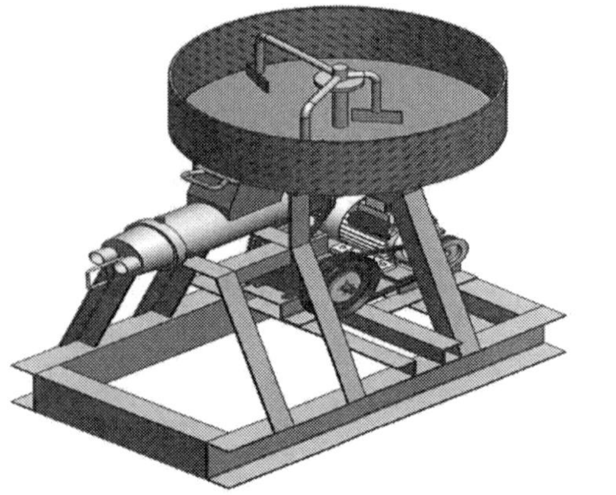

Fig. 1. The screw conveyor machine

3 Results and discussions

3.1 Results

Several activities have been conducted in the production of bio-coal briquettes including the design and manufacture of a machine to form the bio-coal briquettes and testing the characteristics of the briquettes produced. The mixing container has a diameter of 40 cm and a height of 8 cm and it is also equipped with three stirring rods having different lengths. The proximate analyses of raw materials of South Sumatera low rank coals and palm shell charcoals are presented in Table 1.

Table 1. Proximate analyses of raw materials

No.	Samples	Inherent Moisture (%)	Ash (%)	Volatile Matter (%)	Fixed Carbon (%)	Calorific Value (cal/g)
1.	SSC	7.81	3.76	42.60	45.83	6299
2.	PSC	5.41	15.25	36.07	43.27	5550

The proximate analyses of various biocoal briquettes can be seen in Table II.

Table 2. Proximate analyses of samples

No.	Samples	Inherent Moisture (%)	Ash (%)	Volatile Matter (%)	Fixed Carbon (%)	Calorific Value (cal/g)
1.	SSC45-PSC40-B15(1)	2.93	6.95	44.79	45.33	6073
2.	SSC55-PSC30-B15(1)	3.21	6.02	45.20	45.57	6121
3.	SSC65-PSC20-B15(1)	3.26	5.36	45.59	45.79	6139
4.	SSC45-PSC40-B15(2)	2.50	5.55	45.35	46.60	6223
5.	SSC55-PSC30-B15(2)	3.48	4.24	45.56	46.72	6400
6.	SSC65-PSC20-B15(2)	3.68	3.73	45.77	46.82	6438

3.2 Discussions

3.2.1 Effect of blending composition on calorific value (cal/g)

Figure 2 shows the effect of blending composition on calorific value (cal/g) at No. 1 binder type. For each blending composition, the percentage of binder was 15%.

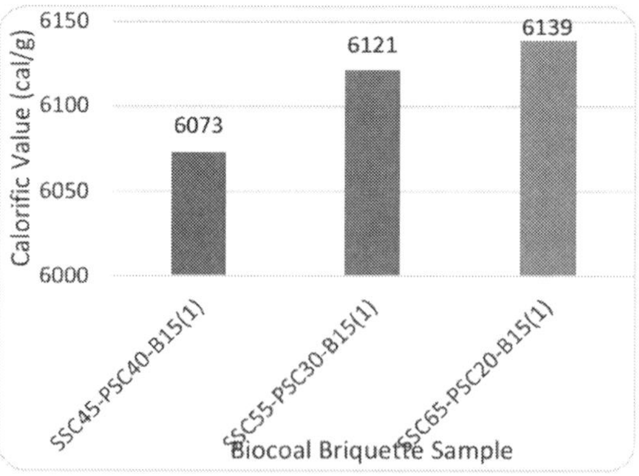

Fig.2. The effect of blending composition on calorific value (cal/g) at No. 1 binder type

The bonding solution was made from a mixture of starch and water with a mass ratio of 1 : 2. It was observed that the bio-coal briquette of SSC65-PSC20-B15(1) has the highest calorific value of 6139 (cal/g) and the bio-coal briquette of SSC45-PSC40-B15(1) has the lowest calorific value of 6073 (cal/g). The calorific value of bio-coal briquette increased from 6073 (cal/g) to 6139 (cal/g) with decreasing proportion of palm shell charcoal from 40% to 20% (by weight). The increase of calorific value may due to the addition of coal material (from 45% to 65%) which has higher calorific value compared to palm shell charcoal material.

Figure 3 below shows the effect of blending composition on calorific value (cal/g) at No. 2 binder type. For each blending composition, the percentage of binder was fixed to15% (by weight).

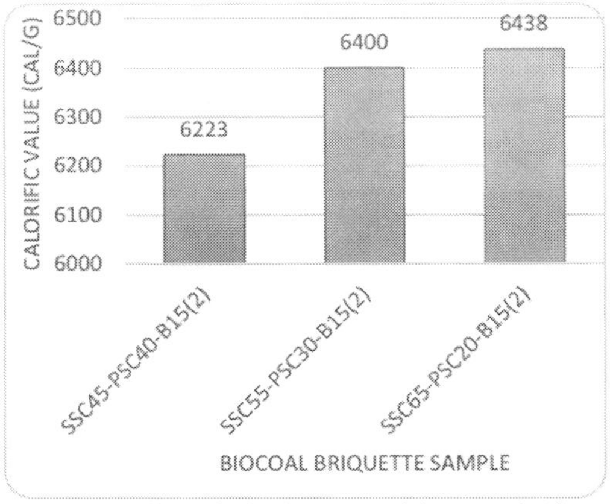

Fig.3. The effect of blending composition on calorific value (MJ/kg) at No. 2 binder type

It was observed that the bio-coal briquette of SSC65-PSC20-B15(2) has the highest calorific value of 6438 (cal/g) and the bio-coal briquette of SSC45-PSC40-B15(2) has the lowest calorific value of 6223 (cal/g). The calorific values of bio-coal briquettes increased from 6223 (cal/g) to 6438 (cal/g) with decreasing proportion of palm shell charcoal from 40% to 20% (by weight). In addition, the increase of calorific value may be due to the addition of coal material (from 45% to 65%) which has higher calorific value compared to palm shell charcoal material.

3.2.2 Effect of blending composition on proxymate analyses

The proxymate analyses of samples were conducted to determine the percentages of fixed cabon, volatile matter, ash and moisture contents. Figure 4 shows the effect of blending composition on proxymate analyses at No. 1 binder type.

It was observed that the bio-coal briquette of SSC45-PSC40-B15(1) has the lowest fixed carbon of 45.33% and SSC65-PSC20-B15(1) has the highest fixed carbon of 45.79%. The percentage of fixed carbon increased from 45.33 to 45.79% with increasing proportion of coal from 45 to 65% (by weight). The increased percentage of fixed carbon may be due to the addition of coal material which has higher fixed carbon compared to palm shell charcoal material. Fixed carbon is a measure of the solid combustible material in solid fuel after the expulsion of volatile matter.

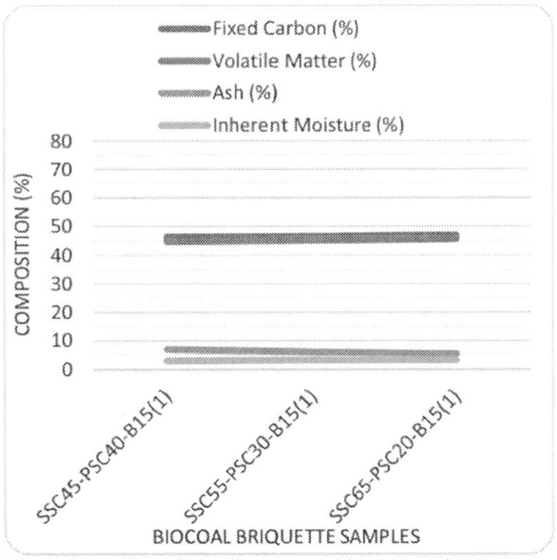

Fig.4. The effect of blending composition on proximate analyses at No.1 binder type

It was observed that the bio-coal briquette of SSC45-PSC40-B15(1) has the lowest fixed carbon of 45.33% and SSC65-PSC20-B15(1) has the highest fixed carbon of 45.79%. The percentage of fixed carbon increased from 45.33 to 45.79% with increasing proportion of coal from 45 to 65% (by weight). The increased percentage of fixed carbon may be due to the addition of coal material which has higher fixed carbon compared to palm shell charcoal material. Fixed carbon is a measure of the solid combustible material in solid fuel after the expulsion of volatile matter.

Fig. 4 also shows that the sample of SSC45-PSC40-B15(1) has the lowest volatile matter of 44.79% and the sample of SSC65-PSC20-B15(1) has the highest volatile matter of 45.59% bound with the binder type of No.1. The volatile matter increased with increasing the proportion of coal material from 45 to 65%.. The increase in volatile matter may be due to the addition of coal material which is higher in volatile matter compared to palm shell charcoal.

In the case of inherent mointure content, it was observed that the sample of SSC45-PSC40-B15(1) has the lowest moisture content of 2.93% and SSC65-PSC20-B15(1) has the highest moisture content of 3.26% (Fig. 4). The percentage of inherent moisture increased with increasing proportion of coal from 45 to 60% (by weight). The increased percentage of moisture content may be due to the addition of coal material which has higher moisture content compared to palm shell charcoal material. Moisture content is a undesirable element because it reduces calorific value and its weight adds to the transportation cost of solid fuel. Nevertheless, the percentages of moisture in all bio-coal briquette are low enough not to have seriously negative impact on the combustibility of the samples when used for household industries.

Fig.5 below shows that the bio-coal briquette of

SSC45-PSC40-B15(2) has the highest fixed carbon of 49.81% and SSC65-PSC20-B15(2) has the lowest fixed carbon of 46.50%.

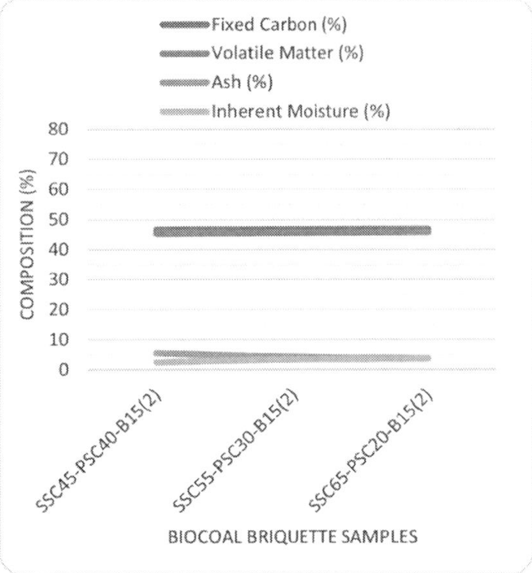

Fig.5. The effect of blending composition on proximate analyses at No.2 binder type

The percentage of fixed carbon at No.2 binder type increased from 46.60 to 49.82% with increasing proportion of coal from 45 to 65% (by weight). The increased percentage of fixed carbon may be due to the addition of coal material which has higher fixed carbon compared to palm shell charcoal material. Fixed carbon is a measure of the solid combustible material in solid fuel after the expulsion of volatile matter. The percentages of fixed carbon of all the bio-coal briquettes bonding with No.2 binder type are higher than of coal material which makes them better as a solid fuel in comparison to South Sumatera low rank coals.

Fig. 5 also shows that the sample of SSC45-PSC40-B15(2) has the lowest volatile matter of 45.35% and the sample of SSC65-PSC20-B15(1) has the highest volatile matter of 45.77% bound with the No. 1 binder type. The volatile matter increased with increasing the proportion of coal material from 45 to 65%. The increase in volatile matter may be due to the addition of coal material which is higher in volatile matter compared to palm shell charcoal. The quantity of volatile matter determine whether the briquette will burn with good flame and whether it will produce smoke. The briquette having higher volatile matter will produce higher level of smoke.

In the case of inherent mointure content, it was observed that the sample of SSC45-PSC40-B15(2) has the lowest moisture content of 2.50% and SSC65-PSC20-B15(2) has the highest moisture content of 3.68% (Fig. 5). The percentage of inherent moisture increased with increasing proportion of coal from 45 to 65% (by weight). The increased percentage of moisture content may be due to the addition of coal material which has higher moisture content compared to palm shell charcoal material. Moisture content is a undesirable element because it reduces calorific value and its weight adds to the transportation cost of solid fuel. Nevertheless, the percentages of moisture in all biocoal briquette are low enough (less than 5%) not to have seriously negative impact on the combustibility of the briquette when used for household industries.

3.2.3 Effect of binde type on calorific value

The following Fig. 6 shows the effect of binder type on calorific value (cal/g) at various compositions of bio-coal briquettes. For each blending composition, the percentage of binder solution was 15%. The bonding solution of No. 1 binder type was made from a mixture of starch (tapioca flour) and water with a mass ratio of 1 : 2. For No. 2 binder type, the binder solution consists of starch, water and caustic soda with the mass ratio of 1 : 1 : 1. It was obvious that the use of caustic soda as part of the binder solution resulted on the increase of calorific value for various compositions of biocoal briquettes. The sample of SSC45-PSC40-B15(2) has the higher calorific value of 6223 (cal/g) compared to the sample of SSC45-PSC40-B15(1) having calorific value of 6073 (cal/g). The increase in calorific value was 2.47%. The highest calorific value was 6438 (cal/g) at the sample of SSC65-PSC20-B15(2) with the increased percentage of 4.87% compared to the sample of SSC65-PSC20-B15(1).

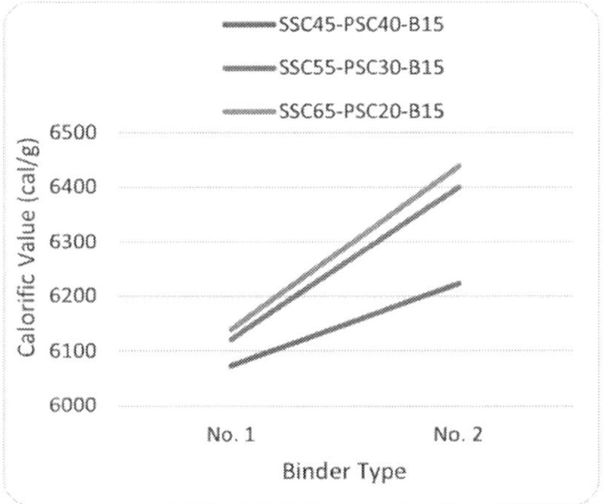

Fig.6. The effect of binder type on calorific value of briquette (cal/g) at various compositions

4 Conclusions

The calorific value of the bio-coal briquettes was greatly influenced by the raw material composition and the binder type. The calorific value increased with increasing proportion of coal in the biocoal briquettes.

The effect of binder type on calorific value was also studied, showing that the briquettes bound with No.2 binder type have higher calorific value campared to that bound with No.1 binder type at the same raw material composition.

The highest calorific value of bio-coal briquettes was

6438 (cal/g) at the composition of South Sumatera low rank coal of 65%, palm shell charcoal of 20% and No. 2 binder type of 15%.

The authors would like to express our sincere and deepest gratitude to Anis Saggaff, the rector of Sriwijaya University, for his support and motivation to present this paper on Sriwijaya International Conference on Engineering, Science and Technology (SICEST) 2016. The authors would also like to thank and express our deepest respect to Subriyer Nasir, the dean of the Faculty of Engineering, Sriwijaya University, for his support and strong encouragement to participate on this SICEST, 2016. Great thanks also submitted to the laboratory officers of Mining and Energy Agency of South Sumatera Province, Indonesia, for providing proximate analyses of the samples.

References

1. J.O. Adekunle, J.S. Ibrahim, and E.I. Kucha, BJAST **7**(1), (2015)
2. H. Kim, G. Lu, T. Li and M. Sadakata, Environ.Sci. Technol. **36**(7): (2002)
3. K. Muthusamy, V.M. Kuma Joshua, C.S. Hoe and L.Y. Ngo David, Drying Technology, **25** (2007)
4. Z. Rao, Y. Zhao, C. Huang, C. Duan and J. He, Prog. in Energy Combust. Sci., **46** (2015)
5. H. Osman, S.V. Jangam, J.D. Lease and A.S. Mujumdar, Drying Technol., **29** (2011)
6. T.U. Onuegbu, N.O. Ilochi, I.M. Ogbu, F.O. Obumselu, and I. Okafor, Curr. Res. Chem., **4**(4): (2012)
7. ASTM D3173, Standard Test Method for Moisture in the Analysis Sample of Coal and Coke, (1987).
8. ASTM D3174-11, Standard Test Method for Ash in the Analysis Sample of Coal and Coke from Coal, ASTM International, West Conshohocken, PA, (2011).
9. ASTM D5142-09, Standard Test Methods for Proximate Analysis of the Analysis Sample of Coal and Coke by Instrumental Procedures, ASTM International, West Conshohocken, PA (2009).
10. ASTM D5865-11a, Standard Test Method for Gross Calorific Value of Coal and Coke, ASTM International, West Conshohocken, PA (2011).

Effect of initial tangential intensity on the fluid dynamic characteristics in tangential burner

Pasymi[1,2], *Yogi* Wibisono Budhi[1], and *Yazid* Bindar[1*]

[1]Department of Chemical Engineering, Faculty of Industrial Technology Institut Teknologi Bandung, Bandung, Indonesia
[2]Department of Chemical Engineering , Faculty of Industrial Technology Bung Hatta University, Padang, Indonesia

Abstract. Swirl turbulent flow is intensively used by the industrial equipments such as combustion, separation and heat transfer equipments. The fluid dynamic characteristics of this flow are influenced by the chamber's geometries and the operating conditions. One of the important operating condition variables which greatly affects the fluid dynamic characteristics is initial tangential intensity (I_{TI}) or often known as swirl number. This study is aimed to quantify the effect of the initial tangential intensity on the fluid dynamic characteristics in a tangential burner. The method of the study is based on the computational fluid dynamic simulation under the Ansys Fluent CFD engine. The fluid dynamic characteristics were modeled using the standard k-ε turbulent model. The simulation results exhibited that the three dimensional flow structure in a tangential burner is dominated by the tangential flow. The fluid dynamic simulations also showed that the effect of the I_{TI} on the mean turbulence intensity and the mean residence time begin to be significant at the I_{TI} values $\geq$ 1.1 and $\geq$ 4.5 respectively, while at low I_{TI} values, its effects on both variables are insignificant. The lowest pressure drop obtained in this study was found on the burner with I_{TI} value of 3.2.

1 Introduction

A tangential burner is a burner that has two inflow orientations, namely axial orientation and tangential orientation. The tangential inflow orientation can be created through a tangential injection or a swirler. This burner type has several advantages over the axial burners. These advantages include increasing the degree of the mixing, extending the residence time of fuel in the burner and assisting the ash separation.

Initially, the development of this burner was intended to reduce coal ash problems in the boiler furnace. But now, a tangential burner is developed for the combustion of other solid fuels, such as biomass fuels. The tangential burners for a solid biomass fuel were developed previously by the names of the McConnell-48 burner [1], the VTS burner [2], and the baggase burner [3].

The performances of a tangential burner are greatly influenced by the burner geometry factors, and the operating conditions. The geometry factors include the shape and the size of the chamber, and the shape and the size the inlet and the outlet. The operating conditions cover the variables of the total mass flow rate and the initial tangential intensity (I_{TI}).

The initial tangential intensity is defined as the ratio between the mass flux entering the tangential inlet to the mass flux entering the burner cylinder. Mathematically, the initial tangential intensity of the flow in the tangential burner is evaluated using the following formula,

$$I_{IT} = (A_c/A_t)\dot{m}_t^2/(\dot{m}_t + \dot{m}_c)^2 \quad (1)$$

The variables A_t and A_c are the cross-sectional area of the tangential inlet and burner cylinder, respectively. While the variables $\dot{m}_t$ dan $\dot{m}_c$ are the mass flow rate through the tangential inlet and the mass flow rate through the burner cylinder [4].

This study is aimed to quantify the effect of the initial tangential intensity to the flow characteristics in a tangential burner. The flow characteristics quantified here are the residence time, the turbulence intensity, the flow structure and the pressure drop. All these flow characteristics are well quantified three dimensionally by CFD display techniques.

1.1 Swirl flow

The swirl turbulent flow is dominated by the tangential velocity component. This typical turbulent flow has many uses in the industrial practices. Therefore, the studies related to this topic obtain great attention from researchers. Both, experimental and numerical methods are commonly used.

Bourgouin [5] conducted the study experimentally and numerically about the influence

* Corresponding author at yazid@che.itb.ac.id

of a swirler geometry on the flow structures inside a gas turbine chamber using the LES approach. They claimed that their predictions agree with the corresponding experimental data. They stated that the flow structures and the precessing vortex core (PVC) frequency are extremely sensitive to the swirler geometry and have a strong effect on the combustion performances [5].

A simulation study on the ability of Unsteady Reynolds Averaged Navier-Stokes (URANS) approach to predict the fluid dynamics inside the swirl burner was reported by Vazquez [6]. He used both k-ε and RSM turbulent models. He reported that both models can predict the values close to the experimental data. For an atmospheric swirl burner, the predictions with k-ε model for the axial and the tangential velocity are better than the RSM model [6].

Escue and Cui [7] investigated the performance of several RANS turbulent models in modeling the developing swirl turbulent flow inside a straight pipe. They reported that although RANS models do not adequate in simulating the developing swirl turbulent flow, but they exhibit a better agreement with the experimental velocity profiles, especially for the low swirl flows. The RSM model even gave an appropriate prediction for higher swirl flows [7].

Another study concerning the flow structure and the combustion performances inside a swirl burner was conducted by Nemoda [8]. They used k-ε turbulent model. Their simulation results satisfactorily agreed with the experimental data. They stated that the flow structures in the swirl burner are influenced by the swirl number or I_{TI} [8].

1.2 Fluid dynamics characteristics

1.2.1 Turbulence intensity

The degree of the mixing is one of the variables that affects the combustion performances. The degree of the mixing inside a chamber can be indicated by the turbulence intensity. It can be determined experimentally or numerically.

According to Nemoda et al. (2005), the flow structure in a burner is influenced by the swirl number. At a high swirl number ($S \geq 2.48$), the recirculation flow structure exists and tends to increase the chemical conversion and the flame stability [8]. In line with the Nemoda results, Li et al. [9] reveals that the turbulence intensity value influences the combustion process and the NO_x formation in a burner [9].

1.2.2 Residence time

The particles movement in the burner can be considered being similar to the air movement. Therefore, the air fluid dynamic in the burner can be used to estimate the pseudo particle residence time.

The residence time of the particles inside the burner is always associated with the particle burning time and the combustion degree that is expected to occur in the burner, whether a partial combustion or a complete combustion. The particle burning time is influenced by the degree of the mixing, the particle shape and size, the oxygen concentration and the temperature. Normally, the greater the particle burning time, the longer the residence time is required.

Momeni et al. [10] conducted the experiments for a combustion process of a single biomass particle in a combustion chamber. Their results showed that the particle combustion process is influenced by the shape and the size of the particle (the particle aspect ratio), the oxygen concentration and the temperature. For a biomass particle with the particle aspect ratio 4, the oxygen concentration 20%, and the burner temperature 1200 ^{0}C, the burnout time of the particle was found at 5.4 seconds, while for the burner temperature 1600 ^{0}C, the burnout time was found at 3.6 seconds [10].

1.2.3 Pressure drop

A pressure drop across the separation and the combustion chamber is directly related to the operating cost. The higher the pressure drop, the higher the cost to drive the carrier fluid across the chamber is. The swirl flow is known to increase the pressure drop in the chamber.

Aydin et. al [11] investigated experimentally the efect of the swirler geometry, the Reynolds number and the cylinder ratio on the pressure drop and the heat transfer performance. They stated that all of the above variables have the effects on the pressure drop and the heat transfer. Unlike the effect of the swirler geometry and the cylinder ratio on the pressure drop which can be ignored, the effect of the Reynolds number on the pressure drop is so significant [11].

1.3 Turbulent model

There are three advanced turbulent approaches often used to solve the fluid dynamic problems. They are Direct Numerical Solution (DNS), Large Eddy Simulation (LES) and RANS-based modeling. The RANS-based modeling is the most frequently used in solving various turbulent flow problems.

In this approach, the turbulent conservation equations are derived by decomposing the Navier-Stokes equations. All decomposed terms are obtained from the Reynolds decomposition method. The decomposed equations are often known as the Reynolds Average Navier-Stokes (RANS) equations.

There are several RANS turbulent models. They are Spalart Allmaras as one equation turbulent model, k-ε and k-ω turbulent models as two equation turbulent model and Reynolds Stress

DOI: 10.1051/matecconf/201710103001

Model (RSM). Each of the above model has its own variants. The $k-\varepsilon$ model has the variants of the standard $k-\varepsilon$, the RNG $k-\varepsilon$, the Realizable $k-\varepsilon$ and the Low N_{Re} $k-\varepsilon$.

The Boussinesq hypothesis is used to close the RANS equations. It is done by introducing the turbulence viscosity, μ_t. The RANS equations are written as follows,

$$\frac{\partial \bar{u}_x}{\partial x}+\frac{\partial \bar{u}_y}{\partial y}+\frac{\partial \bar{u}_z}{\partial z}=0 \quad (2)$$

$$\rho\frac{\partial \bar{u}_x}{\partial t}+\rho\bar{u}_x\frac{\partial \bar{u}_x}{\partial x}+\rho\bar{u}_y\frac{\partial \bar{u}_x}{\partial y}+\rho\bar{u}_z\frac{\partial \bar{u}_x}{\partial z}=-\frac{\partial \bar{p}}{\partial x}+\frac{\partial}{\partial x}\left(\mu_{eff}\frac{\partial \bar{u}_x}{\partial x}\right)+\frac{\partial}{\partial y}\left(\mu_{eff}\frac{\partial \bar{u}_x}{\partial y}\right)+\frac{\partial}{\partial z}\left(\mu_{eff}\frac{\partial \bar{u}_x}{\partial z}\right)+\rho g_x \quad (3)$$

$$\rho\frac{\partial \bar{u}_y}{\partial t}+\rho\bar{u}_x\frac{\partial \bar{u}_y}{\partial x}+\rho\bar{u}_y\frac{\partial \bar{u}_y}{\partial y}+\rho\bar{u}_z\frac{\partial \bar{u}_y}{\partial z}=-\frac{\partial \bar{p}}{\partial y}+\frac{\partial}{\partial x}\left(\mu_{eff}\frac{\partial \bar{u}_y}{\partial x}\right)+\frac{\partial}{\partial y}\left(\mu_{eff}\frac{\partial \bar{u}_y}{\partial y}\right)+\frac{\partial}{\partial z}\left(\mu_{eff}\frac{\partial \bar{u}_y}{\partial z}\right)+\rho g_y \quad (4)$$

$$\rho\frac{\partial \bar{u}_z}{\partial t}+\rho\bar{u}_x\frac{\partial \bar{u}_z}{\partial x}+\rho\bar{u}_y\frac{\partial \bar{u}_z}{\partial y}+\rho\bar{u}_z\frac{\partial \bar{u}_z}{\partial z}=-\frac{\partial \bar{p}}{\partial z}+\frac{\partial}{\partial x}\left(\mu_{eff}\frac{\partial \bar{u}_z}{\partial x}\right)+\frac{\partial}{\partial y}\left(\mu_{eff}\frac{\partial \bar{u}_z}{\partial y}\right)+\frac{\partial}{\partial z}\left(\mu_{eff}\frac{\partial \bar{u}_z}{\partial z}\right)+\rho g_z \quad (5)$$

The standard k-ε turbulent model is supported by modeling μ_t semi-empirically. The formulation of μ_t variable is given by,

$$\mu_t = C_\mu\,\rho\frac{k^2}{\varepsilon} \quad (6)$$

k is the specific turbulent kinetic energy. ε is the dissipation rate of the turbulent kinetic energy. C_μ is the empirical constant.

Two new dependent variables, namely the k and ε variables are modeled by using the conservation equation approach. The concervation equations for these two turbulent variables are expressed as,

$$\rho\frac{\partial k}{\partial t}+\rho\bar{u}_x\frac{\partial k}{\partial x}+\rho\bar{u}_y\frac{\partial k}{\partial y}+\rho\bar{u}_z\frac{\partial k}{\partial z}=\frac{\partial}{\partial x}\left(\frac{\mu_{eff}}{\sigma_k}\frac{\partial k}{\partial x}\right)+\frac{\partial}{\partial y}\left(\frac{\mu_{eff}}{\sigma_k}\frac{\partial k}{\partial y}\right)+\frac{\partial}{\partial z}\left(\frac{\mu_{eff}}{\sigma_k}\frac{\partial k}{\partial z}\right)+P_k-\rho\varepsilon \quad (7)$$

$$\rho\frac{\partial \varepsilon}{\partial t}+\rho\bar{u}_x\frac{\partial \varepsilon}{\partial x}+\rho\bar{u}_y\frac{\partial \varepsilon}{\partial y}+\rho\bar{u}_z\frac{\partial \varepsilon}{\partial z}=\frac{\partial}{\partial x}\left(\frac{\mu_{eff}}{\sigma_\varepsilon}\frac{\partial \varepsilon}{\partial x}\right)+\frac{\partial}{\partial y}\left(\frac{\mu_{eff}}{\sigma_\varepsilon}\frac{\partial \varepsilon}{\partial y}\right)+\frac{\partial}{\partial z}\left(\frac{\mu_{eff}}{\sigma_\varepsilon}\frac{\partial \varepsilon}{\partial z}\right)+C_{\varepsilon 1}\frac{\varepsilon}{k}P_k-C_{\varepsilon 2}{}^{*}\rho\frac{\varepsilon^2}{k} \quad (8)$$

There are six dependent variables contained in the $k-\varepsilon$ turbulent model, those are p, u_x, u_y, u_z, k and ε.

Variables P_k and $C_{\varepsilon 2}{}^{*}$ are the turbulent kinetic energy production rate and the constant. These can be found in the literature [12]. The constant values of the equations above are as follows $C_\mu = 0.085$, $\sigma_k = 0.7194$, $\sigma_\varepsilon = 0.7194$, $C_{\varepsilon 1} = 1.42$ and $C_{\varepsilon 2} = 1.68$ [12].

These conservation equations can be solved simultaneously using the numerical method. All the numerical procedures for solving all equations are blended in a CFD code.

2 Research methodology

This research was conducted using the numerical simulation method with Ansys Fluent as CFD engine. This research is started by designing the burner geometry which includes determining the shape and the size of the burner. The actual designed burner geometry is then built and meshed three dimensionally in the CFD engine.

The CFD simulation is initiated with the problem definitions by assigning all boundary conditions of the specified problem to the meshed boundaries. The next stage is to choose the appropriate turbulent model to simulate the fluid dynamic characteristics in the designed tangential burner. CFD parametric explorations are conducted then to assure that the used CFD parameters produce consistent results. The stage is continued by quantifying the effect of the initial tangential intensity on the fluid dynamic characteristics in the tangential burner using the selected turbulent model.

2.1 Burner geometry

The designed tangential burner in this study has single tangential injection. The tangential injection is installed at the bottom of the frustum that connects the axial inlet and the cylindrical body. The frustum length is about 500 mm. The geometry of the tangential burner is shown in Fig. 1.

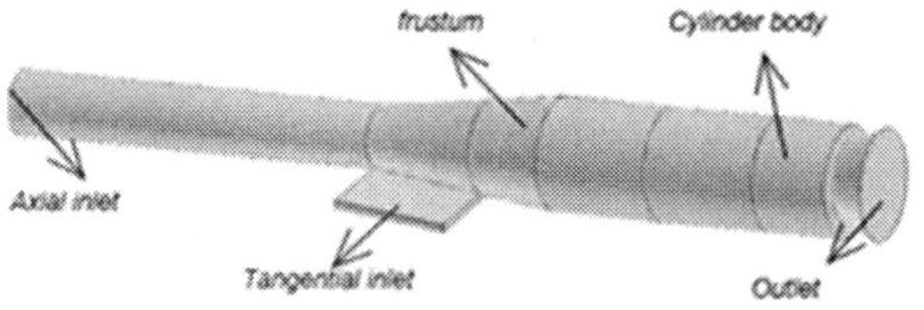

Fig. 1. The geometry of designed tangential burner.

The cylinder diameter is 300 mm with the length of 900 mm. The axial inlet is 172 mm in diameter and 1000 mm in length. The tangential inlet width is 300 mm and the tangential inlet length is 300 mm. The burner inlet aspect ratio is 14.

2.2 Boundary Conditions

The burner has two inlet, namely axial and tangential inlet, and single outlet. The burner is operated at the atmospheric pressure and isothermal condition. The axial and tangential inlet velocities are varied for several I_{TI} values. The variations of I_{TI} values are 0.4, 0.7, 1.1, 1.9, 3.2, 4.5, 6.8, and 10.

This three dimensional fluid flow simulation is set with the stationary walls with no slip condition. The fluid flows near the walls are governed by the standard wall function.

2.3 Model selection

In this study, the simulations are intended for the design purpose where a very high accuracy of predictions does not always become a primary concern. The three dimensional CFD works with complex physical phenomena require very large computational efforts and need a large computer memory. Such this computer facility is very expensive and rarely available for many engineering design companies. To avoid the problems, it is required to use a proper and robust turbulent model and an inexpensive computational technique for the engineering design purposes.

The DNS approach in solving the turbulent flow produces very high accuracy predictions. Therefore the DNS is intended for the validation purpose rather than for the design purpose. The DNS needs a huge computational effort and an excessive computing time. The LES turbulent model still require a very high computational effort and longer computing time although it has a higher accuracy prediction than the RANS turbulent models.

The RANS turbulent models are able to predict the turbulent flow behaviors appropriately and require a relatively low computational effort. The success of RANS turbulence model in modeling the tangential flow has been widely reported by many authors [6], [8]. But, there is no turbulence model that can work well for all geometries and operating conditions. Therefore, an suitable model for an encountered problem must be chosen carefully.

In addition, the current computers and CFD techniques are well established and mature. These allow the CFD simulations to be carried out using an inexpensive computer facility like a small computing server or a simple Personal Computer (PC) or even a laptop computer with acceptable results. Based on the above considerations, the turbulent models used in this study are the RANS-based turbulent models.

2.4 Experimental variables

The numerical quantification on the effect of the initial tangential intensity (I_{TI}) to the fluid flow characteristics was conducted in the range from low to high I_{TI} values. To eliminate the effect of the other variables, the variables such as the size and shape of the geometry, the tangential inlet slope, the inlet aspect ratio, the position of tangential inlet and mass flow rate are kept constant.

The dependent variables that quantified in this study are the turbulence intensity, the residence time and the pressure drop. These variables are chosen because they have significant effects to the overall burner performances. The turbulence intensity as a degree of the mixing and the residence time have a significant influence on the combustion process, while the pressure drop affects the operating cost of the burner.

3 Results and discussion

3.1 RANS models performance

The selection of the turbulent model among the RANS turbulent models is based on the ability of the models to agree with the experimental data of Chen et al. [4]. These data are used because they were measured from the chamber which has similar geometry to the tangential burner used in this study. The RANS turbulent models which are evaluated here are standard k-ε, RNG k-ε, Realizable k-ε, k-ω and RSM model.

The detailed geometry of the Chen's chamber can be found on the original paper of Chen [4]. The model evaluations are done on the case of the high swirl flow (I_{TI} value = 0.8) and with mesh interval size 3.3% of the burner diameter. This mesh interval size is obtained by conducting the mesh size parametric study. The comparisons between our prediction results and the Chen experimental data are shown in Fig. 2.

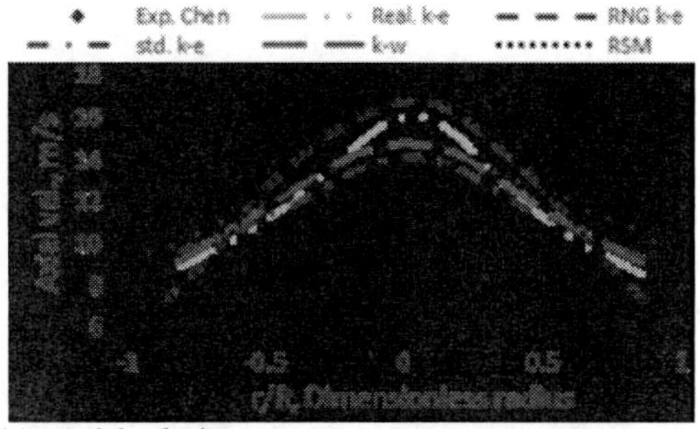

(a) Axial velocity

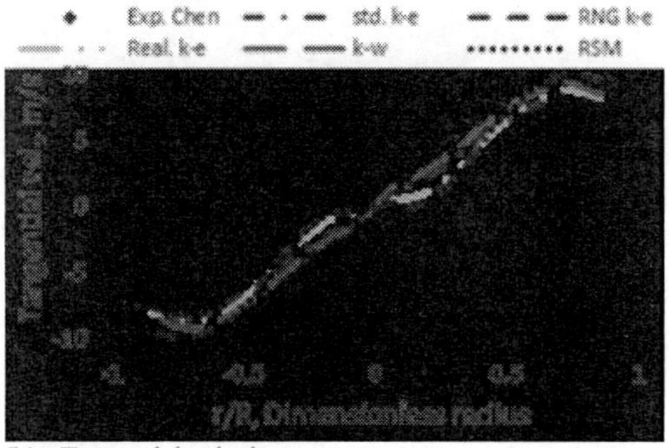

(b) Tangential velocity

Fig. 2. Comparison of some RANS turbulent models performances.

Although with the inherent isotropic turbulent assumption, the k-ε turbulent models are able to arrive with good velocity predictions for a high swirl flow, especially the standard k-ε model. These predictions are more accurate than those given by RSM, although theoretically, the RSM model supposed to provide the better results because it was built under the anisotropic turbulent assumption. These are consistent with statement of Vazquez, who reported that for an atmospheric swirl burner, the predictions of k-ε model for axial and tangential velocity are better than RSM [6].

Quantitatively, the prediction results of standard k-ε model deviate from experimental data about 12% for the tangential velocity and 5% for the axial velocity. In addition, the standard k-ε turbulent model is easy to converge and has relatively low computational effort comparing to other RANS models. Based on the things above, the standard k-ε turbulent model is chosen to quantify the fluid flow characteristics in this study.

3.2 Quantification of the effect of I_{TI} value to the fluid dynamic characteristics

The numerical quantifications in this study were conducted in the tangential burner (Fig. 1) that has 0^0 of the tangential inlet slope, 14 of the inlet aspect ratio and the tangential inlet position at the bottom of the frustum. The simulations were done using the standard k-ε turbulent model at constant total mass flow rate of 0.24 kg/s.

3.2.1 Turbulence intensity

In this simulation, the values of the mean turbulence intensity of the flow are represented by the mean values of the turbulence intensity of 29 fluid flow tracks. The profiles of the turbulence intensity value for each track was shown in Fig. 3.

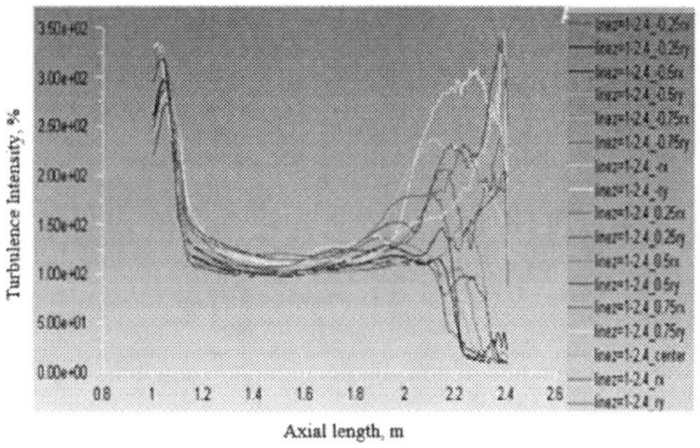

Fig. 3. Turbulence intensity profile along burner at several release points (case for I_{TI} = 10).

The turbulence intensity values near the tangential inlet and the outlet zone are greater than other zones along the burner, as shown in Fig. 3. The high value of the turbulence intensity near the tangential inlet zone (z = ± 1) is caused by the mixing of the axial and tangential inlet flows. The high values of this variable near the outlet (z = ± 2.4) is due to the sudden contraction.

The mean turbulence intensity along the tangential burner is influenced by the I_{TI} value. The higher the I_{TI} value, the higher the turbulence intensity value is. The profiles of the mean turbulence intensities for several I_{TI} values are shown by Fig. 4.

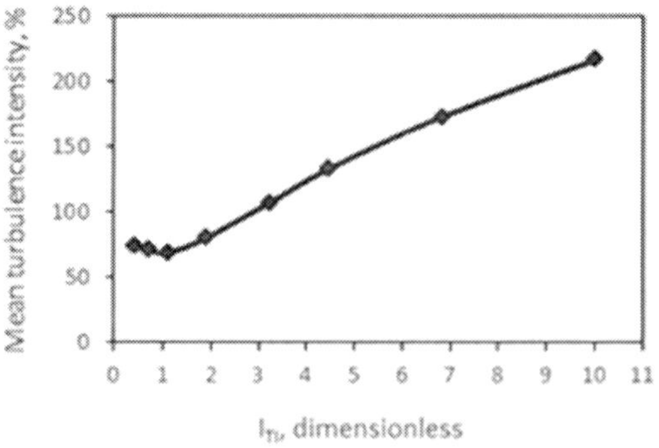

Fig. 4. Mean turbulence intensity profile in a tangential burner for several I_{TI} values.

3.2.2 Residence time

The air residence times in a burner are computed by dividing the path lengths of the air with the magnitude velocities. The air path length is determined by computing the air pathline formed along the burner. The air pathlines at various release points, either from the axial or the tangential inlet, are shown in Fig. 5. The air pathlines in the figures clearly showed that the flow structures throughout the burner are dominated by the tangential flow.

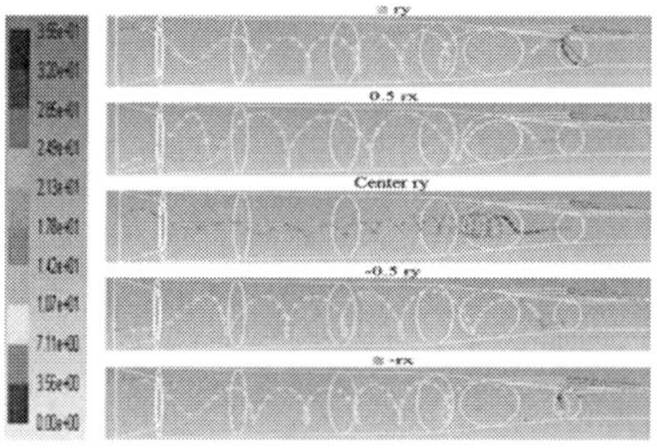

(a) Some release point at axial inlet

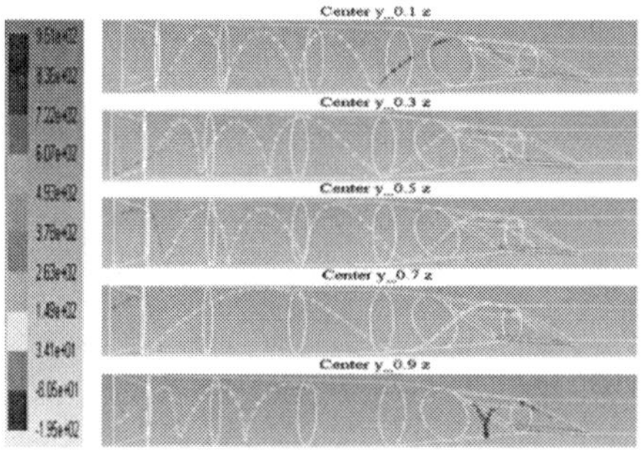

(b) Some release point at tangential inlet

Fig. 5. Air flow pathline for several release points (case for I_{TI} = 10).

A different finding, from those reported by Nemoda about flow structure was obtained in this research. Recirculation flow structure, that is supposed to be produced in high swirl flow, is not found during the simulation, even on a very high swirl flow (I_{TI} = 10). Whereas according to Nemoda [8], the recirculation flow structure will be formed at I_{TI} value ≥ 2.48. This indicates that the formation of the recirculation flow structure is not only determined by the I_{TI} but also determined by the other variables.

The mean residence time along the burner is obtained from the mean air residence time of 20 release points. The profiles of mean residence times in the tangential burner for some of I_{TI} values are given in Fig. 6. The mean residence times along the tangential burner are influenced by the I_{TI} values. The higher the I_{TI} value, the longer the pathline, and the greater the residence time is. Its influence is more significant on the high I_{TI} value.

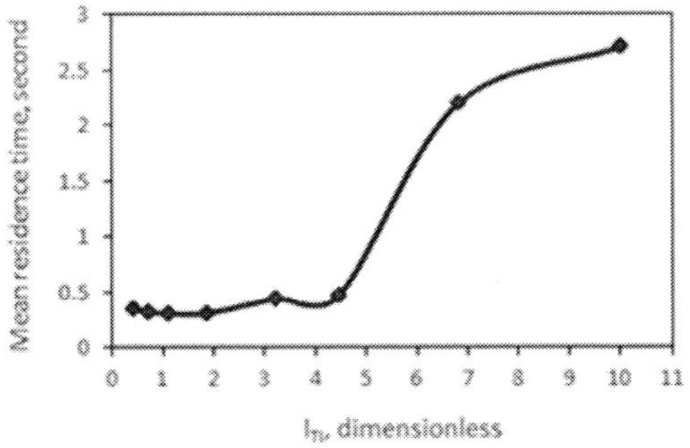

Fig. 6. Mean residence times in a tangential burner for several I_{TI} values.

3.2.3 Pressure drop

The mean pressure drop in this study is calculated from the difference between mean static pressure at the burner inlet and the burner outlet. The simulation results showed that the mean pressure drop across the burner is influenced by the I_{TI} value. Generally, the greater the I_{TI} value, the greater the pressure drop is. This is consistent with the results that are reported by Aydin, where the higher the Reynolds number (it means the higher the I_{TI} value), the higher the pressure drop is [11].

Within the range of I_{TI} values that is investigated here, the lowest mean pressure drop is produced at the I_{TI} value of 3.2. The profiles of the mean pressure drop in the tangential burner at various I_{TI} values are given in Fig. 7.

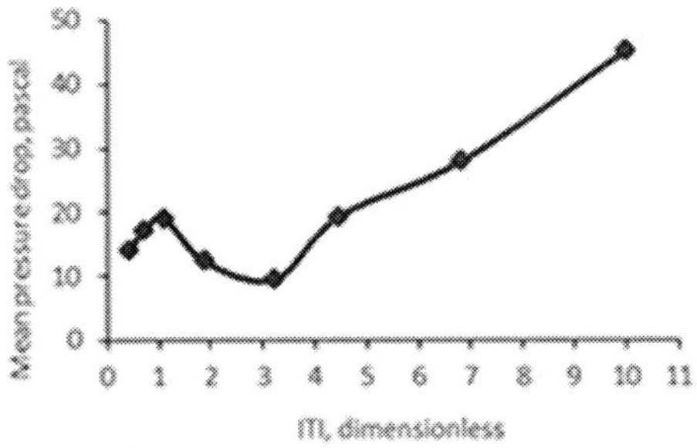

Fig.7. Mean pressure drop in a tangential burner for several I_{TI} values.

The effects of the I_{TI} on the mean pressure drop are more significant at higher I_{TI} value than the lower I_{TI} value. This correlates to the pathline (residence time) of the air along the burner where if the pathlines are longer, the pressure loss that is caused by the friction will be greater. The consistency between the residence time and the pressure drop curves can be seen from Fig. 6 and 7.

4. Conclusion

The numerical quantification of the fluid dynamic characteristics in a tangential burner has been carried out using the standard *k-ε* turbulent model. The simulation results concluded that three dimensional flow structures in a tangential burner are well described as the dominant tangential flows. The numerical quantifications revealed that the fluid dynamic characteristics are greatly affected by the initial tangential intensity.

In general, the higher the I_{TI} value, the greater the mean turbulence intensity, the longer the mean residence time and the higher the mean pressure drop are. At low I_{TI} values, the effect of I_{TI} on the mean turbulence intensity and the mean residence time are insignificant. The effect of I_{TI} on the mean turbulence intensity begin to be significant at the I_{TI} values ≥ 1.1, while its effect on the mean residence time is significant at the I_{TI} values ≥ 4.5. The lowest mean pressure drop in this study is found at I_{TI} value of 3.2.

Our thanks go to the Ministry of Research, Technology and Higher Education of the Republic of Indonesia for the funding support provided to the implementation of this research through the Decentralization Research Scheme.

References

1. S. Paulrud, C. Nilsson, Fuel, **83**, 813-821 (2004)
2. S. M. B. Kops, P. C. Malte, Department of Mechanical Engineering University of Washington, 1-49 (2004)
3. J. H. S. Arnao, D. J. O. Ferreira, C. G. Santos, J. E. Alvarez, L. P. Rangel, S. W. Park, Industrial, Mechatronic and Manufacturing Engineering, **9**, 798-801 (2015)
4. J. Chen, B. S. Haynes, D. F. A. Fletcher, Melbourne, Australia, 485-490 (1999)
5. J. F. Bourgouin, J. Moeck, D. Durox, T. Schuller, S. Candel, C.R. Mecanique, **341**, 211-219 (2013)
6. J. A. R. Vazquez, *Thesis*, University of Zaragoza, Spain, 1-192 (2012)
7. A. Escue and J. Cui, Appl. Math. Model, **34**, 2840–2849 (2010)
8. S. Nemoda, V. Bakic, S. Oka, G. Zivkovic, N. Crnomarkavic, Int. J. Heat Mass Tran., **48**, 4623–4632 (2005)
9. Z. Li, J. Jing, G Liu, Z. Chen, Chem. Eng. Sci., **65**, 1253–1260 (2010)
10. M. Momeni, C. Yin, S. K. Kær, T. B. Hansen, P. A. Jensen, P. Glarborg, Energ Fuel, **27**, 507–514 (2013)
11. Anysy Inc., Ansys Documentation, *Anysy Inc.*, Canonsburg, PA, United States, (2013)

Effect degree of temperature subcooling in the performance of refrigeration with CFC, HFC and hydrocarbons refrigerant

Prayudi Suparmin[1,*], *Vendy* Antono[1], and *Roswati* Nurhasanah[1]

[1]College of Engineering Technology PLN, Mechanical Engineering Department, 11750 Jakarta, Indonesia

Abstract. Global warming is the result of the use of Chloro Fluoro Carbon (CFC) and Hydro Flouro Carbon (HFC) refrigerants. CFC and HFC refrigerants are not environmentally friendly in air conditioning systems. Researchers conducting the research to find an alternative refrigerant with hydrocarbon and the use of liquid suction heat exchanger (LSHX) subcooling in the vapor compression refrigeration system. One kind of LSHX is heat exchanger between tube to tube. Retrofit refrigerant is used to find a replacement refrigerant for R22, R134a, R404A with R600, and mixture refrigerant R290/R600a in the ratio 50:50. From the analysis are conclude that the increases of degree of subcooling temperature significantly affects to the effectiveness of LSHX, increase the refrigeration capacity and also refrigeration capacity index, increase the compressor of work, compressor work of index, COP and COP index. The parameters of the indicators that mixture the refrigerant R290/R600a in the ratio of 50:50 can be used as a substitution for R404A and R22, while the R600 refrigerant can be used as a substitution for R134a in the vapor compression refrigeration system with LSHX.

1 Introduction

In advanced, technology of air conditioning or refrigeration system is influence on human life. There are three major issues associated with the vapor refrigeration system to save energy, ozone layer depletion and also global warming that caused by the use of refrigerant which is not environmentally friendly. Refrigeration or air conditioning systems is generally use electricity. Based on the several studies shows that electricity which is used for refrigeration systems in the building were highest 50%. In the cold storage system and supermarket require energy consumption 50%-60% [1]. Energy power consumption as used to get a big difference temperature between output temperature of the condenser and input of the evaporator, implications increase compressor work in requires the amount of energy that fairly large [2]. One method that can be increase refrigeration capacity is use LSHX subcooling.

Performance refrigeration system with LSHX and using the refrigerant R22, R404A, R134a, R1234ze, and R1234yf influenced by the degree of subcooling in the condenser. Optimum Performance is when using R1234yl. Degree of temperature subcooling maximum depends on the condensation temperature [3].The value of degree of temperature subcooling on vapor compression refrigeration system with LSHX optimal between 2-6°C. Optimal degree of temperature subcooling is strongly influenced by temperature condensation, but it is not temperature evaporation. Low cooling water temperature in the condenser improves performance, heat transfer, and energy destruction [4].

The research for hydrocarbon refrigerant R290, R600, R600a as alternative R134a in vapor compression refrigeration system with subcooling resulted obtain R600 and R600a has the same characteristics as R134a, the saturation pressure is greater than R134a and R290.Performance R600 and R600a in the subcooling refrigeration system better than the R290 and R134a [5]. Incorporation of subcooling heat exchanger in the system, greatly improved the performance of the system, its increases the COP, reduces the input of compressor energy and the specific power consumption of the system. The thermodynamic properties of R432A and R433A are match with R22 and the two refrigerants consistently exhibited better performance than R22 in sub-cooling heat exchanger refrigeration system. Performing of R433A is better than both of R22 and R432A in that the highest RCI, COP, reduction in energy input, and lowest discharge temperature were obtained using R433A in the system [6-7].

Based on the research with three mixture refrigerant hydrocarbon Propane (R290), Butane (R600) and Isobutene (R600a) with the comparison [70:25:5] and [50:40:10] have similar characteristics thermodynamic to R134a. The similarly is a mixture of two refrigerant R290/R600 and R290/R600a with the composition [60.40] have saturated temperature and saturated pressure of each refrigerant this indicates that neither mixture matches closely with HFC-134a. In this case, both of the mixtures may be selected as suitable alternatives [13].

In general, the degree of temperature subcooling is impact on the performance refrigeration system and the

* Corresponding author: prayudi@sttpln.ac.id

majority of research for a replacement refrigerant R22, R134a with hydrocarbons. In this paper, the thermodynamic analysis of vapor compression refrigeration system with LSHX are used working fluid refrigerant type of CFC, HFC and hydrocarbons, while the type of refrigerant are used R22, R134a, R404A, R600 and mixture with refrigerant R290/R600a. The main objective of this paper is to present whether refrigeration R600 and R290/R600a can be used as a substitute refrigerant R22, R134a and R404A in the vapor compression refrigeration system with LSHX.

Cycle vapor compression refrigeration system with subcooling is given in Fig. 1. According to Klein[9], Liquid-suction heat exchangers are commonly installed in refrigeration systems with the intent of ensuring proper system operation and increasing system performance. Specifically, ASHRAE (1998) states that liquid-suction heat exchangers are effective in (1) increasing the system performance; (2) subcooling liquid refrigerant to prevent flash gas formation at inlets to expansion devices; (3) fully evaporating any residual liquid that may remain in the liquid-suction prior to reaching the compressor(s). LSHX is a heat exchanger which is added between the condenser and evaporator in air conditioner. With the aids of LSHX, the refrigerant came out from condenser is further cooled down by the refrigerant from the outlet of evaporator.

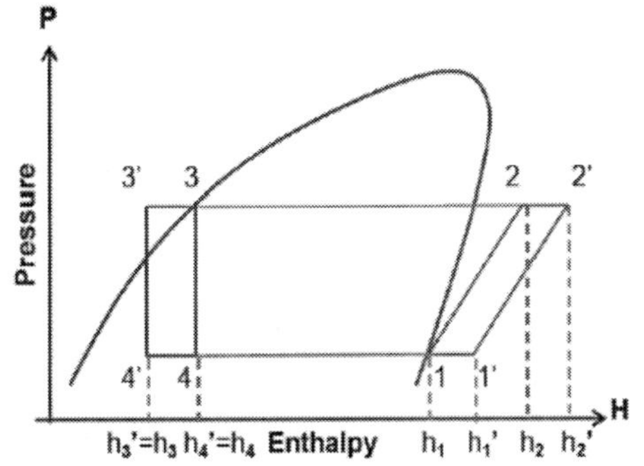

Fig.1. P-H Diagram refrigeration system with subcooling

From Fig. 1, the cycle of vapor compression refrigeration system without subcooling is 1-2-3-4, while LSHX subcooling is 1'-2'-3'-4'. The degree of freedom temperature of subcooling (ΔT_{sub}) is the difference between the temperature of saturation with subcooling fluid temperature before it enters the expansion valve is the difference between T_3-$T_{3'}$.

$$\Delta T_{\text{sub}} = T_3 - T_{3'} \tag{1}$$

Meanwhile, heat transfer on LSHX can be described by the Fig. 2.

The effectiveness on LSHX given by equation [9, 11]

$$\varepsilon_{\text{LSHX}} = \frac{T_{1'} - T_1}{T_3 - T_1} \tag{2}$$

Performance vapor compression refrigeration system with LSHX, influence the degree of temperature subcooling can be reviewed the effect on the refrigeration capacity index (RCI), the compressor work index (WI), and the index of the COP (COP_N) is given by equation [5,12],

$$\text{RCI} = \frac{\text{RC}_{\text{LSHX}}}{\text{RC}_{\text{no-LSHX}}} \tag{3}$$

$$\text{WI} = \frac{\text{W}_{\text{LSHX}}}{\text{W}_{\text{no-LSHX}}} \tag{4}$$

$$\text{COP}_\text{N} = \frac{\text{COP}_{\text{LSHX}}}{\text{COP}_{\text{no-LSHX}}} \tag{5}$$

where RC_{LSHX}, W_{LSHX} and COP_{LSHX} can be seen in the paper [5], and [12].

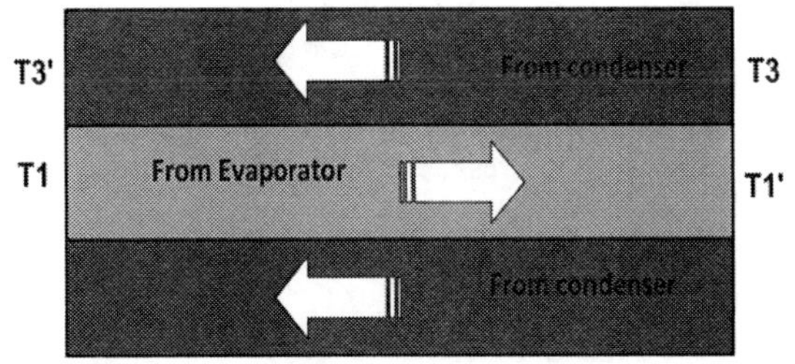

Fig.2. Schematic diagram of heat transfer

2 Experimental setup

Test equipment used is RTU Cold Storage Mini with LSHX tube and tube (Fig. 3), with the cooler box shown in Fig. 4.

Fig. 3. Test equipment RTU cold storage

The system also equipped with measuring devices, accessories and other auxiliary equipment such as pressure gauge, digital thermocouple, oil separator, separator (for refrigerant), shut off valve, sight glass and filter dryer. Number of thermocouples mounted as many as six pieces, and two pieces of a pressure gauge suction and discharge compressor. Refrigerant used CHC type is R22, HFC type is R134a, R404A and hydrocarbon type refrigerant R600 and mixture refrigerant R290/R600a with the ratio 50:50.

DOI: 10.1051/matecconf/201710103002

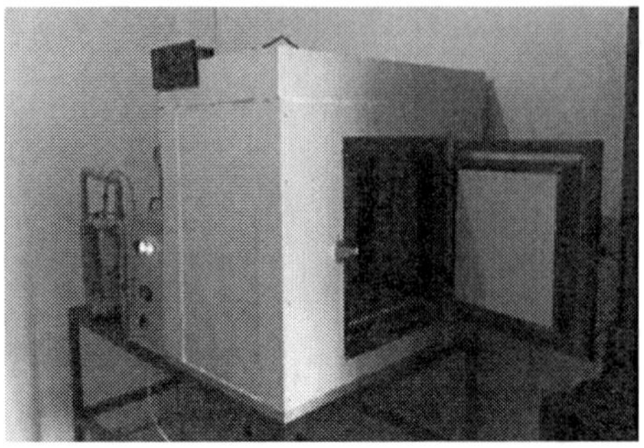

Fig. 4. Box cooling cold storage

The refrigerant is compressed by the compressor hermetic, Tecumseh CAJ9480T power input 0.5 HP, maximum compression pressure of 50 bar. Retrieval of data starting from 0 minutes up to 60 minutes, looking from the degree of temperature subcooling until 7°C according to the results research [5]. The calculation of the value of the enthalpy base on pressure and temperature used software Cooldpack Ver 1.50.

3 Result and discussion

3.1 Effect degree of temperature subcooling on the effectiveness LSHX

Using the lowest temperature of LSHX, in the RTU cold storage with R22 is -14.6°C, R134a is -9.4°C, R404A is -24.4°C, R600 is -5.9°C, and mixture R290/R600a is -21.6°C. From the temperature in the RTU cold storage, the mixture refrigerant R2690/600a has been similar characteristics to R404A. The relationship between the degree of temperature subcooling to the effectiveness LSHX is given in Fig. 5.

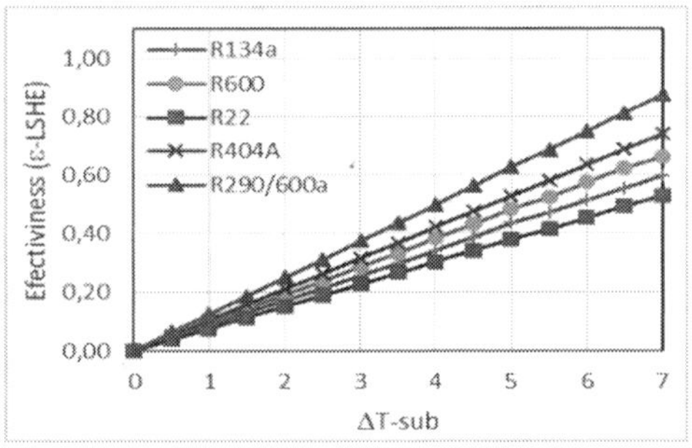

Fig. 5. Effect ΔTsub to effectiveness LSHX

From Fig. 5, its show that if the degree of subcooling temperature increases, the effectiveness LSHX also increased. Lowest effectiveness LSHX with using refrigerant R22 and the highest using mixture refrigerant R290/R600a, ε = 0.88. Average effectiveness LSHX for R22, R134a, R404A, R600 and R290/R600a respectively 0.267, 0.301, 0.370, 0.335, and 0.438. From Fig. 5, the effectiveness of LSHX with a mixture of refrigerant R290/R600a approached characteristics R404A, and effectiveness refrigerant R600 approached characteristics R134a.

3.2 Effect of degree of temperature subcooling to increase the refrigeration capacity.

By using LSHX subcooling, the average of refrigeration capacity for R22, R134a, R404A, R600 and mixed refrigerant R290/R600a respectively are 160.07 kJ/kg, 149.69 kJ/kg, 123.01 kJ/kg, 271.93 kJ/kg and 282.56 kJ/kg. LSHX subcooling lead to increase refrigeration capacity and refrigeration capacity index given by Fig. 6 and Fig. 7. The Fig. show that if the degree of temperature subcooling increase, the refrigeration capacity also increase

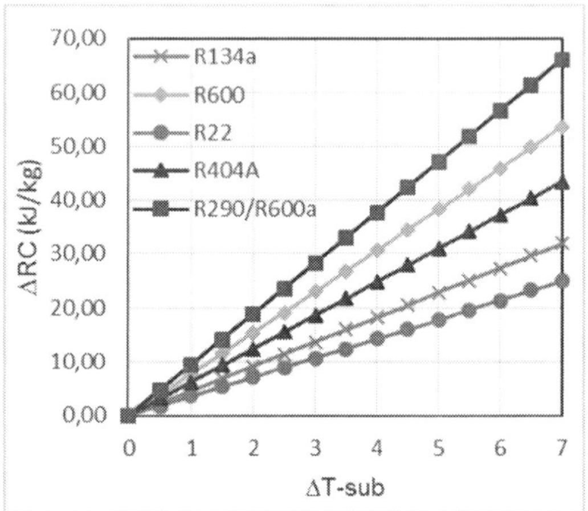

Fig.6. Effect ΔTsub to ΔRC

Fig. 6, its shows that the increase in refrigeration capacity with a mixture refrigerant R290/R600a is greater compared with the others, while the lowest increase refrigeration capacity using refrigerant R22. The average increase in refrigeration capacity than without LSHX for R22, R134a, R404A, R600 and mixture refrigerant R290/R600a respectively is 12.46 kJ/kg, 15.95 kJ/kg, 21.72 kJ/kg, 26.81 kJ/kg, and 33.01 kJ/kg. From indicators the increased refrigeration capacity, the use mixed refrigerant R290/R600a has characteristics similar to R600, and refrigerant R600 has characteristics similar to R404A.

Fig.7 shows the effect the degree of temperature subcooling to the refrigeration capacity index. The Fig.s show that if the degree of temperature subcooling increases, also refrigeration capacity index increased. Usage LSHX and mixture refrigerant R290/R600a has an refrigeration capacity index most large compared with the other, while using refrigerant R22 has the lowest refrigeration capacity index. The average increase in refrigeration capacity index for R22, R134a, R404A, R600 and mixture R290/R600a respectively are 1.09, 1.12, 1.08, 1.11 and 1.13. This means that the average increase largest in refrigeration capacity is used mixture R290/R600a, where percentage in increase refrigeration capacity is 13%.

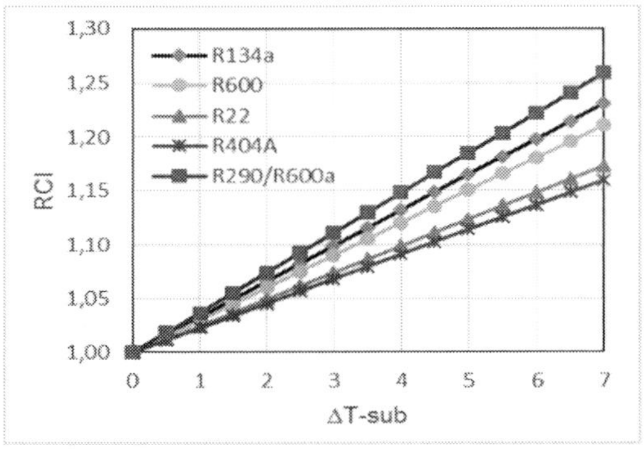

Fig.7. Effect ΔTsub to RCI

With the indicators increase refrigeration capacity (Fig.6), refrigeration capacity index (Fig.7) which function of the degree of temperature subcooling, that mixture refrigerant R290/R600a has a characteristic similar to R134a, and R600 has characteristic similar to R404A. So that the mixture refrigerant R290/R600a can be used as a substitute for R600 and R404A, and R600 can be used as a substitute for R404A in the vapor compression refrigeration system with LSHX. This result relevant to the research [8] and [10].

3.3 Effect degree of temperature subcooling to work compressor.

To see the impact of the installation LSHX subcooling to work may be viewed from the compression ratio between the discharge and suction pressure, increase in work compressor, and work compressor index.

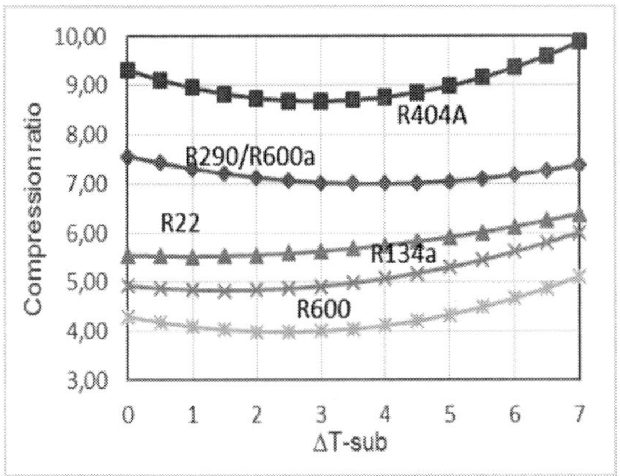

Fig.8. Effect ΔTsub to compression ratio

Fig. 8 shows that the increase pressure ratio work compresso, where the compression work ratio is between 4 to 9 and the minimum value of the compression ratio is at condition degree of temperature subcooling between 3-5 degrees. The average compression ratio for R22, R134a, R404A, R600 and R290/R600a respectively is 5.99, 5.17, 9.04, 4.30 and 7.18. The highest compression ratio using refrigerant R404A, and the lowest use of refrigerant R600. The magnitude of this compression ratio shows the amount of work the compressor performed by the system, which can be seen from the increase in compressor work if the degree of temperature subcooling rise. From indicators work compressor, characteristics mixture refrigerant R290/R600a similar to R404A and, characteristics refrigerant R600 similar to R134a. These results are relevant to papers [13].

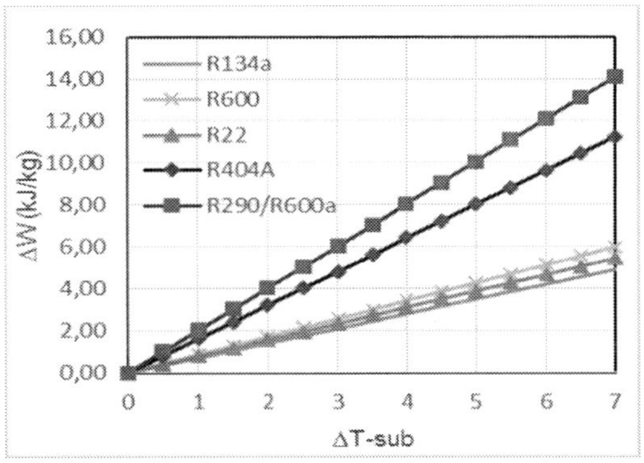

Fig.9. Effect ΔTsub to increase work compression

The Fig.9 as show that if the degree of temperature of subcooling increase, the work compressor work also increased in vapor compression refrigeration system with LSHX subcooling. The average of increasing in work of compressor that use LSHX subcooling for R22, R134a, R404A, R600 and R290/R600a respectively 2.72 kJ / kg, 2.45 kJ/kg, 5.62 kJ/kg, 2.99 kJ/kg and 7.04 kJ/kg. Mixture refrigerant R290/R600a has the largest increase in compressor work. Evaluating from the increase work of compressor, characteristic mixture refrigerant R290/R600a close to characteristics R404A, and characteristic R600 close to R22. These results are relevant to reported of the paper [3], [6] and [9].

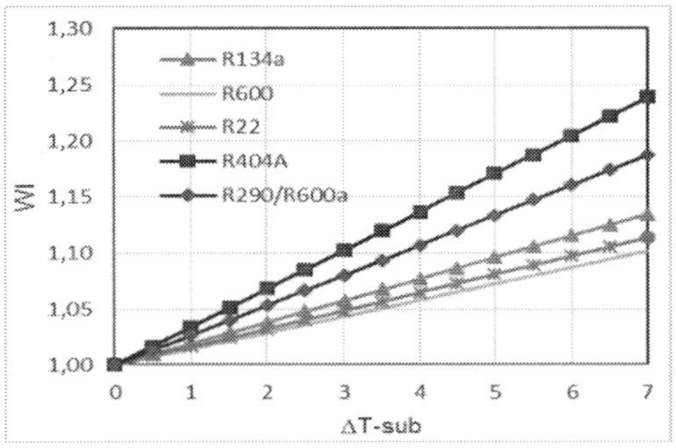

Fig.10. Effect ΔTsub to work compressor index

The work compressor index shows that a rise work compressor as a percentage, which is shows in Fig.10. The average of work compressors index for R22, R134a, R404A, R600 and R290/R600a respectively are 1.06, 1.07, 1.12, 1.05 and 1:09. From Fig. 10, when degrees temperature subcooling increases, the work compressor index also increased. This shows that used LSHX would result in the ratio and the work compressor increasing. The largest work compressor index if used refrigerant R404A and the lowest work compressor index with R600. Meanwhile, are work compressor index R290/ R600a still under R404A and above R134a.

Based on the relationship between the degree of temperature subcooling with a working compressor, work compressor index that mixture refrigerant R290/R600a has close characteristics to R134a and R404A, and characteristics R600 close to R22, so that the refrigerant mixture R290/R600a can be used as substitute for R134a, while the R600 can be used as substitute R22 in vapor compression refrigeration system with LSHX

3.4 Effect degree of temperature subcooling to COP

The average value COP_{LSHX} for refrigerant R22, R134a, R404A, R600 and R290/R600a respectively are 3.31, 4.22, 2.47, 3.59 and 4.85. The Fig.11 show that the value optimum COP_{LSHX} as function of the degree of temperature subcooling between 3-5°C.. This result relevant to research [4]. COP_{LSHX} are greatest using R600, and the lowest use of R404A. This result is relevant that refrigerant R404A has largest compression ratio, while refrigerant R600 has a low compression ratio. The size of the compression ratio will affect the value of COP refrigeration system. Characteristics COP_{LSHX} are relevant to results research [3-4]. Form the value COP_{LSHX}, R290/R600a have closer characteristics to R22 and R600 close to R134a.

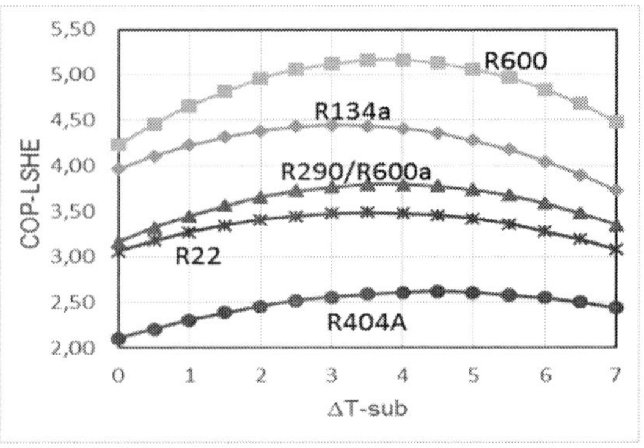

Fig. 11. Effect ΔTsub to COP_{LSHX}

The average COP index (COP_N) for refrigerant R22, R134a, R404A, R600 and mixture refrigerant R290/R600a respectively are 1.03, 1.05, 1.08, 1.06 and 1.09.

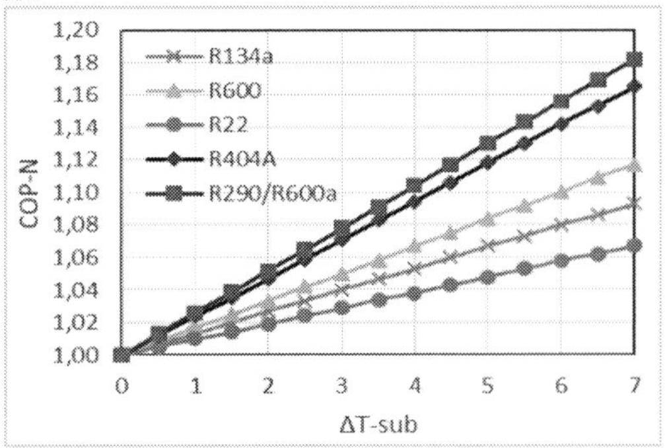

Fig.12. Effect ΔTsub to COP_N

From Fig. 12 shows that increase degree of temperature subcooling, and then the COP index also increased. This shows that there is a significant relationship between the degree of temperature subcooling to COP_N in vapor compression refrigeration system with LSHX. From the value COP_N, its appears that the use of LSHX is impact significantly increase COP_N, particularly COP_N for mixture R290/R600a greater than R404A and COP_N refrigerant R600 close to R134a. In general terms of the parameters COP, mixture refrigerant R290/R600a can be used as a substitute for refrigerant R404A and R600 can be used as substitute R134a in vapor compression refrigeration system with LSHX

4 Conclusion

Based on the above analysis, the installation LSHX on vapor compression refrigeration system, it can be concluded that the increase degree of temperature subcooling significantly influence the effectiveness LSHX, increase refrigeration capacity, increase refrigeration capacity index, increase work compressor, increase work compressor index, COP_{LSHX}, and indexes COPN. From these indicators, that mixture refrigerant R290/R600a with the ratio of 50:50 can be used as a substitute for refrigerant R404A and R22, while refrigerant R600 can be used as a substitution of R134a in vapor compression refrigeration system with LSHX.

This research is part of a research grant funding through the Higher Education Research Lecturer decentralization scheme for "Competitive Research Grant" fiscal year 2016 undur Contract No. 058.A/1/16/PKP/2016. The authors would like to thank Dr. Ir. Supriadi Legino as chairman of STT-PLN and Indrianto, ST, M.Kom. as Chairman of the Research Institute of STT-PLN

References

1. M. Nuriyadi, Sumeru, H. Nasution, Jurnal Teknologi, **76** (11), 57 (2015)
2. B.A.Qureshi, S. M. Zubair, J. Refrigeration **35**, 47, (2015)
3. M.H.Yang, R.H. Yeh, J. Heat Transfer Engineering, **36** (10), 912 (2015)
4. M.H. Yang, R.H..Yeh, J., J. Heat and Mass Transfer, **87**, 1 (2015).
5. B.O. Baloji, Z. Huang, J. Engineering Research, **1**(1), 317 (2013)
6. B.O. Bolaji, J. Science and Technology, **34** (2), 73 (2014)
7. G. Pottker, P.S. Hrnjak, Purdue University, Purdue e-Pubs (2012), paper 1328
8. T.Ren, G. Wu, G. Ding, Y. Zheng, Y. Gao, J. Song, Purdue University, Purdue e-Pubs (2014), Paper 1407.
9. S. A. Klein, D. T. Reindl, and K. Brownell, J. Refrigeration, **23** (8), 588 (2000).

10. L. XU, P. HRNJAK, Conference at Purdue, Purdue University, Purdue e-Pubs, July 14-17, (2014), Paper 2387
11. Prayudi , Roswati N., ARPN Journal of Engineering and Applied Sciences, **11** (2), 901 (2016)
12. Prayudi, Roswati N. and, Retno A.D., AIP Conference Proceedings, (2016) (to be published)
13. S. Wongwises, N. Chimres, J. Energy Conversion and Management, **46**, 85 (2005)

 DOI: 10.1051/matecconf/201710103003

Effect of mixing parameters on the mixing time and density of composite HA/Ti6Al4V feedstock for powder injection molding

Amir Arifin[1,*], and *Abu Bakar* Sulong[2]

[1]Department of Mechanical Engineering, Faculty of Engineering, Sriwijaya University, 30662 Indralaya Indonesia
[2]Department of Mechanical and Material Engineering, University Kebangsaan Malaysia, 43600 Bangi, Selangor, Malaysia

Abstract. Hydroxyapatite is a bioactive material. However, Hydroxyapatite in medical application is limited by its poor mechanical properties. Titanium alloys are biocompatible materials with high specific strength and high corrosion resistance. This combination can successfully be used in biomedical applications. The mixing process is a critical stage in the metal injection molding process because it determines the quality of the metal injection product. The homogeneity of the feedstock depends on the mixing parameters used during the mixing process. An experiment was conducted using Ti6Al4V and hydroxyapatite powder as feedstock at a mixing ratio of 60:40 wt%. This feedstock was mixed with a mixture of polyethylene and palm stearin as the binder. The mixing parameters used in this study were the mixing speed (10 and 30 rpm) and the mixing temperature (130°C and 150°C) at a powder loading of 67%, as determined by the critical powder volume percentage experiment data (69.5%). Increases in the mixing speed and mixing temperature reduced the mixing time to achieve homogeneity. The combined feedstock achieved the highest density at a mixing speed of 30 rpm and a mixing temperature of 150°C.

1 Introduction

Metal injection molding (MIM) is a combination of injection molding and powder metallurgical processes [1]. MIM is one of the near-net-shaped molding processes in manufacturing. The primary advantage of MIM compared with other methods is its low cost for mass-producing small and complex precision parts [2]. The MIM process has four stages: powder and binder mixing, injection molding, debinding, and sintering as shown on the Fig.1. Mixing is a vital stage that determines the homogeneity of feedstock quality in the MIM process. Deficiencies in feedstock quality cannot be corrected by the subsequent process. Failure to distribute powder uniformly on the feedstock can result in molding defects on the final product [3, 4]. Non-homogeneous feedstock can cause defects such as distortion, cracks, and voids in the sintered part. A number of factors affect the feedstock properties, such as the speed and mixing time, geometry of the mixing blades, temperature, particle characteristics, composition, and viscosity [5, 6].

Hydroxyapatite (HA) and titanium are key elements in the development of medical components [7]. HA or [$Ca10(PO4)_6(OH)_2$] is a calcium phosphate ceramics currently used as a bioactive material for dental and orthopedic applications [8]. Its composition and structure is similar to those of bone, but with poor mechanical properties. On the other hand, titanium has high mechanical properties such as specific strength, corrosion resistance. Moreover in term of Young's modulus, titanium alloys has lower value than another metal implant and approaching of Young's modulus of the human bones. The unique properties of both materials can be suitably combined for medical applications. Feedstock have a considerable influence on the quality of products through MIM method. a homogenous feedstock will reduce some of defects such as porosity and cracks.

To obtain a homogeneous feedstocks need to consider the type and composition of the binder used. Generally component of binder can be categorized into two part;

- ✓ Low molecular weight polymer; which has low temperature decomposition.
- ✓ High molecular weight polymer ; which has relatively high temperature decomposition

Moreover a binder system on injection molding process is designed based on several functions such as backbone, surfactant and lubricant. In this work, polyethylene has functioned as the backbone. Palm stearin itself which is fraction of palm oil has fatty acid content play important role as surfactant and lubricant. Palm Stearin is a natural substance that is found in palm oil producing countries in the regions of Asia, Africa and Latin America such as Indonesia, Malaysia, Thailand, Colombia and Nigeria. Palm stearin as natural material will reduce the effects of environmental pollution. moreover the use of palm stearin as a binder can reduce cost incurred in buying commercial binder.

* Corresponding author: amir@unsri.ac.id

DOI: 10.1051/matecconf/201710103003

The objective of this paper is to study the effects of the mixing parameters of a Ti6Al4V/HA composite with a palm stearin and polyethylene binder system on the mixing time and density.

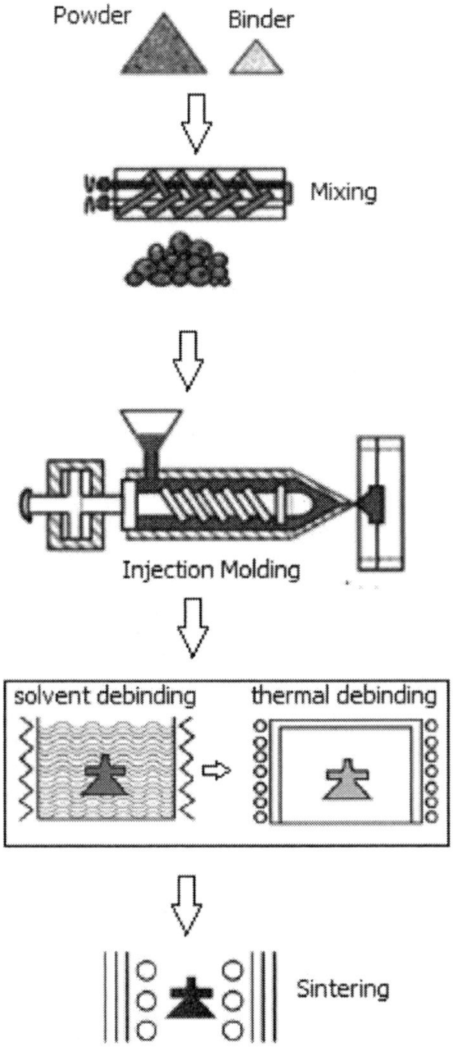

Fig. 1. Injection molding process scheme

2 Experimental

The Ti6Al4V powder used in this experiment was supplied by TLS Technik GmbH & Co. It has an average particle size of 25 μm and is spherical in shape (Figure 2). HA (Sigma-Aldrich Co., Germany) has an average powder size of 5 μm (Figure 3). The HA powder, which is smaller than the titanium powder, HA powder is expected to fill gap between titanium powder.

The binder system consists of 60 wt% palm stearin and 40 wt% polyethylene. Differential scanning calorimetric results for palm stearin and polyethylene (Figures 5 and 6) has showed that the melting temperatures of palm stearin and polyethylene were ~54 and ~125 °C, respectively. Mixing temperature of feed stock was determined by these results. The composite powder has a composition of 60 wt% Ti6Al4V and 40 wt% HA. Before composite powder Ti6Al4V/HA was mixed with palm stearin and polyethylene, premix was performed for Ti6Al4V and HA powder premixed using a Winkworth Sigma Blade Mixer for 0.5 h as shown on Figure 4. Premix stage is conducted to facilitate homogeneity of Ti6Al4V and HA powder.

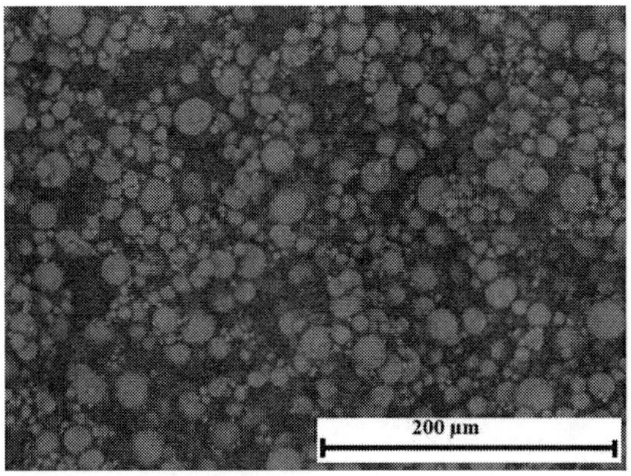

Fig. 2. Scanning electron micrograph of the Ti6Al4V powder.

Fig. 3. Scanning electron micrograph of the hydroxyapatite (HA) powder.

Fig.4. Winkworth Sigma Blade Mixer

 DOI: 10.1051/matecconf/201710103003

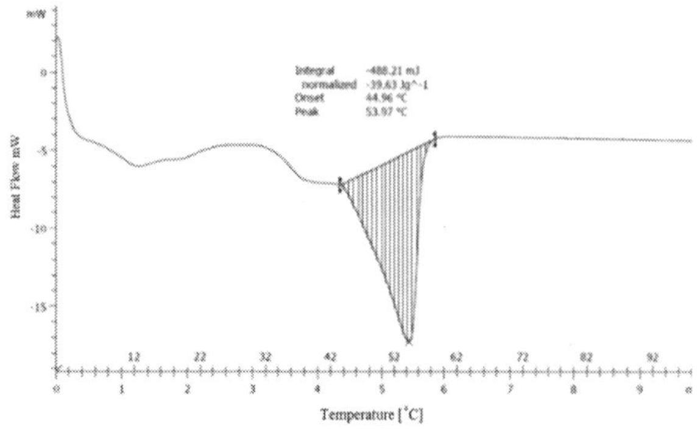

Fig. 5. Differential scanning calorimetric curve of palm stearin.

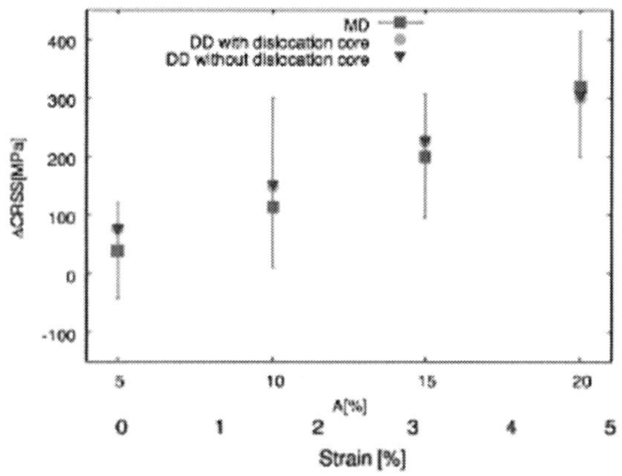

Fig. 6. Differential scanning calorimetric curve of polyethylene

The critical powder volume percentage (CPVP) of the mixture is measured using the oil absorption technique based on ASTM: D-281-12. Powder mixture was mixed with binder system for 30 min at 30 rpm. Firstly, polyethylene was melted completely in mixing machine,then palm stearin and followed by composite powder mixture of Ti6Al4Vand HA. The critical powder volume percentage (CPVP) of was obtain 69.5 vol%, as shown by the maximum torque curves (Figure 8). Brabender W50EHT Measurement and Control Systems was used to analysis CPVP as shown on the Figure 7.

Fig. 7. Brabender W50EHT Measurement and Control Systems

According to German et al 1997, optimum powder loading for MIM ranged from 2% to 5%, which is lower than the critical powder loading. In this study, 67 vol% was determined as the optimum powder loading.

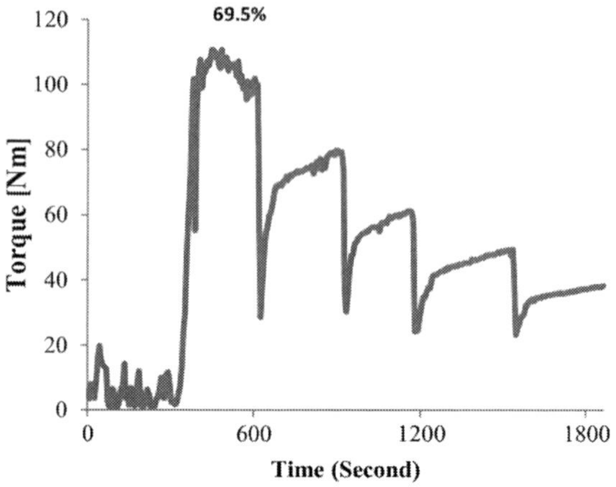

Fig. 8. Mixing process to determine the critical powder volume percentages (CPVP) of Ti6Al4V and HA powder.

All mixing parameters were analyzed using a Brabender mixer that consisted of a mixer backstand, a gear unit, and a mixer bowl. The mixing speed and mixing temperature were set based on the parameters shown in Table 1. The mixing process stops when the homogeneity of the feedstock reaches the value of a steady-state torque on the graph. The density of each sample was measured for comparison with the result of the Brabender mixing.

Table 1. Variations in the mixing parameters.

Identification	Powder Loading vol (%)	Mixing Speed (rpm)	Mixing Temp (°C)
A6710130	67 %	10	130
A6710150	67 %	10	150
A6730130	67 %	30	130
A6730150	67 %	30	150

3 Results and discussion

Figure 9 shows the torque of the mixing process at different speeds and temperatures. In this process, the graph can be divided into three stages. In the first stage, the polyethylene was loaded into the chamber until it fully melted. The graphs remain stable, with a few fluctuations. In the second stage, large fluctuations in the torque values occur when the powder and the palm stearin was added. The number of peaks and the peak heights in the graph depend on the loading process of the powder and the palm stearin, which in turn affects torque. In the final stage, the feedstock reaches homogeneity with the torque and remains at the same value for each sample. Morover some peak of torque was observed on the Fig 9. Initial peaks on the graph was created due to the addition powder on the chamber. The last peak on the graph as indicator homogeneity of feedstock was started until graph constant.

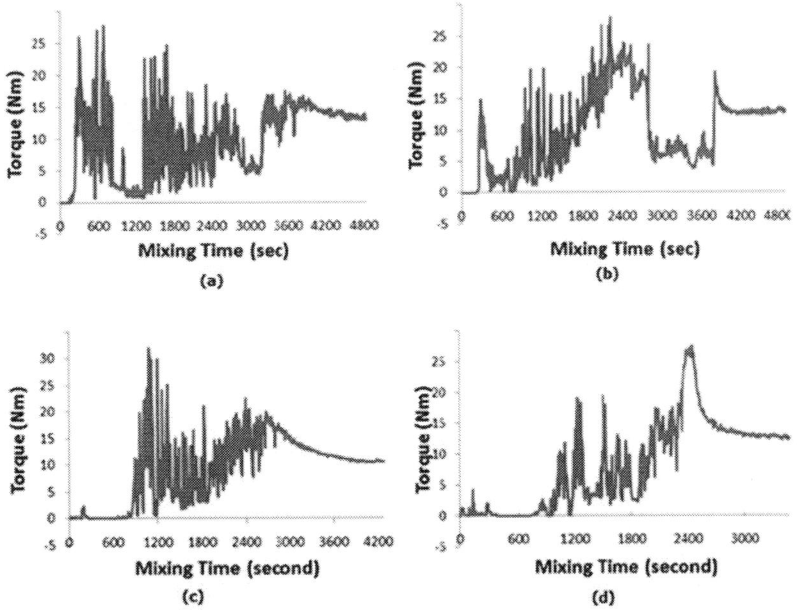

Fig. 9. Torque behavior of the feedstock under various mixing parameters.(a) A6710130, (b) A6730130, (c) A6710150 and (d) A6730150

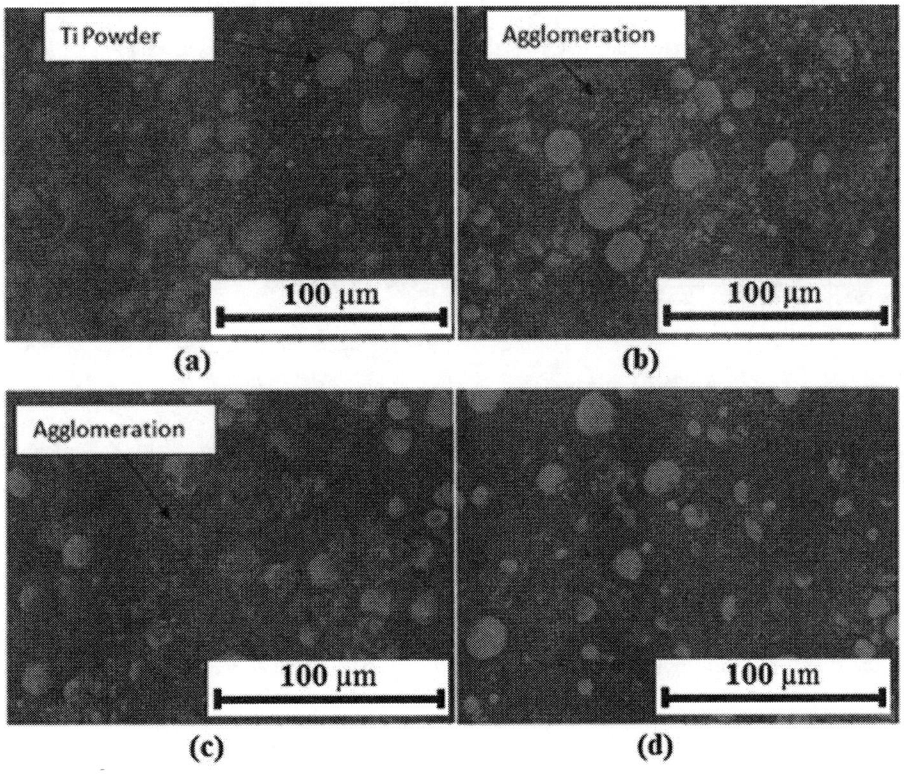

Fig. 11. Scanning electron micrograph of feedstock after mixing; (a). A6710130, (b) A6710150, (c) A6730130 and (d) A6730150

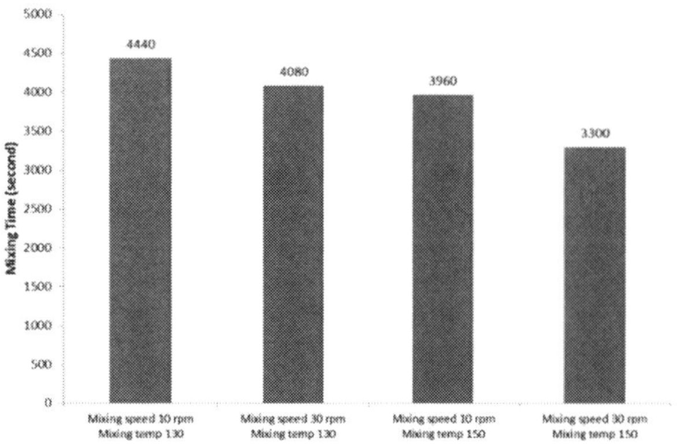

Fig.10. Mixing time of the feedstock at different mixing parameters.

Figure 10 shows that an increase in the mixing speed of the process from 10 rpm to 30 rpm reduced the mixing time required to reach homogeneity from ~4,440 s to ~4,080 s at a mixing temperature of 130°C. At a mixing temperature of 150°C, the reduction was from ~3,960 s to ~3,300 s. Furthermore, an increase in the mixing temperature reduced the mixing time needed to achieve homogeneity. A6710130 showed the longest mixing time among all the composites. However, homogeneity was achieved by A6730150 at the shortest time (Figure 10). This result may be due to the combination of increased mixing temperature. In addition, the amount of input energy that was transferred from the mixer blade during shearing in the mixing process increased with the increase in speed.

Table 2. Bulk density of the feedstock.

Sample	A6710130	A6730130	A6710150	A6730150
Density g/cm^3	2.52 ± 0.02	2.52± 0.04	2.53 ± 0.04	2.54 ± 0.04

Density measurements were performed to determine the density and to investigate the homogeneity of the feedstock. Table 2 shows that each sample has a small standard deviation, indicating that the material achieved homogeneity. At the 150°C mixing temperature, an increase in the mixing speed increased the density. However, the density at the mixing temperature of 130°C appeared stable. A6730150 has the highest density, which is due to the combined increase in the mixing speed and the mixing temperature as a result of the viscosity effect of polyethylene, which has a melting point of 125°C. The viscosity of polyethylene decreased at higher temperatures, which facilitated the mixing.

Agglomeration is the buildup of powder with a powder which is a common phenomenon, especially in the powder at the nanometer scale. Agglomeration is avoided in the powder metallurgy process due to produces a lack of uniformity that results in defects in the final product.

Figure 11 shows, Scanning electron micrograph of feedstock under various mixing parameters. Agglomeration appears on the figure 11 (b) , (c) and (d). Although agglomeration was not occurring on Fig 11 (a) but there were groups of titanium powder tend to close if comparing with other Figure 8 (a,b and c). Titanium powder tend to distribute as shown on Figure 11 (a,b and c) but there were agglomeration especially on Figure 11.(b) and Figure 11(c). Agglomeration tends to decrease with the increasing of mixing speed and mixing temperature. In this term can be concluded that the increase in mixing speed has significant impact on the quality of the resulting feedstock. However high speed on the mixing speed will lead to air trapped in the feedstock. furthermore mixing temperatures can reduce the viscosity of the binder that facilitates the process of achieving uniformity of feedstock. Based on the results of Scanning electron micrograph on feedstock after mixing, agglomeration was found in the composite powder Ti6Al4V / HA. this happens is believed due to differences in size, shape and nature of the powder used.

4 Conclusion

Mixing analysis process for composite powder HA/Ti6Al4V has been performed. Some parameter process of mixing has been analysis such as mixing time and density of feedstock. Experiment results showed that the increasing of mixing speed and mixing temperature tend to reduce the mixing time needed to attain feedstock homogeneity. In addition, the density results show that a mixing speed of 30 rpm and mixing temperature of 150 C are the optimal parameters for obtaining the optimum mixing time and density.

References

1. German, R.M. and A. Bose, Metal Powder Industries Federation.
2. Heaney, D.F., J.D. Gurosik, and C. Binet, Journal of Materials Science, 2005. **40**(4): p. 973-981.
3. Huang, B., S. Liang, and X. QuJournal of Materials Processing Technology, 2003. **137**(1-3): p. 132-137.
4. Liu, Z.Y., et alMaterials Characterization, 2002. **49**(4): p. 313-320.
5. Niespodziana, K., et al., Materials Chemistry and Physics, 2010. **123**(1): p. 160-165.
6. Pramanik, S., et al., Ceramics International, 2007. **33**(3): p. 419-426.
7. Sopyan, I., et al., Science and Technology of Advanced Materials, 2007. **8**(1-2): p. 116-123.
8. Gunawan, et al., Key Engineering Materials, 2012. 531-532: p. 258-261.

* Corresponding author: amir@unsri.ac.id

The current density of hydroxyapatite electrosynthesis by ion exchange resin chambers

Adrian Nur[1,*], *Nazriati* Nazriati[2], *Arif* Jumari[1], *Ega Fitri Novita* Santi[1], and *Shafira Yaumil* Asiffa[1]

[1]Chemical Engineering Department, Faculty of Engineering, Sebelas Maret University, 57126 Surakarta, Indonesia
[2]Chemistry Department, Faculty of Mathematics and Science, State University of Malang, 65111 Malang, Indonesia

Abstract. Electrochemical method has been developed to produce hydroxyapatite (HA) powders. Synthesis of hydroxyapatite particles was carried out from a homogeneous solution. The homogeneous solution was contained of $Na_2H_2EDTA.2H_2O$, KH_2PO_4 and $CaCl_2$ on 0.25/0.25/0.15 M. Water reduction at cathode to form OH^- ion is very substantial in formation of hydroxyapatite. The OH^- ions causes the release of Ca^{2+} ions and change the equilibrium phosphate for the formation reaction of HA. The problem of electrochemical method is OH^- ions move away from the cathode due to the difference in charge. This condition decrease the productivity of the formation of HA because OH^- ions react with H^+ to form H_2O. In this work, we develop new method electrosynthesis of HA with 2 chambers system. The electrolysis use ion exchange resin to separate the OH^- ions and H^+ ions formed. The effect of current density to HA product was studied. HA product analysed with XRD. We conclude that the electrosynthesis hydroxyapatite with ion exchange resin more effective than without ion exchange resin. The higher current density, the more pure HA formed.

1 Introduction

Hydroxyapatite, also called hydroxylapatite, is one of the calsium phosphate based bioceramic material which makes the majority of inorganic components of human bones and teeth. The formula of hydroxyapatite is $Ca_5(PO_4)_3OH$ but usually written $Ca_{10}(PO_4)_6(OH)_2$ to denote that the crystal unit cell comprises two entities.

The calcium hydroxyapatite use as bone substitute materials because of their chemical similarity to the component of bones and teeth of mammalian. Calsium hydroxyapatite also used as fillers and coatings for repairing of large osseous defects [1].

Various synthesis methods have been used for powder hydroxyapatite synthesis, e.g. chemical precipitation method [2], hydrotermal technique [3. 4], emulsion method [5], and sol-gel process [6], mechanochemical method [7]. Chemical precipitation process is the most reported method for preparing hydroxyapatite particles. This process is simple, low cost, and suitable for industrial production but difficult to control and resultant particles have a low quality with a large particle size, wide particle size distribution and a lot of agglomerates. The hydrothermal technique usually gives hydroxyapatite powders a high degree of crystallinity. However, need high temperature and the obtained powders are usually in agglomeration and the size distribution are usually in wide range. The main advantage of emulsion technique is the simplicity of the process, good crystallinity product without high temperature, and favours the formation of small crystallites with a sufficient narrow size distribution. Sol-gel synthesis of hydroxyapatite has many advantages, which include high product purity, homogeneous composition, and low synthesis temperature. When the mass production of high crystalline hydroxyapatite is required, mechanochemical treatment is more suitable. Despite the risk of contamination during milling. Electrochemical deposition is one of the most commonly used method for hydroxyapatite coating at metal. This method is known to be the effective and efficient techniques to assemble fine particle. Reference [8] prepared nanosized powder hydroxyapatite using electrochemical method.

Electrochemical method has been successfully used for the synthesis of hydroxyapatite using constant direct current [9] and pulsed direct current [10]. The development process has also been successfully carried out with the use of EDTA as chelating agent.

References [11, 12] concluded that OH^- ions produced by the reduction of water at the cathode was very substantial in the hydroxyapatite particles formation. The OH^- ions caused the Ca^{2+} ion release from EDTA and the dissociation of phosphoric acid. This condition serve the reactants for particle formation. The problem is OH^- ions move away from the cathode by diffusion and by migration. As a result, the formation of HA becomes ineffective because OH^- ions formed is not fully utilized for the formation of HA.

In this work we have developed 2 chambers system with ion exchange resin to separate the anions (H^+ ions) and cations (OH^- ions). The novelty of this work is used

* Corresponding author: adriannur@staff.uns.ac.id

2 chamber with ion exchange resin to electrosyntesis of hydroxyapatite. Selective cation exchange resin was installed around the cathode and selective anion exchange resin around the anode. Cation exchange resin prevent OH- ions migrate from the cathode while the anion exchange resin prevent the H^+ ions migrate to the cathode. The effect of current density to HA product was studied.

2 Experimental

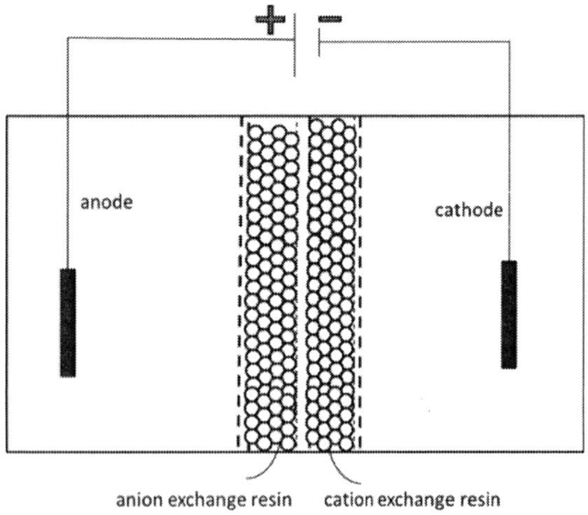

Fig. 1. A schematic presentation of electrochemical cell with ion exchange resin

The solution for the synthesis of HA powders was consisted of $Na_2H_2EDTA.2H_2O$ (Merck, reagent grade), KH_2PO_4 (Merck, reagent grade) and $CaCl_2$ (Merck, reagent grade) with a concentration relationship of $Ca^{2+}/EDTA/PO_4^{3-}$ as 0.25/0.25/0.15 M. All chemicals were used as supplied with no further treatment. The electrolysis was carried out in a cell consisting of two carbon electrodes with dimensions of (2 × 1) cm. A simple two-compartment acrylic cell was used. The electrodes were set parallel with a distance of 3 cm between the electrodes and immersed in the solution at a depth of 2 cm. The electrodes were connected to a DC power supply (Zhaoxin PS-3005D). The ion exchange resin chamber was installed between 2 electrodes (fig.1). The electrosynthesis was performed at 3 hours at 320 to 480 mA/cm^2 under constant stirring at room temperature. The suspension produced was filtered. The particle was washed with demineralized water and dried at 110°C for 2 days. The X-ray diffraction (XRD) pattern of the particles was detected using an X-ray diffractometer (Shimadzu 6000). The morphologies of the particles analysis were observed using scanning electron microscopy (Inspect S40, FEI). Fourier transform infrared spectroscopy (FTIR) (Shimadzu 8400s) was performed to evaluate the functional groups of specimens. The FTIR spectra were obtained over the region 500-4000 cm^{-1}.

3 Result and discussion

The solution was clear at the beginning of the experiment when neither current nor potential had yet been imposed. As electrolysis, the surface of anode and cathode release the gas bubbles. The gas bubbles at anode indicate the formation of oxygen from oxidation of water. The gas bubbles at cathode indicate formation of hydrogen form reduction of water. A white precipitate was obtained.

$$2H_2O + 2e^- \rightarrow H_{2\,(g)} + 2OH^- \quad (1)$$

$$2H_2O \rightarrow O_{2\,(g)} + 4H^+ + 4e^- \quad (2)$$

The phase identification of powder was confirmed by FTIR spectroscopy. Comparison of FTIR spectra of particle with and without ion exchange resin is showed in Fig. 2. There is not significant different of FTIR spectra of particle with and without ion exchange resin. The spectra were collected in the 500 to 4000 cm^{-1} wavenumber range region. The FTIR spectra show that the phosphate and hydroxide absorption peaks for HA were seen. The characteristic bands for phosphate appear at 567 cm^{-1}, 603 cm^{-1}, 960 cm^{-1}, and 1032 cm^{-1}. The broad band of low intensity in the range 3000–3500 cm^{-1} can be attributed to traces of water incorporated to the structure, broad band around 1590 cm^{-1} of H–O–H bending mode. In the OH stretching vibration region, beside the νOH of the hydroxyapatite, surface OH band at 3500 cm^{-1} also appears.

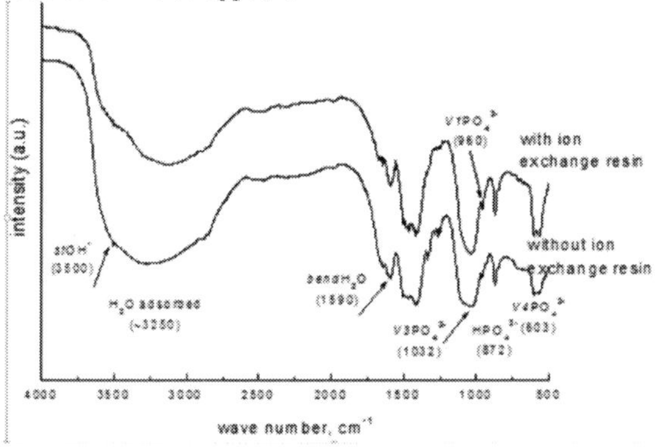

Fig. 2. FT-IR spectra of the HA particles with and without ion exchange resin

The morphology of powders with and without ion exchange resin at high magnification is provided in Fig. 3. The powder of HA without ion exchange resin consist of various shapes. The most of particles is cylinder-like shape. The powders with ion exchange resin consist of agglomerates which are composed of fine crystallites. The SEM micrograph of HA particles show that agglomeration of particles and spherical-like shape. The particles size is about 100 to 300 nm.

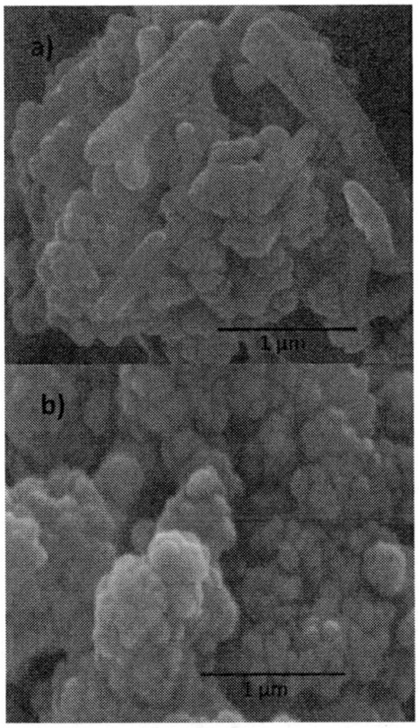

Fig.3. SEM images of the HA particles (a) without and (b) with ion exchange resin

Comparison of the particle with and without ion exchange resin was showed in Fig. 4. The electrosynthesis was performed at 3 hours at 1.5 A under constant stirring at room temperature. The XRD pattern of particle showed that the crystalline phase of particles with and without ion exchange was hydroxyapatite. There are 5 characteristic peaks at 25.9° (002), 29.3° (210), 31.6° (211), 32.2° (300) and 34.1° (202), which match the standard pattern of HA (JCPDS 09-0432). The XRD pattern of particle with ion exchange resin showed that the particle is pure HA, while without ion exchange resin showed they are another peaks. Another peaks for without ion exchange resin are 20.9° and 30.5° which match the standard pattern of brushite (JCPDS 72-0713). Brushite is the intermediate product which rapidly turned into HA. By the time when increase the OH^- ion, the following reaction to take place

$$Ca^{2+} + HPO_4^{2-} \leftrightarrow CaHPO4 \quad (3)$$

$$10CaHPO_4 + 2OH^- \leftrightarrow Ca_{10}(PO_4)_6(OH)_2 + 4PO_4^{3-} + 10H^+ \quad (4)$$

This condition showed that electrosynthesis with ion exchange resin more effective than without ion exchange resin. The ion exchange resin separate electrochemical cell to 2 chambers. Cation exchange resin prevent OH^- ions migrate from the cathode while the anion exchange resin prevent the H^+ ions migrate to the cathode (Fig. 5). The formation of HA becomes more effective because OH^- ions just formed HA particle.

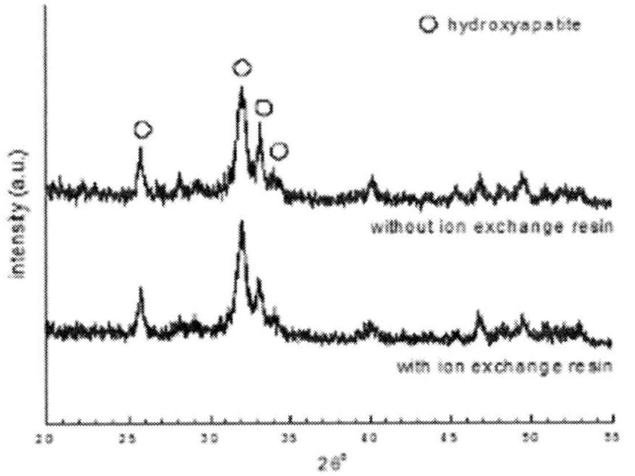

Fig.4. Comparison of XRD pattern of the HA particles with and without ion exchange resin

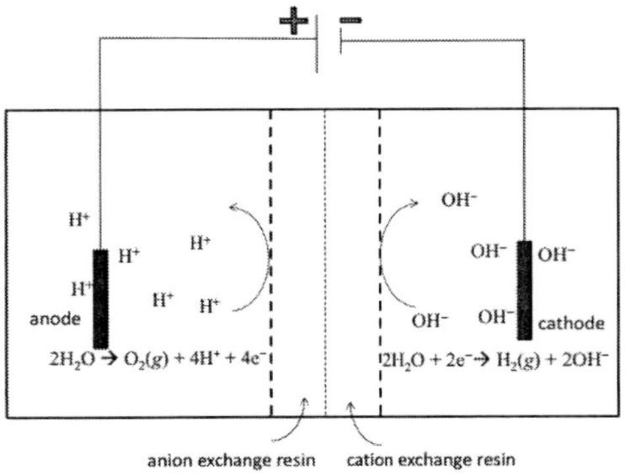

Fig.5. Electrochemical cell with Ion exchange resin

Observation of the solution pH is done all the time electrolysis. Fig.6. shows the solution pH for electrolysis time without resin and with resin (in anode and chatode chamber) (the smoothed line by B-spline). The solution pH without ion exchange resin decreases rapidly than the solution pH with ion exchange resin. As electrolysis, OH^- ions were resulted by water reduction (Eq.1) and H^+ ions were resulted by water oxidation (Eq.2). According Eq.3, OH^- ions react with brushite to form hydroxyapatite and release H^+ ions. The consumption of OH^- ions and the release of H^+ ions causes the solution pH decrease.

The solution pH is very substantial to final product obtained. When the solution pH < 7, the final product is brushite and when the solution pH >7, the final product is HA [14]. The maintaining of the solution pH above 7 during electrolysis is very essential to obtain HA. At electrolysis without ion exchange resin, the solution pH was 8 after 180 minutes and still obtain HA.

At electrolysis with ion exchange resin, there are two chambers i.e the anode and the cathode chamber. Fig.6. shows the solution pH in the anode and the cathode chamber for electrolysis time. In the anode chamber, the solution pH decreases by time electrolysis. The oxidation of water resulted H^+ ions (Eq.2.) due the solution pH decreases.

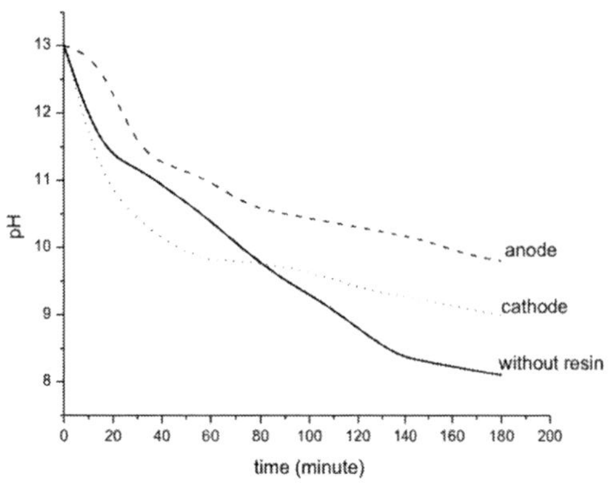

Fig.6. The solution pH for electrolysis time

In the cathode chamber, the reduction of water results OH^- ions (Eq.1.). The OH^- ions are consumed to form HA and resulted H^+ ions. The generation of OH^- ions are smaller than the consumption of OH^- ions. The solution pH in the cathode chamber decreases.

According observation of the solution pH, the electrolysis with ion exchange resin more effective than without ion exchange resin. The solution pH for electrolysis with ion exchange resin decrease more slowly than without resin. Cation exchange resin prevent OH^- ions migrate from the cathode.

Fig. 7 shows the XRD patterns of particles obtained at various the current density at 3 hrs. Hydroxyapatite was produced at the current density of 320 to 480 mA/cm^2. The peaks of hydroxyapatite phase increase in higher the current density. The higher current density, the more pure HA formed. The crystallite properties of the HA particles increase and the peaks of hydroxyapatite phase increase.

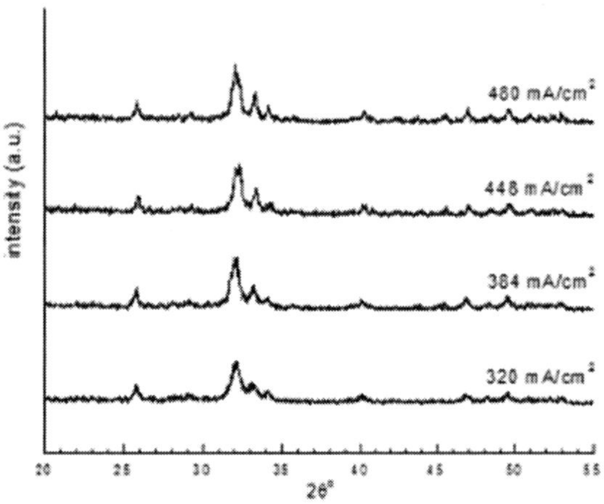

Fig.7. XRD patterns for particles prepared using various the current density

Fig.8. show the weight of precipitate at range of current density 320 to 416 mA/cm^2 for 3 hrs. At electrolysis without ion exchange resin, the weight of precipitate increase rapidly at 320 to 352 mA/cm^2. After current density 352 mA/cm^2, the weight of precipitate increase slowly and tend constant. The higher current density the more weight of precipitate. The higher current density increases the production rate of OH^- ions. The higher concentration of OH^- ions increases the production rate of precipitate.

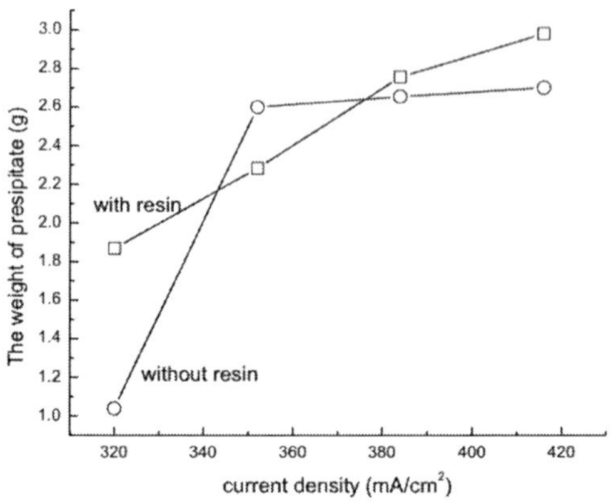

Fig.8. The weight of presipitate at various current density

At electrolysis with ion exchange resin, the weight of precipitate always increases at a range of current density 320 to 416 mA/cm^2. Generally, the weight of precipitate from the electrolysis with ion exchange resin higher than the electrolysis without ion exchange resin except at 352 mA/cm^2. According observation of the weight of precipitate, the electrolysis with ion exchange resin more effective than without ion exchange resin.

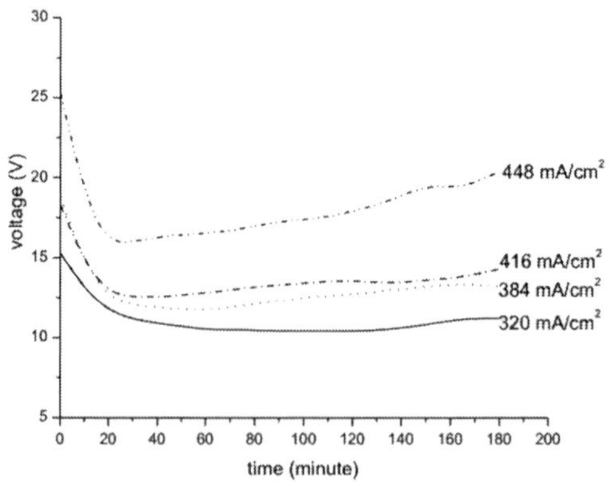

Fig.9. The applied voltage for electrolysis time

The observation of the applied voltage at various of the current density for electrolysis time was done. The result of the observation was showed in Fig.9. The higher current density the higher applied voltage. The applied voltage decreases rapidly from 0 to about 20 minutes and increases after 20 minutes. For the constant current density, the applied voltage depends on the resistance between the anode and the cathode. The higher resistance the higher the applied voltage. In the first 20 minutes, the resistance decreases cause the applied voltage decrease. After 20 minutes, the particles between the both electrodes increase and the resistance increases.

4 Conclusion

The electrosynthesis hydroxyapatite with ion exchange resin more effective than without ion exchange resin. The ion exchange resin prevent OH^- ions migrate to anion chamber. The formation of HA becomes more effective because OH^- ions just formed HA particle. There is not significant different of XRD patterns of particles at various the current density.

The authors would like to thank for KEMENRISTEKDIKTI for Hibah Fundamental 2016.

References

1. S.V. Dorozhkin, Acta Biomaterialia **6** (2010) 715-734.
2. M.A. Martins, C. Santos, M.M. Almeida, and M.E.V. Costa, Journal of colloid and interface science, **318** (2008) 210-216.
3. M. Tomozawa, and S. Hiromoto, Acta Materialia, **59** (2011) 355–363.
4. R. Xin, F, Ren, and Y. Leng, Materials and Design **31** (2010) 1691–1694
5. A. Banerjee, A. Bandyopadhyay, and S. Bose, Materials Science and Engineering C, **27** (2007) 729–735.
6. M.H. Fathi and A.Hanifi, Materials Letters, **61** (2007) 3978–3983.
7. B. Nasiri-Tabrizi, P. Honarmandi, R. Ebrahimi-Kahrizsangi, and P. Honarmandi, Materials Letters **63** (2009) 543–546.
8. M.S. Djosic, V.B. Miskovic-Stankovic, S. Milonjic, Z.M. Kacarevic-Popovic, N. Bibic, and J. Stojanovic, Materials Chemistry and Physics **111** (2008) 137-142.
9. A. Nur, D.L. Martasari, D. Nurwijayanti, S. Affandi, A. Widjaja, and H. Setyawan, Jurnal Teknik Kimia Indonesia, **11** (2013)
10. A. Nur, A. Rahmawati, N.I. Ilmi, S. Affandi, and A. Widjaja, AIP Conferensce Proceedings **1586**(86) (2014)
11. A. Nur, H. Setyawan, A. Widjaja, and I. W. Lenggoro, Bulletin of Chemical Reaction Engineering & Catalysis, **9**(3) (2014) 168-175.

A study of cement additive from varied heating temperature of coconut shell charcoal to increase cement strength

Mursyidah[1], *Novrianti*[1,*], *Adi* Novriansyah[1,2] and *Teguh* Prasetya Utama[1]

[1] Petroleum Department, Faculty of Engineering, Islamic University of Riau, 28284, Indonesia
[2] Department of Energy and Mineral Resources Engineering, Sejong University, Republic of Korea

Abstract. This experimental study introduces an organic wasted material, which hopefully can be another solution besides other expensive chemical-based additives. The study intends to investigate the effect heating temperature on charcoal powder, which is originated from the coconut-shell waste material, related to its ability in enhancing cement strength. Six samples with different additive concentration should be prepared before initiating the strength test. the strength test principally follows the bi-axial loading test procedure, where each sample will be burdened by a certain amount of the load force until failure condition was reached. The recorded pressure where the failure occurs will be treated as maximum load, which is needed for determining compressive strength and shear-bond-strength values. Those parameters should be in refer to respectable standard in the oil and gas industry. Results from the strength test clearly indicate that higher heating temperature yields to a cement with a high strength. The optimal heating temperature has reached for a sample with 700°C charcoal powder. It is a potential opportunity in increasing the economic value of the waste material.

1 Introduction

Well-cementing operation is one of the important aspects of oil and gas drilling activity [1]. The successful cementing job can prolong the oil-well lifetime, which can reduce the workover job frequency. The purpose of the well cementing operation is to adhere the casing into the formation, protected from the mechanical problem such as vibration, prevent from direct contact with the corrosive fluid, and to separate each zone behind it [2]. Failure in cementing operation will bring a serious problem in the hydrocarbon production stage. Therefore, ensuring a good quality cement bond is a serious task.

Wellbore environment leads to a precise effort in designing a cement composition to construct a proper concrete. Such properties shall be considered are fluid loss rate, Waiting-On-Cement (WOC) time, density, and its strength. Those properties can be altered by introducing the additive into the cement slurry without lowering its bond quality [3].

A good quality bonding of cement can be determined from the compressive strength (CS) and shear bond strength (SBS) parameter. CS defines as the ability of cement to resist a geological force from the formation while SBS describes the ability of the cement to bear a vertical force due casing string weight. Cement should have a certain amount of CS and SBS values where cement is strong enough to withstand a casing from the formation pressure. In oil and gas industry, the cement strength should refer to the standard from American Petroleum Institute (API).

Various studies have been conducted to improve the cement bond quality by attaching chemical additives. The compressive strength of the cement can be enhanced by adding pozzolanic material. Moreover, a chemical additive such as silicon oxide (SiO_2), Iron Oxide (Fe_2O_3) and Aluminum oxide (Al_2O_3) are effective additives to accomplish this purpose [4]. Although many studies have proven the effectiveness of chemical additive in enhancing the strength of cement, it will need an extra cost and it will affect the project viability. A low-cost material will be required for a more viable cementing operation.

Our experimental study aims to an introduction of coconut shell charcoal, we called Coconut Shell Carbon (CSC), as an additive to cement. The raw material of CSC can be found easily in Indonesia. With approximately 3.8 Ha, plantation site and produce almost 17 billion coconuts per year, Indonesia is a potential spot for developing this additive. Coconut shell is an organic material which is mostly enriched by carbon-based compounds such as cellulose, lignin, and pentose [5]. The process where the carbon was formed occurs via direct and indirect heating to the carbon-based material. The heating process can be completed in burial-klin-oven, and in open space. The minimum requirement for heating process temperature is 500°C. Works by Basuki in 2008 [6] have confirmed that CSC contains metal ions and water molecule at room condition. If CSC is exposed to heat and the temperature increase up to 100°C, the water molecule will vaporize and CSC turns out as fluid (liquid or gas) absorber. Experiment by Hawley [7]

* Corresponding author: adinovriansyah@eng.uir.ac.id

indicated that charcoal purification and tar creation occur and continue when the temperature reaches 500°C.

Less experimental studies have been investigated the effect of heating temperature to CS and SBS parameter. Work by Novriansyah [8] has approved that CSC additive can enhance cement strength. However, his works utilize CSC-Nanosilica in nanocomposite form and did not inspect the temperature effect. Thus, through this experimental study, we research the effect of heating temperature of CSC to CS and SBS parameter. An optimal heating temperature, a temperature where CSC additive yields the highest CS and SBS, can be obtained. A treatment to produce a representative CSC additive for oil-well cementing will be available.

2 Experimental Design

This experimental study use type-G Portland cement according to API standards (article 10A) [10]. Moreover, CSC in chunks form was purchased from the local industry in Pekanbaru, Indonesia. To attain the powder form of CSC, the chunks should be ground to crack it into a fine form. The fine form of CSC then is sieved with a certain size (in this case, 200 mesh) to ensure all CSC in uniform size. To observe the effect of CSC heating temperature to CS and SBS, each sample should be heated (in this case, the temperature range is 400°C to 900°C with 100°C escalation) by using oven furnace. The physical properties and the chemical composition of CS are listed in Table 1 and Table 2 respectively. From the table 2, The CSC Sample is dominantly composed of SiO_2 mineral, pozzolanic material which is commonly found in sandstone formation, a place where the hydrocarbon was accumulated.

Table 1. Physical Properties of CSC

Physical Properties	Value and Explanation
Color	Black
Specific Gravity	1.12
Density	0.208 gr./cm^3
Ash Concentration	10%
Specific Heat	1 J/g°C
Thermal Conductivity	0.2 W/(mK)
Moisture Content	4%

A hydraulic press apparatus should be prepared to conduct a biaxial loading test. This apparatus consists of a hydraulic pump which is connected to the removable steel plate. The hydraulic air pump supplies a hydraulic force which can push the steel plate. A pressure gauge is installed to apparatus for pressure reading.

Table 2. Chemical composition of CSC

Chemical Compound	Content (%)
MgO	2.5
Al_2O_3	1.2
SiO_2	45.07
P_2O_5	9.77
SO3	3.12
C	4.96
K_2O	6.59
CaO	24.2
Cr_2O_3	0.27
Fe_2O_3	1.8
CuO	0.27
ZnO	0.13
Rb_2O	0.12

2.1 Material and Apparatus

CS and SBS value are obtained from biaxial loading test by using hydraulic press apparatus. Various samples based on the cement compositions (see Table 3) should be prepared before the initiating the loading test. All needed material such as cement powder, de-ionized water, CSC powder and supporting additive were blended together by using a high-speed propeller mixer. The utilization of de-ionized water in the experimental study is due to verify that no chemical interaction between ion in water and other material. The mixer is set at a constant speed (4000 RPM) for 20 seconds before pouring all of the ingredients. The speed then is reduced 1200 RPM for 4 minutes until the homogeneous blends. Each sample is poured into the mold and dried at room condition for 24 hours to confirm all the samples are free from water. Sample cement show at figure 1 and the size of the sample at table 3.

Fig.1. Sample Cement

Table 3. Cement Sample Dimension

Temperature (°C)	Length (cm)	Width (cm)	Height (Cm)
400°C	5.14	5.1	4.7
500 °C	5.14	5.02	4.7
600 °C	5.125	5.1	4.63
700 °C	5.13	5.04	4.42
800 °C	5.13	5.025	4.84
900 °C	5.12	5.1	4.54

Before conducting the biaxial loading test, all of the samples should be removed from the mold after the drying process is completed. A hydraulic press apparatus should be ensured at the standard condition to prevent the error reading when the test is being performed. A certain amount of load will be imposed to the sample by applying pressure from the hydraulic air pump. The pressured air will push the steel plate downwards, and the sample will be compressed until the condition where the failure was achieved. The compressive strength and shear bond strength of the sample can be calculated by dividing the recorded load from the pressure gauge with the sample's cross-sectional area. A Sequential analysis consisting the Scanning electron microscope (SEM) to image the effect of the nanocomposite on cement surface structure, energy dispersive X-ray spectroscopy (EDX) to quantify the chemical compound of cement, and X-Ray Diffraction (XRD) measurements were performed for each slurry sample.

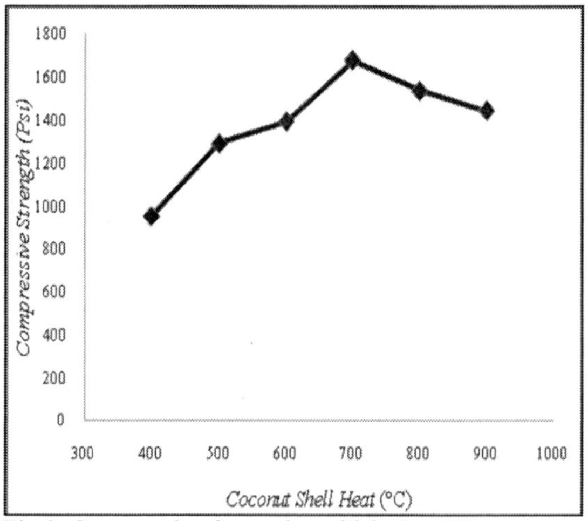

Fig.2. Compressive Strength vs CSC Temperature

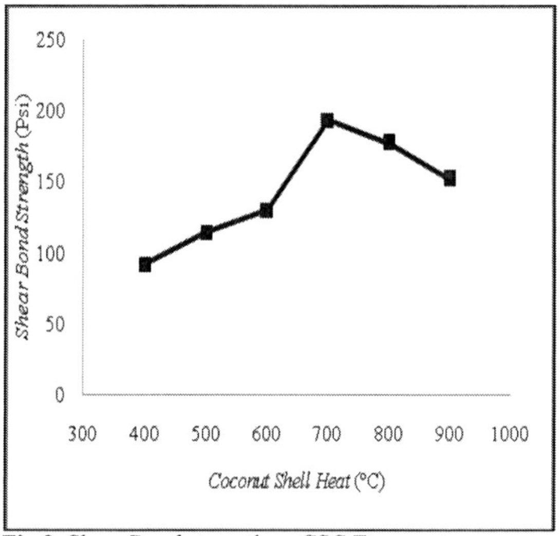

Fig.3. Shear Bond strength vs CSC Temperature

3 Result and discussion

Figure 2 and 3 depicts the plot of CS and SBS for each sample respectively. It is clearly shown in Figure 1 and 2 that CS and SBS indicates a respectable value for oil-well cementing, according to the API standard (article 10A). An increment trend of CS has been revealed when the heating temperature is going up. The CS value rises to around 1675 psi when the heating temperature increased up to 700°C (more than 80% increment, compare to the sample with 400°C heating temperature.A similar phenomenon occurs when investigating the effect of heating temperature on SBS value. The highest SBS value is achieved for the sample with CSC temperature is 700°C, where the increment reaches 100%, compare to the sample with the lowest SBS value in the experiment.

This experimental study has efficaciously proven the effectiveness of CSC additive to enhance the cement strength. Results from the experiment are consistent with Hawley experiment where the CSC undergoes a charcoal purification when the temperature more than 500°C. The CS and SBS are reduced if the heating temperature is higher than 700°C. This is due to the existence of hydrogen gas in the CSC. It reduces the bonding force between cement and CSC. The result from the XRD measurement is displayed in figure 4 for S4 sample. By applying crystal interpellation, the crystal content in the sample with 700°C additive is approximately 66%. Higher crystal content will enhance the inter-particle bonding of cement. The Imaging process from SEM Analysis has demonstrated that cement and CSC have uniformly bonded and distributed (see the blue circle). The result from the SEM measurement is displayed in figure 5 for S4 sample.

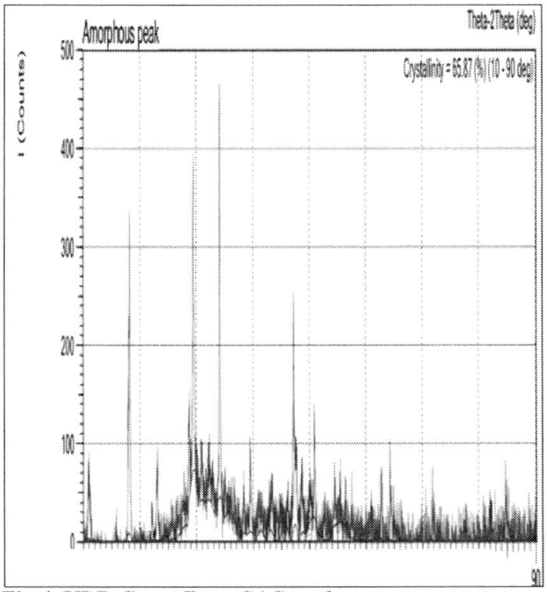

Fig.4. XRD Curve From S4 Sample

Fig.5. Surface Structure from S4 Sample

Comparing the chemical composition of the cement sample (Table 4) has shown that CSC additive can increase silica and aluminum content in the cement. The silica content has shown 113% incremental in form of SiO_2. The similar phenomenon occurs for Aluminum where the content increases up to 90% in the form of Al_2O_3. Other chemical compounds in oxide form show the same trend with the mentioned compound before. However, the contribution is not significant due to the minor content in the cement sample. Nevertheless, the existence of the free oxygen in the cement is not detected in the sample with CSC additive. It is possible because all of the free oxygen will react with those reactive chemical group additive to form an oxide-based compound. Moreover, this existence of the oxide compound can enhance the inter-particle bond of the cement, resulting in a cement with higher strength.

Table 4. Chemical compounds in cement slurry

Element	**Compound Mass (%)**	**Mol (%)**	**Compound**	**Compound Mass (%)**	***Cation***
C	22,06	56,88	C	22,06	0
O	28,38	-	-	-	-
Mg	0,73	0,97	MgO	1,21	0,39
Al	2,22	1,33	Al_2O_3	4,19	1,07
Si	10,35	11,95	SiO_2	22,13	4,81
S	1,02	1,04	SO_3	2,56	0,42
Cl	0,45	0,42	Cl	0,45	0
K	0,74	0,31	K_2O	0,89	0,25
Ca	31,90	25,82	CaO	44,63	10,40
Ti	0,27	0,18	TiO_2	0,45	0,07
Fe	1,89	1,1	FeO	2,43	0,44
Total	**100**	**100**	**Total**	**100**	**17,86**

4 Conclusion

In this study, the charcoal form of the coconut shell has been implemented to enhance the cement strength. Several experiments have been conducted to obtain a representative combination of compressive strength and shear bond strength value. From the experiments, the optimum concentration of coconut shell carbon with 700°C charcoal combustion temperature respectively. Comparison from XRD test, the crystallization value is around 66%. The coconut shell carbon with 700°C charcoal combustion temperature will yield the highest compressive strength value (1676.433 psi) and 192.876 psi for shear bond strength value. Comparison from SEM the cement and CSC have uniformly bonded and distributed.

References

[1] V. Ershadi., et al. J. Environ. Sci. Develop **2:** 128-132 (2011).
[2] A.T Burgoyne, Applied Drilling Engineering, SPE, USA, (1986).
[3] E. B. Nelson, (Ed).Well cementing. Newnes. (1990)
[4] A. Buntoro & R. Rubiandini. Proceeding of The 5thInaga Annual Scientific Conference And Exhibition Yogyakarta(2001).
[5] K.A Widiatama , L. Nuriyah L., D.J. Djoko Herry Santojo., Istiroyah., Masruroh. results.ipi.191833.
[6] K.T. Basuki, Setiawan Budi, Seminar Nasional IV SDM Teknology Nuklir, Yogyakarta (25-26 Agustus 2008).

[7] A. Novriansyah, Mursyidah, S.S.A Putri, Novrianti, and W.S. Bae. *IJAMCE* Vol.**3**, Issue-**2** : 58-61 (2016)
[8] API, RP. 10A. Specification for Cements and Materials for Well Cementing.Twenty- (2002)
[9] Backe, K. R., O. B. Lile, and S. K. Lyomov. Society of Petroleum Engineers (1998)
[10] Kazemian, Sina, Bujang BK Huat, and Maassoumeh Barghchi. Int. J. Phys. Sci 6.**4**: 707-713. (2011)
[11] Kakumoto Masayo, Yoneda Jon, Tenma Nario, Miyazaki Kuniyuki, Aoki Kazuo. Proceeding of the twenty second Intrnational Offshore and Polar Engineering Conference Rhodes, Greece, June 17-22, 77-82 (2002)
[12] Pradipta Imam,. Annual Convention And Exhibition, Indonesian Petroleum Association (2010)
[13] Subhash N. Shah, Ph.D., P.E., Narayan H. Shanker, and Chinenye C. OgugbueWell Construction Technology Center. University of Oklahoma. (2010)
[14] Rubiandini Rudi. New Additive for Improving Shearbond Strenght in High Temperature and Presure cement. IADC/SPE 62750. 1-18
[15] Rubiandini Rudi, Siregar Septoratno, Nursuharcahyo, Efrial Deny. Proc. ITB eng science Vol 37B NO.**1** :29-48. (2005)
[16] Ridha, Syahrir, Sonny Irawan, and Bambang Ariwahjoedi. Journal of Petroleum Exploration and Production Technology 3.**4** :303-311.(2013)
[17] Widjaja A.P, Darjo, S. Making charcoal active with dry destilation II. Komunikasi Balai Penelitian Kimia Bogor. (1980)

The simple fabrication of nanorods mass production for the dye-sensitized solar cell

Dharmanto[1,*], *Hendi* Saryanto[2], and *Darwin* Sebayang[3]

[1,2] Mechanical Engineering Department, Polytechnic of Gajah Tunggal, 15135 Tangerang, Indonesia
[3] Mechanical Engineering Department, Faculty of Engineering, Mercu Buana University, 11650 Jakarta, Indonesia

Abstract. ZnO nanorods were successfully synthesized by the plasma torch with the zinc powder as the source of ZnO. Several testing was conducted to examine the results of ZnO nanoparticles among others are X-ray fluorescence (XRF), Scanning Electron Microscopy (SEM), and X-ray diffraction (XRD). The results of ZnO nanorods diameter vary from 87 nm to 263 nm which can be seen from SEM images and it is length varies from 300 to 1000 nm. It was found from XRD data that a sharp peak occured at 36.253°. It is indicated a good crystal growth and agreed well with the standard data (JCPDF-ICDD card no.: 36-1451). Effects of electrical current variations of plasma torch of 20, 25, and 30 Amperes on the size of ZnO nanorods are indicated from aspect ratio (72.85, 87.42, and 103 nm). The effects of electrical current of plasma torch on the purity of ZnO were about 95.61%, 98.46%, and 96.49% respectively. The performance of the solar cell at the time with the values of Voc, Jsc, FF, and efficiency are 0.466 V, 1.524 mA/cm ^{2}, 51.95%, and 0.37% respectively. ZnO nanorods were successfully synthesized by the plasma torch with the zinc powder as the source of ZnO.

1 Introduction

Dye sensitized solar cells (DSSC) is one application of energy harvesters with low operational cost. Zinc oxide with a similar band gap and comparable electron injection process to that of TiO_2, has been proven to be one of the most promising alternatives photoelectrode materials for application in DSSC. Compared to ZnO or TiO_2 nanoparticle films, single crystals of these materials exhibit significantly faster electron transport and greater electron mobility. The faster electron transport is a result of the high electron diffusion coefficients, which will provide significant advantages to device performance [1,2].

The greater electron mobility is due to the direction and uninterrupted conduction channel, which can improve the efficiency of charge collection and enable the production of optically thick cells which absorb more incident light [3,4]. In spite of the improved charge transport and the larger electron mobility, one-dimensional (1D) ZnO nanowires show a smaller total surface area for nanowire arrays, as compared to the best TiO_2 nanoparticle devices, which adsorb less dye per unit thickness of the cell and thus absorb less light and collect less charge. As a result, the energy conversion efficiency of ZnO nanowire-based DSSC is typically lower [1,5].

To encourage the application of solar cells in real time, various studies have been directed to look for ways to improve the performance of solar cells DSSC. Strategies for reaching the larger currents from solar cells is to simplify and streamline the movement of electrons which of them is to use a semi-conductor single crystalline nanowires, nanofibers, arrays of nanotubes, and core-shell structure [6].

Due to their diverse application in optical, electrical, optoelectronic, photo catalytic, hydrophilic, hydrophobic, pigments, metal compounds, medical ointments and cosmetics, nanometric ZnO such as nanospheres, nanoplates, nanosheets, nanoboxes, nanomallets, nanotripods, nanobelts, nanosprings, nanorings, nanocages, nanoneedles, nanorods, nanotubes, nanopropellers, nanoflowers and nanowindmills have been the attractive research topics in nano science and technology for few decades [7-18].

Various ZnO nanostructures have been extensively investigated for DSSC. In the early reports on ZnO-based DSSC, ZnO nanoparticles were often used as the photoanode prepared by a conventional doctor blade technique [56]. The experimental results of a new solar cell fabricated from an electron-accepting layer of freestanding ZnO nanowires [57]. The nanowire DSSC is an exciting variety of the most successful of the excitonic photovoltaic devices [58]. As an ordered topology that increases the rate of electron transport, a nanowire electrode may provide a means to improve the quantum efficiency of DSSC in the red region of the spectrum, where their performance is currently limited. Raising the efficiency of the nanowire cell to a competitive level depends on achieving higher dye loadings through an increase in surface area [59].

* Corresponding author: dharmanto.mesin@gmail.com

Recently, various techniques have been proposed for ZnO nanoparticles fabrication. It can be classified into either physical or chemical methods [19,20] such as thermal hydrolysis [21], hydrothermal processing [22], sol-gel [23-24], vapor condensation [25], spray pyrolysis [26-27], pulse laser decomposition [28], laser ablation [29], thermal evaporation [30.31], pulse combustion – spray pyrolysis [32], electro-mechanical [33], flame spray pyrolysis [34], direct precipitation [35] and thermal plasma [36,37,38]. More specific, the synthesis of ZnO nanorods has also been performed using various techniques such as hybrid wet chemical route [39], solution process at low temperature [40], physical evaporation [41-43], electrophoretic deposition [44], radio frequency (RF) magnetron sputtering [45], templating against anodic alumina membrane [46] and PTFE capillary tube reaction [47]. Nonetheless, it is important to note that the efficiency and size control still become major research problems in synthesizing ZnO nanoparticles [38].

2 Methodology

ZnO nanorods were successfully synthesized by the plasma torch reactor, as shown in Fig.1 which was operated at atmospheric pressure and less than 3 KW input plasma power. It can be seen that the reactor composed of five main parts among others is a feeder screw as zinc feeder, a plasma torch reactor as main reaction place, a DC plasma as a plasma power source, a filter and a suction blower. Commercial zinc powders from MERCK made in Germany. It has an average particle size of about 45μm was employed in this study. The impurity of raw material has been checked using X-Ray Fluorescence (XRF– PANalytical-Minipal QC) giving results of 98.9% purity of zinc and containing impurities such as Ca, Cr, Fe, and Ni.

A screw conveyor fed zinc powder into the plasma torch reactor at the fix rate of 2 g/min with gas carrier. These having entered the plasma torch reactor then faced plasma zone, then vaporization and oxidation were experienced by the zinc powders immediately forming ZnO nanoparticles.

The time elapsed to form ZnO nanoparticles under this process take about 0.01 s [36]. The results of vapors from burning process were then dropped to silo, then sucked by blower passed through the filter tube so that ZnO powders could be seized by filter membrane. Electrical currents inputted to the plasma poles (cathode and anode) were varied from 20, 25, and 30 Ampere .

Several testing was conducted to examine the results of ZnO nanoparticles among others, are X-ray fluorescence (XRF-PANalytical-Minipal QC), X-ray diffraction (XRD-PHILLIP-XPERT PRO) and Scanning Electron Microscopy (SEM-FEI-Inspect S50), To ensure the purity of synthesized rod-like ZnO nanoparticles, XRF (PANalytical-Minipal QC) operated at 20 kV was used as qualitative and quantitative elemental analysis.

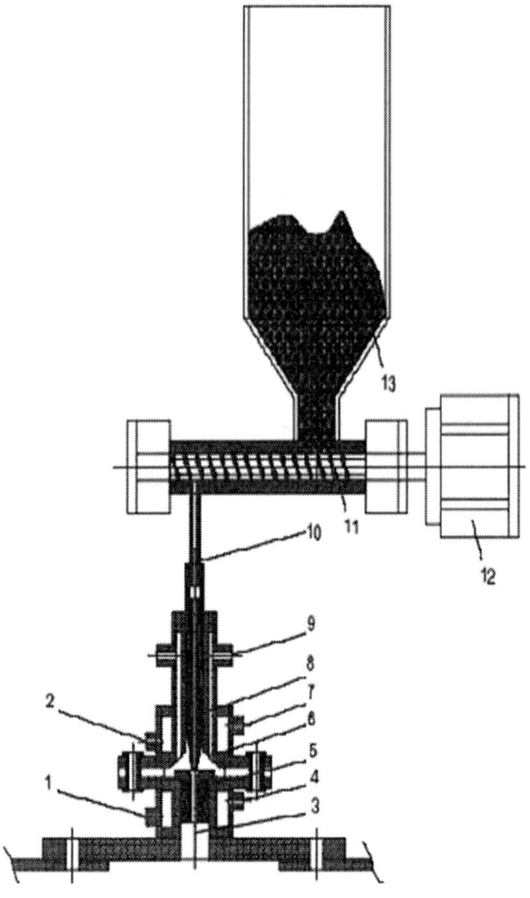

NO	DESCRIPTION	MATERIAL
1	WATER INLET	SUS304
2	WATER INLET	SUS304
3	PLASMA OUTPUT	
4	WATER OUTLET	SUS304
5	BOTTOM GRAPHITE	GRAPHITE
6	COMPRESSION AIR	SUS304
7	WATER OUTLET	SUS304
8	TOP GRAPHITE	GRAPHITE
9	COMPRESSION AIR INLET	SUS304
10	ZINC POWDER INLET	SUS304
11	FEEDER SCREW	SUS304
12	STEPPER MOTOR	
13	ZINC POWDER SILO	SUS304

Fig. 1. The Simple Plasma Torch Machine Design.

In case of crystallite size and phase identification, an examination of the as-prepared ZnO nanorods involved X-ray diffraction analysis by the 2-2θ method. XRD (PHILLIP-XPERT PRO) was equipped with Cu Ka (λ=0.154060nm) radiation at 40 kV and 30 mA. For each step, the scanning step size and collection time were set at 0.02 and 0.5 respectively.

Scanning electron microscopy (FEI-Inspect S50) operated at 20 kV was used to capture the morphology of the synthesized ZnO nanorods.

The DSSC consisted of a conducting glass, the ZnO, dye-absorbing electrode, the electrolyte, and a counter-electrode layer. The fluorine-doped tin oxide (FTO), used as the conducting glass substrate, was prepared using spray pyrolysis [50]. A solution of tin (II) chloride dihydrate (16.12 g, $Cl_2Sn \cdot 2H_2O$, Merck), ammonium fluoride (0.265 g, NH_2F, Merck), and ethyl alcohol (100 ml, 96%, Merck) of 100 ml was mixed and sprayed on the glass surface, followed by heating at 450°C [55]. To prepare the catalytic platinum counter electrode, the FTO glass was first heated at 200°C. Platinum was then chemically deposited from a 5 mM H_2PtCl_6 solution (Merck) onto the FTO glass, followed by heating at 300°C and subsequent cooling [51].

The fabrication of the nanorods ZnO working electrode, a homogenous paste containing nanorods ZnO semiconductor (0.5 g) and ethyl alcohol (0.4 mL, 96%, Merck) was prepared. The paste was then cast onto the FTO glass (1cm × 3 cm) using the doctor-blade method. For a strong bonding between the ZnO film nanorods and the FTO glass [52, 53], the ZnO paste-coated FTO glass was sintered at 450°C for 2 h and subsequently cooled. Afterwards, the ZnO working electrodes were immersed in the respective dye solutions for 24 h at room temperature, prior to the DSSC assembly [53, 54].

The electrolyte solutions for the DSSC were prepared as follows: sodium iodide (3.3 g, 99.9%, Merck) and acetonitrile (30 ml) were dissolved by stirring for 20 minutes, prior to adding iodine (524 g, 99.9%, Merck). The stirring was continued for 24 h to Achieve a homogeneous solution. The electrolyte solution was injected into the assembled DSSC spaced at 50-µm intervals and then sealed. [55]

In this research will be testing the IV characterization tests on solar cells DSSC. The testing uses Xenon lamps under illumination by a lamp of 1000 W/m^2 and solar cell characterization test equipment. For characterization testing of solar cells DSSC several parameters such as J_{sc} Short Circuit Current, Open Circuit Voltage-V_{oc}, and Fill Factor-FF and Efficiency of solar cell DSSC (η) will be tested. This efficiency value which becomes the main reference in determining the quality of performance of a solar cell.

3 Results and Discussion

The diameter of the results of ZnO nanorods varies from 87 nm to 263 nm which can be seen from SEM images with length varies from 300 nm to 1000 nm. Fig.2, Fig.3, Fig.4 shows the morphology and size of the ZnO nanorods. The size of ZnO nanorod range 72 nm - 103 nm. Effects of electrical current variations of plasma torch of 20, 25, and 30 Ampere to the size of ZnO nanorods are also indicated from the aspect ratio of about 72.85, 87.42, and 103 nm.

Table 1 shows the other effects of electrical current variations of plasma torch to the purity of ZnO about 95.61%, 98.46%, and 96.49% respectively.

Table 1. Compound effects of electrical current variations of plasma torch of 20, 25, and 30 Amperes to the purity of ZnO

Ampere	Compound (%)				
	ZnO	CaO	Cr_2O_3	Fe_2O_3	NiO
20	95.61	0.24	0.11	0.15	0.082
25	98.46	0.14	0.098	0.16	0.086
30	96.49	0.55	0.13	0.19	0.13

Fig. 2. Photo of SEM ZnO are formed from zinc powder with the 20 ampere plasma torch process.

Fig. 3. Photo of SEM ZnO are formed from zinc powder with the 25 ampere plasma torch process.

Fig. 4. Photo of SEM ZnO are formed from zinc powder with the 30 ampere plasma torch process

The XRD pattern of resulted ZnO nanorods is performed in Fig.5. In this study, each crystalline structure of ZnO nanorods is a perfect crystal with the growth direction of *c* axis. Three XRD patterns are indexed as hexagonal wurtzite structure of ZnO having space group of P63mc (186) with lattice constant sets of a,b=3.24982 Å and c= 5.20661 Å, which is consistent with the values in the database of JCPDF-ICDD card no.: 36-1451 [49]. Nine peaks appear at $2\Theta = 31.768°$, 34.421°, 36.253°, 47.539°, 56.594°, 62.856°, 66.375°, 67.947°, 69.085°, respective to miller indices of (100), (002), (101), (102), (110), (103), (200), (112), and (201).

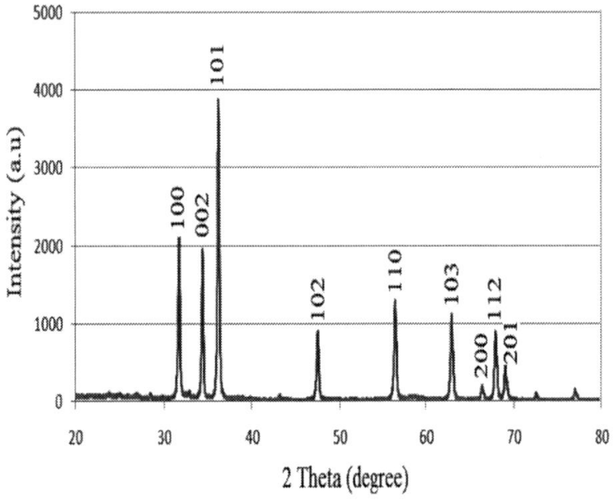

Fig. 5. XRD patterns of ZnO nanorods

Moreover, the strong and narrow intensity diffraction peaks imply the good crystalline nature and size of the synthesized products. Average nano-crystalline sizes (D) obtained from the broadening of XRD peaks are calculated using Scherer's formula as follows [9]:

$$D = \frac{0.94\,\lambda}{B.\cos\theta} \quad (1)$$

Where D is the crystal size, B is the broadening from the sample or its full width half medium (FWHM), λ is the wave length of X-ray, Θ is the Bragg's angle. Calculation results using Equation (1) are listed in Table II.

Table 2. FWHM and crystalline size of ZnO nanorods.

Miller index	2 theta (deg)	FWHM	Crystalline size (nm)
(100)	31.768	0.227	31.7406
(002)	34.421	0.191	37.9835
(101)	36.253	0.226	32.2650

The XRD data shows a sharp peak at 36.253° which indicates a good crystal growth and agrees well with the standard data (JCPDF-ICDD card no.: 36-1451) [49].

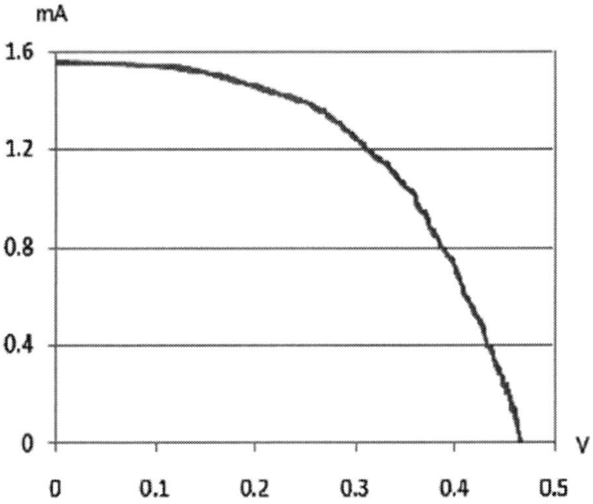

Fig. 6. *J- V* curves of ZnO nanorods

In Fig.6, it has shown the performance of the solar cell at the time with the values of Voc, Jsc, FF, and efficiency generated are respectively 0.466 V, 1.524 mA/cm^2, 51.95%, and 0.37%.

4 Conclusion

ZnO nanorods were successfully synthesized by the plasma torch with the zinc powder as the source of ZnO. ZnO nanorods have a high purity ZnO composition. The plasma torch is a simple fabrication of ZnO nanorods mass production of dye sensitized solar cell. The advantages of this method are simplicity and potential use for future large scale preparation of ZnO nanorods, which is useful for many source of energy applications.

References

1. M. Law, L. Greene, J. Johnson, R. Saykally, P. Yang, Nat. Mater. **4** 455–459, (2005)
2. Y. Hsu, Y. Xi, A. Djuri si, W. Chan, Appl. Phys. Lett. **92** 133507, (2008)
3. V. Noack, H. Weller, A. Eychmuller, J. Phys. Chem. B **106** 8514–8523, (2002)
4. D. Thomas, J. Lander, J. Chem. Phys. **25** 1136–1142, (1956)
5. M. Law, L. Greene, J. Johnson, R. Saykally, P. Yang, Nat. Mater. **4** 455–459, (2005)
6. Kim, I.-D., Hong, J.-M., Lee, B.H., Kim, D.Y., dan Jeon, E.-K., Appl. Phys. Lett, **91** (16) (2007)
7. K. Kakiuchi, E. Hosono, T. Kimura, H. Imai, S. Fujihara, J. Sol-Gel Sci. Technol. **39** 63-72. (2006)
8. S. Mahmud, M.J. Abdullah, IEEE Conf. Emerging Technol.-Nanoelectron 442-446 (2006)
9. L. Shen, H. Zhang, S. Guo, Mater. Chem. Phys. **114** 580-583. (2009)
10. Y. Ding, Z.L. Wang, Micron **40** 335-342. (2009)
11. Z.L. Wang, J. Phys. Condens. Mater. **16** R829-R858. (2004)
12. W. Yu, C. Pan, Mater. Chem. Phys. **115** 74-79. (2009)
13. M.N.R. Ashfold, R.P. Doherty, N.G. Ndifor-Angwafor, D.J. Riley, Y. Sun, Thin Solid Films **515** 8679-8683. (2007)
14. U. Ozgur, I.A. Ya, C. Liu, A. Teke, M.A. Reshchikov,S. Dogan , V. Avrutin, S.J. Cho, H. Morkoc, J. Appl. Phys. **98** 041301. (2005)
15. J. Xie, P. Li, Y. Li, Y. Wang, Y. Wei, Mater. Chem. Phys. **114** 943-947. (2009)
16. M.C. Newton, P.A. Warburton, Mater. Today **10** 50-54. (2007)
17. C. Klingshirn, Phys. Status Solidi B **244** 3027-3073. (2007)
18. L. Yu, F. Qu, X. Wu, Applied Surface Science **257** 7432-7435. (2011)
19. H.-F. Lin, S.-C. Liao, S.-W. Hung, J. Photochemistry and Photobiology A: Chemistry **174** 82-87. (2005)
20. B. Cheng, W. Shi, J. M. Russell-Tanner, L. Zhang, E. T. Samulski, Inorg. Chem. **45** 1208-1214. (2006)
21. H.K. Park, D.K. Kim, C.H. Kim, J. Am. Ceram. Soc. **80** 743. (1997)
22. S.I. Hirano, Ceram. Bull. **66** 1342. (1987)
23. D. Vorkapic, T. Matsoukas, J. Am. Ceram. Soc. **81**. 2815. (1998)
24. Y.X. Li, K.J. Klabunde, Chem. Mater. **4** 611. (1992)
25. V.R. Palkar, Nanostruct. Mater. **11** 369. (1999)
26. C.G. Granqvist, R.A. Burhman, J. Appl. Phys. **47** 2200. (1976)
27. T.T. Kodas, Adv. Mater. **6** 180. (1989)
28. G.L. Messing, S.C. Zhang, G.V. Jayanthi, J. Am. Ceram. Soc. **76** 2707. (1993)
29. J.P. Pollinger, G.I. Messing, J. Aerosol Sci. Technol. **19** 217. (1993)
30. R.D. Vispute, V. Talyansky, S. Choopun, R.P. Sharma, T. Venkatesan, M. He, X. Tang, J.B. Halpern, M.G. Spencer, Y.X. Li, L.G. Salanmanca Riba, A.A. Iliadis, K.A. Jones, Appl. Phys. Lett. **73** 348. (1998)
31. M.A. Gondal, Z.H. Yamani, Q.A. Drmosh, A. Rashid, Int. J. of Nanoparticles. **2** no. 1/2/3/4/5/6 119-128. (2009)
32. K. Triphati, M. Husain, N.A. Salah, S.S. Habib, S.M.A. El-Hamidy, N.Z. Zahed, Z.H. Khan, Int. J. of Nanoparticles **2** no. 1/2/3/4/5/6 148-155. (2009)
33. S.G. Hussain, D. Liu, X. Huang, K.M. Sulieman, S.T. Rasool, Int. J. of Nanoparticles **2** no. 1/2/3/4/5/6 443-450. (2009)
34. I.M. Joni, A. Purwanto, F. Iskandar, M. Hazata, K. Okuyama, Chemical Engineering J. **155** 433-441. (2009)
35. Y. Hames, Z. Alpaslan, A. Kosemen, S.E. San, Y. Yerli, Solar Energy **84** 426-431. (2010)
36. H.-F Lin, S.-C Liao, C.-T Hu, J. Crystal Growth **311** 1378-1384. (2009)
37. H.-F. Lin, S.-C. Liao, S.-W. Hung, J. Photochemistry and Photobiology A: Chemistry **174** 82-87. (2005)
38. T.S. Ko, S. Yang, H.C. Hsu, C.P. Chu, H.F. Lin, S.C. Liao, T.C. Lu, H.C. Kuo, W.F. Hsieh, S.C. Wang, Materials Sience and Engineering B **134** 54-58. (2006)
39. R.N. Gayen, S. Dalui, A. Rajaram, A.K. Pal, Applied Surface Science **255** 4902-4906. (2009)
40. H.-S. Choi, M. Vaseem, S.G. Kim, Y.-H. Im, Y.-B. Hahn, J. Solid State Chemistry **189** 25-31. (2012)
41. L.-Y. Chen, S.-H. Wu, Y.-T. Yin, J. Phys. Chem. C **113** 21572-21576 (2009)
42. D.P. Singh, Sci. Adv. Mater. **2** 245-272. (2010)
43. M.E. Fragala, Y. Aleeva, C. Satriano, J. Nanosci. Nanotechnol. **11** 8180- 8014. (2011)
44. J. Cui, J. Phys. Chem. C. **112** (2008) 10385-10388.
45. C. Li. G. Fang, J. Li, L. Ai, B. Dong, X. Zhao, J. Phys. Chem. C **112** 990-995. (2008)
46. J.-T. Yan, C.-H. Chen, S.-F. Yen, C.-T. Lee, IEEE Photonics Technol. Lett. **22**. 146-148. (2010)
47. S. He, Y. Liu, M. Uehara, H. Maeda, Material Science and Engineering B **137** 295-298. (2007)
48. T.S. Ko, S. Yang, H.C. Hsu, C.P. Chu, H.F. Lin, S.C. Liao, T.C. Lu, H.C. Kuo, W.F. Hsieh, S.C. Wang, Materials Sience and Engineering B **134** 54-58. (2006)
49. R. Verma, B. mantra, Ramphal, A. K. Srivastava, Advanced Materials latters **326** 324-333. (2015)
50. A.V. Moholkar, S.M. Pawar, K.Y. Rajpure, C.H. Bhosale, J.H. Kim, Appl. Surf. Sci., **255** 9358–9364, (2009)
51. P. Calandra, G. Calogero, A. Sinopoli, P.G. Gucciardi, Int. J. Photoenergy, 2010 1-15, (2010)
52. H. Chang, Y.J. Lo, Sol. Energy, **84** 1833–1837, (2010)
53. H. Chang, H.M. Wu, T.L. Chen, K.D. Huang, C.S. Jwo, Y.J. Lo, J. Alloys Compd., **495** 606–610, (2010)
54. K. Wongcharee, V. Meeyoo, S. Chavadej, Sol. Energy Mater. Sol. Cells, **91** 566-571, (2007)
55. Suyitno,T.J. Saputra, A. Purwanto, Z.Arifin, S1386-1425(15)00422-9 (2015)

56. Keis, K.; Bauer, C.; Boschloo, G.; Hagfeldt, A.; Westermark, K.; Rensmob, H. & Siegbahn, H. J. Photochemistry and Photobiology A: Chemistry, **148**, No. 1-3, pp. 57–64, (2002),
57. [Lévy-Clément, C.; Tena-Zaera, R.; Ryan, M. A.; Katty, A. & Hodes, G. Advanced Materials **17**, No.12,pp.1512-1515, ,(2005),
58. Law, M.; Greene, L. E.; Johnson, J. C.; Saykally, R. & Yang, P. (2005). *Nature Materials* **4**, No. 6, , pp. 455-459, (2005)
59. Law, M.; Greene, L. E.; Radenovic, A.; Kuykendall, T.; Liphardt, J. & Yang, P. J. Physical Chemistry B**110**, No. 45, pp. 22652 22663, (2006)

The design of high-temperature thermal conductivity measurements apparatus for thin sample size

Syamsul Hadi[1,*], *Fama* Aqiftiar Falah[1], *Agus* Kurniawan[2], and *Suyitno*[1]

[1]Graduate School of Mechanical Engineering Department, SebelasMaret University, Surakarta, Indonesia
[2]Industrial Machine Engineering Department Polytechnic ATMI, Surakarta, Central Java, Indonesia

Abstract. This study presents the designing, constructing and validating processes of thermal conductivity apparatus using steady-state heat-transfer techniques with the capability of testing a material at high temperatures. This design is an improvement from ASTM D5470 standard where meter-bars with the equal cross-sectional area were used to extrapolate surface temperature and measure heat transfer across a sample. There were two meter-bars in apparatus where each was placed three thermocouples. This Apparatus using a heater with a power of 1,000 watts, and cooling water to stable condition. The pressure applied was 3.4 MPa at the cross-sectional area of 113.09 mm2 meter-bar and thermal grease to minimized interfacial thermal contact resistance. To determine the performance, the validating process proceeded by comparing the results with thermal conductivity obtained by THB 500 made by LINSEIS. The tests showed the thermal conductivity of the stainless steel and bronze are 15.28 Wm-1K-1 and 38.01 Wm-1K-1 with a difference of test apparatus THB 500 are -2.55% and 2.49%. Furthermore, this apparatus has the capability to measure the thermal conductivity of the material to a temperature of 400 ° C where the results for the thermal conductivity of stainless steel is 19.21 Wm-1K-1 and the difference was 7.93%.

1 Introduction

Thermal conductivity is one of the important properties of the material and commonly is depend on the temperature especially in high-temperature condition [1]. The differences of thermal conductivity usually could be ignored and be measured based on the average temperature because the analysis will be complicated if the differences of temperature are involved in the analysis [2]. However, for certain materials particularly for thermoelectric material, the differences of thermal conductivity should not be ignored because thermoelectric operates at high temperature and the Seebeck number as the main parameter of its performance depends on thermal conductivity value [3]. Based on these facts, measurements of thermal conductivity in high temperature need special apparatus in order to obtain an accurate result.

Several methods for measuring thermal conductivity have already proposed such as steady state method using guarded hot plate (GHP) that it shows accuracy about 3% in deviation standard [4] and guarded cut-bar technique which measured thermal conductivity for the *stainless steel* 304 (SS304) was 19.17 $Wm^{-1}K^{-1}$ [5]. Another research using one dimension conduction steady state method and thermal jacket as a protector to minimize heat loss to the ambient was already reported [6]. Other efforts were proposed in collecting, monitoring, and processing data using acquisition Analog to Digital Converter (ADC) Advantech (USM-4718) to the computer [7] and microcontroller ATMEGA16 [8]. All previous efforts are focused on improving new method in order to obtain more accurate measuring apparatus.

In the steady state method for the big samples thermocouples were bored in samples, but for the small samples, this method was not available. A commercial method of measuring thermal conductivity for thin samples usually is a transient method of laser flash diffusivity technique, but this method needs additional test in measuring process. Also, this method has a limitation in high thermal conductivity material and high-temperature measurements [9]. Other proposed thermal conductivity measurements have proposed using point probe method [10] and stamp sensor method [11, 12, 13] that has the capability in in-situ measurements, but this method also has a limitation in high-temperature measurements. The common method in high-temperature thermal conductivity measurements is *axial heat flow* based on the standard of ASTM E1225-99 [14] which use a bored *thermocouple* in the specimen [15,16]. Unfortunately, the last method could not measure for a thermoelectric method which usually thins form and fragile in the machining process.

In this research thermal conductivity measurements, apparatus will be designed and manufactured using axial heat flow steady-state method for thin materials without a boring process to attach thermocouples and capable in various diameter samples. The validity of this apparatus was measured using Stainless Steel and Bronze materials

* Corresponding author: syamsulhadi@ft.uns.ac.id

using ADAM 4018 as data acquisition. Further, the apparatus will be used to measure the thermal conductivity of thermoelectric semiconductor sample until the temperature of 400°C.

2 Research Method

For the first design of thermal conductivity, measurements apparatus was built by using Solidworks computer simulations program. Several designs and applied materials are applied in order to obtain the best solution for one-dimensional thermal transfer, minimal power supply, minimal heat loss, ease of disassembly process, and steady state condition. The design of the apparatus in detail was shown in the Figures 1.

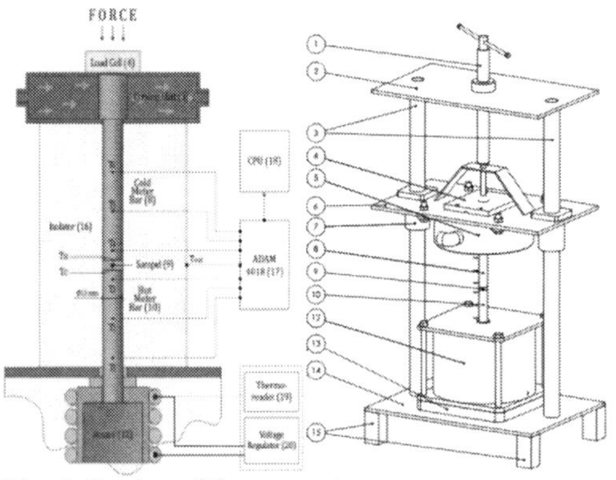

Fig. 1. Design of the apparatus

The frame of apparatus consists of a top base plate (2), middle base plate (6), bottom base plate (14) from mild steel material with a size of 300×180×5 mm and vertical rod (3) from stainless steel material with a height of 500 mm and diameter of 20 mm. In the top base plate, there is a lever (1) giving the pressure to the sample during measurements process. The lever press the load cell (4) then the load cell continuing to the middle base plate which has the capability to move in up and down movement using sliding bearing linear (7) with a vertical rod. In the lower part of the middle base plate, there is a cooling unit (5). The cooling cold meter bar (8) on the one side using water as a cooling medium which circulates by a pump stabilizing the flow rate of the water so the measurements process in the steady state condition. In the bottom of the base plate, a set heater (12) we put and there is an isolator (13) between them avoiding the temperature at the lower part of the base plate too high. This apparatus was designed to be an apparatus which able to measure the sample in high temperature, so the heater element as the heating source of hot meter bar (10) was applied and placed on the lower side apparatus and controlled by a voltage regulator (20). The temperature of the heater was controlled by thermo-reader (19) from box panel. In the space between cold meter bar and hot meter bar, the specimen (9) with a diameter of 13 mm was placed. In each meter bar, there are three thermocouples (11) reading the gradient of temperature from hot meter bar to cold meter bar. Meter bar and specimens were veiled by isolator (16) in order to keep the system in the adiabatic condition. K-type thermocouples were used, and collected data from thermocouples were read by Analog to Digital Converter Advantech ADAM 4018 (17) and were transferred to the computer (18). Furthermore, this apparatus was designed to measure the thermal conductivity of thin materials from 0.02 to 10 mm which not available in bored thermocouples process. Various samples with a different thermal conductivity of 15 $Wm^{-1}K^{-1}$ – 50 $Wm^{-1}K^{-1}$ and sizeable to be measured using this apparatus just only change the meter bar size.

3 Results and Discussion

The proposed thermal conductivity measurements apparatus are intended to measure thermoelectric materials which have middle conductivity in the range of 15 $Wm^{-1}K^{-1}$ to 50 $Wm^{-1}K^{-1}$ [17]. In the validation process materials of stainless steel with a thermal conductivity of ~15 – 20 [18] and bronze with a thermal conductivity of ~50 – 59 [2,19] in range temperature of 50-400^{0}C which have thermal conductivity value between those values were applied. Before validation process both of thermal conductivity materials were measured by using *THB 500 LINSEIS* in ambient temperature to a definite type of metal.

Because the minimum temperature in the apparatus when the apparatus is work was 50°C, several assumptions were used in order to define thermal conductivity of materials. Table 2 shows results of thermal conductivity measurements using test of THB 500 LINSEIS in ambient temperature.

Table 1. Thermal conductivities as results using THB 500 LINSEIS test

Sample Number	Stainless steel		Bronze	
	k ($Wm^{-1}K^{-1}$)	T (°C)	k ($Wm^{-1}K^{-1}$)	T (°C)
1	14.92 ± 2%	31.22	38.89 ± 2%	29.50
2	14.92 ± 2%	31.16	38.12 ± 2%	29.40
3	14.98 ± 2%	31.10	38.15 ± 2%	29.30
4	14.78 ± 2%	31.07	36.41 ± 2%	29.40
5	14.80 ± 2%	31.06	38.50 ± 2%	29.50
Mean	14.88 ± 2%	31.12	38.01 ± 2%	29.42

The validation process for Stainless Steel material in the first step has to check the type of Stainless Stell itself. From THB 500 LINSEIS apparatus test the thermal conductivity of Stainless Stell materials at 31.12°C was 14.88 $Wm^{-1}K^{-1}$. Then it is compared to other references thermal conductivity [15,16,18] which shows thermal conductivity value of Stainless Steel 304 (SS304) was 14.99 $Wm^{-1}K^{-1}$ at temperature 31.12°C and difference about 0,73%. THB 500 LINSEIS test apparatus has capability only at room temperature, so if the test is at 50°C, from the different research thermal conductivity of stainless steel 304 T temperature of 50°C was 15.28 $Wm^{-1}K^{-1}$ [18]. The best-obtained data of thermal conductivity of 304 stainless steel comparing to the literature is -0.64% by 5th number test and the biggest difference is -6.25% on 3rd, see Table 3. Furthermore, from the Table 3, the apparatus demonstrates the proper testability of stainless steel with an average difference from the literature only by -2.55% with a standard deviation is 0.33 Wm-1K-1 or 2.09%.

Table 2. The thermal conductivity of *Stainless steel* at 50°C (k_{ref}= 15.28 $Wm^{-1}K^{-1}$).

Sample number	k ($Wm^{-1}K^{-1}$)	Deviation (%)
1	15.55	-1.73%
2	15.64	-2.35%
3	16.24	-6.25%
4	15.55	-1.74%
5	15.39	-0.69%
Mean	15.67	-2.55%
SD	**0.33**	

The thermal conductivity of commercial bronze of 38.01 $Wm^{-1}K^{-1}$ was measured at a temperature of 29.42°C. From other literature the value of thermal conductivity of bronze was not changed until the temperature of 127°C [2,19], so the thermal conductivity of bronze at a temperature of 50°C can be assumed to be same as result of *THB 500 LINSEIS* test at a temperature of 29.42°C.

Table 3. Thermal conductivities of commercial bronze at a temperature of 50°C (k_{ref}= 38.01 $Wm^{-1}K^{-1}$).

Sample Number	k ($Wm^{-1}K^{-1}$)	Deviation (%)
1	35.97	5.38%
2	38.94	-2.43%
3	36.34	4.40%
4	37.55	1.21%
5	36.54	3.88%
Mean	37.07	2.49%
SD	**1.20**	

Table 4 shows results from five test of thermal conductivity of commercial bronze at a temperature of 50°C with the best result at fourth with the smallest difference only of 1.21% and the biggest difference of 5.38% at the first data. Deviation standard for all five data was 1.20 $Wm^{-1}K^{-1}$ or 3.23%. From the previous validation, good results were shown for all materials at middle thermal conductivity (15 $Wm^{-1}K^{-1}$ - 50 $Wm^{-1}K^{-1}$) at a temperature of 50°C which shown by the difference of within ±3%, comparing to the *THB 500 LINSEIS* test which indicates a fairly accurate of test equipment.The negative value of thermal conductivity difference means greater the testing result than literature value and vice versa. Other than that, of the five-time testing, this tool also showed good repeatability with a standard deviation of less than 3.5%, this indicates that the test tool has made sufficient precision.

Uncertainty analysis also was done to predict uncertainty level for each used tool by using Kline and McClintock method [20].

Table 4. The uncertainty of thermal conductivity.

Material	Thermal Conductivity ($Wm^{-1}K^{-1}$)
Stainless steel 304	15.67 ± 1.61
Commercial Bronze	37.07 ± 2.13

Furthermore, the apparatus was used to obtain thermal conductivity in a high-temperature condition which was conducted at a temperature of 50°C to 400°C, temperature intervals of 50°C, and included uncertainty of ± 9% as shown in Figure 2. The reference value of thermal conductivity of stainless steel 304 at high temperature was also obtained from other research [18] which demonstrate the thermal conductivity of stainless steel 304 alters with the temperature rises, see the continual black straight line. From the Figure 2, it is clear that in the comparative method existence of the thermal contact resistant between surfaces of samples and meter bar has significantly affected by the error [21]. The created equipment has a specification only for thin material testing and does not allow the placing of a thermocouple on the sample material, so to determine the surface temperature of the material or sample (T_H and T_C) only be done by calculation. Meanwhile, from other research which using the same method and allowing the thermocouple to be bored to the material sample, the thermal contact resistance does not significantly affect the calculation of T_H and T_C because those values were measured directly using a bored thermocouple in the sample [15]. Several methods were applied such as pressure application, liquid/gel adding, or minimize the probe for minimizing thermal contact resistance. Application of pressure decreases or minimizes thermal contact resistance, but over valued of pressure damage the specimen and cause deformation of the contact surface [21,22]. For the thin or fragile material, the pressure method is impossible. By ASTM D5470-06 [21], the pressure can be classified based on the hardness of material, for soft materials applied pressure is 0.069 MPa (10 psi) while for the hard material is 3.4 MPa (500 psi). The materials which use as validation process of apparatus were metallic bronze and stainless steel which were be classified as hard materials so that the pressure on the test was 3.4 MPa or 46.0 kg with the cross-sectional area of 1.33×10^{-4} m^2. Then during the pressing process, application of a thermally conducting liquid or so-called thermal grease such as silicone oil to the surface was a logical way for replacing air cavities between both of surfaces. Air cavities have very low thermal conductivity even though on the very smooth surface (0,024 $Wm^{-1}K^{-1}$) or a high thermal contact resistance, [2, 22]. There were two materials used as thermal grease on testing, the HC-131 for the initial validation at 50 ° C and Huskey 2000 Lubricant and Anti-Seize Paste for High Temperature.

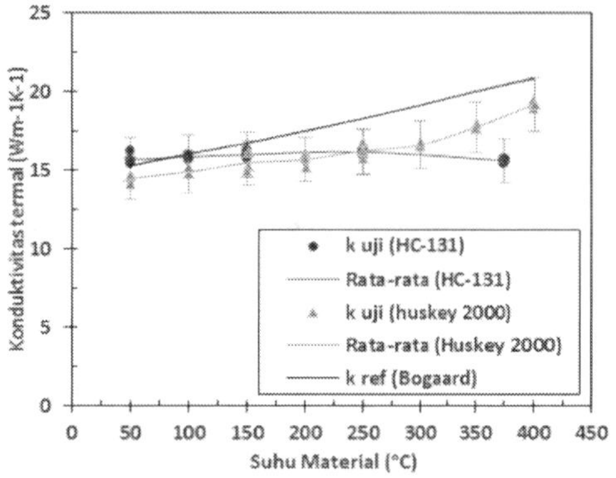

Fig. 2. Thermal conductivity of *stainless steel* at temperature of 50°C - 400°C

Figure 4.1 shows the result of a test using both of thermal greases at the same condition test. At a temperature of 50°C to 200°C for thermal grease of HC-13, the difference of thermal conductivity of material to the literature only by -2.55%. The difference is getting up when the temperature rose from 250°C to 375°C with a maximum difference of 23.6% because HC-131 is only capable in temperature of -30°C to 150°C. At a temperature of 200°C the thermal grease become dry and then air fill the cavities of contact between the sample with the meter bar. In addition to inhibiting heat transfer due to its low thermal conductivity, the air also causes corrosion of the copper used as the material meter bar. Temperatures increase corrosion on copper rapidly so that the copper surface is formed of two layers covering the copper.

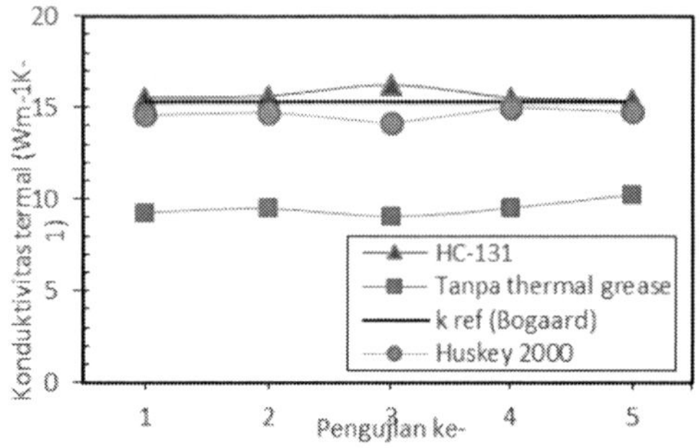

Fig. 3. Comparison of thermal conductivity of *stainless steel using thermal grease HC-131, Huskey 2000* and without *thermal grease* at temperature of 50°C

The inner layer of corrosion that Cu_2O and the outer layer consists of CuO and Cu_2O [23,24]. This layer raises the contact resistance between the meter bar with a bigger sample, resulting in thermal conductivity test results using HC-131 decreases when the temperature further rises.

Meanwhile for Huskey 2000, at a low temperature, the difference of 5.27% was higher than HC-131 because the thermal conductivity of the HC-131 larger (0.62 Wm-1K-1) compared to Huskey in 2000 with its main function as a lubricating oil and anti-seize. However, at a high temperature of 250°C to 400°C, the difference is relatively constant of 7.93% because of the Huskey 2000 capable of operating up to a temperature of 1093.33°C (2000°F). Also, the maximum temperature can be achieved higher at 400°C with the same heater power 858 Watt, exhibiting better thermal conductivity at high temperatures so that the maximum temperature can be accomplished higher also. However, the use Huskey 2000 showed better results compared with air or without using the thermal grease this case is shown in Figure 4.3 the results of comparative testing of the use of thermal grease HC-131, Huskey 2000 and without the use of thermal grease.

4 Conclusions

The thermal conductivity measurements apparatus with axial heat flow method has been manufactured based on the standard ASTM D 5470-06 and has good enough capability in measuring the material at a temperature range of 50°C to 400°C. In the initial validation of the material used stainless steel and bronze by a margin of literature by -2.55% and 2.44% and the standard deviation of the five tests of 2.09% and 3.23% demonstrated accuracy and precision tools are pretty good. As for validation at high temperatures showed good results also by a margin of 7.93% of the literature on the highest temperature of 400 ° C.

References

1. Reif-Acherman, S., J. Heat and Mass Transfer **77**, 542–563 (2014)
2. Cengel, Y.A., *Heat & Mass Transfer: A Practical Approach* (McGraw-Hill Education (India) Pvt Limited, 2007)
3. Kurniawan, A., *Pengembangan Semikonduktor Tipe-P Untuk Modul Termoelektrik Berbasis Material Zno* (Universitas Sebelas Maret, 2014)
4. Xamán, J., Lira, L., and Arce, J., Applied Thermal Engineering **29**, 617–623 (2008)
5. Xing, C., Jensen, C., Folsom, C., Ban, H., and Marshall, D.W., Applied Thermal Engineering **62**, 850-857 (2013)
6. Alrtimi, A., Rouainia, M., and Manning, D.A.C., J. Heat and Mass Transfer **72**, 630-636 (2014)
7. Yunianto, B., *Pengujian Konduktivitas Termal Material Padat Silinder Untuk Kondisi Steady Satu Dimensi Menggunakan Akusisi Data* (Universitas Diponegoro 2008)
8. Permana, B., *Sistem Pengukuran Konduktivitas Panas Pada Logam Berbasis Mikrokontroler* (Universitas Indonesia, 2009)
9. Dasgupta, T., and Umarji, A., Review of Scientific Instruments, **76** (9) (2005)
10. Komiya, A., Mashimo, H., Okajima, J., Takahashi,

I., and Maruyama, S., *The 31st Japan Symposium of Thermophysical Properties*, 158-160 (2010)
11. Hadi, S., Nishitani, M., Fukunaga, T., Kurata, K., and Takamatsu, H., *Proceedings of The Third International Forum on Heat Transfer*, Nagasaki, 089 (2012)
12. Hadi, S., Nishitani, M., Wijayanta, A.T., Kurata, K., and Takamatsu, H., JTST **7** [4] (2012)
13. Hadi, S., Nishitani, M., Wijayanta, A.T., Takanobu Fukunaga, Kurata, K., and Takamatsu, H., J. Heat and Mass Transfer **69,** 256–263 (2014)
14. ASTM E1225-99, E 1225 – 99 (1999)
15. Jensen, C., Xing, C., Folsom, C., Ban, H., and Phillips, J., J. Thermophysics **33** (2), 311-329 (2012)
16. Sweet, J.N., Roth, E.P., and Moss, M., J. Thermophysics **8** (5), 593-606 (1987)
17. Goldsmid, H.J., *Introduction to Thermoelectricity* (Springer Berlin Heidelberg, 2012)
18. Bogaard, R.H., *Thermal Conductivity of Selected Stainless Steels*, pp. 175-185 (1985)
19. Incropera, F.P., Bergman, T.L., and Lavine, A.S., *Fundamentals of Heat and Mass Transfer* (Wiley 2011)
20. Kline, S.J., and McClintock, F., Mechanical Engineering **75** (1), pp. 3-8 (1953)
21. M.Tritt, T., *Thermal Conductivity: Theory, Properties, and Applications* (Kluwer Academic / Plenum Publishers, 233 Spring Street, New York 2004)
22. Holman, J.P., *Experimental Methods for Engineers* (The McGraw-Hill 2012)
23. Tuck, C.D.S., Powell, C.A., and Nuttall, J., Corrosion of Copper and Its Alloys. **3**, 1937-1973, (2010)
24. Wan, Y., Wang, X., Sun, H., Li, Y., Zhang, K., and Wu, Y., J. Electrochemical Science **7** (9) (2012)

Visualization and pressure distribution of gas-liquid flow in ¾ inch plexyglass pipe diameter with the slope orientation upward, horizontal and downward

Dewi Puspitasari[1*], *Marwan* Marwani[1], *Ichsan* Ramdani[1], *Ridho* Mardhan Hadi[1]

[1]Mechanical Engineering Department, Faculty of Engineering, Universitas Sriwijaya, Inderalaya Ogan Ilir, Indonesia

Abstract. The flowing phase in the pipe is affecting on the continuity of fluid transport process, this is due to the interaction when the fluid mixture flowing at a certain superficial velocity. Therefore it will be studied the distribution of pressure and flow pattern of a transparent pipe (plexiglass) ¾ inches, and its influence on the orientation of the pipe 10°, 5°, 0°, -5° and -10° by the experimental equipment. The research was carried out experimentally with the water flow 10 - 20 lpm and air flow 5 - 15 lpm. Results showed the flow pattern was produced on every slope of the pipe is a bubble, elongated bubble, plug, plug-slug transition, plug bubble, bubble slug, slug, stratified wavy and stratified with mixture interfaces. On the upward slope of the bubble flow patterns likely to appear more and result the highest pressure of 1690 N/m2 on Jmix = 9.3 m/s, on the horizontal position bubble slug flow pattern that often appears with the highest pressure drop of 593 N/m^2 on Jmix = 8.23 m/s, while the downward slope of the wavy stratified flow pattern likely to appear more with the highest pressure drop of 685 N/m^2 on Jmix = 5.88 m/s.

1. Introduction

On the industry often encountered two-phase flow, which is part of the multiphase flow. A form or phase of a substance in the two-phase flow can be divided into three: liquid, solid, and gas. Two-phase flow more encountered both in everyday life and in industrial processes, such as boilers, condensers, heat exchangers, nuclear reactors, natural gas liquefaction, and flows in the petroleum industry.

Multiphase flow is divided into several phases' combinations among others the solid-liquid, liquid-gas and solid-gas. It is very important to study the flow pattern from the two-phase flow. The Flow that is changing the flow pattern can cause pressure drop fluctuating. Therefore it is necessary for us to know and identify the flow patterns that occur to predict the pressure drop relationships with the two-phase flow, on a variety of pipe slope orientation.

The result [8] showed that the fluctuation pressure differential of slug flow is very influenced by the superficial velocity of the fluid. [5] describes the visualisation of the air-water mixture flow in a vertical pipe that is heated. The result showed that the heat transfer coefficient of air-water flow increase with increasing air flow and water flow was represented by the superficial velocity of water.

A multiphase flow related to solid, liquid or gas. Multiphase flow is the mixture flow of phases such as bubble flow which shows gas (bubble) in liquid or liquid in gas or a combination of solid, liquid and gas simultaneously in the same pipe. Many engineering and scientific applications of the multiphase, such as gas-liquid flow, solid-liquid (slurry), gas-oil-water (petroleum Industry) and many others applications of the multiphase.

The research about flow regime in horizontal and vertical pipe has been studied by [1] which can be seen in figure 1 and 2.

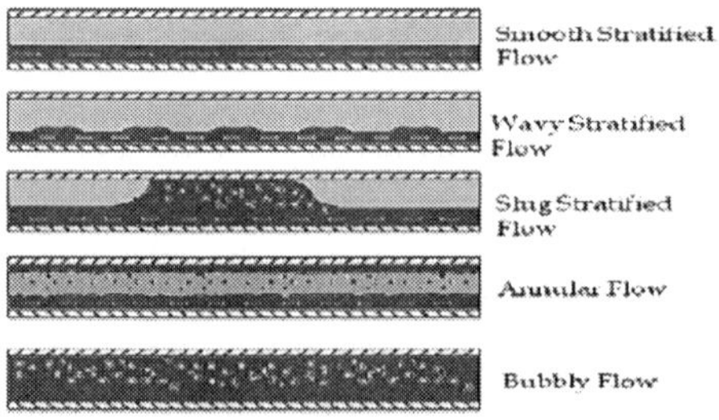

Fig.1. Flow regime in the horizontal pipe [1]

The flow pattern consists of:

a. Bubbles flow (bubble), where the gas bubbles tend to flow at the top of the tube.
b. The flow of gas bag (plug), in which small gas bubbles combine to form gas pockets.
c. Smooth stratified flow, where the interface of the liquid-gas is very smooth, but the flow pattern like this usually does not happen. The interface is almost always wavy (stratified wavy).
d. Stratified wavy occurred when gas velocity increases.

* Corresponding author: dewipuspitasari@unsri.ac.id

e. Liquid plug flow (slug), which are typical large amplitude waves to reach the top of the tube.
f. Flow ring (annular), is same with the vertical tube flow, but film liquid in the bottom is thicker from the top.

The flow patterns in the vertical pipe can be seen in the figure 2.

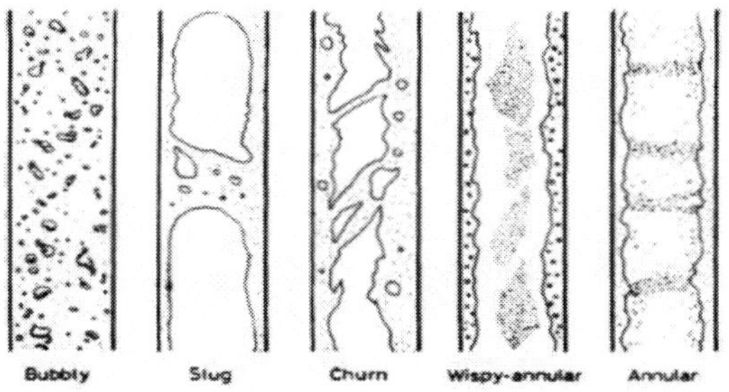

Fig.2. Flow regime in a vertical pipe [1]

The flow pattern consists of:

a. Bubble Flow, where the vapor bubbles are estimated to have the uniform size.
b. Slug Flow, gas flowing to form large bubbles (sometimes distributed small gas bubbles in the liquid)
c. Churn Flow, the oscillating movement occurs so that the liquid becomes unstable.
d. Wispy-Annular Flow, the concentration of droplets in the gas increases and merge to form wisp.
e. Annular Flow, a part of liquid phase as a film in the tube wall, and the other part in the form of droplets distributed in the gas flowing in the middle of the tube.

Hold-up of liquid (α_L) is defined as the ratio between the volume occupied by the liquid phase with overall mixture volume, whereas Hold-up for gas (α_g) is the ratio between the volume occupied by the gas phase with the overall mixture volume.

$$\alpha_L = \frac{V_L}{V_P} \quad (1)$$

$$\alpha_g = \frac{V_g}{V_P} = 1 - \alpha_L \quad (2)$$

where :
α_L = Hold up for liquid
α_g = Hold up for gas
V_L = Volume of liquid filled in pipe (m^3)
V_g = volume of gas filled in pipe (m^3)
V_P = Volume of mixture filled in pipe (m^3)
Parameter flow used in the calculation of multiphase flow is the superficial velocity which is defined as follows:

$$J = Q_{tot} / A.\alpha \quad (3)$$

where :
J = superficial velocity of the phase, m/s
Qtot = total flow rate, m^3/s
A = pipe cross-sectional area, m^2
α = Phase hold up
The superficial velocity of mixture is determined by the equation :

$$J_m = J_L + J_g \quad (4)$$

The pressure distribution in multiphase flow is determined from the measured from the column manometer.

2 Methodology

The study was conducted by using the experimental set up as shown in figure 3.

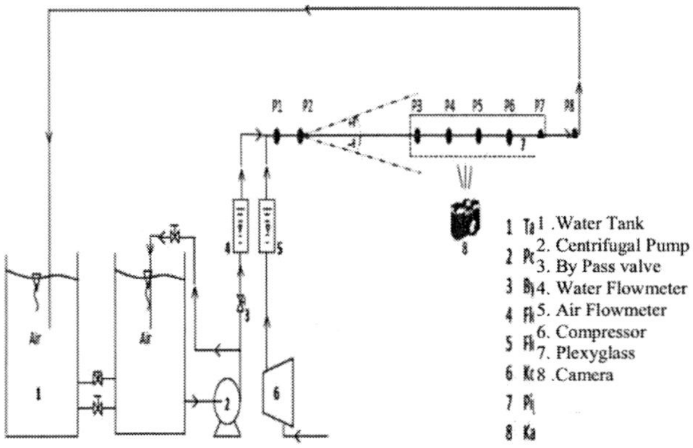

Fig.3. Schematic diagram of experimental set up

Table.1. Properties of work fluid

Work fluid	Density (kg/m³)	Viscosity (kg/m.s)
Water	998	0.001002
Air	1.29	0.000019

3 Results and discussion

From the research results the flow pattern visualisation and pressure distribution according to variation superficial velocity and orientation of pipe slope. There were nine flow pattern results as shown in figure 4 to figure 12.

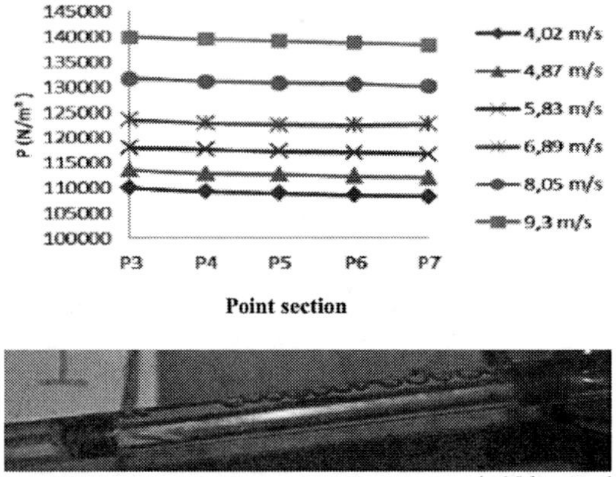

Fig.4. The pressure distribution and flow pattern bubble on the uppward 10°

DOI: 10.1051/matecconf/201710103008

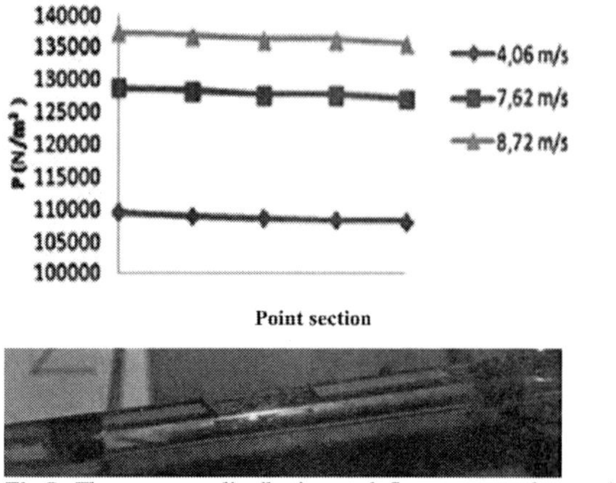

Fig.5. The pressure distribution and flow pattern elongated bubble on the upward 10°

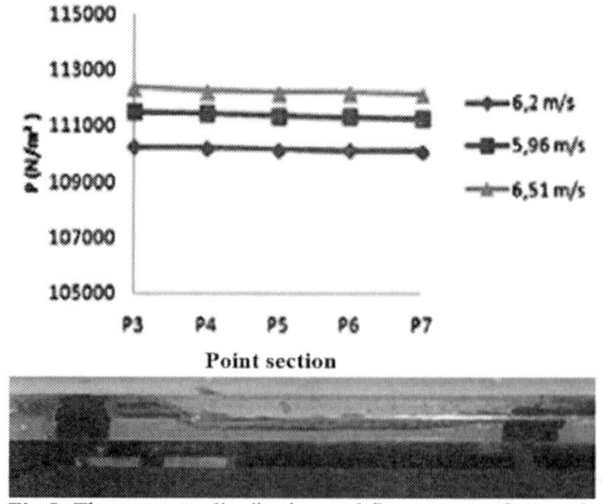

Fig.8. The pressure distribution and flow pattern slug on the horizontal 0°

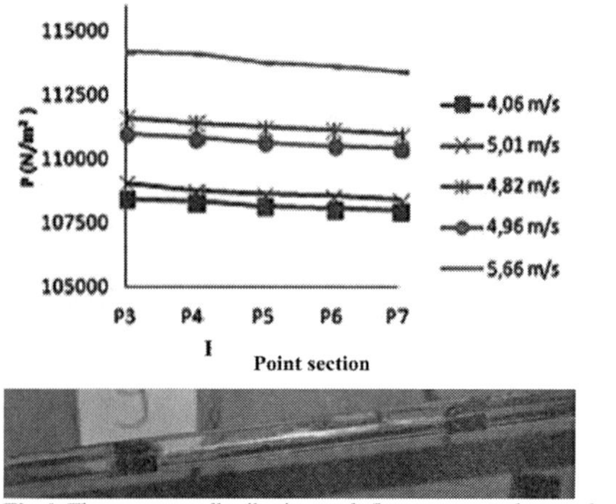

Fig.6. The pressure distribution and flow pattern plug on the upward 5°

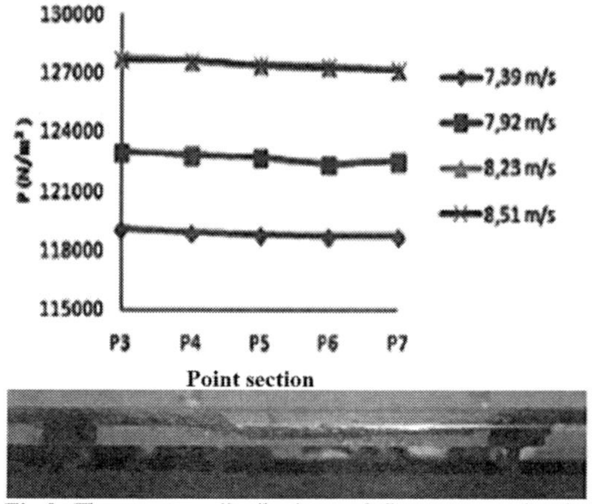

Fig.9. The pressure distribution and flow pattern slug bubble on the horizontal 0°

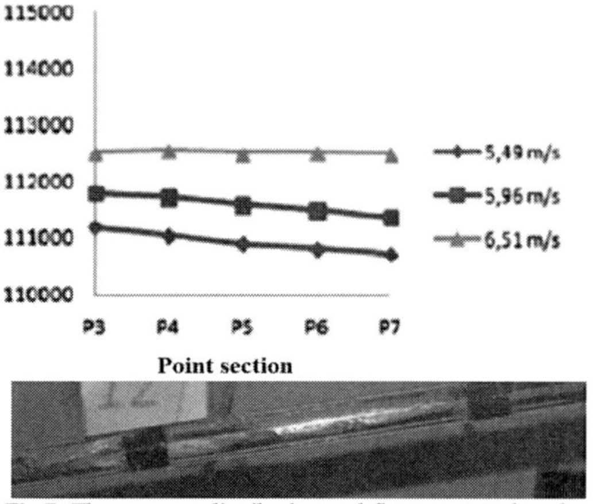

Fig.7. The pressure distribution and flow pattern plug slug on the upward 5°

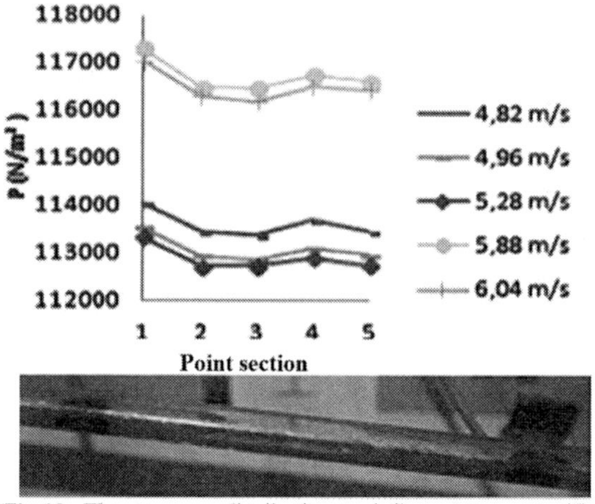

Fig.10. The pressure distribution and flow pattern stratified wavy on the downward -5°

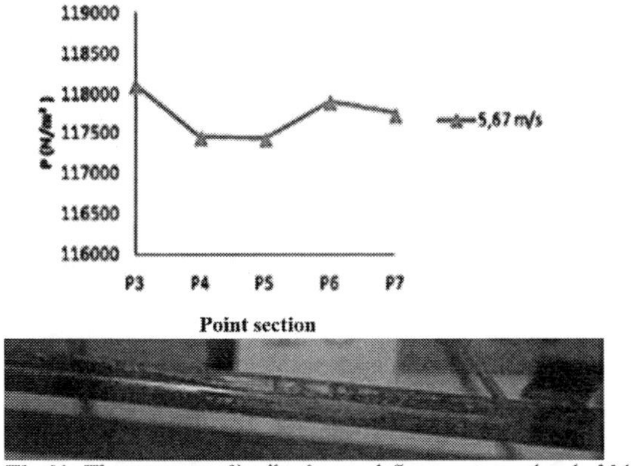

Fig.11. The pressure distribution and flow pattern plug bubble on the downward slope -5°

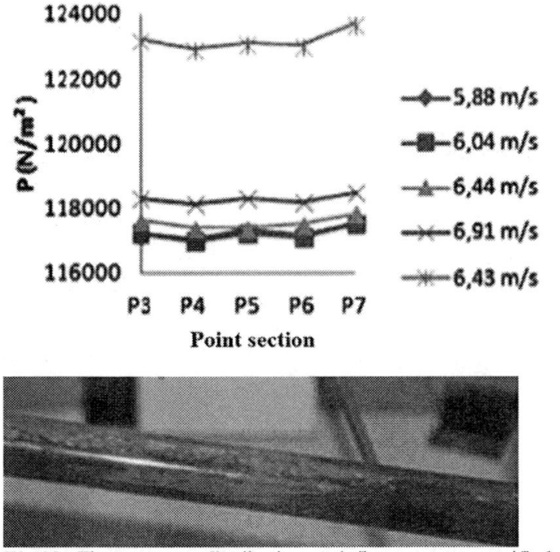

Fig.12. The pressure distribution and flow pattern stratified-mixture interface on the downward -10°

The discussion of the research just focused on the flow patterns that often appear and highly fluctuating pressure distribution.

On the upward slope 10° flow patern form are : bubble, elongated bubble, plug, transition slug-plug, slug, and plug bubble. The lowest pressure drop for bubble flow is 887 N/m^2 on J_{mix} = 6.89 m/s and the highest is 1690 N/m^2 on J_{mix} = 4.02 m/s. For elongated bubble flow the lowest is 1462 N/m^2 on J_{mix} = 4.06 m/s and the highest is 1648 N/m^2 on J_{mix} = 8.72m/s. This phenomenin shows that small gas bubble can combine to elongated bubble when gas flow increase. The formation of flow pattern held by the gravity in this slope, that why the pressure drop is high enought.

On the upward slope 5° flow pattern form are : bubble, plug, slug, transition slug-plug, plug bubble, and the slug bubble. The lowest pressure drop for plug flow is 452 N/m^2 on J_{mix} = 4.06 m/s and the highest is 741 N/m^2 on J_{mix} = 5.66 m/s. For slug flow the lowest is 39 N/m^2 on J_{mix} = 6.51 m/s and the highest is 454 N/m^2 on J_{mix} = 5.49m/s. If gas flow increases the flow pattern can be form to plug, for this condition the amplitude the gas reach the upper pipe and press the water in the bottom with the thin layer, if gas flow is more the flow pattern become plug slug whereas slug in the front section and slug in the rear section, some times was form to combination slug plug. Pressure drop in upward 5° was lower than upward 10°, because an effect of gravity was decreased.

On the horizontal slope, 0° flow pattern form are : elongated bubble, plug, slug, transition slug-plug, plug bubble and slug bubble. The lowest pressure drop for slug flow is 156 N/m^2 on J_{mix} = 6.2 m/s and the highest is 235 N/m^2 on J_{mix} = 5.96 m/s. For slug bubble flow the lowest is 353 N/m^2 on J_{mix} = 7.39 m/s and the highest is 593 N/m^2 on J_{mix} = 8.23 m/s. In this slope the gravity effect was neutral, therefore slug flow becomes slug bubbles with the small area than slug when gas flow increase, but the distribution of pressure become increase very high suddenly, this cause by the existence bubble in the upper of slug layer.

On downward slope -5° flow pattern form are : stratified wavy, plug, bubble, elongated bubble, plug bubble and stratified with mixture interface. The lowest pressure drop for stratified wavy flow is 587 N/m^2 on J_{mix} = 4.82 m/s and the highest is 685 N/m^2 on J_{mix} = 5.88 m/s. For plug bubble flow there is only one flow pattern was found then the pressure distribution is 342 N/m^2 on J_{mix} = 5.67 m/s. Flow pattern stratified wavy become to form plug bubble when gas flow increase, but the pressure drop is lower because gas layer become small and water is dominant in this section. for this slope, effect the gravity play a role again but not significant.

On the downward slope -10° flow pattern stratified wavy, plug, bubble, elongated bubble, plug, transition slug-plug, slug, atnd stratified with mixture interface. The lowest pressure drop for stratified with mixture interface flow is 184 N/m^2 on J_{mix} = 6.91 m/s and the highest is 1690 N/m^2 on J_{mix} = 4.02 m/s. For elongated bubble flow the lowest is 450 N/m^2 on J_{mix} = 6.43m/s. For this slope, the effect of gravity was significant, that why flow pattern become stratified wavy with mixture interface and pressure drop higher same with upward 10°.

4 Conclusion

1. The flow patterns were found throughout the study for each slope are : bubble, elongated bubble, transition plug-slug, plug-slug, plug bubble, slug bubble, stratified wavy, and stratified with mixture interface.
2. The pressure drop for each flow patterns in the variation of the slope is different. For upward slope 10° highest pressure drop is achieved under conditions bubble flow with a drop of pressure of 1690 N/m^2 and J_{mix} = 4.02 m / s. The lower of J_{mix} the higher of pressure drop.
3. On the upward slope of 5° highest pressure drop occurs in the Plug flow of 741 N/m^2 with J_{mix} = 5.66 m/s. In the horizontal position (0°) highest pressure

drop occurs in slug bubble flow of 593 N/m^2 on J_{mix} = 8.23 m/s.

4. On the downward slope of the highest pressure drop occurs in conditions the Stratified wavy flow of 685 N/m^2 with J_{mix} = 5.88 m / s.

We acknowledge for the University of Sriwijaya as my institution, that has provided research grants PNBP 2016. My beloved family who always supports me to write the scientific paper more and more.

References

1. Boyun, Guo. *Offshore Pipelines* (University of Louisiana at Lafayette, 2005)
2. I Kadek EHW, J. mechanical engineering Bali State Polytechnic, Bali (2015)
3. Korawan, A.D., Yuwono, T. J. Mechanical Engineering Institute of Technology (2015)
4. Livia A et al, *The 13th Brazilian Congress of Thermal Science and Engineering* (2010)
5. Kunarto, J. Mechanical Engineering University of Bandar Lampung (2011)
6. Munson, B. R., Young, D. F., and Okiishi, T. H., *Fluid Mechanics (Translation by Dr. Ir. Harinaldi and Ir. Budiarso, M.Eng), Volume 1, Edition 4. Jakarta, Erland*, 2003)
7. Priyo H A, J. Mechanical Engineering, State University of Surabaya (2010)
8. Santoso, Budi., et all., J. Mechanical and Industrial Engineering University of Gadjah Mada (2012)

Optimization high vortex finder of cyclone separator with computational fluids dynamics simulation

Caturwati Ni Ketut, *Dwinanto,* Attegar

Mechanical Engineering Department , University of Sultan Ageng Tirtayasa, 42435 Cilegon - Banten, Indonesia

Abstract. Cyclone separator is an equipment that separates particles contained in the fluid without using filters. The dust particles in the flue gases can be separated by utilizing centrifugal forces and different densities of particles, so that the exhaust gases to be cleaner before discharged into the environment. In this paper carried out a simulation by Computational of Fluids Dynamics to determine the number of particles that can be separated in several cyclone separator which has a ratio body diameter against vortex finder high varied as : 1:0.5 ; 1:0.75 ; 1:1 ; 1:1.25 and 1:1.5. Fluid inlet are air with antrachite impurity particles that are commonly found in the exhaust gases from tire manufacturers with inlet velocities varied as: 15 m/s and 30 m/s. The results of simulation show the fluids with 15 m/s of inlet velocity is generate particle separation value is higher than the fluids with 30 m/s inlet velocity for ratio of body diameter and height vortex finder a: 1:0.5 and 1:1.5. For both of inlet velocities the best ratio of body diameter and height vortex finder is 1:1.25, where it has the highest values of percentage trapped particles about 86% for 30 m/s input velocity and also for 15 m/s input velocity.

1 Introduction

Cyclone separator is a device for separate solid or liquid particles from carrier gases by centrifugal force wich exists in a stationnaries mechanical device. Cyclones separators are the most widely used for air pollution reducted devices in industries or agricultural processing , without the use of filters The cyclone separator is a highly effective and strong equipment due to the absence of moving parts, so that it can be operated in any environments. The efficiency of separator systems is determined by the size of particles to be separated and the velocity of flow entering devices. [1].

The cyclone separators advantages compared with the other devices, are :

- Capital investment and maintenance cost is low.
- No moving parts.
- Very strong.
- The particles are separated remains dry and can be used again.
- Very compact to be operated.
- Can be formed from a variety of materials needed or suitable such as : casting metals, plate steel, aluminium , alloys, ceramics, plastics, etc.
- Can be equipped with coating material to resist corrosion and erosion such as Teflon.
- With appropriate design can separate either liquid particulates , solid particulates or even both in combination.

Besides the favorable things that there are some things less well among :

- Fluid through cyclone separator has higher pressure loss than the other separator such as scrubber or bag filters.
- Low efficiency values for very small particulate size when operated under small amount particle in the fluids.
- Subject to fouling for the sticky particulate and to erosive wear if particulates being processed are abrassive.
- If the cyclone designed and operated are not properly then product is below expectations.

Figure 1. shows a schematic of a cyclone separator. The tangential inlet gas flow enters cyclone separator at near the top, which gives rise to an axially descending spiral of gas and a centrifugal force field that causes the incoming particles to concentrate along the inner walls of the cyclone separator, and spiral down. Denser or larger size of particles in the rotating stream have big inertia to follow the tight curve of the stream then it strike the outside wall, fall to the bottom of the cyclone where they can be colected. The rotational radius of the stream is reduced in a conical system, so the rotating flow moves towards the narrow end of the cyclone, thus separating smaller and smaller particles

The maximum collection efficiencies is quantified by W.B.Faulkner and B.W.Shaw for the inlet velocities obtained for 1D3D (cylindrical section height = 1 x body diameter ; conical section height = 3 body diameter) and 2D2D (cylindrical section height = 2 x body diameter ; conical section height = 2 body diameter) cyclones and demonstrated that for large particles,

* Corresponding author: n4wati@untirta.ac.id

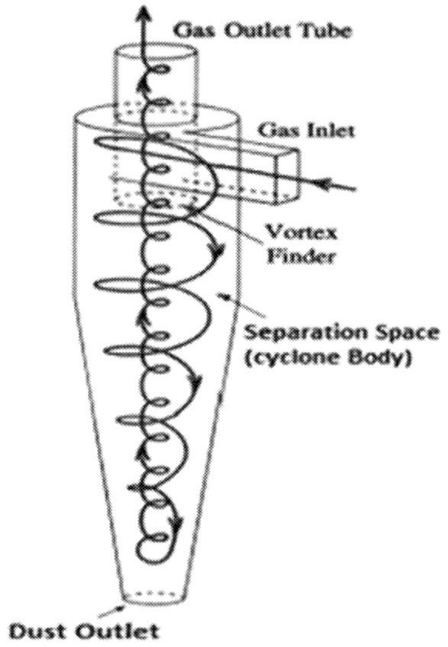

Fig. 1. Schematic of a cyclone separator

the collection efficiency is similar for inlet velocities from 10.16 m/s up to the design velocity of six inch diameter 1D3D and 2D2D cyclones [2]. While they have been considered low values efficiency colectors. Recent studies have shown that cyclones efficiencies can reach exceeding 99% for particle size larger than 5 μm.[3].

By experimental 1D3D and 2D2D cyclones also evaluated for effect of standard air density on cyclone performance, The results indicate that optimal cyclone design velocities for 1D3D cyclones is 16 m/s and for 2D2D cyclones is 15 m/s [4].

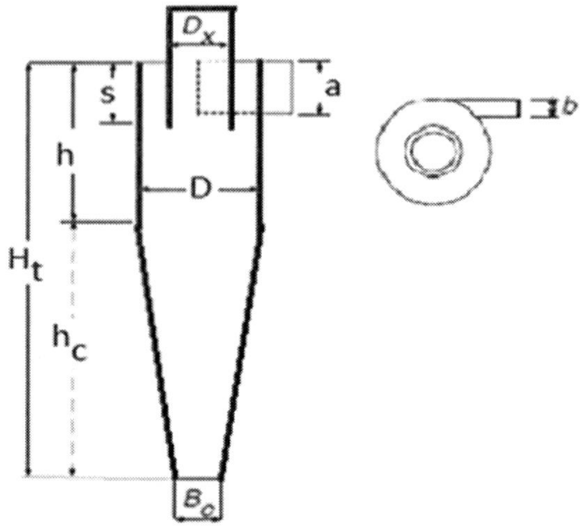

Fig. 2. Geometrical notation of a cyclone separator.

The geometrical notation is indicated in Figure 2.
Where :

D : body diameter
H_t : total height of the cyclone
D_x : vortex finder diameter
S : vortex finder length
a : inlet height
b : inlet width
h_c : conical section height.
h : cylindrical/barrel section height.
B: cone-tip diameter

The performance of cyclone separator and flow field pattern has been investigated by computational for the effect of cyclone height. Increasing the cyclone height cause decrease values of maximum tangential velocity, increasing the cone height cause considerably change in axial velocity , and increasing the barrel height makes the axial velocity change small. The changes in the performance at constant cone height for beyond h/D = 1.8 are small, whereas the performance at constant barrel height improvement stops after h_c/D = 4.0 (H_t/D = 5.5).[5]

Type of a cone-shaped cyclone separator with a slot type fluid inlet and reverse flow fluids is a most popular cyclone. Some of relevant literature show simulation results theoretically, experimental and even simulation by computational fluid dynamics (CFD to determine the performance of the cyclone with a slot type inlet for the influence of the geometry and operating conditions. This study also used to develop the design parameters in order to improve the effectiveness of particulate collected or effectiveness of separation. For all types of cyclones separators show general performance of the cyclone separator is determined based on the type and amount of solid particles in fluids inlet. [6], [7], [8], [9].

Another experiment showed an increase effectiveness of particulate collection by adding coarse particles in the cyclone with the coarse particles carrying fine particles move toward the cyclone wall region and down to collector. [10],[11],[12].

The principles of cyclone separator is dust-containing gas is circulation moving around so that the dust particles bounced to the outside wall of cyclone body and then the particles move down to collected.

Swirling flows types can be distinguished of :

1. Free vortex flow : fluid move swirly without friction so that the tangential velocity in the vortex has a value such that a uniform momment of momentum at the same radii.
2. Forced vortex flow : fluid move swirly with friction fluids so that the values of tangential velocity as same a rotating body.

Actual swirling flows has behavior between the two types of swirling flows. For liquids which have an infinitive viscousity (as solid materials) so no sliding motion between the liquid layer. So all liquids at different positions has uniform angular velocity,V_θ, which equals to the tangential velocity,V_τ , devide to radii,r. So for solid body rotation or forced vortex flow, tangential velocity can be expressed as :

$$V_\tau = V_\theta . r \qquad (1)$$

In the other words, the movement of fluid element in the swirling fluid without viscosity, the movement of each element is not influenced by neighboring elements where are in the radius of the greater or the radius of the smaller. While the values of the tangential velocity increases to radii smaller to maintain constant value of the moment of momentum.

For frictionless vortex , multiple of tangential velocity and radii is constant, $V_\tau \cdot r = C$, where C is a constan number, so that :

$$V_\tau = \frac{C}{r} \tag{2}$$

An actual Swirling flow generally has a loss- free rotation which area is near the center of solid-body rotation.

The theory for increased of cyclone separators characteristics have been studied by many researchers but no has studies found the influence of vortex finder length dimensions. Therefore, in this paper studied about ratio of body diameter comparison to vortex finder heights in relation to increased efficiency of cyclone separator.

2 Experimental details

This study conducted a simulation by a computational fluids dynamics software with geometric parameters of cyclone separator are given in Table 1.

Table 1. Geometric Parameters

No.	Parameter	Symbol	Dimension (mm)
1.	*Body Diameter*	D	1000
2.	*Inlet height*	a	500
3.	*Inlet width*	b	200
4.	*Vortex finder diameter*	Dx	500
5.	*Vortex Finder length*	S	varies
6.	*Cylindrical section height*	h	1500
7.	*Conical section height*	hc	2500
8.	*Cone tip diameter*	B	400

The size of vortex finder length is varied from 500 mm , 750 mm , 1000 mm , 1250 mm to 1500 mm to make ratio of body diameter against vortex finder heights varied as : 1 : 0.5 , 1 : 0.75 , 1 : 1 , 1 : 1.25 and 1 : 1.5 .

So we have some type of cyclone separator to make the simulation as :

Type a : vortex finder length S = 500 mm
Type b : vortex finder length S = 750 mm
Type c : vortex finder length S = 1000 mm
Type d : vortex finder length S = 1250 mm
Type e : vortex finder length S = 1500 mm

Figure 3 shows the model geometry of each type of cyclone separator.

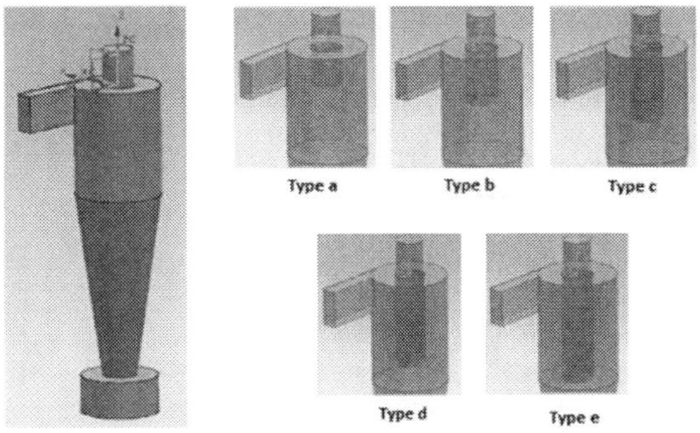

Fig. 3. Model geometry of each type of cyclone separator

Conditions and boundary conditions for make simulation by CFD's are :

Meshing size is 10 mm.
Fluids is air with density : 1.125 kg/m^3 and viscosity : 1.7894 x 10^{-5} kg/m.s
Impurities particle is anthracite with density 1,550 kg/m^3.
Inlet velocity are : 15 m/s and 30 m/s

3 Results

The simulation result show maximum pressure in the cyclone separator for velocities inlet 15 m/s and 30 m/s as showed at Table II and Figure 4. Maximum pressure in cyclone separator as increased as velocities inlet and vortex finder length, S.

Table 2. Maximum Pressure

Ratio S/D	P_{max} (Pa)	
	V_{in} = 15 m/s	V_{in} = 30 m/s
0,50	424,58	1697,11
0,75	428,95	1715,74
1,00	437,56	1747,97
1,25	450,40	1802,54
1,50	462,31	1854,84

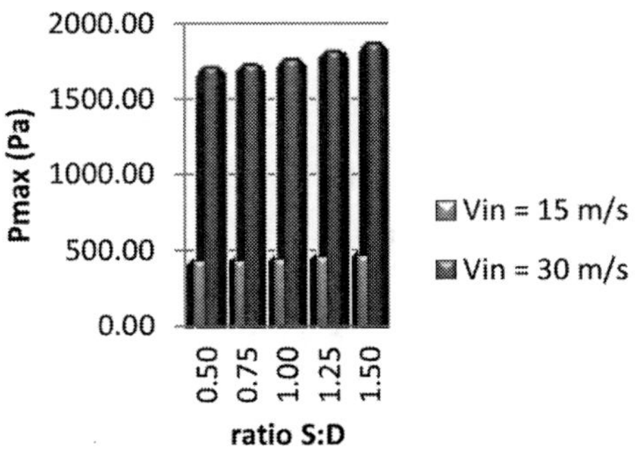

Fig. 4. Maximum pressure in cyclone separator.

The maximum velocity in cyclone separator generated from simulation are shown in Table III and Figure 5. Special conditions happen to the value vortex finder length (s) equal to body diameter (D) which has a minimum value for the maximum speed of the fluid in cyclone separator.

Table 3. Maximum Velocity

Ratio S/D	V_{max} (m/s)	
	V_{in} = 15 m/s	V_{in} = 30 m/s
0,50	18,97	38,07
0,75	19,12	38,41
1,00	18,90	37,94
1,25	19,03	38,22
1,50	19,41	38,88

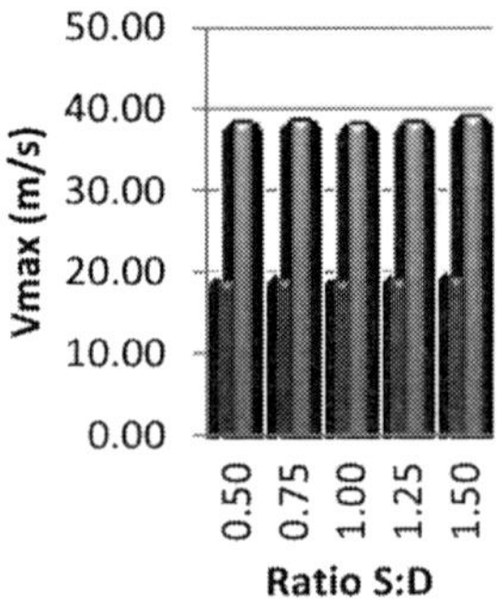

Fig. 5. Maximum velocity in cyclone separator

Besides values of maximum pressure and maximum velocity resulted from the simulation, the most important thing to consider is the separated effectiveness of particles. Which is presented by precentation of traped particles. Table 4. and Figure 6 shows the precentation of traped particles.

Table 4. Percentation Of Trapped Particles

Ratio S/D	Traped Particles (%)	
	V_{in} = 15 m/s	V_{in} = 30 m/s
0,50	82,14	78,73
0,75	81,24	83,39
1,00	80,55	83,51
1,25	85,71	85,95
1,50	86,50	84,58

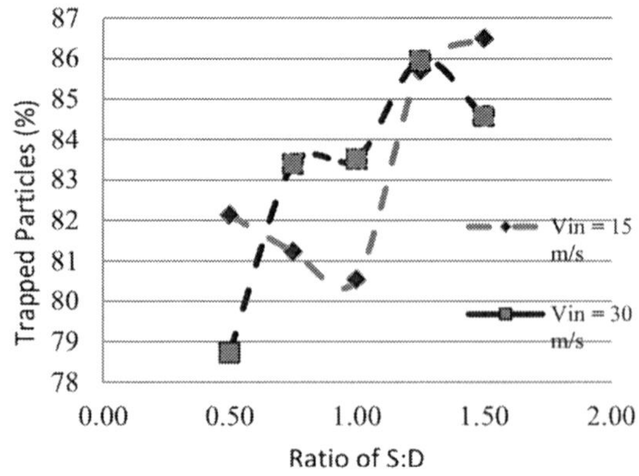

Fig. 6. Percentage of trapped particles

The simulation results for the fluids with velocity input greater (30 m/s) has maximum value of trapped particles are in the vortex finder length is 1250 mm. Contrary to the fluid with lower velocity input (15 m/s) has a percentage trapped value as length as vortex finder.

4 Conclusions

Five cyclones with different vortex finder length have been simulated by computational fluids dynamics methods. Each cyclone imposed on two input values fluid velocity of 15 m/s and 30 m/s. The simulation results are maximum pressure in cyclone separator as increased as velocities inlet and vortex finder length.

Besides that the simulation results show the cyclone separator with vortex finder length is equal with body diameter (S = D) has the lowest value of the maximum fluid velocity.

The effectiveness of particle separation for input flow rate of 30 m/s occur in the value of S/D = 1.25. As for the input flow rate of 15 m/s the value effectiveness of particle separation occur in the value of S/D greatest. Conclusions from the simulation results show cyclone separator by S/D is equal to 1.25 provides the best value as a particle separator compared to the others.

References

1. Wang, L.,C.B.Parnell, and B.W. Shaw. Proc. Of the 2000 Beltwide Cotton Conferences. San Antonio, Tex.: National cotton Council (2000).
2. W.B.Faulkner and B.W.Shaw, Applied Engineering in Agriculture Vol. **22**(1) : 155-161American Society of Agricultural and Biological Engineers ISSN 0883-8542 (2006)
3. Cooper, C.D., and F.C. Alley, A Design Approach, 3rd ed. Prospect Heights, III, Waveland 2002.
4. LingJuan Wang, Texas A&M University (2004)
5. Khairy Elsayed, Department of Mechanical Engineering, Vrije Universiteit Brussel, (2011)
6. A.C. Hoffmann and L.E.Stern, *Gas Cyclones and Swirl Tubes*, Springer (2002)

7. G.B. Sakura, Andrew Y.T.Leung, International Journal of Environmental Science and Development, **Vol. 6**, No. 3 March (2015)
8. W. Barth, Brennstoff-Warme-Kraft, **vol. 8** (1956)
9. C.J.Stairmand Trans Inst. Chem. Eng. **Vol 29** pp 356-383 (1951)
10. A.J.T. Linden, Proc. The Institution 0f Mechanical Engineers, **vol.160**.pp 233-251 (1949)
11. A.Stem, K Caplan, and P. Bush, Division of Refining, American Petroleum Institute, New York, **vol 67** (1955)
12. Z.Ji, Z. Xiong, X.Wu, H. Chen and H.Wu, Powder Technology, **vol.191**, pp. 254-259 (2009)

Engine optimization grate multipurpose analysis method with quality function deployment

Nukman[1], *Irsyadi* Yani[1] *and Firdaus*[1]

[1]Mechanical Engineering Department, Faculty of Engineering Universitas Sriwijaya, 30139 Palembang, Indonesia

Abstract. The object that will be developed in this study is a simple technology shredding machine design, artificial and has been patented by Andi Patolla (1988). which is often used by people specifically to grate the coconut. The methods used for product development using QFD. This method is needed to analyze and evaluate the level of consumer desires in order to develop products grate machine. The subject matter of the research outlined in the distribution table discussion. The research instrument used when data collection is questionnaire technique, which is supported by the observation technique, documentation, and interviews. From the discussion that has been done, it can be concluded that the priority consumers are very stressed to do a modification component or an additional tool in the machine grater multipurpose namely Cover Safeguard Fan Belt, Exhaust, Priority Motor Driving Fuel Gasoline, layout position Mover, Motor Starter, Plastic cleaning brush Roll Grate, and Plate Cap. 2.

1 Introduction

Increasing customer satisfaction is one of the actions undertaken to determine the value added to a product. Improved product quality is a measure to improve the quality of products to provide the best for consumers. Reference [14] said that the development process has meaning, how to develop in order to be developed, good or perfect. Innovation or creative ideas play a major role in achieving marketing success in order to maintain and increase the superior quality of high competitiveness so that the product become acceptable in a market.

To identify and evaluate problems in order to develop a product, then use the analysis method of Quality Function Deployment (QFD) in order to know exactly the wishes of consumers. In accordance with the statement of the experts, Ref [17] "QFD is a methodology, the which establishes a relationship between product and customer, determines the product's sales ability and carry out it to a high level during the process", Then by Bernal [2] "QFD is a system with the aim of translating and planning the "voice of the customer" into the quality characteristics of products, processes and services in order to reach customer satisfaction".

The object that will be developed in this study is a simple technology shredding machine design, artificial and has been patented by Andi Patolla [11]which is often used by people specifically to grate the coconut. The methods used for product development using QFD. This method is useful to analyse and evaluate the level of consumer desires in order to develop products grate machine. The subject matter of the research outlined in the distribution table discussion.

Manufacturing development study using QFD method that has prospects for development approaches that lead to the development of services, to improve services to the public which deals with the design, models and product selection has been done by several authors [2, 6, 8, 12, 15]. Development in the field of manufacturing, with respect to the development of manufacturing strategy based on fuzzy, Integration and value engineering in the Furniture Manufacturing Industry, as well as the picture of the future quality of the product for future research are discussed by [1,5,10]. Meanwhile other works by Zarei [18] was discussed on food supply chain leanness, and [7] on designing a sustainable maritime supply chain. Supply chain is a series of link that is often used in the manufacturing process. The development made by both literatures is the aspect of wear and the design of supply chain [4]. The Assessment of Aircraft Maintenance Education, process Machining, Concurrent engineering were set on approach in the development of composite [9]. There was a different product development discussed on the literatures. Each of them was develop products related to the assessment of aircraft maintenance, machining processes, and composites. Basically each literature has a goal to develop better products.

Reviewing the literature over from year to year the diverse areas of study and products developed by some experts. Over time, the QFD method is always used in a variety of product development, in 2016. This method was also applied to the development QFD Engine Optimization Multipurpose shredders. Applying QFD in this study is a right step in meeting consumers' desire to get new innovations to grate multipurpose machine.

* Corresponding author: ir_nukman2001@yahoo.com

2 Research method

Questionnaire dates were made from a predetermined sample of 40 people. It consists of four regions population research sites, then integrated in a distribution table traders overall assessment questionnaire, ranging from OI region, Muara Enim, OKI, and Palembang as shown in Table 1.

Table 1. Questionaire for Assesment

No	Type of Problem	Total Samples (people)	Rating (*) SA	A	D	SD	Persentase	Category
1	Do you think the engine size coconut grater you need a large place ?	40	140	15			$\frac{155}{160}$x100 = 96.8 %	Strongly agree
2	Is your coconut grater machines are too complicated when the operator of the initial (start the engine) ?	40	136	18			$\frac{154}{160}$x100 = 96.2 %	Strongly agree
3	Coconut grater machine type do you use ?	40					$\frac{158}{160}$x100 = 98.7 %	Strongly agree
	a. Type Electric Motor							
	b. Type Motor Gasoline		152	6				
	c. Type Diesel							
4	Is cocanut grater machines you use too wasteful of energy?	40					$\frac{69}{160}$x100 = 43.1 %	Disagree
	a. Type Electric Motor (*kWh*)							
	b. Type Motor Gasoline (*fuel*)			3	35	11		
	c. Type Diesel (*Diesel fuel*)							
5	If you are given the opportunity to choose , you prefer a machine that uses coconut grater ?	40					$\frac{154}{160}$x100 = 96.2 %	Strongly agree
	a. Type Electric Motor		4	6				
	b. Type Motor Fuel		132	12				
6	Is it difficult to perform maintenance coconut grater machines ?	40					$\frac{51}{160}$x100 = 31.8 %	Disagree
	a. Clean				2			
	b. Lubricate				18	31		
7	Is the tire cord (Fanbelt) pully intrusive security ?	40	160				$\frac{160}{160}$x100 = 100 %	Strongly agree
8	Is it hard to put the raw material into the roller grated coconut ?	40			44	18	$\frac{62}{160}$x100 = 38.7 %	Tidak Setuju
9	Are all the raw materials with refined coconut terparut ?	40	156	3			$\frac{159}{160}$x100 = 99.3 %	Strongly agree
10	Is it hard to do the cleaning on the roller grate , when not in use ?	40	128	24			$\frac{152}{160}$x100 = 95 %	Strongly agree
11	Is coconut grater machine too noisy ?	40					$\frac{160}{160}$x100 = 100 %	Strongly agree
	a. Engine vibration noise							
	b. Sounds kanalpot		160					
12	Whether , component (Spare Parts) are difficult to obtain in the market ?	40					$\frac{45}{160}$x100 = 28.1 %	Disagree
	a. worn-out Bearing							
	b. Dead Spark Plug							
	c. Roller grate broken			3	6	36		
	d. Tire cord pulley (fanbelt) breaks							
*I	How old is the use of coconut grater machine you now ?	40					$\frac{156}{160}$x100 = 97.5 %	Strongly agree
	a. 1 to 6 months							
	b. 6 to 12 months			3				
	c. 1 to 2 months				6			
	d. 2 to 5 months					147		
*II	Do you think the usefulness of coconut grater machines already help people ?	40	160				$\frac{160}{160}$x100 = 100 %	Strongly agree
*III	Is the machine uses coconut grater has a major role in improving your economy ?	40	160				$\frac{160}{160}$x100 = 100 %	Strongly agree

Dates obtained from the questionnaire for each area, further data processing with statistical dates analysis techniques described referring to issues based consumer desires. The results in the table above and then poured into House of Quality QFD (House Of Quality).

Fig1. Phase of QFD

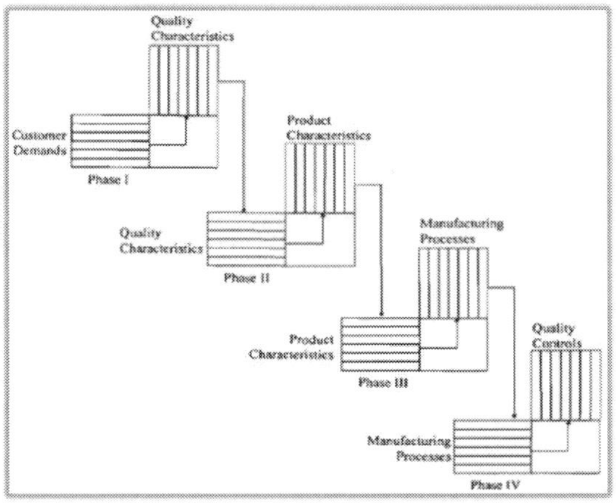

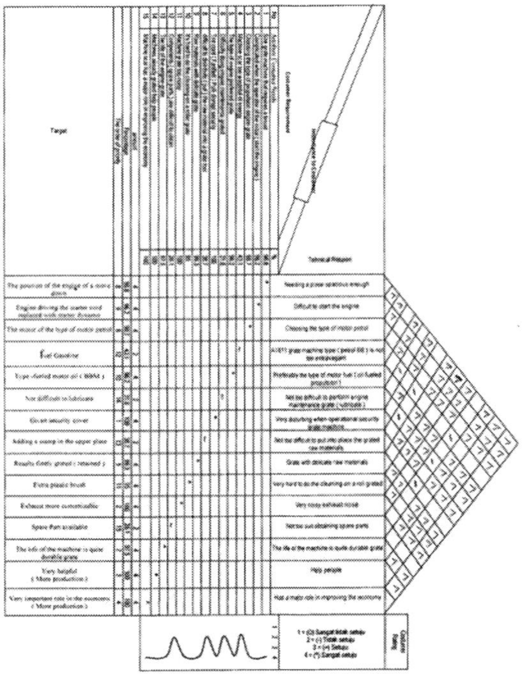

Fig 1. House Of Quality (HOQ)

HOQ house of quality the can be seen Priority needs are desired by consumers. The order of priority in the HOQ above can in description the following caption:

1. Fan Belt interfere with the safety of an operation and a grater machine too noisy then the solution by adding components aluminum cover plate that protects the tires Ropes (fanbelt) and modified the exhaust forms from simple to more customized and equipped with a silencer.
2. The machine uses a greater role in helping people and machines grate in improving the economy. Questionnaire dates showed consumers still want the existence of a grater machines, essentially consumers still need a grater machines as a means to create jobs by selling services grated or shredded results, and become involved in economic principles so that producers do not have to worry mengsa market to customer needs.
3. Is the raw material refined gratering? and what type grater machines using? ", dates obtained most consumers stated that the results gratering fine grater. as well as questionnaire dates and field observations the average trader using the engine driving type Gasoline Engine.
4. Age use a grater machine and engine size grater, the average merchant traders had undergone profession for 2 years and above, as well as those reported to have never replace the engine, it is proven that the machine age grate quite durable. as well as dates showed that, the engine size grater, it is not suitable. Then the solution by moving the position of the engine of the machine tool gets under grater, so their size is slim.
5. Elaborate current operator of the initial (start the engine) and If you had to choose, would prefer the

machine what type of motor grater? "Operation consumers complained of a beginning. is it hard to use the starter rope, then the solution by adding the electric starter system, as well as the majority of merchants to choose the type of motor fuel than other types of electric motors, because it can increase operational costs to prepare electricity.

6. It's hard to do the cleaning roller grate, when not in use and energy-intensive grater machines. Analysis of statistical data showed 95% explained that it is difficult to do the cleaning roll grater (eye roll grated), so the solution must be available grater machine brushes, as well as In accordance with the problem to 3, the majority of consumers use an engine of types Motor Gasoline. As well as scores on this subject showed 43.1% (* disagree) it can be concluded that the grater machines they use less wasteful of energy.

7. It is difficult to enter into the roller grated raw materials and difficult to perform engine maintenance grater, questionnaire data showed 38.7% (* disagree). But assessed the level of security of the observation space, a trader presses the raw material used pieces of wood or fingers. Then add Aluminum Cap Plate is suitable to prevent the remaining pieces of the raw materials did not toss out, without having to press it with a finger / timber. As well as dates showed 31.8% (* do not agree) if such treatments lubricate.

8. Components (Spare Parts) are difficult to obtain in the market. The data shows 28.1% (* disagree), but sometimes the quality is somewhat different

Fig 2. Engineering design grater [11]

The resut found may explains the priorities, the function and purpose of any problems in accordance with the expectations and desires of consumers leading to operation security, maximizing the function of the tool, as well as the addition of new tools in the development of multipurpose grater machines.

3. Result and Discussion

Distribution of the questionnaire merchant value (Muara Enim) Distribution of the questionnaire merchant value (Muara Enim)

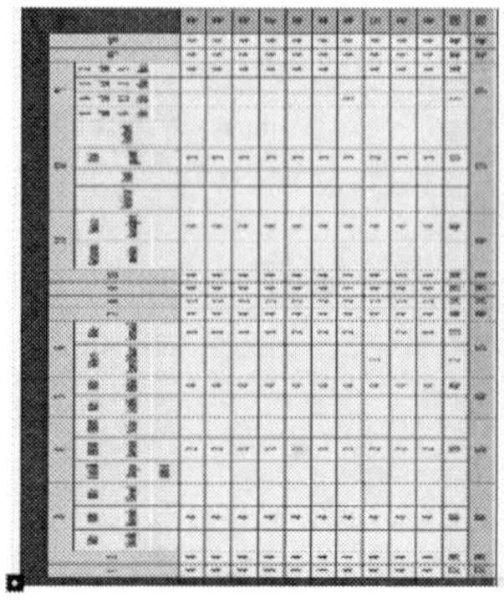

Distribution Frequency

No	X	F	FX	X^2	FX^2
1	51	1	51	2601	2601
2	49	8	392	2401	153664
3	47	1	47	2209	2209
	∑	10	490	7211	158474

Mean $(\bar{x}) = \sum \frac{fx}{N} = \frac{490}{10} = 49$

Varian $(S^2) = \frac{n \sum fx^2 - (\sum fx)^2}{n\,(n-1)} =$

$\frac{10\,(158474) - (490)^2}{10x9} = 14{,}93$

standard Deviation (SD) = $\sqrt{S^2} = \sqrt{14{,}93} = 3.86$

Modus (Mo) = 49

Normality test (Km) = $\frac{\bar{x}-Mo}{S}$ = (49-49)/3.86 = 0

Normal distribution of data for value Km = 0

located in reception area $-1 \leq Km \leq 1$

To determine the interpretation of the level of consumer desires of achievement scores obtained from the value of questionnaire data vendors machine user of grated coconut are required in order to interpret the results of the assessment with a score value (n) the average number of samples 49 and (N) total score of base 60, then percentage of calculation process are :

$P = \frac{n}{N} \times 100\%$

then :

$P = \frac{49}{60} \times 100\% = 81.66\%$

From the data contained in the questionnaires for the district of Muara Enim, with the calculation of distribution table, frequency scores questionnaire data vendors, then calculate the Mean = 49, Varian = 14.93, standard deviation = 3.86, mode = 49, Normality Test = 0 and interpersentase score level consumer desire with the score value achieved was 81.66%. Then the instrument data from the questionnaire scores interpretation value questionnaire merchants engine users grated achieve a score of 81.66%, can be matched with a conversion table in the form of a graduated scale model of which is contained in Table No. 1 by weight 4 results definition "Strongly Agree". Where the scores achieved are in the range of 76% -100% criteria. So we can conclude that for an assessment based on an overall assessment criteria for the desires of consumers (Muara Enim district) to the problems that exist "Strongly Agree" to undertake a development of the engine multipurpose grater.

Distribution of the questionnaire merchant value (OKI)

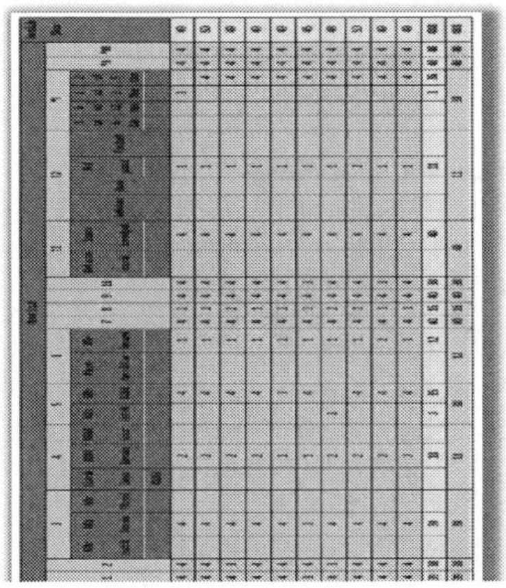

Distribution Frequency

No	X	F	FX	X^2	FX^2
1	50	2	100	2500	10000
2	49	6	294	2401	86436
3	48	1	48	2304	2304
4	46	1	46	2116	2116
	$\sum$	10	488	9321	100856

$$\text{Mean } (\bar{x}) = \sum \frac{fx}{N} = \frac{488}{10} = 48.10$$

$$\text{Varian } (S^2) = \frac{n \sum fx^2 - (\sum fx)^2}{n\,(n-1)} =$$

$$\frac{10\,(100856) - (488)^2}{10x9} = 8.55$$

Standard Deviation (SD) = $\sqrt{S^2} = \sqrt{8.55} = 2.92$

Modus (Mo) = 49

$$\text{Normality test (Km) (Km)} = \frac{\bar{x} - Mo}{S} = \frac{48.10 - 49}{2.92} = -0.3$$

Normal distribution of data for value Km = - 0.3

located in reception area $-1 \leq \text{Km} \leq 1$

To determine the interpretation of the level of consumer desires of achievement scores obtained from the value of questionnaire data vendors machine user of grated coconut are required in order to interpret the results of the assessment with a score value (n) the average number of samples 48.1 and (N) total score of base 60, then percentage of calculation process are :

$$P = \frac{n}{N} \text{ x } 100\%$$

then :

$$P = \frac{48.10}{60} \text{ x } 100\% = 80.16\%$$

From the data contained in the questionnaire instrument for OKI district, with the calculation of the distribution table, frequency scores traders questionnaire data, then calculate the Mean = 48.10, Varian = 8.55, standard deviation = 2.92, mode = 49, = -0.3 and Normality Test interpersentase score level consumer desire with the score value achieved was 80.16%.

So from the instrument data interpretation value questionnaire scores merchants of grated machine user reached a score of 80.16%, can be matched with a conversion table in the form of a graduated scale model of which is contained in Table No. 1 by weight 4 results definition "Strongly Agree". Where the scores achieved are in the range of 76% -100% criteria. So we can conclude that for an assessment based on an overall assessment criteria for the desires of consumers (district OKI) against the existing problems "Strongly Agree" to undertake a development of the engine multipurpose grater.

Distribution of the questionnaire merchant value (OI)

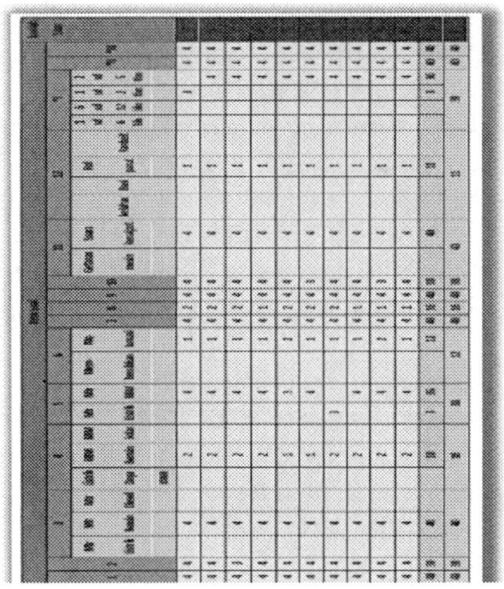

Distribution Frequency

No	X	F	FX	X^2	FX^2
1	50	2	100	2500	10000
2	49	7	343	2401	117649
3	40	1	40	1600	1600
	$\sum$	10	483	6501	129249

Mean $(\bar{x}) = \sum \frac{fx}{N} = \frac{483}{10} = 48.3$

Varian $(S^2) = \frac{n \sum fx^2 - (\sum fx)^2}{n\,(n-1)} =$

$\frac{10\,(129249) - (483)^2}{10x9} = 11.76$

Standard Deviation (SD) = $\sqrt{S^2} = \sqrt{11.76} = 3.42$

Modus (Mo) = 49

Normality test (Km) = $\frac{\bar{x}-Mo}{S} = \frac{48.3 - 49}{3.42} = -0.2$
Normal distribution of data for value Km = -0.2 s located in reception area $-1 \leq Km \leq 1$.

To determine the interpretation of the level of consumer desires of achievement scores obtained from the value of questionnaire data vendors machine user of grated coconut are required in order to interpret the results of the assessment with a score value (n) the average number of samples 48.3 and (N) total score of base 60, then percentage of calculation process are :

$P = \frac{n}{N}$ x 100%

then :

$P = \frac{48.3}{60}$ x 100% = 80.5%

Data generated from the questionnaires contains with the calculation of the distribution table, frequency scores traders questionnaire data. Afterward, it found a calculation of Mean=48.3, Varian= 11.76, standard deviation = 3.42, mode = 49, = -0.2 and normality test interpersentase score level consumer desire with the score value achieved was 80.5%.

From the instrument data, the interpretation value questionnaire scores, it was reached a score of 80.5%. It can be matched with a conversion table in the form of a graduated scale model of which is contained in Table 1 by weight 4 results definition "Strongly Agree". Where the scores achieved are in the range of 76% -100% criteria.

Therefore, it can be concluded that for an assessment based on an overall assessment criteria for the desires of consumers (district OI) to the problems that exist "Strongly Agree" to undertake a development of the engine multipurpose grater.

4 Conclusion

From the discussion, it can be concluded that the priority consumers are very stressed to do a modification component or an additional tool in the machine grater multipurpose namely Cover Safeguard Fant Belt, Exhaust, Priority Motor Driving Fuel Gasoline, layout position Mover, Motor Starter, Plastic cleaning brush Roll Grate, and Plate Cap.

The conclusion is based on data obtained through questionnaires, and supported by observation and open interviews. Conclusions is generated from the data within the desires and expectations of consumers as user's grater multipurpose machine. The results of this study is expected to enhance an action of development to create new innovations using the latest ideas in optimizing the function of the multipurpose grater engine.

References

1. Annappa, C., & Panditrao, K. (2013).. IJITEE, 2(6), 2278-3075.
2. Bernal, L., Dornberger, U., Suvelza, A., & Byrnes, T. Report Universitat Leipzig. (2009).
3. German, M., Mengoni, M., & Peruzzini, M. Computers in Industry 63, 12–29 (2012)
4. Jen, C., Cheng, Y., Shen, H., & Yeong, L. ICEE2004/Proceedings/Papers (2015)
5. Jia, G., & Bai, M. (2011). Computers & Industrial Engineering 60, 445–454.
6. Kuo, H. A. (2011). ICIC International, 7, 253-268.
7. Lam, J. Transportation Research Part E., (2014).
8. Na, L., Xiaofei, S., Yang, W., & Ming, Z. International Conference on Complexity Science and Information Engineering, 243 – 251. (2012).
9. Nukman. Jurnal Rekayasa Sriwijaya, Vol. VI, No. 1, April (1999).
10. Oke, S. Literature Review and Future Trends. DOI, 17(3) (2013)
11. Pattola, A. Indonesian Patent No. 10333. (1988)
12. Rajesha, G., & Malliga, P. Proceedings International Conference On Design and MAnufacturing IConDM 2013), 1283 – 1292. (2013).
13. Sapuan, S., & Mansor, M. Materials and Design 58, 161–167 (2014).
14. Sudiharto, P. Oration (2009).
15. Tian, Y. Report. Aalto University. (2011)
16. Tunggal, A. Report PT Rineka Cipta. (1998)
17. Yilmaz, H. Thesis ed. Izmir Institute of Technology. (2009)
18. Zarei, M., Fakhrzad, M. B., & Paghaleh, M. J. Journal of Food Engineering, 25-33. (2011)
19. Luigi T. De Luca, *Propulsion physics* (EDP Sciences, Les Ulis (2009)
20. F. De Lillo, F. Cecconi, G. Lacorata, A. Vulpiani, EPL, **84** (2008)

Experimental study on a hybrid loop heat pipe

Iwan Setyawan, *Imansyah* Ibnu Hakim, *Nandy* Putra*

Applied Heat Transfer Research Group, Department of Mechanical Engineering, Universitas Indonesia, 16424 Jawa Barat, Indonesia

Abstract. A conventional loop heat pipe two-phase heat transfer device of passive system often can no longer meet the challenging cooling needs due to the inherent limitations of the capillary pumping which can lead to dry out. This study aims to create a *loop heat pipe* uses capillary wick copper sintered with centrifugal casting method. The stressing effort to overcome the dry-out by adding a diaphragm pump to accelerate the fluid transportation from the condenser to the evaporator (*hybrid loop heat pipe, HLHP*), where the pump is equipped with a reservoir and both installed on the liquid line. In testing the performance of HLHP also varying the *filling ratio, FR*: 50%, 60%, and 80%. The pump will be activated when the dry-out took place, by the piezo electric diaphragm pump with temperature controller installed in the evaporator that was set to activate the pump to work. From the results of the experimental, the pump successfully prevented the occurrence of dry out, and reduced the temperature of the evaporator from 130°C to 80°C, owing the pump distributed the working fluid from the condenser to the evaporator efficiently. The result indicated the best performance of HLHP was filling ratio, FR of 60%.

1 Intoroduction

Nowadays electronic devices tend to have increasingly higher power densities, that certainly would require a heat transfer large amounts as well to maintain their performance and reliability [1]. Since the high heat flux and large heat dissipation operation, many commercial electronics and military weapon systems require high performance cooling. With the rise of hybrids and electric vehicles, next generation power electronics that are known to reach heat fluxes of 100 W/cm^2 are growing sources needing high-performance cooling. The military has been developing various direct energy weapons such as railguns and high power laser systems requiring large amounts of cooling and high heat flux dissipation that are critical for their efficient operation and likely achievable only by two-phase heat transfer [2]. In this case the heat pipe is expected ideal applied to the devise that demanding high reliability and performance as a two-phase cooling system [3, 4]. As known that the two-phase heat transfer has higher heat transfer than heat transfer using a single-phase sensible heat of fluid. Latent heat of vaporization of water for instance, is two orders greater than sensible heat in the cooling system of one phase of the same size. This makes the two-phase heat transfer is superior in terms of heat transport capacity. In addition, the effect of boiling gives the possibility of increased absorption of heat per unit volume of liquid and boiling also produces heat transfer coefficient is much larger ranged 22,5-100 kW/m^2-K [5-9].

The two-phase cooling technologies used for aerospace applications include *heat pipes, loop heat pipes* (LHP) and *capillary pumped loops* (CPL) which are all passive cooling system, thus very reliable devices relying on only capillary pumping. However, the passive devices could not meet future challenging cooling demand because of the inherent limitation of the capillary pumping in terms of heat flux, transport distance and multiple heat source capabilities [10-12]. Meanwhile, high heat flux in advanced defense devices is found in radars, directed-energy laser and microwave weapon systems, and avionics. These devices follow the trend of escalating power density of commercial electronics, heat fluxes from defense devices are now projected to exceed 1000 W/cm^2. This level of heat dissipation exceeds the capabilities of today's most advanced dielectric liquid cooling systems, single-phase or two-phase, which points to an urgent need to develop new powerful cooling solutions[13]. The limitations of the above seem to be resolved with the successful integration of mechanical pump into the exhaust heat system of the Rover Exploration Mars, NASA which also helped create the current technical trends with increasing use of active cooling design for space applications [10]. This active cooling system is known as a hybrid loop cooling system that combines mechanical pumping liquid active with capillary pumping liquid passive. Thus, a hybrid two-phase loops (HTPL) combines the capabilities of the high heat flux and robust operation of conventional pumping two-phase loop with the simplicity and reliability of the

*Corresponding Address: nandyputra@eng.ui.ac.id

capillary mechanism [7, 14]. Then the hybrid system also has been observed for other applications, no longer confined on the space applications[6, 7, 9, 10, 14-16]. The results of their research represented a significant advance over the state-of-the-art in hybrid loop heat pipes.

Based on the limitation in the two-phase cooling technologies, hence in this study, a loop heat pipe has been designed where is modified with installing a pump mechanism in LHP, that is called *hybrid loop heat pipe, HLHP*. The concept of design is slightly different from other research concepts, which in this study of the loop heat pipe works as passive cooling when the heat and other parameters are entered did not cause the appearance of dry out phenomenon, and if dry out happen, then the pump will be activated. Thus, the HLHP can operate with two different conditions.

2 Methodology

In this study, HLHP is developed by the integrating hybrid method. The study was conducted through several stages namely manufacture of wick capillary, design and manufacture and performance testing of HLHP.

2.1 Manufacturing of the capillary wick

The manufacturing process of the wick, i.e., first of all make preparation the copper powder. Then mixed with a binder, where a mixture ratio is 45wt% binder and 55wt% copper powder. The slurry was then mixed until homogeneous. Furthermore slurry poured into the pipe that will be used as evaporator, where in this study was made of copper pipe with lenght of 65 mm and inner diameter of 23 mm. In order that slurry can cling with thickness evenly on the evaporator, inner wall of the pipe, then the process was continued in stage of centifugal casting, by giving a constant rotation with connecting to the 12 volt DC motor. Then proceed to the final stage, where the product through the sintering process by entering into vacuum furnace with the sintering temperature was set at 800 ° C, the temperature rate was set at 20 ° C / min and the heating time was set for 3 hours.

2.2 Design and manufacture of HLHP

The evaporator is designed in cylinder form and made of copper pipe with 65mm in length, an inside diameter of 23 mm and a wall thickness of 1.2 mm which merges with the capillary wick, having a thickness of 2 mm that was made previous. The condenser is also made of copper tubing with dimensions of 70 mm long, 23 mm in diameter. For ensuring the availability of liquid, that will be pumped if it is needed, HLHP is equipped with a reservoir made of pipes with dimensions of 23 mm in diameter and 65 mm in length. In the mid of liquid line from the condenser is designed, where a straight pipe to the evaporator and a short pipe deflected into the reservoir which then the outlet pipe on the bottom of reservoir connected to the piezoelectric pump (**fig.1**).

The addition of the piezo electric pump was used to pump liquid into the evaporator when required, where the hybrid circulation system is expected to help circulate the working fluid from the condenser to the evaporator .

Fig. 1. Piezo electric diaphragm pump

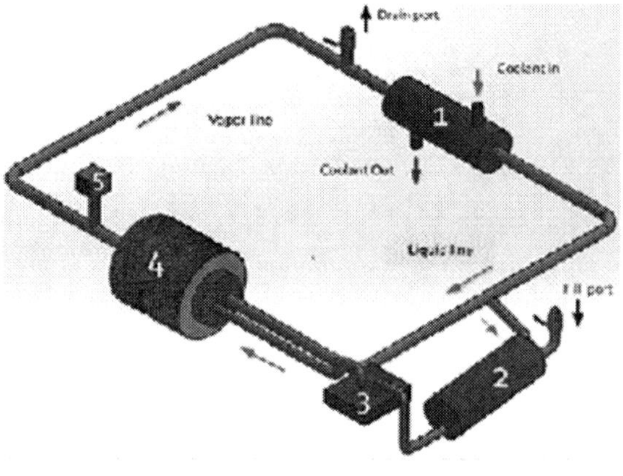

1 : Condenser 4: Evaporator
2 : Reservoir 5: Pressure sensor
3 : Piezo Electric Pump

Fig. 2. Design of hybrid loop heat pipe

The principle of operation of HLHP is depicted in Figure 2. it has evaporator, condenser, reservoir and piezoelectric pump. The liquid reservoir provide enough space to accommodate the liquid to be used when needed. So, the liquid from the condenser flow through the evaporator in two ways: by simply capillary pumping or the combined of capillary pumps and pump piezoelectric.

The manufacture of the evaporator is depicted in Figure 3. The evaporator wick has thicness 2 mm and made from powder copper sintered with 100 μm particle diameter. To prevent the flow of vapor to the liquid line, then at the entrance pipe side of the evaporator inserted a wick with 20 mm lenght that made from the powder copper sintered with 300 μm particle diameter.

In contrast to the loop heat pipe (LHP) conventional works passively, the Hybrid Loop heat pipe (HLHP) which will be made of this, when the conditions in the evaporator reaches the temperature set point, which has been set on the temperature controller, the pump becomes active that will diliver liquid from the reservoir to the evaporator. Thus, the total liquid supplied to the evaporator equal to that produced by the pump plus liquid due to capillary pressure on the wick.

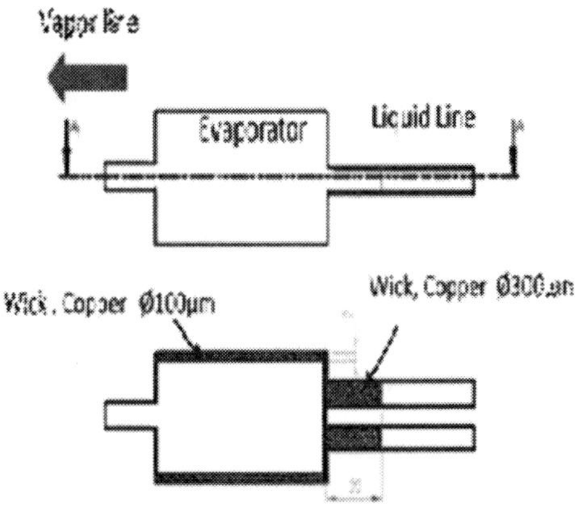

Fig. 3. Design of the HLHP evaporator

2.3 Performance testing of HLHP.

2.3.1 Working fluid filling procedure on HLHP

The leakage on HLHP should really be checked at each connection before charging the working fluid. The vacuum pump was used to look for any leaks to as low as -99 kPa which is the limit value of the pump maximum capacity we used. After that, HLHP was left for a few time to ensure no leaks in any part. Then charging the HLHP by filling the distilled water into HLHP through the charging port (fill port) in the reservoir fluid by using an equipment that has been made specifically.

Distilled water was used as the working fluid HLHP with variying the filling ratio (FR) i.e, 50%, 60%, and 80%. The maximum vacuum was done on the HLHP during charging to ensure that there were no non-condensable gases in the system[17].

In the charging process, although it has been done with the maximum effort to avoid the incoming ambient air as well as into HLHP when the charging process, but the amount of air which dissolves in the water can not be avoided, therefore HLHP requires degas procedure before continuing to testing. On the removal procedure of trapped air, the authors followed the method of Park et al. [15], namely operate HLHP with low heating load was enough to make vapor flow in the system. The air is removed from the water, eventually collected in the liquid reservoir. Then the collected gas be withdrawn by a vacuum pump that was connected to the top of the liquid reservoir.

2.3.2 Schematic of performance testing on HLHP

In this expriment, thermocouples K type are installed at some points on the HLHP as shown in the **fig.4.**. Some thermocouples are also mounted on the inlet and outlet of the evaporator, vapor line, inlet and outle condenser, outlet and inlet coolant from CTB, liquid line, and reservoir, then the measured data will be recorded using the NI cDAQ-9174 and NI-9213 which then processed by using Labview software. The HLHP working pressures during the test was measured by using a pressure transmitter PSA-C01, and the NI 9203.

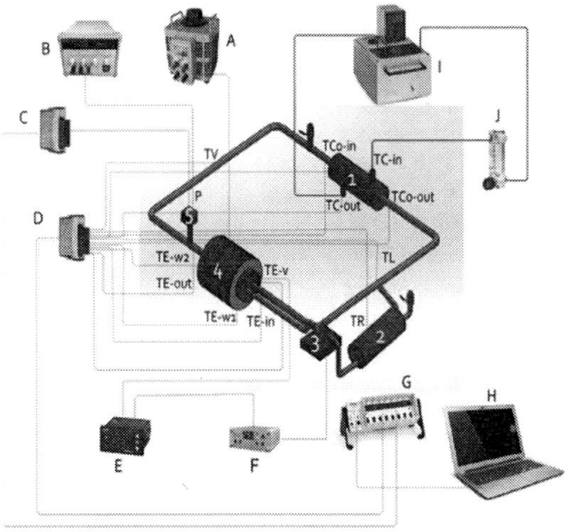

A	: AC power supply	I	: Circulating Thermostatic bath (CTB)
B	: DC power supply	J	: Rotameter
C	: NI- 9203	1	: Condensor
D	: NI- 9213	2	: Reservoir
E	: Themperature Controller	3	: Piezo Electric Pump
F	: Pump Voltage Controller	4	: Evaporator
G	: NI cDAQ- 9174	5	: Pressure control
H	: Computer		

Fig. 4. Schematic of testing on HLHP

3 Results and discussion

In the Figure 5, shows a graph when the experement is conducted at 20W heat load and filling ratio, FR 50%. Form the figure, it can be seen that the boiling condition in the evaporator chamber occurred after 2,500 seconds of operation. It was indicated by the increasing temperature difference between the evaporator wall and its vapor followed by a fall of temperature in the vapor line suddenly. At the same time, temperature at the inlet of the condenser increased, this indicated the increasing of the flowrate of working fluid in HLHP. The steam generated in the evaporator is not sufficient to generate adequate pressure to force the fluid flow more smoothly in HLHP. Since the heat in vapor cannot released in the condenser perfectly, and condesat from condeser is not suficient to supply cooling liquid to the evaporator, it caused the evaporator temperature increaseed rapidly which will lead to dry-out phenomenon. But in this study, dry-out condition has to be prevented by seting the controller to activate the pump before the evaporator temperature reaches 130°C.

In the Figure 5 also presented the phenomenon or temperature distribution on HLHP, when the pump is activated. It can be seen, that the vapor temperature Tv and wall evaporator temperature Te-w decreased drastically after piezoelectric pump is turned on. This is a contrast with rising extrem temperature momentary on the condenser inlet. This phenomenon was probably caused by the alternation of the status of two-phase flow due to the vapor-liquid fluid from the pump supply which suddenly actived is resulting in instantaneous evaporator temperature drops drastically. With the volume of the evaporator mostly liquid will stop the evaporation and furthermore the supply of fluid from the pump will force the working fluid circulation which causes the temperature dramatically increased momentary to whole HLHP except evaporator section. After a certain time of turning on the pump, the temperature on the evaporator reduced into stable condition at 80 ° C .

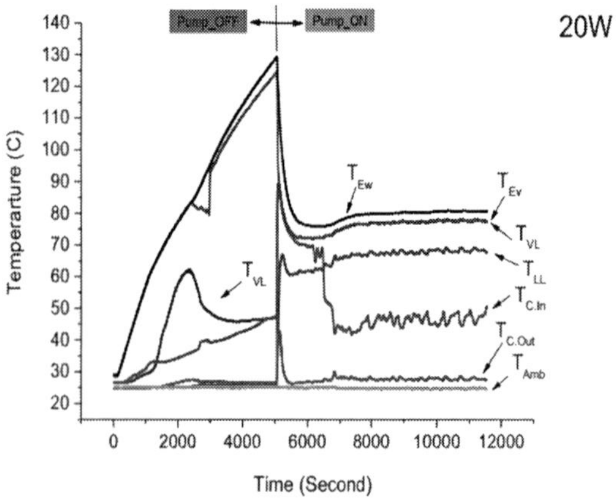

Fig. 5. HLHP experimental result, test at 20W

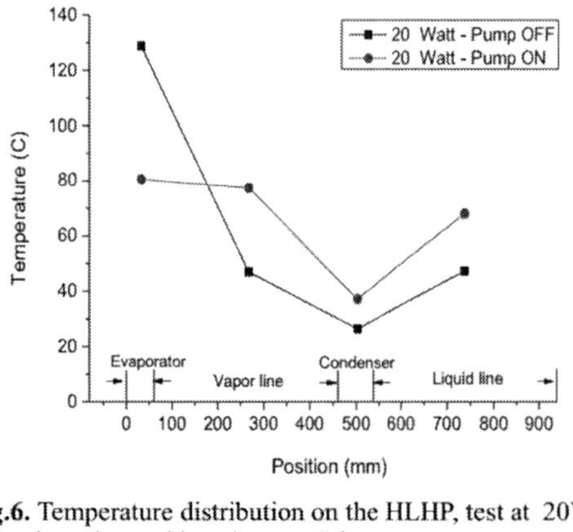

Fig.6. Temperature distribution on the HLHP, test at 20W with pump in active and inactive condition.

In the Figure 6, shows the temperature distribution in HLHP with a heat load on the evaporator 20W with pump in active and inactive condition. It is clear that HLHP can reduce the temperature of the evaporator and can prevent the LHP from the occurrance of dry-out conditon.

In the Figure 7, shows the temperature distribution in HLHP, with 60W heat load on the evaporator by varying the filling ratio, FR 50%, 60% and 80%. In the figure can be seen, that for all cases when it reached steady state condition, the temperature distribution has the same trend, the temperature decreased when it entered the vapor line and then through the condenser and to the liquid line except in cases FR 50% when the pump was activated, after exit condenser the temperature slightly increases when entering the liquid line.

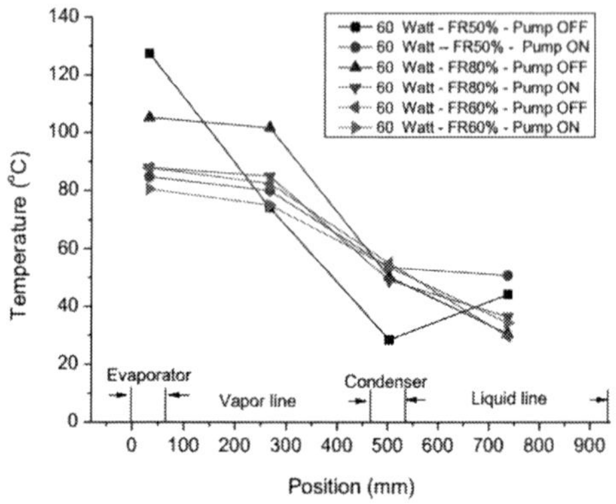

Fig.7. Temperature distribution on the HLHP, test at 60W with pump in active and inactive condition (Filling ratio, FR: 50%, 60%, and 80%).

Furthermore, it can be seen, when the pump inactive, the temperature of the HLHP with FR 60% and FR 80%, HLHP successfully achieved steady state when the evaporator temperature about 88°C and 105°C for FR 60 % and FR 80% respectively. On the graph can be seen the condition with the FR 50% after pump is switched on, the evaporator temperature successfully decreased from 130°C to be steady at 82°C or 48°C temperature difference. However with FR 60%, when pump in HLHP is activated when temperatur on the evaporator reached 88°C, after that temperature on the evaporator decreased and reaches constant at 80°C or differences only 8°C. Similarly, for FR 80% when temperature of evaporator reached 105°C then pump is turned on, the temperature of evaporator reduce and to be constant at 85°C or it decreased 20°C. A possible reason to explain this is by FR FR80% and 60% although pump has not been activated but HLHP already operated properly.

4 Conclusions

The experiment on HLHP has been conducted with the filling ratio, FR : 50%, 60%, and 80% at heat load 20W and 60W. Instalated piezoelectric pump in modified LHP system was able to prevent the occurance of dry-out in the evaporator. With this system, it will more working fluid flow from condenser to evaporator. From the experiment, the condition of FR 60% has the best performance. The system of HLPH can be considered for the system with high heat flux.

The authors would like to thank DRPM, Universitas Indonesia, for funding this research.

References

1. N. Putra, I. Ibnu Hakim, I. Setyawan, and M. Zayd AI, *APCEAS*. **484**, pp. 353-360, August 27 (2015)
2. M. Crepinsek and C. Park, *Applied Thermal Engineering* **38**, pp. 133-142 (2012)
3. J. A. Weibel, S. V. Garimella, and M. T. North, Int. J. Heat and Mass Transfer **53**, pp. 4204-4215 (2010)
4. J. A. Weibel and S. V. Garimella, Int. J. Heat and Mass Transfer **55**, pp. 3498-3510 (2012)
5. W. Anderson, P. Dussinger, S. Garner, J. Hartenstine, and D. Saraff, *ASME 2009 Heat Transfer Summer Conference collocated with the InterPACK09 and 3rd Energy Sustainability Conferences*, pp. 497-503 (2009)
6. C. Park, J. Zuo, P. Rogers, and J. Perez, *Hybrid Loop Thermal Bus Technology for Vehicle Thermal Management (*DTIC Document, 2004)
7. C. Park, A. Vallury, and J. Perez, *Advanced hybrid cooling loop technology for high performance thermal management* (Defense Technical Information Center, 2006)
8. D. Sharar, N. R. Jankowski, and B. Morgan, *Review of two-phase electronics cooling for army vehicle applications* (DTIC Document, 2010)
9. C. Park and M. Crepinsek, J. Thermophysics and Heat Transfer **25**, pp. 572-580 (2011)
10. C. Park, A. Vallury, J. Zuo, J. Perez, and P. Rogers, *ASME/JSME 2007 Thermal Engineering Heat Transfer Summer Conference collocated with the ASME 2007 InterPACK Conference*, pp. 911-916 (2007)
11. N. Putra, I. Setyawan, and D. Raditya, *Applied Mechanics and Materials*, pp. 575-579 (2016)
12. M. H. Kusuma, N. Putra, A. R. Antariksawan, Suryadi, and F. A. Imawan, Nuclrear Engineering and Technology (2016)
13. M. K. Sung and I. Mudawar, J. Electronic Packaging **131**, p. 021013 (2009)
14. D. B. Sarraf and W. G. Anderson, *High Temperature and High Heat Flux Thermal Management for Electronics*
15. C. Park, A. Vallury, and J. Zuo, J. Thermal Science and Engineering Applications **1**, p. 022004 (2009)
16. C. Jiang, W. Liu, H. C. Wang, D. D. Wang, J. G. Yang, J. Y. Li, *et al.*, *Applied Thermal Engineering* **71**, pp. 581-588 (2014)
17. D. Wang, Z. Liu, J. Shen, C. Jiang, B. Chen, J. Yang, *et al.*, *Experimental Thermal and Fluid Science* **57**, pp. 157-164 (2014)

A probability approach in cans identification

Yulia Resti[1,*], *Amrifan* S.Mohruni[2], *Firmansyah* Burlian[2], *Irsyadi* Yani[2], and *Ali* Amran[1]

[1]Department of Mathematics, Faculty of Mathematics and Natural Science Universitas Sriwijaya, 30162 Inderalaya, Indonesia
[2]Department of Mechanical Engineering Faculty of Engineering Universitas Sriwijaya, 30162 Inderalaya, Indonesia

Abstract. The objective of this study is to identify can waste into three types based on the images by using a probability approach of trinomial distribution in term regression. Predictor variables considered are the color intensity of red, green, and blue of the images taken at the top, down, and side pose successively. From an independence test between each of the predictor variable and can waste type noted that only the color intensity of red which the image taken at top pose that does not correspond to the can waste types. Based on the Nagelkerke value is found that the variance of the predictor variable data in identifying the can waste type is able to explain the variance of the types of 59.1 percent. The final model show that the significant predictor variables are the colors intensity of green and blue which the image taken at the top pose, the color intensity of red which the image taken at down pose, and the color intensity of red, green and blue which the image taken at side pose successively. The model can identify cans waste into three types based on the images correctly by 73.13%.

1 Introduction

Aluminum cans are the most widely recycled solid waste in the world. A lot of energy is taken to make them, throwing away one aluminum can wastes as much energy as pouring out half a can of gasoline, and if aluminum cans are thrown in a landfill, it takes between 100-500 years for just one aluminum can to decompose [1]. In recent years, solid waste management, especially recycling based on automatic sorting system is a major challenge for large cities in developing countries [2].

A cans is a metal container designed to hold a fixed portion of liquid such as carbonated soft drink, fruit juice, herbal tea and the like. Cans are made of aluminium (75% international manufacture) or tin-plated steel (25% international manufacture). Worldwide production for all beverage cans is approximately 52 billion units. An aluminium can is a container for packaging, made primarily of aluminum. It is commonly used for foods and beverages but also for products such as oil, chemicals, and other liquids. A common 12-ounce-size can weigh 15 grams when empty.

The identification process of cans waste is needed in an automatic sorting system of cans recycling. Several approaches have been applied to identify cans waste either the parametric such as discriminant analysis and non-parametric approach such as back propagation algorithm. The application of discriminant analysis to identify cans waste can be found in [4], while the implementation of back propagation algorithm can be found in [3].

The identification system research using automatic or semi-automatic techniques was started in the 1960s and intensified in the last two decades. With the increased availability of powerful computer, it has received significant attention from the research community. There are currently automatic systems that perform well when the object images are captured under uniform and controlled conditions. However, these technologies have several limitations due to the fact that they are difficult to develop, the quality of the image captured, and lighting problem.

The identification system is generally based on the selection of the color space to detect the color identification methods. Object identification is a feature based approach in which face geometry is taken which includes shape and other features like size, edge, color etc. The algorithm requires 2-D images whose threshold values of intensities are taken into consideration in the measurement of the number of the pixels to get the entire object feature area.

Trinomial logistic regression is also the parametric approach which often considered in many analysis because it does not assume normality, linearity, or homoscedasticity, whereas discriminant analysis requires these assumptions. [10] applied trinomial logistic regression to classify the health level of BUMN is based on the financial ratios, [9] classify the severity of traffic accident victims based on the characteristics of victims, types of vehicles, the characteristics and time of collision, while [8] applied trinomial logistic regression in another name as polynomial logistic regression to classify the risk of motor insurance for the third party. Application of trinomial logistic regression in other fields can be found in [11], [12], and [13]. In this paper, the trinomial logistic regression is applied to identify the cans waste based on the some images into three types.

*Corresponding author: yulia_resti@mipa.unsri.ac.id

2 Material and methods

2.1 Data collection and description

We collected three types of cans waste; 100% recyclable (type 1), not 100% recyclable (type 2), and unrecyclable (type 3). The proportion of each of cans waste types is presented in Table 1.

Correct identifying and sorting of aluminium cans from non-degradable materials in solid waste requires many facets of intelligent sorting system. First, for the item to identify itself, it should minimize the number of information carried or remembered on the automatic sorting system. For example, it must be able to distinguish between aluminium and non-aluminium cans and add a new object for automatic sorting in the future. This is to ensure that the best and optimal methods can be chosen in order to produce significant and efficient results.

Table 1. Proportion of cans waste for each type.

Type	% populations
1	50.00%
2	31.25%
3	18.75%

We have collected 3 pictures for each cans. In order to store the feature values, a database with two tables was created using Excel. One of the tables was used to store raw sample data images and another was used to store the template data.

An edge detection algorithm based on the Sobel function was applied to extract the contour of the object after eliminate noise of gray image. Figure 4.4 provides the edge of beverage cans after filtering the global edge.

Each of them is placed on a conveyor belt, which connect to a webcam. The images of them are taken using the webcam for three phases; top, down, and side poses. We save it in JPEG format with a resolution of 320 x 240 pixels, and then extract the color intensity of red, green and blue of each pose using a Matlab comment. Of this population, 50% is taken as a database or for learning, and the rest for testing.

The image capture contained image acquisition and image pre-processing steps. The feature extraction contained cropping and extraction steps, where the cropping consisted of template matching and window feature. On the other hand, the extraction comprised edge and color extraction, thus similarity measurement is to measure the similarity between the measured features and the values stored in the database.

Fig 1. object on the conveyor was detected by a webcam

Some factors, such as lighting conditions, speed of the conveyor belt, camera gain, contrast and brightness, height of the camera above the object, may strongly influence the quality and sharpness of the final image; experiments were carried on in order to find an optimum configuration.

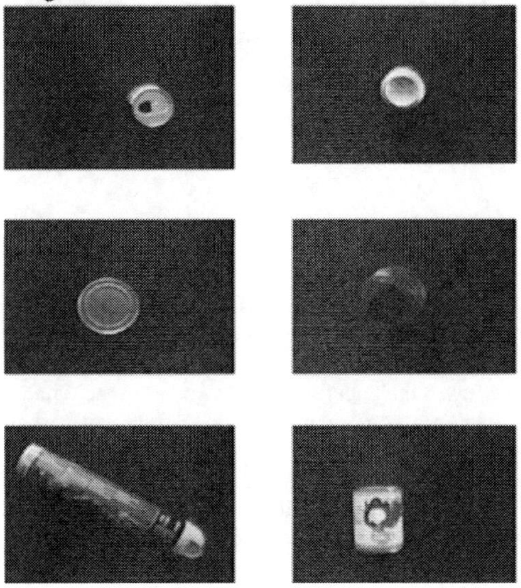

Fig 2 The standard pose of the cans

In feature extraction phase also considered both color and gray scale images. For color image each of the three color components RGB are considered separately. For gray scale image, standard grayscale transformation is obtained from the original RGB image. Processing with the gray-level of one image is called point processing. Once the raw image has been provided by the webcam and grabbing routine, the edges are extracted and the shape of the cans is identified. This procedure has been designed to also work in conditions of bad lighting, high image noise although the experimental condition may not require all the following image processing steps. Processing with the gray-level of one image is called point processing. The manipulation of image is meant for extracting pixel value and pixel position of each pixel of the image. The image is loaded in a form or picture box using load Picture function. Then, using the point method, pixel values of the pixels of an image are obtained. Next, the red, green and blue values for the pixel values are separated. After the necessary operations, the new pixel values are placed in the appropriate positions in a form or in a picture box using pset method and RGB function. All these steps will result in image manipulation.

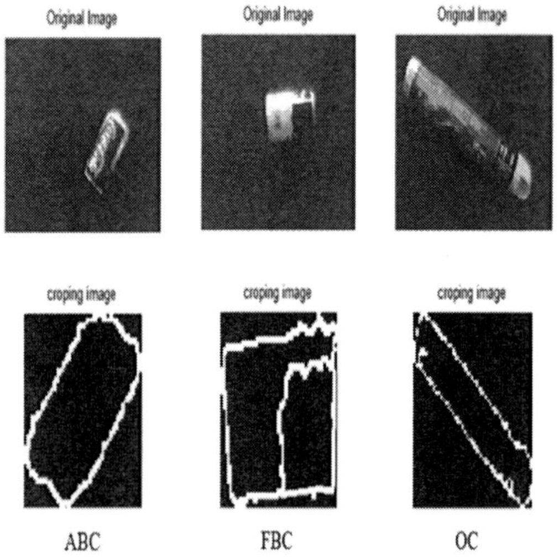

Fig 3. Cropping image

The identification process was handled by the system, after each can was automatically detected by the webcam. The features of the captured object were then matched with the data for the cans waste in the database. For cans waste identification, the rule based classification was developed using the energy distribution with different quantization level.

The main process in develop the database system consisted of image pre-processing, feature extraction and classification. Precision is defined as the ratio of genuine records in the template database retrieved by the identification system and the total number of templates retrieved. Recall is defined as the ratio of the genuine records in the template database retrieved by the identification system and the total number of genuine records in the template database.

The identification process is undertaken by comparing the value of the image with the values in the database to find the closest values. To determine the accuracy of the identification system, the can images are stored in the database with a standard pose.

2.2 Trinomial logistic regression

Consider an image data of cans waste with n observations $(i = 1,2,\cdots,n)$, three cans types $(j = 1,2,3)$, the number of sample in j-th type is denoted n_{ij} and $n = \sum_{j=1}^{3} n_{ij}$. Let N_{ij} be the random variable for i-th can waste with probability into j-th type, $\pi_j(x_i)$, $\mathbf{x}_i = (x_{i0}, x_{i1}, \cdots, x_{ip})^T$ be the predictor variables for i-th can waste. If N_{ij} follow trinomial logistic regression model, the probability mass function is

$$P(N_{ij} = n_{ij}|\mathbf{x}_i) = \frac{n}{\prod_{j=1}^{3} n_{ij}!} \prod_{j=1}^{3} \pi_j(\mathbf{x}_i)^{n_{ij}}$$

(1)

This probability mass function in equation (1) implied that probability of n_{i1}, n_{i2}, n_{i3} fall into can type 1, 2, 3 respectively given a set of predictor variables $\mathbf{x}_i$, where $\sum_{j=1}^{3} \pi_j(\mathbf{x}_i) = 1$. The trinomial logistic regression model was formed with stated $P(N_{i1} = 1|\mathbf{x}_i) = \pi_1(\mathbf{x}_i)$, $P(N_{i2} = 1|\mathbf{x}_i) = \pi_2(\mathbf{x}_i)$ and $P(N_{i3} = 1|\mathbf{x}_i) = \pi_3(\mathbf{x}_i)$ as the following logistic functions,

$$\ln \frac{\pi_j(\mathbf{x}_i)}{\pi_3(\mathbf{x}_i)} = \beta_j^T \mathbf{x}_i$$

(2)

where we chose the cans waste type 3 as reference category in this research. Using the logit link function, we have

$$\pi_1(\mathbf{x}_i) = \frac{g_1}{1 + g_1 + g_2} \quad (3)$$

$$\pi_2(\mathbf{x}_i) = \frac{g_2}{1 + g_1 + g_2} \quad (4)$$

and

$$\pi_3(\mathbf{x}_i) = \frac{1}{1 + g_1 + g_2} \quad (5)$$

where

$$g_1 = \exp(\beta_1^T \mathbf{x}_{i1}) = \sum_{k=0}^{p} \exp(\beta_{1k} x_{i1k}) \quad (6)$$

$$g_2 = \exp\left(\beta_2^T \mathbf{x}_{i2}\right) = \sum_{k=0}^{p} \exp\left(\beta_{2k} x_{i2k}\right) \qquad (7)$$

The regression parameters, β_k, can be estimated from (1) using maximum likelihood estimation, where the log likelihood function is (Agresti, 2013 in Thorsen, 2014),

$$\mathrm{Ln}\left(\beta_{ij}\right) = \sum_{i=1}^{n} \sum_{j=1}^{3} n_{ij} \ln \pi_j\left(\mathbf{x}_i\right)$$

$$= \sum_{i=1}^{n} \sum_{j=1}^{3} n_{ij} \ln \left(\frac{\exp\left(g_j\left(\mathbf{x}_i\right)\right)}{1 + \sum_{j=1}^{2} \exp\left(g_j\left(\mathbf{x}_i\right)\right)} \right)$$

$$= \sum_{i=1}^{n} \sum_{j=1}^{3} n_{ij} g_j\left(\mathbf{x}_i\right) - n_{ij} \ln\left(1 + \sum_{j=1}^{2} \exp\left(g_j\left(\mathbf{x}_i\right)\right)\right)$$

$$= \sum_{i=1}^{n} n_{ij} \left(\sum_{j=1}^{2} \sum_{k=1}^{p} \beta_{j0} + \beta_{jk} x_{ijk} \right)$$

$$- n_{ij} \ln\left(1 + \sum_{j=1}^{2} \exp\left(\sum_{k=1}^{p} \beta_{j0} + \beta_{jk} x_{ijk} \right)\right)$$

3 Results and discussion

From the results of extracting color intensity of red, green, and blue of the images of cans waste types, we obtain the value of intensity of red, green, and blue is quite varied. They both mean and standard deviation of the values range of color intensity of each cans waste type can be seen in Table 2.

Table 2. Mean and standard deviation of values range of color intensity.

Type	Mean	Standard Dev
1	33.78	12.41
2	25.78	2.33
3	9.11	1.45

The cans waste for type 1 has the biggest mean and standard deviation of values range, whereas for type 3 has the smallest mean.

For identification cans waste based on the colors intensity of red, green and blue, we employed the probability approach; trinomial distribution in term regression as described in Sec.2, where predictor variables X_1, X_2, X_3 denote color intensity of red, green, and blue which the image is taken at top, X_4, X_5, X_6 denote color intensity of red, green, and blue which the image is taken at down, and X_7, X_8, X_9 denote color intensity of red, green, and blue which the image is taken at side successively.

We employed an independence test between each of the predictor variable and cans waste type at first. Based on the chi-square test ($\alpha = 5\%$) is revealed that only the color intensity of red which the image is taken at top pose (X_1) that does not correspond with the can waste type. Then the variable is not involved in the trinomial regression modeling to identify the type. By using *G*-statistic, the significance of predictor variables simultaneously in the trinomial model denotes that at least one predictor variable of eight predictor variables that influence in identifying the type cans waste ($\chi^2 = 235.74 > \chi^2_{(0.05;14)} = 23.69$). The significance of predictor variables partially based on the Wald test ($\alpha = 5\%$) notes that some variables have p-value $> \alpha$. By using the backwardstepwise procedure, we have the significant variables as presented in Table 3.

Table 3. The significant variables.

Predictor Variable	p-value
X_2	0,00
X_3	0,02
X_4	0,03
X_7	0,02
X_8	0,00
X_9	0,01

The final trinomial regression model parameters obtained given in Table 4, where type 3 as the reference category. The variance of the predictor variables data in identifying the cans waste type are able to explain the variance of the types by 59.1 percent (based on Nagelkerke value), while the rest is explained by the other predictor variables than this model.

From Table 4, we known that for cans waste type1, the higher the color intensity of the green of the image taken at top pose (X_2), the tendency of an image to be identified as type 1 is 1.31 times more than in type 3. The higher the color intensity of the blue of the image taken at top pose (X_3), the tendency of the image to be identified as type 1 is 1.06 times more than type 3, etc, while for type 2, the higher the color intensity of the green of the image taken at top pose (X_2), the tendency of the image to be identified as type 2 is 1.75 times more than in type 3. The higher the color intensity of the color of the image taken at top pose (X_3), the tendency of the image to be identified as type 2 is 0.68 times more than type 3, etc.

The color intensity of red of the image taken at side pose (X_7) has the highest tendency to be identified as type 1 whereas the color intensity of blue of the image taken at side pose (X_9) has the lowest tendency. The color intensity of blue of the image taken at side pose (X_9) has the highest tendency to be identified as type 2 whereas the color intensity of blue of the image taken at top pose (X_3) has the lowest tendency.

Table 4. The final model of trinomial regression.

Can waste type		β	Wald	exp (β)
1	intercept	-1,67	2,59	-
	X_2	0,27	2,83	1,31
	X_3	0,06	0,13	1,06
	X_4	-0,50	7,67	0,61
	X_7	0,79	12,63	2,21
	X_8	0,62	8,26	1,87
	X_9	-0,56	6,46	0,57
2	intercept	-6,49	26,12	-
	X_2	0,56	7,15	1,75
	X_3	-0,39	3,74	0,68
	X_4	-0,32	2,14	0,73
	X_7	0,23	0,80	1,25
	X_8	0,27	1,20	1,30
	X_9	1,64	35,68	5,15

The correct result percentage of testing to identify the can waste type is presented in Table 5.

Table 5. Percentage of identification result.

Identification to Cans Waste Type	True Cans Waste Type		
	1	2	3
1	43.13	6.88	9.38
2	6.56	22.81	2.19
3	0.31	1.56	7.19
% correct	43.13	22.81	7.19
Total	73.13		

The detection algorithm was simplified for cans classification. When a can is moved on the belt conveyor through the detection area, it will automatically be identifying.

The higher the level correctly of identification cans waste will denote the higher the level accuracy which will indicate the better the sorting system performance. We found that the approach can identify correctly more than 70%. It is enough for a good sorting system.

This method is a useful procedure for analysing and comparing three types of cans. These results are a definite improvement on all other published results so far which quote a sorting accuracy of only.

4 Conclusion

Employing the probability approach of trinomial distribution in term regression, the predictor variables that affect the identification of cans waste type based on the images are the color intensity of green and blue which the image is taken at top pose, the color intensity of red which the image is taken at down pose, and the color intensity of red, green and blue which the image is taken at side pose successively.

The model can identify cans waste into three types based on the images correctly by 73.13%. The level correctly which more than 70% is enough for a good sorting system. This method is a useful procedure for analysing and comparing three types of cans. but the other approaches can be tried to improve performance sorting system.

It is estimated that more investigations on the modern system are necessary. Although knowing how each different component works is a step in the right track, complete combination will give a better sense of the identification system's overall cycle time, failure rates, and problem areas.

This research was funded by Kementerian Riset, Teknologi, dan Pendidikan Tinggi Republik Indonesia (Penelitian Unggulan Perguruan Tinggi, Anggaran DIPA Direktorat Penelitian dan Pengabdian kepada Masyarakat, 2016). The author are grateful to the Universitas Sriwijaya. The author are also many thanks to reviewers for providing valuable comments that help to improve this manuscript

References

[1] http://triaa.com/environment-society/recycling/tri-arrows-recycling/just-kids, accessed on Nov 28, (2015).
[2] S. Cointreau, Seminar on Solid Waste Primers and Lessons Learned from World Bank Projects, (2005)
[3] Irsyadi Y, Edgar S, M. A. Hannan, Djuraidah AW, Hassan B, in Proc. 3rd Int. Conf. on Soft Computing Technology in Civil, Structural and Environmental Engineering, pp.1-11, (2013).
[4] Yulia R, in Proc. 1st Int. Conf. on Education, Technology and Science, pp. 414-417, (2014).
[5] Yulia R, J. of Physics: Conference Series, doi:10.1088/1742-6596/622/1/012052, pp. 1-6, (2015).

[6] Agresti, A., New York. John & Son, (2002).
[7] Hosmer, D.W. &Lemeshow, S., New York. John & Son, (2000).
[8] Yulia R, Prosiding Seminar Nasional Sains dan Teknologi 2015, Bandung, Indonesia, (2015).
[9] Ambarwati, A.N. , H. Kuswantodan I. Zain, Prosiding Seminar Nasional Penelitian, Pendidikan dan Penerapan MIPA, Fakultas MIPA, Universitas Negeri Yogyakarta, (2012).
[10] Buana, I.S., Mahendrawathidan N. Irawan, Seminar Nasional Informatika 2010, UPN Veteran, Yogyakarta, ISSN: 1979-2328, (2010).
[11] Abdalla El-Habil, Pakistan Journal of Statistics and Operation Research. Vol.VIII No.2, pp.271-291, (2012).
[12] Mahdi Khodadadzadeh, Jun Li, Antonio Plaza, and José M. Bioucas-Dias, IEEE Geoscience and Remote Sensing Letters, Vol. 11, No. 12, pp. 2105-2109, (2014).
[13] Erik Thors´en, Mathematical Statistics Stockholm University Bachelor Thesis, (2014).

Vibration analysis of rotary cement kiln using finite element method

Hasan Basri*, *Irsyadi* Yani, and *Akbar* Teguh Prakoso

Department of Mechanical Engineering Faculty of Engineering University of Sriwijaya, 30162 Inderalaya, Indonesia

Abstract. . In this research, the implementation of shell of kiln problem has been discussed. The results are analysed in detail in this research for fatigue life for the shell of a kiln. In this work, the shell of the kiln has been modelled by Solid Works. This simulation showed how the most relevant aspects of the developed work presented in this paper can contribute to the state-of-the-art of the analysis of fatigue life of rotary cement kiln technique with innovative ideas and strategies. It also reviews that the obtained results achieve the proposed objectives. Based on the FEA the transfer matrices and overall transfer equation are developed to calculate natural frequencies, and response overall motion equation is established for response analysis. Due to the dimensionality of the problem addressed, the research specification has to set limits to the applicability of the research by selecting only mechanical load problems in rotary cement kiln tasks and goal-seeking to predict the fatigue life simulation investigated. From the simulation, model and boundary conditions are defined. Crack growth behaviour in the rotary kiln was predicted.

1 Introduction

A kiln is basically an industrial oven, and although the term is generic, several quite distinctive designs have been used over the years, such a kiln in PT. Semen Baturaja made about 1,200.000 tonnes of clinker per year. Rotary kiln shell is a large scale welded structure with 4.5 m in diameter and 75 m in length, and produced by welding thin cylindrical steel plate one by one.

Padded plates are directly soldered to the shell in the supporting rollers places to reduce their concentrated stress. Crack are often initiated at these welded joints, and the over long circumferential crack are prevailing at welded joints near the supporting rollers.

However, Kikuchi et.al, (2010) has predicted of two interacting surface cracks of dissimilar sizes by FEA. The simulations were performed for fatigue crack growth experiments and the method validity was shown on this research. It was shown that the offset distance and the relative size were both important parameters to determine the interaction between two surfaces of crack; the smaller crack stopped growing when the difference in size was large. It was possible to judge whether the effect of interaction should be based on the correlation between the relative spacing and relative size. In 2014, Fatigue crack growth simulation in heterogeneous material using finite element method has generated by Kikuchi et.al. Kikuchi have developed a fully automatic fatigue crack growth simulation system using FEM and applied it to three-dimensional surface crack problems, in order to evaluate the interaction of multiple surface cracks, and the crack closure effects of surface cracks. The system is modified to manage residual stress field problems, and the stress corrosion cracking process is simulated.

The prediction of crack propagation under thermal, residual stress fields using S-Version FEM (S-FEM), Kikuchi was employed to solve a crack growth problem by combining with the auto-meshing technique, this re-meshing process of the local mesh becomes very simple, and modelling of three-dimensional crack shape becomes computationally easy. On the other hand, in 2004, Irsyadi has developed visualization of finite element analysis in 3D (C, C++, under Linux/Fedora), with this system, analysis for extra-large problems such as fatigue life predictions becomes easy and fastly. Irsyadi, Kikuchi, and Kanto employed numerical analysis of 3-D Surface Crack in 2006, and then, Irsyadi and Kikuchi was developed a numerical analysis in the low carbon steel by finite element method and experimental method under fatigue loading. In this research they were predicted fatigue life of material under stresses.

The prediction of fatigue life of rotary cement kiln welded shell is not completely understood, therefore it should be investigated. For rotary kiln shell where the vibration occurs with high displacement, cracks can grow with a complex overloading conditions for over thousands of tons, and then results in premature shell failure. The affecting conditions crack growth include material characteristics, initial crack size, service stresses, and stress concentration due to overheated in hot spot area, all these conditions are random. The fatigue life of the welded shell during crack growth need to be predicted numerically by using finite element analysis and experimentally. In relation to the problems

* Corresponding author: hasan_basri@unsri.ac.id

above, we propose the research topic of vibration analysis and fatigue life analysis of rotary cement kiln (case study in PT. Semen Baturaja).

The primary objective of this study is to investigate the area of vibration with high displacement and fatigue life of the crack growth analysis in a rotary cement kiln in PT. Semen Baturaja.

2 Research method

This research has the overall goal of identifying the characteristics of crack growth that determine fatigue life and hence the risk posed to rotary cement kiln (welded shell). To accomplish this goal, this collaboration research brings together a multi-disciplinary team with expertise in Tribology, Fracture Mechanics, Computational Mechanics and Smart Engineering. It includes two research projects: 1). Dynamic modelling and analysis of the large-scale rotary machine with multi supporting; and 2). Fatigue life of rotary kiln, along with supporting cores.

Separate report follow for each of the projects. In each section, we discuss: a). the research performed and results generated in this year, and b). the challenges encountered in the research and proposed actions.

Research Project 1: Dynamic modelling and analysis of the large-scale rotary machine with multi supporting.

The analysis of dynamic modelling of the kiln shell is shown in Figure 1 and Figure 2

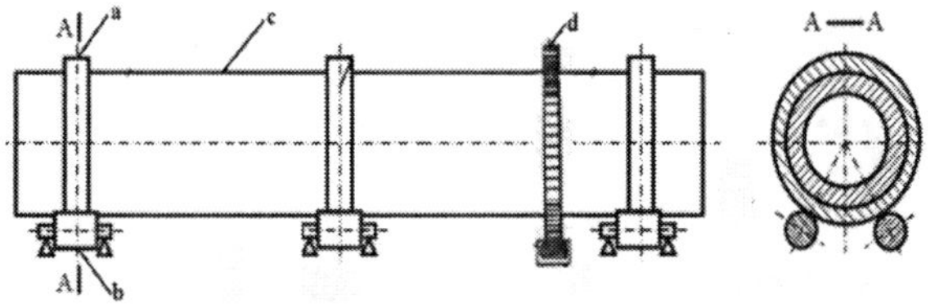

Fig. 1. The sketch of the large-scale rotary machine with multi-supporting. (a: tyre, b: roller, c: body, d: kiln drive)

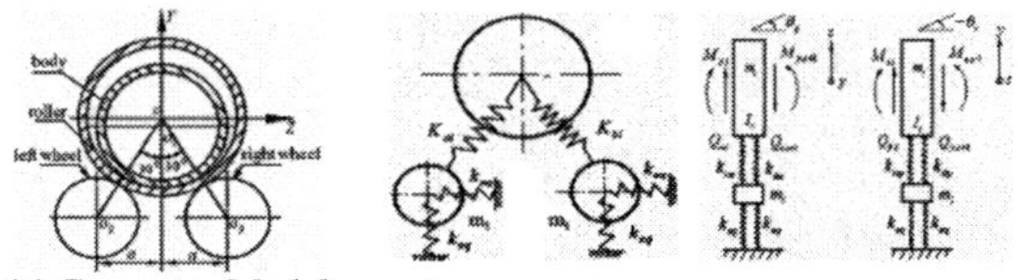

(a) Geometry Model (b) Mechanic model (c) The Sketch in y and z plane

Fig. 2. The geometry and mechanic model of wheel bearing supporting structure

Data Input:
Mass of rotary kiln shell = 55.92 ton.
The weight of rotary kiln shell = 54,86 x 104 N

The data of mass, material the rotary kiln and the boundary conditions are shown in Figure 3, Table 1 and Figure 4 respectively.

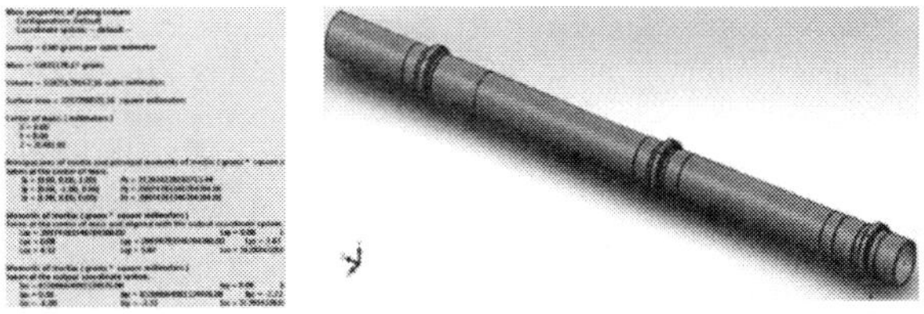

Fig. 3. Mass of rotary kiln shell is calculated by Autodesk 2016

Table 1. Material of a kiln shell

No	Specification of ASTM 516 Grade 70	Value
1.	Tensile strength	485 MPa
2.	Yield strength	260 MPa
3.	Density	7.85 gr/cm^3
4.	Poisson ratio	0.29
5.	Modulus of elasticity	200 GPa

Boundary Conditions:

- Type of analysis: Natural Frequency with (modal) load stiffening.
- Clinker weight is neglected.
- Speed of rotary kiln = 3.5 rpm.
- Ambient temperature = 37^{0}C.
- The kiln shell inclination (deviation) = 3.5^0
- Rotation Axis be based on the position of kiln drive

X = - 57149. 77 mm
Y = 29605.84 mm
Z = 27150. 75 mm

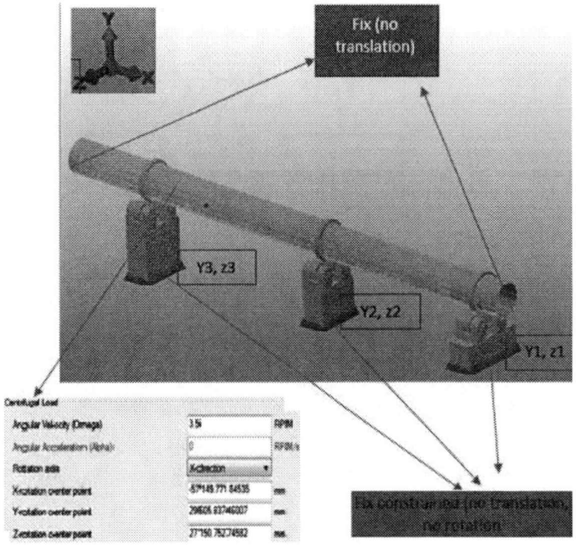

Fig. 4. Boundary condition

3 Results and discussion

We computed the natural frequencies of the body system of a rotary kiln, as well as the response with axis line deflection of the supporting structure. The rotary kiln is shown in Fig 1. The body length and radius are 75 m and 4.5 m respectively, the total weight is 54,86 x 10^4 N. The kiln has 3 supporting structures, located 4 m, 31 m, and 62 m from kiln head respectively. The kiln shell is divided into various elements according to body structure as shown in Figure 1 (b: roller). Any subsystem is composed of rigid disks, elastic shafts, and linear springs.

Anisotropy stiffness of the supporting structures in y and z directions are given as follows (see Figure. 5);

Direction	K Value
Y	K_{yy1} = 6.5 x 10^4 N/mm, K_{yy2} = 36 x 10^4 N/mm, K_{yy3} = 39.08 x 10^4 N/mm
Z	K_{zz1} = 3.76 x 10^4 N/mm, K_{zz2} = 20.78 x 10^4 N/mm, K_{zz3} = 20.56 x 10^4 N/mm

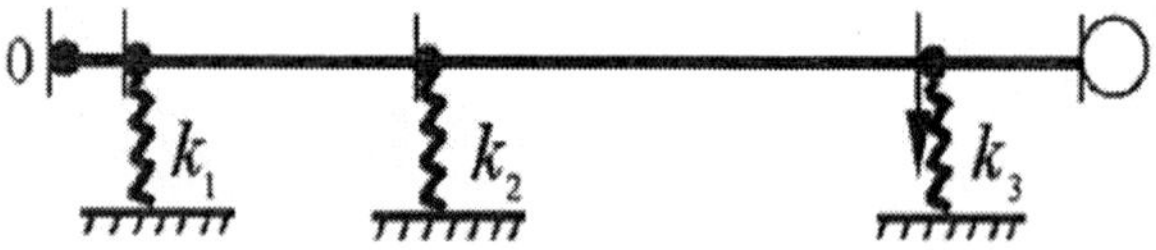

Fig. 5. The sketch of rotary kiln and dynamics model of the body system.

Table 2. Order of natural frequencies of a kiln shell. β_k (for k=1, 2, 3, ... 150)

Number of frequency	Circular frequency (rad/sec)	Frequency (Hertz)	Priode (sec)
1	1136600	180900	552,80
2	1157100	184160	543,01
3	1756500	279550	357,71
4	1785700	284200	351,86
5	1814300	288760	346,31
...	...	...	...
143	15188000	2417200	41,37
144	15284000	2432500	41,11
145	15429000	2455700	40,72
146	15453000	2459400	40,66
147	15566000	2477400	40,36
148	15642000	2489500	40,17
149	15673000	2494400	40,09
150	**15715000**	**2501100**	**39,98**

Natural frequencies of the rotary kiln.

The results of the natural frequencies β_k (k = 1, 2, 3, ... 150) of the body system are shown in Table 2. Displacement value at first of 150 number of frequencies are shown in Figure 6, Figure7 and Figure 8. It can be seen clearly from Table 2 that the results of natural frequencies obtained by FEM are shown in Figure 9.

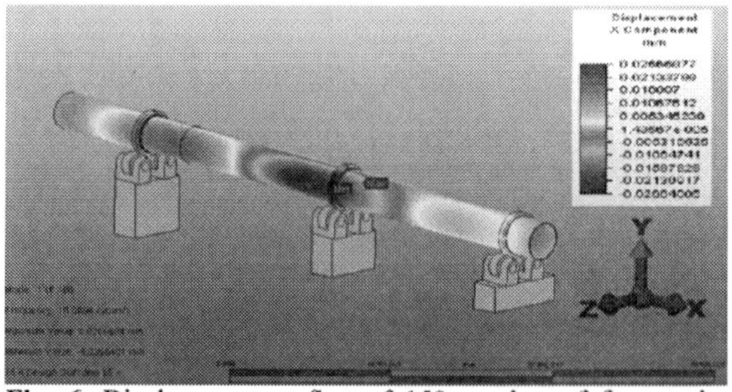

Fig. 6. Displacement at first of 150 number of frequencies along x-axis

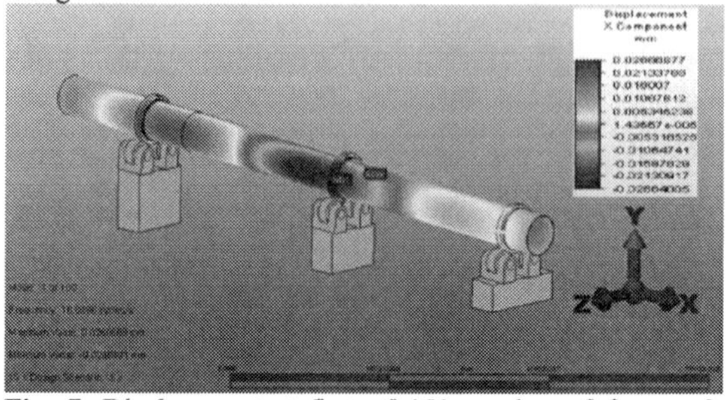

Fig. 7. Displacement at first of 150 number of frequencies along y-axis

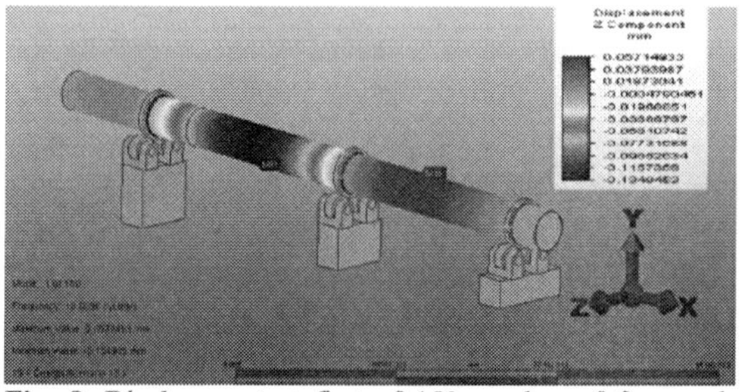

Fig. 8. Displacement at first of 150 number of frequencies along Z-axis

Research Project 2: Fatigue life of rotary kiln along with supporting cores.

Considering a finite element analysis, characteristics of a kiln shell are a pretension and a mating part contact. The pretension can generally be modeled with static loading, thermal deformation, a constraint equation, or an initial strain. For a thermal deformation method, the pretension is generated by assigning virtual different temperatures and thermal expansion coefficients to the shell, clinker and temperature of gases. In this work, in order to generate a finite element model for the kiln shell with clinker and temperature of gases, two kinds of models are introduced. All the proposed models are taken into account above primary characteristics such as a pretension effect and a contact behavior between shell, clinker and temperature of gases. The prediction of the crack propagation are considered on crack emanating from contact surface between kiln shell and clinker and welded joint on the kiln shell surface. On this finite element model, the final mesh consisted of 40,856 elements and 45,245 nodes. Based on vibration analysis and thermal analysis, the elements along the direction of

DOI: 10.1051/matecconf/201710103013

crack advance had a length of 13-23 m and 22-24 m respectively. Following the stress and deformation simulation, fatigue crack growth was modelled by repeated loading (mass of clinker and thermal gases), unloading, advancing the crack and then the loading again.

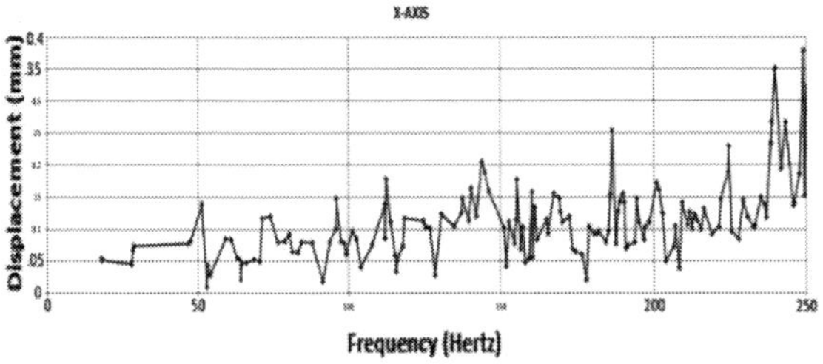

(a) X - axis

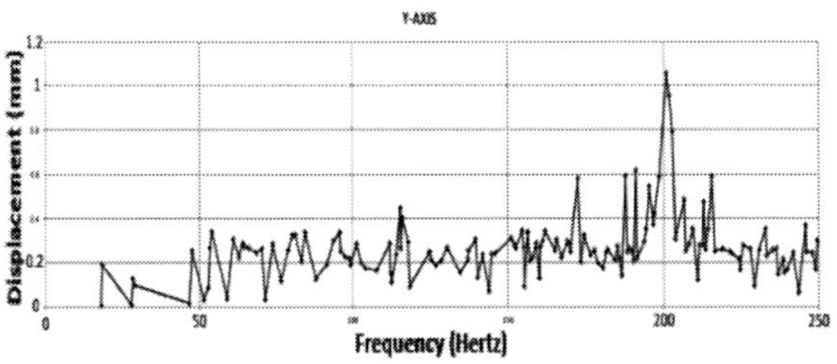

(b) Y - axis

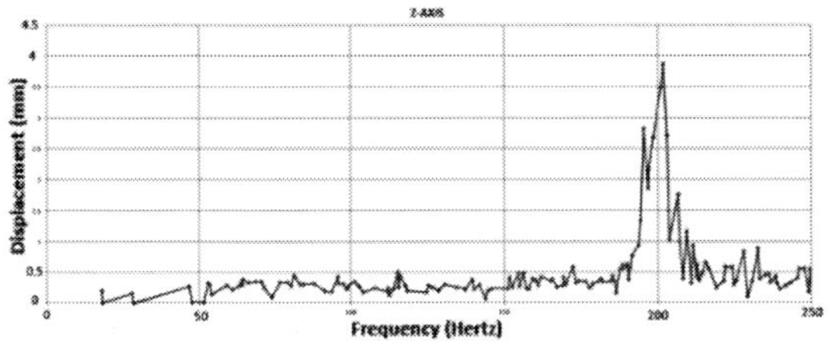

(c) Z - axis

Fig. 9. The results of natural frequencies by FEM

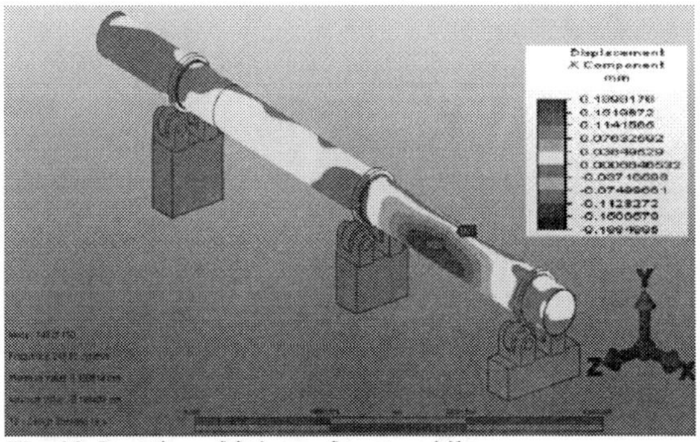

Fig. 10. Location of fatigue of a rotary kiln

4 Conclusion

The dynamics of the large scale rotary machine with multiple wheel bearing structures is investigated in this research. Based on the finite element method, the transfer matrices and overall transfer equation are developed to calculate natural frequencies, and response overall motion equation is established for response analysis.

Structures are simplified as linear springs, and their anisotropy equivalent stiffness are deduced. Taken a rotary kiln as an instance, natural frequencies, modal curves, and response vibration are obtained. The body vibration modal curves illustrate the cause of dynamical errors in common axis line measurement methods. The displacement response can be used for further measurement dynamical error analysis and compensation. The response overall motion equation could be applied to predict the body motion under abnormal mechanics condition, and provide theory guidance for machine failure diagnosis.

The main goal of the work presented in this research is to propose fatigue life analysis algorithm in the shell of a kiln using finite element analysis. Due to the dimensionality of the problem addressed, the research specification has to set limits to the applicability of the research by selecting only mechanical load problems in rotary cement kiln tasks and goal-seeking, to predict the fatigue life simulation investigated.
From the simulation, model and boundary conditions are defined. Crack growth behavior in rotary kiln was predicted. As the crack grows, the speed of the crack depth increase.

References

1. ABP, *ASME boiler and pressure vessel code II - material properties* (The American Society of Mechanical Engineers, 1995)
2. ABP, *ASME boiler and pressure vessel code VIII division 2 -alternative rules, Appendix 5. Design based on fatigue analysis* (The American Society of Mechanical Engineers, 1995)
3. Coz Díaz JJ, Mazón FR, García Nieto PJ, Suárez Domínguez FJ. **39**(1):17–42, (2002).
4. *Energy Efficiency Component EU-China* (A reference book for the industry. Beijing, 2009)
5. Frederic O., *The kiln book; materials, specifications and construction* (Chilton Book Company, 1997)
6. Irsyadi Yani, *Seminar PPI Jepang Tengah*, Toyohashi University of Technology, Japan, (2003)
7. Irsyadi Yani, Masanori Kikuchi, *Proceeding of The 5th International Conference on Numerical Analysis in Engineering (NAE)* (2007)
8. Irsyadi Yani, Masanori Kikuchi, Yasuhiro Kanto, Int. Conference of Computational Mechanics and Numerical Analysis (2006)
9. Irsyadi Yani J. Rekayasa Mesin, Universitas Sriwijaya (2005)
10. Kikuchi M., Int. J. Fatigue **58**, (2014)
11. Kikuchi M., Engineering Fracture Mechanics **77**, 3120–3131 (2010)
12. Kikuchi M., Int. J. Pressure Vessels and Piping 90-91 (2012)
13. Liu XY., Specht E., Process Intensification; **49**(2):147–50 (2010)

14. Merk Blatter A., *Technical rules for pressure vessell* (Verlag KG, Carls Heymanns, 1997)
15. Xuejun Li, Yiping Shen and Songlai Wang, *Hunan Provincial Key Laboratory of Health Maintenance for Mechanical Equipment, Hunan, China* (publisher Hindaw)
16. Basri, H., *Proceeding of 5th International Conference on Numerical Analysis in Engineering (NAE 2007)*, ISSN: 979-96139-6-7, pp. 8.1-8.7, (2007).

Influences of pitch-length louvered strip insert on thermal characteristic in concentric pipe heat exchanger

Indri Yaningsih[1,*], and *Agung* Tri Wijayanta[1]

[1] Mechanical Engineering Department, Faculty of Engineering, Sebelas Maret University, 57126 Surakarta, Indonesia

Abstract. The effects of pitch-length louvered strip inserts with a forward arrangement on heat transfer and friction factor characteristics in a concentric pipe heat exchanger were experimentally investigated. Louvered strip insert was installed inside the tube. For comparison, the experiment without louvered strip (plain tube) was conducted. All cases were tested under steady state condition for Reynolds number ranging from 5300-17,500. The results show that the utilization of louvered strip insert produced a higher heat transfer rate compare to a plain tube. Nusselt number, friction factor, and heat transfer coefficient ratio increased with decreasing pitch length. The Nusset number for louvered strip insert with S of 40, 50, and 60 mm increasing in the value of 67% - 77%; 48% - 53%; and 21% - 24% than plain tube, respectively. While the tube fitted with louvered strip insert by S of 40, 50, and 60 mm, friction factor increased 2.91 - 3.84; 2.41 - 2.92; and 1.88 - 2.08 times greater than a plain tube. The addition of louvered strip insert with S of 40, 50, and 60 mm producing heat transfer coefficient ratio in the range of 1.02 - 1.12; 1.03 - 1.07 and 1.03 - 1.04.

1 Introduction

Enhanced heat transfer is a method to increase the thermal performance system by enhancing heat transfer and flow characteristics. In the last years, enhanced heat transfer technology has been developed and extensively applied to heat exchanger applications; such as cooling devices, automotive, chemical industries and others. Some investigations have been done to increase the heat exchanger performance with the aim to diminish the size and cost of a heat exchanger, which can lead to more economical design. This economical design can help to make energy saving related to heat exchange process. The required pumping power influences energy saving in this case.

The significant effect of using heat transfer enhancement technology is the rise of pressure drop that leads to the increase of required pumping power. Pressure drop in the tube flow occurs due to the frictional losses with the solid surface. Therefore, while selecting the enhancement method should be optimal with the consequence of the rising heat transfer coefficient and the pumping power because of the rising pressure drop.

Among three enhancement methods, passive, active, and compound methods, the passive methods are mostly used to achieve the highest heat transfer rate. There are many kinds of mechanism of enhanced heat transfer by using passive methods such as modified surfaces or geometries, extended surfaces, and insertion. The insertion considered as the most advantageous passive method. There are several advantageous of the use of insertion (1) can be fitted on the smooth tube; (2) can be formed in various shapes; (3) easy to install and operate. Researchers have been conducted a large numbers of experimental investigation using insertion methods. Including louvered strip, twisted tape, wire coil and helical wire coil to examine the thermal characteristics on each insertion.

The experiment related to a modification of twisted tape insert, Yaningsih et al. [1] experimentally investigated the heat transfer and flow characteristics of heat exchanger by employing perforated twisted tape inserts with a different axial pitch ratio. At the given Reynolds number, heat transfer rate, friction factor and thermal performance factor increases with decreasing axial pitch ratio. Sibel et al. [2] developed a study by using equilateral triangle cross-section coiled wire employed inside the tube. The use of insertion guided to a significant increase in heat transfer and pressure drop over the smooth tube. Mohammed [3] and Fan [4] studied numerically on the convective heat transfer characteristics of heat exchanger fixed with louvered strip insert with various slant angle and pitch length. In the tested range, the Nusselt number increases with the increasing slant angle and tend to decrease with rising pitch length. Eswaraiah [5] studied numerically on heat transfer enhancement using louvered strip inserts in the oil cooler. Numerical modelling was using computational fluid dynamics (CFD). These experiments were carried to compare the theoretical correlations and CFD simulated result. The result shows that a good agreement was found between the theoretical correlations and CFD simulated result. Eimsa-ard [6] experimentally studied the influenced of louvered strip insert with backwards and forward arrangement on the

* Corresponding author: indriyaningsih@staff.uns.ac.id

characteristics heat transfer and friction factor of concentric pipe heat exchanger. The results reveal that the using of louvered strip insert yielded the higher heat transfer rate compared with a plain tube. Pethkool [7] experimentally researched thermal characteristics of concentric pipe heat exchanger fitted with louvered strip insert with the different inclination angle. The results show that the Nusselt number and friction factor tend to increase with increasing inclination angle. Sarada et al. [8] using louvered square leaf inserts inside a horizontal tube heat exchanger to estimate the enhancement heat transfer rate for air in the presence of insert. The experimental investigation carried out by combination different arrangement of the direction of the airflow and inclined angle. The results show that Nusselt number and pressure drop increased. The overall enhancement ratio is highest for the highest inclined angle. Variation of the geometry louvered strip insert includes pitch and slant angle were investigated by Huisseune et al. [9]. They investigate the thermal and hydraulic performance of such a compound heat exchanger. They found that the combination between small fin pitch and large slant angle cause the best performance of the compound heat exchanger. Other experiment conducted by Raut and Farkade [10], they studied convective heat transfer enhancements in a tube with three arrangements of louvered strip insert (forward, semi-forward, semi-backward, and forward-backward arrangements). The results show that the that the different arrangement has a various value of Nusselt number and friction factor.

The literatures above show that the using of insertion can increase the thermal performance of the heat exchanger. From literatures [3-10], there are many aspects that influence the use of louvered strip insert (geometry, the arrangement to the flow direction, and shape of louvered strip insert). The main mechanism of heat transfer enhancement due to insertion include the reduction of the hydraulic diameter and induced swirling flow makes a better fluid mixing [11]. Based on the literature this experiment will be carried out to investigate the thermal characteristics of concentric pipe heat exchanger installed with louvered strip insert. The innovative design and new materials of louvered strip insert were present in this experiment.

2 Experimental details

The schematic diagram of test rig is shown in Fig. 1.

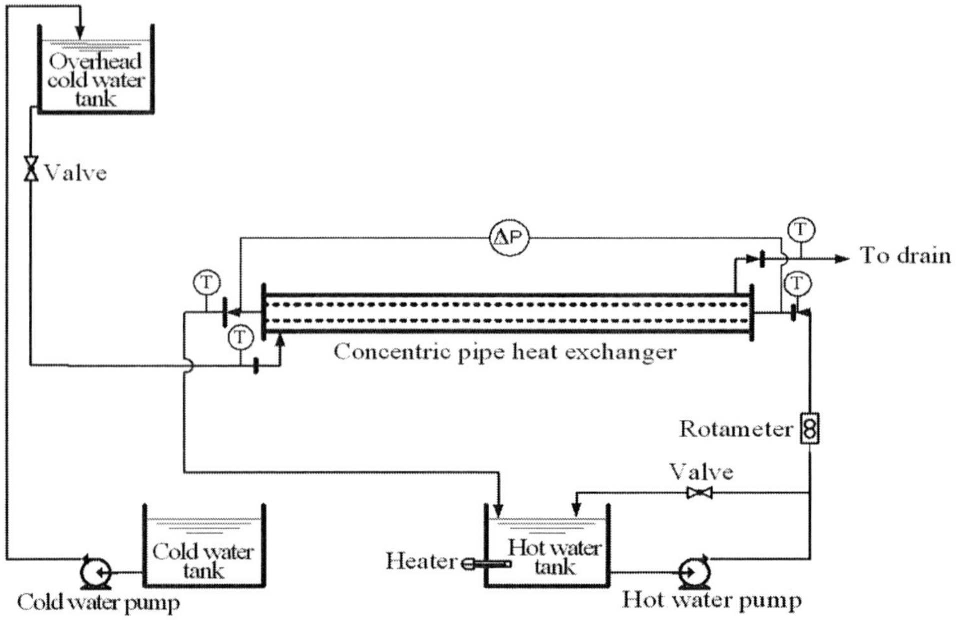

Fig. 1. Schematic diagram of concentric tube heat exchanger rig

The experimental concentric tube heat exchanger rig consists of concentric tube heat exchanger, cold water circulation system, hot water circulation system and instrumentation system to measure cold water and hot water flow rate, pressure and temperature. Concentric tube heat section made from the straight aluminum pipe consists of an outer pipe and inner pipe. The dimensions of the inner and outer tubes of the heat exchanger are: d_i = 14.3 mm, d_o = 15.8 mm, L_i = 2500 mm, D_i = 23.4 mm, D_o = 25.4 mm and L_o = 1950 mm. The heat exchanger is oriented horizontally, and the fluids inside the pipe are in counter flow directions. Hot water was passed through the inner tube, while cold water was flowing through the annulus. The outer wall of the pipe is wrapped with the glass wool as insulation to minimize the heat losses to ambiences.

Cold-water circulation system consists of an overhead cold-water tank, cold-water tank, and cold-water pump. The overhead water tank is used for supplying cold water using gravity flow with constantly flow rate at 0.103 kg/s. The inlet cold-water temperature was the ambient temperature that relative constant at ± 28°C. After passed through the system, cold water exposed to the surrounding. The hot water circulation system consists of hot water tank, flow regulation valves, heating system, and hot water pump. The heating system consists of a water tank with eight electric heaters, 4000 W for each electric heater. The inlet hot water temperature was kept constant at 60°C. Rotameter was used to measure the flow rates. The test runs are done at the hot water mass flow rates ranging between 0.033 kg/s and 0.099 kg/s. Data were collected at steady state conditions.

Temperature measurement was done by K-type thermocouples. K-type thermocouples were used for the measurement of inlet and outlet temperatures of cold and hot water fluids. Similar to the fluid, ten thermocouples were used to measurement of inner pipe outer wall. The temperatures were read by using a multi-channel digital thermometer. Pressure loss of the inner pipe side was determined by using U-tube manometer. The working fluid inside manometer was water. At present work, it is intended to find the changes in the convection heat transfer coefficients of the inner pipe side turbulent flow by affecting the regions near the wall of the pipe flow. For this objective, louvered strip insert was installed on the inner pipe side of the concentric double pipe heat exchanger as a tubulator. For comparison, this study also tested inner pipe without insert (plain tube).

The louvered strips that used in this experimental study were shown in Figs. 2 and 3.

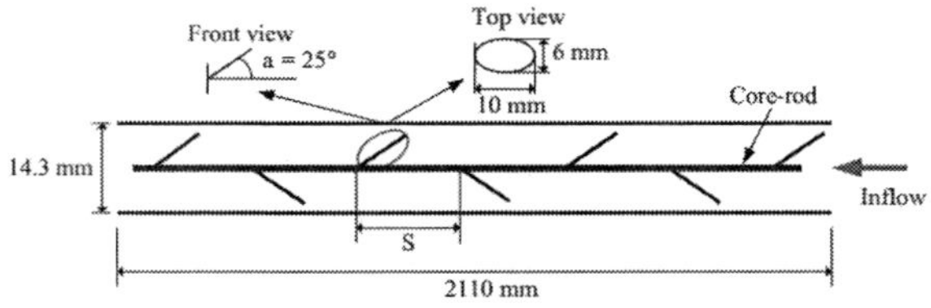

Fig. 2. Louvered strip with forward arrangements

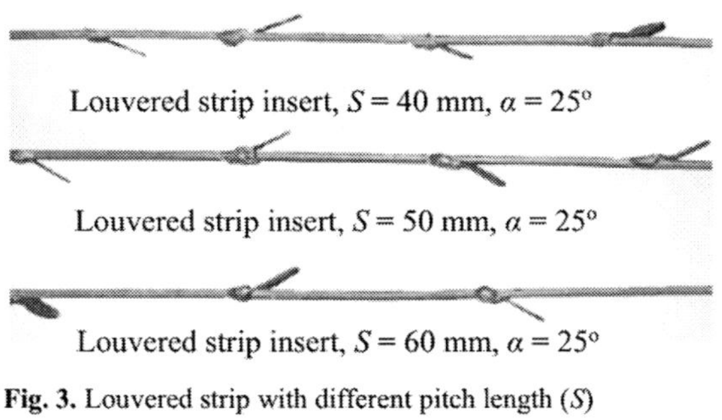

Fig. 3. Louvered strip with different pitch length (*S*)

Louvered strip inserts made from mild steel, 1 mm in thickness with an elliptical shape. The dimension of elliptical louvered strip insert was 6 mm, 10 mm respectively for minor and mayor axis. The louvered strip insert was connected to a central rod of 2 mm diameter as shown in the Fig. 2. The influenced of different pitch length (S = 40, 50, and 60 mm) are studied in the present study. The slant angle (α) was kept constant at 25°. Louvered strips insert is on a forward arrangement with the direction of fluid flow.

Steady state values of the fluid and wall temperatures of the experiments were obtained for given mass flow rate of the hot water fluid. These values were used to compute heat transfer rate to the fluid flowing inside a tube. Nusselt number and friction factor were also calculated to know the effect of louvered strip inserts on heat transfer and friction characteristics. The following expression were used for calculation of heat transfer rate of hot and cold water (Q), heat transfer coefficient (h), Nusselt number (Nu), friction factor (f), Reynolds number (Re) and thermal performance factor (η). Heat transfer rate from hot water in the inner tube:

$$Q_h = \dot{m}.C_{p,h}.(T_{h,i} - T_{h,o}) \tag{1}$$

Heat transfer rate from cold water in the annulus:

$$Q_c = \dot{m}.C_{p,c}.(T_{c,o} - T_{c,i}) = h_o.A_o.(\overline{T}_{w,o} - T_{b,o}) \tag{2}$$

where

$$T_{b,o} = \frac{T_{c,i} + T_{c,o}}{2} \quad \text{and} \quad \overline{T}_{w,o} = \frac{\sum T_{w,o}}{10} \tag{3}$$

$\overline{T}_{w,o}$ is the average of the outer wall inner tube temperature. There are ten point measurements located on the outer wall inner tube. $T_{b,o}$ is the bulk cold temperatures, computed from the average of inlet and outlet cold water temperature. The differences between equations (1) and (2) indicate the convective heat loss, which may be assumed to be negligible:

$$Q_{loss} = Q_h - Q_c \tag{4}$$

An average heat transfer rate between hot and cold water,

$$Q_{ave} = \frac{Q_h + Q_c}{2} \tag{5}$$

Overall heat transfer is:

$$U_i = \frac{Q_{ave}}{A_i.\Delta T_{LMTD}} \tag{6}$$

while an average convection heat transfer coefficient, hi, obtained from the overall thermal resistance consist of three resistances in series: convective thermal resistance of the inner tube, conductance thermal resistance of the inner tube wall, and convective thermal resistance on the annulus side:

$$\frac{1}{U_i.A_1} = \frac{1}{h_i.A_i} + \frac{\ln(d_o/d_i)}{2\pi k_p L} + \frac{1}{h_o.A_o} \tag{7}$$

Thus h_i can be computed from equation (7) as:

$$h_i = \frac{1}{\left[\frac{1}{U_i} - \frac{d_i . \ln(d_o/d_i)}{2k_p} - \frac{d_i}{d_o . h_o}\right]} \quad (8)$$

The mean Nusselt is expressed as follows:

$$Nu_i = \frac{h_i . d_i}{k_i} \quad (9)$$

The friction factor is defined as:

$$f = \frac{\Delta P}{(\rho u^2/2)(L/d_i)} \quad (10)$$

The Reynolds number can be calculated as:

$$Re = \frac{\rho . u . d_i}{\mu} \quad (11)$$

The fluid thermo-physical properties of the water are determined at the mean bulk fluid temperature (T_b).

Thermal performance factor (η) is defined as the ratio of the heat transfer enhancement ratio to the friction ratio at the same pressure drop. The thermal performance factor can be written as:

$$\eta = (Nu_t / Nu_p) / (f_t / f_p)^{1/3} \quad (12)$$

The present experimental data will be compared with standard correlations for Nusselt number and friction factor under a turbulent flow regime. The Nusselt number obtained from the present plain tube are compared with those from the proposed correlations by Petukhov (13) and Gnielinski (14) [12] and friction factor proposed correlation by Blasius (15), [13]. Nusselt number correlation from Petukhov is:

$$Nu = \frac{(f/8) Re . Pr}{1.07 + 12.7 . (f/8)^{1/2} . (Pr^{2/3} - 1)} \quad (13)$$

for $10^4 < Re < 5 \times 10^6$.

Nusselt number correlation from Gnielinski:

$$Nu = \frac{(f/8) . (Re - 1000) . Pr}{1 + 12.7 (f/8)^{1/2} . (Pr^{2/3} - 1)} \quad (14)$$

for $10^3 < Re < 5 \times 10^6$.

Friction factor correlation from Blasius:

$$f = 0.3164 . Re^{-0.25} \quad (15)$$

for $4 \times 10^3 < Re < 10^5$

3 Results and discussion

3.1 Validation

The present data (Nusselt number and friction factor) of plain tube were compared with the empirical equations (13-15) to confirm the accuracy and reliability of the experimental setup. Figure 4 demonstrated the comparison between experimental data and the standard equation for Nusselt number.

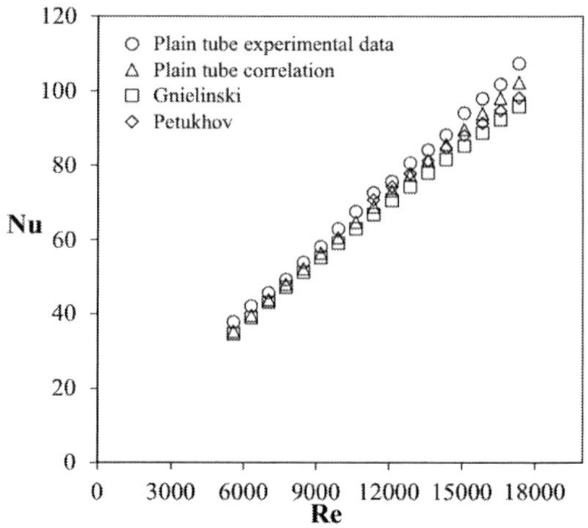

Fig. 4. Nusselt number validation for plain tube

From Fig. 4, it appears that the Nusselt number of the present data deviated from those of Dittus-Boelter and Pethukov empirical expression within ±2.78% and ±7.05% respectively. Figure 5 shows the comparison between friction factors from the present data with Blasius correlation.

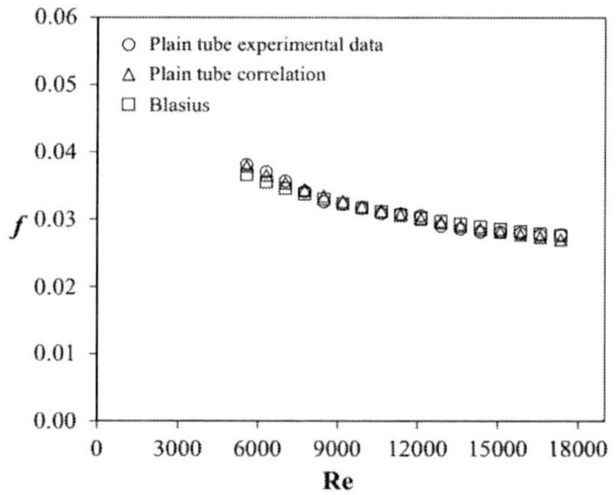

Fig. 5. Friction factor validation for plain tube

The friction factor deviated from the proposed correlation by Blasius within ±2.46%. The comparisons imply that the present data accord well with the correlations, validating the reliability of the present experimental setup and method.

The present results are correlated with the Nusselt number and friction factor for the plain tube as follows:

$$Nu_i = 0{,}007\ Re^{0{,}946}\ Pr^{0{,}3} \quad (16)$$

$$f = 0{,}495\ Re^{-0{,}298} \quad (17)$$

Equations (16) and (17) are found to represent the experimental data within ±6.6% for Nusselt number and ±2.6% for friction factor, as shown in Figs. 3 and 4 respectively.

3.2 Heat transfer characteristic

Figure 6. depicted the heat transfer characteristics of the tube in term of Nusselt number (*Nu*) with Reynolds number (*Re*).

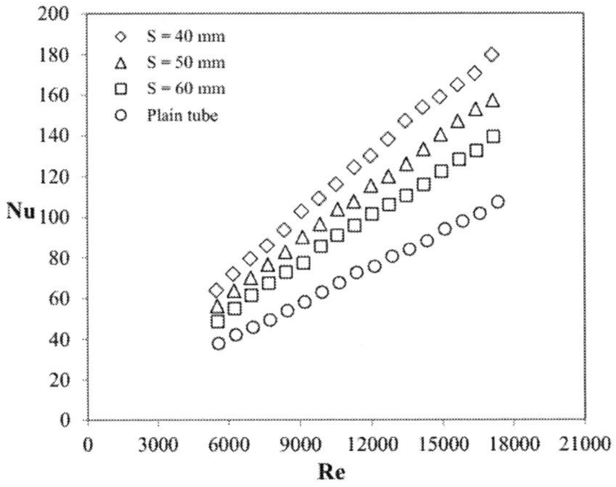

Fig. 6. Nusselt number vs Reynolds number

For all cases, the results present that the Nusselt number (*Nu*) rises with the increase of Reynolds number. The results are similar with previous experiment [1]. For different pitch-length (40, 50, 60 mm), the Nusselt number tend to increase with decreasing the pitch-length. The experimentation showed that pitch-length 40 mm provided the maximum value. However, pitch-length 60 mm has the lowest value of Nusselt number. The reasons why the excellent heat transfer was produced by smallest pitch-length is with the smaller pitch-length value resulting the higher compactness of the leaves of the louvered strip insert. The effects of the higher compactness are there was more streamlined pattern breaks up of fluid flows, therefore the flow between the elements of the leaves produces higher turbulence intensity due to rapid fluid mixing. Particularly on a pitch that was getting closer [3]. At given Reynolds number, the comparison between the plain tube and louvered strip insert with different pitch-lengths the Nusselt number found to be 67% - 77%; 48% - 53%; and 21% - 24%, respectively.

3.3 Friction factor

Friction factor characteristic is shown in Fig. 7. In term of friction factor with Reynolds number. The friction factor is inversely proportional to Reynolds number for all conditions. These results are in accordance with previous experiment [1]. The friction factor becomes greater for smaller pitch-lengths and decreases for the higher pitch length. The highest value of friction factor was found at pitch-length 40 mm, while the plain tube gave the minimum. Concentric tube fitted with louvered strip insert, the values of friction factor are decreasing with the rise of pitch-length [4]. In the range of 5.300 < *Re* < 17.500 friction factor of the inner tube fitted with louvered strip insert with different pitch length (40, 50, 60 mm) were 2.91 – 3.84; 2.41 – 2.92; and 1.88 – 2.08 times greater than the plain tube.

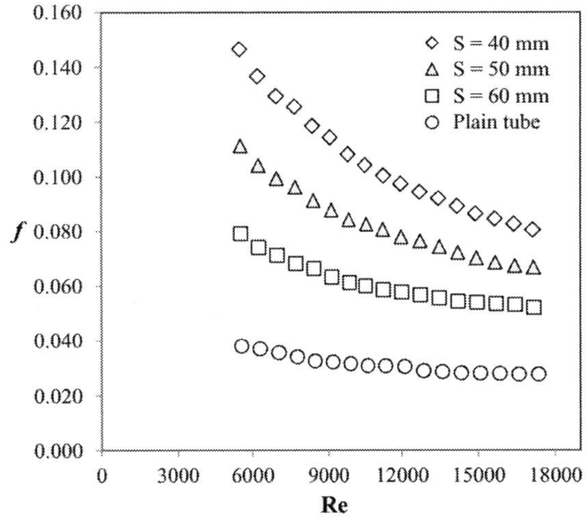

Fig. 7. Friction factor vs Reynolds number

3.4 Thermal performance factor

Thermal performance factor is used to evaluate the quality of the enhancement technique used. The thermal performance factor gradually increased with the rise of Reynolds number as shown in Fig. 8.

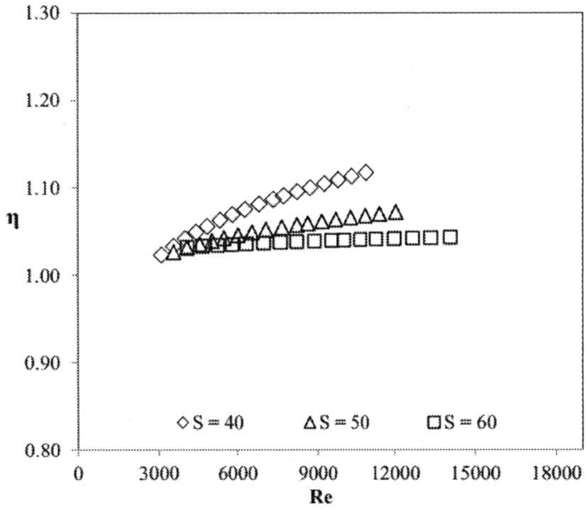

Fig. 8. Thermal performance factor vs Reynolds number

Thermal performance factor is observed highest for pitch-lengths 40 mm. The values of thermal performance factor of the inner tube with louvered strip insert at various pitch-lengths (40, 50, 60 mm) were found 1.02 – 1.12; 1.03 – 1.07 and 1.03 – 1.04 times greater than the plain tube.

4 Conclusions

The experimental study is conducted to enhance the performance of concentric pipe heat exchanger by installing louvered strip insert with various pitch-lengths. The results can be concluded as follows:

1) Louvered strip enhanced the thermal performance of the concentric tube heat exchanger. Nusselt number, friction factor and thermal performance factor tends to be higher than the plain tube.
2) The influenced of the different pitch-length on the thermal performance, the smallest pitch length gave the higher Nusselt number, friction factor and thermal performance factor.
3) The 40, 50, 60 mm of pitch lengths augmented average heat transfer 67% - 77%; 48% - 53%; and 21% - 24%, respectively at Reynolds number 5300 – 17,500 compared to the plain tube.

Nomenclature

A_i	inner surface area of inner pipe (m^2)
A_o	outer surface area of inner pipe (m^2)
C_p	specific heat capacity (J/kg.°C)
d_i	inner diameter of inner pipe (m)
d_o	outer diameter of inner pipe (m)
f	friction factor (dimensionless)
f_p	friction factor of the plain tube (dimensionless)
f_t	friction factor of the inner pipe with louvered strip insert (dimensionless)
h_i	average convective heat transfer coefficient of the inner pipe side (W/m^2.°C)
h_o	average convective heat transfer coefficient of the annulus side (W/m^2.°C)
h_p	average convective heat transfer coefficient of the plain tube (W/m^2.°C)
k_i	thermal conductivity of hot water (W/m.°C)
k_p	thermal conductivity of the inner pipe material (W/m.°C)
L_i	length of the inner pipe (m)
L_o	length of the outer pipe (m)
Nu_i	average Nusselt number of the inner pipe side (dimensionless)
Nu_p	average Nusselt number of the plain tube (dimensionless)
Nu_t	average Nusselt number of the inner pipe side with louvered strip insert (dimensionless)
Pr	Prandtl number (dimensionless)
Q_h	heat transfer rate from the hot water in the inner pipe (Watt)
Q_c	heat transfer to the cold water in the annulus (Watt)
Re	Reynolds number (dimensionless)
S	pitch length
T_b	bulk water temperature (°C)
T_c	cold water temperature (°C)
T_h	hot water temperature (°C)
T_w	outer wall inner pipe temperature (°C)
u	velocity of hot water in the inner pipe (m/s)
U_i	overall heat transfer coefficient based on the inner pipe inside surface area (W/m^2.°C)

Greek symbol

α	slant angle (°)
ρ	density of hot water (kg/m^3)
μ	dynamic viscosity of hot water (kg/m.s)
η	thermal performance factor (dimensionless)
ΔP	pressure drop across the inner pipe (Pa)
ΔT_{LMTD}	logarithmic mean temperature difference (°C)

References

1. I. Yaningsih, T. Istanto, A.T. Wijayanta, *Proceedings of International Conference and Exhibition Sustainable Energy and Advanced Material (ICESEAM), AIP Conference Proceedings* **1737**, 030012-1 – 030012-10 (2015)
2. G. Sibel, O. Veysel, B. Orhan, Exp. Therm Fluid Sci. **34** 684-691 (2010)
3. H.A. Mohammed, Husam A. Hasan, M.A. Wahid, Int. Commun. Heat Mass Transfer **40** 36-46 (2013)
4. A.W. Fan, J.J. Deng, A. Nakayama, W. Liu, Int. J. Heat Mass Tran. **55** 5205-5213 (2012)
5. D. Eswaraiah, S.S. Naga, IJERT **1** (7) 1-9 (2012).
6. S. Eimsa-ard, S. Pethkool, C. Thianpong, P. Promvonge, Int J Heat Mass Tran **35** 120-129 (2008)
7. S. Pethkoll, S. Eimsa-ard, A. Ridluan, P. Promvonge, *The 2nd International Conference on Sustainable Energy and Environment (SEE)*, A-028, Bangkok (2006)
8. N. Sarada, R. Reddy, G. Ravi, IJTAE **3** (8) 420-424 (2013)
9. H. Huisseune, C. T'Joen, C., P. De Jaeger, B. Ameel, S. De Schampheleire, M. De Paepe, Int. J. Heat Mass Transfer **57** 58-72 (2013)
10. R. Raut, H.S. Farkade, Int. J. Eng. Res. Appl. **2** (4) 01-04 (2014)
11. C. Zhang, D. Wang, K. Ren, Y. Han, Y. Zhu, X. Peng, J. Deng, X. Zhang, Renew. Sust. Energy Rev. **53** 433-449 (2016)
12. Y.A. Cengel, *Heat and Mass Transfer 5th ed.* (McGraw-Hill, New York, 2008)
13. Frank M. White, *Fluid Mechanics 7th ed.* (McGraw-Hill, New York, 2011)

Performance of palm oil as a biobased machining lubricant when drilling inconel 718

Erween Abd Rahim[1], *Hiroyuki* Sasahara[2]

[1] Precision Machining Research Center (PREMACH), Faculty of Mechanical and Manufacturing Engineering, UniversitiTun Hussein Onn Malaysia, 86400 Parit Raja, Johor, Malaysia
[2] Department of Mechanical System Engineering, Tokyo University of Agriculture and Technology, 2-24-16 Naka-cho, Koganei-shi, 184-8588 Tokyo, Japan

Abstract. Metalworking fluid acts as cooling and lubrication agent at the cutting zone in the machining process. However, conventional Metalworking fluid such mineral oil gives negative impact on the human and environment. Therefore, the manufacture tends to substitute the mineral oil to bio-based oil such as vegetables and synthetic oil. In this paper, the drilling experiment was carried out to evaluate the efficiency of palm oil and compare it with minimal quantity lubrication technique using synthetic ester, flood coolant and air blow with respect to cutting temperature, cutting force, torque and tool life. The experimental results showed that the application of palm oil under minimal quantity lubrication condition as the cutting fluid was more efficient process as it improves the machining performances.

1 Introduction

Drilling process can be divided into either short hole drilling or deep hole drilling. A short hole is defined as a hole with a small ratio of depth to a diameter. Typically, it includes holes up to 30 mm having a depth of not more than five times the diameter. Meanwhile, for holes greater than 30 mm in diameter, the depth is usually more than 2.5 times hole diameter. Drilling performance can be measured in terms of thrust force, torque, power, tool life, dimensional accuracy and surface quality. Each of these measures is affected by various factors such as the tool and workpiece materials properties, the tool geometry, the cutting conditions, cutting fluid and machine tool. Drilling can be considerably the most difficult while milling and turning were the easiest in the machining process. Many researchers have studied the machinability of aeroengine alloys in the past, especially in the turning and milling process. Although extensive investigation reports have been published, considerable progress is being made and reported on the drilling of these alloys.

Ezugwu *et al.* [1] stated that approximately 50% of the total consumption in the manufacture of an aircraft engine is nickel-based alloys. These refractory materials may come in wrought (bar, sheet, large forging) and casting. Nickel-based alloys can be divided into three sub-categories namely nickel-copper, nickel-chromium-iron and nickel-molybdenum-chromium. Many of the industrial nickel-based alloys contain alloying elements, including chromium, aluminium, titanium, molybdenum, tungsten, niobium, tantalum and cobalt. It has an excellent track record for providing corrosion resistance, high strength at high temperatures and aesthetic beauty in a wide range of applications. In most mentioned cases, Inconel 718 meets the requirements for many of the components in the aerospace engine. A major used of nickel-based alloys is in the manufacture of aerospace turbine blades whom the operating temperature is exceeded 500°C. An analysis on the chemical content of nickel-based alloys revealed that machining problems could be occurred due to the relatively high percentage of nickel, cobalt and chromium content.

Basically, work hardening, low thermal conductivity, abrasiveness, high strength level and high heat generated were the dominant reasons for the difficulty in machining aeroengine alloys. Excessive heat could damage the cutting tool. Furthermore, lack of rigidity when holding the tool holder with cutting tool and workpiece can also shorten the tool life. Prolong machining also caused severe chipping and fracture of the tool edge. Cracks on the cutting tool and fracture of the entire cutting edge were mainly observed when machining aeroengine alloys.

Chen and Liao [2]summarized that wear can be divided into four stages; coated layer is abraded-off, then followed by flank wear and chipping at the outer cutting edge, formation of chipping and lastly the wear rate increases drastically and the cutting action is now replaced by extrusion. Azarhoushang and Akbari [3] developed a ultrasonic-assisted drilling method (UAD) in order to improve the hole quality. They found that

* Corresponding author: erween@uthm.edu.my

the conventional drilling method suffered from catastrophic failure where the drills broke at the cutter exit. In addition, the UAD shows significant improvement in drill's lives. The study on the effect of minimal quantity lubrication (MQL) and biobased lubricant was conducted Rahim and Sasahara [4-8]. They found that the biobased lubricant made from palm oil significantly improved the drilling performance in terms of tool life and surface integrity under the MQL condition.

Unlike flood machining, previous researches showed that the machining of nickel based alloys is still limited especially when the effectiveness of the MQL approach has taken into consideration. In addition, much of the previous work done on MQL machining has been focused on the application of turning and milling process, however little research has been completed on drilling. In addition, published literature contains few articles on the performance of vegetable oil, in fact it can be an alternative to aforementioned cutting fluid. The main objective of this research is to highlight and investigate the drilling performance under various coolant-lubricant conditions, as well as a different type of MQL fluids namely synthetic ester and palm oil.

2 Experimental details

2.1 Drilling tests under external coolant supply

Drilling tests were performed using a Mazak Nexus 410-A machining center. These tests were conducted on the Inconel 718 with a diameter and thickness of 50 mm and 20 mm, respectively, which was mounted on a piezoelectric dynamometer (Kistler 9365) as shown in Fig. 1. The dynamometer was connected to the charge amplifier and data logger system to measure and congregate the thrust force and torque data. All drills were AlTiN coated indexable carbide (Mitsubishi TAWNH1400T) with a diameter of 14 mm. The point and helix angles were 130° and 30°, respectively. The tool shank was collet mounted with an 80 mm overhang. A type-K thermocouple wires were embedded in the workpiece and located 9 mm (TC1) and 17 mm (TC2) from the top of the workpiece surface. The distance between the thermocouple wire to the hole wall is approximately 1.25 mm.

In this experiment, the cutting speed of 30 m/min and feed rate of 0.05 mm/rev were employed. The through hole was drilled under the action of the external air blow, minimum quantity of lubricant (MQL) and flood coolant conditions (water soluble type). Synthetic ester (MQLSE) and palm oil (MQLPO) were used as the lubricant for the MQL condition. The physical characteristics of the MQL lubricants are summarized in Table 1.

2.2 Drilling tests under internal coolant supply

A series of MQL drilling tests under internal through-coolant supply were performed on the same machining center as described earlier. Fig. 2 shows the experimental setup with a rotating 20 mm diameter of Inconel 718 bar and a stationary 14 mm diameter of AlTiN coated indexable carbide drill (Mitsubishi TAWNH1400T). The overhang of the workpiece and the drill were kept constant at 5 mm and 75 mm, respectively. The thrust force during the drilling tests was measured with a Kistler 9257B dynamometer. The force signals were acquired by a data acquisition system connected to the personal computer. The type-K thermocouple wires were embedded in the cutting tool and located near to the cutting edge, denoted as TC1 and TC2. The cutting speeds were 30, 40 and 50 m/min at two feed rates of 0.05 and 0.1 mm/rev. A commercial available MQL unit was used to deliver the mist spray via an internal through-coolant. The lubricants used in this experiment are synthetic ester (MQLSE) and palm oil (MQLPO).

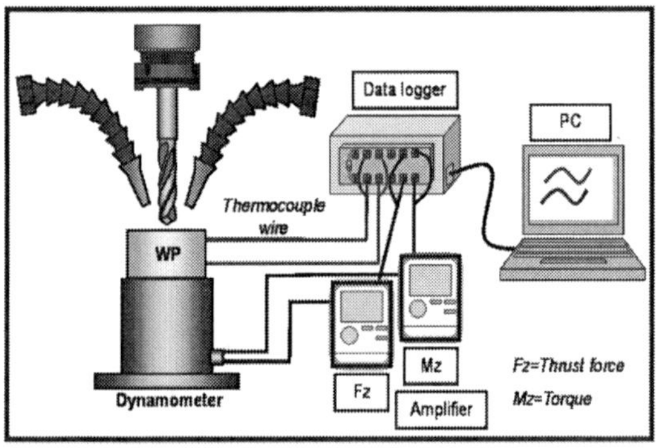

Fig. 1. Set up for external coolant supply

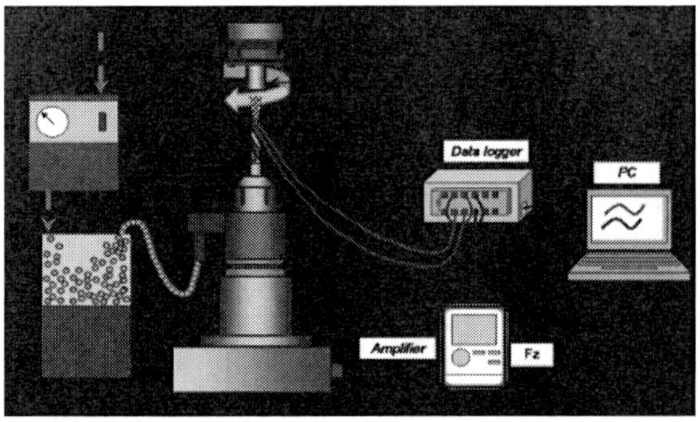

Fig. 2. Set up for internal coolant supply

3 Result And discussion

3.1 Performance of various coolant conditions under external supply

Fig. 3 shows the results of thrust force and torque at a constant cutting speed and feed rate of 30 m/min and 0.05 mm/rev. It can be seen that MQLPO yielded significantly lower thrust force than the MQLSE, flood and air blow conditions. The MQLPO condition yielded a mean thrust force of 2638 N, while the MQLSE condition yielded a mean thrust force of 2886 N. Furthermore, the air blow condition recorded the highest thrust force value (3699 N) followed by the flood condition (3443 N). These results indicated that the thrust force was reduced by 40% (MQLPO), 28% (MQLSE) and 7% (flood) relative to the air blow condition. The values of torque obtained while drilling Inconel 718 depicted that the MQLPO gave the lowest torque, whereas MQLSE and flood yielded slightly lower torque values than the air blow condition. The results indicate improved performance of the MQLPO condition up to 24%, followed by MQLSE (23%) and flood (7%) relative to the air blow condition.

The results show that thrust force and torque are considerably less with the application of coolant (flood, MQLSE and MQLPO) compared to the air blow condition. This performance is due to the ability of coolant-lubricant to both reduce the cutting temperature and the friction thus preventing seizure at the tool-chip interface. However, the flood condition exhibits slightly higher thrust force and torque values than the MQL condition. The application of the MQL condition (MQLSE and MQLPO) substantially affected the thrust force and torque results. Furthermore, the MQLPO condition recorded the lowest thrust force and torque values. This suggests that MQLPO was a more effective lubricant than MQLSE. A vegetable oil such as palm oil consists of a high percentage of unsaturated fatty acid in the carbon chain. Unsaturated fatty acids contain double bonds which are easily broken with high temperature [9]. This enhances the formation of high lubrication strength film on the tool-chip or tool-workpiece interface.

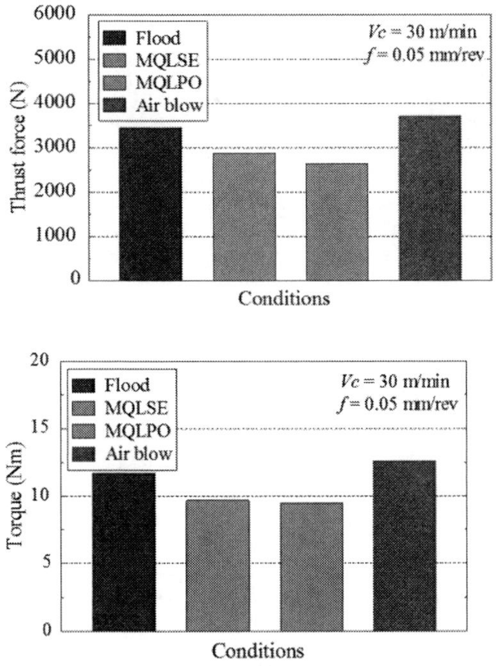

Fig. 3. Results of thrust force and torque

Fig. 4 shows the maximum workpiece temperature obtained under different coolant-lubricant conditions when drilling Inconel 718 at the cutting speed and feed rate of 30 m/min and 0.05 mm/rev, respectively. In this experiment, the air blow condition yielded the highest workpiece temperature at TC1

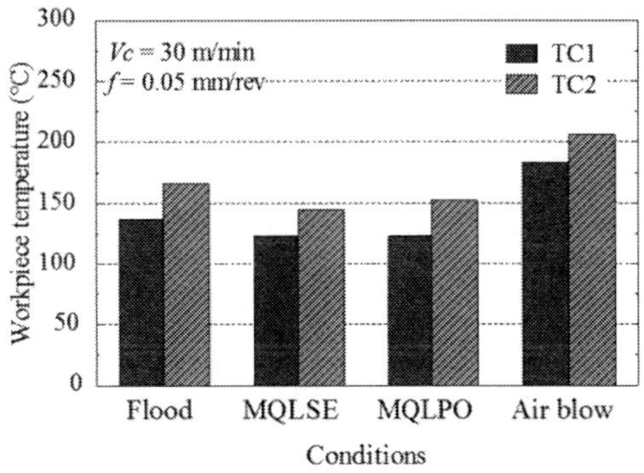

Fig. 4. Results of temperature

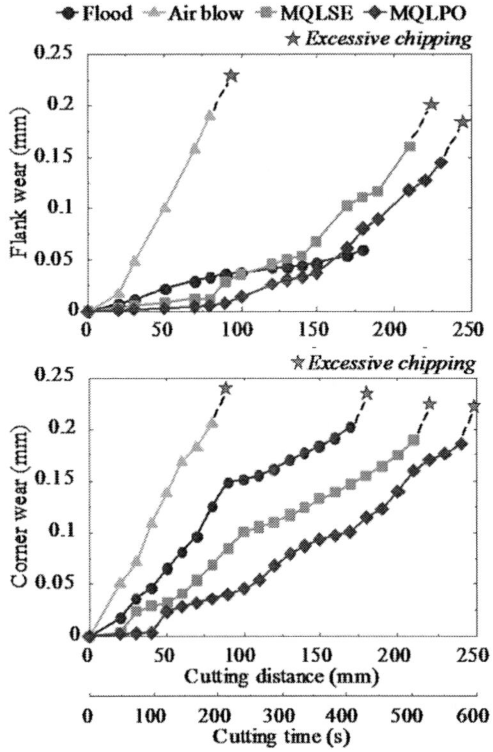

Fig.5. Tool wear progression

and TC2 with the value of 183°C and 209°C, respectively. This was due to the poor cooling effect performed by the air blow. Also, it was observed that both MQL conditions yielded slightly lower temperatures than the air blow and flood conditions. At TC1 and TC2, the workpiece temperature of the flood condition indicated a reduction of 25% and 19%, respectively, as opposed to the air blow condition.

Furthermore, the MQLSE condition exhibited a reduction of 33% and 30% at TC1 and TC2 at both locations. MQLPO, however, showed a comparable performance with MQLSE with a reduction of 33% and 26%.

Fig. 5 illustrates the variation of flank wear and corner wear at the cutting speed of 30 m/min and feed rate of 0.05 mm/rev under various coolant-lubricant conditions. Drilling of Inconel 718 under the air blow conditions revealed a rapid flank and corner wear progression. The cutting tool under this condition suffered from excessive chipping after 90 seconds. Additionally, a large amount of workpiece material was observed adhering to the cutting tool. The flank wear progression curves under the flood, MQLSE and MQLPO conditions were almost the same. However, at the corner region, the wear rate for the flood condition accelerates faster than the MQLPO followed by the MQLSE condition. A tool life of 414, 507 and 578 seconds was recorded under the flood, MQLSE and MQLPO conditions, respectively.

It was found that the application of MQL (MQLSE and MQLPO) can efficiently cool and lubricate the tool-chip interface, thus improving the tool life. The considerable improvement in the MQL condition is due to the high velocity of small droplets of lubricant that penetrate into the tool-chip interface more effectively than in the flood condition. Accordingly, the cutting temperature and cutting force are expected to be reduced and, consequently, the tool life improved.

3.2 Performance Of Various Coolant Conditions Under Internal Supply

Fig. 6 shows the variation of thrust force for MQLSE and MQLPO conditions via internal through-tool supply when drilling Inconel 718. In general, thrust force decreased with an increase in cutting speed and decrease in the feed rate. Furthermore, thrust force under MQLPO condition is lower than under MQLSE condition at all tested cutting speeds and feed rates. This figure shows that drilling at the feed rate of 0.05 mm/rev decreased the thrust force of MQLSE and MQLPO conditions by 25% and 24% respectively, when the cutting speed increases from 30 to 50 m/min. Meanwhile, MQLSE and MQLPO conditions recorded the thrust force reduction by 6% and 12% respectively, at the feed rate of 0.1 mm/rev. Furthermore, drilling at the lower feed rate of 0.05 mm/rev reduced the thrust force of MQLSE condition by 17% for the cutting speed of 30 m/min, 24% for the cutting speed of 40 m/min and 38% for the cutting speed of 50 m/min. In addition, MQLPO condition recorded the reduction of 18% for the cutting speed of 30 m/min, 22% for the cutting speed of 40 m/min and 32% for the cutting speed of 50 m/min.

The observation revealed that high thrust force at lower cutting speeds is believed to be due to the presence of a built-up edge on the cutting edges, thus increasing the tool-chip friction as well as thrust force. Moreover, thrust forces were lowest at the higher cutting speed. This reduction is due to the decreased of tool-chip contact length and corresponding decrease in the friction of tool-chip interface. The phenomenon of lower thrust force by increasing the cutting speed can also be elucidated by the temperature influence. Higher temperature generates at high cutting speed thus the lower resistance against chip formation causes a lower thrust force and results in the thinner chip. It can be seen that the thrust forces also increased in feed rate. This is because the higher feed rate leads to a increase in the volume of material removed and more deformation energy is required for chip formation. It is evident from Fig. 6 that the thrust forces are lower under MQLPO condition as compared with MQLSE condition. This is due to the reduction of the friction at the tool-chip interface under MQLPO condition.

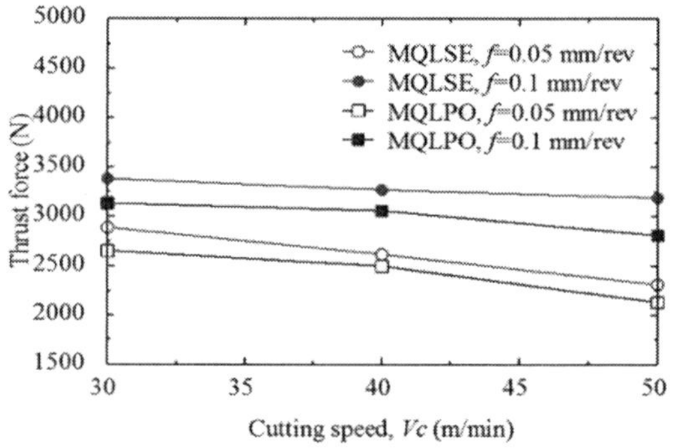

Fig. 6. Thrust force for MQLSE and MQLPO

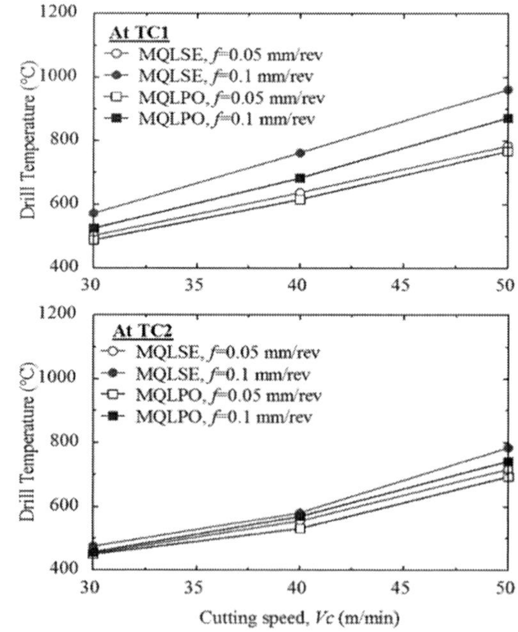

Fig. 7. Maximum drill temperature at TC1 and TC2for MQLSE and MQLPO

Fig. 7 illustrates the maximum temperature on the cutting edges measured at TC1 and TC2, respectively. An examination of these figures revealed that the drill temperature is affected by the main effects of the cutting speed, feed rate and lubricants. Increasing the cutting speed and feed rate significantly increases the drill temperature. For instance, under the MQLSE condition, the tool temperature at TC1 and TC2 goes up from 502°C to 784°C and 453 °C to 717 °C at the feed rate of 0.05 mm/rev when the cutting speed change from 30 to 50 m/min, while the tool temperature goes up from 572 °C to 962 °C and from 475 °C to 784 °C at the feed rate of 0.1 mm/rev. Furthermore, MQLPO condition indicates an increment of tool temperature at TC1 and TC2 from 488°C to 767 °C and from 450 °C to 694 °C at the feed rate of 0.05 mm/rev. In addition, the drill temperature also rises from 525 °C to 872 °C and 457 °C to 742 °C at the feed rate of 0.1 mm/rev.

Fig. 7 also revealed that the drill temperature generated for feed rate of 0.05 mm/rev was lesser than the drill temperature generated for the feed rate of 0.1 mm/rev. For instance, for the thermocouple positioned at TC1, the recorded drill temperature values at the cutting speed of 50 m/min for MQLSE condition were 784 °C (feed rate of 0.05 mm/rev) and 962 °C (feed rate of 0.1 mm/rev). Under the same cutting speed, MQLPO condition recorded 767 °C (feed rate of 0.05 mm/rev) and 872 °C (feed rate of 0.1 mm/rev). The result indicates the reduction of 23% and 14% for MQLSE and MQLPO conditions, respectively. It is believed that the reduction of drill temperature was due to the reduction of contact time at the tool-chip interface during the machining process, subsequently reduces the friction. High feed rate also generates higher stress at the cutting tool edge thus promotes localized heat built-up in the cutting tool. The drill temperature is reduced in the presence of MQLPO condition compared to that when drilling with MQLSE condition. As stated in previous section, the viscosity of palm oil (MQLPO) was higher than that of synthetic ester (MQLSE). This suggests that the friction was reduced as the viscosity was increased, thus improved the lubrication condition.

4 Conclusions

Based on the analysis results, the evaluation of MQLPO and MQLSE was determined. The following conclusions can be drawn from this study;

MQLPO outperformed MQLSE and other coolant techniques in terms of cutting force, torque, cutting temperature and tool life. It can be suggested that the vegetable oil such as palm oil consists of a high percentage of unsaturated fatty acid in the carbon chain. This enhances the formation of high lubrication strength film on the tool-chip or tool-workpiece interface.

Acknowledgments

The authors thank the Ministry of Education Malaysia and UniversitiTun Hussein Onn Malaysia for the financial support via Fundamental Research Grant Scheme (Vot 1467).

References

1. E. O. Ezugwu, Z. M. Wang, A. R. Machado. Journal of Material Processing Technology, **86** (1999), 1-16.
2. Y. C. Chen, Y. S. Liao, Y. Study on wear mechanisms in drilling of Inconel 718 superalloy, Journal of Materials Processing Technology, 140 (2003), 269-273.
3. B. Azarhoushang, J. Akbari. **47** (2007), 1027-1033.
4. E. A. Rahim, H. Sasahara. An Analysis of Surface Integrity when Drilling Inconel 718 using Palm Oil and Synthetic Ester under MQL Condition. Machining Science and Technology, **15** (1) (2011) 76-90.
5. E. A. Rahim, H. Sasahara. Surface Integrity in MQL Drilling Nickel-Based Superalloy. Key Engineering Materials, 447-448 (2010) 811-815.
6. E. A. Rahim, H. Sasahara. Surface Integrity in MQL Drilling Nickel-Based Superalloy. Key Engineering Materials, 447-448 (2010) 811-815.
7. E. A. Rahim, H. Sasahara. Surface Integrity when Drilling Nickel-Based Superalloy under MQL Supply. Key Engineering Materials, 443 (2010), 365-370.
8. E. A. Rahim, H. Sasahara. Application of Minimum Quantity Lubrication when Drilling Nickel-Based Superalloy at High Cutting Speed. Key Engineering Materials, 407-408 (2009)612-615.
9. N. Talib, E. A. Rahim. *Procedia* CIRP, (2015) 346-350.

Evaluation of cutting force and surface roughness in high-speed milling of compacted graphite iron

Mohd Azlan Suhaimi[1]*, *Kyung-Hee* Park[2], *Safian* Sharif[1], *Dong-Won* Kim[3], *Amrifan* Saladin Mohruni[4]

[1]Faculty of Mechanical Engineering, Universiti Teknologi Malaysia, 81310 Johor Bahru, Malaysia
[2]Manufacturing System R&D Division, Korea Institute of Industrial Technology, Republic of Korea
[3]Department of Industrial and Information Systems Engineering, Chonbuk National University, Republic of Korea
[4]Department of Mechanical Engineering, Universitas Sriwijaya, 30162 Inderalaya, Indonesia

Abstract. Compacted Graphite Iron, (CGI) is known to have outstanding mechanical strength and weight-to-strength ratio as compared to conventional grey cast iron, (CI). The outstanding characteristics of CGI is due to its graphite particle shape, which is presented as compacted vermicular particle. The graphite is interconnected with random orientation and round edges, which results in higher mechanical strength. Whereas, graphite in the CI consists of a smooth-surfaced flakes that easily propagates cracks which results in weaker and brittle properties as compared to CGI. Owing to its improved properties, CGI is considered as the best candidate material in substituting grey cast iron that has been used in engine block applications for years. However, the smooth implementation of replacing CI with CGI has been hindered due to the poor machinability of CGI especially at high cutting speed. The tool life is decreased by 20 times when comparing CGI with CI under the same cutting condition. This study investigates the effect of using cryogenic cooling and minimum quantity lubrication (MQL) during high-speed milling of CGI (grade 450). Results showed that, the combination of internal cryogenic cooling and enhanced MQL improved the tool life, cutting force and surface quality as compared to the conventional flood coolant strategy during high-speed milling of CGI.

1 Introduction

Grey Cast Iron (CI) has been widely used in the manufacturing of engine blocks and cylinder heads for so many years due to its mechanical and physical properties. However, the repetition of start-up and shut-down of engines can cause high mechanical loading and may lead to localized crack at the engine blocks and heads as a result of thermo-mechanical fatigue [1, 2]. From this aspect, researchers have found that Compacted Graphite Iron (CGI) is the best candidate in replacing CI. The graphite structure of CGI is interconnected with random orientation and round edges exhibits better mechanical strength than that of CI. Whereas graphite in the CI presents smooth surface flakes that easily propagates cracks resulting in weaker and brittle properties when compared to CGI as shown in Fig. 1. Table 1 shows the mechanical and physical properties of CGI in comparison to CI [3]. The tensile strength and fatigue strength of CGI are almost double than CI, the elastic modulus is 38% higher than CI, and the thermal conductivity is 8% smaller than CI. Owing to the improved properties, CGI is considered a suitable candidate to substitute the grey cast iron in various components for diesel engine; such as engine blocks, cylinder heads and cylinder sleeves [4].

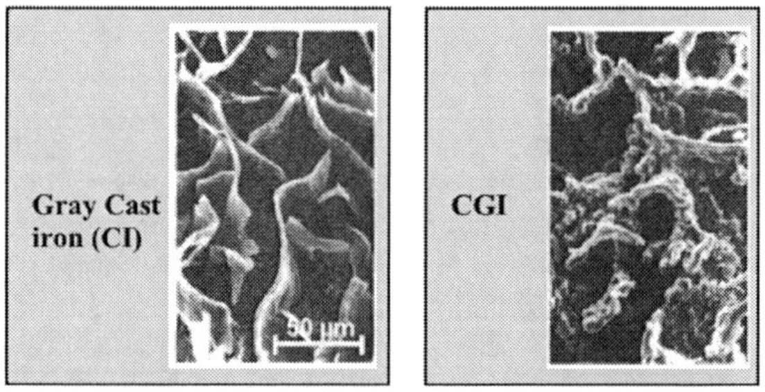

Fig. 1. Graphite particle shape of gray iron and compacted graphite iron [5]

Furthermore, the outstanding mechanical strength of CGI material enables the CGI-based engines to operate at higher temperature and cylinder pressure, which improved fuel consumptions, hence produce lower levels of emissions. In addition, CGI is lighter than CI as confirmed by Hyundai Motor Company whereby their CGI-based engine of 1.8L I-4 Diesel is 22 % lighter compared to the CI-based engine of the same kind [3].

Yang *et al.* (2015) studied the remanufacturing capability of automotive products by considering four different materials, used as engine block, namely Grey Cast Iron (CI) ASTMA48 Class 40, Aluminum A356-t6, Magnesium AMC SC1 T6 and CGI ASTM A482 Grade 450. The remanufacturing performance was ranked

* Corresponding author: azlans@fkm.utm.my

based on material's durability, clean-ability, restorability, environmental health and safety, material cost and material density. The authors found that CGI is more superior compared to CI in terms of remanufacturing performance. However, the poor machinability of CGI is the major barrier to the automotive industries in producing high performance engines with effective tool life, especially, when high-speed machining is applied to bore engine cylinders in an automated transfer line. According to [7], the tool life is decreased approximately by the factor of 20 when changing to CGI from CI for the same cutting condition. Whereas, [8] reported that when machining CGI the tool life reduced to 10% as compared to CI for both turning and milling process using PCBN and carbide tools. Da Silva *et al.* (2011) performed a tool life comparison between CGI and CI for high speed face milling. They found that, the flank wear was 100 % greater for CGI compared to CI at cutting speed of 600 m/min.

Table 1. Mechanical and physical properties of CGI (GJV) compared to conventional grey cast iron (GJL).

Properties	Units	GJV 450	GJL 250	GJL 300
Ultimate tensile strength	MPa	450	250	300
Rotating-bending fatigue (20 °C)	MPa	210	110	125
Rotating-bending fatigue (225 °C)	MPa	205	100	120
Elastic modulus	GPa	145	105	115
Elongation	%	1-2	0	0
Thermal conductivity	W/m.K	36	46	39
Thermal expansion	µm/m.K	12	12	12
Density	g/cc (CGS	7.7	7.1	7.1
Brinell hardness	BHN 10-3000	215-255	190-225	215-255

It is well known that, the decreased in tool life is due to the absence of Manganese-Sulfide (MnS) layer, which acts as a protective surface to avoid adhesion at the tool-chip interface in the case of machining CI [5, 10]. The most significant difference between CI and CGI is the sulfur content. CGI consists of almost 90% less sulfur content compared to CI. In machining CI, sulfur reacts with manganese to form MnS inclusions that forms a thin layer and acts as lubrication on the cutting edge surface. The MnS layer may also reduce the cutting temperature, hence decreases the oxidation and diffusion rates [10]. To demonstrate this effect, [8] compared the tool life by using different cutting speeds for machining both CGI and CI. As cutting speed increased to 400 m/min, a built-up layer containing MnS began to form on the cutting edge and the thickness of the layer increased as the cutting speed was raised to 800 mm/min. In contrast, with the same cutting condition applied to CGI, the tool suffered from abrasive flank wear with increasing cutting speed. Bazdar *et al.* (2009) attempted to increase the sulfur content on the graphite aspect ratio of a commercial CGI in order to improve the machinability. However, the tensile properties of the modified CGI decreased as compared to the commercial CGI.

The main purpose of this study is to investigate the effectiveness of various cooling/lubrication strategies in high-speed milling of CGI. The performance of uncoated carbide tool was tested at three different cutting speeds (400, 600 and 800 m/min) and two feed rates (0.1 and 0.2 mm/tooth). The progression of tool wear under various cooling/lubrication was recorded and analyzed. In order to evaluate the effectiveness of each different cooling/lubrication strategy, cutting force and surface roughness were measured at all conditions investigated. First, an experiment with conventional flood coolant (WET) was employed and the tool wear progression was used as the benchmark for other cooling/lubrication strategies, such as; Minimum Quantity Lubrication (MQL) with added nano-particles (nMQL), nano-MQL with combination of external cryogenic cooling (nMQL + ExtCryo), and nano-MQL with combination of internal cryogenic cooling (nMQL + IntCryo). Then, the respective cutting forces were acquired in real-time and surface roughness was measured at the end of each experiment. Finally, the tool wear was measured using a confocal laser microscope. The worn tools were analyzed under a Scanning Electron Microscope (SEM) and Energy Dispersive X-ray to examine the predominant type of wear mechanisms and chemical characterization, respectively.

Cooling and lubrication play an important role in metal cutting process for transferring excessive heat at the tool/workpiece engagement region and reducing tool wear. In addition, these applications are able to reduce the power consumption and at the same time improve the surface quality and productivity. In a conventional machining of most metals, a flood coolant system is preferable due to the effective cooling and ease of the application. However, flood coolant is less effective when working with difficult-to-cut and low thermal conductivity materials such as CGI. To deal with extremely high cutting temperature when machining these materials, several studies were conducted by employing cryogenic cooling with liquid nitrogen (LN2). It has been proved that cryogenic cooling results in a better tool-life [12, 13] and lower cutting force [14] when machining titanium alloys. Due to the low temperatures below -180 °C, liquid nitrogen has been used in a wide variety of application for cooling purposes. In addition, LN2 is odorless and more environmental friendly compared to the conventional flood coolant [15]. Nevertheless, reports on the

DOI: 10.1051/matecconf/201710103016

performance of cryogenic cooling when machining of CGI are very limited. It was very common that an external spray nozzle is used to supply LN2 to the cutting tool as shown reported by many researchers since year 2001 until now [13, 16-20]. By this way, the surface hardness of the workpiece material may increase due to the excessive LN2 spray. Fig. 2 shows the hardness of CGI Grade 450 with respect to testing temperature, and the effect of the excessive cryogenic cooling using conventional external spray. It is clearly showed that the hardness can be easily increased to 16%, due to the effect of excessive LN2 of the external cryogenic cooling. Similarly, [21] reported that the hardness of Ti-6Al-4V increases from 32HRc to 42HRc due to the exposure of LN2. Therefore, in order to solve the excessive spray effect, a specially designed tooling kit (as shown in Fig. 3) was used in this study which enables the supply of liquid nitrogen to the cutting tool internally. The performance of the newly designed tooling kit was evaluated accordingly.

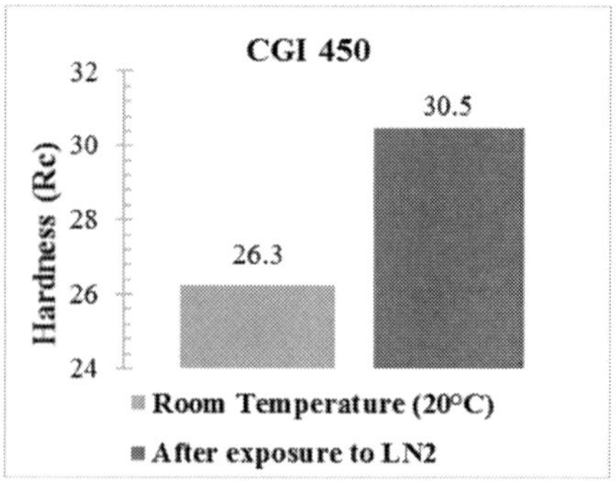

Fig. 2. The effect of the excessive cryogenic due to the conventional external spray of CGI 450

Minimum quantity lubrication (MQL) is well known for its capability to cool and lubricate the cutting tool simultaneously. However, MQL is less effective under high-speed machining as the sprayed oil simply evaporates as soon as the spray contacted the high temperature tool. In this study, a nano-particle is added with the vegetable-based oil, similar to that study conducted by [22]. It their study, it has been proved that, the enhanced exfoliated nano-graphene (xGnP) vegetable-based oil performed better in terms of reducing the tool wear of a ball-milling process compared to the traditional MQL vegetable oil. The added nano-particles provide additional lubrication to the cutting tool even when the MQL spray already evaporates. This is due to the nano-particles is arranged in a multiple-layer structure. With the same interest, [23] did a comparative study on the effectiveness of added nano-particle MQL by xGnP and Hexagonal Boron Nitride (hBN). They reported that, the mixture of hBN particles reduced the flank wear and central wear of a ball-milling process better than xGnP mixture.

This study investigated the effectiveness of MQL with added nano-particle using MQL with added hBN nano-particle when machining CGI. In addition, the combination of both external and internal spray of cryogenic cooling and enhanced nano-hBN lubricant was also performed.

2 Experimental methodology

The experiments were performed on a three-axis vertical milling center, Mori Seiki NVD-4000-DCG-HSC with Fanuc CNC controller. The cutting tool used in the machining test was an indexable end-milling cutter with designated code R390-016A16L-11L with two cutting edges. The tool diameter was 16 mm. The insert used was uncoated carbide (grade H13A) from Sandvik-CoroMill 390. Its helix angle was 21° and the corner radius was 0.8 mm.

The workpiece material is a commercially CGI (grade GJV 450) with size of 100 x 100 x 300-mm block from SinterCast. The mechanical and physical properties of the workpiece are shown in Table 1. Table 2 presents the chemical composition of the workpiece according to the material producer. The cutting parameters was defined according to the standard production line and also based on previous published work. The machining conditions are shown in Table 3.

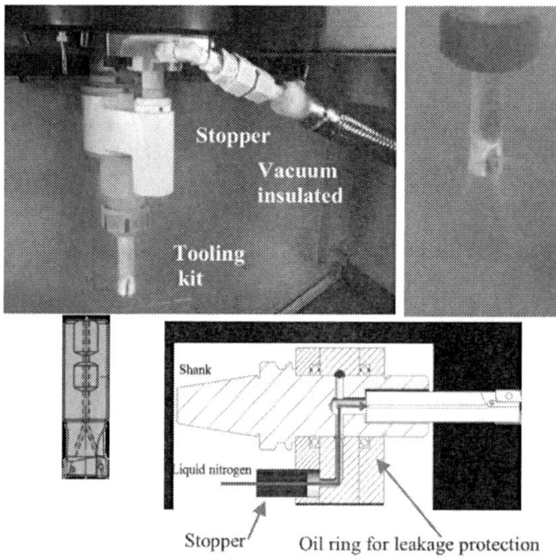

Fig. 3. Custom designed of internal cryogenic cooling tooling kit.

For each experiment, a 3D laser scanning microscope (CLSM), Model: Keyence VK-X200 with a magnification of 10X, has been used to capture the image of corresponding tool wear. Measuring software, known as 'VK Analyzer', has been used to measure the average flank and crater wear from the captured image. The worn tools were analyzed in a scanning electron microscope (SEM) and energy dispersive X-ray spectroscopy (EDX) in order to identify the predominant type of wear mechanisms and to measure the chemical weight percentage adhere on the tool.

Due to the irregular shape of the workpiece stock from the as-cast effect, a layer of approximately 2 mm was removed by face milling before the actual experiment starts. A total of 24 experiments were conducted which consists of different 6 combinations of cutting parameters for each cooling/lubrication method.

Table 2. Chemical composition of CGI, GJV 450.

Component	wt. %
C	3.55
Si	2.25
Mn	0.40
S	0.0065
Cr	0.029
Cu	0.81
Mg	0.009
Sn	0.075
Ti	0.009

Table 3. Machining conditions

Condition	Description
Cutting speed, Vc	400, 600 & 800 m/min
Feed per tooth, fz	0.1 & 0.2 mm
Axial depth of cut	2 mm
Radial depth of cut	5 mm
Cutting length	375 mm

3 Results and discussion

Fig. 4 presents the main effects of cutting force, surface roughness and average flank wear with respect to different cutting speed (Vc), feed-per-tooth (fz) and cooling/lubrication methods. Results show that, there is no significant effect on the cutting force for the change of Vc from 400 to 600 m/min. But, cutting force is decreased as Vc increased to 800 m/min, similar to the findings by Oxley [24], cutting force decreased as the cutting speed increases until at certain point before increased back due to material's strain characteristics. However, cutting force increased significantly with increase in feed rate (fz), due to the double increment of material removal rate from 7.95 cm^3/min to 15.91 cm^3/min. Meanwhile, WET condition, consumed the highest cutting force compared to other cooling and lubrication methods. This is due to the inefficient cooling of the conventional flood coolant method. For that reason, WET strategy appears to suffer from the worst tool life as shown in Fig. 4(c). Fig. 5 reveals the EDX analysis of the tool flank face under WET strategy. Due to inefficient cooling, the EDX spectra reveals that, the tool material for WET strategy exhibits a strong adhesion of workpiece material. A highest concentration of iron, carbon, silicon, manganese and copper was observed under all cutting conditions using WET strategy as cooling/lubrication method. These chemical elements are the main element of the workpiece material. As an example, for Vc = 800 m/min and fz = 0.2 mm (Fig. 5), the weight percentage of iron for WET are 86% followed by 84%, 78% and 76% for NanoMQL, NanoMQL+ExtCryo and NanoMQL+IntCryo, respectively. Perhaps, the cutting tool from WET strategy suffers thermal cracks at the tool cutting edge. These cracks are caused by cycling variation of temperature and frequently happen perpendicular to the cutting edge. The thermal cracks damage only appears after machining at cutting speed of 600 and 800 m/min under WET strategy as reported by Silva et al. [9]. Severe cracks may cause frittering hence causing poor surface finish and excessive flank wear as shown in Fig. 6.

The results for surface roughness (Fig. 4 (b)) for the whole set of experiments were fund to be below 1.0 micron, which is within the acceptable range for rough milling process. The main effects diagram showed that at Vc = 600 m/min, the finest quality was recorded below the average surface quality for both MQL and MQL + Internal Cryogenic cooling, compared to MQL + ExtCryo cooling and WET which produced the worst surface roughness value. There was no significant effect on surface roughness for the change of feed, fz.

The main effect for average flank wear (Fig. 4 (c)) reveals that flank wear is at smallest amount when cutting speed was at 600 m/min. The average flank wear is improved when higher feed was applied. More importantly is the effects of cooling/lubrication methods. The results show that WET, Nano-MQL and NanoMQL+ExtCryo has the worst tool life compared to NanoMQL+IntCryo. As mentioned before, under WET condition, the tool suffered from thermal cracks and excessive adhesion of workpiece material. As for Nano-MQL, the tool flank wear recorded the worst roughness value due to inability of MQL to penetrate outside the region of slipping, between the primary and secondary zone. This also may be due to effect of higher cutting speed and higher feed employed.

As shown in Table 4, the burn mark increased as cutting speed increased for both feed rates under MQL strategy. For NanoMQL+Ext Cryo, even though cryogenic cooling promises an excellent cooling capability, however due to the excessive amount of liquid nitrogen being exposed to the workpiece, it causes the hardness of the workpiece to increase, similar to the study reported by Park et al. [25]. It was evident that cryogenic cooling using the proposed tooling kit devise is able to control the cryogenic spray from spreading to the workpiece hence increases the hardness. As a result, NanoMQL+IntCryo was found to outperform other cooling/lubrication methods in improving the tool life. Furthermore, similar result was achieved in previous study [26] that was conducted on machining Ti-6Al-4V. It was evident that proper application of cryogenic is able to provide 93% reduction in tool wear compared to dry machining when utilizing the same research apparatus as in this study.

Since the main effect only shows the effect of one independent variable on the response variables by ignoring the effects of other independent variables, the potential of making error conclusion may occur. Hence, Fig. 7 shows the interaction effects for the cutting force and surface roughness. A statistical interaction occurs when the effect of one independent variable on the dependent variable changes depending on the level of another independent variable.

DOI: 10.1051/matecconf/201710103016

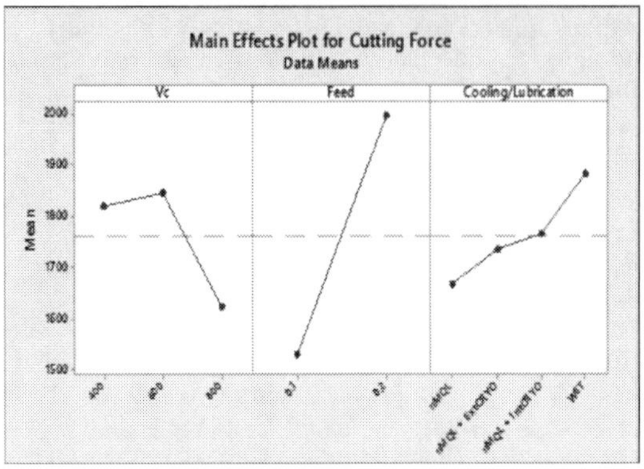

a) Cutting force

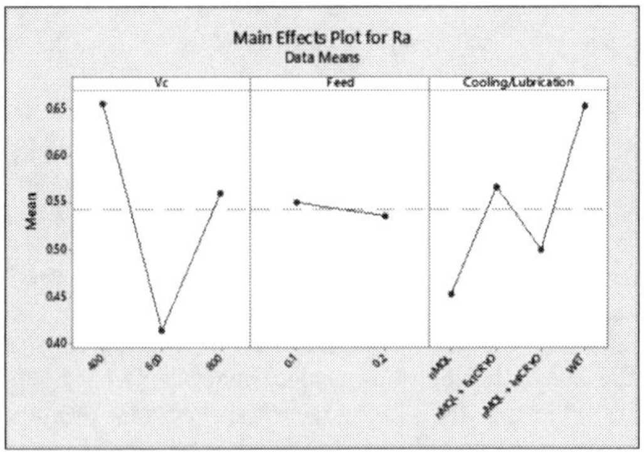

b) Surface roughness

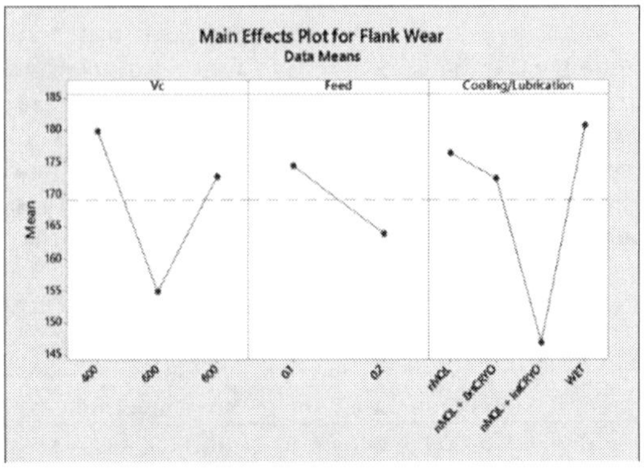

c) Average flank wear

Fig. 4. Main effects of cutting force, surface roughness and average flank wear

From Fig. 7, it can be concluded that, there is no interaction for all the main effects for cutting force (Fig. 7 (a)). As for the surface roughness (Fig. 7 (b)), there is no interaction effect except for Vc and fz. Instead, a cross over interaction effect was observed for Vc and fz on surface roughness, which implies both variables have no influence on surface roughness. So, the variation of surface quality is solely dependent on the application of cooling/ lubrication methods.

Table 4. Tool flank face captured for average flank wear measurements

Cutting Condition		**Coolant/lubrication Methods**	
Vc	Fz	Nano-MQL	Nano-MQL+IntCryo
400	0.1		
600			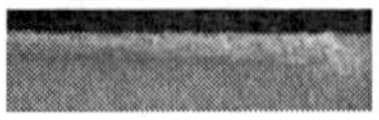
800		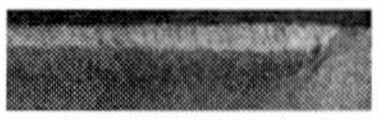	
400	0.2	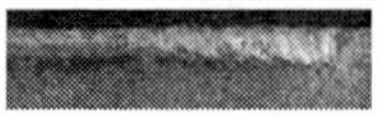	
600		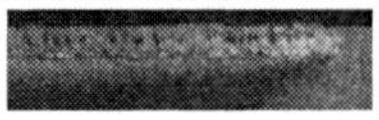	
800		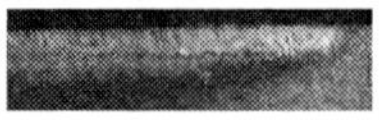	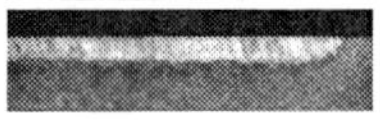

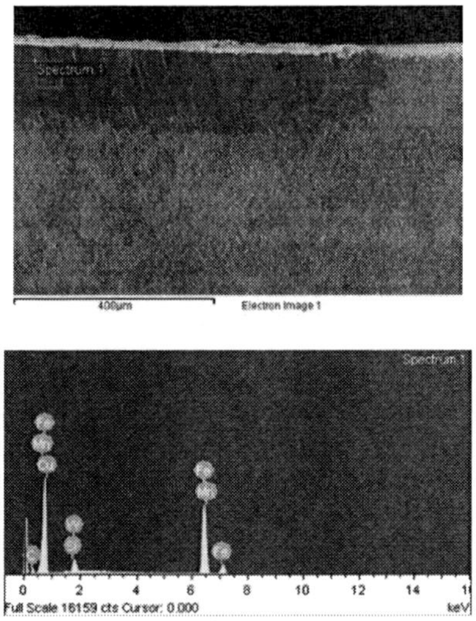

Fig. 5. EDX at the flank face after machining CGI at Vc = 800 m/min and fz = 0.2 mm with WET strategy

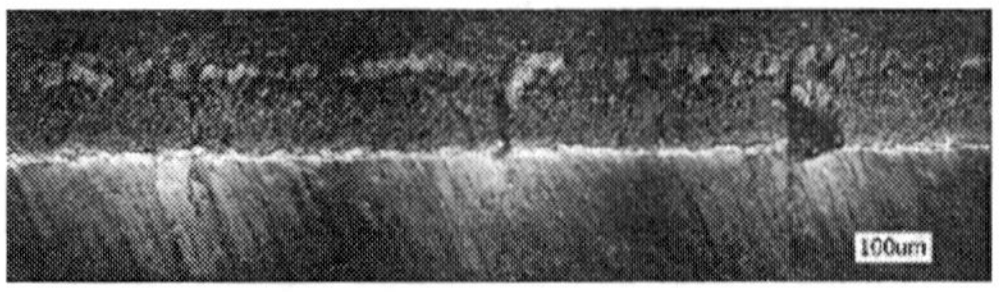

Fig. 6. Tool cutting edge for WET strategy for cutting coondition of Vc = 600 m/min and fz = 0.2 mm

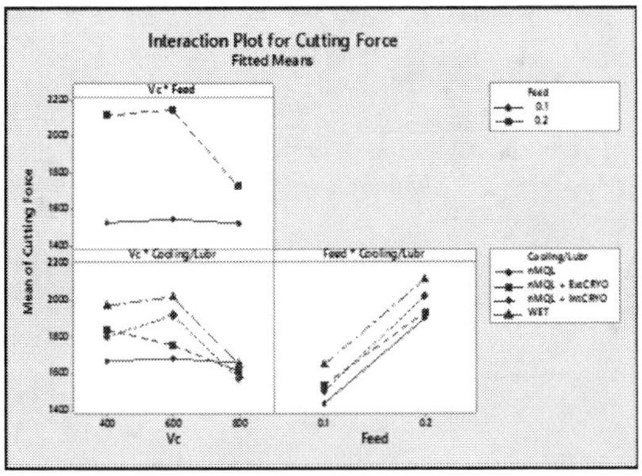

a) Cutting force

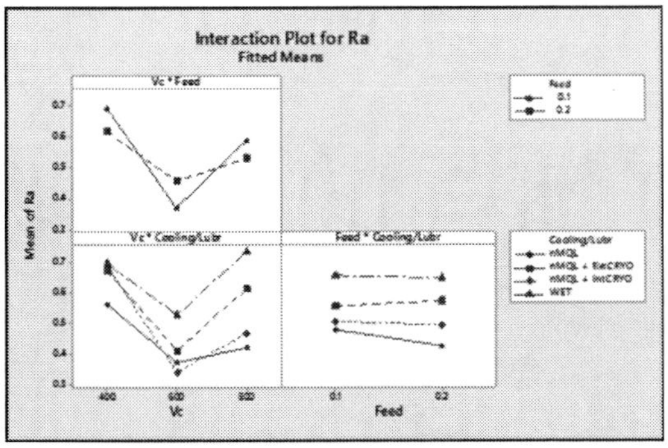

b) Surface roughness

Fig. 7. Interaction effects of cutting force and surface roughness

Results indicated that MQL and MQL + Internal Cryogenic cooling were able to produce the best surface quality while WET and external cryogenic cooling produced the worst surface quality. From the overall performance it can be concluded that, increase in cutting speed and feed do not affect the surface roughness, due to the acceptable range of surface roughness reading obtained for all cooling/lubrication methods which is below than 1.0 micron.

4 Conclusions

From the investigation conducted, it can be suggested that:

1. Cutting force is inversely proportional to the cutting speed. In addition, cutting force increased significantly when higher feed is employed.
2. NanoMQL has the worst tool life even though consumed lowest cutting force. This effect was due to thermal softening of the tool and workpiece. The combination of heat and shear stress at the primary and secondary zones resulted in high temperature and stress generation at the cutting zone.
3. The proposed tooling kit was able to efficiently transfer the cryogen to the cutting zone and effectively cool the tool without increasing the workpiece hardness. Hence, the tool life for Nano-MQL+Internal Cryogenic method outperformed other cooling/lubrication strategies.
4. Cutting speed and feed per tooth do not influence the surface roughness. The surface roughness was significantly affected by the application of cooling and lubrication methods. However, all experiments present acceptable surface roughness values of below 1.0 micron.

References

1. M. Riedler, H. Leitner, B. Prillhofer, G. Winter, and W. Eichlseder, *Meccanica* **42**, pp. 47-59 (2007)
2. V. Norman, P. Skoglund, D. Leidermark, and J. Moverare, Int. J. Fatigue **80**, pp. 381-390 (2015)
3. S. Dawson and F. Hang, *China Foundry* **6**, pp. 241-246 (2009)
4. L. L. Myagkov, K. Mahkamov, N. D. Chainov, and I. Makhkamova, pp. 370-408e (2014)
5. E. Abele, A. Sahm, and H. Schulz, *CIRP Annals - Manufacturing Technology***51**, pp. 53-56 (2002)
6. S. S. Yang, N. Nasr, S. K. Ong, and A. Y. C. Nee, J. Cleaner Production (2015)
7. M. Heck, H. M. Ortner, S. Flege, U. Reuter, and W. Ensinger, Int. J. Refractory Metals and Hard Materials **26**, pp. 197-206 (2008)
8. S. Dawson, I. Hollinger, and M. Robbins, *Detroit, SAE Technical Paper* **409**, pp. 4-16 (2001)
9. M. B. Da Silva, V. T. G. Naves, J. D. B. De Melo, C. L. F. De Andrade, and W. L. Guesser, *Wear* **271**, pp. 2426-2432 (2011)
10. M. Gastel, C. Konetschny, U. Reuter, C. Fasel, H. Schulz, R. Riedel, *et all.*, Int. J. Refractory Metals and Hard Materials **18**, pp. 287-296 (2000)
11. M. Bazdar, H. R. Abbasi, A. H. Yaghtin, and J. Rassizadehghani, J. Materials Processing Technology **209**, pp. 1701-1705 (2009)
12. C. Courbon, F. Pusavec, F. Dumont, J. Rech, and J. Kopac, *Tribology International* **66**, pp. 72-82, (2013)
13. X. Huang, X. Zhang, H. Mou, X. Zhang, and H. Ding, J. Materials Processing Technology **214**, pp. 3169-3178 (2014)
14. H. Safari, S. Sharif, S. Izman, H. Jafari, and D. Kurniawan, *Materials and Manufacturing Processes* **29**, pp. 350-356 (2014)
15. Y. S. Hong and Z. Zhao*Clean Products and Processes* **1**, pp. 107-116
16. S. Y. Hong and Y. Ding, Int. J. Machine Tools and Manufacture **41**, pp. 1417-1437 (2001)
17. B. D. Jerold and M. P. Kumar, J. Manufacturing Science and Engineering **135**, pp. 1-8 (2013)
18. T. C. Yap, N. S. M. El-Tayeb, and P. Von Brevern, J. Brazilian Society of Mechanical Sciences and Engineering **35**, pp. 11-15 (2013)
19. Aramcharoen and S. K. Chuan, *Procedia CIRP* **14**, pp. 529-534 (2014)
20. Shokrani, V. Dhokia, and S. T. Newman, *Second International Conference on Sustainable Opus : University of Bath Online Publication Store* (2015)

21. Shokrani, V. Dhokia, P. Muñoz-Escalona, and S. T. Newman, Int. J. Computer Integrated Manufacturing **26**, pp. 616-648 (2013)
22. K.-H. Park, B. Ewald, and P. Y. Kwon, J. *Tribology*, **133**, p. 031803 (2011)
23. T. Nguyen, I. Do, and P. Kwon, Int. J. Precision Engineering and Manufacturing **13**, pp. 1077-1083, (2012)
24. P. L. B. Oxley, *The mechanics of machining : an analytical approach to assessing machinability* (Chichester [England]; New York: E. Horwood ; Halsted Press, 1989)
25. K.-H. Park, G.-D. Yang, M. A. Suhaimi, D. Y. Lee, T.-G. Kim, D.-W. Kim, *et al.*, J. Mechanical Science and Technology **29**, pp. 5121-5126 (2015)
26. G.-D. Y. Kyung-Hee Park, M. A. Suhaimi, Dong-Yoon Lee, Tae-Gon Kim, Seok-Woo Lee, *Recent Advances in Manufacturing, Machine Design and Tribology* (2015)

Tandem bike design for apartment residents as an idea to reduce air pollution

Bambang Iskandriawan[1], *Jatmiko*[2], *Eri* Naharani Ustazah[1] and *Firman* Hawari[3]

[1]Department of Industrial Design, Faculty of Civil Engineering and Planning, Institut Teknologi Sepuluh Nopember, Indonesia
[2]Department of Interior Architecture, Faculty of Technology and Design, Universitas Ciputra, Indonesia
[3]Department of Interior Design, Faculty of Civil Engineering and Planning, Institut Teknologi Sepuluh Nopember, Indonesia

Abstract. Vertical housing is recently popular in urban areas since the availability of land is very limited. A strong preference for living in the city is mainly due to the home to work travel that is relatively close for those who work in the city. While residing in the vertical housing (i.e. apartment) can be an advantage, several accommodation facilities are often provided at their minimum standards. As such, a special design of bicycle for apartment residents may an effective solution. Multi-function bike, ridden either by one or two riders and its aesthetic part of interior element can be an attractive offer. Identification of marketing objectives and observation of comparable products were a crucial step in the initial process of tandem bike design. The use of bicycle in urban areas can overcome a number of motorized vehicles operating wherever the using of them may create issues such as traffic jam and air pollution. This study aims to adjust the tandem bike which can be produced commercially for the apartment residents. Stated in the research problems, the design process was successfully fulfilled. Fabrication and product testing were positively completed. A concise and comfortable bike to carry from the lobby to the apartment unit, or vice versa, was successfully created.

1 Introduction

In 2012, around 134 million of new middle class society grew in Indonesia, five times larger than Malaysian population and twenty five times larger than Singaporean citizens [1]. More than 50 % of this number lives in urban areas. This factor causes a very high population density in the city and makes the house prices more expensive. Vertical housing (i.e. apartment) offers an affordable accommodation for those working in urban areas. Several years ago, living in an apartment was viewed as an elegant lifestyle. Nowadays, however residing in an apartment is a common lifestyle for any class of society.

As the size of an apartment unit is normally undersized (Figure 1), household furnishings and electronic equipment have to be smallest dimension as well. The selection of air conditioning type for apartment model, for example, has to fulfil the criteria of the most cost-effective type [2]. It could be concluded, which one the variant of furniture arrangement along with the best working based on the value of Cooling Coefficient of Performance (COP_c) [3]. In addition, the use of bicycle around the city is becoming more popular since the government allows cyclists to ride in the main streets on *car-free* days. This trend encourages apartment residents to use bike. Unfortunately, the design of the bike is often problematic in such a vertical housing. Thus, a special design of the bike is required to be able to adapt to the characteristics of apartment units. The bike features needed include its function, used both as a tandem bike and as a single bike; its efficiency, a bike use depends on rider's needs; and its dimension, consuming less space in a room when not used. Such considerations are beneficial for apartment residents [4] (Figure 2).

Careful attention to notable ideas in the phases of design process results in a successful design [5]. Correlation amongst bicycle design, material, and art was investigated comprehensively by Ashby [6]. An updated analysis of bike based on either functions or conceptions is discussed comprehensively in the literature [7 and 8]. Bike design must fulfil the ergonomic rules [9, 10 and 11] directly from working trials to the finished products. The correlation amongst seat-tube angle, pedalling efficiency, and the main function of bike is crucial [12]. Construction efficacy aspect is considered to be significant aspect on designing a bike [13].

The principal question of this research is how to create tandem bike design especially for the apartment residents in flexible and efficient concepts, as well as safe and appealing to urban society. The increasing use of tandem bike is able to reduce the number of vehicles used by apartment occupants expectantly; therefore the amount of air pollution in the apartment area can also be minimized.

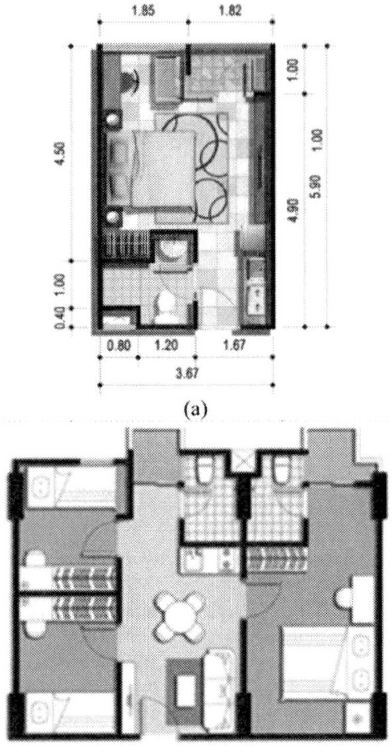

Fig. 1. An illustration of an apartment unit plan which has limited dimension: studio (a) and 3 (three) bed room type (b).

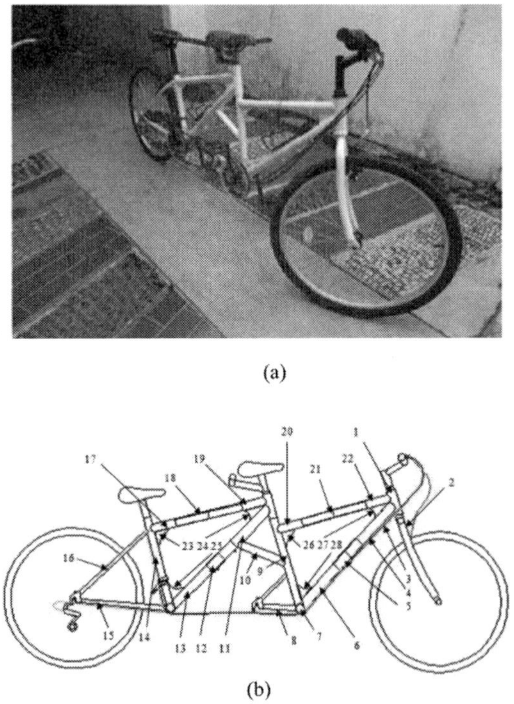

Fig. 2. Tandem bike prototype on ITS LPPM Competency Research Program (a, 2011), while still in patent process (number: P.00201200855) [3] and the technical drawing (b).

2 Method

The prototype development of multi-function tandem bicycle for apartment residents is well-programmed. Its development in this study is explained in the flow chart below (Figure 3.)

It is very often that problem/sub-problems/sub-solutions/solution has symmetrical relationships [4]. If there is a possibility, the result of the development process will give feedback to the prior development process.

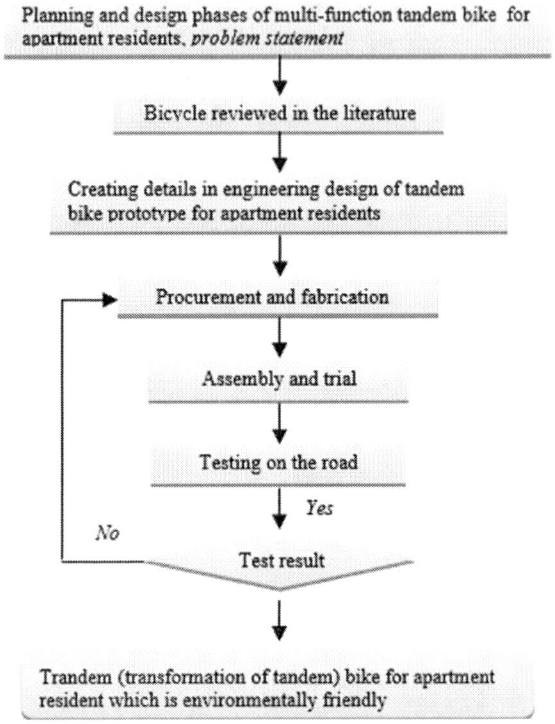

Fig. 3. Flow chart of the development of tandem bike design for apartment residents.

3 Results and discussion

In this study, the researchers who engaged in interior design disciplines focused more on designing and exploring the uniqueness of the apartment, while other researchers who majored in product design discipline conducted the tandem bike design. The results of analysis in this development process are compiled and discussed concurrently in the following part of this paper.

Identification of the marketing objectives

The identification of product users including the product environment where this product will be used should be clearly identified (Figure 4). The planning situation where the product will be stored can be seen in Figure 1. To be in line with the title of this study, the marketing objectives of the product users are the apartment residents, mainly the single and young couple.

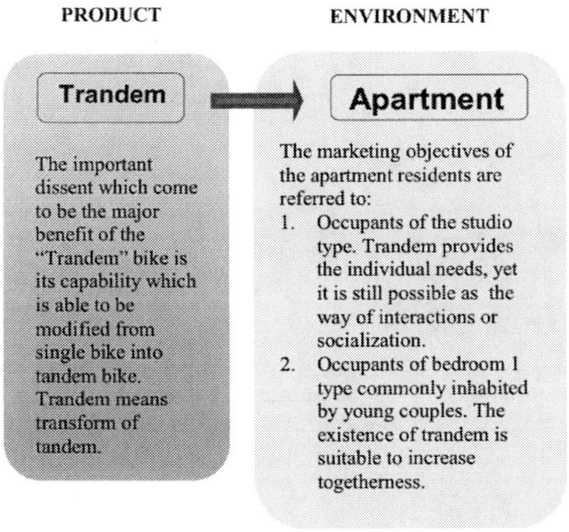

Fig. 4. Product identification and the target of trandem market segmentation. The meaning of "Trandem" is the transform of tandem.

Observation of the product settings

Commonly, there are 2 (two) kinds of bicycle parking area in the apartment (i.e. in Figure 5); outdoor area and (semi) indoor area. Both kinds of parking area have several advantages and disadvantages. Below is the illustration of the kinds of the parking areas

Itinerary of tandem bike from parking area to the apartment unit

The route of tandem bike starting from parking area to its placement inside the apartment unit can be seen in Figure 6 (illustration 1 to 6). For each location, it needs several instructions concerning the unique condition of all areas mentioned.

Product storage techniques in an apartment unit

Alternative positions of tandem bicycle placement for storage techniques can be seen in Figure 7. Based on the result of apartment location analysis, as well as the problem statements, the idea schemes can be developed as can be seen in Table 1.

Designing process

Designing process phase of tandem bike for apartment residents produced several sketch ideas, including the 3D technical drawing as can be seen in Figures 8 to Figure 12.

It is crucial to create initial idea by producing several sketches of bike. Through brainstorming amongst the team members, bicycle designs were selected from the most excellent sketches related to both single and tandem bike model for apartment residents. The main ideas of selection process were whether the design could work comfortably and whether there is a possibility to develop other bicycle frame designs for future studies.

Outdoor parking area

It provides bike racks for security reason. Also, it serves appropriate spaces to organize and to make it tidy. To protect the bike from hot weather and raining, the users use bike cover.

Indoor parking area

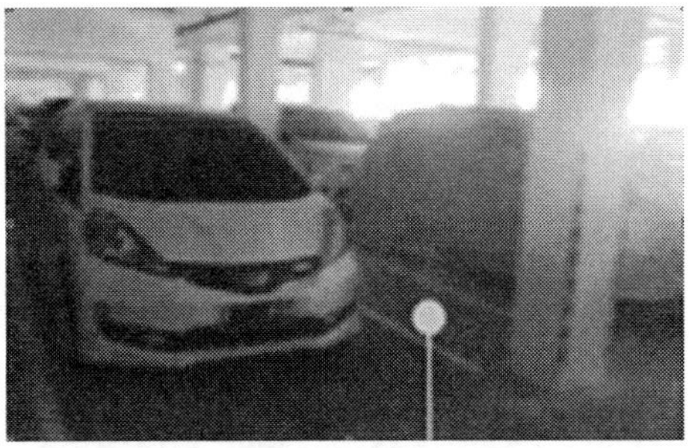

This kind of parking area provides spaces for two-wheel vehicle, motorcycle and bicycle. However, this facility commonly limits the vehicle numbers; one car and one motorcycle spaces parking area for one resident of the apartment unit.

Fig. 5. Bicycle parking area in the apartment; (semi) outdoor and indoor.

After determining its functions, bike component design which was easy to operate and uncomplicated fabrication was taken into account. To fulfil the process of fabrication and assembly, it needed detailed engineering design from technical drawing. 3D drawings were eminent to validate the works of skilled labourers since it provided closer to the representativeness of specific object than technical drawing. In addition, 3D drawings could visualize how the bike was mounted on the wall as one of possibilities to store the bike in an apartment unit (Figure 12).

Fabrication process and components assembly

Fabrication and assembly process phase (Figure 13) constitute to complicated process and as a matter of fact, the choice of the process is decided by employing *trial and error* approach. In the sketch process, technical or operational drawing can be imagined, yet in the process of execution, some difficulties can be found during fabrication.

One of the significant findings in this study is the tandem bike can be transformed into trolley. By transforming into trolley, the bike could be carried from the lobby to apartment unit or vice versa uncomplicatedly. Figure 14 shows the model of it. The bike is possible changed to another function such as a table when it is not to be used as the bike (Figure 15).

DOI: 10.1051/matecconf/201710103017

From parking area to the lobby

From parking area to the apartment building, the bike can be ridden.

At the front of the lobby, the bike must be folded and then put in the bag entirely or some parts with the remains are carried.

To keep the lobby clean, the bike has to be packaged in the bag, then pulled like a trolley, or brought like a backpack, when the biker is ready to enter the elevator in the lobby.

From the lobby to the apartment unit

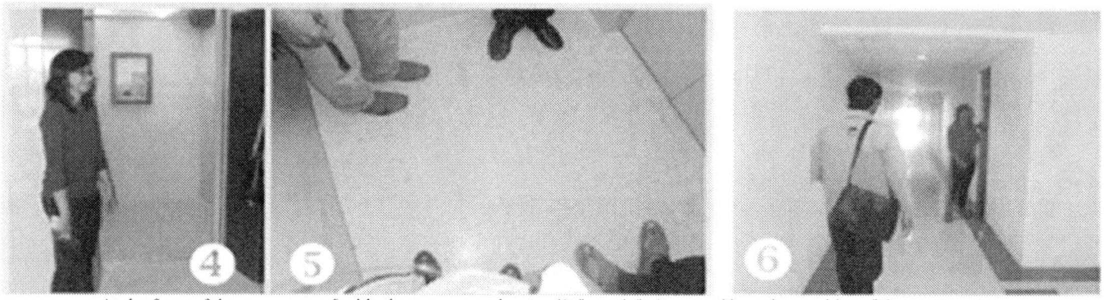

At the front of the elevator, the techniques in carrying the bike should be considered to avoid banging people when they enter or leave the elevator

Inside the passenger elevator (1.5m x 1.5m), the safe packaging and carrying techniques can prevent hitting other passengers. Since the bike needs proper space, the bike orientation should be vertical.

Since the corridor of the apartment must be clean from any object, the bike must be stored or parked inside the apartment unit.

Fig. 6. Tandem bike carrying scenarios in the apartment (1st to 6th steps).

Behind the main door. It is vertically stored above the shoes rack.

On the kitchen set or under the additional fence. It is possible to place the bike in a special storage.

Beside the refrigerator. The gap between the wall and the refrigerator can serve as the bike protector.

Wall mounted. The bike attached to the wall space in the living room possibly becomes an interior decoration.

Suspended at the ceiling. The opening space between the walls which are equipped with a hoist is potential to function as a garage.

Fig. 7. The possible storage techniques for tandem bike in an apartment unit (1 to 5).

Table 1 Ideas scheme (concept)

Parking method	Additional frame function	Carrying method	Storage method
All bike parts parked at the basement	They are only as the component (part)	Back-pack	Behind the main door of an apartment unit
All bike parts are carried and stored in an apartment unit	They are functioned as the support for the bike (part)	Carried	On the kitchen set Beside the refrigerator
Some bike parts parked at the basement and the remains stored in the apartment unit	They are functioned as a tool which is useful for user activities, i.e. table, partition, container, and hanger.	By pushing or pulling (as a trolley)	Wall mounted Hung at the ceiling

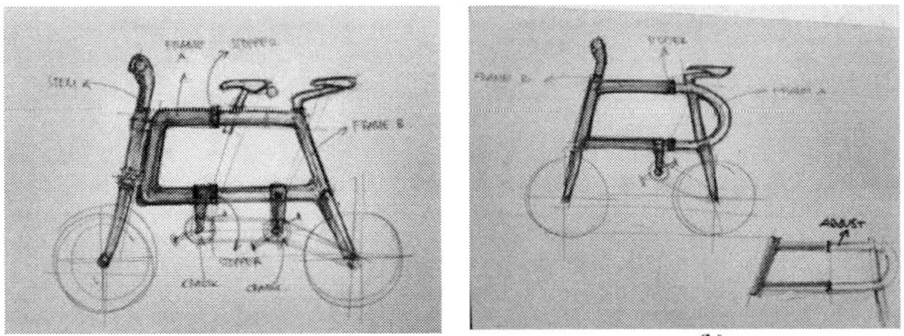

(a) (b)

Fig. 8. Tandem bicycle sketches for the apartment residents as a tandem (a) and as a single bike (b).

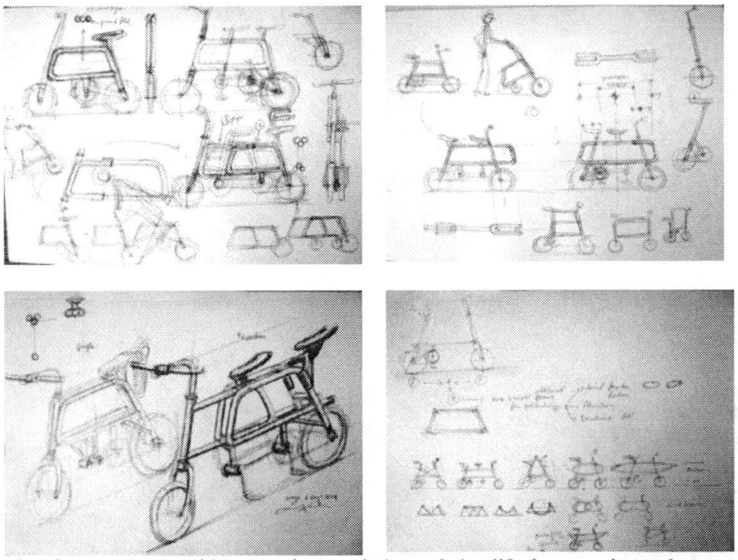

Fig. 9. Tandem bike sketches for apartment residents, made to optimize and simplify the operation and to accomplish the design purpose.

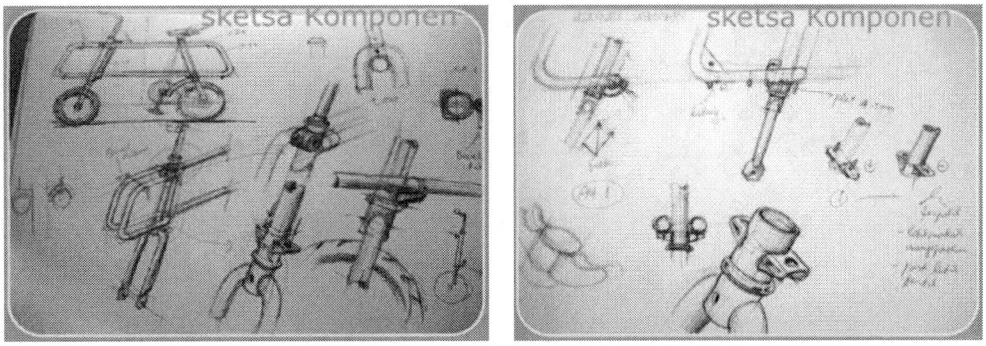

Fig. 10. Tandem bike component sketches for apartment residents, made to obtain simple operation and fabrication.

DOI: 10.1051/matecconf/201710103017

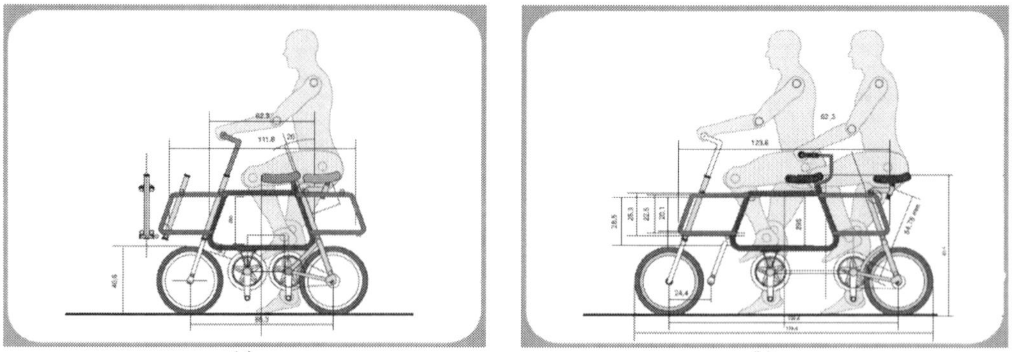

(a) (b)

Fig. 11. Tandem bike technical drawing results for apartment residents: as a single bike (a) and as tandem bike (b).

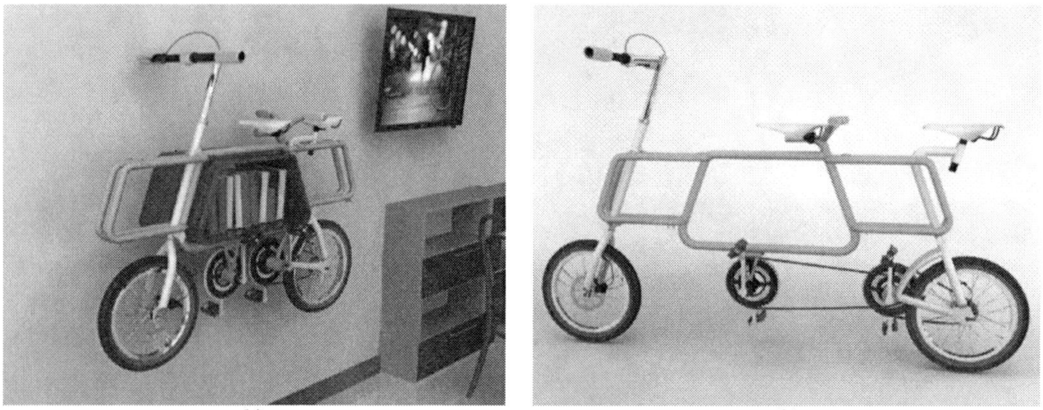

(a) (b)

Fig. 12. Placement alternatives results of tandem bike in a single bike form as the book shelf in apartment unit (a), and the bike form which is ready to be used as a tandem bike (b).

Fig. 13. Fabrication and components assembly process results of tandem bike for the apartment residents.

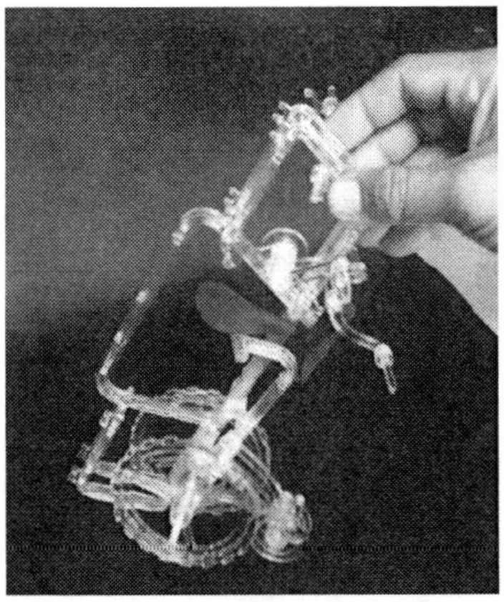

Fig. 14. Carrying method study result of bike from parking area to the lobby and to the apartment unit as a trolley.

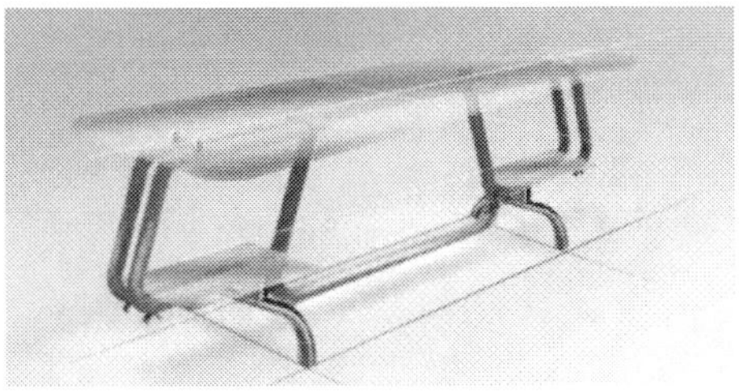

Fig. 15. Another function study result of the apartment bike: table.

4 Conclusions

Currently, there is a possibility to construct tandem bike prototype for apartment residents. Although the present research has produced some findings, there are some other related areas that need to be investigated to improve both the design and the function of tandem bike in the future research.

The purpose of this research is to enrich the design process theory in which it is viewed as specific tools that are likely to be used by designers. However, those specific tools can be altered on the users' purpose.

Since the fabrication process of tandem bike was created manually, the bike dimension was not perfectly shaped. The main frame, for instance, was dissimilar. In the future manufacturing process, however, it can be improved.

Concerning the test results, the tandem bike was strong enough to support two adult people at the time it was used. This try-out was also conducted on the bumpy road to examine the strength of the fork and the main frame.

By using tandem bike, the quality of social life around the apartment can be increased. Unlike the car users, the bike users are most likely to be able to interact with others comfortably. Furthermore, the escalating number of cyclists possibly reduces the number of vehicle driven by apartment occupants. More importantly, pollution around the apartment can be lessened.

The researchers are aware that the present study has some limitations. Therefore, further research on more innovative and precise tandem bike for apartment residents is required.

This work was supported by the Ministry of Higher Education, Research, and Technology, Republic of Indonesia.

References

1. *Apartment Guide*, Edition 01/I/July 2012, (Publisher: Kompas Gramedia, 2012)
2. B. Iskandriawan, IJAER **9**, 19, 6277-6290 (2014)
3. B. Iskandriawan, ARPN JEAS **11**, 2, 851-856 (2016)
4. B. Iskandriawan and Jatmiko, AMM **607**, 920-925 (2014)
5. N. Cross, Engineering Design Methods, *Strategies for Product Design*, Second Edition (John Wiley & Sons Ltd, Baffins Lane, Chichester, West Sussex, PO19, England, 1998)
6. M. Ashby, K. Johnson, *Materials and Design, The Art and Science of Material Selection in Product Design* (Butterworth Heinemann, Oxford, UK, 2007)
7. M. Embacher, Foreword by Paul Smith, *Cyclepedia A Tour of Iconic Bicycle Designs* (Thames & Hudson, ISBN 978-0-500-51558-7, 2011)
8. D. Wiyancoko, *Desain Sepeda Indonesia* (KPG (Kepustakaan Populer Gramedia) Jakarta, ISBN: 978-979-91-0243-0, 2010)
9. E. Grandjean, *Fitting the Task to The Man: An Ergonomics Approach* (Philadelphia: Taylor & Francis, 1986)
10. H. Christians and A. Bremmer, Applied Ergonomics, **29**, 3, 201-211 (1998)
11. PC Donkers, HM Toussaint, JF Molenbroek and LP Steenbekkers, Applied Ergonomics, 24, 2, 109-118 (1993)
12. C.Hsiang Chen, Y. Hao Huang and T. Yuang Shiang, J. Science and Cycling, **4**, 1, 28-32 (2015)
13. R.L. Mott, *Machine Element in Equipment Design*, Volume 1 and 2 (Publisher: Andi, 2009)

Numerical model for pollutant dispersion in the Dumai estuary

Mubarak[1], *Sigit* Sutikno[2,*], *Rena* Dian Merian[3]

[1]Marine Science Department Faculty of Fisheries, University of Riau, 28293 Pekanbaru, Indonesia
[2]Civil Engineering Department University of Riau, 28293 Pekanbaru, Indonesia
[3]Environmental Science Department Post Graduate Program, University of Riau, 28293 Pekanbaru, Indonesia

Abstract. The development of industry and the urban population in Dumai city potentially increases the amount of domestic sewage into the Dumai River, which is gradually deposited in the Dumai estuary. This condition has an impact on the environment and public health because of the higher concentration of pollutants. Knowing the dispersion of pollutants is very important in order to manage and monitor the water quality in the Dumai estuary. This research simulated a numerical model that was developed using MIKE 21 to study pollutant dispersion in the Dumai estuary as an impact of domestic sewage from Dumai city. Three important parameters of pollutants were discussed in this study, such as ammonia, nitrates and nitrites. The data used for modeling included winds, tides, bathymetry, and the pollutant concentration. The result of the numerical simulations shows that the concentration of all pollutant parameters tends to increase in the Dumai estuary. The concentration of ammonia, nitrates, and nitrites increase to 0.4 mg/L, 0.15 mg/L, and 0.14 mg/L respectively. Those concentrations exceed the standard maximum value that has been appointed by Indonesian regulations. It means the domestic sewage from Dumai city should be treated before it is discharged into the river.

1 Introduction

Dumai is the capital city in the Riau province on the island of Sumatera, Indonesia, which has an area of 1,727.38 km² and had 280,129 inhabitants in 2014 [1]. As a result of the geographic location, Dumai has an important role in transport and trade both regionally and internationally, especially to Malaysia. Dumai also has an important role in the oil industry both for fuel and palm oil, which generates economic activity. The development of industry and the urban population of Dumai city potentially increase the amount of domestic sewage into the Dumai River, which is gradually deposited in the Dumai estuary. This condition has an impact on the environment and public health because of the higher concentration of pollutants. Knowing the dispersion of pollutants is very important in order to manage and monitor the water quality in the Dumai estuary.

The water quality in the river should be managed and monitored periodically in the whole river, from upstream to downstream. This can be done by collecting and analyzing the time histories of water quality data of the river. By collecting and analyzing of those data, the pattern of pollutant dispersion and the potential impact because of the pollutants can be predicted. Those historical data can be obtained through continuous monitoring activities. However, these activities require substantial funds and long-term observation. To overcome these problems, an alternative solution can be found with a numerical model simulation. Numerical model simulation can be used to predict the dispersion of pollutants for various scenarios of input data both in the current conditions and in the future conditions. Numerical model simulation provides a powerful and unique capability for engineering studies of complex problems occurring under realistic field conditions [2].

To study pollutant dispersion in the Dumai estuary as impact of domestic sewage from Dumai city, this research simulated a numerical model which was developed using MIKE 21 as in many similar studies [3], [4], [5], [6]. MIKE 21 is a model frequently used for coastal simulations, developed by DHI (Danish Hydraulic Institute Environment). The hydrodynamic module in MIKE 21 is a general numerical modeling system for the simulation of water levels and flows in estuaries, bays and coastal areas; it simulates unsteady two-dimensional flows in one layer.

2 Methods

2.1 Study area

This study focused on the Dumai estuary, which is located in Dumai Kota sub-district, Dumai city, Riau province. Dumai is a capital city in Riau province on the

* Corresponding author: ssutiknoyk@yahoo.com

island of Sumatera, Indonesia as shown in Figure1. Dumai River is the main river that flows through the city and disembogues into the Rupat Straits. The river potentially brings many kinds of pollutants as a result of human activity in Dumai city.

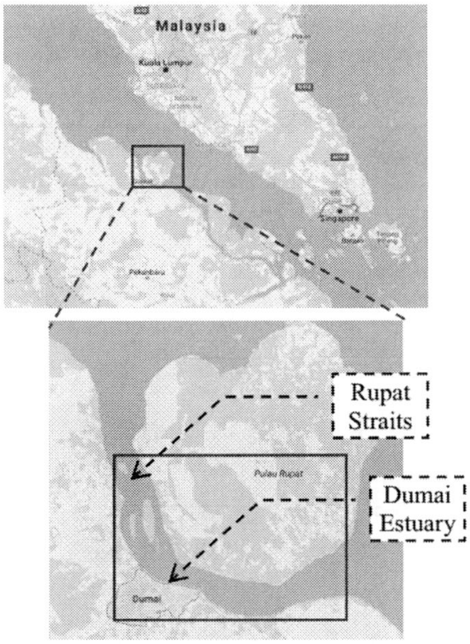

Fig. 1. Study area of this research.

2.2 Field survey

For numerical calibration, this research needs ground measurement data. The field survey was conducted on May 13-16, 2016 to get the measurement data. The position of sampling points was decided based on the purposive sampling method. The coordinate position of sampling points in the research area was used to guide the certain location for data measurements as shown in Figure 2.

Water quality parameters, which were measured in the sampling points, are ammonia, nitrates, and nitrites. Ammonia, a chemical compound with the formula NH_3, is a chemical parameter to determine the quality of water based on the Ministry of Environment's regulation No. 51, 2004. These compounds are usually found in the form of gas with a pungent odor that is characteristically called 'ammonia odor'. Ammonia is a nitrogen compound that is important and widely produced. Although ammonia has contributed significantly to the presence of nutrients in the earth, ammonia itself is a caustic compound that can damage health. Ammonia is toxic to fish and is also produced from the secretions or feces of fish.

Nitrate is a polyatomic ion with the molecular formula NO_3 that, in freshwater or estuarine systems close to land can reach high levels that can potentially cause the death of fish. While nitrate is much less toxic than ammonia, levels over 30 ppm of nitrate can inhibit growth, impair the immune system and cause stress in some aquatic species [7]. However, in light of inherent problems with past protocols on acute nitrate toxicity experiments, the extent of nitrate toxicity has been the subject of recent debates. The nitrite ion, which has the chemical formula NO_2 is the toxic by-product of the nitrifying bacteria (Nitrospira) in a filter or substrate consuming ammonia. It is only mildly less toxic than ammonia but it still can kill aquatic animals if its levels get too high.

These parameters were measured at a distance of 25 m, 50 m, and 100 m from the river mouth towards the Rupat Straits and were performed at high tide and low tide. The analysis of ammonia, nitrate and nitrite concentrations were conducted at the Marine Chemistry Laboratory, Faculty of Fisheries and Marine Sciences, University of Riau. The method and working principle refers to the Indonesian National Standard (SNI) 06-6989.30-2005, SNI 06-6989.9-2004, and SNI 06-2480-1991 for the measurement of ammonia, nitrites, and nitrates respectively.

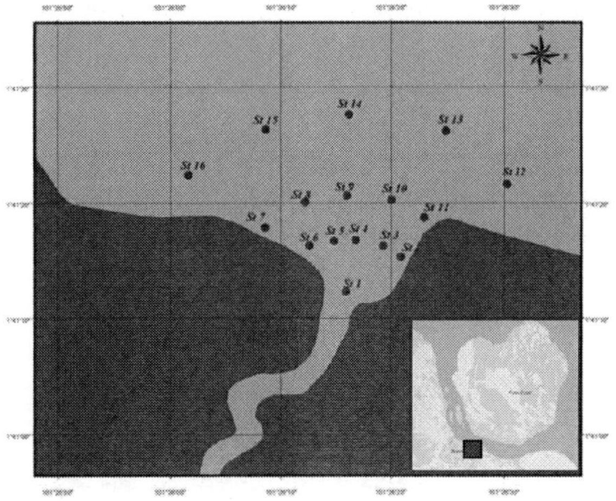

Fig. 2. Sampling points for field survey.

2.3 Numerical model

With the increase in computational technology, many numerical models and software programs have been developed for various engineering practices. Numerical modeling has been used extensively in industries for both forward problems and inverse problems. Forward problems include simulation of space shuttle flight, ground water flow, material strength, earthquakes, and molecular and medication formulae studies. Inverse problems consist of non-destructive evaluation (NDE), tomography, source location, image processing, and structure deformation during loading tests.

The hydrodynamics are represented through the shallow water equations, which must be solved numerically [6], [8]. Software Mike 21, from DHI Water

& Environment, was used. The model domain was the whole area of the Rupat Straits, as shown in Figure 3.

A non-structured triangular elements grid was built, with a higher density of nodes within the problem zone. The spatial step ranged from about 100 to 2000 m, for a total of about 75,000 nodes. The bathymetry of the sea bottom was built based on data reported in the nautical charts of the Hydro-Oceanographic Agency, Indonesian Army as shown in Figure 4. The bathymetry of Rupat Straits shows that the depths ranging from 2.5 to 35 m. Regional marine waters that have a depth of about 5 m are usually not so far from the shoreline. The depth of the waters around the Strait Rupat troughs is relatively steep. The depth of the waters of east, south and north Rupat range from 10 to 20 m, while the depth of the waters in the northeast and east range from 30 to 35 m.

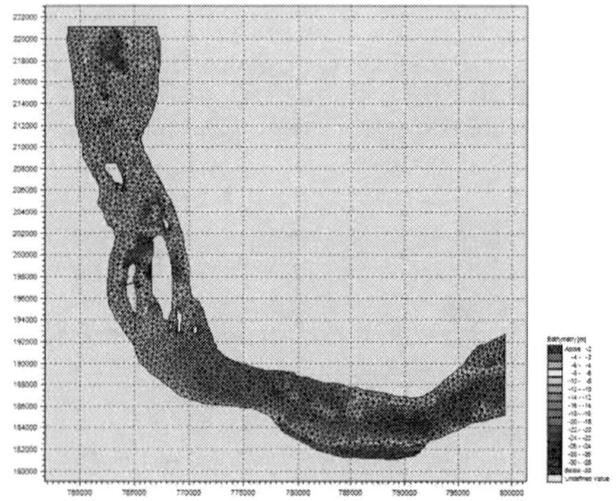

Fig. 3. Domain area of the numerical model.

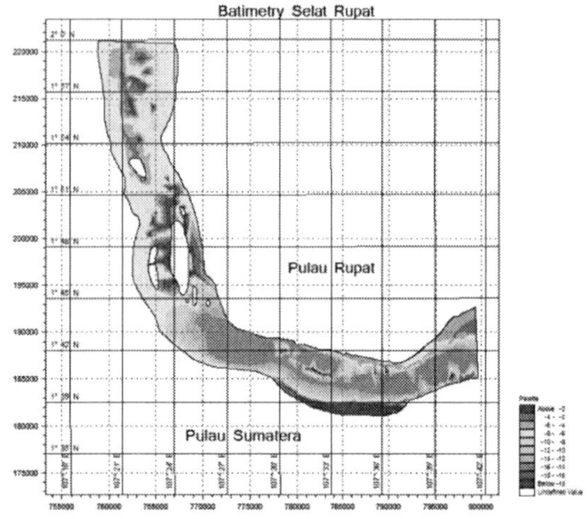

Fig. 4. Bathymetry map of the research area.

The numerical model is based on the mass conservation of pollutants, that leads to a system of 2D advection–diffusion equations, including source (or sink) terms to represent transformation processes through which coupling among them exists. This system must also be solved numerically. In order to achieve this, Mike 21 software was used. Three water quality parameters modeled were ammonia (NH_3), nitrates (NO_3), and nitrites (NO_2).

2.4 Model calibration

Two parameters were calibrated, which dominantly affect the hydrodynamics of the estuary, such as roughness coefficient of the Manning value (n) and viscosity turbulent or eddy viscosity (ϵ). Those parameters were adjusted so the predicted tide was as close as possible to the measured tide [9]. Figure 5 shows the comparison between the predicted tide and the measured tide as a result of the calibration process. The mean relative error (MRE) for this condition is 34.02%, and the Manning value and eddy viscosity are 32 m $^{(1/3)}$/s and 0.28 m^2/s respectively.

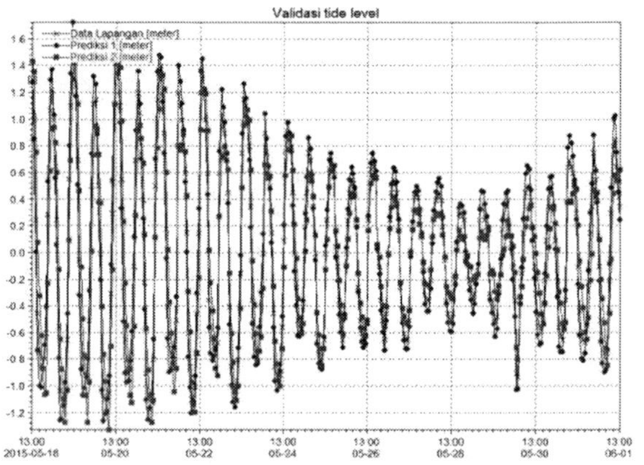

Fig. 5. Comparison between predicted tide and measured tide for the calibration process.

3 Results and discussions

3.1 Currents pattern of the rupat strait

The currents pattern in the Dumai estuary is greatly affected by the tide phenomenon because during a low tide the sea water is quickly attracted to the ocean causing sea levels to drop dramatically and quickly. Otherwise, in the case of high tide, sea water will get into the estuary area and bathe low-lying areas. This phenomenon has formed a pattern of movement of currents in the Dumai estuary. It means the tidal conditions are a major factor in the movement of aquatic pollutants.

The Dumai estuary experiences semidiurnal tide, which means it has two high tides and two low tides each day. The two high and the two low tides are nearly the same height. Currents in the Dumai estuary are generated by the long-wave movement of tides that creep from the Malacca Strait. During high tide, the current travels from the north to the south and east, as shown in Figure 6. In contrast, during low tide, the current moves from the east to the west, turns north, and exits in the

Malacca Strait, as shown in Figure 7. The magnitude of the current velocity is generally in the range of 0.2 to 0.45 m/s.

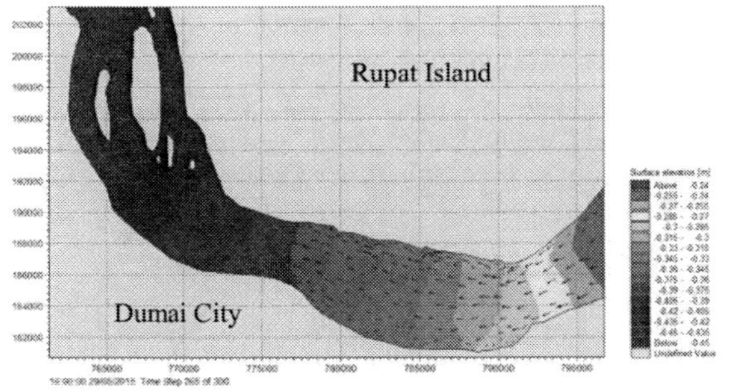

Fig. 6. The current pattern of Rupat Strait during low tide.

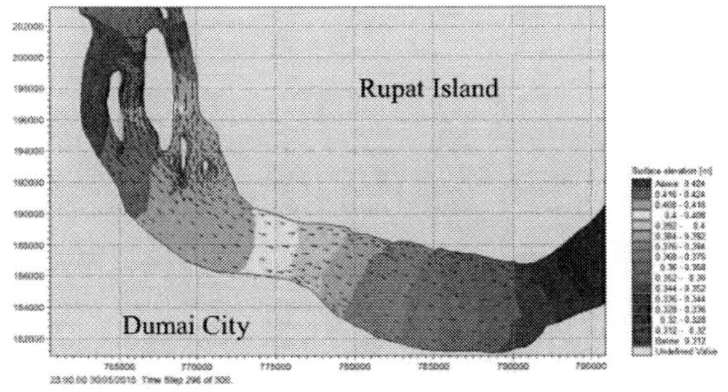

Fig. 7. The current pattern of Rupat Strait during high tide.

3.2 The dispersion of pollutant

The results of the current pattern from simulation of the hydrodynamic model were used as the basis for the simulation of the pollutant transport model. The simulations were carried out by the data input concentration of each pollutant that flows from Dumai River. The concentration of ammonia, nitrate, and nitrite in the river are 0.912 mg/L, 0.352 mg/L, and 0.115 mg/L, respectively. Meanwhile, the concentrations of ammonia, nitrates, and nitrites in the waters of the Rupat Strait are 0.433 mg/L, 0.096 mg/L, and 0.103 mg/L, respectively. Those concentrations were used as the initial condition value for computation.

3.2.1 Ammonia dispersion

The distribution of ammonia tends to spread to follow the movement of currents. During low tide, the ammonia dispersion spread to the northwest, as shown in Figure 8. On the other hand, during high tide, the ammonia dispersion spread to the southeast as shown in Figure 9. Because of the Dumai River which contains ammonia concentration of about 0.912 mg/L, the concentration of ammonia around the estuary increases from 0.32 mg/L to 0.40 mg/L in the high tide and from 0,012 mg/L to 0.0135 mg/L in the low tide, as shown in Figure 8 and Figure 9, respectively. In the high tide condition, the concentrations exceed the standard of maximum values that have been appointed by Indonesian regulations (0.30 mg/L).

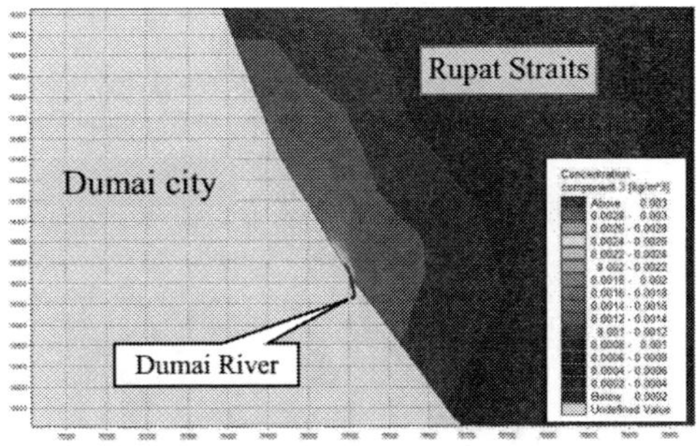

Fig. 8. Ammonia dispersion during low tide.

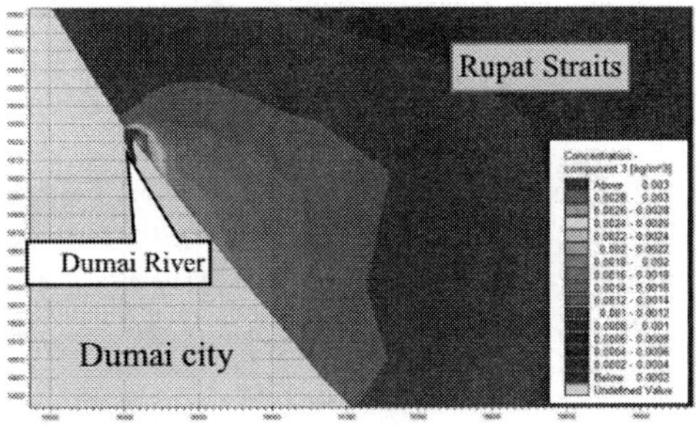

Fig. 9. Ammonia dispersion during high tide.

3.2.2 Nitrate dispersion

Similar with ammonia dispersion, the dispersion pattern of nitrates also tends to spread to follow the movement of currents but the rate of dispersion is smaller, as shown in Figure 10 and Figure 11. Nitrate pollutants originate from the Dumai River at a concentration of 0.352 mg/L, thus increasing the nitrate concentration around the estuary from 0.105 mg/L to 0.15 mg/L during low tide and from 0,048 mg/L to 0.056 mg/L during high tide. The concentrations of nitrates exceed the standard maximum value that has been appointed by Indonesian regulations (0.008 mg/L) in both high and low tide conditions.

3.2.3 Nitrite dispersion

Nitrite is a form of nitrogen that is contained in water. Nitrogen in the form of nitrite is the form between nitrate and ammonia, both in the process of oxidation of ammonia to nitrate and in the reduction of nitrate to nitrite [10]. This is the reason why the nitrite content in water is very limited.

The concentration of nitrite tends to spread to follow the movement of currents. In the low tide conditions, the nitrite dispersion spreads to the northwest, as shown in Figure 12. In the high tide condition, the nitrite spreads to the southeast, as shown in Figure 13. Because of the Dumai River which contains nitrite concentration of about 0.115 mg/L, the concentration of nitrite around the estuary increased from 0.1 mg/L to 0.14 mg/L in the high tide and from 0,01 mg/L to 0.012 mg/L in the low tide as shown in Figure 12 and Figure 13, respectively. The nitrite concentrations do not exceed the standard of maximum value that has been appointed by Indonesian regulations (0.06 mg/L) in both high and low tide conditions.

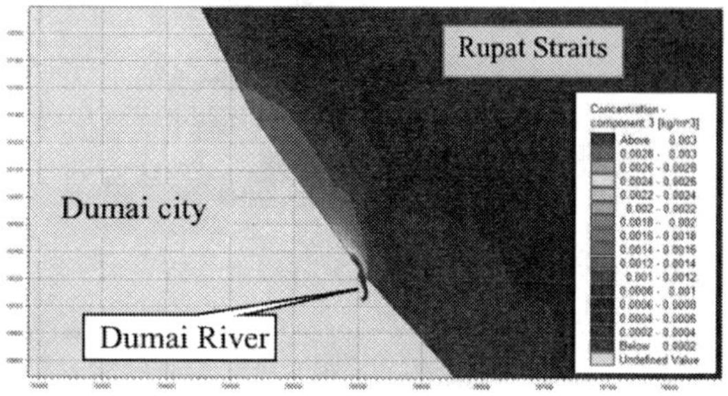

Fig. 10. Nitrate dispersion during low tide.

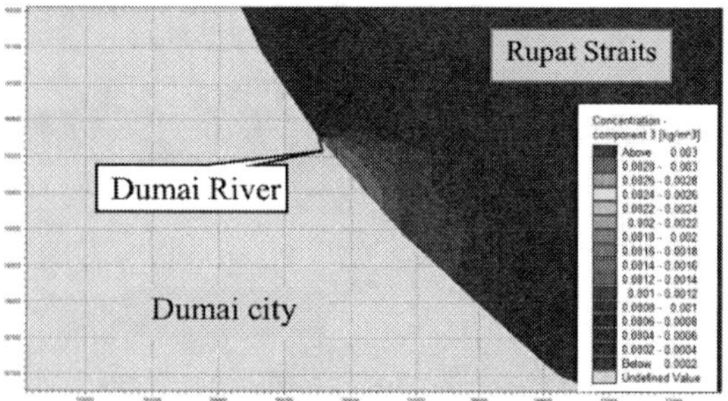

Fig. 11. Nitrate dispersion during high tide.

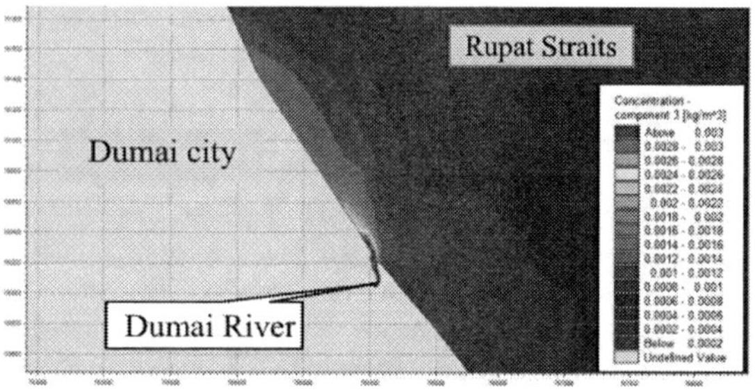

Fig. 12. Nitrite dispersion during low tide.

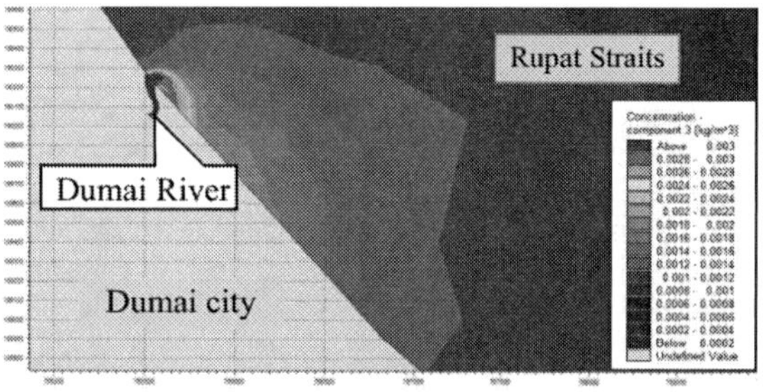

Fig. 13. Nitrite dispersion during high tide.

4 Conclusions

This study applied a numerical model to predict pollutant dispersion in the Dumai estuary using the MIKE 21 tool in order to manage and monitor water quality. The result of the numerical simulations showed that the concentration of all pollutant parameters tends to increase in the Dumai estuary. The concentration of ammonia, nitrates and nitrites increase to 0.4 mg/L, 0.15 mg/L, and 0.14 mg/L, respectively. Those concentrations exceed the standard of maximum value that has been appointed by Indonesian regulations. It means the domestic sewage from Dumai city should be treated urgently before discharged into the river.

The author would like to thank the Directorate General of Higher Education (DIKTI) - Ministry of Research, Technology and Higher Education for its research grants program.

References

1. *Dumai Dalam Angka (Dumai in Figure) 2015*, vol. **1**. 2015.
2. S. Sutikno, K. Murakami, D. P. Handoyo, and M. Fauzi, "Calibration of Numerical Model for Shoreline Change Prediction Using Satellite Imagery Data," *Makara J. Technol.*, vol. **19**, no. 3, pp. 113–119, 2015.
3. a. N. Menendez, N. D. Badano, M. F. Lopolito, and M. Re, "Water quality assessment for a coastal zone through numerical modeling," *J. Appl. Water Eng. Res.*, vol. **1**, no. 1, pp. 8–16, 2013.
4. C. Paper, M. Ben, M. Politecnico, and M. M. Politecnico, "Numerical results of the pollutant spreading offshore Taranto," 2009, no. August 2016.
5. D. Vouk, G. Lončar, and D. Bekić, *Tech. Gaz.*, vol. **6**, pp. 255–262, 2010.
6. S. Davies, H. Mirfenderesk, R. Tomlinson, and S. Szylkarski, "Hydrodynamic, Water Quality and Sediment Transport Modeling of Estuarine and Coastal Waters on the Gold Coast Australia," *J. Coast. Res.*, vol. 2009, no. 56, pp. 937–941, 2009.
7. N. Romano and C. Zeng, "Acute toxicity of sodium nitrate, potassium nitrate, and potassium chloride and their effects on the hemolymph composition and gill structure of early juvenile blue swimmer crabs(Portunus pelagicus Linnaeus, 1758) (Decapoda, Brachyura, Portunidae).," *Environ. Toxicol. Chem.*, vol. **26**, no. 9, pp. 1955–1962, 2007.
8. J. Jakeman, "ON NUMERICAL SOLUTIONS OF THE SHALLOW WATER WAVE EQUATIONS," 2006.
9. M. Abbaspour, a Javid, P. Moghimi, and K. Kayhan, "Modeling of thermal pollution in coastal area and its economical and

environmental assessment," *Int. J. Environ. Sci. Technol.*, vol. **2**, no. 1, pp. 13–26, 2005.
10. M. K. Pigue, "Changes in Dissolved Oxygen, Ammonia, and Nitrate Levels in an Extended Aeration Wastewater Treatment Facility When Converting From Counter Current to Disc Diffuser Aeration," 2013.

Analysis of erosion hazardous level and sedimentation in Manna Watershed, Bengkulu Province Indonesia

Khairul Amri[1,*], *Muhammad* Faiz Barchia[2], *Yuzuar* Aprizal[3]

[1,3]Civil Engineering Department Faculty of Engineering, University of Bengkulu, 38371 Bengkulu, Indonesia
[2]Soil Science Department Faculty of Agriculture, University of Bengkulu, 38371 Bengkulu, Indonesia

Abstract. The research purposeis to analyze erosion hazardous level and sedimentation on Manna watershed, Bengkulu Province. The research area geographically lies between 102°51'38.2"-103°10'57.8" East., and 4°0'39.6"-4°29'38.0" South covering 72,071.9 ha. Potential erosion yield and level were calculated by applying Universal Soil Loss Equation, and spatial analysis of the erosion was done using ArcGIS. The potential erosion in Manna watershed was classified as moderate hazardous level of 12,442,368.12tons yr^{-1} or similar to170.68tons $ha^{-1}yr^{-1}$. Moderate to high level Soil erosion covered areas of 47,359.00ha, or 65.19% of Manna watershed, andhigh risk of potential erosion with veryhigh sedimentation rate amounted to 248.851,1 tons yr^{-1}. In short, land cover in Manna watershed was generally in high environmental pressures, therefore, and in future, water resources would face environmental problems.

1 Introduction

A watershed as catchment areas plays important functions in human ecosystems [1]. In this ecosystem, human activities cultivated natural resources such as lands, water, vegetation and other biodiversities. In last decades, unwise usage on natural resources in watershed caused acceleration soil erosion hazard, degrading productivities on forest and agricultural land and sedimentation on river and lowland [2]. Depletion of watershed productivities today is mostly caused by mis management and poor coordination amongs take holders in controlling and managing natural and human resources within the watershed [3]. Manna watershed covers about 72.054,7 Ha (720.54 km^2), consisting of upstream areas of 503 km^2, middle-stream areas of 168.5 km^2 and down-stream areas of 87,924 km^2. The watershed lies from upstream in Lahat District, South Sumatra Province to downstereams in South Bengkulu.

Landscapes of Manna watershed mostly are rolling topographic with moderate to highly steep slope especially in the upstream areas having slope of 25–45 %, or more than 45% [4]. Manna water shed also faces environmentally depletion caused by human activities.

Manna watershed facessome problems in relation to encroachment onprotected forest areas such as illegal logging, and traditional agricultural activities for coffee plantation. Agricultural activities on agricultural land on steep slope with coffee, rubber, and oil palm plantations are also done by traditional farming. Cultivation activities on Manna watershed, which neglected environmental aspects had caused severe land degradation widespread from upstream to down-stream areas as indicated by soil erosion on the top soil and river bank distributed along the Manna river bank. Furthermore, land degradation with high potential erosion decreased water quantity and quality on Manna River. In future, pressureon Manna watershed due to population increase and land demandcould threaten the balancing ecosystem for human life on this area such as risks of sedimentation, flooding, landslide, etc.

Soilerosionis carried by upstream run-off flowing to river systems, thensoil an-organic particles and organic materialis deposited on riverbed and suspended on the water [5]. The deposited materials causess hallowing drainage system on the river, therefore high intensity rainfall maycause river over flow . [6]. Because of these, flooding in the middle and down-stream occurred, and people living in the down-stream areas might face environmental disaster at anytime [7]. Based on the above problems, study of potential erosion hazard and sedimentation on Manna watershed was conducted.

The objective of this study is to analyze the erossion hazard level occuring on Manna river Basin and total sedimentation within the river.

2 The materials and method

This study was conducted on Manna watershed located between 102°51'38.2"-103°10'57.8" East., and 4°0'39.6"-4°29'38.0" South, covering area of about 72.054,7 ha as shown in figure 1.

* Corresponding author: khairulftunib@yahoo.com

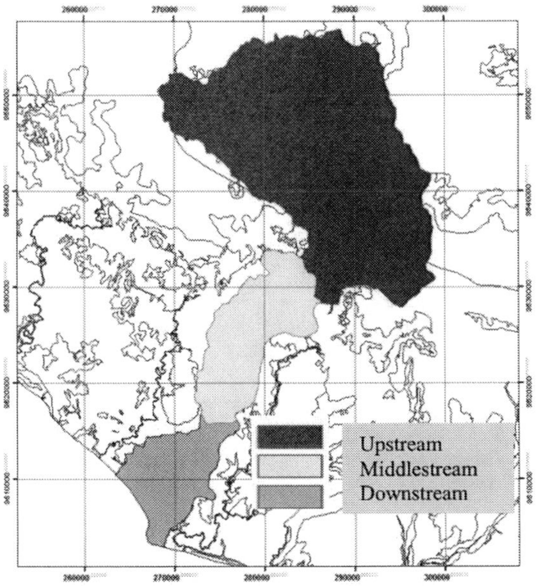

Fig. 1. Study location in Manna Watershed

Figure 1 shows the watershed comprising upstream, middle stream and downstream area.

Soil survey was conducted from May to July 2016 in order to collect soil samples, both disturbed and undisturbed soil samples. The soil samples were analyzed in Soil Science Laboratory, University of Bengkulu. The analysis of Soil characteristics include soil organic matter contents, soil textures, soil structures, and permeability.

Secondary data involved the last 10 year rainfall from The River Management Agency (Balai Wilayah Sungai) Sumatra VII, spatial data of land cover, land unit, topographical land slope, etc from Ketahun Watershed Management Office (BPDAS Ketahun), Bengkulu, Ministry of Environment and Forestry, Republic of Indonesia, and digital ALOS photos from Geospatial Information Agency, Republic of Indonesia.

Erosion is a process of soil removal or transport from one place to another one. The soil may be transported and eroded by wind or water.

Then, the transported soil is deposited in a river or stream . Erosion causes top soil loss which decreases soil absorbtion capacity .

There are a lot of models to predict erosion rate andUniversal Soil Loss Equation(USLE) is commonly used model to estimate erosion on a watershed.

Potentialsoilerosionwasestimatedusingempirically-based USLE formula [9] :

$$E= R\ K\ LS\ C\ P$$

E = Potentialsoilerosion (tons $ha^{-1}yr^{-1}$)
R = Erosivity index
K = Erodibility index
LS = Length and Slope indices
C = Crop Management Index
P = Land Conservation Practice Index

Erosivity index (R) wascalculatedbased on the following equations [10];

$$R = 2.21.\ CH^{1.36}$$

Inwhich CH = rainfall.

Soilrodibility index (K)was calculated using formula as follows [11]:

$$100K = 1.292\ [2.1M^{1.14} \times 10^{-4} \times (12 - a) + 3.25\ (b - 2) + 2.5\ (c - 3)]$$

Inwhich
M = soil texture index,
M = [(% silt) + (100 - %clay]
α = organicmatter content,
α $= (\frac{100}{58} \times$ C-organic content)
b = soil structure
c = soilpermeability

LS factor was predicted according to Rules of Directorate General for Land Rehabilitation and Social Forestry, Ministry of Forestry, Republic of Indonesia No.P.4/V-SET/2013 [12]. Crop Management Index and Soil Conservation Practice Index followed published indices [13]. Visualisation and calculation of areas of soil erosion degrees was done by overlaying digital map of erosivity index (R), erodibility index (K), length and slope degree (LS), crop management (C), and soil conservation practice (P) using ArcGIS software.

GIS package has been used for carrying out geographic analysis. The GIS has beeen used for data analyses of Aereal photo data for deriving land cover characteristics.

Tolerable Soil Loss (TSL) was determined according topublished criteria. Total TSL was calculated using next equation:

$$\text{Total TSL} = \text{TSL x A}$$

In which
A = watershed areas (km^2)
The concept of Sediment Delivery Ration (SDR) is the process of transfering erosion to sedoment yield.

SDRis ratio between values of sedimentflowing to river and values of soilerosionyield onwatershed. Based on proposed formula [14], The concept of SDR was used to dertmine total sediment yield which can be expressed as the following :

$$SDR= -0.02 + 0.385.\ A^{-0.2}$$

In which
A = watershed areas (km^2)

Then, sediment values (Y) was determined using next formula [15]:

$$Y = E\ (SDR)\ A$$

In which
E = Potential soil erosion (tons $ha^{-1}yr^{-1}$)
The values of factors in USLE equation are usually computed according to Agricultural Handbook. The Rainfall Erosivity Index (R) is calculated using an annual simulation of rainfall data using rainfall energy 30 minute duration.

 DOI: 10.1051/matecconf/201710104002

The yield of sediment is then estimated using Sediment delivery Ratio (SDR) which as function of Erosion and Watershed areas. The SDR concept is used to route surface erosion from catchment to the basin outlet.

3. Results and discussion

2.1 Potential soil erosion level

Erosivity index which wascalculated based on 10 year rainfall data over Manna watershed was 1,515.38. Soil lying on Manna watershed was catagorized into fairly high to very high level of erodibility class, therefore the soils wereprone to deteriorate when vegetative land cover wasdisclosed (Fig.2). Agricultural practices in Manna watershed should consider conservation values to avoid soil destruction on the land covering the watershed.

On the upstream of Manna watershed, landscape of the watershed was dominated by steep to very steep landslope (Fig. 3). Therefore, poor watershed management accelerated soil erosion on the upstream, and soil particles from the soil erosion would be deposited and shallowed riverbed on the down-stream of the Manna watershed.

Land use on the upstream of Manna watershed wasdominated by traditional agricultural activities with coffee plantation, and some parts on the upstream were covered by shrub and rubber plantation. Forest land coverwasfound only in some small areas in the border or in peak of hilly areas of the upstream Manna watershed.

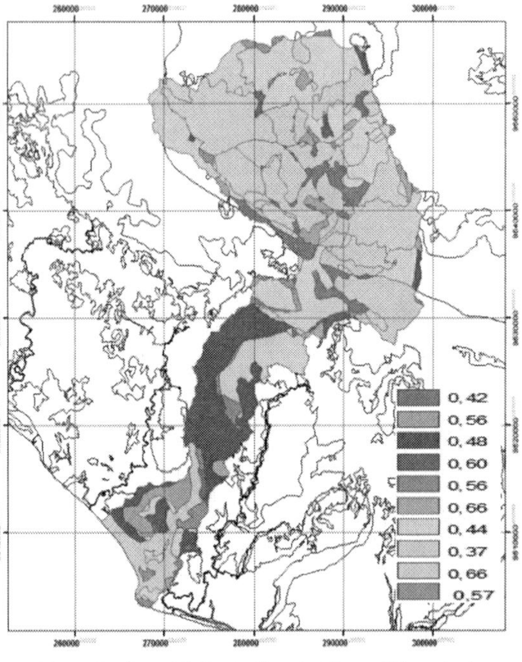

Fig. 2. Map of soil erodibility indices in Manna Watershed

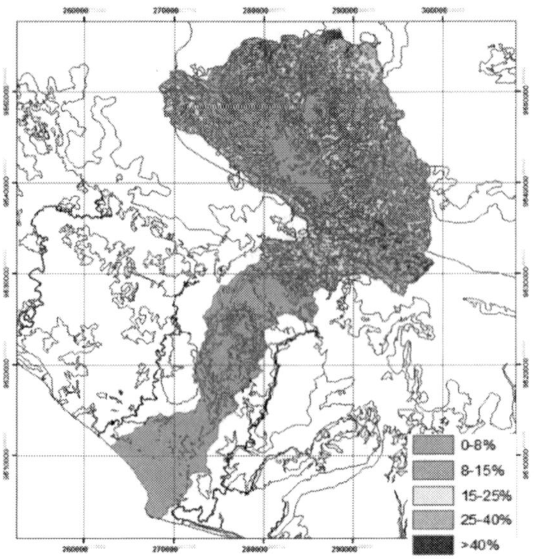

Fig. 3. Map of slope length and degree in Manna Watershed

Land use on the upstream of Manna watershed wasdominated by traditional agricultural activities with coffee plantation, and some parts on the upstream was covered by shrub and rubber plantation. Forest land cover was only found in some small areas in the border or in peak of hilly areas of the upstream Manna watershed (Fig. 4). Land management in the watershed was under poor conservation practices, especially in upland areas (Fig. 5). Some areas of protected forest in the upland today werecultivated as traditional perennial crops particularly coffee plantation.

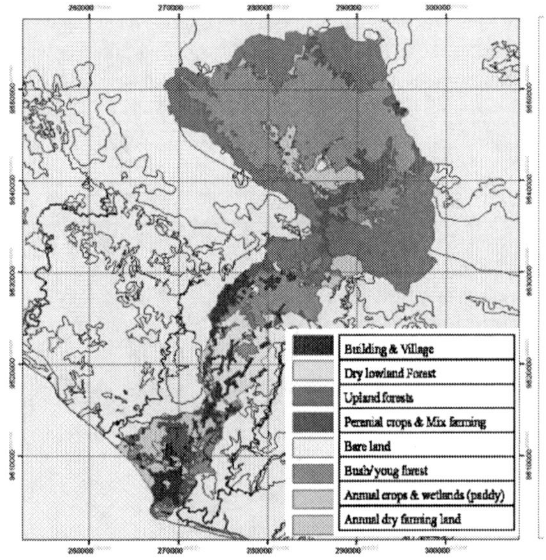

Fig. 4. Map of crop management in Manna Watershed

 DOI: 10.1051/matecconf/201710104002

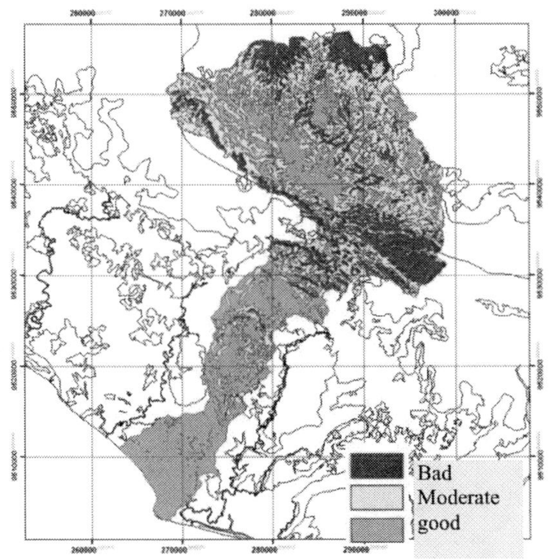

Fig. 5. Map of conservation practicesin Manna Watershed

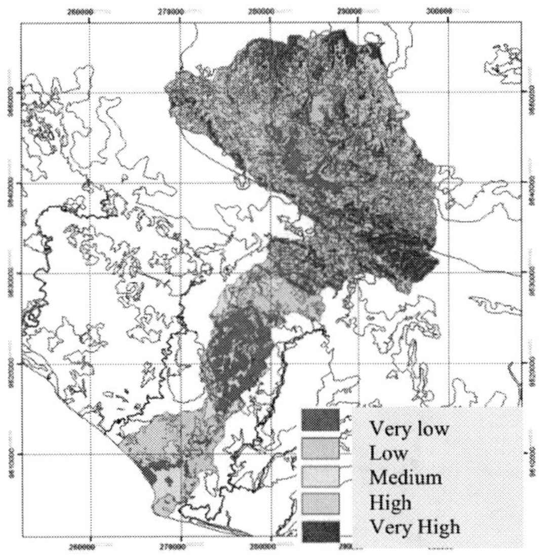

Fig. 6. Map of potential soil erosion level in Manna watershed

Potential soil erosion which occurred in Manna watershed amounted to 12,442,368.12tons yr^{-1}or to average170.68tons $ha^{-1}yr^{-1}$. In average, the potential soil erosion on Manna watershed was classified asmoderate erosion level. Potential soil erosion with very low potential erosion hazard, lower than 15 tons yr-1 occurred in the areas of 6,036.29ha or 8.28% of the Manna watershed. Also, low erosion level of 15-60 tons yr^{-1} covered the area of 19,255.04 ha or 26.41% of the watershed (Table 1).The low levels of erosion hazard covered only about 34.69%. Therefore Manna watershed today undergoes deteriorated function for supporting human ecosystems.

The potential soil erosion in the level of moderate to very high covered about 47,608.57 ha or 65.31%. These hazardous levelwasdistributed mostly on upstream of the watershed (Fig. 6).

Table 1. Areas of the erosion hazard level in Manna watershed

No	Erosion Level	Areas (ha)	(%)
1	Very Low	6.036,29	8,28
2	Low	19.255,04	26,41
3	Moderate	19.655,14	26,96
4	High	19.497,94	26,75
5	Very High	8.455,49	11,60
Total		72.899,9	100

Considering sustainable function of ecosystem for human life in this area, the high potential soil erosion in the watershed is going to threaten sources of human life such as decreasing land productivities, water supplies for irrigation, and water qualities for water biotic ecosystems, etc.

2.2 Sedimentation

Sediment involved organic materials transported by water run-off and deposited on the riverbed and suspended on water flow. From the whole areas of Manna watershed, the SDR index was 0.02. This index was very small comparing to 1, it means that whole potential sediments from lands flowed to Manna River, and was deposited to river bed or flowed to Indian Ocean. The total potential sediment flowing into Manna River was of 248.851,1 tons yr^{-1}, and this value was categorized ashigh level of sedimentation.

Tolerable sediment based on the value of total TSL and the value of SDR on the Manna watershed was about 19.610,07tons yr^{-1}. Comparing the total potential sediments to the tolerable sediments, the potential sediments arebigger than permitted ones , therefore cultivation and management on Manna watershed especially on the upland should consider environmental values both land characteristics and water behavior as human ecosystems.The function of protected forestwhich wascultivated as traditional land agriculture for coffee plantation and bush has to be improved as protected forest areas. Collaborative land management with traditional villagers implementing social forestry might be as a wise solution to improve land productivities in Manna watershed.

4 Conclusions

Based on analyses result and discussion, it can be concluded:

1. High Potential soil erosion in Manna watershed in high values is prone to deteriorate to be critical ecosystems.
2. Sedimentation flowing from upland polluting downstream of Manna River decreases water qualities in the whole downstream ecosystems.
3. Considering Manna watershed as a fragile ecosystem, agricultural practices should considerManna watershed as sustainable human ecosystems.

References

1. [1] H.Ramdan, *Prinsip Dasar Pengelolaan Daerah Aliran Sungai.* Forest Ecology Lab. Faculty of Forestry WinayaMukti University. Jatinangor, West Java.Indonesia (2006).
2. [2]H.Santoso. KebijakanPengelolaan DAS Terpadu.Director General of Watershed Management and Social Forestry, Ministry of Forestry.Coordination Meeting on Integrated Watershed Management.Batam. Indonesia. March 23(2011).
3. [3]I.E.Soegiri. *PeranPengelolaan DAS Terpadudalam Pembangunan Wilayah.*Director for Planning and Evaluation of Watershed Management, Ministry of Forestry.Coordination Meeting on Integrated Watershed Management.Batam. Indonesia. March 23 (2011).
4. [4] BPDASKetahun, 2007, *Karakteristik Daerah Aliran Sungai Manna Bengkulu.* Watershed Management Agency Ketahun Bengkulu. Ministry of Forestry.Republic of Indonesia.BPDAS Ketahun, Bengkulu (2007).
5. [5] Sucipto. AnalisisErosi Yang Terjadi Di Lahan Karena Pengaruh Kepadatan Tanah.*Wahana Teknik Sipil***12**, 1 pp 51-60 (2007).
6. [6] R. Setiaji. *Pengaruh Kepadatan Tanah Terhadap Erosi yang terjadi di Lahan,* Magister Thesis. Postgraduate Program.Gadjah Mada University. Yogyakarta. Indonesia(2006).
7. [7] B.A. Kironoto. *KonservasiLahan*, GadjahMada University Press, Yogyakarta. Indonesia (2000).
8. [8] BWS Sumatera VII. *Karakteristik Air Manna Bengkulu Selatan.*River Management Agency (Balai Wilayah Sungai) Sumatera VII.Bengkulu.Indonesia (2012).
9. [9] W.H. Wischmeier and D.D. Smith. *Predicting rainfal erosion losses, A guide to conservation planning.*USDA. Agric. Handbook. 537. Washington DC (1978).
10. [10] J. Lenvain, M.DeBoodt and Suwardjo. An effective way in fighting soil erosion: promoting growth of young trees through plantpit treatment with soil conditioners. Mededelingen van de FaculteitLandbouwwetenshappenRijksuniversiteit Gent.**41**, 4 pp 141-157 (1976).
11. [11] W.H. Wischmeier. C.B. Johnson, and B.V. Cros. A soil erodibilitynomograph for farmland and construction sites.*J. Soil and Water Conservation*, pp.189-193. Sept–Oct (1971).
12. [12] Directorate General of Watershed Management and Social Forestry. *PetunjukTeknisPenyusunan DataSpasialLahanKritis,* Ministry of Forestry. 2013. www.dephut.go.id accessed in August 19 (2015).
13. [13] S. Arsyad. *Konservasi Tanah dan Air*, InstitutPertanian Bogor Press, Bogor.Indonesia (2010).
14. [14] K. Auerswald. Sensivitat erosions best immenderFaktoren. In: *Wasser und Boden*, 34-38. Verlag Paul Parey (1987).
15. [15] C. Asdak. *HidrologidanPengelolaan Daerah Aliran Sungai*, GadjahMada University Press, Yogyakarta.Indonesia (2002).

Preliminary study of the human activities and perception on the groundwater conservation in lowlands of Semarang

Novie Susanto[1,*], *Thomas* Triadi Putranto[2], and *Dian* Agus Widiarso[2]

[1]Industrial Engineering Department Faculty of Engineering, Diponegoro University, 50275 Semarang, Indonesia
[2]Geological Engineering Department Faculty of Engineering, Diponegoro University, 50275 Semarang, Indonesia

Abstract. This paper aims at investigating the human activities and perception of the residents in lowland areas of Semarang city regarding the groundwater conservation. As the first step, classifying and rating land use in the study area is conducted as the interpretation of human activities. A map of human perception using the adapted Protocol Perception Monitoring is then configured to evaluate the deviation between the expected and observed behavior of the community in an integrated ecosystem. The result of preliminary study shows that about 68% areas in lowland have score 8 (built-up area, irrigated field crops), 25% has score 7 (industries and reservoirs), 1% has score 5 (plantation), 3% has score 2 (non-irrigated field crops) and about 3% has score 1 (forest, park and open areas). It means that more than 90% lowland areas in Semarang city have been intervened by the human activities. Based on the result of the perception mapping, it is found that there is a high deviation (about 40-90%) between expected and observed behavior in critical areas, including the understanding of area situation, utilization of groundwater and water resources (wells/PDAM) as well as the community comprehension of government regulation on groundwater conservation.

1 Introduction

Groundwater as natural resources and the representation of valuable drinking water resources has been affected particularly by the increasing population in urban cities [1,2,3,4,5,6,7]. An enormous stress has been established on this resources caused by the increasing water demand from urban and industrial development both in quantity and quality aspects.

Semarang is an urban city in the central north of Java Island which is facing a rapidly increasing population as well. This phenomenon leads to the increasing water demand from groundwater for domestic and industrial purposes. The groundwater exploitation affects some environmental problems. Therefore, to solve the problems and achieve a sustainable ecosystem, there are some steps that should be considered. They are a mapping of the conservation area in Semarang as technical aspect assessment as well as mapping of human activities and perceptions regarding groundwater conservation as human factors evaluation.

A preliminary study of groundwater quality in Semarang coastal areas [8] showed that there were four groundwater conservation zones in Semarang based on the Electrical Conductivity (eC) value assessment. The area classifications were secure (1.1%), vulnerable (24.5%), critical (43.6%) and damage (30.9%) as seen in Figure 1, while based on lowering groundwater level, there were 68,8% secure zone, 28.5% vulnerable zone, 1.1% critical zone and 1.6% damage zone in lowland area of Semarang city. This study is focused on the critical class as the largest area in conservation classification.

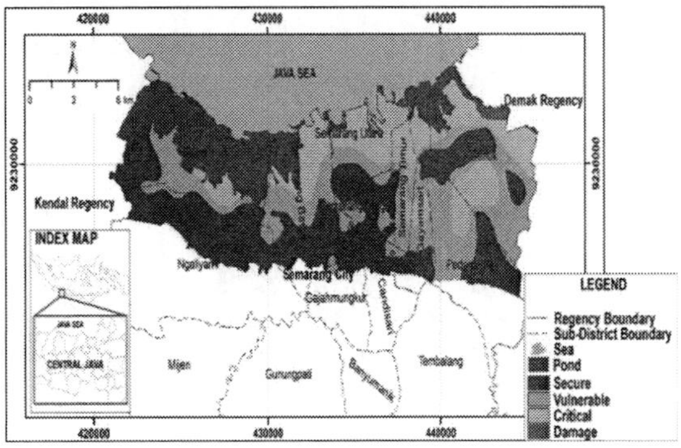

Fig. 1. Map of groundwater conservation zones in lowland areas of Semarang city based on electrical conductivity value [8]

Based on these results, preliminary results of human factor can be taken into account. The evaluation of human factor cover a land use map that explains the impact of human activities on groundwater state [9]. Besides, the assessment of human perception using adapted Protocol Perception Monitoring [10] for each classification also taken into account. Water resources

* Corresponding author: nophie.susanto@gmail.com

and their management are critical to human security and are essential for achieving sustainable development [11].

Recent global analyses have revealed that almost no area of the world is untouched by human impacts [12]. Thus, an understanding of the location and intensity of human impacts on the ecosystems is an essential component of informed and successful coastal management. Mapping of human activities and behaviour allows an assessment of human activities impacts on the health of these ecosystems [13].

A recent study of the human factor in groundwater management is conducted by [14]. They design an ergo-hydrogeology system to model groundwater based on zonation method. However, this study only considers technical system in groundwater management including mapping of the groundwater utilisation and zonation. Ergonomic aspect only adopted on the concept of the study using SHIP (Systemic, Holistic, Interdisciplinary, and Participatory) approach.

Therefore, this study aims at exploring the human factor in the groundwater management to achieve a balance and an integrated design of groundwater system both on the technical and human aspects.

2 Literature reviews

2.1 Human activities

Human activities can affect the groundwater quality and increase its pollution risk. Using the land use/cover map of the study area, a classification of human activities can be mapped and analysed. Classification of the land use covers six classes: urban, agriculture, natural vegetation, water, evaporation pond and bare land [9] and the rating for each class will be referred to the categorization of land use by Secunda *et al.* [15].

Considering human as an integrated part of the ecosystem and a role holder of environment stresses regarding urbanisation and industrial development, human activities that may impact the condition of groundwater should be taken into account. Data of human activities will be obtained based on different land use or cover map. Industrial activities, traffic, septic tanks and sewer systems within urban areas and fertilisers, used in agriculture, are considered as 'hazards' [16] and categorised under the human activities parameter. These hazards can affect the groundwater quality and increase its pollution risk.

Therefore, there will be a classification of land use/cover including urban, agriculture, natural vegetation, water, evaporation pond, and bare land. The detailed ratings for each class are given in **Error! Reference source not found.**.

2.2 Perception monitoring

There are many studies about perception monitoring in environmental topic, i.e., Protocol Perception Monitoring [17,18,19] applied in Marine Protected Areas in Indonesia. These studies aimed at highlighting the trends in the measured perceptions, and look at differences in these perceptions between the different villages that took part in the surveys. Perception monitoring protocol includes the structured questionnaire, the interviewer selection procedure, sampling procedure and interview procedure, is an adaptation of the MPA monitoring standard developed by Bunce and Pomeroy [20].

This study adapted perception monitoring protocol to obtain perception data of the communities around the lowland areas of Semarang city. Deviation of the human perception is defined as the gap between expected and observed behaviour. Therefore, this study designed the questionnaire based on the necessity of perceived data of the communities to be compared with the actual data.

Table 1. Land Use [15].

Land Use Category	Rating
Site-specific Land Usage	
Toxic-waste disposal	9
Oil spillage	8
Industries	7
Solid-waste disposal (regional)	6
Domestic-waste disposal (local)	5
Effluent irrigated fields	4
Effluent reservoir	3
Extensive Land Usage	
Cotton	10
Built-up area	8
Irrigated field crops	8
Greenhouses/tomatoes	8
Reservoirs	7
Citrus orchards	7
Orchards of other fruits	6
Pasture or other land unsuitable for agriculture use	5
Uncultivated land	5
Temporary uncultivated land	5
Vineyards	5
Olives	5
Quarries	5
Non-irrigated field crops	4
Avocados	2
Forests	1
Natural areas or reserves	1
Dune sands – open areas	1

3 Research methods

Methods in this research were applying into two stages including land use analysis and human perception analysis. Land use analysis was focused on lowlands of Semarang city and mapped during collecting water samples in a preliminary study 7.

The human perception was analysed using the deviation between expected and observed behaviour of the residents. The respondent was selected based on the location of residence. It also considered the proportion of the population compared with the area study [21]. As the preliminary study, there were in total 94 respondents in lowlands of Semarang city. These respondents were distributed based on the area of each zone of conservation as seen in Table 2.

Table 2. Distribution of respondents

No	Conservation zone of the area	Number of respondents
1	Secure	37
2	Vulnerable	28
3	Critical	21
4	Damage	8

This stage was using a questionnaire to obtain human perception data regarding the groundwater conservation. The questionnaire contained the understanding of area situation (V1) (6 points), utilisation of groundwater (V2) (7 points) and water resources (well (V3) with 12 points and/or PDAM (V4) with 5 points) as well as the community comprehension of government regulation on groundwater conservation (V5) (7 points).

Actual human perception data (Observed Value/OV) was compared with ideal condition (expected value = EV) of human perception regarding groundwater conservation. The formula can be seen in Equation (1). Expected value was obtained based on the multiplication of the maximum points and number of respondents for each variable, while the observed value was generated from the total right point for each variable.

$$\text{Deviation (D)} = \frac{EV-OV}{EV} x100\% \quad (1)$$

An analysis was then conducted to evaluate the human perception in lowland areas of Semarang city.

4 Result and discussion

4.1 Land use

The result of the study showed that the largest area of Semarang city lowland was covered by the built-up areas and irrigated field crops (96.9 km^2 or 68% of total area). Besides, industry and reservoir were placed in a second largest area (35.8 km^2 or 25.1%). Detail of the study result can be seen in Table 3 and Figure 2.

With these results, it should be further examined whether the communities aware of the situation of the land cover. The utilisation of the groundwater should be considered due to the least area of forest and open areas (only about 2% of total area in lowlands of Semarang city).

Table 3. Land use cover

Land use	Score	Area (km^2)	%
Forests, Park, open areas	1	3.4	2.4
Non-irrigated field crops	2	4.7	3.3
Plantation	5	1.7	1.2
Industries, reservoirs	7	35.8	25.1
Built-up area, irrigated field crops	8	96.9	68.0
	Total	**142.4**	**100**

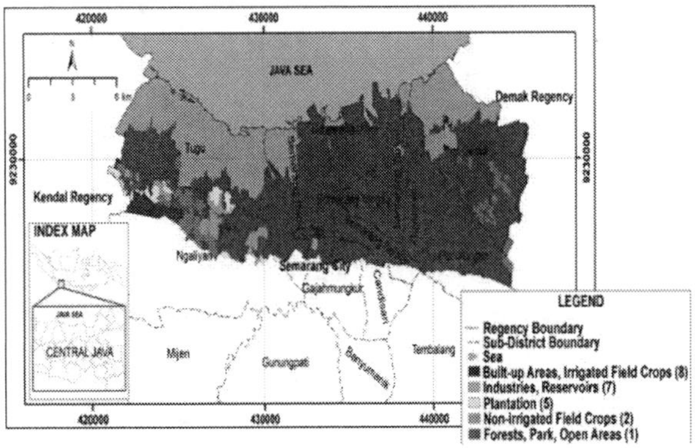

Fig. 2. Map of land uses in lowland areas of Semarang city.

4.2 Human perception monitoring

Based on the questionnaire result, it can be concluded that the communities in lowland areas of Semarang city had a high deviation value considering expected and observed behaviour in groundwater utilisation. The recapitulation of the study results for each classification of conservation zone can be seen in Table 4. Detail deviation for each variable in a different zone of conservation can be seen in Table 5. The deviation value can be different for each variable depended on the number respondents. The number of respondents using the well and/or the PDAM can be varied for a different zone of conservation area.

Table 4. Deviation of human perception.

No	Conservation area	EV	OV	D
1	Secure	210	114.5	45.5%
2	Vulnerable	160	73.3	54.8%
3	Critical	107	54.3	49.9%
4	Damage	47	20.8	53.5%

Table 5. Deviation of human perception for each variable

	Secure			Vulnerable		
	EV	**OV**	**D (%)**	**EV**	**OV**	**D (%)**
V1	222	97	56.3	168	73	56.5
V2	264	152	42.4	204	83	59.3
V3	95	54	43.2	80	36	55
V4	259	155	40.2	196	101	48.5
V5	147	75	49	196	32	83.7
	Critical			**Damage**		
	EV	**OV**	**D (%)**	**EV**	**OV**	**D (%)**
V1	126	62	50.8	48	16	66.7
V2	60	28	53.3	36	26	27.8
V3	95	49	48.4	40	18	55
V4	147	78	46.9	56	35	37.5
V5	147	34	76.9	56	4	92.9

Based on these results, it can be further analysed that community comprehension regarding government regulation obtained the highest deviation in vulnerable (83.7%), critical (76.9%) and damage (92.9%) areas. It means that community in damage areas do not conform and obey the government regulation regarding groundwater system in their areas, whereas they should be aware of the risk and danger of living the vulnerable areas especially related to the utilisation of water and groundwater resources. The result also confirmed that understanding regarding area situation of the residents in damage area was in low level compared with other areas. Therefore, even the damage area only cover less than 2% lowland area, the intervention for the residents should be actively implemented.

5 Conclusions

Based on the result of the research, it can be concluded that more than 90% lowland areas in Semarang city have been intervened by the human activities. It covered by the built-up areas and irrigated field crops (96.9 km^2 or 68% of total area), industry and reservoir (35.8 km^2 or 25.1% of total area), non-irrigated field crops (4.7 km^2 or 3.3 of total area), forests, park, open areas (3.4 km^2 or 2.4% of total area), and plantation (1.7 km^2 or 1.2% of total area). The result of human perception conducted a low level of community understanding regarding government rule and area situation, especially for residents in vulnerable (deviation = 54.8%) and damage areas (deviation = 53.9%). The further studies can be developed research topics into ecological behaviour, the design of intervention for the community as well as mapping the human perception in the groundwater conservation.

The authors thank Diponegoro University who funded the project "Desain intervensi kognitif untuk meningkatkan indeks kesiapsiagaan individu di daerah rawan longsor" within "Riset Pengembangan dan Penerapan (RPP)" scheme, Contract Number: SP DIPA-042.01.2.400909/2016, December 7th, 2015, with funding resource: DIPA PNBP of Diponegoro University 2016.

References

1 D. Lerner, A.S. Issar, and I. Simmers. Groundwater recharge: A guide to understanding and estimating natural recharge. *International contribution to hydrogeology: Vol.8*, Verlag Heinz Heise GmbH and Co. KG, Hannover Germany, (1990).

2 R. Healy and P. Cook. Using groundwater levels to estimate recharge. *Hydrogeology Journal, 10* (1) (2002) 91-109.

3 D. Lerner. Identifying and quantifying urban recharge: a review. *Hydrogeology Journal, 10*(1), (2002) 143-152.

4 E. Khazaei, A. E.F. Spink, and J. W. Warner. A catchment water balance model for estimating groundwater recharge in arid and semiarid regions of south-east Iran. *Hydrogeology Journal, 11*(3), (2003) 333-342.

5 S. S. D. Foster and P. J. Chilton. Downstream of downtown: urban wastewater as groundwater recharge. *Hydrogeology Journal, 12*(1) (2004). 115-120.

6 M. Kumar, M. S. Rao, B. Kumar, and A. Ramanathan. Identification of aquifer-recharge zones and sources in an urban development area (Delhi, India), by correlating isotopic tracers with hydrological features. *Hydrogeology Journal, 19*(2) (2011) 463-474.

7 E. Kruse, E. Carol, M. Mancuso, P. Laurencena, M. Deluchi and A. Rojo. Recharge assessment in an urban area: a case study of La Plata, Argentina. *Hydrogeology Journal, 21*(5) (2013) 1091-1100.

8 T. Putranto, N. Susanto and D. A. Widiarso. Assessment of Groundwater Quality to Achieve Sustainable Development in Semarang Coastal Areas. Proceeding of International Conference Coastal Planning for Sustainable Maritime Management (CITIES) Surabaya. (2016).

9 A. Al-hanbali and A. Hondoh. Groundwater vulnerability assessment and evaluation of human activity impact (HAI) within the Dead Sea groundwater basin, Jordan. *Hydrogeology Journal* 16 (2008) 499–510.

10 D. N. Pada, A. Y. Fauzan, dan W. Sulstyaningrum. Persepsi Masyarakat di Kawasan Konservasi Laut Daerah Kabupaten Kaimana, Indonesia. Conservation International Indonesia, Kaimana Program. (2011).

11 UN-Water. Water security and the global water agenda: a UN-Water analytical brief. Available at http://www.unwater.org/downloads/watersecurity_analyticalbrief.pdf. [Online Access: 02.11.2015] (2013).
12 B. S. Halpern, S. Walbridge, K. A. Selkoe, C.V. Kappel, F. Micheli, C. D'Agrosa, J.F. Bruno. A Global Map of Human Impact on Marine Ecosystems. Science, 319 (2008) 948-952.
13 InVEST Documentation., Habitat Risk Assessment. Retreived from: http://ncpdev.stanford.edu/~dataportal/investreleases/documentation/currentrelease/habitat_risk_assessment.htm.[Online access 12.03.2015]. 2015
14 L. Sudiajeng, I.G.L. Parwita, M. Mudhina, I.W. Arya, M. Sudiarsa, K.W. Andayani, I.I. Wiraga, A. Andayani. Kajian Teknis Pengelolaan Air Tanah Kota Denpasar. Politeknik Negeri Bali. Report. (2013)
15 S. Secunda, M.L. Collin and A.J. Melloul. Groundwater vulnerability assessment using a composite model combining DRASTIC with extensive agricultural land use in Israel's Sharon region. *Journal of Environmental Management,* **54** (1998) 39–57.
16 F. Zwahlen. Vulnerability and risk apping for the protection of carbonate (karst) aquifers, final report (COST action 620). *European Commission, Directorate General XII Science, Research and Development*, Brussels (2004) 297.
17 A. Halim, P.J. Mous, and J. Schoemaker. General Protocol for The Implementation of Perceptions Monitoring Program at SEACMPA's Marine Conservation Sites in Indonesia. The Nature Conservancy, (2005).
18 H. Widodo., E. Carter, T. Soekirman, and A. Halim. Community Perceptions of Marine Protected Areas in Indonesia – Phase II. The Nature Conservancy, (2009).
19 H. Widodo, D.N. Pada, A.Y. Fauzan, dan W. Sulistyaningrum. Protokol Pelaksanaan Monitoring Persepsi Kawasan Konservasi Laut Daerah (KKLD) Kabupaten Kaimana, Papua Barat. Conservation International Indonesia dan The Nature Conservancy, (2010).
20 L. Bunce. and B. Pomeroy. Socioeconomic Monitoring Guidelines for Coastal Managers In Southeast Asia: SOCMON SEA. World Commission on Protected Areas and Australian Institute of Marine Science, Townsville, Australia (2010) 85.
21 M. Nazir. Metode Penelitian. Ghalia Indonesia, Jakarta. (2003).

Modelling of phosphate enrichment and trophic status of Sipin Lake Jambi using TSI Carlson method

Ira Galih Prabasari[1,*], *Hutwan* Syarifuddin[2], and *Damris* Muhammad[1]

[1]Engineering Faculty, Jambi University, 36361 Jambi, Indonesia
[2]Graduate Program of Environment Science, Jambi University, 36361 Jambi, Indonesia

Abstract. Floating fishery net (keramba) activity in Sipin Lake Jambi has increased significantly in the last decade. This intensive fishery results in enrichment of phosphate due to the application of fish feeding and induces eutrophication. The objectives of this study are to determine phosphate enrichment, the trophic status using TSI Carlson method and to propose a model of phosphate enrichment in the lake water. TSI Carlson is a method to determine the trophic status condition in the water body. Composite water samples were collected at up-, downstream and middle of the lake. Total phosphate content, chlorophyl-a, and brightness were measured. The results show that the average of total phosphate, chlorophyl-a, and brightness of the lake water are 0.0505 mg/L, 5.86 µg/L and 58 cm, respectively. These figures indicate that fishery net activity has induced eutrophication with phosphate enrichment of 9.65 T/yr. A proposed TSI Carlson model estimates the TSI value would be 61.79 within next ten years with phosphate enrichment of 12.04 T/yr. Application of double fishery nettings (keramba) would be expected to reduce these figures and the risk of eutrophication in the lake.

1 Introduction

Developing aquaculture in a lake such as fishery activity using floating fishery nets potentially improves social welfare. However, this is usually followed by enrichment of phosphate in water body as a result of the application of fish feeding and induces algae bloom/ eutrophication. Sometimes, phosphate concentrations in lake water exceed eutrophication level of 0.035–0.1 mg/L [1]. This affects physical appearances of lake water such as muddy and stinks thereby disrupting local community activities.

Sipin Lake located at Jambi city of Jambi Province is surrounded with a quite dense population of three kelurahan (Legok, Buluran Kenali, and Sungai Putri) with about 2,500 people per square km. There are offices, hospital, and residence area in the south of the lake and Batanghari River in the north. In the east, there is a hotel and residence areas and only residence area in the west. The lake area about 227 Ha and water depths of 2–6 meters are an ideal place for fishery activity. Three tributaries flow into the lake [4] with only one output to Batanghari River. Phosphate originating from these flows is ignored in this study but floating fishery net is the major source of phosphate in the lake water [5]. Based on the data reported by Local Environmental Institution (BLH) within the last two years (2013 and 2014) showed that the average of total phosphate concentration in the lake water is above eutrophication limits [4]. This indicates that the increasing number of floating fishery nets in the lake results an eutrophication.

In year 2013 annual fish production are 900 tons and number of floating fishery net are 4,348 units. With 10% increase in the fishery net and 5% fish production, it is estimated an increase of 100-ton annual fish production within next 5 years [3]. This means that fish feeding material released to the lake system also increases. It is known that 150-200 kg feeding material is needed for every 1000-fish for every 3-month. Since the feeding materials contain 1.2% of phosphate [2] and 41% of it is released into water, it is estimated that approximately 35 tons of phosphate are released into Sipin Lake water each year.

High concentrations of phosphate in the water affect the trophic status of the lake such as water productivity, nutrient enrichment, brightness and some biological activities in the water. The objectives of this study are to determine phosphate enrichment and the trophic status of the lake and to propose a model of phosphate enrichment in the lake system. The latter is expected to be useful for the local government as an anticipative policy of phosphate enrichment in Sipin Lake later. A model of phosphate enrichment within next 10-year was developed using Powersim 2.5.

2 Research methods

This study was conducted in January to April 2015. The samples were collected at up-, downstream, and the middle of the lake on 28th January, 6th February, and 17th March. Water sample collection and brightness measurements were conducted on summer of January

* Corresponding author: iragalih@gmail.com

and February, but on March the rain is higher than the previous ones. Although this study does not provide data for the whole year, however, the data could be used as an estimation of the whole year as no significant different in phosphate concentration and brightness of these two-period sampling ($p>0,05$). This is also supported by a report from Local Environmental Institution (BLHD) shows there is no significant variation of phosphate and brightness of Sipin lake water [4]. Sampling location is shown in Figure 1.

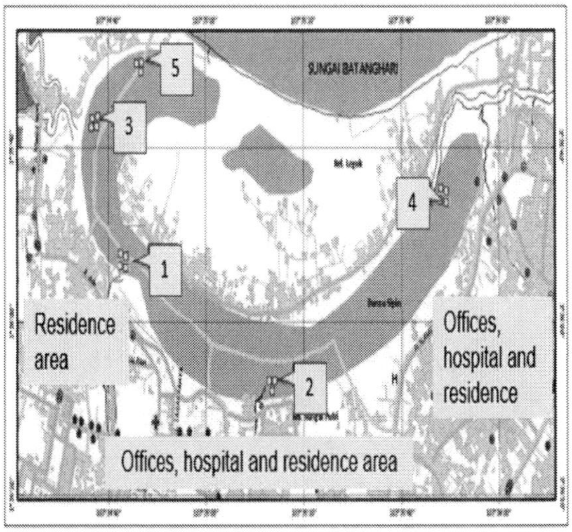

Fig. 1. Sampling point location

1. Kambang River – Upstream (01°35'51.4"S and 103°34'39.6"E)
2. Putri River – Upstream (01°36'09.6"S and 103°35'13.9"E)
3. Kenali Besar River – Upstream (01°35'40.4"S and 103°34'34.9"E)
4. Downstream (01°35'34.1"S and 103°36'02.6"E)
5. Pandan Island - Middle (01°36'02.5"S and 103°35'03.2"E)
6. Buluran – Middle (01°36'31.1"S and 103°34'37.7"E)

Total phosphate concentration was analyzed using spectrophotometer UV-Vis Pharo 300 with λ = 880 nm, intercept 0.0331 and slope 0.6752 with regression line of Y = 0.6752 + 0.0331. Water brightness was measured using Secchi Disk and chlorophyl-a was calculated based on total phosphate concentration using equation 5. Chlorophyl-a is an indicator of the biological condition in the lake water. All measurements in the field were conducted in three replications.

Trophic status of the lake was calculated using TSI Carlson classification given in Table 1. Trophic Status Index (TSI) of the lake was calculated by measuring brightness of the lake water using Secchi Disk (TSISD), biological condition using chlorophyl-a indicator (TSICHL) and total phosphate concentration (TSITP) measured in the lake water.

Table 1. TSI Carlson index.

TSI	Status Trophic	Brightness (m)	TP (mg/m3)	Chl-a (mg/m3)
1	Oligotrophic	64	0.75	0.042
10	Oligotrophic	32	1.5	0.012
20	Oligotrophic	16	3	0.34
30	Oligotrophic	8	6	0.94
40	Oligotrophic	4	12	2.64
50	Mesotrophic	2	24	6.4
60	Eutrophic	1	48	20
70	Eutrophic	0.5	96	56
80	Eutrophic	0.25	192	154
90	Eutrophic	0.12	384	427
100	Eutrophic	0.062	768	1183

As Mamatha reported [6] the classification of TSI Carlson index based on the following equations:

$$TSID = 60 - 14.41\ ln(SD) \quad (1)$$

$$TSICHL = 9.81\ ln(CHL) + 30.6 \quad (2)$$

$$TSITP = 14.42\ ln(TP) = 4.15 \quad (3)$$

$$TSI = (TSID + TSICHL + TSITP)/3 \quad (4)$$

Where TSISD is TSI for Secchi Disk (m); TSICHL is TSI for klorofil-a (µg/l) and TSITP is TSI for Total phosphate (µg/l).

Chlorophyl-a was calculated using total phosphate concentration as Beveridge equation [7] as below.

$$CHL\text{-}a = 0.416(P)^{0.675} \quad (5)$$

In the water phosphorus originating from fish feeding is transformed into phosphate. The mathematical model for phosphate enrichment is given below.

$$Phosphate\ enrichment = PU + PF - PD \quad (6)$$

$$Phosphate\ from\ ffishery\ net = PW + PFc \quad (7)$$

Where:
PU = Phosphate from up stream
PF = Phosphate from floating fishery net
PD = Phosphate from downstream

PW = Phosphate release to water
PFc = Phosphate in fish feces

Approximately 37.7% of phosphate is absorbed by fish, the remaining is released to water as residual feed and feces [7].

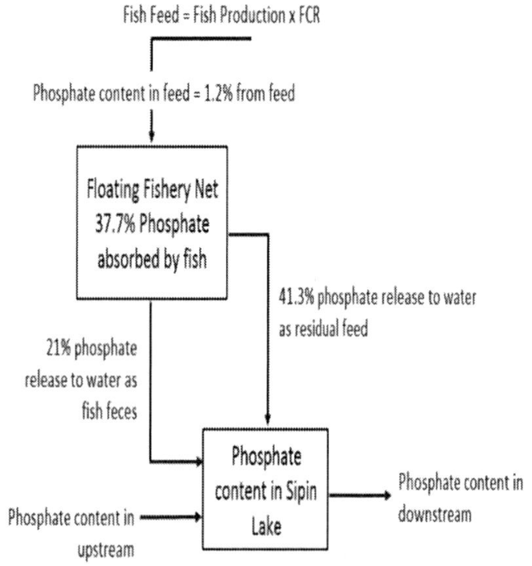

Fig. 2. Phosphorous enrichment empirical model.

3 Result and discussion

The analysis results of total phosphate shown in Table 2. From t-test, we know that the samples from each sampling point are not significantly different.

Table 2. Sampling location and total phosphate analysis result in mg/l.

Sampling point	Sampling date		
	28-01-2015	06-02-2015	17-03-2015
1	0.166	0.166	< 0.01
2	0.150	0.136	0.022
3	0.062	< 0.01	< 0.01
4	0.110	0.108	< 0.01
5	0.056	< 0.01	< 0.01
6	0.045	< 0.01	< 0.01

,

Phosphate enrichment calculation based on total phosphate analysis, empirical model, and 2013 total fishery production. This calculation showed that phosphate enrichment in Sipin Lake is 9.65 T/yr

Table 3. Phosphate enrichment calculation.

Source of phosphate	T/yr
Phosphate from floating fishery net	8.71
Phosphate from upstream	3.37
Phosphate from downstream	2.43
Phosphate enrichment	9.65

There are 2 areas for development of aquaculture in Sipin Lake, Pulau Pandan and Buluran area, the samples compiled from the two areas. Trophic status index calculation is shown in Table 4 and Table 5.

Table 4. Trophic status of Buluran area.

Parameter	Data	TSI Carlson
Total phosphate	45 μg/l	59.04
Brightness	0.4 m	73.20
Chlorophyl-a	4.43 μg/l	47.20
TSI Carlson Index		60.00

TSI Carlson index of Buluran area is 60 which indicates the eutrophic level in the lake.

Table 5. Trophic status of Pandan Island.

Parameter	Data	TSI Carlson
Total phosphate	56 μg/l	62.19
Brightness	0.76 m	63.95
Chlorophyl-a	6.29 μg/l	48.64
TSI Carlson Index		58.26

TSI Carlson index of Pulau Pandan is 58.26, indicates the eutrophic level in the lake.

To figure out the trophic status condition and phosphate enrichment for next 10 years, a modeling system is developed using Powersim 2.5. The modeling system has 3 sub-models.

Sub-model 1: phosphate enrichment. The input of this sub-model is fishery production and the output is Sipin Lake phosphate enrichment.

Sub-model 2: concentration of total phosphate. the output of sub-model 1 is input for sub-model 2. The conversion of phosphate enrichment in sub-model 1 into total phosphate concentration was calculated using morphological condition as depth, upstream, downstream, area, and volume of the lake. The output of this sub-model is total phosphate concentration in mg/l.

Sub-model 3: trophic status. the output of sub-model 2 is input for sub-model 3. In this sub-model, trophic status of the lake is calculated using total phosphate concentration from sub-model 2. Total phosphate concentration is used to calculate chlorophyl-a. The output of this sub-model is status trophic index.

All sub-model will build a modeling system of Sipin Lake's status trophic as shown in Figure 3.

From the modeling system, the figures of phosphate enrichment, total phosphate concentration and trophic status of Sipin Lake for next 10 years could be estimated as given in Table 6. The target proposed by agriculture and fisheries department of Jambi province fish production would increase approx. 2% per year in 2015. This means that total phosphate enrichment and trophic status of the Lake also increase as shown in Figure 4. Table 6 and Figure 4 show that in the year 2023 phosphate enrichment, total phosphate concentration and trophic status index of the Lake increase 18.94%, 15.64% and 61.79, respectively. The latter indicates that the Lake would be in eutrophic condition.

Table 6. Sipin lake estimation condition for next 10 years.

Year	Fish prod. (T/yr)	TP (mg/l)	Phosphate enrichment (T/yr)	Trophic Status Index
2014	909	0.0507	9.76	60.59
2015	934	0.0518	10.01	60.73
2016	959	0.0528	10.27	60.88
2017	984	0.0539	10.52	61.01
2018	1010	0.0549	10.77	61.15
2019	1034	0.0560	11.02	61.28
2020	1060	0.0571	11.28	61.41
2021	1084	0.0581	11.53	61.54
2022	1109	0.0592	11.78	61.67
2023	1135	0.0601	12.04	61.79

The model was validated using MAPE, MAD, MSD, and Durbin-Watson test. From the test, the value of MAPE, MAD, and MSD are lower than 5%, indicates that the model doesn't have significant error. Durbin-Watson test indicates that the model doesn't have auto-correlations from each equation.

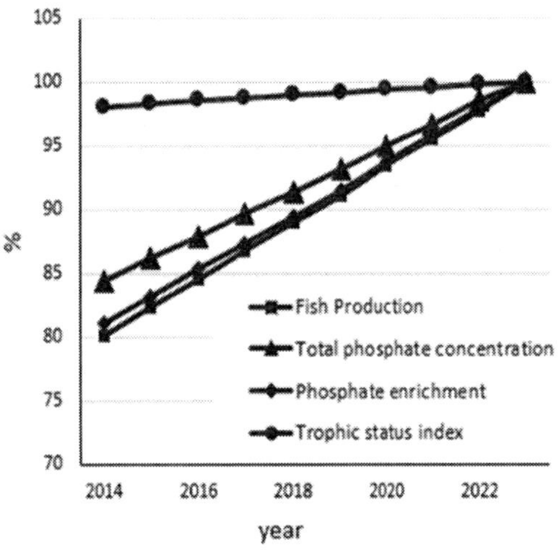

Fig. 4. Sipin Lake estimation condition for next 10 years

Some alternative policies that may be undertaken by the Local Government to reduce the risk of phosphate enrichment, total phosphate concentration and the trophic status increase in the Sipin Lake are discussed as follows.

1. Application of double floating fishery net. This system can cut nutrient in the water more than 90% since the system is combined with the fish culture outside the net. Application of double netting only reduces the nutrient to approximately 50% as reported in Waduk Sermo in 2009 [8]. Figure 5 shows reduction estimation in percent of phosphate enrichment, total phosphate concentration, and trophic status of Sipin Lake by applying double fishery netting. Figure 5 shows that in year 2015, total phosphate concentration and phosphate enrichment in lake water will be reduced 13.32% and 16.38% after 8 years application of double netting and trophic status of the Lake is mesotrophic level.

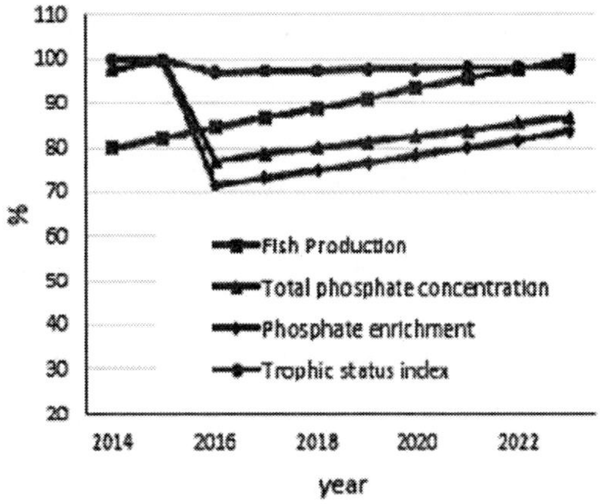

Fig. 5. Sipin Lake estimation condition by applying double floating fishery net

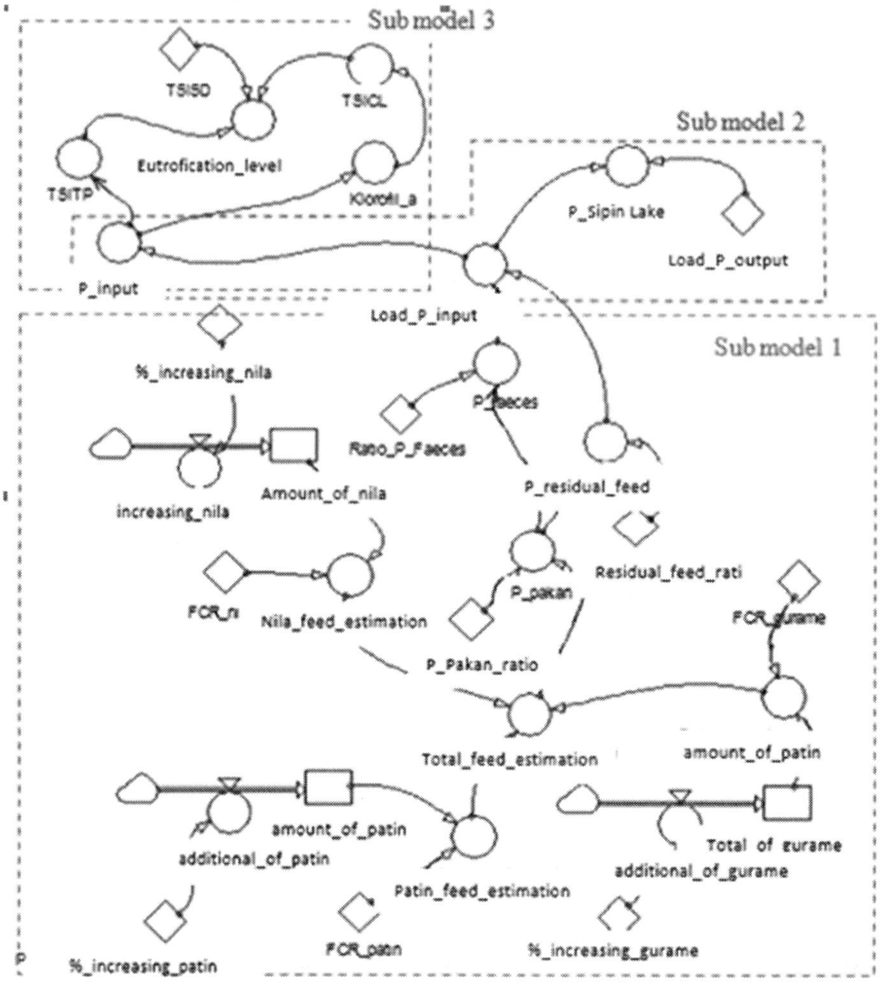

Fig. 3. Sipin Lake trophic status modelling

2. Application of 'pelet' (food) additive on fish feeding to break down the phytic acid. Phytic acid is a compound in fish feces. It will turn phosphate to the water and enrich phosphate concentration. Pelet additive usually used is Ronozyme-P with the application of 0.1 gr in 46.42 gr fish meal, 22 gr rebon flour, and 20.58 gr lamtoro seed flour, it will reduce 0.29% phytic acid content in fish feces [9]. It is estimated that phosphate enrichment, total phosphate concentration and trophic status of the Lake are also reduced as shown in Figure 6. Figure 6 shows that application of food additive also increases of phosphate enrichment, total phosphate concentration and the trophic status of the Lake. However, all those parameters will be lowered since nothing to in the Lake system. The total phosphate concentration, phosphate enrichment would be 0.16% and 0.41%, respectively. The trophic status is still eutrophic level.

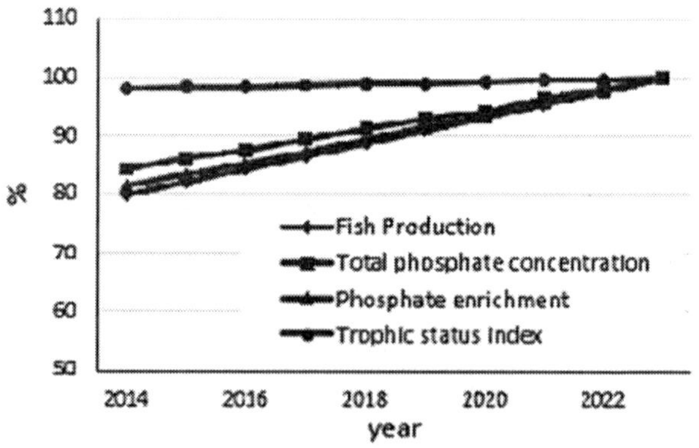

Fig. 6. Sipin Lake estimation condition by applying food additive

3. Reduction of phosphate from river streams. Phosphate in upstream of Sipin Lake probably comes from community activity using detergent, agricultural and farming activities. Sodium tripolyphosphate

4. content in detergent increases phosphate enrichment in lake water up to 20 – 30% [10]. If this policy applies in Sipin Lake the reduction of total phosphate concentration and phosphate enrichment is shown in Figure 7.

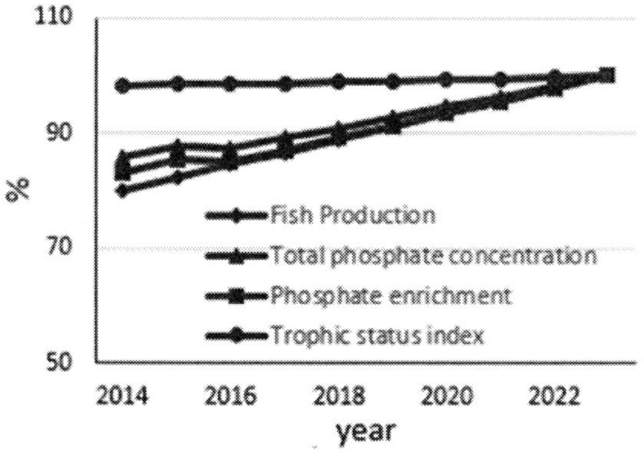

Fig. 7. Sipin Lake estimation condition since no detergent discharge into upstream river

Figure 7 shows that total phosphate concentration is reduced 0.34% in 2015. After 8 years total phosphate concentration and phosphate enrichment will be 1.8% and 2.5%, respectively. However, it is lower than if the system does not enforce anything.

4 Conclusion

Floating fishery net in Sipin Lake has triggered of increasing phosphate concentration in the lake. Trophic status index of the Lake is at eutrophic level. The increase of phosphate and trophic status of Sipin Lake has been identified by TSI Carlson method, there is estimated that within next 10 years, there is an increase of phosphate and total phosphate concentration of 18.94% and 15.64%, respectively and trophic status of the Lake is at eutrophic level. From the simulation, application of double floating fishery net is able to reduce total phosphate concentration and the risk of eutrophication in the lake higher than other policies. So, using double fishery net is a recommendation for the government to anticipate the future condition.

References

1. L. Cargnelli and M. Doyle. *Lake Erie Bi-national Nutrient Management Strategy*. Canada: Lakewide Management Plan. 3-5 (2011)
2. W. Gunawan, Zahidah and D. Mulyanti. *Model Eutrofikasi untuk Merancang Kebijakan Pengelolaan Waduk yang Berkelanjutan Melalui Pendekatan Sistem Dinamis*. Bandung: Faculty of Fisheries and Marine Science Padjajaran University. 1-18 (2007)
3. Dinas Pertanian, Peternakan, Perikanan dan Kehutanan. *Profil Perikanan Budidaya Kota Jambi* (2014).
4. Badan Lingkungan Hidup Daerah. *Status Lingkungan Hidup Daerah* (2014).
5. M. Everard. *Taking a system-oriented view of phosphorus enrichment in fresh waters- A System-Oriented View of Phosphorus Enrichment. England*: Gloucestershire GL50 1QB 35-54 (2011)
6. T.S.Mamatha, M.R.Nandini, T.N.Parinitha, U.Umesha, P.Tuppar, B.M.Kumar and B.M.Krisna. *Water quality monitoring and trophic status classification of Karanji Lake* Mysore J. Current Engineering and Technology ISSN 2277-410, 43-48 (2013)
7. Lukman. *Evaluasi Keseimbangan Fosfat di Danau Toba*. Jakarta: Prosiding Seminar Nasional Limnologi, **VI**, 429-430 (2012)
8. Rustadi. *Eutrofikasi nitrogen dan fosfat serta pengendaliannya dengan perikanan di Waduk Sermo* J. Manusia dan Lingkungan, **16**, 176-186 (2009)
9. D. Rachmawati and J. Hutabarat. *Efek Ronozyme P dalam pakan buatan terhadap pemanfaatan pakan dan pertumbuhan ikan kerapu macan (Epinephelus fuscoguttatus)* J. Ilmu Kelautan, **11**, 193-200 (2006)
10. H. Horth, E. Glennie, L.J. Davis, P. Jones and O. Tortolea. *Recommendation for the Reduction of Phosphorous in Detergents*. UK: WRc plc, 15-20 (2006)

WindSat and RAMA Buoy: a comparison of ocean-atmosphere data

Ayu Agustin[1], *Wijaya* Mardiansyah[1], *Dedi* Setiabudidaya[1] and *Iskhaq* Iskandar[1,2,*]

[1]Physics Department Faculty of Mathematics and Natural Sciences, Sriwijaya University, 30662 Inderalaya, Indonesia
[2]Environmental Research Center, Sriwijaya University, 30139 Palembang, Indonesia

Abstract. Ocean-atmosphere data deployed along the equatorial Indian Ocean were used for validating the ocean-atmosphere data recorded by the WindSat satellite. Three RAMA Buoys mounted at 67 °E, 80.5°E and 90 °E along the equator were used to evaluate the sea surface temperature (SST), wind speeds and precipitation obtained by the WindSat satellite, except for precipitation in the eastern equatorial Indian Ocean. The SST from WindSat shows significant correlation with that from RAMA Buoy, with coefficient correlations of 0.95, 0.88 and 0.83, respectively. The root mean square errors (RMSEs) of the WindSat SST relative to RAMA Buoy data were 0.28, 0.29 and 0.32 °C, respectively. In this study, the quality performances of the WindSat wind speed were also satisfactory, with coefficient correlation of 0.64, 0.7 and 0.58 in each RAMA buoy. The corresponding RMSEs of the wind speeds were 1.46, 1.37 and 1.63 ms^{-1}. Meanwhile, the WindSat precipitations reveal weak correlation with RAMA Buoy data, in particular in the eastern equatorial Indian Ocean. Overall, the WindSat satellite data for each measurement parameters can be used to replace in-situ observation data after taking into account the measurement errors.

1 Introduction

The Indian Ocean is the smallest of all oceans. It has a north-south extent of 9600 km from antartica to the inner Bay of Bengal and spans 7800 km in east-west direction between southern Africa and western Australia. Without its southern ocean part it covers an area of 48.10^6 km^2. If the southern ocean part is included, the area increases to 74.10^6 km^2 [1].

The Indian Ocean has unique characteristics and different from the ocean and atmospheric circulation in the Pacific Ocean and Atlantic Ocean. Wind circulation in the Indian Ocean is dominated by monsoon winds that change direction every six months [2]. Meanwhile, ocean circulation especially in the Indian Ocean around the equator characterized by the flow Wyrtki [3]. Flow Wyrtki a surface current that flows from west to east along the equatorial Indian Ocean. Wyrtki currents have an important role in distributing the mass of water and heat in the Indian Ocean.

So far, the data in-situ measurement result in the Indian Ocean is still very limited and far behind when compared with the other two ocean. One of the multinational cooperation program to perform in-situ measurement in the Indian Ocean is a Research moored Array for African-Asian-Australian Monsoon Analysis and Prediction (RAMA) buoy network. RAMA program is part of a multi-national coorperation activities for the in-situ measurements in the tropics known as the Tropical Atmosphere Ocean (TAO) project. Measurement parameters include parameters dynamics of the atmosphere and ocean dynamics parameters.

2 Data

2.1 WindSat satellite

The WindSat Polarimetric Radiometer was developed by the Naval Research Laboratory (NRL) Remote Sensing Division and Naval Center for Space Technology for the U.S. Navy and the National Polar-orbiting Operational Environmental Stellite System (NPOESS) Integrated Program Office (IPO).

WindSat data product include the standard set of measurements (sea surface temperature, wind speed, water vapor, cloud liquid water, and rain rate) and differ from other RSS radiometer products in that we can include additional measurements:

- WDIR: Because of its polarimetric channels, WindSat can measure surface wind direction. WDIR is only provided if the wind speed exceeds 3 m/s and the rain rate is below 15 mm/hr.
- WSPD_AW: We have made significant progressin improving the retrival of wind speeds in the presence of rain. We can retrieve WindSat surface wind speed in both rain-free and rainy atmospheres. We combine various algorithms, including a global wind speed through rain algorithm that works under all rain conditions and H-wind algorithm that has been

* Corresponding author: iskhaq@mipa.unsri.ac.id

specially trained for tropical cyclones. The final all-weather wind speed, WSPD_AW, provided in our binary files is a smooth blend between the standard wind speed obtained in non-raining conditions WSPD_LF, the global wind speed through rain and the H-wind derived wind speed.

Gridded data files are organized according to observation date. All date and time are Coordinated Universal Time (UTC), also known as Greenwich Mean Time (GMT), Zulu Time (Z), universal Time (UT), and World Time.

2.2 RAMA Buoy

The Tropical Moored Array Programs consist of three different TAO Buoy Arrays. They are TAO/TRITON in Pacific, Prediction and Research Moored Array (PIRATA) in Atlantic, and RAMA in the Indian Ocean. The Research Moored Array for African-Asian-Australian Monsoon Analysis and Prediction (RAMA) Buoy network over the Indian Ocean, part of the Tropical Atmosphere Observation (TAO) program helps to fill this data void to good extend.

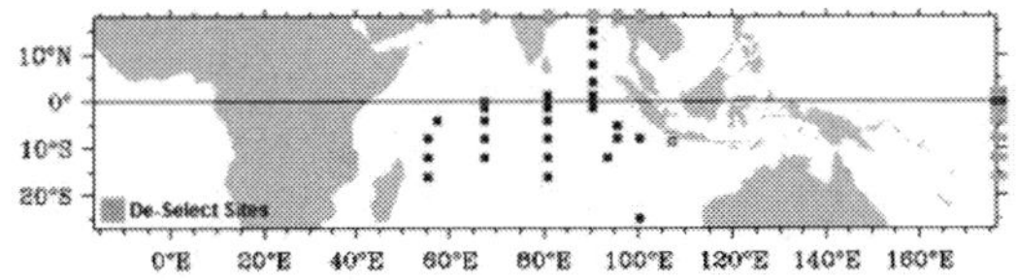

Fig. 1. Location of RAMA Buoy

The measurements on the RAMA surface floats include the meteorological parameters like air temperature, relative humidity, and wind velocity at height of 3-4 m above the mean sea level (MSL). Daily averages of all data and several hourly samples per day of most meteorological variables are transmitted to shore in real time via Service Argos [4]-[6]. Parameter and period of this study data contained in the Table 1.

Table 1. WindSat satellite data and RAMA buoy data

Parameters	Period
Sea Surface Temperature	1/1/2015-31/12/2015 (0˚LU,67˚BT) 9/7/2013-17/8/2014 (0˚LU,80.5˚BT) 9/7/2013-31/12/2015 (0˚LU,90˚BT)
Wind Speed	1/1/2015-31/12/15 (0˚LU,67˚BT) 9/7/2013-16/8/2014 (0˚LU,80.5˚BT) 17/5/2013-30/8/2015 (0˚LU,90˚BT)
Rainfall	1/1/2015-31/12/15 (0˚LU,67˚BT) 31/10/2012-31/12/15 (0˚LU,80.5˚BT) 21/5/2006-18/2/2009 (0˚LU,90˚BT)

In this study using data from 2015 due to the data of the three parameters in 2015 are of better quality than in other years.

Measurement data be seen in the graph below

(a)
(b)
(c)

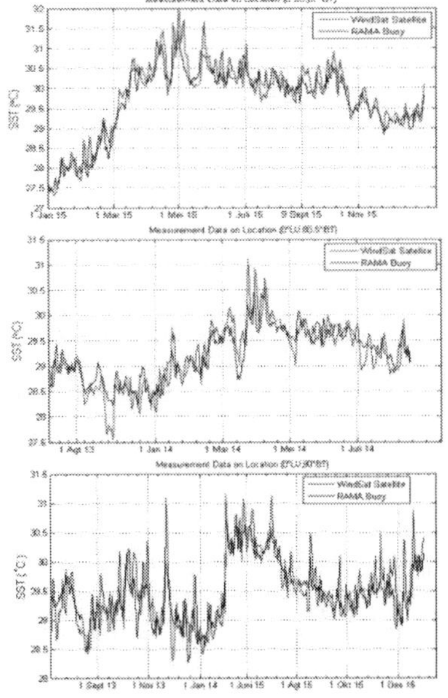

Fig. 2. The Graph comparison of sea surface temperature WindSat satellite and RAMA Buoy on location (a) 0˚LU,67˚BT (b) 0˚LU,80.5˚BT and (c) 0˚LU,90˚BT

(a)
(b)
(c)

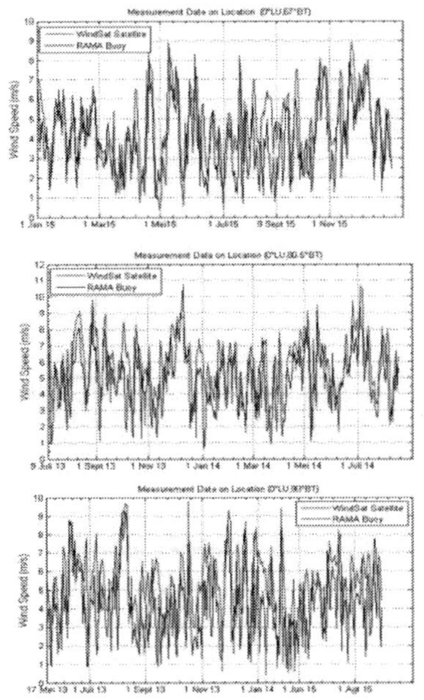

Fig. 3. The Graph comparison of wind speed WindSat satellite and RAMA Buoy on location (a) 0˚LU,67˚BT (b) 0˚LU,80.5˚BT and (c) 0˚LU,90˚BT

(a)
(b)
(c)

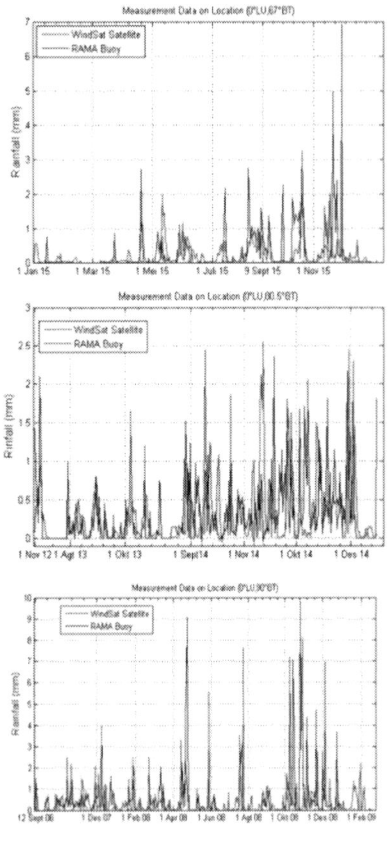

Fig. 4. The Graph comparison of rainfall WindSat satellite and RAMA Buoy on location (a) 0°LU,67°BT (b) 0°LU,80.5°BT and (c) 0°LU,90°BT

3 Methodology

WindSat satellite data downloaded from Asia-Pacific Data-Research-Center (APDRC). This data has a daily temporal resolution and a spatial resolution of 0.25° x 0.25° and the RAMA Buoy data downloaded from Tropical Atmosphere Ocean (TAO) project. As for the in-situ measurement data that will serve as reference data is the measurement RAMA Buoy data at three location along the equatorial Indian Ocean, each at 0°LU,67°BT; 0°LU,80.5°BT and 0°LU,90°BT. This study will calculate the level of accuracy WindSat satellite measurement by comparing them with the data in-situ measurement from RAMA Buoy. Sea surface temperature, rainfall and wind speed parameters were used in this study.

Sea Surface Temperature (SST) is a measure of energy caused by the movement of molecules on the surface of the sea. SST measurement can be made directly using a measuring instrument which generally is Conductivity Temperature Depth (CTD). If the term of the dynamics of the atmosphere, the water of the Indian Ocean has the unique and complex nature. This condition causes Asian-Australian Monsoon and Indian-African Monsoon. This will cause the circulation of currents especially sufficient surface complex [3],[7]. Wind over the Indian Ocean is dominated by annual circulation pattern (seasonal). Wind circulstion pattern in the semiannual period is a characteristic of wind circulation in the equatorial Indian Ocean which distinguishes it from the other two oceans, Pacific Ocean and Atlantic Ocean [8]. Equatorial rainfall pattern has the characteristic two peaks of rainfall on an annual cycle [9].

This research were used statistical analysis and analysis of standard errors to calculate the level of accuracy satellite measurement.

3.1 Linier correlation

The correlation is a way of determining how well two (or more) variables co-vary in time or space. For two random variables. The correlation coefficient can be written:

$$r_{xy} = \frac{1}{N-1}\sum_{i=1}^{N}\frac{(xi-\bar{x})-(yi-\bar{y})}{s_x s_y} \tag{1}$$

Where s_x and s_y are the standart deviations for the two data record. For r = ±1, the data points (x,y) cluster along a straight line and the sampels are said to have a perfect correlation (+) for "in-phase" fluctuations and minus (-) for 180° 'out-of-phase"fluctuations). For r≈0, the points are scattered randomly on the graph and there is little or no relationship between the variables [10].

3.2 Linier regression

Linier regression analysis was used to determine the shape of the relationship between variables. Intended use this analysis is to estimate the value of a avariable in relation to the other variables that are known through the regression line equation [11]. The general form of linier regression is:

$$y = a + bx \tag{2}$$

where a are intercept and b are slope or gradient of the line. y is a dependent variable and x is a independent variable. The constants a and b are calculated used equation :

$$a = \frac{(\sum y \sum x^2)-(\sum x \sum xy)}{N(\sum x^2)(\sum x)^2} \tag{3}$$

$$a = \left(\frac{\sum y}{N}\right) - b\left(\frac{\sum x}{N}\right) \tag{4}$$

$$b = \frac{N(\sum xy)-(\sum x \sum y)}{N(\sum x^2)(\sum x)^2} \tag{5}$$

DOI: 10.1051/matecconf/201710104005

3.3 Analysis standart errors

3.3.1 Mean bias error (MBE)

MBE value indicated systematic error estimation data (satellite). If a positive value means the estimation data is greater than the measurement data. This means that the data was biased low, if a negative value meansthe data estimation less of the measurement data.

$$MBE = \frac{1}{N}\sum_{i=1}^{N}(xi - yi) \qquad (6)$$

3.3.2 Root mean square error (RMSE)

RMSE is used to indicate random error of the measurement, which in this research is data from satellite.

$$RMSE = \sqrt{\left(\frac{1}{N}\sum_{i=1}^{N}\{xi - (MBE) - yi\}^2\right)} \qquad (7)$$

3.3.3 Mean absolute error (MAE)

MAE is used to measure the accuracy of estimate with the average error in unit of the same size as the original.

$$MAE = \frac{1}{N}\sum_{i=1}^{N}|xi - yi| \qquad (8)$$

Where xi is the estimated value (satellite data), yi is the value of in-situ measurement (RAMA Buoy), and N is the number of data used.

4 Result and discussion

4.1 Sea surface temperature

The result of calculation using the equation 1,6,7 and 8 show in Table 2.

Table 2. The correlation coefficient and error of sea suface temperature

Parameter	0°LU,67°BT	0°LU,80.5°BT	0°LU,90°BT
r	0.95	0.88	0.83
MBE	0.09	-0.02	0.07
RMSE	0.28	0.29	0.32
MAE	0.23	0.23	0.24

The results of Figure 2 measuring in all of located showed sea surface temperature data obtained from WindSat satellite that have high value than results measuring of RAMA buoys. That caused sea surface temperature was recorded by WindSat satellite is temperature in layers of sea surface, meanwhile sea surface temperature in-situ is sea surface temperature in deepth 1-2 m in below surface of the sea. As known that sea surface temperature was influenced by meteorology conditions, like as rainfall and velocity of wind that have high variations. This allegedly cause temperature was recorded by RAMA bouys that also have high variation.

(a)
(b)
(c)

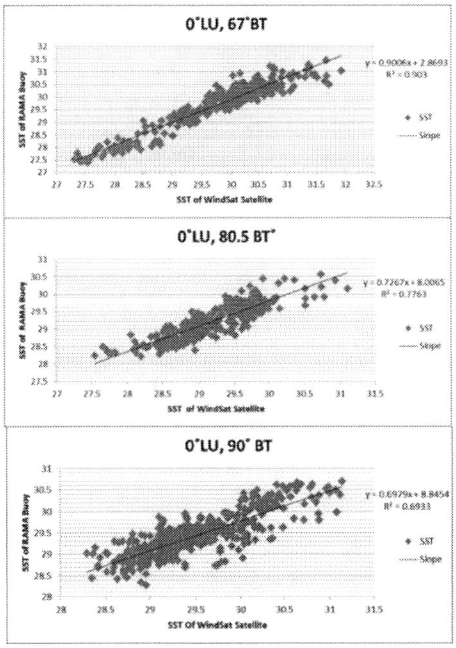

Fig. 5. The Graph linier regression of sea surface temperature (a) 0°LU,67°BT (b) 0°LU,80.5°BT and (c) 0°LU,90°BT

The highest correlation coefficient was shown on location 0°LU,67°BT i.e r=0.95. The correlation coefficient r≈1, the points are scattered regularly on the graph (show in figure 5) and there is perfect relationship between the variables.

The correlation coefficient obtained due to the values of systematic error is small. This is reinforced by the result of the linier regression.

4.2 WindSpeed

The result of calculation using the equation 1,6,7 and 8 show in Table 3.

Table 3. The correlation coefficient and error of wind speed

Parameter	0°LU,67°BT	0°LU,80.5°BT	0°LU,90°BT
r	0.64	0.7	0.58
MBE	0.59	0.98	1.24
RMSE	1.46	1.37	1.63
MAE	1.23	1.34	1.64

The results of measuring WindSat satellite couldn't caught signal with short time variation. That is showed in Fig.3, where the result of the graph measuring WindSat satellite more smooth than the result of RAMA

DOI: 10.1051/matecconf/201710104005

Buoys measuring. As known WindSat satellite caught signal in high variation. This allegedly cause wind speed was recorded by RAMA Buoys that also have high variation.

(a)
(b)
(c)

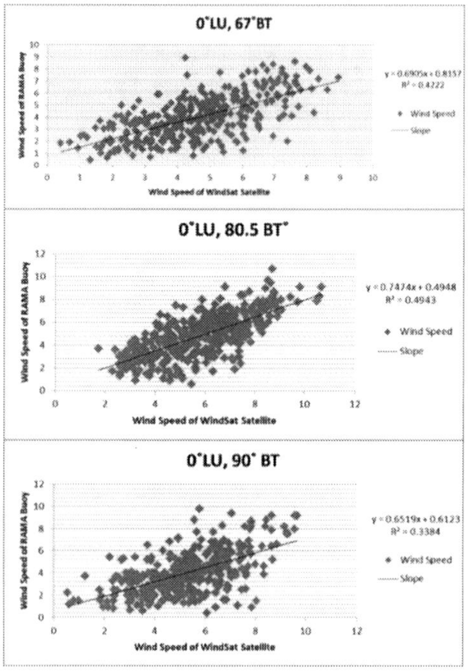

Fig. 6. The Graph linier regression of wind speed (a) 0°LU,67°BT (b) 0°LU,80.5°BT and (c) 0°LU,90°BT

4.3 Rainfall

The highest correlation coefficient was shown on location 0°LU,80.5°BT i.e r=0.7. The correlation coefficient the points are scattered regularly on the graph (show in figure 6) and there is good relationship between the variables.

The result of calculation using the equation 1,6,7 and 8 show in Table 4.

Table 4. The correlation coefficient and error of wind speed

Parameter	0°LU,67°BT	0°LU,80.5°BT	0°LU,90°BT
r	0.41	0.27	0.2
MBE	-0.10	-0.02	-0.15
RMSE	0.61	0.52	1.18
MAE	0.25	0.31	0.04

The correlation coefficient was shown the points are scattered randomly on the graph (show in figure 7) and there is no relationship between the variables.

The results of statistic test of rainfall measure showed correlation value between WindSat satellite data and RAMA bouys data have small coefficien correlation and high errors. That caused spacial resolution data of WindSat satellite (0.25x0.25) couldn't caught signal detail in smaller location than resolution.

(a)
(b)
(c)

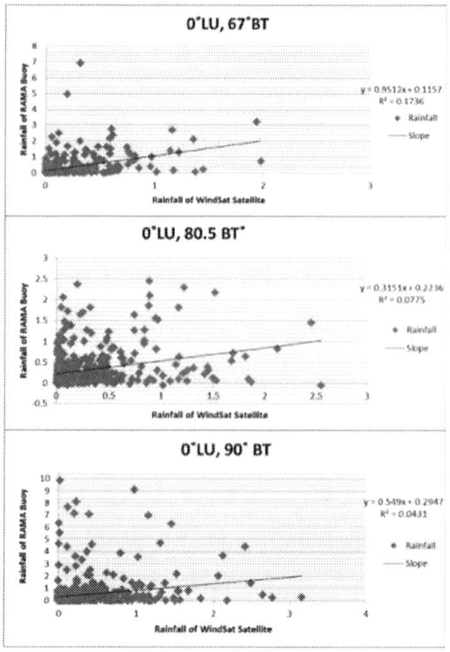

Fig. 7. The Graph linier regression of rainfall (a) 0°LU,67°BT (b) 0°LU,80.5°BT and (c) 0°LU,90°BT

The main obstacle to search information atmosphere-Ocean data (sea surface temparature, wind speed, and rainfall) from radiometric measure is correction atmosphere. Cloud made some problem to measure. Because make sea surface randomly. Part of satellite data was contaminate by effect of cloud that must be identification and elimination from processing procedure.

5 Conclusion

The greater the systematic errors in measurement data, the smaller the correlation coefficient resulting in less efficient estimation data (satellite) to replace the in-situ data. This is reinforced by the result of the linier regression on every measurement parameter used in this research.

Authors are grateful to the Physics Department, Faculty of Mathematics and Natural Sciences and Environmental Research Center, Sriwijaya University for the encouragement.

References

1. M. Tomzack, and J.S. Godfrey, *Regional Oceanography: an Introduction* (2011)
2. F.A. Schott, P. Julian, Jr. McCreary, The Monsoon Circulation of the Indian Ocean, Progress in Oceanography, **51**, 1-123 (2001)
3. K.Wyrtki, An Equatorial Jet in the Indian Ocean, Science, **181**, 264-264 (1973)
4. S.I. Rani, and M.D Gupta,

5. Oceansat-2 and RAMA Buoy Wind : A Comparison, **122** (J.Earth Syst.Sci, 2013)
6. S.C.M. Prakash, R.M. Gariola, S. Pokhrel, Surface Freshwater Flux Estimation Using TRMM Measurements over the Tropical Oceans, **1**, 225-234 (Atmospheric and Climate Science, 2011)
7. M.J. McPhaden, G. Meyes, K.. Ando, Y. Maumoto, V.S.N. Murty, M., Syamsudin F. Ravichandran, J. Vialard, L. Yu and W. Yu, RAMA the Research Moored Array for African-Asian-Australian Monson Analysis and Prediction, **90**, 459-480 (Bull.Am. Meteor.Soc, 2009)
8. Martono, *Karakteristik dan Variabilitas Bulanan Angin Permukaan di Perairan Samudera Hindia* (2009)
9. I. Iskandar, *Diktat Matakuliah Hidrodinamika Atmosfer dan Laut* (2015)
10. Bannu, D.A. Suriamiharja, N. Takeuchi., H. Kuze, Impact of the ENSO and IOD Phenomena : Long-Term Analysis in Indonesian Region, 3-7 (36th COSPAR Scientific Assembly, 2003)
11. W.J. Emery, R.E. Thomson, Data Analysis Metods in Physical, 233-238 (Elsevier, 2004)
12. Spiegel, R. Murray, Theory and Problems of Statistics, Schaum's Outline Series : McGraw-Hill, 311-313

* Corresponding author: iskhaq@mipa.unsri.ac.id

Assessment of indoor air quality in comparison using air conditioning and fan system in printing premise

Nazirah Ramlan[1], *Siti* Nurhalimatul Husna Ahmad[1], *Eeydzah* Aminuddin[2], *Hazrul* Abdul Hamid[1], *Siti* Khalijah Yaman[1] and *Abd* Halid Abdullah[1]

[1] Faculty of Civil and Environmental Engineering, Universiti Tun Hussein Onn Malaysia, 86400 Batu Pahat, Malaysia
[2] Faculty of Civil Engineering, Universiti Teknologi Malaysia, 81310 Johor Bahru, Johor, Malaysia

Abstract. Printers contribute to various emissions consist with chemical contaminants. High concentration of the particulate matter can cause serious health problems. This study focuses on the indoor air quality in printing premise unit in Universiti Tun Hussein Onn, Malaysia. Field testing involving air sampling methods were taken from 900 hours to 1600 hours, for every 30 minutes using physical measurement which is Multi-Channel Air Quality Monitor (YESAIR), E-Sampler and Ozone Meter. Air sampling was recorded based on one sampling point and most suitable point for production. A comparison based on different ventilation using fan and air-conditioning were also taken and results is being compared based on OSHA and NIOSH standards. Besides that, the statistical analysis is being conducted in order to predict the effect on number of printers. From the result, the O_3 concentrations show, 10% reduced for printing premise using fan ventilation compared to air-conditioning but remain the same value for PM2.5. The concentration of O_3 increased when the number of printers decreased, while the concentration of PM2.5 increased the increase of printers number. Overall, the use of fan in printing premise is more suggested since the level is slightly lower than the printing premise using air-conditioning.

1 Introduction

Air is a precious resource for all humans and living things. Indoor air quality is significant to the human since most of the time spent in the buildings. According to Mohd Dzulkifli S.N. et al. (2014), poor level of indoor air quality since indoor air pollutants is not easy to dilute compared to outdoor pollutants. There are many factors that damage the air quality from rapid development such as increasing number of vehicles, factory's waste, dusts from construction works, industrial sources and materials use for building. Some of the electrical components used in the building are computers, printers, copiers and other electronic equipment [1]. The Environmental Protection Agency (EPA) defines Indoor Air Quality (IAQ) as the temperature, humidity, ventilation and chemical or biological contaminants of the air inside a building [2].

Nowadays, printing services are very popular and it has been widely used in Malaysia. It became one of the most important services for people in their daily lives. Printing services have been used by many people include the workers and students from school or university. The services provided by printing premises are photocopy, scanning, binding and lamination, booklet and nametag. It is a well-known the fact that in the printing process hazardous substances occur in the air, threatening human health [3]. There are few potential source from indoor pollutants from printers [4,5] whereby the main pollutants identified are from volatile organic compound [6], ozone [7] and fine particles [8]. One of important factor in indoor environment is fine particle due to the effect to health and also for how inhabitants perceive the environment [9].

According to Kim et al. (2008), a fuser or also called as a toner fixation unit released a particles and addition to it, different condition of fuser from the operation of printer may contributed to different size of particle distributions. Damanhuri *et. al.*, 2015 report that the emission rates of compound are significantly influence by printer speed and the coverage of printing job that affected the quality of indoor environment.

Kiurski, J. S. *et al.*, 2013 added that the printer that were used in printing premises for printing services is one of the items that emit the air pollution and it will lead to poor air quality. Some illnesses such as cough, headaches, eyes irritation or dizziness were caused by poor air quality. In a report by World Resources Institute 1998), indoor air pollution has been recognize as a global environmental problem. These pollutants include airborne particles and particulate matter such as gases, toxic chemicals and volatile organic compounds (VOCs). Based on a studies by U.S. Environmental Protection Agency (EPA), there is a cases where level of pollutants for indoor environment are higher than outdoor environment. EPA is concern about particles

* Corresponding author: khalijah@uthm.edu.my

that are 10 micrometers in diameter which are "inhalable course particles" and below 2.5 micrometers which are "fine particles". Due to Fisk WJ (1987), '*buildings were assumed to shelter occupants from outdoor pollutants and little thought was given to pollutants generated or trapped indoors*'. Table 1 shows the list of indoor air contaminants and acceptable limits based on ASHRAE standard 62-2004 and DOSH.

Therefore, this paper investigates the indoor air quality in printing premises. The investigation is directly to monitor the main potential pollutants with is the VOCs Ozone and fine particle. Besides that the IAQ parameters such as the temperature and relative humidity were also being recorded.

2 Analysis method

2.1 Mapping

Figure 1 shows the position of the printers inside the premise. The arrangement of furniture is being organized based on one cubicle and u-shape desk.

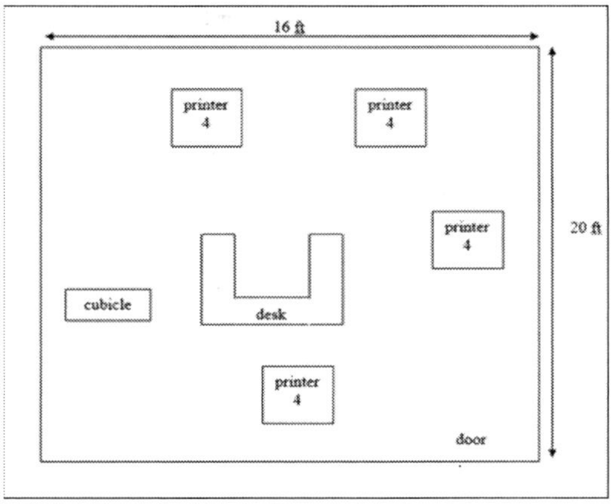

Fig. 1. Position of printer in the premise.

Table 1. List of indoor air contaminants and the acceptable limits

Indoor Air Contaminant	Acceptable limits		
	ppm	mg/m³	cfu/m³
Chemical contaminants			
a. Carbon monoxide	10	-	-
b. Formaldehyde	0.1	-	-
c. Ozone	0.05	-	-
d. Respirable particulates	-	0.15	-
e. Total volatile organic compounds (TVOC)	3	-	-
Biological contaminants			
a. Total bacterial counts	-	-	500*
b. Total fungal counts	-	-	1000*
Ventilation performance indicator			
a. Carbon dioxide	C1000	-	-

2.2 Sampling

The space for printing premise is about 16ft by 20ft with closed door as to control the contaminants value. The location situated in printing premises unit in University of Tun Hussein Onn Malaysia. The field testing involve are from air sampling methods which were taken from 0900 hours to 1600hours, for every 30 minutes using physical measurement which is Multi-Channel Air Quality Monitor (YESAIR), E-Sampler and Ozone Meter. The room temperature ranging in between 26°C-28°C with relative humidity in the range of 50-60 %.

2.3 Descriptive analysis

The descriptive statistics parameters that were used for analyses the data collected are the number of data, mean, mode, median, maximum number and minimum number. From the descriptive statistics, the simpler interpretation of the data can be obtained and the data was presented in a more meaningful way. In order to determine the outliers of O_3 and $PM_{2.5}$, the box plot was used in the analysis. Outlier is an observation point that is distant from other observation. From the box plot obtained, the skewness of data can be obtained.

2.4 Correlation analysis

In order to achieve the objective for this research, the correlation analysis method was used. The correlation analysis was used to describe the linear relationship between two continuous variables which is between the numbers of printers operated with the level of contamination indoor air. The correlation measures the strength and direction of the linear relationship. This method can show whether and how strongly pairs of variables are related.

3 Result and discussion

3.1 Descriptive statistics

The collected data were analysed and the values of mean, mode, and median were determined and tabulated in the Table 2. The results show that the mean of ground level ozone and $PM_{2.5}$ are still comply the standard which is 0.05 and 150 $\mu g/m^3$ respectively. For comparison purposes, the mean of O_3 for printing premise using air-conditioning is slightly high compared to printing premises using fan and based on the value of mean, mode and median, the distribution of monitoring records are almost symmetry or normally distributed.

Based on the monitoring records, the data of O_3 and $PM_{2.5}$ for air-conditioning and fan have been transfer into boxplot. From the information that was obtained from the boxplot, the skewness of the data can be known. The boxplot of the data can be seen in Figure 2, Figure 3, and Figure 4. Both boxplot for O_3 and $PM_{2.5}$ are negatively skewed and positively skewed respectively. Negatively skewed shows that the most frequent data to occur indicate a larger value and as positively skewed means

 DOI: 10.1051/matecconf/201710104006

that the most frequent data that were collected are smaller than median.

Table 2. Descriptive statistics parameters

	Air-conditioning		Fan	
	O_3 (ppm)	$PM_{2.5}$ (µg/m^3)	O_3 (ppm)	$PM_{2.5}$ (µg/m^3)
N	141	141	141	141
Mean	0.020	11	0.019	11
Mode	0.021	8	0.020	7
Median	0.020	9	0.020	7
Maximum	0.023	169	0.022	106
Minimum	0.012	1	0.015	2

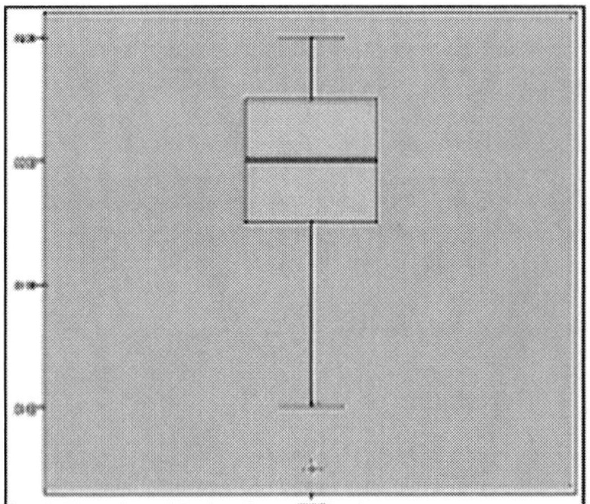

Fig. 2. Boxplot of O_3 in air-conditioning system

3.2 Correlation analysis

From the data collected, the relationships between the numbers of printers operated with the level of contamination indoor air $PM_{2.5}$ and O_3 were analysed. The correlation analysis method was used in this study in order to analyse the relationship. Table 3 shows the correlation coefficient for concentration of O_3 and $PM_{2.5}$.

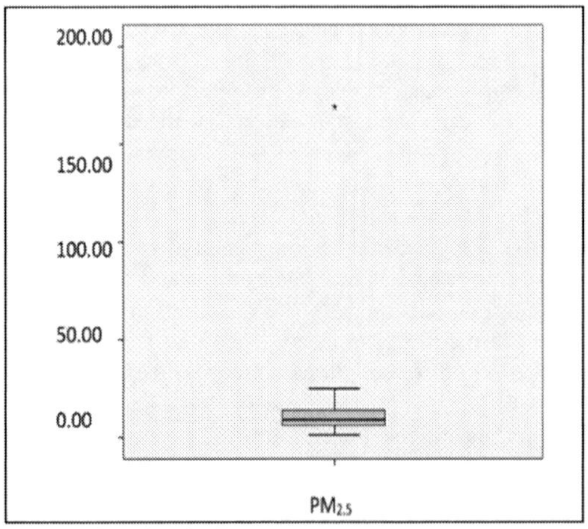

Fig. 3. Boxplot of O_3 in fan system

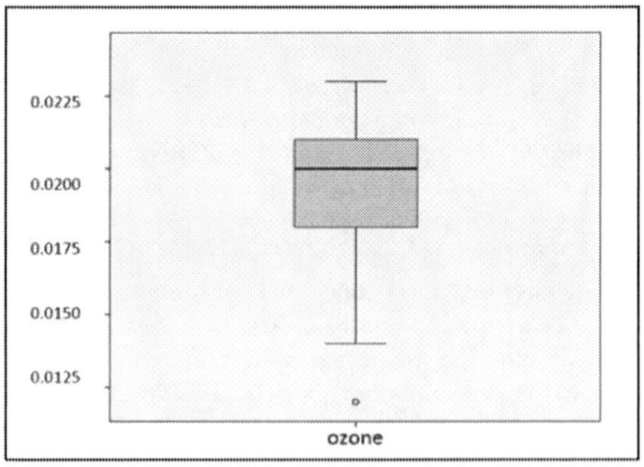

Fig. 4. Boxplot of $PM_{2.5}$ in air-conditioning system

Table 3. Correlation coefficient of O_3 And $PM_{2.5}$

Parameters	Correlation of air-cond.	Correlation of fan
O_3	-0.41	-0.18
$PM_{2.5}$	0.07	-0.09

Based on the Table 3, the correlation coefficient of O_3 in air-conditioning and fan system is -0.41 and -0.18 while correlation coefficient of $PM_{2.5}$ is 0.07 and -0.09 respectively. The relationship for O_3 is known as the negative relationship while the relationship for $PM_{2.5}$ is known as positive relationship. Otherwise, the relationship of $PM_{2.5}$ for fan is known as the negative relationship. The strength of the correlation can be analyzed based on the guide that Evans (1996) suggests that were tabulated in the Table 4.

Table 4. Correlation guideline: absolute value of pearson's coefficient, R

Strength	Coefficient, r
Very weak	0.0-0.19
Weak	0.2-0.39
Moderate	0.4-0.59
Strong	0.6-0.79
Very strong	0.8-1.0

From the Table 4, the result of correlation of O_3 was in the range of $0.4 < r < 0.59$ which means the correlations are moderate. Meanwhile, the correlation coefficient of $PM_{2.5}$ was in the range of $0.0 < r < 0.19$ which means very weak. From the result, the concentration of O_3 increased when the number of printers decreased. Meanwhile, the concentration of $PM_{2.5}$ increased when the number of printers increased. Otherwise, the result of correlation of O_3 and $PM_{2.5}$ are in the range of $0.0 < r < 0.19$. This indicates that the relationship in between $PM_{2.5}$ and O_3 is very weak.

Based on the result from Brown et al. (1999), there are different results for contaminated indoor air when the number of printer use is different. From the recent reports, it has been identified and measured that the emissions of ultrafine particle were exist during operation of laser printers [4, 15] and copiers [16]. Further studies are needed to better understanding on the type of printer used and the types of printing activity such as fax, colour printer, copier, and scanner.

3.3 Graphical analysis

Figure 5 shows the graph of O_3 against time while Figure 6 shows graph of $PM_{2.5}$ against time. The figure below shows the graph of O_3 and $PM_{2.5}$ against time for ten (10) days which are from 28th February 2016 until 13rd March 2016. The guidelines for both O_3 that cannot be exceed which is 0.05 ppm. Meanwhile, the acceptable limit based on Department of Occupational Safety and Health (DOSH) for $PM_{2.5}$ is 150 μg/m³. The critical result that was obtained for O_3 is 1st March 2016 and for $PM_{2.5}$ is 3rd March 2016. Based on the graph, the reading of O_3 rose from 0.018 to 0.022 ppm starting from 12.15 until 2.30pm because on that time the printers are not working or at rest condition. Meanwhile, when the printers were at rest, both contaminations were volatile in indoor air. Also the surrounding was lack of airflow from an outdoor and the air in surrounding was caught at one place. This is the critical part where both contaminant emission from printers. A by-product from printing processes of laser printers and photocopiers is ozone. The fluctuate values of ozone is depending on airflow around printer, ozone diffusion in air, or the characteristic of ozone monitor used in this study [4].

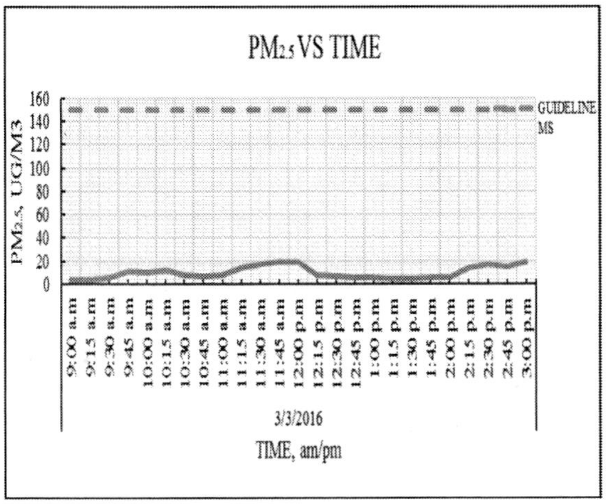

Fig. 5: Graph of O_3 vs. time

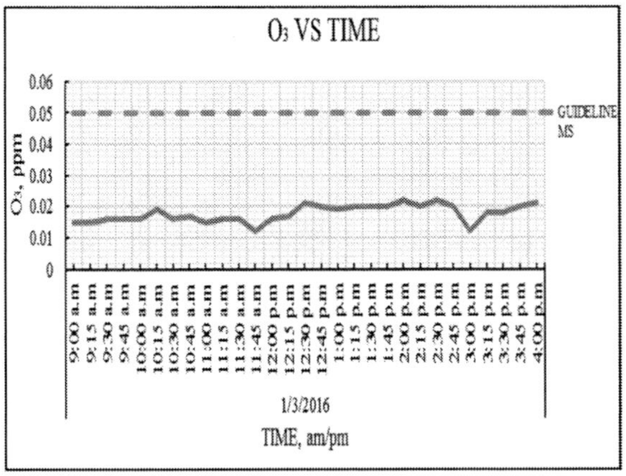

Fig. 6: Graph of $PM_{2.5}$ vs time

4 Conclusion

Based on the discussion and the analysis, IAQ inside printing premise unit was found safe to occupants where the levels of contaminants inside the building are low. The result for air temperature, relative humidity, O_3, VOCs and $PM_{2.5}$ shows a positive sign for good IAQ and below DOSH standard and permissible limit by DOSH. Besides that, the relationships between numbers of printers operated with the level of contamination shows that using air-conditioning system, the relationship between the number of printers with the level of contaminate air for O_3 is moderate (negative), indicated when numbers of printers operated increase, the concentration of O_3 decrease. Meanwhile, the relationship for $PM_{2.5}$ is weak (positive) indicated numbers of printer increases will cause increasing in concentration. Unlike O_3 and $PM_{2.5}$ the relationship shows a weak results when fan is applied.

The authors wish to sincerely thank the managers of the copy center whose generosity made this study possible. This investigation was made possible by Short Term Grant, U338, UTHM, Malaysia.

References

1. Jeong, H. B. & Jang W. K., Particle Emission from Laser Printers with Different Printing Speeds. *Atmospheric Environment. (2012)*
2. Fiedler N., Laumbach R, Kelly-McNeil K, Lioy P., Fan Z.H., Zhang J., Ottenweller J., Ohman-Strickland P., and Kipen H., Health effects of a mixture of indoor air volatile organics, their ozone oxidation products, and stress. Enviro health Perspect, 113(11):1542-8. (2005)
3. Kiurski, J. S., Marić, B. B., Aksentijević, S. M., Oros, I. B., Kecić, V. S. & Kovačević, I.M., Indoor Air Quality Investigation from Screen Printing Industry. *Renewable and Sustainable Energy Reviews* (2013)
4. Kagi N., Fujii S., Horiba Y., Namiki N., Ohtani Y., Emi H., Tamura H., Kim Y.S., Indoor air quality for chemical and ultrafine particle contaminants from printers, Building and Environment, **42**(5), 1949-1954. (2007)
5. The World's Worst Pollution Problems; assessing health risks at hazardous waste sites, Blacksmith Institute, p.9,32, 43.(2012)
6. Destaillats H., Maddalena R.L., Singer B.C., Hodgson A.T., McKone, T.E., Indoor pollutants emitted by office equipment: A review of reported data and information needs, Atmospheric Environment, **42**(7), 1371–1388. (2008)
7. Wang H., He C., Morawska L., Mcgarry P., Printer-generated ultrafine particles : the role of ozone, 10th International Official Conference of the International Society of Indoor Air Quality and Climate, Brisbane, Australia. (2012)
8. Saraga D., Pateraki S., Papadopoulos A., Vasilakos C., Maggos T., Studying the indoor air quality in

three non-residential environments of different use: A museum, a printery industry and an office. Building and Environment, **46**(11), 2333–2341. (2011)
9. Gudmundsson, A., Löndahl, J., Bohgard, M., Methodology for Identifying Particle Sources in Indoor Environment. *Journal of Environmental Monitoring* (2007).
10. Kim, Jae Hoon, Jin Hee An, Young Soo La, Jin Su Jung, Han Mo Jeong, Suck Man Kim, Nam Gu Moon, Byung Wha Lee, Young Ho Yoon, and Young Il Choi. *Journal of Industrial and Engineering Chemistry* 14 (2): 194–201 (2008)
11. Damanhuri A.A.M., Leman A.M., Abdullah A.H., Hariri A., Effect of toner coverage percentage and speed of laser printer on total volatile organic compound (TVOC), *Chemical Engineering Transactions*, **45**, 1381-1386 (2015)
12. WRI [World Research Institute]. In collaboration with the United Nations Environment Programme, the United Nations Development Programme, and the World Bank. World Resources Report 1998-99 New York: Oxford University Press. (1998)
13. Evans, J. D., Straightforward Statistics for The Behavioral Sciences. Pacific Grove, CA: Brooks/Cole Publishing (1996)
14. Brown SK. Assessment of pollutant emissions from dry-process photocopiers. Indoor Air. 9-25 (1999)
15. He, C., Morawska, L. & Taplin, L., 2007. Particle Emission Characteristics of Office Printers. *Environment Science Technology.* **41**(17): 6039–6045
16. Lee, C. W. & Hsu, D. J., Measurements of Fine and Ultrafine Particle Formation in Photocopy Centers in Taiwan. *Atmospheric Environment.* 41:6598–6609. (2007)
17. DOSH [Department of Occupational Safety and Health], 2005. Code of Practice on Indoor Air Quality. *Ministry of Human Resources Malaysia*
18. EPA, 1997. An Office Building Occupant's Guide to Indoor Air Quality. Washington, DC (20460)
19. S.N. Mohd Dzulkifli, Abdullah A.H and A.M. Leman. Indoor Air Quality of Museum Building Environment in a Tropical Climate: Proposed Study. *3rd Scientific Conference on Occupational Safety and Health.* (2014)

* Corresponding author: khalijah@uthm.edu.my

Surface-flow wetland for water reclamation at Batamindo Industrial Park

Chris Salim[1,*], *Andita* Rachmania[1], and *Rahma* Dewi[2]

[1]Surya University Department of Environmental Engineering, 15810 Tangerang, Indonesia
[2]PT. Batamindo Investment Cakrawala, Wastewater Treatment Plant & Sewage Treatment Plant, 29433 Batam, Indonesia

Abstract. The reclamation of wastewater as clean water resource is essential in the concept of water conservation. In industries, this will also lead to overall plant operational cost reduction. In this study, a pilot-scale surface-flow constructed wetland system filled with water hyacinth was used to treat effluent from existing sewage treatment plant at Batamindo Industrial Park. The sewage treatment plant effluent with quality fulfilling the regulation of Indonesian Ministry of Environment No.5/2014/XLVII-Group I was treated further to meet the raw water quality standard of Class I defined by Indonesian Government Regulation No.82/2001 as potable water resource. The system was able to maintain high chemical oxygen demand removal efficiency of around 80% from initial value of 100 mg/L in sewage treatment plant effluent despite the change of flow rate from 10 to 40 m^3/day (corresponding to hydraulic retention time of 2 to 0.5 days in constructed wetland respectively). The constructed wetland could successfully increase dissolved oxygen concentration, remove coliform bacteria and improve clarity (reduce turbidity) of water while maintaining neutral pH and low TSS values. Some additional post-treatments such as simple aeration and ultraviolet irradiation may improve the water quality further to meet the Class I Standard of potable water resource.

1 Introduction

Constructed wetland has been studied and used extensively in treatment of various kind of wastewater, not only those containing organic compounds or plant nutrients such as municipal sewage and agricultural wastewater but also wastewater containing heavy metals or other hazardous compounds such as metallurgic industry, chemical industry, landfill leachate and so on [1]. Guidelines from United States Environmental Protection Agency (US-EPA) for municipal wastewater treatment using constructed wetland have been issued as early as 1988 [2]. A handbook of constructed wetlands for agricultural wastewater, domestic wastewater, coal mine drainage, and storm water was provided by United States Department of Agriculture cooperating with US-EPA and Pennsylvania Department of Environmental Resources [3]. During the years, other guidelines related to constructed wetland for treatment of specific wastewaters have also been published by cities and other agencies around the world [4-7]. It is fairly safe to say that wetland treatment is an already established method to treat various kind of wastewater.

As in natural wetland system that is well known for the water purification capability, constructed wetland is actually comprised of several units such as plants, microorganisms and soil that take part in water purifying process physically, chemically and biologically. Plant absorbs both organic and inorganic compounds in water into its biomass. Microorganism can accelerate the decomposition of organic compounds due to its metabolism. Some pathogens existing in water can also be removed when the water comes into contact with natural microorganisms living in the wetland. Soil can serve as filter to remove solid particles, as ion exchanger to trap cations or anions in water, and also as media to facilitate many kinds of chemical and biological reactions occurring in water. Because of these characteristics, wetland treatment system is capable to process almost any kind of wastewater removing both the organic and inorganic contaminants as long as the toxicity of the contaminants is within tolerance of the ecosystem inside the wetland. With sufficient nutrients in wastewater, no chemical addition is necessary to operate the wetland treatment system. Constructed wetland is also generally constructed to achieve gravity flow so that no energy is required for pumping system inside the wetland. The only disadvantage of wetland treatment system is the relatively slow treatment process that requires large area to achieve large-scale water or wastewater treatment.

According to the position of water layer, constructed wetland can be categorized as surface-flow wetland with exposed water layer flowing above soil layer and subsurface-flow wetland with water layer flowing beneath the surface of soil layer. Both types have their own advantages and disadvantages. Surface-flow wetlands may be easily aerated due to photosynthesis processes and direct contact of water with the atmosphere, while subsurface-flow wetlands may be suitable for heavier load of contaminants due to additional filtration system of soil. Subsurface-flow

* Corresponding author: chrissalim@gmail.com

wetlands may require larger area than surface-flow wetlands due to the volume occupied by soil media that could reach almost 50% of total volume.

Until now, most of the studies and practices relating to wastewater treatment using constructed wetland were aimed at safe disposal of treated effluent to the environment or reuse of the treated effluent as a non-potable water resource such as water for irrigation that will eventually return to the environment. Some cities such as Orlando utilize constructed wetlands that are large enough to reclaim municipal wastewater as non-potable water resource [8-9]. The quality of water coming out of a wetland depends on the treatment capacity of the wetland that could be affected by several factors such as the wetland area, type of vegetation, hydraulic retention time and so on. Therefore, it is considered possible to integrate constructed wetland into wastewater treatment system to provide effluent with quality high enough for reuse in water treatment system that provides potable water.

In this study, pilot-scale of surface-flow wetland was built to treat effluent from sewage treatment plant (STP) of Batamindo Industrial Park (BIP) as initial phase of BIP water reclamation project scheme to reduce water intake from surrounding environment while increasing water recycling and reuse. The objective in this study was to evaluate the wetland performance in improving water quality of STP effluent into raw water resource for potable use. Performance of the constructed wetland was observed at several different operational conditions that include water depth, flow rate and soil layer. Water quality of the effluent from wetland was monitored and compared to the water quality of raw water resource for potable use defined in regulation of Indonesian Government No.82/2001 Water Standard Class I (PP 82/2001/Class I) [10].

2 Experimental methods

2.1 Constructed wetland design

Flow diagram of STP and wetland system is shown in Figure 1. BIP domestic wastewater being treated at STP using sequential batch reactor (SBR) was flowed into intermediate collection tank prior to entering the wetland to achieve constant flow rate. After that, effluent from wetland was collected at the outlet for sampling and disposal.

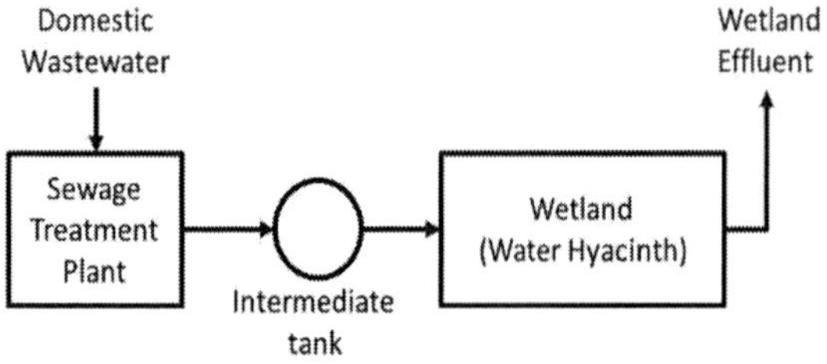

Fig. 1. Flow diagram of STP and wetland system

Two open concrete ponds with dimension of 5 m x 10 m x 1 m (width x length x height) were used as the vessels to contain pilot-scale surface-flow wetlands operated with different conditions. Soil layer was laid out in the bottom with height of 0.4 m. The outlet piping was arranged specifically to maintain water level at 0.2 or 0.4 m above the soil layer even during the raining period. Flow rate of STP effluent were adjusted to hydraulic retention time (HRT) of 0.5, 1 or 2 days. Plant type use in the wetland was *Eichhornia crassipes*, known commonly as water hyacinth. This type of aquatic plant is known for its resiliency and its survivability in contaminated water, and it is even classified as an invasive aquatic plant that often poses a threat in an open aquatic ecosystem.

2.2 Sampling and analysis

Influent and effluent of wetland were sampled daily and analyzed for several parameters, i.e., chemical oxygen demand (COD), total suspended solid (TSS), dissolved oxygen (DO) and turbidity. COD was measured using colorimetry with potassium permanganate method, while TSS and turbidity was measured using spectrophotometry (Spectroquant Nova 60, Merck). Dissolved oxygen was measured using portable DO meter (Cyberscan DO 300 Series). Periodically, full water quality analyses of STP and wetland effluents were carried out according to parameters in regulation of Indonesian Environmental Ministry (PerMenLH 5/2014/XLVII-Group I) [11] (for domestic wastewater) and PP 82/2001/Class I (for potable use raw water resource), respectively.

Plant growth in the wetland was also observed visually. The plant number was counted manually starting from the initial number of 30 units in the 50 m^2 pond area.

3 Results and discussion

3.1 Plant growth

Figure 2 shows the increase of plant count of water hyacinth in the wetland during the experiment. The initial number was 30 units and increased significantly up to 680 units in 36 days. This significant growth of water hyacinth was sustained only by nutrients existing in STP effluent without any additional nutrients from external source. Maximum growth rate calculated from this data is about 0.5 unit of water hyacinth per square meter of wetland per day.

The number of plant did not show any significant increase afterwards due to saturation in the 5 x 10 m^2 wetland pond. Visual observation showed that some plants became submerged and withered.

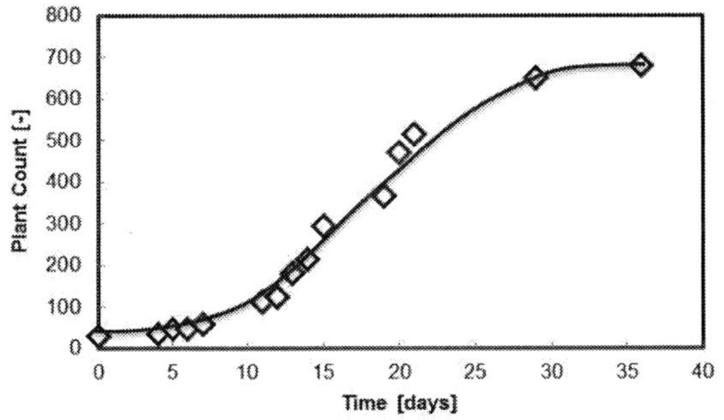

Fig. 2. Plant count of water hyacinth in the wetland

3.2 pH, DO, TSS and turbidity measurements

The results of pH, DO, TSS and turbidity measurements are shown in Figure 3-6. The water quality data during plant growth period (0-29 days) showed some fluctuation (data not shown) due to a few internal and external factors such as growth of other type of plants (lemna and pistia) and disturbances to water in pond during observations. The results shown here is those from steady-state period after the number of water hyacinth reached saturation (>29 days) and the pond is fully covered only by water hyacinths.

The dashed lines in Figure 3-5 indicate the regulation standard limits for class I raw water resource with pH value between 6 and 9, DO value of 6 mg/L or higher and TSS value of 50 mg/L or higher.

pH value was maintained near neutral throughout the experiment (Figure 3). Although a spike of pH value (data not shown) reaching nearly 10 was observed during plant growth period (0-29 days), the value decreased and stabilized quickly after the plants' number has reached saturation. Further study is necessary to confirm and reveal the mechanism of pH increase during plant growth period.

DO value of wetland influent showed some fluctuation that could be caused by the fluctuation in the STP at Batamindo which is utilizing two SBR units for the treatment of domestic wastewater (Figure 4 and 5). In fact, the full water quality analyses carried out periodically for both SBR units at the same time showed quite different values for those parameters (Table 1). Despite the fluctuation, the wetland can improve and stabilize the DO concentration from one as low as 2 mg/L up to around 5-6 mg/L. During the treatment, organic compounds contributing to BOD and COD values were absorbed and/or degraded, thus reducing the oxygen demand in the water and increasing the DO concentration. A spike of DO concentration was also observed (data not shown) during plant growth period (0-29 days), although the mechanism is yet to be studied. The DO value is actually balanced between several processes in the water that includes photosynthesis processes and oxygen transfer from atmosphere that increase DO, and aerobic metabolisms of microorganisms and reaction of oxygen with reduced compounds such as sulfides that decrease DO. During the experiment, the DO value in the effluent could stabilize at around 6, the minimum value required to meet the raw water quality standards for potable use. With simple aeration method such as cascade aerator, the DO value can be improved further up to nearly the saturation value of around 8 mg/L at 25°C.

Despite the fluctuation, TSS value and turbidity were also able to be lowered and maintained at a minimum level (Figure 5 and 6), thus improving the clarity of the wetland effluent. The contributor of TSS and turbidity in the water was mainly the solid biomass sludge residue after settling and clarifying processes in SBR. Given enough treatment time inside the wetland, the biodegradable solid residue from SBR can settle down to the bottom or be dissolved and utilized in metabolism processes. Although turbidity is not included in the water quality standards, it may affect the effectiveness of some post-treatment processes of the wetland treated water such as disinfection process. The turbidity is also one of the aesthetic parameters in water quality.

3.3 Effect of hydraulic retention time

The flow rate of influent at Pond 2 was adjusted from 10 to 20 m^3/day and then to 40 m^3/day on day 29th and 56th, respectively. This corresponds to decreasing HRT from 2 days to 0.5 day. The decrease of HRT (or the increase of influent flow rate) did not reduce the water quality in terms of pH, DO, TSS and turbidity. With a lower HRT (or a higher flow rate), the treatment capacity of the wetland is increased, meaning that smaller wetland area is necessary to treat certain amount of wastewater, or in other words, larger amount of wastewater can be treated using a certain area of wetland.

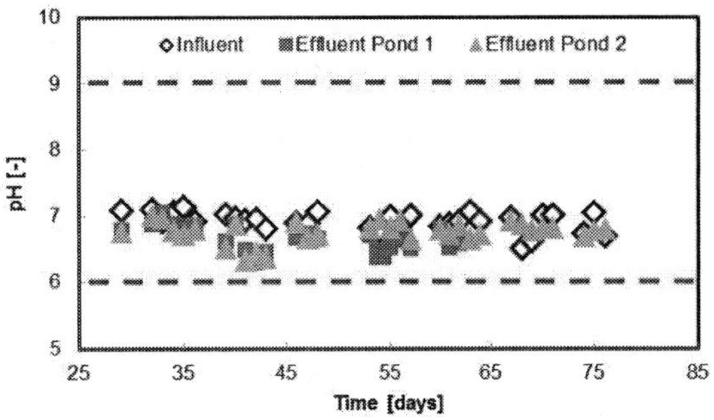

Fig. 3. pH value of influent and effluent in wetland

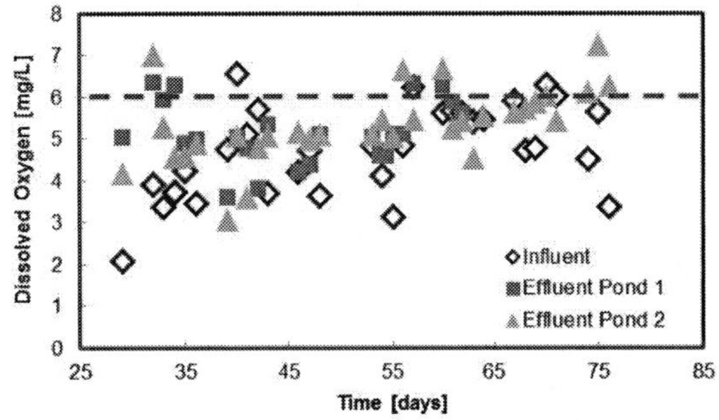

Fig. 4. DO value of influent and effluent in wetland

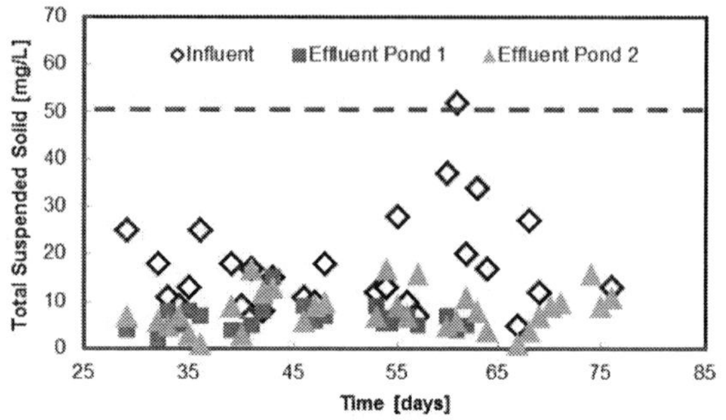

Fig. 5. TSS value of influent and effluent in wetland

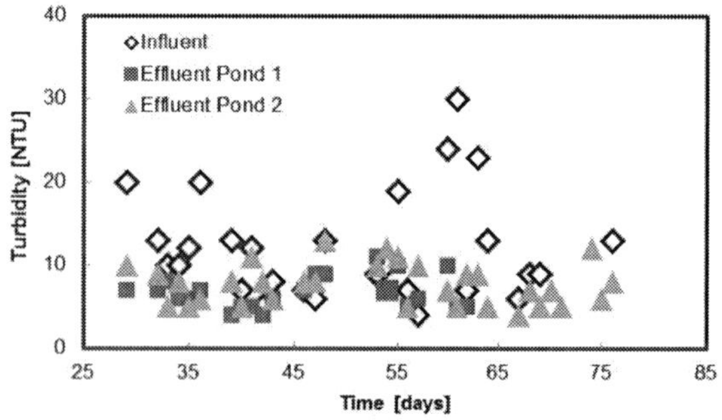

Fig. 6. Turbidity of influent and effluent in wetland

Table 1. Water Quality of SBR Effluents

Parameters	SBR 1	SBR 2
TSS [mg/L]	16	65
Turbidity [NTU]	3	25
BOD [mg/L]	48	27
COD [mg/L]	112	63
Ammonia [mg/L]	3.10	3.24
E. coli [MPN/100 ml]	2,000	2,000
Total Coliform [MPN/100 ml]	2,400	2,400

3.4 Comparison of effluent quality with water quality standards

Table 2 shows the comparison of wetland effluent quality and the standards in PP 82/2001 for class I raw water resource for potable use. Comparison of Table 1 and 2 shows that water quality of wetland effluent was improved considerably where BOD were lowered significantly from 48 mg/L to a range of 5-10 mg/L while COD were lowered from 112 mg/L to a range of 11-34 mg/L. The wetland could also suppress the number of pathogenic microorganisms as shown by the minimization of the number of *Escherichia coli* and Total Coliform.

There are, however, some parameters that still require improvements in order for the wetland effluent to reach Class I raw water resource standards, i.e., BOD, COD, ammonia and hydrogen sulfide (H_2S) concentrations. It is possible that BOD and COD values were fluctuating at higher levels than standards and could not be reduced further because of metabolisms of organisms inside the wetland that not only consume organic compounds contributing to BOD and COD values but also produce them as a part of the metabolisms.

Table 2. Water Quality of Wetland Effluent

Parameters	Effluent	Class I raw water resource standards
BOD [mg/L]	5-10	≤ 2
COD [mg/L]	11-34	≤ 10
Ammonia [mg/L]	0.65	≤ 0.5
H_2S [mg/L]	0.050	≤ 0.002
E. coli [MPN/100 ml]	3	≤ 100
Total Coliform [MPN/100 ml]	15	≤ 1000

Although the levels were very low, the existence of ammonia and hydrogen sulfide in water may indicate the possibility of an anaerobic condition occurring inside the wetland. Since the whole pond surface was fully covered by water hyacinth and no aeration system installed in the wetland, it is possible that anaerobic condition prevails at bottom layer of the sediment. The activity of microorganisms at this condition will certainly produce reduced compounds such as ammonia, hydrogen sulfide and methane.

These parameters may be improved further using post-treatment methods such as simple aeration to remove excess ammonia, H_2S, and other volatile organic compounds contributing to BOD, COD levels. Photolysis using ultraviolet irradiation or photocatalysis such as UV/TiO_2 system [12] may also be an option to improve these parameters.

3.5 Operational conditions and maintenance

The optimum operational conditions of the wetland in this study are yet to be investigated further. For example, DO levels during the fast growing period between day 10 and 30 (Figure 1) showed high values that were close to saturation (around 8 mg/L at 25°C). This could be caused by wider exposure of water surface to the surrounding air when the surface was not yet fully covered by water hyacinth and also by photosynthesis activities happening beneath the surface, both providing more oxygen into the water. Usual practice of surface-flow constructed wetland will leave a part of the surface exposed to air and sunlight, although the portion is yet to be determined according to some factors such as vegetation type, optimum plant count per unit area (planting density), water depth, etc.

By controlling the plant density, it is also possible to maintain an optimum consumption rate of organic compounds that will reflect on higher removal of BOD, COD, ammonia, nitrites, nitrates and other contaminants that serve as nutrients for the plant and other aquatic organisms. On the other hand, constructed wetland using water hyacinth or similar plant that grow and multiply very fast will certainly face a problem in the disposal of the by-product, i.e., the excess number of plants. For example, simple calculation using the maximum plant growth rate of around 0.5 unit per square meter per day shows that the treatment of 1000 m^2/day of STP effluent using 1000 m^2 of wetland (HRT=0.5 day) would produce 500 units of water hyacinth daily. Therefore, it is also necessary to find a suitable way to utilize this by-product. Some studies related to utilization of water hyacinth proposed methods such as biogas production [13], fiber extraction [14] and animal fodder [15].

4 Conclusions

The evaluation of surface-flow wetland treatment performance on sewage treatment plant effluent at Batamindo Industrial Park using water hyacinth has been carried out and the ability of this wetland treatment method to improve water quality of STP effluent was confirmed. Post-treatment using simple aeration and ultraviolet irradiation may be necessary to improve further the water quality of wetland effluent so that it can fulfill the requirement standards of PP 82/2001/Class I raw water resource for potable use. Excess water hyacinth in this wetland treatment can be a benefit if suitable industry can be implemented for its utilization, whether as energy or material resources.

The next phase of this study will be focusing on treatment of effluent from industrial wastewater treatment plant at Batamindo Industrial Park. Due to its characteristics, the WWTP effluent may prove to be more challenging than the sewage treatment plant effluent discussed in this paper, in which case a more efficient wetland treatment method will be necessary to achieve the expected result.

This pilot study of surface-flow wetland for water reclamation at Batamindo Industrial Park was fully supported by PT. Batamindo Investment Cakrawala as initial phase towards full-scale implementation on water reclamation of domestic and industrial wastewater treatment plant effluents.

References

1. J. Vymazal, Water, **2**, 530-549 (2010)
2. U.S. EPA, EPA/625/1-88/022 (1988)
3. U.S. EPA, EPA 843B00005 (1995)
4. U.S. EPA, *Manual*, EPA/625/R-99/010 (2000)
5. UN-HABITAT, *Constructed Wetlands Manual* (2008)
6. Melbourne Water, *Constructed Wetlands Guidelines* (2010)
7. City of Saskatoon, Wetland Design Guidelines (2014)
8. S. Glass, J. Thullen, J. Sartoris, R. Roline, General Technical Report RM-GTR-272, 243-252 (1995)
9. K. Rothfeld, M. Sees, *Orlando Easterly Wetlands Annual Report 2013* (2013)
10. Regulation of Indonesian Government PP 82/2001
11. Regulation of Indonesian Ministry of Environment PerMenLH 5/2014
12. M.N. Chong, B. Jin, C.W.K. Chow, C. Saint, Water Res, **44**, 2997-3027 (2010)
13. P. Njogu, R. Kinyua, P. Muthoni, Y. Nemoto, EPE, **7**, 209-216 (2015)
14. N.F. Ramirez, Y.S. Hernandez, J.C. de Leon, S.R.Vasquez Garcia, L.D. Lvova, L.G. Gonzalez, Fibers Polym, **16**, 196-200 (2015)
15. C. Polprasert, N. Kongsricharoem, W. Kanjanaprapin, Waste Manag. Res., **12**, 3-11 (1994)

Lessons learned in developing a green environment at the Engineering Faculty, University of Indonesia

Gabriel Andari Kristanto[1,*], *Cindy* Priadi[1], *Nyoman* Suwartha[1], *Erly* Bahsan[1], and *Arief* Udhiarto[2]

[1]Civil Engineering Department Faculty of Engineering Universitas Indonesia, 16424 Depok, Indonesia
[2]Electrical Engineering Faculty of Engineering Universitas Indonesia, 16424 Depok, Indonesia

Abstract. Today, institutions of higher education play increasingly active roles in helping achieve a sustainable society. This study aims to demonstrate the lessons learned in the development of a green environment at the Faculty of Engineering, University of Indonesia. This paper reviews the development of the necessary supporting infrastructure and discusses opportunities and challenges in establishing a green environment. The availability and condition of groundwater sources are identified, and the process by which canteen wastewater is treated by phytoremediation is described. Additionally, composting and recycling of solid wastes are encouraged in communities, and a model for an energy-efficient classroom is discussed. Three lessons learned are discussed: a focus on defining the program, continuous education and monitoring, and sources of funding. The implications of this study for relevant practices are also identified.

1 Introduction

Worldwide economic growth has caused the amount and volume of waste generated and energy consumed to increase exponentially. As a result, developing a sustainable society is a major challenge for many countries as they strive to balance green issues with economic development [1]. Increasing demands for energy and clean water coupled with a limited amount of resources mean that every human activity has a resulting impact on the environment. The term "sustainable development" was first defined in the Brundtland Report in 1987 as development that satisfies the needs of the present without diminishing the abilities and choices of future generations [2]. Although a general environmental awareness is more prominent among developed countries, many universities from emerging countries with visionary academic leaders are expanding their roles to solve global and environmental challenges [3, 4]. In Indonesia, several universities are starting to play increasingly active roles in helping the country achieve sustainability, as reflected in the increasing number of participating universities in Indonesia in the Green Metric World University Ranking (from 22 in 2010 to 25 in 2011). This ranking system was developed by the University of Indonesia (hereinafter called UI) in 2010; it applies specifically to sustainable and green campuses according to applicable conditions. The evaluations are based on qualitative and quantitative approaches and employ the Berlin Principles [5].

In line with the Research Agenda for Higher Education in Indonesia and Millennium Development Goals (MDG), the Faculty of Engineering at the University of Indonesia (hereinafter, FTUI) developed a research theme for the period 2010–2014 entitled "Integrated Design in Urban Eco-technology for Quality of Life and Humanity". Consistent with this theme, FTUI aims to become a role model for building a community with a strong orientation toward a green and sustainable environment. Hence, an integrated program was launched on the FTUI campus with the theme of "FTUI Toward a Green and Sustainable Society." The program has several sub-programs consisting of solid waste, water and wastewater management, and energy savings. Research on waste management focuses on solid waste management to encourage the "3Rs"—reduce, reuse, and recycle—and research on water management focuses on protecting a sustainable water supply in the neighborhood by the FTUI campus. Through this program, a classroom was developed as a model for energy efficiency. Subsequently, all research and results have been promoted throughout the FTUI community to increase awareness and encourage participation in developing a green and sustainable society.

2 Framework

The University of Indonesia (UI) with 14 faculties and numerous research centers and laboratories, is located in the city of Depok, West Java, Indonesia. Detailed descriptions of FTUI are presented in Table 1. As one of the largest faculties, FTUI is challenged to become one of the best engineering faculties in Indonesia; additionally, it aims for high quality in campus environmental conditions. The increasing demands for

* Corresponding author: gakristanto@gmail.com

 DOI: 10.1051/matecconf/201710104008

water and energy, the generation of a large amount of solid wastes, and the low quality of wastewater are some of the issues that must be resolved.

Table 1. Faculty of engineering, University of Indonesia (FTUI) key statistics in 2014.

Description	Number	Unit
Total area	10	Hectares
No. of departments	7	
No. of students	5,300	People
No. of academic lecturers	350	People
No. of admin. personnel	125	People

To improve the quality of the environment, it is important to design a program that reflects consideration for the specific needs of the individual area. Each university environment is unique in terms of water and wastewater quality and quantity, solid waste composition, and energy consumption; therefore, defining a suitable program is important, especially when funding is limited. The availability of superior human resources, a reliable baseline environmental condition, and a strong commitment from campus management are some of the key factors that need to be considered when such a program is initiated. Additionally, setting up priorities and appropriating them into several sub-programs will enable achievement of the main goals.

The two-year development of the sustainable society program at FTUI is summarized in Figure 1, followed by a detailed description of each related activity. In the first year, the program focused on developing the system and infrastructure to support the development of a sustainable society. In the second year, emphasis shifted to increasing awareness and encouraging stakeholder participation.

2.1 Water supply

To maintain the supply of clean water at FTUI, two activities were developed: employing an artificial recharge system and monitoring water quality and groundwater levels throughout the year. The artificial recharge of groundwater is achieved by capturing rainwater in a shallow well so that the water infiltrates the soil and moves downward to recharge aquifers. The benefits of artificial recharges are well-documented and include reducing seawater intrusion or land subsidence, storing water, improving the quality of water through soil-aquifer treatment or geo purification, using groundwater in a water conveyance system, making groundwater out of surface water (for drinking) [6, 7, 8], and helping to reduce flooding. During this program, along with the development of artificial recharge wells, five piezometers were installed to monitor the level of groundwater.

The location of the three deep wells (SP01 to SP03), 27 artificial recharge wells, and piezometers wells (DB01 to DB05) on the FTUI campus are shown in Figure 2.

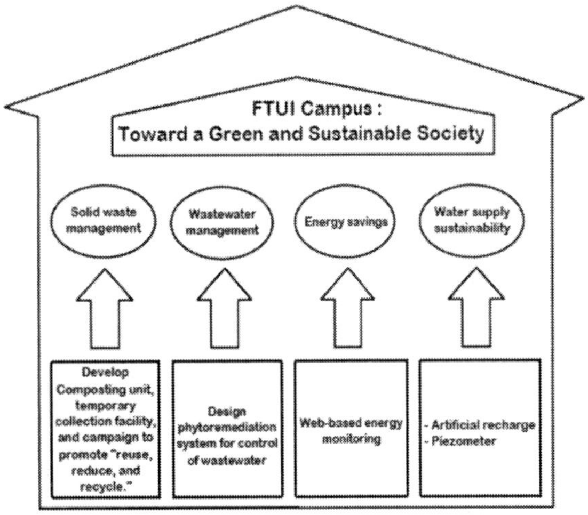

Fig. 1. Framework of FTUI green campus.

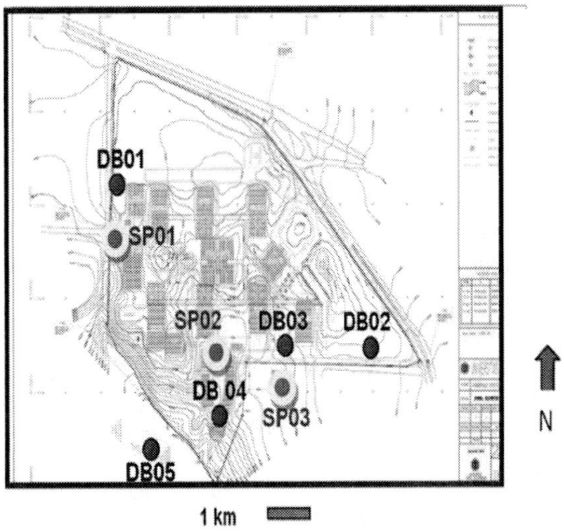

Fig. 2. Location of deep wells (red dots) and piezometric wells (blue dots) around the Engineering Faculty, University of Indonesia (FTUI) campus.

Twenty-seven of the shallow artificial recharge wells were designed by directing the rainwater into 0.8×0.8×1.1 m3 wells located in drainage intersections; hence, water penetrated directly into the aquifer. The groundwater quality was monitored by periodically sampling it from three deep wells near the artificial recharge wells. Monitoring groundwater level and water quality to maintain the water supply throughout the year, monitoring of groundwater levels and water quality was initiated through this program. Table 2 shows the laboratory test results for groundwater quality from each sampling point.

Table 2. Water quality monitored at FTUI.

Parameters	Quality standard *	Quality standard **	SP01	SP02	SP03
Color, Pt-Co	15	-	127	4	4
Chloride (Cl), mg/L	250	-	5.3	7.7	11.7
Turbidity, NTU	5	-	39.2	0.36	1.12
TDS***, mg/L	500	1,000	88.3	147	258
pH	6.5–8.5	6–9	5.32	5.8	6.4
Iron (Fe), mg/L	0.3	0.3	1.11	0.04	0.23
Manganese (Mn), mg/L	0.4	0.1	0.3	0.1	0.1

*[9]
**[10]
****Potentiometry method*

Most of the tested water parameters were within safe limits (i.e., below the maximum level/threshold quality standards issued by both the Indonesian Health Ministry [9] and the Indonesian Government [10]. The results may have been potentially affected by the proximity of sampling locations to the artificial recharge wells established around the main Faculty building. In addition, the results may have been influenced by the dilution effect of excessive rainfall during the sampling period (May–June). On the other hand, the pH showed that the groundwater had a relatively high acidity, especially at SP01 (i.e., about 5.3). This result is in line with the previous results, probably because of the amount of CO_2 in the soil that significantly affects acidity of the groundwater [11, 12]. Contact between CO_2 and water will form H_2CO^3, which causes the groundwater to become acidic. Meanwhile, the results of laboratory tests on water samples at SP02 and SP03 showed a similar quality issue in turbidity and Fe. In both samples, turbidity and Fe content showed fairly low values; however, the influence of drilling a new deep well (SP01) may have caused the fine materials and clay fragments to remain suspended in the water samples. In Figure 3, it shown the existing condition of groundwater levels recorded at each *piezometric* well from May 2013 to February 2014.

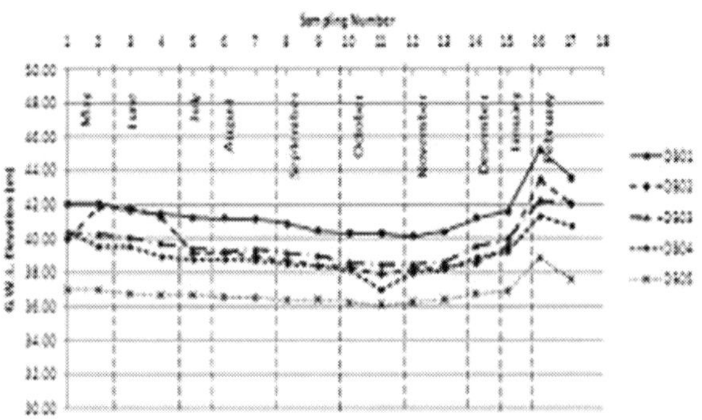

Fig. 3. Fluctuation of groundwater level at each piezometric well

During this program, several sustainable building concepts were analyzed. The main challenge for effective application of the concept on the FTUI campus was the fact that existing buildings are relatively old, having been built more than 15 years ago with limited implementation of energy-efficient designs. During the first year of this program, the energy-saving policy at FTUI was carried out in stages. The first stage included developing a policy cycle, meaning that all policies needed to continuously adhere toward achieving the overarching objectives. Policies included the UI's Rector Policy on Climate Change Mitigation at the University Campus, UI's Rector Policy on Energy Saving, and the Ministry of Energy's Decree on Electricity Saving. The second stage was to establish energy efficiency indicators. Baseline indicators, such as energy intensity for every room (including classrooms and administrative offices) and energy intensity for every student were applied and used to evaluate energy savings at year's end. To set up data for the baseline indicator, real-time electricity consumption was measured in one of the classrooms as a prototype of the smart energy building at FTUI. Data was gathered and converted to IP data and sent to the Green Campus server by LAN or WLAN. Using the proxy server in the Electrical Engineering Department (FTUI), data were made accessible via smartphones, tablets, or computers. Data were beneficial for analyzing electricity usage in real time and tracking energy and cost movements with each and every flip of the switch [7]. Such information can help faculty stakeholders understand which activities consume the most energy so that their consumption patterns, behaviors, and composition of appliances can be amended to reduce their electricity bills and carbon footprints.

2.2 Energy savings

Energy is fundamental to human life. As the engine of the economy, it has become an integral part of society. For example, it is used to power cities, provide fuel for various types of vehicles, and set home and indoor temperatures. In other words, everything people do is connected to energy in one way or another. Since the Industrial Revolution, fossil fuels have been the main source for supporting the energy needs of mankind. Unfortunately, fossil fuels are non-renewable and emit different kinds of pollutants to the atmosphere. Therefore, there is a dire need to find options to improve energy efficiency in multiple sectors, including reducing the energy demands of buildings.

2.2 Wastewater treatment

Currently, centralized wastewater treatment in Indonesia is very limited. Only 1% of domestic wastewater generated by the urban population is collected by municipalities and conveyed to a wastewater treatment plant. Most domestic wastewater is treated on-site (in home environments) using septic tanks (64%). Thus, wastewater treatment is often regarded as a personal responsibility, as is the case for academic institutions.

Although most of the black water at FTUI is treated using septic tanks, gray water is usually untreated and conveyed to the nearest body of water through the drainage system. This issue needs to be addressed based on the increasing quantities of wastewater and pollutants. During this project, it was identified that around 17% of the wastewater produced by FTUI originated from the faculty canteen. Although the amount is relatively small compared to other sources, its impact becomes more significant when wastewater quality is taken into account (Table 3). All parameters revealed that the effluent limit was exceeded by 2–10 times. The concern is not only exceeding the limit of waste but also its discharge, which pollutes the artificial recharge lakes around the UI campus.

The ability to treat wastewater high in organics has been studied extensively during the last decades [15, 16, 17, 18]. Several engineering designs were analyzed to determine a viable technology to treat canteen wastewater with a design flow rate of 1 L/s [19, 20]. The combination of biological and phytoremediation is considered to be the most economical method; it requires low maintenance and zero energy. The plants patch in the phytoremediation process serves as a demonstration of landscape-friendly wastewater treatment. Prior to implementing the biological and phytoremediation processes, individual grease traps were installed in each sink. Afterwards, water passed through a bar screen for transport downstream to a 6 m3 anaerobic tank using an attached growth biological system. This treatment was followed by phytoremediation using Vetiveria zizaniodes—known as vetiver—with tall, tufted, perennial, scented grasses; long narrow leaves; and an abundant network of roots. It grows easily in the FTUI landscape.

2.3 Solid waste management

Developing solid waste management (SWM) programs is a key to achieving institutional sustainability [21, 22]. Effective SWM requires a complete understanding of the composition of waste streams as well as the activities that determine their generation in the first place [23, 24]. During the first year of this program, FTUI solid waste generation was characterized; further, a recycling system and compost houses were designed [25].

Characterization of FTUI waste was the first essential step toward greening the campus. This analysis helped decision makers at FTUI identify relevant opportunities in reducing and recycling wastes. The nature of the wastes indicated that among the recovery options, composting would be most appropriate as it would reduce more than 30% of the waste volume from the waste stream when providing compost for the faculty garden. The program was designed to use garden and food wastes as compost feedstock. During the first year, garden wastes were composted by the open windrow system; during the second year, a combination of food and garden wastes were treated in the in-vessel composting system.

Table 3. Effluent of the Engineering Faculty Canteen.

Parameters	Unit	Canteen wastewater quality		Wastewater regulation	
		2013	2012	Quality standard*	Quality standard**
pH		4.2	4.5	6–9	6–9
TSS	mg/L	286	498	100	50
Phosphorus	mg/L	194	-	-	-
COD	mg/L	2016	488	-	80
BOD	mg/L	159	185	100	50
Oil and grease	mg/L	-	115	10	20

*[13]
**[14]

3 Lessons learned

Based on literature reviews and findings for developing a green environment, several key lessons are highlighted as follows; Lesson No. 1, Focus on designing a program, development of a green environment can be accomplished through several initiatives, such as providing the following: facilities and clean water systems, wastewater treatment, waste management, and clean energy. Each initiative has considerable options in technology and program selection; therefore, one should assess the potential applications of technology and programs based on characteristics of each initiative. Selection criteria should be prioritized with care. Technology applications—from the simple to the advanced—have pros and cons; thus, a "cost-benefit analysis based on need" must be performed and assessed thoroughly. The availability of reliable suppliers and resources for the various technologies is essential for supporting the continuity of the initiatives.

Lesson No. 2, Ensure continuous education and monitoring, involving faculty stakeholders in a community-based program is a key to the success of building a sustainable society [26]. Students, lecturers, administrative staffs, and cleaning service workers participated in this program in planning, collecting and analyzing data, designing, and reporting results. Their active participation was immediately beneficial for stakeholders in terms of establishing practices for growing and maintaining a green and sustainable society at the university. Several strategies were developed to increase stakeholder participation in this program. Increasing awareness and participation of stakeholders by providing sufficient information and education was prioritized to stimulate the adoption of a green campus program.

Incentives and rebates were created for participants to recycle according to each group's interests. However, monetary rewards could only be utilized to initiate new recyclers until they were able to sustain the habit of recycling [27]. In the future, the performance levels of recycling programs and participation need to be assessed [28, 29]. Assessments will be beneficial for addressing certain factors that influence the recycling program and achieving more sustainable approaches to SWM.

Lesson No. 3, identify sources of funding, most of the funding available for universities in Indonesia is research-oriented, whereas funding for community awareness is limited in amount and availability. It is important to encourage each institution to establish a vision and mission on green infrastructures and allocate funding for the development of a green environment. The infrastructure serves as a start-up, but the operation, maintenance, and further development of a green environment must be consolidated to involve both the stakeholders and program team. For example, the polluter pays, meaning that the cost of pollution is borne by those who cause it. Essentially, the cost is the integration of water, wastewater, energy, and solid waste treatment. Some of the inherent expenses are offset by research grants and meal prices in the canteen. Ultimately, this system could be self-funded, especially as awareness among all stakeholders increases.

Implications of the findings for actual practice. The findings of this study demonstrate that defined programs, continuing education and monitoring, and sources of funding were critical factors in developing a green environment at FTUI. Two main limitations of this study were identified. First, it was based on conditions at UI; although some of the conditions related to other universities, care should be taken in making decisions regarding waste treatment and energy consumption in contexts outside UI. Second, regarding the lesson learned from actual practices in developing a green environment, improvements will not be realized without addressing other organizational issues such as culture, time, and sense of value. It is very important to undertake a comprehensive assessment before making decisions and understand the reasoning behind the methods and approaches used.

4 Conclusion and recommendations

To enhance the role of the university as an agent of change in Indonesia, FTUI aimed to be a pioneer in managing its campus environment by focusing on water, wastewater, SWM, and energy savings. Implementing the program resulted in building rainwater harvesting and artificial recharge systems, installing canteen wastewater treatment units, providing for SWM, designing energy-smart buildings, and involving stakeholders in every related activity. Support from all academic members and stakeholders through the effective dissemination of information and continuous education were keys for developing a successful and green environment at FTUI.

The authors would like to express deep gratitude to the Indonesia Ministry of Higher Education for supporting this research with funding under BOPTN 2013 Contract Number 1613/h2.PPK/PPM.00.01/2013.

References

1. K. Zaman, A.B. Abdullah, A. Khan, M.R.D. Nasir, T.A.A.T. Hamzah, S. Hussain, Dynamic Linkages among Energy Consumption, Environment, Health and Wealth in BRICS Countries: Green Growth Key to Sustainable Development, *Renewable and Sustainable Energy Reviews*, **56**, 1263-1271 (2016)
2. World Commission on Environment and Development: Our Common Future. Available at: http://www.un-documents.net/our-common-future.pdf, accessed July 5th (2015)
3. D.N. Norfadillah Derahin, H.S. Hashim, N. Ali, S.A. Abdul, G. Aziz, UKM's Staff Perspective on Sustainability and Its Contribution Towards a Sustainable University, *Procedia—Social and Behavioral Sciences*, **59**, 376-381 (2012)
4. Y. Wang, H. Shi, M. Sun, D. Huisingh, L. Hansson, R. Wang, Moving Towards an Ecologically Sound Society? Starting from Green Universities and Environmental Higher Education. *Journal of Cleaner Production*, **61**, 1-5 (2014)
5. N. Suwartha, R.F. Sari, Evaluating UI Greenmetric as a Tool to Support Green Universities Development: Assessment of the Year 2011 Ranking. *Journal of Cleaner Production*, **61**, 46-53 (2013)
6. S. Patel, Water Management Conservation, Harvesting, and Artificial Recharge, New Age International, New Delhi (2008)
7. R.B. Thomas, M.J. Kirisits, D.J. Iye, K.A. Kinney, Rainwater Harvesting in the United States: A Survey of Common System Practices. *Journal of Cleaner Production*, **75**, 166-173 (2014)
8. S.-H. Moon, J.-Y. Lee, B.-J. Lee, K.-H. Park, Y.-J. Jo, Quality of Harvested Rainwater in Artificial Recharge Site on Jeju Volcanic Island, *Korea. Journal of Hydrology*, **414–415**, 268-277 (2012)
9. Decree of the Indonesia State Ministry of Health Ministry No. 492/MENKES/Per/IV/2010 on drinking water quality. Available at: http://www.litbang.depkes.go.id/sites/download/regulasi/permenkes/Permenkes_492-th-2010.pdf, accessed February 20th (2014)
10. Government of Indonesian Law No. 82/2001 concerning management of water quality and control over water pollution. Available at: http://www.minerba.esdm.go.id/library/sijh/PP8201_KualitasAir.pdf, accessed February 20th (2014)
11. M. Gusnisar, The effect of recharge wells on the quality of groundwater in the Faculty of Engineering. Final project for bachelor's degree, Environmental Engineering Department, Faculty of Engineering, University of Indonesia, Jakarta (not published)
12. P. Freeze, J. Cherry, Groundwater, Prentice Hall, New Jersey (1979)
13. Decree of the Indonesia State Ministry of the Environment No. 112/2003 on effluent quality standards. Available at: http://www.iec.co.id/pdf/Kepmen%20LH%20No.%

2011203%20%28Baku%20Mutu%20Air%20Limbah%20Domestik%29.pdf, accessed June 3t[h] (2014)
14. Jakarta Governor Regulation No. 112/2005 on domestic wastewater management. Available at: http://bplhd.jakarta.go.id/peraturan/pergub/pergub_122_2005.pdf, accessed June 3t[h] (2014)
15. S.S. Andra, R. Datta, D. Sarkar, S.K.M. Saminathan, C.P. Mullens, S.B.H. Bach, Analysis of Phytochelatin Complexes in the Lead Tolerant Vetiver Grass [Vetiveria zizanioides (L.)] Using Liquid Chromatography and Mass Spectrometry. *Environmental Pollution*, **157**(7), 2173-2183 (2009)
16. Y. Chen, Z. Shen, X. Li, of Soils Contaminated with Heavy Metals. *Applied Geochemistry*, **19**(10), 1553-1565 (2004)
17. M. Hammer, Water and Wastewater Technology, John Wiley and Sons, Singapore (1986)
18. J. Vymazal, Horizontal Sub-surface Flow and Hybrid Constructed Wetlands Systems for Wastewater Treatment. *Ecological Engineering*, **25**(5), 478-490 (2005)
19. G. Moshiri, Constructed Wetlands for Water Quality Improvement. Lewis Publishers, London (1993)
20. Z. Li, H. Xiao, S. Cheng, L. Zhang, X. Xie, Z. Wu, A Comparison on the Phytoremediation Ability of Triazophos by Different Macrophytes. *Journal of Environmental Sciences,* **26**(2), 315-322 (2014)
21. Y. Geng, K. Liu, B. Xue, T. Fujita, Creating a "Green University" in China: A Case of Shenyang University. *Journal of Cleaner Production*, **61**, 13-19 (2013)
22. K. Foo, A Vision on the Role of Environmental Higher Education Contributing to the Sustainable Development in Malaysia. *Journal of Cleaner Production*, **61**, 6-12 (2013)
23. C.A. de Vega, S.O. Benitez, E.R. Bareto, Solid Waste Characterization and Recycling Potential for a University Campus. *Waste Management,* **28**(Supplement 1), S21-S26 (2008)
24. G. Tschobanoglous, H. Thiessen, S. Vigil, Integrated Solid Waste Management: Engineering Principles and Management Issues, McGraw-Hill Inc., Singapore (1993)
25. G.A. Kristanto, C.K. Banaget, I.G. Sofyan, A Baseline Study of Solid Waste Characterization and Recycling Potential at University of Indonesia, Proceedings of the Conference for Civil Engineering Research Networks, 62-67 (2014)
26. J.A. Fadamiro, A. Bobadoye, Managing the Building Design Process for Sustainabilty and Improved Quality. *Civil Engineering Dimension*, **8**(1), 1-7 (2006)
27. Y.C. Moh, L.A. Manaf, Overview of Household Solid Waste Recycling Policy Status and Resources, *Conservation and Recycling*, **82**, 50-61 (2014)
28. S. Suttibak, V. Nitivattananon, Assessment of Factors Influencing the Performance of Solid Waste Recycling Programs, *Resource, Conservation, and Recycling*, **53**, 45-56 (2008)
29. A. Disterheft, S.F. Caeiro, R.M. Ramos, U.M. Azeiteiro, Environmental Management Systems (EMS) Implementation Processes and Practices in European Higher Education Institutions—Top-down Versus Participatory Approaches, *Journal of* The Use of Vetiver Grass (Vetiveria zizanioides) in the Phytoremediation *Cleaner Production*, **31**, 80-90(2012)

Allocation of potential value socio-ecology post tin mining Kolong in Bangka Island

Wahyudi Himawan[1,2,*], *Fachrurrojie* Sjarkowie[3], *Indra* Yustian[4], and *Ardiyan* Saptawan[5]

[1] Doctor Programs of Environmental Study Postgraduate University of Sriwijaya, 30139 Palembang, Indonesia,
[2] Education and Training Department of Bangka Belitung Province, Indonesia
[3] Deptartment of Agribusiness Faculty of Agriculture, University of Sriwijaya, 30662 Inderalaya, Indonesia
[4] Deptartment of Biology Faculty of Mathematics and Natural Sciences, University of Sriwijaya, 30662 Inderalaya, Indonesia
[5] Deptartment of Public Administration Faculty of Social & Politics, University of Sriwijaya, 30662 Inderalaya, Indonesia

Abstract. Kolong is a water basin as an impact of post tin mining in Bangka Island. Kolong should be managed properly to minimize some impacts, improve the quality of socio-ecological, and the income of local citizen. The purpose of this study was to analyze the potential allocation of socio-ecology value of the post tin mining. The case studies are Kolong in the Central Bangka Regency. This research was used resources valuation methodology, ABAD model classification, and observations. To determine the relationship of social ecology of the Kolong using SEM WarpPLS. The result shows that revenue of economic value of kolong in Bangka Island is IDR 17.9 Billion/year from utilization of Water Supply System, raw water and aquaculture. The diversity Kolong utilization as production tool of social needed in Bangka Island are 6 % as Water Supply System, 9 % as aquaculture, 0.2 % as tourism, and almost 84.8 % unused and owned by local government and private companies. The changing of kolong system influence the changing of social systems. The social variables that are most affected are the technology and local population. Utilization of Kolong as the economic value depends on its classification which based on the ABAD model.

1 Introduction

Exploration and exploitation activities of tin ore inBangka Island had big impact to physical change, spatial structure, social, and environment use. Its social impact should not be underestimated, the interaction between mining industry and the citizen could cause various of social problems, economic, culture, and public policy. The economic wellbeing enhancement is the hope of local citizen from mining industry in their area [1].

Exploration and exploitation tin ore in Bangka Island had left a post mining area which is flooded and called Kolong. Kolong's management not only related to environment aspects and economic, but also social, politics, and public policy [2]. The environment restoration that has been done through partial ecologic-technical approach gradually lost its legitimacy and interest [3,4]. Kolongs that was spread around Bangka have a big potential to be managed and used for citizen well-being and regional development. Regional government has tried to develop post mining area with various functions [5,6]. These functions are for tourist attraction, community economic production facilities, raw water source, new settlement area, etc to preserve and optimize economics value, and post mining area ecologists.

For planning and implementing those policies, it is important to know and valuate economics and socio-ecology values. Valuation of potential value is used for classification and allocating Kolong of post mining potential utilization. The value of economic potential allocation and socio-ecology are used to make a strategy of management policy and Kolong of post mining utilization. The focus of research on Central Bangka Regency is believed to represents the conditions and characteristics of kolong on Bangka Island. The purposes of this research are to do economic valuations of kolong's utilization, socio-ecology valuations of utilization and management of Kolong, classify Kolong of post mining on Bangka Island and counts economic potential allocation, social, and ecology of Kolong on Bangka Island.

The aims of this study:

1. Conducting economic valuation and socio-ecological utilization of Kolong;
2. Classify the post tin mining of Kolong in Bangka Island;
3. Calculating the allocation of the potential economic, social, and ecological of the Kolong on Bangka Island.

* Corresponding author: wahyudi.himawanbabel@gmail.com

2 Methodology

2.1 Economic Valuation

Valuation method was used based on Ministry of Environment Regulation Number 15 year 2012. The total economics value reflects monetary value which is produced by ecosystem function of Kolong, both extractive and non-extractive, calculated by:

$$TEV = \text{Use Value} + \text{Option Value} \quad (1)$$

$$TEV = [(DUV+IUV) + (OV+QOV+BV+EV)] \quad (2)$$

Where :

TEV : Total economic value
DUV : Direct use value
IUV : Indirect use value
OV : Option value
QOV : Quasi option value
BV : Bequest value
EV : Existence value

The productive effort of Kolong resource utilization was counted by unit rate calculation, market rate minus production cost minus proper profit, formula is given by:

$$UR = (MR - APC) - PPIR \quad (3)$$

where:

UR : Unit rent
MR : Market rate
APC : Average production cost
PPIR : Proper profit as investment retribution

2.2 Classification of Kolong

Classification of kolong was done by ABAD (Access, Broad, Age, and Depth) model as shown in Table 1 and Table 2.

Table 1. Classification criteria according to social economic perspective

No	Access (km)	Broad (ha)	Age (year)	Depth (m)
1	Good < 0.5	Wide 4>	Old 20>	Deep 5>
2	Fair 0.5-1	Fair 1-4	Fair 5-20	Fair 2-5
3	Bad >1	Narrow <1	Young <5	Superficial <2

Table 2. ABAD model classification

No	Classification base	Value		
		Low	Fair	High
1	Access	A_1	A_2	A_3
2	Broad	L_1	L_2	L_3
3	Age	U_1	U_2	U_3
4	Depth	K_1	K_2	K_3

where's 1: low, 2: fair, and 3: high

2.3 Socio-ecology allocation

Socio-ecology potential allocation methodology is summing all of benefits and function of Kolong either extractive or non-extractive. Socio-ecology analysis was done by using five fundamentals in ecology, i.e: (1) interaction, (2) interdependency, (3) diversity, (4) harmonization, and (5) sustainability [7,8,9]. To determine the variables in the social system both formative and reflective, use Structural Equation Modelling Analysis with SEM Warp PLS Program [10].

3 Results and discussion

3.1 Economic allocation

Based on the identification and inventory by doing a closed area map overlay, IUP (mining permit) map, administration map, PIPIB moratorium map, and Central Bangka local government year 2011-2013, where has 227.991 ha area, is known:

- Mining permit (IUP) = 91.971 ha or 40 % of the total area.
- IUP with expired permit up to 2015 = 2,.936.14 or 28% of the total area
- Area that becomes Kolong = 1,614,64 ha or 0.7% of the total area.
- Number of identified Kolong = 423
- Number of inventoried Kolong = 276 or 56 % of which are identified
- Assumed water volume = 18,128,301 m^3
- Potential allocation = IDR 17.9 Billion/year

Based on ABAD model classification, among 276 of inventoried kolong, there are 27 pcs for first class or 10%, 193 pcs for second class or 70% and 56 pcs for third class or 20%. As shown in the map **Fig.1** below:

3.2 Classification of kolong

Based on NEA(the economic value of water) Study, Optimum water allocation for PDAM (local water company) and BBI(fish breeding centers) Sungailiat, in such condition that the sums of water economic value of Kolong Dam-3 Pemali for each corresponding 1.36 Billion/Year [11]. Based on ABAD model classification, among 276 of inventoried Kolong at Central Bangka Regency, there are 27pcs for first class or 10%, 193 pcs for second class or 70% and 56 pcs for third class or 20%, after economic value was done, both of extractive or non-extractive. Regarding to this valuation, economic value of kolong on Bangka, calculated by the equation (1) and (2) was IDR 9.5 billion/year for Water Supply System utilization and fish cultivation, and IDR. 17.9 billion/year for the Kolong itself.

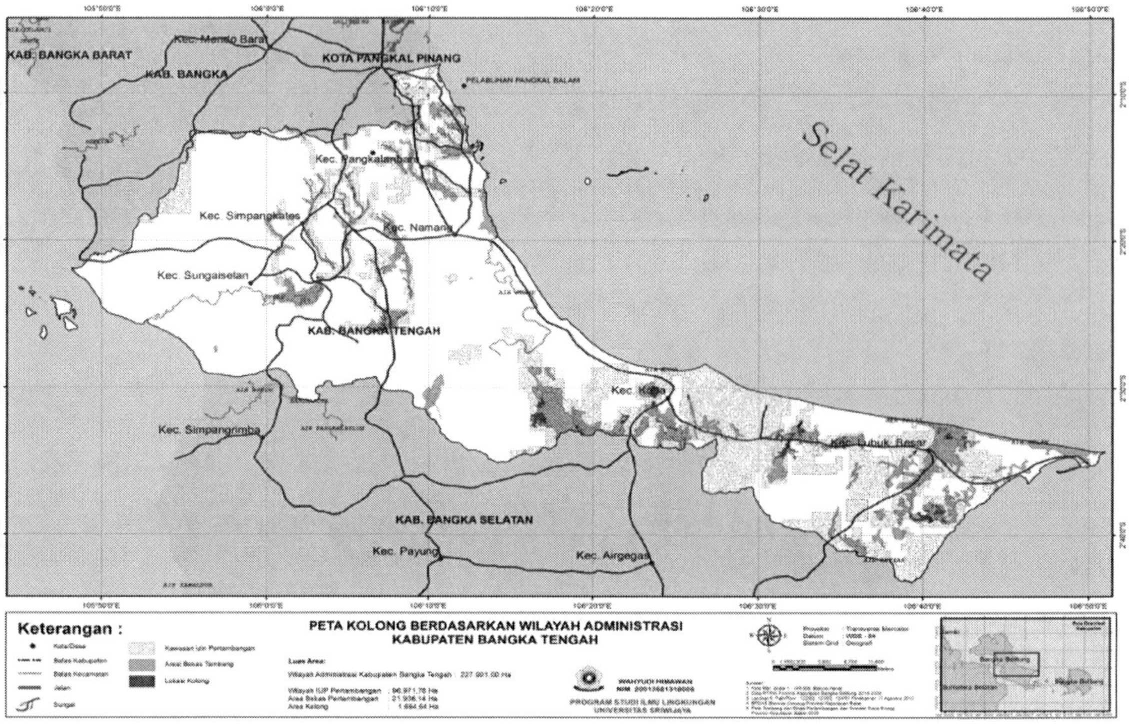

Fig.1. Map of Kolongs in Central Bangka Regency

3.3 Socio-ecology allocation

According to the perception of stakeholders, the utilization type of Kolong, the type of tourism utilization from of a tourist village and artificial edutourism [5]. In this research, according to the perception of stakeholders, the utilization types of Kolong were as Water supply system, raw water for citizen, local asset, community settlement, and water resource for agriculture and industry. The reflective analysis of SEM WarpPLS as shown in Figure 2.

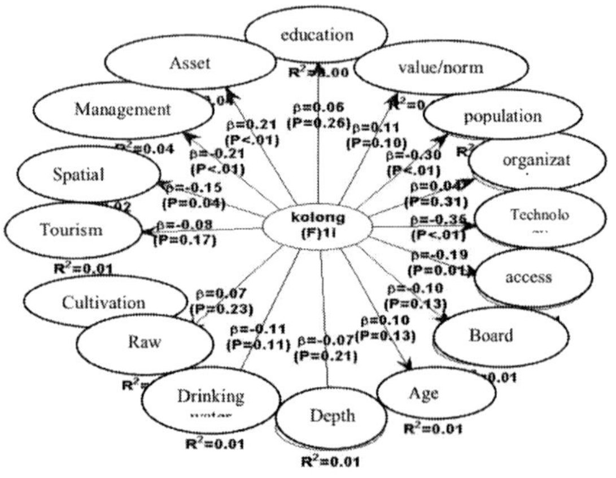

Fig. 2. Analysis kolong reflective contsruct

Table 3. Reflective analysis of kolong's socio-ecology

No	Component	Formative		Coefficient
		βr	P_r	*βr*^{2}
	Social			
1.	Education	0.060	0,20	0,49
2.	Value/Norm	0.110	0,49	0,36
3.	Population	0.300	0,37	0,09
4.	Organization	0.040	<0,01	9,00
5.	Technology	0.350	<0,01	12,25
	Influence	0.860		**88,36**
	Potential			
1.	Access	0.190	0.010	3.610
2.	Broad	0.100	0.120	1.000
3.	Age	0.100	0.130	1.000
4.	Depth	0.070	0.210	0.490
	Influence	0.460		21.160
	Economic use			
1.	Drinking Water	0.110	0.11	1.210
2.	Raw Water	0.070	0.23	0.490
3.	Cultivation	0.110	0.10	1.210
4.	Tourism	0.080	0.17	0.640
	Influence	0.370		13.690
	Policy			
1.	Spatial	0.150	0.04	0.006
2.	Management	0.210	0.01	0.002
3.	Asset	0.210	0.01	0.002
	Influence	0.570		32.490

DOI: 10.1051/matecconf/201710104009

3.3.1 Socio-ecologi fungtion

Based on the reflective analysis of SEM WarpPLS generated that Kolong system changes significantly influence 88.6% to changes the social system, as shown in Tabel 3. While based on the analysis of the construct of formative changes in the social system has no significant effect on the changes of kolong.

The social systems component which influenced by the kolong's changing can be observed in reflective analysis as shown in **Fig. 3** below.

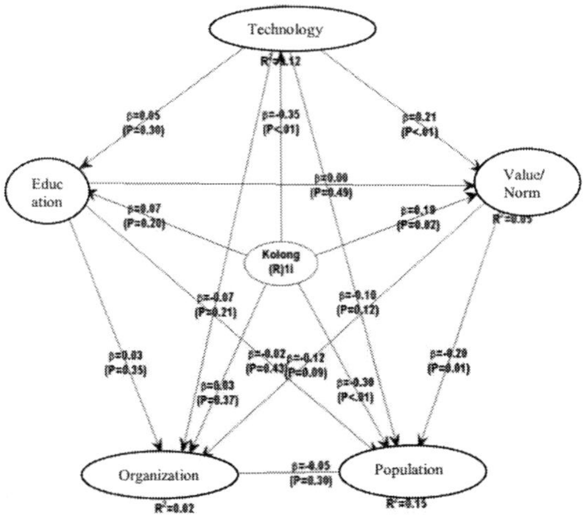

Fig. 3. Social component influenced convertion used of Kolong.

From the analysis above, the function kolong the assumption of the general equation is:

$$K = f(aS + bP + cK + dE) \quad (1)$$

From the reflective analysis result above, the equation (1) can be simplified as follows:

$$K_r = f(aS + bP) \quad (2)$$

Where:

Kr : Reflective kolong (Kolong effect on external valuable change)
a : Coefficient on social influent
b : Coefficient on potential influent
c : Coefficient on influent of policy
d : Coefficient on influent of economic use
S : Social variable
P : potential variable
K : Variable of policy
E : Variable of economic use

It can be concluded that changes in the Kolong for the management and utilization has very significant effect (88.36%) on the social variables. So that the model equations that apply are:

$$Kr = f(aS) \quad (3)$$

Based on the equation (4) above means that: technology and local population in communities are influenced by kolong utilization changes fungtion. Components of the social system variables reflective affected by the changes of kolong system based on the relationship rows are: technology, organizations.

3.3.2 Ecological principle analysis

Based on ecological principles, it can be explained as follows:

1. Interaction has occured between social system and environment system for utilization of Kolong. The social utilization system has a format that fit to environment system by using local knowledge and technology from citizen to use Kolong as an economic production tools.
2. Interdependency between social system and environment of Kolong has occured with an adaptation format of social system which fit to environment system.
3. Diversity of Kolong utilization as socio-ecology value were classified and used to Water Supply System and raw water for citizen, aquaculture and recreation. Class II of Kolong was used as water source for industry and agriculture. In that, Kolong's class III was left as it was because a Kolong can only be used for economic production activities after 5-10 years.
4. Harmonization The key of harmonization of socio-ecology system are ownership asset and technology control.
5. Sustainability Kolong's utilization will happen if Kolong's ownership status and economic value have various benefit in economic, social, and environment (Class I and Class II based on ABAD Model).
6. The Construction Strategy of Socio-ecology must consider several factors as follows; Stability(socio-psychology), Eficiency (socio-ecology), Productivity (socio-economic) and Sustainability (socio-culture)[12].

4 Conclusion and suggestion

Based on the results and discussions, it can be concluded that:

1. Potential allocation of the economic value of Kolong in Bangka Island is IDR 17,9 billion/ year.
2. Potential allocation of the socio-ecology benefit value has various allocations utilization and management as:
   a. Water supply system
   b. Raw water for citizen
   c. Local assets

   d. Community settlement
   e. Tourism
   f. Water source for agriculture and industry
3. The main factor as an indicator for utilization and management of Kolong, based on socio-ecology perspective, are ownership assets, technology controls, and Kolong's management as a production source.
4. Adaptation process of social system is balance to ecosystem change pattern of Kolong.

References

1. S. Sujitno, *Dampak Kehadiran Timah Indonesia Sepanjang Sejarah.* PT. Timah, Jakarta (2010)
2. Bapedalda Provinsi Kepulauan Bangka Belitung. *Kondisi Kerusakan Lingkungan Hidup di Provinsi Kepulauan Bangka Belitung.* (2007)
3. F.H. Buttel., R C. Humphrey. Sociological Theory and the Natural Environment. pp. 33–69 in *Handbook of Environmental Sociology* edited by R.E. Dunlap and W. Michelson., C.T. Westport. Greenwood Press (2002)
4. F.H .Buttel *Ecological Modernization as Social Theory*. Pergamon Journal Madison. USA. (2000)
5. C. Henny.. "KOLONG" Bekas Tambang Timah di Pulau Bangka: Permasalahan Kualitas Air dan Alternatif Solusi Untuk Pemanfaatanya, *JOLI* **37**(1): 119-138 ISSN 0125-9830. (2011)
6. L. Meyana, M. Sudadi, B. Tjahjono.. JPSL. **5**(1) 51-60 : 20086-4639 (2015)
7. O. Sumarwoto, *Ekologi Lingkungan Hidup dan Pembangunan* ISBN 979-428-285-5 (1997)
8. E.P. Odum, *Basic Ecology* (1983)
9. G.T. Miller, *Living in the Environment: Principle, Connection, and Solution* (2006)
10. M. Sholihin, D. Ratmono, Analisis SEM-PLS dengan WarpPLS, ISBN 978-979-29-4267-5 (2013)
11. F. Sabri, *Nilai Ekonomi Air Kolong DAM-3 Pemali*, Gadjah Mada University, Yogyakarta (2008)
12. F. Sjarkowie, *Agro Ekosistem Lahan Basah Lestari.* Palembang (2014)

The Impact of tin mining in Bangka Belitung and its reclamation studies

Eddy Nurtjahya[1,*], *Jennifer* Franklin[2], Umroh[3] and *Fournita* Agustina[4]

[1]University of Bangka Belitung, Biology Department, 33172 Merawang, Indonesia
[2]University of Tennessee, Forestry, Wildlife and Fisheries Department, 274 Ellington Plant Sciences Building, TN 37996, USA
[3]University of Bangka Belitung, Fisheries Department, 33172 Merawang, Indonesia
[4]University of Bangka Belitung, Agribusiness Department, 33172 Merawang, Indonesia

Abstract. Tin mining in Bangka Belitung has been exploited for hundred years. The province is the second largest tin producer in the world. Secondary data from studies which took place in all four regencies in Bangka Island were discussed to show the impact of mining activities and its reclamation studies. In order to add plant selection criteria for regevetate mined soil, the greenhouse and laboratory experiment was carried out with fourteen herbs and grass species in Tennessee. The mining activities increase the wealth of the community, but the other hand they change and decrease the environmental stability, and cause horizontal conflicts. Offshore mining reduced water quality, change sea bed caused the change of biodiversity. Onshore mining activity reduces biodiversity and causes floods and damages infrastructure. While the more economic species are demanded, planting local tree species is challenging. An evaluation with local tree species concluded that best adapted species based on anatomical and physiological measurements was not those that showed the best performance in the field. The greenhouse and laboratory findings indicate that some physiological characteristics i.e. plant height and cover, transpiration rate, and foliar pigments may be used to select plant adaptability to mined soil.

1 Introduction

Bangka Belitung Islands produced approximately 106,000 t of tin in 2013 or more than one third of global tin supply, and the majority is exported [1]. Tin mining is the most significant economic-driver in the province, taking place inland and offshore, including in protected forests and marine ecosystems [1].

Following the issuance of a 1999 Ministry of Trade and Industry decree that tin is not an export item to be monitored and regulated, the Bangka regent issued a decree in 2001 giving permission for the people to mine tin.

There were 80 dredges and nearly 4000 floating tin mines off the shore of Bangka Island in 2013 [2], and are up to 50,000 artisanal small scale mines (ASM) and approximately 30 independent smelters [1]. Tin production from ASM contributes up to 80% of Indonesian tin exports [3].

Tin mining activities increase the wealth of the local people, but most of the activities neglect good mining practices, safety and land reclamation [3]. The most accidents at inland mine sites are due to landslides, and non-standard diving devices [4]. The mining activity decrease the environmental stability, causes pollution, and cause horizontal conflicts. Offshore mining reduced water quality, change sea bed caused the change of biodiversity.

While the more plant species planted in mined soils in the last decade, economic species such as rubber, oil palm, and some fruit species are demanded. Planting local tree species is challenging. There is an additional challenge in cases where reclaimed soils are contaminated with metals, are highly acidic, or coarsely textured. Natural succession takes a long time [5]. Soil amendments and land preparation are the major costs [6].

An evaluation of ten native tree species concluded that four of ten species showed highest survival rates and cover development [7] but measurement of anatomical and physiological measurements of five year saplings of those species, best adapted species were not those that showed the best performance in the field (Table 1) [8].

The identification of traits that can be used to identify species for potential use in mine reclamation would greatly speed the search for appropriate native species. The early growth, pigment content, and transpiration rates of 14 herbaceous species are studied to determine whether these traits can be used to predict ground cover success on mine reclamation sites.

Anatomy and physiology parameters: stomatal density, leaf thickness, palisade thickness, sponge thickness, upper epidermal thickness, lower epidermal thickness, upper cuticle thickness, lower cuticle thickness, xylem diameter, xylem bundles, root diameter, root conductity ratio. Morphology parameters: survival rate and cover area (1 year old), and height, stem diameter, and cover area (5 year old)

* Corresponding author: eddy_nurtjahya@yahoo.com

Table 1. The most adaptive local tree species on sandy tin tailing at one and five year old based on anatomy and physiology, and morphology measurements.

	1 year old [7]	5 year old [8]
Anatomy and Physiology	not measured	*Vitex pinnata* *Calophyllum inophyllum* *Syzygium grande*
Morphology	*Hibiscus tiliaceus* *Ficus superba* *Calophyllum inophyllum* *Syzygium grande* *Vitex pinnata*	*Syzygium grande* *Calophyllum inophyllum* *Vitex pinnata*

2 Methods

Bangka Island, with a population of nearly one million, is located off the eastern coast of South Sumatra Island [9]. The island has a surface area of 11,900 km^2 and is mainly lowland below 50 m; its climatic differences within the island are small. Its climate belonging to the Af-type Köppen-Geiger climate classification [10], with an average temperature of 26.3°C, average humidity of 61.7% and average annual rainfall of approximately 2,400 mm [7].

The authors gathered secondary data from studies which took place in all four regencies in Bangka Island (Figure 1), and in the green house and laboratory in the University of Tennessee, Knoxville, USA to indicate that some physiological characteristics may be used to select plant adaptability to mined soil.

The experiment was carried out with fourteen herbs and grass species whose seeds are widely available commercially within the southeastern United States. Germination test was conducted in the laboratory, and the plants were grown in the green house. Plant height and cover were measured at weeks 2, 4, 6, and 8 after seeding. Transpiration rate was measured beginning eleven weeks after planting. All analysis was completed using SPSS © Statistical Software (version 18.1, SPSS Inc., Chicago, USA) [28]. Leaf tissue was extracted in 80% (v÷v) acetone for chlorophylls a and b [11], and equation of [12] for carotenoids.

3 Socio-economic impacts

3.1. Positive impacts

The positive impact of tin mining is economic [13-15]. The increase of income is shown by the number of motorcycles and cars, and from 1999 to 2011, motorcycle and car taxes increased 15-fold. The number of people who perform the hajj pilgrimage increased almost 10% from 2001 to 2012 [16].

The income percentage for tin miners compared to overall income per month of people in Lubuk Kelik, Bangka; for ex-pepper farmers in Silip, Bangka; and for ex-rubber farmers in Bencah, Central Bangka are 90% and above. Pepper and rubber plantations contribute less than 3 % each of overall monthly income [17]. The net monthly income of fishermen in Rebo and Bubus beaches, Bangka, is just about one-third of the income of their colleagues working in tin mining [18].

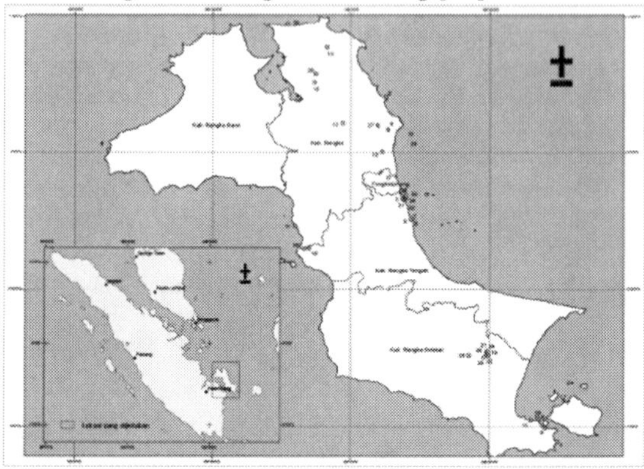

Fig. 1. Inland and offshore study sites in Bangka Island from secondary data [19]

3.2 Negative impacts

Tin mining also causes societal conflicts at mining sites. Most of the conflict in both inland and offshore mining is between locals and immigrants [16; 4]. Attitude changes and conflicts are reported in the hamlets and villages of the studied area [19].

The drop-out rate from elementary to senior high school has increased. In 2011, the province of Bangka Belitung had the second-largest student drop-out rate in the country because of children's involvement in mining or following their parents when they move to new mining sites [16].

In some areas, fishermen and farmers have changed their professions to become miners. Fishing boats are modified to become mobile floating dredges in Bangka [18]. Rubber plantations and pepper plantations have been mined in some areas in Central Bangka and South Bangka [17].

Flooding in many areas of the province is believed to be caused by tin mining. The original small stream channels have been changed by the mining activity.

4 Water qualities and offshore biotas

Offshore tin mining has reduced water quality. This is shown by a 40% total soluble solid (TSS) increase, a 75% sedimentation rate increase, a 25% water pH decrease and a 50% dissolved oxygen (DO) increase [20] (Figure 2). It is reported that, owing to tin mining, the concentrations of lead (Pb) (0.223 ppm) and TSS in solution offshore at Batu Belubang (705 ppm) were above the ministerial regulations of (Kepmen LH No. 51 tahun 2004) 0.008 ppm and 400 ppm respectively [21].

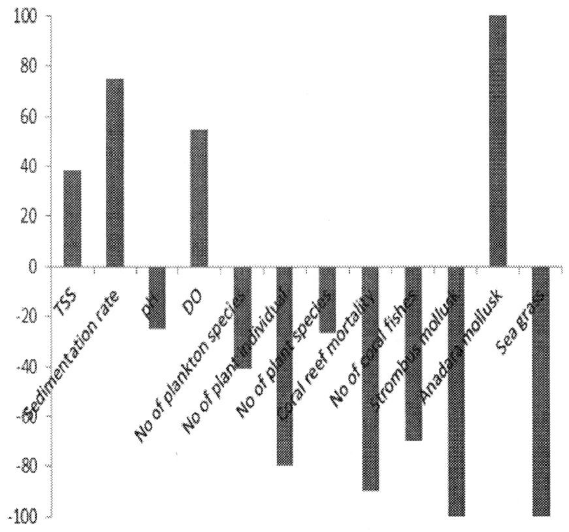

Fig. 2. Increase and decrease percentage of water quality and offshore biotas [20].

In another study, offshore mining was found to cause a 40% reduction of the number plankton species [18]. The number of species of seagrass in mined water was about 70% of the number in less mined water [18]. The dominant substrate in mined water was sand and rubble, in contrast to macroalga *Halimeda* sp. in less mined water [18].

The number of coral reef–associated fish in mined water was 30% of that in less mined water [18]. Coral reef life coverage was less than 25% in mined water compared to more than 90% in less mined water [18]. However, the growth rate of the coral reef species *Acropora digitata* transplanted to Teluk Limau Beach, Bangka, was 2.2–2.4 mm/month [22].

Because of floating small-scale artisanal tin mining units (*TI apung*), the number of fish caught has decreased, causing some fishermen not to go fishing, and no need to go further with no guarantee of a good catch [23]. The habitat changes have caused the economic benthic mollusc species *Laevistrombus canarium* L. (*siput gonggong*) of the family Strombidae to be replaced by the bivalve species *Anadara granosa* [24]. Small pelagic and demersal fish production decreased at three offshore mined sites in three regencies over the period 2009–2010 across the island, from 10 – 70% [19].

5 Soil degradation and inland biotas

Alluvial tin deposits - cassiterite (SnO_2) - were revealed after stripping the vegetation above the upper soil and removing the non-tin deposit overburden. The extraction is done by pouring a large volume of highly pressured water over the sediment. Heavy tin ore separated from light material by gravity. Non-tin sediment settles in a lower area with acidic pH.

Dredger exploits tin deposits located offshore up to 70 m depth with bucket wheel dredging. ASM use small traditional gravel pumps to pump tin-ore deposits to floating dredge units (*TI apung*) or modified small fishing boats.

Inland mining decreases soil properties, changing sand faction up to 97% (Table 2). The concentrations of phosphate, potassium and sodium in undisturbed land are higher than in mined soil, and are gradually decreasing as the site is abandoned [5]. C-organics are less than 2%, and the cation-exchange capacity (CEC) of tin-mined lands is very low (0.4–3.9 units) [5]. The soil temperature may reach 45°C during the day [25], and evaporation on sandy tailings may reach 4 L/m^2/day or double than of undisturbed soil [26].

Table 2. Soil properties of 0-, 7-, 11-, 38- year old tin-mined land, and riparian forest [7].

Sites	Sand	Silt	Clay	pH	C	N	P_2O_5	K_2O	CEC
	(%)			H_2O	(%)		(mg/100g)		
0	94	2	4	4.8	0.2	0	2	3	0.4
7	94	4	3	4.8	1	0.1	49	3	3.3
11	83	5	13	4.9	0.2	0	11	4	2
38	96	2	2	5.1	0.3	0	5	2	1
forest	78	13	10	4.7	1.6	0.2	22	5	5.8

C (Walk & Black); N (Kjeldahl); Cation-exchange (NH4-Acetate 1 N, pH 7); CEC (Ca+Mg+K+Na)

The water and the sediment from the washing process bring acidic material, which may reach the pH below 3. The acidity negatively affects soil flora and fauna [5]. A river that receives tin sedimentation has nearly 30% less fish species compared to a river free from tin mining [27].

Mining activity changes the vegetation structure and composition. The vegetation structure after 38 years of natural succession on old tin-mined land was less than 2%, similar to that of a riparian forest on Bangka Island [5] (Table 3). The number of arbuscular mycorrhizal fungi (AMF) spores increases with the abandonment of tin-mined land, and the number of phosphate solubilising bacteria (PSB) shows different readings with the period of abandonment [5] (Table 4).

Land recovery and coral reef transplantation are costly. The revenue from tin through land function change is lower than for non-mining land uses: protected forest, rubber plantation and beach [20]. The expenditure to convert one hectare of previously tin-mined site into rice field is estimated at Rp. 31 million, with the land preparation component representing the major portion, and almost half of the costs are for soil amendment [6].

Table 3. Number of individuals (A), species (B), and families (C) of 0-, 7-, 11-, 38-year old tin-mined land and riparian forest [7].

A. Number of individual per hectare

Sites	Number of individuals / ha				
	seedlings	saplings	poles	trees	total
0	0	0	0	0	0
7	890	0	0	0	890
11	1675	45	0	0	1720
38	2125	55	0	0	2180
forest	2665	4155	305	170	7295

B. Number of species

Sites	Number of species				
	seedlings	saplings	poles	trees	total
0	0	0	0	0	0
7	6	0	0	0	6
11	7	2	0	0	8
38	15	1	0	0	16
forest	42	66	24	11	85

C. Number of families

Sites	Number of families				
	seedlings	saplings	poles	trees	total
0	0	0	0	0	0
7	4	0	0	0	4
11	4	2	0	0	5
38	12	1	0	0	13
forest	24	30	14	8	44

Table 4. Number of phosphate solubilizing bacteria (PSB) and arbuscular mycorrhizal fungus (AMF) of 0-, 7-, 11-, 38-year old tin-mined land and riparian forest [7].

Sites	PSB (10^5 c/ g soil)	AMF spores/ 50 g soil	AMF genera	*Glomus* spp. (%)
0	5	1	1	100
7	10.3	69	4	67
11	6	87	4	59
38	3.2	372	3	95
forest	7	30	4	57

6 Physiological characters

Reclaimed mined land is a highly variable and often challenging environment for the establishment of plants. Although sites differ greatly depending on climate, local geology and reclamation methods there are several characteristics that are commonly encountered on reclaimed mines: relatively low water holding capacity due to coarsely textured soils with little or no organic content, and low nutrient availability.

Species that have been widely adopted for mine reclamation are likely to be ones that are tolerant of a wide range of environmental conditions, and in particular, those that are tolerant of low water and nutrient availability [28].

For water stressed environments, the most drought–adapted species at the individual plant scale has the lowest daily transpirational water consumption [29]. Transpiration reduction also means increasing water use efficiency [30] which may show better adaptability in unfavorable soil conditions. Other species may adapt to drought through morphological changes such as leaf size [31] or possibly anatomical characters [8] rather than through physiological mechanisms.

Although rapid growth of vegetation is often cited as a desirable characteristic to control erosion on reclaimed sites, we found that the most frequently used species tended to have a moderate growth rate. Only *Trifolium pratense* L. showed higher values that are within the typical reported range for that species [32] (Table 5).

Better able to endure the stressful, low nutrient environments of mine sites, species of moderate growth rate may have been favored for reclamation due to their ability to persist and spread. This finding provides a starting point for further testing.

Table 5. Summary of traits potentially desirable in reclamation ground covers in Eastern United States [28].

	A	B	C	D	E	F	G	H
T. pratense	√	√	√	√	√	√	√	√
H. esculentus	√	√	√			√	√	√
T. repens	√	√	√			√	√	√
L. multiflorum		√	√		√	√	√	
L. corniculatus			√	√		√	√	√
B. napus	√	√				√	√	
B. perviridis	√	√	√				√	
L. perenne	√		√	√				√
P. virgatum			√	√		√	√	
V. unguiculata		√		√		√	√	
D. glomerata			√	√		√		
P. fagopyrum		√	√			√		
S. scoparium			√			√		√
S. *nutans*	√		√			√		

A= germination >30%, B= cover > 50% within 2 weeks, C= number of germination x pH, D= number of cover x soil type, E= number of height x soil type, F= number of transpiration x soil type, G= low transpiration rate, H= foliar pigments > 0.4 mg/g

7 Conclusions

The mining activities increase the wealth of the community, but the other hand they change and decrease the environmental stability, and socio economic impact stimulate horizontal conflicts. Offshore mining reduced water quality, change sea bed caused the change of biodiversity, and increase the mortality of coral reefs and their associated fishes. Inland mining activity reduces biodiversity and causes floods and damages infrastructure.

While the more economic species such as rubber, oil palm, and some fruit species are demanded, planting local tree species is challenging. An evaluation with local tree species concluded that best adapted species based on anatomical and physiological measurements was not those that showed the best performance in the field.

The work of finding physiological characteristics to predict ground cover success on mine reclamation sites has being conducted. Plant height and cover, transpiration rate, and foliar pigments may be used to select plant adaptability to mined soil. Species most widely used in reclamation tended to be perennials of moderate rate.

The first author gratefully acknowledges the funding of this research by Directorate General of Higher Education, Republic of Indonesia (Hibah Bersaing 092/SP2H/PP/DP2M/III/2007;086/SP2H/PP/DP2M/III/2008), and to thank Fulbright for funding the research (68150141). The first author would like to thank Department of Forestry, Wildlife and Fisheries, University of Tennessee kindly provided facilities to undertake the research. University of Bangka Belitung is thanked for its permission allowing the first author to conduct research.The authors acknowledge Mr. A. Akbar and Ms. W.E.A. Putri, Dr. Z.Q. Liu, and Mr. J. Seaton for the valuable data and help.

References

1. IDH Tin working group communiqué - To share recent study findings and consult Indonesian stakeholders about their sustainability views and priorities for action (2013).
2. Bangka Pos, 2013, Idil: arogansi bupati akan dibayar mahal. Bangka Pos 20 Maret 2013, viewed 19 April 2013.
3. International Tin Research Institute (ITRI) Indonesia's position in the global tin industry. *Proceedings ITRI Indonesia Tin Forum*, 11 December 2013, Pangkalpinang (2013).
4. Walhi - Friends of the Earth Indonesia, Mari buat Bangka Belitung lebih baik. *Proceedings ITRI Indonesia Tin Forum,* 11 December 2013, Pangkalpinang, Indonesia.
5. E. Nurtjahya, D. Setiadi, E. Guhardja, M. Muhadiono, Y. Setiadi, Succession on tin-mined land in Bangka Island, Blumea, **54** (2009), no. 1–3, pp. 131–138.
6. E. Nurtjahya, M.M. Nur, E. Mulyono, Rice field cultivation on tin-mined land in Bangka Island, Indonesia. *Proceedings of Mine Closure 2009*, A.B. Fourie, and M. Tibbett (eds), September 2009, Perth,2009 Australian Centre for Geomechanics, Pe rth, pp. 549-560.
7. E. Nurtjahya, Revegetasi lahan pasca tambang timah dengan beragam jenis pohon lokal di Pulau Bangka, Dissertation, Bogor, Institut Pertanian Bogor, 163 pp, (2008).
8. E. Nurtjahya, R. Robika, D. Dorly, Can anatomical and physiological characters predict plant adaptation on tin-mined land in Bangka Island? *Proceedings of the 6th International Conference on Mine Closure 2011*, A.B. Fourie, and M. Tibbett and A. Beersing (eds), September 2011, Alberta, Canada,2011 Australian Centre for Geomechanics, Perth, pp. 75-83.
9. Badan Pusat Statistik (BPS), *Bangka Belitung dalam Angka*, Pangkalpinang (2012), Indonesia.
10. PT Timah Tbk, ANDAL, RKL, dan RPL. *Kegiatan penambangan timah dan pasir laut di perairan Pulau Bangka Kabupaten Bangka, Propinsi Sumatera Selatan* (1997).
11. Z.D. Sesták, J. Čatský, P.G. Jarvis, Determination of chlorophylls a and b, in *Plant photosynthetic production: Manual of Methods*, Z. Sesták, J. Čatský, and P.G. Jarvis, eds., Junk NV, The Hague (1971), pp. 672–701. B.H.

12. Davies, Analysis of carotenoid pigments. In: T.W. Goodwin (ed.) *Chemistry and Biochemistry of Plant Pigments*, pp. 489-532. Academic Press, New York, (1965).
13. L. Juniarti, *Disharmonisasi keluarga pada pekerja tambang inkonvensional (TI) di Dusun Parit 19 Kecamatan*, Dissertation, Jurusan Sosiologi, Universitas Bangka Belitung, Pangkalpinang, Indonesia, (2014).
14. Indra, *Tambang inkonvensional dan implikasinya terhadap perilaku sosial masyarakat di Desa Tempilang Kabupaten Bangka Barat*, Dissertation, Jurusan Sosiologi, Universitas Bangka Belitung, Pangkalpinang, Indonesia, (2013).
15. B. Romeo, *Tambang inkonvensional dan perubahan perilaku ekonomi masyarakat Desa Semulut Kecamatan Jebus*, Dissertation, Jurusan Sosiologi, Universitas Bangka Belitung, Pangkalpinang, Indonesia, (2011).
16. E. Erman, Dampak penambangan timah dan respon masyarakat local. *Proceeding ITRI Indonesia Tin Forum*, 11 December 2013, Pangkalpinang, Indonesia, (2013).
17. E. Nurtjahya, F. Agustina, W.A.E. Putri, Neraca ekologi penambangan timah di Pulau Bangka Studi kasus pengalihan lahan di ekosistem darat, Berkala Penelitian Hayati, **14**(1) (2008), pp. 29–38.
18. E. Nurtjahya, F. Agustina, A. Akbar, *Kajian manfaat sosial ekonomi penambangan timah inkonvensional dan kerusakan lingkungan dan keanekaragaman hayati yang ditimbulkannya di Pulau Bangka*, Laporan Penelitian Hibah Bersaing, Universitas Bangka Belitung (2008), 70 p.
19. E. Nurtjahya, F. Agustina, Managing the socio-economic impact of tin mining on Bangka Island, Indonesia — preparation for closure. *Proceedings Mine Closure 2015*, A.B. Fourie and M. Tibbett, L. Sawatsky and D. Van Zyl (eds), Juni 2015, Vancouver,Canada,Australian Centre for Geomech anics, Perth, pp. 817–826.
20. E. Nurtjahya, U. Umroh, F. Agustina, Impact of tin mining on the biota of Bangka Island, Indonesia – a proof to convince the tin supply chain of smartphone companies. *Proceedings Mine Closure 2014*, A.B. Fourie, M. Tibbett, and I. Weiersby (eds), October 2014, Johannesburg, SouthAfrica,Australian Centre for Geomechanics, P erth (to be published).
21. H. Wahyuni, S.B. Sasongko, D.P. Sasongko, Kandungan logam berat pada air, sedimen dan plankton di daerah penambangan masyarakat Desa Batu Belubang Kabupaten Bangka Tengah, in *Proceedings Seminar Nasional Pengelolaan Sumberdaya Alam dan Lingkungan* (2013), 489–494.
22. H. Sodikin, *Pertumbuhan Acropora digitata pada transplantasi karang di pantai Teluk Limau Sungailiat Provinsi Kepulauan Bangka Belitung*, Dissertation, Pangkalpinang: Jurusan Perikanan Universitas Bangka Belitung (2011), 36 p.
23. P. Pratama, *Alih profesi nelayan menjadi penambang timah di Lingkungan Nelayan 2 Sungailiat*, Dissertation, Jurusan Sosiologi, Pangkalpinang: Universitas Bangka Belitung (2014).
24. F. Yulianda, F. Achmad, A.H. Armin, H. Sri, K. Kusharjani, *Ekologi ekosistem perairan laut tropis*, PUSDIKLAT Kehutanan - Departemen Kehutanan RI, Bogor: SECEM - Korea International Cooperation Agency (2009).
25. E. Nurtjahya, D. Setiadi, E. Guhardja, M. Muhadiono, Y. Setiadi, Revegetation of tin-mined land using various local tree species in Bangka Island, Indonesia. *Proceedings of the 2008 National Meeting of the American Society of Mining and Reclamation New Opportunities to Apply Our Science*, R.I. Barnhisel (ed), June 2008, Lexington, USA, ASMR, pp. 739-755.
26. E. Nurtjahya, Ex tin-mined research - a local tree species search (2010) (unpublished).
27. K. Muslih, E.M. Adiwilaga, S. Adiwibowo, Pengaruh penambangan timah terhadap keanekaragaman ikan sungai dan kearifan local masyarakat di Kabupaten Bangka. *Prosiding Pertemuan Ilmiah Tahunan MLI I* (2013).
28. E. Nurtjahya, J. Franklin, Some physiological characteristics to estimate species potential as a mine reclamation ground cover. Int. J. Min. Reclam. Environ (2016) (to be published).
29. C. Tong, J.-Z. Gong, R. Marrs, I. Zhang, W.-Q Wang, Pattern of transpiration of four shrub species and water consumption from shrub stands in an eco–reclamation catchment in northwest China, Arid Land Res. Manag. **22** (2008), pp. 242–254.
30. A.M. Gonzalez-Rodriguez, Z. Baruch, D. Palomo, G. Cruz-Trujillo, M.S. Jimenez, D. Morales, Ecophysiology of the invader Pennisetum setaceum and three native grasses in the Canary Islands, Acta Oecol. **36** (2010), pp. 248–254.
31. F.M. DaMatta, Exploring drought tolerance in coffee: a physiological approach with some insights for plant breeding, Braz. J. Plant Physi. **16** (2004), pp. 1–6.
32. L. Simova-Stoilova, K. Demirevska, A. Kingston-Smith, U. Feller, Involvement of the leaf antioxidant system in the response to soil flooding in two Trifolium genotypes differing in their tolerance to water logging, Plant Sci. **183** (2012), pp. 43–49.

* Corresponding author: eddy_nurtjahya@yahoo.com

DOI: 10.1051/matecconf/201710104011

Olistostrome and the mesozoic tectonic of the bantimala complex, South Sulawesi

Kaharuddin MS[1*], *A.M.* Imran[1], *Chalid* Idham Abdullah[2], and *Asri* Jaya[1]

[1]Geology Engineering Department Faculty of Engineering, Hasanuddin University, 90245 Makassar, Indonesia
[2]Geology Engineering Department Faculty of Earth Science and Technology, Bandung Institute of Technology, 40116 Bandung, Indonesia

Abstract. The study aims to determine the presences and spreads of the olistostrome as one components of the tectonic complex of Bantimala area. The basement rocks of Bantimala area is a metamorphic rocks, which are unconformably overlain by Balangbaru Formation and radiolarian chert. Beneath in between cherts and basement rocks are presence breccia schists which give rise to various presumptions and interpretations of the environment and conditions of formation as submarine deposit, ideally cherts in the area underlain by oceanic crust.Schist breccias presence underneath of cherts in the Bantimala Complexes were suggested an olistostrome deposit. It was characterized by poorly sorting, unfoliated, shows deformed textures and composed olisthtolits which are embedded in sandy matrices, and in the cherts are presence layer sandstones and schist fragments. Olistostrome is sedimentary deposit as preserved in the trench, they will give us an interpretation that prior to formed cherts in the Bantimala area, initially tectonic subduction activity which are deforming and brecciation of the basement rocks subsequently as the constituent material of olistostrome.

1 Introduction

The olistostrome of Bantimala Tectonic Complexes is likely related to Western Pacific plate has subducted to Asian Continent during Mesozoic, the interaction both of the plates were caused deformation, metamorphism and olistotrome deposits. These situation is reasonably allow to occur a debris fall due to steep slope in the part of trenches as suggested by [1], [2], their fragments possibly are eroded debris materials from continent and oceanic plates, they have been mixed as a chaotic sedimentary deposits, its immediately covered by deep marine sediments of radiolarian cherts.

The oldest rocks were exposed in the Bantimala Complexes are Triassic metamorphic rocks composed of glaukofan schists, mica-hornblende schists, eclogites, granulits, quartz phyllite and quartzite [3]. On the top of this basement rocks is overlain by breccias, schists (olistostrome), sandstones and radiolarian cherts indicating Jurassic - Cretaceous age [4].

Our side observations that schist breccias in the study area is an olistostrome deposits characterized by present vary of olistolith schists, stratigraphically it overlain by radiolarian cherts. So far, the study of olistotrome in the area has never been studied, but previous studies it was indicated a mélange [5] thought a chaotic sedimentary processing which is more appropriately considered as olistostrome.

2 Methods

This study emphasizes field observations and petrographic analysis. Field observations was performed stratigraphic detail by measuring section and cross section along Patteteyang river, observing the relationship between the lithologic unit and indication deformation structures in the field and rocks sampling. Further observations were made followed by petrographic analysis to determine the texture and rock composition for all components of rocks, especially for olistolith and matrices.

3 Regional geology and tectonic setting

Tectonically the Bantimala Complexes is still contentious, including presence of rocks faragments and interrelationship each other such as presence of mélanges, schist basement rocks, rocks continent affinity, radiolarian cherts and Paleogene volcanics.

3.1 Geological setting

The basement rocks were exposed in the Bantimala area is part of shelf margin of East Kalimantan which was separated since Miocene coincided with the occurrence of Makassar Strait. The basement rocks is called the Bantimala Tectonic Complex where composed of: Triassic metamorphic rocks namely glaucophane schists,

* Corresponding author: kaharuddin_geounhas@yahoo.com

hornblende-mica schists, eclogites, granulites, phyllitesand metaquarzites [3]; Jurassic-Cretaceous of mélange composed of quartz schists, metacherts, metabasalts. Cretaceous sedimentary rocks which composed of fine-siliceous shales, sanstone, mudstone and radiolarian cherts.

Ophiolite blocks consisting of harzburgites and serpentinites where occur through out tectonic obduction which covered or structural contact with the Tertiary rocks in the area, whereas sedimentary type of continental shelf margin are Cretaceous of typical flysch sedimentary rocks of Balangbaru-Paremba Formations. They were unconformity overlain by quartz sandstones of Mallawa Formation and Paleocene-Eocene volcanics. They are covered by Eocene-Miocene limestone of Tonasa Formation and Upper Miocene-Pliocene breccias and tuffs of Camba Formation [6] Fig 1.

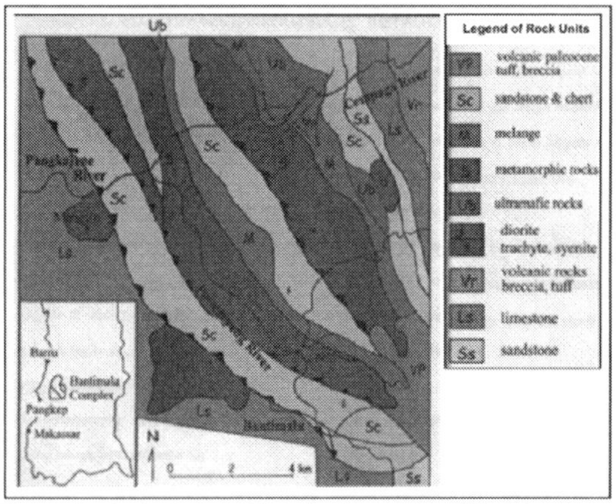

Fig. 1. Geological map of Bantimala area (modified from Sukamto, 1986).

3.2 Tectonic setting

The Bantimala Complex is formed in two tectonics phases, the first is subduction system of oceanic plate that took place during the Mesozoic to Tertiary. The second is ophiolite obduction system that took place during Tertiary to Quaternary. The presence of high grade of metamorphic rocks associated with system during pre-Cretaceous (Triassic - Jurassic), it may cause by the West Pacific plate subducted to the East Kalimantan continent, which was produced accretion complex on the eastern shelf of Asian Continental plates. During Jurassic to Cretaceous the accretion complex has deformed to produce a mélange rocks [7-8]. During the Early Cretaceous occurred brecciation in the accretion complexes formed breccia - schist [5] or here called olisostrome [9], while continuing deposited chert and contemporaneously with deposition in the trench of terrigenic sediment of Balangbaru (Figure 2).

More over the Cretaceous of Bantimala Complexes have been situated by some of previous study such as [10] and [11], according to them that the establishment of a high-pressure metamorphic rocks associated low-grade metamorphic rocks, mélange and ultramafic in this region are the results of system subduction of oceanic crust into the shelf margin of continental plates in the Jurassic to the Early Cretaceous (114 to 132 Ma). Based on calculations of pressure - temperature of rocks garnet - glaucophaneresulted a temperature of 580 – 640 °C and pressures 18-24 kbar, occurs at depth 65-85 km [5]. Furthermorethe breccia of Bantimala Complexis unconformably overlain by the Radiolarian chert which ranging from Albian to Cenomanian in age (± 100 Ma) [12].

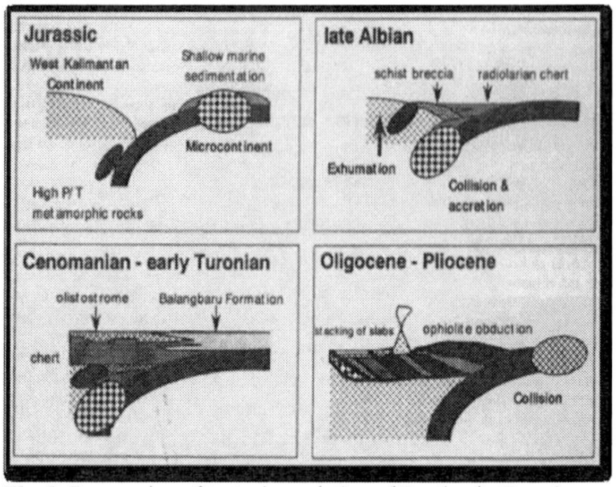

Fig. 2. Progressive of tectonic evolution of Bantimala Complex, during Mesozoic to Tertiary (Wakita et al., 1996)

4. Physical properties of Olistostrome

Lithologic characterized of olistostrome of Bantimala Complex could be described into two aspects i.e. based on components and sedimentary types.

4.1. Component Fabrics

Olistostrome component in the Bantimala Complex shows various clast of material component swhich is indicated difference origin (polymictic), it composed by deformed and reworked materials derived from subduction activity which consisting of schist, quartzite and gneiss, serpentinite, metachert, the components are vary in grain size ranging from centimeter to several meter, poorly sorting and angular to subangular. Fragments floating in reddish chert matrices suggesting retransported material within a trench, deformed textures indicated by lenses, fracturing, shows pseudofoliation with slipped surface (Figure 9 and 10).

The following is results of petrographic analysis of olistoliths of olistostrome, schists, cherts, and sandstoneas follows:

1. Schists and Quartzite Olistoliths:
- Schist olistoliths of schist muscovite quartz (OL, 5C), lepidoblastic textures, composed by muscovite 40%, quartz 30%, chlorite 20%, actinolite 5%, mineral opak 5% (Fig. 3).
- Quartzite Olistolith (OL, 5D), composed byquartz 75%, muscovite 12%, and biotite 10%, mineral opak 3%, granoblastic textures (Figure 4).

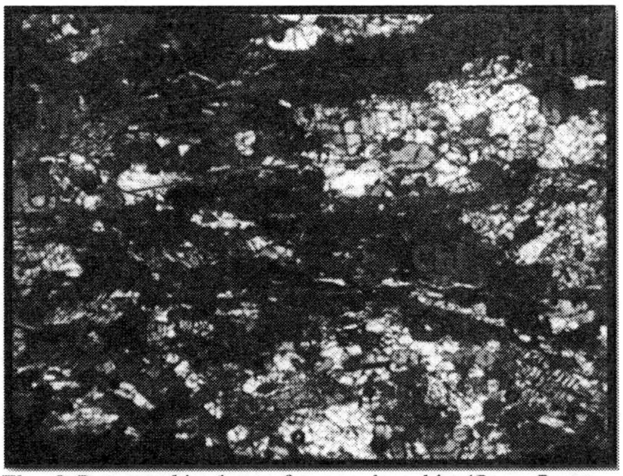

Fig. 3. Petrographic photo of muscovite schist (Qtz = Quartz ; Chl = Chlorite ; Ms = Muscovite)

Fig. 4. Petrographic photo of quartzite (Qtz = Quartz ; Bt = Biotite ; Ms = Muscovite)

2. Schist and greenschist

- Olistolith schist (OL, 2B), lepidoblastic textures composed by muscovite 65%, quartz 20%, chlorite 5%, albite 6%, and ore mineral 4%(Fig. 5).

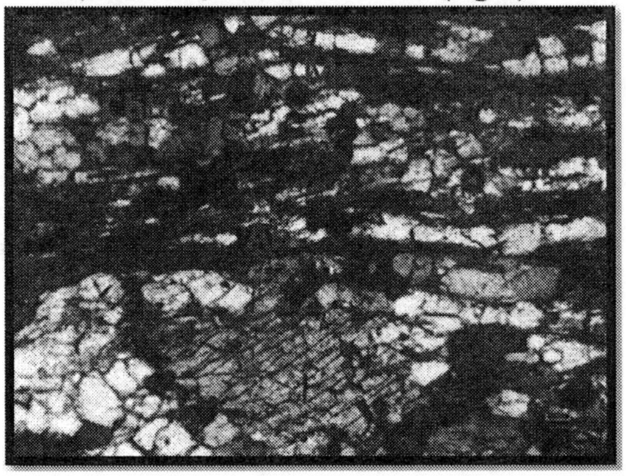

Fig. 5. Petrographic photo of olistolith schist (Qtz = Quartz ; Ab = Albite ; Chl = Chlorite ; Ms = Muscovite)

- Green schist/ actinolite schist (ML, 3B), lepidoblastic textures, composed by actinolite 90%, quartz 5%, chlorite 3%, ore mineral (Fig. 6).

Fig. 6. Petrographic photo of actinolite schist (Act = Actinolite)

3. Chert (C, 4E)

Chert composed by radiolaria 40%, quartz 56%, ore mineral 4%, (Fig. 7).

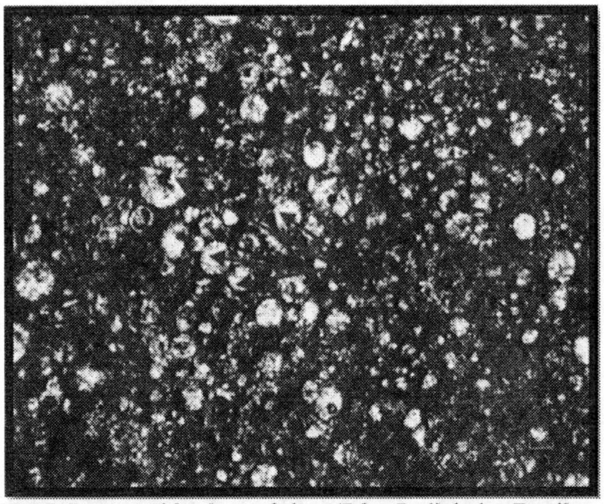

Fig. 7. Petrographic photo of chert (Rfs = Radiolarian Fossil ; Qzt = Quartz)

4 Sandstone (ST. 13A)

Sandstone composed by muscovite 70%, quartz 25%, ore mineral 5%, (Fig.8).

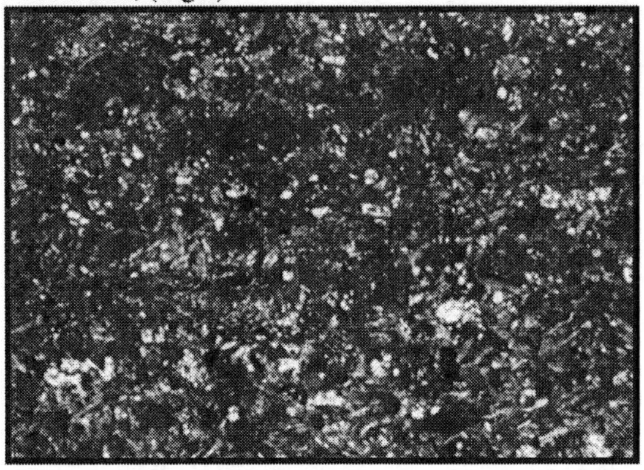

Fig. 8. Petrographic photo of actinolite schist (Act = Actinolite)

Fig. 9. Outcrop of olistostrome (fluxo) with tectonit textures, schist olistolith floating on matrices, Pateteyang River, Bantimala.

Fig. 10. Outcrop of olistostrome in the eastern part are show proxima deposit, Pateteyang River, Bantimala.

4.2 Type and physiography sediment

4.2.1 Deposition type Olistostrome

Quartz muscovite schist in one complex with glaucophane schists, granulits, eclogites and radiolarian cherts, as an indication of sediment avalanche or slope slumping at a relatively steep slope suggested rocks that preserve from subduction between the continental plates with oceanic plates, where olistostrome interbedded in part the bottom of the radiolarian cherts [13]. To mixing materials in chert clast and fining upwards structures, cobble – granule in grain sizes, indicating that they were preserve in a deep marine or trenches on subducting tectonic conditions, based on the above determination so that the olistostrome of Bantimala Complex is categorized as subduction type.

4.2.2 Sediment Physiography

The appearance of interbedded between the sandstones and radiolarian chert as the top of the olistostrome, gradational structure components in the western part of the study area including fluxo deposit. Based on deposit type in the eastern part and the contact between the radiolarian chert coarse grainsof olistolith indicated proksima type deposit (Fig. 9 and 10). Based on the spread and characteristics of layering can be interpreted the depositional direction of olistostrome from west to east. The presence of serpentinite – jadeite blocks of as exotic block in olistostrome is separate phenomenon to be determine in the future work. Body of olistostrome in Bantimala Complex suggested as a lensis in radiolarian cherts unit.

5. The occurrences of olistotrome

The process of subduction of the Pacific plate along continental margin of Kalimantan during Jurassic, initially the process of forming olistostrome of Bantimala. During that period waspreserved tectonic deformation, brecciation and metamorphism of the two plates subducted each other, which is accompanied by the formation of deep trenches as the depositional environment. Plate continental and oceanic plates that have undergone at low grade metamorphism and brecciated forming blocks as olistoliths that shows deformation textures.

Based on experiments [14] and [15] on the slopes of the critical condition allow submarine landslide occurred in the trench, the material falling as debris flow or slumping and widespread deposited on the ocean floor which its mixture with clast of continental and oceanic components, [16] and [17], (Fig. 11).

Based on their appearance in the field, olistostrome layer at Pateteyang River (Bantimurung Village), there are at least four cycle of avalanche or debris deposition of material inter bedded with cherts, namely:

- The first is preserve debris avalanches and slumping, forming as rock blocks with a thickness of ± 340 m olistolith sizes between 1-150 cm, angular and boudinage. Components consist of chlorite schist, mica schist, amphibolite schist, and quartzite gneiss.
- The second, before the second deposition is preceded by the deposition of a thin layer chertsof 20 cm thick and mixed with coarse grains of schist material followa debris avalanche of gravel-blockof schist clasts (2-40 cm), relatively smoother than the first avalanche, with a thickness ± 150 cm.
- The third is also inter bedded by chert layers 60 cm in thick. The third layer is composed of schist clast with 25 cm thick.
- The fourth, deposition above of layer chert (120 cm thick), a thin layer of schist clast ± 20 cm thick and

DOI: 10.1051/matecconf/201710104011

covered by layers cherts that show the body olistostrome of Bantimala (Fig. 12). The composition and structure of the sediment grain size olistostrome shows fining upwards as debris flow sediment or turbidity (Table 1).

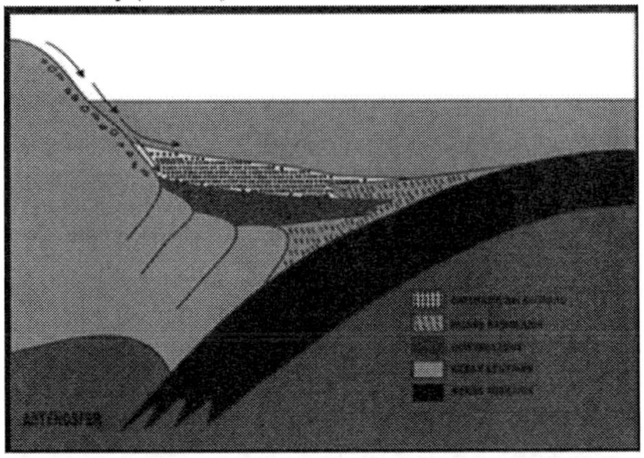

Fig. 11. Cross section of formation of olistotrome of Bantimala Complex, subduksi lempeng Oceanic Pasific plate (black) subducted to Asean Continental plate (orange) during Early Cretaceous.

Fig. 12. Outcop olistostrome (X) with radiolarian chert (Y) in the Pateteyang River, Bantimala

6 Discussion

Synsedimentary deformation and tectonic implications, field data shows deformation of soft-sediment on cherts with showing small folds (antiform and synform) (Fig. 13), brecciation, dragfold, microthust, boudinage olistolith surrounded by a fine grained schist matrix, slump fold and slump ball (Fig. 14), indicated that during sedimentation contemporaneous with active tectonic condition.

The stratigrphy position of olistostrome by Wakita et al. (1996) is above the wedge with Radiolarian chert and intercalated with Balangbaru sandstone, and there are also schists interbedded chert on the bottom. Breccia schist by Wakita et al. (1996) and the authors include as olistostrome deposit. Stratigraphy position hardly visible in the field but by outcroping thin layer of ± 50 cm (Figure 10) olistostrome interbedded in chert in the eastern part of the area of research, it can be interpreted that the (1996) that is deposited in the same time with the slump sedimentation system on the bottom layer of chert.

Thicknesses (cm)	Lithologic Units	Description
> 1000		Folded Radularian Chert
20		Reddy Sandstone (4)
120		Folded Chert (antiform and sinform)
25		Coarse Sandstone (3)
60		Folded Radularian Chert
150		Polylitic Breccia (2), Gradation Structure Components : Schist, Quartzite, Metachert
20		Schist Clasts In chert
> 5000		Polylitic Breccia (1), Chaotic & Proorly Sorted Components : Schist, Quartzite, Metachert, Gneiss

Fig. 13. Tabel of stratigraphic column of olisostrom in Bantimala Complex

The stratigraphy relationship breccia schist in this case olistostrome is unconformable overlaying cherts (Wakita, et al., 1994), conversely based on the age of the fossil in radiolarian chert, breccia or sandstone is conformable. It is also supported by the appearance of the field that showed gradational from bottom-top with fining upward structures and almost breccias composed by schist clast indicated same environment and sedimentary sources suggested that the relationship between radiolarian chert and breccia schists is conformable.

7 Conclusion

a. Geological conditions and tectonic Bantimala Complex composed by pre-Tertiary components (Schists, olistostromes, radiolarian cherts and sandstones of Balangbaru Formation) and Tertiary components (Volcanic Paleocene, sandstones of Mallawa Formation, limestones of Tonasa Formation, Camba volcanic and basic-intermedite-acid igneous intrusions. Some of them bounded by thrust faults.
b. Characteristics of olistostrome in the Bantimala Coplex domostrated by heterogeneous components with polymictic breccias consisting of schist, gneiss,

quartzite and metachert components, composed of three layers, namely breccia polymictic, coarse sand - granule and sand. They including in olistostrome subduction type that formed in Lower Cretaceous (Albian).

c. Stratigraphycally the position of olistostrome or breccia schist conformably overlaying bv radiolarian chert.
d. The olistostrome component is characterized by deformation textures and interrelationship with radiolarian chert where it was deformed during of post-sedimentation of the Jurassic – Cretaceous.

Fig. 14. Antiform (X) and synform (Y) structures in the cherts, Pateteyang River, Bantimala

Fig. 15. Outcrop of radiolarian cherts deformed, Pateteyang River, Bantimala.

References

1. Hall, R., Ophiolite Emplacement and the Evolution of the Taurus Suture Zone, Suthern Turkey, The Geological Society of America, Benchmark Paper in Geology/**66**, p. 275 – 285 (1976)
2. Abbate, E., Bortolotti, V., Passerini, P., , Olistostromes and Olistoliths, Benchmark Papers in Geology/**66**, pp.86 – 110 (1970)
3. Sukamto, The Structure of Sulawesi in the Light of Plate Tectonics, Proceeding Regional Conferention Geology Mineral Resources SE Asia (1975)
4. Sukamto, R., Geologi Lembar Pangkajene dan Watampone Bagian Barat, Sulawesi, P3G, Departemen Pertambangan dan Energi Bandung (1982)
5. Wakita, K., Sopaheluwakan, J., Miyazaki, K., Zulkarnaen, I., Munasri, Tectonic of the Bantimala Complex, South Sulawesi, Indonesia, Tectonic Evolution of Southeast Asia, London (1996)
6. Sukamto, Tektonik Sulawesi Selatan dengan awan Khusus, ciri-ciri himpunan batuan daerah Bantimala, Disertasi S3 ITB, Bandung, 363p, unpublished (1986)
7. Kaharuddin, M., Geotectonic of the Sulawesi Region, Toyama University, Japan (1995)
8. Kaharuddin, Perkembangan Tektonik dan Stratigrafi Kompleks Bantimala, Sulawesi Selatan, Prosiding Hasil Penelitian Fakultas Teknik Unhas, vol.**4**, hal.TG 5-1-TG5-9 (2010)
9. Kaharuddin, Jaya, A., Sirajuddin, H., 2015, Olistostrome dan Batu Mulia Kompleks Tektonik Bantimala Kabupaten Pangkajene dan Kepulauan, Prosiding TPT XXIV dan Kongres IX Perhapi, Jakarta (2015)
10. Zulkarnain, I., Cretaceous Tectonic Events of the Bantimala Area, South Sulawesi – Indonesia : Evidence from Rock Chemistry, Jurnal Teknologi Mineral No. 2 – Vol. VI (1999)
11. Maulana, A., Mamma, K., Imai, A., Petrological Characteristic and Metamorphic Evolution of the Bantimala Complex, South Sulawesi Indonesia (2010)
12. Wakita, K., Munasri, Sopaheluwakan, J., Zulkarnain, I., Miyazaki, K., Early Cretaceous Tectonic Events Implied in the Time – log Between the Age of Radiolarian Chert and its Metamorphic Basement in Bantimala Area, South Sulawesi, Indonesia, Research Article, Bandung, Indonesia (1994)
13. Shuib, M.K., The Olistostrome in the Bentong Area, Pahang and their Tectonic Implications, Department of Geology, University of Malaya, Kuala Lumpur, Malaya (2000)
14. Yamada, Y., Yamashita, Y., Yamamoto, Y., Submarine Landslide at Subduction Margins : Insights From Physical Models, Tectonophysics, Elsevier (2009)
15. Festa, A., Pini, G.A., Dilek, Y., Codegone, G., Melanges and Melange-forming Processes : A Historical and New Concepts, International Geology Review, Vol. **52**, Nos. 10 – 12, October – December 2010, 1040 – 1105 (2010)
16. Elter, P., Trevisan, L., Olistostromes in the Tectonic Evolution of the Northern Apennines, Benchmark Papers in Geology/66, pp.111 – 124 (1973)
17 Leonov, M.G., Olistostromes and their Origin, Benchmark Papers in Geology/**66**, pp.125-134 (1978)

DOI: 10.1051/matecconf/201710104012

Collaboration of high activity soil and geological structure factors in Pagelaran soil creep occurrence, Indonesia

Salam Ahmad F.[1], *Kristyanto* T. H. W. [2,*], and *Asriza* [3], *Syahputra* Reza[2], *Tempessy* Albert S.[2], *Indra* Tito L.[2]

[1]Resevoir Geophysics Graduate Program FMIPA Universitas Indonesia, 16424 Depok, Indonesia
[2]Geology Study Program FMIPA Universitas Indonesia, 16424 Depok, Indonesia
[3]Geophysics Study Program FMIPA Universitas Indonesia, 16424 Depok, Indonesia

Abstract. Soil creep is a landslide that moves slowly but it does not mean harmless. Soil creep that succeeded in forcing the citizens to migrate occurred in Pagelaran by wiping out 13 houses. This paper aims to know the contribution of geological structures and physical properties of soil toward the creeping occurrence. The methods used in this research consist of geological structure mapping; soil sampling; laboratory works; and analysis of geological structures and soil activity. Geological structures analysis indicates that the main force worked predominantly in NE-SW direction, parallel with creeping. It expresses that the creep was influenced by the orientation of them. Additionally, their distribution can be used as access for rainwater to infiltrate and add to the burden of the slopes. While laboratory results show that, the soil is classified into MH and CH. Calculation of activity numbers of soil indicates that it has a high – very high activity with value between 0.705 – 1.4931 (Skempton, 1953). It means that the soil will swell easily if contaminated by water.

1 Introduction

Soil creep is one type of landslide that has unique characteristic. It is the slow movement of soil materials down slopes under the influence of gravity. This type of land movement moving slowly, but it does not mean harmless [1,2]. One of soil creep that succeeded in forcing the population to migrate en masse occurred in one of the villages in Pagelaran, Cianjur, West Java. This landslide wiped out 13 houses and collapsed road along 280 meters [3].

Soil creeps usually occurs during and after rain. The amount of creep has a positive correlation with the amount of rain and increase in soil moisture. Soil creeps, mostly, are triggered by soil contained with active clay minerals, such as kaolinite or montmorillonite. This kind of mineral cause the soil has high activity number [4]. It means that the slope, which is formed by high activity soil, can be prone to creep [5]. On the other hand, landslide, including soil creep, also can be influenced by geological structures [6,7]. Their orientation can affect the slope stability if it has the same or inline orientation with the orientation of the slope [8].

Geological structures, as discontinuities, also can be access for water to flow inside the slope and affect the properties of soil [9]. Therefore, this paper aims to know the contribution of physical properties of soil, which is represented by its activities, and geological structures that develop at the site of research on the occurrence of the soil creep.

2 Geological setting

2.1. Geomorphology

Geographically, Pagelaran, Cianjur located in 7°08'48.9"S 107°12'05.3"E which is part of Southern West Java Mountain. This zone, triangular facet highland, is culminating in Bandung and extending from West to East. Overall, geomorphology of the zone is slightly sloping geanticline. The geomorphology of this area was affected by rejuvenation process (peneplain), lifting and re-sedimentation of erosion material. Thus in some area, the surface is not plateau, but narrow and deep valleys shaped. Van Bemmelen [10] concluded that not all of the Southern West Java Mountains were drown by sea level when late Miocene transgressions occur. Geological processes during the formation process, differences in the hardness and type of rock, and geological structure that occur in this area [11] influence morphology of this area.

2.2. Stratigraphy

Bedrock that formed sliding surface of this area, based on Sindangbarang Geological map and its Stratigraphic column [12], is Bentang Formation (Late Miocene). This formation containing less compressible tuff sandstone, crystal tuff and pumice tuff with globigerina clay, silt stone, breccia (andesite) and conglomerate. Then Koloberes Formation was deposited above the Bentang Formation. Koloberes Formation tends to be

* Corresponding author: twin.hosea@hotmail.com

DOI: 10.1051/matecconf/201710104012

similar with the Bentang Formation, which consists of well-layered sandstone, tuff breccia, and brownish grey sandstone.

2.3. Tectonics

Geological structures that developed in research area was faults, lineament, folds, and joints, which can be found in Oligocene – Miocene rocks and Quaternary one. Faults that spread out in research area are in the forms of NW-SE strike-slip faults and N-S normal faults. Folds that found there are SW-NE anticline and SW-NE syncline. Lineament that lies in study area is interpreted as fault with orientation NW-SE and SW-NE. Joints can be commonly found in Oligo-Miocene rocks [12].

3 Methods

3.1. Field Works

Field works that conducted in this research are geotechnical drilling and geological structure mapping. Geotechnical drilling was done to get subsurface information, i.e. soil lithology, and undisturbed soil samples for laboratory analysis. Figure 1a shows the activity of drilling in research area.

a b

Fig. 1. (a) Drilling activity and (b) joint orientation measurement

Geological structure mapping was done by investigating outcrop and measuring the orientation of geological structures. Measurement of it also conducted towards cracks on the surface of slope. Figure 1b shows the measurement activity of cracks and joints.

3.2. Laboratory analysis

3.2.1 Grain size analysis

The purpose of this analysis is to determine grain size distribution of the soil sample. This method using sieve test equipment to classify soil type based on USCS classification [13]. Table 1 and 2 show the results of grain size analysis and the classification of the soil based on USCS.

3.2.1 Atterberg Limit Test

Atterberg limits are a basic method to measuring PI (plasticity index), plastic limit (PL), and liquid limit (LL) of the fine-grained soil (silt and clay) with water content changes. If we know the value of this parameter, then generally soil properties can be known.

3.3. Studio analysis

Studio analyses are in the form of geological structure, and physical properties (Atterberg Limits, Plasticity Index, and Activity Number) analyses. Geological structures analysis is done to know the orientation of them and compare with the orientation of slope.

Physical properties analyses were using Casagrande method [14], Gillot Method [15] and the Method of Seed et al. [16], and method of William and Donaldson [17]. Casagrande method [14] was done to know the properties and classification of fine soil regarding with its plasticity.

Gillot method [15] and William and Donaldson method [17] was done to know the relationship between clay particle content and PI. While, Seed et al. [16] method was done to know the relationship betwen clay particle content and acitvity number of soil. Three of them were done to predict the activity number of soil that represents the swelling potential of soil.

4 Results and discussion

4.1. Physical properties

The results of laboratory test are shown in Tab. I and II in Appendix. These results are tested from undisturbed soil samples. Laboratory tests that were conducted consisted of Atterberg Limits and Plasticity Index.

4.2. Casagrande method (1948)

Based on the results of laboratory tests using Casagrande method ([14], there are two types of soil in research area, they are Clay High Plasticity (CH) and Silt High Plasticity (MH). Figure 2 show the graphic of the results of this test.

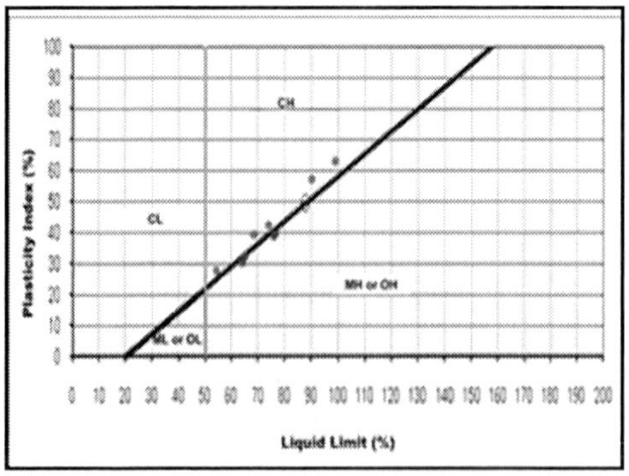

Fig. 2. Diagram of Casagrande method [14].

Table. 1. Physical properties of soil in toe of slope

Sample Depth (m)	Soil Description	Laboratory Test Result							
		Atterberg Limit Test			Grain Size Distribution				USCS
		LL (%)	PL (%)	PI (%)	Gravel (%)	Sand (%)	Silt (%)	Clay (%)	
0.00 – 0.50	Silty Clay, Brown, High Plasticity	90.24	34.50	55.69	0.00	18.00	32.40	49.60	CH
N/A	N/A	N/A	N/A	N/A	N/A	N/A	N/A	N/A	N/A
1.00 – 1.50	grades with sand	54.81	57.60	27.23	0.00	21.80	34.60	43.60	CH
1.50 – 2.00	Same as above	68.79	29.60	39.16	0.00	21.80	32.90	45.30	CH
2.00 – 2.50	Same as above	74.48	32.10	42.41	0.00	6.00	43.00	51.00	CH
2.50 – 3.00	Same as above	99.31	36.00	63.31	0.00	16.40	36.20	47.40	CH

Table. 2. Physical properties of soil in top of slope

Sample Depth (m)	Soil Description	Laboratory Test Result							
		Atterberg Limit Test			Grain Size Distribution				USCS
		LL (%)	PL (%)	PI (%)	Gravel (%)	Sand (%)	Silt (%)	Clay (%)	
0.00 – 0.50	Silty Clay, Brown, High Plasticity	90.24	34.50	55.69	0.00	18.00	32.40	49.60	CH
N/A	N/A	N/A	N/A	N/A	N/A	N/A	N/A	N/A	N/A
1.00 – 1.50	grades with sand	54.81	57.60	27.23	0.00	21.80	34.60	43.60	CH
1.50 – 2.00	Same as above	68.79	29.60	39.16	0.00	21.80	32.90	45.30	CH
2.00 – 2.50	Same as above	74.48	32.10	42.41	0.00	6.00	43.00	51.00	CH
2.50 – 3.00	Same as above	99.31	36.00	63.31	0.00	16.40	36.20	47.40	CH

Blue dots represent sample from the base of the slope and the red one represents the top part of the slope.

4.3. Activity number analysis

Based on the Atterberg Limits Analysis and grains size distribution analysis, soil that formed the slope has a number of activities ranging between high to very high. Figure 3 shows the plot diagram between % of clay particle and plasticity index. It is confirmed with the results of activity number analyses based on Skempton [18] that shown in Tab. 3 and 4. Its results show that soil samples in research area has range of activity number (A) from 0.705 – 1.4931 (medium to very high).

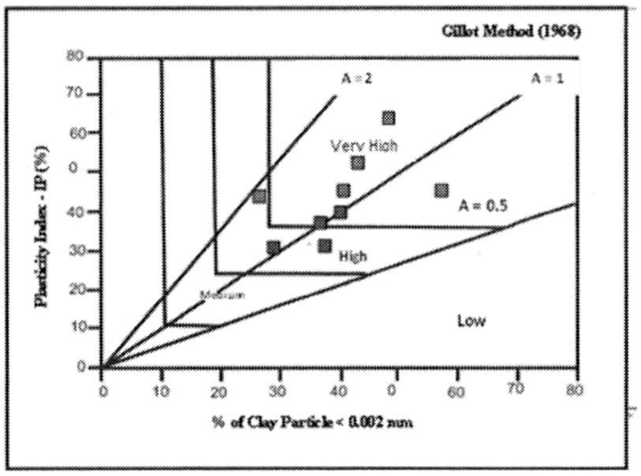

Fig. 3. Diagram of Gillot Method [15].

DOI: 10.1051/matecconf/201710104012

Table 3. Activity number noil at toe of the slope

Sample	Depth (m)	Activity Number (A)	Activity
UDS 1	0.00 – 0.50	1.099	High
N/A	N/A	N/A	N/A
UDS 2	1.00 – 1.50	1.268	Very High
UDS 3	1.50 – 2.00	1.004	High
UDS 4	2.00 – 2.50	1.2566	Very High

Table 4. Activity number soil at top of the slope

Sample	Depth (m)	Activity Number (A)	Activity
UDS 1	0.00 – 0.50	1.248	High
N/A	N/A	N/A	N/A
UDS 2	1.00 – 1.50	0.705	Medium
UDS 3	1.50 – 2.00	0.9717	High
UDS 4	2.00 – 2.50	0.922	Very High
UDS 5	2.50 – 3.00	1.4931	Very High

Figure 4 shows the relationship between activity number of soil and percentage of clay particles. It shows that soil samples derived from slopes have range of activity from medium to very high. It is in line with previous method, i.e. Gillot Method [1] and Skempton [2].

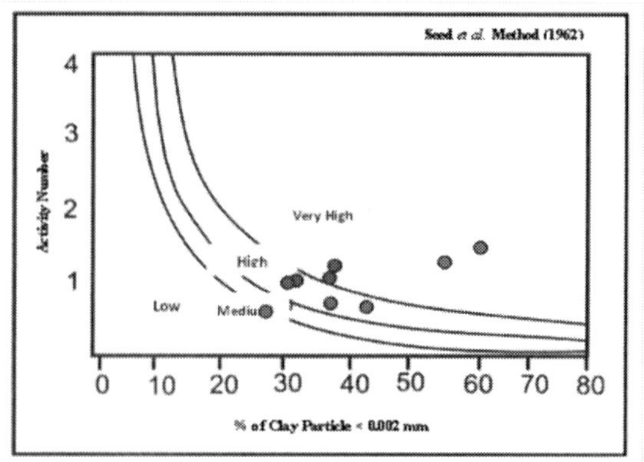

Fig. 4. Diagram of Seed et al. Method [16].

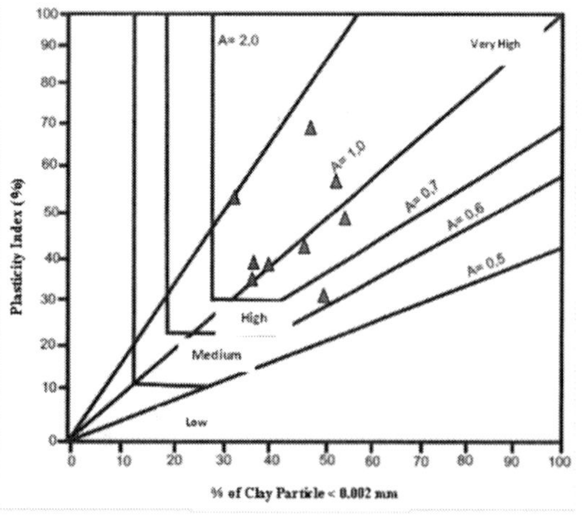

Fig. 5. Diagram of William and Donaldson Method [17].

Activity number (A) analysis using William and Donaldson Method [3] also shows that the activity distribution of soil in research area is ranging from high to very high activity (Fig. 5). It is, once more confirm that soil in research area has high to very high activity.

Based on Casagrande [14], Gillot [15], Seed *et al.* [16], and William and Donaldson [17] method, it can be linked that creep occurrence in Pagelaran was influenced by the high activity of soil that formed the slope. This high activity can be caused by active clay mineral, e.g. Montmorillonite or Smectite.

4.4. Geological structures orientation

Geological structures in research area are the manifestation of forces that worked inside of the Earth. In Creep location there was not found outcrops that indicate the apppearence of faults or folds. Nevertheless, it was found a lot of joints and cracks on the outcrops of rocks and soil. It was measured the orientation of strike and dip of joint plane using Geological Compass. The data resulted from it then analysed using Dipsto become Rosset Diagram to determine the orientation of main force ($\sigma 1$).

The main force ($\sigma 1$) that obtained from geological structures analysis of research area is show the orientation of NE-SW as shown in diagram Rosset of Fig. 4a and 4b. This orientation is in line with slope orientation. It can be interpreted that the slope and the landslide occurrence were affected and influenced by geological structure.

It was also found many cracks in the. These results also confirmed with dominant force that worked in Java Island. Based on Martodjojo (1984), there are three dominant structural lineaments in Java Island, and one of them is Meratus Pattern with NE-SW orientation body of slope. Joints or cracks in it can be access for rainwater to flow inside the slope. It then can change the properties of soil by decreasing the shear strength of soil.

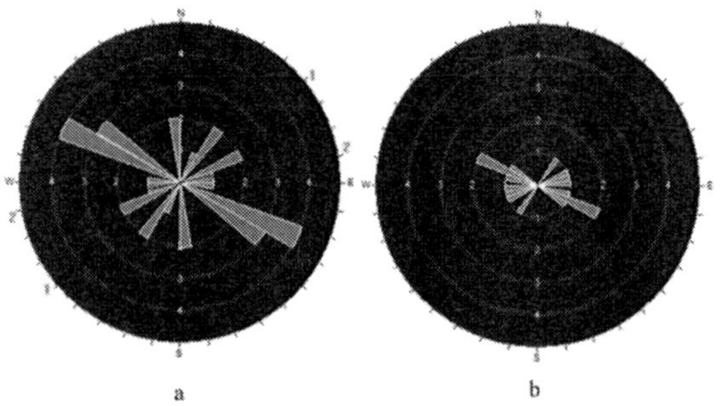

Fig. 6. Joint orientation of rock outcrop (a) and joint orientation measurement of cracks in the body of slope (b)

5 Conclusion

Based on Atterberg limits and activity number analyses, it can be concluded that creep occurrence in Pagelaran was influenced by high activity of soil that formed the slope. It can be caused by high activity clay mineral. It can be analysed using XRD test for further detail. It is also can be concluded that the creep also influenced by geological structure, especially joints that spread out on and inside the slope. It expressed by same orientation between joints and the slope, i.e. NE-SW. The joints also can be as access for rainfall to flow inside the slope and reduce its stability

Authors want to say thanks Universitas Indonesia for providing funding to conduct this research. A lot of thanks also we give for Nabila Arief who has provided literature for this research. Finally, yet importantly is for Dr. Supriyanto, M.Sc., the Head of Geology Study Program, Universitas Indonesia who always gives valuable support for authors.

References

1. B. Fatahi, T.M. Le, M.Q. Le, H. Khabbaz. Geomech. Geoeng. (2012) 1-18.
2. D.H. Kang, T.S. Yun, Y.M. Lau, Y.H. Wang. Comput. Geotech. (39) (2012) 98-106.
3. Centre of Volcanology and Geological Hazard Mitigation, Geological Agency (2015).
4. Md. Anisuzzaman, D. Muslim, Arifuzzaman, N. Sadhrina. Proceeding of the 3rd International Conference on Ecological, Environmental, and Biological Sciences, Singapore (2013) 188-191.
5. Ratna Husain, Thesis, Universitas Hasanuddin, Makassar (2015).
6. R.W. Fleming and A.M. Johnson. Eng. Geol., (27) (1989) 39-114.
7. Saro Lee, Ueechan Chwae, Kyungduck Min. Geomorphology, (46) (2002) 149-162.
8. S.A. Dunning, C.I. Massey, N.J. Rosser. Geomorphology, 103 (2009) 17-29.
9. Deying Li, Kunlong Yin, Chin Leo. Environ. Earth. Sci., (60) (2010) 677-687.
10. R.W. van Bemmelen. The Hague. (1949) 442.
11. S. Martodjojo. PhD Thesis, Institut Teknologi Bandung, Bandug (2003).
12. M. Koesmono, Kusnama, N. Suwarna. Geological Agency, (1996).
13. Braja M. Das, Book. Jakarta: Erlangga (1993).
14. Casagrande. The U.S. Army Corps of Engineer and the U.S. Bureau of Reclamation. (1948).
15. Gillot. Clay in engineering geology. Elsevier Publishing Company (1968) 234.
16. H.B. Seed, R.J.W. Ward, R. Lundgren. Highway res. Board Bull. (1962) 12-39.
17. A.A.B. William and G. Donaldson. Proceeding of 4th International Conference on Expansive Soil, Denver (1980) 834-840.
18. A.W. Skempton.. Geothecnique (4) (1953).

Environmental value losses as impacts of natural resources utilization of in coal open mining

Restu Juniah[1], *Rinaldy* Dalimi[2], *M.* Suparmoko[3], *Setyo* S Moersidik[4] ,and *Harry* Waristian[1]

[1]Department of Mining Engineering Faculty of Engineering Sriwijaya University, 30139 Palembang, Indonesia
[2]Regional Developments Studies, Electrical Engineering, University of Indonesia, 16424 Jakarta, Indonesia
[3]Environmental Economics, University of Budi Luhur, 12260 Jakarta , Indonesia
[4]Regional Developments, Environmental Science, University of Indonesia, 16424 Jakarta, Indonesia

Abstract. Mining industry provides benefits to the state in mining as a source of state revenue and foreign exchange earner, producing raw materials for the industrial, infrastructure and socio-economic facilities, and employment providers. Utilization of natural resources in openly coal mining in the forest did to get these benefits, but on the other hand cause environmental harm. Land clearing activities in coal mining, open forest vegetation is causing the loss of economic benefits of forest. Exploitation of coal caused the shrinking amount of coal (coal depletion). Although revegetation activities have been conducted on coal mined land, those would still cause harm to the environment. The research conducted at the site of coal mining PT Bukit Asam (PTBA) South Sumatra Indonesia using quantitative methods of Net Present Value (NPV) aims to assess the value of environmental damages incurred as a result of the utilization of natural resources in coal strip mine in forest areas. The calculation of forest benefit in and after the forest recovery period is using time dimensions that has innovated in this study. The results found for the environmentalloss duringlife of mine for a period of time 1997-2023 PV 2009 amounted to Rp 73.98 trillion,-

1 Introduction

Environmentally sound development becomes an important requirement for any nation that wants the preservation of natural resources. Therefore, natural resources need to be preserved and maintained for the survival of human kind in the present, and for the next generations. Forest destruction has led to reduction in the value of forest benefits.

Damage of the forest will also damage its ecosystem and its surrounding areas, as well as destroying all living systems. According to Reference [1] a lot of forest ecosystem services value clearly can not be known its monetary value. The impact of these policies affect the forest use, including for open pit coal mining activities more consider the financially benefits, so that the rate of degradation become faster.

Constanza *et al.* [1] stated that the increased of degradation threatens the value of the benefits of forest ecosystems. Use of natural resources by open mining coal land clearing activities and coal exploitation activities pose a negative externality on the function of forest ecosystems, such as forests function as a provider of natural resources such as timber resources and coal resources contained in the forest.

The loss of this function results in loss of economic benefit to the forest so that the resulting loss of forest and coal depletion. Revegetation activities on coal mined lands to forests provide economic benefits in the form of the benefits of revegetation. But unfortunately coal mining activity in Indonesia including PT Bukit Asam South Sumatera not account for the losses suffered as a result of the loss of forest and coal depletion. Similarly, the value of the environmental benefits derived from the activities of revegetation.

In accordance with this study seem it necessary to assess economic losses of natural resources and the environment due to the use of natural resources in coal mines open in logged-over secondary forests on coal stock loss and the loss of value of the economic benefits of forests.

2. Material and methods

2.1 Study area

PTBA founded on 1981, now the company is the 6th biggest mining company in Indonesia. Mining systems used in PTBA to exploit the natural resources of coal using open pit methods. Mining area of PT. Bukit Asam (Persero), Tbk. PTBA's coal mining consists of three mining blocks namely Tambang Air Laya (TAL), Tambang Bangko Barat (TBB), and the Tambang MuaraTiga Mine Besar (MTB). This study focused on

* Corresponding author: author@e-mail.org

DOI: 10.1051/matecconf/201710104013

TAL Block, located in District Lawang Kidul Muara District Enim, South Sumatra Province, Indonesia.

2.2 Methodology

Based on field investigation, the research conducted is environmental filed. Research conducted by surveys directly to the field of TAL PTBA to verify the secondary data obtained by instasional studies, and direct observation conducted to examine the effects of externalities of coal mining activities on the environment. Instational Survey conducted by visiting institutions associated with this study.

The population of study is a block of the mines in the coal mine site PTBA which consists of (1) Tambang Air Laya (TAL), (2) Tambang Bangko Barat (TBB), (3) Tambang Muara Tiga Besar (MTB). Samples were taken in Tambang Air Laya block. Sample determination is conducted by purposive sampling [8]. Determination based purposive sample on consideration of: (1) how long the PTBA coal mining blocks has been mined and reclaimed; searched the first block mined and reclaimed, and to know the influence of the distribution of the effects of these activities on the environment.

The study is a quantitative method to calculate the net present value (NPV) of the benefits and costs associated with the externalities of the utilization of natural resources in the open pit coal mining to the environment.

3. Economic loss assessment of natural resources and environmental impacts of natural resource utilization of open pit coal mining

The utilization of natural resources of coal in the open pit is one of the activities that alter land conversion and forest ecosystems can be a threat to the economic value of natural resources and environment. The conceptual framework of the research is shown in figure 1.

Broadly speaking, the calculation of depletion and degradation of natural resources has been done in previous studies.

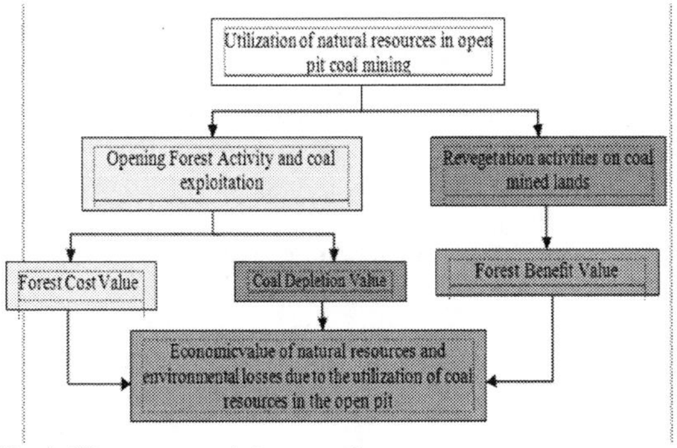

Fig .1. The conceptual framework

In this study, the calculation of depletion of natural resources is still being done on coal resource depletion calculation and standing timber logged-over secondary forests. The development of the previous studies is to provide an economic assessment of forest loss, the study calculated the dimension of time is when the forest opened up the forest revegetation approach requires its original estimated recovery time for 15 years.

The economic value of each component (the value of coal depletion, loss of forests, indigo benefits of revegetation) next calculated the present value (present value) in 2009. Present value to the value of each component are summed to obtain a total economic value (NET) loss of natural resources and the environment due to the use of natural resources in the coal mine open.

Calculations are made up of natural resource depletion and environmental losses, the benefits of revegetation on coal mined lands and forest benefits to n after the end of revegetation activities on coal mined lands (during the recovery period of 15 years), with the calculation of the time dimension is a unitary concept of determination environmental losses due to natural resource use coal strip mine the state of the art in this research complements a lack of previous research studies lack.

Value of forest loss is not taken into account only in the dimension of space, but also with the calculation of time dimensions (15 years), adjusted to the time of recovery (restoration) secondary forest logged over to approach its original state, and the mature eucalyptus plants are most abundant in the location TAL PTBA and can be extracted as eucalyptus oil, as well as the loss values for each component and forest benefits over the life of the mine and the benefits of forests to t years after the end of the forest restoration activities carried out calculations to obtain the present value of the total economic value as the novelty of this study are as follows:

a. Value of coal depletion

$$VD_c = \sum_{p=1}^{p=q} (D_c) \frac{1}{(1+r)^t} \qquad (1)$$

where:

VD_c = value of coal depletion (Rp)
D_c = coal depletion (Rp)
P = 1 = first year of exploitation= 1997
= 1,2,3, ... p=q
Q = last year of explitation= 2023
r = interest rate (%)
t = periods of coal depletion time since the first calculation of present value (year 2009)

b. Value of logged over secondary forest loss

$$VFL_{slo} = \sum_{p=1}^{p=q} (FL_{slo}) \frac{1}{(1+r)^t} \qquad (2)$$

where:

VFL_{sl} = value of forest loss (Rp)
FL_{slo} = forest loss (Rp)
p = 1 = first year of exploitation= 1997
= 1,2,3, ... p=q
q = last year of explitation = 2023

r = interest rate (%)
t = period of forest loss time since the first calculation of present value (year 2009)

with, the forest loss use the equation below:

$$FL_{slo} = FAO_{slo} \, x VF_{slo} \quad (3)$$

Source: Restu Juniah, 2012
where:

FL_{slo} = forest loss (Rp)
FAO_{slo} = forest area opened (ha)
VF_{slo} = value of secondary forest logged over /ha/year (Rp)

c. Value of logged-over secondary forest benefits

$$VFB_{slo} = \sum_{p=1}^{p=q} (FB_{slo}) \frac{1}{(1+r)^t} \quad (4)$$

Source: Restu Juniah, 2012
where:

VFB_{slo} = value of logged over secondary forest/revegetation benefit (Rp)
FB_{slo} = benefit of logged over secondary forest/revegetation benefit (Rp)
p = 1 = first year of exploitation= 1997
= 1,2,3, ... p=q
Q = last year of explitation = 2023
r = interest rate (%)
t = period of forest benefit time since the first calculation of present value (year 2009)

d. Result

Utilization of natural resources by open pit coal are conducting clearing forests through land clearing and revegetation on coal mined lands as an externality impacts on natural resources and the environment.

4. Impacts of forest clearing and revegetation of coal mined lands.

The forest clearing activities at the site of TAL PTBA conducted in order to clear mined lands for the excavation of coal. After the coal excavation accomplished the area there is an empty land called coal mined lands. With the end of coal excavation at the least after 6 months, then the coal mined lands should be started with revegetation (Government of RI Regulation No. 78 year 2010). The forest area and opened in locations revegetated of TAL PTBA from1997 until 2023 is Area of forest revegetated (1,374.5 Ha), and Area of forest opened (3,106.59 Ha).

4.1 Calculation methode of value of forest benefits

Valuation losses of natural resources and environmental externalities arising from the use of natural resources in the open pit coal use calculation methods that have been modified from previous studies [10] and those found in this study. Method of calculation used to determine the percentage of forest benefits revegetation approach for the 15-year secondary forest recovery. Percentage revegetation approaches do given the plant to become mature through the process of growth each year. 15-year time horizon used is based on eucalyptus plants most dominating in the area of revegetationat TA PTBA can be extracted to be taken advantage at the ripe age of 15 years, and according to [7] it would take 15 years for the recovery of logged-over secondary forests in the TAL PTBA site.

Value of environment benefits obtained from each year revegetation percentage for 15 years time horizon. Similarly for the year to 16 and so on for the operational life of the mine will continue to value the environmental benefits obtained in a sustainable manner. If the value of the benefit during forest restoration continues to increase each year in accordance with the process of plant growth up to 15th year (the printed image and his broken red line), then for n years after the end of the recovery period (16th year an so on) a decline in the value of benefits cumulative environmental (printed lines and blue arrows). It is assumed that with the increasing age of the plant after the mature will decrease the ability to be productive. The diagram of growth process and counting revegetation of forest benefits during forest restoration, and year after the recovery period) in TAL Location PTBA is shown in Figure 2.

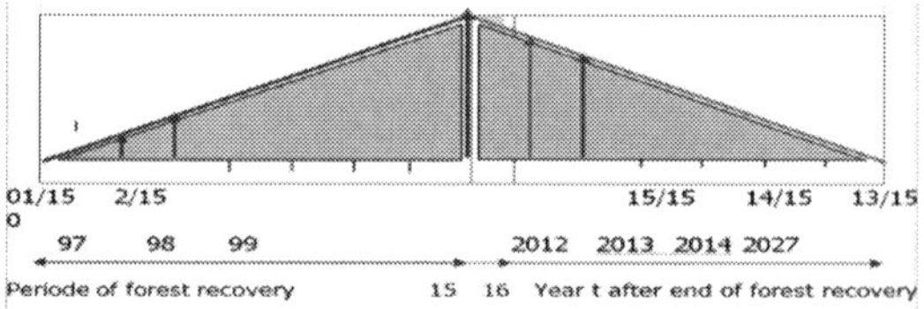

Fig.2. Diagram of revegetation growth process and counting for the forests benefit and forest restoration into the year t recovery on tal ptba location.

Forest benefits use the following equation:

$$FB_{slo} = \frac{M - m}{M} \, x \, RFA_{slo} \, x \, VF_{slo} \quad (5)$$

Persentage of revegetation

Source: Restu Juniah, 2012
where:

FB_{slo} = benefits of secondary forests looged over (Rp)
RFA_{slo} = forest area revegetated (ha)
VF_{slo} = value of secondary forest logged over /ha/year (Rp)
M = periode of forest recovery = 15 years
m = year n of forest recovery

Calculation of value of forest benefits year of t after period of forest recovery15 year use the equitation 6 .

$$FB_{t,T+1} = FB_{t,T} - (M\text{-}t/M \; x \; FB_{t,T}) \quad (6)$$

where:

$FB_{t,T+1}$ = forest benefit at the end of recovery period (Rp)

$FB_{t,T}$	=	forest benefit at the last year of recovery (Rp)
T	=	last year of recovery period
T	=	1,2 ..., M-1
M	=	periode of forest recovery = 15 years

4.2 Assessment of natural resources and environmental losses due to natural resource utilization activities of coal open mining.

Assessment of natural resources and environmental losses due to the utilization of natural resources in coal open mining to assess and determine losses to the environment as a result of the utilization of natural resources in coal strip mine that caused the depletion of coal and forest loss. While the total economic value ((TEV) is obtained by subtracting the value of the loss (the value of coal depletion and loss of forests) to the value of forest benefits obtained through the benefits of revegetation on coal mined lands. Assessment was conducted on the following components of the value of coal depletion, loss of value, value of forest benefits.

a. Value of coal depletion

Coal mining activities with the exploitation of coalcausedcoal depletion, coal reserves shrinking resulting inthe loss of ecosystem servicesto provide thenatural resourcesof coal. Calculation of coal depletion annually using the equation:2. While the economic rent of coal obtained Rp 242,325,

Based on the calculations, value of coal depletion in 1997 - 2023 PV 2009 logged over TAL PTBA location is Rp 73,474,494,298,342, - or 73.5 trillion rupiah. Thus, this study found that coal depletion occurred in coal mines during mine life TAL PTBA years 1997-2023 PV 2009 as externalities of natural resource utilization activities of coal open mining valued at 73.5 trillion rupiah.

b. Value of forest loss

The forest clearing activities at the TAL PTBA site since 1997 has been done in stages in accordance with the stages of mining. The need to internalize external costs into the cost of production due to the cost required to address the environmental damage is much greater than the cost of prevention. The results of Sudirman (2011) found no environmental damage costs incurred as externalities PT Indominco Indonesia coal mining in protected forests. Sudirman assess the cost of environmental damage caused by coal mining in protected forests, whereas in this study examined the effect of external costs of coal mining activities secondary forest areas. Costing forest at TAL PTBA location uses equations: 4. Based on calculations that have been made to known the economic value of forests per ha per year in TAL PTBA coal mining is Rp. 39,082,000, -. Using equation 3 by assuming the interest rate of 7% (the average rate of Bank of Indonesia, 2011) and remains valid for a year, the value of forests costs over the life of the mine in 1997-2023 PV 2009 for the 15-year restoration of logged over forest in TAL PTBA site is Rp 834,001,622,078.003, - or 834 billion rupiah. Thus, this study found the loss suffered by secondary logged-over forests in the coal mines during mine life of TAL PTBA years 1997-2023 PV 2009 as externalities arising from the utilization of natural resources in the open pit coal worth 834 billion rupiah.

c. Value of forest benefit.

Forest is a natural resource which is very important and useful for life and living either directly or indirectly, that contains the value of services for the life of the environment service. Previous research by [9] stated to get the value of 1,835.83 U.S. $ / ha/ year as value of secondary forest services in Indonesia. Similarly, research in Thailand by [4] suggests a value of USD 756.16/ha/year obtained from externalities reforestation with type *Eucalypthuscamaldulensia*. Compensation for people who lost access to the forest of USD 673,000 per year and the value of farmers' losses due to the flood of USD 71,000 per year in Madagascar shows the value of environmental services of forests [3]. Value of the external benefits in this study was obtained from the benefits of revegetation at the TAL PTBA site. Revegetation activities at TAL PTBAsite are performed each year in accordance with the annual plan of environmental management of TAL PTBA. Revegetation activities is calculated by assuming that (1) the economic value of plants emerged a yearafter the implementation of revegetation activities initiated by 15-year recovery period, (2) After 15 years of restoration at the TAL PTBA site, revegetation activities not yet done, and no logging activity, but the forest still preserved until coal post-mining, so in year 16 and post-mining coal revegetation activities still providing value of benefits.

The calculation of the benefits of forest each year in TAL PTBA coal mines using equation 6. Using equations 7, assuming the rate of interest 7% (the average rate of Bank of Indonesia, 2011) and remains valid for a year, and the economic value of forests per ha per year in coal mines TAL PTBA is Rp 39,082,000, - discovered the value of forest benefits are environmental benefits over the life of the mine in 1997-2023 PV 2009 with a 15-year restoration of logged forest in PTBA TAL site of Rp 324,233,514,093, - or 324 billion rupiah.

Thus, this study found that the benefits arising from the revegetation on coal mined lands in the coal mines during mine life TAL PTBA years 1997-2023 PV 2009 as externalities of natural resource utilization activities of coal open mining worth 324 billion dollars. Value of coal depletion, value of forest cost, value of forest benefits, and value loss in the year 1997-2023 PV TAL PTBA's coal mines in South Sumatra Province, Indonesia.

Land conversion activities that alter forest ecosystems such as the utilization of coal resources in open pit mines pose externalities on natural resources and the environment that cause harm to the environment. Environmental losses that resulting from the utilization of open pit coal mining as shown in the Table 1 above as a sum of 73.98 trillion rupiah. Revegetation activities on coal mined lands are expected to recover the loss of environmental benefits can only benefit the forest worth 324.23 billion rupiah. Value of this benefit is very small compared with the losses suffered by the environment worth 73.98 trillion rupiah.

Table 1. Environmental losses

Year	Forest Loss PV 2009 (a)	Depletion of Coal PV 2009 (b)	Forest Benefits PV 2009 (c)	Environmental Losses (a+b)-c
1997	32,385,212,266	763,104,135,591	-	795,489,347,856.97
1998	32,898,428,785	741,156,209,096	21,300,027,319	752,754,610,562.09
1999	33,328,908,303	894,950,539,672	19,017,286,295	909,262,161,679.71
2000	33,949,385,774	792,599,074,362	18,708,592,124	807,839,868,012.19
2001	34,487,389,043	781,868,058,132	18,404,911,091	797,950,536,084.43
2002	17,797,496,341	862,635,485,539	18,106,159,460	862,326,822,419.51
2003	30,865,683,990	985,458,196,909	10,475,303,648	1,005,848,577,250.44
2004	36,325,138,652	864,656,381,795	18,484,621,823	882,496,898,625.05
2005	36,045,560,895	908,395,918,328	20,474,485,923	923,966,993,299.14
2006	48,384,868,936	958,621,294,118	19,972,191,292	987,033,971,762.37
2007	39,899,562,069	974,517,243,680	28,221,939,119	986,194,866,629.74
2008	5,122,695,421	1,061,285,503,770	13,553,425,005	1,052,854,774,186.16
2009	8,641,994,298	1,279,476,000,000	10,357,937,753	1,277,760,056,545.28
2010	3,552,435,842	2,544,124,427,664	10,389,688,458	2,537,287,175,047.11
2011	1,403,861,859	2,747,654,381,877	11,638,498,935	2,737,419,744,800.64
2012	13,058,023,230	2,967,466,732,427	10,858,414,843	2,969,666,340,814.18
2013	3,486,788,579	3,204,864,071,021	10,110,677,323	3,198,240,182,277.53
2014	0	3,461,253,196,703	9,393,170,245	3,451,860,026,457.78
2015	13,958,524,170	3,738,153,452,439	8,703,915,916	3,743,408,060,693.01
2016	9,013,150,696	4,037,205,728,634	8,041,066,026	4,038,177,813,304.02
2017	0	4,360,182,186,925	7,402,893,182	4,352,779,293,743.06
2018	78,310,010,140	4,708,996,761,879	6,769,096,030	4,780,537,675,988.71
2019	73,186,925,365	5,085,716,502,829	6,158,075,278	5,152,745,352,915.68
2020	68,398,995,668	5,492,573,823,055	5,568,340,880	5,555,404,477,843.22
2021	63,924,295,017	5,931,979,728,900	4,998,500,271	5,990,905,523,645.46
2022	59,742,331,792	6,406,538,107,212	4,447,251,988	6,461,833,187,014.87
2023	55,833,954,945	6,919,061,155,788	3,913,379,711	6,970,981,731,022.71
Total	834,001,622,078	73,474,494,298,342	324,233,514,094	73,984,262,406,325.80

Assessment of natural resources and environmental losses due to natural resource utilization activities of coal open mining can be applied by: 1) Government as policy makers should internalize the environmental cost into coal mining feasibility study using open pit methods. The Government, should implieda policy to internalizethe external cost (environmental cost) in the feasibility study of coal mining activity with open pit method, 2) Openly coal mining business in general should determine the value of losses incurred as externalities of coal open mining activities.

Conclusion

The calculation of forest benefit in and after the forest recovery period is using time dimensions has been innovated in this study. The results found for the environmentalloss during life of mine for a period of time 1997-2023 PV 2009 amounted to Rp 73.98 trillion,. The results of this study indicate that such large losses suffered as a result of the utilization of the open pit coal mining apart due to the loss of forest resources worth 834 billion rupiah also caused the loss of coal stock in the nature worth 73.47 trillion rupiah. With the findings of this study should coal mining operation account for losses arising as environmental externalities of natural resource utilization activities of coal open mining.

References

1. R. Costanza, , R. D'Arge, ,R. de Groot, S. Farber, , Grasso, M., Hannon, B., Limburg,K., Naeem, S., O'neill, R., Paruelo, J., Raskin, R., Ssutton, P., & van den Belt, M. Nature, 387 : 253-260 (1997)
2. Directorate General of Mineral and Coal. Annual Report Management &Environmental Monitoring Tambang Air LayaTanjungEnim- Sumatera Selatan (2012)
3. Kramer, RA., N. Sharna., and M. Munasinghe. World Bank Environment Paper. Number 13. The World Bank, Washington DC (1995)
4. A. Niskanen, Value of External Environmental Impact of Reforestation (1998)
5. In Thailand. Ecological Economic Journal No. 26 (1998).
6. Government Regulation Number. 78 Year 2010 about Reclamation and Mine Closing. Directorate General of Mineral. Ministry of Energy and Minaral Resource Republic of Indonesia (2010)
7. A. Stenger, P. Harou, & S. Navrud, *Journal of Forest Economics*, 15, 1–14 (2009)
8. Sukandarrumidi, Research Methodology.3rd Edition, Gajahmada University Press, Yogyakarta, ISBN: 979-420-508-7 (2006)
9. Suparmoko, Journal of Development Economic (Economic Study on Developing Countries) Vol. 13 No.3 December 2008 Pages:217-229. (2008).
10. Sudirman, Valuation econict of Mining Impact" disertation, PSIL UI (2011)

Marine ecology conditions at Weda Bay, North Maluku based on statistical analysis on distribution of recent foraminifera

Anis Kurniasih[1], *Septriono* Hari Nugroho[2], and *Reddy* Setyawan[1]

[1]Geological Engineering Diponegoro University, 50275 Semarang, Indonesia
[2]Deep Sea Research Centres-LIPI, 16003 Bandung, Indonesia

Abstract. Analysis of foraminifera in geology, usually being used to find the age of rocks/ sediments and depositional environment. In this study, recent foraminifera was used not only to determine the sedimentary environment, but also to estimate the ecological condition of the water through a statistical approach.Analysis was performed quantitatively in 10 surface seabed sediment samples in Weda Bay North Maluku. The analysis includes dominance (Sympson Index), diversity and evenness (Shannon Index), and the ratio of planktonic - benthic. The results were shown in the plotting diagram of M-R-T (Miliolid-Rotalid-Textularid) to determine the depositional environment. Quantitative analysis was performed using Past software (paleontological version Statistic 1:29).The analysis result showed there was no domination of certain taxon with a moderate degree of evenness and stable communities and considerably a moderate diversity. The results of this analysis indicated that research area had a stable water conditions with the optimum level of carbonate content, oxygen supply, salinity, and temperature.The ratio of planktonic and benthic indicate the relative depth, which was deeper the water increased the percentage of planktonic foraminifera. Based on M-R-T diagram showed the distribution of sediment deposited on exposed carbonate (carbonate platform) environment with normal saline.

1 Preliminary

Analysis of the waters condition is very important because Indonesia is an archipelagic country that most of its territoryis covered by the sea. Ecological condition of a water body can become a reference about the usage of the waters for the fishing industry. The ecological condition of waters can be measured directly on water samples including the levels of carbonate, salinity, and temperature. However, this measurement is less precise to describe the ecological conditions in a long time scale. Ecological condition of waters can also be known by observing the organisms that live in the waters, which is to determine the correlation between the components of biotic and abiotic. One of the living organisms that is abundant in the sea is foraminifera.

This research was conducted by applying the method in seabed sediment samples taken from Weda Bay, North Maluku. The goal is to determine the recentecological condition of the water based on the distribution of foraminifera contained in the sediments. The benefit of the study is to determine the ecological conditions in a long time scale as the basis for the usage of the waters in fishing industry and so on.

Within the scope of geosciences, foraminifera are generally only used as a marker of age and depositional environment of rocks, but in this study,statistical methods was used in recent foraminiferacommunity in sediments to estimate the ecological condition of waters. Previous research on ecological analysis based on distribution of foraminifera had been done by [1] on the sediment surface in the Makassar Strait. The purpose of this study was to determine the distributional pattern of foraminifera, which include uniformity, diversity, and dominance of the individual as well as knowing the ecology of the waters at this time (Recent) based on the distribution of foraminifera contained in the sediments.

2 Research method

2.1 Research site

The research site was the territorial waters of Weda Bay, North Maluku (Fig.1). From all the sediment samples were taken, 10 of them performed a quantitative analysis of foraminifera, that was sample B2, B3, ST1, ST2, ST5, ST6, ST7, ST9, ST10, and ST11.

2.2 Sampling method

Sediment sampling was conducted by using a grab sampler. Sample of the sediment was weighed and then washed using a 100-mesh sieve with openings. Sample of the sediment was analyzed, mainly the foraminifera from 1 gram of sediment leftovers (washed residue). Thedistribution of foraminifera was quantified by the maximum number of 300 individuals.

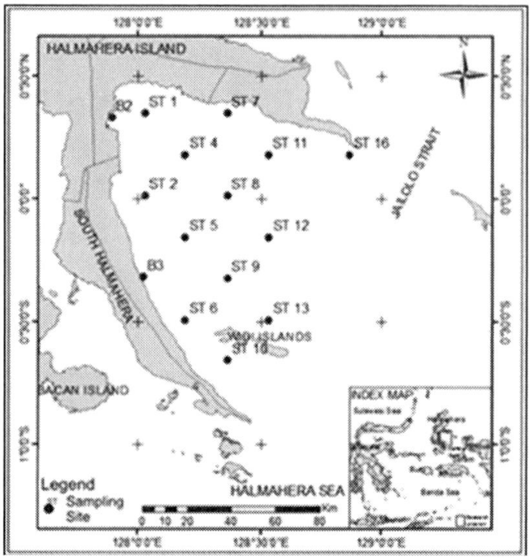

Fig. 1. Location of the study[2].

2.3 Analysis of recent foraminifera

2.1.1 Qualitative and Quantitative analysis

Qualitative analysis wasused to determine the taxonomic classification of foraminifera. Planktonic foraminifera taxonomy based on [3]andfor the benthic foraminifera was based on [4, 5, 6]. The taxonomic classification of foraminifera was the basis data to get the age and the depositional environment of sediment or rock samples.

Beside of qualitative analysis, this research used the quantitative analysis calculatedthe number and types of foraminifera that exist in each samples.The distribution of foraminifera was established in the distribution table, which then was used to conduct statistical analysis.

2.1.2 Community Structure of Foraminifera

Foraminifera community structure expresses in uniformity index (similarity), the diversity index (diversity), and dominance index. The variability or diversity is the number of species or taxon presents in a community. Diversity index in this study was performed using the index Shannon – Wiener [7], calculated as the formula below (1).

$$H = -\sum \frac{ni}{n} ln\left(\frac{ni}{n}\right) \quad (1)$$

Where:

H = Shannon-Wiener index

ni = Number of individuals of a taxon

n = total number of individuals

Diversity index criteria are divided into three categories [6]:

H <1 = low diversity

1 <H <3 = moderate diversity

H> 3 = high diversity

In addition of using the index of Shannon - Wiener diversity can also be determined using the Fisherdiagram[8] via the alpha index calculation, as the following equation (2).

$$S = \alpha \ln (1 + n/ \alpha) \quad (2)$$

S = the number of taxa

n = total number of individuals

α = total number of individuals

Uniformity value is the distributional pattern of each foraminiferataxonis equal or not, if the uniformity index value is relatively high the existence of any kind of foraminifera in the waters isunder equitable conditions. Uniformity index stated in Evenness Index (e), in accordance with the following formula (3) based on [7].

$$e = \frac{H}{logS} \quad (3)$$

e = evenness / uniformity index

H = index of Shannon - Wiener

S = the number of taxa

Value e <0.3 indicates uniformity is low, e = 0.3 - 0.6 classified as moderate and e> 0.6 classified as high [7].

Dominance index is used to get information about taxon dominance in certain communities. Dominance index value ranging from 0 - 1. If the dominance index value close to 1 then the community is dominated by certain taxon of foraminifera and if the dominance index close to 0 then no taxon dominates [7]. Here is the calculation formula dominance index created by [9] (4).

$$D = \sum \left[\frac{ni}{n}\right]^2 \quad (4)$$

D = Dominance Index

ni = Number of individuals of a species I

n = total number of individuals

Community structure of foraminifera is one of the biotic parameters, which show the ecologic condition of waters, whether it is inhabitable by such organism, or not. This method is used to be a reliable method to determine the existence of disturbance in an area inhabited by a specific organism. The disturbance can occur by the cause of water pollution and or sedimentation.

2.1.3 The individual presence ratio of planktonic and benthic foraminifera

The individual presence ratio of planktonic and benthic foraminifera was used as an indicator of the depth of a water body [10].

2.1.4 Triangular diagram

Triangular diagram was arranged by the abundance of three types of foraminifera distinguished by the composition of foraminifera shell whether it wasagglutinate, porcelain, or hyaline (Fig.2). Triangular diagram parameter can be used as indicator of the carbonate, dissolved oxygen and salinity levels [11].

3 Results and discussion

3.1 Foraminifera Community Structure

Based on the quantitative analysis of 10 sediment samples taken from several locations in the study area (Fig. 1), obtained distribution of foraminifera which include uniformity, diversity, and their particular taxon dominates. Prior to the analysis, identification of each appearing taxon was performed and its population was counted. After that, all data was inserted in the program PAST [12] to determine the pattern of its spread (Table1).

Based on analysis of structural community in the foregoing discussion, it is known that the value of Shannon - Wiener diversity index of sample ST1, ST2, ST5, ST6, ST7, ST9, ST10, and ST 11ranged from 1-3, which means that the level of diversity is moderate. Fisher - Alpha index value of those samples show the level of diversity in the open ocean with normal salinity. This fits in with the results of the triangular diagram plotting showing the distributional pattern that characterizes the open sea environment with salinity, levels of carbonate and oxygen levels conditions were normal (normal marine continent).

Based on the dominance index value that is almost entirely close to 0 can be determine that there is no type of foraminifera that dominates from each community in the whole sample. Degree of uniformity or evenness of each total sample ranged from 0.3 to 0.6, which means the level of uniformity moderate.

A conclusion from the analysis of the structural community is that the condition of the waters where the foraminifera were living was in optimum condition so that both planktonic and benthic organisms can thrive maximum. This can be seen from the absence of one type of foraminifera that dominates, meaning that each type of foraminifera can live and thrive. In addition, the level of diversity and uniformity were moderate, meaning that all types of foraminifera that live in the water has the same chance to thrive, or in other words, competition is low.Sample B2 and B3 were exceptional, because those samples were taken in transition between marine and terrestrial area, which not allowed foraminifera to live. Moreover, foraminifera that were occurred in those two samples might be derived from different environment. In other word, foraminifera that were contained in sample B2 and B3could not be used to indicate the structural community.

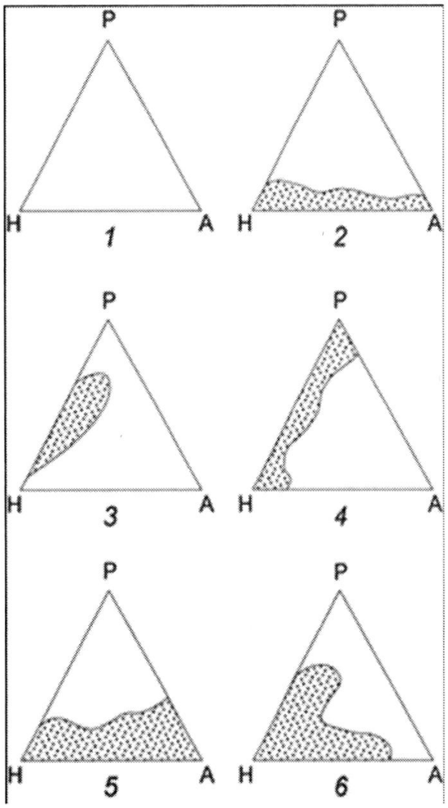

Fig. 2. Triangular diagram showing the abundance of foraminifera in several different environment [11], note: 1. Fresh water; 2. Hypo saline lagoon; 3. Normal marine lagoon and carbonate platform; 5. Shelf; 6. Normal marine continental; P-porcelain; H-hyaline; A-agglutinated.

3.2 The presence ratio of planktonic and Benthic Foraminifera

Based on the ratio of planktonic and benthic foraminifera presence in the research area, nearly all samples had percentages of planktonic foraminifera presence far greater than benthic foraminifera, e.g. on the samples ST1, ST2, ST5, ST6, ST7, ST9, ST10, and ST 11. Planktonic foraminifera live by floating in the water and when it dies, the shell will be swept away. The higher the percentage of planktonic foraminifera presence indicates that the water allowed the shells of foraminifera to be deposited alongside other material. Thus, the place where the samplesweretakenpossibly has a low current or far away from the beach.

Meanwhile, there are two samples that have very lowpercentageof planktonic foraminiferaappearance compared to benthicforaminifera, the samples are B2 and B3. Based on observations in the microscope, sample B2 contains many pellets. Pellets are formed by

Table 1. Calculation value of dominance index, uniformity Index, and diversity index using PAST software.

	B2	B3	ST1	ST2	ST5	ST6	ST7	ST9	ST10	ST11
Taxa S	12	11	21	16	12	18	15	14	16	15
Individuals	77	216	295	240	296	237	287	238	250	240
Dominance D	0.1682	0.3791	0.1035	0.2378	0.2525	0.2165	0.1441	0.2858	0.3007	0.1801
Shannon H	2.054	1.369	2.532	1.913	1.676	2.008	2.124	1.652	1.667	2.025
Simpson 1-D	0.8318	0.6209	0.8965	0.7622	0.7475	0.7835	0.8559	0.7142	0.6993	0.8199
Evenness e^H/S	0.6499	0.3574	0.5988	0.4234	0.4452	0.414	0.5576	0.3725	0.3309	0.5048
Fisher alpha	3.984	2.45	5.171	3.859	2.512	4.526	3.365	3.25	3.811	3.547

chemical reactions that formed oval-shaped like material and contain many organic materials, probably derived from invertebrate organism's feces. Pellets are common in carbonate mud, as the sample B2 is based on granulometric analysis have mud dominant material. Water conditions that are dominated by mud do not allow foraminifera to live. The presence of foraminifera in the samples B2 likely derived from ocean currents that carry the shells of foraminifera from the sea towards the mainland. Meanwhile, sample B3 contained many plant particles, which indicates a high influence of land material so it does not allow foraminifera to live too. The presence of foraminifera in the samples B3 possibly formed during the transportation of sediment grain from the ocean.

The ratio of planktonic and benthicforaminifera individual presence can be seen in Table 2.

Table 2. Appearance ratio of foraminifera planktonic and benthic.

Code	Total Plankton	Total Benthic	Total Individual	*percentage plankton*	*percentage benthic*
B2	21	56	77	27.3	72.7
B3	36	180	216	16.7	83.3
ST1	217	78	295	75.6	26.4
ST2	200	40	240	83.3	16.7
ST5	274	22	296	92.6	7.4
ST6	180	57	237	76	24
ST7	184	103	287	64.1	35.9
ST9	226	12	238	95	5
ST10	228	22	250	91.2	8.8
ST11	219	21	240	91.25	8.75

3.3 Triangular diagram

Triangular diagram was arranged based on the abundance of three types of foraminifera, distinguished by the composition of the shell wall, which is agglutinated, porcelain, and hyaline. The type of foraminifera that has an agglutinated shell wall are *Bathysiphon sp.* and *Ammobaculites sp.,* while the type of foraminifera that has porcelain shell wall are *Quinqueloculina sp.* and *Pyrgosp.*Foraminifera which has a hyaline shell wall are *Nodosaria sp.*, *Elphidium sp.*, *Uvigerinasp.,Cibicides sp., Obesopleustomella sp., Oridorsalisumbonatus, Bolivina sp., Bullimina sp., Lagena sp., Amphistegina sp.*, and *Baculogypsinasp.*The presence of these groups of foraminifera in each sample is plotted in the triangular diagram, as shown in Figure 3.

The resulting from plotting in triangular diagram of the entire sample (Fig.3) shows the distributional pattern that characterizes the open sea environment with salinity conditions, levels of carbonate and oxygen levels were normal (normal marine continent).

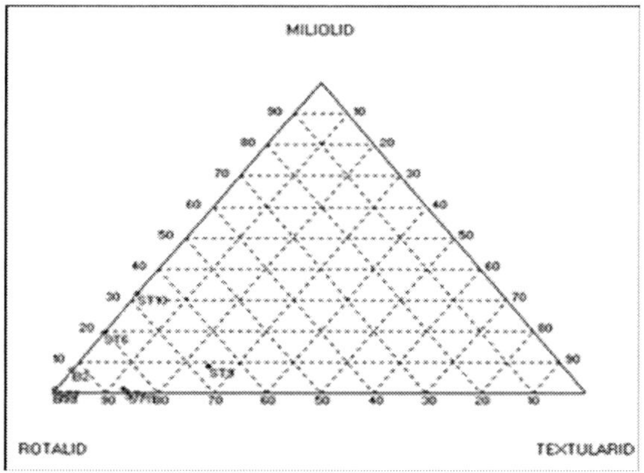

Fig. 3. Plotting a triangular diagram of all samples from the study area (Note: P = porcelain; H = hyaline; A = agglutinated).

4 Conclusion

Based on the stages and methods of research that has been done, it can be concluded as follows:

1. Based on analysis of structural community it is known that the value of Shannon - Wiener diversity index of all samples ranged from 1-3 which means that the level of diversity is moderate.
2. Fisher - Alpha index value shows the level of

diversity in the open ocean with normal salinity.

3. The resulting from plotting in triangular diagram shows distributional pattern that characterizes the open sea environment with salinity conditions, levels of carbonate and oxygen levels were normal (normal marine continent).
4. Based on the dominance index value that is almost entirely close to 0 can be determine that there are no type of foraminifera that dominate from each community in the whole sample.
5. Degree of uniformity or evenness of each total sample ranged from 0.3 to 0.6, which means the level of uniformity is moderate.
6. Based on the analysis of the community structure it can be seen that the condition of the waters where the foraminifera were lived in conditions of salinity, levels of carbonate and oxygen levels that optimum.

The authors thank to Abdul Basit, M.App.Sc as coordinator of the activities that have provided an opportunity for the retrieval and use of sediment samples from Weda Bay expedition in 2013 organized by UPT BKBL LIPI Ambon research funding sources DIPA 2013. The thanks also go to parties who have been willing to provide corrections, criticisms, suggestions and input, so that the research and writing ofthispapercan be resolved

References

1. S. M. Natsir, A. Firman, I. Riyantini, and I. Nurruhwati. J.IlmudanTek. KelautanTropis**7**,671-680 (2015).
2. S. H. Nugroho and A. Basit. J. IlmudanTek. KelautanTropis 6 No. 1,229-240 (2014).
3. J.A. Postuma. *Manual of planktonik foraminifera. Amsterdam*. Elsevier Publishing Company (1971) 252 – 420.
4. E. Boltovskoy. Marine Geo.**26**139-175 (1978).
5. A. R. Loeblich and H. N. Tappan. Foraminiferal genera and their classification. New York. Springer (1998).
6. F. P.C.M. van Morkhoven, W. A. Berggren, and A. S. Edward. *Cenozoic cosmopolitan deep-water benthic foraminifera*. Pau. Elf-Aquitaine (1986)
7. E. P. Odum. *Fundamentals of ecology Third Edition*. W.B. Philadelphia. Saunders Company (1971) 143 – 154.
8. J.W. Murray. *Ecology and paleoecology of benthic foraminifera*. Essex. Longman Harlow (1991).
9. E.H. Simpson. *Measurement of diversity*. Nature (1949) 163.
10. H. A. Armstrong and M. D. Brasier. *Microfossils second edition*. Cornwall. Blackwell Publishing (2005) 156.
11. B. Valchev. *50 Years University of Mining and Geology "St.IvanRilski" Annual, Geology and Geophysics*, Sofia, 46 (2003) 189-194.
12. Ø. Hammer, D.A.T. Harper, and P.D. Ryan. *PAST: Paleontological statistic software package for education and data analysis*. (2004)

Effect of socioeconomic status and institution of the environmental concern level

Zuber Angkasa Wazir[1,*]

[1] Department of Architecture. Muhammadiyah University of Palembang, 30263 Palembang, Indonesia

Abstract. Development in developing countries should be implemented in a sustainable and responsible. The responsibility not only by the government but also the Society. Society must have a high environmental concern. Therefore, the sociological factors that affect the level of environmental concern must be identified. This study used socio-economic status and social institutions as potential variables affect the level of environmental concern. The study was conducted in three housing in the suburbs, where the land conversion happens on a large scale in the city of Palembang. The level of environmental concern identified by NEP scale which had Cronbach Alpha value of 0.658. Socio-economic status is represented by the last educational variables and the average monthly income while institutions are represented by profession occupants. Total respondents were 103 heads of households with the majority of high school educated, the average income of IDR 4.3 million per month, and 53 respondents as a trader. Data analysis using linear regression. The analysis showed that the negative effect of institutions while the socio-economic status did not affect the level of environmental concern. These results show the importance of sociological intervention of institutions that positively affects the level of environmental concern.

1 Introduction

In developing countries, the discourse of responsible and sustainable development is an important discourse in an effort to meet the needs of the present without compromising the ability of future generations to meet their own needs [1]. Even so, it is difficult to achieve because of human behavior in developing countries experiencing a transition between state-oriented environment (agrarian or maritime) to the industrial countries. New emerging society moved from an era that basically is already ongoing, but the offer of consumerism and industrialization irresponsible provide hedonic desire in an atmosphere that is still sustainable environment. Consequently, the reluctance to adopt a sustainable development that seeks to be more careful in the exploitation of natural resources and behavioral development in general.

Their behavioral deficits such as these require corrective steps are involving elements of the government in order to reach the public in a large scope. Hopefully, with the gap being filled public awareness can lead to the creation of sustainable behavior so that development can be more certain in the future.

In line with this, studies in psychology and sociology neighborhood went on efforts to assess aspects of the physical-biological process of the physical environment of nature, increasing complexity of spatial-physical for their aspects of socio-physical that need to be considered in the study, as well as their paramount concern in behavior and actions in everyday human environment in relation to natural resources or processes in nature considered ecologically sustainable [2]. The present study attempted to assess the factors of socio-economic and social institutions in their influence on one of the indicators in environmentally friendly behavior which is the level of environmental awareness.

The study was conducted in Palembang, Indonesia. Palembang is a city that is growing rapidly, characterized by the growth of housing in large quantities. Residential neighborhood that continues to grow is to produce a large-scale land conversion. Originally, the land around and in Palembang itself is so marshy that practically the main transportation is unreliable public transport streams [3]. Since the 20th century, large-scale changes to the conversion of wetlands into housing in the suburbs and in the city give the effect of flooding. Indeed so, the City government strives to increase public awareness of the environment in a multi-ethnic city. Various physical programs conducted as forming a city park or clean-up activities on a regular basis. Even so, the psycho-sociological aspects still less untouched. therefore, this study sought to touch this aspect by highlighting the factors that can affect the level of public environmental concern in Palembang.

2 State of the art

In 1978, Dunlap and Van Liere filed New Environmental Paradigm Scale (NEP scale) to measure the belief that humans are part of nature that they should preserve it

* Corresponding author: zuberpalembani@gmail.com

and avoid exploitation of natural resources [2]. NEP scale is a belief system embedded and difficult to change on an individual regarding his perspective on the environment. NEP scale measure the ecological core beliefs as their ecological limits and the importance of maintaining the balance of nature [2].

In its development, the NEP scale was found significantly associated with pro-environmental behavior [2]. Therefore, we can approach the NEP scale as a form of human values guide to describe their behavioral tendencies pro-environment.

Previous research has identified several demographic variables that influence individual attitudes toward the attitude towards the environment [4]. These variables are generally weak and sometimes inconsistent between populations. Among them, the respondents were young and highly educated tend to be more positive about environmental issues [4;5]. Women tend to be more pro-environment than men [6;7]. Certain ethnic groups in care more about the environment than other ethnic groups [4]. High socioeconomic status and ethnic origin majority (elitism) tend to have a positive attitude towards the environment [4]. Meanwhile, other studies have shown that social context influence the behavioral expression of an attitude, such as between ethnic groups with environmentally friendly behavior [8].

The social institution itself is a means of transmitting norms, values, beliefs, habits, and so on between generations [9]. That is, social institutions is an effective tool to raise a certain world view, including new ecological paradigm. Social institutions themselves structure the availability of social and economic incentives in the community for a certain behavior or attitude [10].

Changes in society created mainly through the support of social institutions that contain systems of values stored in social institutions. Social institutions increasingly important role in modern society is a society of risk. Risk society characterized by a fragmented view and there is competition between institutions and values. It can be said, containing risk society institutions that pro-environment, are vying for dominance. Such a society is going to be tough when pro-environmental institutions able to control the entire system in a holistic and integrated [11]. This is because social institutions, especially if tied to a spatial context, will be a significant impact on the lives of community members [12].

Indeed so, the literature is still saying that there is still much to be learned from the relationship between a social institution with human individuals [12]. This can include the relation between the social institutions in influencing new ecological paradigm embraced by the community. A close relationship should signify that social institutions that surround the community have been effective in encouraging the development of pro-environmental paradigm. Conversely, if there is a close relationship, it can be said that the spatial context that surrounds the pro-environmental regulation cannot support the development paradigm of the environment in the community.

As the description above, this study will examine the relationship between socioeconomic characteristics and social institutions of the community with new ecological paradigm scale (referred to as Environmental Concern Level) that embraced by the community

3 Data collection

This study collected data from three housing in the city of Palembang, namely Taman Sari Kenten (TS), Talang Kelapa (TK), and Taman Ogan Permai (TOP). Housing projects have been selected because they exhibit the same characteristics in locations on the outskirts of Palembang. In addition, residential housing is classified as having the occupants of diverse backgrounds. The housings are shown by the following data:

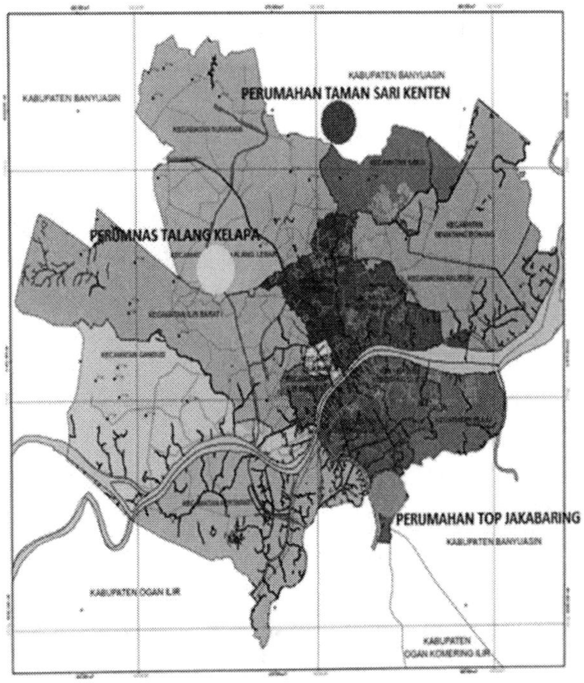

Fig. 1. Research sites on the city of Palembang
Source : [13], with modifications

The data source is the residents of the house in this residential area. One unit of data representing the randomly selected homes by housing blocks. It can be seen to be representative because it avoids some individuals represent one family. The total sample was 103 homes.

4 Methodology

4.1 New environmental paradigm scale

NEP scale applied in this study after a pass from the validity and reliability. Validity and reliability of data using SPSS software version 20 executed after data collected from residents by questionnaires. The options are Strongly Agree, Agree, Do not Know, Disagree and Strongly Disagree. NEP scale instruments shown in Table 1.

Table 1. NEP scale instruments

Descriptive	Dimensions	Item No. Question
Limits on Humans Growth	We are approaching the limit of the number of people the Earth can support	1
	The Earth has plenty of natural resources if we just learn how to develop them	2
	The Earth is like a spaceship with very limited room and resources	3
The dominance of humans in nature	Humans have the right to modify the natural environment to suit their need	4
	Plants and animals have as much right as humans to exist	5
	Humans were meant to rule over the rest of nature	6
The balance of nature	When humans interfere with nature it often produces disastrous consequence	7
	The balance of nature is strong enough to cope with the impacts of modern industrial nations	8
	The balance of nature is very delicate and easily upset	9
Humans Exemptionalism	Humans ingenuity will ensure that we do not make the Earth unlivable	10
	Despite our special abilities, humans are still subject to the laws of nature	11
	Humans will eventually learn enough about how nature works to be able to control it	12
Ecocrisis	Humans are seriously abusing the environment	13
	The so-called "ecological crisis" facing humankind has been greatly exaggerated.	14
	If things continue on their present course, we will soon experience a major ecological catastrophe	15

4.1.1 Validity

Validity is determined by Pearson correlation (r). The results of Pearson correlation of the sample is shown as follows:

Table 2. Calculation result validity nep scale questionnaire

Item	r	Significant	Conclusions
1	0,188	0,116	Not valid
2	0,046	0,706	Not valid
3	0,331	0,005	Valid
4	0,428	0,001	Valid
5	-0,151	0,210	Not valid
6	-0,038	0,752	Not valid
7	0,568	0,001	Valid
8	0,161	0,179	Not valid
9	0,434	0,001	Valid
10	-0,023	0,851	Not valid
11	0,491	0,001	Valid
12	0,451	0,001	Valid
13	0,564	0,001	Valid
14	0,411	0,001	Valid
15	0,421	0,001	Valid

The results indicate that the item is a valid item number 3, 4, 7, 9, 11, 12, 13, 14 and 15. Therefore, these items are then taken as an item to be used for the calculation of reliability.

4.1.2 Reliability

Calculation of reliability using Cronbach's Alpha parameter. The results of calculation of reliability on items that have valid (9 items) show the reliability of 0.658. This value is deemed reliable because the limit of reliability is 0.6 [14]

4.2 Socio-economic status and institution

Organizations or socio-economic status is an organized way of the urban population according to social stratification, political systems, and economic systems [15]. Socio-economic status was approached with a total income of the household and the highest education of the household. The parameters used are the logarithm multiplication between the level of education and income levels. The level of education consists of seven levels: Elementary (1), SMP (2), High School (3), Diploma (4), S1 (5), S2 (6), and S3 (7) which refers to the highest level of education held family members. Income alone is numerical data that shows the total amount of earnings of all family members in rupiah unit. Measurements are made through a questionnaire.

Institutions in this study were approached by their family members who work in the fields of education, health, leisure, and commerce. If there is, then the sample is given the index 2 while no indexed 1. Value institution itself is the multiplication of educational status, health, and comfort divided by trade status. This is because the status of education, health, and comfort should be oriented pro-environment because of their obligation to promote sustainability or a special focus on aspects of sustainability. Meanwhile, the trade is viewed, Therefore, if a house has four professions, the total score is 2x2 x 2/2 = 4. If a house has a profession of education, health, and comfort, but no trade profession, then the total score is 6. it is obtained through a questionnaire.

5 Result

Table 3 shows the descriptive statistical study respondents. The majority of families have the highest level of high school (43 of 103 respondents). The average income of respondents was IDR 4,297,000. 13 respondents working in education, four work in the field of hospitality, 8 working in the health sector, and 53 working in the field of trade.

Table 3. Descriptive statistics respondents

Last Education	
Primary school	3
Junior high school	4
Senior high school	43
Diploma	11
Bachelor	35
Postgraduate	7
Income	
Mean	4.297.000
Standard Deviation	3.583.500
Institutions	
Educational institutions	13
The hospitality institutions	4
Health institutions	8
Trading institutions	53

The potential of the relationship between socio-economic status and pro-environmental institutions with environmental concern level, tested through regression analysis using SPSS software version 20. Figure 2 shows the normality of residual regression results is a prerequisite for regression analysis. Observations in Figure 3 shows that the results of the regression residuals clear that normal distribution so that the linear regression model can be accepted as a method of data analysis.

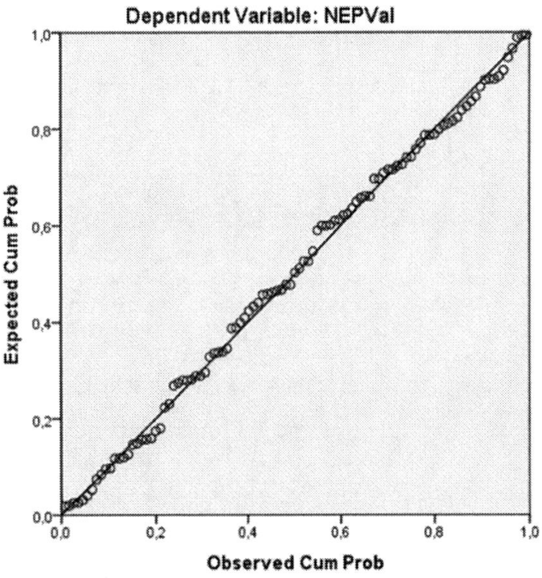

Fig. 2. Graph normality of regression results

Figure 3 shows a scatter plot of the regression residuals. Spread of points do not form a clear pattern, so it can be inferred that the regression model having no problem with heteroscedasticity

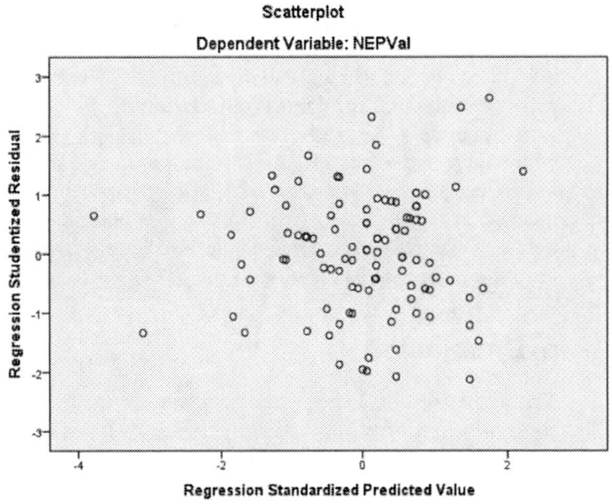

Fig. 3. Scatter plot regression residual

To see the relationship between variables, we calculate regression by putting variable socioeconomic status as a single variable that is the natural logarithm of the level of income and education level of respondents. This is done because the amount of income has a very great value compared to other indicators, as calculated in units of currency. The regression results are shown in Table 4. It appears that the effect weak institutions to environmental concerns while socioeconomic status had no significant effect on environmental awareness. But the influence of institutions on environmental concerns actually negative. That is, the person becomes a member of a pro-environmental institution, the more he behaves not care about the environment.

Table 4. Calculation results of multiple regression socioeconomic status and institution of the nep scale

Coefficients

Model		Unstandardized Coefficients		Standardized Coefficients	t	Sig.
		B	Std. Error	Beta		
1	(Constant)	24,693	1,391		17,747	,000
	Institution	-1,214	,680	-,175	-1,785	,077
	Ln (Education x Income)	-,738	,486	-,149	-1,519	,132

a. Dependent Variable: NEP Scale

The above results show that the institutions that developed among the people in the three housing destructive rather than constructive for the environment. Education does not give a positive impetus on environmentally friendly behavior. In contrast, higher education justifies one's behavior to the consumer and not environmentally friendly behavior, such as in educational institutions. There is no awareness to encourage students and teachers that are environmentally friendly. Health institutions are also not concerned with efforts to encourage patients and health care to behave environmentally friendly. Meanwhile, institutions are not sustainable comfort and are consumptive. Instead, institutions trade show's positive attitude. Orientation pro-environmental regulation of trade can be understood as respondents generally a small enterprise (SMEs) which have stalls with low environmental impact and has little environmental impact rising (eg. garbage in large quantities in the store), will affect the economic value obtained traders seek profit. Therefore, they will try to keep the environment location.

6 Conclusion

The above results indicate that to increase the level of environmental concern, sociological intervention is required. In this study, it was found that variable institutions affect the level of environmental concern was negative and weak. This is because institutions housing residents in developing countries have consumer behavior and less concerned with the environment. In other words, Institution as one form of social organization has a negative role in environmental concerns, therefore in order that the institutions have a positive impact, it must do some activities as follows:

– The government should strengthen environmental education curriculum so as to encourage the younger generation to love the environment and strive to reduce environmental impact at their residence. This is an easy program because there are no sociological barriers in urban communities. Urban communities are relatively liberal and scientific. Would be different if we did in the marginal communities that have a conservative and fatalistic attitude.

– Health institutions more easily integrated with the environmental concerns, because the quality of meteorological would have a direct impact on human health. Therefore, the health center should be warned about the danger of pollution or destruction of the environment and the importance of healthy behavior.

– Hospitality institutions should use components that are more natural in every activity, for example by extending the trees or by replacing manufacturing materials with natural materials. of course, these activities are not attractive to local travelers who are accustomed to tropical life, but it will attract foreign tourists are more interested in the local characteristics. They come to Indonesia not to find something that can be found in his own country. They want the typical situation of Indonesia and this is indicated by a tropical environment in harmony with nature Indonesia. For example, the guest house can be made with the concept of traditional architecture including the typical environment of Palembang city in the past.

– Trade is the most difficult institutions to be triggered environmental changes for the better because the essence of trade is the exploitation of natural resources into goods of human needs. exploitation caused environmental pollution, not least in the form of smoke from a factory or waste in traditional markets. The rational step that can be done to make this trade regulation as the trigger for the better is to expect the public awareness and efforts to forcefully through trade laws more restrictive.

References

1. R. Kutter, W. Information Strategies in Websites for Sustainable Consumption: An Exploration of the EcoRunner Tool, (2009).
2. R. Bechtel & Churchman, A. (Eds.). (2003). *Handbook of environmental psychology*. John Wiley & Sons, (2003).
3. R. Schefold., Nas, P., & Domenig, G. (Eds.). *Indonesian Houses*: Tradition and transformation in vernacular architecture. NUS Press, (2004).
4. S. Clayton., Myers, G. *Conservation Psychology*: Understanding and Promoting Human Care for Nature. John Wiley and Sons, (2015).
5. M. Bodur. Sarigollu, E. Environmental Sensitivity in a Developing Country: Consumer Classification and Implications. J. Environment and Behavior, **37** (4), 487-510, (2005)
6. T. Milfont, L. & Duckitt, J. J. Environmental Psychology, **24,** 3, (2004)
7. L. Zelezny, C. Chua, P. P. & Aldrich, C. J. Social issues, **56,** 3, (2000)
8. M. Floyd. Race, Ethnicity, and Use of the National Park System. *Social Science Research Review*, 1(2), (1999)
9. A. Mareck, F. *Acquiring Biospheric Literacy*: Discursive Tools, Situated Learning, and the

10. Rhetoric of Use. In Goggin, P.N (ed), Rhetorics, Literacies, and Narratives of Sustainability. Routledge, (2010)
11. R. Godfrey, J. Risky. *Messages to Whom?*: Understanding Our Toleration for Risk Through an Examination of Values. ProQuest, (2008)
12. M. Pelling. Resilience and Transformation. In Pelling, Mark, David Manuel-Navarrete, and Michael Redclift, eds. *Climate change and the crisis of capitalism:* a chance to reclaim, self, society and nature. Routledge, (2012).
13. J. Porter, R & Howell, F. M. (2012). *Geographical sociology*: Theoretical foundations and methodological applications in the sociology of location (Vol. 105). Springer Science & Business Media, (2012)
14. Pemerintah Kota Palembang. Peta Kota Palembang, (2012)
15. J. Hair, F. Anderson, R. E., Tatham, R. L., & William, C. (1998). *Multivariate data analysis*. (1998)
16. R. Palen. Emergence of Cities, (2006) http://www.bolender.com/Sociology/Palen/Palen%202.htm. Accessed on July 12nd 2016.

Source rock formation evaluation using TOC & Ro log model based on well-log data procesing: study case of Ngimbang formation, North East Java basin

Yosar Fatahillah[1], *Widya* Utama[1], *Kukuh* Suprayogi[2], *Anik* Hilyah[1], and *Iqbal* Maulana[1]

[1] Department of Geophysical Engineering, Institut Teknologi Sepuluh Nopember, 60118 Surabaya , Indonesia
[2] Pertamina Upstream Technology Center, Pertamina, 10110 Jakarta, Indonesia

Abstract. Ngimbang Formation is known as one major source of hydrocarbon supply in the North Eastern Java Basin. Aged Mid-Eocene, Ngimbang is dominated by sedimentary clastic rocks mostly shale, shaly sandstone, and thick layers of limestone (CD Limestone), with thin layers of coal. Although, laboratory analyses show the Ngimbang Formation to be a relatively rich source-rocks, such data are typically too limited to regionally quantify the distribution of organic matter. To adequately sample the formation both horizontally and vertically on a basin–wide scale, large number of costly and time consuming laboratory analyses would be required. Such analyses are prone to errors from a number of sources, and core data are frequently not available at key locations. In this paper, the authors established four TOC (Total Organic Carbon Content) logging calculation models; Passey, Schmoker-Hester, Meyer-Nederloff, and Decker/ Density Model by considering the geology of Ngimbang. Well data along with its available core data was used to determine the most suitable model to be applied in the well AFA-1, as well as to compare the accuracy of these TOC model values. The result shows good correlation using Decker (TOC) Model and Mallick-Raju (Ro- Vitrinite Reflectance) Model. Two source rocks potential zones were detected by these log models.

1 INTRODUCTION

Ngimbang is one of 3 proven major supplier of organic rich sourch rocks in the petroleum system of the North Eastern Java (NEJ) Basin [1]. Known as the oldest formation of sedimentary deposits in the basin, Ngimbang lays on top of the basement of the NEJ basin. The formation is dominated by sedimentary clastic rocks mostly shale, sandstone, thick layers of limestone (CD Limestone), and thin layers of coal [2].

North East Java Basin was formed through northward plate collision between Austro-Indian plate and Sunda plate. Characterized with low gravity anomaly in collision area and high anomaly in south Java mountain range. The meeting of ocean crust of India and continental crust from sunda plate forms islandic arch system [3].

One of the biggest concern in determining the TOC value is the cost of cutting or coring analysis. Adding to that, the result of laboratory analyses are most of the times very limited and not available at key locations/depth and the possibility of errors from many external factors is very high [4].

In this study, well log data was used to observe the characteristics of vertical distribution of TOC in the Ngimbang Formation. These wells are SH-1, SH-2, SH-3, and AFA-1. However well AFA-1 is the only well where no TOC core data is available. Therefore, this well was used as comparison well to prove whether the TOC in the well based on determined applicable model from well SH-1, SH-2, and SH-3 is considered to be having proper TOC value.

The identification of the total organic content of the source rocks in the subsurface, its type and levels of maturity are very important aspects in the field of petroleum exploration. This work aims to provide preliminary information of hydrocarbon source rock potential of the Ngimbang fomation, using well log data [5]. Classification of TOC Quality and Formation Maturity can be seen in Table.1 and Table.2.

Method of determining organic richness from well log data to predict the TOC value offers a practical alternative to laboratory analysis of core or cuttings. The well log data derived method has several advantages over laboratory analyses including but not limited to [4][6]:

- The well-log data provides continuous measurement of the formation, reducing statistical uncertainties of limited and sometimes random picked spot sampling.

*Corresponding Author email: yosar94@gmail.com

- Well log data are more common and more readily available than core cuttings.
- Working with well logs data is simpler and less costly than laboratory procedures.

Table 1. Organic Richness

Quality	Critetia		
	TOC	**S1**	**S2**
Poor	0 – 0.5	0 – 0.5	0 – 2.5
Fair	0.5 – 1	0.5 – 1	2.5 – 5
Good	1 – 2	1 – 2	5 – 10
Very Good	>2	>2	>10

Rock-Eval Pyrolysis

Rock-Eval pyrolysis is the most widely used pyrolisis technique in geological analyses. It measures organic richness and maturity level of source rocks. In a pyrolysis analysis, the organic content is pyrolyzed in the absence of oxygen, then combusted. The amount of hydrocarbons and carbon dioxide released is measured. (see Table. 1 and Table. 2).

Table 2. Maturity Level

Quality	Critetia		
	Ro	**LOM**	**Tmax**
Immature	0 – 0.65	<9	<420
Mature	0.65 – 1.4	9-11	435-445
Post Mature	>1.4	>11.5	>470

Rock-Eval data are expressed as mg/g of rock and include three basic parameters [7][8]

- S1 represents the quantity of free hydrocarbons present in the rock and is roughly analogous to the solvent extractable portion of the organic matter;
- S2 represents the quantity of hydrocarbons released by the kerogen in the sample during pyrolysis;
- T-max is the temperature at which the maximum rate of generation (of the S2 peak) occurs and can be used as an estimate of thermal maturity.

2 METHOD

The methods used in this paper is based on proven models that has previously been tested in various basins. However the main objective of this paper is to review the assumptions and methodology of the approach and establish the validity and limitations and inspect further observations of possible improvement of the method, by comparing log derived data to laboratory analyses result. In this paper the authors established 5 TOC log Model:

- Schmoker-Hester Model (1983)
- Meyer-Nederloff Model (1984)
- Passey/ DlogR Method (1990)
- Decker Model (1992)
- Improved Decker Model (2016)
- Mallick-Raju for Ro (1995)

Since TOC core data in the wells are mostly taken on random sampling without certain interval picks, this paper used deviation method where each TOC value taken from log model is summed with the core TOC value taken from laboratory analyses. High deviation low value of correlation value describes poor agreement between TOC model Log and TOC result of laboratory analyses. Thus, will further be considered as unsuitable to be applied in the formation or just on certain depth intervals area.

Schmoker-Hester Model (1983) based on four major components. The content of pyrite in the formation, interstitial pore fluid of the formation, rock matrix, and organic matter [4]. These four components was fitted to the Bakken Formation. Thus follows the empirical state of the Bakken geological background. However, the final formula given by Schmoker-Hester (1983) was one of the simplest TOC calculation form.

$$TOC(wt.\%) = \left(\frac{154.497}{\rho}\right) - 57.261 \quad (1)$$

Where ρ is the formation density taken from density log. TOC calculated using this method shows poor qualitative result. Therefore, enormous deviation value on quantitative analysis was found. Average value of deviation on the experiment of well SH-1 is 1.77, Deviation on SH-2 is 3.38, and SH-3 is 3.39

***Decker Model* (1993)** the organic matter density in shale (1.03-1.10 g/cm3) is significantly lower than the surrounding rock matrix (the density of clay skeleton is 2.30-3.10 g/cm3), resulting in lower density of high-quality source rocks [9]. There is a certain relationship between TOC and total formation bulk density of Ngimbang formation source rocks. Therefore, it is possible to determine TOC through this relationship, as the formation bulk density can be accurately measured by density log data [10]. The mathematical relation for calculating TOC by using data is written as:

$$TOC\ (wt.\%) = A + B\rho \quad (2)$$

Where, ρ is bulk density (g/cm3); a and b are Shale regional empirical factors. Based on the experiment data of 12 core samples in Well SH-1 taken from shalestone of Upper Ngimbang Clastics calculated are:

a= -2.970, b= 8.034, cc = 0.867

Only applicable in Shale formation, with relatively same Maturity – Ro (Vitrinite Reflectance) or LOM as well as other geological background (facies, and other mineral content indicator) see (Figure 1).

This relatively good correlation (cc>0.8), is resulted from minimum variation range in the TOC value with respect to the change (increasing, or decreasing) density value. Therefore, formed linear line that have rather poor distribution point throughout the linear regression line.

Linear regression process is also applied to gain correlation between the Limestone and the formation density. This relation is written in mathematical notation as written eq. 2 with formation constant as written bellow:

c=2.59, d= -6.08, cc= 0.764

Where, c is gradient constant, and d is the intercept

constant taken from limestone of Ngimbang formation. Applicable in Limestone formation with same formation condition as Decker-Shale.

Meyer & Nederloff Model **(1984)**, uses discrimination of Source Rocks from Non-source Rocks Meyer and Nederlof [11] developed an approach to distinguish source rocks from non-source rocks using well log readings. Based on the resistivity log values cross plotted with either the sonic or density log values. The resistivity values used in these plots must be standardized to 75°F using Arp's formula:

$$R75° = R(T-7)/82 \qquad (3)$$

Where, R is the resistivity of rocks corrected to a 75 degree standard temperature of °F (24°C) and T is formation temperature at the concerned depth.

The formation temperature is derived from gradient temperature using bottom temperature of the various logging runs. Two linear equations, was first introduced by Meyer and Nederlof. For the discriminant score (D), are used on the basis of log combination R- ρ and R-Δt:

$$D = -6.906 + 3.186\log(\Delta t) + 0.487\log(R75°) \qquad (4)$$

$$D = 2.278 - 7.324\log(\rho) + 0.387\log(R75°) \qquad (5)$$

If D is positive, the rock is a probable source rock; if D is negative, the rock is probable barren;

Passey Model **(1990)**, Passey developed a technique used for both carbonates and clastics using sonic-resistivity overlay. They also introduced the term "ΔlogR" which is linearly related to TOC and is a function of maturity. In this technique, sonic travel time "Δt" and true formation resistivity"Rt" are scaled such as a ratio of 164 μs/m to one resistivity cycle. The separation between two curves (Δt to left and Rt to right) defined Δlog R that can be calculated from the following equation [12]:

$$\Delta\log R = \log\left(\frac{R}{Rbaseline}\right) + 0.02(\Delta t - \Delta tbaseline) \qquad (6)$$

In this paper we also utilize the density log as an alternative solution in finding TOC value, using the following equation:

$$\Delta\log Rden = \log\left(\frac{R}{Rbaseline}\right) + 2.5(\rho b - \rho baseline) \qquad (7)$$

Then, the TOC is calculated using the following relation:

$$TOC(wt.\%)\Delta\log R\ x\ 10^{(2.297-0.1688LOM)} \qquad (8)$$

The value of LOM (Level of Metamorphism) needed to use Passey Model was determined by using crossplot between DlogR and TOC from core data. Using the crossplot we determine the value of LOM in the range of 11 to 12. However, Passey method is only limited to LOM value of 10.5. Therefore, we used 10.5 instead of 11.5. However, we expect erroneous result from this calculation. Applicable in rocks with Δt ranges from 262 to 460 μs/m, And LOM ranges from 6-10.5, with Ro- 0.5 to 0.9 [8].

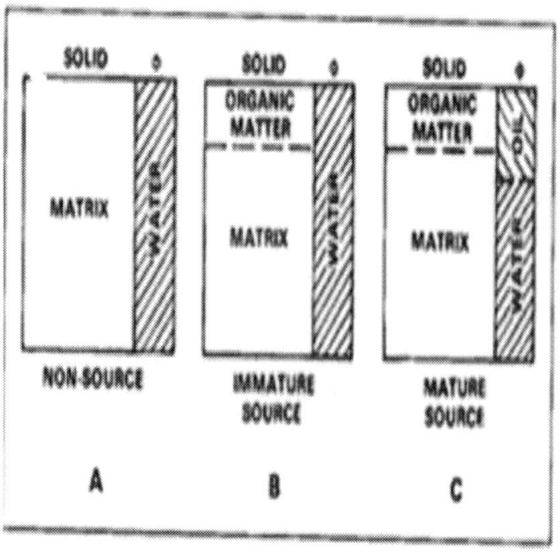

Fig.1. Solid and fluid materials of source rock and non-source rock (Passey et al, 1990)

Mallick-Raju Model (1993) was used to determine the source-rock maturity level using Vitrinite Reflectance (Ro). It is a measure of the thermal maturity of a sedimentary rock containing kerogen. Also used as an indicator of whether a source rock has been heated enough to produce oil, oil and gas, or gas only. [13] Pointed out that the measured quantitiy of Ro has a linear correlation toward the sonic log. Where, certain maturity level correspond to higher or lower sonic transit time. Thus, gives a simple equation as written below:

$$Ro = A - \Delta t/B \qquad (9)$$

Where, A and B is empirical constant of Ngimbang Formation, and BHC is borehole compensated log (sonic transit time). Applying this method for Ngimbang formation gives equation (10)

$$Ro = 3.24 - 0.0082(\Delta t) \qquad (10)$$

With total corralation number is 0.769. Note the equation (10) is a simplified equation which was derived from (9).

3 RESULT AND DISCUSSION

From the abovementioned method, the TOC calculation model for the Ngimbang Formation was conducted. The result describes the lack of accuracy of Schmoker-Hester Model. In contrast, the Passey and Decker Model gives satisfactory result with very small amount of deviation from core data TOC (Figure.2)

Decker-Shale model is applicable in the shale zone of the Ngimbang Formation. Since, only shale data are taken into account during least-square processing. There's no proven correlation between the formation densities to the TOC of Limestone by using this method. Thus, Low

DOI: 10.1051/matecconf/201710104016

deviation in Limestone are only considered as incidental (see Figure.2).

Since, **Decker Shale Model** is only applied to the Shale interval, similar approach with Decker model is needed to be applied in the limestone intervals. The result shows this method is also considered to be applicable, but shall be used prudently, as erroneous result is expected. Most of the erroneous result occur on the Lower Ngimbang Limestone. We concluded that this was resulted by difference in geologic background, as indicated by available Ro data. Unlike the case of Upper Ngimbang Shale (Clastics) which has minimum difference in maturity level, Lower Ngimbang has various maturity level ranging from Immature (Ro<0.65), Mature (Ro>0.65), and Post Mature (Ro>1.4) with no linear correlation to depth of Interval.

DlogR (Passey) Model both Sonic and Density derived Model shows both promising quantitative result. However, as Passey has pointed out that the method is only optimally applicable when the LOM of the formation ranges from 6.0 to 10.5. The determined LOM value of 11.5 is considered to be "non Passey". Moreover, This model is valid only in Δt (Sonic Log) ranging from 262 to 460 µs/m. Error in TOC calculation can be expected at extreme low or high Δt due to the proposed scale is not valid at this situation [8]. Combination of both LOM (above the range) and Sonic Log (below) range in the formation often gives negative separation, and gives erroneous TOC calculation result. Thus, this method shall be used more prudently. However, applying Passey method gives satisfactory result in the application on some of limestone intervals, as estimated TOC value

Since, **Schmoker-Hester Model** is based on Shale Source Rocks experiment, there's no correlation between the models with the application on Limestone. Thus, the equation is only considered usable in shale zone with high pyrite component, and specific ratio of organic matter to organic carbon relatively around range of 1.33 with density of organic matter of 1.01 g/cm3. Low pyrite component and undetermined OM (organic matter) to (OC) Organic Carbon ratio exacerbates the accuracy of predicting the TOC value. The result showed that this model is unsuitable to be applied both in Shale and Limestone of Ngimbang Formation. Good quantitative result in Limestone are only considered to be incidental. The equation used in this method shall be recalibrated, especially in Pyrite component, and Organic Matter-Organic Carbon Ratio.

Mallick-Raju Model, was used to determine the level of maturity of the formation. Using eq. 10 the result shows a very good correlation between predicted to the core Ro data, where cc=0.92 was obtained. Thus, this model is considered to be applicable in the Ngimbang Formation. Applying this model suggests that the Ngimbang formation in the study area has maturity level ranging from 0.44 – 1.8 or immature to post mature level. However, this model require several Ro core or laboratory data to be applied (See Figure.2 bottom right).

Using the aforementioned result of log model selection, we concluded that Decker (especially in Shale intervals), Passey-Density for limestone intervals (shall be used more prudently and erroneous result is expected as R square is lower than 0.8) and Malick-Raju are the only applicable models to be used in determining the organic richness and maturity level in the Ngimbang Formation. Hence, these models can be applied in the well AFA-1. Application of Meyer-Nederloff model using Eq.4 gives negative result throughout the entire Ngimbang formation interval detected through well-bore data. This hypotheses is also being supported by the fact that applying Decker Model indicates the lack of organic richness where the average TOC% throughout the entire interval is 0.22 TOC wt.% with maturity level ranging mature to post mature ranging

Zone A is located at depth 2240 meters to 2243 meter Measured Depth. The main lithological component of zone A is limestone. By using Passey-Density model, the maximum TOC value obtained is 5.676, and the minimum value 0.318. Decker Model shows the maximum TOC value of 1.261 and 0.546 low.

Table 3. Zone A (Limestone) Well AFA-1

Model	Critetia		
	Min	Max	Quality
Passey TOC	0.318	5.676	Good
Decker TOC	0.546	1.261	Fair-Good
Mallick Ro	1.524	1.821	Post-Mature

From the results of the matching process, the most reliable TOC value of zone A is predicted by Decker (bulk density) method. This is due to the maturity value detected above LOM 10.5 or Ro>0.9. Hence, can be concluded zone A is potential source rock with a value of TOC ranging from 0.546 to 1,261 (Model Decker).

Zone B is located at depth interval 2244-2251 meter (MD), which mainly consist of shale. The model indicated by the results of TOC Passey density shows maximum TOC value in the depth interval zone B is 1,813 and a minimum of 0.758. Decker model shows maximum TOC value of 0.523 with thin layer of poor organic richness level. However, as zone B are mainly post mature zone with Ro above 0.9, then only Decker model givies the closest TOC prediction to original value- in accordance with the principles of the Passey model and test results on the model of the well SH-1, SH-2 and SH-3.

Table 4. Zone B (Shale) Well AFA-1

Model	Critetia		
	Min	Max	Quality
Passey TOC	0.758	1.64	Fair-Good
Decker TOC	0.314	0.523	Poor-Fair
Mallick Ro	1.502	1.821	Post-Mature

Using the Waples theorem (1985) zone A is considered as good source rock. Whilst, zone B is mostly categorized as poor source rock. However, the insignificant thickness of both layer, zone A and B reduces the potential of possible charge towards the reservoir. For further analyses, seismic data with exceptional resolution is needed in order to identify the lateral distributions of Zone A.

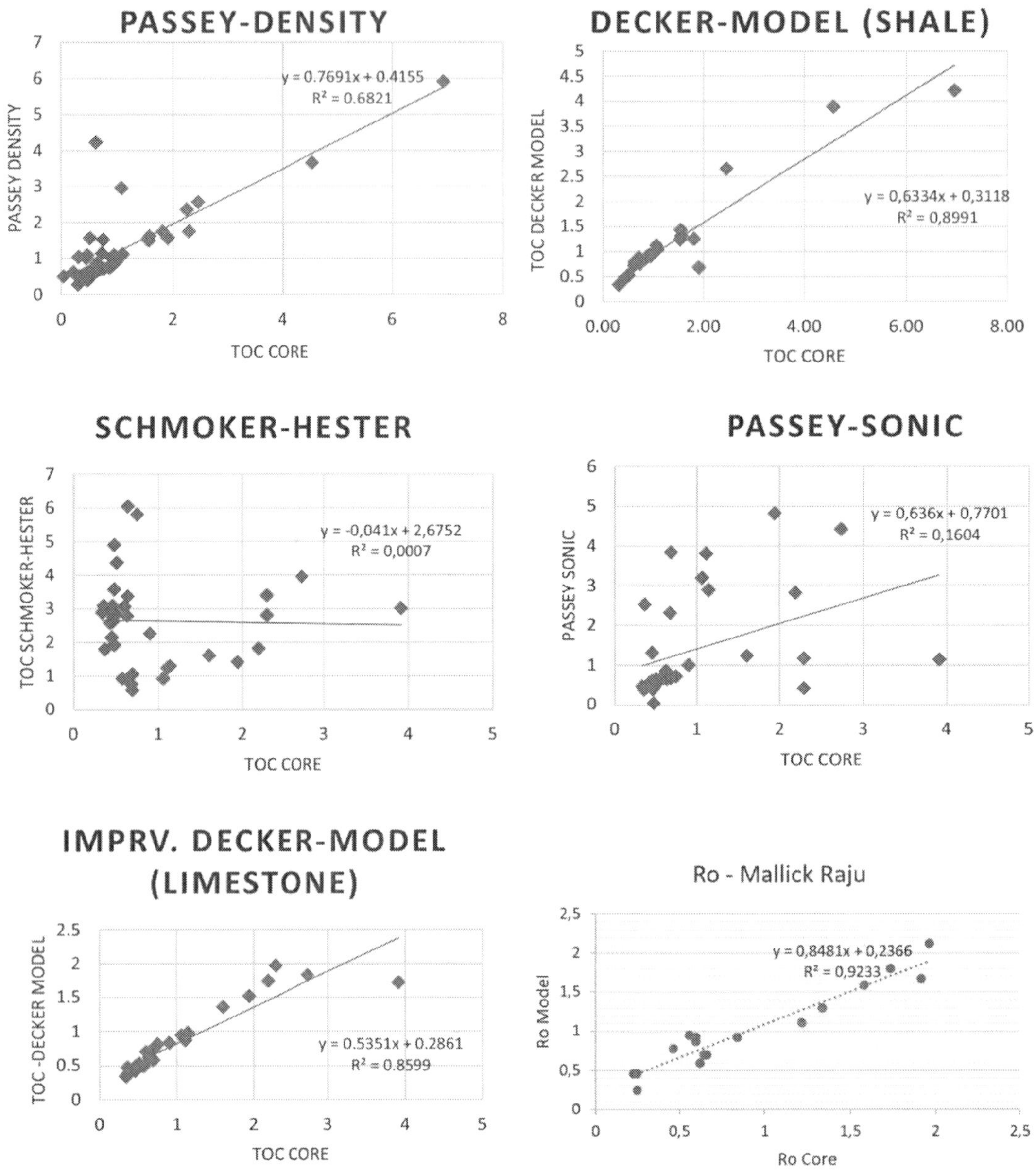

Fig.2. Relationship between TOC Core Data and TOC Predicted using TOC Log Model. The square error as indicator of log model accuracy.

4 CONCLUSIONS

Based on the application and results of each model on the Ngimbang Formation, we concluded:

- Decker-Shale method is applicable in the shale zone of Ngimbang Formation. Since, only shale data are taken into account during least- square processing. There's no proven correlation between the formation densities to the TOC of Limestone by using this method. Thus, Low deviation in Limestone are considered as incidental.
- Decker-Limestone method is applicable, but should be used carefully, as erroneous result is expected.
- Since, the qualitative and quantitave analysis has shown fair-good result (slightly below standard correlation accepted <0.8), Passey Density method shall be used more prudently. As most of the Ngimbang formation intervals are considered to be "non-Passey" when the LOM is beyond the range of 6-10.5.
- Passey-Sonic Method- combination of both LOM (above the range) and Sonic Log (above and sometimes below)

- range in the formation often gives negative separation, and thus gives erroneous TOC calculation result.
- Schmoker-Hester Model gives poor accuracy to predict TOC both in Shale and Limestone of Ngimbang Formation. Good quantitative result in Limestone are only considered to be incidental. Application of this method shall be recalibrated mostly from the Pyrite component, and Organic Matter-Orcanic Carbon Ratio.
- Mallick-Raju Ro model is applicable in the Ngimbang Formation.
- Applying Decker, Passey-Density, and Mallick-Raju Model, detected several good source-rock potential in the well AFA-1, namely Zone A and Zone B, with layer thickness ranging from 3 to 7 meters.

The authors would like to thank Pertamina Upstream Technology Center, Pertamina Persero, for their assistance and support in providing the needed data and advisory upon completing this research.

REFERENCES

1. Saultan Panjaitan. *Buletin Sumber Daya Geologi* **5** No **3** – (2010)
2. Mudjiono dan Pireno, *Proceedings Indonesia Petroleum Association 28th Annual Convention and Exhibition, Jakarta.* (2001)
3. Satyana A.H and Purwaningsih, M.E.M. *Proceedings IPA* ***29th*** *Convention and Exhibition, Jakarta.* (2003)
4. Hester. T & Schmoker J.W. *Determination of Organic Content From Formation-Density Logs,Devonian-Mississippian Woodford Shale, Anadarko Basin, Oklahoma.* Denver, Colorado: Department of Internal U.S Geological Survey. (1987)
5. Ali Younis Ahmed Abdel-Rahman. *Source Rock Evaluation and Maturity Levels of the Rudeis Formation,Northern Gulf of Suez, Egypt, Using Well Log Analysis and Rock-Eval Pyrolysis.* Giza, Egypt: Department of Geophysical Science, National Research Center, (2013)
6. Schmoker JW. *AAPG Bull*;**65(7):1285e98**. (1981)
7. Peter, K.E and Cassa, M.R, *AAPG Memoir* **60**. Tulsa, Oklahoma. (1994)
8. Mohamed S. El Sharawy, Gamal R. Gaafar. *Journal of Applied Geophysics* **80** 129–143 (2012)
9. Decker AD, Hill DG, Wicks DE. *The Low Permeability Reservoirs Symposium. New York: SPE;* 1993. **26e28** April 1993. Denver, Colorado, USA. (1993)
10. Huang Renchun et al. *Research Institute of Sinopec Exploration Southern Company, Chengdu, Sichuan* 610041. (2015)
11. Meyer, B.L., Nederlof, M.H, *AAPG Bulletin* **68**, 121– 129. (1984)
12. Passey, Q.R., Moretti, F.U., Stroud, J.D. *AAPG Bulletin* **74**, 1777– 1794. logs II: Geol. Soc. London, Spec. Publ., 65, pp. 369– 376. (1990)
13. Mallick, R.K., Raju, S.V. *Organic Geochemistry* **23 (10)**, 871– 879. (1995)

Iron ore deposits model using geoelectrical resistivity method with dipole-dipole array

Adree Octova [1,*], and *Dedi* Yulhendra [1]

[1]Mining Engineering Department Faculty of Engineering, State University of Padang, 25131 Padang, Indonesia

Abstract. Mining industry is an industry with very high risk (losses). In order that mining activities can be run well, then the potential of the Earth's resources must be known for sure. one of the Earth's resources of high economic value is the iron ore. Iron ore is rarely found in a free state in nature, it is usually associated with other minerals and exposed randomly. With these properties, iron ore needs to be modeled before doing mining activities in order to avoid large losses. Iron ore deposits can be modeled with geoelectrical resistivity method. Dipole-dipole array will produce good imaging both vertically and laterally. From the measurement results of geoelectrical resistivity with dipole-dipole array will be obtained the value of measuring the current and potential difference. This value will generate into 2D and 3D model of the cross section of the iron ore deposits. One of the areas in West Sumatra has the potential for iron ore. Five lines were applied in this area. The result of cross section got the iron minerals associated with quartzite at 30 meters depth below the surface.

1 Introduction

Iron is one of the natural resources that most abundant in nature and most use in life. Iron found in nature in form of compounds, such as hematite, magnetite, pyrite and siderite. From iron ore minerals, magnetite is a mineral with the highest Fe but there is a small amount. While hematite is the main ore mineral needed in the iron industry. One of the uses of iron ore is as steelmaking material. Basically, there are a lot of iron ore mineral as well as contact metasomatik secondary sludge buried and exposed randomly.

In practice, the exploration activities carried out by utilizing the properties of physics and chemistry of rocks, soil, elements and minerals, such as: magnetism, density, electricity, radioactivity, and mobility elements. There are several methods that can be used to describe these properties, one of them is indirect exploration method. These method is applied at the early stages of exploration, which time and cost are quite limited.

Geoelectrical method is one of indirect exploration method that is often used in mining exploration. In addition to the cost of exploration costs, this method is also pretty well describes the rock layers below the surface.

Geoelectrical resistivity has several electrode arrangement in providing a picture of the earth's subsurface, one of them is dipole-dipole array. The arrangement of dipole-dipole electrode array is different from the Wenner and Schlumberger array. For the dipole-dipole array, current and potential electrodes are arranged separately, longer line will be further away. While the Wenner and Schlumberger array will provide potential electrode arrangement in which current electrode arrangement in conjunction with the long line.

The differences of array will give the different appearance of subsurface models. Based on [3][5], Schlumberger array will provide a better picture for the vertical model with a deeper current distribution. Wenner array will provide a better picture laterally with more shallow current distribution than Schlumberger array [6][7]. While the dipole-dipole array provides better picture vertically and laterally with deeper current distribution than two electrode array mentioned above[8][11][13].

See the random distribution of iron ore, it will be easier to interpret the iron ore by modeling with dipole-dipole array. The result model of the dipole-dipole array can be modeled in 2D and 3D display. The cost of this array is not expensive when compared with Wenner and Schlumberger. This configuration can also be used for mapping, ie measurement focuses laterally results.

The purpose of this study was:

- Provide an overview of the subsurface using geoelectrical resistivity method with dipole-dipole array.
- Provide information of iron in the study area in the form of 2D and 3D models of geoelectrical resistivity method with dipole-dipole array.

2 Electrical properties of rocks and minerals

The electrical properties of rocks is characteristic of rocks when electric current flows into it. Electric current

* Corresponding author: adree.octova@gmail.com

can be derived from nature caused by the presence of the atoms making up the earth's crust that interact with each other due to the charge imbalance, or electric current accidentally put into it. Some of the electrical properties of rocks that are useful in the geoelectrical exploration particularly in resistivity method is the natural electrical potential, electrical conductivity, and dielectric constant.

Based on the value of resistivity, rocks and minerals can be classified into: a good conductor with resistivity value of 10^{-6} Ωm to 1 Ωm, bad conductor with resistivity value of 1 Ωm to 10^{7} Ωm, and isolator with resistivity value of greater than 10^{7} Ωm [1]. It means conductive material will has smaller the resistivity value, and isolator material has greater the resistivity value.

Earth material has no fixed resistivity value, in other word has a specific range. Resistivity material depends on the state of the geology and rocks structure at each location [4]. The range of resistivity value of some materials shown in Table 1 [9].

Table 1. Resistivity of rocks and minerals

Rock/mineral	Resistivity (Ωm)
Topsoil	50 - 100
Loose sand	500 - 5000
Gravel	100 - 600
Clay	1 - 100
Weathered bedrock	100 - 1000
Sandstone	200 - 8000
Limestone	500 - 10000
Greenstone	500 - 200000
Gabbro	100 - 500000
Granite	200 - 100000
Basalt	200 - 100000
Kuarsite	100 - 2500000
Graphitic schist	10 - 500
Slates	500 – 500000
Pyrite (ores)	0.01 - 100
Phyrotite	0.001- 0.01
Chalcopyrite	0.005 - 0.1
Galena	0.001 – 100
Sphalerite	1000 – 1000000
Magnetite	0.01 - 1000
Cassiterite	0.001 - 10000
Hematite	0.01 - 1000000

3 Geoelectrical resistivity

Geoelectrical method is one of the geophysics methods that studies the current distribution properties in subsurface. The response is based on the measurement of potential field and current on the surface that occur either naturally or due to the injection of the current into the earth. Some of geoelectrical method are: self potential method (SP), Telluric currents, magnetotelluric, electromagnetic, induced polarization (IP) and resistivity method.

Resistivity method is one of geoelectrical method that used to study current distribution based on the differences of rock resistivity. The basic principle of this method is injected the current into the earth using two current electrodes, then measure the potential difference through two other electrodes on the surface of the earth. The current will flow through the rock layers below the surface, and generate data potential difference whose value depends on the resistivity of rock [13]. Based on the value of resistivity, the subsurface structure and constituent materials of the earth can be known.

The resistivity measurement is generally related to the size of the saturation and pore spaces of water. Water has a low resistivity, so that the current will follow the flow of the smallest resistivity [2]. Saturation, salinity of groundwater, and porosity rocks increasing will reduce the resistivity value. Increasing soil or rock compaction unit will chase away water and effectively increase the value of resistivity. Air has a high resistivity value, resulting in the exact opposite response to water when filling pores. As the result, the presence of water will reduce the resistivity value, whereas the presence of air in pores increase the resistivity value below the earth's surface.

Generally, the simple approach of earth current properties is thinking the earth as a homogeneous isotropic medium. Assuming the current field from a source point in the earth is spherical symmetries [13]. Fig.1. shows the current distribution on the earth as a homogeneous medium.

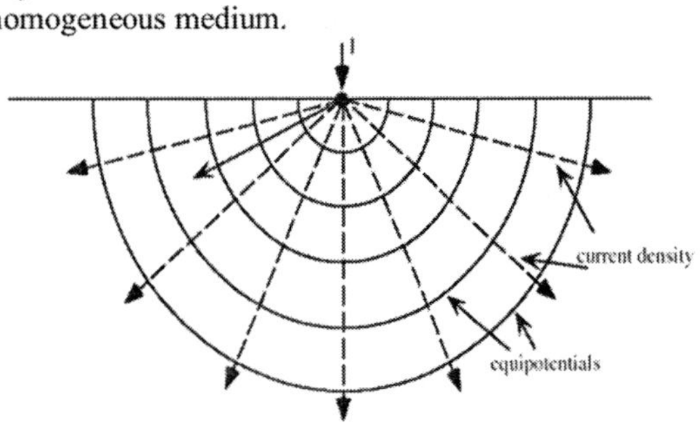

Fig. 1. Current distribution on earth as homogeneous medium [13]

The equation to calculate the resistivity of rock is:

$$\rho = \frac{V}{I} 2\pi r \qquad (1)$$

ρ is resistivity material (Ωm), r is radius (m), V is potential difference (v), I is current (A).

Equation (1) can be simplified to

$$\rho = k\frac{\Delta V}{I} \qquad (2)$$

K is geometric factor. The K value will depend on electrode array.

4 Apparent resistivity

Basic assumption of geoelectrical resistivity method is the earth with homogeneous isotropic medium that provides the same resistance values for the same electrode array [12]. But in fact, the Earth is composed of layers with different resistivity value. The measured potential difference is the influence of these layers [1].

The measured resistivity is the value of pseudo (apparent) on various rocks [10], this resistivity is called the apparent resistivity. Apparent resistivity is the resistivity of a fictitious homogeneous medium equivalent with layered medium that reviewed for one layer.

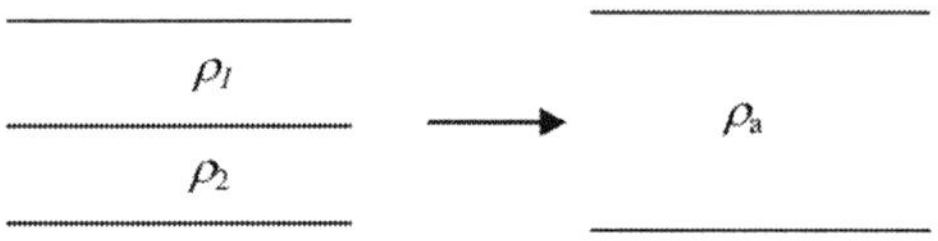

Fig. 2. Apparent resistivity concept [1]

Fig.2. shows the concept of apparent resistivity with layered medium assumption that consists of two different resistivity layers (ρ_1 and ρ_2). This medium is considered as one homogeneous medium layer which has one resistivity value that is measured apparent resistivity ρ_a. Based on this explanation, (2) can be written as:

$$\rho_a = K\frac{\Delta V}{I} \quad (3)$$

with ρa is apparent resistivity. The data analysis of apparent resistivity will produce actual resistivity of Earth.

5 Dipole-dipole array

The data of measured resistivity is plotted at corresponding points to the value of n (n = 1,2,3,4,5,6) with the apparent depth level, so that it can be made pseudodepth section contours of resistivity variations to lateral direction toward the direction of the apparent depth.

The results of measurement using more spaces between current and potential electrode will give deeper information on subsurface structure. Thus, the dipole-dipole configuration can be effectively in mapping, both laterally and vertically (Fig.3).

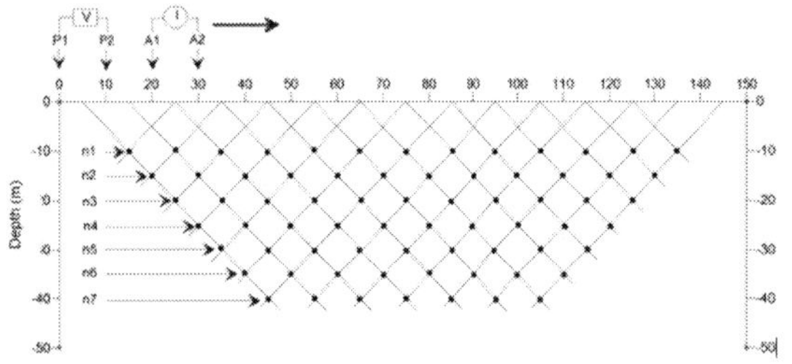

Fig. 3. Electrode array of dipole-dipole

In dipole-dipole array, both current and potential electrode sare separated by distance "a". While the current and the electrode potential inside are separated section as far as "na", where n is an integer. Variations of "n" is used to get a certain depth, greater "n" will obtain bigger depth. The level of range sensitivity of the dipole-dipole array is affected by the value of "a" and "n" variation [5].

6 Design of measurement

Regionally, the research area is situated on the path of magmatic. As an active path magmatic, then along this zone has potential mineralization (Fig.4). The liquid magma rises to the top interacting with groundwater forming rich mineral hydrothermal fluids, rise up through the fault as a weak zone and settles on fractures or porous of rock.

Lithologies contained in the investigation area from old to young, as follows, namely:

- Quartzite and quartz sandstones, shales, conglomerates, regionally included in the lithologies of the Lower Formation Kuantan Member in Perm aged.
- Slate, black shale, regionally included in the Triassic Formation in Tuhur aged.
- Granite, structure ranges from leucogranit to quartz monzonite, Triassic aged, gray-green, massive structures, mineral composition of quartz, plagioclase, hornblende.

Based on observations in the research area obtained a few chunks of rock containing iron ore found as a result of transportation that expected not far from its original location seen from the angle of that angled boulder. Mineralization of iron ore deposits are found in the form of boulder of quartzite and granite gravel containing mineral magnetite spots that found along the river

Measurements were taken at the block X on one of iron ore mining in West Sumatra. Design of measurement is performed with 5 lines with 50 meters distance between the lines and 300 meters long line. Measurements were made from South to North direction (Fig.5).

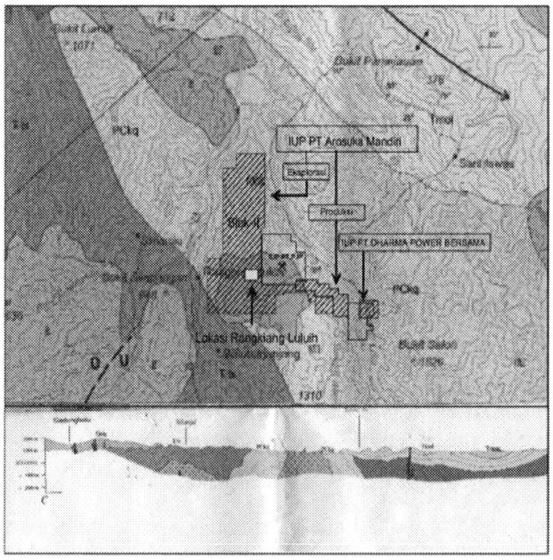

Fig. 4. Regional geological of research area

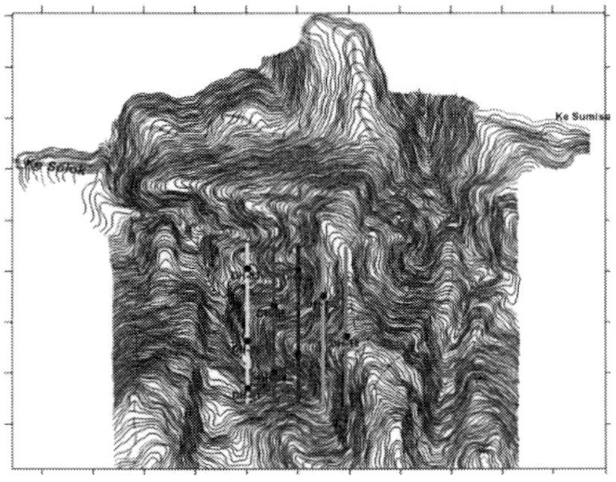

Fig. 5. Design of measurements

7 Results and discussion

Data obtained from geoelectrical measurements are potential difference, electrical current, and geometric factor. The data is then processed to find apparent resistivity using (3) and will be obtained 2D cross-section using geolectical resistivity software. The measurement results are illustrated in geoelectrical cross-section line 1 to line 5. Geoelectrical measurements was using 5 meters, 10 meters and 20 meters electrodes space along 300 meters line. With the following results:

Line 1

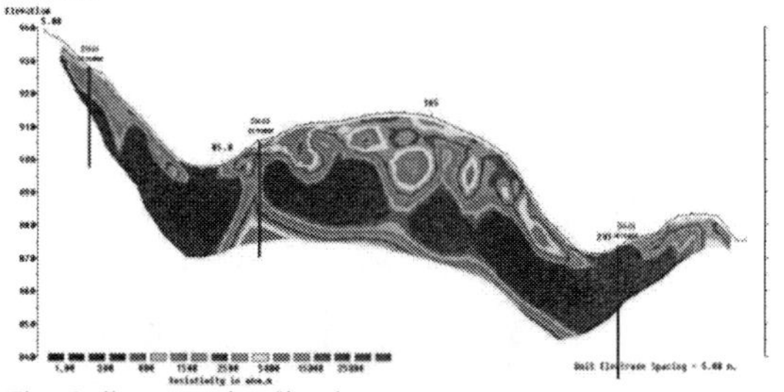

Fig. 6. Cross-section line 1

Resistivity value in line 1 has range 18.56 Ωm to 18855.5 Ωm (Fig.6). Based on resistivity value, line 1 has three layers, namely:

- The first layer is a soil layer in the form of clay. Yellow to orange color supposedly is a boulder of quartzite rocks were transported from elsewhere.
- The second layer is dominated by clay, a dark blue color is suspected of clay containing water. The blue color that is closely to the third layer presumably of iron, because the drilling results has weathering of quartzite containing little iron at a depth of 18 meters to 30 meters. The weathering is thought same with line 2 because iron also found in this the second layer. While the blue color in the centre of cross section is a layer of clay and schist. This is reinforced by the discovery of a layer of clay and schist in the drilling point. In the end of line 1 is clay that contains lots of pyrite. While the light blue color on the boundary layer is the boundary between clay, schist and quartzite rock intrusions.
- The third layer is an intrusion of quartzite rock, the drilling results obtained quartzite rock layers that contain iron minerals at a depth of 30 meters to 35 meters.

Line 2

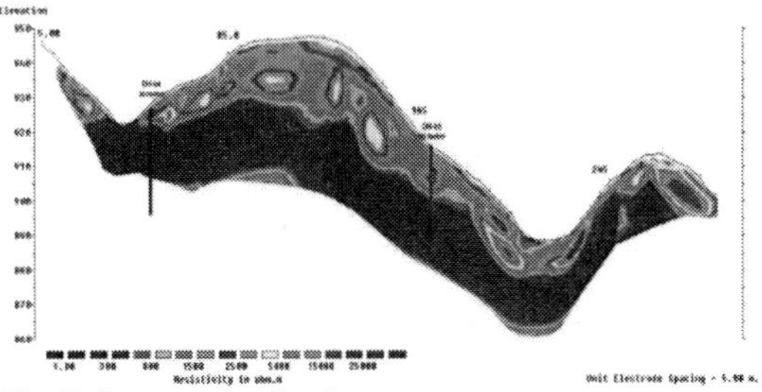

Fig. 7. Cross-section line 2

Resistivity value in line 2 has range 1.19 Ωm to 75429.4 Ωm (Fig.7). Line 2 has three layers:

- The first layer is a layer of soil in the form of clay. Green color to yellow is boulders of quartzite rock. Yellow to purple color at the end of line is granite lifted upwards due to fracture of the rock nearby.
- The second layer has the similar rock properties with the line 1.
- The third layer is an intrusion of quartzite rock, the drilling results obtained quartzite rocks that contain iron minerals in 15 meters depth. Quartzite layer containing iron were also found in line 1.

Line 3

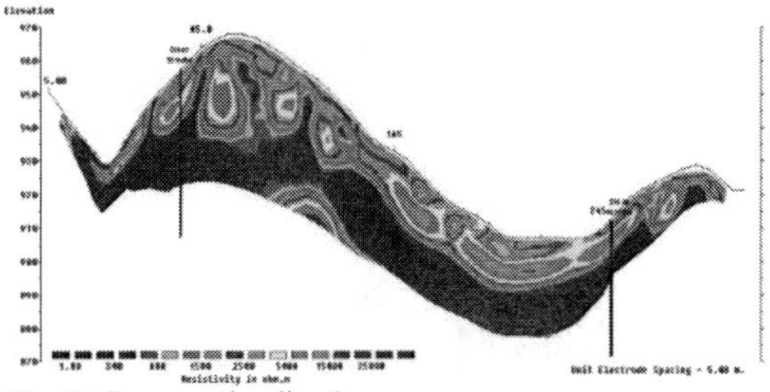

Fig. 8. Cross-section line 3

Resistivity value in line 3 has range 2.28 Ωm to 58567.3 Ωm (Fig.8). There are three layers that identified in this line.

- The first layer is a layer of clay that there are many weathering of quartzite. Light blue color in this layer is sand as the result of quartzite weathered rocks that contain iron and has low resistivity value.
- The second layer below the point of 0 to 100 meters from the geoelectrical measurements is a layer of clay that contains lots of water as it is above the layer there is a stream. The iron indication caused of the possibility of iron in line 1 and 2 are still a part of line 3 in the second layer. While at 100 to 200 meters is a layer of clay and schist, it is obtained from the drilling results. Layers of schist and clay in line 3 is believed to be the unity of connecting layer from line 1 to line 3. At the point of 200 to 300 meters is a mixed layer of clay and pyrite.

- The third layer is an intrusion of quartzite rock. The drilling results obtained quartzite rock layers that contain iron minerals.

Line 4

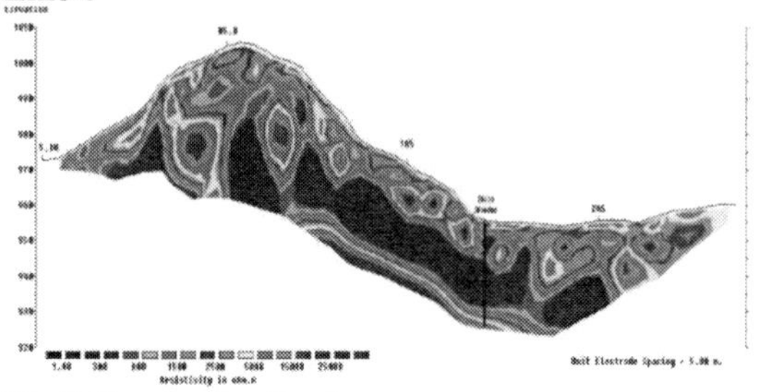

Fig. 9. Cross-section line 4

Resistivity value in line 4 has range 24.66 Ωm to 89475.3 Ωm (Fig.9). The contain layer in this cross section:

- The first layer of this line almost similar with other layers in previous lines. The layer contain soil in form of clay. There is also quartzite rocks containing iron with small resistivity value.
- The second layer below point 0 to 100 meters from geoelectrical measurements is weathering of quartzite and containing iron. At the point of 100 to 300 meters has a clay containing pyrite.
- The third layer is quartzite intrusion. Because of resistivity value in this layer is almost equal with quartzite resistivity value in previous lines, so there is connection of this layer with previous layer. However, the high resistivity value of this layer make possible the iron contained on quartzite intrusion measly or no at all.

Line 5

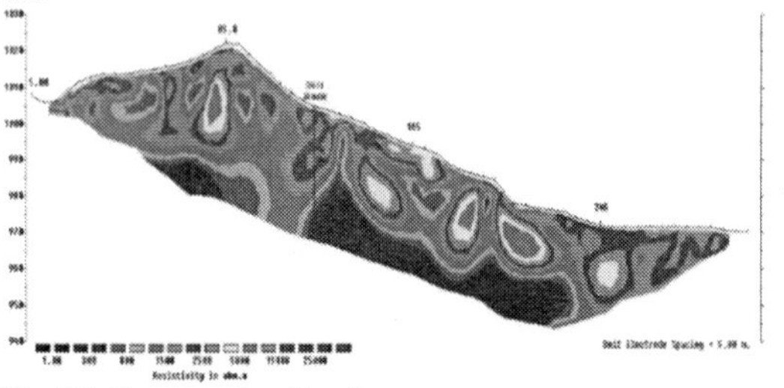

Fig. 10. Cross-section line 5

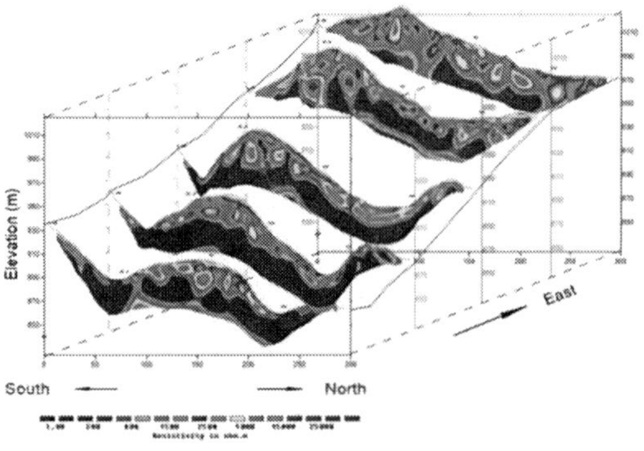

Fig. 11. Combination of 5 cross-sections

Resistivity value in line 5 has range 56.3 Ωm to 10181.3 Ωm (Fig.10). There are only two layers in this line.

- The first layer is clay with many boulders of quartzite. The presence of blue color on this line is probably the clay fill the gaps between the quartzite boulders.
- The second layer of this line is clay with 20 meters thickness.

The results of geological mapping and outcrop on the field, the presence of iron in the research area are in quartzite and granite rocks (Fig.11). However, not all of quartzite and granite rock contains iron minerals due to discontinuous the nature of the iron mineralization. Thus, in interpretation of geoelectrical result, the resistivity value is thought to be the body of quartzite and granite rocks.

Based on the combined cross-sectional imaging and the results of previous 2D cross-sectional images, can be concluded:

a. Overburden in every line generally chunks clay of quartzite.

b. The second layer at points 200 to 300 meters of line 1 to 5 allegedly is a layer of clay that contains pyrite. This is certainly due to the continuity of the blue color in the area.

c. The second layer at the point of 0 to 200 meters of line 1 to 4 are suspected of quartzite layer containing iron and estimated blue color formed is a continuity of the iron.

d. The third layer is an intrusive rocks containing iron quartzite and this layer connects from the line 1 to 4.

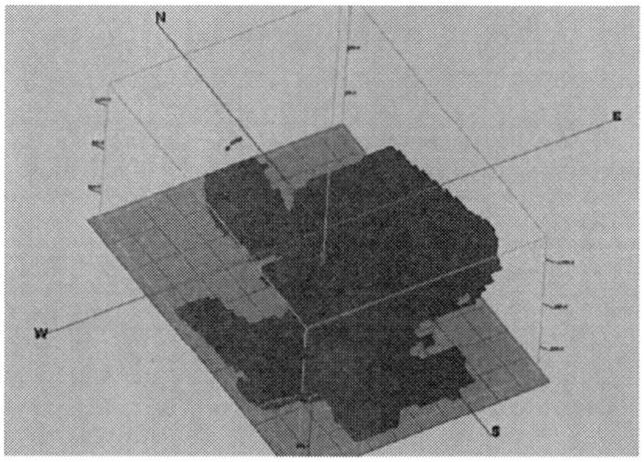

Fig. 12. 3D Resistivity Modeling

From the results of the 3D model (Fig.12), the iron is seperated in 30 to 40 meters below the surface. The concentration of the iron is more legible on the midpoint of the geoelectrical measurement, it caused midpoint has deeper current penetration.

8 Conclusions

Geoelectrical resistivity method with dipole-dipole array can provide a representative picture of the iron ore in the subsurface. It is seen from the cross section produced by the iron minerals associated with quartzite at 30 meters depth below the surface.

We thank to PT. Arosuka Mandiri West Sumatera for the permission in using the data.

References

1. Akmam. *Elastisitas Zat Padat dan Batuan Serta Aspek Geofisiknya dan Metoda Geolistrik Tahanan Jenis*. Jurusan Fisika. Fakultas Matematika dan Ilmu Pengetahuan Alam. Universitas Negeri Padang. (2002).
2. Cardimona, Steve. *Electrical Resistivity Techniques for Subsurface Investigation*. Department of Geology and Geophysics, University of Missouri-Rolla, Rolla, MO (2000).
3. Dahlin,T. and M.H. Loke. Resolution of 2D Wenner Resistivity Imaging as Assessed by Numerical Modelling, J. of App. Geophy. (1998).
4. Keller, George V. *Electrical Properties of Rocks and Minerals,* Cambridge University Press. (1990)
5. Loke, M.H. *Electrical Imaging Surveys for Environmental and Engineering Studies, A Practical Guide to 2-D and 3-D Surveys*. 5, Cangkat Minden Lorong 6, Minden Heights, 11700 Penang, Malaysia. (1999).
6. Loke, M.H. *Electrical Imaging Surveys For Environmental And Engineering Studies*, Penang, Malaysia. (2000).
7. Loke, M.H. *2D And 3D Electrical Imaging Surveys*. Penang, Malaysia. (2002).
8. Loke, M.H. *Rapid 2D Resistivity & Ip Inversion Using The Least-Square Method*, Penang, Malaysia. (2004).
9. Milsom, John. *Field Geophysics*. England. Willey. (2003)
10. Ocvianti, Marini A, I.Suyanto, E.Hartantyo. *Pengolahan Data Resistivitas Mapping Menggunakan Program Probabilitas Tomografi*. Prosiding Himpunan Ahli Geofisika, Pertemuan Ilmiah Tahunan ke-25. Bandung. (2000).
11. Reynolds, J.M. *An Introduction To Applied And Environmental Geophysic*, John Welly & Sons. New York. (1997).
12. Sunaryo, A.Rachmansyah, D.Sisinggih. *Penentuan Lapisan Aquifer dengan Metode Geolistrik Resistivitas di Desa Tempuran, Jatilangkung dan Awang-awang, Kec. Pungging, Kab. Mojokerto*. Proceedings of Joint Convention Jakarta 2003. The 32nd IAGI and The 28th HAGI Annual Convention and Exhibition. (2003).
13. Telford, W.M, L.P.Geldart, R.E.Sheriff. *Applied Geophysics Second Edition*. Cambridge University Press. (1990).

DOI: 10.1051/matecconf/201710104018

The study of ore minerals parageneses in Ponorogo area, East Java

Endang Wiwik Dyah Hastuti[1,*]

[1]Geological Engeneering Department Faculty of Engeneering, Sriwijaya University, 30128 Palembang, Indonesia

Abstract: The present study was undertaken in the Southern Mountain Range of East Jawa, Ponorogo District. Tectonically, the region extends along the Magmatic Sunda-Banda Arc, which comprises predominantly volcanic rocks of Mandalika Formation, sedimentary rock units of Arjosari Formation, and intrusive sequences such as dacite, andesite and diorite. Structurally, the region is controlled by faults striking NE-SW, NW-SE and N-S. Mandalika Formation and Arjosari Formation have an interfingering relation and are Oligo-Miocene. Results of the field observation and analyses of petrography and mineragraphy on outcrops reveal that the region has commonly undergone alteration and mineralisation. The principle ore minerals occurring in the Ponorogo area are pyrite and sphalerite with abundant mineral assemblages of chalcopyrite, magnetite, hematite, galena, covelite, bornite, and limonite. Mineralisation occurs in argillic zone and subprophyllitic zone. Based on textures, structures and ore mineral assemblages, mineralisation in the study area can be devided at least into two stages. The earlier stage was present in relation to hypogene processes, and resulted in pyrite–sphalerite–chalcopyrite–magnetite–galena. The subsequent stage took place due to supergene enrichment processes, and yielded pyrite–sphalerite–covelite–bournite–limonite. Such mineral assemblages suggest that they are formed at temperatures of about 100–360° C.

1. Introduction

The study area is situated in the southern section of East Java. Physiographically, the region forms part of the Southern Mountain Range of East Java (Fig. 1) [1]. Stratigraphically, it is included in the Old Andesite Formation [2]. The region consists predominantly of rhyolitic and basaltic volcanic rock successions. Geology of this area has been mapped and resulted in the 1:100,000 geological map of the Ponorogo sheet [6] (Fig. 2).

Rock sequences constituting the region consist from oldest to youngest units of Formation Mandalika, comprising volcanic breccias and tuffs, and Arjosari Formation, consisting of sandstones alternating with breccias. These formations show an interfingering relation, and are Oligo-Miocene in ages. The sequence was intruded by andesite, dacite and basalt which resulted in local alteration and mineralization. In addition, the study area has also influenced by volcanic activities during Late Miocene, indicated by the presence of andesitic lava that has experienced weak alteration and mineralization. Figure 2 shows the regional stratigraphic correlation proposed by several workers.

Ore deposits seem to take place in the likely geological controls such as the presence of heat source in the form of magmatism, channel way that fasilitated permeability of rocks, and structures determining the occurring ore pattern [7-9]. The ore mineralization has been controlled by a strike slip fault extending to the NW-SE, NE-SW, and N-S direction (Fig 3). In the mineralized zone, quartz veins are present as veinlet or stockwork and disseminated crystals in both fissure vein and wall rock.

The wall rock mineralization is generally present in volcanic sequences, the Oligocene-Middle Miocene sedimentary rocks, and the Lower-Middle Miocene extrusive successions [10]

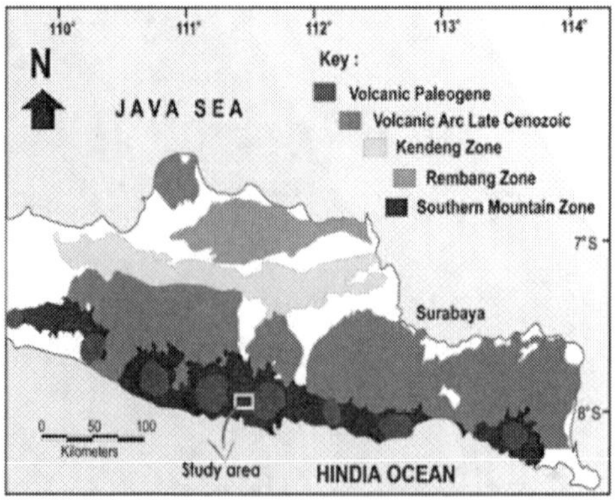

Fig. 1. A map showing physiography of East Java [1]

[*]Correspoding author: endanghastuti59@gmail.com

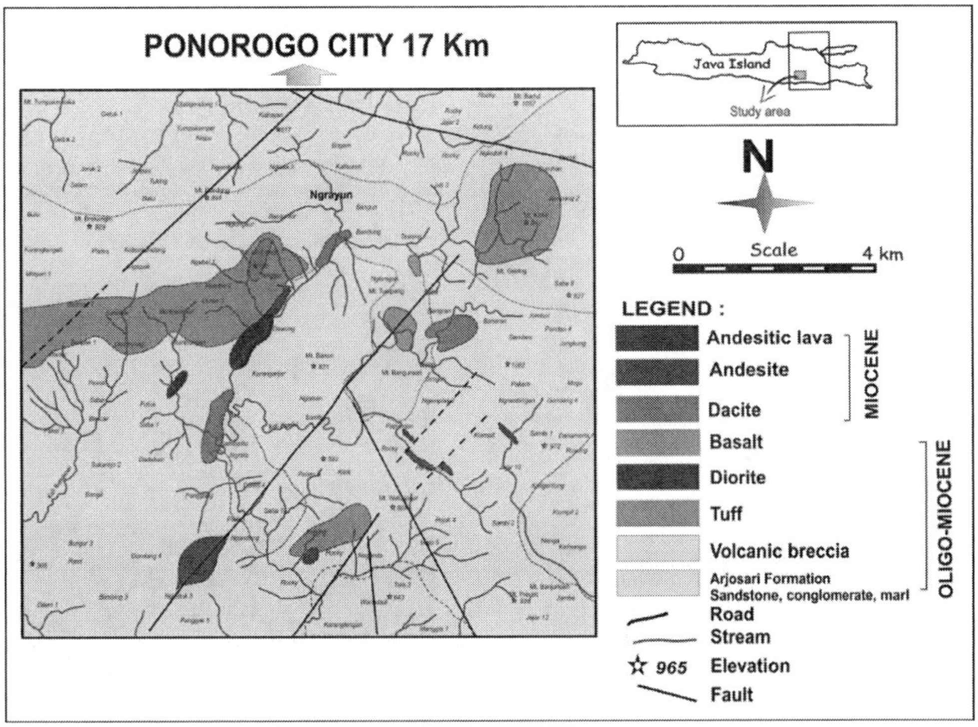

Fig. 2. Geological map of the study area (modified from [3-5]

2. Samples and methods

Mineral analysis done by macroscopic and microscopic observation. Determination of mineral in fine grained sizes is done with analysis polishes section. The polished rock thin sections analysis is performed to determine the type of mineral, texture, the relationship between minerals and microscopic characteristics related to the occurrence of ore mineralization as well as paragenesis interpretation. Data polish section done by microscope Zeis, 2000.

Texture formed in mineral deposits vary considerably depending on the characteristics of the fluid, chemical and physical characteristics of the wall rock and the way its deposition [8]. From the appearance of the texture will be able to assist in interpreting the chronologies set precipitated minerals (paragenesis), environment of formation (type deposit) and the way its deposition.

The analysis was performed on 30 samples in the present study were collected from outcrops which have undergone alteration and mineralization. The rocks samples Analysis of the polished rock thin sections was undertaken at one of accredited laboratories available in Bandung.

3. Mineral characteristics

Ore minerals recognized in the study area include pyrite, sphalerite, galena, and minor chalcopyrite, magnetite, covellite, and bornite (Fig. 3). Gangue minerals such as feldspar, quartz and clay minerals are common, but calcite, chlorite and muscovite are few, and adularia, actinolite, epidote and pyrophilite are dispersed [10].

Pyrite is the most abundant sulphide mineral in the study area. The occurrence of this mineral can be divided into two generations. The earlier phase resulted in euhedral-subhedral grains ranging from 0.03 to 0.5 mm in size, generally formed zoning, and associated with sphalerite. The later episode yielded euhedral-anhedral pyrites with sizes between 0.01 to 0.2 mm, generally associated with sphalerite and galena in fractures.

Results of petrographic analysis show that pyrite is yellow to white, cubic or euhedral, smooth or rough anhedral, fractured, and brecciated (fragmental). This suggests that the formation of the crystal has a relatively long range. They are commonly present with chalcopyrite and quartz. Under the microscope, mineral pyrite is partially replaced by covelite forming a blanket at the edge of pyrite (Fig. 4A).

Pyrite formed at the early stage or the first magmatic generation. The crystals are subhedral-anhedral, and there appears corrosion on the edges, cracking (cataclastic textures), filled in with chalcopyrite, sphalerite, and quartz, and often brecciated with drag patterns. Pyrite is interpreted to be generated prior to deformation of the rock section.

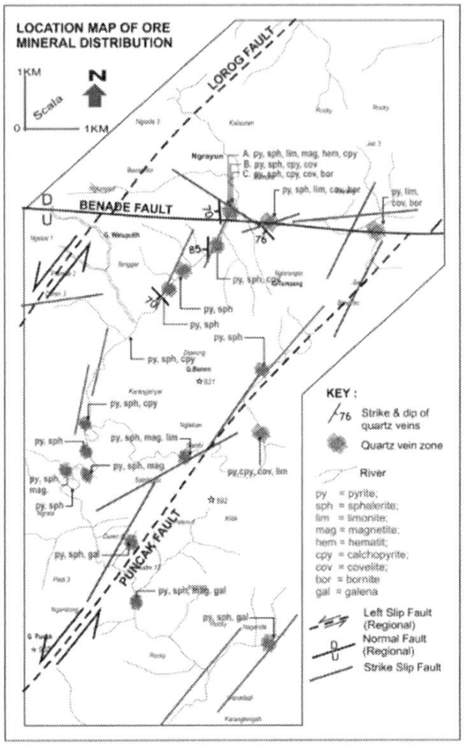

Fig. 3. Location map of ore mineral distribution and structural trend in the study area

Chalcopyrites in the field area is encountered in a moderate amount, present in quartz veins and wall rocks around the intrusive body, and shows subhedral – anhedral crystals. This mineral often occurs associated with pyrite and galena (Fig. 4B). Under microscope, chalcopyrites are anhedral in shape with visible cleavage and alteration around the edges. Chalcopyrites are generally existed as crack filling in quartz veins altogether with pyrite and arsenopyrite. Some chalcopyrites grow together with galena, chalcopyrite and quartz. Chalcopyrite in the samples analyzed is wide spread distributed and shows cleavages. The crystal and cleavage of chalcopyrites have undergone alteration into covelites, which have blue-gray color (Fig. 4D) and sphalerite inclution.

Galena is found in small quantities, present in the wall rock and quartz veins, anhedral–euhedral in shape, light gray-white in color, very fine-moderate (<0.01-0.2 mm) in size, and appears to be associated with pyrite, chalcopyrite and arsenopyrite. Microscopic appearance on a number of samples shows that galenas are present as inclution in pyrite and chalcopyrite (Fig. 4E), and show textures of triangular facets (Fig. 4F).

Covelite and Bornite, these minerals replace some of pyrite and chalcopyrite, and fracture-filling in pyrite, but it is also present as exsolusion on chalcopyrite. Covellite is the altered chalcopyrite (Fig. 4B). In the field, the crystal is easily observed, blue-gray in color, usually occurs on the margin of chalcopyrite and associates with malachite and bornite. Whereas, bornite is commonly present in quartz veins together with covelite and malachite, brownish-red, and subhedral - anhedral.

Magnetite veins are present, distributed and partly replaced by hematite. Some magnetites are dissected by pyrite. This feature indicates that the magnetite was formed ahead of pyrite.

Sphalerite occurs with pyrite and pyrite partially envelop, locally associates with galena (Fig. 4G & H). It is dark gray in color, fine to medium-sized grain (0.01 - 0.2 mm), subhedral–anhedral in shape, and partially shows the texture of emulsion.

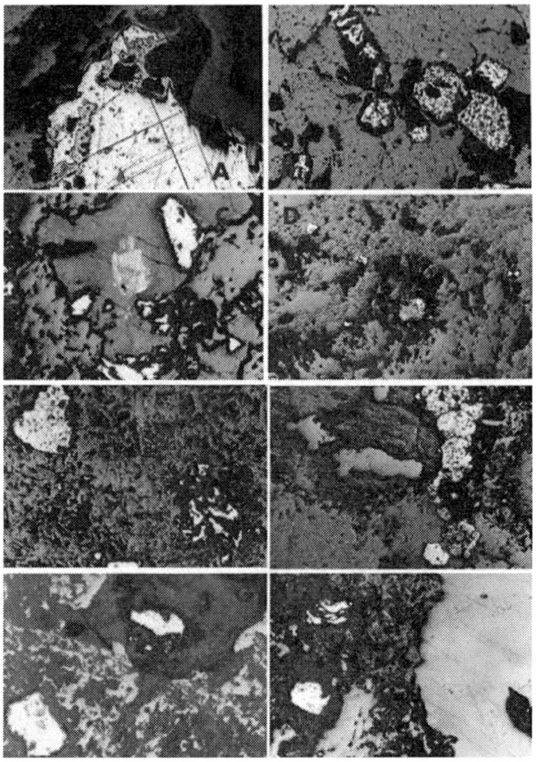

Fig. 4. Photomicrograph of ore minerals poles petrography in the study area showing A. Sphalerite (gray) enveloped by pyrite, displaying texture replacemen; B. Pyrite is partly replaced by covelite (bluish) and bornite (brownish); C. Arsenopyrite shows euhedral in shape and most visible feature is that of corrosion on the edges. D. Sphalerite (gray) largely replaced by chalcopyrite (yellow); E. chalcopyrite (yellow, top left) altered partly into covelit (blue-gray); F. Quartz veins filled by pyrite, and chalcopyrite particularly intruded by sphalerite; G. Sphalerite is an inclusion of pyrite (bottom left); H. galena is present within pyrite (bottom left)

Diginite and chalcocite, both minerals are very rare, form as secondary minerals and replace pyrite and chalcopyrite. Chalcopyrite is mostly covered by digenit.

Gold in the study area is found in quartz veins and wall rock in the form of electrum. In thin section, the mineral occurs together with galena forming emulsion texture and is usually found in fractures and disseminated on quartz and quartz breccia. Gold is often found in quartz filling fault lines and is very hard to see under thin section. This is only found from the essay analysis (0.2–2.7 ppm).

Limonite is an alteration or a secondary mineral, generally comes from pyrite and pyrhotite. The occurrence of the mineral in in the field is relatively easy to be treced through its brownish pink color. Limonites are often found in quartz breccia fractures, reddish yellow in color, usually resulted from dissolved minerals within ore. Under microscopic observation, most of limonite is a kind of goethite that has a cubic shape and shows colloform textures, and contains in the gap between quartz crystals. Limonite is formed due to the oxidation process.

Hematite, under megascophic appearance, it is present in fracture and associated with pyrite. It is estimated that the occurrence of hematite is formed by the oxidation of pyrite. On microscopy observation, hematite is partially replaced by some magnetite.

4. Fluid inclusion analysis

Fluid inclusions are a small volume of fluid, microscopic trapped during the formation of a mineral. Based on the genesis, fluid inclusions are divided into two, namely primary and secondary inclusions. Primary inclusions trapped during the formation of a primary crystal, while the secondary inclusions trapped in the fracture-tectonic fractures.

Fluid considered homogeneous phase during the process of entrapment. When the temperature decreases, there is a difference between the shrinkage that traps inclusions and crystal, in which the existence of this shrinkage resulting inclusion becomes more homogeneous with the onset of vapor bubbles and may also appear a crystal in fluid inclusions. Inclusion consisting of fluid and vapor-called biphasic, while containing fluid, vapor and crystals called trifase. The composition of the fluid inclusions could be CH4, CO2, H2O, hydrocarbons and so on

Analysis of the formation temperature quartz veins using methods mikrotermometri research areas, namely the inclusion heat until the steam disappears. In general, the method used is divided into two parts, namely thermometry which include heating fluid inclusions to obtain the homogenization temperature (Th) which is used to determine the formation temperature and kriometri in the form of cooling fluid inclusions for measuring the fusion temperature (Tf) which is used to infer the composition. Analysis of fluid inclusions conducted at the Laboratory of Physics Mineral Research Center for Geotechnology LIPI, assuming that the temperature is measured in quartz veins more or less the same as the minimum temperature at the time of the fluid trapped during the process of the fluid in a state of homogeneous, then the temperature calculated at the time the loss of steam (Th).

The results of measurements of fluid inclusions in the study area indicate that the quartz veins formed at a temperature ranging 164.1°C–293.2°C with an average temperature 216.1°C.

Mikrotermometri measurement is performed only on the primary, measuring more than 5µm, subhedral-euhedral shape (Tabel 1).

Tabel 1. Mikrotermometri measurement and calculation results of fluid inclusion analysis

Parameter	Kisaran	Rata-rata
Liquid temperatures (Tm)	-2.1– -0.5 °C	-1.3 °C
Homogenisation temperatures (Th)	164.1–293.2 °C	216.1 °C
Salinity Na Cl	0.9–3.9 %WT	2.4 %WT
Depth [11]	62.83–913.71 m	228.82 m
Pressure [11]	6.85–76.24 bar	21.28 bar

Based on the results of the measurement and calculation of temperature conditions of quartz vein, So that quartz vein in the satudy area form and grow at a depth of 62.83 m from 913.71m and form at temperatures from 293.2 °C up to 164,1°C, or included in the epithermal zone [11].

5. Paragenesis

Paragenesis is a time sequence during mineral deposition; how many times a mineral deposition has formed [7-9]. Chronology of mineral deposition, called the sequence paragenesa. This study used the term paragenesis to discuss the sequence of mineralization.

Observation on the structure of ore mineral deposits present in stockwork, brecciations, veins, mineral composition and microscopic appearance of texture such as exolusion, charging, replacement, suggest that there are at least two episodes of mineralization. The early episode relates to the mineralization hypogene processes (Tabel 2), while the subsequent episode results in supergene enrichment processes (Tabel 3).

Tabel 2. Temperature range of mineralization in the study area (a range of temperatures taken from [11])

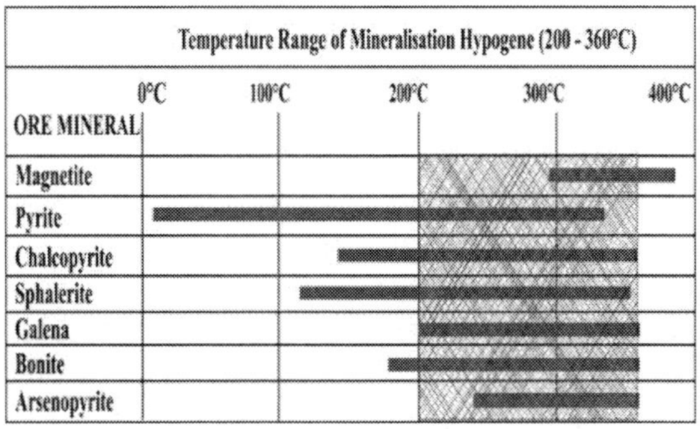

Tabel 3. Temperature range of mineralization in the study area (a range of temperatures taken from [12])

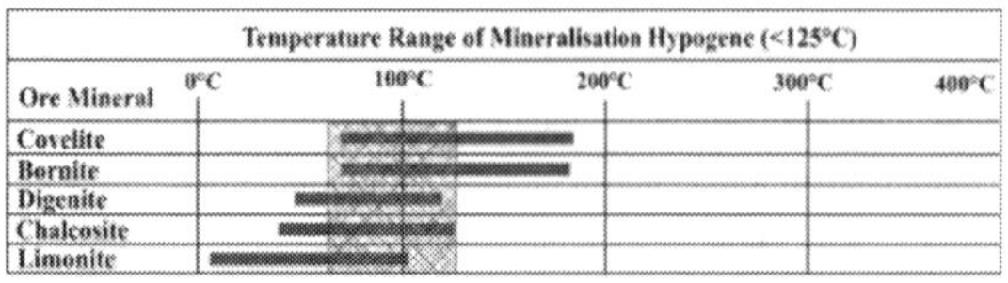

In the hypogene zone, ores mineralization formed as a product of rocks reaction with hidrothermal fluid due to oxidation [13]. Therefore, the formation of hypogene mineralization is likely formed at temperatures about 260°–360° C (Tabel 2). The chronology of hypogene mineralization in the study area is as follows: started by the formation of magnetite that forms magnetite veins and dessimination, followed by other metal sulphide mineralization (pyrite, sphalerite, galena, and chalcopyrite) prior to the formation of magnetite and chalcopyrite. Subsequently, mineralization of sphalerite, galena and chalcopyrite continued until to the end of this episode in conjunction with the formation of bornite.

Enrichment Supergen in the study area is characterized by mineralization digenit-bornite-covellite that replaced part of chalcopyrite and pyrite. Pyrite is the dominant sulphide mineral, suggesting the presence of dissolving agent. According to [8-9] the occurrence of dissolving agent might have formed ferric sulfate, which then reacted with sulfides in the vicinity. Under water $CuSO_4$ would react with primary sulfide and form sulfide supergen.

Ferric hydroxide precipitate changes to form limonite in the oxidation zone. Limonite contained in the study area is largely formed on the fracture. Based on mineral assemblage, supergene mineralization takes place at temperatures less than 125°C (Table 4).

Tabel 4.: Paragenesa in the study area

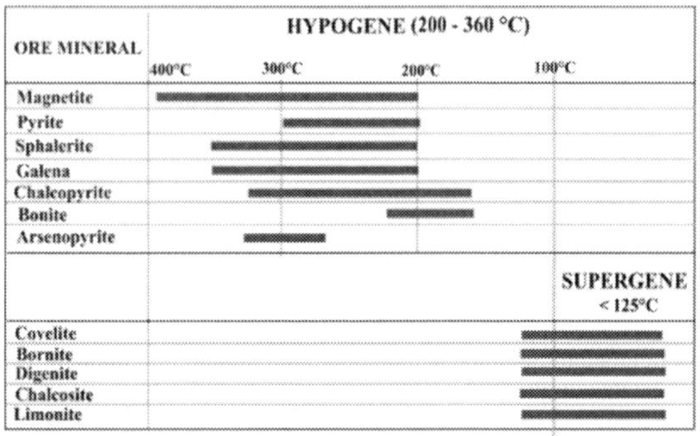

6. Conclusion

Igneous rocks in the study area are generally exposed to follow the pattern of structures. Igneous rocks in the region suffer alteration and mineralization. The event was due to intrusion of igneous rocks which have brought hydrothermal solution. Quartz veins are generally found in old rocks of Oligocene to Lower Miocene, especially those that have undergone argillic alteration. This indicates that the mineralization occurs from the first episode of magmatic intrusion. Hence, it can be interpreted that the gold was formed at the time of the second breakthrough.

In the study area there are at least two episodes of mineralization. The early episode wes related to the hipogene mineralization, characterized by the formation of magnetite, followed by sulphide mineralization (pyrite, chalcopyrite, sphalerite, galena, bornite and arsenopyrite). The final episode is characterized by supergene enrichment, characterized by the formation of digenit, chalcocite, bornite and covelite. These minerals are formed due to charging and replacement textures as well as little exsolusi.

Base on the fluid inclusion analysis mineralisation in the study area formed at a depth from 913.71 m to 62.83 m and at temperatures from 293.2 °C to 164,1°C,

The presence of mineral sericite, chlorite, adularia, calcite, chalcopyrite, galena, sphalerite indicates that the possible mineralization environment in the study area is a low sulphidation system.

The author is grateful to the University of Sriwijaya for reseach fund. We also thank my colleagues Edy Sutriyono, who has been involved in finishing this paper especially in editing

References

1. Sribudiyani; Muhsin, N.; Ryacudu, R.; Kunto, T.; Astono, P.; Prasetya, I.; Sapiie, B.; Asikin, S.; Harsolumakso, A.H.; Yulianto, I. *The collision of the East Java microplate and its implication for hydrocarbon occurrences in the East Java Basin.* Proceedings Indonesian Petroleum Association.
2. Van Bemmelen, RW. *The Geology of Indonesia.* Govt. Printed Office, the Haque (1949), **V. I** p. 545-675. & **V. II** p. 92-174
3. Samodra, H.; Gafoer, S.; & Tjokrosapoetro, S. *Peta geologi lembar Pacitan. Jawa. Sekala 1 : 100.000.* Puslitbang Geologi. Bandung (1992).
4. Sampurno dan Samodra, H. *Peta geologi lembar Ponorogo. Jawa. Sekala 1 : 100.000.* Puslitbang Geologi (1997). Bandung.
5. Hartono, U.; Baharuddin; dan Brata, K. *Peta geologi lembar Madiun. Jawa, Sekala 1 : 100.000.* Puslitbang Geologi (1992). Bandung.
6. Sudarto, T. dan Prapto, A.S. *Laporan penyelidikan logam dasar dan logam mulia di daerah Nawangan, Pacitan dan Ngrayun, Ponorogo, Jawa Timur.* Direktorat Sumberdaya Mineral (1995). Bandung
7. Corbett, G.J. & Leach, T.M. *Southwest Pacific Rim gold-copper Systems: Structure, alteration and mineralization.* Sociaty of Economic Geologists (1995.). **Special Publication 6**
8. Guilbert, J.M. and Park, C.F., JR. 1986.The Geology of Ore Deposits. W.H. Freeman and Company, New York, p. 55-209

9. Pirajno, F. *Hydrothermal Mineral Deposit*. Berlin Heiderberg, Springer – Verlag (1992)
10. Dyah Hastuti, E.W., Suparka, E., Asikin, S., Harsolumakso, A.H. *Miocene Volcanism Related to Hydrothermal Alteration in Ponorogo, East Java, Indonesia*. Procceding **8th** International Congress on Pacific Neogene Stratigraphy, Chiang Mai (2003), Thailand, p. 418-425
11. Haas, J.L. Jr.. The effect of salinity on maximum thermal gradient of a hydrothermal system at hydrostatistic pressure. Economic Geology (1984) 66 : 940 -946.
12. Anonim. *Important Hydrothermal Minerals and Their Significance*. Sixth Edition, Kingston Morrison Limited (1995), 65 p.
13. H. Leal-Mejía, J. C. Melgarejo, and Draper, Ore Mineral Paragenesis of the Gramalot Gold Deposit, Colombia. Macla, revista de la sociedad española de mineralogía (2008), no. 9. 139-140

MATEC Web of Conferences **101**, 04019 (2017)
SICEST 2016
DOI: 10.1051/matecconf/201710104019

Geothermal hot water potential at Parangwedang, Parangtritis, Bantul, Yogyakarta as main support of Geotourism

KRT. Nur Suhascaryo[1,*], *Hadi* Purnomo[1], and *Jatmika* Setiawan[1]

[1] Faculty of Mineral Technology, UPN Veteran Yogyakarta, 55281 Yogyakarta, Indonesia

Abstract. The paper aims to determine the condition of Parangwedang as hot spring source in Parangtritis, Bantul, Yogyakarta and provide a guidance to develop Parangwedang as one of tourism destinations by controlling geological factor. The study is limited to examining the physical condition in the form of color, turbidity, odor, temperature and chemical condition (pH), compositions of calcium (Ca), sodium (Na), silica (SiO_2), magnesium (Mg), bicarbonate (HCO_3), sulfate (SO_4) and chloride (Cl) and water source debits of Parangwedang hot springs as part of geohydrology research. The methodology used in the paper is divided into three steps. Firstly, the methodology was based on orientation and survey location. Then, it examined mapping the hot water temperature distribution. Lastly, it was implementing laboratory analysis of rocks and water. As a result, the paper portrays that there are potential water of hot of spring which meets the standards as clean water and the heat capacity can be utilized to support as geological tourism at Parangwedang, Bantul, Yogyakarta, Indonesia.

1 Introduction

Natural resources are being popular among tourism industry, for example in Bantul, Yogyakarta, which offers several beaches such as Parangtritis, Samas, Baron, Kukup, Krakal and Glagah Beach. Those places are well known for their eye catching view of the nature that portrayed in the area. Geothermal phenomenon is one example of natural resources that has attractiveness in tourism industry. This condition is commonly known as geotourism.

Geotourism as a natural attraction is being promoted alongside with the beautiful landscape of the beach. It uses geothermal fluid consisted of water, steam, and fumaroles that are carried out of the surface in the tourism activities. The fluid that appears on the surface is commonly used for bathing, swimming-pools of hot water, and medical therapy. The development of geotourism can be an added value that increases local revenue and the welfare of the population.

Geotourism is implemented in Parangtritis area in Bantul, Yogyakarta. Parangtritis is located in ± 30 km in south of central of Yogyakarta. The location is a coastal tourist spot of Indian Ocean. Based on the geographical position, Parangtritis is located in front of the path of active subduction zone of southern Java, namely the collision between the Indian Ocean and Australian plate under the Eurasian Plate. It leads to tertiary volcanic pathways that can lead to the emergence.

In addition, there is a hot water spring area in Parangtritis area, namely Parangwedang. The average fluid temperature in Parangwedang is at an average <50 C and it can be used for baths and hot tubs. Reviewing from its utilization, hot fluid distributions in Parangwedang are not fully utilized. It seems that Parangwedang needs huge exposure of its potential attraction in terms of geotourism.

Nevertheless, it is not an easy task to promote Parangwedang with its geotourism concept as tourism attraction because up to recent condition it has not been popular yet among tourist visitors although local government of Bantul has been managed tourism in Parangtritis area pretty well. As an illustration, Depok beach which offers a common attraction such as culinary is more popular than Parangwedang which offers limited attraction such as hot water spring area. Its visitors that come intensely can be counted few people. Their purposes are experiencing bathing, relying on tourist overflow from Parangtritis, doing a skin disease treatment by soaking in hot water which contains sulfur, until coming for spiritual purposes. This circumstances occurred in Parangwedang were attracted the researchers to conduct this research.

In relevance, the research aims to portray geotourism condition and opportunity that can be enhanced in Parangwedang and suggest a guidance on how to develop Parangwedang for tourism attraction, especially by controlling the geological factor. This geological research focuses on knowing the physical condition in the form of color, turbidity, odor, temperature and chemical condition (pH), compositions of calcium (Ca), sodium (Na), silica (SiO_2), magnesium (Mg), bicarbonate (HCO_3), sulfate (SO_4) and chloride (Cl) and water source debits of Parangwedang hot springs as part

*Corresponding author: suhascaryo@yahoo.com

DOI: 10.1051/matecconf/201710104019

of geohydrology research. This research refers to science and technology development.

2 Geological review

Parangtritis is located on the topographic map sheets 47/XLIII-c, sheet 47/XLII-c and sheet 47 / XLII-b, with a map 1:25,000 scale. Administratively it is located between 8° South Latitude and 110° East Longitude. The study area is in the Mancingan, Kretek District, Bantul (see fig.1).

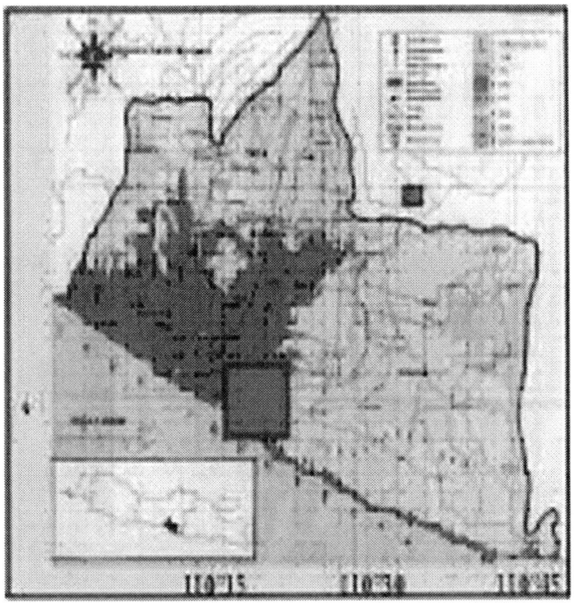

Fig.1. Parangtiritis Topographic Map

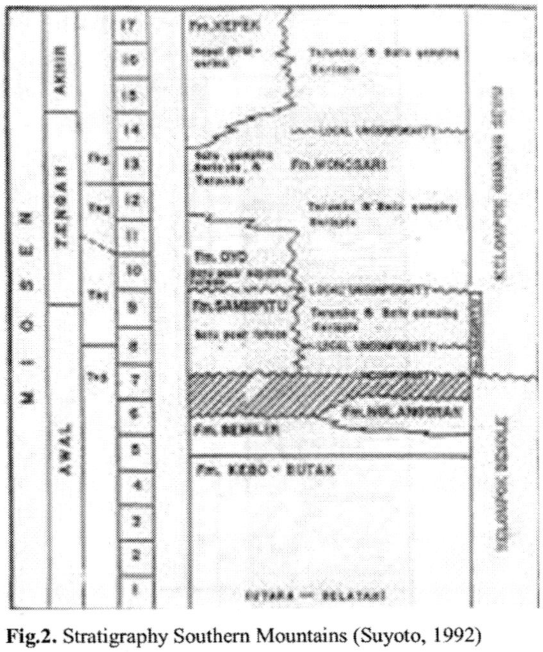

Fig.2. Stratigraphy Southern Mountains (Suyoto, 1992)

There are several places surrounded Parangtritis area. In the north is bounded by the Opak river and Oyo river, the southern part is bounded by the Indian Ocean, the eastern part is bounded by Sewu Karst mountains and the western part is bounded by Opak river downstream. Geologically, Parangtritis is located in Central Java Province South Mountains Zone and the coastal alluvial plain zone (Bemmelen, 1949). According to Rahardjo, et al (1977) and Surono, et al (1992), the oldest rocks in the South Mountains in the form of Pre-Tertiary metamorphic rocks were exposed in the Jiwo Hills, Bayat, Klaten regency, Central Java. In the western part of South Mountains, above metamorphic rocks were deposited sequenced Gamping-Wungkal Formation, Kebo-Butak, Semilir, Nglanggran, Sambipitu, Oyo, Wonosari, Kepek and alluvium sediments. According to Suyoto (1992), he suggested the structure of the stratigraphy in the Southern Alps as seen in fig. 2.

3 Methodology details

There were three stages implemented in conducting the research. They are:

Stage 1, literature and the provision of basic map (preparation) was used to obtain information related to the study. Secondary data were needed in the form of a topographical map, a map in such a world, analysis and interpretation, geological maps regional concerned areas, the study of reports/articles/papers in good geological condition, geochemistry, hydrothermal research of previous investigators, and information from the Internet.

Stage 2, phase study of area or field observation was direct research activities in the field, by data recording of geological phenomena, landscapes and environment observation, and also water sampling at Parangwedang. The findings were analyzed in the laboratory.

Stage 3, the analysis in the laboratory was testing to determine the geological and environmental aspects which include the analysis of physical characteristic of hot springs include color, turbidity, odor, and temperature. Meanwhile, hot springs chemical analysis was by testing pH, hardness, and nitrite. Then, inundated condition analysis was carried out in two places, namely in the field and Industrial Engineering Laboratory of UPN Veteran Yogyakarta.

4 Result and discussion

Morphology research areas included in Southern Mountains zone in Central Java and Yogyakarta were found undergoing a process of the sinking. It was covered by alluvial sediments in the form of broad coastal plain, from Parangtristis to Kebumen and appeared in the Karangbolong Hills and Nusakambangan Island (Bemmelen, 1970). Then, the top of South Mountain was largely composed of Miocene limestone, which had undergone a process of dissolution form karst topography, such as reefs hills, limestone caves in which rivers flow underground. Meanwhile, there are three subzones in the Southern Mountain, namely Northern subzone or better known as Complex Batur-Grand (Baturagung range), Midsection subzone or better known Wonosari and Southern subzone or better known by the name of Thousand Mountains (Sewu Mountains).

Sewu Mountains has elevation between 0-400 meters above sea level, which is composed of material that is readily soluble limestone. Sewu Mountains are divided into three zones, the North, Central and South region. In physiographic, Parangtritis and surrounding areas that lie in the South Central Java Plato zones which are formed by karst mountains are bordered by Baturagung Strip, and beside around Opak river estuary is a coastal alluvial plain (Pennekoek, 1949).

Parangtritis area and surrounding area stratigraphy are structured by Tertiary rocks composed by clastic volcanic sedimentary rocks, volcanic rocks and clastic carbonate sedimentary, as well as surface-old sediment quaternary (Idral et al, 2008). Clastic volcanic rocks are structured by clastic volcanic rock units which are in Semilir Formation. It consists of an ash-sized interlude tuff to lapilli and Miocene. These rock units are waterproof and found has become bedrock in the research area. In addition, volcanic rocks which are constituent rocks are composed by andesite breccia and andesite lava were found included in Nglanggran formation. This andesite breccia extended in the north and west of research areas with the general slope elevation to the south-southeast. Furthermore, igneous andesite rocks were found in Grogol Village and Bibis Village in outcrop conditions which were still fresh with a stocky structure of shear zones. Andesite is an igneous rock that is contained in the surface or near surface which typically lava forms, dike and sill. Next finding is located in Parangkusumo area nearby Parangwedang. There was found common black basalt lava gray-green with lava flows and fractured structure with parallel pattern or a hefty plate.

Clastic sedimentary carbonate rocks are constituent lithologies which consist of limestone reefs and kalkarenit included in Wonosari Formation aged between Middle Miocene and Under Pleistocene (Wartono-Rahardjo et al, 1977). The position of the rock layers is generally between N50^{o}E- N70^{o}E with 4°-28° elevation to the south. Clastic sedimentary carbonate deployment was found located on the western and eastern of research areas with the karst hills appearance.

Sediment surface is deposition of alluvium which consists of clay to gravel shuffling results of weathering and leaching surrounding rock and shore sediment that composed of fine-sized till coarse-sized sand, sandbanks forming crescent type along the Parangtritis coast until Depok Beach. According to Idral, et al (2008), the pattern of the geological structure in Parangtritis region is mostly structured by the symptoms associated with tectonic that had ever happened in the Java Trench and the formation of mountain systems in the south of Java. Structures in Parangtritis area is normal fault (Bantul, Bambang Lipuro and Mudal), Horizontal fault (Parangkusumo, Soka Nambangan and Siluk) and stocky-built Fault, with the orientation of the fault SE-NW and NE-SW. Parangkusumo faults trending N300^{o}W sub ducting 80° to the southwest, is an important fault that controls the appearance of Parangwedang hot water spring.

Hydro geological of Parangtritis area defined by the climate and geology or geomorphology of the area and divided into 3 units of hydrogeology; (1) The area traversed by Oyo river and Opak river fluvial sediment with porous material, (2) Regional alluvial beaches and dune sand, where most rain seep into the porous soil into groundwater and (3) The topography area above the Karst Sewu Hills with specific appearance, such as developing its dissolution cavities, the water flow that suddenly disappear and underground drainage system.

In addition to the hydro-geological characteristics above, the hot springs in Parangwedang has founded which is hot spring fissure due to the influence from Parangkusumo fault and the groundwater heated by a heat source (magma) or associated with symptoms of post-volcanic. The hot rocks are thought to result igneous intrusions in the surrounding LYSIS areas. Based on observations in the field there are three hot springs that emerge with gas bubbles and accommodated in storage ponds of hot water which has an area of 8 x 9 meters, the color of the water in the ponds is green, due to achieved by high amount of algae (fig. 3).

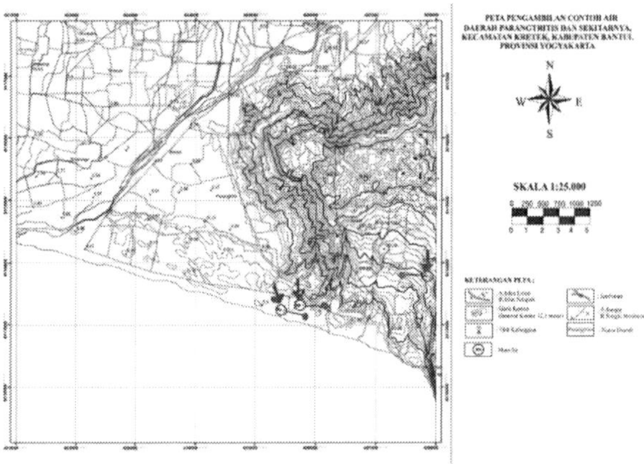

Fig.3. Sampling Hot Water Spring in Parangwedang

Based on physical analysis, chemical and limited microbiological performed on water samples analysis, it can be argued on the analysis of physical results, chemical and micro-biological limited to Parangwedang hot spring (sample is checked in the location the temperature is 44°C, pH is 7.8, $CaCO_3$ is 1,795 mg/l hardness, sample is checked in the laboratory temperature is 29°C, pH 7.5, $CaCO_3$ is 3,385 mg/l hardness which has an amount of dissolved substances (TDS) about 1.283 mg/l, Color 2 PtCo, 0.002 mg/l nitrite, 8 MPN/100 ml MPN Coliform and 5 MPN/100 grams fecal coli MPN). See at table 1.

From the table, it portrays several findings, such as the concentration of a MAP dissolved chemical compound 1 and 3 in Parangwedang are significant in mg/L units is among Cl = 7291.06; Na = 2470.59; Ca = 2450.98; SiO_2 = 62.25; Mg = 11.62; NH_4 = 5.1; B = 7.71, F = 2, while Al, Fe and As are undetected. Then MAP Parangwedang 2, the concentration of dissolved

Table 1. Result and Sample Analysis

Samples Analyzed in The field						
No.	Results of the Analysis			Water Quality Standards		
	oTemp.(C)	pH	Hardness(CaCO)mg/l	oTemp.(C)	pH	Hardness(CaCO3) mg/l
1.	40	7.6	1,790	3	6.5-8.5	500
2.	30	7.5	1,790	3	6.5-8.5	500
3.	44	7.8	1,795	3	6,5-8,5	500

Samples Analyzed in the Laboratory			
	Parameter	Results of the Analysis	Quality Standards
1	Temperature	29°C	3°C
2	pH	7.5	6.5-8.5
3	Hardness ($CaCO_3$)	3,385mg/l	500 mg/l
4	Total of Dissolved Substances (TDS)	1,283mg/l	500mg/l
5	Color	2 PtCo	15 PtCo
6	Nitrite	0.003 mg/l	1.0 mg/l
7	MPN Coliform	8MPN/100ml	50MPN/100ml
8	MPN Coli Tinja	5MPN/100 gram	10MPN/100 gram

and Fe = 0.04, while Al and As are undetected. From the limited physical, it refers that chemical and microbiological analysis to the findings above which are the hardness of $CaCO_3$ at 3,385 mg/l has been exceeded from quality standards which is formed by Ministry of Health Decree No. 416/Menkes/Per/IX/1990 which is supposed to be equal to 500 mg/l.

In addition, the result of the dissolved substance amount (TDS) of 1,283 mg/l is quite high, and has exceeded the quality standard by The Minister of Health No. 416/Menkes/Per/IX/1990 that has to be equal to 500 mg/l. Meanwhile in microbiological analysis, researcher found that MPN Coliform and MPN Fecal Coli were under quality standard. Concluding from all above, the data were found that Parangwedang hot spring already met the standards as clean water, so it can be used for hot spring.

The literature review which examines the results of the integrated geological, geochemical and geophysical investigation for geothermal Parangtritis regions (Idral et al, 2008), indicates that there is a potential for geothermal utilization in Parangwedang because it has low enthalpy with the temperature (T) below the surface which is smaller than 125° C. This condition presents that Parangwedang is safely can be used for a hot water bath.

Moreover, the analysis of the triangle diagram Cl-SO_4-HCO_3 based on literature review about the type of hot water which mentioned the type of chloride or Cl (Nicholson, 1993), is in line with the presence of chloride found which is generally indicated by the high permeability reservoir zone (breccia fault on the pipe eruption). The phenomenon emerged by the condition of Parangwedang hot spring which is formed by Parangkusumo across fault.

The further discussion in this research is that the reserves quantity of groundwater in the research area is large enough to warrant the formation of a natural steam and hot water. Perhaps, it is occurred because of meteoric water which can infiltrate through the karst topography of the area as well as upland areas Oyo river and Peel river, then it is turned form the groundwater reservoir rocks of andesite breccias, and also because of the presence hot springs which has high Chloride inside Parangwedang that can be useful as disinfectant and skin bleaching. Consequently, Parangwedang has high probability to offer hot water spring with good quality in Parangwedang to be used as geotourism.

Nevertheless, current condition of Parangwedang hot spring as mentioned before has to be developed to support the idea of making Parangwedang as profitable geotourism. If the location does not well maintain, the potential means nothing than only natural phenomenon. The researchers suggest that local government collaborated with local people in Bantul to improve its promotion and facility. It can be started from improving the extent of cleanliness, public needs in the area, comfortable access to reach the area, conducting various activities, or even by building a well designed interior and decoration in the location. The government may alliance with several sponsors to support this idea. Researchers believe that Parangwedang can be one of popular geotourism in Bantul and attract huge visitors if it is well developed and maintained. In order to support this idea, the researchers also suggest that further research will also useful to examine deeply in supporting the proof that Parangwedang has the potential as geotourism. The future research may explore in the field of hot springs discharge research, implementation of engineering technology, or economical and management development in this case.

5 Conclusion

Below are the conclusions after researchers conducting the research:
First, the rocks in the Parangtritis area are structured by Tertiary rocks which are composed by volcanic clastic sedimentary rocks, volcanic rocks and clastic sedimentary carbonates, as well as surface sediment Map Quaternary aged. Parangwedang appear in alluvium rocks that are controlled by the northwest-southeast Parangkusumo trending fault. The existence of this fault is also supported by magnetic anomalous, gravity, resistivity and head-on.

Second, geothermal expression in Parangtritis area such as hot springs, which appears in Parangwedang 1 and 3, each with a temperature of 40°C and 44°C with a normal pH. Parangwedang hot spring has chloride type due to contamination with sea water, while it is below the surface temperature and estimated about 115°C.

Third, the main structure of this area is trending to northwest-southeast which opens towards the south, where is an area of hydrothermal alteration characterized by the presence of mineral chlorite and sericite. Source rocks are not exposed on the surface and it is estimated as residual magma rock due to subduction zone in South Java and predicted as heat source in investigation area. Meanwhile, reservoir of geothermal area is estimated to be at 600-700 meter depth.

Last, according to the analysis of geological, the limited physical, chemical and microbiological analysis, Parangwedang in Parangtritis area, has a potential of groundwater as a source of hot water spring and it needs to be overlooked by local government and local people, or even sponsor, while further research in this field is also still available to support the idea such as the fields of hot springs discharge research, implementation of engineering technology, or economical and management development in this case.

6 Further research

As mentioned earlier, there are several topics available for further research related to this current research. They are:

a. Hot springs discharge research

b. Implementation of engineering technology research

c. Economical and management development research (DCF, real option, marketing, etc)

References

1. Anonymous. *The Minister of Health Quality Standard Number 416/Menkes/Per/IX/1990.* Ministry of Health: Jakarta, Indonesia. (1990)
2. Anonymous. *The Southern Mountains Field Test Guidelines.* Geological Engineering of UPN Veteran Yogyakarta: Yogyakarta, Indonesia.
3. Chasin, M. *Inventory of Appearance of Symptoms Geothermal.* Yogyakarta Regional Directorate of Volcanology: Yogyakarta, Indonesia. (1974)
4. Geothermal Regional Integrated Team Adum. *Unpublished Report of the Regional Integrated Research Geothermal Adum.* Lembaga District: NTT: NTT, Indonesia. (2002).
5. Idral, A., Suhanto E, Sumardi, E., Kusnadi, D and T. Situmorang. *Integrated Research Geology, Geochemistry and Geophysics Geothermal Regional Parangtritis, Subdit Geothermal.* ITB Central Library: Bandung, Indonesia. (2008)
6. Kadar, D. *Biostratigraphy of Neogene Planktonic Foraminiferal The Area South Central Java, Indonesia.* Special Publication **No. 5**. Geological Research and Development Centre: Bandung, Indonesia. (1986)
7. Komazawa..M.,et al. *Gravity Anomalies of the Central Flores Island, Indonesia. page 211 – 223.* Special Publication: Indonesia-Japan Geothermal Expl. Project: Flores Island, Indonesia. (2002)
8. Kusumadinata, K., Alzwar, M. *Mapping the spread of geothermal manifestations in Java Island, Directorate of Volcanology.* Bandung. (1969).
9. Nicholson, K. *Geothermal Fluids: Chemistry and Exploration Technique page 262.* Springer-Verlag: Berlin, Germany. (1993)
10. Pannekoek, A.J. *Outline of the geomorphology, of Java, Geological Survey.* T.A.G.:The Netherlands. (1949).
11. Telford, W.M. et al. *Applied Geophysics.* Cambridge University Press: Cambridge, United Kingdom. (1982)
12. Thornburry, W.D. *Principle of Geomorphology 2nd Edition.* Mc Graw Hill: New York, USA. (1969).
13. Van, Bemmelen, R.W. *The Geology of Indonesia Vol.* ***1B***. Government Printing Office: The Hague, Netherlands. (1949).
14. Wartono, R., Sukandar, R., and Rosidi, H.M.D. *Geological Map Sheet Yogyakarta, Java.* **Issue 2**. P3G: Bandung, Indonesia. (1995).
15. Wohletz, K., Heiken, G. *Volcanology and Geothermal Energy.* University of California Press: Los Angeles, USA. (1992).

Pleistocene reef development in Bulukumba, South Sulawesi

Andi Muhammad Imran[1*], *Ratna* Husein[1], *Meutia* Farida[1], and *Afdan* Prayudi[1]

[1]Geological Department, Hasanuddin University, Kampus Unhas Tamalanrea, Makassar 90245, Indonesia

Abstract. Quaternary reefs are commonly studied right now to explain climate change during that time. They act as a good archive of climate change, because their development is influenced by climate condition. The research area is located in the southern tip of Bulukumba Regency, South Sulawesi. The objective of this research is to define the development of the reef. Methods applied in this research are field survey of 4 line transects along reef cliff. Laboratory work is mostly on petrographic and biofacies analyses in order to reconstruct the reef development. Four reef biofacies have developed in this study namely 1) *Coralgal framestone - wackestone*, 2) Massive coral *framestone facies*, 3) Platy-like coral Bindstone facies, and 4) Branching Coral Bafflestone facies. Based on the facies association and organism accumulation, the reefs are interpreted to be developed within a reef complex in a shallow marine environment.

1 Introduction

Pleistocene coral reefs in eastern part of Indonesia are abundantly found along the coast. Coral reefs well developed in this area due to a wide shallow marine environment. The reefs exposed by an uplifting and form terraces. Kennedy and Woodroffe [1] reported that the tropical reefs develop in a complex variety of ways and commonly form fringing reefs. The principal factor that appears to determine the growth and morphology of fringing reefs is the available accommodation space controlled by sea level.

Coral reef is an important archive of climate change, because coral is a sensitive organism toward climate changes. Experts believe that some climate change events in the past such as El-Nino are well preserved within the coral skeleton as long as they do not experience a diagenesis. These proxy climate indicators provide paleoclimatic records which are important for the assessment of the global climate system and the mechanisms of decadal climate variability [2]. It is clear that the study on climate change in the past could help to predict the climate changes in the future.

Bira area consists of Plio-Pleistocene reef, Pleistocene reef and even modern reef. They are a member of the Selayar Limestone, Walanae Formation [3], [4]. The rocks are exposed in several terraces indicating sea level fluctuations and/or tectonic activities during their development. The study area experienced the rapture, forming terraces of limestone reefs [5] indicated by the presence of notches. The purpose of this study is to determine biofacies and depositional environment based on reef developments.

2 Methods

The study focus on Lower Terrace unit which has Pleistocene age. The study area locates at Bira and P. Liukang Loe, Bulukumba region (Fig. 1). Method applied in this study were field survey of 5 line transects, laboratory analysis and data processes. Facies data was taken along line transect with perpendicular and parallel to the shoreline. Laboratory work was done for petrography and paleontology analyses.

3 Geological setting

Pleistocene reef is a younger unit in the carbonate series of Selayar Limestone [5]. The rock is a member of the Walanae Formation formed within the Walanae Depression [3], {4]. The Plio-Pleistocene event was accompanied by general uplift of the region and triggered most of Quaternary deposit to be exposed including Walanae Formation. The local appearance of thin coal layers at the upper part of the Walanae Formation may indicate the beginning of this uplift. It continued to the raised Pleistocene coral reefs in the southern of Walanae Depression. The reef unconformably overlies siliciclastic rock of Walanae Formation [6].

Stratigraphy of the area (Table 1) is composed by1) lower Late Miocene foraminiferal limestones, 2) upper Late Miocene to Pliocene coral reef, 3) Pliocene coralgal reef and 4) Pleistocene coral reef [5]. Each unit is characterized by terrace showing oldest rock exposes landward with high altitude and youngest one seaward

* Correspondent author: mudhimran@gmail.com

with lower altitude. Lower terrace has been described as a lower Pleistocene reef [5] and in Selayar Island was dated as very early Pleistocene or 1.6 to 1.4 Ma [7].

Selayar Limestone, which crops out in the Bulukumba region shows a reef complex dominated by coral reef. Based on fossil contents such as coral, larger foraminiferal assemblage and lithologic characteristics, the Selayar Limestone is informally divided into four reef biofacies have developed in this study namely 1) Coral gal framestone – wackestone, 2) Massive coral framestone facies, 3) Platy-like coral Bindstone facies, 4) Branching Coral Bafflestone facies. Based on the facies association and organism accumulation, the reefs are interpreted to be developed within a reef complex in a shallow marine environment.

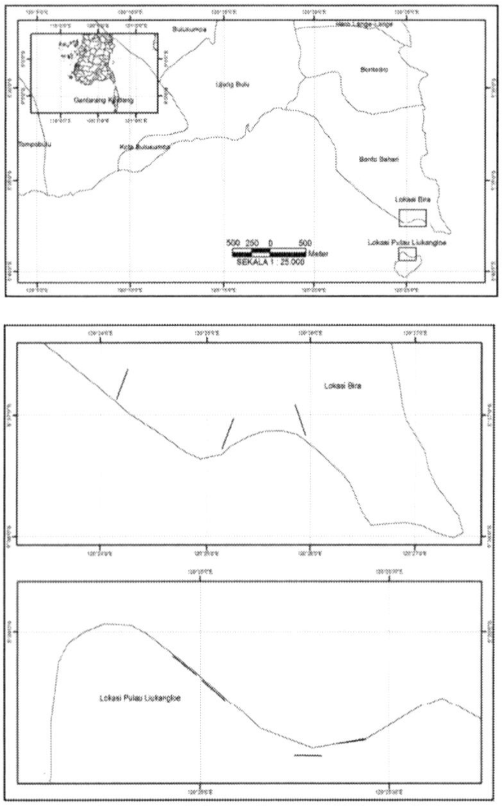

Fig. 1. Location map of the study area of Bira and Liukang Loe area

4 Pleistocene reef

Pleistocene reef is a raised coral reef [8] cropping out at the lower terraces of the Selayar Limestone [5]. The rock exposes in the southern tip of Bulukumba region and in the east cost of the Selayar Island. This study issubjected to the reef in the Bulukumba region. TheThe Pleistocene reef in Bulukumba composes four reef biofacies namely 1) Coralgal framestone – wackestone facies, 2) Massive coral Framestone facies, 3) Platy-like coral Bindstone facies, 4) Branching Coral Bafflestone facies. Based on the facies association and organism accumulation, the reefs are interpreted to be developed within a reef complex in a shallow marine environment.

Pleistocene reef unconformably overlies volcanic rock of the Walanae Formation [4] in Bira area. On the other hand the Walanae Volcanic does not expose in Pulau Liukangloe (small island closed to Bira).

Table 1. Stratigraphy of South Sulawesi and Selayar Limestone, modified from several authors [9, 3 & 10].

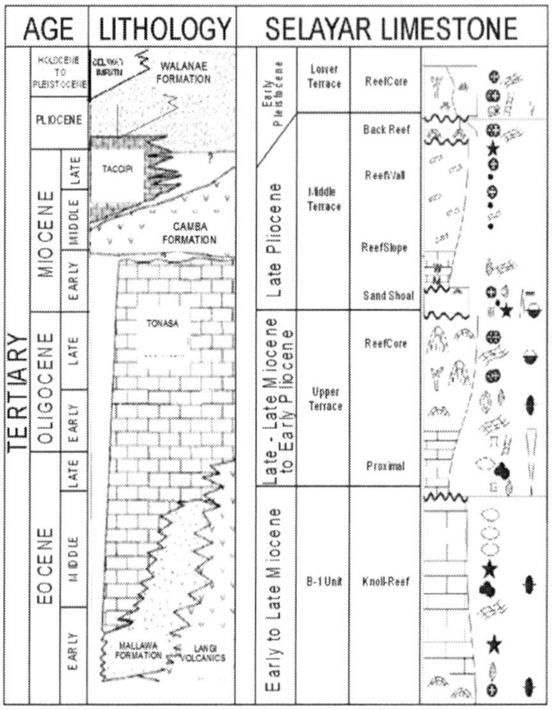

4.1 Coralgal framestone – wackestone facies

The coralgal framestone-wackestone facies crops out just above present sea level as high as 80 - 160 cm at P. Liukangloe and about 2-3 m at Bira area. It composes mainly of massive coral (Fig. 2a) and mud supported matrix. In a certain place, *Tridacna* sp. is present ingrowth position with up to 119 cm in diameter (Fig. 2B).The facies also consists of a rhodholithic lens with globular and encrusting habit at the Bira area. This encrusting algae has 1-30 mm in diameter. The lens forms graded bedding structure with rudstone texture. Fragments of red algae (Fig. 2C) and lithoclasts are common present in this facies. They are interpreted to be transported from upper part of the reef and accumulated in the fore reef [11]. Wackestone texture in which the grains commonly originated from coral indicates a low energy environment. Based on the bio-physical data, the facies is interpreted to develop in front reef in the open marine environment. (for example in accordance with the guidelines in the use of the font size)

4.2 Massive coral framestone facies

The facies consists of massive coral bed overlaying rhodholithic rudstone lens. The corals are predominantly of family Faviidae and Poritidae which have 1,5 m high and 10-45 cm diameter (Fig. 3A). To the west the coral gradually turns into a branching coral of the family Acroporidae. *Tridacna* sp. is commonly found at the basement of the facies (Fig. 3B) associating with small pelecypods and gastropods.

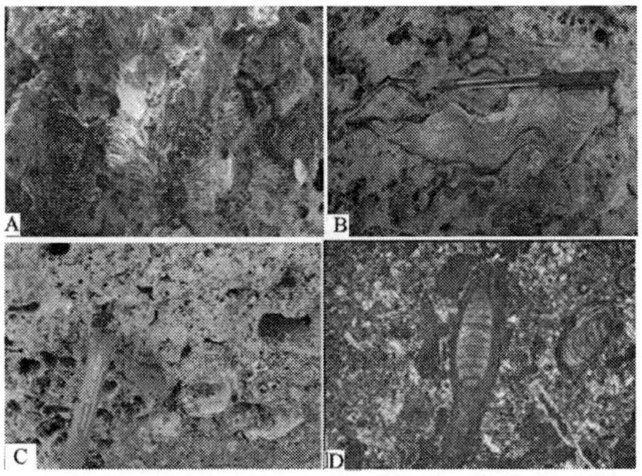

Fig. 2. Shows massive coral (A) and *Tridacna* sp. (B), rhodolit(C) and thin section of matrix showing red algae fragments (D) as components of massive coralgal framestone – wackestone facies.

Massive corals as a dominant organisms in growth position indicate a high energy environment with sufficient temperature and light [12]. The presence of *Tridacna* sp. with reference of modern sea of Spermonde Island, indicates a water depth <10 m with good water circulation environment [5]. The condition mention above represents a reef front zone to reef flat [11], [12]. By the same characteristics, the facies is interpreted to be developed in reef core toward fore reef environment with a good circulation and sufficient sun light.

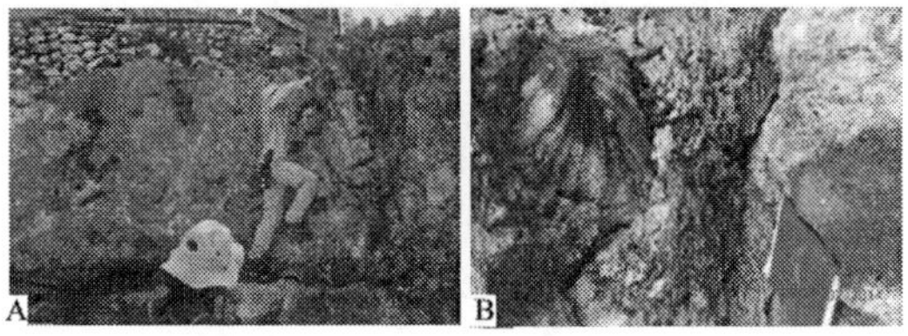

Fig. 3. Outcrop of massive coral framestone showing massive coral in growth position (A) and *Tridacna* sp exposed on the lower part of the facies (B).

4.3 Platy-like coral Bindstone facies

The facies is dominated by platy coral and coralline algae. The facies exposes at the western part of Bira beach and at the west side from Massive coral framestone facies. The corals are predominantly of Acropora and genus Montipora in growth position. The coral dimension shows a different in size, where in P Liukangloe is larger than that in Bira area coral has (5 – 80 cm in length) (Fig. 4). Matrix of the facies is locally found and consist predominantly of grain from coral and red algae (Fig. 4C). The organism association indicates the depositional environment of the facies is a protected reef in lagoon.

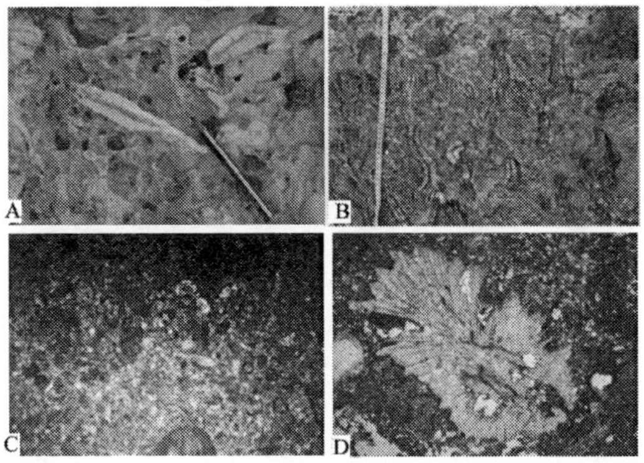

Fig. 4. Flatty coral in growth position at Bira area (A) and P Liukang Loe (B) and thin section of matrix (C) and coral fragment (D).

4.4 Branching coral bafflestone facies

The facies has been highly abraded and partly refined by travertine deposits. The facies locates at cliff as high as 7 meter to the northern part and slightly flat at P Liukangloe. It consists commonly of delicate branching coral from Acropora robusta with 1-3 cm in diameter (Fig. 5A &B). The texture of the matrix (Fig. 5C) shows skeletal grain support consisted of foraminifera, algae, echinoderms and gastropods. The association of the organism indicates a slightly high energy in reef front at 20-60 m depth [11], [13]. The organism and textural association suggests a shallow reef environment and is exposed to strong wave action.

Fig. 5. Field view of robust branching coral at P Liukangloe (A) and Bira (B). Thin section photograph of matrix showing red algae (C) and *Halimeda* fragments (D).

 DOI: 10.1051/matecconf/201710104020

5 Fringing Reef Of Bira Area

Fringing reef growth attaches to a landmass and form without or very narrow lagoon. It has variation in morphology and size. The reef produces a range of reef structures controlled by sea-level regional storm Climate [1]. It is possible to describe the variation between fringing reefs based on the nature of the antecedent surface that they are established over as well as the proportion of framework and detrital sediments within the reef structure [14].

The characteristics of the Pleistocene reef which is attached to the basement of shoreline, allow the reef to grow vertically and form reef flat during the still stand of sea level. This type of reef corresponds to the simple type of narrow fringing reef, proposed by Kennedy & Woodroffe [1]. The reef growth responds to the sea level change by catching-up in shallow water environment (Fig. 6). This type of reef corresponds to the reef model A of Kennedy & Woodroffe [1]. The reef consist of reef front, core reef and back reef with small lagoon. It is widespread surround the older reef where its lagoon is narrow in the north and reef front in the south.

6 Conclusion

The reef model proposed in this study is based on geomorphology and facies association of the reef. This model shows that both areas (Bira and P Liukangloe) form a narrow fringing reef. The Bira area developed from reef front to back reef. The reef was growing vertically by catching-up sea level and finally formed reef crest. The reef facies are constrained by pseudo bedding plane and they are interpreted to develop in different period. The first period was the establishment platy-like coral bindstone, coralgal framestone - wackestone, and massive coral framestone facies. The second period was Branching coral bafflestone facies.

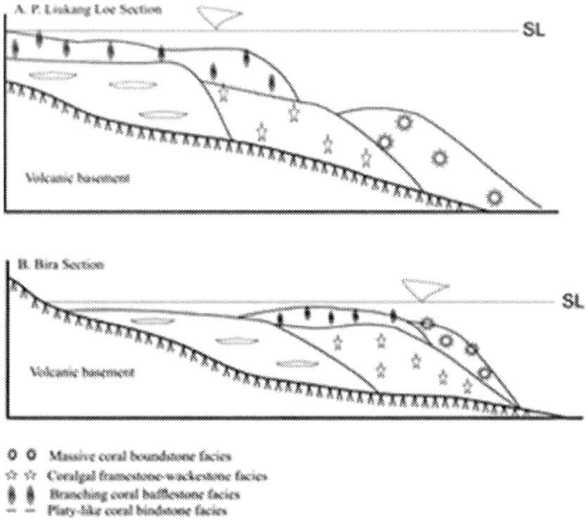

Fig. 6. Idealized of fasies development at the Bira Reef Complex (not to scale).

The author would like to thank to DIKTI for the financial support by research grant year 2015 – 2016.

References

1. D.M. Kennedy and C.D, *Elsevier*, vol. **57**. 255–277. (2002)
2. T. Felis and J. Pätzold, Eds., M. Shiyomi et al., *TERRAPUB* (2004) pp. 205–227., (2004).
3. R. Sukamto and S. Supriatna, *Geological Research Center General Directorate of Energy and Mining Department*. (1982).
4. A. M. Imran, M. Farida, M. F. Arifin, R. Husain and A. Hafidz, *Indonesian Journal on Geoscience* **3** No. 1 53-66. (2016)
5. A. M. Imran. Microfacies and Diagenesis of the Tertiary Selayar Limestone (Walanae Formation), South Sulawesi, Indonesia. Dissertation, (2000). Universitat Erlangen-Nuermberg, Jerman. (unpublished)
6. T. M. van Leeuwen. *The Geology and Tectonics of Eastern Indonesia, Geological Research and Development Center,* Spec. Publ. No.2, (1981). 277-304.
7. K. Bromfield, *Bulletins of American Paleontology*, no 387 Des. (2013). 1-60.
8. R.W. Van Bemmelen. The geology of Indonesia: General Geology. (Government Printing Office, The Hague) Batavia,Indonesia. (1949). 40-441.
9. R. Sukamto, *J. Geologi dan Sumberdaya Mineral*. Bandung (Pusat Penelitian dan PengembanganGeologi). (1975) **2**/1, 3-8.
10. A. M. Imran and R. Koch. *In: 35th PIT IAGI Riau*. Riau: IAGI. Code S-28. (2006). 8pp.
11. N.P. James. *Reef Environment*. in Schole, P. A., D. G. Bebout, & C. H. Moore (Eds.), *Carbonate Depositional Environments*. Oklahoma: AAPG Memoir 33. (1983).
12. J. Veron. New Species Described In Corals of the World. Melbourne: Australian Institute of Marine Science Monograph Series, **11**. 207 pp. (2002)
13. J.M. Webster, L Wallace, E. Silver, D. Potts, J.C. Braga, W. Renema, K. Riker-Coleman, C. Gallup. *Marine Geology* **204** (1–2), 59– 89. (2004)
14. D. Hopley, B. Partain. *In: Baldwin, C.L. (Ed.), Fringing. Reef Workshop: Science, Industry and Management. Great Barrier Reef Marine Park Authority, Townsville*, pp. 13– 33. (1986)

Development of erosion risk map using fuzzy logic approach

Manyuk Fauzi[1,*], *Imam* Suprayogi[1], *Sigit* Sutikno[1], *Ari* Sandhyavitri[1] and *Eko* Riyawan[1]

[1]Civil Engineering Department, Universitas Riau, 28293 Riau, Indonesia

Abstract. Erosion-hazard assessment is an important aspect in the management of a river basin such as Siak River Basin, Riau Province, Indonesia. This study presents an application of fuzzy logic approach to develop erosion risk map based on geographic information system. Fuzzy logic is a computing approach based on "degrees of truth" rather than the usual "true or false" (1 or 0) Boolean logic on which the modern computer is based. The results of the erosion risk map were verified by using field measurements. The verification result shows that the parameter of soil-erodibility (K) indicates a good agreement with field measurement data. The classification of soil-erodibility (K) as the result of validation were: very low (0.0–0.1), medium (0.21-0.32), high (0.44-0.55) and very high (0.56-0.64). The results obtained from this study show that the erosion risk map of Siak River Basin were dominantly classified as medium level which cover about 68.54%. The other classifications were high and very low erosion level which cover about 28.84% and 2.61% respectively.

1 Introduction

Siak river region is one of the major rivers that is included in the national strategic river criteria (Minister Regulation No. 04/PRT/M/2015). The criteria of national strategic has the following parameters: a) the potential of water resources in the relevant river area is bigger or equal to 20% of the potential water-resource in the province; b) the number of resources sectors in the relevant-river basin is at least 16 (sixteen) sectors; and the number of population in the basin is at least 30% (thirty percent) of the total province population; d) the negative impact due to the destructive force of water to economic growth resulting in economic losses is at least 1% of gross regional domestic product of the province.

In 2003, the palm-oil plantation area dominated the land cover in upstream of Siak river basin (Tapung River Basin), which had the area of 76,249.6 ha or 44.80%. However the area of forest was only 15,624.36 ha or 9.18% of the total land area. In 2007, the forest area decreased to 7,352.64 ha or 4.32%. There were additional forest area in 2012 to 12,186.32 ha, or 7.16%. If the land use change is allowed continuously, this activity will have many impacts on the water quality and quantity of the Siak River Basin, especially in the upper watershed (Tapung Basin). In addition to this problem, the impact of the land use changes also contributed to the potential soil erosion. Thus, effort is needed to build a map of potential erosion under conditions of existing land use. The application of soft computing approach is very powerfull to develop erosion risk map based on geographic information system [1].

In the last decade, the soft computing model as a branch of artificial intelligence was introduced as a forecasting tool such as knowledge-based systems, expert systems, fuzzy logic, artificial neural network and genetic algorithms [1]. Applications of fuzzy-logic soft computing models particularly (fuzzy logic) in water resource management efforts have been carried out in various countries, such as research on the fluctuation of groundwater flow simulation in India [2], the study of erosion by using GIS and fuzzy logic in Turkey [3], the study of optimal water network using fuzzy logic in Singapore [4], and the research about development of erosion risk map using fuzzy logic method in combination with geographic information system (GIS) [5]. The input data for this model were soil maps, topographic maps and density-flow data. The data processing were conducted by decision-making system using the fuzzy-logic Mamdani type models with triangular representation. The research results proved that the model provided an effective description of erosion-risk map.

2 Method

2.1 Research location

Siak River basin is located in Riau Province, Indonesia with the area of 1,481,499.25 ha. Geographically the Siak river basin is in the position between 100^0 28'-102^0 12' East Longitude and $0^0$20'-$1^0$16' North Latitude.

2.2 Fuzzy logic

The system of fuzzy membership function has a very important role to present a problem [6-7]. Membership

* Corresponding author: manyu_fauzi@yahoo.com

function is a curve that shows the mapping of points of input data into membership values (often called the degrees of membership) which has the interval between 0 and 1. One way that can be used to get the value of membership is through the function-approach. There are several functions that can be used include:

a. Linear representation

In the linear representation, mapping input to the degree of membership described as a straight line. This is the simplest form and be a good choice to approach a concept that is less clear. The linear state of fuzzy set consists of two state: linear-rise and linear-down. In the linear-rise, hike begins at the set of domain values that have a membership degree of 0 (zero) moves to the right towards the domain value that has a higher degree of membership with membership functions:

$$\mu[x] = \begin{cases} 0 & x \le a \\ (x-a)/(b-a) & a \le x \le b \\ 1 & x \ge b \end{cases} \quad (1)$$

While the linear down, straight line starting from the value of the domain with the highest degree of membership on the left side, then move down to the domain values that have lower degrees of membership with membership functions.

$$\mu[x] = \begin{cases} 1 & x \le a \\ (b-x)/(b-a) & a \le x \le b \\ 0 & x \ge b \end{cases} \quad (2)$$

b. Triangle representation

Triangular membership function is characterized by the presence of three (3) parameters {a, b, c} which will determine the x coordinate of the three angles. This curve is basically a combination of the two lines (linear). The equation for this triangular shape is:

$$\mu[x] = \begin{cases} 0 & x \le a \text{ atau } x \ge c \\ (x-a)/(b-a) & a \le x \le b \\ (c-x)/(c-b) & b \le x \le c \end{cases} \quad (3)$$

c. Trapezoidal-representations

Trapezoidal curve is essentially like a triangular shape, with some points that have a value of membership 1. The equation for a trapezoid curve are:

$$\mu[x] = \begin{cases} 0 & x \le a \text{ atau } x \ge d \\ (x-a)/(b-a) & a \le x \le b \\ 1 & b \le x \le c \\ (d-x)/(d-c) & c \le x \le d \end{cases} \quad (4)$$

Fuzzy inference system is a process of formulating a mapping from a given input to an output using fuzzy logic. This system involves the fuzzy logic and the components therein such as membership functions, logic operations and if-then rules. In general, there are two kinds of fuzzy inference systems that can be implemented, namely the Mamdani Type and the Sugeno Type. According to [8]; [9], there are four phases to obtain output in the type of Mamdani Fuzzy Inference System, namely:

1. Comparing the input variables with membership functions in the antecedent ("because" part) to obtain the membership value of each linguistic variables. This process can be established by comparing the input variables and membership functions to obtain the membership value of each linguistic variables.
2. Combine all input variables by applying the t-norm. T-norm is sliced operations on fuzzy sets. System rules are used is min with liaison "AND". Operating results with the operator "AND" is expressed as α-predicate.
3. Generate consequences ("due" part) qualified or each rule based on weight. The set composition of the fuzzy-set rules is a process that states the output of each rule to be combined together into a fuzzy set. Each rule can be written in the form of the following languages: "If the antecedent then the consequent" or it can also be defined as follows: If x1 is A1 AND...AND xn is An THEN y is B. "If-Then" form is a form of expression based-on reasoning system that consists of numbers (r) of fuzzy rules.
4. Aggregation of all parts of the consequences. Forms part of the consequences of the process of aggregation is called "defuzzification". "Defuzzification" is a convertible model of the form of fuzzy-value into the amount / value more precise. One of the method is "the centroid method" with a mathematical formulation in the following equation:

$$z = \frac{\int_a^b x\mu_A(x)dx}{\int_a^b \mu_A(x)dx} \quad (5)$$

where z defuzzification-value; x fuzzy members set A; and uA (x) the degree of membership of an element x in a set A. Input from "defuzzification" process is a fuzzy-set obtained from the composition of fuzzy-rules, whereas the resulting output is a fuzzy set of the domain.

3 Result and discussion

3.1 Analysis of soil erodibility value (K)

Based on geological and soil unit maps, the upstream of Siak River Basin consists of several orders of soil; such as inceptisol, ultisol, oxisol, entisols and histosol. Inceptisol order consists of several types namely dystrudept, dystropepts, and endoaquepts. Ultisol order consists of great group (equivalent) of land that is hapludults, kandiudults, tropodults. Order Oxisol soil consists of one type of soil that is hapludox. Land entisols order consists of one type of soil that is tropofluvents. Then ground histosol order consists of several types of ground (great group) namely haplosaprits, troposaprits and tropohemists. The description soil maps of upstream Siak river basin is presented in Fig. 1.

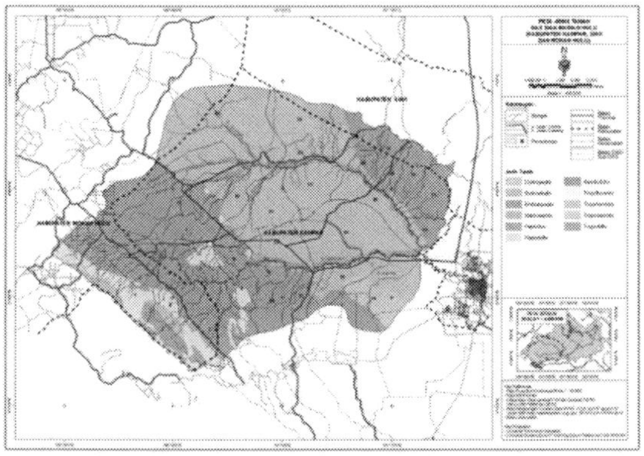

Fig. 1. Soil maps of upstream Siak Basin

Table 1 shows that the erodibility of the upstream section of Siak river basin is included in the classification between low (0.11 to 0.20) and medium (0.21 to 0.32). In principle, soil properties affecting soil erodibility are soil properties that affect the infiltration rate, permeability, soil's capacity to hold water, and soil properties that affect resistance of soil structure against dispersion and attenuation beads of runoff. The soil properties are texture, structure, organic matter, soil nature and fertility.

Table 1. Value of weighted erodibility (K)

ID Number of Tapung Kiri Sub Basin	Rated K	ID Number of Tapung Kanan Sub Basin	Rated K
1	0.254	17	0.248
2	0.251	18	0.238
3	0.238	19	0.221
4	0.258	20	0.271
5	0.265	21	0.222
6	0.227	22	0.233
7	0.231	23	0.211
8	0.259	24	0.222
9	0.258	25	0.239
10	0.251	26	0.244
11	0.251	27	0.249
12	0.221	28	0.223
13	0.244	29	0.251
14	0.219	30	0.251
15	0.110	31	0.249
16	0.239	32	0.243

3.2 Slope analysis (LS)

The length and slope were obtained from the [10], then were multiplied with the slope comprehensive classification.

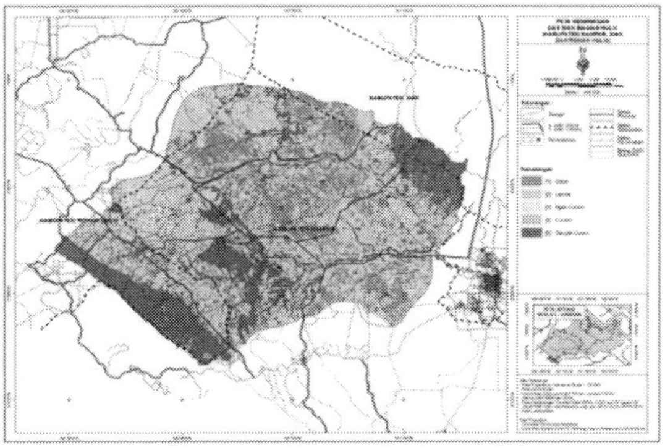

Fig. 2. Slope of upstream Siak River Basin

Table 2. Classification of slopes

Class	Slope (%)	Class Description
I	0 – 8	Flat
II	8 – 15	Sloping
III	15 – 25	Somewaht steep
IV	25 – 45	Steep
V	> 45	Extremely steep

Table 3. Percentage area of class slopes

Class	Slope (%)	Percentage (%)
I	71,359,556	15.36
II	153,523,425	33.04
III	71,775,121	15.45
IV	53,231,838	11.45
V	114,775,492	24.70
Total	464,665,432	100.00

Table 4. Value of weighted slope (LS)

ID Number of Tapung Kiri Sub Basin	Rated LS	ID Number of Tapung Kanan Sub Basin	Rated LS
1	5.726	17	4.035
2	4.620	18	2.514
3	7.390	19	3.615
4	4.790	20	3.171
5	6.459	21	2.425
6	6.488	22	2.573
7	6.652	23	1.358
8	7.346	24	2.357
9	8.318	25	2.427
10	5.202	26	2.910
11	5.786	27	7.654
12	2.459	28	2.787
13	3.353	29	6.133
14	2.569	30	8.039
15	2.239	31	7.764
16	2.470	32	3.440

The classification of slope can be seen in Table 2, which were classified in to 5 (five) classes of slope. Table 3 shows that the slope on upstream Siak river basin is dominated as ramps (33.04%). To grade the next slope is very steep (> 45%) with an percentage 24.70%. Small slope will give more opportunities to rainfall for infiltration. In conjunction with the large loss of land, the small slope will reduce runoff, so the ability to erode and sediment transport to be low.

3.3 Density index (Dd)

Results of the map data processing in upstream Siak river basin section were used to determine the flow or drainage density which was calculated from the ratio of the total length of the river network. The density of the flow of the river is an index number that indicates the number of tributaries in a watershed. Description density maps of upastream Siak river basin and the value of weighted density index are presented in Fig. 3 and Table 5 respectively.

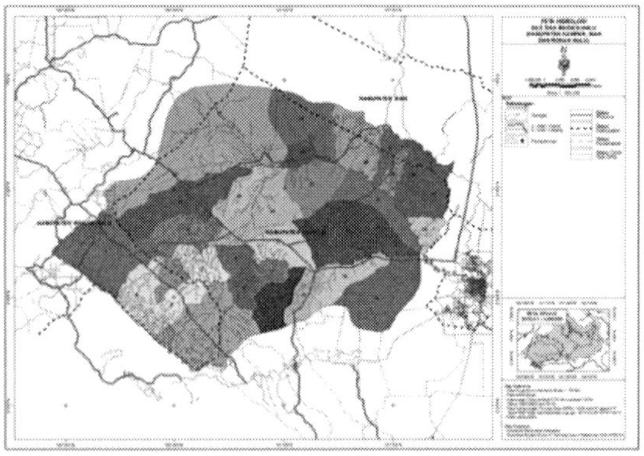

Fig. 3. Density index maps upstream of Siak Basin

The density of the river flow that illustrates the storage capacity of the surface water in basins such as lakes, marshes and rivers that flow in a watershed. The higher density level of the river flow, the more water can be absorbed in the river bodies.

Table 5. Classification of flow density index

Class	Flow Density Index (km/km^2)	Class Description
I	< 0.25	Low
II	0.25 - 10	Moderate
III	10 - 25	High
IV	> 25	Very High

3.4 Fuzzy inference system algorithm

The method used in this study was the fuzzy inference system, which the stages of these methods are as followings.

1. Establish the variable input and output variables
   Input variables used are erodibility (K), length and slope (LS) and current density (Dd), while the output variable is the index of the risk of erosion.
2. Establish a fuzzy set. Fuzzy set was formed for each variable as follows:
   a. The Variable Erodibility forming three fuzzy set, i.e. low, medium, and high.
   b. The Variable Slope of forming three fuzzy set, i.e. low, medium, and high.
   c. The Variable Density Flow forming three fuzzy set, i.e. low, medium, and high.
3. Establish a set of discourse universe
   The universe of discourse of each variable is the range of possible values of the input variable data. Universe of discourse based on [5] can be shown as in Table 6.

Table 6. Fuzzy variables

Functions	Variables	Classification
Input	Erodibility	Low [0-0.5] Moderate [0.25-0.75] High [0.5-1]
	Slope/Tilt	Low [0-0.5] Moderate [0.25-0.75] High [0.5-1]
	Density Flow	Low [0-0.5] Moderate [0.25-0.75] High [0.5-1]
Output	Erotion Risk Index	Very Low [0-0.25] Low [0-0.5] Moderate [0.25-0.75] High [0.5-1]

3.5 Erotion risk map of fuzzy logic result

Results that were obtained from the decision making process was erosion risk level classification of each macro basin (ID Number of Sub Basin). The classification described very low, medium and high grade. By using those erosion risk levels, the erosion risk can be mapped. The results of this mapping were shown in Fig. 4.

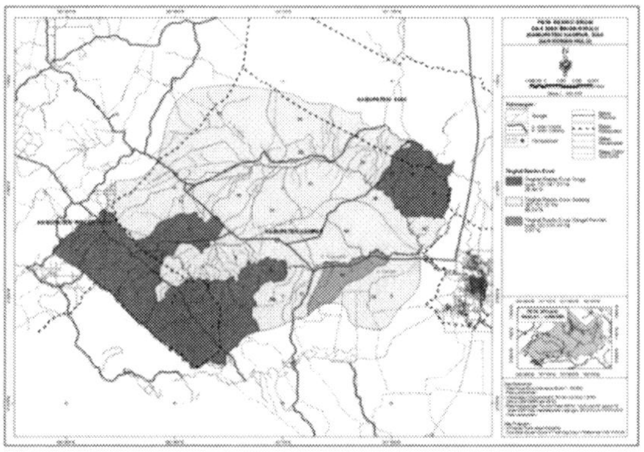

Fig. 4. Risks erosion maps in upstream of Siak Basin

Fig. 4 shows that the erosion risk in upstream of Siak river basin is dominated with the moderate level of erosion risk which is about 321,151,320 ha (68.54%). For high level of risk erosion with extensive 135,147,910 ha (28.84%) is more predominant in the Tapung Kiri sub basin. While, the attrition rate is very low with an area of 122,434,440 hectares (2.61%). More details about the area of erosion risk class can be seen in Table 7.

Table 7. The percentage of erosion risk class area

Erosion Risk Class	Area (Ha)	Percentage (%)
Very Low	12,143,440	2,61
Moderate	318,497,176	68,54
High	134,024,999	28.84
Total	464,665,432	100.00

According to [5] the input variables such as soil erodibility, slope and stream density is the decisive factor for the risk of soil erosion. Based on the results of erosion risk map and the results of validation studies the risk of erosion in upstream of Siak basin indicates that the variable erodibility, slope and stream density showed as factors that influence the risk of soil erosion.

Overall the rate of erosion is caused and influenced by a factor of five factors, including the type of soil (erodibility), topography (slope), climate, vegetation, and soil management factors. To determine the amount of erosion on the surface of the watershed can be performed quantitatively in several ways, one using the empirical method USLE [10].

4. Conclussion

This study applied a fuzzy logic approach to develop erosion risk map based on geographic information system. The result of the erosion risk map was classified as very low to height. While the value of very low index was stated in between 1.01-4.0 and the height value was in between 4.01–10. It was identified that the dominant area prone to erosion risks was 68.54% of the total watershed area. The level of high erosion risk classification was approximately 28.84 % of the watershed area and very low level was 2.61%. These results matched the soil erodibility classes (K). Hence, this result finding is considered valid. The classification of the soil erodibility (K) were as follow, very low (0.0
10. I.W. Sutapa, Jurnal SMARTek, **8**, 169-181 (2010)
–0.10), moderate (0.21-0.32), height (0.44-0.55) and very high (0.56 – 0. 64).

The author would like to thank the Directorat General of Higher Education, Ministry of Research, Technology and Higher Education for funding this research through grants postgraduate

References

1. M. H. Purnomo, *Prospect and Apllication of Artificial Intelligent Knowledge* (Inauguration speech for Professor, Institute of Sepuluh November, Indonesia, 2004)
2. D.C.S Bist, M. Pohan, M. C. Joshi,. J. Comp Modelling and New Technologies, **13**, 16-23 (2009)
3. Z. Akyurek, O. Kıvanç, 7th International Symposium on Spatial Accuracy Assessment in Natural Resources and Environmental Sciences, Turkey, 2010.
4. M. Amirabdollahian, A. Keyvan and M. R. Chamani, *International Conference on Information and Intelligent Computin,* IPCSIT, **18**, Singapore, 2011
5. T.H. Gournellos, N. Evelpidou, A, Vassilopoulos. J. Natural Hazards, **31**, 63-83 (2004)
6. S. Kusumadewi, *Analyses and Design in Fuzzy utilizing Toolbox Matlab*, (Indonesia, 2002)
7. Sutojo, Mulyanto, Suhartono, *Arificial Intelligent.* (Indonesia, 2011)
8. S.A. Asklany, K. Elhelow, I.K. Youssef, M.A. El-Wahab, Journal Atmospheric Research, 228-236 (2011)
9. D.R. Navianti, I.G. Usadha, F.A. Widjajati. Jurnal Sains dan Seni ITS, **1**, (2012)

Drought mitigation participatory at Pinang Luar Lowland, West Borneo (Indonesia)

Henny Herawati[1,*], *Stefanus* B.Soeryamassoeka[1]

[1]Civil Engineering Department, Tanjungpura University, 78124 Pontianak, Indonesia

Abstract. Indonesia is a tropical country with 2 seasons, the rainy season and the dry season. In the rainy season, flood can occur causing a disaster due to excess water. But in the dry season it happens to prolonged drought cannot fulfilll the society needs as decent life. In the areas of the downstream of Kapuas River, the impact of the drought is the intrusion salinity. Where as agricultural land as Pinang Luar lowland agitated of drought that leads to crop failure. This study will be conducted by qualitative descriptive analysis of survey results at Kapuas downstream area which is Pinang Luar tidal lowland. As the result of this research is mitigation of drought can be made bycommunity participatory. For agricultural cultivation, the community has to get involved in some activities such as: (1) conservation ofland and water (2) maintain the irrigation network (3) developing water-saving agriculture, and (4) setting the cropping pattern. And for the mitigation efforts in participatory forest fires can be done by making the channels controlling groundwater levels. To complying the water needs of the community in the dry season can be solve by rain storage ponds, so that the stored water can be used in drought times.

1 Introduction

Indonesia is a tropical country with 2 seasons, the rainy and the dry season. The seasons do not always occur in the same month every year. Similarly,the amount of water available. The water available is determined by the amount of rainfall that occurred. That was not always happened the same amount of rainfall every season [1]. West Kalimantan is one of provinces in Indonesia that located on the equator line with significant water resources, the average annual rainfall reaches 2,500-3,000 mm almost around this region butWest Kalimantan have experienced excess water and water supply shortage.

Kapuas River is the longest river in Indonesia that located in West Kalimantan. During the rainy season flooding can occur that spread everywhere, especially on the residential areas, roads and the agriculture areas. All human activity and productivity do not run properly.

However in the dry season the opposite occurs, prolonged drought can not fulfill clean water, water available for agriculture, plantation and fishery. The impact of drought, cause the forest fire, andsalinity intrusion in the zone that located in the lower part (near the Kapuas River estuary such as Pontianak City).

Pinang Luar is one of the areas that was developed as agricultural land is in the Kapuas River estuary. Pinang Luar area is an area that affected by tides. In the rainy season, agricultural land in area of Pinang Luar have waterlogging. Whereas in the dry season the area is experiencing drought and crop failure, and in certain circumstances can cause forest fires.

With optimal tidal lowland water management the area will produce high agricultural production [2]. The amount and distribution of rainfall in the rainy season which is enough for one growing season will give high agricultural production with proper water management [2]. Potential rainfall around the tidal lowlandof Pinang Luar is very high. The average annual rainfall reaches 3,000 mm and the average monthly rainfall reaches 250 mm [3]. The amount of monthly precipitation with a probability of 80 % greater than the evapotranspiration that occurs in a year (except in February and August) [3].

Based on the description above, in the area of West Kalimantan in general potentially have excessed water during the rainy season and otherwise have drynessed in the dry season. The tidal lowlandof Pinang Luar is an area that was developed as one of land for cultivation. But as an area that affected by tides, that potential is not always reliable to irrigate agricultural land in Pinang Luar, because the availability of water also affected by the rainy season and the dry season. Problems that have been created by all of aspects above, therefore, it is necessary to analysis how to anticipate the water excess and how prevention (mitigation) dryness in tidal lowlandof Pinang Luar.

This paper have purposed to identify the potential of water in tidal lowlandof Pinang Luar and determined the concept of drought mitigation. This research needs to determine the efforts to mitigate drought in this area, so

* Corresponding author: hennyherawati@civil.untan.ac.id

the water can be provided as needed, as a basic need for people and for the agriculture.

2 Methodology and study area

The study was conducted in tidal lowlandof Pinang Luar. Geographically, Pinang Luar is at 0 ° 16 '- 0 ° 18' South Latitude and 109 ° 18 '- 109 ° 23' East Latitude. This area is located at Kapuas River estuary and periodically influenced by tides. Land elevation in Pinang Luar area is relative flat that is 1-3.5 m (refer to the MSL). This area was developed as an agricultural cultivation area for the community. Local communities take advantage of the movement of the tide to fulfill their water needs.

This study will be conducted by qualitative descriptive analysis of field survey results. The datas are secondary and primary data which can be describe the potential, constraints and existing problems in tidal lowland of Pinang Luar. Secondary data consist of rainfall data and topographic data. The data of rainfall obtained from the closest station to the study area, BMKG Station Supadio Pontianak. Land conditions and channel system in tidal lowlandof Pinang Luar can be seen in Figure 1.

Primary data is a tidal measurement data in the study area. Tidal observations conducted at the Kapuas River, thetidal lowland of Pinang Luar for 15 days, taken byobservations on April 15 to 28, 2014.

Land elevation displayed in contour map of ground surface elevation. By using the Global Mapper Software v.17.0, the water fluctuation and groundelevation will be analysis.The overlay of tidal measurement data and land elevation, can be describe the changing of inundation level when the water level in highest and the lowest oftides. The next step is finding the location of rainfall water storage (pond) that can be built for drought mitigation.

(a) Tertiary channels filled with sediment and grass

(b) The control gates are not functioning

(c) Channel tertiary treated by society in a participatory manner

Fig. 1. Land conditions and channels in tidal lowlandof Pinang Luar

3 Result and discussion

3.1 Rainfall potential in study area

The research area is located at Kapuas River estuary. Kapuas River estuary is a lowland areas that are affected by tides. The study area elevation ranges from 1-3.5 m (refer to the MSL). This causes the water level can not reach land, both during neap tides and spring. So,the paddy cultivation can only be done on a small portion in Pinang Luar, and can only be planted once a year. Because the water level is not enough to irrigate the land, the land was allocated to plant paddy replaced by crops. And in some other parts, changed as oil palm plantations.

DOI: 10.1051/matecconf/201710104022

Based on rainfall data in BMKG Supadio Station,years 1970-2015 describe the potential of rainfall in the study area is high, with an average of 3,130 $mm.year^{-1}$ or an average of 261 $mm.month^{-1}$. The amount of rainfall that occurred during the last 45 years tend to decrease. This can be seen in Figure 2. The number of days of rain events tend to decrease that resulted in the drought tends to rise. This condition is not favorable for the farmers to make the cultivation of agriculture [1].

3.2 Tidal potential in study area

From the observation and forecasting, the elevation tides that occur in Pinang Luar area is 152 cm, the important value elevations of subsequent observations made forecasting as shown in Table 1. Elevation's tidal effect on the availability of water in Pinang Luar area. By using the tidal water movement in the Kapuas River, water can irrigate Pinang Luar agriculture area. This can occur onrainy season.

The hydro-topographical conditions in the tidal lowlands as the starting point for the analysis of land suitability is determined from ground elevation than the elevation of the river or canal. From the analysis of the high tides that occur, in generally Pinang Luar never have water logged 4-5 times during one tide cycle, both spring and neap tides are referred to as hidrotopography A. In the rainy season and the highest tide only a few parts which overflow. At low tide conditions, water can only reach the channel. This will worsen the dry season. Comparison of water level and overflow areas in Pinang Luar can be seen in Figures 3 and 4. The map is based on the water level of the Kapuas River in Pinang Luar, precisely located in the intake of primary channel. Discharge from upstream assumed in normal condition.

In the dry season, the discharge from upstream is decrease. The impact is fresh water that flows to downstream also decrease. So the water level in of Pinang Luar is also decrease, especially at low tide conditions. Water will drainfrom the channels ofPinang Luar water management network system. If the management do not control the water level, in the dry season will occur over drain in the channel. Suppletion water will be decreaseand ground water at low tide will be ran out into the channel. It is critical for the occurrence of forest fires.

Table 1. Important tidal water level elevation observations at Pinang Luar.

Name	Simbol	Elevation (m)
Highest Water Spring	HWS	0.80
Mean High Water Spring	MHWS	0.65
Mean High Water Level	MHWL	0.33
Mean Sea Level	MSL	.00
Mean Low Water Level	MLWL	-0.32
Mean Low Water Spring	MLWS	-0.62
Lowest Water Spring	LWS	-0.72

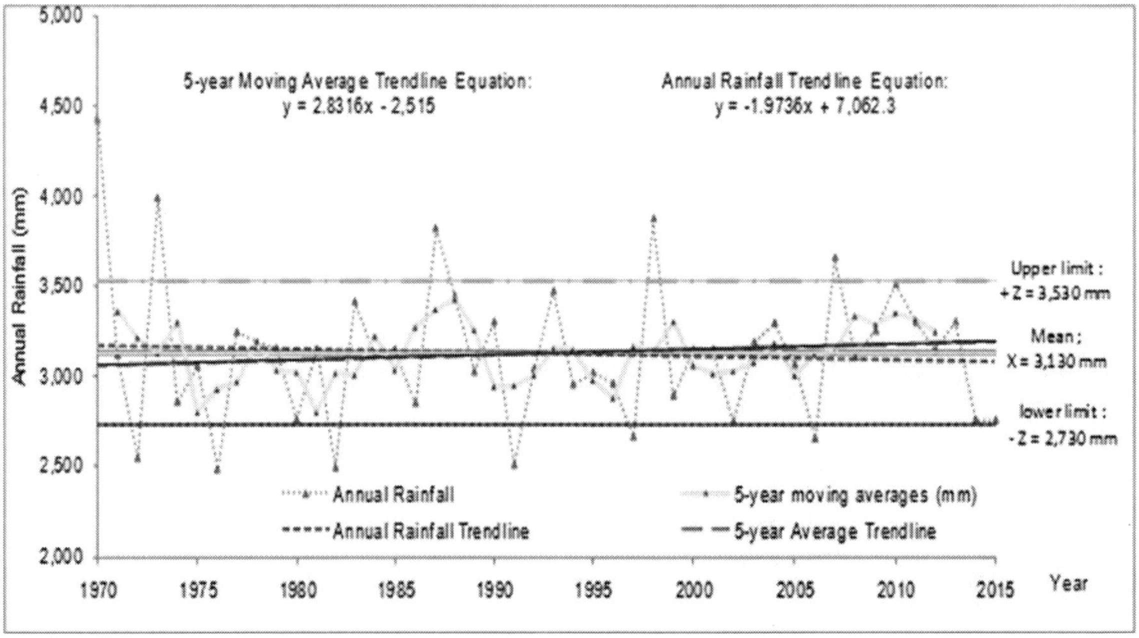

Fig.2. The annual rainfall for 45 years (1970-2015), BMKG Stations Supadio, Pontianak [1,4]

 DOI: 10.1051/matecconf/201710104022

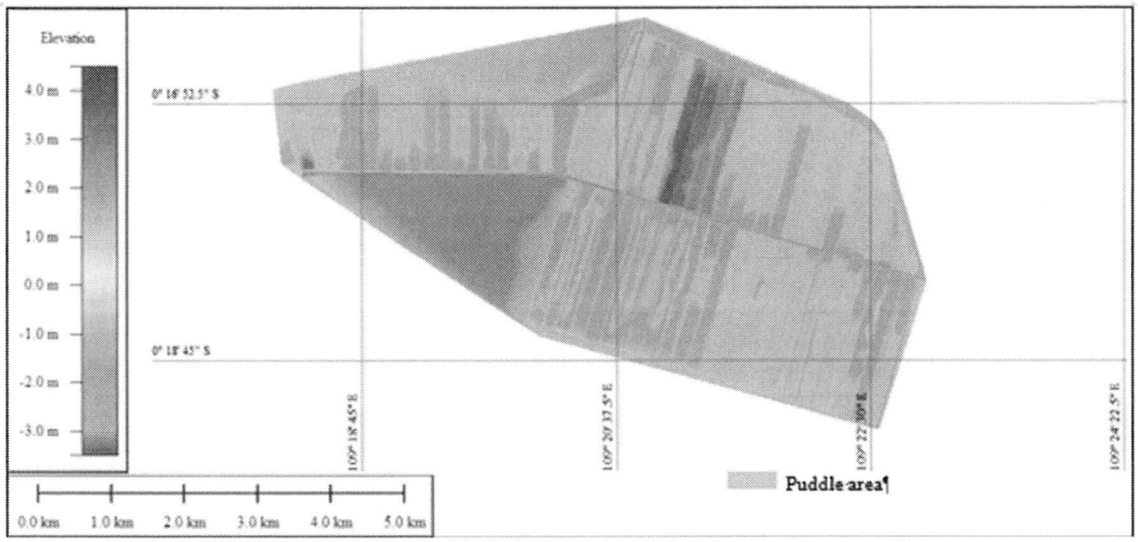

Fig. 3. The land surface elevation and a puddle due to the influence of the highest water spring level (HWS) is 0.8meters at the Kapuas River estuary

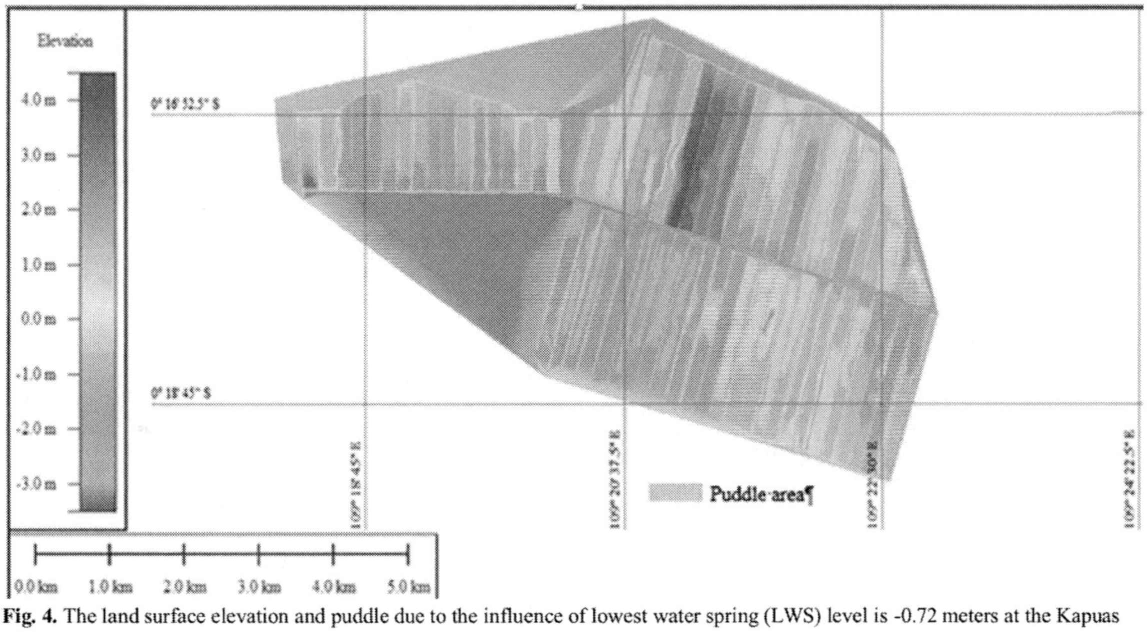

Fig. 4. The land surface elevation and puddle due to the influence of lowest water spring (LWS) level is -0.72 meters at the Kapuas River estuary

4 Constraints on the field

Research area already constructed network system of water tidal management. If the facility of water system is functioning properly, the water management system can be set optimally so the plants can be grown with high productivity. Operations and maintenance will support the efforts of agricultural cultivation. Based on the observations, the availablity offacilities have not been fully utilized. This can be seen in Figure 1. Many channels are not well maintained. Tertiary channel capacity is reduced due to a lot of sediment and vegetation in the middle of the channel which makes the shallow ofchannel.

In the Pinang Luar consists of the primary channel, secondary channels and tertiary channels. At each end of the tertiary channels equipped with sluice gates. Based on the observations, the control gates have not functioned well to control the water level. Then, the flat area will be difficult to release the water to the network system. The discharge is highly depend on the Kapuas Riverwater level, which also detracted from the tide.

Moreover the operation and maintenance have not been apply, Pinang Luar have not storage and freshwater reservoir. Fresh water depend on rainfall and the tides from Kapuas River. Therefore, when water gets into the channel will be ran out into the river at low tide. As low land area, Pinang Luar area is a marshland that has a high void ratio: 0.6-0.7. So the ability to store water is poor when the ground water level is low.

5 Mitigation efforts

Water management strategies should be focus on maintaining the ground water level at the appropriate level for soil, water, hydro-topography conditions and land use [2, 5, 6]. Water management strategy aims to fulfill the water needs of plants although for community. In addition management strategy is also intended to reduce the adverse effects on the environment [7]. Wetlands can be used as a source of irrigation water [8].

One of the constraint is no infrastructure which can savethe fresh water. The fresh water from rainfall and the tides wasted through the channel. With the potential rainfall and soil conditions, Pinang Luar is very needed to build a pond as a water reservoir to fulfill the irrigation needs.

Based on the analysis of Pinang Luar, drought mitigation and land as well as to prevent forest fires can be done various efforts. The effort is a way to control the availability of water through the control of water level in the soil and in the channel. The efforts are:

1. Provide knowledge of ideal water level in the channel and on the land to farmers.
2. Assign or provide operational and maintenance jobs to farmers in a participatory manner.
3. Performs the channel utilization as control water levels on the farm.
4. Install the door controllers to the tertiarychannels or quarter which directly flow to the land.
5. Make some ponds as water reservoir.

The efforts to control the groundwater table above can be done by the participation of community. By the team work called *gotong royong* (mutual aid society), has ability to perform operations and maintenance and optimal control. Community involved to clean the channel(Figure 1 (c)). So the channel can function optimally. Channels that function well, will be as long storage to reserves of water that can be used at low tide conditions.

Ponds/storages construction can be built on the land with the elevation above 3 m as shown in Figures 3 and 4. At high tide the water will be flow into ponds and when the low tide the doors will be close. So the water will be collected in ponds. Stored water will be release when the water in channelsis less to fulfill the water needs for irrigation.

To increase the agricultural production, the appropriate cropping patterns is needed and the selection of adaptive seeds with tidal land. If these attempts to do so will increase agricultural productivity. Based on field data, the paddy productivity about 2 ton.hectare^{-1}. Most of agricultural land still dependent to rainfall.

6 Conclusion

Based on the results of this study concluded that drought mitigation efforts can be done in a participatory manner, the role of the community actively to mitigate the drought. Drought mitigation can control the water level in the channel and land. Controling of water level aims to set the level of the water table to fit the needs of the plant. Good control of water level also can prevent land dryness as to prevent the forest fires and can improve the quality of the land for agriculture. Mitigation measures are participatory forest fires can be done by controlling groundwater levels.

For agricultural cultivation, the community has to get involved in some activities such as: (1) conservation of land and water (2) maintain the irrigation network (3) developing water-saving agriculture, and (4) setting the cropping pattern. To fulfill the needs of household water in the dry season can made ponds rain reservoir, so the stored water can be used in times of drought. In addition to rainwater, ponds also can store fresh water into the channel at high tide. And shut the door controllers at low tide, so the water will not be wasted during low tide.Creating a rainfall reservoir ponds is a new concept in the area of tidal marshto address the needs of irrigation water.

This research was conducted with financial support to authors through Fundamental Research financed by the Directorate General of Higher Education Ministry of Research Technology and Higher Education through a grant letter of number 0299/E3/2016 dated January 27th, 2016. The authors would like to thank to the reviewersfor the valuable comments and suggestions in order to improve the quality of the paper.

References

1. Agency of Meteorology, Data Curah Hujan Stasiun Supadio Pontianak, Pontianak Clim. and Geoph., (2016)
2. H. Herawati, Suripin. Suharyanto. Procedia Engineering **125**, 185-192 (2015)
3. F. X. Suryadi, Soil and Water Management Strategies for Tidal Lowlands in Indonesia, Netherlands (1996)
4. H. Herawati, IRJES **4**, Issue 9, 01-06 (2015)
5. F.X. Suryadi, E. Schultz, Proc.Workshop on the Agricl. Based Dev. of Tidal Swamps and Estuaries and Env. Cons. Seoul-Korea, (2001)
6. M. S. Imanudin, E. Armanto, R. H. Susanto, S. M. Bernas, J. Trop Soils **15**, No. 3, 277-282 (2010)
7. S.K. Hong, C.H. Koh, R. R. Harris, J.E. Kim, J.S. Lee, B.S. Ihm, Land Use in Korean Tidal Wetlands: Impacts and Mgt Strategies, Env. Mgt **45**, 1014-1026 (2010).
8. Y. Wang, Y. Yao, M. Ju, Wise Use of Wetlands: Current State of Prot. and Utilz. of Chinese Wetlands and Rec. for Impr., Env. Mgt **41**, 793-808 (2008).

Heave induced reduction of friction capacity of pile embedded in clays

Gogot Setyo Budi[1]*, *Gondo* Wibowo Tantri[2]

[1]Civil Engineering Department, Petra Christian University, 60236 Surabaya, Indonesia
[2]PT Teno Indonesia, 60116 Surabaya, Indonesia

Abstract. Installation of new piles may cause heave which influence friction capacity of existing piles. The heave can be observed from the difference in the elevation of existing pile heads recorded before and after the installation of new piles or through load-settlement diagram from Static Load Test data. This paper presents the study of bearing capacity of hollow cylindrical concrete piles with diameter of 800 mm from two projects. The piles at Project I and Project II were hydraulically jacked into depths of 46 m and 42 m, respectively. The bearing capacity of the piles was determined from the pressing force required to install the instrumented indicator-piles, and Static Load Test performed after the installation of the new piles. The indicator-piles, which were instrumented using Vibrating Wire Strain Gauges to observe the distribution of transferred load along the piles. The pressing force required to install the piles, generally defined as mobilized bearing capacity of piles, was smaller than the bearing capacity of instrumented indicator-piles. The result shows that, the driving of the new piles reduced the bearing capacity of the existing piles to as low as 34% and 39% in Project I and Project II, respectively.

1 Introduction

Increasing demand for housing and offices in urban area drives investors to build high rise buildings, which is generally supported by pile foundations. In the past ten years, the use of hydraulically jacked-in machine has been popular in Indonesia to install pile foundation, due to its several advantages such as the absence of noise and vibration. However, driving relatively close spacing of piles generates problems such as heaving of ground surface and previously-installed piles [1, 2, 3, 4, 5, 6], increasing pore water pressure [7, 8], and lateral deformation of surrounding soil that might affect the neighbouring structures [1, 8].

Installation of several piles in a group triggers upward movement of soil within the group and it decreases gradually in lateral direction up to the distance of the pile length. In other words, the installation of a pile affects an area of surrounding soil starting from the pile tip spreads upward to the ground surface with the angle of 45° [4]. However, according to Cooke and Price [9], the heave of ground surface due to piling installation of relatively shallow piles disappears at the distance about five times pile diameter.

In general, there are several factors influence the soil heave due to pile driving, which include volume of displaced soil, thickness and depth of clay layer, pile installation procedure, existing vertical stress, etc. [3]. Meanwhile, Hagerty and Peck [1] stated that heaving of piles depends on the ratio between volume of penetrated piles and the volume of soil enclosed by a group of piles.

The installation of piles in a group not only generates heaving of soil and previously installed piles but also increases pore water pressure. The maximum pore water pressure occurs in the area around pile shaft and decreases with the distance from the installed pile [2, 7]. Pore pressure build-up in the sensitive marine clay subjected to pile driving reached 1.5 to 2 times of that hydrostatic state and it dissipated in about seven to eight months [2]. Based on the analysis conducted by Broms and Hansbo [10], the increase of pore pressure at the pile tip embedded into the marine clay was about five to seven times of its undrained shear strength. In some cases, the increase of pore water pressure surpasses the over burden pressure. The effect of pore water pressure due to pile driving disappears at the distance about twenty times pile diameter [7].

Basu et al. [11] stated that the build-up of pore water pressure due to pile driving reduces the shear strength of clays. According to Broms [12], the disturbance of soil surrounding installed pile spread out up to one diameter from the periphery of the installed pile. It was also stated that the shear strength of soil immediately around the pile during driving is reduced to the shear strength of the remoulded soil, and it recovered in one to six months.

Bozozuk et al. [2] conducted research to investigate the effect of marine clay disturbance due to pile driving. The sensitivity of marine clay used in the research was in the range between five and twenty. It was reported that the shear strength of marine clay decreased to range between 70% and 85%. The increase of soil strength was noticed in the first five days after pile driving, but it did

* Corresponding author: gogot@petra.ac.id

not increase in the period up to three months. While Orrje and Broms [13] described that the disturbed soil due to installation of concrete piles regained its undrained shear strength in nine months.

The objective of the research is to study the bearing capacity of heaved piles due to driving of piles. The research was carried out in two different projects, namely Project I and Project II. Three data of bearing capacity of piles were collected. The first data were collected from instrumented piles using Vibrating Wire Strain Gauge (VWSG) that were installed prior to the projects, second data were gathered from driving records during piling, and the third were collected from Static Load Tests that were conducted after the piling activities had completed.

The piles used in both projects were hollow cylinder prestressed-concrete spun pile with diameter of 800 mm. In order to minimize the lateral deformation of surrounding soil and structures, the open-end (bottom) piles were used. The piles were installed using hydraulically jacked-in machine with total capacity about 900 ton. The length of piles tested statically at Project I and Project II was 46 m and 42 m, respectively.

2 Soil data

The soil stratigraphy in both projects is dominated by very soft to stiff clays. The soil layers at the two locations are quite similar, and there is no bearing layer up to the depth of 50 m. The soil profile at Project I and Project II is summarized in Table I and Table II, respectively. The value of Standard Penetration Test (N_{SPT}) with depth is presented in Fig. 1 for Project I and Fig. 2 for Project II.

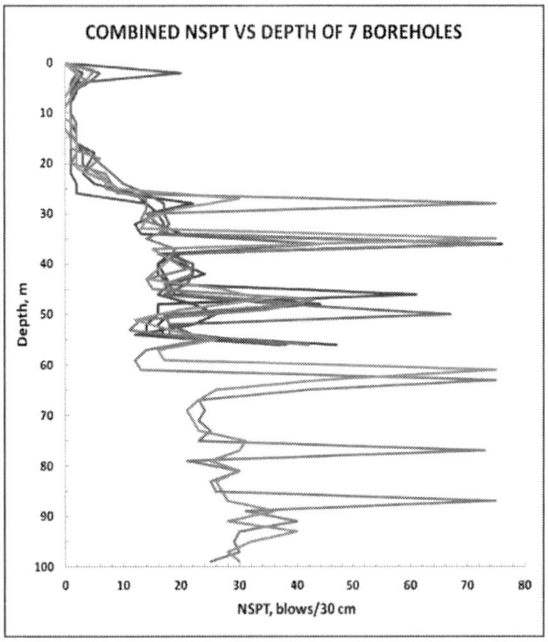

Fig. 1. N_{SPT} Vs depth of project I

Table 1. Soil profile at project I

Depth, m	Soil Classification	N_{SPT}
0 – 3	Fill	-
3 – 22	Very soft clay	0 – 2
22 – 27	Soft clay	2 – 9
27 – 29	Silt	30
29 – 33	Stiff clay	14 – 19
33 – 37	Sandy silt	14 – 75
37 - 45	Stiff clay	14 – 18
45 - 49	Sandy silt	14 – 34
49 - 53	Stiff clay	12 – 20
53 - 60	Silty clay	16 – 26

Table 2. Soil profile at project II

Depth, m	Soil classification	N_{SPT}
0 - 2	Fill	-
2 - 10	Very soft clay	1 - 2
10 - 15	Fine sand	17 - 18
15 - 23	Stiff clay	11 - 20
23 - 25	Fine sand	20
25 - 31	Stiff clay	18 - 24
31 - 35	Silty clay	21 - 23
35 - 37	Sandy clay	25
37 - 45	Silty clay	22 - 26
45 - 50	Stiff clay	22 - 25

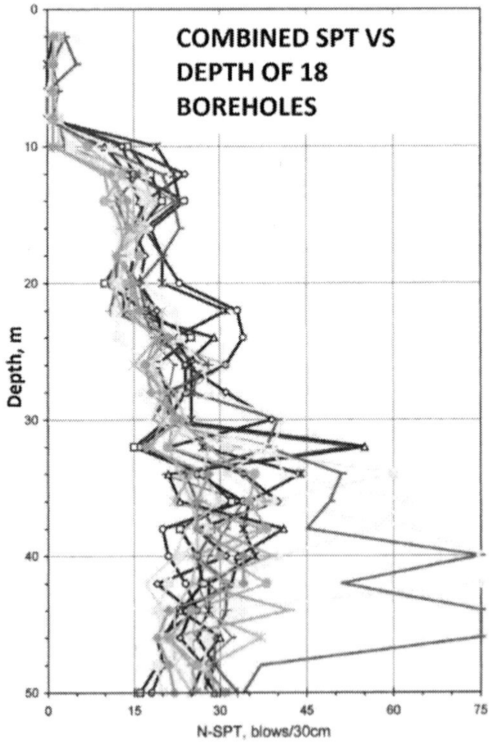

Fig. 2. N_{SPT} Vs depth of project II

DOI: 10.1051/matecconf/201710104023

3 Testing program

Bearing capacity of the tested piles was collected from the static load test on the indicator piles that were instrumented with Vibrating Wire Strain Gauges (VWSG), driving records, and Static Load Test on heaved piles. The instrumented piles were installed prior to the driving of main piles so that the soil was considered undisturbed. The piles were called as indicator piles. The driving records were gathered from the pressure gauges indicated in the hydraulic jack-in machine compartment during driving, so that the bearing capacity of the installed piles was in mobilized condition. The static load tests (Quick load) were conducted to the tested piles after the main piling activities had completed so that bearing capacity of the tested piles were very likely affected by disturbance of surrounding soil due to installation of main piles.

The statically load tested piles at Project I, which were conducted after the piling activity, were located about 20m from the area of piling areas, as presented in Fig. 3. Static Load Test on piles at Project I was carried out about three months after piling activity had concluded. While Static Load Test on piles at Project II was carried out about 20 days after the pile installation had finished. The location of Statically Load tested piles is presented in Fig. 4.

Fig. 5 and Fig. 6 present the load transferred distribution along the piles recorded from VWSG data at Project I and Project II, respectively. The data of tested piles instrumented by VWSG is summarized in Table III. The length of piles that were tested after the completion of piling of main piles was 42 m. Load settlement diagrams constructed from the records of the tested piles at Project I and Project II are presented in Fig. 7 and Fig. 8, respectively.

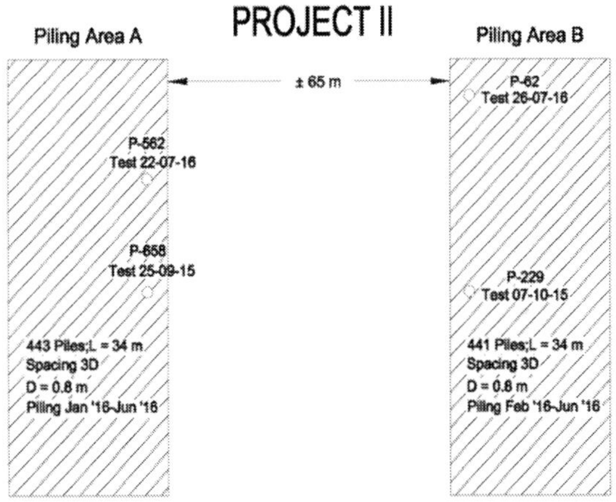

Fig. 4. Location of tested piles at project II

Table 3. Data of instrumented piles

Subject	Project I	Project II
Pile diameter, mm	800	800
Length of pile, m	47.5	43.5
Applied load, ton	700	920
Load transferred at pile tip, ton	112	71
Load transferred at 46 m depth, ton	118	-
Load transferred at 42 m depth, ton	-	100
Friction resistant along 46 m, ton	582	-
Friction resistant along 42 m, ton	-	820

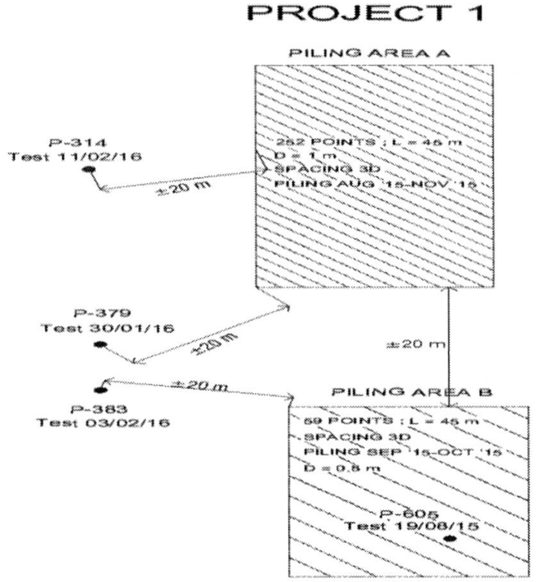

Fig. 3. Location of tested piles at project I

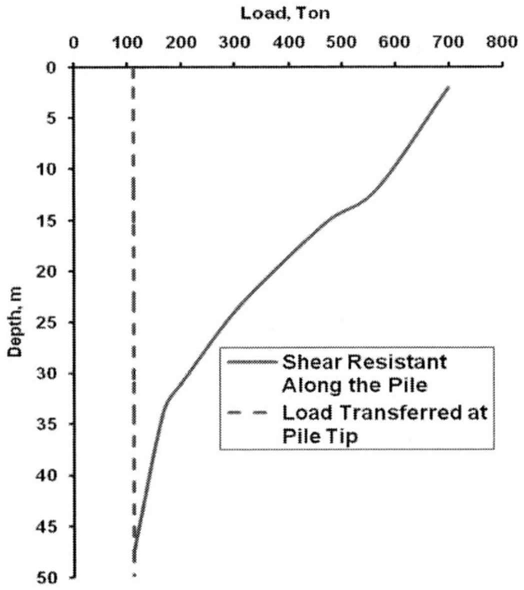

Fig. 5. Load transferred distribution at project I.

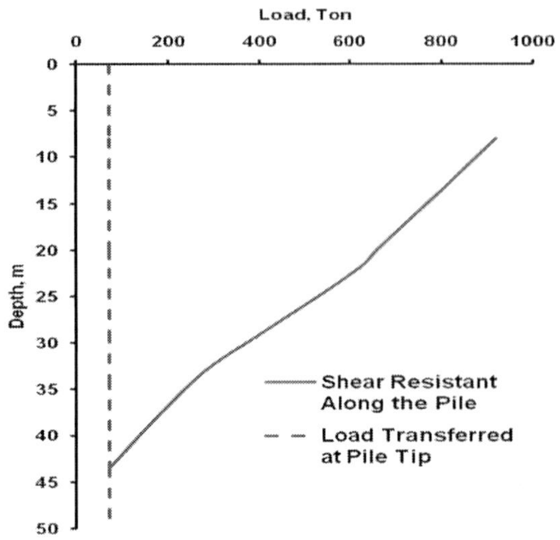

Fig. 6. Load transferred distribution at project II.

4 Analysis

Four piles at Project I and three piles at Project II were tested using Static Load. It shows that in general, the maximum loads can be applied to the piles were less than those recorded from the instrumented piles indicators. Fig. 7 shows that piles experienced heave about 20 mm to 25 mm. It can be seen that the bearing capacity of heaved piles, which is presumably contributed by friction capacity only, in the order between 200 ton and 400 ton compared to 582 ton of the friction capacity of instrumented pile (about 34% to 68%). The applied load increased to about 550 ton to 700 ton at the corresponding settlements of 45 mm to 50 mm.

Very significant reduction of bearing capacity of tested piles at Project II was observed, as shown in Fig. 8. The maximum load can be applied to the both tested piles was only about 320 ton. This load was determined just before the piles experienced excessive settlement. In other words, the maximum friction capacity of tested piles conducted 20 days after completion of piling activity was only 320 ton. Compared to the 820 ton friction capacity of the instrumented pile, the resistance of piles was about 39%.

The bearing capacity reduction of tested heaved piles at both projects very likely caused by the soil disturbance and pore-water pressure built-up during driving of neighbouring piles. The reduction of bearing capacity of piles at Project I, which were located about 20 m from the piling area, was smaller than that of piles in Project II, which were located in the pile group and determined in about 20 days after the piling activity had finished.

The load required to penetrate the tested piles up the depth of 42m at Project I was in the range between 300 ton and 400 ton, as depicted in Fig. 9. In addition, the force needed to re-drive the piles in the following day (about 12 hours later) was more than 700 ton. While in Project II, the applied load to penetrate the tested piles up to the depth of 42 m were in the range between 500 ton and 700 ton, as presented in Fig. 10.

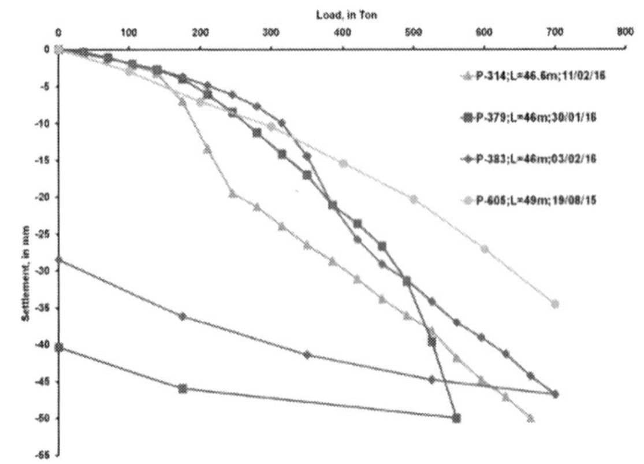

Fig. 7. Load-settlement diagrams of the tested piles at project I

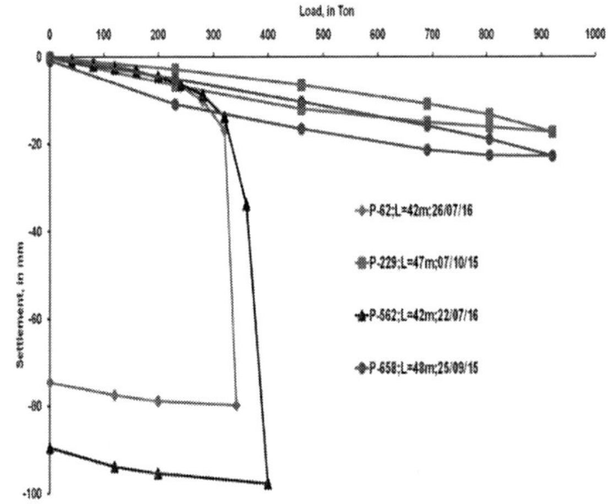

Fig. 8. Load-settlement diagrams of the tested piles at project II

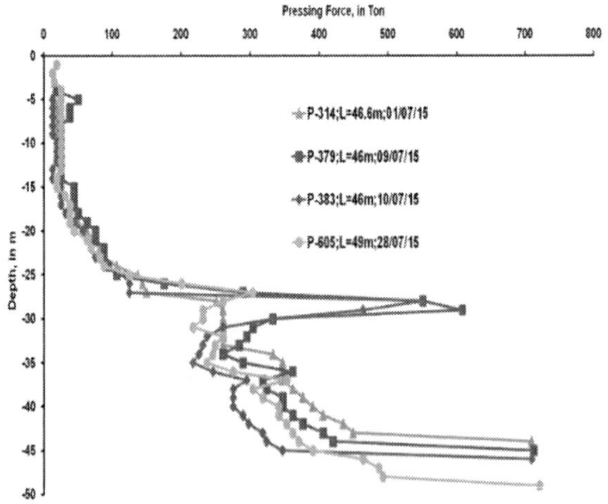

Fig. 9. Pressing force vs depth at project I

DOI: 10.1051/matecconf/201710104023

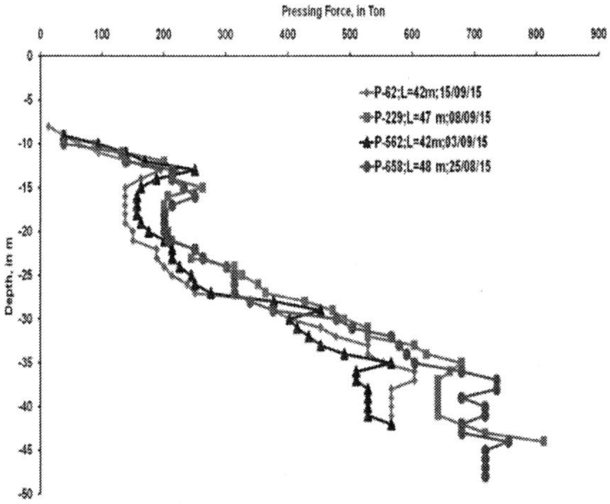

Fig. 10. Pressing force vs depth at project II

5 Conclusions

The study provides insight information regarding the effect of driving new piles to existing piles.

Soil heave resulted from driving the new piles induced a reduction on the friction capacity of the existing piles, thus the pressing force required to install the piles was smaller than that of the bearing capacity of indicator piles embedded in undisturbed soil, which were installed prior to the driving of the main piles.

The bearing capacity of existing piles at Project I and Project II decreased to as low as 34% and 39%, respectively due to the installation of the new piles.

The authors would like to express their gratitude and appreciation to PT. Teno Indonesia for giving permission to access all valuable data required in this research.

References

1. D. Hagerty, R. Peck. J. of The Soil Mech. and Found. Eng. ASCE, Vol. **97**, SMII 1513-1531 (1971).
2. M. Bozozuk, B.H. Fellenius, L. Samson. Can. Geotech. J., Vol. **15**, No. 3, 340-361 (1978).
3. J. P. Dugan, D. L. Freed. *International Conference On Case Histories in Geotechnical Engineering*, Paper 28. (1984)
4. C. Olsson, G. Holm. *Palgrundlaggning*. AB Svenk Byggtjanst Och Statens Geotekniska Institut. (Stockholm, Sweden, 1993)
5. C. Sagaseta, A. J. Whittle. J. of Geotech. and Geoenv. Eng. ASCE. Vol. **127**. Pp 55-66 (2001)
6. C. Wërsall, K. R. Massarsch, Geotechnical Special Publication (GSP) 230 ASCE. Pp 481-499 (2013)
7. H. G. Poulos, Davis. *Pile Foundation Analysis and Design*. John Wiley & Sons Inc., Canada (1980)
8. T. Hørlykke, B. Dedenroth. *Proceeding of The 17th Nordic Geotechnical Meeting: Challenges in Nordic Geotechnic*. 25 – 28 May 2016, NGM, Reykjavik (2016)
9. R. W. Cooke, G. Price. *Proceedings of the 8th International Conference on Soil Mechanics and Foundation Engineering*, Moscow, Vol. **2**, No. 1, 53-60 (1973)
10. B. B. Broms, S. Hansbo. Foundation on Soft Clay. *Soft Clay Engineering: Developments in Geotechnical Engineering 20*. Elsevier Scientific Publishing Company. Pp 417-424 (1981)
11. P. Basu, M. Prezzi, R. Salgado, T. Chakraborty. J. of Geotech. and Geoenv. Eng., ASCE, Vol. **140**, No. 3, 04013026 (2014)
12. B.B. Broms. *Precast Piling Practice*. Thomas Telford Ltd., London (1981)
13. O. Orrje, B. B. Broms. J. of the Soil Mech. and Found. Div., Vol. **93**, No. 5, 59-73 (1967)

Options for land conservation practices based on land use and land degradation in upland Luas Watershed Bengkulu Indonesia

Muhammad Faiz Barchia[1,*], *Khairul* Amri[2], *Friski* Namura[2]

[1]Soil Science Department, Faculty of Agriculture University of Bengkulu, 38371 Bengkulu, Indonesia
[2]Civil Engineering Department Faculty of Engineering University of Bengkulu, 38371 Bengkulu, Indonesia

Abstract. This study is objected to determine the options for land conservation based on degraded land in upland of Luas watershed, Bengkulu, Indonesia. This work was conducted on the upland from November 2015 to February 2016. The location was geographically lying on 103°20'47" - 103°58'50" Latitude., and on 04°21'10" - 04°37'12" longitude, in Kaur District, Bengkulu Province. The total area is 55,045.23 ha where the forest covers 52,432.95 ha or 95% of the upland of Luas watershed. The spatial analysis for land degraded level was based on overlaying from attributes of some parameters such as; land cover, slope, erosion hazardous level, land productivity and management. The analysis is conducted by converting analytical geographical coordinates into Universal Transverse Mercator (UTM). The upland of the watershed is currently under environmentally pressures which around 24% in poor conditions and 23% of the area are degraded. The Limited Timber Forest areas of Air Kinal, Kaur Selatan, and the protected forest area of Bukit Raja Mendara can potentially be rehabilitated by applying concepts of social forestry with introducing agro-forestry cultivation models. All areas of agriculture cultivation in this study can be effective using land conservation tillage and management which concerning all environmental values for the sustainable productivities.

1 Introduction

Watershed in a hydrological concept is an important element in the water management system. Watershed as a component in the systems transforms rainfall to be run-off and other water characteristics [1]. Watershed plays important rules on the ecosystem sustainability, considering interactions within, and surrounding of it through of human life [2]. Sustainable ecosystems are maintained when interactions whole components in the systems well run and stabilized by each other [3]. Threaten to one of the ecosystem elements influence one or some other components with effects of continuous destabilization human ecosystems [4].

Nowadays, there are 458 of watersheds in Indonesia are reported under critical conditions. It is estimated that 60 of the watersheds are in high priority for rehabilitation. Otherwise, a number of 222 watersheds are in the second priority due to poor environmental values, and 176 of the Indonesia watersheds are in the third priority. The degraded lands covered watershed with forest functions in numbers of 19,506,487 ha, and of 10,690,311 ha in agriculture lands [5]. Increasing degraded land productivities within forest areas and agriculture cultivated land is not only in the down- and middle-stream areas but also continuous encroachment to upland watershed. It is known, that the watershed degradation have been tirggered by increasing exploitation of natural resources due to rapid population, economic growth, lack of experiences in watershed management and practices [6]. In the management planning and integrated conservation practices, it requires some important strategies and innovative actions in order to improve and maintain the land productivities and watershed surrounding values [7].

The landforms and watersheds in Bengkulu area has different characteristics than others. The distance between upstream and downstream are relatively short. It is around 10 km with narrow boundaries [8]. Some watersheds in Bengkulu, as like in Luas watershed, have been lost its natural and social function. Agriculture land and forest areas on the upland Luas watershed, are under intensive cultivation practices by both traditional farmers and coming from out the area. The cultivation practices on the upland Luas watershed have not been applied well by considering good agricultural practices. It was applied by un-integrated planning for whole functions of ecosystem components of the upland Luas watershed. In the future, the integrated management planning and technical conservation practices should be focus on the improvement of the upland of Luas watershed. Therefore, empowerment of the natural resources on the Luas watershed could optimally harvested in peak of productivities and in long running for sustainable environmental values.

* Corresponding author: faiz_barchia@yahoo.com

DOI: 10.1051/matecconf/201710104024

2 Materials and method

The upland of Luas watershed lies on 103⁰20'47" - 103⁰58'50" Latitude., and on 04⁰21'10" - 04⁰37'12" Longitude., in Kaur District, Bengkulu Province with covering area of 55,119.37 ha (Fig. 1).

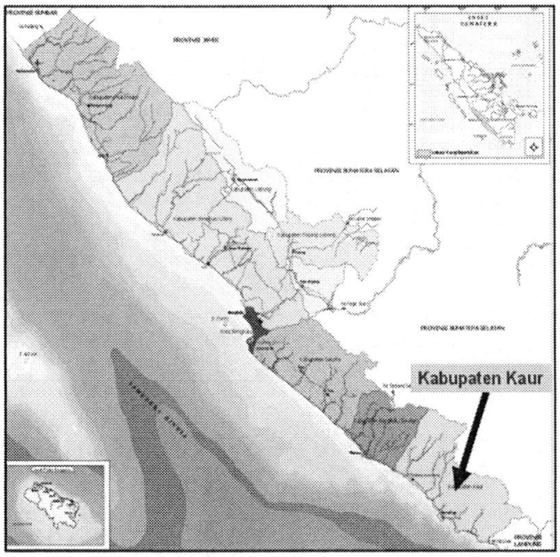

Fig. 1. Map of Kaur District, Bengkulu Province, Indonesia

This research is related to previous conducted research as reported in [9]. It consisted the analysis in general view for the whole degraded land conditions on upland Luas watershed. In this work, the most concerns are in determining the options of land conservation practices based on land use and environmental values management.

The land degradation degrees on different land use based on Rules of Directorate General for Land Rehabilitation and Social Forestry, Ministry of Forestry, Republic of Indonesia No. P.4/V-SET/2013 [10]. The spatial analysis for land degraded level was based on overlaying from attributes of land cover, slope, erosion hazardous level, land productivity and management. The spatial analysis was converted from analytical geographical coordinates to Universal Transverse Mercator (UTM) [11]. Secondary data collected related with this research were validated through ground check in some location in field to perform the data. Options for land management practices were different between agriculture area and forestry area [12; 13], and other related references.

In orger to figure out the options of land conservation and management practices based on land use and functions, it required a map of Indonesia landforms and digital ALOS photos released from Geospatial Information Agency, Republic of Indonesia, spatial data of upland Luas watershed from Ketahun Watershed Management Office (BPDAS Ketahun) Bengkulu, Ministry of Environment and Forestry, Republic of Indonesia.

3 Results and discussion

The Luas watershed is commonly provided some functions such as forest protection on Bukit Raja Mendara for water recharge, limited timber forest of Air Kinal and, Kaur Selatan, Bukit Barisan Selatan National Park. It also protect the traditional agriculture cultivation areas. The land function on the upland Luas watershed are mostly in form of forest with land cover of 52,432.95 ha, or 95.25% of the upland, meanwhile the protected forest covers of 47,948.56 ha, or of 87.11% of total area. The agricultural cultivated lands are 2,612.28 ha or only 4.75% of the upland area.

The forest land in the upland Luas watershed are in intensive pressure due to agricultural cultivated land uses by traditional farmers. It caused a fast land deterioration in the upland of Luas watershed. When the watershed in a degraded condition, so all components to form and related to the watershed are suffered. The levels of degraded land condition on the upland of Luas watershed are more widespread. It is not only on traditional agriculture area, but it also exterminated until area of Bukit Barisan Selatan National Park, limited timber forest, and forest protected as shown in Table 1. The distribution critical lands based on degradation levels for the whole land functions on the upland Luas watershed are shown in Figure 2.

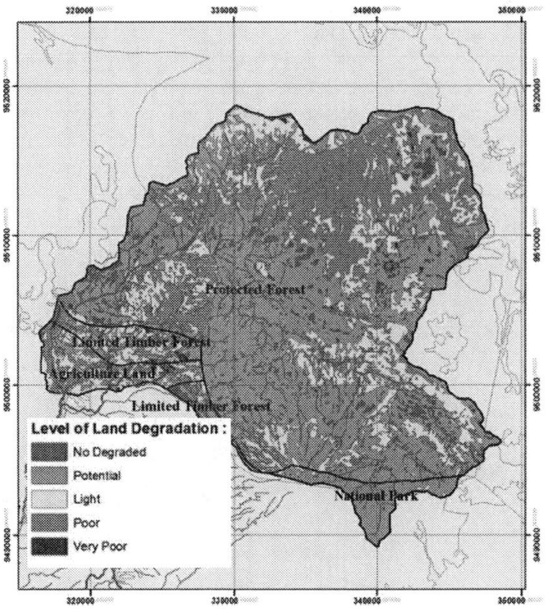

Fig.2. Degraded land in the upland Luas watershed

As shown in Table 1 and Figure 2, the level of degraded lands was performed from the minimum to the poorest area. It very poor area reached to 26,326.93 ha, or of 47.83% from the upland Luas watershed. At the same time, the areas were threated due to unwise cultivation practices with the potentially deteriorated level reached to 26,430.39 ha, or 48.01%.

Table 1. Wide areas of degraded land on upland Luas watershed Bengkulu Indonesia

Degraded Level	Wide Area of Forests (Ha)			Agriculture Area	Total	%
	Protected Forest	Limited Timber Forest	Bukit Barisan Selatan National Park			
No Degraded	1,985.91	249.92	21.40	30.68	2,287.91	4.15
Potential Degraded	23,558.96	814.07	1,751.90	305.46	26,430.39	48.01
Light Degraded	10,864.72	841.11	94.67	1,127.62	12,928.12	23.48
Degraded	11,423.12	582.09	0.66	1,060.38	13,066.25	23.74
Poor Degraded	115.85	128.57	0	88.14	332.56	0.60
Σ	47,948.56	2,615.76	1,868.63	2,612.28	55,119.37	100.00

Based on forest and land functions, the watershed should be performed as service areas for water interception and re-charge, biodiversity sources, and other forest functions, options for conservation practices. In order to restore forest productivities in the upland Luas, it should be designed and developed in same concept by considering the traditional farmers. The better options for conservation practices over the protected forest areas are by applying some developed models of social forestry with introducing agro-forestry cropping systems. There are several choices of plants in the agro-forestry model which commonly using multi purposes plant species. Those options can potentially lead farmers to harvest only non-wood forest products such as various fruits, root-crops, sap, resin, etc. Crops choice for this model involved durian, rubber crops, jackfruit, bread-fruit, candle-nuts, etc. The protected forest area should be rehabilitated in width of 22,403.69 ha. The visualization of the protected forest should be re-planted with the social forestry model as seen on Figure 3.

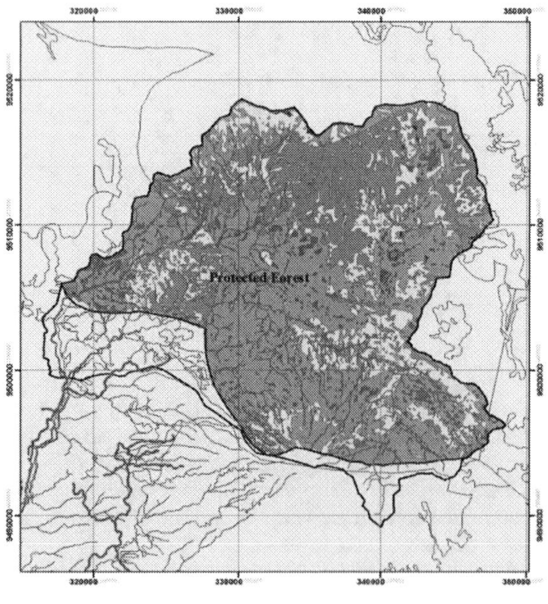

Fig. 3. Agro-forestry, option for conservation practice on protected forest

The limited timber forests on the upland Luas watershed was 2,615.76 ha. The widest area of limited timber forest was located on the middle of Luas watershed. The limited timber forests also cultivated by traditional farmers for coffee plantation. Recently, the land productivities are decreased due to land degradation in around of 1,551.77 ha or 59.32% of the limited timber forest. The models for forest restoration on the timber forest could be conducted with engagement through social forestry program, therefore the surrounding societies in the forest will take an advantage by harvesting non wood forest products. The asre should be be restored in term of its functions as timber forest through the social forestry. The distribution areas are revealed in Figure 4.

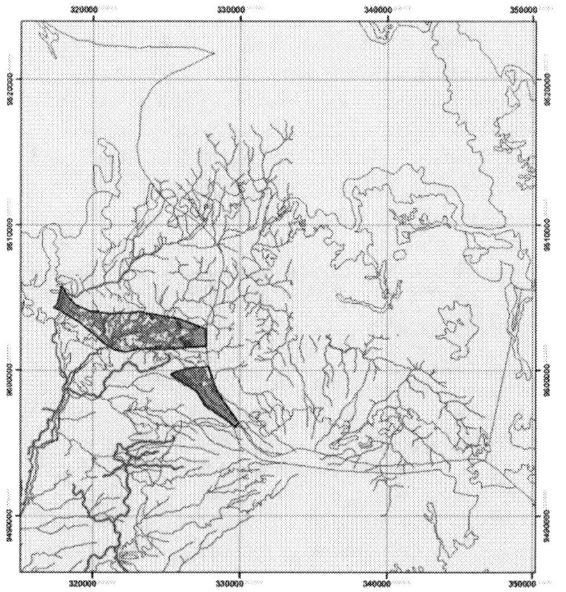

Fig. 4. Agro-forestry, option for conservation practice on limited timber forest

Some forest encroachment were found for land cultivation of agricultural activities in the upland Luas watershed. It also happened to Bukit Barisan Selatan National Park with small area of 95.33 ha or 5.10% of

DOI: 10.1051/matecconf/201710104024

the National Park. However, it indicated a threaten for sustainability in natural sources on the forest functions on the National Park in the future. The distribution area that should be re-vegetation at Bukit Barisan Selatan National Park on the upstream of Luas Watershed are shown on Figure 5.

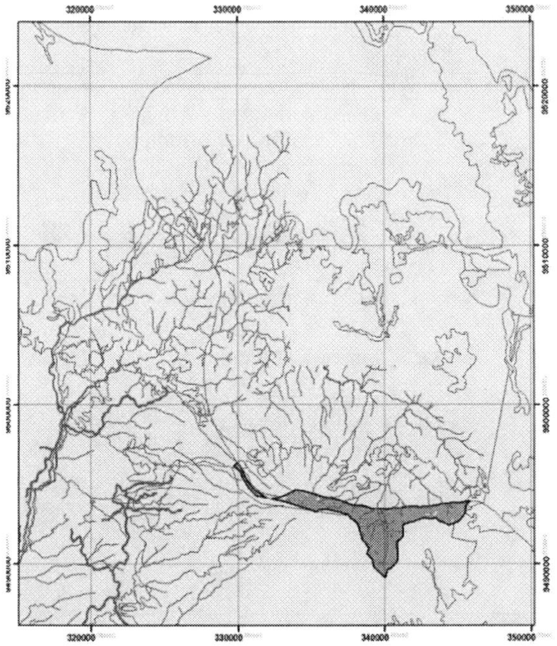

Fig. 5. Reforestry, option for conservation practice on Bukit Barisan Selatan National Park

Some efforts are necessary to conduct in order to protect the Bukit Barisan Selatan National Park from encroachment. The degraded land on the National Park should be rehabilitate by plantation using local plant seeds. All agricultural activities and people living in the Bukit Barisan Selatan National Park need to re-settle out of the location. In this park, it was found huge biodiversities in the ecosystems, therefore it should be protected carefully from any potential disasters.

The area of agricultural cultivated land on the upland of Luas Watershed are 2,612.28 ha, or only covers 4.74%. Meanwhile, from all the agricultural land area, there are 2,276.14 ha, or 87.13% of the cultivated lands were under deteriorated conditions. It is found in low degree to the highest level of land degradation. This condition indicated that when a decrease in productivities occurred on the agricultural cultivated lands, there will be an expansion of encroachment in the forest areas due to incoming of new traditional farmers from surrounding or out-side Kaur District. For this reason, empowerment of local societies is required to improve the land productivities using appropriate technologies. There are several options in conservation practices on agricultural land, such as; minimum tillage, bench terrace, and agro-forestry. Options for conservation practices on agricultural cultivated land in the upland of Luas watershed are shown on Figure 6.

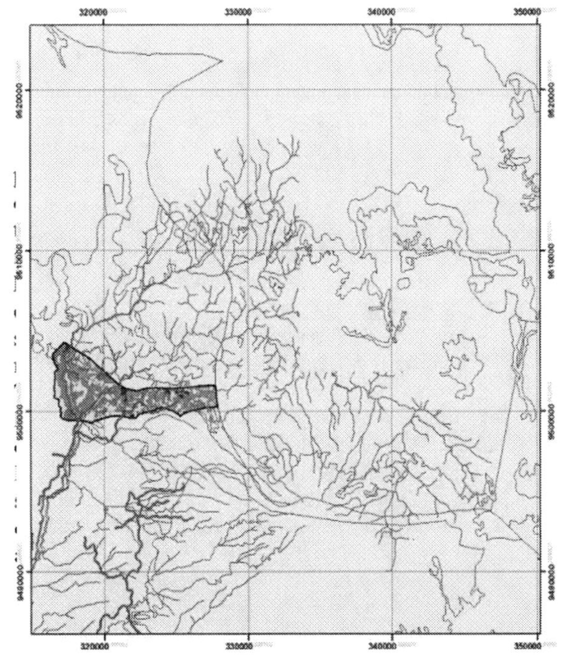

4 Conclusions

The upland of Luas watershed mostly covers by forest area of 52,432.95 ha, or of 95.25% of total area in the watershed. It consists of protected forest of Bukit Raja Mendara, Limited Timber Forest of Air Kinal and Kaur Selatan, and the National Park of Bukit Barisan Selatan. On the area of Bukit Barisan Selatan National Park, it is found area of 95.33 ha that urgently to be restored. The options for land conservation and management practices may be proceed with reforestation using local plant species with starter seeds. Agricultural and social activities on the National Park must be persuasively deported from this area.

The forest protection with area of 22,403.69 ha and the limited timber forest with area of 1,551.77 ha are detected as land degradation. The area distribution are varied from light to heavy level of degradation. Furthermore, this condition led to a program of land rehabilitation in order to increase the land productivity. Those areas could be developed as social forestry through empowerment society around the forest by planting multi-purposes tree species.

The agricultural cultivated land on the upland of Luas Watershed are covers only of 2,612.28 ha, but the productivity of 87.13% of the land was depleted. The degraded agriculture land should be cultivated by considering environmental values synergistically with the long run of improvement in land productivities and sustainable agriculture. The options for land management practices on the agriculture area can be applied by planting cover crops with minimum tillage, with credit and/or bench terraces.

References

1. I.G.Tunas. *Jurnal Smartek*. **3**, 3 pp 137-145 (2005).
2. Hoff, Holger. *Proc 5th Rosenberg Int Forum on Water Policy, Banff, Alberta, Canada. September*. (2006).
3. C. Asdak. *Book*. Gadjah Mada Univerisity Press. Yogyakarta (2014).
4. D.I.D Arini, L.B. Prasetyo, and O. Rusdiana. Media Konservasi. **12**, 2 (2007). http://journal.ipb.ac.id index.php/konservasi/article/view/2985
5. H. Santoso. Report of Coordination Meeting on Integrated Watershed Management. Batam. Indonesia. March 23 (2011).
6. K. Amri., A. Halim., Ngudiantoro, and M.F. Barchia. APCBEE Procedia. **3**: 235-249 (2014).
7. I.E. Soegiri. *Report*. Coordination Meeting on
8. Integrated Watershed Management. Batam. Indonesia. March 23 (2011).
9. A. Farianto., K. Amri and M. Ali. Report. repository.unib.ac.id /4239 (2012).
10. F. Namura. Thesis. University of Bengkulu. Indonesia (2016).
11. Directorate General of Watershed Management and Social Forestry. Guide Book. www.bpdasps.dephut.go.id, accessed on November 22 (2015).
12. Sadikin., S. Nurmuin and M. F. Proceedings. Seminar Geomatika. Balai Penelitian Geomatika. Geospatial Information Centre. Cibinong. Indonesia. May 9 (2012).
13. Djaenuddin. Technical Report No. 7 ver. 1.0 Centre for Research and Development on Soil and Agro-climate. Bogor. (1994).
14. Purwowidodo. Book. Institut Pertanian Bogor Press. Bogor. (1999).
15. A. Dariah, Book. Ministry of Agriculture. Bogor. Indonesia (2004).

Analysis on the influence of rainfall and mine water ratio against pH in East pit 3 West Banko coal mine

Neny Rochyani[1*], *Ngudiantoro*[2], *Helmi* Harris[3]

[1]Chemical Engineering Department, Faculty of Engineering PGRI University, 30116 Palembang, Indonesia
[2]Mathematics Program Faculty of Mathematics and Natural Sciences Universitas Sriwijaya, 30128 Palembang, Indonesia
[3]Chemical Engineering Department, Faculty of Engineering PGRI University, 30116 Palembang, Indonesia

Abstract. In the coal mining area, the pH of mine water is found tend to low and acids. In order to increase the pH, it is important to consider the treatment of acid mine drainage using lime, due the indicators of pollution. This work is focused on the influence of rainfall volume on the pH of acid mine drainage. This research conducted using a ratio of mine water and rainfall water that varies in the 9 (nine) conditions, respectively: 1: 1, 1: 2, 1: 3, 1: 4 and 1: 5 and 5: 4, 5: 3 , 5: 2 and 5: 1. The results were then measured and tested with statistical analysis. The ratio of rainfall and mine water showed a significant effect on the pH. The higher of the rainfall lead to increase pH. This condition will affect the water neutralization process using lime where there are some possible differences on dose of lime needed to neutralized the acid mine drainage in the rainy season and dry season.

1. Introduction

The main problem faced in the coal industry is a pollution due to the presence of acid mine drainage (AMD). It is commonly formed by the oxidation of sulfide minerals, mainly pyrite that produces sulfuric acid [1]. With a high acidity level, coal mine water can dissolve other minerals and releasing cations such as Fe, Mn, Al, Cu, Zn, Cd, Ni and Hg. When the coal mine water flow into water sources, it can reduce biological productivity that was existed in the water system. In severe conditions, the water becomes dangerous for consumption and for other purposes [2].

Acid mine drainage become major concern, where the volume of this water was cames from rainfall and the water which produces from exploration and mining activities at the mine front [3]. The Accumulation of mine water may produce pollution caused by the mining activities that bring pollutants material.

The input factor on producing mine water is the climate, especially the wind that brought mine material, dust, catchment areas, slopes and basin that will be a carry out the water to settling ponds that directly or indirectly will bring mine materials. In addition, the water from the front mines that are the result of excavation and exploration activities will be accumulated with rain water which in certain conditions will produce acid mine drainage.

Mine water management system in the mining area of West Banko using a pumping series system, which means that there are several pumps are placed in certain spots to remove water from the surface / mine. This is because the mining area that has been excavated is located at a depth of ± 100 m, thus allowing a puddles which if not removed would disrupt mining activities[4].

The mine water was pumped from the mine front to the drainage system which located on the surface of mine and then flow into the settling pond. In many cases, mining companies generally use lime to neutralize mine water before the water flows into the settling ponds. This is intended to allow the pH of the water that goes into the pool increases. According to Hakim [5], lime is material that containing calcium which can be given to raise the pH. The increase in pH can occur because the hydrogen ions (H +) in solution disabled.

According to Setyamidjaja [6] calcification, in general is supplying materials for the purpose of increasing the pH. Lime reacts with acids to produce a near-neutral pH value of about 6-7. The degree of neutralization and the end result of this reaction is not known with certainty. When lime is given to the water, probably it will react with water containing CO 2 and soil colloids. The reaction is as follows:

$$CaO + H_2O \longrightarrow Ca(OH)_2$$
$$CaCO_3 + H_2O \longrightarrow Ca_2^{+} + HCO_3^{-} + OH^{-}$$

The reaction with H_2CO_3 is as follows:

$$CaCO_3 + H_2CO_3 \longrightarrow Ca\ (HCO_3)_2$$
$$Ca\ (OH)_2 + 2H_2CO_3 \longrightarrow Ca(HCO_3)_2 + 2H_2O$$

Lime generally used because the conditions of water are too acid. Therefore, The pH of the water should be increased so that the nutrients can be absorbed by plants

* Corresponding author: nenyrochyani@yahoo.com

DOI: 10.1051/matecconf/201710104025

and Al toxicity can be avoided. Usefulness of granting lime is to increase the pH, Mg and Ca. Adding lime could increase the availability of P and Mo, eliminating the element content of Al, Fe and Mn, and improve the life of the organism. [7].

The existence of acid mine drainage can reduce the acidity (pH) and the content of total suspended solids (TSS), Fe and Mn, TSS in relatively high amounts. It will reduce the penetration of light / sunlight into the water thus affecting the regeneration of oxygen in photosynthesis. Fe content that exceeds quality standards could endanger aquatic organisms and may cause corrosion to the equipment that metal made, while Mn at high concentrations are toxic. The pH value outside normal range may affect to the organisms in the water such as fish and other animals.

The experimental results in previous studies of acid mine drainage in the sump without rainfall water known that the optimum dose of lime that effectively increasing the pH of acid mine water is at a dose of 0.8 g / liter with operating time of 50 minutes, the pH obtained is 7,32. based on the results of these measurements , it can be known that to raise the pH of 1, the dose of lime needed of 0.24 g / l [8].

Lime requirements to conduct the treatment of acid mine drainage needed a valid data, where lime treatment performed in the third pond. Therefore, precise data is required on the ratio of precipitation and acid mine water or needed to know acid mine drainage debit and rain water discharge then the amount of lime dose required can be known [9].

2. Method

2.1. Laboratory Studies

Laboratory experiment conducted to determine the effect of the ratio of the rainfall volume and mine water to pH. mainly analysis on the use of lime to increase the pH on various conditions (ratio of rainfall volume and mine water).

2.2. Variable Research

1.Mine water
2.Rainfall water

2.3. Effect of rainfall valume and mine water on pH

Measurement on the effect of the ratio of the volume of rainfall and mine water on pH is based on optimum conditions obtained in the analysis of the effect of the operating time and lime dose on pH, . The Variations of ratio on rainfall volume and mine water and its effect on pH was : 1: 5, 2: 5, 3: 5, 4: 5 and 5: 5.

The effect of the ratio of rainfall volume and mine water on pH to be analyzed using ANOVA test and further tests such as real difference test and Tukey method to determine whether a significant effect between variables.

3 Result And Discussion

1. The Influence of Rainfall Volume on the pH

One of the factors that generate acid mine drainage is rainfall. In order to determine whether there is influence of rainfall on the pH of acid mine water, further testing conduct in the laboratory using a variety of volume of rainfall and acid mine water from Settling pond on East pit 3 West Banko Coal mine againts the pH.

From the laboratory test results, it can be seen that the pH levels will be increase if the volume of rain water also increasing, it can be concluded that the volume of rain water would give effect to the pH levels, thisis due to the concentration of acid mine water mixed with rain water will reduce the acidity of the water and then, on the high level of rainfall volume, the higher of the pH levels generated. To test and analyze the effect of rainfall, then the testing on hypothesis as the basic for this research in order to obtain valid results in the conclusion of this experiment. Tests conducted to determine whether there is singnificant effect between the volume of rainfall on pH and whether there is any real difference between each treatment in this trial.

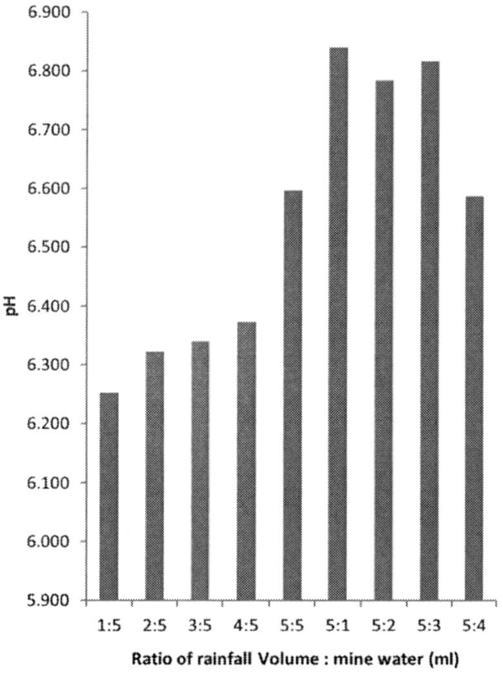

Fig 1. The Influence Of Rainfall Volume and mine Water on pH

As a base of test, then using a hypothesis to the effect of rainfall volume on the pH is as follows.
Ho: $\alpha_1 = \alpha_2 = \alpha_3 = 0$, meaning that there is no difference between the mean treatment of rainfall volume on the pH
H_1:at least there is one $\alpha_i \neq 0$, i = 1, 2, 3

For the above hypothesis, the decision-making that reject Ho if $t_{count} > t_{ref}$, or otherwise accept Ho if $t_{count} < t_{Ref}$ [10].The test results influence the volume of rainfall volume on the pH of acid mine drainage as shown in Figure 1. From the calculation and analysis show that there was a significant difference between the volume of rainfall water and pH, because the value of significant level is less than 0.01.

Table.1. Variance Analysis

Source	Sum	df	Average Squares	F	Sig.
Rainfall Volume	610 259	5	122. 052	43486.39 9	.000
Error	.028	10	.003		
Total	287	15			

Statistical result

The changes in the pH of acid mine drainage caused by the volume of rainfall. The magnitude coefficient of determination or the closeness of the variable volume of rainfall effect on the pH of acid mine water, known is 0,99 or 99%, which shows the significant level on influencing the pH of acid mine drainage.

The test results further by using the Tukey method is known that the volume of rainfall water affect with a significant level on pH.Variations in rainfall volume shows that in ratio 5: 5 is the most significant level on increasing the pH. This is shown in the table where the condition of five (5: 5) until the condition 9 (5: 4) were significantly different from the conditions 1, 2, 3, and 4. It can be concluded that the higher volume of rain water will increase the pH.

Table 2 ..Comparisons of rainfall volume

(I) rainfall	(J) rainfall	Mean Difference (IJ)	Std .Error	Sig.	95% Confidence Interval	
					LowerBound	Upper Bound
1:00	2:00	-.0700	.04514	.562	-.2260	.0860
	3:00	-.0867	.04514	.379	-.2426	.0693
	4:00	-.1200	.04514	.148	-.2760	.0360
	5:00	-.3433 *	.04514	.000	-.4993	-.1874
2:00	1:00	.0700	.04514	.562	-.0860	.2260
	3:00	-.0167	.04514	.995	-.1726	.1393
	4:00	-.0500	.04514	.799	-.2060	.1060
	5:00	-.2733 *	.04514	.002	-.4293	-.1174
3:00	1:00	.0867	.04514	.379	-.0693	.2426
	2:00	.0167	.04514	.995	-.1393	.1726
	4:00	-.0333	.04514	.941	-.1893	.1226
	5:00	-.2567 *	.04514	.003	-.4126	-.1007
4:00	1:00	.1200	.04514	.148	-.0360	.2760
	2:00	.0500	.04514	.799	-.1060	.2060
	3:00	.0333	.04514	.941	-.1226	.1893
	5:00	-.2233 *	.04514	.007	-.3793	-.0674
5:00	1:00	.3433 *	.04514	.000	.1874	.4993
	2:00	.2733 *	.04514	.002	.1174	.4293
	3:00	.2567 *	.04514	.003	.1007	.4126
	4:00	.2233 *	.04514	.007	.0674	.3793

Statistical Result

The analysis results also concludes that acid mine drainage conditions will be different during the rainy and dry seasons, in which the pH during high rain water tends to be higher and closer to the environmental quality standard, while in the dry season the water will contain lower pH.

Along with it also the amount of lime dose that is given there will be a difference on dose, in the rainy season, the lower dose of lime due to the pH of acid mine water is already approaching the environmental quality standard, while the dry season when rainfall is low, then the dose of lime will be the higher to increase the pH due to acid mine drainage is low.

2. Effect of Rainfall on the use of lime Dose

The estimation amount of precipitation plan that occurred in the region of East Pit 3 West Banko coal mine must be supported by their maximum daily rainfall data over the last few years were obtained from the Civilian Planning and Hydrology on PT. Bukit Asam. This analysis was conducted for the last 10 years ranging from 2005 to 2015. Based on data obtained and analyzed by the Planning and Hydrology section of PT Bukit Asam gained the highest maximum daily rainfall of 165 mm / day and the lowest was 76.4 mm / day.

In Figure 2, note the pattern of rainfall monthly average in the region of West Banko where conditions are the highest in December amounted to 366.09 mm / month and the lowest in August amounted to 97.76 mm / month. Based on the pattern of rainfall monthly averages also showed that the monthly rainfall of high intensity occurs from November to April, while the monthly rainfall at low intensity occurred in the month of May to October. It can be concluded that the rainfall will always be different on each month in one year, it is certainly going to affect the condition of acid mine water, where there will be mixing with rain water levels will be different in each month

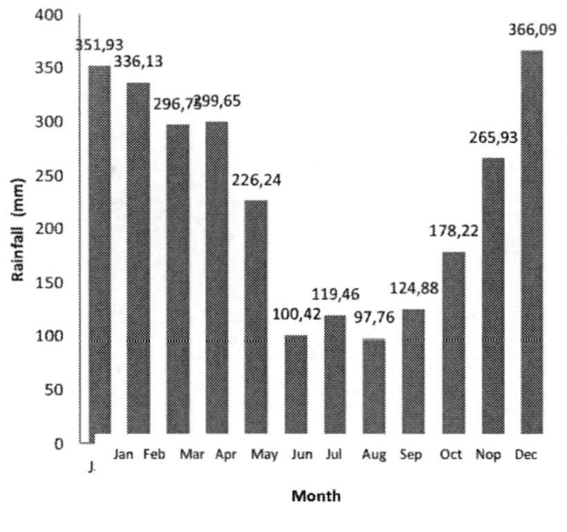

Fig 2.1. Average monthly rainfall on West Banko

Based on the data and the measurement result, it is known that the volume of acid mine drainage is equal to

1.75176 million liters / day. While the volume of rainfall for the east pit 3 West Banko coal mine area as following table.

Table 3. Rainfall Volume

Month	The volume of rainfall for Period					
	2 years	5 years	10 years	25 years	50 years	100 Years
Jan	203,244,064.97	261 636 728	301 704 783	352 323 156	365 021 297	399 869 120
Feb	197,450,329.86	259 587 125	299 339 987	349 562 516	386 821 449	423 750 474
March	188,946,816.62	244 355 895	281 777 086	329 052 432	364 125 242	398 887 460
April	189,887,913.60	252 968 904	291 708 303	340 650 521	376 959 477	412 947 021
May	165,666,191.35	209 066 965	241 092 481	281 542 150	311 550 958	341 294 078
June	120,997,411.20	155 473 820	179 283 763	209 363 178	231 678 606	253 796 477
July	109,149,597.92	143 786 194	165 805 503	193 623 765	214 261 578	234 716 533
Aug	141,075,597.74	190 669 748	219 868 882	256 757 458	290 709 597	311 249 377
Sept	98,656,779.37	129 962 584	149 865 124	175 008 915	193 662 766	212 151 340
Oct	159,444,721.66	210 655 636	242 920 459	283 670 628	313 906 802	343 874 789
nop	174,220,115.36	228 109 732	262 987 688	307 751 591	340 554 085	373 066 081
Der	199,282,486.48	264 529 275	305 039 224	356 217 800	394 186 016	431 818 127

PT. Bukit Asam (2015)

According to table 4 it is known that the volume of acid mine drainage and rain water volume will be different in each month, so the use of lime for acid mine water management differently.

Relating with the use of lime dos , the ratio of the rainfall volume and acid mine drainage is important to know to obtained effectiveness lime treatment monthly pattern on acid mine drainage. The effect of the ratio of rainfall volume and acid mine drainage on the pH of water known based on the trial and interpolation data as follows.

Table 4.1 Effect of Ratio Rainfall and Acid Mine on the pH

RV / AMD	pH	RV / AMD	pH
1.0: 1	5.6	2.7: 1	6.11
1.1: 1	5.64	3.0: 1	6.2
1.2: 1	5.66	3.3: 1	6.29
1.3: 1	5.69	3.4: 1	6.32
1.4: 1	5.72	3.9: 1	6.47
2.0: 1	5.9	4.0: 1	6.5
2.3: 1	5.99	4.1: 1	6.53
2,5: 1	6.05	5.0: 1	6.83

Description: RV: Rainfall Volume, AMD: Acid Mine Drainage

For monthly lime dose, using the data in Table 3 and 4, it can be seen that the monthly dose of lime used is as follows.

Application of lime doses were conducted to determine the effectiveness treatment in order to achieve

Table 4.2 Lime Dose Monthly

Month	ratio RV: AMD	pH	pH ENV	pH Needs	Dose/ 1 pH	Lime dose
January	3.9: 1	6.47	7	0.53	0.24	0.1
February	4.1: 1	6.53	7	0.47	0.24	0.1
March	3.3: 1	6.29	7	0,71	0.24	0.2
April	3.4: 1	6.32	7	0.68	0.24	0.2
May	2,5: 1	6.05	7	0.95	0.24	0.2
June	1.1: 1	5.64	7	1.36	0.24	0.3
July	1.3: 1	5.69	7	1.31	0.24	0.3
August	1.1: 1	5.64	7	1.36	0.24	0.3
boss	1.4: 1	5.72	7	1.28	0.24	0.3
October	2.0: 1	5.9	7	1.1	0.24	0.3
nop	3.0: 1	6.2	7	0.8	0.24	0.2
December	4.1: 1	6.53	7	0.47	0.24	0.1

Description: RV: Rainfall Volume; AMD = Acid Mine Drainage

the environmental quality standards, in which the use of lime will be less in the rainy season and more during the dry season.

Based on precipitation data for the area East Pit 3 West Banko, it is known that the conditions of high rainfall during the period from 2005 to 2015 is from November to April so that the use of doses of lime was low, while in the low rainfall that on May to October the use of lime doses will be much more.

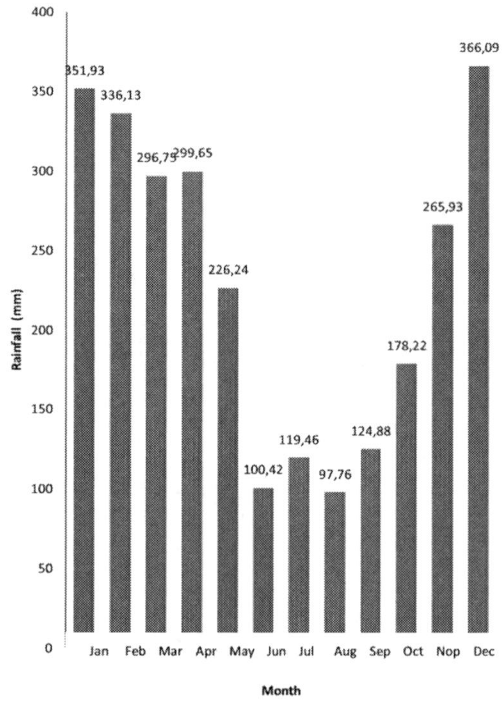

Fig 2.2 Rainfall Volume on East Pit 3 West Banko

4 Conclusion

Based on the analysis and discussion in the previous section, it can be concluded as follows:

a. Laboratory test results shows that the pH levels will be higher if the volume of rain water high, it can be concluded that the volume of rain water effect to the pH levels where the concentration of acid mine water is mixed with rain water and it will reduce the acidity. then, then, the higher the volume of rainfall, the higher the pH.
b. The rainfall volume affect to thepH water due to the process of dilution by rainfall water. It will decreasing H^+ there by increasingt he pH of the mine water.
c. The difference in the volume of rainfall as a result of different seasons will affect the dose of lime that given to raise the pH. Then ,there will be differences in the dose of lime to the dry season and the rainy season.

Acknowledgments

This research was made possible through a permission from the management of PT. Bukit Asam coal mine that would give us an opportunity to research and work in coal mine field at the west Banko. And special thanks to all of the employees on coal mine field that providing support and helping us every equipment that we all needs.

References

[1]. J.G. Skousen, *Proc. Reclamation of Drastically Disturbed Land, American Society of Agronomy And the American Society for Surface Mining and Reclamation*, **41** 1- 42 (2000).
[2]. J.P. Widowson, *Biological Resources Utilization Local For Acid Mine Drainage Passive Treatement*, (2007).
[3]. J. Fripp, J. ERDC TN-EMRRP-SR-14, 1-7 (2000)
[4]. N. Rochyani, E. Ibrahim, M. Faizal, And Ngudiantoro. Unsri.E J., 158-164 (2014)
[5]. M. Hakim, *Fundamentals of Soil Science*. (1986)
[6]. D. Setyamidjaja. *Fertilizers and Fertilization (*1986)
[7]. R.S. Hedin. Mine Water Env. J., **29** 165-167 (2009).
[8]. N. Rochyani, E. Ibrahim, M. Faizal, And Ngudiantoro. Int. J. Environ Sci. Develop. 430-434 (2015)
[9]. N. Rochyani, E. Ibrahim, M. Faizal, And Ngudiantoro. Int. J. Adv. Sci. Eng. Inform. Technol., 169-172 (2014).
[10].KA. Gomez, *Statistical Procedures for Agricultural Research* .(2010).

The influence of Musi river sedimentation to the aquatic environment

Achmad Syarifudin[1,*]

[1]Civil Engineering & Environmental Department Faculty of Engineering, Bina Darma University, 30264 Palembang, Indonesia

Abstract. Musi river is the largest river in Palembang with a length of more than 750 kilometers. The sedimentation in the Musi river is commonly existed due to the high level meeting of velocity between the Musi river and the ocean in the *Bangka* Strait. The silting condition of Musi river is getting more severe due to the sludge growth that reach about 40 cm per month. In fact, the volume of sludge could reach 2.5 million m^3. There are 13 silted points along the river shipping channel from the *Boom Baru* port to Bangka Strait. Four points are vulnerable due to the siltings reach up to 4 m. This research was conducted by using the approach of software models MIKE 21 Flow Models. The results obtained from this study are the greatest speed direction occurred, elevation of the outer surface and total depth of water will contribute and effect to the free area for ships to perform well and safe movement. The other side will effect of the amount of sedimentation and the aquatic environment of the swamp area.

1 Introduction

Musi river is the largest river in Palembang with a length of more than 750 kilometers and an average width of 540 meters where the maximum width of 1,350 meters is located around the Kemaro island and the minimum width of 250 meters is located around the Musi II Bridge. Musi river has two islands, namely Kembaro (Kemaro) and Kerto islands. The three other major rivers are the Ogan river with an average width of 236 meters and the Keramasan river with an average of 103 meters [1].

The movement of sediment shows profound significance to river. The process of sediment transport and deposition can change the topography of the river bed [2]. The sediment deposition is a key factor to limit the river development and management [3]. The problem of sedimentation will have significant impacts on the operation lifetime of the river, the flood control capacity and the transportation.

Musi river is not only used by people around the the river, but also by big companies located along the Musi river. They use the Musi river to deliver the products and bring in the raw material through the vessel. There are many big ships and even giant pacing in the Musi river. Some companies located along the Musi river are PT. Pertamina, PT. Pupuk Sriwidjaja (fertilizer industry), Wilmar Group and the Port of Boom Baru, and Port at 35 ilir. The role of the Musi river is very vital and this is a pulse of Palembang city is now preoccupied by various problems. One of the problems is the river silting which continue to increase every year. This is surely become some disadvantageous to the Government of South Sumatra Province, especially at this time of South Sumatra province being intensively attracts investors to invest in the business sector. Basically, sedimentation in the Musi river does include a high level of sedimentation caused by the current meeting between Musi river and ocean currents in the Bangka Strait.

Major sources of sludge in the country resulted from soil erosion and sedimentation are due to agricultural land occupation [4]. Common definition mentioned that siltation or siltification is the pollution of water by particulate terrestrial clastic material, with a particle size dominated by silt or clay.

Currently, the silting condition of Musi river is getting more severe because of the sludge that reaches about 40 cm per month. In fact, the volume of sludge could reach 2.5 million meter³. There are 13 silting points along the Musi river shipping channel from the Boom Baru port to Bangka Strait. Four points are very vulnerable because due to the silting reaches up to 4 meters. The location which is quite compress is from the northern Payung Island to the estuary, while the location of the worst silting among others is on the outside verge of the Jaran strait and the water groove of the sedimentation in the southern part of the Payung island reaches 7 km, so the boat that crosses the groove of Musi river should be guided against the tide that occurs.

2 Methodology

This study was carried out with the modeling and simulation of flow patterns and sediment transport in water area of the Bangka Strait in position 2.07° - 2.38°

* Corresponding author: syarifachmad6080@yahoo.co.id

DOI: 10.1051/matecconf/201710104026

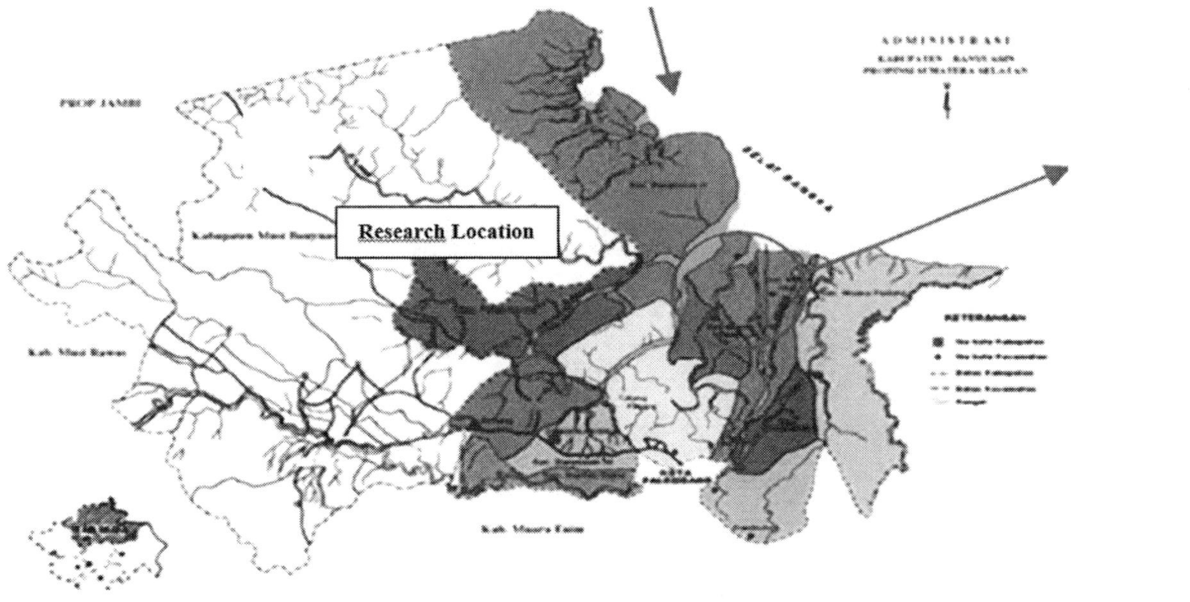

Fig. 1. Map of Research Location [5]

of South Latitude and 104.85° - 105.17° of East Longitude. The research location is presented in Figure 1.

2.1. Hydro-climatological conditions

Natural state of Palembang is relatively valley an tropical area, with an average temperature of most areas of Palembang 21°- 32° Celsius, rainfall of 22-428 mm per year. Based on data from the Meteorological Station in 2003 the average air temperature range between 23,9°-32° Celsius, 24,04°-32,60° Celsius (2004), 22,44°-33,65° Celsius (2005), 26,4°- 28,9° Celsius (2006) and 21,2°-35,5° Celsius (2007). In 2007, the heaviest rainfall fell in April by the amount of rainfall is 540 mm. While the air humidity in 2007 an average of 80%, an average wind speed of 20 km / hours with the largest direction of the northwest, as well as the average air pressure at sea level at 1009 mbar and in the mainland amounted to 1007.5 mbar [4].

2.2. Geological conditions

Based on geological conditions, Palembang has a diverse relief consists of the ground in the form of a layer of alluvial and sandy loam. In the southern part of the city, the rock in the form of impermeable clay sand, north rocky sand impermeable clay, while the west rocky clay gravel, sand clay is impermeable to water-resistant. From the data obtained on the geological structure of the Jakabaring area folding structure is in the form of a number of anticline has a NW-SE trending axis. Their multiplicity - the multiplicity leading to a muscular structure. Theoretically muscular structure will consist of four sets as follows:

- Perpendicular axis Sumatra Island
- Perpendicular direction of flow of the river Ogan
- Ogan or nearly parallel to the axis of Sumatra Island
- Directions NNE – SSW
- Direction NNW – SSE

2.3. Hydrometric measurement

From the results of measurements performed by consultant, the data obtained are:

- Fluctuations maximum tidal water level during spring tide in the estuary of the Ogan is 2.20 m and 2.40 m Ogan River estuary. In the period of spring tide ebb and flow of nature is a single daily, and gradually became mixed, eventually became semi - diurnal during the neap tide.
- Damping along the tidal section of the Ogan River estuary to the canal Java (14 km) is relatively small ± 10 cm and on segment Ogan River from the mouth to the canal Java (28 km) ± 40 cm.
- The current in the river is always moving downstream Ogan without ever reversed despite a speed variation is from 0.05 to 0.06 m / s and large discharge is 60-500 m^3 / sec.
- In the Ogan River occurs reversing flow, the discharge variation between 110-120 m^3 per secon. There is a transfer nearly all upstream river discharge into the river Ogan Ogan through the canal Java.
- The cross section of the Ogan River estuary to km 14.5 (Java canal) has a wide variation between 100-200 m with an average depth of 10 m. A cross section Ogan River has a wide variation between 60-330 m with a depth of the average between 2.2 -7.8 m
- The content of the sediment in the river Ogan ranged between 51-166 ppm at low tide and 16-56 ppm at high tide. The value in Ogan River is 36-116 ppm at low tide

and 52-136 ppm at high tide

- Bed material Ogan river and consists of silty sand, fine sand fraction to moderate with the d50 to the river Ogan ranged from 0.07 to 0.15 mm and for the river Ogan 0.15 to 0.32 mm.[4]

The tools used in this research are as in Table 1 below.

3 Tool and materials

Tabel 1. List of tools used in the study

No	Name of Tools	Number of Uses
1	Stationery 2 pieces	Writing to recording data
2	Computer (RAM 2 GB) 1 unit	Performing general modeling
3	Printer 1 unit	Displaying form of report
4	Software MIKE-21 Model, MS-Excel	Performing modeling and data processing
5	Dongle (licensed program) 1 pieces	Activating the software MIKE-21 FM
6	Laptop and Printer 1 pieces	Assisting to preparing reports

4 Mike-21 flow model Simulation

MIKE-21 Hydrodynamic Module (HD Module) is a mathematical model to calculate the hydrodynamic behavior of the water against a wide variety of styles functions. Particular wind conditions and water levels are specified in the open model of the boundary. HD module simulates the water level and the current differences in the various styles function in lakes, estuaries, and beaches.[5]

5 Results and discussion

5.1 Output model hydrodynamics

Therefore the boundary condition is discussed in the modeling, the results of global model output on the outside verge of the Musi river are as in Figure 2, 3 and 4. The greatest speed direction occurred on the Musi river bank ranges between 300-350 degrees.

The data used in the analysis are as shown in Table 2.

Table 2. Types and sources of data required

No	Data properties	Data type	Sources	Units
1	Tidal	Primary	Field	m
2	Bathymetry	Secondary	Bakosurtanal / Pelindo	m
3	The direction and speed of current	Primary	Field	(°) & m/s
4	River capacity	Primary	Field	m^3/s
5	The direction and speed of wind	Secondary	BMG	(°) & m/s
6	Base of Sediment	Primary	Field	mm or ø
7	The concentration of suspended sediment	Primary	Field	mm/l
8	Discharge of sediment from the river	Primary	Field	gram/s

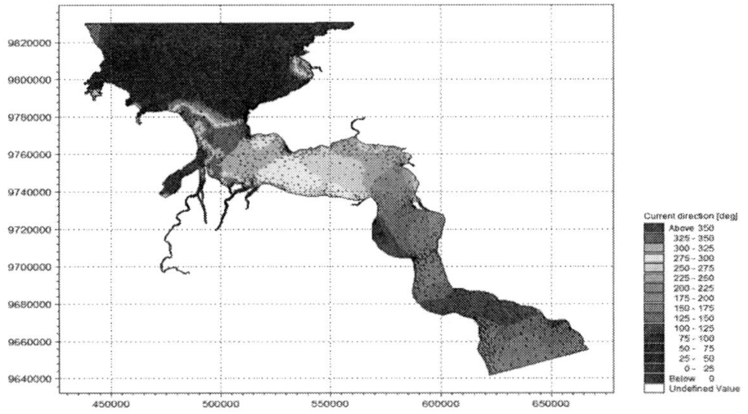

Fig. 2. The hydrodynamic model of the river water velocity direction.[5]

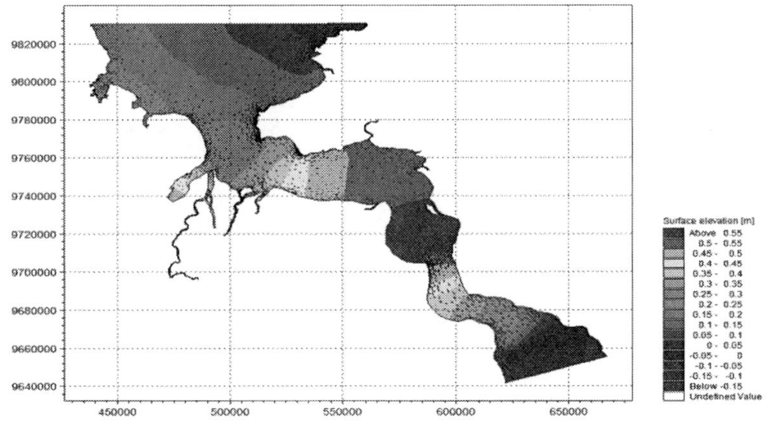

Fig. 3. Model hydrodynamic surface elevation Watersheds [5]

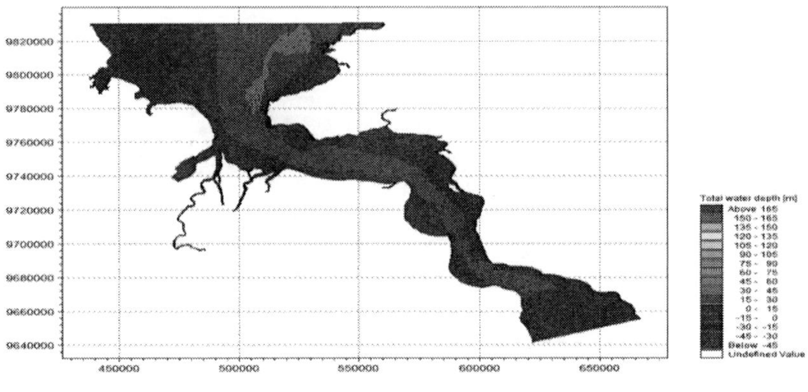

Fig. 4. The total depth of the hydrodynamic model of river flow [5]

Thresholds elevation of the outer surface of the Musi river ranges between 0.20 m - 0.40 m will affect the free area for ships to perform well and safe movement.

The total depth of water occurs at 15-30 m and therefore contributes to the flow of the ship movement into the Musi river as a whole.

6 Conclusions and recommendations

From the analysis of this study, we can conclude that the greatest speed direction occurred, elevation of the outer surface and total depth of water will contribute and effect to the free area for ships to perform well and safe movement. The other side will effect of the amount of sedimentation and the aquatic environmental of the swamp area [4,9].

I would like to say thank you to Prof. Ir. H. Bochari Rachman, M.Sc, Rector of Bina Darma University, Prof. Dr. H. Zainuddin Ismail, MM, Vice Rector I and Dr. Sunda Ariana, MM, M.Ed, Vice Rector II of Bina Darma University who are pleased to give permission and assistance to the author, especially in the preparation of the data so that the paper can be completed.

References

1. AKNOP, Report. Department of Public Works Directorate General of Water Resources (2011)
2. S. J. Chen, Y. X. Yan, and Y. K. Li, "Spatial and temporal variations of suspended sediment deposition in the alluvial reach of the upper Yellow River from 1952 to 2007," Catena, vol. 92, pp. 30-37, 2012.
3. Wanshun Zhang, Yanhong Xu, Yanru Wang, and Hong Peng, Modeling Sediment Transport and River Bed Evolution in River System, Journal of Clean Energy Technologies, Vol. 2, No. 2, April 2014
4. Fulazzaky, Mohamad Ali. "Challenges of integrated water resources management in Indonesia." *Water* 6.7 (2014): 2000-2020
5. BMKG, Hydro-climatology of Palembang (2007)
6. DHI Software, MIKE-21 Reference Manual, Danish Hydraulic Institute, Denmark (2007)
7. D G Wetlands and Beach, Report, Directorat General of Water Resources, Department Public Works (2006)
8. H. Sumarjo, R.H. Susanto, B. Schult, and F.X. Suryadi, Proceedings of the 9th Inter-Regional Confrence on water environment. Enviro water, Concept for Water management and multifunctional land uses in lowlands, Delft, the Netherlands (2006)
9. L. Hayde, Canal Designs, Lecture notes, IHE. Delft, The Netherlands (2007)
10. Land and Water Management Tidal Lowlands (LWMTL) South Sumatra Province, June 2004. Operation and Maintenance Network with water user associations (P3A), Rikjkswaterstaat, UNESCO-IHE, ARCADIS-Euroconsult in Cooperation with Regional Infrastructure, Ministry of Agriculture, Sriwijaya University and Local Government South Sumatra, Indonesia (2004)
11. F.X. Suryadi. Report: Regional Development Experience Swamp and O & M Telang I. Land and Water Management Tidal Lowlands (2004)
12. A. Syarifudin. The 2nd International Conference on Informatics, Environment, Energy and Applications (IEEA 2013), Bali, Indonesia, March 16-17, 2013, JOCET (Journal of Clean Energy and Technology) Journal ISSN : 1793-821X Vol. **2**, No. 1, January 2014. (2013)
13. A. Syarifudin, 5th ICIBA International conference on information technology & Engineering Application, Palembang (2015)

Industrialized building system - an innovative construction method

Anis Saggaff [1*]

[1]Civil Engineering Department, Faculty of Engineering, Universitas Sriwijaya, 30662 Indralaya, Sumatera Selatan, Indonesia

Abstract: Industrialized building system (IBS) has been introduced in construction industry as early as 1960's. The challenges facing our construction industry are due to shortage of skill labour and high cost of construction. Therefore, the concept of construction using IBS is proposed to reduce our dependency on intensive labor works and to reduce the cost of construction. However, the process of changing the traditional construction method to IBS has been delayed due to the reluctant to accept the changes and even resistant to changes. Therefore this paper presents the concept of Industrialized Building System, the scope of IBS, challenges and way forward to inculcate the use of IBS construction in Indonesia.

1 Introduction

Housing needs continue to rise very rapidly in Indonesia, particularly in urban areas. Back log on housing needs up to 2016 has reached up to 7.6 million units, while demand for new homes every year has reached up to 800,000 units [1]. There are about 1,350,000 units gap that must be resolved in order to fulfill the housing need in Indonesia [1]. One of the ways to speed up the construction of these affordable housing is to use Industrialized Building System (IBS) in construction. Industrialized Building System is usually associated with sustainable construction which is described as the ability of the construction system to consider the environmental impact of a building over its entire lifetime, while optimizing its economic viability and the comfort and safety of its occupants [2]. Typical standard building practices are guided by short term economic considerations; while IBS construction is based on best practices which emphasize on long term affordability, quality and efficiency. At each stage of life cycle of the building, it increases comfort and quality of life, while decreasing negative environmental impacts and increasing the economic sustainability of the project [3]. The concept of IBS usually associates with the preservation of the environment which also concern on related issues such as the efficient use of resource, continual social progress, promising economic growth, and improve standard of living. The main goals of IBS construction are to meet present day needs for housing, working environments and infrastructure without compromising the ability of future generations to meet their own needs in times to come. In Indonesia, typical building construction is usually associates with construction system based on typical reinforced concrete design where formwork is formed using timber and plywood while the concrete is cast in-situ. As IBS system is quite new in Indonesia, a way forward to motivate and promote the system is a big challenge to the Indonesia construction industry.

This paper generally discussed on issues facing our construction industry and current researches related to the use of Industrialized Building System (IBS). One of the main problems facing our construction industry is a serious shortage of skill construction workers. Therefore, the introduction of Industrialized Building System (IBS) by Indonesia Government is a good alternative to reduce the dependency on skill workers. However, the use of IBS as an alternative to conventional system (reinforced concrete) has not widely accepted by construction industry players. Some of negative perceptions given by construction industry players on the use of IBS are listed as not ready towards new concept, insufficient information and lack of knowledge to understand changes in IBS. Hervas revealed that construction sector is known as traditional sector that can be identified as not willing to accept changes or even resist changing [4]. This paper also discusses the current implementation of IBS, its shortcomings and the way to enhance the system forward. It also elaborates the current issues by focusing on stability, build-ability and cost efficiency and research done in IBS.

2 Definition and types of IBS

IBS is defined as a construction system which components are manufactured in a factory, on or off site, positioned and assembled into structure with minimum additional site work [5]. Dietz defined IBS as total integration of all sub-system and components into

* Corresponding author: anissaggaff@yahoo.com

overall process fully utilizing industrialized production, transportation and assembly techniques [6]. According to Parid (1997), IBS is a system which use industrialized production technique either in the production of component or assembly of the buildings or both [7]. Lessing et al defined IBS as an integrated manufacturing and construction process with well-planned organization for efficient management, preparation and control over resources used, activities and results supported by the used of highly developed component [8].

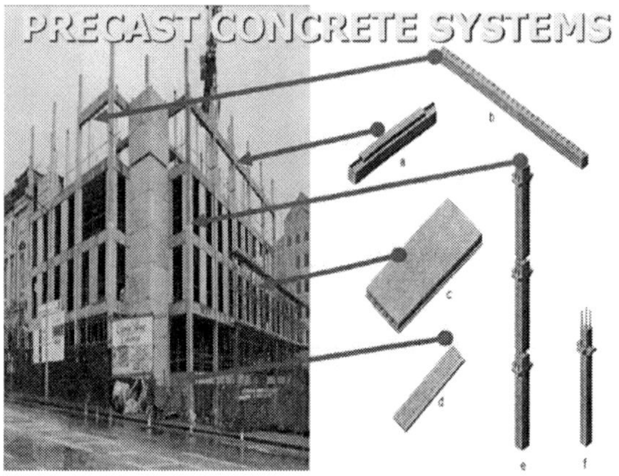

Fig. 1. Pre-cast concrete system

Nowadays, there is an increase in demand for housing due to an increase in population. In most countries including Indonesia, the demand is exceeding the supply. In order to solve the inadequacies of housing, Industrialized Building System (IBS) is introduced to replace conventional construction method. Basically, CIDB has classified IBS into five groups [3] which are

- Precast Concrete System (see Fig. 1)
- Steel Formwork Systems (see Fig. 2)
- Steel Framing Systems (see Fig. 3)
- Prefabricated Timber Framing Systems (see Fig. 4)
- Block Work Systems.

These systems are used due to several advantages. It saves a lot of time as the casting of precast structure can be done in factory and installed the construction components on site.[9] Besides, the delay in construction can be minimized as the construction components are fabricated in the factory with controlled environment [9].

Fig. 2. Steel formwork system

Fig. 3. Steel framing systems

3 Advantages of IBS

Awareness and knowledge are the main important barriers to the implementation of IBS in Malaysia [10]. Industrialized building system has the benefits when compared to the conventional construction method in the aspects of cost savings, less relied on labor, quality control, environmental friendly and less dependent on weather problem.

3.1 Cost

The implementation of the IBS in construction will reduce the construction time. CIDB [3-4] stated that the casting of the precast element in the factory and foundation at site can be carried out simultaneously. As the construction time decreases, the overall cost of the construction will also be reduced. Besides, the repetitive use of system formwork made up of steel and scaffolding can save the cost [11].

Fig. 4. Pre-fabricated timber frame system

3.2 Labour

Since IBS on mechanical means to produce prefabricated elements and only minimal in-situ construction, it reduces the requirement of labors for prefabrication of element and erection at site. Skilled and semi-skilled labors are needed for IBS and unskilled workers are mostly not needed. Countries like Malaysia where the construction industry is the host to a large number of foreign workers from Indonesia, Bangladesh and Pakistan. It is important to stress on that the labors are still have to send for training for the skills appropriate with the IBS. Although the training may cause some cost, it is believed that trained labor would be more quality than the untrained labor.

Fig. 5. Block work system

3.3 Quality control

The quality of the prefabricated concrete components in IBS is much better than the conventional method due to the quality in terms of the selection of materials, manufactured under controlled environment and inspection before the components send to the site. The materials used are based on the quality standard. Besides, the prefabricated concrete components are normally carried out in the factory which is indoor environment and it will not affected by the raining season in Malaysia. Inspection will be done to make sure the quality of the components is achieving the standard or the requirement needed by the clients.

The overall advantages of the using IBS system be summarized as shown in Fig. 6.

Fig. 6. Advantages of IBS

4 Roadmap towards IBS

Malaysia has earlier initiated the use of IBS by publishing IBS Roadmap 2003-2010 which prescribed the needs and requirements of Malaysian Construction industry [12]. The roadmap listed five main focus areas that are needed to drive forward the use of IBS in our construction industry in Malaysia. The five areas are listed as Manpower, Materials, Management, Monetary, and Marketing [13]. IBS Steering Committee was formed to enhance further the concept of IBS in the construction system [14]. The functions of the Steering Committee were to develop standard drawings for common use, to provide expertise in the process of development and production of IBS components, to carry out test and evaluation and to provide assistance in analysis and design procedures [14].

4.1 Challenges and delays

After all efforts done by the steering committee, shortcomings and delays of IBS implementation can be listed as follows [15]:

- IBS is not popular choice by consultant and developers due to lack of knowledge on the performance of the IBS system.
- The needs to change the mind-set of construction industry player that for long run IBS is much better than the traditional construction.
- Insufficient push factors from governments and policy makers such as Public Works Department and local authority.
- Lack of technical know-how from production to erection.
- Volume and initial cost of production are also issues to be addressed in the supplier chain.

Besides all the shortcomings and negative remarks mentioned above the way forward to implement the use of IBS are very much needed in order to benefit the advantages developed from IBS. A series of workshops to enhance awareness have been carried out by CIDB to all construction industry players. The awareness in using IBS has also been initiated by consultants, universities, and companies produce the IBS components.

4.2 Planning and implementation

The establishment of steering committee by Indonesia government should be seen as a very good step to inculcate IBS system in construction industry. One of the ways forward to enhance the effective planning and implementation of IBS is to promote research and development (R&D) of IBS in all related parties in construction. The R&D and workshops organized by Indonesia Construction Board should be carried out to the public so that the awareness of the IBS can be well accepted and understood. The initiatives to accept the use of IBS should be done by all parties such as contractors, consultants, universities, developers and research institutes. The main objectives of promoting the use of IBS is to ensure that factors such as improve the performance, quality, and minimize the dependency of unskilled workers and reduced cost can be well explained and understood. A complete and comprehensive study on IBS solutions which take into consideration the role of the entire value chain should be emphasized. A technology transfer model from other parts of developed countries should be done through smart collaboration.

5 Some experimental works

5.1 Influence of longitudinal spaced bolted shear connectors on composite beam integrated with concrete and cold-formed steel

[16] have experimented composite construction with conventional hot rolled steel (HRS) sections has been known to perform much better than Cold-formed steel (CFS) sections for decades. However, the composite action of CFS with an in-situ concrete, especially Self-Compacting Concrete (SCC) and bolted shear connector 'haven't been investigated'. This study attempted to investigate the behavior of bolted shear connector used with SCC and CFS as 'shear connector' to form a composite beam system at designated longitudinal spacing. Push-out test specimens of shear connector longitudinal spacing of 300 mm, 250 mm and 150 mm with bolted shear connector of grade 8.8.

Steel-concrete composite beam has being in use as a structural member in building and bridges for decades. The shear connection between steel and concrete slab in composite beams is inherently significant, as it resists separation between the two components as well as enhance longitudinal shear transfer.

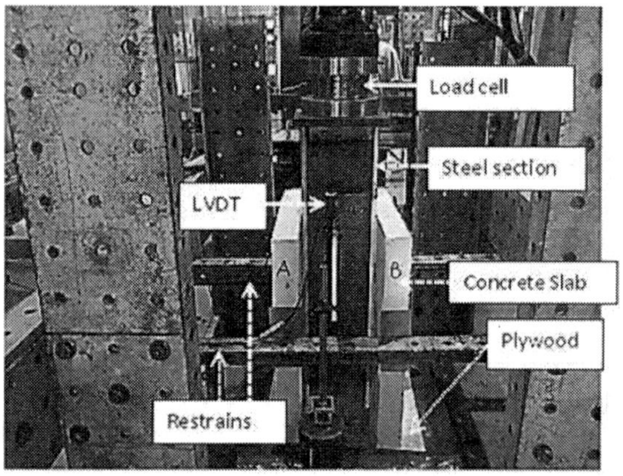

Fig. 7. Test of shear composite

The paper has come to summary that the effect of shear connector longitudinal spacing was investigated, and its influence on the ultimate strength capacity of the shear connector was established. The connection was classified as ductile shear connector since it attained a characteristic slip capacity. The longitudinal spacing of bolted shear connector had shown the influenced the strength capacity of the shear connector.

5.2 Performance of varying bolted shear connectors in cold-formed steel composite beam

Anis Saggaff study a suitable bolt and nut type of shear connection mechanism has been proposed. The Experiment has presented the possibility of using a type of bolt and nut as shear connector. The experimental tests results were conducted to ascertain the performance of the proposed connectors. The aim of the research was to investigate the structural performance of bolted shear connectors in providing composite action between Self-compacting concrete (SCC) and CFS section integrated as composite beam system.

Fig. 8. Push-out test

The push-out test results have shown the load-slip relationships of the tested specimens. The specimens with M14 and M16 bolt diameters could be attributed to the failure of CFS section by flange local buckling. Perhaps high resistance of the applied load by the concrete slabs and bolted shear connectors could be responsible for it. The failure due to steel buckling occurred at the top shear connector position of specimens which was close to the load application position. The failure then extended upwards from the initial position where it had occurred to the part where the CFS was not covered by the concrete slab.

The Remarkable shear resistance and slips at ultimate loads were attained by the specimens with all the bolted shear connectors. It has proved that the maximum design capacity of the specimens with M14 and M16 bolted shear connectors is based on the tested value obtained due to CFS failure.

The study has shown the influence of shear connectors. The shear strength capacity of bolted shear connectors was noted as follows:
40.6% between M12 and M14
41.0% between M12 and M16 and
0.3% between M14 and M16.
It was concluded that the shear connectors with single embedded nut and washer influence on the ultimate strength capacity of bolted shear connectors. The bolted shear connectors used in this study possessed good shear resistance capacity.

DOI: 10.1051/matecconf/201710105001

5.3 Behavior of pad footing with cold-formed lipped C-Channels steel section as permanent form-work

Anis Saggaff has done the experimental work using light concrete and cold-formed steel with wire-mesh as reinforcement. The experimental investigation of pad footing designed using cold-formed lipped C-channel steel section as permanent formwork. The tested specimens assembled as square and rectangular pad-footing where cold-formed steel of lipped C-channel section.

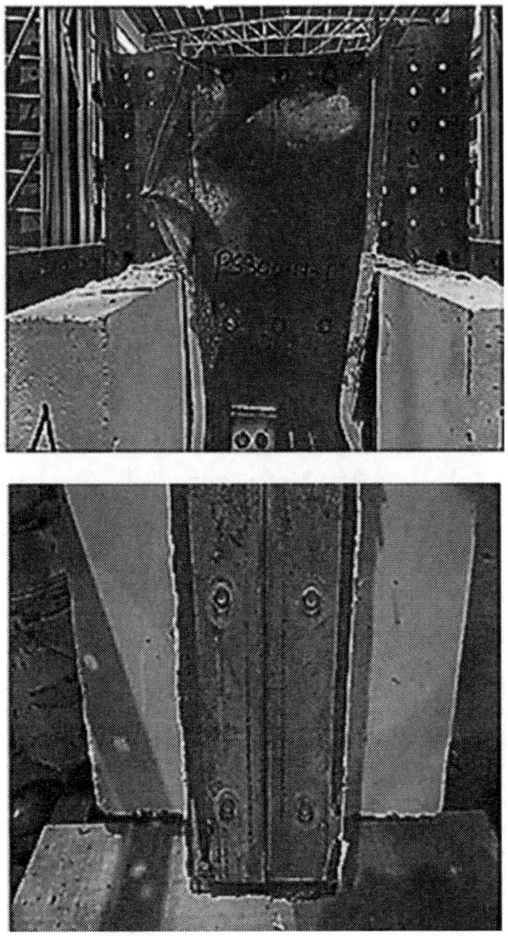

Fig. 9. Failure of shears

The formation of the formwork for each specimen was done using the lipped cold-formed steel section with stiffeners positioned at right angle to shape the footing as a boxed type specimen and to avoid changes in shape during concreting. The reinforcement bars used in the LCCFSS specimens were A10 which was bent downward about the length of the cover needed for the footing and sit on the LCCFSS specimen as shown in. This was purposely done so that no spacer is needed underneath the A10 wire mesh to meet the requirement of the concrete cover. For typical reinforced concrete pad footing the A10 wire mesh was bent upward which result to the need of the spacer underneath the reinforcement to meet the requirement of concrete cover. Test results has shown the maximum axial load applied until failure based on the theoretical calculation from BS 8110 part 1:1997 [12]. The load capacity of the footing also depends on the length of the footing besides the thickness of the footing. This study has maintained the thickness is kept constant. The load capacity of the footing has been reduced by up to 30% as the length of the footing increased from 1000mm to 1750mm. The reduced in load capacity is very much related to the failure of the footing which is shifted from punching shear failure to bending failure.

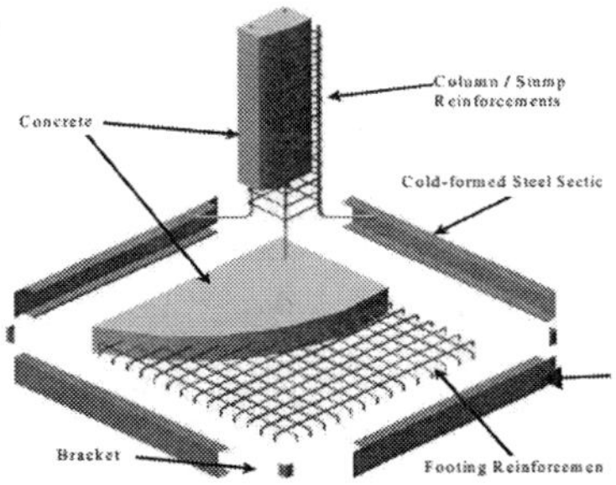

End of A10 wire mesh bending about concrete cover

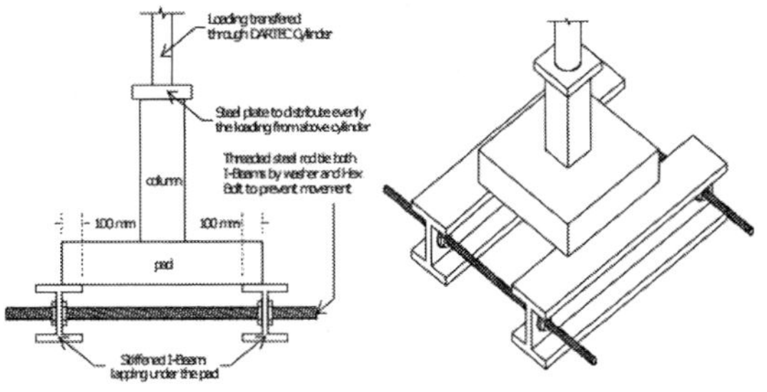

Fig. 10. Layout of test setting

6 Conclusions

The use of IBS system needs all construction players to adopt or willing to accept the way forward in construction industry. It is hoped that the finding addressed in this paper could highlight the benefits of using IBS in actual construction. It is also hoped that the IBS can be used to solve the problem of housing needs in Indonesia. The delays in implementation should not hinder the way forward in construction as the benefits of using IBS are too many to be ignored. Indonesia government should take a drastic measures and regulation in order to prosper the success of IBS in Indonesia.

The author of this paper would like to acknowledge Specials thanks addressed to Construction Research Center UTM, Prof Ir. Dr Mahmood Md Tahir., and Structure and Construction Research Center also Sriwijaya University that make this paper

References

1. Pupera, *The Ministry of Public Works and Housing Report* (PUPERA, Indonesia, 2016)
2. CIDB Malaysia, *Strategic recommendations for improving environmental practices in construction industry* (CIDB Publisher, Kuala Lumur, 2007a)
3. CIDB Malaysia, *Guidelines for implementing environmental management system in the construction industry* (CIDB Malaysia, Kuala Lumpur, 2007b)
4. Hervas, *Open Building Manufacturing – Core Concept and Industrial Requirements* (Manu BUILD Consortium, Holland, 2007)
5. National IBS Survey, *Construction Industry Development Board (CIDB)* (CIDB, Malaysia, Kuala Lumpur, 2003)
6. A. G. H. Diets, *Global Trends in Research Development and Construction Proceeding of the International Conference of the International Conference on Industrialised Building System* (IBS CIDB Malaysia, Kuala Lumpur, 2003)
7. W. Parid, *Global Trends in Research Development and Construction Proceeding of the International Conference of the International Conference on Industrialised Building System* (IBS, CIDB Malaysia, Kuala Lumpur, 2003)
8. J. Lessing, A. Ekholm, and L. Stehn, *13th International Group for Lean Construction* (Sydney, Australia, 2005)
9. T. M. Toole, J. CEM, 124, **4**, (1998)
10. M. M. Nuruddin, Buletin Bulanan Jurutera (Institute of Engineers Malaysia, 2003)
11. CIDB Malaysian, *Construction Industry Technology Foresight Report* (CIDB Malaysia, 2000)
12. Construction Industry Master Plan 2006 – 2015 (CIDB, Kuala Lumpur, 2007)
13. S. S. Gue, *Bridging the Gap between R & D and Construction Industry*", (CIDB, PWTC, Kuala Lumpur, 2007)
14. J. Hussein, *Industrialized Building System: The Challenge and The Way Forward* (CIDB PWTC, Kuala Lumpur, 2007)
15. A.H. Zuhari, K.G. Mohd, H.A.R. Ahmad and A.M.K. Kamal, *Industrialised Building Systems (IBS): Current Shortcomings and the Vital Roles of R& D* (JURUTERA, Malaysia, 2008)
16. M. M. Lawan, M Tahir, J. Mirza, and A. Saggaff, *Influence of Longitudinal Spaced Bolted Shear Connectors on Composite Beam Integrated with Concrete and Cold-Formed Steel* (2008)
17. A. Sagaff, M. M. Lawan, M Tahir, and J. Mirza, *Performance of Varying Bolted Shear Connectors in Cold Formed Steel Composite Beam* (2008)
18. A. Saggaff, M. Tahir, W. K. Leong, and S. P. Ngian, *Behavior of Pad Footing with Cold-formed Lipped C-Channels Steel Section as Permanent Form-Work* (2016)

* Corresponding author: anissaggaff@yahoo.com

Track access charge for Indonesian railways using full cost method: improving industry competitiveness

Mohammed Ali Berawi[1,*], *Perdana* Miraj[1,2], *Abdur* Rohim *Boy* Berawi[1], *Nahry*[1], and *Jachrizal* Sumabrata[1]

[1]Center for Sustainable Infrastructure Development, Faculty of Engineering, Universitas Indonesia, 16424 Depok, Indonesia
[2]Civil Engineering Department, Faculty of Engineering, Universitas Pancasila, 12640 Jakarta, Indonesia

Abstract. The government of Indonesia currently focuses on improving the public railway services to be more effective and efficient by formulating suitable railways infrastructure charges. This paper examines a possibility to determine the track access charge (TAC) for Indonesia railways infrastructure based on a benchmark to other countries. Compared to existing charge that used basic price and product factor, the alternative proposed concept also considers multiplicative surcharge and load component. In-depth interviews will be conducted with related stakeholders from government bodies, operator, and academics to gain more insights and further to validate the research outcomes. The result shows that the suitable TAC value for the railway's operator ranging from 56% to 66% from total proposed payable TAC, while government subsidy will cover the rest of 34% to 44% of determined track access charge to liberalize railways industry in Indonesia.

1 Introduction

Railway infrastructure plays a significant role in increasing national transport connectivity. Many benefits have been identified by many scholars ranging from reliability, speed, and accuracy compared to other modes of transportation such as a bus or personal vehicles [1-3]. The government of Indonesia through the Ministry of Transportation currently focused on improving the public transportation system to be more effective and efficient by increasing the quality and quantity of railway services.

Currently, the management of rail services in Indonesia held by PT Kereta Api Indonesia (KAI) – a state-owned enterprise – that acts as sole operator for railway sector. They annually receive subsidies from the government to operate railway services. In 2012, government supported the operator through a subsidy that amounted for about US $ 79 million and the budget keep increasing every year [4]. Despite improved performance and earnings of PT KAI in the past few years, there are still many aspects in term of technical aspect to user services that can be improved.

The government continues to increase the reliability of rail services by balancing the reduction of costs and subsidies. Encouraging the private sector to get involved in creating efficiencies and innovation is one of the solutions to improve rail services. Firstly, the government needs to create Track Access Charge (TAC) mechanism – access fees for using railway infrastructure that provided by the government.

However, TAC system in Indonesia has not properly formulated to accommodate private sector investment. It can be seen from the framework that shown an unclear formula which plausible for dispute due to undefined components. In the 2000s, TAC mostly equal or higher than infrastructure maintenance and operation or IMO [5]. IMO is the cost of railway infrastructure both for maintenance (e.g. Tracks, ballast) and operation (e.g. human resource) [6]. Considering TAC equal to IMO, thus operator budget from government support to maintain the infrastructure unable to cover the cost of the use of infrastructure. This condition makes a significant decline in term of service for users.

Other developed countries have been successfully improving their respective railway services by implementing proper rail infrastructure charge such as Germany, France, UK and others [7-8] as well as Australia [9]. Considering the needs to create better services in railway infrastructure in Indonesia, this research will examine the conceptual framework of track access charge by evaluating components in current TAC scheme.

It is expected that this research will produce a robust framework that can be used by decision makers to increase the quality services for public interest and also to attract more private sector investment in the Indonesia's railway sector.

2 Experience from other countries

Western Europe countries such as United Kingdom, France, and others, mostly use Marginal Cost (MC+) method for their rail infrastructure charge. In contrary, Full Cost (FC) method commonly implemented by Eastern Europe Countries such as Latvia, Hungary, and

* Corresponding author: maberawi@eng.ui.ac.id

others. In the western part of Europe, only two countries that use this system; Germany and Italy. Both of them will be included in the case study to gain more insight in determining suitable FC-method for Indonesia's rail infrastructure charge.

2.1 Germany

Deutsche Beteiligungs AG (DBAG) is one of Germany's leading private companies that run railway networks in the country. They oblige to pay for track access charge which covers 60% from railway infrastructure cost. Germany track access charge components consist of following items: basic price, product factor performance – based component and load component. More detail can be seen in figure 1 [10].

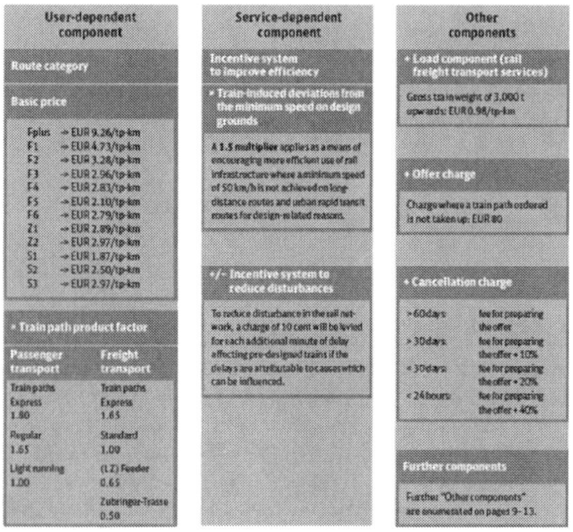

Fig. 1. Structure of the train path pricing system in Germany

Basic price is categorized by using long distant lines (Fplus to F6), feeder lines (Z1 and Z2) and urban lines (S1 and S2). The value varied from 1.87 up to 9.26 €/train km. Nonetheless, F5 has a limitation in the speed for only maximum 120 km/h that runs for inter – regional area with lower traffic. S1 is an exclusive track where S-Bahn train is operated. Compared to that, S2 is a network that connects S-Bahn Berlin and Hamburg [10]. For further calculation, the basic price will then multiply by product factor that considers whether the train categorized into passenger or freight train [11].

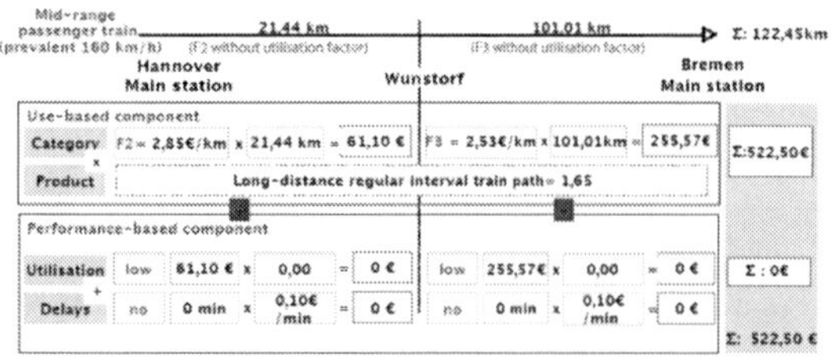

Fig. 2. Cost calculation of medium passenger train in Hannover – Bremen route

In term of performance – based component, it was formed through sub – categories including utilization, and probability of delays. Last, load factor is mainly used for the freight train. 0.92 €/train km will be charged in the condition where freight train surpasses maximum tonnage threshold in 3,000 tons. A sample of track access charge calculation in Germany can be seen in figure 2 [12].

2.2 Hungary

The Hungarian State Railway (MAV Co.) managing Hungarian railways which serve 7 million train passenger and 15 million freight train. Track access charge formula in this country is categorized into three travel distances from long distance passenger (LDP), short distance passenger (SDP) and freight transport (F). The operator also mandates to pay the cost for the quality of service in particular line sections. It comprises of three categories:

- 1st-category (advanced) lines [international routes (corridors)]
- 2nd-category (normal) lines [domestic mainlines]
- 3rd-category (local) lines [regional (branch lines)]
- The basic charge formula for track access charge in Hungary can be seen in table 1 [13]

Table 1. Hungary's basic charge formula for TAC

Line category	1st category International lines (il)	2nd category Domestic lines (dl)	3rd category Regional lines (rl)
Long distance passenger traffic (ldp)	$BC_{ldp,il} = BC_{fixed,il} + BC_{variable;ldp,il}$	$BC_{ldp,dl} = BC_{fixed,dl} + BC_{variable;ldp,dl}$	$BC_{ldp,rl} = BC_{fixed,rl} + BC_{variable;ldp,rl}$
Short distance passenger traffic (sdp)	$BC_{sdp,il} = BC_{fixed,il} + BC_{variable;sdp,il}$	$BC_{sdp,il} = BC_{fixed,dl} + BC_{variable;sdp,dl}$	$BC_{sdp,rl} = BC_{fixed,rl} + BC_{variable;sdp,rl}$
Freight traffic (f)	$BC_{f,il} = BC_{fixed,il} + BC_{variable;f,il}$	$BC_{f,dl} = BC_{fixed,dl} + BC_{variable;f,dl}$	$BC_{f,rl} = BC_{fixed,rl} + BC_{variable;f,rl}$

Table 2. Simulation for Budapest – Austria passenger train

Item	Amount	Unit	Price	Total
Path Allocation	1	Path	23.2	23.2
Station Usage	1	Event	27.	55.6
Stops	4	Stop	13	52
Train Running	270	Train.km	2.3	431.48
Overhead Catenary Usage	270	Electric. train.km	0.5	93.8
Total				€ 656.08

Furthermore, following simulation will visualize the implementation of infrastructure charge that has been conducted in Hungarian Railways Infrastructure [14].

2.3 Italy

Italy has separated its railway in a vertical way from infrastructure manager (*Rete Ferroviaria Italiana SPA* (RFI) and operators (5 passenger operator companies and ten freight operator companies). Railway in Italy differs from node and network. Network consist of trunk network and complementary/secondary network (secondary, scarce traffic and shuttle lines).

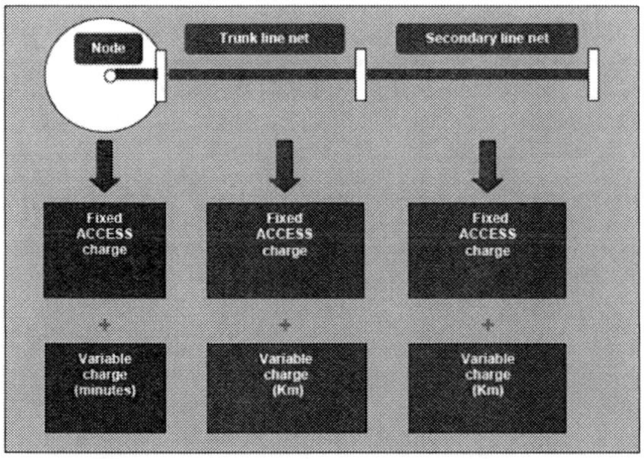

Fig. 3. Track access charge of network and node in Italy

Table 3. TAC comparison among Germany, Hungary and Italy

Countries	Germany	Hungary	Italy
% Total fees charged	60	80	16
Method	FC-	FC	FC-
Contribution	Federal or Local support for investment	Direct support from the state	State pay for investment to operation. User pays for smaller part of service cost
TAC Approach	Simple	2 part tariff	2 part tariff
Categorization between Passenger and Freight	Y	Y	Y
Traffic considered in calculation	Y	N	Y
Categorization through speed	Y	Y	Y
Capacity is being charged	N	Y	N
Delays	N	N	N

Every node and network will have a fixed access charge and variable charge. In node, the variable charge will consider time (minutes) as a multiplying factor. It divided into three categories of time: 22.00 – 06.00, 06.00 – 09.00 and 09.00 – 22.00. Moreover, variable charge in the network will consider the travel distance and expressed by km. More detail about track access charge in Italy can be seen in Figure 3 [15].

Based on previous explanation about the comparison of track access charge system in three countries (Germany, Hungary, and Italy). Therefore a comparison can be derived as follows.

3 Current Indonesia's track access charge

The payment method of existing rail infrastructure in Indonesia has been regulated through Minister of Transportation Decree No 62, 2013 that contain a calculation of railway infrastructure usage. It has been agreed by both government and KAI – a state-owned enterprise that acts as sole the operator.

Several aspects have been considered to determine the payment method including operational and maintenance; the quality of the infrastructure, and operator capability. The formula expressed as follows:

$$TAC = (IO + IM + Depreciation) \times f \quad (1)$$

where:

TAC = Railway infrastructure charge

IO = Infrastructure Operation cost

IM = Infrastructure Maintenance cost

f = Coefficient

Infrastructure depreciation value is annualized according to its standard life cycle. Meanwhile, Coefficient value in the above formula determined based on government policy for basic use of infrastructure. The magnitude of this factor ranging from 0.6 for a passenger train to 0.9 for a freight train. The detail of priority factors are classified using the following type of classes:

- Economic class: 0.6
- Business class: 0.7
- Executive class: 0.9
- Freight train: 0.9

According to reference [16], one of the reasons why Indonesia's government yet to produce a solid framework of railway infrastructure charge is because the coefficient value justification remains unclear. No standardized coefficient makes track access charge value tend to exceeds infrastructure maintenance cost. Therefore, the operator was at disadvantage position and possible to leads into company shut down due to declining revenue.

4 Proposed rail charge framework

Proposed rail charge framework will evaluate existing Indonesia's rail charge by adopting principles and formula of TAC in Germany. It will use a full cost (FC) principle where all infrastructure cost that being used must be paid by the operator that runs the network. However, despite using single tariff through a train – km approach, proposed rail charge will use a single tariff with a gross – ton calculation.

This approach is more favorable to be applied because freight train playing a significant role in degrading rail infrastructure in Indonesia compared to the passenger train. According to the reference [17], a gross ton-km approach is a right choice for a country that

DOI: 10.1051/matecconf/201710105002

still develops its railway infrastructure formulation into a robust framework. The formula of proposed rail charge can be seen as follows:

TAC = {Base charge x product factor x multiplicative surcharge} + Load Component (2)

Base charge will consider the value of station maintenance, rail infrastructure maintenance, bridge, signal and telecommunication, planning and monitoring cost, maintenance and operating personnel cost, general cost, and depreciation, weight, and length of the track.

The length of corridor derived from all active railway network for every Daop/Divre (kilometer) and also put aside the classification of speed and rail traffic. The type of railway divided into passenger train (executive, business, and economy) and a freight train.

Meanwhile, the multiplicative surcharge will be involved in the calculation to evaluate the use of diesel locomotive in Indonesia railway network. The coefficient expected to reduce pollution and as a means to introduce more sustainable locomotive such as an electrical train. Multiplication among base charge, product factor, and a multiplicative surcharge will produce basic track access charge.

On the other hand, the variable will consider load component that applied for trains that exceed 3,000 ton. Every gross – ton kilometer (GT-KM) will charge around 15,514 rupiah or equal to US$ 1.07. It calculated through a benchmark process with converted currency exchange. This charge aims to cover additional maintenance cost in railway infrastructure from freight train excessive load which occurs multiple times.

Compared to other countries that implement rail infrastructure charge using full cost method, the proposed components for Indonesia's track access charge exclude several items such as incentive system, offer charge and cancellation charge.

Offer charge is a condition where the operator decline to use the track that given by the infrastructure provider. Since Indonesia's still use a single operator, this variable yet to be used. However, it may be possible to be applied for new railway track to produce competition among railway operators.

The following table is summarizing the components of proposed track access charge for Indonesia railway transportation.

Furthermore, considering table 4 and TAC formulation, following flowchart will summarize the proposed calculation of track access charge for Indonesia railway transportation.

Table 4. Proposed TAC components for Indonesia railway

Component	Applied in Indonesia
Product Factor	✓
Multiplicative Surcharge	✓
Incentive System	×
Load Component	✓
Offer Charge	×
Cancellation Charge	×

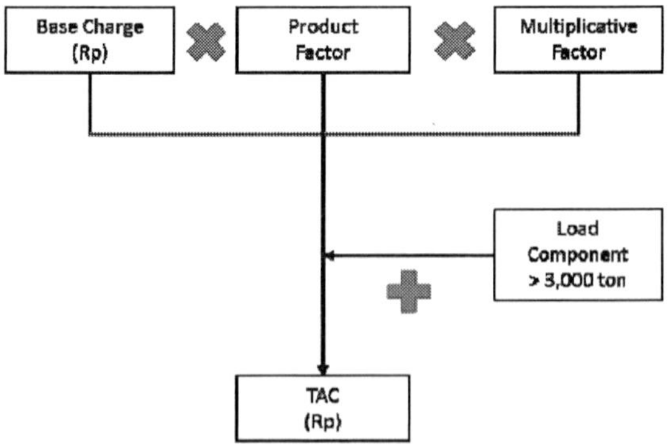

Fig. 4. Flow chart of TAC calculation

Previously, TAC – IMO in Indonesia for 2011 period are about Rp 2,046,095,743,034.44 and Rp 1,354,687,243,847.58 respectively. Using the interpolation from the increasing of TAC – IMO during the past five years (5%), track access charge for Indonesia in 2017 is estimated around 2.7 trillion rupiahs.

5 Government support

Since the proposed track access charge requires an enormous amount of cost to be paid by the operator, the government needs to support with a subsidy to produce a competitive ticket price for users. Compared to other countries, their rail charges are lower than its infrastructure maintenance and operation due to government support by reducing its TAC.

For instance, Germany's operator pays for 60% from total TAC; Italy pays for 16%, Slovenia around 13% and Romania with 52% from total TAC. However, there are also countries where TAC excluded from the government support such as Latvia and Estonia.

Using the figure above where the value is ranging from lowest percentage of TAC around 20% (Belgium) to 100% (Latvia and Estonia), it concludes that there is no single formulation about the government support in reducing the cost of access rail charge. It depends on various factors from the availability of data, economic growth, railway competitiveness, political factors and others.

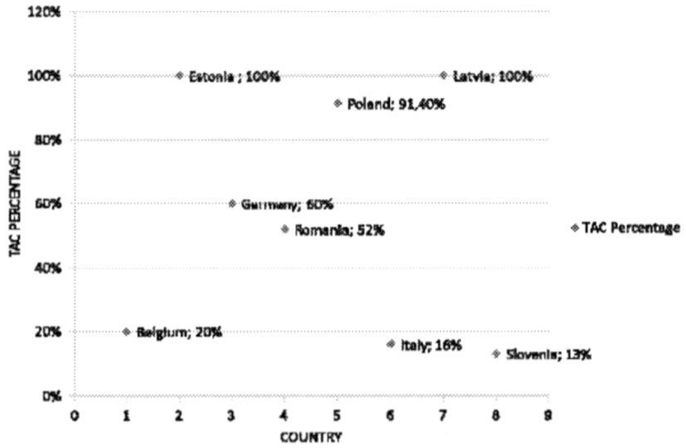

Fig. 5. TAC percentage in European countries

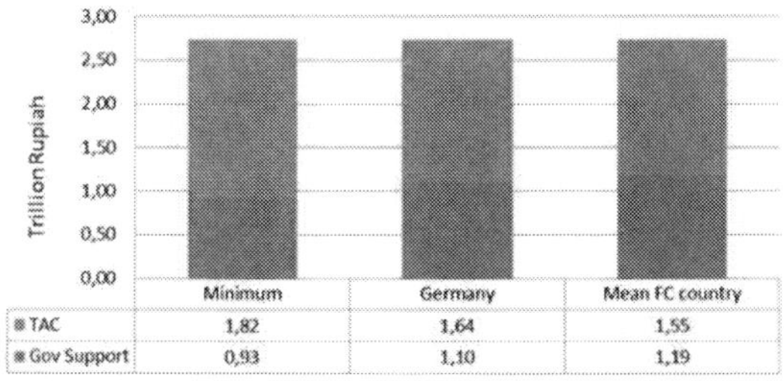

Fig. 6. TAC and government support for Indonesia railway industry

As the comparison between TAC and government support from Germany and other countries that use FC concept, the suitable TAC value for the railway's operator ranging from 56% to 66% from total proposed payable TAC, while government subsidy will cover the rest of 34% to 44%.

6 Conclusion

This research produces alternative track access charge calculation for Indonesia's railway infrastructure. It uses a modified formula from Germany track access charge by considering the local context in Indonesia. The components will involve the base charge, product factor, multiplicative surcharge and load component.

The track access charge by using this approach produce a TAC and IMO values for about 2,046,095,743,034.44 rupiahs and 1,354,687,243,847.58 rupiahs respectively. Furthermore, to generate a competitive ticket price, the operator shall pay the TAC ranging from 56% to 66%. The other will be supported by the government for around 34% to 44% of the overall TAC price.

This research is fully supported by research grant from Ministry of Research and Education, Republic of Indonesia.

References

1. M. Jordi, J. Transp. Geogr. **26,** 126-138 (2013)
2. N. Olsson, A. Økland, S. Halvorsen, Transport Policy **22**, 29-35 (2012)
3. J. Heather, F. Moura, T. Domingos, Procedia Soc. Behav. Sci. **111**, 400-409 (2014)
4. Ministry of Transportation Republic of Indonesia, Infrastructure Maintenance and Operation in Indonesia (2014)
5. A. Samosir, Railway Prospect: PSO-IMO-TAC (2011)
6. C. Nash, Research in Transportation Economics **22,** 1 (2008)
7. A. Zoeteman, European Journal of Transport and Infrastructure Research **1**, 4 (2001)
8. B. Szekely, International Journal of Sustainable Economy **1**, 2 (2009)
9. P. Mees, Transport Rev **25**, 4 (2005)
10. Deutsche Beteiligungs AG, The Train Path Pricing System of DB Netz AG (2013)
11. H. Link, International Journal of Transport Management **2**, 1 (2004)
12. P. Ronai, Rail charging and accounting schemes in Europe–Case studies for six countries, CER/EIM Brussel, 40 – 49 (2008)
13. K.Tánczos, F. Gyula, Periodica Polytechnica Transportation Engineering **28**, 1 (2000)
14. A. Dénesfalvy, G. Farkas, P. Rónai, Acta Technica Jaurinensis **2**, 3 (2009)
15. S. Impastato, M. Vivaldi, 9th Conference on Competition And Ownership In Land Transport, 15 (2005)
16. I. Muthohar, T. Sumi, Journal of Sustainable Development **3**, 3 (2010)
17. European Conference of Ministers of Transport, Railway Reform and Charges for the Use of Infrastructure. Paris: OECD (2005)

* Corresponding author: maberawi@eng.ui.ac.id

Study on effectiveness of flood control based on risk level: case study of Kampung Melayu Village and Bukit Duri Village

Mohammad Farid[1,*], *Harni* Harumi Pusparani[2,3], *Muhammad* Syahril Badri Kusuma[1], and *Suardi* Natasaputra[2]

[1]Water Resources Engineering Research Group, Faculty of Civil and Environmental, Institut Teknologi Bandung, 40132, Bandung Indonesia
[2]Professional Master Program in Water Management, Institut Teknologi Bandung, 40132 Bandung, Indonesia
[3]Ministry of Public Works and Housing, 40393 Bandung, Indonesia

Abstract. Jakarta, the capital city of Indonesia, experiences flood which causes activities disruption and losses almost every year. Many studies have been done to mitigate the impact of flooding. Most of them focus on reducing the inundated area as an indicator of the effectiveness of flood control. This study aims to evaluate the effectiveness of several flood control alternatives based on risk level reduction. The case study is located in Kampung Melayu Village and Bukit Duri Village which are densely populated with several economic area and almost every year experiencing severe flooding in Jakarta. Risk level analysis was carried out by the method based on guidelines issued by the Head of National Disaster Management Agency (BNPB) No. 2/2012 with few modifications. The analysis is conducted on five alternatives of flood control which are diversion channel to East Flood Canal (KBT), diversion channel to the West Flood Canal (KBB) in Pasar Minggu, river improvement, capacity enhancement on all components, and capacity enhancement focusing on one component. From the results, it is showed that enhancing capacity which focus on preparedness component by two levels are the best in terms of investment value to risk level reduction.

1 Introduction

Flood is identical with disaster due to its loss impact especially in urban areas. Special Capital Region (DKI) Jakarta is one of the most flood-prone area in Indonesia. In 2002, Jakarta experienced the first severe flood in the 21st century which caused widespread damage and affected more than 75% of the total population of the city [1]. Furthermore, based on the report from National Development Planning Agency (Bappenas) in 2007 [2], damages and losses caused by floods that occurred in 2007 were estimated to reach 5.16 trillion rupiah which almost 70% from this number was in economic sector. While on January 15, 2013, major floods in Jakarta and West Java had caused at least 40 people died, 45 thousand refugees, and considerable economic losses [3].

One alternative way in determining priorities and action of flood management is conducting risk analysis [4]. There are differences of risk determinant factors level of each area, whether it is the threats, vulnerabilities, or capacities. By developing the risk mapping of an area the various demands of many users including policy makers, can be fulfilled. Furthermore, variety of purposes such as for awareness building and for communication about the local hazard situation, are also served by the maps [5].

Recently, studies of flood management effectiveness based on risk level is not as much as studies that based on inundation (flood volume), whereas the inundation reduction is not necessarily proportional to the reduction of risk and loss in a region. The aim of this study is to analyse the effectiveness of several flood control efforts based on risk reduction by taking the case study in Kampung Melayu Village and Bukit Duri Village, Jakarta, which has predominance spatial function as settlements area with several points of economic and service.

2 Methodology

Flood events used as input in this study was obtained from previous study [6]. According to Jakarta river plan, design flood discharge is for return period of 100 years which is equal to 690 m^3/s.

Risk analysis was carried out according to General Guidelines for Disaster Risk Assessment issued by National Agency for Disaster Management (BNPB) [7] with some modifications considering the location of study. They are the accuracy level of analysis which is village (RW) based, flood classification, and weighting parameter of capacity index.

In this study, except for inundation map, the data sets are collected from institutions and field survey conducted on 9 to 22 November 2015. Some interviews with local people, local government, and the provincial government, represented by the secretary of the village chief and staff BPBDs DKI, are also conducted. The data sets are mainly used as an input to the analysis of

* Corresponding author: mfarid_si99@yahoo.com

population exposure index, losses index, and capacity index. Those data sets are demographic data, buildings and public facilities data, economic condition, capacity condition, and infrastructure plans for flood management in the study area [8-10].

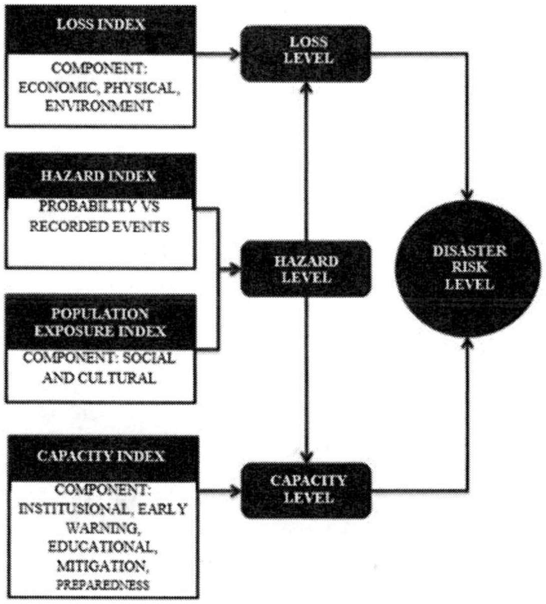

Fig. 1. Methodology of disaster risk assessment

Furthermore, the effectiveness of flood control alternatives were analysed by comparing the investment to risk reduction with also considering area location from the river. Investment which is taken into account is the cost for implementation phase without the cost for planning and maintenance phase. Investment for structural approaches includes the cost for construction and land acquisition, while the investment which are taken into account for non-structural approaches includes the procurement of goods and services, training, and system improvement/development.

3 Analysis

3.1 Hazard analysis

Hazard parameter indicated by hazard index is determined based on depth and area of inundation. Flood classification used in this study is following classification suggested by Sarminingsih in 2014 [11] which uses a combination of depth and speed of the flow, based on velocity data from BBWS Ciliwung Cisadane as in Table 1.

From the analysis of the hazard index that can be seen in Figure 2 that there are 13 RWs that have hazard index on 100 years return period flooding scenario, while there is no hazard index in RW 2, 3, 4, 5, 6, 7 and 8 Bukit Duri with the same scenario. The colour green, yellow, and red show the low, moderate, and high hazard index respectively.

Table 1. Flood hazard classification

Indicator	Flood hazard classification		
	Low	**Moderate**	**High**
Depth	D < 0,8 m	0,8 m < D < 1,2 m	D > 1,2 m

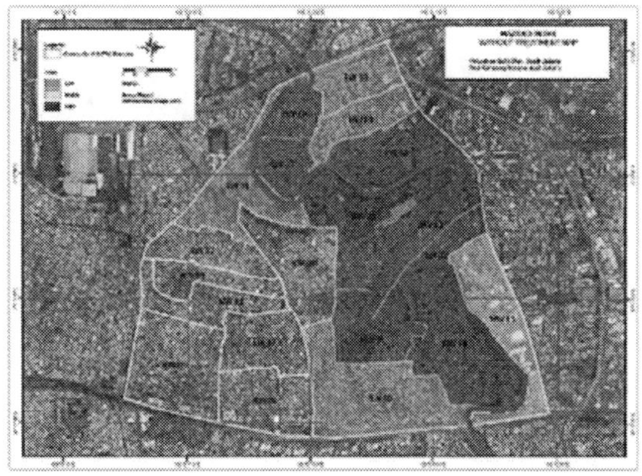

Fig. 2. Hazard index without any action in Kampung Melayu Village and Bukit Duri Village

3.2 Vulnerability analysis

Vulnerability parameter described in two indices that are exposed population index determined by social condition and losses index determined by physical condition and economic condition.

Social vulnerability variables that are considered in the analysis are population density, gender ratio, number of children and elderly citizens, and number of disabled/ill citizens. Physical vulnerability that counts is the density of houses, either permanent or non-permanent, buildings, public and critical facilities consist of religious facilities, educational facilities, medical facilities, electrical facilities, telecommunication facilities, and government/public service offices. Economic vulnerability variables that are taken into account in the analysis is trade sector (i.e. traditional market, mini market, and shop) and industrial sector.

In this analysis, classification of indices and weight factor for each vulnerability components are determined according to the guidelines issued by BNPB in 2012 [7]. While the reference for value per unit variables used in physical and economic vulnerability analysis are obtained from Damage and Loss Assessment, and Recovery Needs due to Jakarta Flood Occurred in January 2013 Report compiled by the Regional Disaster Management Agency (BPBD) Jakarta in 2013 [12]. For some values that are not listed in the report, they were determined from field survey.

Figure 3 and 4 show the analysis of exposed population and loss index. The colour green, yellow, and red show the low, moderate, and high of the index respectively. It can be seen that the exposed population index and losses index in most of the study area are in the moderate and high class. It confirmed the fact that the study area is a dense residential areas with a number of trade and industries points, both small and medium scale.

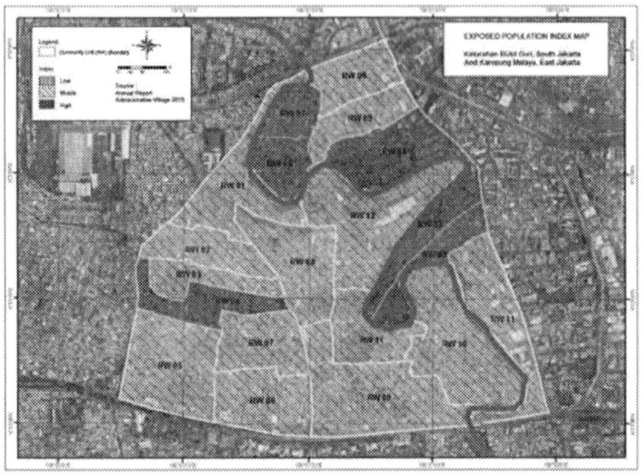

Fig. 3. Existing exposed population index in Kampung Melayu and Bukit Duri

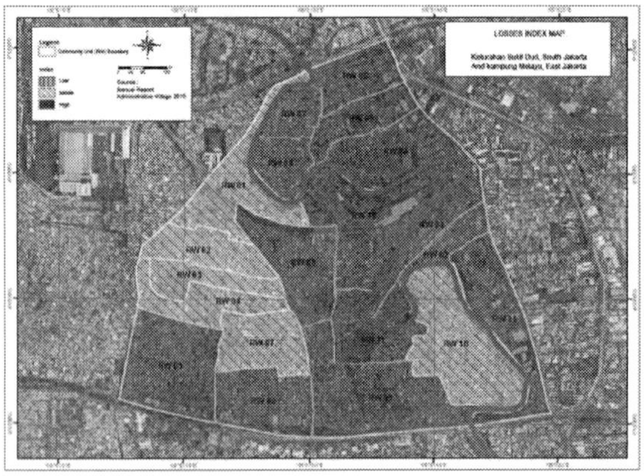

Fig. 4. Exsisting losses index in Kampung Melayu and Bukit Duri

3.3 Capacity analysis

Capacity index is determined by five components consisting of 22 achievement indicators based on Hyogo Framework for Action (HFA), according to the BNPB Head Regulation number 2 in 2012 [7]. Those five components are legal and institutional, early warning system and disaster risk assessments, disaster education, efforts for reducing basic risk factors, and preparedness development on all levels.

There are five levels of capacity depends on achievements and supporting system in order to reducing losses due to flooding. Table 2 shows the region capacity indicator levels.

A number of variables used in the assessment of the region capacity, e.g. availability and implementation of legal and master plan, flood control infrastructure condition, disaster response officer, early warning system, training and simulation, land use and spatial plan, access road condition, drainage, SOP for disaster response and also inventory of casualties and damage.

From the analysis of the capacity index that can be seen in Figure 5, the capacity index of Kampung Melayu is higher than Bukit Duri. The colour green, yellow, and red show the low, moderate, and high of the index respectively. Capacity index in the study area is mainly reinforced by early warning system, mitigation system at local level and citizens who are familiar with flood.

Table 2. Region capacity indicator levels

Level	Indicator			Capacity index
	Achievement	Legal policy	Commitment	
1	Small	No	No	Low
2	Sporadic	No	No	
3	Enough	Yes	Low	Moderate
4	Good	Yes	Enough	High
5	Comprehensive	Yes	Good	

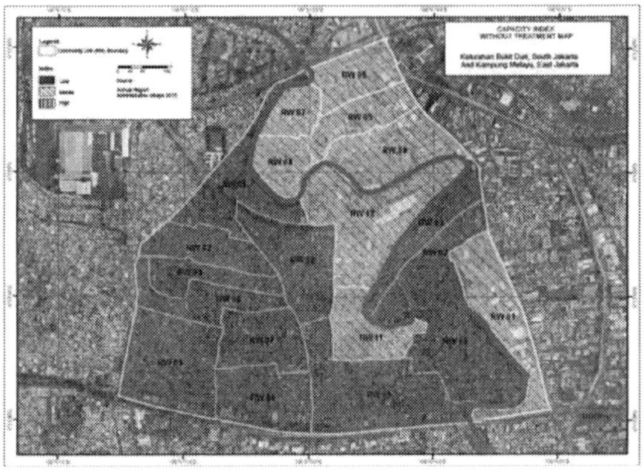

Fig. 5. Existing region capacity index in Kampung Melayu and Bukit Duri

3.4 Risk level analysis

Next process is analysing hazard level by taking hazard index and exposed population index into account. Then it will be input in losses index and capacity index analysis which product of these two indices is called risk level.

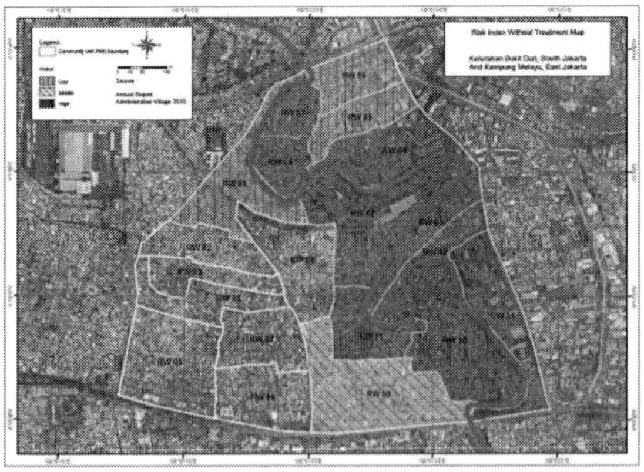

Fig. 6. Flood risk level without any action in Kampung Melayu and Bukit Duri

The risk level of low, moderate, and high is represented in the colour of green, yellow, and red respectively. From the analysis shown in Figure 6, it can be found that from 13 RWs in study area that has potential of being flooded, there are 9 RWs that has high risk level, which six of them were in Kampung Melayu, while the other three were in Bukit Duri. This condition were effect of the high hazard index and loss index due to the characteristics of the area which was densely covered by buildings.

3.5 Effectiveness analysis of flood mitigations

Considering the engineering process, action of flood control alternatives analysed in this study were those which aiming to hazard index reduction or capacity index enhancement. Those five alternatives to be analysed in this study consists of:

- floodway to East Flood Canal (KBT) with capacity design is about 90 m^3/s
- floodway from Pasar Minggu to West Flood Canal (KBB) with capacity design is about 260 m^3/s
- Ciliwung River improvement with the capacity plan of the channel is about 500 m^3/s
- enhancing capacity on all components by one point, and
- enhancing capacity on one component of parameters, i.e. preparedness, by two points

Floodway to KBT can reduce the inundation in Kampung Melayu and Bukit Duri by 5% and this quite affected the hazard index in five RWs [6]. However, at the end of the analysis it can be seen in Figure 7 that this scenario can only decrease risk level in two RWs, i.e. RW 1 Kampung Melayu and RW 11 Bukit Duri. Risk level in both RWs have decreased by two levels, from high to medium, while the rest remained in the same condition.

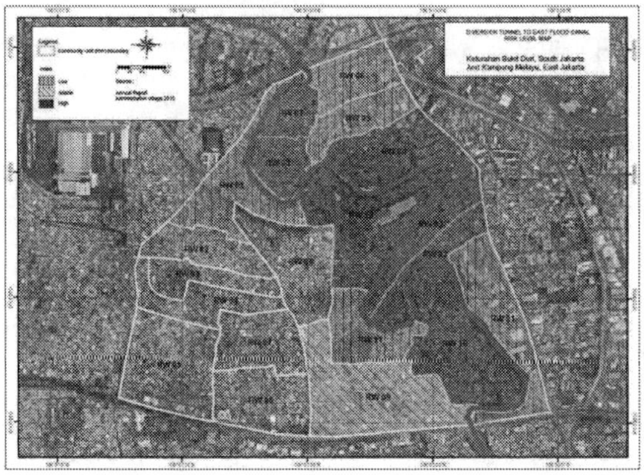

Fig. 7. Risk level in Kampung Melayu and Bukit Duri with floodway to KBT

In other hand, floodway to KBB can reduce inundation by 10%, based on the inundation map [6], two times bigger than floodway to KBT. But apparently this is not enough to reduce the risk level better than floodway to KBT does as can be seen in Figure 8. There are only two RWs that its risk level can be reduced while the rest remained the same.

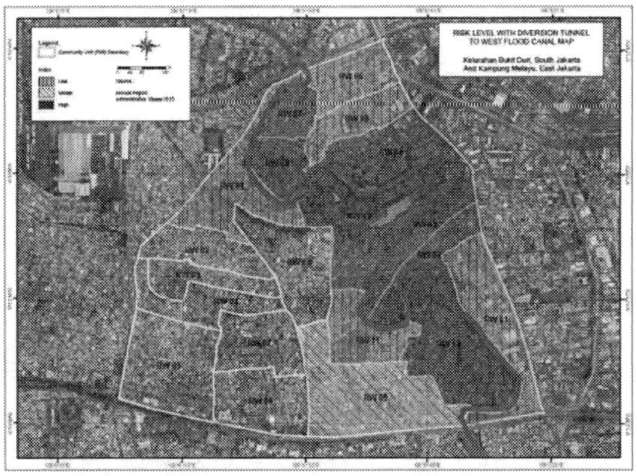

Fig. 8. Risk level in Kampung Melayu and Bukit Duri with floodway to KBB

River improvement as the third alternative gave better result than the previous two. From inundation map [6], it is known that there are 61% inundation reduction in the study area. This scenario could reduce the hazard index significantly as in Figure 9 which affected to risk level reduction. The analysis result showed that there are four RWs which their risk level decreased from high to low, i.e. RW 1 Kampung Melayu and RW 10, 11 and 12 Bukit Duri.

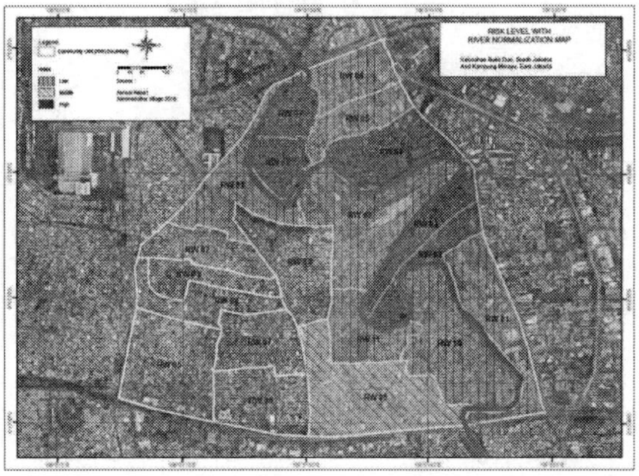

Fig. 9. Risk level in Kampung Melayu and Bukit Duri with river improvement

For capacity enhancement scenario, enhancing all capacity indicators by one point would affected the average value of capacity index. In the study area, as can be seen in Figure 10, after the enhancement of all capacity indicator, the capacity index of all RW's increased, from low to medium, or medium to high. However, it turns out that there are only two areas which the risk level can be reduced with this scenario, i.e. RW 1 Kampung Melayu and RW 9 Bukit Duri.

Last scenario alternative in this study was enhancing one component of capacity index by two points. The component that be enhanced is preparedness. With this scenario, there is a change in the capacity index of seven

RWs located in riverside. From the analysis as in Figure 11, it is known that there are only two RWs that the risk level could be decreased, i.e. RW 1 Kampung Melayu and RW 9 Bukit Duri. Meanwhile, there are still eight RWs with high risk level.

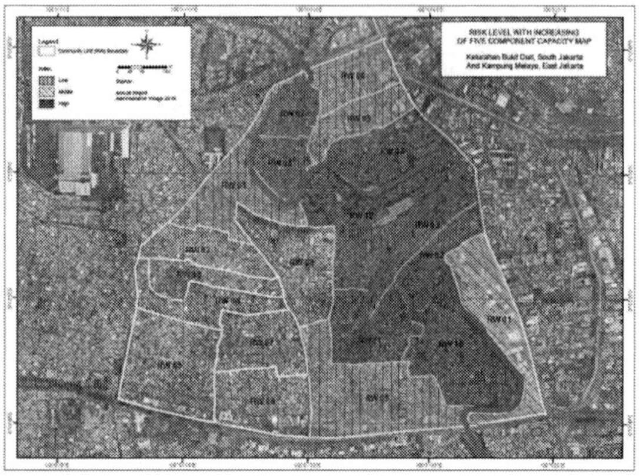

Fig. 10. Risk level in Kampung Melayu and Bukit Duri with enhancing capacity on all components by one point

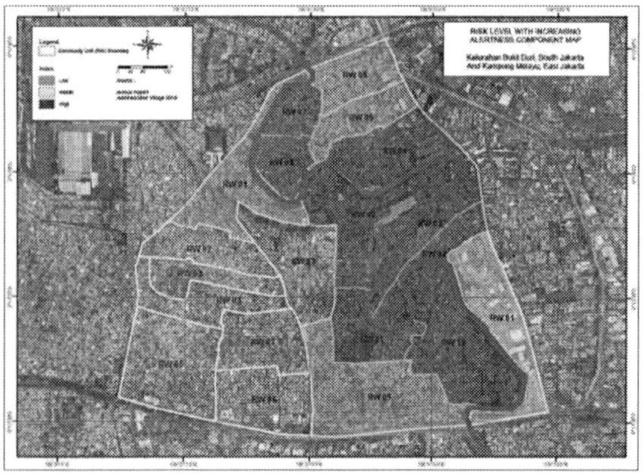

Fig. 11. Risk level in Kampung Melayu and Bukit Duri with enhancing capacity on preparedness components by two points

The effectiveness of each alternative is analyzed by comparing the investment to risk level and considering the location from the river. The investment value of floodway to KBT and river improvement are obtained from BBWS Ciliwung Cisadane River Agency (BBWS) in 2008 [13], the investment value of floodway to KBB is assumed to be proportional to investment of floodway to KBT by the channel length, while investment of capacity enhancement was referring to BPBD DKI [12], that was determined by Rp. 10 billion/level/component / year for 5 years of implementation.

The effectiveness analysis is also done by evaluating each region depends on the risk level reduction and location from the river. From these results as in Table 3, it is known that the value for flood management by river improvement could reduce the risk level of a region by 43.24%, most than other scenarios. Scenario with floodway to KBT and floodway to KBB gave the same result, and it was more than result the capacity enhancement scenarios could give. These results also indicate that, in general, the effect of capacity enhancement, whether on all components or focusing on one component, was not very much different one to another. In general, those five alternative scenarios did not give a very significant impact in the study area. Five of six RWs that had high risk level remained in the same conditions due to the loss index and population exposed index in these areas were high, so that more effort was needed to reduce the risk level. One alternative for this was to combining structural and non-structural scenario in the same time.

Table 3. Summary of investment cost, risk level reduction, and investment value for each scenario of flood control in Kampung Melayu and Bukit Duri

Scenario	Investment (Rp million)	Risk reduction	Investment to risk reduction value (Rp million)
Floodway to KBT	865,000.00	21.62%	4,000,625
Floodway to KBB	681,500.00	21.62%	3,151,938
River improvement	1,860,000.00	43.24%	4,301,250
All capacity enhancemet	300,000.00	5.41%	5,550,000
Preparedness improvement	120,000.00	5.41%	2,220,000

From these results, it can be concluded that even though scenario with river improvement gave the biggest effect both on inundation and risk level reduction, scenario to improve capacity on preparedness components by two levels gave the most effective results, considering its investment value.

4 Conclusions

Simulation using 100 years return period flood at the rate of 690 m^3/s in condition without addressing any action causes 62% of the Kampung Melayu and 31% of Bukit Duri area are inundated. This causes eight RWs in Kampung Melayu and Bukit Duri had hazard index in high category, one RWs in moderate category and four RWs in the low category.

The vulnerability of the area indicated by population exposed index and loss index in most of study area are in the category of moderate and high. The capacity index in order to cope with floods in eight RWs located in the riverside are in moderate category, while the other five are in the low category. The capacity index of other seven RWs located relatively far from the river are in the low category. There were 13 RWs in Kampung Melayu and Bukit Duri that had flood risk level. These 13 RWs were located in the riverside. From those 13 RWs, three RWs were in low level, one in moderate level and nine RWs in high level.

The impact of flood control management on inundation reduction were proportional with risk level reduction but not directly. The same results for non-structural scenario, which capacity index enhancement did not directly caused risk level reduction of an area. Preparedness component improvement scenario by two points were known to be the most effective of all scenarios analysis in this study due to its investment to risk level reduction, even though river improvement scenario gave the biggest impact on risk level reduction. In the case of regions with high vulnerability, in areas with high hazard index, scenario to reduce the inundation would give better effect to the risk level reduction. In the other hand, in regions with relatively low hazard index, non-structural scenario will give more significant risk level reduction than structural scenario. Flood control management by only targeting on one risk parameters would not give significant results. Because of that, for area with vulnerability that is difficult to changed, scenarios which were targeting on other two risk parameters, i.e. hazard and capacity components, were needed. For further development of this study, analysis which use vulnerability factor as function of inundation, whether depth, area, velocity, or duration, would give a more thorough result. Deeper analysis related to the weighting of each component of the capacity indicator is also needed as review for the BNPB Guideline number 2/2012 about regional capacity assessment guidance for disaster management. Furthermore, benefit analysis of a disaster management needs to take all factors that are included in investment and profits into account, such as maintenance and cycle life of the infrastructure. In areas with high vulnerability like in study area, non-structural scenario which directly connected with citizens live in the area, e.g. to increase awareness and preparedness of society needs to be an alternative for flood management, especially during structural action still has not functioned yet.

Authors would like to express their gratitude to the Ministry of Public Works and Housing Republic of Indonesia and Water Resources Engineering Research Group of Institut Teknologi Bandung for their support in this study.

References

1. M. Farid, A. Mano, K. Udo, J. Disaster Res., **7(5)**, 554-559 (2012)
2. Badan Perencanaan Pembangunan Nasional (Bappenas), *Keterangan Menteri Negara PPN/Kepala Bappenas tentang Laporan Awal Penanganan Bencana Banjir di Wilayah Jakarta dan Sekitarnya*, Bappenas, Jakarta, Indonesia (2007)
3. S. Kure, M. Farid, Y. Fukutani, A. Muhari, J.D. Bricker, K. Udo, A. Mano, J. Japan Society of Civil Engineers Series G, **70(5)**, 211-217 (2014)
4. M.S.B. Kusuma, H.P. Rahayu, M. Farid, M.B. Adityawan, T. Setiawati, R. Silasari, Jurnal Teknik Sipil ITB, **17(2)**, 123-134 (2010)
5. M. Hagemeier-Klose, K. Wagner, Nat. Hazards Earth Sys., **9**, 563-574 (2009)
6. S. Ginting, *Tesis Magister Pengelolaan Sumber Daya Air: Kajian dan Efektivitas Pengendalian Banjir di DKI Jakarta*, Fakultas Teknik Sipil dan Lingkungan ITB, Bandung, Indonesia (2015)
7. Badan Nasional Penanggulangan Bencana (BNPB), Pedoman Umum Pengkajian Risiko Bencana, *Peraturan Kepala*, **2** (2012)
8. Biro Pusat Statistik (BPS) Kota Administrasi Jakarta Selatan, *Statistik Daerah Kecamatan Tebet 2014*, BPS, Jakarta Selatan, Indonesia (2015)
9. Biro Pusat Statistik (BPS) Kota Administrasi Jakarta Timur, *Jatinegara dalam Angka 2014*, BPS, Jakarta Timur, Indonesia (2015)
10. Biro Pusat Statistik (BPS) DKI Jakarta, *Jakarta Dalam Angka 2014*, BPS, Jakarta, Indonesia (2015)
11. A. Sarminingsih, *Disertasi Doktor Teknik Sipil: Model Spasial Analisa Resiko sebagai Alternatif Penentuan Kriteria Sistem Pengendali Banjir: Studi Kasus DAS Citarum Hulu*, Fakultas Teknik Sipil dan Lingkungan ITB, Bandung, Indonesia (2014)
12. Badan Penanggulangan Bencana Daerah (BPBD) DKI Jakarta, *Penilaian Kerusakan dan Kerugian, serta Kebutuhan Pemulihan Pasca Banjir di Jakarta pada Januari 2013*, BPBD, Jakarta, Indonesia (2013)
13. Balai Besar Wilayah Sungai (BBWS) Ciliwung Cisadane, *Laporan Akhir Pekerjaan Perencanaan Teknis dan Detail Desain Penataan Sungai Ciliwung dari Pintu Air Manggarai ke Hulu*, BBWS, Jakarta, Indonesia (2008)

* Corresponding author: mfarid_si99@yahoo.com

DOI: 10.1051/matecconf/201710105004

Saturated and unsaturated stability analysis of slope subjected to rainfall infiltration

Nurly Gofar[1,*] and *Harianto* Rahardjo[1]

[1]School of Civil Engineering, Nanyang Technological University, 50 Nanyang Avenue, 639798 Singapore

Abstract. This paper presents results of saturated and unsaturated stability analysis of typical residual slopes subjected to rainfall infiltration corresponds to 50 years rainfall return period. The slope angles considered were 45° and 70°. The saturated stability analyses were carried out for original and critical ground water level commonly considered by practicing engineer. The analyses were conducted using limit equilibrium method. Unsaturated stability analyses used combination of coupled stress–pore-water pressure analysis to evaluate the effect of rainfall infiltration on the deformation and transient pore-water pressure on slope stability. Slope stability analyses were performed at some times during and after rainfall infiltration. Results show that the critical condition for slope made by sandy material was at the end of rainfall while for clayey material was at some specified times after the rainfall ceased. Unsaturated stability analysis on sandy soil gives higher factor of safety because the soil never reached saturation. Transient analysis using unsaturated soil concept could predict more critical condition of delayed failure of slopes made up of clayey soil.

1 Submitting the manuscript

Many residual slopes in tropical region with very deep ground water table stand at a steep angle. This is due to the strength of unsaturated soil which is governed by two independent stress-state variables: matric suction (negative pore-water pressure) and net normal stress [1]. The stability of the slope could be disturbed by rainfall infiltration through the slope surface. Rainwater infiltration into the unsaturated soil decreases the matric suction, hence affecting the shear strength properties and thus the probability of slope failure. Furthermore, heavy downpour may cause raising ground water level (GWL), development of a perched water table and positive pore-water pressure in the soil slope. This condition will further decrease the shear strength of soil and eventually make the slope increasingly susceptible to failure.

Although the contribution of soil suction to the shear strength of unsaturated soils has been increasingly accepted, the inclusion of soil suction in the long term slope stability analysis is still not a common practice [2]. Practicing engineers found that it is impractical and risky to perform such rigorous and time consuming analysis. Conventional slope stability analysis is usually performed for saturated condition, hence the effect of suction is not considered. Effect of rainfall is considered by assuming raise in GWL to the assumed most critical condition. In Singapore, the most critical condition is considered by bringing up the actual GWL below crest by 1 m, positioning the GWL at ground surface (GS) at the toe and connecting the two lines by a straight line. This practice is based on the ground water fluctuation in Singapore which is about 1 m [3].

The stability of slopes is generally assessed using limit equilibrium method where Factor of Safety (FoS) is estimated as the ratio of the soil strength and driving stress acting along the potential failure surface. Many computer programs for slope stability analysis (e.g. SLOPE/W) were developed based on limit equilibrium approach and are available in the market. The limit equilibrium method treats the soils as rigid plastic materials. The method does not take into account the effect of slope deformation induced by stress. Hence, finite element software (e.g. FLAC, PLAXIS, SIGMA/W) have been utilized by practicing engineers and researchers to assess the stability of the slope through deformation analysis and the FoS was obtained as strength reduction factor [4-6]. Similarly, pore-water pressure variation induced by rainfall infiltration is not considered in the current practice using limit equilibrium method. Rainfall infiltration influences the strength; thus soil resistance is not constant throughout the process of rainfall infiltration. The reduction of shear strength due to wetting trigger the shear deformation of slopes [7]. The effect of strength and modulus reduction due to rainfall infiltration could be considered if the concept of unsaturated soil was applied in the analysis [1].

Both effect of stress and pore water pressure should be incorporated in the slope stability analysis. The analysis should be carried out for pore-water pressure condition during and after rainfall because researchers [8-9] have suggested that delayed failure could occur in slope made of clayey soil due to low permeability.

* Corresponding author: nurlygofar@ntu.edu.sg

This study was aimed to show that, for some conditions, analysis using unsaturated soil concept could predict more critical condition to the slope. In this study, the transient analyses were performed using coupled stress – pore-water pressure analysis using SIGMA/W [10] to obtain deformation and pore-water pressure conditions at different times during and after a rainfall event. Slope stability analyses were performed using SLOPE/W [11] at some specified times during and after the rainfall event. FoS obtained from slope stability analysis were plotted with respect to time and compared to the factor of safety obtained from conventional analysis considering the original GWL and the assumed critical GWL.

2 Properties of unsaturated soil

Limit equilibrium method is widely used in practice for slope stability analysis due to its reliability for most practical cases. The FoS of a slope is assessed by the ratio of the resistance force (quantified by the shear strength of the soil) to the mobilized force. Thus, the shear strength of soil is an important parameter for the slope stability analysis. The shear strength is derived from Mohr-Coulomb failure criterion and Terzaghi's effective stress concept, defined as:

$$\tau_{ff} = c' + (\sigma_n - \mu_w)\tan\phi' \quad (1)$$

where, τ_{ff} is the shear stress at failure; c' is the effective cohesion and $\phi'$is the effective angle of shearing resistance with respect to net normal stress ($\sigma_n - \mu_w$).

For unsaturated soil, the water phase occupies only parts of the pore volume, while the rest is filled by air. Therefore, the main difference between shear strength of saturated soils and unsaturated soils is the definition for effective normal stress. A practical formulation for shear strength of unsaturated soil was proposed by [12]:

$$\tau_{ff} = c' + (\sigma_n - u_a)_f \tan\phi' + (u_a - u_w)_f \tan\phi^b \quad (2)$$

Where ϕ^b is the angle of frictional resistance with respect to matric suction ($\mu_a - \mu_w$). Eq. (2) shows that the shear strength of an unsaturated soil has a close relationship with pore water pressure (μ_w) and pore air pressure (μ_a) because the shear strength is governed by the wetted contact area of the soil particles. Fig. 1 shows Mohr-Coulomb representation of shear strength of unsaturated soil.

The ratio between the pore water and the pore air pressures is controlled by the amount of water in soil (water content) [13]. A fundamental property that relates the suction and water holding ability of soil is the soil water characteristic curve (SWCC) as shown in Fig 2. The SWCC curve can be divided into three regions i.e. boundary effect zone, transition zone and residual zone. The rate of desaturation with respect to an increase in matric suction is the greatest in the transition zone which occurs between the air-entry point (AEV) and the residual suction (θ_r). The variation of shear strength with respect to matric suction in the transition zone significant [1, 7], hence the definition of AEV and residual matric suction is very crucial in the estimation of shear strength of unsaturated soil. Theoretically, the unsaturated soil with matric suction less than AEV is deemed to behave like a saturated soil. Thus, the ϕ^b under this condition can be approximated as equal to ϕ' [7] Beyond the AEV, the ϕ^b of fine grained soil decreases to a lower value. When laboratory data is not available ϕ^b between 0.5 ϕ' and ϕ' could be used for analysis [14].

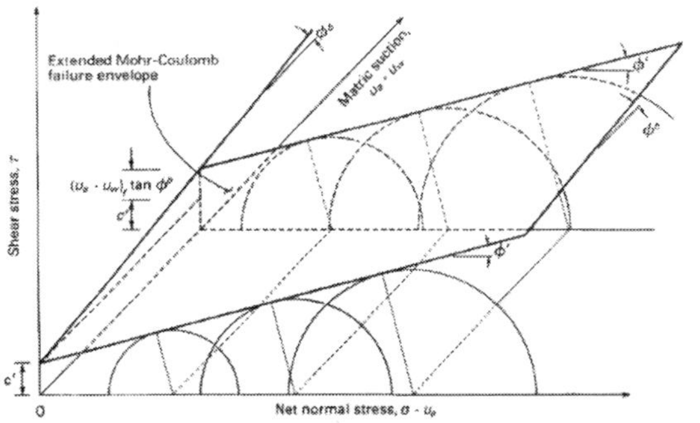

Fig. 1. Extended Mohr-Coulomb failure envelope for unsaturated soils [1]

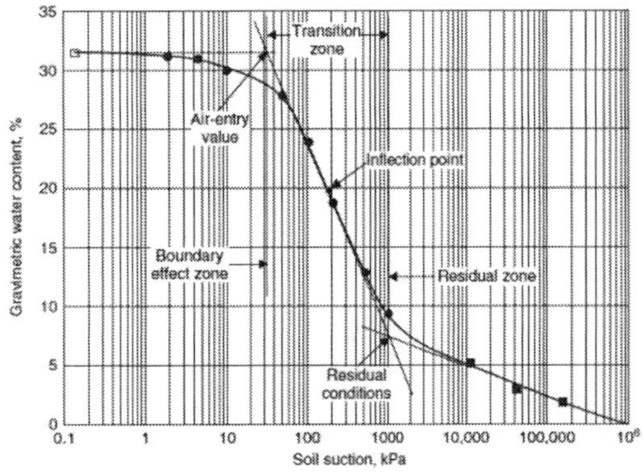

Fig. 2. Typical SWCC curve [1]

The elasticity form of the volumetric strain constitutive relationship for an unsaturated soil under plain strain condition as presented in [1] as:

$$d\varepsilon_v = \frac{2(1+\mu)(1-2\mu)}{E} d(\sigma_{ave} - u_a) + 2\frac{1+\mu}{H} d(\mu_a - \mu_w) \quad (3)$$

where $d\varepsilon_v$ is volumetric strain increment, $d(\sigma_{ave} - \mu_a)$ and $d(\mu_a - \mu_w)$ are the net normal stress and matric suction increment respectively. E is the elastic modulus with respect to change in net normal stress while H is the elastic modulus with respect to change in suction. The relationship between E and H for saturated soil in given in the following equation whereby ν is the Poisson's ratio [1]:

$$E / H = (1 - 2\nu) \quad (4)$$

DOI: 10.1051/matecconf/201710105004

The effect of suction on the *E/H* ratio was studied by [15] for Bukit Timah Granite in Singapore. The relation was obtained in this study:

$$E / H = (1-2v)\,[1-0.003(\mu_a - \mu_w)] \qquad (5)$$

3 Methodology

3.1 Slope model and analysis

In order to investigate the effect of infiltration on slope stability and to compare the results of unsaturated slope stability analysis with those of conventional analysis, a typical slope was selected in this study as shown in Fig. 3. Two slope angles and two types of soil were considered. The slope angles (α) were 45° and 70° while the soils were sandy and clayey soils represented by their shear strength, elastic properties, SWCC and permeability function. The slope was 4 m high. Slope of 45° was selected because a 1:1 slope is considered as stable for residual soil slope in Singapore, while slope of 70° is the maximum inclination of ground to be analyzed as a slope. The initial ground water level (GWL) condition is as shown in Fig. 3.

Conventional (saturated) stability analysis was carried out using SLOPE/W with ϕ^b equal to zero. The analyses were performed for the original GWL and critical condition by raising the GWL below crest by 1 m and the GWL at the toe side coincide with ground surface (GS).

The effect of rainfall infiltration on the slope stability was evaluated by applying a rainfall of 22 mm/h with duration corresponding to 50 year return period (RP) in Singapore i.e. 8 hour [16, 17]. The infiltration process of the rainwater into the soil was analyzed using stress-deformation finite element software SIGMA/W. The SIGMA/W is capable of performing seepage analysis coupled with stress/deformation analysis. Slope stability analyses were carried out using SLOPE/W for several conditions during and after rainfall.

The model adopted in this study was two-dimensional plane strain. The boundary conditions applied to the finite element analysis are illustrated in Fig. 3. The distance between the slope and the sides of the slope model was set to three times the height of the slope to eliminate the effect of boundary conditions. No flow boundary were simulated by assigning a nodal flux Q equal to zero at the bottom and sides of the slope model above GWL. A constant total head h_w was applied on each side boundary below GWL. Rainfall was applied to the slope as flux boundary q. Ponding was not allowed to occur on the slope surface. This meant that if a flux greater than the permeability of the soil was applied, the seepage model will not allow pore water pressure greater than 0 kPa to built-up. This simulates the actual condition whereby the excess rainwater on the slope would become surface flow or runoff. The initial condition of the slope was taken based on the position of the initial GWL as in Fig. 3. Unsaturated slope stability analyses were performed for seepage conditions during with an interval of 2 h and after rainfall with interval of 1 day.

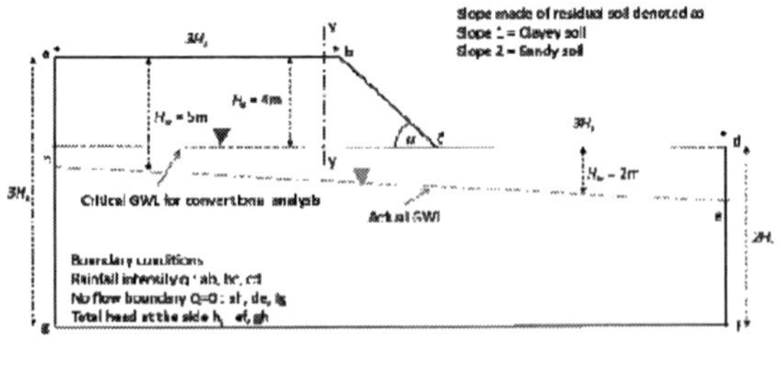

Fig. 3. Slope geometry and boundary conditions adopted in this study

3.2 Material properties

The slope was made of residual soil. Two types of soil were selected to represent sandy and clayey soils. The typical shear strength properties of Singapore residual soils as shown in Table 1 were selected for limit equilibrium analysis using SLOPE/W. The properties of clayey soil was based on [18] while for sandy soil, typical values were adopted.

Table 1. Soil properties

Soil	γ_b (kN/m³)	c' (kPa)	ϕ'	ϕ^b	*E* (kPa)	ν
Clayey	20	10	26	17	10,000	0.40
Sandy	20	5	33	23	15,000	0.33

SWCC and permeability functions are two primary soil properties used in the deformation-seepage analysis. The SWCC and permeability function used in this study was based on [18] to represent clayey ($f_{100,-6}$) and sandy soil ($f_{10,-4}$). The SWCC was plotted based on Fredlund and Xing (1994) with $c(\psi)$= 1 [19] while the permeability function was based on [20]. The SWCC and permeability function of the residual soil is shown in Fig. 4. The SWCC and permeability function parameters adopted in this study are given in Table 2.

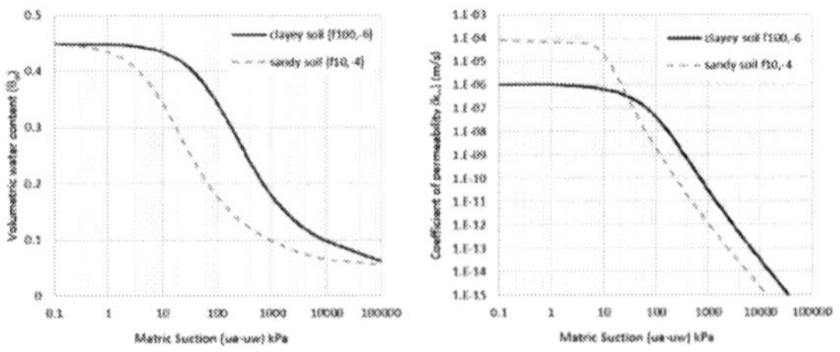

Fig. 4. SWCCs and permeability functions used in this study

Table 2. SWCC and permeability function parameters

Soil	θ_s	*a*	*n*	*m*	k_s(m/s)	*p*
Clayey	0.45	100	1	1	10^{-6}	4
Sandy	0.45	10	1	1	10^{-4}	4

4 Results and discussion

4.1 Unsaturated slope stability analysis

Unsaturated slope stability analysis was performed on Slope 1 and Slope 2 in three stages. First was the establishment of insitu stress for the slopes using SIGMA/W. Second was the application of 50 year RP rainfall (22 mm/h for 8h) using coupled stress – pore-water pressure analysis using SIGMA/W. Finally, slope stability analyses were carried out using SLOPE/W for stress and pore-water pressure conditions during and after rainfall.

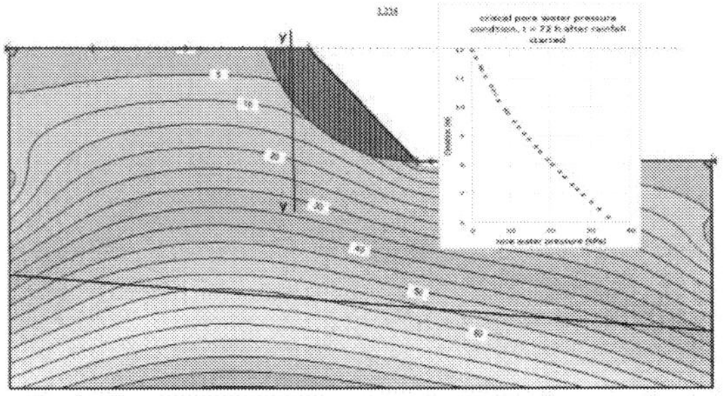

(a) Slope 1 (clayey soil)

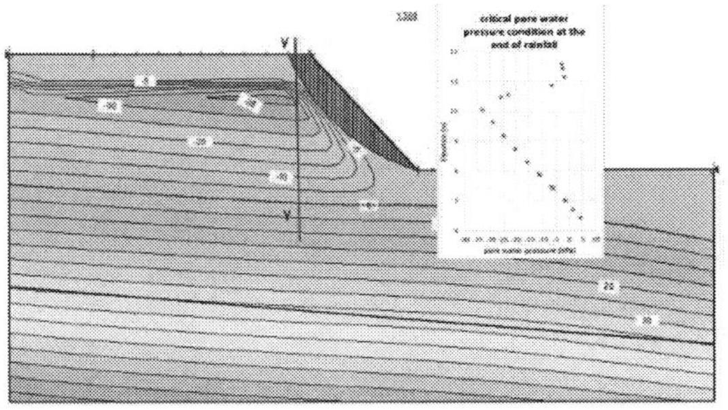

(b) Slope 2 (sandy soil)

Fig. 5. Critical pore water pressure variation due to rainfall infiltration and factor of safety for slope angle 45°

Analysis of Slope 1 (Clayey soil) for 70° and 45° slopes showed that the critical condition occurred 48 h and 72 h after rainfall respectively. In these times, the GWL has raised up to the GS. On the other hand, for Slope 2 (Sandy soil), the critical condition occurred at the end of rainfall and the GWL only raised slightly from the original location. The critical pore water pressure distribution along cross-sections y-y (in Fig. 3) are plotted in Fig. 5a and 5b for Slope 1 and Slope 2 respectively. The pore-water pressure contours shown in Fig. 5 indicate the position of GWL at critical condition.

The FoS of Slope 1 and 2 with slope angle of 45° and 70° are plotted with time in in Figure 6a and 6b respectively. It can be seen that the FoS of the slopes decreased during rainfall. For slope made of Sandy soil, the FoS reached minimum at the end of rainfall and increase again as the pore water pressure started to redistribute due to high coefficient of permeability. On the other hand, the FoS of the slope made of Clayey soil continued to decrease as the GWL continued to raise.

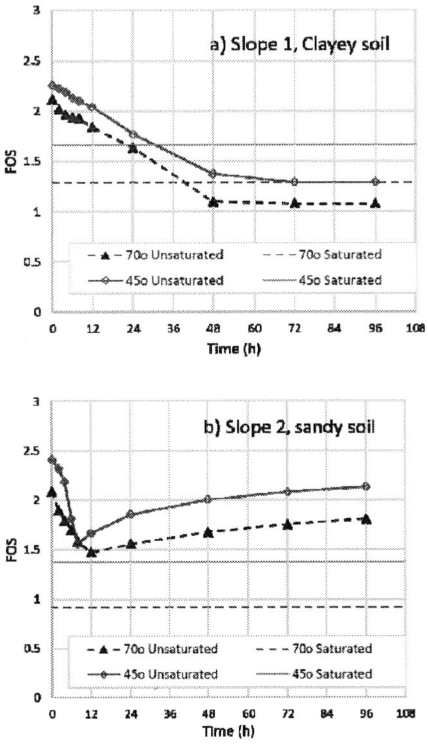

Fig. 6. Change in FoS with pore-water pressure and stress

4.2 Conventional slope stability analysis

The conventional stability analysis was carried out using SLOPE/W for Slope 1 and Slope 2 with original GWL as shown in Fig. 1 and critical GWL 1 m above the original GWL. Since both assumed GWL did not coincide with slip surface, the FoS obtained from both analyses are equal i.e. 1.661 and 1.285 for Slope 1 with slope angle 45° and 70° respectively. For Slope 2 with slope angle 45° and 70°, the FoS are 1.372 and 0.917 respectively. These values are also plotted in Fig. 6.

Table 3 Summary of FoS obtained from different methods

Slope	Saturated/conventional analysis for GWL			Unsaturated analysis	
	Original	Raise by 1m	at GS	Initial	Min FoS. Time in bracket
Clayey Slope 1, 45°	1.661	1.661	1.083	2.54	1.289 (72h)
Slope 1, 70°	1.285	1.285	0.732	2.12	1.075 (48h)
Sandy Slope 2, 45°	1.372	1.372	NA	2.409	1.668 (8h)
Slope 2, 70°	0.917	0.917	NA	2.090	1.458 (12h)

For slope 1, conventional slope stability analysis was also performed for condition where GWL as at the GS and the FoS were 1.083 and 0.732 for slope angle 45° and 70° respectively. Comparisons in the FoS obtained from different analyses are summarized in Table 3.

4.3 Effect of suction on shear strength and FoS

Slope stability analysis on four cases evaluated in this study indicated that unsaturated concept offer a more realistic estimate of the stability of slope. Rainfall infiltration does not only cause the raise in GWL as assumed in the conventional analysis, but also change in strength due to pore water pressure variation and deformation of soil.

The variation of shear strength with matric suction is illustrated in the following example for clayey soil. Assume a soil has a unit weight of 20 kN/m^3; shear strength parameters of c' = 10 kPa and ϕ' = 26° and ϕ^b = 2/3 ϕ' = 17°. Assuming the GWL is at 3 m depth, the hydrostatic shear strength at depth of 1 m based on Eq. (2) is equal to 25.9 kPa. If rainfall infiltration caused the GWL to rise to GS, then the shear strength decreased to 11.8 kPa. For the same case, Eq. (1) will give the initial shear strength of 19.7 kPa and final shear strength of 14.9 kPa. This shows that use of Eq (2) is not only results in higher shear strength due to suction but could also results in lower shear strength; hence lower factor of safety of the slope. Table 3 also shows that assuming GWL at GS for conventional stability analysis on Slope 1 gives a lower FoS as compared to the minimum FoS using unsaturated stability analysis even though the GWL was at the same position assumed in conventional analysis.

5 Conclusions

For slope made of coarse grained material, unsaturated stability analysis always give higher FOS than conventional analysis. The most critical condition is achieved a while after the end of rainfall.

For cases of delayed failure of slope made of clayey soil, unsaturated stability analysis could be more critical as compared to conventional analysis considering GWL raises about 1 m from original position (current practice). Conventional slope stability analysis using GWL coincide with GS give lowest FoS, however this consideration is over-conservative and not realistic, especially for sandy material whereby the effect of rainfall on GWL is minimum.

Unsaturated stability analysis of slope is a more realistic approach because it considers the actual GWL position as well as the change in shear strength due to deformation and variation of pore-water pressure variation.

The first author would like to thank Er. Lim Shiyi for discussion on state of practice on slope design in Singapore.

References

1. D.G.Fredlund, H. Rahardjo, *Soil mechanics for unsaturated soils* (John-Wiley & Sons Inc. New York,1993)
2. A. Kassim, N. Gofar, N.Mokhtar, M.L. Lee,*Proc. international conference on slope Malaysia*,1-12, 2006.
3. H. Rahardjo, A. Satyanaga, E.C. Leong, Y.S. Ng, ASCEJournal of Geotechnical and Geo-environmental Engineering, 136(11), 1555–1564(2010)
4. J.M. Duncan, ASCE Journal of Geotechnical Engineering, 122(7) 577-596(1996)
5. H. Jitno, N. Gofar,Malaysian Journal of Civil Engineering. 17 (1) 1-12(2005)
6. H. Khabbaz, B. Fatahi, C. Nucifora, *Proceedings of the 11th Australia New Zealand Conference on Geomechanics* (ANZ 2012) 1293-1298(2012)
7. S.G. Goh, H. Rahardjo, E.C. Leong, ASCE Journal of Geotechnical and Geo-environmental Engineering. 136(4) 594-606(2010)
8. P.R. Vaughan, H.J. Walbancke, *Géotechnique* 23(4) 531-539(1973)
9. E. Conte, A. Troncone, *Geotechnique* 62(1) 87-91(2012)
10. Geo-Slope International Ltd. SIGMA/W Stress deformation modelling. Calgary, Alberta, Canada (2012a).
11. Geo-Slope International Ltd. SLOPE/W for Slope Stability Analysis. Calgary, Alberta, Canada (2012b).
12. D.G.Fredlund, N.A. Morgenstern, R.A. Widger,Canadian Geotechnical Journal, 15, 3131-321(1978)
13. A.S. Muntohar, *Proc. 10th Indonesian Geotechnical Conf. and 19th Annual Scientific Meeting* (2015)
14. E.C. Leong, H. Rahardjo, S.K. Tang, *Proc. International Workshop on Characterization and Engineering Properties of Natural Soils.* Vol. 2. 1279-1304(2002)
15. H. Rahardjo, F. Melinda, E.C. Leong, R.B. Rezaur, *Stiffness of a Compacted Soil.* Engineering geology. 120, 60-67(2011)
16. Public Utilities Board. Code of practice on surface water drainage. Singapore Drainage Department (2009)
17. Q. Zhang, J. He, A.M. Hee, The IES Journal Part A: Civil & Structural Engineering, 5(1)62-64(2012)
18. H. Rahardjo, T.H. Ong, R.B. Rezaur, E.C. Leong, ASCE Journal of Geotechnical and Geo-environmental Engineering, 133(12)1532-1543(2007)
19. D.G. Fredlund, A. Xing, Canadian Geotechnical Journal. 31(3) 521-532(1998)
20. E.C. Leong, H. Rahardjo, ASCE Journal of Geotechnical and Geo-environmental Engineering. 123(12) 1118-1126(1997)

DOI: 10.1051/matecconf/201710105005

Parametric study on the effect of rainfall pattern to slope stability

Fathiyah Hakim Sagitaningrum[1,*] and *Erly* Bahsan[1]

[1]Department of Civil Engineering, Universitas Indonesia, Depok 16424, Indonesia

Abstract. Landslide in Indonesia usually occurs during the rainy seasons. Previous studies showed that rainfall infiltration has a great effect on the factor of safety (FS) of slopes. This research focused on the effect of rainfall pattern on the FS of unsaturated slope with different slope angle i.e.: 30°, 45°, and 60°. Three different rainfall patterns, which are normal, advanced, and delayed were considered in the analysis. The effects of low or high hydraulic conductivity of the soil are also observed. The analyses were conducted with SEEP/W for the seepage and SLOPE/W for the slope stability. It is found that the lowest FS for gentle slope is reached under the application of advanced rainfall pattern and the lowest FS for steep slope is reached under the application of delayed rainfall pattern. Reduction of FS is known to be the largest for gentle slope rather than steep slope due to negative pore water pressure reduction and the rising of ground water level. The largest FS reduction caused by rainfall was achieved for gentle slope under advanced rainfall pattern.

1 Introduction

Slope stability failure in Indonesia usually occurs mainly during the rain. BNPB (National Board for Disaster Management) mentions that out of 493 landslides occurred in Indonesia in 2011-2014, 286 of them were caused short duration, high intensity rainfall, and long duration rainfall. The rainfall infiltrated the slope through soil cracks, saturating the soil and induced landslide or slope failure [1].

The rainfall infiltration cause the reduction in matric suction. As a factor of an increase in shear strength. The reduction of matric suction decrease the shear strength and reducing the stability of a slope [2-3].

The stability of slope usually represented by the factor of safety (FS) value. For example, [4] mention that FS $\geq$ 1.5 can be used to represent that a slope is stable. On the contrary, FS $<$ 1.5 means that the slope will probably fail. Even though there is no such a unique correlation between the FS value and probability of failure [5], FS value is still favorable to predict the stability condition of slope by practical engineers.

This study will focus on the effect of different rainfall patterns to the factor of safety of slope. Nevertheless, the effect of slope's angle is also considered. Rainfall patterns will be taken form the higher five-days cumulative rainfall from Faculty of Engineering, Universitas Indonesia rain station from July 2014 until March 2015 data. The rainfall is then divided into three patterns, named normal, delayed, and advanced. Slope Geometry in this study is divided to three angles which are steep, gentle, and an angle between.

A finite element software from Geostudio, SEEP/W is used to analyse the effect of rainfall pattern to the groundwater level conditions. Another product from Geostudio, SLOPE/W, a limit equilibrium-based program, is used to analyse the factor of safety value of the slope regarding the groundwater conditions and the geometry of the slope.

The purpose of this study is to observed the effect of slope geometry and rainfall pattern to the slope's factor of safety and effect of pore water pressure change for slope's factor of safety.

2 Theoretical overview

2.1 Unsaturated soil

Unsaturated soil is a concept which stated that soil is not in a fully saturated condition. This concept explains that above the ground water level, the value of pore water pressure is not positive, but negative. The reason behind the negative value is due to the existence of matric suction where cohesion-adhesion reaction fills in between the soil particle because the void is filled with both air and water at the same time [3, 6-7].

Unsaturated soil consists of two components, which are matric suction and osmotic suction. While osmotic suction is a function of how many salts dissolved inside the soil void's fluid as pressure, matric suction is a deviation of air pressure inside the void with water pressure inside the void. This suction will then change the formulation of shear strength, seepage, and volume change of soil above the ground water level [7-8].

* Corresponding author: fathiyah,hakim@gmail.com

One of the distinct feature about unsaturated soil mechanics is the Soil-Water Characteristic Curve [9-10] which connects the function of volumetric water content and hydraulic conductivity with soil suction (matric suction).

2.2 Infiltration

Infiltration is related to rainfall, which usually defined as rainfall intensity (mm/hr). The factors that affect the infiltration rate are degree of saturation, compression of soil void, clogging void due to fine soils, vegetation, topography, and rainfall intensity [11].

Water will seep through soil and analysed as steady-state and unsteady-state, with the difference between the two analysis types is the change of hydraulic head and hydraulic conductivity because of time. Steady-state is an analysis which doesn't change through time and have a constant condition and vice versa for unsteady-state.

2.3 Factor of safety of slope

In slope stability typically factor of safety is formulated as the ratio between the resisting shear strength to the driving shear stress. In the unsaturated soil condition, the factor of safety can be written as [12]:

$$F_s = \frac{\tau_f}{\tau_m} = \frac{c' + (\sigma - u_a)\tan\phi' + (u_a - u_w)\left(\frac{\theta - \theta_r}{\theta_z - \theta_r}\right)\tan\phi^b}{\gamma_t z_w \sin\alpha\cos\alpha} \quad (1)$$

With Fs is the factor of safety, τ_f is the average shear strength, τ_m is the average shear strength with the potential to fail, α is the slope's angle, z_w is the vertical depth of unsaturated soil, γ_t is the unit weight of the soil, c' is the cohesion, σ is the total normal pressure, $(\sigma-u_a)$ is the effective normal pressure, $\emptyset'$ is the effective friction angle, and $\emptyset_b$ is the increasing friction angle due to the presence of matric suction.

2.4 Related studies

To know the differences between this study and the previous ones, a parametric study from [2] is taken as a reference. The study focuses on a typical Singapore slope geometry behaviour at three different rainfall patterns (normal, advanced, and delayed) at two different conductivity value, Low (1 x 10^{-6} m/s) and High (1 x 10^{-5} m/s).

At Low Conductivity (LC), the greatest reduction of FS is achieved at advanced pattern with the average of all patterns reduction about 40-45% from the initial FS. Different results shown at High Conductivity (HC) with the average reduction of 10-13% and the greatest reduction is achieved at delayed pattern. It is also known from this research that FS reduction is caused by increasing value of pore water pressure which decrease matric suction and rising ground-water level.

3 Research method

3.1 Soil parameters and slope geometry

The soil material in this study is taken to be a homogeny silty clay layer in an unsaturated condition of 80% water content. The parameters of the soil are assumed based on some literatures [13-14], shown in Table 1.

Table 1. Silty clay soil properties

Soil type	Silty clay
γ_{sat} (kN/m^3)	18
γ_d (kN/m^3)	15
c (kN/m^3)	25
φ (°)	20
Water content	70-80%
K_{sat} (m/s)	10^{-6}

The soil is at an unsaturated condition, which in SEEP/W, needed to input the property of volumetric water content and hydraulic conductivity function for matric suction. The function used in SEEP/W is Fredlund-Xing estimation from Silty Clay material inside SEEP/W itself. Without any residual water content, the relationship between volumetric water content and hydraulic conductivity with matric suction is as Fig.1.

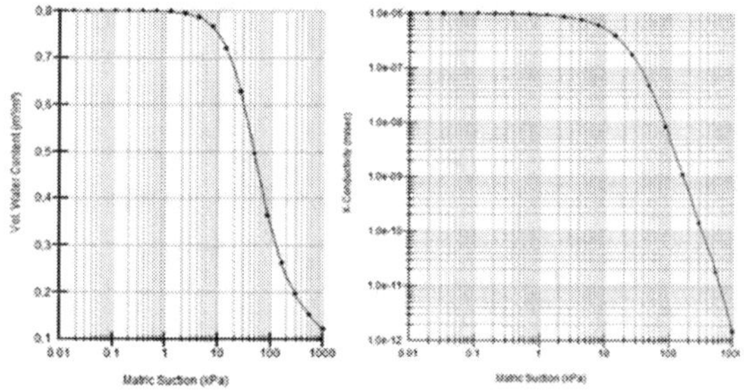

Fig. 1. Volumetric water content and hydraulic conductivity versus matric suction function SEEP/W

The slope height is taken to be 10 meters. The ground water level is assumed to be at the elevation of the slope's foot and the bedrock is at the elevation of 3 times the height under the slope's foot elevation, which is 30 meters under the slope's foot elevation. The geometry of the slope is varied by the slope's angle, which are 60° that represent the steep slope, 30° as the gentle slope, and the 45° consider as the in between.

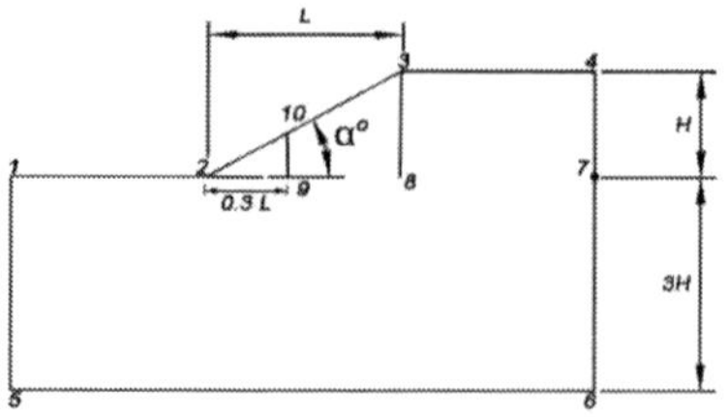

Fig. 2. Slope geometry

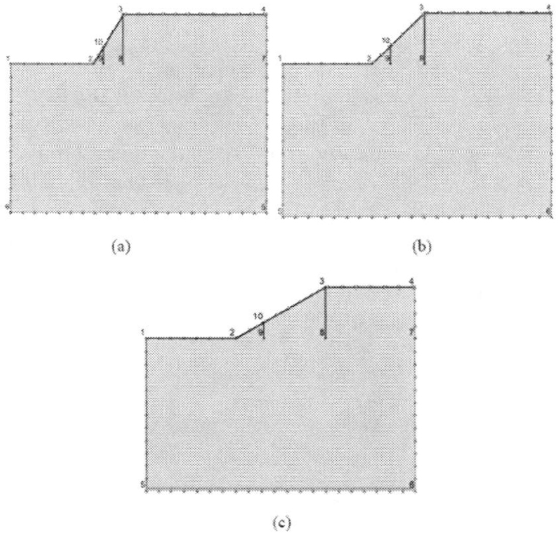

Fig. 3. Slope model for (a) 60°, (b) 45°, and (c) 30° angles

3.2 Boundary conditions

The geometry and soil property is then applied to the program. For SEEP/W, there are two types of boundary condition. From Fig.2, there are several boundary conditions applied for steady-state analysis in order to achieve the initial condition of the slope, which are the 1-5-6-7-4 line as the zero total flux that shows no water that will flow through the boundary, the 2-10-3-4 line as the unit flux boundary with potential seepage review so that the pore water pressure at the surface will stay as zero throughout the analysis to avoid any ponding when the rain falls at the transient analysis, and line 1-2 and point 7 as the zero pore water pressure boundary that shows the elevation of the ground water level. For the transient analysis, the boundary conditions are only the unit flux boundary and zero pore water pressure boundary because the initial condition for the slope is sustained throughout the analysis from the result of the steady-state analysis.

As for the SLOPE/W, the analysis taken is Bishop analysis with the entry-exit method boundary condition.

3.3. Rainfall parameters

The rainfall pattern parameter is taken from Faculty of Engineering, Universitas Indonesia rain station from the period of July 2014 – March 2015. From all the data taken at the period, the hourly rainfall for the month is then divided into five-days rainfall. This is according [2] and the data of the rain station which provides a 5-day rainfall occurring at one of the month.

The 5-days rainfall then divided into 15 intervals. The number of intervals taken because an interval with long duration of rainfall will make the rainfall intensity too small that would make it hard to differentiate the rainfall patterns. On the contrary, if the interval time is too short, the rainfall pattern will be too much so that it would be hard to categorize. The rainfall patterns are taken from January 2015's rainfall. The total rainfall from each of the interval is then divided by the total rainfall of the 5-days rainfall to get the percentage of every interval.

From the actual percentage of each interval, idealization is made from random numbers generated so that the patterns which is like normal distribution is called normal, while the other two patterns which are like normal distribution with skewness are called delayed and advanced that were differentiate by the top of the intensity. The idealized percentage is then multiplied by the total 5-days rainfall in November 2014 with the total intensity of 225 mm. This value is used because the 5-days rainfall intensity in January 2015 is not as high as the 5-days rainfall intensity in November 2014. From the value, there are three rainfall patterns as shown in Fig. 4.

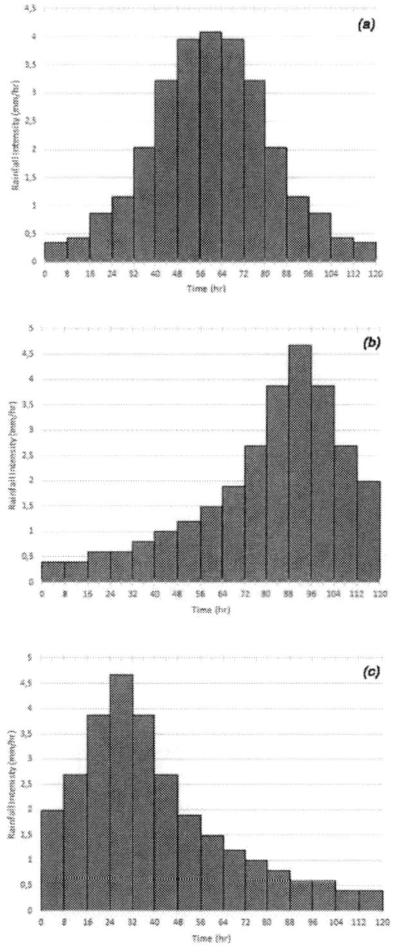

Fig. 4. Rainfall patterns (a) normal, (b) delayed, and (c) advanced

3.4 Scenarios

There are two scenarios conducted in this study are shown in Fig. 5.

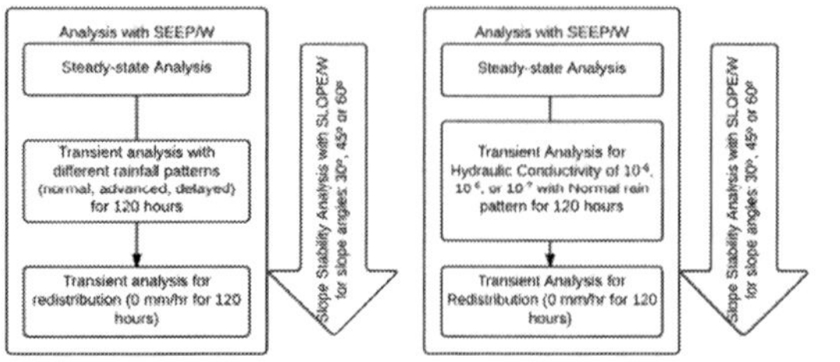

Fig. 5. Scenarios (a) Differentiates rainfall pattern and (b) Differentiates hydraulic conductivity value

The first scenario (Fig.5a) shows that the first analysis is a steady-state analysis with SEEP/W, which is a steady analysis that doesn't change through time so that the initial condition of the slope can be obtained. After that, the rainfall is applied to the slope for 120 hours for each of the rainfall patterns; normal, delayed, and advanced with transient analysis that can change the pore water pressure's profile through time and obtained different behaviour for every rainfall patterns applied. Lastly, the slope is applied to a redistribution step with transient analysis that doesn't have any rainfall for 120 hours to distribute the infiltrated water evenly throughout the slope. Every step of analysis performed is accompanied by SLOPE/W analysis to know the FS of each of the conditions and applied to each of the angles; 30°, 45°, and 60°.

The second scenario (Fig.5b) is almost the same as the first, but the rainfall pattern used in this scenario is the normal rainfall pattern with the variation is the hydraulic conductivity of the soil, which are differentiate as 10^{-6} m/s, 10^{-5} m/s, and 10^{-7} m/s to analyse the effect of different hydraulic conductivity for the change in FS.

4 Results and discussion

The results obtained from SEEP/W and SLOPE/W for each case are then compiled for three different discussions.

4.1 The effect of rainfall pattern to FS for different slope angles

With different rainfall patterns, such as normal, advanced, and delayed for different slope angles, which are 30°, 45°, and 60°, the results are shown in Fig.6.

From Fig.6, it is shown that the initial FS value for each slope is 2.111 for 30°, 1.898 for 45°, and 1.850 for 60°. From Fig.6a and Fig.6b, it is shown that for angle 30° and 45°, the smallest FS is achieved at advanced pattern with the value of 2.085 for 30° and 1.883 for 45°. As for 60° angle, the smallest FS is achieved at delayed pattern with the value of 1.828.

From Fig.6, it is also shown that FS reduction has the same pattern as the location of peak intensity of each rainfall pattern. For example, the advanced rainfall pattern that has the peak at the beginning of the pattern resulting in faster FS reduction at the beginning of the rainfall. As for the 60° angle, the least FS is achieved at delayed rainfall due to insufficient time for water to infiltrate the slope and recover the FS before the rainfall stops, which give the least FS at delayed pattern rather than at the advanced and normal pattern.

As a comparison from [2], the least FS for LC is at 30° slope angle at advanced rainfall pattern on the 96-hr with 40-45% reduction and 450 mm rainfall for 120 hours while this research only has 225 mm rainfall for 120 hours with the least FS is at advanced pattern at t = 120-hr and reduction of 1.29%. The difference between the previous study with this study is the FS reduction percentage which is caused by the different rainfall intensity for the two studies.

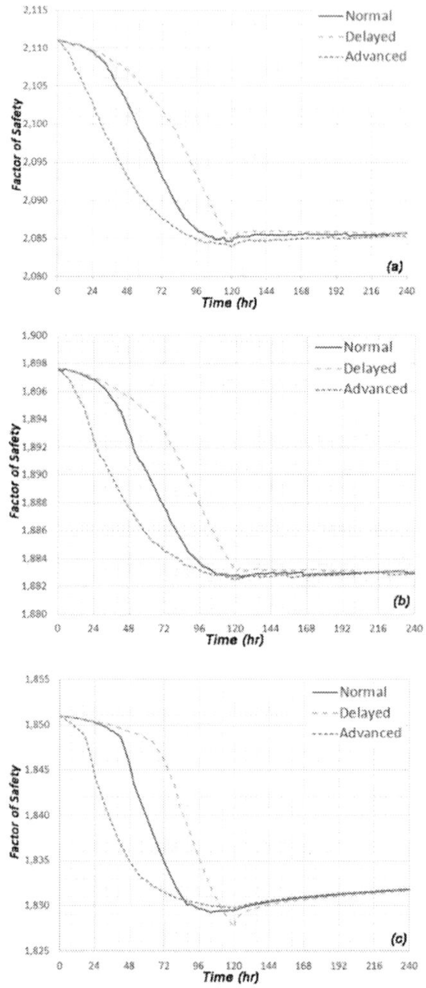

Fig. 6. FS vs time graph on different rainfall pattern for (a) 30°, (b) 45°, and (c) 60°

4.2 The effect of different slope angles to FS for different rainfall patterns

The comparison between FS and time for different slope angles and different rainfall pattern were observed with normalized FS. The normalized FS is used to see the different behaviour for each slope angle at a certain rainfall pattern at the same initial value. The results are shown in Fig.7.

From Fig. 7, it is shown that the largest FS reduction for each rainfall pattern is achieved at 30° angle, followed by 60°, and the least is 45°. The behaviour that is shown on the three figures are likely because the steeper the slope, the less water will be infiltrated and the rest will turn into runoff [6, 15].

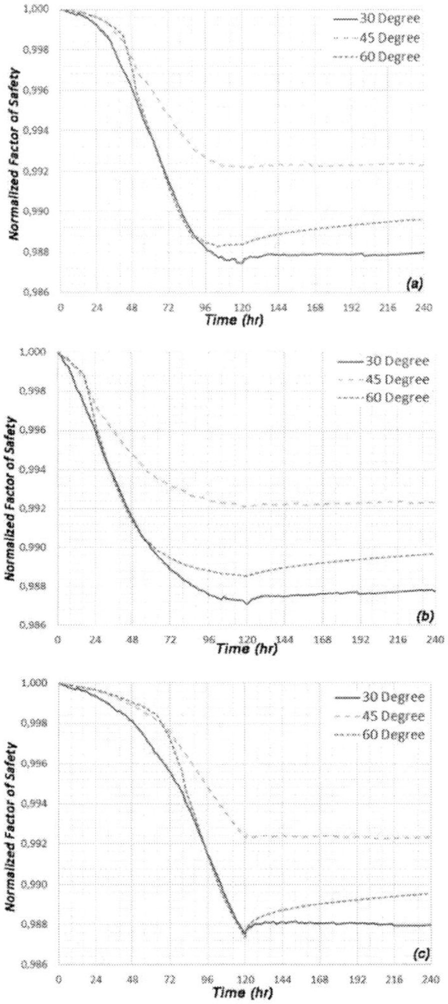

Fig. 7. Normalized FS vs time for (a) Normal, (b) Advanced, and (c) Delayed pattern

The 30° angle slope has a smoother FS reduction pattern compared to the ones from the 45° and 60° slopes. The 45° and 60° slopes have a similar slower FS reduction at the beginning of the rainfall, while the 60° slope FS will drastically decreased after that, following the pattern of FS from the 30° angle slope. Considering the slope model in the 2D finite element, it is suspected that the extended horizontal surface behind the 60° slope is allowing the rainfall infiltration like the gentle slope surface as in the 30° slope model (see Fig.3.).

As for the redistribution step, for the three rainfall patterns, the change of FS for 60° angle is much more significant than the two other angles because of the less infiltrated water is not much affecting the redistribution process.

4.3 The effect of different hydraulic conductivity to FS for each slope angles

From the comparison between FS's with different slope angles, it is known that the increase of FS on redistribution step is not significant with the applied hydraulic conductivity. To know whether different hydraulic conductivity affect the change of FS on normal rainfall pattern at 30° slope angle, there will be three different hydraulic conductivities used in this section. The three different hydraulic conductivities noted as Initial (1×10^{-6} m/s) which represents the hydraulic conductivity used on the beginning of this study, Low (1×10^{-7} m/s), and High (1×10^{-5} m/s). The Initial conductivity represent the value in between the Low and High conductivities. The results are shown in Fig.8.

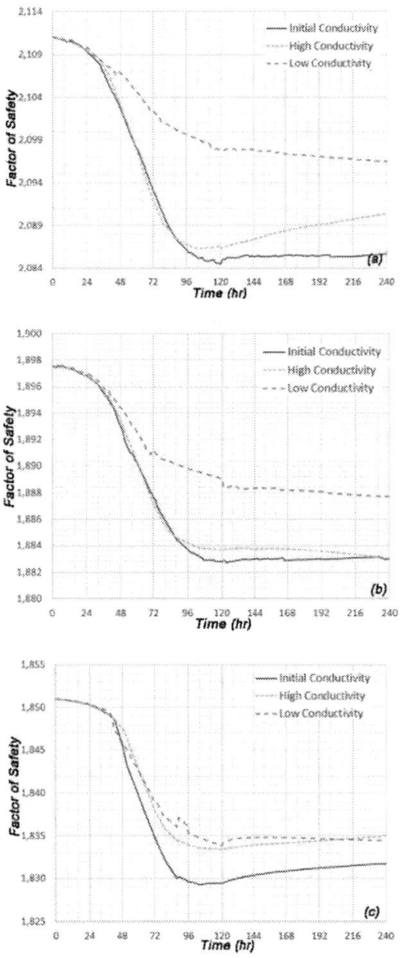

Fig. 8. FS vs Time on different hydraulic conductivity for (a) 30°, (b) 45°, and (c) 60°

From Fig.8, it is shown that for the three slope angles, 30°, 45°, and 60°, the Low conductivity always gives the least FS reduction. The largest difference of FS

DOI: 10.1051/matecconf/201710105005

reduction between the Low conductivity and the Initial or High conductivities is shown at the 30° slope. This means that the effect of infiltration is significant to the reduction of FS since the low conductivities slowing down the infiltration of rainfall through the slope.

Compared to [2] at 30° slope angles for different hydraulic conductivity, which is High (1 x 10^{-5} m/s) and Low (1 x 10^{-6} m/s), it is known that the least reduction of FS is obtained at High. As for this research, the same behaviour corresponds at Initial (1 x 10^{-6} m/s) and High (1 x 10^{-5} m/s) for the normal rainfall pattern.

4.4 The change of pore water pressure during the rainfall

There are two cross sections observed for pore water pressure profiles (Fig.5), which are 3-8 near the top of the slope and 10-9 near the toe of the slope. Both pore water pressure profiles are observed to see the behaviour of changing pore water pressure profile and rising ground-water level due to the two scenarios.

For the first scenario, the changing pore water pressure (PWP) are observed for different slope angles at each of the rainfall pattern. The results shown that the behaviour of different slope angles for every rainfall pattern is similar, as an example, the results for the changing PWP at advanced rainfall pattern are shown in Fig.9 (Section 3-8) and Fig.10 (Section 9-10).

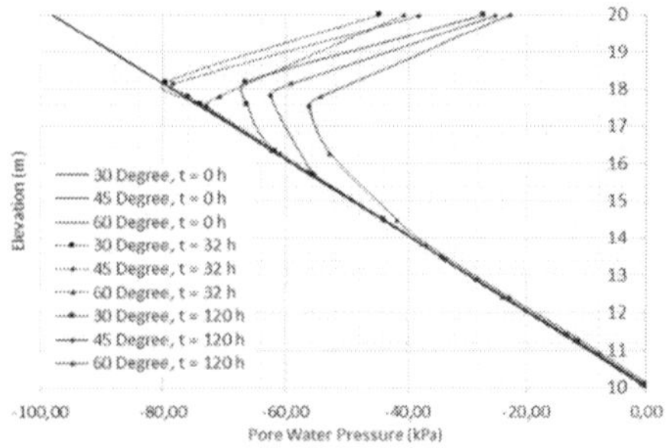

Fig. 9. PWP profile change during advanced rainfall pattern at section 3-8

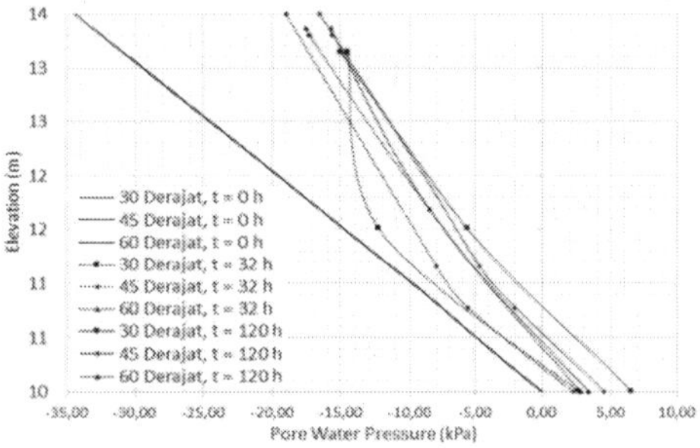

Fig. 10. PWP profile change during advanced rainfall pattern at section 9-10

From Fig.9 and Fig.10, it is shown that the PWP increases as the duration of the rainfall increase which means that the rainfall is infiltrated into the slope and lowered the FS. This behaviour corresponds to the behaviour shown in Fig.7 which shows that the least FS is at the end of the rainfall.

From Fig.9, at a position near the top, the 60° slope has the least value of negative PWP than the two other angles which indicates that the slope will have the least FS value. This behaviour corresponds for the initial FS of the three angles where the 60° slope angle have the least initial FS than the two others. While from Fig.10, at a position near the toe, the 30° slope has the biggest value of positive PWP than the two other angles which indicates the greatest rise of the ground-water level due to rainfall infiltration. The behaviour of rising of ground-water level at the 30° slope corresponds to the behaviour where 30° slope have the greatest FS reduction.

Compared to [2], for 30° slope LC soil, the increased value of PWP for every rainfall is the same even if at different time. At this research, the increased value of PWP is not the same but are achieved at the end of the rainfall. Even though the value is different, both research exhibits the same behaviour.

At the second scenario, the PWP profile at every angle is observed with different hydraulic conductivity. The behaviour of every angle is similar, as an example, the 30° PWP Profile at normal rainfall is shown in Fig.12 (Section 3-8) and Fig.13 (Section 9-10).

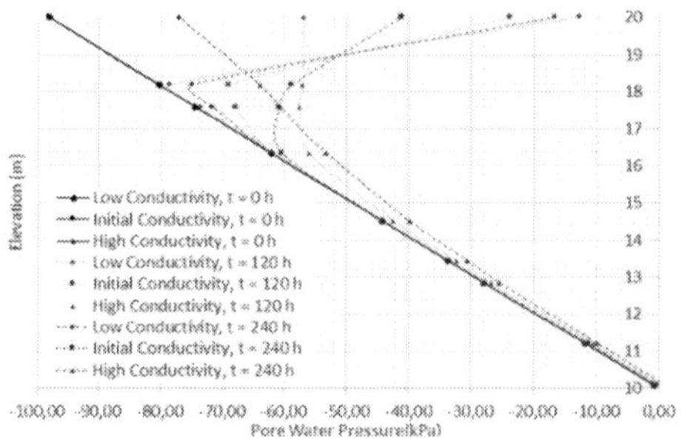

Fig. 11. PWP profile change for normal rainfall for 30° slope at section 3-8

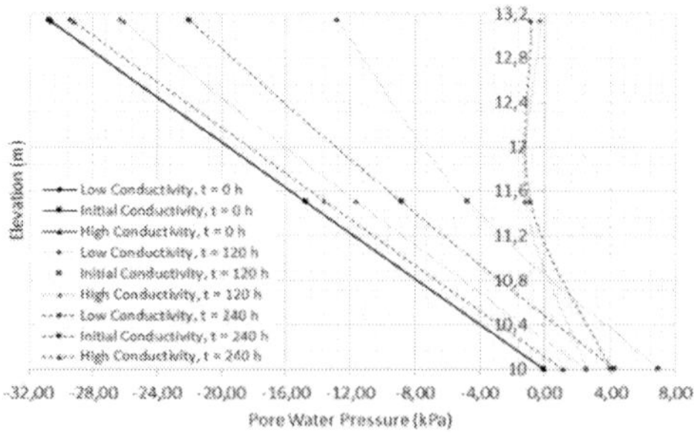

Fig. 12. PWP profile change for normal rainfall for 30° slope at section 9-10

From Fig.11 and Fig.12, it is shown at the end of the rainfall, the greatest PWP is achieved at Low Conductivity PWP near the top while the greatest PWP near the toe is achieved at Initial Conductivity. But, at elevation of 18-m, inside the slope's body, the Low Conductivity PWP is the lowest than the other conductivity which leads to higher FS. This corresponds to the behaviour of which the Low Conductivity FS redistribution step, the Low Conductivity's PWP is greater than the other conductivity which corresponds to the behaviour of which the FS at redistribution keeps decreasing due to the slow infiltration rate.

From Fig.11 and Fig.12, it is also shown that the PWP for High Conductivity is the smallest and at the end of the distribution step, the PWP decreases which leads to the behaviour of increasing FS at the end of the redistribution and great infiltration rate.

5 Conclusion

This study tries to analyse the effect of different rainfall pattern to FS value of slope. The effect of the slope's angle and hydraulic conductivity also being observed. The following conclusions can be drawn from this study:

The steeper the slope, the smaller the initial FS of the slope. When rainfall occurs, FS reduction in gentle slope is found to be higher than the FS reduction in steep slope at any type of rainfall pattern. Gentle slope tends to have the least FS after advanced rainfall pattern while steep slopes have the least FS after delayed rainfall pattern.

The FS reduction is mainly caused by the rising of the ground-water level during rainfall which builds positive PWP for the slope.

The higher the hydraulic conductivity, the smaller the FS reduction during the rainfall and the greater the increase of FS during redistribution which is caused by the infiltration rate of the hydraulic conductivity that enable the PWP that rises during the rainfall to decrease after the redistribution.

Some factors related to the slope model in 2D finite element may contribute to the results of this study, such as the effect of horizontal surface behind the slope. This issue should be considered for the next study.

The authors would like to thank Civil Engineering Department, Universitas Indonesia, especially Soil Mechanic Laboratory staffs and lecturers for the spirit and input of this research.

reduction is the smallest. But, at the end of the

References

1. H. Rahardjo, T.H. Ong, R.B. Rezaur, E.C. Leong, J. Geotech. Geoenv. Eng., **133:12**, 1532-1543 (2007)
2. A. Rahimi, H. Rahardjo, E.C. Leong, J. Geotech. Geoenv. Eng., **137:5**, 483-491 (2011)
3. T.T. Lim, H. Rahardjo, M.F. Chang, D.G. Fredlund, Can. Geotech. J., **33(4)**, 618-628 (1996)
4. J.M. Duncan, S.G. Wright, *Soil Strength and Slope Stability*. (John Wiley & Sons, Inc, 2005)
5. F. Silva, W. Lambert, W.A. Marr, J. Geotech. & Geoenv. Eng. V **134**, 12 (2008)
6. B.B.K. Huat, F.HJ. Ali, T.H. Low, Geotech. & Geo. Eng. V **24** 1293-1306 (2006)
7. D.G. Fredlund, *The emergence of unsaturated soil mechanics* (Fourth Spencer J. Buchanan Lecture, Texas A&M, 1996).
8. D.G. Fredlund, H. Rahardjo, M.D. Fredlund, *Unsaturated Soil Mechanics in Engineering Practice*. (John Wiley & Sons Inc, 2012)
9. D.G. Fredlund, J. Geotech. Geoenv. Eng., **132:3**, 286-321 (2006)
10. D.G. Fredlund, A. Xing, Can. Geotech. J., **31(4)**, 521-532 (1994).
11. D.M. Gray, D.I. Norum, *Proceedings of Hydrology Symposium, No.6* (1967)
12. H.F. Yeh, C.C. Lee, C.H. Lee, J. Environ. Eng. Manage. **18(4)**, 261-268 (2008)
13. M. Budhu, *Soil Mechanics and Foundations* (John Wiley & Sons Inc, 2011)
14. B. Look, *Handbook of Geotechnical Investigation and Design Tables* (Taylor & Francis, 2007)
15. M.P. Purba, *Besar Aliran Permukaan (Run-off) pada Berbagai Tipe Kelerengan dibawah Tegakan Eucalyptus spp. (Studi Kasus di HPHTI PT. Toba Pulp Lestasi, Tbk. Sektor Aek Nauli)* (Universitas Sumatera Utara, 2009)

 DOI: 10.1051/matecconf/201710105006

The study of the usage of coral and limestone aggregates as asphaltic layer on coastal structures

Adelia Dwidarma Nataadmadja[1,*], *Oki* Setyandito[1], *Fiona* Maida Basrian[1], and *Michael* Grashinton Kurniawardhani[1]

[1]Civil Engineering Department, Bina Nusantara University, 11480 Jakarta, Indonesia

Abstract. Indonesia is a maritime country with the second longest coastline in the world. Any structures constructed along the coastline, such as pavement layer on sea dike, must be able to resist the hydrodynamic forces from the waves, such as the overtopping wave. It is necessary to find an aggregate to be used as surfacing layer for sea dike and/or flexible pavement that can withstand these (compressive) forces. This research analyses the feasibility of two natural aggregates, namely the coral and limestone aggregates, to be used as an ingredient for asphalt surface layer on coastal structures. Asphalt mix samples were prepared by using these two aggregates with various percentage of asphalt. The Marshall stability, flow, and the compressive force (F_{lab}) of the prepared samples were tested in the laboratory. The stability and the flow of asphalt samples were tested to verify that the asphalt mix satisfies the requirements specified by *Standar Nasional Indonesia* 06-2489-1991. Moreover, this research also calculates the overtopping waves by varying the height and the period of the waves. The results were then converted as a compressive force (F_{wave}) and compared to the F_{lab} to see whether the prepared samples could resist the calculated F_{wave}.

1 Introduction

Indonesia is a maritime country with the second longest coastline in the world, and hence, it is important to ensure that the infrastructure built around the coastline is resistant to the hydrodynamic forces from the waves, such as the overtopping wave. One of the structures that directly face the sea wave is the sea dike, which is built with the aim to protect coastal areas from the water pressure. It has sloping sides and watertight protection surface layer, which can be made from concrete, asphalt, asphaltic concrete or synthetic material [1-4].

Asphalt layer has been commonly used as the surfacing layer on the dyke and research project has been undertaken to analyse the relationship between the dyke slope and the percentage of void in asphalt mixture [5]. It was found that the higher the dyke slope, the higher the percentage of void in asphalt due to the increasing difficulty in the compaction process will be. Additionally, the stiffness of the protection layer significantly affects the capacity of the layer to resist sea wave. More studies are required to be conducted to assess the capability of asphalt mixture to be used as protection layer of sea dike as the wave in Indonesia's coastal area might be different to other countries, and hence, adjustment in design might be needed to suit the local condition.

Additionally, due to the increasing construction cost, it is necessary to utilize locally available material (wherever possible) to minimize the transportation cost of materials. In Indonesia, both coral and limestone aggregates are available in abundance and in a close proximity. Therefore, this paper focuses on studying the feasibility of two natural aggregates, coral and limestone aggregates, as an ingredient of asphalt mixture to be used as the protection layer of sea dike structure against the overtopping wave.

2 Limestone and coral aggregates

There are two natural aggregates used in this research, coral and limestone aggregates. Both of them are categorized as organic sedimentary rock and mainly consist of calcium carbonate ($CaCO_3$) [6]. However, their forming process is different which results in diversity in the rock form. Limestone aggregates are crystallized rock, while the coral aggregates are more porous [7].

Limestone aggregates have been commonly used as construction materials, including as filler in asphalt mixture, and have been proven to have some advantages for pavement performance, such as reduces the potential of asphalt to deform at high temperatures and alters the plastic properties of clay fines to improve moisture stability and durability [8]. They have high affinity for asphalt and low affinity for water, so that they can bond well with asphalt. Additionally, the usage of certain limestone aggregates used in the coarse aggregate fraction of asphalt concrete wearing course can increase mix stability and resistance to moisture damage [9, 10].

Coral aggregates have been used as construction material for road pavements, especially in Pacific islands [11-13]. In Indonesia, a research project has shown that coral aggregates have also been used in asphalt mixture, although it was found that additive has to be added to

* Corresponding author: anataadmadja@binus.edu

improve the quality of the bond between the asphalt and the coral aggregates [7].

3 Wave run up and overtopping

The wave run-up height is defined as the vertical difference between the highest point of wave run-up and the still water level [14]. Due to the stochastic nature of the incoming waves, each wave will give a different run-up level. If only 2% of the waves reach the crest of a coastal structure, the crest and inner slope do not need specific protection measures. Wave run-up had previously been the focus in research projects. However, in the past decades, the design or safety assessment has been changed to allowable overtopping instead of wave run-up.

Wave overtopping is one of the most important hydraulic responses of a breakwater. It occurs when the top of the structure is lower than the run-up level during maximum wave [15]. Wave overtopping significantly affects its functional efficiency and to a lesser extent, its structural safety [16]. It has also been the major cause of the inner slope of sea dikes [17-18]. Therefore, it is important to ensure that the design of sea dikes can resist the compressive force generated by this wave overtopping.

4 Experimental design

4.1 Material

There are two aggregates used in this research, coral and limestone aggregates as shown in Figures 1 and 2. Six asphalt samples were constructed for each type of aggregates and for each percentage of asphalt content (4.5%, 5.5%, and 6%). Three samples were used for Marshall Stability and Flow Test and the other three were used for Compressive Strength Test. The aggregates were crushed into four sizes, as shown in Table 1. The coarsest aggregates are Aggregate I and the finest aggregates are Aggregate IV. The asphalt samples were then constructed by using the asphalt mix design used in [19], as shown in Table 2. The samples were compacted 75 times on each side.

It is important to note that the limestone aggregates used in this research were covered by a thin layer of particles, although they did not interrupt the adhesion between the aggregates and the asphalt.

Table 1. Grading of aggregates

Aggregate	Passed Sieve No.	Retained by Sieve No.
I	3/4" (19,1 mm)	3/8" (9,6 mm)
II	3/8" (9,6 mm)	#8 (2,4 mm)
III	#8 (2,4 mm)	#16 (1,2 mm)
IV	#16 (1,2 mm)	#200 (0,075 mm)

Table 2. The composition of asphalt mix design

Asphalt Content		Weight of Aggregate (gr)			
%	gr	I	II	III	IV
4.5	52.87	242	193	181	476
5.5	66	242	193	181	476
6.0	71.62	242	193	181	476

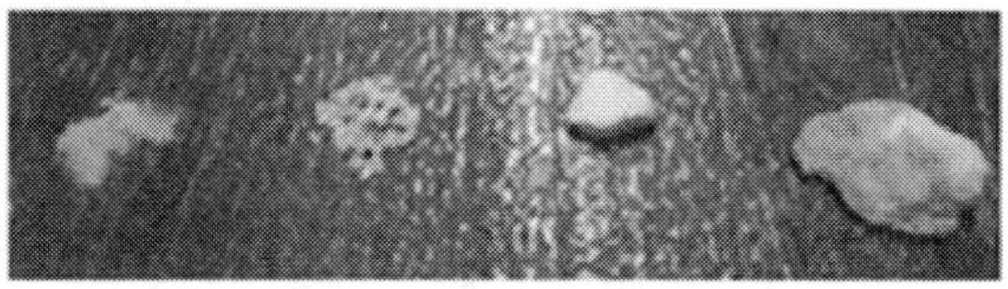

Fig. 1. Coral aggregates

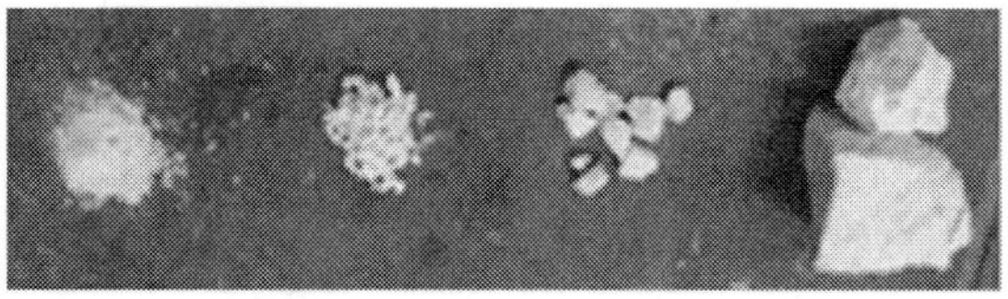

Fig. 2. Limestone aggregates

4.2 Laboratory tests

There are a number of laboratory tests conducted for this research, namely the Marshall Stability and Flow Test, which was conducted according to *Standar Nasional Indonesia* (SNI) 06-2489-1991, and Compressive Strength Test, which was conducted according to *SNI* 03-1974-1990. Each test was conducted three times. The compressive strength test result was denoted as F_{lab} and the device used is shown in Figure 3. It is important to note that the device and the standard used to test the compressive strength of the asphalt mix in this research are usually used for concrete testing. However, several steel plates were placed underneath the asphalt samples so that the asphalt samples could fit into the device.

Fig. 3. Compressive strength test device

4.3 Overtopping calculation

In calculating the wave overtopping to be used in this research, there are some assumptions made, including:

- the wave overtopping is assumed to fall perpendicular to the protection layer;
- the wave condition used in this research is based on the wave occurred at Southern Beaches of Java Island, Indonesia, including using three wave period (T) namely, 8, 10 and 12 seconds and using the incoming wave height were varied between 0.5 and 5 meters, with 0.5 meter interval;
- the wave overtopping hits against a plain vertical wall;and
- three cress freeboard (R_c) heights were calculated, namely 0.7, 1, and 2 meter.

The overtopping debit is the volume of water that fall onto the inner slope of sea dike. In this research, there are two wave overtopping conditions considered, which are the impulsive and the non-impulsive conditions. A non-dimensional parameter, h*, was used to determine whether the wave overtopping was impulsive or non-impulsive and was calculated as follows:

$$h^* = 1.3\frac{h_s}{H_{m0}}\frac{2\pi h_s}{gT_{m-1}^2} \quad (1)$$

where:

h_s = water depth at the toe of the structure

H_{m0} = wave height at the toe of the structure

$T_{m-1.0}$ = wave period

If the h* parameter is more than 0.3, the wave was categorized to be non-impulsive, and if the h* parameter was less than 0.2, the wave was categorized to be impulsive.

Based on the equations introduced by [20], for non-impulsive wave, the wave overtopping debit was calculated based on the equation as follows:

$$\frac{q}{\sqrt{gH_{m0}^3}} = 0.04e^{\left(-2.6\frac{Rc}{H_{m0}}\right)} \quad (2)$$

where:

q = average overtopping discharge

R_c = cress freeboard

For impulsive wave, the wave overtopping debit was calculated based on the equations as follow:

$$\frac{q}{h_*^2\sqrt{gh_s^3}} = 1.5\times10^{-4}\left(h_*\frac{R_c}{H_{m0}}\right)^{-3.1} \quad (3)$$

$$\frac{q}{h_*^2\sqrt{gh_s^3}} = 2.7\times10^{-4}\left(h_*\frac{R_c}{H_{m0}}\right)^{-2.7} \quad (4)$$

Equation 3 was used when $0.03 < h_*\frac{R_c}{H_{m0}} < 1.0$ and equation 4 was used when $h_*\frac{R_c}{H_{m0}} < 0.02$.

The calculated wave overtopping debit data was verified by using an online calculator [21], which was constructed based on the EurOtop Manual [22]. The debit was then converted to obtain the volume of water (V_{bar}) by using an equation introduced by [16], which is:

$$V_{bar} = 0.8\frac{q_{m-1}N_w}{N_{cw}} \quad (5)$$

where:

N_w = number of incoming wave

N_{ow} = number of overtopping wave

Lastly, the V_{bar} was then converted as compressive force (F_{wave}), which was calculated by using the following formula:

$$F_{wafe} = V_{bar} \times \rho \times g \quad (6)$$

where:

ρ = density of water = 1000 kg/m^3

g = gravity = 9.81 m/s^2

The F_{wave} was compared to the F_{lab} to see whether the prepared samples could resist the calculated F_{wave}. F_{wave} calculated from Equation 6 was in kN. To compare it with F_{lab}, which was in MPa, the F_{wave} was also converted to MPa by dividing the value by the surface area of asphalt sample, which was 8167.14 mm^2.

5 Results and discussion

5.1 Characteristics of asphalt mixture

Tables 3 and 4 show the characteristics of asphalt mixture made of limestone and coral aggregates, respectively. The characteristics measured including specific gravity, Void in Mix (VIM), Void in Mineral Aggregate (VMA), Void Filled with Asphalt (VFA), stability, flow, Marshall Quotient (MQ), and the compressive strength (F_{lab}). The values listed in Tables 3 and 4 are the average of the results of the three samples tested.

Table 3. Characteristics of asphalt mixture with limestone aggregates

Characteristics	Asphalt Content		
	4.5%	**5.5%**	**6%**
Specific Gravity	2.343		
VIM (%)	3.97	2.08	3.27
VMA (%)	12.54	10.3	14.12
VFA (%)	69.82	78.67	76.81
Stability (kg)	3220	2208	2568
Flow (mm)	2.73	2.33	3.53
MQ (kg/mm)	1182	963	727
F_{lab} (MPa)	62.41	126.46	97.90

The data in Tables 3 and 4 shows that the coral aggregates have a higher percentage of VMA and VFA than the limestone aggregates, regardless of the percentage of asphalt content. This data suggests that the coral aggregates are more porous than the limestone aggregates. It can also be seen from the specific gravity values of both aggregates, which represent the density of the aggregates. The specific gravity of coral aggregates is approximately 36% lower than the specific gravity of limestone aggregates, which also suggest that the coral aggregates have more pores than the limestone aggregates.

Table 4. Characteristics of asphalt mixture with coral aggregates

Characteristics	Asphalt Content		
	4.5%	**5.5%**	**6%**
Specific Gravity	1.499		
VIM (%)	3.71	3.04	5.46
VMA (%)	24.83	28.74	25.21
VFA (%)	84.95	89.62	82.10
Stability (kg)	1692	1890	2137
Flow (mm)	3.40	4.60	3.33
MQ (kg/mm)	490	411	644
F_{lab} (MPa)	77.51	83.63	85.67

After the samples were compacted, the percentages of VIM of all samples were similar, although the percentages of VIM of the asphalt samples made of limestone aggregates are slightly higher than the asphalt samples made of coral aggregates for all variety of asphalt content.

From the results presented in Tables 3 and 4, it can be seen that all samples have met the requirements stated in [23]. By analysing the stability, flow, and MQ values of all samples, it can be seen that asphalt samples made of limestone aggregates have a higher stability, a lower flow, and a higher MQ values than the asphalt samples made of coral aggregates, which suggest that the asphalt mixture with limestone aggregates is more stable and more resistant to permanent deformation than the asphalt mixture with coral aggregates.

5.2 Compressive strength test (F_{lab}) results

The Compressive Strength Test results of the asphalt samples made of both aggregates are presented in Tables 3 and 4. There is a significant difference in the compressive strength of asphalt samples with limestone aggregates with the addition of asphalt. Increasing the asphalt content from 4.5% to 5.5% doubled the compressive strength of the asphalt samples with limestone aggregates, while increasing the asphalt content from 4.5% to 6% increased the compressive strength by approximately 57%. However, this phenomenon was not found in the asphalt samples with coral aggregates. Increasing the asphalt content by 1% and 1.5% increases the compressive strength by approximately 8% and 11%.

Furthermore, in general, the asphalt samples made with limestone aggregates with 5.5% and 6% asphalt content can resist more compression force than the asphalt samples made of coral aggregates, except the asphalt samples with 4.5% asphalt content. This could be related with the mineralogy of the aggregates. Structurally, coral aggregates have more pores than the limestone aggregates, which make the coral aggregates are not as strong as the (more) compact limestone aggregates. During the compaction process in the asphalt sample making, some of the coarse coral aggregates (aggregates I and II) might be crushed into finer particles. Adding more asphalt to the mixture will not help increasing the compressive strength by much as the asphalt would have already seeped through the void between aggregates and/or between the minerals. On the other hand, limestone aggregates are generally more compact than the coral aggregates, and hence, adding more asphalt to the asphalt mixture will improve the adhesion between the asphalt and the aggregates. However, adding too much asphalt to the mixture will have an adverse effect on the compressive strength. Too much asphalt will reduce the compressive strength as the force would be partially resisted by the asphalt, which is not as strong as the aggregates. In short, the more compact the mineralogy of an aggregate, the higher the force that could be resisted by the asphalt mixture made of those.

5.3 Comparison between F_{lab} and F_{wave}

Based on the wave overtopping calculation explained in Section 4, both F_{lab} and F_{wave} for R_c of 0.7, 1, and 2 m were plotted in Figures 4, 5, and 6, respectively. The data points plotted in those figures include both non-impulsive F_{wave} (blue points) and impulsive F_{wave} (green) for various height and period combinations and the magnitude is displayed in x-axis. The division between impulsive and non-impulsive waves was made based on the parameter h* (equation 1). The black lines (straight and dashed) represent the F_{lab} for asphalt samples with limestone aggregates and the red lines (straight and dashed) represent the F_{lab} for asphalt samples with coral aggregates.

From the data shown in Figures 4, 5, and 6, it can be seen that the magnitudes of F_{wave} depending on the height and the period of the incoming waves. The higher the height (H) of the incoming wave, the more likely it is for the wave to become an impulsive wave, and the impulsive waves created higher compressive forces than the non-impulsive waves. The non-impulsive F_{wave} data ranges between 0.002 MPa and 0.665 MPa for R_c of 0.7, 1, and 2 m. Additionally, it was found that the higher the cress freeboard, the lower the impact due to the force generated by the overtopping wave onto the protection layer of sea dike, which can be seen from the decreasing value of F_{wave}. For the sea dike with R_c of 0.7 m, the highest (impulsive) F_{wave} was above 250 MPa, while for the sea dike with R_c of 2 m, the highest (impulsive) F_{wave} was (only) approximately 17 MPa.

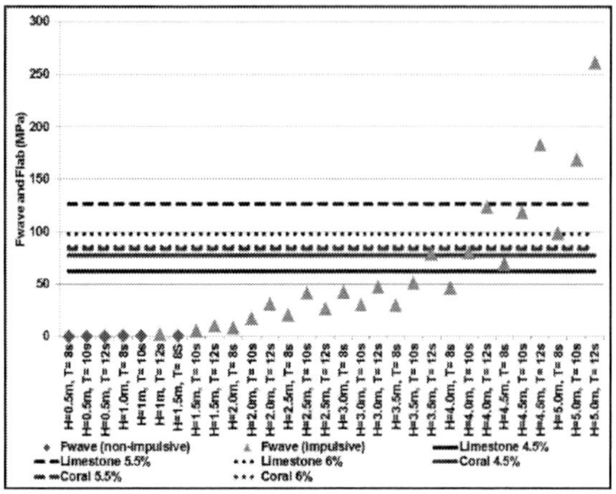

Fig. 4. The comparison between F_{lab} and F_{wave} for $R_c = 0.7$ m

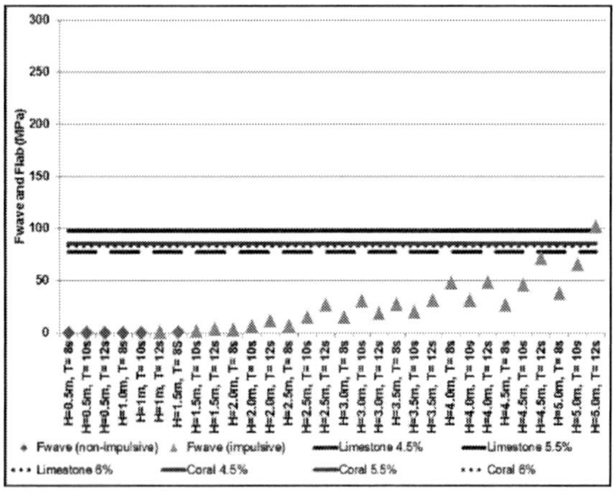

Fig. 5. The comparison between F_{lab} and F_{wave} for $R_c = 1$ m

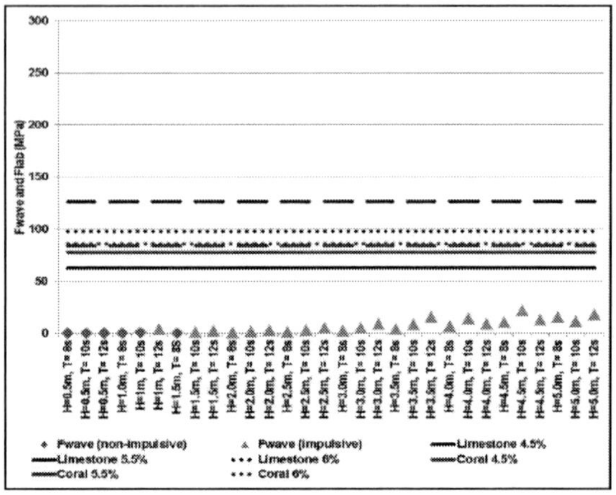

Fig. 6. The comparison between F_{lab} and F_{wave} for $R_c = 2$ m

All asphalt mixtures with either limestone or coral aggregates could resist the F_{wave} generated for sea dike with R_c of 2 m, as seen in Figure 6. However, if the R_c is 1 m and the wave height is 5 m, only asphalt mixture with 5.5% asphalt content and limestone aggregates that could resist that wave.

5 Summary and conclusions

It is important for Indonesia as a maritime country to ensure that the infrastructure built around the coastline is resistant to the hydrodynamic forces from the waves. Due to the increasing construction cost, it is desirable to use locally available material, including coral and limestone aggregates. The main question of the paper is to analyse the feasibility of two natural aggregates, coral and limestone aggregates, as an ingredient of asphalt mixture to be used as the protection layer of sea dike structure against the overtopping wave.

In this research, asphalt samples with limestone and coral aggregates have been constructed to assess the compressive strength of the asphalt mixture. A calculation has also been made to predict the force due to wave overtopping that could occur. From the laboratory and calculation results, it can be concluded that:

- asphalt mix with 5.5% asphalt content and with limestone aggregates have the highest compressive strength;
- both limestone and coral aggregates can be used as an ingredient of asphalt mixture, depending on the magnitude of the F_{wave} needs to be resisted;
- it is important to study the characteristics of the incoming wave to decide the height (and the type) of the sea dike to be constructed; and
- building a higher vertical wall could help in reducing the magnitude of the F_{wave} needs to resisted, and hence, the protection layer could be built with, for example, coral aggregates.

The authors would like to thank Bina Nusantara University for the research grant provided to conduct this research.

References

1. I. Jassim, S. Tan, P. A. Vermeer, M. A. Hicks, *Conference of the Association for Computational Mechanics in Engineering ACME* (2012)
2. K. W. Pilarczyk, *Dikes and Revetments: Design, Maintenance and Safety Assessment* (1998)
3. S. Nandlal, *Lightweight semi-flexible dike* (1985)
4. J. van der Meer, *Seawalls, dikes and revetments*, ed. K.W. Pilarczyk (1998)
5. Committee on The Compaction of Asphalt Revetments of Dyke Slopes, *Asphalt Revetments of Dyke Slopes* (1977)
6. R. Devnita, Jurnal Geologi Indonesia **4**, 1, p. 19-29 (2009)
7. A. Yamin, *Jurnal Jalan - Jembatan* **28**, 3 (2011)
8. O. M. Ogundipe, *Transportation Research Procedia*, **14** p. 685-693 (2016)
9. D. Manglorkar, P. S. K. Kandhal, F. Parker, *Evaluation of Limestone Coarse Aggregate in Asphalt Concrete Wearing Courses* 1991, Auburn University Highway Research Center Alabama, United States of America (1991)

10. F. Chairuddin, M. W. Tjaronge, M. Ramli, J. Patanduk, *IACSIT International Journal of Engineering and Technology* **8**, 3 (2016)
11. N. A. Bowers, Engineering News Record, **July** p. 80-85 (1944)
12. Overseas Centre, *A Guide to the Structural Design of Bitumen-Surfaced Roads in Tropical and Sub-Tropical Countries*, Transportation and Road Research Laboratory (1993)
13. J. Palmer, *Use of Coral as an Engineered Construction Aggregate in Pacific Runways* (2013)
14. H. Schlüttrumpf, J. van der Meer, A. Kortenhaus, T. Bruce, L. Franco, Handbook of Coastal and Ocean Engineering, ed. Y.C. Kim (2010)
15. J. W. van der Meer, J.P.F.M. Janssen, Wave Forces on Inclined and Vertical Wall Structures, ed. N. Kobayashi and Z. Demirbilek (1995)
16. L. Franco, M.d. Gerloni, J.W. van der Meer, *Coastal Engineering* p. 1030-1045 (1994)
17. M. R. A. van Gent, *28th Conference on Coastal Engineering* (2002)
18. G. J. Steendam, J. W. van der Meer, B. Hardeman, A. van Hoven, *International Conference on Coastal Engineering ICCE (2012)*
19. W. Wati, *Pengaruh Penambahan Bahan Alami Lateks (Getah Karet) Terhadap Karakteristik Campuran Hot Rolled Sheet Wearing Course (HRS-WC)* (2015)
20. T. Bruce, J.W. van der Meer, T. Pullen, W. Allsop, *Wave Overtopping at Vertical and Steep Structure* (2009)
21. HR Wallingford Ltd. *Calculation Tool*, Available from: http://www.overtopping-manual.com/about-this-site.html (2007)
22. T. Pullen, N. W. H. Allsop, T. Bruce, A. Kortenhaus, H. Schlüttrumpf, J. W. van der Meer, *EurOtop Wave Overtopping of Sea Defences and Related Structures: Assessment Manual* (2007)
23. Direktorat Jenderal Bina Marga, *Dokumen Pelelangan Nasional - Penyediaan Pekerjaan Konstruksi (Pemborongan) Untuk Kontrak Harga Satuan* (2010)

* Corresponding author: anataadmadja@binus.edu

Influence of pore water pressure to seepage and stability of embankment dam (case study of Sermo Dam Yogyakarta, Indonesia)

Undayani Cita Sari[1*], *Sri* Prabandiyani Retno Wardani[1], *Suharyanto*[1], and *Windu* Partono[1]

[1]Civil Engineering Department, Faculty of Engineering, Diponegoro University, 50275 Semarang, Indonesia

Abstract. Today, water requirements are not analogously with its availability. Therefore, build water reservoir dam is one of the simple concepts for fulfilling water requirements. Embankment dam is one of the most popular dam build in Indonesia. Stability of embankment dam should be taken into account to evaluate the safety condition in retaining water. Safety against seepage is one of the most important steps for checking the possibility of failure of embankment dam. PLAXIS is alternative software that can be used for evaluating safety of embankment dam due to seepage condition. This paper explains the study of pore water pressure and seepage effects against Sermo Dam safety using PLAXIS and compares it with field measurement data from piezometer and V-Notch monitoring. The study was conducted during dam operational conditions. The analysis was carried out using Mohr-Coulomb model to calculate discharge of seepage and settlement. The value of pore water pressures within the model, upstream and downstream, positions are then reviewed with the real dam. The results of model analysis shows that the more higher water level within the dam reservoir the more higher pore water pressure will be happened that affects in increasing discharge of seepage and settlement of dam.

1 Introduction

Construction of the dam is needed so that water as a fundamental natural resource, can be long in the land before flowing back into the sea, as well as allow time to absorption the ground water. In order for the benefits to be sustainable, it needs to know the safety, especially when the dam operates. More than 70% dam in Indonesia is an embankment dam as stated by Soedibyo [1]. In this type of embankment dam, it should be taken into account in order to ensure its stability in retaining water.

One of the principal criteria for dam safety is safe against failure due to seepage [2]. Seepage through dam body and foundation can cause erosion. The transport of material through dam by seepage flow may result in fractures, which can reduce the capacity of dam. In addition, excessive seepage can cause reed erosion increasingly and potentially cause landslide until the dam collapse. Sudardja [3] stated that the seepage that occured within embankment dam is affected by water level height.

In addition, with the change in water level can also affect the increase in pore water pressure. Pore water pressure can trigger embankment slope instability and deformation. Morton et al. [4] study on the high walls suggested that the increase in pore water pressure will lead to a reduction in stress, while decreasing pore water pressure will lead for an increasing in stress. When the stress on land increases, it can increase strength of the ground to against instability resulting landslide on the slope. On the other way, if the stress on the ground is reduced, the more easily on slopes in landslide.

Deformation due to increased water levels studied on the Shuibuya Dam, China which is rockfill dam with impermeable layer of concrete in the upstream (Concrete Faced Rockfill Dam) [5]. Observation instrument for monitoring done at WS01 to WS11 which measures the deformation in the dam. Observations monitoring shows that when the water level increases, the settlement in dam also increases. In addition, the observation was also made by using Interferometric Synthetic Aperture Radar (InSAR), which has result that consistent with the result from the monitoring measurements. The combination of the results between monitoring in-situ and InSAR measurements indicate that the settlement is affected by gravity and the water level of reservoir at the dam, so that the largest settlement occurred at the peak of the dam.

The influence of dam water level with settlement is also given using a laboratory model of the embankment dam [6]. The study was conducted at three variations of the upstream slope of the dam, 1:1; 1:1.5; and 1:2. The result shows, when water level height increases, then the deformation is getting bigger. In addition, with the fill on the upstream slope steeper makes sliding on the downstream slope of the dam models more quickly occur.

* Corresponding author: undayanicita@yahoo.com

Based on these problems, it is necessary to study the seepage and stability analysis (in this case to be reviewed is settlement), to analyze the security of the dam caused by the influence of pore water pressure. Seepage and stability analyzes performed on operational condition so that it can be seen the influence of pore water pressure in these conditions. This study took place in the Sermo Dam located in the Hargowilis village, Kokap, Kulon Progo, Yogyakarta, Indonesia.

2 Research method

The research method is an important thing which is used as reference studies for research stage. In this study, the first stage is carried out based on the field data collection of water level from monitoring Peil Schaal and discharge of seepage monitoring V-Notch in the field at zero rainfall.

Sermo Dam modeling done by using PLAXIS in analyzing every operational conditions reviewed in the study. There are three cross sections on Sermo Dam, which are sta 12, sta 15, and sta 18, but in this study the cross section is reviewed in sta 15. Further analysis discharge of seepage, settlement, and the pore water pressure are based on modeling using PLAXIS and the explanation can be seen as follows.

2.1 Field data collection

Water levels are needed to determine the conditions used in this research analysis. In this study, determined the water level used is at the zero rainfall. Then graphed the relation between water level and discharge of seepage measurement with V-Notch, so it can be determined the discharge of seepage for each condition water levels is analyzed. Graph relation between water level and discharge of seepage can be seen in Figure 1, which uses 120 data field from 2000 to 2014 based on Departemen Pekerjaan Umum [7]. The graph's correlation value (R^2) is 0.666.

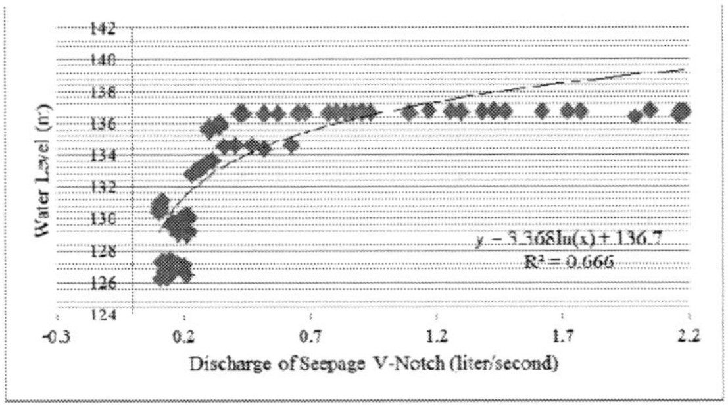

Fig. 1. Correlation between water level and discharge of seepage without rainfall condition

Water levels reviewed in this study analysis are 126.33 meters; 133.47 meters; 136.60 meters; and 136.71 meters. The selection of water level was adjusted to the availability of data piezometers in the field which is recorded every seven days. This is because before analysing this research study, then also do the data validation checks between the value of pore water pressure monitoring with piezometers in the field and value of pore water pressure modeling simulation results using PLAXIS.

Based on the graph in Figure 1 and four water level that have been determined for analysing in this study, it can be obtained the new discharge of seepage V-Notch that can be seen in Table 1. Standard error of 120 data is 0.05, whereas the standard deviation is 0.537, so that can be determined the uper value limit and lower value limit of the new discharge of seepage. The uper value limit and lower value limit for the new disharge of seepage can be seen in Table 1.

Table 1. New discharge of seepage data through V-Notch

Water Level (m)	Discharge of Seepage V-Notch (liter/sec)	Uper Limit (liter/sec)	Lower Limit (liter/sec)
126.33	0.05	0.58	0.00
133.47	0.38	0.92	0.00
136.60	0.97	1.50	0.43
136.71	1.00	1.54	0.46

2.2 Modeling of Sermo Dam

Sermo Dam is included in zonal dam, consisting of permeable zone (shell), impermeable zone (core), and a filter zone (classification zone based on Sosrodarsono and Takeda [8]). Sermo Dam itself is a rockfill dam zonal types with clay core.

Figure 2 shows a cross-sectional Sermo Dam which is reviewed in this study, namely the sta 15. Figure 3 shows a cross section Sermo Dam on the sta 15 which is modeled using PLAXIS. It should be noted that at this Sermo Dam given chamber of 2.142 meters, from a height of 141.00 meters to 143.142 meters.

This study is conducted using modeling Mohr-Coulomb on PLAXIS, which is applying the finite element method by dividing the dam structure into small dots related with nodal (meshing). The finer meshing is done, then the result will be more accurate. In this study, using very fine type of meshing with 15-node. Very fine type of meshing produces about 1000 elements [9].

Parameters of material at the dam determined based on data from soil investigation by Caturbina Guna Persada [10] and field testing data from Departemen Pekerjaan Umum [11]. The data's parameters materials bedrock based on design data from Departemen Pekerjaan Umum [12].

Determining the value of permeability coefficient is from the result discharge of seepage measurements using PLAXIS modeling and calibrated with discharge of seepage from monitoring V-Notch in the field at water level 135.20 meters. The result in this discharge of seepage monitoring with V-Notch is 1.055 liters/second [7].

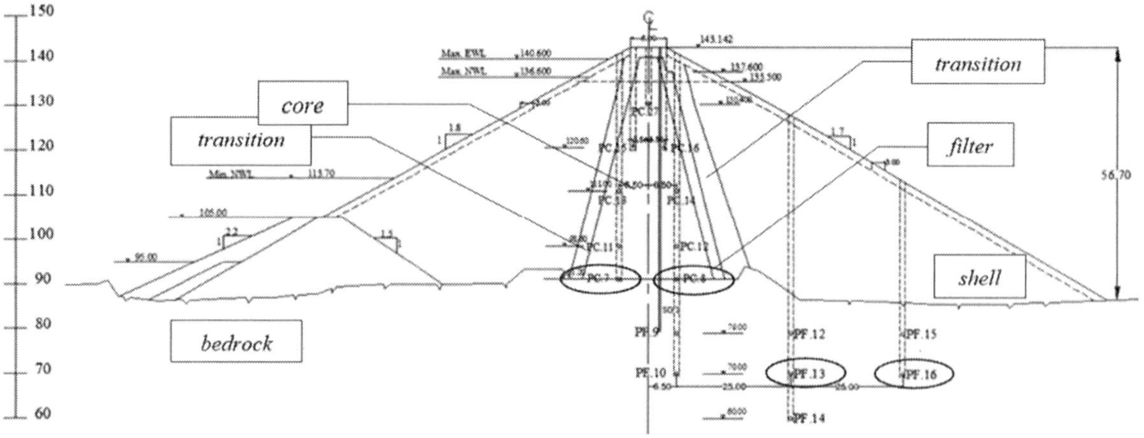

Fig. 2. Location of piezometer within the cross section of Sermo Dam [13]

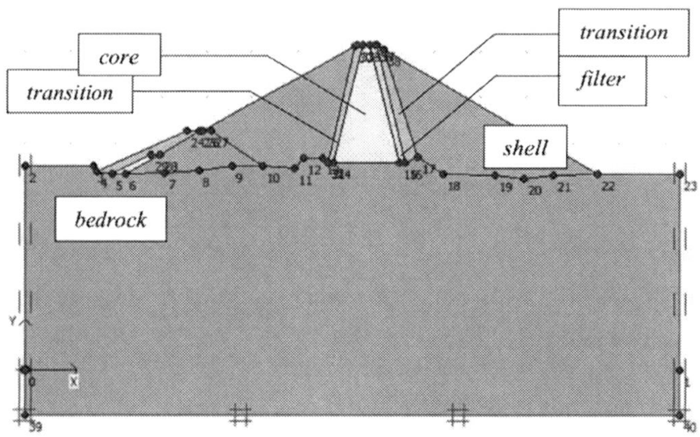

Fig. 3. PLAXIS modeling overview of Sermo Dam

Table 2. PLAXIS parameters of material used in the model of Sermo Dam

Zone of Dam	Parameter			
	$\gamma_{saturated}$ (kN/m^3)	$\gamma_{unsaturated}$ (kN/m^3)	Cohesion (c) (kN/m^2)	Angle of Friction (^{0})
Core	18.74	15.41	8.3	32
Filter	21	18.5	0.001	35
Transition	21.7	21.6	0.001	35
Shell	21	20	0.001	43
Bedrock	21.82	20.12	680	55

Table 3. PLAXIS parameters of material used in the model of Sermo Dam

Zone of Dam	Parameter		
	Coefficient of Permeability (*k*) (m/day)	Young's Modulus (E) (kN/m^2)	Poisson (*v*)
Core	2.32 x 10^{-4}	5500	0.45
Filter	134.10	6000	0.30
Transition	14.292	8000	0.30
Shell	12154.69	200000	0.35
Bedrock	6.00	210000	0.45

Calibration the parameters material is performed at three station locations within Sermo Dam, which are sta 12, sta 15, and sta 18, along 190 meters as the length of the peak of dam, so expect the results is obtained in accordance with the real situation on the field. Neither, the parameter values of Young's modulus and Poisson value figures on core zone, which are determined using back analysis method by calibration with settlement gauge. Thus parameter of materials values shown in Table 2.

2.3 Discharge of seepage analysis

Discharge of seepage calculation by analyzing the resultant discharge of the dam cross-section modeling using PLAXIS, which is calculated from the total discharge of seepage along 190 meters as the length of the dam. Discharge of seepage calculation results using modeling PLAXIS can be compared with measurements of the new discharge of seepage from V-Notch which has been calculated in Table 1, for each condition is analyzed.

Pore water pressure is measured at each stress point in PLAXIS modeling that has nearly same coordinates with the coordinates of the piezometer in field under review. Piezometer's reviewed in this study (see Figure

2) are on the upstream side of the PC-7 and PC-8 (core zone). Meanwhile, the downstream are in PF-13 and PF-16 (bedrock zone). Coordinates PC-7 and PC-8 are located at 143.85; 91.20 and 156.85; 91.20. While coordinate PF-13 and PF-16 are located at 181.85; 70.0 and 206.850; 70.0. In this study do a review of the settlement, as one aspect of stability. Analyse settlement is to look at the maximum settlement that occurred in the Sermo Dam using PLAXIS modeling.

3 Results and discussions

Based on the analysis performed, the results discharge of seepage using PLAXIS modeling in each waters levels are analyzed in this study shown in Table 3. From Table 3 it can be concluded that the higher water level then also higher the discharge of seepage. This is according to research that have been done before [3], which examined the effects of fluctuations in water level in the dam seepage using laboratory models. The stability in this study to evaluated is settlement. Large drop settlement can lead to instability of the dam. The relation of water level and the maximum settlement in the core zone is shown in Table 4.

Table 4. Discharge of seepage analysis on Sermo Dam using PLAXIS

Water Level (m)	Discharg e of Seepage V-Notch (lt/sec) *	Uper Limit (lt/sec)	Lower Limit (lt/sec)	Discharge of Seepage Using PLAXIS (lt/sec)
126.33	0.05	0.58	0.00	0.58
133.47	0.38	0.92	0.00	0.69
136.60	0.97	1.50	0.43	0.71
136.71	1.00	1.54	0.46	0.83

* Based on Figure 1 and Table 1

Table 5. Maximum settlement on core zone of Sermo Dam using PLAXIS modeling

No	Water Level (m)	Maximum Settlement in the Core Zone (m)
1	126.33	1.10
2	133.47	1.14
3	136.60	1.16
4	136.71	1.16

Table 4 shows that in operational condition, the higher water level so does the settlement. This is relevant with research that have been done before [5], which examines monitoring settlement in Shuibuya Dam, China. The monitoring results show that when the water level increases, the settlement in dam will also increase. Additionally, there is also a study that studied the effect of fluctuations in water level in the dam deformation in laboratory model which is giving relevant conclusion [6].

According to the tolerable resulting discharge of seepage should not exceed 0.56 liters/minute/meter, for high dams greater than 40 meters [14]. The sta 15 of Sermo Dam which is analyzed in this study, has a height of 56.70 meters (Figure 2). Meanwhile, based on Table 3, the maximum discharge using PLAXIS is in the water level 136.71 meters with a discharge of 0.83 liters/sec or equal to 0.26 liters/minute/meter. The discharge value is smaller than 0.56 liters/minute/meter, so it is still less than the tolerable discharge requirements. Meanwhile, when seen from the settlement, the value of the maximum settlement result from numerical modeling using PLAXIS is 1.16 meters in water level of 136.71 meters (Table 4). The tolerance settlement at the end construction is 2% of dam height [15]. It is approaching with the value of settlement result from modeling using PLAXIS.

In this study, also given the relation between the water level and pore water pressure result using PLAXIS modeling. As already explained, that piezometers are reviewed on the upstream and downstream, they are PC-7, PC-8, PF-13, and PF-16 (Figure 2). The value of pore water pressure from PLAXIS modeling is obtainabled by taking the stress point in PLAXIS which have similiar or approaching coordinates (in x-axis and y-axis) with the coordinates in each piezometer that is reviewed. Relation between water level and pore water pressure in PC-7 and PC-8 modeling using PLAXIS can be seen in Table 5. Meanwhile Table 6 shown the relation between the water level and pore water pressure in PF-13 and PF-16.

Table 6. Relation between water level and pore water pressure on the PC-7 and PC-8 using PLAXIS

Water Level (m)	Pore Water Pressure			
	(PC-7)		(PC-8)	
126.33	34.57 kPa	126.32 m	16.83 kPa	126.67 m
133.47	41.20 kPa	133.46 m	33.32 kPa	130.97 m
136.60	54.30 kPa	135.59 m	45.88 kPa	132.25 m
136.71	66.19 kPa	136.50 m	61.13 kPa	135.25 m

Table 7. Relation between water level and pore water pressure on the PF-13 and PF-16 using PLAXIS

Water Level (m)	Pore Water Pressure			
	(PF-13)		(PF-16)	
126.33	6.79 kPa	88.46 m	4.30 kPa	86.67 m
133.47	9.50 kPa	87.66 m	9.47 kPa	85.59 m
136.60	14.26 kPa	87.01 m	10.54 kPa	85.28 m
136.71	16.17 kPa	86.97 m	11.54 kPa	85.37 m

Table 5 and Table 6 show that the higher water level, the higher the pore water pressure, and at a position downstream of the dam (bedrock zone in PF-13 and PF-16), the pore water pressure will be smaller than in upstream of the dam (core zone in PC-7 and PC-8). In addition, it can be concluded that when the water level increases, the pore water pressure will also increases. Increasing water level can cause an increase pore water pressure, this is because the material is in the saturated condition resulting from the flow of water seepage into the body of the dam, as a consequence of the water level height. Saturated material will increase pore water pressure, especially with the long drainage because the material in the core of Sermo Dam is a cohesive soil (as a location of PC-7 and PC-8). So that, in PC-7 and PC-8 have value of pore water pressure greater than in the PF-13 and PF-16. From Figure 2, is known that PF-13 and PF-16 are located in the bedrock zone or foundation of the dam, which is composed of andesite breccia rocks making the material easier for draining water seepage.

The increase in pore water pressure that occurs may lead to increases the discharge of seepage, which can be seen in Table 3. The discharge of seepage could cause settlement that occurred also increased significantly (Table 4). Increasing the settlement occured because of water seepage in the dam body causes the material changes in saturated condition. Then, the value of the materials will changed. So that, the stress will reduced. The stress reduction will impact on increasing the settlement and reducting the stability of dam in retaining water. This is also according to research conducted on a high wall [4], which is an increasing pore water pressure will lead to a reduction in stress, while decreasing pore water pressure will lead to an increasing in stress. When the stress increases, so that can increase the soil strength to against instability.

4 Conclusion

Based on the existing problem in this study is about the influence of pore water pressure on the seepage and stability with a case study on Sermo Dam, could be concluded as follows:

1. Analysis of the PLAXIS modeling shows that water levels affect the increase of pore water pressure. The higher the water level, the higher the pore water pressure. With the increase in pore water pressure, the discharge of seepage that occurs will also increase.
2. Increasing the pore water pressure due to change of water level height led to increased discharge of seepage and settlement. This is due to water seepage causes the materials become saturated and the value of materials will changed and the stress will reduced. Reduction of stress will impact on the instability of dam structure.

This research supported by Work Unit of The Central River Region (BBWS) Serayu Opak for helping to collect data Sermo Dam.

References

1. Indonesian National Committee on Large Dams. (Buletin No.38-39-40 TH.XII **Kwartal I/II/III – 2009**, 16, 2010)
2. Departemen Pekerjaan Umum. *Pedoman Pengendalian Rembesan pada Bendungan Urugan*. (Jakarta: Direktorat Jenderal Sumber Daya Air, Direktorat Sungai dan Waduk, 2005)
3. H. Sudardja, Civil Engineering Forum, **XXI/1**, 1199 – 1207 (2012)
4. K.L. Morton, M.C. Muresan, F.R. Debswana, *Importance of Pore Pressure Monitoring in High Walls*. The Southern African Institute of Mining and Metallurgy, 225 – 238 (2008)
5. W. Zhou, S. Li, Z. Zhou, X. Chang, *Remote Sensing of Deformation of a High Concrete-Faced Rockfill Dam Using InSAR: A Study of the Shuibuya Dam, China*. Remote Sens **8**, 255, 1 - 15, (2016)
6. N. Meisyara. *Pengaruh Fluktuasi Muka Air Reservoir terhadap Deformasi dan Stabilitas Bendungan Tanah (Uji Model Laboratorium)*. (Tesis Universitas Gadjah Mada, 2012)
7. Departemen Pekerjaan Umum. *Bendungan Sermo*. (Satuan Kerja Balai Besar Wilayah Sungai Serayu Opak, Yogyakarta, 2015)
8. S. Sosrodarsono and K. Takeda. *Bendungan Type Urugan*. (Jakarta: Pradnya Paramita, 1977)
9. R.B.J. Brinkgreve. *Manual PLAXIS 2D Version 8*. (Delft University of Technology & PLAXIS b.v, Belanda, 2007)
10. Caturbina Guna Persada. *Pekerjaan Studi Faktor Lingkungan dan Stabilitas Lereng Bendungan Sermo Tahun 2005*. (Laporan Studi Teknik, Yogyakarta, 2005)
11. Departemen Pekerjaan Umum. *Laporan Kajian Pelaksanaan Pembangunan Waduk Sermo D.I. Yogyakarta* (Direktorat Jenderal Pengairan, Unit Keamanan Bendungan, Yogyakarta, 1996)
12. Departemen Pekerjaan Umum. *Pekerjaan Desain Detail Proyek Bendungan Sermo*. (Direktorat Jenderal Pengairan, Direktorat Irigasi II, Proyek Irigasi Kali Progo, Yogyakjarta, 1985)
13. Departemen Pekerjaan Umum. *Laporan Ringkas Waduk Sermo*. (Direktorat Jenderal Pengairan, Proyek Pembangunan Waduk Sermo, 1994)
14. B. G. Look. *Handbook of Geotechnical Investigation and Design Tables*. (Taylor & Francis Group, London, 2007)
15. C. Kutzner. *Earth and Rockfill Dams, Principles of Design and Construction*. (A.A. Balkema, Rotterdam, Brookfield, 1997)

* Corresponding author: undayanicita@yahoo.com

Rainwater quality improvement using zeolite, activated carbon, limestone and preheated 400°C limestone

Meilani[1,*], *S.* Syafalni[2], *Yulianto* Santoso[1], and *Gorga* Green Malau[1]

[1]Civil Engineering Department, Bina Nusantara University, 11480 Jakarta, Indonesia
[2] Civil Engineering Department, National Institute of Sciences and Technology, 12640 Jakarta, Indonesia

Abstract. Resources of clean water have decreased significantly in recent years, one of the reasons is water pollution. Utilization of rainwater as clean water resource is among the best option, especially in big cities like Jakarta. Excessive rainwater during wet season can be saved and used as the source of clean water during dry seasons. However, rainwater contains some pollutants, thus suitable treatment method should be utilized to improve its quality as clean water for daily needs. This study compared the effectiveness of zeolite, limestone, activated carbon and preheated 400°C limestone as additive to improve the rainwater quality. Results show that both limestone and preheated 400°C limestone improve all the parameters observed, meanwhile both activated carbon and zeolite had shown to have detrimental effect on water conductivity.

1 Introduction

Although the earth surface is covered almost 70 % by water, but only approximately 3 % of water can be used [1]. People in Indonesia, mostly in large city like Jakarta, using ground water as one of the water resources to fulfill their daily needs [2]. The rainwater has not been an option yet for people in Jakarta while there are many advantages that can be obtained from rainwater [3].

Rainwater in Indonesia contains acid and the pH level is 5.6 [4]. It needs treatment so it can be used for daily needs.

Water conductivity, pH and TDS are the water characteristics to be observed. Water conductivity shows the ability of water to transmit electricity [5]. Water pH is indicator of level of acidity and basicity of water [6]. Water TDS is the total amount of ions, including minerals, salts or metals dissolved in the water [7].

2 Experimental details

The first stage of experiment was to determine the material used as filter. Zeolite, activated carbon, limestone [8,14-15] and preheated 400°C limestone were used to improve the rainwater quality.

Zeolite was used as an absorbent for reducing water hardness[9]. Activated carbon removed water odor and taste, absorbed water chlorine and purified water [10]. Limestone purified water and stabilized water pH. Limestone (CaCO3) was heated to produce Calcium Oxide (CaO) to reduce the water hardness [11].

Zeolite was soaked using 1 N NaCl for 24 hours.Activated carbon was soaked using 1 N NaOH for 24 hours. Then both were dried using an oven with a temperature of 115°C for 24 hours [12,13].

Limestone was heated gradually. The temperature was increased by 100°C every 15 minutes, starting from 200°C and ended at 400°C [14].

Each of materials were categorized into 3 different sizes: large size (materials passing sieve number 4 and retained by sieve number 8), medium size (materials passing sieve number 8 and retained by sieve number 10) and small size (materials passing sieve number 10 and retained by sieve number 16) [14,15].

Each variation of size divided to 10 weight of variation (0.5 g, 1 g, 3 g, 5 g, 7 g, 10 g, 15 g, 20 g, 25 g, 30 g). This experiment was conducted 3 times for each variation.

3 Results

3.1 Conductivity

Zeolite increased the water conductivity as shown in Fig.1. The results of water conductivity using zeolite varied from 0.01 to 0.25. Medium sized zeolite resulted the highest conductivity compared to the other sizes. Large sized zeolite and small sized zeolite resulted the similar effect on water conductivity. The addition of zeolite weight into the water increased the value of water conductivity.

As shown in Fig.2, the water conductivity increased significantly because of activated carbon. The results of water conductivity using activated carbon varied from 0.03 to 1.29. The results showed the size of activated carbon affected the water conductivity. Large sized

* Corresponding author: meilani@binus.edu

activated carbon resulted the highest conductivity compared to the other sizes. Small sized activated carbon resulted the lowest conductivity compared to the other sizes. The addition weight of large sized activated carbon increased the value of water conductivity. When the weight of medium sized activated carbon were 0.5 g, 1 g, 3 g, 5 g, 7 g, 10 g, 15 g and 20 g, the value of water conductivity gradually increased. But when the weight of medium sized activated carbon was 25 g, the value of water conductivity decreased. The value of water conductivity increased again when the weight of medium sized activated carbon was 30 g. The water conductivity of small sized activated carbon kept increasing until the weight of 25 g, and it decreased when the weight was 30 g.

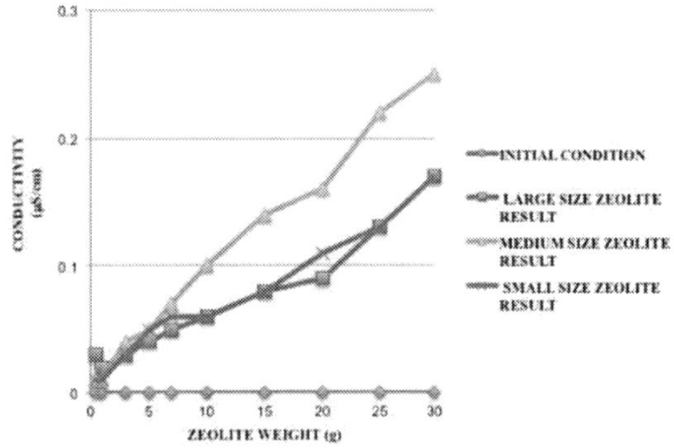

Fig. 1. Water conductivity using zeolite

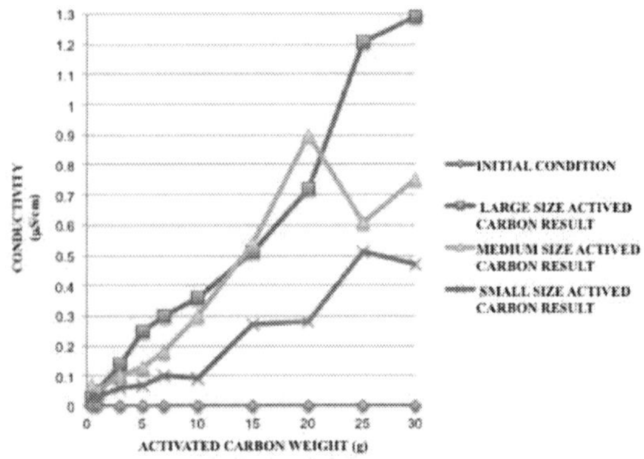

Fig. 2. Water conductivity using activated carbon

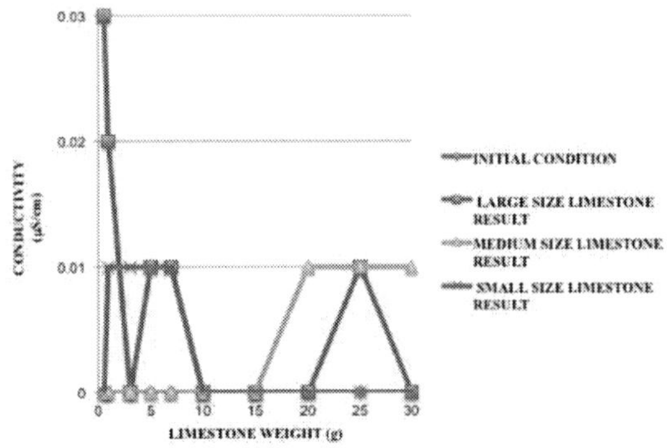

Fig. 3. Water conductivity using limestone

Both preheated 400°C limestone and limestone, didn't affect much on the water conductivity. It can be seen at Fig.3 and Fig.4. As shown in Fig.3., the results of water conductivity using limestone varied from 0.00 to 0.03. The large sized limestone with weight less than 3 grams resulted the highest conductivity compared to the other sizes and weights.

As shown in Fig. 4, the results of water conductivity using preheated 400°C limestone varied from 0.00 to 0.02. The small sized preheated 400°C limestone with weight less than 3 grams resulted the highest conductivity compared to the other sizes and weights.

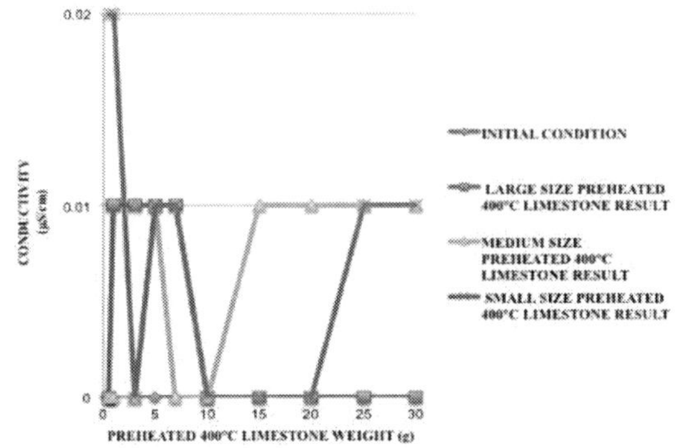

Fig. 4. Water conductivity using preheated 400°C limestone

3.2 Water pH

The increasing results of water pH using zeolite varied between 0.37 and 2.31 point higher from the initial condition. Large sized zeolite and medium sized zeolite resulted the similar effect on water pH but the large sized zeolite resulted the highest water pH compared to the other sizes, as shown in Fig.5. The addition weight of large and medium sized zeolite increased the value of water pH.

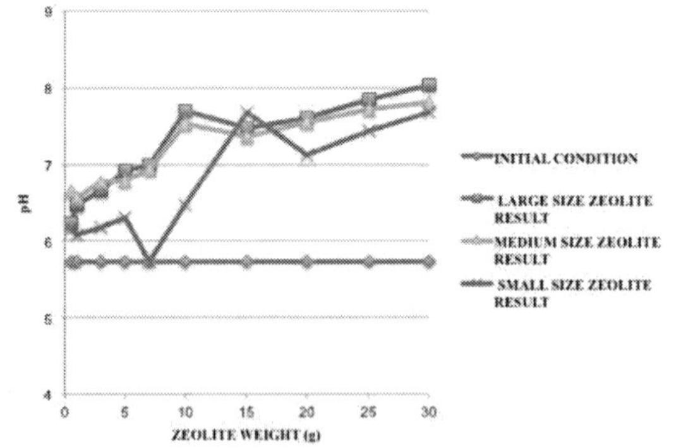

Fig. 5. Water pH using zeolite

The increasing results of water pH using activated carbon varied between 0.58 and 2.32 point higher from the initial condition. The large sized activated carbon resulted the highest water pH compared to the other sizes, as shown in Fig.6.

 DOI: 10.1051/matecconf/201710105008

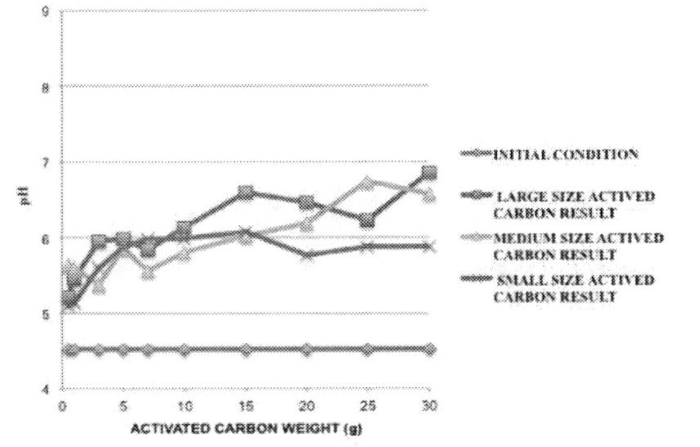

Fig. 6. Water pH using activated carbon

The increasing results of water pH using limestone varied between 1.01 and 2.18 point higher from the initial condition. The small sized limestone resulted the lowest water pH and the large sized limestone resulted the highest water pH compared to the other sizes, as shown in Fig.7.

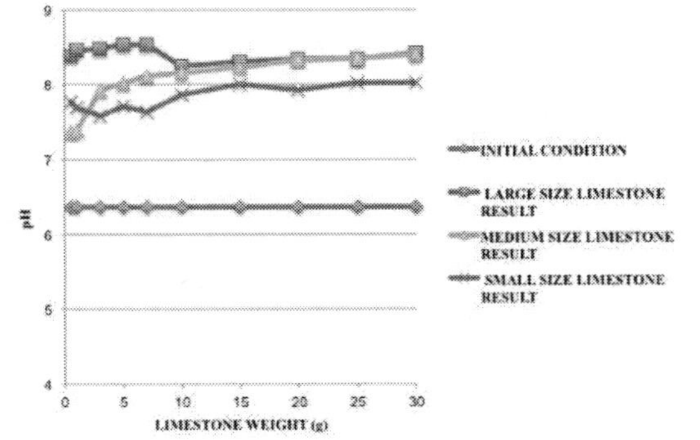

Fig. 7. Water pH using limestone

The increasing results of water pH using preheated 400°C limestone varied between 1.06 and 2.64 point higher from the initial condition. The large sized preheated 400°C limestone resulted the lowest water pH and the small sized preheated 400°C limestone resulted the highest water pH compared to the other sizes, as shown in Fig.8.

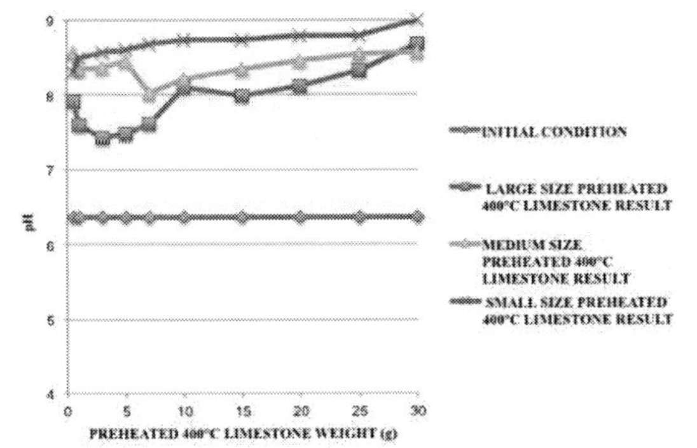

Fig. 8. Water pH using preheated 400°C limestone

3.3 Water TDS

As shown in Fig.9, the results of water TDS using zeolite varied between 0.00 and 0.12. Medium sized zeolite resulted the highest TDS compared to the other sizes. Large and small sized zeolite resulted the similar effect on water TDS.

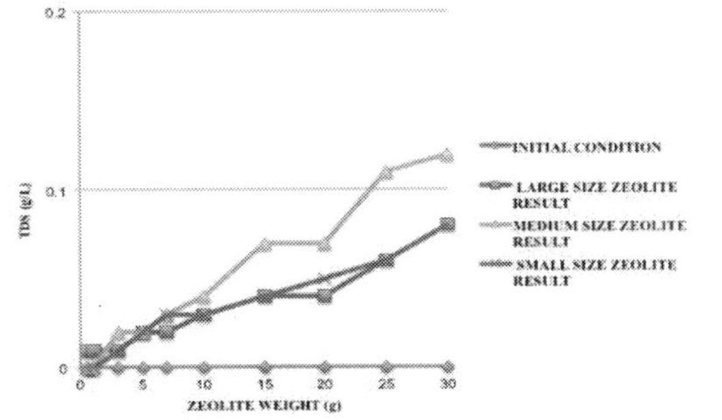

Fig. 9. Water TDS using zeolite

As shown in Fig.10, the water TDS increased because of activated carbon. The results of water TDS using activated carbon varied from 0.01 to 0.64. The large sized activated carbon resulted the highest water TDS compared to the other sizes. The small sized activated carbon resulted the lowest water TDS compared to the other sizes.

The results showed the size and the weight of activated carbon affected the water TDS. The addition weight of large sized activated carbon increased the value of water TDS. When the weight of medium sized activated carbon were 0.5 g, 1 g, 3 g, 5 g, 7 g, 10 g, 15 g and 20 g, the value of water TDS gradually increased. But when the weight of medium sized activated carbon was 25 g, the value of water TDS decreased. The value of water TDS increased again when the weight of medium sized activated carbon was 30 g. The water TDS of small sized activated carbon kept increasing until the weight of 25 g, and it decreased when the weight was 30 g.

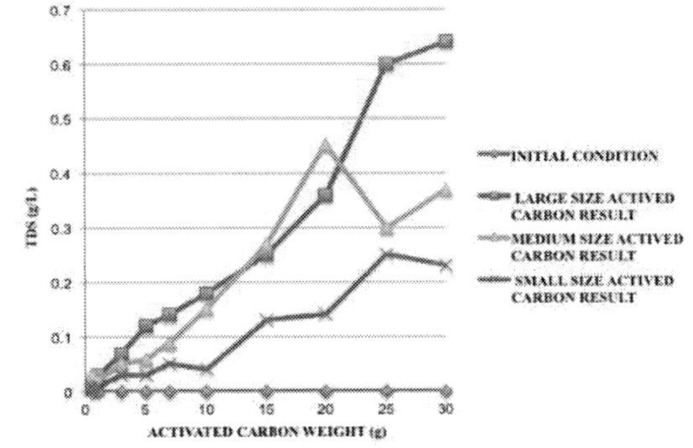

Fig. 10. Water TDS using activated carbon

As shown in Fig. 11 and Fig. 12, both preheated 400°C limestone and limestone, didn't affect much to the water TDS. The results of water TDS using limestone and preheated 400°C limestone varied from 0.00 to 0.01. The

DOI: 10.1051/matecconf/201710105008

large sized preheated limestone with weight less than 3 grams resulted the highest TDS compared to the other sizes and weights. The medium and small sized limestone had shown no effect on the water TDS, as shown in Fig. 11.

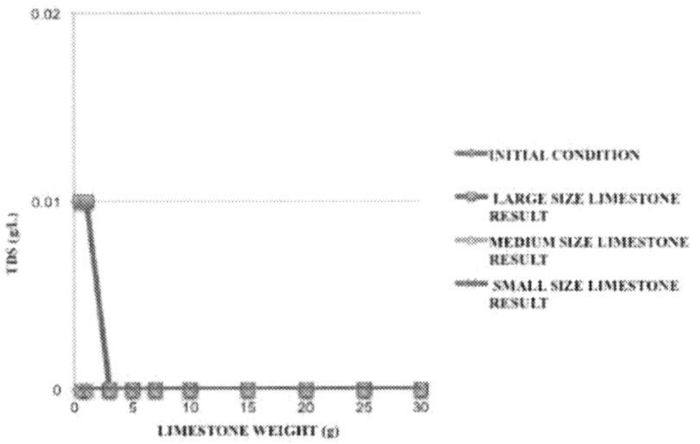

Fig. 11. Water TDS using limestone

The small and large sized preheated 400°C limestone had shown the similar effect on the water TDS. The small and large sized preheated 400°C limestone with weight less than 3 grams resulted the highest water TDS compared to the other sizes and weights. The medium sized preheated 400°C limestone had shown no effect on the water TDS, as shown in Fig. 12.

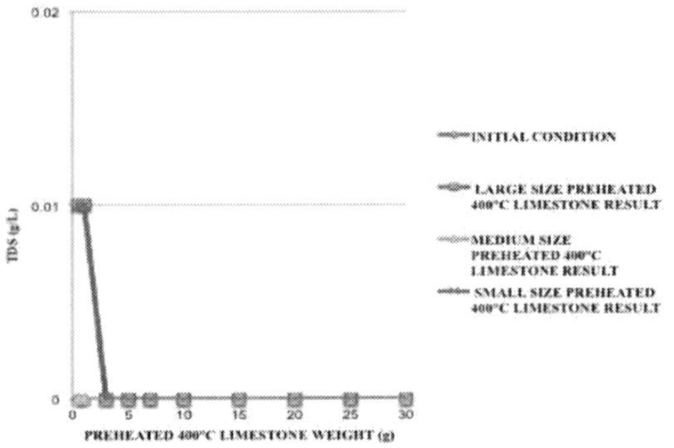

Fig. 12. Water TDS using preheated 400°C limestone

4 Conclusion

From the results, it concluded that both limestone and preheated 400°C Limestone were found to improve all parameters observed.

The small sized preheated 400°C Limestone was found to be the best material to obtain low water conductivity, higher pH and lower water TDS values.

Both activated carbon and zeolite had shown to have detrimental effect on water conductivity.

Activated carbon had shown similar effect on water conductivity and TDS.

This research was funded by Ministry of Research, Technology and Higher Education of the Republic of Indonesia (contract number: 791/K3/KM/SPK.LT/2016).

References

1. Information from http://savethewater.org/ education- resources/water-facts (August 2016)
2. Information from http://www.ipsnews.net/2000/02/ environment-indonesia-drowing-in-water-woes (August 2016)
3. Information from http://rainharvesting.com.au /knowledgecenter/ reasons-for-using-rainwater (August 2016)
4. Information from http://www.bmkg.go.id/BMKG_ Pusat/Kualitas_Udara/Informasi_Kimia_Air_Hujan. bmkg (August 2016)
5. Information from http://www.lenntech.com/ applications/ultrapure/conductivity/water-conduct ivity.htm (November 2016)
6. Information from http://water.usgs.gov/edu/ph.html (November 2016)
7. Information from http://www.water-research.net/index.php/water-treatment/tools/total-dissolved-solids (November 2016)
8. S. Syafalni, Meilani, O. Setyandito, F. Han Yang, IJAES, **9**, 1811-1826 (2014)
9. G.B. Gholikandi, M.M. Baneshi, E. Dehghanifara, S. Salehi, A.R. Yari, IJOT, **3**, 302-310 (2010)
10. Information from http://www.hc-sc.gc.ca/ewh-semt/pubs/water-eau/devices-dispositifs-eng.php (November 2016)
11. Information from http://www.randwater.co.z a/waterandinfastructuremanagement/pages/waterpur ification.aspx (November 2016)
12. S. Syafalni, I. Abustan, I. Dahlan, K.W. Chan, G. Umar, MAS, **6**, 37-51 (2012)
13. S. Syafalni, I. Abustan, A. Brahmana, S.N.F Zakaria, R. Abdullah, MAS, **7**, 39-49 (2013)
14. G.G. Malau, *Perbaikan Kualitas Air Hujan Untuk Air Minum (Menggunakan Karbon Aktif, Batu Kapur Alami, Batu Kapur yang Dipanaskan 400 C dan Zeolite), Bachelor Thesis* (Bina Nusantara University, 2014)
15. Y. Sentoso, *Perbaikan Kualitas Air Hujan Menggunakan Karbon Aktif, Batu Bata,Limestone dan Zeolit untuk Air Minum, Bachelor Thesis* (Bina Nusantara University, 2014)

* Corresponding author: meilani@binus.edu

Cracking of open traffic rigid pavement

Chatarina Niken[1,*], *Yudi* Siswanto[1], Widodo[1], and *Elly* Tjahjono[2]

[1]Civil Engineering Department, Faculty of Technology, University of Lampung, Indonesia
[2]Civil Engineering Department, Faculty of Technology, Universitas Indonesia, Indonesia

Abstract. The research is done by observing the growth of real structure cracking in Natar, Lampung, Indonesia compared to C. Niken's et al research and literature study. The rigid pavement was done with open traffic system. There are two main crack types on Natar rigid pavement: cracks cross the road, and cracks spreads on rigid pavement surface. The observation of cracks was analyzed by analyzing material, casting, curing, loading and shrinkage mechanism. The relationship between these analysis and shrinkage mechanism was studied in concrete micro structure. Open traffic make hydration process occur under vibration; therefore, fresh concrete was compressed and tensioned alternately since beginning. High temperature together with compression, cement dissociation, the growth of Ca^{2+} at very early age leads abnormal swelling. No prevention from outside water movement leads hydration process occur with limited water which caused spreads fine cracks. Limited water improves shrinkage and plastic phase becomes shorter; therefore, rigid pavement can't accommodate the abnormal swelling and shrinking alternately and creates the spread of cracks. Discontinuing casting the concrete makes both mix under different condition, the first is shrink and the second is swell and creates weak line on the border; so, the cracks appear as cracks across the road.

1 Introduction

Rigid pavements generally are made from concrete, with or without reinforcement. Rigid pavement lays over sub base and sub grade as shown in Figure 1. Sub base can be made from lean concrete or stone structure. Over the past few years, the reliance has shifted more to rigid pavement because of its low maintenance cost, long service life, and smoother riding surface [1]. Besides its of the advantages, rigid pavement is expensive due to the high cost of concrete and reinforcement. The economic consideration are carried out for the pavement design of a section by using the results obtained by design methods and their corresponding component layer thickness [2]. Because of low initial cost, flexible roads are preferred to concrete road pavement, but bituminous roads deteriorate during the rainy season and maintenance becomes costly [3]. In concrete pavement, a single concrete mixture design and structural surface layer are selected to resist mechanical loading without an attempt to affect pavement shrinkage, ride quality or noise attenuation adversely [4]. It is important to prevent water evaporation right after casting because micro cracks of rigid pavement can avoided. The micro cracks can grow and accumulate due to applied load and by climate. The accumulation of cracks can create deterioration in the concrete; therefore, rigid pavement performance is decreased and its service life becomes shorter. These circumstances can change into serious problems because accumulated cracks make the structure unsafe. Rigid pavement or concrete pavement needs high attention during planning, preparing, placing, curing and load scheduling. The faults in every step of producing rigid pavement leads to concrete deterioration, micro cracks and wider cracks.

There are many types of cracks such as the tearing of concrete through paver, plastic shrinkage cracking at pavement surface, map cracking (craze cracks), transfer and oblique cracks within the panel, random longitudinal cracks within the panel, corner cracks (break at panel corner), random transverse cracks at or near transverse joints, random longitudinal cracks at or near longitudinal joints, and cracks in front of saw during joint cutting [5]. Variation of crack width in Texas was studied [6]. They mentioned that an immediate decrease in crack width was observed after rain. There was clear evidence that moisture variation in concrete has substantial effect on concrete volume change and crack widths [6]. Locally calibrate the transfer cracking and IRI (smoothness) performance model for newly constructed and rehabilitated rigid pavement has been published [7].

Cracks are also caused by flexural fatigue, and non-homogenous deflection in sub-base or sub grade. The non-homogenous deflection has been led by erosion of the sub-base or sub-grade. Erosion of the sub-base occurs if it is structural stone; thus, this erosion is not similar in every point in the sub-base. Non-homogenous erosion makes deflection different at any point of rigid pavement. The swelling of roadbed soil also causes deformation of its upper layer [8]. Erosion and flexural crack types, generally begin from the bottom to the top, and crack by swelling of roadbed soil begin from the top. Top-down cracking has been found to be a predominant mode of asphalt pavement distress in Florida [9]. Deflection, erosion, and swelling always occur because of climate change; so, the base layer of rigid pavement

* Corresponding author: chatarinaniken@yahoo.com

has dynamic properties. Field observation has demonstrated that there is an increase in reflecting cracking of PCC (Portland Cement Composites) slabs placed over relatively a stiff base layer [10].

Foundation layers in finite element analysis of jointed pavement concrete system were done with appropriate idealization to predict the thermo mechanical response of jointed PCC slabs accurately [10]. Many researchers have also studied rigid pavement response using the finite element method [11-13]. Significant changes in the slab curvature shape (curl/warp) also occurs through thermal gradient, from positive to negative and vice versa. Even, critical cases in the stress were found in the combination of wheel load and the positive thermal gradient [14-17]. Statements as mentioned above lead particles movement to any direction. The horizontal movement of any point is almost uniform, but transverse movements are different [18].

From the above mentioned, it can be seen that there are complex mechanisms which cause deformation due to induced cracks such as the raw materials of concrete, mix design, casting system, treatment right after casting, curing system, loading time, loading system, fatigue, weather, erosion in sub-base/sub grade, and swelling of roadbed soil. Therefore, the study of cracking rigid pavement can involve a very wide field. The objective of this study is to understand the cause of rigid pavement cracking which occurs in Natar, Lampung Province, Indonesia.

2 Experimental details

The experimental details were divided into 2 parts: Natar, Lampung, Indonesia rigid pavement and Niken, Elly, and Supartono's research of rigid pavement (un-published).

The research was done through a case study in one place in Natar Lampung Province, Indonesia. The research was done by observing cracks which appeared in Natar. The observation included the age of lean concrete (LC) as sub-base and rigid pavement when the cracks appeared, the growth of cracks, crack width, quality of concrete, treatment right after casting, curing system, weather, and loading system. The rigid pavement was built with open traffic system: half part was done longitudinally, while another part was used as traffic.

Rigid pavement in Natar uses concrete quality of 30 MPa. The rigid pavement (RP) lays over the lean concrete (LC) with a compressive force (fc') of 15 MPa. Rigid pavement in Natar was build with the cross section as shown in Figure 1.

The embankment in Fig.1 is similar to roadbed soil according to AASHTO 1993. The lean concrete function is to obtain homogenous support for rigid pavement and also for rigid pavement plate work. There is no permeable sheet in the base and all sides of the lean concrete and rigid pavement. Right after casting, the lean concrete and rigid pavement were un-covered resulting in contact between the surface of lean concrete or rigid pavement and the surrounding weather occurred.

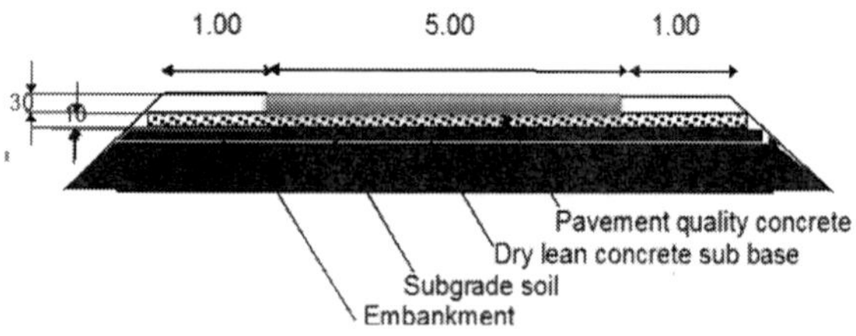

Fig.1. Cross section of the rigid pavement

Curing was done by wet sacks which were sprayed with water 3 times daily without de-molding the side part of both concrete layers. These sacks covered the plate surface for 1 until 7 days. Because of the lack of understanding and control, the sacks were not always wet during curing time. After that, the pavement came into direct contacts with the surrounding weather. The rigid pavement was cast when the lean concrete was 2 days of age.

3 Observation results

3.1 Natar Lampung Indonesia Rigid Pavement

When the lean concrete and rigid pavement was at the age of 1 day, many fine cracks appeared on the surface and all sides of the both. After side molds were released, there also were many cracks in the side part of the lean concrete and rigid pavement and some of them showed more crack width than general fine cracks (Fig.2).

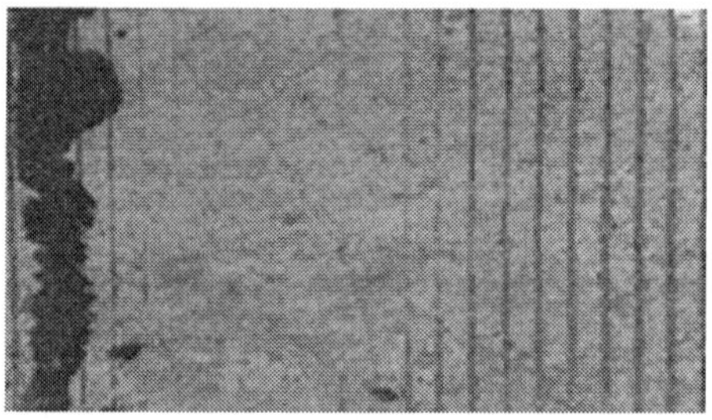

Fig. 2. Fine cracks over the surface of rigid pavement

Fig. 3. Single crack lying a cross the rigid pavement with fine cracks

Figure 3 and 4, show crack that is a single crack a cross the road. The crack divided the road into two parts. Besides the main crack, there are many fine cracks in the lean concrete and rigid pavement layer as shown in Fig.3 and 4.

DOI: 10.1051/matecconf/201710105009

Fig. 4. Single cracks with chain

3.2 Niken's et al research

There search of Niken et alwhich wereused a full scale plate (3m x 2m) with 130mm thickness to observe deformation behavior of plates.Analysis also involved the change of weather. Indonesian weather was observed over 3 years from November 2009 until December 2011 in Jakarta, Indonesia as shown in Fig. 5 and 6 [19].

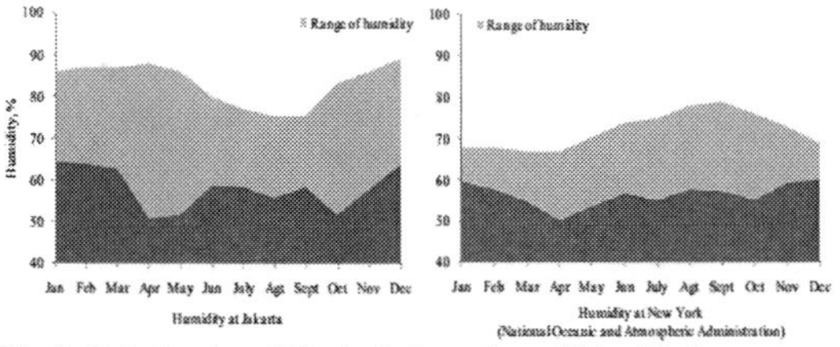

Fig 5. Relative humidity in Indonesia and NewYork

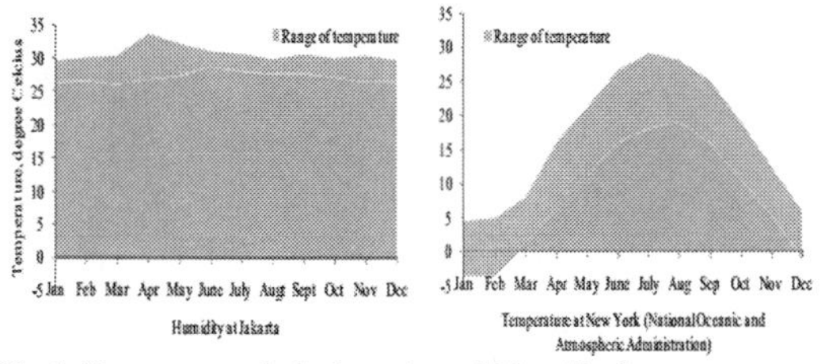

Fig 6. Temperature in Indonesia and New York

The difference between maximum and minimum humidity is about 25% (Fig. 5), while the temperature in Indonesia is always high throughout the year (Fig. 6). This research also used high range water reducer and silica fume. The compression strength of plate is 60 MPa. The concrete quality of Niken's*et al* research was high performance concrete (HPC); thus limited water and much OPC (ordinary Portland cement) were used (Table 1). The mix design of Niken et al's research, and Natar rigid pavement is displayed in Table 1.

The water to cementitious ratio (w/cm) is 0.264 (Table 1). To obtain the accuracy of w/c, HRWR/c, and w/cm, an electric scale was used. Coarse and fine aggregate are under saturated surface condition (SSD). Right after casting, the plate is covered with styrofoam to eliminate water evaporation. After the mold are removed, wet cure is done carefully for 1 to 7 days using sacks by continual monitoring and spreading water on sacks which are beginning to dry. Deformation and concrete temperature are obtained from the vibrating wire embedded strain gauge(VWESG). The VWESG has the ability to detect the strain up to 3000 με with an accuracy of about 0.025% and concrete temperature between -80°C and 60°C with about 0.5% accuracy. Observation was performed righ after pouring with the following schedule: 0-24 hours, every 15 minutes; 24-48 hours, every 60 minutes; days 3-7, every 2 hours; and one time each day using a read out.

Table 1. Material composition

Material	Natar		C Niken (kg/m³)
	LC (kg/m³)	RP (kg/m³)	
OPC	280	380	500
Silica fume	0	0	40
Coarse aggregate	1067	1059	935
Fine aggregate	819	754	800
HRWR	1.12	1.52	7.6
Water	187	183	142.6
w/c	0.668	0.482	0.285
HRWR/c	0.4%	0.4%	1.52%

w/c : water to cement ratio
HRWR/c : HRWR to cement ratio

Crack observations wereperformed using a loop 50 times larger displayed in Fig.7.

Fig. 7. Surface of concrete plates in C. Niken's et al research

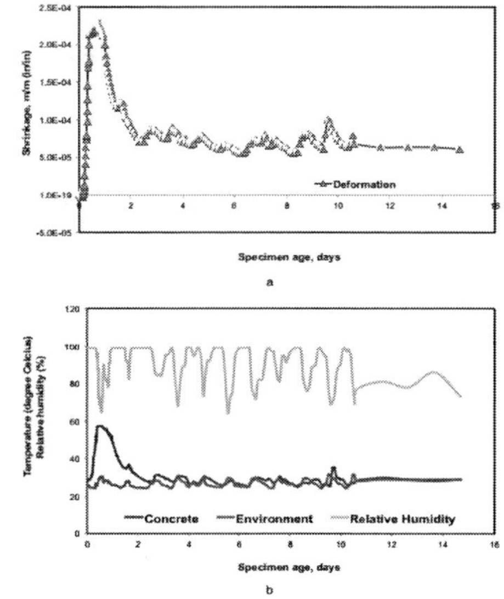

Fig. 8. Deformation of HPC rigid pavement, concrete and environmental temperature and relative humidity

No cracks were found on the surface and the side of the rigid pavement although limited water was used. The relationship between deformation and time, relative humidity, environment and concrete temperature during observation are displayed in Fig. 8.

Although the concrete quality of Natar is different, the relationship between the two types is available, such as the relationship between deformation and the age of concrete, relative humidity, environment and concrete temperature (Fig.8). By using supporting literature study, the cause of cracks could be obtained.

4 Discussion

Natar rigid pavement cracks occur 1 day after casting and before applied loads. Because there is no preventive mechanism to handle water movement outside the rigid pavement, environment in the surrounding weather completely influences the inner process of rigid pavement since the beginning. Therefore, the internal concrete mechanism together with the influence of the surrounding weather and open traffic as a unity is one factor. The first process is solution. This process is a transition from fresh concrete to hard concrete. In this phase the repulsive/dispersion force is dominating which makes the volume swell. Together with this process, the hydration process is also beginning. The hydration process is a chemical reaction, and in every chemical reaction, there is a change in heat, volume, and adhesive power. Chemical reaction results in free energy which causes particles to move [20]. This energy change is caused by pure thermal processes. The increase of heat is the evidence of an increase in chemical reaction;therefore, the hydration process occurs since casting (Fig.8).

The open traffic system makes the hydration process occur under vibration. Vibration makes the hydration process under compression and the tension condition alternately. Vibration makes coarse aggregate down to the bottom; this action inflicts compression. At the same time high temperature occurs in the concrete. The combination of compression and high temperature make volume swell abnormally [21]. In the tense condition, volume may shrink. A long-term plastic phase of fresh concrete was needed to accommodate both conditions which occur with high fluctuation according to the traffic. This condition could not be met because of the reasons below.

Vibration causes coarse aggregate to be placed at the bottom; hence, matrix becomes filled at the upper part of the rigid pavement, so that most of the hydration process occurs at the area. Plate has a surface which is more spacious than beam or column, without covering this surface, the influence of the surrounding weather is larger. It is clear that the top of rigid pavement is more sensitive to crack than the deeper part.

The hydration process consumes internal water; thus, it can be approached as progressive conversion of free water in the capillary pores to structural water in the solid hydration product. All hydration products contain chemically bonded water. Free water influences ionic transportation, and shrinkage. The condition in which there are no outside influences is called autogenous. No outside influence is needed, so that the amount of calculated water for the hydration process has not changed. Autogenous deformation is a change of volume purely through inner processes occurring right after casting. The maximum rate of autogenous deformation happens during the first 3-12 hours. Limited water causes increasing autogenous shrinkage. Autogenous shrinkage can be more than drying shrinkage [22]. Concrete temperature at this phase can reach 55^{0} C, at the age less than 1 hour (Fig. 8). Until the age of 2 days concrete temperature is higher than the surrounding temperature (Fig.8).

At this critical condition, Natar rigid pavement and lean concrete have no prevention from outside influences right after casting. Casting was done at day time; therefore, outside temperature is high and inside of concrete temperature is also high. Mosa *et al*, 2015, was explains that all crack types are caused by temperature [5].

Internal water which was needed for the hydration process moved outside rigid pavement; in addition to this, high temperature of concrete leads inner water to evaporate; thus, the hydration process occurs with water shortage. High temperature of the surrounding area leads surface water to evaporate. The large surface area makes much water evaporation.

Tension caused by vibration also increases the early shrinkage. At this age, capillary pores was dominating. Until the age of 14 days, the number of concrete pores decreases significantly (Fig.9). Pores with 10-20 cm in diameter in the C-S-H cluster are available [23]. These pores are medium capillary pores. Water in these pores can evaporate even in high humidity such as Lampung, Indonesia, as this water has strong menisci.

Strong menisci make capillary pores and surface forces increase. The plate surface area is larger than another; thus, surface force is also larger. Internal water decreases again because of this force. All these things cause early shrinkage to increases very rapidly.

The loss of water as mentioned above causes the loss of plasticity of rigid pavement also to occur very rapidly. The loss of plasticity is not accompanied with strong particles bond because silica bridges have not formed. Plastic losses and weak bond causes matrix unable to accommodate the change of abnormal swelling and shrinking ; as a result, the particles bonds break and cracks occur. Because top area is more sensitive than the deeper, a lot of fine cracks spread over the surface appear immediately as shown in Fig.2.

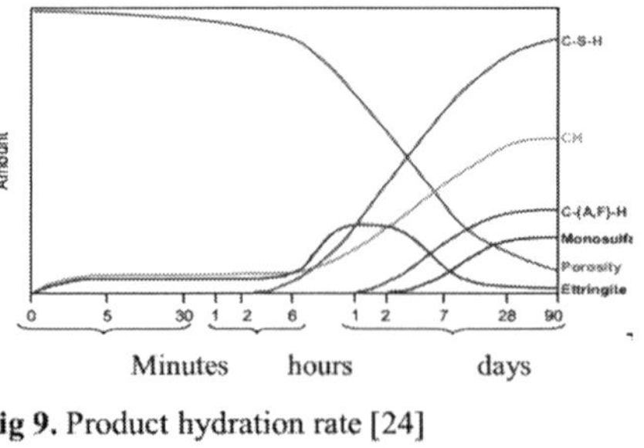

Fig 9. Product hydration rate [24]

Lacking of water at the early stages, can cause delayed ettringite formation or DEF. Ettringite formation causes expansion. If the ettringite formation happens during the plastic phase, expanded space becomes available automatically, but if the ettringite is formed at the beginning or during the hardening of the rigid pavement, deterioration will occur. Inner sulfat attacks concrete or the rigid pavement [25]. The deterioration of concrete will accelerate the growth of existing cracks.

Besides stresses as mentioned above, there are also product hydration forces, and disjoining pressure forces, caused by hidden water in the concrete. Product hydration force will increase according to the product hydration rate (Fig.9). All of stress in microstructure is called microprestress [26]. All stresses both at the level of microstructure and outside influences, such as vibration or surrounding weather should be accommodated by the rigid pavement.

Accumulation of all stresses will break the weakest of bonds and the stress vines and breaks other weak bonds. The position of the weak bond depends on the casting system. Casting of Natar rigid pavement was done from the end of the cross section of road to another end. Waiting time more than 1 hour (depending on the traffic weight) to the next casting makes the concrete character has a significant difference. The more weight traffic, the more compression on the fresh concrete, and the larger the abnormal swelling becomes. The difference is the first casting in the shrinkage phase, while the next casting is in abnormal swelling. The difference makes the connection line becomes weaker than another part. Cracks appear as a single line as shown in Fig.3.

New rigid pavement is cast over lean concrete at the lean concrete age of 2 days; therefore, lean concrete cracks have been growing. The cracks penetrate the young concrete of the rigid pavement, and the rigid pavement becomes cracked. The penetrate force from lean concrete is increases because of the increases of product hydration growth, capillary force, disjoining pressure and also vibration force. The similar forces also occur in rigid pavement. All of the forces hit the relative stiff layer (lean concrete); hence, the forces will be reflected. The reflected force will improve the breaking bond and cracks occur more and more. The circumstances are in accordance with the observation of Ashtiani *et al*, 2016 as mentioned above. To avoid these cracks, the new layer should be cast when the rate of the hydration product is low. According to Fig.9, the rate of product hydration growth declines at the age of 7 days. Besides this, to eliminate reflecting forces, the reflective layer damping force can be retrofitted. Rigid pavement or lean concrete length and width should also be restricted in order that the shrinkage over the width can still be handled.

Surrounding humidity is accompanied by concrete deformation until the concrete age is 11 days. The rhythm of deformation is inverted with the rhythm of humidity even in curing conditions. The highest deformation occurs at the lowest humidity. The peak of deformation occurs when the humidity is the lowest (Fig.8).

5 Conclusion

Based on field observations, in comparison to the research of C. Niken's*et al*, rigid pavement cracks have been studied. The causes of open traffic rigid pavement cracks are:

a. Vibration by open traffic causes coarse aggregate to move down, and concrete matrix fills in the upper part of rigid pavement, as surface stress becomes larger.
b. Vibration makes the hydration process occur under compression and tension alternately.
c. High temperature at the early hydration process together with compression leads to concrete swelling abnormally.
d. The absence of prevention from outside water movement makes the hydration process occur with limited water, and early shrinkage increases.
e. The long time plastic phase is needed to accomodate the both conditions.
f. Lacking of water at early age shortens plastic time; therefore, it can not accomodate the alternate of abnormal swelling and shrinking.
g. Product hydration bonds are not strong enough to sustain accumulated stresses, such as compression, tension, microprestress, and the influence of surrounding weather.
h. The bonds cracks and the stress continue to vine and breaks other weak bonds.
i. Discontinuing casting the concrete makes both mixes under different conditions: the first is shrink and the second is swell. This condition creates a weak line on the border of the second mix; thus, the cracks appear as a single crack a cross the road.
j. The lack of water by the absence of prevention for moving water inside the rigid pavement, high concrete temperature,high surrounding temperature, and delayed ettringite formation makes spread fine cracks appear over the surface of concrete.
k. Microprestress and stress from outside concrete such as vibration, and surrounding weather could be reflected by the refative base layer and penetrate the rigid pavement, and cracks occur more

Suggestions:

To avoid cracks, the suggestions are:

1. Materials conditions are similar to the mix design condition.
2. Sub grade/sub base should in stable condition.
3. If the sub base/ sub grade is lean concrete, lean concrete should free of cracks. If the lean concrete cracks, it is necessary to recover the cracks before casting the rigid pavement.
4. Lean concrete should be strong enough to sustain the rigid pavement.
5. Surrounding weather with high humidity (evening) is better time to cast.
6. All of molds should be covered with an impermeable sheet.
7. Planning of cast should consider: material and equipment readiness, anticipation in change of

weather and location of discharge casting. Casting should be done continuously.
8. Cover all of surface right after casting by an impermeable sheet.
9. Cover all of surface right after casting by an impermeable sheet and wet sacks if the rigid pavement is cast with the open traffic system.
10. Cover the stiff base layer with a damper sheet.

I would like to thank all contributors for their generous support to this paper. It is my pleasure to thank Gatot Eko Susilo, PhD., Meutia and Hanindya for their advices and help. We are grateful to the Department of Civil Engineering, the Faculty of Technology, University of Lampung and Universitas Indonesia for their support, spontaneous response, and generous assistence.

References

1. S. Singh, Sachdeva SN, Journal of Mechanical and Civil Engineering, 52-57 (2015)
2. S. Jain, Y.P. Joshi, S.S. Goliya. Int. Journal of Engineering Research and Application, **3** (5), 119-123 (2013)
3. J.D. Supe, M.K. Gupta. International Journal of Scientific & Engineering Research, **5** (11), 46-57 (2014)
4. J.Roesler, G.Paulino, C.Gaedicke, A. Bordelon, K. Park. Journal of the Transportation Research Board, 2037 **04** (2014)
5. A.M. Mosa, N.N. Ismail, N. I. MdYusoff, M.A Mubarak, N.A. Memon, M.R Taha, M. R. Hainin. Jurnal Teknologi (Science & Engineering), **76** (14), 105-119 (2015)
6. J.H Nam, D.H Kim, S.C. Choi, M. Won. Journal of the Transportation Research Board, 2037 **(01)** (2014)
7. S. W. Haider, W. C. Brink, N. Buch. Proceeding T & DI Congress, 100-110 (2014)
8. AASHTO: Guide for design pavement structure. American Association of State Highway and Transportation Official, Washington, II-37-II-66 (1993)
9. J. Wang, B. Birgisson, R. Roque. Journal of the Transportation Research Board, 2037 **(08)** (2014)
10. A.Z. Ashtiani, C. Tirado, C. Carrasco, S. Nazarian. International Journal of Pavement Engineering, 17 (10) (2016)
11. M.I. Khan, M.A Qadeer, A.B.Harwalkar. Journal of Mechanical and Civil Engineering, **11** (2), 90-107 (2014)
12. M. Gibigaye, C.PYabi, I. E. Alloba. *Dynamic response of rigid pavement plate base on inertia soil.* International Scholarly Research Notices, 2016, 1-9 (2015)
13. I. Harik, P. Jianping, H. Southgate, D. Allen. Technical papers, ASCE, **1** (127) 127-143 (1994)
14. A.E.A El-Maaty, G.M.Hekal, Eman M. S. El-Din. Civil Engineering Journal, **2** (2), 38-51 (2016)
15. H. Ceylan, S. Kim, K. Gopalakhrisnan, K. Wang. Journal of the Transportation Research Board, 2037 **(03)** (2014)
16. Y. Qin, J. E. Hiller. Journal of Construction and Building Materials, Elsevier, (2011)
17. A. Guclu, H. Ceylan, K. Gopalakhrisnan, S. Kim. ASCE Journal of Transportation Engineering, **135** (8), 555-562 (2009)
18. S. Gill, D.K. Maharaj. International Research Journal of Engineering and Technology, **2** (3), 1251-1261 (2015)
19. C. Niken, T. Elly, F.X. Supartono. Journal of Civil and Environmental Research, **3** (2) (2013)
20. J.W. Gibbs. Methods of geometrical representation of the thermodynamic properties of substances by means of surface, Trans Conn, 4 Akad 2, 382-404 (1873)
21. L.H. Van Vlack. *Materials science for engineers.* Addison Wesley,4, USA, (1973)
22. M. Larson, J.E. Jonasson. Journal of Advanced Concrete Technology, **1**(2), 172-187 (2003)
23. V. Morin, F. Cohen-Tenoudji, A. Feylessoufi, P. Richard. Cem.Con. Res. Journal, **32,** 1907-1914 (2002)
24. K. Kurtis. *Portland cement hydration.* School of civil Engineering, Georgia Institute of Technology, Atlanta, Georgia (2009)
25. B.S. Moffat. *Shrinkage compensating concrete: An investigative study.* ASCE licence, 1-11 (2005)
26. Z.P. Bazant, Fellow, ASCE, A.B. Hauggaard, S. Baweja, F.J. Ulm. Journal of Engineering Mechanics, Nov, 1188-1194 (1997)

* Corresponding author: chatarinaniken@yahoo.com

MATEC Web of Conferences **101**, 05010 (2017) DOI: 10.1051/matecconf/201710105010
SICEST 2016

Development of site class and site coefficient maps of Semarang, Indonesia using field shear wave velocity data

Windu Partono[1,*], *Masyhur* Irsyam[2], and *Sri* Prabandiyani Retno Wardani[1]

[1]Civil Engineering Department, Diponegoro University, 50275 Semarang, Indonesia
[2]Civil Engineering Department, Bandung Institute of Technology, 40132 Bandung, Indonesia

Abstract. The new Indonesian National Code for seismic resistance design (SNI-03-1726-2012) issued recently utilizes seismic response spectra for the whole area of the country. Site class and site coefficient are two parameters needed for designing response spectra. Site class can be estimated using average standard penetration test (N-SPT), average shear wave velocity (Vs) and average un-drained shear strength (Su) of top 30 meter soil deposit. Site coefficients can be predicted using probabilistic seismic hazard analysis (PSHA) by implementing total probability theorem. To perform PSHA, Vs30 is a parameter needed for calculating ground motion at bedrock elevation. This paper presents the results of PSHA and site class analysis using Vs30 values estimated based on N-SPT results collected from 265 boring locations in Semarang. Seismic data in a radius of 500 km from Semarang were collected for PSHA. Site class and site coefficient maps are then developed for the whole study area.

1 Introduction

Indonesia has implemented the new seismic design code for building and non-building structure SNI-03-1726-2012. The code contains maps of Maximum Considered Earthquake Geometric Mean Peak Ground Acceleration (MCE_G) and Risk-Targeted Maximum Considered Earthquake (MCE_R). The MCE_R maps were developed by conducting 1% probability of collapse in 50 years at short-period (0.2 second) spectral response acceleration, S_S, and long period (1 second) spectral response acceleration, S_1, at bedrock elevation. Site class and site coefficient are two parameters needed for designing response spectra at ground surface. Based on [1] site class can be estimated using average standard penetration test (N-SPT), average shear wave velocity (Vs) and average un-drained shear strength (Su) of top 30 meter soil deposit. An average shear wave velocity in the upper 30 meter soil deposit is referred to as Vs30. Site coefficients can be predicted using PSHA by implementing total probability theorem [2]. This paper presents the results of site class analysis and PSHA using real Vs30 values. The shear wave velocity (Vs) for each soil layer for Vs30 calculations is estimated using N-SPT values. The Vs values were predicted using three empirical equations proposed by [3, 4 and 5]. The objective of this study is to get complete information of Vs30, site class and site coefficient distributions for the whole area of Semarang city for developing site class and site coefficient maps. This study is part of a comprehensive seismic microzonation research of the study area. Seismic microzonation study of Semarang is still on-going as part of seismic mitigation study following the work conducted by the Team for Revision of Seismic Hazard Maps of Indonesia 2010 (TRSHMI-2010).

Semarang is the capital city of Central Java Province, Indonesia, and located at the center of the Northern part of Central Java Province. The city has an area of about 374 square kilometers and spans for 22 km long and 22 km wide and elongated in North-South and East-West directions. The study area can be separated into two different land-slopes, coastal plain area in the northern part with maximum slope of 5% and hilly area in the center and southern parts with maximum slope of 33%.

Seismic sources that are significantly influence Semarang are the Java subduction zone and shallow crustal faults [6, 7]. Three big earthquakes due to subduction zone were reported by [6] including 7.9 Ms (1903), 7.2 Ms (1937) and 7.9 Ms (1977). Asrurifak [2] stated that for the last 300 years no earthquakes occurs with magnitude of more than 8.0 Mw. Tectonic condition for Semarang is quite similar to that of Yogyakarta, there is an active fault near both cities. The latest 2006 Yogyakarta earthquake with 6.3 Mw was a destructive earthquake that caused thousands of casualties in Yogyakarta Province [8]. Salahuddin [7] stated that the 2006 Yogyakarta earthquake was caused by Yogya Fault. Learning from Yogyakarta earthquake, Semarang with Lasem Fault required a comprehensive seismic microzonation study for hazard mitigation and disaster preparedness.

Development of site class and site coefficients maps of Semarang was performed using five following procedures: 1) collecting and processing recorded earthquake data for entire Indonesian region from 1900

* Corresponding author: windu_bapake_dila@yahoo.com

until 2013, 2) modeling seismic source zones, 3) determining seismic parameters which include a-b values, maximum magnitudes, and slip-rates, 4) selecting appropriate attenuation functions and 5) conducting the PSHA.

2 Experiment details

2.1 Seismic sources and seismic models

According to [2, 9] the study area is significantly influenced by seismic sources which can be classified into subduction zone, transform zone and diffuse seismicity zone. Following the work conducted by TRSHMI-2010 for developing national seismic hazard maps, seismic sources were divided into subduction, shallow crustal fault, and background sources. The subduction zone at South of Java, the Java segment of the Sunda arc, extends from Sunda Strait on the west to Bali Basin on the East. Old oceanic crust is relatively converging in a direction essentially normal to the arc at the rate of about 60 mm/year in the west Java trench and 49 mm/year in the east Java trench [9]. The Benioff seismic zone along the Java segment dips approximately 50° and extends to depths of about 600 km and the gap in seismicity exists in the segment between a depth of 300 and 500 km [9].

Java transform zones occurred on well identified shallow crustal faults in Java Island such as Cimandiri, Lembang, Lasem, Pati and Yogya faults. According to [6, 10] no faults of regional extent similar to the Sumatra faults are present in Java Island. For Semarang the closest distance fault, Lasem fault, is considered as the main shallow crustal source that can significantly influence the hazard of the city. Investigation of Lasem fault is therefore very important for microzonation study.

Comprehensive data collection was performed to identify every potential source capable of generating significant ground motion to Semarang. The earthquake data recorded by national agency (Indonesian Agency for Meteorology Climatology and Geophysics/BMKG), international agencies (National Earthquake Information Center and U.S. Geological Survey/NEIC-USGS, International Seismological Summaries/ISS) and personal catalog [11] were collected for determining seismic hazard parameters for PSHA. Based on the collected earthquake data from 1900 to 2013 with minimum 5 Mw and maximum magnitude (Mmax) 8.3 Mw, all earthquake data are then clustered within radius 500 Km from the study area. All epicenter data are then distributed into subduction, shallow crustal fault, and background sources. Background seismicity used to model random earthquakes on unmapped faults and smaller earthquakes on mapped faults [12]. The background seismicity was divided into four different depth intervals (shallow background 0 – 50 km, deep background 50 - 100 km, deep background 100 – 150 km and deep background 150 – 300 km).

The a-b values from Gutenberg Richter (GR) are required as input parameters for PSHA. The parameters are important because they can significantly affect hazard value from PSHA. Based on independent earthquake events, the Guttenberg-Richter parameter (a-b parameter) was calculated and conducted for subduction and background sources.

Faults that have been well identified and influencing Semarang in a radius of 500 km are Cimandiri, Lembang, Lasem, Pati and Yogya. Parameters of fault required for input of PSHA include fault traces, focal mechanism, slip-rate, dip, length of fault, and maximum magnitude. Location of each fault was determined based on the previous study conducted by TRSHMI-2010 and [2]. Table 1 shows required seismic parameters for PSHA for fault, subduction, and background sources.

2.2 Seismic hazard analysis

PSHA was conducted to obtain probabilistic peak ground acceleration (PGA) and spectral acceleration for short periods (0.2 seconds) and long periods (1.0 seconds) at bedrock elevation. Spectral response accelerations at bedrock were calculated for 2% probability of exceedance in 50 years (2500 years return period of earthquake). This method of analysis was conducted based on the total probability theorem proposed by [13].

Table 1. Seismic sources parameters in a radius of 500 km from Semarang.

Seismic Sources	Name	Mechanism	Mmax (Mw)	Length (km)	Slip Rate (mm/year)	GR Parameter	
						a	b
Fault	Cimandiri	Strike-Slip	7.2	62.2	4	-	-
	Lembang	Strike-Slip	6.6	34.4	1.5	-	-
	Yogya	Strike-Slip	6.8	31.6	2.4	-	-
	Lasem	Strike-Slip	6.5	114.9	0.5	-	-
	Pati	Strike-Slip	6.8	51.4	0.5	-	-
Subduction	Java Megathrust	Reverse	8.3	-	-	5.36	1.0307
Background	Shallow Background (0-50 km)	-	6.8	-	-	7.04	1.3549
	Deep Background (50-100 km)	-	8.3	-	-	7.62	1.4116
	Deep Background (100-150 km)	-	6.6	-	-	5.73	1.0608
	Deep Background (150-300 km)	-	7.5	-	-	7.27	1.3974

This method of analysis was conducted using three-dimensional seismic source model. Geological and seismological data used by TRSHMI-2010 were also conducted for PSHA. According to [13], the probability that a particular level of strong ground motion will be exceeded (λa*) is calculated using equation (1) where the value "V", "m" and "r" represents the mean rates of exceedance, seismic magnitude and seismic distance respectively.

$$\lambda a^* = V \iint P(a>a^*|m,r) P_m(m) P_r(r) dr dm \quad (1)$$

Other important step in seismic hazard analysis is selection of attenuation relationships. Due to inadequate ground motion records to develop attenuation function in Indonesia, attenuation functions used in this study were adopted from other countries and following the work proposed by TRSHMI-2010. The attenuation functions were selected considering the source types. For shallow crustal faults and shallow background sources three attenuation functions used in this study were proposed by [14, 15 and 16]. Three attenuation functions proposed by [17, 18 and 19] were used for subduction sources. And finally attenuation functions proposed by [17, 18] were used for deep background sources.

Following the work proposed by [2] and TRSHMI-2010, two PSHA models were implemented using different Vs30 values. The first model (Model-1) was implemented using Vs30 = 760 m/s hence the second model (Model-2) was implemented using real Vs30 values. PSHA for both models were conducted at 265 locations which represents 265 boring locations. Fig.1. shows the location of boring investigations. The real Vs30 values for all boring locations were calculated using estimated shear wave velocity (Vs) values at 30 meter top soil deposit by using equation (2). The shear wave velocity values for every soil layer (Vs_i) were estimated using three empirical correlation equations proposed by [3, 4, 5]. The value "d_i" on equation (1) represents the thickness of each soil layer.

$$Vs30 = 30/[\Sigma(d_i/Vs_i)] \quad (2)$$

3 Results and discussions

Based on the Vs30 values calculated at 265 boring locations the distribution of Vs30 can be seen on Fig. 2. The distribution of Vs30 as shown in this figure shows that lower Vs30 values are observed along the seaside zone of the town in between 100m/s until 200m/s, hence higher Vs30 values observed along the middle and southern part of the study area between 200m/s until 450m/s. Fig. 3 shows contour of site class for the whole area of the city. The site class SE, SD and SC were predicted using Vs30 values and following the same method proposed by [1]. Based on the distribution of site class as shown in Fig. 3, site class SE is observed along the northern part of the study area, hence the SD and SC site class are observed along the middle and southern part of the city.

According to [1] site coefficients are divided into FPGA, Fa and Fv which are represents site coefficients for PGA, spectral 0.2 seconds and spectral 1 seconds respectively. FPGA, Fa and Fv are predicted by comparing the PGA, short periods (0.2 seconds) and long periods (1.0 seconds) ground acceleration values calculated using Model-2 and Model-1. Fig.4 shows the distribution of site coefficients FPGA for the study area. Fig.5 shows the distribution of site coefficients Fa for the study area.

The distributions of FPGA and Fa for the whole study area are quite similar. The highest values of both site coefficients are observed in the northern part of the study area, hence the lowest values are observed in the middle and southern part of the study area. The FPGA values are distributed between 1.11 until 1.41 and the Fa values are distributed between 1.14 until 1.30.

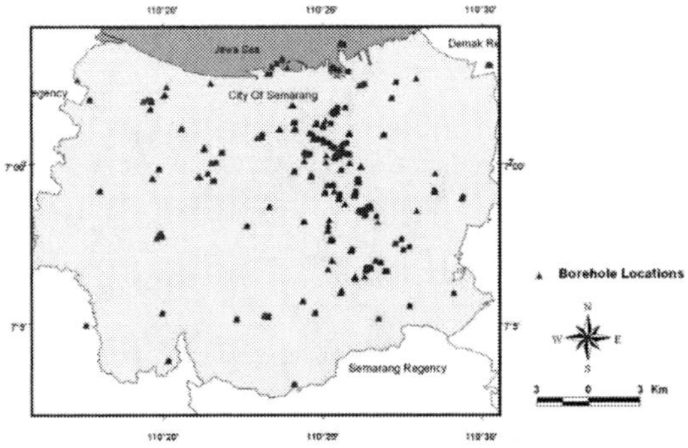

Fig. 1. Location of boring investigations in Semarang

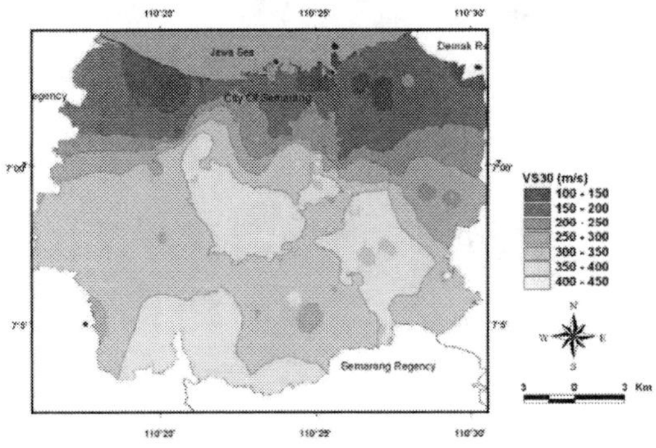

Fig. 2. Contour map of Vs30 in Semarang.

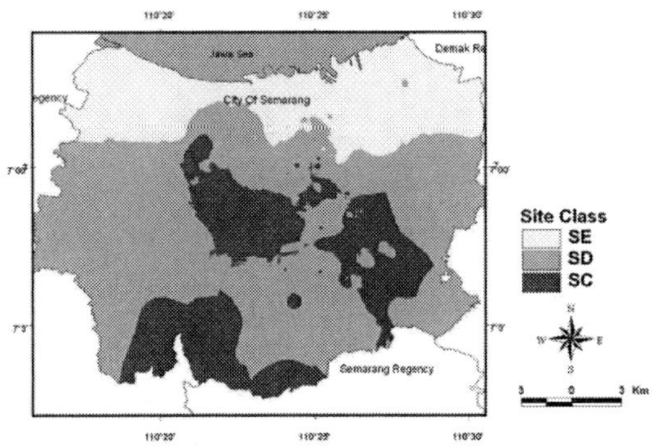

Fig. 3. Contour map of site class in Semarang

* Corresponding author: windu_bapake_dila@yahoo.com

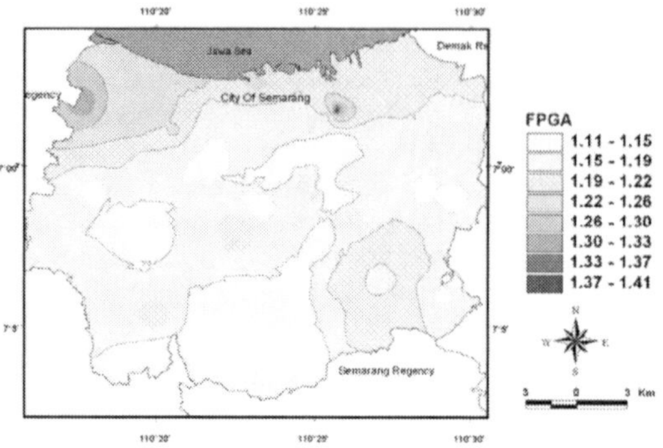

Fig. 4. Contour map of FPGA in Semarang

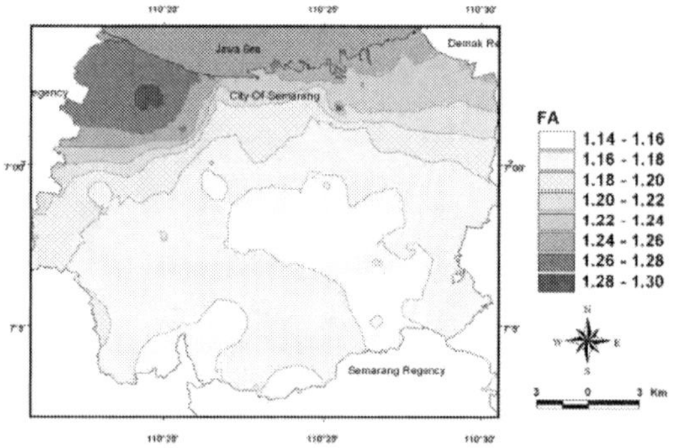

Fig. 5. Contour map of Fa in Semarang

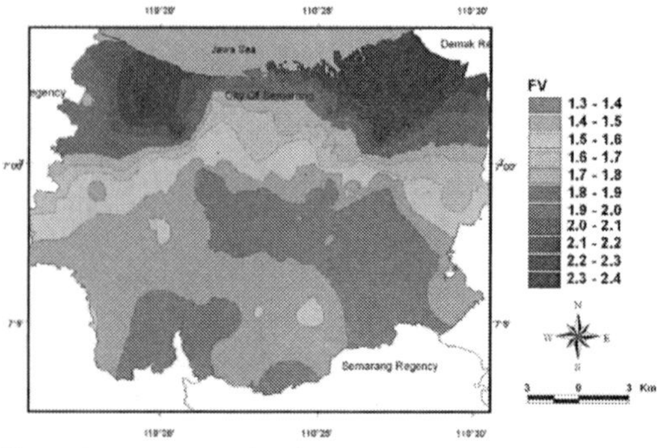

Fig. 6. Contour map of Fv in Semarang

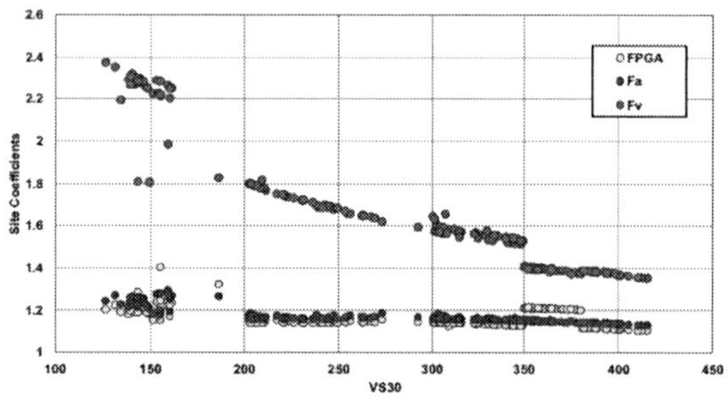

Fig. 7. Graphical plots of Fa, Fv and FPGA in terms of Vs30 in Semarang.

Fig.6 shows the distribution of Fv for the study area. As can be seen in this figure the highest Fv values is observed along the northern side of the city and distributed between 1.6 until 2.4, however the lowest Fv values is observed in the middle and southern part of the study area, between 1.3 until 1.6.

Fig.7 shows the FPGA, Fa and Fv plots in terms of Vs30 calculated from 265 locations. Based on site coefficient analysis results calculated from 265 boring investigations it seems that there is a good correlation between increasing of FPGA, Fa and Fv values with decreasing Vs30 values.

Table 2 shows the minimum and maximum spectral accelerations and site coefficients calculated at 265 locations. As shown in this table the PGA values are observed between 0.24g until 0.28g, the spectral 0.2s values are observed between 0.50g until 0.59g and the spectral 1s values are observed between 0.16g until 0.18g where "g" is gravitational accelerations. The corresponding site coefficients FPGA are observed between 1.11 until 1.41, site coefficients Fa are observed between 1.14 until 1.30 and site coefficients Fv are observed between 1.36 until 2.38.

Table 2. Spectral accelerations and site coefficients calculated at 265 locations

		PGA		**Spectra 0.2s**		**Spectra 1s**	
		min	**max**	**min**	**max**	**min**	**max**
SE	accel.(g)	0.24	0.27	0.50	0.56	0.16	0.17
	site cof.	1.16	1.41	1.19	1.30	1.99	2.38
SD	accel.(g)	0.24	0.28	0.52	0.58	0.16	0.18
	site cof.	1.13	1.33	1.15	1.27	1.40	2.33
SC	accel.(g)	0.26	0.28	0.55	0.59	0.16	0.18
	site cof.	1.11	1.22	1.14	1.17	1.36	1.58

Table 3. Average spectral accelerations and site coefficients calculated at 265 locations.

Site Class	**References**	**Site Coefficients**	
		Fa (0.2s)	**Fv (1 s)**
SC	[1]	1.012	1.472
	[2]	1.14	1.53
	[20]	1.12	1.63
	This study	1.15	1.422
SD	[1]	1.112	1.744
	[2]	1.35	2.29
	[20]	1.35	2.12
	This study	1.173	1.669
SE	[1]	0.936	2.687
	[2]	1.55	3.56
	[20]	1.58	3.29
	This study	1.248	2.262

Table 3 shows average Fa and Fv values derived from 265 locations at the study area and compared it with the same site coefficients proposed by [1, 2 and 20]. The average values of Fa and Fv observed in this study area are quite similar with the same site coefficients proposed by [1, 2 and 20].

4 Conclusions

Site class and site coefficients maps of Semarang were implemented based on the Vs30 values predicted at 265 locations. The site coefficients were implemented using PSHA by conducting two PSHA models. The Vs30 values used in this study were predicted using N-SPT data. Maximum Vs30 values were observed in the middle and southern part of Semarang while the lowest Vs30 values were identified at the northern part of the study area. The distribution of Vs30 values has a good correlation with the distribution of site class. By using Vs30 distributions for the entire study area, site class SD and SC are identified along the middle and southern part of the city, hence site class SE is identified along the northern part of the study area.

Distribution of site coefficients has a good correlation with the distribution of site class. The lowest FPGA, Fa and Fv site coefficients are identified in the middle and southern part of the study area, hence the highest FGPA, Fa and Fv are identified in the northern part of the study area. Based on two difference models of PSHA it can be seen that the value of FPGA, Fa and Fv can easily be predicted by comparing peak ground acceleration and spectral acceleration calculated using PSHA by conducting real Vs30 of site and Vs30 = 760 m/s.

The authors express their sincere gratitude to the National Disaster Management Agency (BNPB) for financial support for this research, Team for Revision of Seismic Hazard Maps of Indonesia 2010 for providing seismic data and technical assistances and USGS for supporting free PSHA software PSHA.

References

1. Standar Nasional Indonesia, *Tata Cara Perencanaan Struktur Bangunan Gedung dan Non Gedung* (SNI-03-1726-2012) (2012)
2. M. Asrurifak, *Peta Respon Spektra Indonesia untuk Perencanaan Struktur Bangunan Tahan Gempa dengan Model Sumber Gempa Tiga Dimensi dalam Analisis Probabilitas*, Disertasi Doktor Teknik Sipil, Institut Teknologi Bandung (2010)
3. Y. Osaki and R. Iwasaki, JSSMFE, Vol. **13**, N.4, pp. 59-73, Dec. (1973)
4. Y. Ohta and N. Goto, Earth. Eng. Struc. Dyn., Vol. **6**, pp.167-187 (1978)
5. T. Imai and K. Tonouchi, *Proceeding of Second European Symposium on Penetration Testing*, Amsterdam, The Netherlands, pp. 67-72 (1982)
6. K.R. Newcomb and W.R. McCann, J. Geophys. Res., Vol. **92**, pp. 421-439 (1987)
7. Salahuddin, Jurnal Fakultas Geologi UGM, Yogyakarta (2007)
8. A. Elnashai, S. J. Kim, G.J. Yun, D. Sidharta, *The Yogyakarta Earthquake of May 27, 2006*, (Mid-America Earthquake (MAE) Center Report, No 07-02, 2007)
9. M. Irsyam, D.T. Dangkua, Hendriyawan, D. Hoedajanto D.B.M. Hutapea, E.K. Kertapati, T. Boen and M.D. Petersen, J. Earth Syst. Sci., Vol. **117**, S2, pp. 865-878 (2008)
10. Hamilton, *Tectonics of Indonesian Region*, (U.S. Government Printing Office, Washington D.C., U.S. Geological Survey Professional Paper, 1078, 345, 1979)
11. E.R. Engdahl, A. Villasenor, H.R. DeShon and C.H. Thurber, Bull. Seismol. Soc. Am., Vol **97**, S43-S61 (2007)
12. M. Petersen, S. Harmsen, C. Mueler, K. Haller, J. Dewey, N. Luco, A. Crone, D. Lidke and K. Rukstales, *Documentation for the Southeast Asia Seismic Hazard Maps*, (Administrative Report September 30, 2007)
13. R.K. McGuire, Bull. Seismol. Soc. Am., Vol. **85**, No. 5, pp 1275-1284 (1995)
14. D.M. Boore and G.M. Atkinson, Earthq. Spec., Vol. **24**, No. 1, pp. 99-138, Earthq. Eng. Res. Inst. (2008)
15. K.W. Campbell and Y. Bozorgnia, Earthq. Spec., Vol. **24**, No. 1, pp. 139-171, Earthq. Eng. Res. Inst. (2008)
16. B.S.J. Chiou and R.R. Youngs, *NGA Model for Average Horizontal Component of Peak Ground Motion and Response Spectra,* (PEER 2008/09, Pacific Engineering Research Centre, College of Engineering, University of California, Berkeley, 2008)
17. R.R. Youngs, S.J. Chiou, W.J. Silva and J.R. Humprey, Seismol. Res. Lett., **68**, pp. 58-73 (1997)
18. G.M. Atkinson and D.M. Boore, Bull. Seismol. Soc. Am., Vol. **93**, No. 4, pp. 1703-1729 (2003)
19. X. Zhao, J. Zhang, A. Asano, Y. Ohno, T. Oouchi, T. Takahashi, H. Ogawa, K. Irikura, H. Thio and P. Somerville, Bull. Seismol. Soc. Am., **96**, 898 (2006).
20. International Codes Council Inc., *International Building Code* (IBC) (2009)

* Corresponding author: windu_bapake_dila@yahoo.com

MATEC Web of Conferences **101**, 05011 (2017) DOI: 10.1051/matecconf/201710105011
SICEST 2016

Port performance evaluation tool based on microsimulation model

Jzolanda Tsavalista Burhani[1,*], *Febri* Zukhruf[1], and *Russ* Bona Frazila[1]

[1]Institut Teknologi Bandung, Civil Engineering Department, 40132 Bandung, Indonesia

Abstract. As port performance is becoming correlative to national competitiveness, the issue of port performance evaluation has significantly raised. Port performances can simply be indicated by port service levels to the ship (e.g., throughput, waiting for berthing etc.), as well as the utilization level of equipment and facilities within a certain period. The performances evaluation then can be used as a tool to develop related policies for improving the port's performance to be more effective and efficient. However, the evaluation is frequently conducted based on deterministic approach, which hardly captures the nature variations of port parameters. Therefore, this paper presents a stochastic microsimulation model for investigating the impacts of port parameter variations to the port performances. The variations are derived from actual data in order to provide more realistic results. The model is further developed using *MATLAB and Simulink* based on the queuing theory.

1 Introduction

Maritime transportation plays a major role in national and international trade, and it has a significant impact to national economic growth [1]. For instance, more than 90% of international cargo move through seaports, and 80% of seaborne cargo ships use containerized vessels, which has increased by 3.4% and 5.3% annually [2]. In response to this growth, the container terminal (CT) operators thus strive to stay competitive by offering competitive services and land transportation services [3]. In addition, in order to meet the container demands effectively and efficiently, the port performance evaluation tool becomes overly essential. It functions to answer several issues relating to the port performances: to know the effectivity and efficiency of its operation; to compare the current performance and the past performance; to adjust its targets for future periods; etc. [4].

CT operations have been considered as complex systems which involve the interactions of many resources, entities, and activities, namely, the ships arriving, the multiple quay cranes (QC) and the rubber tyred gantry cranes (RTGC) loading/unloading, the trucks delivering, and the security gates open/closing [5]. Due to the complexity and scale of operations, it is relatively challenging to develop the mathematical model for analyzing the whole CT system, and hence, the existing research mostly considers the single aspect of CT [6]. The simulation modelling (SM) is then prospectively used as a tool for demonstrating, evaluating, and optimizing the system performance [7], which is practically developed using the computer program as it is implemented in this paper.

SMs have been widely used and applied in the modelling, planning, management, and analysis of CTs [6, 8, 9, 10, 11, 12]. Started in 1960s, SM remains popular as a decision-making tool for various purposes, especially for complex systems that cannot be represented by the analytical models [13]. Dragović *et al.* [14] and Angeloudis *et al.* [15] provided the complete reviews about the SMs in ports and CTs. The investigation of CT performance has been treated using various simulation models, which are coded in the different languages: *SIMPLE++* [16], *ARENA* [5, 17, 18, 19], *Matlab* [20], *AweSim* [21], *AutoMod.* [22], *PORTSIM* [23]. Mainly, previous researches present the simulation models for analyzing and evaluating the terminal performance at current configuration. In addition, preceding papers also show that the simulation models can be built from a sequence of operational procedures for determining the port performances in the various environment and heterogeneous cases within the different points of view. Relating to the CT operations, Carteni *et al.* [24] proposed a discrete-event simulation model and applied it to the Salerno CT to investigated the effects of different hypotheses on handling equipment models.

Practically, SMs can be divided into some categories: microscopic, macroscopic, stochastic, and deterministic models. The microscopic (operational) model takes into account the detail behaviour of vehicles and the traffic control infrastructure in order to replicate the dynamic systems of terminal. In contrast, the macroscopic (strategic) model, which has a lower fidelity, is developed based on the aggregate information [15]. Furthermore, the stochastic model considers the stochastic features by generating random numbers to

* Corresponding author: tjzolanda@ymail.com

capture the uncertainty of real-world systems. The rest of SM, which does not contain any probabilistic components, are called deterministic models [8, 13].

This paper then develops a microsimulation model for evaluating the CT performance. For capturing the stochastic phenomena of CT, this paper also conducts several experiments with different initial seed value using *MATLAB and Simulink* software. The rest of the paper is organized as follows. In the following section, the system and modelling framework is described. In the third section, the experimental study is conducted for investigating the applicability of model. Finally, in the fourth section, the methodologies, results, and analyses in the paper are summarized.

2 The simulation model development

2.1 System description

The simulation model developed in this paper considers the whole process of import container movements in the CT. The process starts from the ship unloading process and ends in the administration process in the terminal gate. The container is firstly discharged from a ship, and it is sequentially headed to the berth using QCs. Each ship is practically served by 2 or 3 QCs, where the time required by the QC for serving the container is called as the QC service time. There are various types of QC that possibly influence its service time, however this paper only involves a Panamax QC (P-QC), which is generally used in Indonesia.

The container is then transferred to the truck trailer unit (TTU), which is utilized for transporting it to the container yard (CY) through the horizontal transport area. The TTU service time, which is defined as the time required for transferring container to CY, is evaluated as its performance indicator. Arriving at CY, the container is further stacked by the first RTGC (F-RTGC) until the consignees fulfil the requirements for taking it. The time spent by the container within the CY area is named as the storage service time. If the consignees can pass the requirement, the last RTGC (L-RTGC) is employed for transporting and heading the container to the external truck (XT). The time required by the L-RTGC to serve a container is called L-RTGC service time. This process indicates the completion process at the terminal.

In the next step, the container is transported by XT for leading out of the terminal area through a gate. The time required by the XT to transports the container is called as XT service time. In the final stage, the XT faces the administration process in the gate, which is indicated by the gate service time, before it is distributed to their hinterland area. The total time spent by the container within the CT area, namely from the P-QC until the gate, is defined as the dwelling time.

The rich picture-based procedures are invoked for constructing the CT system in the simulation model. The procedures contain three main components, namely the structure, process, and each process relationship [25]. Structural aspects describe the body structure of model, which can be illustrated as follows:

1. *Entity*, is an object that is served in the system: container.
2. *Generator*, is a unit that produces entities at any given time period: ship.
3. *Servers*, are units that serve entites in a certain period of time: P-QC, TTU, F-RTGC, L-RTGC, storage area, XT, and gate.
4. *Queues*, are places for entities to wait before they are served: P-QC queue and storage queue. Queue has its own capacity for serving entities.
5. *Attributes of entity*, are properties which are owned by the entity: inter-arrival rate (which represent the period of time between the arrival of the entity), waiting time, P-QC service time, TTU service time, F-RTGC service time, storage service time, L-RTGC service time, XT service time, gate service time, and dwelling time.
6. *Attributes of system*, are properties which are owned by the system: capacity of P-QC queue, capacity of storage queue, length of P-QC queue, and length of storage queue.

The second aspect of procedure is related to the process which occurs in the CT system, such as:

1. The arrival process of the ship carries containers.
2. The waiting process of container for the P-QC service.
3. The transporting process of container from the ship to the TTU using P-QC.
4. The transporting process of container by the TTU along the horizontal transportation area and the transferring process to the F-RTGC at CY.
5. The stacking process of container by F-RTGC in CY.
6. The waiting process of container before being taken by the consignee.
7. The servicing process of container by L-RTGC.
8. The transferring process of storage container to the XT.
9. The transporting process of container to the gate area by the XT.
10. The gate services for XT.
11. The time spent by the container within the CT area since served by P-QC until out of gate.

The relationship among processes is then explained as follows:

1. P-QC service time is affected by the process of transporting container from the ship to the TTU.
2. TTU service time is affected by the transporting and transferring process of container from the TTU to the F-RTGC in CY.
3. F-RTGC service time is affected by the stacking process of container from the TTU to the F-RTGC.
4. Storage service time is affected by the waiting and servicing process of container to be served by L-RTGC.
5. L-RTGC service time is affected by the transferring process of container to the XT.
6. XT service time is affected by the transporting process of container from the CY to the gate.
7. Gate service time is affected by the XT service process.

8. The length of P-QC queue is influenced by the P-QC service time.
9. The length of storage queue is influenced by storage service time.
10. Dwelling time is an acculumation of all existing attributes.
11. Dwelling time is affected by the whole process for handling the import containers.

2.2 Input and output variables

There are two types of input variables that affect the system: controllable and uncontrollable input. Controllable input variables are generally described as decision variables in the system, for instance the number of P-QC, the number of TTU, the number of F-RTGC, the number of XT, and the number of gate. Uncontrollable input variables are related to the input values that cannot be controlled, for example, the inter-arrival rate of a container, the capacity of P-QC queue, the P-QC service time, the TTU service time, the F-RTGC service time, the number of storage, the storage service time, the capacity of storage queue, the number of L-RTGC, the L-RTGC service time, the XT service time, and the gate service time.

The outputs are then illustrated by the performances of CT operation, for instance, the number of containers served, the average number of containers which are waiting in each queue, the average time of containers for waiting in each queue, the average service time of each server, the dwelling time of system, and the utilities of equipment.

2.3 Assumptions of model

Although the model is developed for representing the actual condition of CT, there are several assumptions used, which can be described as follows:

1. There is no warming up time on the model, and hence a continuous time is assumed.
2. In the beginning of the simulation, it is assumed that there are no containers in the system, and thus the queue is empty and the server is not used (idle).
3. The port is assumed to be operated 24 hours in a day.
4. The replacement of the vessel process is not considered.
5. The queue priority follows the principle of First In First Out (FIFO).
6. The server is always available during the simulation process.
7. The overbrengen process in the container handling is not considered.
8. Empty container handling processes are not considered.
9. The import customs process is not considered.

2.4 Model transformation

The model further will be transformed into a programming language i.e. *MATLAB and Simulink* using *SimEvents* blocks [26]. Here are brief explanations of the existing subsystems created in the SM.

1. *Container generator subsystem*, is a subsystem that regulates containers which is produced in a specific time. Some important parameters in this subsystem, namely, the distribution parameter and the initial seed parameter.
2. *Set service time subsystem*, is a subsystem that organizes the attributes of the service time. Some important parameters in this subsystem i.e. distribution parameter that controls the distribution of random numbers that we want.
3. *Output switch subsystem*, is a subsystem that organizes the entites output port selection criteria for the entities departure. Some important parameters in this subsystem, such as, the switching criterion parameter.
4. *Enabled gate subsystem*, is a subsystem that organizes the entrance arrangement for the entities.
5. *Server queue subsystem*, is a subsystem that organizes the entities waiting to be served in specified or unspecified time.
6. *Single server subsystem*, is a subsystem that organizes the service for entities in a certain period of time.
7. *Read timer after server subsystem*, is a subsystem that organizes the statistic data information on the timer which associated with the arrival of entities.
8. *Path combiner subsystem*, is a subsystem that organizes the incorporation of entities pathway.
9. *Containers sink subsystem*, is a subsystem that organizes the acception or rejection of entites.

3 Experimental study

In order to test the model's applicability, the experimental study is conducted by considering the actual data from Koja Container Terminal (KCT) relating to the facilities performance (i.e., berth facilities, CY, container handling equipment). The KCT can handle containers up to 7560 TEUs for import and 7696 TEUs for export. Operating 5 blocking parallel layouts of CY, the KCT also uses the combination of RTGC and TTU for handling such operations. In addition, KCT is also equipped with 7 QCs, 48 head trucks, 60 chassis, 25 RTGCs, 3 reach stackers, and a fire fighting car.

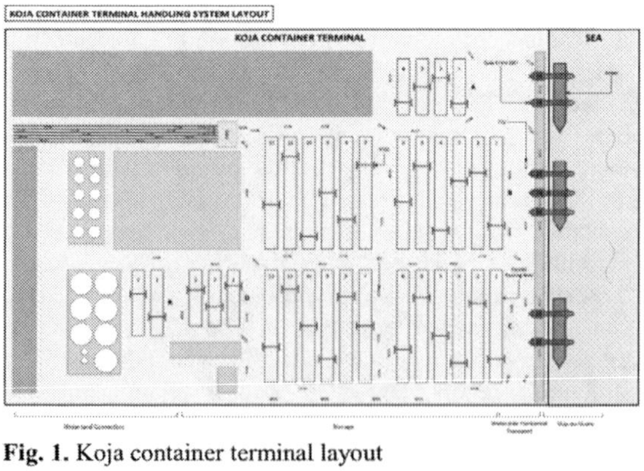

Fig. 1. Koja container terminal layout

The actual data, which is derived from the KCT operational reports, is then utilized as the simulation inputs: the inter-arrival rate of entities, the capacity of server queue, the number of equipment, and the equipment service time. The characteristics of simulation inputs are summarized in Table 1, in which the random number is generated by following the triangular distribution.

Table 1. The simulation input

Parameters	Units	Min	Max	Mode	Mean	Stdev
Inter-Arrival Rate of Entities	TEUs/min	1.00	2.00	1.00	1.00	0.00
Capacity of P-QC Queue	TEUs	-	inf	-	-	-
P-QC Service Time	min	2.00	3.00	2.50	0.00	0.00
TTU Service Time	min	24.00	43.00	29.00	31.00	5.00
F-RTGC Service Time	min	5.00	8.00	7.00	6.00	-
Capacity of Storage Queue	TEUs	-	630.00	-	-	-
Storage Service Time	min	70.56	180.00	93.60	11.00	-
L-RTGC Service Time	min	5.00	8.00	7.00	6.00	-
XT Service Time	min	94.00	191.00	106.00	130.00	24.00
Gate Service Time	min	1.00	3.00	2.00	2.00	-

Since the container practically needs to spend around 12 days for passing the terminal gate from the ship unloading, this paper then simulates 12 days of unloading operation in KCT. The model implementation is conducted with 20 attempts of different initial seed value for capturing the natural stochastic phenomena of CT. In other words, the experiment tries to imitate 240 days of stochastic terminal operations (see Figure 2).

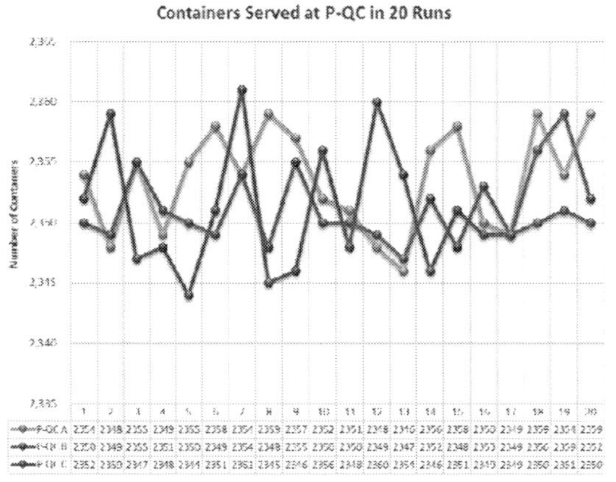

Fig. 2. Number of containers served by P-QC

As the important indicator of port performance, the container queue is firstly elaborated. The experimental results show that the longest queue occurs before the P-QC server, which cause the queue for 698 containers (see Table 2). A similar result also occurs in the storage area, which involve 409 containers in queue. The long queue is probably caused by the lack number of equipment compared to the container arrival rate. In contrast, the results show that there is no queue before F-RTGC. It indicates that the number of F-RTGC and F-RTGC service time is on adequate condition.

Table 2. Number of containers in queue

Parameters	Results (containers)
Total number of queue at P-QC	2094
Average number of queue at P-QC	698
Total number of queue at storage	3674
Average number of queue at storage	409
Total number of queue at F-RTGC	0
Average number of queue at F-RTGC	0

Furthermore, the utilization ratio of equipment is also explored. The interesting result is laid on the P-QC, TTU, and storage, where they have a utilization ratio of near one (i.e. 0.999, 0.988 and 0.997, respectively). This result describes that those servers already face their highest performance for servicing high workloads. On the contrary, the F-RTGC, L-RTGC, XT, and gate results in a low utilization ratio (i.e., 0.691, 0.058, 0.563, 0.039, respectively). This condition may indicate the server's ability of service is too fast compared to the inter-arrival rate of containers. In addition, it can also be caused by the long deposition on the previous server, which causes in an idle and unproductive server.

4 Conclusion

This paper presents a microsimulation model for analyzing the operational performances of CT. The model is developed by utilizing the rich picture-based procedures for imitating the whole of import CT system. For investigating the applicability of model, the actual data of CT is considered in the model. The simulation is conducted by generating several different numbers of initial seed in order to describe a real world system. The experimental results show that the model can identify the several operational problems of CT, which is potentially used as a tool for evaluating the CT performance as well as for managing the CT operation.

References

1. De Monie, G. *Measuring And Evaluating Port Performance And Productivity. UNCTAD monographs on port management, No. 6* (United Nations, 1987). doi:GE.87-56444/0295G
2. Regina *et al. Review of Maritime Transport 2015.* (United Nations, 2015).
3. Alderton, P. & Saieva, G. *Port Management and Operations.* (Informa, 2008).
4. Esmer, S. Performance Measurements of Container Terminal Operations. **10,** 238–255 (2008).
5. Kotachi, M., Rabadi, G. & Obeid, M. F. *Procedia Comput. Sci.* **20,** 229–234 (2013).
6. Huang, S. Y., Hsu, W. J., Chen, C., Ye, R. & Nautiyal, S. *Int. J. Comput. Appl. Technol.* **32,** 246–253 (2008).
7. Law, A. M. & Kelton, W. D. *Simulation Modeling and Analysis 3rd Edition.* (McGraw-Hill, 2000).
8. Dragovic, B. & Zrnic, N. D. A Queuing Model

Study of Port Performance Evolution. *Analele Univ. 'Eftimie Murgu' Resita Fasc. Ing.* **2,** 65–76 (2011).
9. Gudelj, A., Krčum, M. & Twrdy, E. Models and Methods for Operations in Port Container Terminals. *PROMET - Traffic & Transportation* **22,** 43–51 (2010).
10. Legato, P., Mazza, R. M. & Trunfio, R. Simulation-Based Optimization for Discharge/Loading Operations at a Maritime Container Terminal. *OR Spectr.* **32,** 543–567 (2010).
11. Peng-fei, Z., Zi-jian, G. & Xiang-qun, S. *International Conference on Management Science and Engineering* 177–182 (IEEE, 2006). doi:10.1109/ICMSE.2006.313837
12. Ramani, K. V. An Interactive Simulation Model for the Logistics Planning of Container Operations in Seaports. *Simulation* **66,** 291–300 (1996).
13. Kelle, P. & Jin, M. *A Simulation Model for Intermodal Freight Transportation in Louisiana.* (2015).
14. Dragović, B., Tzannatos, E. & Park, N. K. Simulation Modelling in Ports and Container Terminals: Literature Overview and Analysis by Research Field, Application Area, and Tool. *Flex. Serv. Manuf. J.* 1–31 (2016). doi:10.1007/s10696-016-9239-5
15. Angeloudis, P. & Bell, M. G. H. A Review of Container Terminal Simulation Models. *Marit. Policy Manag.* **38,** 523–540 (2011).
16. Yun, W. Y. & Choi, Y. S. *Int. J. Prod. Econ.* **59,** 221–230 (1999).
17. Kulak, O., Polat, O. & Guenther, H.-O. *3rd German-Korean Workshop on Container Terminal Management* (2008).
18. Kyu Park, N., Dragović, B., Zrnić, N. & S-H Moon, D. Simulation Approach of Container Terminal Modelling.
19. Park, N.-K. & Dragović, B. A Study of Container Terminal Planning. *FME Trans.* **37,** 203–209 (2009).
20. Ansorena, I. L., Orive, A. C., Cancelas, N. G. & Orive, A. C. *Marit. Transp. V. Technol. Innov. Res.* 448–460 (2012).
21. Nam, K.-C., Kwak, K.-S. & Yu, M.-S. *J. Waterw. Port, Coastal, Ocean Eng.* **128,** 126–132 (2002).
22. Lau, Y. K. H. & Lee, M. Y. N. Simulation Study of Port Container Terminal Quay Side Traffic. *AsiaSim* **5,** 227–236 (2007).
23. Nevins, M. R., Macal, C. M., Love, R. J. & Bragen, M. J. Simulation, Animation and Visualization of Seaport Operations. *Simulation* **71,** 96–106 (1998).
24. Cartenì, A. & Luca, S. de. Tactical and Strategic Planning for A Container Terminal: Modelling Issues Within A Discrete Event Simulation Approach. *Simul. Model. Pract. Theory* **21,** 123–145 (2012).
25. Daellenbach, H. G. *System and Decision Making: A Management Science Approach.* (John Wiley & Sons, Ltd., 1996). doi:10.1002/bdm.3960080407
26. *MATLAB & Simulink: SimEvents 3 User's Guide.* (The Mathworks, Inc.).

* Corresponding author: tjzolanda@ymail.com

DOI: 10.1051/matecconf/201710105012

Bamboo reinforced concrete slab with styrofoam lamina filler as solution of lightweight concrete application

Ari Wibowo[1,*], *Indradi* Wijatmiko[1], and *Christin* Remayanti Nainggolan[1]

[1]Department of Civil Engineering, Faculty of Engineering, Brawijaya University, 65149 Malang, Indonesia

Abstract. Energy resilience is becoming more important nowadays especially in the field of building sustainability. Some implementations can be carried out including using recycled materials instead of nonrenewable materials such as steel. Hence, one of the investigation conducted in this paper is replacing steel reinforcement with bamboo bars and using recycled materials such as Styrofoam with the aim of producing a concrete element structure that is lighter and more economical. In this research stage, flexural strength test on bamboo reinforced concrete slab with Styrofoam lamination filler was conducted. The results showed that the flexural strength of specimens decreased by 15% but with the weight advantage of 20% less compared with those of normal reinforced concrete slab with the same dimension. It is considered good performance in practical design context, since the nominal flexural capacity of RC slab when designed with minimum reinforcement are usually much higher than the required moment.

1 Introduction

Bamboo reinforcements become increasingly considered as substitutes for steel reinforcements since being cheaper especially in developing countries where bamboo can be obtained more easily. The production of bamboo also does not endanger environment and hence it is environmentally friendly. However, material properties of bamboo and its application as reinforced concrete materials still need further investigation. Many research studies on bamboo reinforced concrete members have been conducted in the last few years. Ghavami [1] has invesigated the characteristics of bamboo as an engineering material (including mechanical properties, durability, water absorption effect, bonding strength), as replacement of steel bars in concrete members and also flexural strength analysis of concrete elements. The study showed that bamboo can substitute steel reinforcement satisfactorily. However, more research studies need to be undertaken to further determine international norms for bamboo reinforcement design applications. Yamaguchi et al [2] studied on applying bamboo as main rebars and stirrups replacement for reinforced concrete beams. The experimental tests demonstrated that the collapse mechanism of the beams were affected by the rupture strength of bamboo bars, whilst the stiffness of beams was inluenced by bond slip strength between bambo rebars and concrete. Further, Terai and Kinami [3] focused on bamboo bond slip strength capacity as their main study by conducting tests using various bonding agents, application methods and treatments. It was observed that the bond strengths of coated bamboo were found in between those of plain and deformed steel bars. Also the difference shrinkage rate of bamboo and concrete caused the deterioration of the bond stress, and hence, the choice of proper curing treatment can be significant.

To further improve the sustainability of structural elements, styrofoam can be used as panel filler inside the concrete element which is also useful to minimize the self-weight of concrete member. Styrofoam is regarded as environmentally friendly material since it is fully recyclable and reusable made of plastic. Moreover, by providing lightweight concrete member, the total weight of the building can be lowered and in turn reduces the lateral loads imposed by the earthquake excitation as well. However, previous reconnaissance study in damage area due to earthquake disaster observed that many lightweight concrete members did not have adequate strength and deformation capacity to withstand such energy [4]. Therefore, many further investigations are required to address the issue.

In this study, the applications of bamboo as reinforcement and styrofoam as lightweight filler for concrete member are combined and investigated. The members studied in this test were bamboo reinforced slab using styrofoam panel filler as shown in Figure 1.

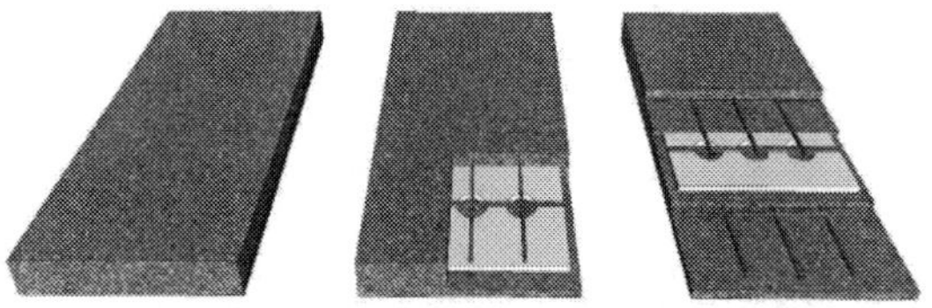

Fig. 1. Bamboo reinforced concrete with styrofoam panel filler

* Corresponding author: ariwibowo@ub.ac.id

The main aim of this study was to study the ultimate strength capacity and collapse mechanism of such slab system so that it can be applied in concrete design and construction practices.

2 Experimental test program

2.1 Pull-out test on bamboo bars

Bamboo can expand while absorbing water and shrink while losing its water content. The shrinkage volume and speed of bamboo are larger and faster compared to those of concrete, and hence bamboo bars need to be coated for preventing water absorption and also strengthening the bond strength between bamboo and surrounding concrete. Therefore, varnish were brushed on the bamboo bar surface; and while the coating was still viscous, bamboo bars were rolled onto fine aggregates to create rough surface and to improve the friction between concrete and bamboo bars.

Pull-out tests were conducted on 6×6 mm sand coated bamboo bars which embedded in concrete cubes (fc' = 20 MPa) with embedment length of 220mm as shown in Figure 2. The test results showed that the pull-out force was about 1.5kN, which equivalent to pull-out strength of 0.3 MPa.

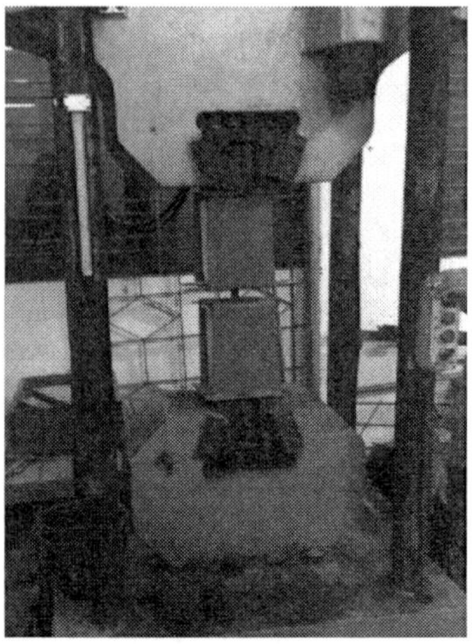

Fig. 2. Bamboo bar pull-out test setup

2.2 Slab specimens

The specimens consisted of two types of bamboo reinforced concrete slabs i.e.: Spesimen S series (with Styrofoam panel filler) and specimen TS series (without Styrofoam panel filler). All slab specimens had the same dimension (width × length × thickness) of 40×80×5 cm as shown in Figure 3. Main reinforcements were manufactured by splitting Petung bamboo (botanical name: *Dendrocalamus Casper,* giant bamboo) bars with dimension of 6 × 6mm and placed on both longitudinal and transversal directions of slab. The presence of Styrofoam panel alters the load-displacement mechanism from regular flexural member to composite system. And hence, galvanized wire shear connectors were placed via 3cm diameter holes connecting top and bottom longitudinal-transversal bar joints. The clear concrete cover was 8mm, whilst the specified concrete compressive strength and bamboo yield stress were 20MPa and 190 MPa respectively.

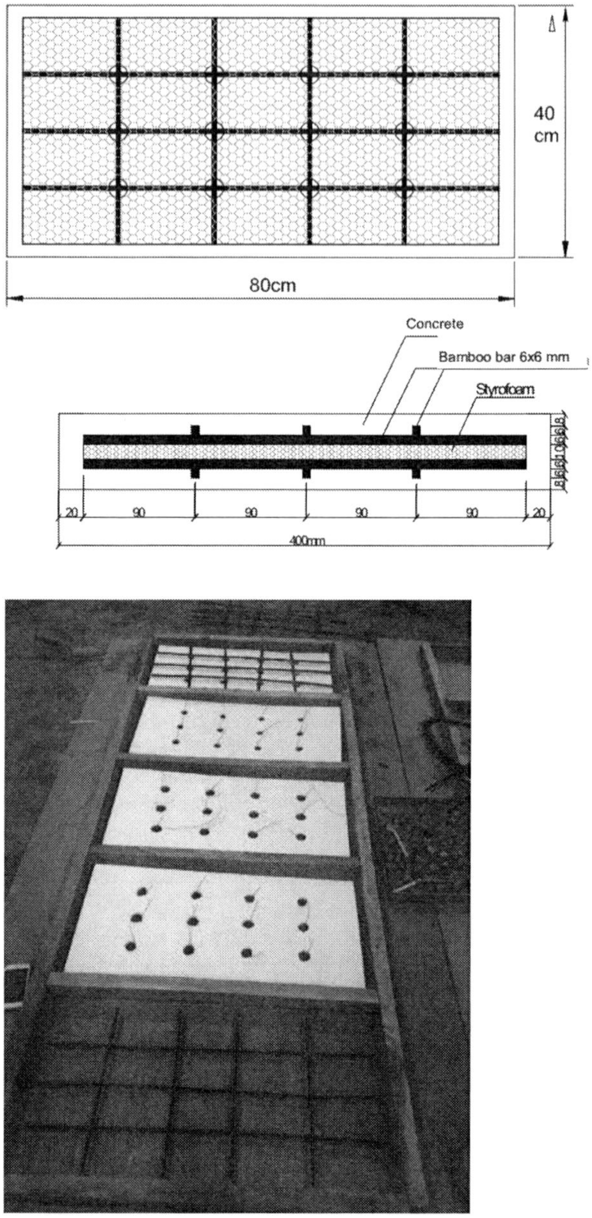

Fig. 3. Detail of slab specimens

The slab specimens were tested as one-way slab system spanning 75cm between two pin supports (refer to Figure 4). Concentrated line loading was applied at mid span of slab under load control mechanism at increment of 20 kg until peak load and then followed by displacement-controlled load until failure.

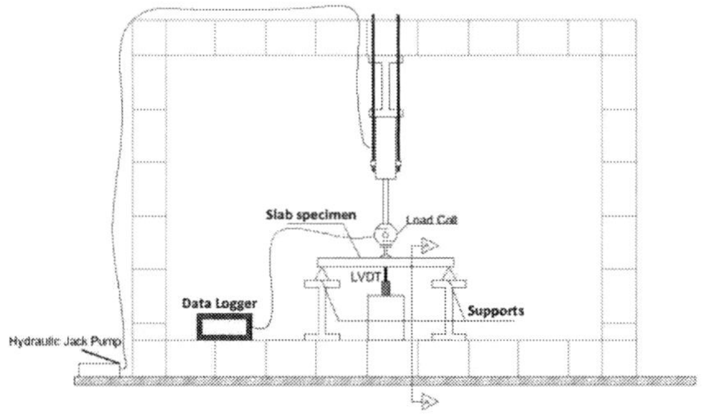

Fig. 4. Experimental test setup

3 Experimental test results

All specimens were set as one-way slab member, where the LVDTs were initialized for subsequent readings. The subsequent load-controlled were then applied. The peak loads obtained for specimens without Styrofoam TS1 and TS2 were 6.7 KN, 6.3 KN, whilst the line loads of specimens S1, S2 and S3 (with Styrofoam) peaked at 5.5 KN, 5.7 KN and 5.2 KN respectively; which correlated with 15% reduction of ultimate strength capacity. Correspondingly, the ultimate deflections at mid span for specimens TS1 and TS2 were about 23 mm and 21 mm respectively, whereas the deflections for specimens S1, S2 and S3 reached 31 mm, 21 mm and 22 mm; which generally showed no significant decrease in deflection at failure point. All test results are shown in Table 1.

Table 1. Peak load and displacement test results for specimens TS and S (without and with Styrofoam respectively)

Specimen	P_{cr} (KN)	P_u (kN)	Δ_{cr} (mm)	Δ_u (mm)
TS1	4.0	6.7	1.45	23
TS2	4.0	6.3	1.35	21
S1	4.3	5.5	5.2	31
S2	4.6	5.7	1.2	21
S3	4.0	5.2	4.7	22

Note Pcr = Cracking Load, Pu = Peak Load, Δ_{cr} = cracking displacement Δu = Displacement at Failure

The load-deflection behavior for all specimens are shown in Figure 5. Interestingly, two different type of initial stiffness were observed on the specimens. Specimens TS1, TS2 and S2 had a very similar initial stiffness which are steeper than those of specimens S1 and S3 (refer Figure 6). It shows that the Styrofoam panel did not directly contribute to the initial stiffness of the slabs. The reason for this behavior might be attributed to the different bond slip strength between bamboo bars and concrete, since the sand as coating for bamboo could not be applied homogeneously over the bamboo length. The use of Styrofoam also reduced the thickness of concrete layer surrounding the bamboo bars, and hence reduced the bond slip resistance; which in turn, increased the possibility of slippage of bamboo bars. From the tests, it was obvious that specimens with Styrofoam tended to have lower initial stiffness up to the first crack compared to those of slabs without Styrofoam.

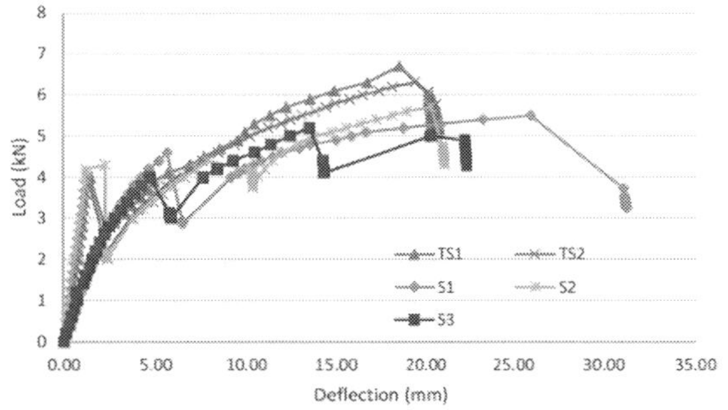

Fig. 5. Load-deflection behavior for all specimens

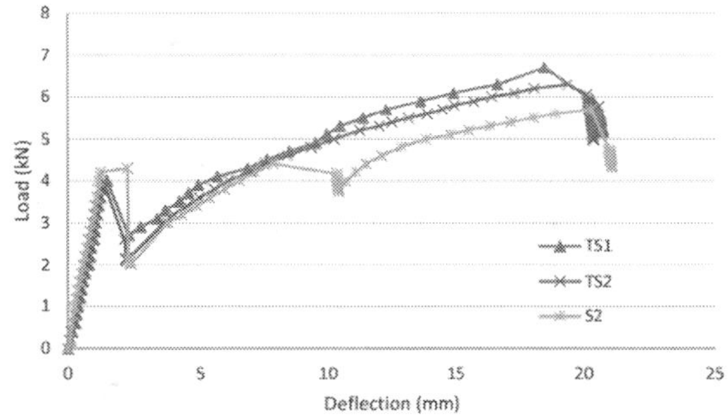

a. Load-deflection behavior for specimens with higher initial stiffness

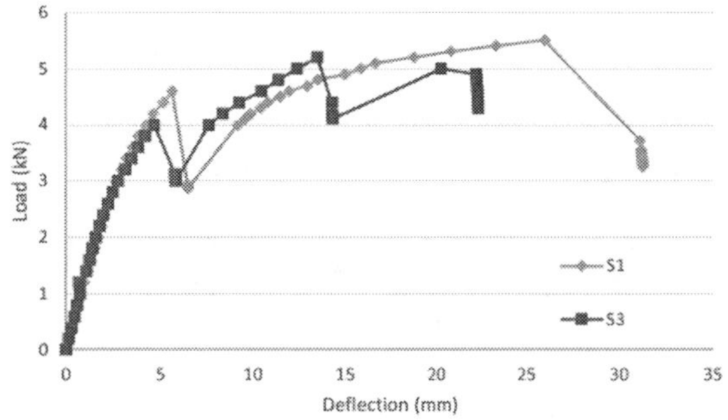

b. Load-deflection behavior for specimens with lower initial stiffness

Fig. 6. Detail of load-deflection behavior of specimens

The load-displacement behavior for specimens S2 and S3 showed two peak loads which correlated with the occurrence of two subsequent main flexural cracks, whereas the flexural cracks of specimen TS1, TS2 and S1 occurred simultaneously. Interestingly, instead of only one crack at maximum moment location as commonly found at one-way slab, those two cracks were observed propagated from tension extreme fiber below the lateral bamboo bars close to mid span. The thin concrete cover underneath lateral bamboo bars reduced the cracking moment capacity compared to that at mid span (the area in-between two lateral bamboo bars) and hence shifted the cracks further from mid span. This showed that concrete cover provided significant effect on cracking moment compared to the gross area of concrete especially on thin element structures such as slabs.

The collapse of concrete slabs occurred with two different behaviors for each type of element as shown in Table 1. Concrete slabs with Styrofoam failed due to bond slip mechanism between Styrofoam and compression concrete since the bond slip strength between concrete and Styrofoam was relatively small. In contrast, classical flexural failures were observed at concrete slabs without Styrofoam. However, the ultimate moment capacity between those two failure mechanisms were quite similar due to the short level-arm moment of concrete slab on both types of slabs

Table 1. Crack patterns for specimens S1, S2 and TS1

Crack pattern	**S1**	**S2**	**TS1**
under side of slab	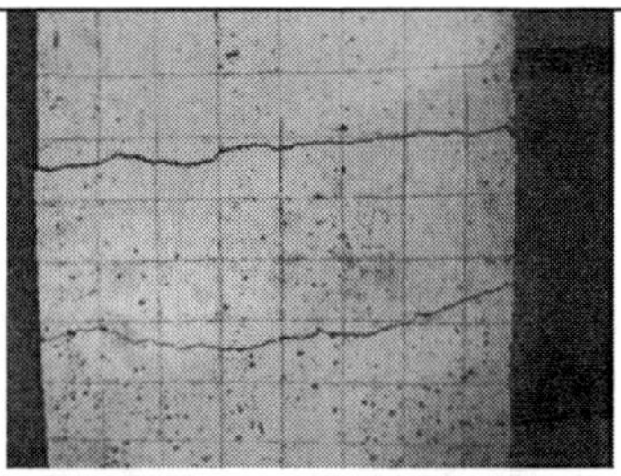	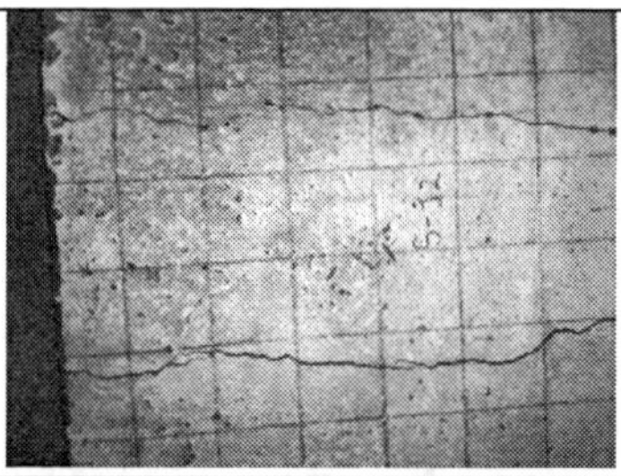	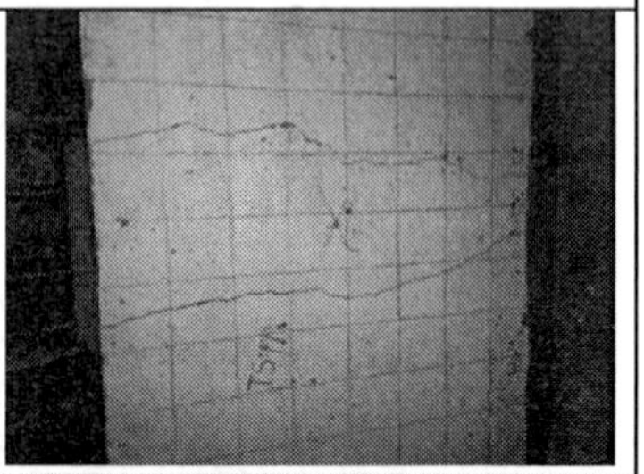
Outer edge of slab		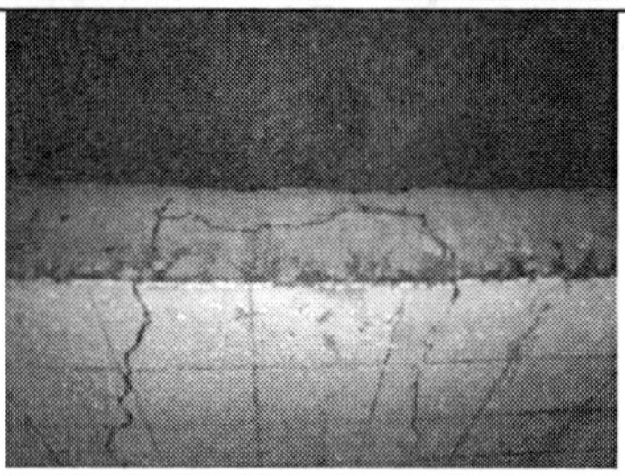	

4 Analytical results

Flexural displacement prediction of a slab has been developed in this study based on the fiber section method by Park and Paulay [5] to obtain a moment-curvature relationship. Further, from the obtained moment and curvature, the load and deflection of slab can be calculated.

The moment-curvature analysis used the assumption that plane section remains plane as shown in Figure 7. Hence, the strain of concrete is equal to the strain of reinforcement bamboo at the same depth level and consequently, stresses of concrete and bamboo can be calculated from the stress-strain relationship of each material. The moment (M) and curvature (φ) relationship is defined as follows:

$$EI = \frac{M}{\varphi} \tag{1}$$

$$\varphi = \frac{\varepsilon_c}{c} = \frac{\varepsilon_c + \varepsilon_b}{d} \tag{2}$$

where c is neutral axis depth, d is slab effective depth, and ε_c and ε_b are concrete and bamboo strain respectively.

The stress-strain curve for the concrete cover is assumed to have the same strain-stress relationship as unconfined concrete for strains up to fracture, whilst for larger strains, the concrete cover is assumed to have spalled and is ignored.

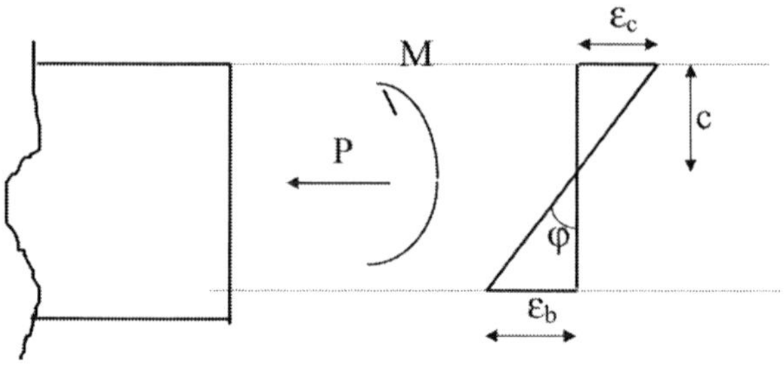

Fig. 7. Moment-curvature relationship

For simplicity, buckling is assumed not to occur on the compressive longitudinal reinforcement. Consequently, in the analysis the longitudinal reinforcement will not lose the compressive strength when the surrounding concrete cover spalls.

A computer program [6] was developed to establish the moment-curvature relationship by dividing the cross section area into '*n*' layers as follows (refer Figure 8):

1. Fix increment of compressive concrete strain ε_{cm},
2. Guess an initial value of neutral axis, $c = kd$.
3. Calculate concrete and bamboo forces:
   a. Divide the cross section area into *n* horizontal layers. This program uses 2000 layers.

b. For each depth y_i, calculate the strain that directly proportional to the strain ε_{cm} at the extreme fiber
c. Calculate concrete stress f_{ci} for segment-*i* based on modified-Kent & Park stress-strain relationship diagram.
d. Calculate force at each layer, $P_i = b(h/n)f_{ci}$
e. Calculate moment about the extreme fiber,

$$M_i = P_i\, d_i$$

f. Total concrete force: $P_{tot} = \sum_{i=1}^{n} P_i \ (i=1,n)$

g. Total moment: $M_{tot} = \sum_{i=1}^{n} M_i \ (i=1,n)$

4. Check equilibrium equation:

$$P = 0.85\ f_c'\ b\,kd + \sum_{i=1}^{n} f_{bi} A_{bi}$$

where:

b	= the column width
d	= the effective depth of a column
kd	= the neutral axis depth
f_{bi}	= bamboo stress of each main rebars
A_{bi}	= cross section area of each main rebars

Iterate steps 2-4 with new value of neutral axis (*kd*), until equilibrium equation is satisfied and then calculate the moment capacity and the related curvature.

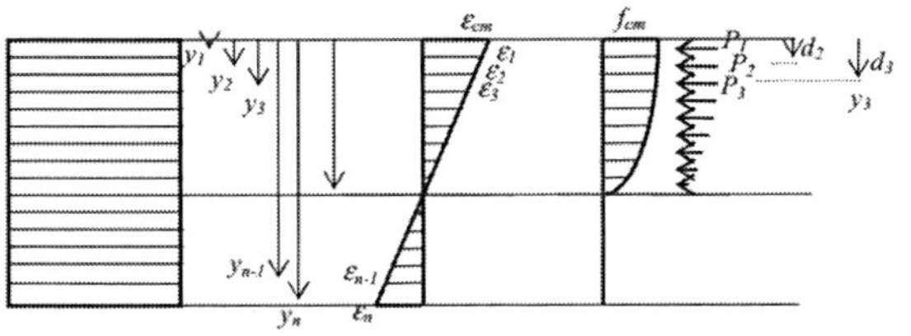

Fig. 8. Fiber section analysis

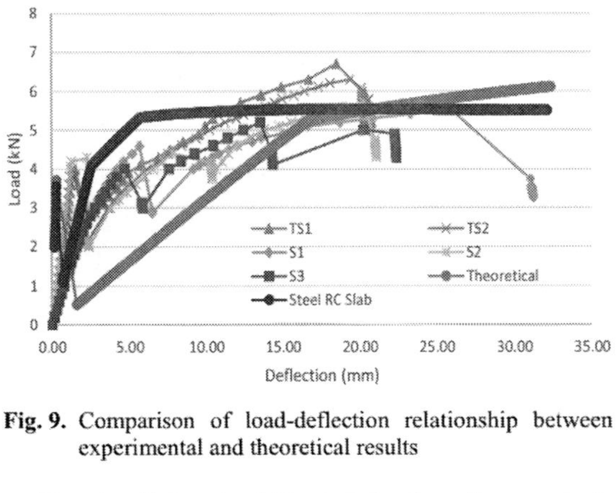

Fig. 9. Comparison of load-deflection relationship between experimental and theoretical results

For simplicity, the bilinear behavior of stress-strain relationship for bamboo derived from experimental data was used. Whereas, the Modified-Kent & Park model as stress-strain relationship of concrete was applied into the model.

The theoretical model using moment-curvature analysis describe above showed good agreement with the experimental results as shown in Figure 9. The cracking load, ultimate load and the corresponding deflections of theoretical model were also in good agreement with experimental data (refer Table 2).

Table 2. Comparison of average load and deflection of experimental data and theoretical model for specimens TS and S (without and with styrofoam respectively)

Specimen	P_{cr} (KN)	P_u (kN)	Δ_{cr} (mm)	Δ_u (mm)
TS	4.0	6.5	1.40	22
S	4.3	5.5	1.2-4.7	21-31
Theoretical	3.8	6.1	0.9	32

As comparison, the load-deflection of steel reinforced concrete slab with identical dimension and concrete properties as bamboo reinforced concrete slab has also been obtained. The steel reinforced concrete used the three steel bars with dimension of 6mm (corresponding to ϕ6-100) and yield stress of 240 MPa. Interestingly, as shown in Figure 10, the performance of bamboo reinforced concrete slab was relatively similar but with less stiff and less ductile to those of steel reinforced concrete slab.

5 Conclusions

Five similar bamboo reinforced slabs have been tested with three slabs were designed with Styrofoam panel as filler and the other two without the Styrofoam panel. The theoretical modeling based on moment-curvature relationship has also been conducted which showed good agreement with experimental test data. Several outcomes can be observed as follows:

1. The ultimate load of slabs without Styrofoam panel were 15% higher but 20% heavier compared with those of Styrofoam panel slabs with the same dimension It is considered good performance in practice design context, since the nominal flexural capacity of RC slabs when designed with minimum reinforcement are usually much higher than the required moment.
2. However there was no significant difference in ultimate deflection capacity, since the deflections at failure point for all specimens were roughly similar.
3. Bamboo reinforced concrete slabs with Styrofoam panel were adequate as a substitute for normal reinforced concrete slabs.

We acknowledge the support of Higher Education Excellent Research Grant from Ministry of Research, Technology and Higher Education of the Republic of Indonesia.

References

1. K. Ghavami. Cem. Concr. Res. ***27***(6), pp.637-649. (2005)
2. Yamaguchi, M., Murakami, K. and Takeda, K., 2013. *Proceedings of Third International Conference on Sustainable and Technologies*. (2013)
3. M. Terai, K. Minami. *Proceedings of 15th World Conference on Earthquake Engineering, Lisbon (Portugal*. (2012).
4. A. Wibowo, B. Kafle, A.M. Kermani, N.T.K. Lam, J.L. Wilson, E.F. Gad. *Proceedings of Australian Earthquake Engineering Society Conference*, Ballarat, Australia, 21-23 November (2008)
5. R. Park, T, Paulay, *Reinforced Concrete Structures*. John Wiley & Sons, New York. **769** pp. (1975).
6. A. Wibowo. Seismic performance of insitu and precast soft storey buildings. *Doctoral Thesis*, Swinburne University of Technology. (2012)

* Corresponding author: ariwibowo@ub.ac.id

Regression models for compressive strength of concrete under different curing conditions

Kolawole Adisa Olonade[1], *Heni* Fitriani[2] *and Olutobi* Toluwalase Kola[1]

[1]Department of Civil Engineering. Obafemi Awolowo University, 220005 Ile-Ife, Nigeria
[2]Faculty of Engineering, University of Sriwijaya, 30662 Palembang, Indonesia

Abstract. Effect of different curing methods on the compressive strength of concrete was investigated. A total of 69 concrete cubes of size 150 mm was cast and cured under five different conditions (immersion in water, sprinkling with water, moist sand covering, polythene covering and air curing) for 3, 7, 14 and 28 days. At the expiration of curing ages, compressive strength was determined and regression analysis was conducted. The concrete cured with these techniques was equally subjected to water penetration test. The results showed that Polythene covering method produced concrete specimen with the highest compressive strength of 23.41 Nmm^{-2} followed by Immersion in water (22.86 Nmm^{-2}). Regression models formulated for each condition indicated that the strength development was dependent on methods of curing. From the results, it was concluded that the compressive strength of concrete depends on the medium in which they were cured and there exist a positive correlation between the compressive strength of concrete and curing age.

1 Introduction

Moisture is essential to the performance of cement-based products. When water is added to a mixture containing cement, hydration reaction begins. The extent to which this reaction continues has great influence on both mechanical and durability of concrete. Though, freshly mixed concrete contains more water than required for hydration reaction but due to the fear of loss of water through evaporation, which could inhibit further hydration reaction. Thus, there is need to keep the conrete matrix saturated or near saturated as much as possible to facilitate anhydrous portion of cement to take up water through curing.

Curing usually has immediate effect on the surface zone, the depth of which varies depending on the type of curing techniques adopted and ambient environmental conditions. It is estimated that the depth at which the curing directly affect concrete is about 20 mm in temprate climate region and up to 50 mm in extreme arid conditions [1]. Unfortunately, those regions beyond these depths are not significantly affected by normal curing technique (wetting). For this reason, effectiveness of curing technique on the strength properties could partly be linked with the size of concrete element. Slender elements would be more affected than bulk ones.

According to [2], cement needs about 23% of water of its weight for hydration to proceed and 15% for filling the space or voids of gel, while [3] reported that 0.22 grams of water is required to fully achieve hydrated cement of about 1 g. If the required amount of moisture is not available for complete hydration, there would be region of anhydrous substance that could be a window for attack. It is equally identified that improper curing entails insufficient moisture and this has been found to produce cracks, compromise strength, and reduce long-term durability [4]. On the other hand, proper curing of concrete is essential to obtain maximum durability of concrete [5].

A number of curing techniques have been identified such as water ponding, spraying of water, or covering with wet burlap and plastic sheet [5]. However, there is still dearth of information as to which curing condition could be considered as the best. Meanwhile, effectiveness of any curing technique is a function of extent of time it can maintain moist condition and also temperature [7]. The rise in temperature speeds up the hydration reaction while water facilitates the reaction.

Reference [8] studied the effect of immersing concrete in water for curing on compressive strength of concrete. They found that the strength of specimens cured by immersion in water was about 30 percent higher than that cured in air. Many other researchers reported similar findings [9-10]. The work reported by [11] suggested that moist sand curing method produced specimens with the highest compressive strength when compared with sprinkling methods. However, [10] opines that sand curing method could cause patchy discoloration especially, if the sand contains calcium chloride, which could negatively affect performance. Reference [12] reported that wrapping curing is more efficient than air curing as it results in greater

* Corresponding author: olonade1431ah@gmail.com

compressive strength. Membrane-forming curing compounds, self curing agents as well as accelerated curing are some other techniques that are found effective [13-15].

A careful perusal of the literature pointed that quite a number of curing techniques have been studied with comparison of three or four methods. Subsequently, there appears to be dearth of information on comparison for wider range of techniques. The thrust of this present work, therefore, is to compare more techniques with a view to rating the performance of each technique on the compressive strength of concrete. In addition, the paper also formulates a regression model that could be used to predict the compressive strength of concrete for different curing techniques.

2 Materials and method

2.1 Materials

Ordinary Portland cement of grade 32.5 was used as binder. Fine and Coarse aggregate used was river sand and crushed granite of maximum nominal size of 3.48 mm and 12.5 mm respectively. Potable water was used for mixing the concrete. Concrete of mix ratio 1:2:4 was batched by weight. A water cement ratio of 0.56 was used to keep the slump in the range of 65±15 to maintain relatively constant consistency [16]. Thereafter, 69 concrete cubes of size 150 mm were cast.

2.2 Curing of specimen

Six curing techniques were considered in this study. The choice of the selected techniques was deliberate. Apart from accelerated curing, others techniques are commonly found in construction sites especially in most developing countries like Nigeria.

- **Immersion in water:** The concrete cube specimens were submerged in water for the required number of days until testing (Figure 1a).
- **Sprinkling with water**: Water was sprinkled on the concrete cube specimens daily (Figure 1b).
- **Polythene covering**: The specimens were wrapped/covered with polythene membrane (nylon) for the required age of curing (Figure 1c).
- **Air Curing**: No active form of curing was carried out. The specimens were left exposed in the Laboratory (Figure 1d).
- **Moist Sand**: The entire concrete cube specimens were buried in sand which was kept moist by wetting with water on a daily basis. Sand cover of approximately 150mm was provided (Figure 1e).

2.3 Determination of material properties

The chemical composition of the cement used in this study was determined using XRF technique, while the gradation parameters of the aggregates were determined in accordance with [17].

2.4 Determination of compressive strength

Compressive strengths of the concrete cubes cured using the techniques identified were determined at the expiration of 3, 7, 14 and 28 days; and after 6, 9 and 12 hours for accelerated curing. The procedure highlighted in [16] was adopted in determining the compressive strengths. Average of three values was determined for each specimen. The 28-day strength of the accelerated curing was obtained using the model proposed by Regressions Models for Predicting Compressive Strength under different curing techniques. Regression analysis was conducted on the compressive strength values obtained. Using MS Excel software, equation that represents the line of best fit was established. Pearson's correlation coefficient was determined for each model.

(a) Immersion in water (b) Sprinkling with water

(c) Polythene covering d) Moist Sand Curing

(e) Air Curing

Fig. 1. Curing of concrete under different conditions

2.5 Determination of efficiency of curing techniques

In order to determine effectiveness of each of the curing techniques, curing efficiency was determined for each using equation (1) as proposed by [17]

$$E = \frac{k_1 - k_2}{k_1 - k_3} \times 100 \quad (1)$$

where:
E = Efficiency (%)
k_1 = the compressive strength of concrete cured in air (N/mm^2)
k_2 = compressive strength of concrete cured in the medium being evaluated (N/mm^2)
k_3 = compressive strength of concrete cured in water (N/mm^2)

2.6 Water absorption capacity of concrete cured under different techniques

Effect of the curing techniques considered in this study was evaluated by determining water absorption and depth of water penetration in concrete cured by the techniques. Concrete cubes were cured using different techniques for 28 days. Thereafter, they were dried in oven for 24 hours at temperature of 60°C.

The temperature chosen was to minimize damage to microstructure of the hardened concrete. For penetration test, about 5 mm of the surface of the concrete cube was held in contact with water (Figure 2) and rise in water by capillary was monitored. This was noticeable by change in colour of the layer of the concrete soaked with water. The depth of water rise in the concrete was recorded at every 30 minutes until no physical rise in water was noticed. In water absorption test, the dried concrete cubes were weighed before they were soaked in water for 24 hours, after which their weights were measured. Water absorption capacity was determined using Equation (2).

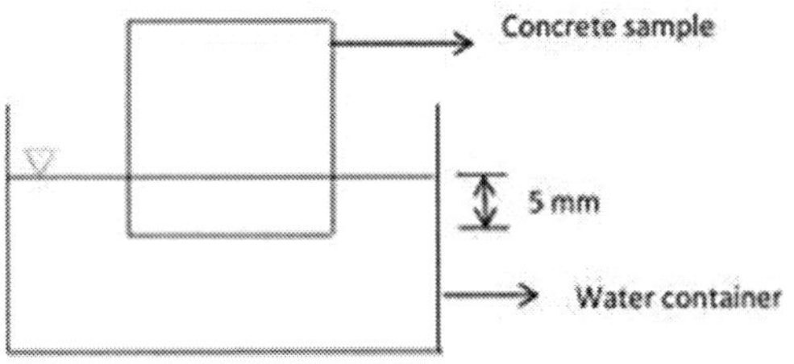

Fig. 2: Penetration depth set-up

$$WAC = \frac{W_d - W_w}{W_d} \times 100\% \quad (2)$$

where:
WAC = water absorption capacity (%)
W_d = weight of dried concrete (kg)
W_w = weight of concrete after soaked in water for 24 hours (kg)

3 Results and discussion

3.1 Material characterization

The chemical composition of the ordinary Portland cement (OPC) used, as determined from XRF technique, is presented in Table 1. Also shown in the table are the mineral contents obtained, using Bogue's equations. The values indicated that the oxides contents were within the stipulated limits given by [18]. Thus, the cement could be categorized as CEM I (general purpose cement). Fig. 3 shows the particle size distribution of the aggregates used in the study. The pattern is in accordance of materials accepted for concrete works [19].

Table 1. Chemical and mineral composition of the cement in comparison with bs en 196

Oxides	Cement	BS EN 196-2 [18]
SiO_2	20.48	18 – 24
Al_2O_3	5.02	2.6 – 8.0
Fe_2O_3	3.15	1.5 – 7.0
CaO	62.45	61.0 – 69
MgO	1.92	0.5 -4.0
Na_2O	0.61	-
K_2O	0.72	0.2 – 1.0
SO_3	0.18	0.2 – 4.0
LOI	4.76	-
Minerals		
C_2S	59.74	
C_3S	13.36	
C_3A	7.98	
C_4AF	9.59	

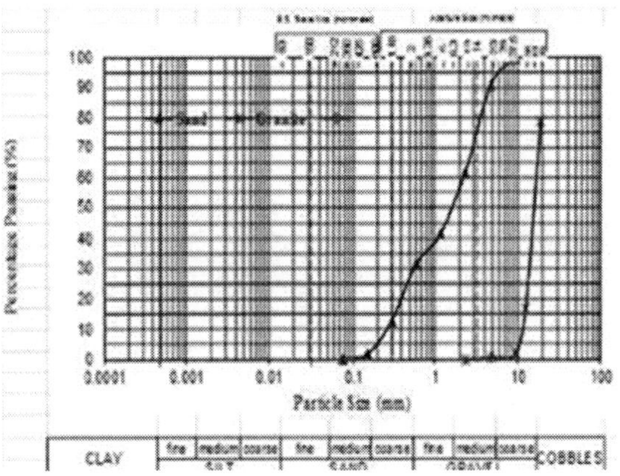

Fig. 3. Particle distribution curves of the aggregates

3.2 Effect of curing methods on compressive strength

In all the curing methods, the compressive strength of the concrete increased with increasing ages of curing (Figure 4). The increase in strength might be associated with production of various silicate hydrates as hydration reaction progressed. It is a known fact that silicate hydrates contribute to increase in strength by filling the voids in concrete matrix. As long as these voids are sufficiently filled with hydration products, the quality of concrete will be maintained at a satisfactory level [19].

DOI: 10.1051/matecconf/201710105013

Compressive strength obtained varied as the curing methods varied, indicating that curing condition influences compressive strength (Figure 1). For example, at age 28 days, the compressive strength of concrete cured by polythene covering was highest (23.41 Nmm^{-2}), while air curing gave lowest strength of 13.94 Nmm^{-2} (60% of polythene covering). Strengths obtained from immersion in water and burying in sand were about 2% and 10% respectively lower than that of the polythene covering, while that of sprinkling with water was about 20% lower. These results are comparable to those reported by [14].

The relative low strength obtained from air curing could be partly attributed to loss of moisture from the concrete through evaporation, subsequently slowing down or retarding hydration of cement; hence low strength [20]. Though, air curing may likely produce higher strength if the relative humidity is maintained at a value greater than 80% [21] as against 70% in this case. Another possibility of low strength in air curing is carbonation, which would interfere with hydration process. Conversely, better perform was given by Polythene covering, where moisture in the concrete matrix was not allowed to escape into the atmosphere and was fully utilized during hydration process.

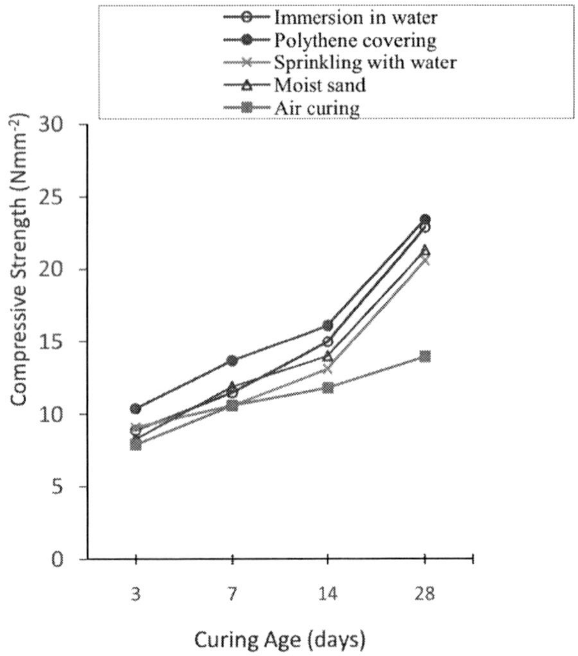

Fig. 4. Compressive strength of concrete under different curing conditions

For total immersion, the strength obtained was in contrary to what [9] obtained in their study, where it was reported that immersion in water gave the best strength. The reason may not be readily known but experimental handling, type of water used, difference in ambient environmental conditions and material constituents could be responsible. Similarly, results from this study indicated that strength of concrete cured by immersion in water was about 60% higher than that of air curing as against 30% obtained by [8]. Though concrete cured by layer of moist sand gave relative higher strength (Figure 3), the method lead to patchy coloration of concrete element, and could have exposed to acidic attack especially, if the soil contains deleterious substances.

3.3 Regression models for compressive strength under different curing methods

Relationship between compressive strength (f_c) and curing ages in days (t) was established using regression analysis, and (1) to (5) were obtained for the different curing techniques considered. These equations further affirmed that there was strong correlation between the compressive strength and curing ages as the Spearman's correlation coefficients were close to 1 for all the equations (0.87 - 0.98). These coefficients, as they approached unity, suggested that the models could be used to predict compressive strength for each of the curing techniques at any given age, all other factors being equal. Nevertheless, the rate of strength development as indicated by the gradients of the equations varied from one technique to the other. The Polythene covering, for instance, had rate of 4.54 (2) followed by immersion in water (1), while air curing had the least (4). The relative low rate of strength development seen in (3) for sprinkling with water could be partly attributed to partial exposure of the concrete to the atmosphere, leading to evaporation of moisture; hence slow hydration process.

$$f_c = 4.14t + 3.22 \quad (R = 0.93) \quad (1)$$
$$f_c = 4.54t + 5.55 \quad (R = 0.94) \quad (2)$$
$$f_c = 3.68t + 4.11 \quad (R = 0.87) \quad (3)$$
$$f_c = 1.93t + 6.23 \quad (R = 0.98) \quad (4)$$
$$f_c = 4.11t + 3.60 \quad (R = 0.95) \quad (5)$$

3.4 Efficiency of curing techniques method

Efficiency of a curing technique is a measure of its effectiveness in influencing concrete properties. The results of efficiency computated suggested that sprinkling of water technique was not as efficient in improving compressive strength of the concrete when compared with immersion in water because its efficiency was less that 100% (74.4%). Moist sand was equally less efficient than water with efficiency of 80.1%. It is interesting to note that polythene covering had about 106.1% efficiency, indicating that it was more effective than water curing. The reason for this performance has could be attributed to preserving mixing water from evaporation. As for curing in air, the efficiency is 0% while that of immersion in water was 100%. If the provision of [23], which stipulated minimum efficiency of 85%, then only polythene covering and immersion in water methods of curing med the requirement, while

DOI: 10.1051/matecconf/201710105013

buried in moist sand was close to the requirement (80.1%).

3.5 Effect of curing techniques on water absorption and penetration

Water penetration depth was found to be the highest in concrete cured in air, while polythene covering had the lowest penetration depth (Fig. 5). The results seem to suggest that the concrete sample cured in air was more porous than other, which made water to travel through it. This assertion is supported by the water absorption capacity of the concrete sample cured in air (Fig. 6).

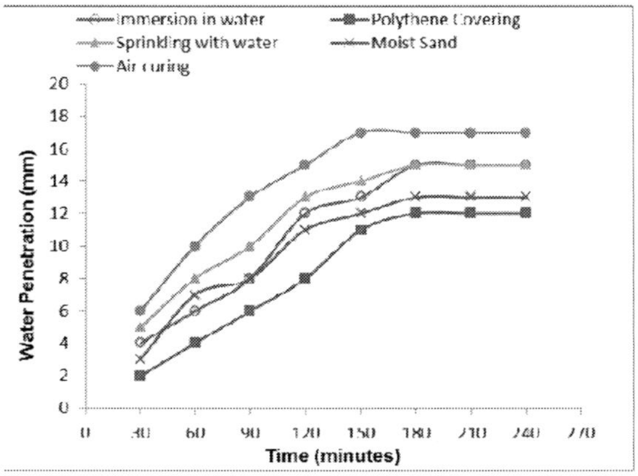

Fig. 5. Water penetration depth for concrete cured using different techniques

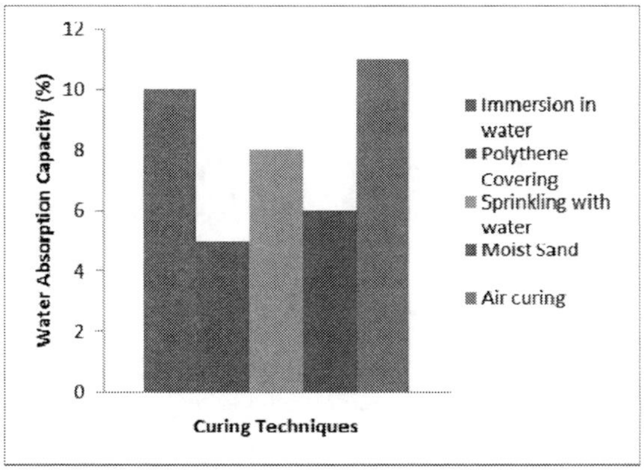

Fig. 6. Water absorption capacity of concrete cured using different techniques

It is possible that the hydration of anhydrous cement was not fully completed due to insufficient moisture, which most of which might be evaporated as against polythene concering and other techniques. The guiding rule is that the more a curing technique could retain moisture required for complete hydration within the concrete matrix the more anhydrous cement reacts with water forming cementitious compound . As more water is consumed the calcium silicate hydrate formed tends to increase volume and hence filling the internal voids. This makes the concrete more compact and resist permeability of water to some extent. These results further confirmed that air curing technique shoud be avoided in practice, especially where strength matter most. On the contrary, polythene covering appeared to perform excellently.

4 Conclusions

Effect of curing methods on compressive strength was studied experimentally and regression models were established for each of the methods. The following conclusions were drawn from the study:

i. Curing of concrete is essential in attaining the desired strength of concrete.
ii. Variations in compressive strengths were influenced by the curing techniques depending on their ability to prevent moisture movement from the concrete.
iii. Polythene covering was found to be the most efficient method of curing as compared to the other methods employed.
iv. Regression models established could be used to predict compressive strengths of concrete for different curing techniques considered.

References

1. CIRIA. Project Report 49, On-site curing – Influence on the durability of concrete: A review (1997)
2. B.L. Gupta, A. Gupta, *Concrete Technology*. A.K. John Publishers for Standard Publishers Distributors, Nai Sarak, Delhi, India (2004)
3. M. Wanielista, M. Chopra, *Compressive Strength of Previous Concrete Pavements*. Final Report. FDOT project BD521-02 University of Central Florida. Orlando Florida (2007)
4. G.S. Wojcik, D.R. Fitzjarrald, Journal of Applied Meteorology, 40 (2001)
5. J.B. Chougale, M. Bhagat, S.M. Nagargoje, International Journal of Advance Foundation And Research In Science & Engineering, 1(2015)
6. M.F. Zain, Y. Matsufuji, Proceedings of the Fifth International Conference on Concrete Engineering and Technology, Kuala Lumpur, Malaysia, (1997)
7. CIP. Concrete in Practice, National Ready Mixed Concrete Association (2000)
8. R. Preetha, S.R. Kishore, C. Sundaramurthy, P.C. Sivathanu, A.K. Laharia, Concrete research letters, 5 (2014)
9. T. James, A. Malachi, E.W. Gadzama, V. Anametemoka, Nigerian Journal of Technology 30 (2011)
10. N.R. Kholia, A.V. Binita, T.G. Tank, International Journal of Advanced Engineering Technology, 4 (2013)
11. A.A. Raheem, A.A. Soyingbe, A.J. Emenike, International Journal of Applied Science and Technology. 3(2013)
12. S.N. Safuddin, Raman, M.F. Zain, Australian Journal of Basic and Applied Sciences, 1 (2007)

13. Y. B. Raghavendra, M.U. Aswath, International Journal of Advance Scientific Research and Technology, 3 (2012)
14. T.S. Nagesh, "Assessment of the Properties of the Self-Cured concrete", Civil Engineering Portal, http://www.engineeringcivil.com/assessment-of the properties- of-self-cured-concrete.html. (2012). Accesed: 18th September, 2015.
15. S.H. Kosmatka, B. Kerkhoff, W.C. Panarese, "Design and Control of Concrete Mixtures", Fourteenth Edition, Portland Cement Association, USA (2003)
16. BS 1881 – Parts 108 & 116. "Method for making test cubes from fresh concrete & Method for determination of compressive strength of concrete", British Standard Institution (BSI), London, (1983)
17. J.G. Cabrera, N. Gowripalan, P.J. Wainwright, Magazine of Concrete Research, 1989, Vol. 41, No. 149, pp.193-198. 1989.
18. BS EN 196-2, Methods of testing cement – Part 2: Chemical analysis of cement', British Standards Institution (BSI), London (1995)
19. BS 1377. "Method of test for soil for civil engineering". British Standard Institution (BSI), London, (1990)
20. D. Ye, D. Zollinger, S. Choi, and M. Won, Literature Review of Curing in Portland Cement Concrete Pavement 2nd edn. (2006) 1 – 80.
21. M.A. Rashid, A. Rahman, S. Chakma, M. Rahman, Duet Journal 21 - 15 (2014)
22. Neville A. M. *Properties of Concrete, Fourth and Final Edition*. John Wiley and Sons, Inc., New York, USA (1996)
23. ACI Committee 318. Building Code Requirements for Structural Concrete and Commentary, 2014

* Corresponding author: olonade1431ah@gmail.com

Performance of rainwater harvesting system based on roof catchment area and storage tank capacity

Imroatul C. Juliana[1], *M.* Syahril Badri Kusuma[2], *M.* Cahyono[2], *Hadi* Kardhana[3], and *Widjaja* Martokusumo[4]

[1]Faculty of Civil and Environmental Engineering, Bandung Institute of Technology, Indonesia
[2]Water Resources Development Center, Bandung Institute of Technology, Indonesia
[3]Center for Research on Infrastructure and Regional, Bandung Institute of Technology, Indonesia
[4]School of Architecture, Planning, and Development Policy, Bandung Institute of Technology, Indonesia

Abstract. Increasing population growth has created problems in water resources. Natural water resources become progressively more expensive and difficult to develop. In addition, it is also becoming increasingly polluted and difficult to obtain. Many countries shown a resurgent interest in the use of rainwater harvesting (RWH) technique to overcome these problems. There are several factors that will influence the RWH performance, such as the rainfall, catchment area, storage tank capacity, and water demand. The performance parameter determines by the volumetric reliability, time reliability, and yield. The RWH system used in this study is a simple RWH system that utilizes roof as a catchment area, pipes as a distribution system and tank as a storage. An analysis is carried out to investigate the effect of altering the large of the catchment area and storage tank capacity to the RWH system performance parameters. A suitable behavioral model based on the water balance method is implemented to evaluate the inflow, outflow, and the storage volume. Results demonstrate that with up to 15 years daily rainfall data in 15 cities in Indonesia, the most influential parameters on the performance of RWH system is the time reliability.

1 Introduction

Human population growth for the last few decades has increased exponentially. Increasing human population growth lead to the emergence problems in the water resources either directly or indirectly. In terms of quality or quantity, the usefulness of water will reduce, and water damage will be even greater. Furthermore, the problem of water resources becomes complicated as the lack of management of water resources. Accordingly, if the water resources optimally use and manage, such problems can be avoided.

Rainwater is one of the potential water resources that is often overlooked. The potential rainfall-yet- unused rainwater often thrown out into a run-off. Whereas the rainwater can be collected and afterward processes in raw water that is ready to use [1].

Rainwater harvesting is a simple ancient technology and can vary from minor and basic to major and complex. Previous studies have shown that the consumption of rainwater provides economic and environmental benefits. Rainwater that has been collected can be used for nonpotable water use, thereby it could reduce the consumption of groundwater and provide financial benefits [2].

Similar studies corroborate that the rainwater harvesting system also provides the advantage from an economic standpoint [3-9]. The rainwater harvesting concept also bestows significant impact on the surrounding environment [10]. The usage of rainwater harvesting system on a large scale is taken into consideration as an adaptive strategy in dealing with climate change, related to water shortages [11].

The use of rainwater harvesting system (RWH) had declined in many developed countries including Indonesia. There are several factors that contribute to that e.g. the public perception on the installation of RWH system is expensive and does not provide significant benefit in terms of economic or environmental. Therefore, there should be a study comprehensively on the RWH system performance in order to analyze the feasibility of the system implementation. Several parameters, such as the volumetric reliability, time reliability and yield could define the RWH system performance.

This study aims to investigate the contribution of the roof catchment area and storage tank capacities to the alteration of RWH system performance parameters. Thus, it can analyze which performance parameter is the most affected. This would be outstandingly beneficial for the design of the RWH system.

2 Methodology

This study uses a simple RWH system that used the roof as a catchment area, pipes as a distribution system and tank as a storage. The storage tank capacity is a vital component as it will determine the entire system performance and cost.

The characteristic of the catchment area, potential rainfall, and water demand will contribute to the RWH system performance [1]. The daily rainfall data from 29

* Corresponding author: icjuliana76@gmail.com

DOI: 10.1051/matecconf/201710105014

rainfall stations in 15 cities in Indonesia from Sumatera to Sulawesi were used. The length of the data is varying from 7 years to 15 years, according to the data availability. The cities were divided into 4 categories according to the annual rainfall range.

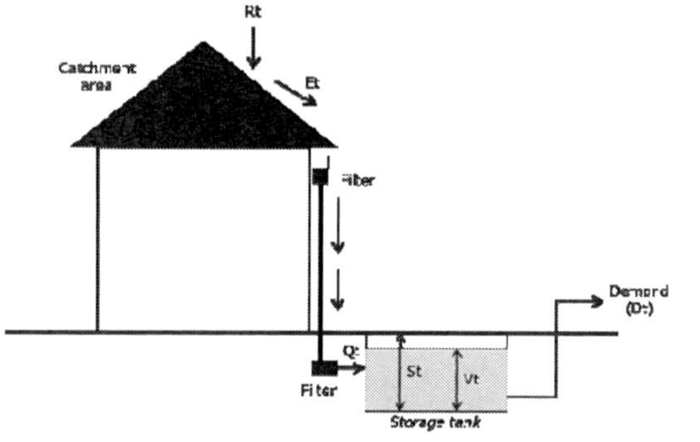

Adapted : Roebuck, 2010

Fig. 1. Schematic representation of the simple rainwater harvesting system

The effective rainfall volume obtains by multiplying the rainfall with the catchment area and runoff coefficient.

$$ER = R_t \,.\, A \,.\, C \tag{1}$$

Where ER is the effective runoff (m^3), R_t is daily rainfall (mm), A is catchment area wide (m^2) and C is the ratio of the rainwater that could be caught in the catchment area.

The water balance method is used to describe the simulation of the system. The yield before spillage (YBS) algorithm used to describe the storage tank behavior.

$$V_t = V_{t-1} + Q_t - D_t - E_t - L_t \tag{2}$$

V_t is storage volume at time t (m^3), V_{t-1} is the storage volume at time t - 1 (m^3), Q_t is the volume that enters the storage in a time interval t (m^3), D_t is a demand release in lapse of time t (m^3), E_t is evaporation (mm), L_t is the losses (mm) and S is the storage capacity (m^3). In this study, the effect of evaporation and losses is disregarded since the storage tank is below ground, so that:

$$V_t = V_{t-1} + Q_t - D_t \tag{3}$$

where $0 \le V_t \le S$

The RWH system performance expressed with the reliability parameter. The units of volume and time could express the reliability of the storage tank capacity [12]. Liaw and Tsai defined volumetric reliability as the total volume of harvested rainwater supplied divided by the total water demand (VR) and time reliability is the fraction of *time* that demand is fully met divided by total time under consideration (TR) [13].

$$VR = \frac{\sum_{t=1}^{T} Y_t}{\sum_{t=1}^{T} D_t} \times 100 \tag{4}$$

$$TR = \frac{\sum_{t=1}^{T} YT_t}{\sum_{t=1}^{T} T_t} \times 100 \tag{5}$$

where VR is the volumetric reliability, TR is the time reliability, Y_t is yield in time t (m^3), D_t is total demand in time t period (m^3), YT_t is the total days that the water need fulfill with RWH (days), and T_t is total time under consideration (days).

The sensitivity analysis carried out in order to determine the level of variation in predicting RWH system performance to changes in the catchment area and storage tank capacities. The alteration in the catchment area and storage tank capacity defined by:

$$C_c = \frac{CA_n - CA_b}{CA_b} \times 100\% \tag{6}$$

$$S_c = \frac{S_n - S_b}{S_b} \times 100\% \tag{7}$$

where C_c is the alteration of the catchment area, CA_n is the catchment area at n, CA_b is the benchmark of the catchment area, S_c is the alteration of storage tank, S_n is the storage tank at point n, S_b is the benchmark of the storage tank.

Percentage changes in RWH system performance, such as VR, TR and yield were selected as the metric for judging the sensitivity. By performing this analysis, the possible consequences of such changes can be evaluated. Sensitivity analysis performed for two conditions variation : the large of the catchment area and the storage tank capacity. The graph gradient will show the extremely affected parameter. The steepest gradient generates, the greater the effect of these parameters on the RWH system performance.

3 Results and analysis

3.1 Analysis on the roof catchment area alteration

The analysis conducted on alteration of the roof catchment area from 10 m^2 to 1200 m^2. The 70 m^2 catchment area and 2 m^3 storage tank capacity are set as a base case result. The water uses divided into all water use and nonpotable water use. A constant occupancy of 1 occupant with total water use is 150 L/person/day [14, 15]. Two storage tank capacities : 2 m^3 and 5 m^3 was assumed. The results of the sensitivity analysis are in the following graphs. In each case, the results from all rainfall ranges were plotted on the same chart to make direct comparisons easier.

DOI: 10.1051/matecconf/201710105014

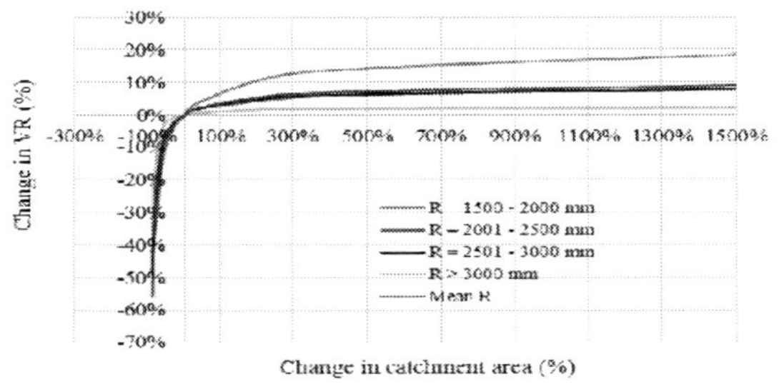

(a) tank 2 m^3 – all water use

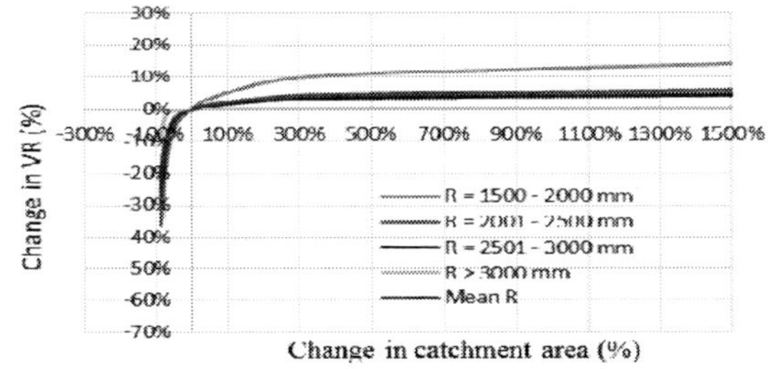

(b) tank 2 m^3- nonportable water use

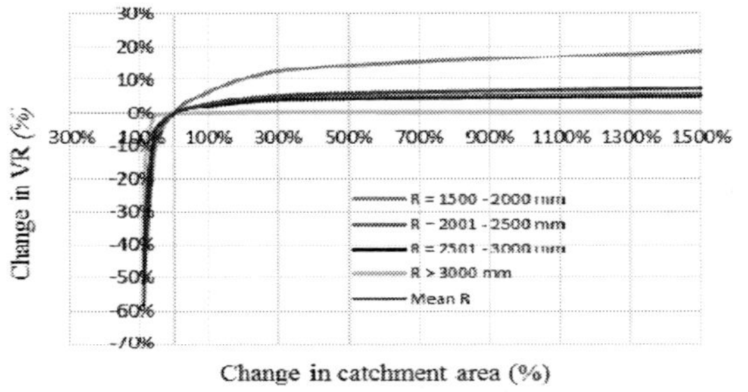

(c) tank 5 m^3 – all water use

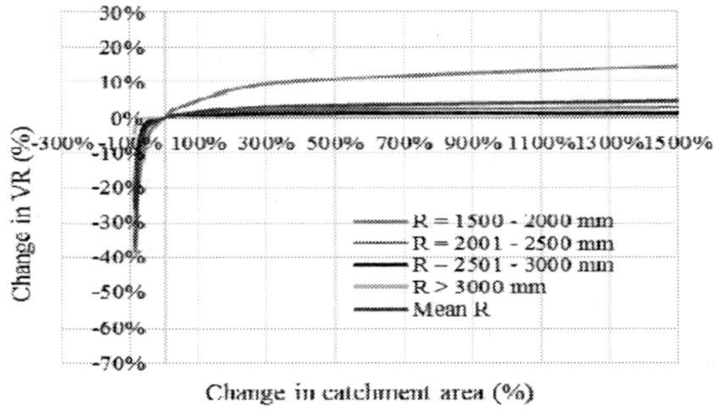

(d) tank 5 m^3- nonportable water use

Fig. 2. Sensitivity to alteration in catchment area of volumetric reliability (VR)

Figures 2 to 4, shows the average differences between the results for eight catchment areas. Each tank displayed similar trends with regards to the influence of the catchment area and water uses. Analysis of the results showed that there is a nonlinear relationship between the increasing in the catchment area with the system performance and annual yield. Furthermore, there is a positive gradient indicating that the increased in the catchment area would lead to an increase in system performance.

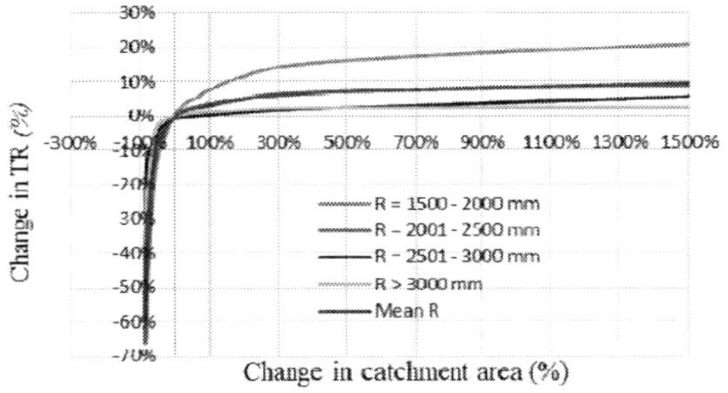

(a) tank 2 m^3 – all water use

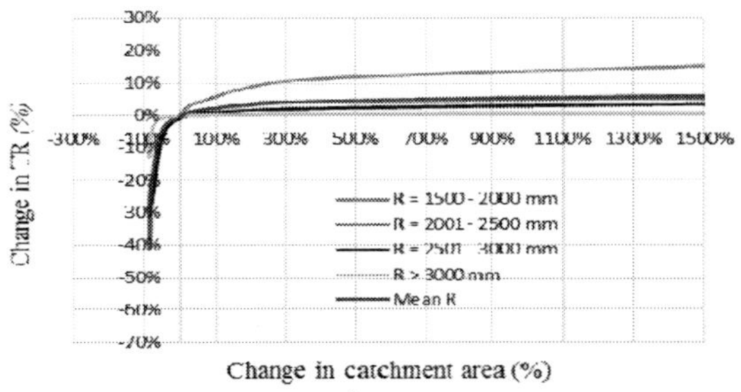

(b) tank 2 m^3- nonpotable water use

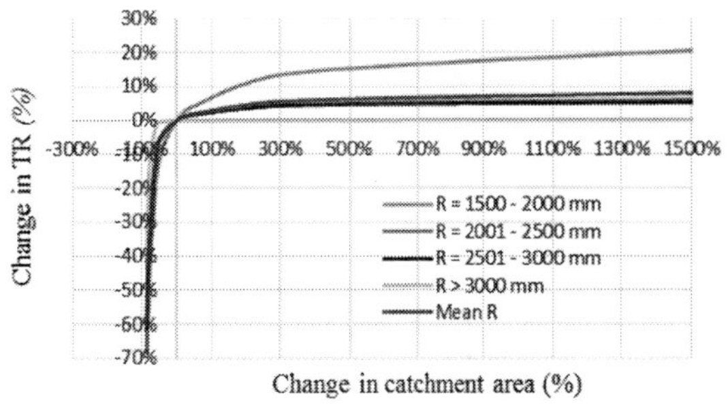

(c) tank 5 m^3 – all water use

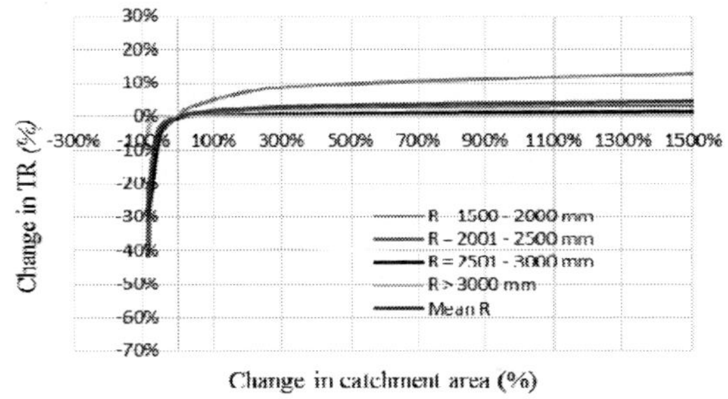

(d) tank 5 m^3- nonpotable water use

Fig. 3. Sensitivity to alteration in catchment area of timemetric reliability (TR)

 DOI: 10.1051/matecconf/201710105014

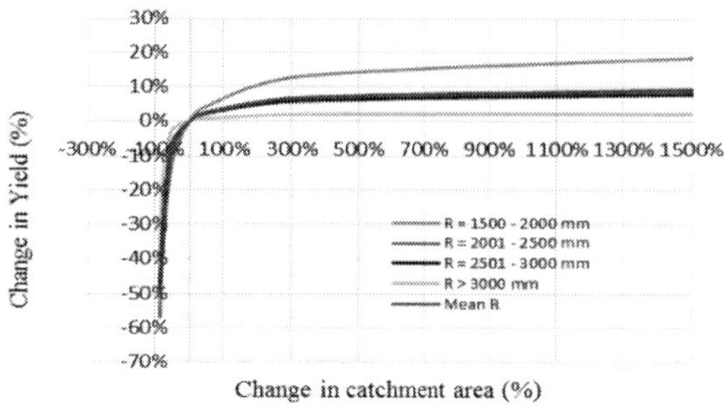

(a) tank 2 m^3 – all water use

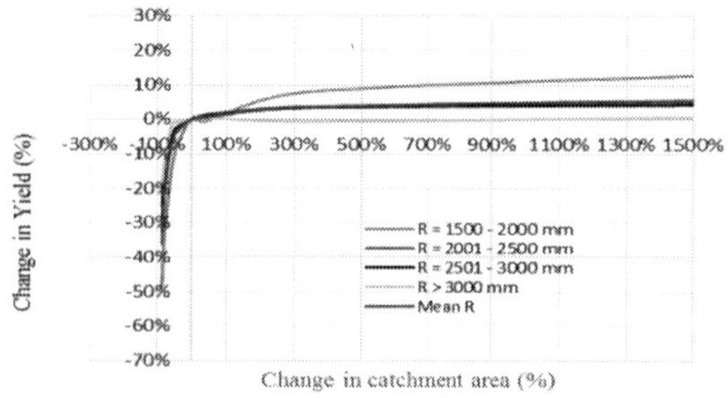

(b) tank 2 m^3- nonpotable water use

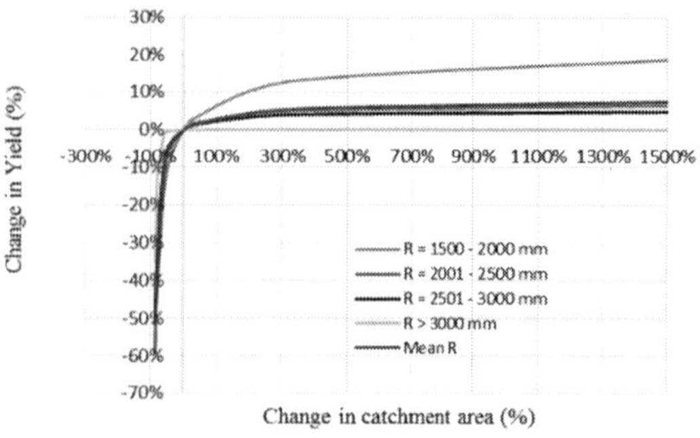

(c) tank 5 m^3 – all water use

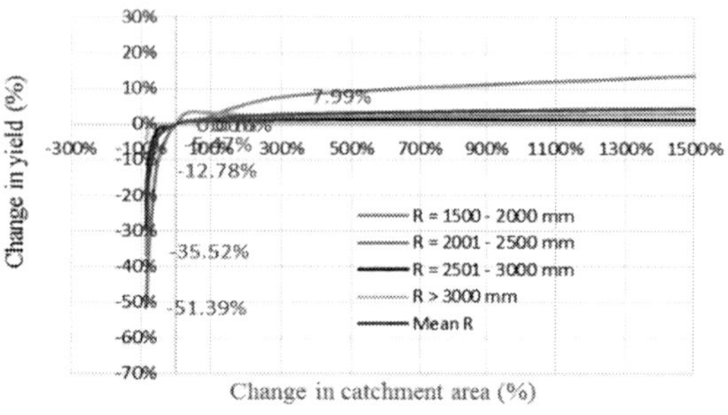

(d) tank 5 m^3- nonpotable water use

Fig. 4. Sensitivity to alteration in catchment area of yield

The gradient of the graph progressively ramps, especially in the wide of catchment area over 300 m^2. The rate of graphs gradient increasingly sharps for the catchment area less than 70 m^2. Therefore, this indicates that the increase in the catchment area below 70 m^2 extremely influential on the performance of RWH systems.

Rainfall also affects system performance. The lesser the annual rainfall, the increase of the catchment area gives the extra impact on the system performance and vice versa. Moreover, the increase in the catchment area is influential for annual rainfall 1500 mm - 2000 mm/year. For annual precipitation above 3000 mm/year, the rate of changes was relatively small, especially for the catchment area above 70 m^2.

Table 1. The graph gradient for each parameter to alteration in the catchment area

Catchment area (m^2)	Mean VR gradient		Mean TR gradient		Mean yield gradient	
	2 m^3	**5 m^3**	**2 m^3**	**5 m^3**	**2 m^3**	**5 m^3**
10	0.44	0.44	0.52	0.52	0.47	0.47
15	0.27	0.26	0.31	0.32	0.32	0.29
30	0.11	0.10	0.15	0.12	0.14	0.11
45	0.07	0.06	0.10	0.07	0.10	0.07
70	**Bench mark**					
100	0.04	0.03	0.04	0.03	0.03	0.03
300	0.01	0.01	0.02	0.01	0.02	0.01
1200	0.00	0.00	0.00	0.00	0.00	0.00

Table 1 shows the graph gradient for each parameter. The gradient of 2 m^3 and 5 m^3 storage tank capacity in each parameter approximately the same particularly for the larger catchment area. From the results, it can be concluded that the most affected parameter by the alteration in catchment area is the time reliability parameter. Despite the differences tend to be diminutive, not more than a few percentage. As a result, the differences from each parameter are not further than 50%. Therefore, the alteration in storage tank capacity also does not perform a significant increase in system performance. The larger the catchment area, the alteration in parameter is close to (zero).

3.2 Analysis on the alteration of storage tank capacity

The analysis conducted under 4 storage tank capacities alteration, which are 1 m^3, 2 m^3, 5 m^3, and 10 m^3. The 2 m^3 storage tank capacity assumed as a benchmark. The water uses divided into all water use and nonpotable water use for 1 occupant. The catchment area is defined as 70 m^2. The sensitivity on the alteration of storage tank capacity can be seen in figure 5 to 7.

Figure 5 to 7, shows the effects of varying annual rainfall on the RWH system performance. A range of water uses and storage tank capacity were used in order to cover a wide range of possible scenarios. From the graphs, the lesser the rainfall, there is a linear relation between the upsurge in storage tank capacity and the system performance. For storage tank capacity below 2 m^3, the relations tend to be nonlinear with fairly steep gradient, conversely, compared with storage tank capacities above 2 m^3.

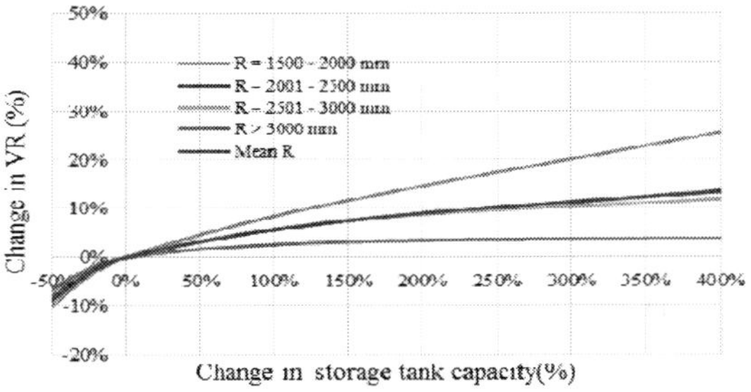

(a) all water use

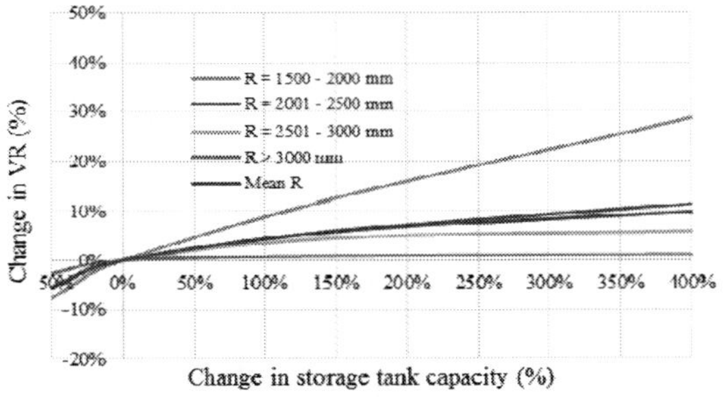

(b) nonpotable water use

Fig. 5. Sensitivity to alteration in storage tank capacity of volumetric reliability (VR)

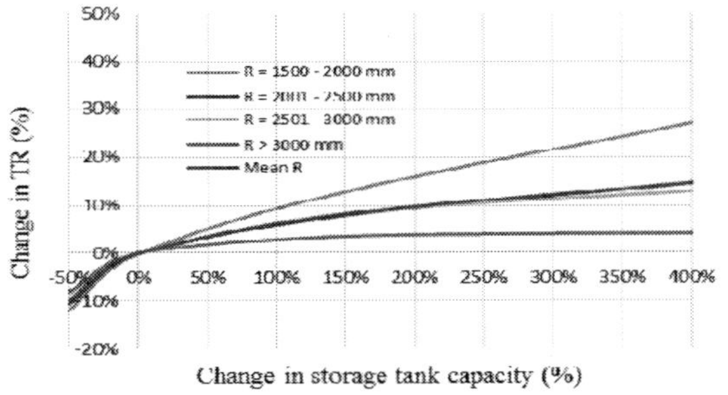

(a) all water use

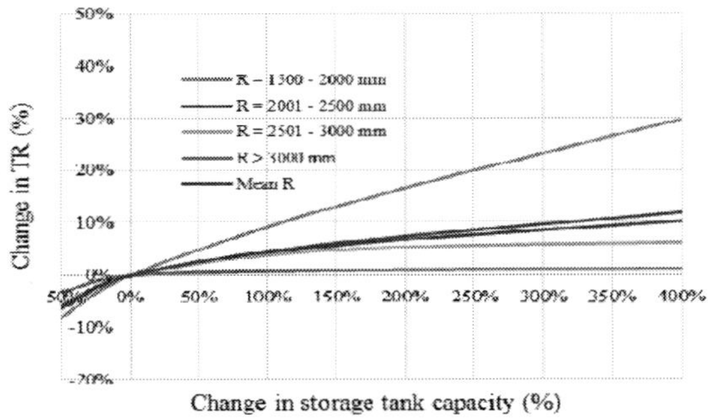

(b) nonpotable water use

Fig. 6. Sensitivity to alteration in storage tank capacity of time reliability (TR)

The graph gradient is positive which indicates that the increase in storage tank capacity will lead to the increase of the system performance. Ordinarily, the larger the rainfall, the gradient becoming declivous. Especially for the rainfall above 3000 mm/year, the increase in storage tank capacity will not significantly affect the system performance. However, the increase in storage tank capacity will deliver the vast effect for a rainfall with a range 1500 mm/year - 2000 mm/year. This matter can be seen from the sharp gradient and far enough span with the annual rainfall 2001 – 3000 mm/year. The increase of RWH system performance is almost equal for annual rainfall of 2000 - 2500 mm/year and 2501- 3000 mm/year. It is perceived from the merging gradient lines between the two ranges primarily for the more water use.

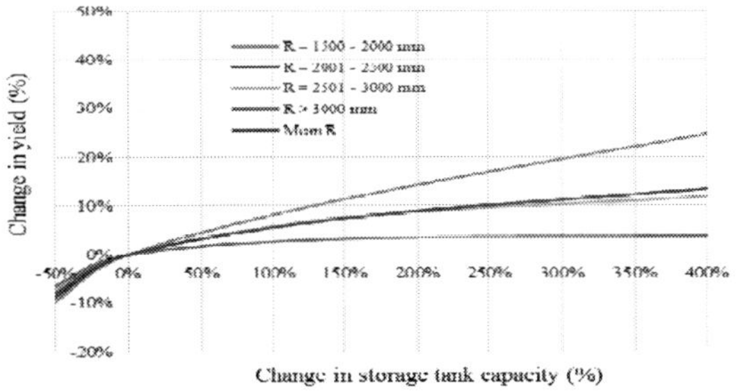

(a) all water use

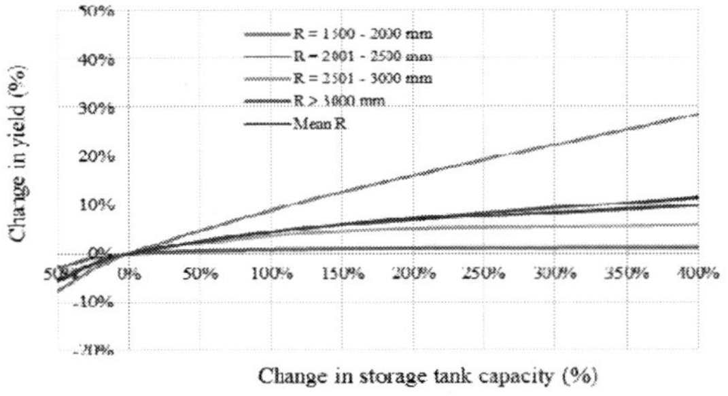

(b) nonpotable water use

Fig. 7. Sensitivity to alteration in storage tank capacity of yield

Table 2. The graph gradient for each parameter to alteration in storage tank capacities

Storage tank capacity (m^3)	Mean VR gradient	Mean TR gradient	Mean yield gradient
1	0.135	0.156	0.134
2	Benchmark		
5	0.045	0.048	0.044
10	0.031	0.033	0.031

Table 2 shows the graph gradient associated with each of the key parameters. The gradient of VR, TR and yield for each storage tank capacity are approximately the same particularly for the bigger storage tank capacity. From the results, it can be concluded that the most sensitive parameter by the alteration in storage tank capacity increase is the time reliability parameter. Therefore, the alteration in storage tank capacity also does not perform a significant increase in system performance. Whilst, the performance for up to 5 m^3

storage tank capacity is decreased. This indicated that the saturation point of storage tank capacity for the range of rainfall is 5 m^3.

4 Conclusions

The predicted RWH system performances depend overwhelmingly on the annual rainfall, storage tank capacity, and catchment area. From the analysis, it can be concluded that the most sensitive parameter within the alteration in the catchment area and storage tank capacity is the time reliability even though the difference with the other parameters is insignificant. The graphs also demonstrate that the smallest annual rainfall encounters the biggest influence of the alteration.

The authors would like to express the highest gratitude to Engineering Faculty, Sriwijaya University for the initial funding for this research. Our sincere thanks go to Water Resources Development Center, Center for Research on Infrastructure and Regional, Bandung Institute of Technology, for its supports during this publication process. The authors would also like to gratefully the BMKG Online for providing the secondary data.

References

1. Frasier, Gary, Myers, *The Handbook Of Water Harvesting*, US Department of Agriculture, Washington DC, (1983)
2. R.M. Roebuck, Ashley, *Proceeding 7th International Conference of Urban Drainage Modelling*, (2006)
3. Yan Zhang, Donghui Chen, L. Chen, L., S. Ashbolt, Journal of Environmental Management **91**, p : 222-226, (2009)
4. Vivian Tam, Zeng S.X., Journal of Resources, Conservation and Recycling **54**, p : 178-186, (2010)
5. Anthony M. Ryan, Clive Spash, Thomas G. Measham, Journal of Hydrology **379**, p : 164-171, (2009)
6. A. Rahman, J. Dbais, M. Imteaz, American Journal of Engineering and Applied Sciences **3 (1)**, p : 73-82, (2010)
7. Ataur Rahman, Joseph K, M. Imteaz, Journal of Resources, Conservation and Recycling **61**, p : 16-21, (2012)
8. R. Farreny, X. Gabarrell, J. Rieradevall, Journal of Resources, Conservation and Recycling **55**, Netherlands, p : 686-694, (2011)
9. N. Nagaraj, U. Pradhani, U., Chengappa, Agricultural Economics Research Review **24**, India, p : 217-223, (2011)
10. Sara Angrill, Farreny, Ramon, Gasol, International Journal of Life Cycle Assess, DOI 10.1007/s11367-011-0330-6, (2011)
11. K. E. Trenberth, P.D. Jones, P. Ambenje, R. Bojariu, D. Easterling, A. Klein Tank, D. Parker, F. Rahimzadeh, J.A. Renwick, M. Rusticucci, B. Soden, and P. Zhai, *Observations: Surface and atmospheric climate change*. Cambridge University Press, Cambridge, (2007)
12. A. Fewkes, D. Butler, *Proceedings of the 9th International Rainwater Catchment Systems Conference*, (1999)
13. C. Liaw, Y. Tsai, Journal of the American Water Resources Association **40 no. 4**, p : 901-912, (2004)
14. Dian Mangiring, *Kajian Pola Konsumsi Air Bersih Rumah Tangga di Kelurahan Setiamanah Kota Cimahi sebagai Masukan Bagi Upaya Konservasi, Master Thesis*, ITB, Bandung, (2007)
15. Direktorat Penataan Bangunan dan Lingkungan, *Peraturan Perunandg-unandgan Bidang Bangunan Gedung and Lingkungan,* Kementrian Pekerjaan Umum, Direktorat Jenderal Cipta Karya, (2010)

* Corresponding author: icjuliana76@gmail.com

Hydrologic modelling using TRMM-based rainfall products for flood analysis

Sigit Sutikno[1,*], *Yohanna* Lilis Handayani[1], *Manyuk* Fauzi[1], *Fitriani*[1], *Ariani* Kurnia[1], and *Rinaldi*[1]

[1]Civil Engineering Department, University of Riau, 28293 Pekanbaru, Indonesia

Abstract. The low resolution and the limitation of data are the main problems for flood analysis, so that the flood prediction often becomes inaccurate. This led to the decision policy for flood disaster mitigation often becomes ineffective. Currently the available of satellite-estimated rainfall data is gradually becoming a cost-effective source of input for flood prediction under a variety of circumstances. This study developed a hydrologic modelling using Tropical Rainfall Measuring Mission (TRMM) satellites data and IFAS (Integrated Flood Analysis System) tool for flood analysis. Upper Rokan watershed with the AWLR (Automatic Water Level Recorder) station of Lubuk Bendahara and Pasir Pangaraian in Riau Province were picked up as study area of this research. The study areas are located in Rokan Hulu regency which is the most vulnerable regency against flood disaster in Riau Province. The results show that the hydrologic modelling using satellite data can be used as an alternative for flood analysis in the area with limited ground data. The flood discharge prediction shows an acceptable error with the ground measurement data. The wave shape error, the volume error, and the peak discharge error for Lubuk Bendahara station are 9.58%, 17.28%, and 10.81%, and for Pasir Pangaraian station are 33.62%, 8.60% and 5.88% respectively.

1 Introduction

Floods are among the most recurring and devastating natural hazards, impacting human lives and causing severe economic damage throughout the world. It is understood that flood risks will not subside in the future, and with the onset of climate change, flood intensity and frequency will threaten many regions of the world [1]. The main problem for flood analysis in Indonesia and other developing country is the low resolution and the limitation of data, so that the flood prediction often become inaccurate [2, 3]. This led to the decision policy for flood disaster mitigation often becomes ineffective.

Currently the available of satellite-estimated rainfall data is gradually becoming a cost-effective source of input for flood prediction under a variety of circumstances. One such satellite rainfall product, the National Aeronautics and Space Administration (NASA) Tropical Rainfall Measuring Mission (TRMM) has been used extensively and provides quasi global (50°S–50°N) precipitation analyses at 3-hourly, 0.25° latitude–longitude resolution, with all satellite estimates calibrated or adjusted to the information from the TRMM satellite itself, which carries both a radar and passive microwave sensor [4, 5].

This study developed a hydrologic modeling using TRMM satellites data and IFAS (Integrated Flood Analysis System) tool for flood analysis. IFAS was developed by Hydrologic Engineering Research team, International Centre for Water Hazard and Risk Management (ICHARM), Japan. The application of IFAS which includes interface for utilisation of satellite-based rainfall as the input is suitable for the catchment in developing country which the availability of ground rainfall stations is limited. The design concept in developing the flood analysis model are the software can utilize both satellite and ground-based rainfall as an input; implementation of runoff analysis engine; implementation of a model creation and a parameter estimation function; visualisation of flood result and free distribution [3, 6-8].

Upper Rokan watershed with the AWLR (Automatic Water Level Recorder) station of Lubuk Bendahara and Pasir Pangaraian in Riau Province were picked up as study area of this research. The study areas are located in Rokan Hulu regency which is the most vulnerable regency against flood disaster in Riau Province. The applicability of the IFAS in Indonesia has been studied in some cases [9-11]. However, those study were still limited for low flow hydrologic analysis. The applicability of IFAS for flood analysis also have been used in wide range around the word [12]. The applicability of IFAS for high flow analysis in case of flood in upper Rokan watershed, Indonesia was further evaluated in this study.

2 Methods

2.1 Study area

This study is focused in the Upper Rokan watershed with the AWLR station of Lubuk Bendahara and Pasir

* Corresponding author: ssutiknoyk@yahoo.com

Pangaraian in Riau Province as shown in Fig. 1. Lubuk Bendahara and Pasir Pangarayan Sub-watershed have catchment area about 3,196 km^2 and 1,595 km^2 respectively. The topography of study area is quite varied ranging from 100 meters to over 1,000 meters above sea level, which is about 80% of its altitude in range between 100 and 500 meters. The upper Rokan watershed generally is a tropical climate with an average maximum temperature of 31°C-32°C. The climate is strongly influenced by changes in wind direction, which the rainy season usually fall on October to April.

Rokan Hulu regency wich is located in upper Rokan watershed had experienced flood diaster almost every year since 2002. The most severe damage was occured in 2014. It was reported that 69 villages in 13 sub-districts in Rokan Hulu regency was inundated by the flood, 18 people were killed, 17,816 houses were damaged, and 53,725 people were evacuated [13].

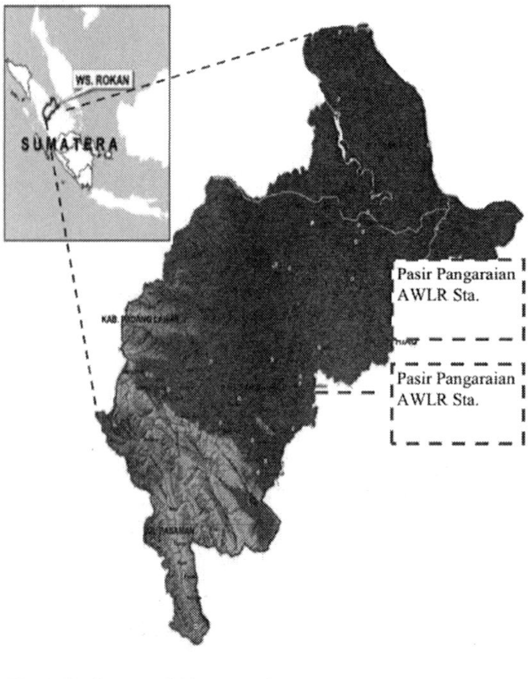

Fig. 1. Study area of this research

2.2 Basic structure of IFAS

IFAS which was developed by ICHARM) Japan, employs the Public Works Research Institute Distributed Hydrological Model (PDHM) as its runoff simulation model. The conceptual structure of PDHM is shown in Fig. 2. Surface flow, rapid subsurface flow, infiltration, slow subsurface flow and base flow are calculated by the equations from (1) to (5). In the process of river routing, the Kinematic Wave Model is employed. Also, IFAS can import some kinds of satellite-based rainfall data such as GSMaP and 3B42RT for insufficient ground observation area [2].

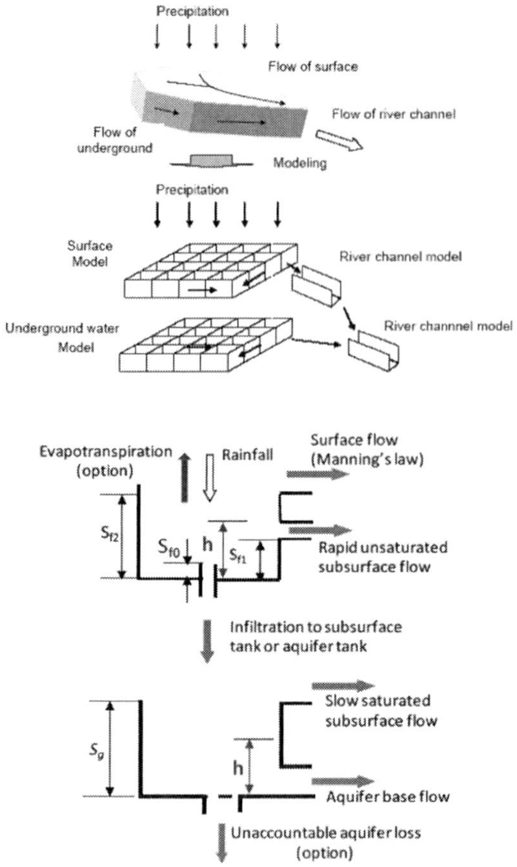

Fig. 2. Conceptual structure of IFAS

$$Q_d = L\frac{1}{N}(h - S_{f2})^{\frac{5}{3}}\sqrt{i} \quad (1)$$

$$Q_{ri} = \alpha_n A f_o \frac{(h - S_{f1})}{S_{f2} - S_{f1}} \quad (2)$$

$$Q_0 = A f_0 \frac{(h - S_{f0})}{S_{f2} - S_{f0}} \quad (3)$$

$$Q_{g1} = A_u^2 (h - S_g)^2 A \quad (4)$$

$$Q_{g2} = A_g h A \quad (5)$$

where Q_{sf}: surface flow (m^3/s), L: mesh length (m), N: manning's roughness coefficient (m$^{-1/3}$/s), h: water height for the tank (m), S_{f2}: height from which surface flow occurs (m), i: slope with the adjacent cell, Q_{ri}: rapid unsaturated subsurface flow (m^3/s), A: mesh area (m^2), S_{f1}: height from which rapid unsaturated subsurface flow occurs (m) Q_0: infiltration to aquifer tank (m^3/s), S_{f0}: height where ground infiltration occur (m), Q_{g1}: slow saturated subsurface flow (m^3/s), S_g: height from which slow saturated subsurface flow occurs (m), Q_{g2}: base flow (m^3/s),α_n ,f_0, A_u, A_g: coefficients.

2.3 Data for hydrologic model setup

The data used in the study consisted of satellite data and the discharge measured data. The discharge measurement data was needed for calibration process. The remote sensing data sets enabling the implementation and testing of a distributed hydrologic model in the Rokan waterhed include Digital Elevation Model (DEM), rainfall, land-use / land cover, and soil parameters.

2.3.1 Digital Elevation Model (DEM)

Digital Elevation Model (DEM) is determined from the Global Data which which can be downloaded from site (http://edcftp.cr.usgs.gov/pub/data/gtopo30/global/) directly from IFAS. Global data sets relevant to flood analysis which can be used to create run-off analysis model and estimates parameters for the river basin. The elevation within the catchment area can be determined and the model will then generate the catchment boundary. Fig. 3 shows the river basin and sub-catchment area of AWLR Lubuk Bendahara in upper Rokan watershed.

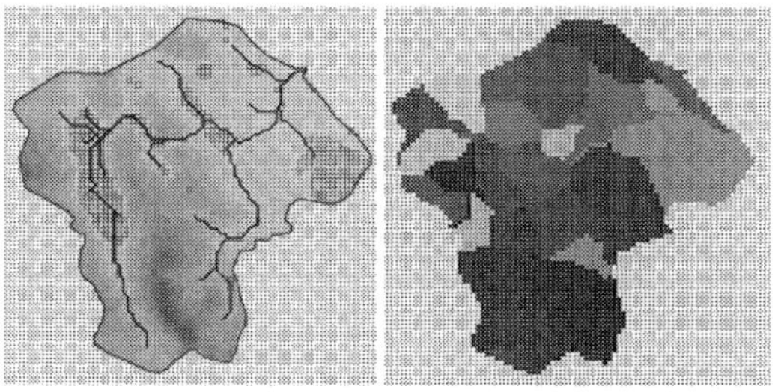

Fig. 3. (i) Lubuk Bendahara river basin, (ii) Sub catchment generates from IFAS.

2.3.2 Rainfall input

Two types of rainfall data are used as input data to the IFAS model which are ground and satellite-based rainfall data. IFAS can read rainfall point data (like rain gauges data) if saved in CSV format where location of the station and time of the measurements are specified. For the satellite rainfall data, GSMaP (Global Satellite Mapping of Precipitation) was was used in this study. GSMaP data can be downloaded directly from IFAS Rainfall Data Manager.

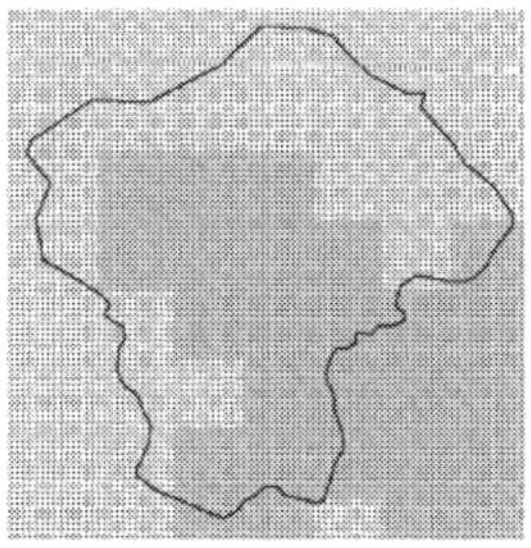

Fig. 4. GSMap rainfall data 2012

IFAS has a built-in GSMaP import function which uses the Microwave radiation device. Furthermore, 3B42RT is also one of the rainfall product provided by NASA (National Aeronautics and Space Administration, US) that can be downloaded and imported by IFAS through the Rainfall Data Manager. Fig. 4 shows the GSMaP rainfall data 2012 of catchment area of AWLR Lubuk Bendahara in upper Rokan watershed.

2.3.3 Land-use and soil parameters

Soil parameterse data is ulilized from United Nations Environment Programme (UNEP) GNV25 Soil Water which can be downloaded from site (http://www.grid.unep.ch/data/data.php). GNV25 data is the data of soil water holding capacity, which have been selected for commonly used for hydrological modeling. Fig. 5(i) shows the soil parameters data of river basin AWLR Lubuk Bendahara in upper Rokan watershed.

Land Use data is ulilized from GLCC (Global Land Cover Characterization) which which can be downloaded from site (http://edcftp.cr.usgs.gov/pub/data/glcc/globe/latlon/) directly from IFAS. This data is a database of characteristics of land created by the USGS, University of Nebraska-Lincoln (UNL), and the European Commission's Joint Research Centre (JRC) with a resolution of global 1 km. Fig. 5(ii) shows the land-use data of river basin AWLR Lubuk Bendahara in upper Rokan watershed.

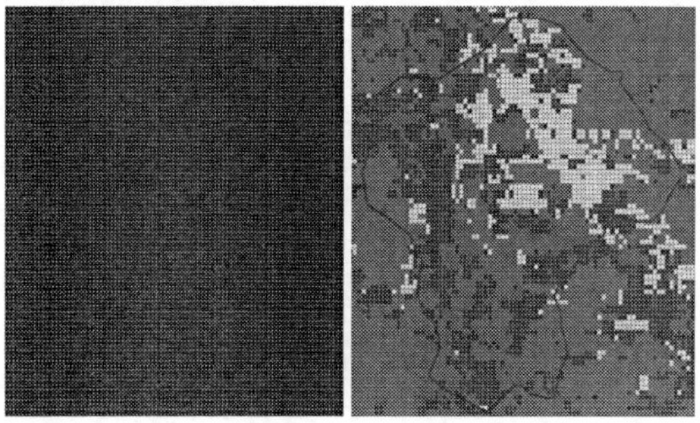

Fig. 5. (i) Soil data (UNEP) and (ii) land-use data (GLCC) of study area

2.3.4 Model calibration and validation

Model calibration is a process of optimizing or systematically adjusting the model parameter values to obtain a set of parameters that provide the best estimate of streamflow were observed. In other words, the process of optimizing the value of the parameter to improve the coherence between the hydrological response observed and simulated. In this research, the IFAS system has some parameters that can be calibrated using a reference of regional hydrological data were observed (measured data). There are eleven hydrological parameters are calibrated on this modeling. The sensitivity of these parameters on the response of hydrological model calibration phase has been examined in study area [11].

The performance of the IFAS model is evaluated by three indices like wave shape error, volume error and peak discharge error which are defined by the Japan Institute of Construction Engineering (JICE). The each and every indicator can be described as shown in the Table 1.

Table 1. Indicators for the error analysis of IFAS

Wave Shape Error	Volume Error	Peak Discharge Error
$E_w = \frac{1}{n}\sum_{i=1}^{n}\left(\frac{Q_{O(i)} - Q_{C(i)}}{Q_{O(i)}}\right)^2$	$E_v = \frac{\sum_{i=1}^{n} Q_{O(i)} - \sum_{i=1}^{n} Q_{C(i)}}{\sum_{i=1}^{n} Q_{O(i)}}$	$E_p = \frac{Q_{OP} - Q_{CP}}{Q_{OP}}$

where, E, E_W, E_v and E_p represent Error, Wave Shape Error, Volume Error and Peak Discharge Error respectively; n: The number of calculating time; $Q_{O(i)}$: Measured run-off at time I; $Q_{C(i)}$: Calculated run-off at time I; Q_{OP}: Calculated maximum run-off; Q_{OP}: Measured maximum run-off.

3 Results and discussions

3.1 Initial simulation

At the beginning of the simulation, hydrological parameters used for the simulation was the default parameters specified by IFAS based on satellite data input. This simulation was carried out for a period of two weeks during flood event in 2012. The simulation results of this model is the form of discharge hydrograph along those period as shown in Fig 6. and Fig. 7 for Lubuk Bendahara and Pasir Pangarayan station respectively. Comparison between the simulation and measured hydrograph show the similar trend, but the magnitude is not acceptable.

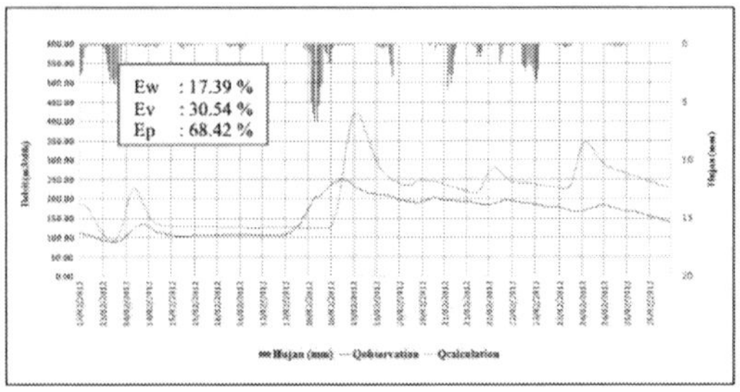

Fig. 6. Discharge hydrograph for initial simulation in Lubuk Bendahara station in 2012

As shown in Fig. 6, the peak of simulation discharge of Lubuk Bendahara sub-basin is still significantly under estimate compared with the measurement peak discharge. The magnitude of simulation peak discharge about 250 m^3/sec, while the measurement peak discharge about 420 m^3/sec. The value of wave shape error, volume error and peak discharge error are 17.39%, 30.54% and 68.43% respectively. The peak of simulation discharge of Pasir Pangarayan sub-basin is quite close with the measurement peak discharge as shown in Fig.6. The magnitude of simulation peak discharge about 250 m^3/sec, while the measurement peak discharge about 270 m^3/sec. The value of wave shape error, volume error and peak discharge error are 43.52%, 9.92% and 5.94% respectively. To improve the accuracy and the correlation, it is necessary to calibrate the hydrological parameters.

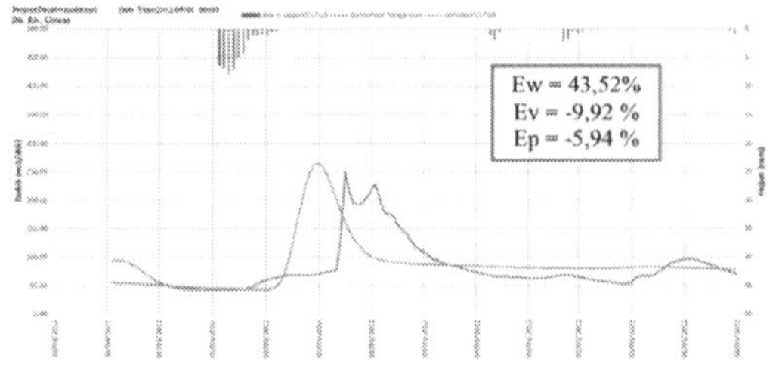

Fig. 7. Discharge hydrograph for initial simulation in Pasir Pangarayan station in 2012

3.2 Model calibration

Calibration process was done by trial on hydrological parameters so that the output discharge from the model as close as possible with measured discharge. After calibration process, the simulation hydrograph showed better corelation with measured hydrograph as shown in Fig. 8 and Fig. 9 for Lubuk Bendahara and Pasir Pangarayan sub-basin respectively. For Lubuk Bendahara sub-basin, the value of wave shape error, volume error and peak discharge error are 9.58%, 17.28% and 10.81% respectively, and for Pasir Pangarayan sub-basin, the value of wave shape error, volume error and peak discharge error are 9.58%, 17.28% and 10.81% respectively.

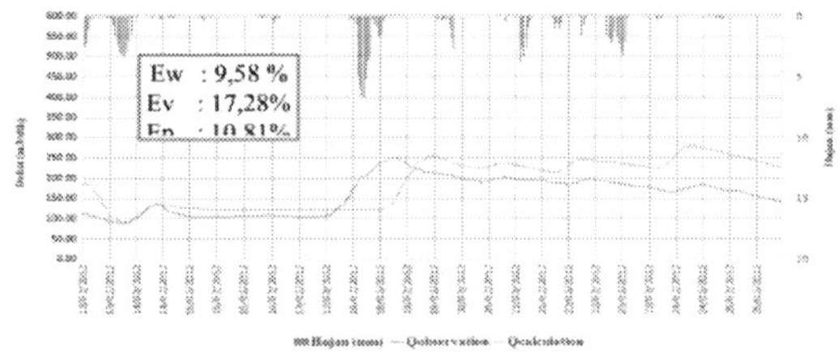

Fig. 8. Discharge hydrograph after calibration in Lubuk Bendahara station in 2012

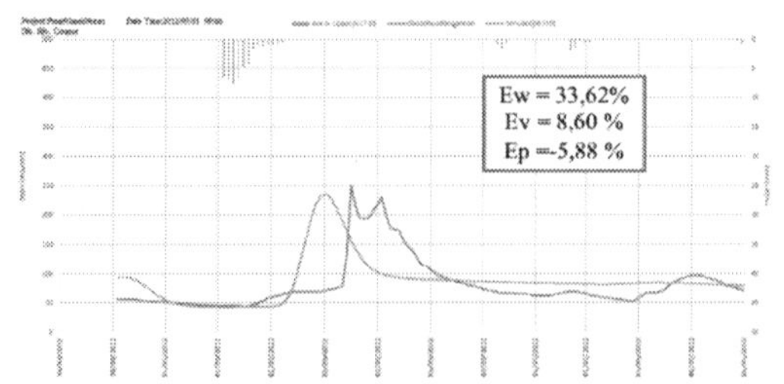

Fig. 9. Discharge hydrograph after calibration in Pasir Pangarayan station in 2012

As shown in Fig. 8 and Fig. 9, the peak discharge generated from models already approaching the magnitude of the peak discharge measurement results in the field, with a peak discharge error (Ep) amounted to 10.81% and 5.88% for Lubuk Bendahara and Pasir Pangarayan sub-basin respectively.

3.3 Flood event simulation

After the model was calibrated, then it was used to simulate a flood event in Lubuk Bendahara village which was occured on February 15-17, 2013. The model was simulated using GSMap_NRT rainfal data on those period and the calibrated parameters. The simulation result shows that flood event was occured on February 15-17, 2013 as shown in Fig. 9. It means that the simulation result has a good correlation with the data from flood historical record in Rokan Hulu regency. The maximum discharge from the simulation on that flood event is about 760.57 m^3/sec. The measured peak discharge in the same time periode is 690.57 m^3/sec. It means that The flood discharge prediction show an acceptable error with the ground measurement data.

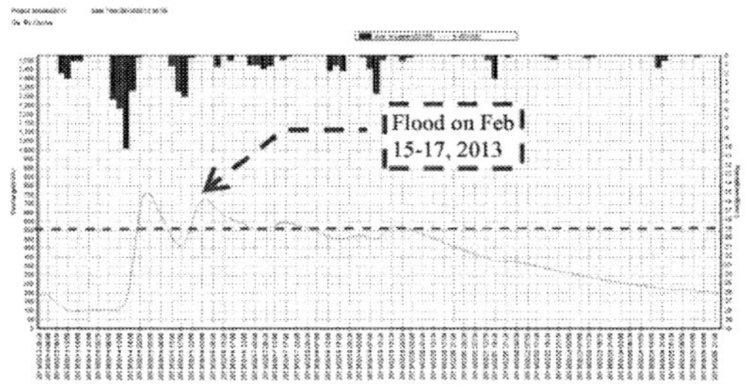

Fig. 10. Flood event in Lubuk Bendahara Sub-basin, Rokan Hulu regency on February 13-15, 2013

4 Conclusions

This study applied a method on hydrologic modeling using TRMM satellites data and IFAS tool for flood analysis. The results of this study show that the hydrologic modeling using satellite data can be used as an alternative for flood analysis in the area with limited ground data. The flood discharge prediction show an acceptable error with the ground measurement data. The wave shape error, the volume error, and the peak discharge error for Lubuk Bendahara station are 9.58%, 17.28%, and 10.81%, and for Pasir Pangaraian station are 33.62%, 8,60% and 5,88% respectively

The author would like to thank the Directorate General of Higher Education (DIKTI) - Ministry of Research, Technology and Higher Education for its research grants program.

References

1. S. N. Jonkman and J. K. Vrijling, *J. Flood Risk Manag.*, vol. **1**, no. 1, pp. 43–56, (2008).
2. K. Fukami and Y. Shiraishi, *Public Work* (2009).
3. T. Sugiura, *Proceedings of 7th ISE & 8th HIC, Chile*, pp. 1–6, (2009).
4. H. Wu, R. F. Adler, Y. Hong, Y. Tian, and F. Policelli, *J. Hydrometeorol.*, vol. **13**, no. 4, pp. 1268–1284, (2012).
5. A. Harris, S. Rahman, F. Hossain, L. Yarborough, A. C. Bagtzoglou, and G. Easson, *ISSN 1424-8220*, vol. **7**, pp. 3416–3427, (2007).
6. I. Hafiz, N. D. M. Nor, L. M. Sidek, H. Basri, M. N. Hanapi, and L. Livia, *IOP Conf. Ser. Earth Environ. Sci.*, vol. **16**, p. 012128, (2013).
7. K. Fukami, S. Nabesaka, M. Miyamoto, and A. Sugiura, *International Centre for Water Hazard and Risk Integrated Flood Analysis System (IFAS)*, pp. 1–15 (2009).
8. K. Fukami, *International Centre for Water Hazard and Risk Integrated Flood Analysis System*. (2009)
9. Y. Hendra, M. Fauzi, and S. Sutikno, *Proceedings ACES (Annual Civil Engineering Seminar)*, vol. **1**. pp. 61–72, 14-Dec-(2015).
10. S. Sutikno, M. Fauzi, and Hamiduddin, *Seminar Nasional Teknik Sipil X-2014*, 2014, p. 721 (2014).
11. S. Sutikno, M. Fauzi, and D. M. Mardhotillah, *Pertemuan Ilmiah Tahunan (PIT) HATHI XXXI*, 2014, pp. 481–492 (2014).
12. D. D. Chinh, N. Thi, T. Thuan, P. T. Van, T. N. Thanh, and V. Van, *Proceedings of the 28th EnviroInfo 2014 Conference, Oldenburg, Germany*, 2014, no. 1 (2014).
13. Bappeda, Dokumen Kajian Risiko Bencana Kabupaten Rokan Hulu, I. Pasir Pangarayan, (2015).

DOI: 10.1051/matecconf/201710105016

The application of backpropagation neural network method to estimate the sediment loads

Taufik Ari Gunawan[1,*], *M.* Syahril Badri Kusuma[2], *M.* Cahyono[2], *Joko* Nugroho[2]

[1]Civil Engineering Departement, Faculty of Civil and Environmental Engineering, ITB, 40132 Bandung, Indonesia
[2]Water Resources Development Center, Bandung Institute of Technology, 40132 Bandung, Indonesia

Abstract. Nearly all formulations of conventional sediment load estimation method were developed based on a review of laboratory data or data field. This approach is generally limited by local so it is only suitable for a particular river typology. From previous studies, the amount of sediment load tends to be non-linear with respect to the hydraulic parameters and parameter that accompanies sediment. The dominant parameter is turbulence, whereas turbulence flow velocity vector direction of x, y and z. They were affected by water bodies in 3D morphology of the cross section of the vertical and horizontal. This study is conducted to address the non-linear nature of the hydraulic parameter data and sediment parameter against sediment load data by applying the artificial neural network (ANN) method. The method used is the backpropagation neural network (BPNN) schema. This scheme used for projecting the sediment load from the hydraulic parameter data and sediment parameters that used in the conventional estimation of sediment load. The results showed that the BPNN model performs reasonably well on the conventional calculation, indicated by the stability of correlation coefficient (R) and the mean square error (MSE).

1 Introduction

A number of researchers across the globe have developed various procedures and theory to predict sediment load. Formulations of bed load were first developed and introduced extensively by Du Boys (1879) with the concept of shear stress approach [1]. Furthermore, several approaches were developed such as graphics solution, the concept of probability, stream power concepts and the methods of multimodal characteristics. The method that most frequently used by researchers is the regression method [2]. So far, the existing theories have not been able to accurately predict the actual sediment load when implemented on particular rivers [3]. From various measurement data test through the sediment flume experiments in the laboratory and natural river measurement data, there are limitations in the application of available theories [4]. The results show that there is a big difference between the rates of sediment transport estimation with the sediment discharge measurement results.

The development of the sediment load estimation formulation in the last decade has been developed by [5] for large rivers in terms of the universal stream power is compared with selected bed-material load equations using large and medium river data. The next development, total bed-material discharge in alluvial channels [6], a comparative study of some of the sediment transport equations for an alluvial channel with dunes [3], fractional transport of sediment mixtures [7] and sediment concentration prediction and statistical evaluation for annual load estimation [8] as well as some computer programs developed by [9] about comparison of sediment transport computations using hydrodynamic versus hydrologic models, estimating the sediment transport capacity using HEC-RAS [10] and software for evaluating sediment-induced stratification in open-channel flows [11]. Final conclusions indicate that these formulas generally remain localized, so that this method is only suitable for certain river typology and cannot be used for any random river.

The magnitude of sediment load tends to be non-linear with respect to the hydraulic parameters and parameter accompanying sediment. The character of non-linear influenced by turbulence, where the turbulence flow velocity vector direction of x, y and z are influenced by the morphology of the water body in 3D on the vertical cross section and horizontal [12]. Whilst, all sediment load estimation formulations does not take into account this and are usually very difficult to retrieve data relating to a third of the velocity vector.

The past few years, researchers chose an alternative approach by developing artificial neural network (ANN) models. In some cases, the ANN has proven to be a

* Corresponding author: tagz.unsrir@gmail.com

better alternative for nonlinear modeling [13]. This is indicated by one of the features on ANN i.e. the ability to extract the relationship between input and output without reviewing the physical process. The network was able to do a mapping of the multivariate space with one another, even apply to a data set that contains errors. In some studies, ANN method is suitable to solve the problem of estimation and prediction on hydrology field [14].

The following applications in various calculations of ANN in sediment load or transport can be developed by [15] for prediction of suspended sediment using ANN GA conjunction model with Markov chain approach at flood conditions, combining deterministic modelling with ANN for suspended sediment estimates [16]; integrative neural networks model for prediction of sediment rating curve parameters for ungauged basins [17]; a review sediment load change [18]; stream flow discharge and sediment rate relation using ANN [19]; evaluation of transport formulas and ANN models to estimate suspended load transport rate [20]; daily suspended sediment load prediction using ANN and support vector machines [21]; estimate sediment load in ungauged catchments using ANN [22]; suspended sediment modeling using genetic programming and soft computing techniques [23]; estimation of daily suspended sediments using support vector machines [24] and prediction of bed material load transport using neural network [25].

This study aims to demonstrate the non-linear properties of the hydraulic parameters data and sediment parameter with the sediment load measurement data. This fact can be seen clearly on the calculation results by applying some conventional calculation method i.e. Yang method, Ackers and White method, Engelund and Hansen method, and Laursen method. The result of this conventional calculation will be compared with the results on the application of the backpropagation neural network (BPNN) schema. The sediment load of the hydraulic parameter data was projected with this schema. The stream data in the United States and data from laboratory testing were used for this purpose.

2 Methodology

In this study, a description of estimation methods only serves basic formulas and calculation results. Furthermore, the main focus is on the application of BPNN to estimate sediment load.

The study adopted a simple methodology to prove that the model BPNN are able to map the data input-output regardless of the physical processes that accompany it. Variable data are used as input data matrix and concentration measurement data sediment as a target matrix. Before conducting simulation model, the first data set is identified carefully, this is to avoid extreme values into the training process in order to produce a natural output. According to the rules in the modeling BPNN input data and targets for training as much as 80% and the remaining 20% is used for testing.

The advantages of BPNN is the use of the hidden layer algorithm. The function of hidden layers is to reduce the error output through the propagation control information errors on output layer unit to the hidden layer. This method is one way to optimize the neural network. The differences of hidden layer numbers are compared to obtain the optimal estimation. Variations in the amount of BPNN hidden layers scheme in the implementation of the MATLAB program application can be seen the figure 1.

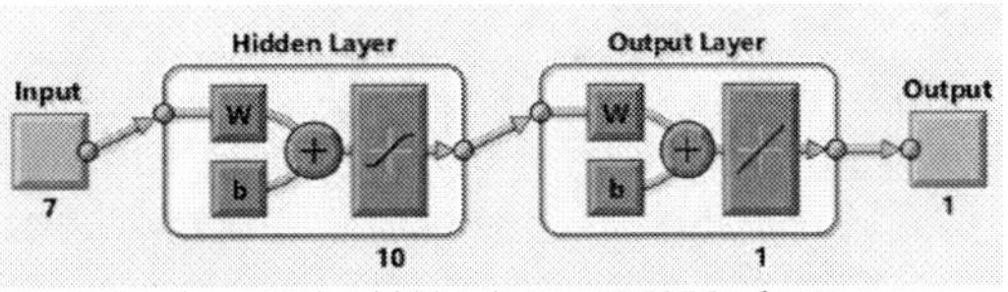

Fig. 1. Implementation hidden layer at BPNN schema

3 Data collection

In order to get the best results, a vast amounts of data are required. This study used 806 hydraulic and sediment data collected from 18 sets of laboratory data streams in the United States. This secondary data was obtained from the database of the Civil Engineering Department, Diponegoro University, Indonesia. The existing data set covering; name of the researchers, measurement date, water discharge (Q_w), flow velocity (V), flow depth (h), channel width (w), slope (S), water temperature (T), the mean bed diameter (d_s) and the concentration of sediment load (C_m). The concentration is converted to a form sediment load using the formula:

$$q_m = \frac{Q_w}{w} C_m \times 10^{-3} \quad (1)$$

Table 1. Data collection

Researcher / Sources	**Number of data**
Barton & Lin (1955)	30
Brooks (Vanoni & Brooks, 1957)	21
Franco (1968)	19
Guy, Simons & Richardson (1966)	290
Kalinske & Hsia (1945)	9
Kennedy & Brooks	9
Laursen (1958)	24
Meyer, Peter & Muller (1948)	55
Nomicos 1 (Toffaleti, 1968; Vanoni & Brooks, 1957	12
Nomicos 2 (Vanoni & Brooks, 1957)	26
Onishi, Jain & Kennedy (1976)	14
Stein (1965)	57
Straub (1954 & 1958)	24
Taylor & Vanoni (1972)	6
Vanoni & Brooks (1957)	15
Vanoni & Hwang (1967)	16
Williams (1970)	83
Willis, Coleman & Ellis (1972)	96
Total data	**806**

Table 2. Statistics parameter data

Statistics parameter	average	st-dev	minimum	maximum
Q_w [m³/s]	0.31	0.67	0.003	4.614
w [m]	1.25	0.80	0.267	2.438
h [m]	0.19	0.14	0.027	1.092
V [m/s]	0.83	0.50	0.188	2.879
d_s [mm]	1.72	5.71	0.011	28.65
S_w [m/m]	0.00337	0.00369	0.00015	0.03310
T [°C]	20.36	5.32	1.70	48.00
C_m [ppm]	4198	10206	1	11099
q_m [(kg/s)/m]	0.82	1.88	4E-05	18.33

4 Estimation using formulation methods

In this study, the data were tested using a variety of estimation methods of total sediment load. The value of transport sediment concentration measurement data is the total value of the sediment load. This study limited to the use of 4 estimation of total sediment load methods that represents the concept of different formulation development. The relationship between the calculation parameter formula for each method, presented as follows :

A. Yang method (stream power's concept)

$$\log C_t = 5.435 - 0.286\log\frac{\omega d_s}{v} - 0.457\log\frac{u_*}{\omega} + \left(1.799 - 0.409\log\frac{\omega d_s}{v} - 0.314\log\frac{u_*}{\omega}\right)\log\left(\frac{V\cdot S_w}{\omega} - \frac{V_{cr}\cdot S_w}{\omega}\right) \quad (2)$$

B. Ackers and White Formula method (regresion's concept)

$$C_t = \frac{G_{gr} d_s\left(\frac{\gamma_s}{\gamma}\right)}{h\left(\frac{u_*}{V}\right)^n}\cdot 10^6 \quad (3a)$$

where:

$$F_{gr} = \frac{u_*^n}{\left[g d_s\left(\frac{\gamma_s}{\gamma_w} - 1\right)\right]^{1/2}}\left[\frac{V}{\sqrt{32}\log\left(\frac{\alpha h}{d_s}\right)}\right]^{1-n} \quad (3b)$$

$$G_{gr} = C\left(\frac{F_{gr}}{A_1} - 1\right)^m \quad (3c)$$

C. Engelund and Hansen method (stream power's concept)

$$q_t = 0.05\gamma_s V^2\left[\frac{d_{50}}{g(\gamma_s/\gamma - 1)}\right]^{1/2}\left[\frac{\tau_0}{(\gamma_s - \gamma)d_{50}}\right]^{3/2} \quad (4)$$

D. Laursen method (stochastic's concept)

$$C_t = 0.01\gamma\left(\frac{d_s}{h}\right)^{7/6}\left(\frac{\tau'}{\tau_c} - 1\right) f\left(\frac{u_*}{\omega_i}\right) \quad (5)$$

The formulation is converted to a form sediment load using the formula:

$$q_t = \gamma_w h V\left(C_t \times 10^{-6}\right) \quad (6)$$

The final result of each method in the form of sediment load estimation is presented in Fig.2 for the data distribution and Fig.3 for the comparison of measurement data.

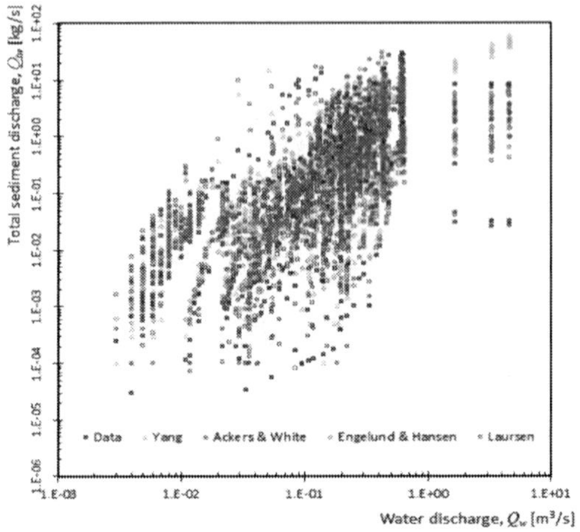

Fig. 2. The distribution of measurement data and calculation results

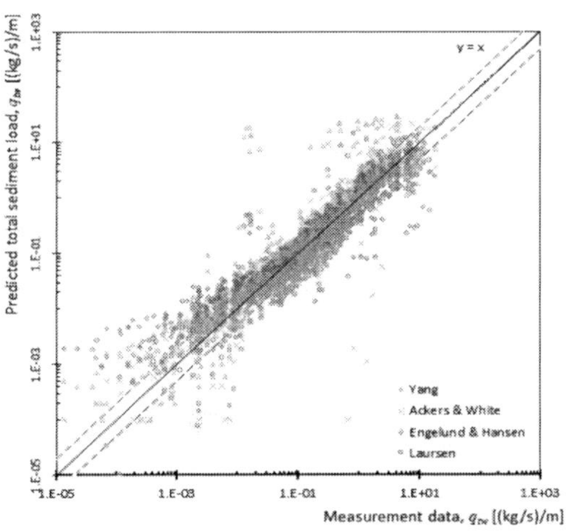

Fig. 3. Predicted against measured sediment load

5 Back propagation Neural Network (BPNN)

BPNN is a multi-layer device and it's able to overcome the complex non-linear problem since there are interconnections between neurons in algorithms. BPNN algorithm using delta rule generalization (Widrow-Hoff) applying the gradient descent method to minimize the total squared error at the output of the network and transferring non-linier derivative function [26]. The uniqueness of the backpropagation algorithm is able to map the input-output process through the feed forward and back forward. Input and output vectors are used to train the network until it can approach the appropriate function. BPNN has the built-in capability i.e. the adaptation of synaptic weights to changes in the surrounding environment through pattern recognition for adaptive information. The contextual information introduction will produce output that is natural since every neuron in a neural network may be potentially affected by the global activity against all the other neurons in the network. Furthermore, the analysis of neural network system performance smoothly degraded since the inherent fault tolerance. All ANN calculations in this study using a MATLAB program application because it has a very strong ability to solve problems in the form of a matrix calculation. The simple BPNN program used a command in the window command and create a program listing in the file editor. MATLAB has provided NNTool to train and test input data, targets and testing. Some types of neural network algorithms e.g. hebb, adaline, perceptron and probabilistic neural network.

Matrix that contains the input data and the target is used in the program listings, wherein the input matrix and the target should be in the format transpose. Before processing the data, a process of normalization and preprocessing conducted, so that the value of each data can be calculated with smaller values without losing the characteristics of the data itself. Hereinafter, BPNN network coding created to determine the number of hidden layers, the number of neutrons and the number of the output layer. Bipolar sigmoid activation function (tansig), unipolar (logsig) or function identity (purelin). In order to BPNN learning methods, the method used: *trainlm*.

The next step is to set the parameters that contain; the number of epoch (iteration) maximum learning = 1000, minimum number of errors = 0.001, the number of learning rate = 0.5, the number of the epoch in which the appearance of the graph will be updated = 1000, error and the amount of momentum = 0.7. The next process is learning and evaluating the network output simulation performance, simulated with the target and training. Data need to be trained to determine which the most appropriate smoothing factor is. At the end of the training, the individual smoothing factors can be used as a sensitivity analysis tool: the bigger factor for a given input, the more important that the input is a model at least as far as the test set is concerned.

6 Results and analysis

From the calculation results using conventional sediment load estimation method, the data distribution almost coincides with the measurement data. The number of calculation results data that in the discrepancy ratio range was 71%, 75%, 67% and 69% respectively for Yang, Ackers and White, Engelund and Hansen method, and Laursen method. This results considered generally accepted since the discrepancies ratio has been above 60% (Fig. 2 and Fig. 3).

In the BPNN model, the input parameters that used in each simulation is in the same value, the differences is in the number of hidden layers. The results of simulations and calculated indicate that the performance of the BPNN model is good enough. However, from the distribution of the measurement data and the calculation results, it can be seen that the result of its spread still accumulate in the data center as seen in Fig. 4, 6 and 8. For more details, the result of the comparison expressed by a discrepancy ratio of 0.5 - 2, can be seen in Fig. 5, 7, and 9. The percentage of the amount of the data that comes into the discrepancy ratio range are respectively 61%, 63%, and 64% for 2, 5, and 10 hidden layers. The more hidden layers applied, the more accurate the results obtained. From these results it can be seen that modeling with BPNN suitable to predict the sediment load. The inaccuracies of the results caused by a large range between sediment load distribution data (target). Likewise, the range of each parameter input is also uneven. Correlation coefficient (R) training nearly to 1, and mean square error (MSE) nearly to 0. To obtain the good simulation results, consider the training, testing, validation and overall regression data were already close to 1. The value of R for each hidden layer could be seen in table 3.

Table 3. Correlation coefficient (R)

Hidden layers	Correlation coefficient (R)			
	Training	Validation	Testing	All
2	0.961	0.943	0.936	0.953
5	0.955	0.983	0.963	0.964
10	0.963	0.971	0.972	0.965

From table 3, it can be conclude that the number of hidden layers will influence the R value, although the differences is insignificant. When the data were already close to 1, then the simulation is finished and fittings are considered reliable. Conversely, if it is not, then the network should be retrained.

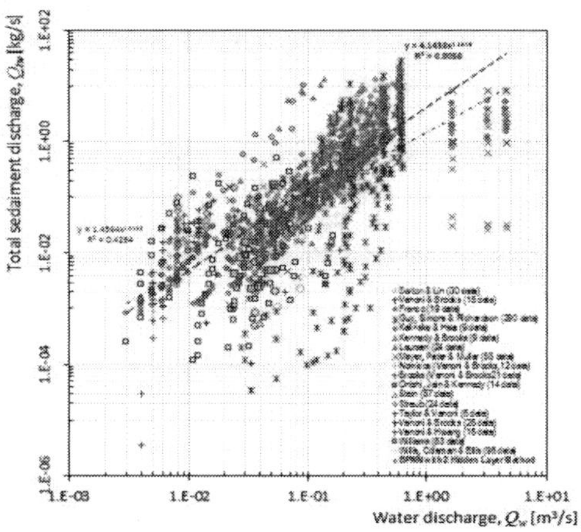

Fig. 4. The distribution of measurement data and calculation results with 2 hidden layers

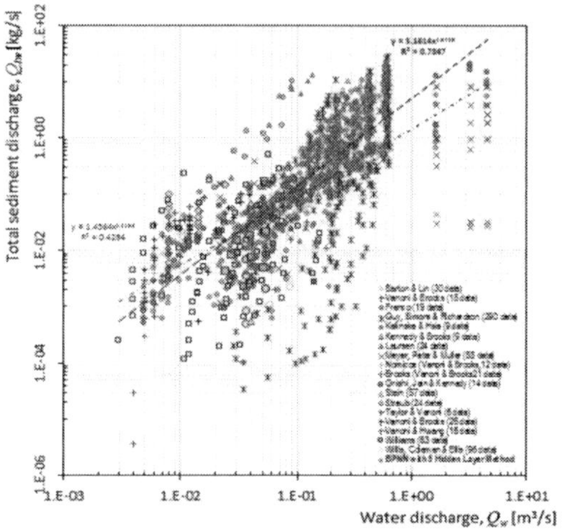

Fig. 6. The distribution of measurement data and calculation results with 5 hidden layers

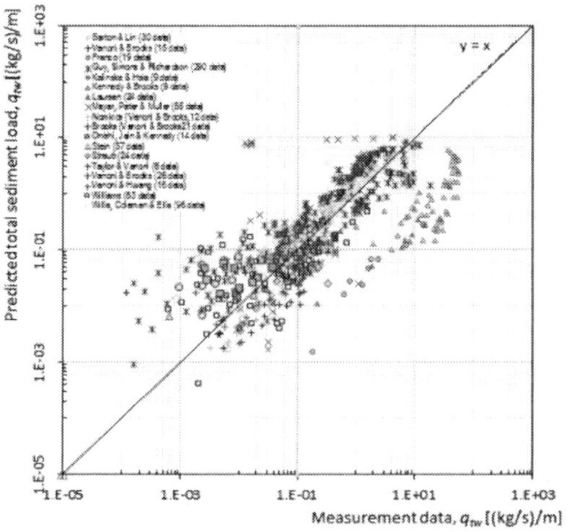

Fig. 7. Predicted against measured sediment load using BPNN with 5 hidden layers

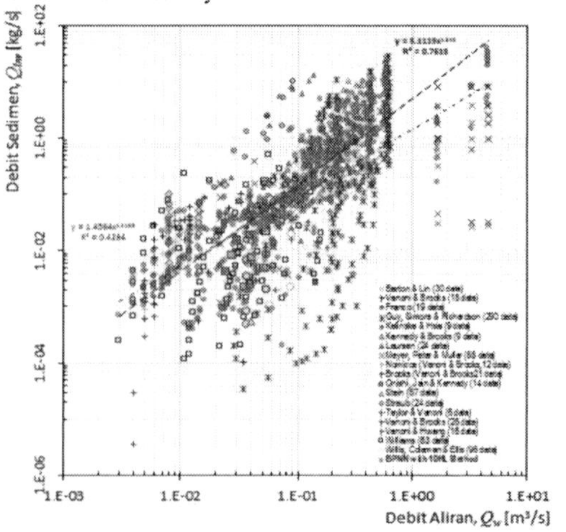

Fig. 8. The distribution of measurement data and calculation results with 10 hidden layers

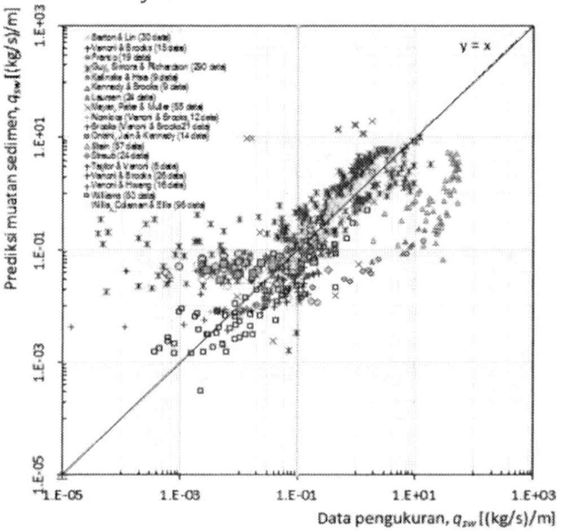

Fig. 9. Predicted against measured sediment load using BPNN with 10 hidden layers

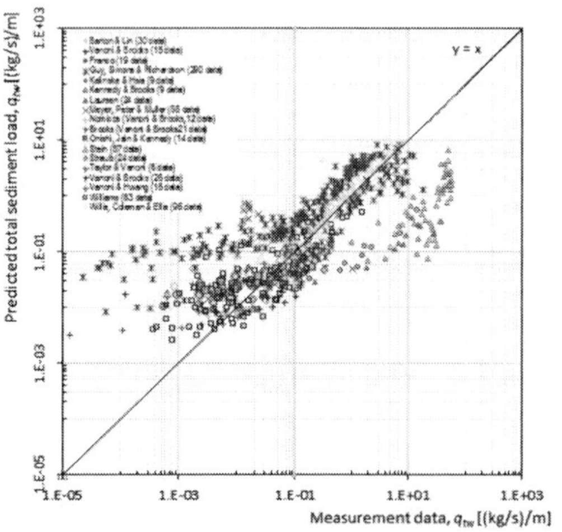

Fig. 5. Predicted against measured sediment load with 2 hidden layers

7 Conclusions

The BPNN model can be used in the estimation of sediment load in rivers. The results obtained with the BPNN model is not satisfying compare with the results from conventional methods. Better outcomes will be obtained by increasing the number of hidden layers and variation parameters calculation contained in MATLAB models. The model predicted approximately 61% to 64% accuracy of the data for 2 to 10 hidden layers. This gives a clear robustness of the model for use in sediment prediction for rivers with different hydraulic and sediment characteristic.

The authors gratefully acknowledged Prof. Iwan Kridasantausa from Bandung Institute of Technology and Dr. Robert J. Kodoatie from Diponegoro University for the brief discussion and the support of secondary data. The authors would also gratefully Engineering Faculty, Sriwijaya University for the funding during this study period. Sincere thanks to Water Resources Development Center, Bandung Institute of Technology for the support during this publication process.

References

1. Haddadchi A, Dehghani AA., *Journal of Hydrodynamics*, Ser B. **24,** p :145-51, 1 (2012)
2. Junaidah A, Nurashikin AK, Muhamad Syahreen Sa, Mohd Nasir T, Suhaimi A-T, Aminuddin A-G, *Journal – The Institution of Engineers* **69**, 3 (2008).
3. Srinivasan VS, Cavalcante RG, Santos CA, *Journal of Urban and Environmental Engineering (JUEE)* **2**, doi: 10.4090/juee. 2008. v2n1. 028032, (2009)
4. Habibi M, Sivakumar M, *11th Australasian Fluid Mechanics Conference Hobart*, 1992.
5. Molinas A, Wu B., *Journal of hydraulic research*, **39,** p :135-46, 2 (2001)
6. Kiat CC, Ghani AA, Wen L, *2nd International conference on managing river in the 21st century-Solutions toward sustainable river basins*: Kuching, Sarawak, Malaysia, (2007)
7. Baosheng W, Van Maren D, Lingyun L. *International Journal of Sediment Research.* **23,** p :283-98, 4 (2008)
8. Wang Y-G, Tian T, *Journal of Hydrology* **482**, p :69-78, (2013)
9. Van Griensven A, Popescu I, Abdelhamid M, Ndomba PM, Beevers L, Betrie GD., Physics and Chemistry of the Earth, Parts A/B/C **61**, p :12-21, . (2013)
10. Talreja J, *Global Research Analysis* **2**, 4 (2013)
11. Yeh T-h, Parker G., *Computers & Geosciences* **53**, p : 94-104, (2013)
12. Kodoatie RJ., *Hidrolika Terapan*, Andi Yogyakarta, p : 17 – 18, (2002)
13. Kumar M. *Evapotranspiration modeling using artificial neural networks,* PhD thesis, Indian Institute of Technology, Kharagpur India. PhD thesis. 2003.
14. Sudheer K, Gosain A, Ramasastri K. A, *Hydrological processes* **16**, p :1325-1330, 6 (2002)
15. Adib A, Mahmoodi A., *Journal of Civil Engineering*, p : 1-11, doi : 10.1007/s12205-016-0444-2, (2016)
16. Makarynskyy O, Makarynska D, Rayson M, Langtry, *Applied Soft Computing* **35**, p : 47-56, 2 (2015)
17. Atieh M, Mehltretter S, Gharabaghi B, Rudra R., *Journal of Hydrology* **531,** p :95-107, 10 (2015)
18. Dai S, Lu X., Geomorphology **215**, p :60-73, (2014)
19. Vafakhah M., *Arabian Journal of Geosciences* **6,** p : 3-18, 8 (2013)
20. Haddadchi A, Movahedi N, Vahidi E, Omid MH, Dehghani AA., *Journal of Hydrodynamics* Ser B. **25**, p :459-70, 3 (2013)
21. Lafdani EK, Nia AM, Ahmadi, *Journal of Hydrology* **478,** p : 50-62, (2013)
22. Heng S, Suetsugi T., *Journal of Water Resource and Protection* 5, p : 111 -123, (2013)
23. Kisi O, Ozkan C, Akay B., *Journal of Hydrology* **428**, p :94-103, (2012)
24. Çimen M., *Hydrological Sciences Journal* **53**, p : 656-66, 3 (2008)
25. Kumar B., *Hydrological Sciences Journal* **57**, p : 956-66, 5 (2012)
26. Nazari J, Ersoy OK. *Implementation of back-propagation neural networks with MatLab*: Purdue.

Automatic gate design model from wood & tire for farmers

Ivan Indrawan[1], *Nursyamsi*[1], *Sayed* Iskandar Muda[1]

[1]Department of Civil Engineering, Engineering Faculty, University of Sumatera Utara, 20155 Medan, Indonesia

Abstract. Indonesia is one of the potential paddy farming area in Southeast Asia, and North Sumatra Province is one of many provinces that provides it. Yet, Indonesia is still importing rice from foreign country, eventhough today the government has been willing to supply its own need. Almost 10% irrigation areas in Indonesia are connected to sea current, which means it must have a system to manage the circulation of fresh water and block the seawater from entering the irrigation area through the irrigation channel. Many systems use gates to control the water management, and most of them are using automatic sluice gate because the gates are usually positioned far from village, this makes the manual operating become difficult. Unfortunately, not all farmers can use this kind of gate due to its accessibility and cost. This research was done to design the automatic gates, which are easy to build, user friendly, low cost and dependable. In the future, poor farmers or farmers who do not have connection to government, can make this gate by themselves. The research was conducted in laboratory, using flume, pumps, reservoir, and gate prototype.

1 Introduction

Over 9.136.028 hectares (*Permen PUPR No.14, 2015*) of paddy field in Indonesia, some of them did not have a technical irrigation [1]. The irrigation infrastructure should keep the paddy production sustainable. When the production went down, most of the farmers gave up and turned into other crops such as palm which requires less water and infrastructure. This condition would be the opposite of the willing of government to become self-sufficient rice country. One of many problems that the farmers have to deal with is the irrigation gate. This gate should be an instrument which holds the water during ponding stage and drains the water when its flooding stage. The gate must be built by private company or with help from the government which is so difficult to process. This paper will answer the challenge to create a gate that is affordable, easy to install and operate, without ignoring the basic principles of fluid mechanics and hydraulics theory. The gates are made from wood and rubber tire (bicycle tire), and formulated from Buoyancy and Flotation theory.

This paper will give the result of the gate experiment which shows the design of gates, specification of gates, and installation-operational manual. The basic principle of buoyancy and flotation was first discovered and stated by Archimedes which states that a body which is floating or submerging in a fluid is buoyed (lifted) upward by force equals to the weight of the body that would be displaced by fluid. The force is known as buoyant force. It shows that the floating body displaces sufficient volume of fluid to balance its own weight. The point which the buoyant force act occurred is called the centre buoyancy, it is located at the centre of gravity of the displaced fluid [4,7].

If the weight of an object is less than the displaced fluid, the object rises, as in the case of a block of wood that is released beneath the surface of water or a helium-filled balloon that is let loose in air. An object heavier than the amount of the fluid it displaces, will sink when released, it has an apparent weight loss equal to the weight of the fluid displaced. In fact, in some accurate weighings, a correction must be made in order to compensate the buoyancy effect of the surrounding air. The buoyant force, which always opposes gravity, is nevertheless caused by gravity [10]. Fluid pressure increases with depth because of the (gravitational) weight of the fluid above. This increasing pressure applies a force on a submerged object that increases within depth. The result is buoyancy [3].

2 Experiment details

2.1 Automatic gate model

The dimension of the gate is 40 cm x 40 cm x 15 cm, made from first class of wood (*Damar Laut*), 1 inch thickness

* Corresponding author: ivanindrawan76@gmail.com

DOI: 10.1051/matecconf/201710105017

[2]. The gate designed as a pocket filled with rocks to have significant weight. The gate was coated with oil-paint. The upstream side of the gate has been installed with a rubber bicycle tube tire (Fig.1). All the materials of the gate was meant to be simple and available everywhere, such as plank wood used from ex-building material as long as it meets the gate's dimension and the bicycle tire, which we assume most of the farmer in Indonesia have it. The bicycle tire attached to wooden gate is made to give buoyancy to the gate at significant water level. The wood and rubber tire will manage the water circulation from upstream to downstream automatically. The main principle is to keep the gate closed at W weight and opened at water level, y, by adding air to the tire rubber as pressure to the gate, FU, until the gate opened at a significant water level. The schematic of instrument shown as figure 4.

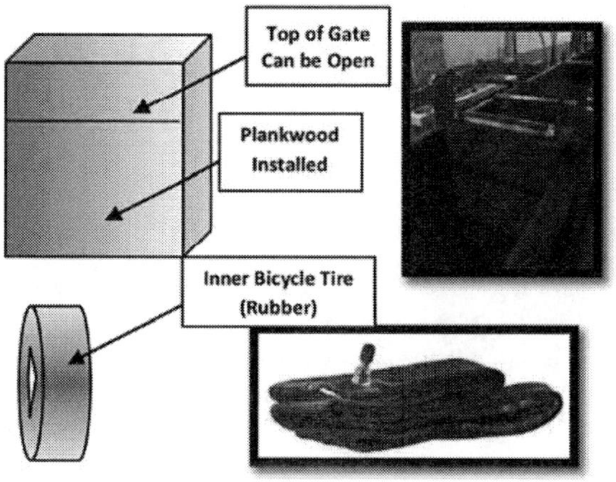

Fig. 1. Material of gate

2.2. Experiment procedures

The experiment measures the height of water at upstream (hu), at the gate (hp) and at downstream (hi). The measurement is conducted at different flow velocity from 0.4 m/s to 0.1 m/s, with a constant weight of wooden gate, where the pressure at rubber tire set from zero to maximum. This figure below explains where the point of measurement is taken.

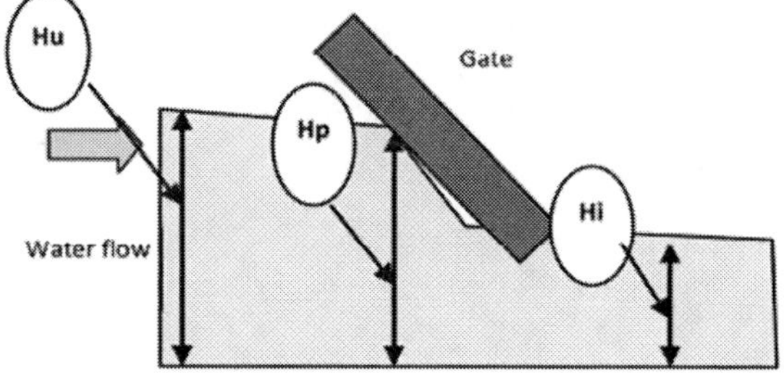

Fig.2. Point of measurement

Procedure of experiment conducted as follow:

1. The top of the gate open and filled with rock to make significant weight so the gate remain closed until the purposed water level is achieved. Make sure the tire tube is attached to gate.

2. The water flows into the flume. The gate must be filled again with rock to add the weight in order to make the gate remain closed.

3. When the gate is opened, it is expected that the water level on downstream is lower than 1/10 height of gate. Otherwise, the weight of the gate should be added. Then, the tire will manage the gate for purposed height of water on downstream.

4. The air volume inside the tube tire (inner tube) gives bouyant force to lift the gate. The adding of air volume uses air hand pump which formerly used for bicycle tire. The height water at downstream can be adjusted by adding the volume of air.

5. To close the gate when needed, the airinside the tire tube can be released by simply opening the tire tube valve. All the methodology of the experiment are repeated and recorded in experiment's form for analysis.

This figure below shows how the experiment was conducted.

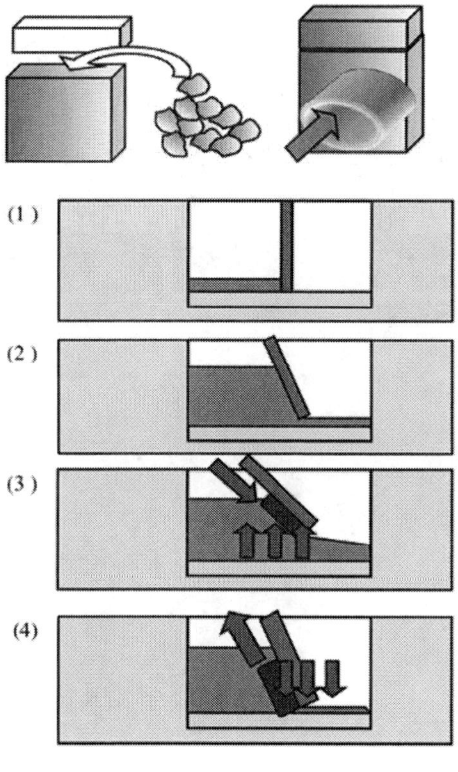

Fig. 3. Experiment procedure

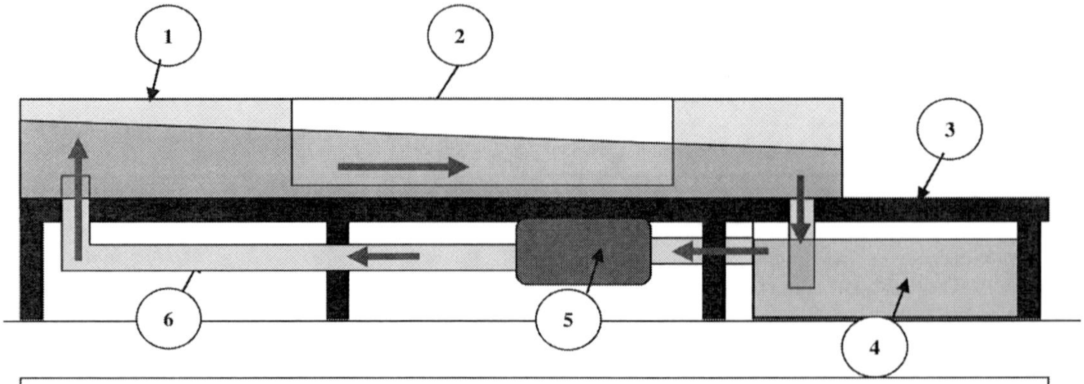

Annex:
1. Flume, for water flow, made from plywood 9 mm, size 40 cm x 40 cm x 1500 cm
2. Glass, for observation view
3. Steel framework, for flume supporting and leveling.
4. Reservoir, for water circulating
5. Pump, for pumping and sucting water, with 10 litre per-second capacity
6. Pipe, for water flow, PVC 2" diameter

Fig. 4.(a) Instrumentation of experiment

Fig. 4.(b) Instrumentation of experiment

3 Result and discussion

After trial and error procedure to set up the constant weight by filling the gate with rock gradually, the optimum result for the weight is 41 kg. This trial and error procedure was done at maximum velocity and zero pressure. After the weight of gate is set, then the experiment was carried on with the following procedure above. The result of experiment is shown in table below.

Fig. 5. Condition of gate at pressure = 0

DOI: 10.1051/matecconf/201710105017

Table 1. Result at pressure = maximum

No.	Velocity (m/s)	Hu (mm)	Hp (mm)	Hi (mm)	Opened Gate (mm)
1	0.40	281	281	21	241
2	0.35	279	279	19.5	238
3	0.30	276	276	17.5	235
4	0.25	272	272	16	232.5
5	0.20	269	269	15	229
6	0.15	266	266	14	217
7	0.10	263	263	12.5	214
8	0.05	262.0	262.0	0.0	0.0
9	0.03	261.0	261.0	0.0	0.0

Table 2. Result at pressure = 0

No.	Velocity (m/s)	Hu (mm)	Hp (mm)	Hi (mm)	Opened Gate (mm)
1	0.40	270.0	270.0	33.0	235
2	0.35	267.0	267.0	31.5	229
3	0.30	263.5	263.5	29.5	225
4	0.25	260.0	260.0	27.5	222
5	0.20	257.0	257.0	26.0	216
6	0.15	255.0	255.0	24.5	213
7	0.10	253.0	253.0	23.0	208
8	0.05	252.0	252.0	0.0	0.0
9	0.03	251.0	251.0	0.0	0.0

From the experiment, it is shown that the weight required to make the gate remain closed is 41 kg. At zero pressure in tire tube, to start opening the gate, requires 253 mm height of the water at upstream point. Meanwhile, the gate is opened about 23 mm height of water at downstream point. At maximum pressure in tire tube, to start opening the gate, requires 263 mm height of the water at upstream point. Meanwhile, the gate is opened about 12.5 mm height of water at downstream point.

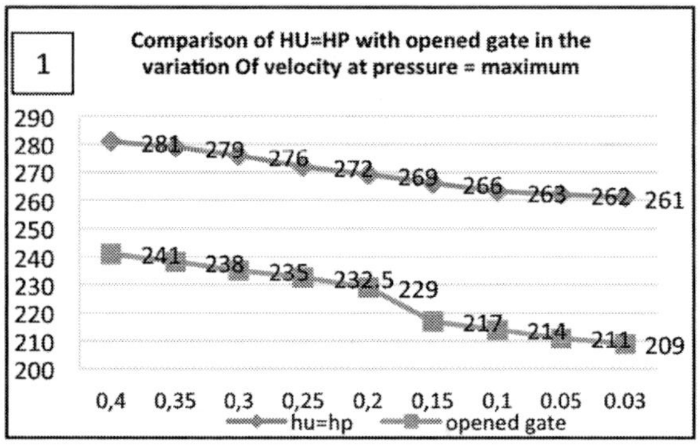

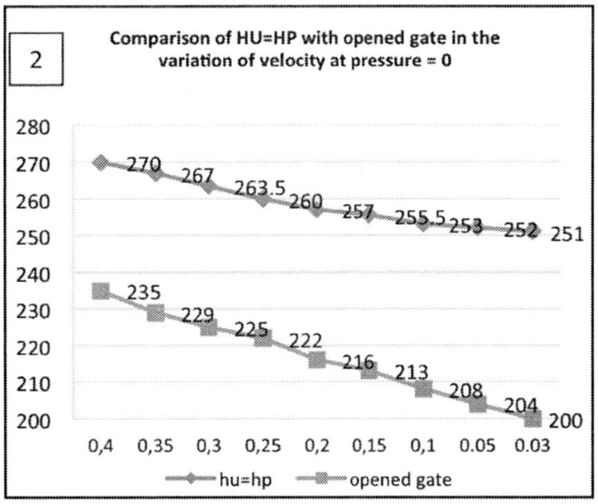

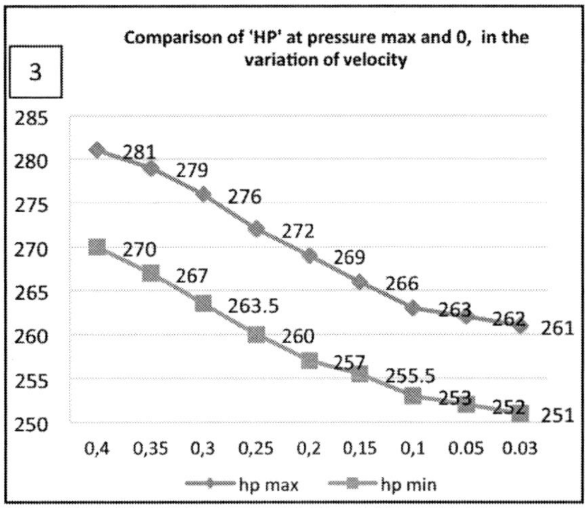

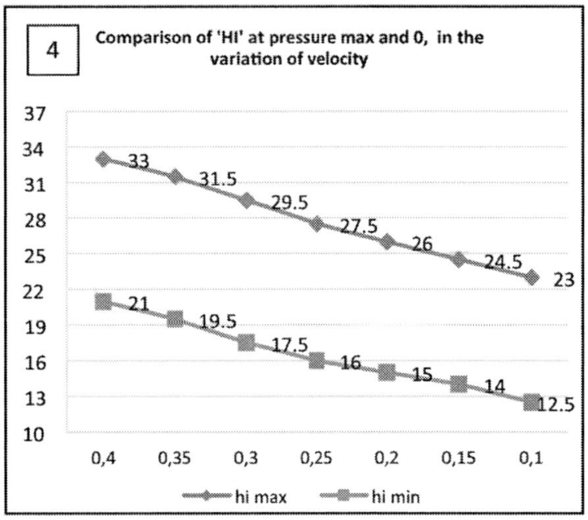

 DOI: 10.1051/matecconf/201710105017

Fig. 6. Relationship between various water level, pressure and velocity

4 Experiment result – analysis

This chapter is explaining the math to analyse behaviour of the gate as it is sub-merged [5, 9].

Assuming the wooden gate is a plane of area as figure bellow;

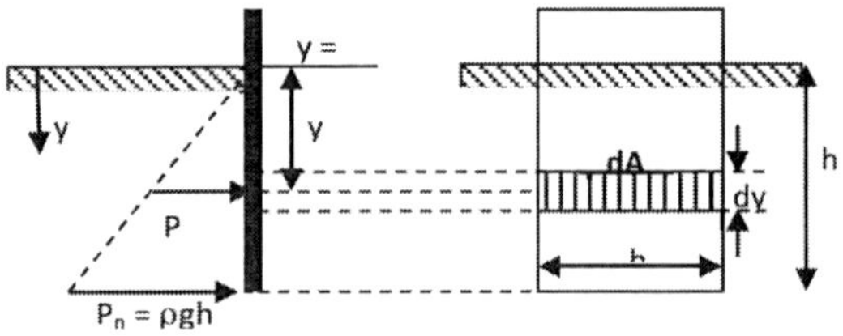

Fig. 7 Forced on a submerged plate

Force on plate work on element dA, where,

$$dF = p.dA$$

Since,

$$p = \rho.g.y$$

So,

$$dF = \rho.g.y.dA$$

Then, total force on plate can be define as

$$F_R = \int dF$$

Which mean,

$$F_R = \int \rho.g.y.dA$$

It is same as,

$$F_R = \rho.g \int y.dA$$

Where the equation fullfill the boundary of 0 to h. So the equation is,

$$F_R = \rho.g \int_o^h y.dA$$

Since, *A* is area of *b* and *y*, so integral of *y.dA* result is

$$\int_o^h y.db.y = b.[\tfrac{1}{2}y^2]_0^h = ½\ b.h^2$$

So,

$$F_R = \rho.g.1/2.b.h^2 = ½.\ r.g.b.h^2$$

Besides the pressure from water flow, the force acting on gate is the weight of gate itself and the floating force from rubber tire [6], that is,

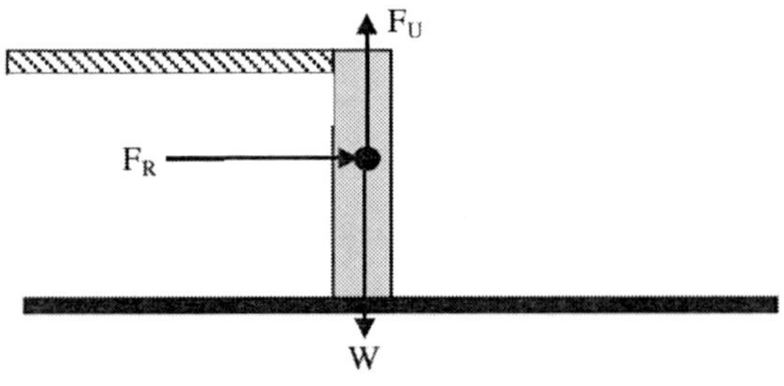

Fig. 8. Forced on a submerged plate

To make **the gate remain close**, we need balance equation, which is,

$$F_R - W + F_U = 0$$

So,

$$F_R = W - F_U$$

During the experiment, we use zero floating force in order to have a weight of the door at closed.

$$F_R = W - F_U$$

Then, we add floating force using the rubber tire to make **the gate opened**.

$$F_R - W + F_U > 0$$

During the experiment the water flow start to push the gate to open until it reach its maximum height, *Hp* = maximum. The problem is when the water level is at low point, 0< Hp<2/3Hp the gate is still open. Eventhough, this event can be applied as water flow management at the field, it should reach the optimumwater level point, *y*, before the gate open. In order to find the optimum point, we must calculate the centre of pressure.

Refering from figure 7 the equilibrium point can be described as figure below,

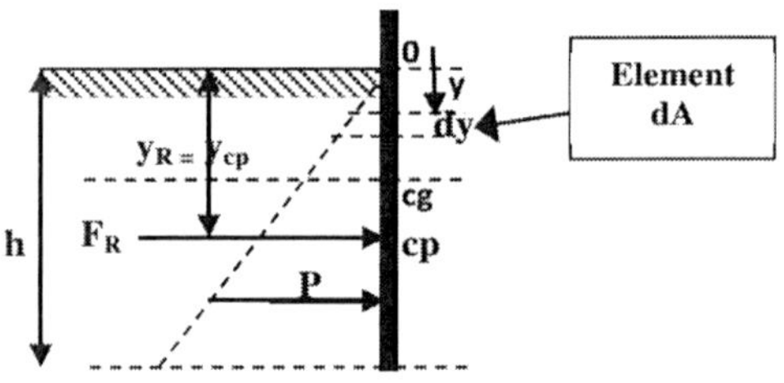

Fig. 9 Equilibrium point

For a plate in equilibrium we can take moments about any point, such as "0". Sum of moments on all elements = moment of resultant force.

So,

$$\int_0^h \rho.g.y.dA.y$$

$$= \rho.g \int_0^h y^2.dA$$

$$= \rho.g \int_0^h y^2.d(b.y)$$

$$= \rho . g \int_0^h by^2 dy$$

$\int_0^h by^2 dy$, called as "second moment of area"

Then,

$$\rho g [1/3by^3]_0^h$$

$$= 1/3 \rho g b h^3$$

Since, $\int_0^h \rho . g . y . dA . y = F_R . y_{cp}$ to reach equilibrium

Then,

$$1/3 \rho g b h^3 = \frac{\rho g h}{2} . (bh) . y_{cp}$$

So the result is,

$$y_{cp} = 2/3h$$

Using h as hp in experiment, the result is that the gate is remain closed at W , y, F_U at 0 < Hp < 2/3Hp. And then the gate start to open at Hp >2/3Hp. This value will give an automatic control for farmer, so they don't have to control the gate as day to day activity. When the farmer needsy_1(where $y_1 > y_{cp}$) water level, F_U should be increased by pumping air intotire tube. When the farmer need y_2 (where $y_2 < y_{cp}$) water level, F_U should be decreased through tire tube. Moreover, when the farmer keeps the water by closing the gate, the farmer should simply release air from tire tube. In Table 3 the point of experiment are described.

Table 3. Experimental Result Analysis

Water level Condition at 0 - 30 cm height (cm)	Gate Level Condition at 25 kg weight (cm)	Rubber Tire Condition (psi)
0	Close	Fu = 0
0 < hp < 15,4	Open, hi = 1,7	Fu = 0
15,4< hp < 18,5	Open, hi = 3,1	Fu = 0
18,5< hp < 20,0	Open, hi = 3,2	Fu = Maximum

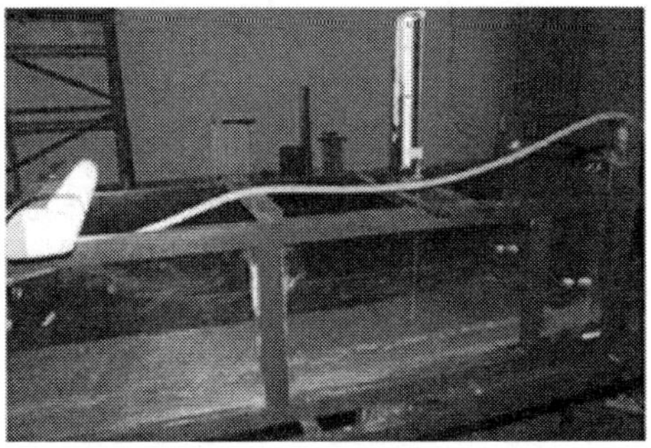

Fig. 10.a. The gate *closed* at water level 0

Fig. 10.b. The gate start to *open* at water level hp<2/3y

Whether the result is not fullfiled in condition at equilibrium, as optimum level, this experiment can be used for farmer asa starting effort and for further research. Every possible options can be excecuted using the same model until we can build the prototype to be used by the farmers all of around Indonesia.

5 Conclusion and recomendation

Conclussion from this experiment is;

1. The gate with 41 kg weight was opened below at proposed point at 2/3 H (since H = 400 mm, so 2/3H is 266 mm). Its 281 mm the gate was open or it is 67% water level height require. So, the gate succeed toopened below at proposed point at 2/3 H.
2. Using Buoyancy and Floting Method, as described at experiment result analysis, the gate should be close at W , y, F_U at 0<Hp<2/3Hp. And then the gate start to open at Hp>2/3Hp. This value will give an automatic control for farmer, so they do not have to control the gate as day to day activity. When the farmer needs$y_1$water level, F_U should be increased by pumping air into the tire tube. When the farmer needsy_2 water level, F_U should be decreased throughtire tube. And when the farmer keeps the water by closing the gate, the farmer should simply release air from tire tube.

Recomendation for completing the experiment;

1. The gate filler material must be changed so the weight required is fullfilled. Otherwise, the gate will function only in certain condition. And we are expecting the gates to be functional in all conditions
2. Sedimentation is necessary to be tested in this model

3. since this problem eventually emerged in the field
4. Various velocity category must be tested to make the gates more useable.

References

1. Regulation of the Minister of Public Works and Public Housing *Republic of Indonesia* No. 14/2015 (2015)
2. Department of Public Works Republic of Indonesia, Regulation of Wood Constructions *NI-5, PKKI-1961,* (1961), SNI 7973-2013 (2013)
3. Reed, B. Cameron. *Physics education* 39.4 (2004)
4. Ranald, V.Giles, Fluid Mechanics and Hydraulics, Schaum Outlines, McGraw-Hill (1995)
5. Triatmodjo, Bambang, Hidrolika I, Beta Offset, Yogyakarta (2001)
6. Kojima, Haruyuki, Takeshi Ijima, and Akinori Yoshida. *Coastal Engineering 1990.* 1228-1241. (1991).
7. Chen, Wai-Fah, and Lian Duan. Bridge Engineering Handbook: Superstructure Design. CRC press, 2014.
8. Eades, Andrew, and Jeffrey Penno.. U.S. Patent No. 7,037,432. 2 May (2006)
9. Groves, Michael F. U.S. Patent No. 5,716,041. 10 Feb. (1998)
10. Rempel, A. W., J. S. Wettlaufer, and M. G. Worster. *Physical review letters* 87.8 (2001)

Analysis characteristic of corruption in construction project in Indonesia

Felix Hidayat[1,*], *Sherly* Mulyanto[1]

[1]Civil Engineering Department, Faculty of Engineering, Parahyangan Catholic University, 40141 Bandung, Indonesia

Abstract. The competition among providers of construction services often leads corruption. The purpose of this study was to examine the characteristics of corruption in construction projects in Indonesia. The analysis of 15 (fifteen) cases are conducted by grounded theory method using coding techniques. The results showed that (a) the characteristics of construction projects involved in corruption is dominated by projects which, are owned by government that hire private company as the contractor; happen on infrastructure project; located in Java and Sumatra; (b) the vulnerability characteristics of construction projects involved in corruption are: occurs at construction phase due to contractual violation; sequential acts of corruption is begin with the difference between the volume of work that resulted in payments, which is not in accordance with progress reports; and the parties involved is Committing Officer (Pejabat Pembuat Komitmen/PPK); (c) the characteristics of corruption in construction project are: cause losses of 16-33% of the project value; and prison sentence of 44.8 months with the demands of a fine of USD. 10.660,- (d) the characteristic of the impact of corruption in construction project are: cause construction failure; 73% had an appeal to the High Court level and 33% had an appeal to the Supreme Court.

1 Introduction

Due to the need for construction which continue increase every year, it drives the hard competition among providers of construction services. The competition often leads corruption in Indonesia. According to the Corruption Perception Index 2014 released by Transparency International, Indonesia is ranked 107th of 175 countries, Singapore is ranked 7th, and Malaysia is ranked 52nd. The number of corruption in Indonesia is still higher compared to other ASEAN countries, such as Singapore and Malaysia.

Corruption occurs repeatedly leave a significant impact both on people and the government. Corruption results in infrastructure development is hampered because of the budget should be utilized for the development are being exploited by some specific parties. Besides corruption also led arises a sense of distrust of the government (Mc Mullan, 1961). In the aspect of its own development, corruption can affect the design and construction has been planned so that the construction can lead to failure.

Corruption is a problem that is often faced by the construction project in Indonesia. Corruption in construction projects that occur repeatedly lead to a pattern of how the corruption occurred. However, the value of the corruption that is large enough with corruption is difficult to be supervised both by society and the government led to corruption in construction projects in Indonesia are still common, especially in government-owned projects. So that's quite interesting if the characteristics of corruption in construction projects in Indonesia to be analyzed. The purpose of this study was to examine the characteristics of corruption in construction projects in Indonesia.

2 Corruption on construction project

Characteristics of corruption in construction projects in this study is an extension of previous research on the causal relationship between cause and vulnerability to corruption in the construction sector in China. A causal relationship between cause and vulnerability to corruption in the construction sector in China (Le et al., 2014) can be seen in Figure 1, with parameters that can be seen in Table 1.

Corruption in construction projects that occur repeatedly led to a pattern of corruption. The pattern that eventually developed into the characteristics of corruption in construction projects. Based on previous research on the causal relationship between the causes and vulnerabilities of corruption in the construction sector in China, developed a new pattern of cases of corruption in construction projects, particularly in Indonesia. Vulnerability described in previous studies led to corruption in construction projects in Indonesia. The cases of corruption in construction projects in Indonesia impacts due to corruption in construction projects in Indonesia. So when do the translation of the anatomical form of the structure and mechanisms of corruption in construction projects in Indonesia can be

* Corresponding author: hidayat@unpar.ac.id

described with an anatomic characteristics of corruption in construction projects in Indonesia.

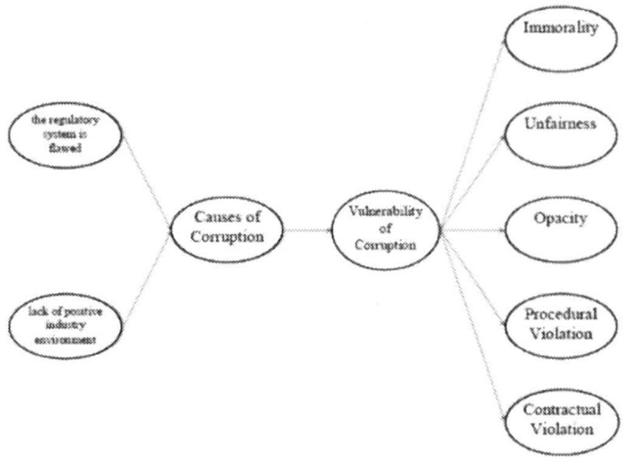

Fig. 1. Causal relationships between causes and susceptibility to corruption in the construction sector in China (Le et al, 2014)

Table 1. Parameter vulnerability to corruption in the construction sector in China (Le et al., 2014)

Construct	Code	Measurement Item
Immorality	IMM1	The work is not executed as per original design accorded
	IMM2	Work is executed without the availability of funds for the said purpose
	IMM3	The changes, especially in abnormally high rated and high value items are not properly monitored and verified
	IMM4	contractors provide false certificate in bidding
	IMM5	Subsitution of unqualified materials in construction
	IMM6	Site supervisor neglects duties for taking bribe from contractor
Unfairness	UNF1	The consultant is not appointed after proper publicity and open competition
	UNF2	The criteria adopeted in prequalification of consultant are restrictive and benefit only few consultant
	UNF3	The selection of consultant not done by appropriate authority
	UNF4	The criteria for selection of contractor are rstrictive and benefit only few contractor
	UNF5	The conditions and spesifications are relaxed in favor of contractor to whom the work is being awarded
	UNF6	Confidential information of bidding is disclosed to a spesific bidder
Opacity	OPA1	Adequate and wide publicity is not given to tender
	OPA2	Adequate time for submission of tender or offer not given
	OPA3	The evaluation of tenders is not done exactly as per the notified criteria
	OPA4	The negotiation on tender not done as per laid down guidelines
	OPA5	A large project should have called for bids split into several small projects and contracted without bidding
Procedural Violation	PRV1	Administrative approval and financial sanction not taken to execute the work
	PRV2	Lack of the sanctioned financial provision from the
	PRV3	Work is not executed for the same purpose for which the
	PRV4	The proper record of hindrances is not being maintained
Contractual Violation	COV1	Escalation clause is not applied correctly for admissible
	COV2	Compliance with conditions regarding deployment of
	COV3	The work order or supply order is not placed with justified

Anatomy of the characteristics of corruption in construction projects in Indonesia can be formed by a process system of corruption , which starts from the causes of corruption lead to vulnerability of corruption, then the process of corruption that occurred on the construction project, until the impact of corruption, where the system is influenced by the characteristics of the construction project. Anatomy of the characteristics of corruption in construction projects can be seen in Figure 2.

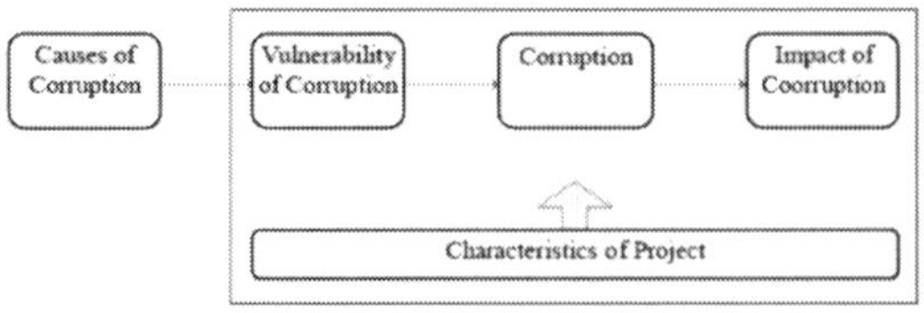

Fig. 2. Characteristics anatomy of corruption in construction projects

3 Research method

Data collection techniques used in this research is secondary data. Data used in the form of a Supreme Court ruling against corruption involving construction projects. Source of data used comes from the domain of the Supreme Court, namely www.putusan.mahkamahagung.go.id.

Referring to the characteristics of the data, the research method used is a combination of qualitative and quantitative research methods. Qualitative research methods used in this research is qualitative research with grounded theory approach. The analysis technique used in this research using coding techniques. According Strauss and Corbin, analysis using coding technique is divided into three stages namely open coding, axial coding, and selective coding.

1. At this stage of open coding is done categorization of data based on questions or variables that have been made with a grounded theory approach. In this study, the use of parameters and indicators of previous studies. Parameters and indicators of vulnerability in the form of corruption in construction projects, which in this study was developed back in the vulnerabilities that occur at each stage of construction. If the question explorative discovered new parameters and indicators that did not exist in previous studies, it will be created a new parameter based on the discovery in exploratory questions.
2. The next stage is the stage of axial coding. At this stage the data is new collected after analyzed by open coding. From the collected data obtained four categories that can be analyzed. The categories are vulnerability of corruption in construction projects, corruption in construction projects, the impact of corruption in construction projects, and the characteristics of the construction project. The last stage is the stage of selective coding.
3. Phase selective coding is done by seeking to harmonize between open and selective coding that has been done. So that the parameters will be included in a category that is relevant and of a series of stories. In order to be a story that is consistent, can be made a fishbone diagram. The anatomy of the skeleton to be

 DOI: 10.1051/matecconf/201710105018

used as a basis for making a fishbone diagram as a result of the analysis in this study.

4 Data analysis

4.1 Characteristics analysis of construction projects

The 15 cases analyzed was obtained in the form of construction project characteristics of the dominant service provider is private while the dominant service users are the government, the dominant construction project site is Java and Sumatra. The dominant types of construction projects are infrastructure projects, with an average contract value USD. 195.000.000,- (USD. 1 = Rp. 13.500,-) and the average execution time is 166 days. While the type of tender is most often done is through an auction.

4.2 Characteristics of vulnerability analysis corruption on construction project

Vulnerability corruption in the construction project is divided into three stages, namely the pre-construction phase, construction phase and maintenance phase. Vulnerability analysis of corruption in construction projects on coding techniques included in axial coding phase. Of the 15 cases analyzed, sequential most common is the difference in the volume of work that resulted in payments not in accordance with the progress and news reports a fictitious event. It is suspected of corruption in the form of the difference in the volume of work more easily identified if found non-compliance with the contract documents, while another sequential as bribes to get the project could not be found immediately if it does not receive the report. The results of the analysis can be seen in Figure 3. While the majority of the parties involved in corruption in construction projects is the Committing Officer, allegedly because the Commitment Officer is responsible for reports of physical work and payment approval. On the external factors that affect the public intervention seen the construction process so that the implementation of construction projects is not in accordance with applicable regulations and lead to corruption. Of the 15 cases analyzed, the public intervention is more common in construction projects in remote locations and outside Java. This is apparently due to the inland location of indigenous and influence is very strong compared to more modern locations such as in big cities.

4.3 Analysis on characteristics corruption on construction projects

Characteristics of corruption in construction projects in the form of loss of value of construction projects involved in corruption and the prosecution of the cases of corruption in construction projects. Analysis of the characteristics of corruption in construction projects in coding techniques included in axial coding phase. Based on sequential corruption in construction projects can be seen that this kind of corruption is most often done is the difference between the volume of work. This means that there is a mismatch between the work done in the field with the work that should be done in the agreement contract. The difference of the volume of work resulting in differences in payments that are not in accordance with the work progress report. The payment difference is the value losses of corruption in construction projects.

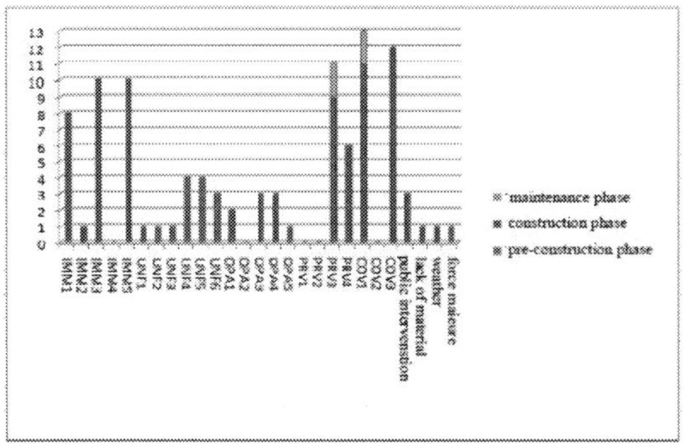

Fig. 3. Corruption vulnerability

Based on 15 cases analyzed, it can be concluded that the difference in the volume of work is a kind of corruption most often occurs in the construction project. Values project losses are in the range of 16.71% - 33.42% of the contract value of the project. While the demands on the criminal cases of corruption in construction projects had an average demand of 44.8 months imprisonment with the demands of average fine of USD. 10.716.

4.4 Characteristics analysis on impact corruption construction projects

The impact of corruption in construction projects can be divided into two, namely the technical and legal implications. On the technical impact is more focused on the effects of corruption in the construction of the building after the maintenance period runs out. While the impact of the law is focused on the decision at the level of district courts, high courts and the supreme court.

Technical impact in the case of corruption in construction projects such as the failure of the construction and the design life of the building is not reached. The impact of construction failure present in 4 cases out of 15 cases analyzed. Construction failure that occurs in the form of collapse of the building, as well as damage to roads. While the impact of the age of the building plan is not achieved only in one case out of 15 cases analyzed. Through this analysis it can be said corruption in construction projects construction prone to failure. The impact of the law in cases of corruption in construction projects is a decision at the level of district courts, high courts and supreme court. Decision against corruption in construction projects can be found guilty or not guilty.

On the characteristics of the impact of corruption in construction projects, it can be concluded that the

DOI: 10.1051/matecconf/201710105018

technical impact of the most common is a construction failure. While the impact of the law are most numerous in the district court verdict. A judicial decision can be found guilty or not guilty. Decision of the Supreme Court level, only 1 case was found not guilty, while four other cases convicted. Imprisonment verdict at the Supreme Court level had an average of 33 months with an average fine of USD. 9.260. As for compensation damages ruling by the Supreme Court level there is only one case, which amounted to USD. 15.618,-. Most of the value of money substitutes nominal damages was not mentioned in the decision of the Supreme Court Level. The cases that have not reached the level of the Supreme Court still has the possibility to appeal to the Supreme Court.

4.5 Summary characteristics corruption construction project

Characteristics of corruption in construction projects most often occur in the implementation phase of construction with this type of vulnerability is most often done is a vulnerability in the form of contractual violation. Activity corruption is the most frequently performed work orders are not carried out in accordance with applicable regulations. The vulnerabilities that do give rise to corruption in the form of project losses and prosecution of corruption criminal. Based on this analysis, much as 16.71% to 33.42% of the nominal value of the project contract losses due to corruption.

The impact of corruption in construction projects in the form of the technical and legal implications. On the technical impact, construction failure is the most common. But of the 15 cases analyzed not all cases are experiencing technical impact. While the impact of the law, all cases have been affected by the law. Legal effect in this study experienced a change rather than the demands put forward in corruption.

Through the analysis of 15 cases of corruption in construction projects, it appears that the impact of the law is not appropriate when compared to the value in corruption. Characteristics of corruption that occur pose a groove pattern or acts of corruption. Based on the analysis in this study sequential corruption in construction projects most often is the difference between the volume of work in the form of changes in the quality and quantity of materials resulted in progress in the field does not correspond to the minutes, so there is a discrepancy between the payment by progress in the field. Sequential dominant corruption can be seen in Figure 4.

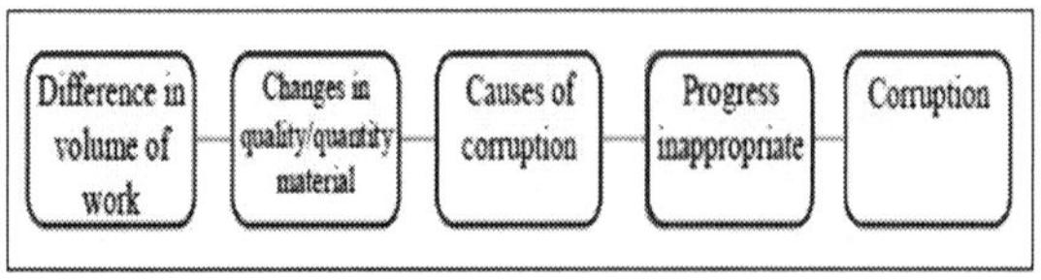

Fig. 4. Sequential corruption in construction projects (dominant)

5 Conclusion and recommendation

Based on the analysis performed on the data that has been collected, it can be obtained the following conclusions:

1. On the characteristics of the construction project:
   a. Of the 15 cases analyzed, the type of ownership of government service users and the type of ownership of private service providers is a type of ownership that experienced corruption in construction projects.
   b. 60% of the 15 cases analyzed, the type of construction project that involved corruption is an infrastructure project.
   c. 53.85% of the 15 cases analyzed, the project involved the construction of corruption have an average contract value of Rp. USD. 195.000.
2. At vulnerability anatomy of corruption in construction projects:
   a. In addition to the parameters and indicators of vulnerabilities that have been there on previous research on the causal relationship between cause and vulnerability to corruption in the construction sector in China (Le et al., 2014), found a new parameter in the form of external factors with indicators of intervention of society, weather, emergencies, and unavailability of materials. In Indonesia, the division of the project site greatly affect the implementation of the project, which at inland locations and customs of the community intervention influence is still very strong compared to the big cities. In addition, limited infrastructure affect the availability of material required in the field.
   b. Of the 15 cases analyzed, all cases are susceptible of corruption during the implementation phase of construction. This is apparently due to the implementation phase of construction irregularities committed corruption is easier to detect. Addition of 15 cases analyzed, it was found that the vulnerability in the form of contractual violation be repeated frequently. This is presumably due to vulnerabilities in the form of contractual violation is easier to detect by comparing the execution in the field with the agreed work contracts.
   c. Of the 15 cases analyzed it was found that sequential corruption most often occurs is the difference between the volume of work that resulted in payments not in accordance with progress reports with the parties most often involved is the Committing Officer. This is presumably due to the difference in the volume of work more easily identified by comparing the implementation of a contract of employment. In addition, the Commitment Officer is the party most responsible for project progress reports and payment approval of projects based on project progress report.
3. In the anatomy of corruption in construction projects:
   a. Of the 15 cases analyzed, it was found construction projects with total losses amounting

to 16.71% up to 33.42% of the most frequently experienced corruption. However, in the area of Nusa Tenggara Islands, and Papua discovered a greater loss value project up to 80% of the contract value of the project. This is due to the influence of external factors that are large enough in these locations, so that the payment received is not in accordance with the progress that is implemented on the ground.

b. Of the 15 cases analyzed, the demands of imprisonment against corruption in construction projects had an average value during the 44.8 months with the value of the average penalty for USD. 10.716,-. While demand restitution damages only some cases including the same in the Supreme Court ruling. It is suspected that cause corruption in construction projects is often done because of the demands form of imprisonment and fined not comparable with values in corruption.

4. In the anatomy of the impact of corruption in construction projects is 73.33% of the 15 cases analyzed has reached a verdict in the High Court level and 33.33% of the 15 cases analyzed has reached a verdict in the Supreme Court.

Suggestions for further research are:

1. Future studies should focus on each type of construction project, in order to take the tendency of corruption in each type of construction project.
2. Future studies should focus on corruption in construction projects by years of corruption in construction projects.
3. Further research should be done further analysis regarding the location of the construction projects involved in corruption in order to know the relationship between the location of construction projects with the trend of corruption.

We would like to express our appreciation to Civil Engineering Department, Faculty of Engineering, Parahyangan Catholic University, for their support and cooperation.

References

1. Barrie, Donald S. dan Paulson, Bayd C. (1992), Professional construction management: including C. M, Design-construct, and general contracting Ed.3, McGraw-Hill, New York:
2. Bogdan, R.C., dan Biklen, S.K., (1982). Qualitative Research for Education: An Introduction to Theory and Method. Boston, MA: Allyn and Bacon, Inc.
3. Creswell, John W. (2010), Qualitative Inquiry & Research Design: Choosing Among Five Approaches, London.
4. Grisham, T. (2013). Journal of Construction and Management Engineering, Vol. 5, No. 1, 13-21.
5. Le, Y., Shan, M., Chan, A., and Hu, Y. (2014). Journal of Construction and Management Engineering, Vol. 140, No. 9
6. Le, Y., Shan, M., Chan, A., and Hu, Y. (2014). Journal of Construction and Management Engineering, Vol. 30, No. 4
7. Strauss, A. L., dan Corbin, J. (1990), Basic of Qualitative Research: Grounded Theory Procedures and Techniques. New Park, CA: Sage.

MATEC Web of Conferences **101**, 05019 (2017) DOI: 10.1051/matecconf/201710105019
SICEST 2016

Accuracy analysis of SRTM usage for upper Citarum River flood modeling

Riza Inanda Siregar[1,*]

[1]Civil Engineering Department, Sumatera Utara University, 20155 Medan, Indonesia

Abstract. Natural channel cross section is generally very irregular and varied, for complex cross-sections and natural cross sections no specific formula presents to express the elements of the geometric cross section of the channel. The study uses HEC-Geo RAS with Arc GIS 10 and HEC-RAS 4.1.0 for modeling and 1D hydrodynamic. The surveyed cross sections modeling is calibrated and validated with observed daily discharge. Then the surveyed data cross sections are replaced by the cross sections extracted from SRTM. Accuracy is suitability between the surveyed data cross sections with data cross sections which extracted from SRTM. Accuracy is expressed by the magnitude of the error. The ratio of simulated discharges between two models by surveyed data cross sections and SRTM data cross sections have index of performance indicator for Nash–Sutcliffe Index (NI) is 0.95 and index of Root Mean Square Error (RMSE) is 0.41. Index of Performance indicator for comparison of simulated discharges between models with Nash–Sutcliffe Index (NI) is 0.95 and index of Root Mean Square Error (RMSE) is 0.41. The results show that cross sections from SRTM can be an alternative to extract data cross sections for Upper Citarum River flood modeling.

1 Introduction

Natural channel sections are in general very irregular from approximate parabola to approximate trapezoid. The river which frequent floods, the channel may consist of main channel section carrying normal discharges and one side channel sections to release the overflow [1].

Hydraulic river modeling needs cross sections to geometric data analysis. In generally, river cross sections can be obtained by surveying river geometric. Geometric data such as data cross sections of river is complicated to obtain the data information, so we need an alternative to acquire data easily without field surveys. The alternative aims to get the river data cross sections that can be used in river flood modeling. The modeling is expected having results as accurate as using the surveyed river cross section [2].

In this study, there are two different sources data of river cross sections. The first river modeling uses source data of river cross sections from surveyed cross sections and the second uses alternative data. The source of alternative data uses SRTM (Shuttle Radar Topography Mission) with the resolution is 90 meters for the entire surface of the world. The SRTM is generated to release river cross sections.

DEMs are used for 2D hydrodynamic flood modeling and simulation algorithm. The data is used for predictions which are compared with published flood maps and observed flood conditions [3].

A method of approach to generate river data cross sections from the ASTER DEM with low resolution and the inadequate vertical accuracy of such as global data present differentiating features of hydraulic importance, which necessitate processing of the DEMs before use for river flood modeling [4].

1-dimensional modeling is modeling that has a one-way flow is the main flow along the flow direction. Hydraulic River modeling uses software HEC-RAS to simulate daily discharge based on source data from surveyed cross sections and generated SRTM. The model with two source data will be compared to analyze the accuracy of SRTM usage for Upper Citarum River Modeling.

1.1 Upper Citarum River

The Upper Citarum watershed is at West Java with a height ranging between 625-2600 meters above sea level. The Upper Citarum is geographically located at 6°43'21,8" - 7 °19'38,1" South Latitude and 107°32'2" - 107°53'51,6" East Longitude. The upstream basin is surrounded by mountains and hills.

Annual rainfall varies from an average of 1000 mm in the coastal areas and increases to 4000 mm in the mountains, at the top of the watershed [5].

Upper Citarum watershed's area is about 1771 km^2. Upper Citarum River flow is affected by the thirteen tributaries. It flows through in South Bandung to the Saguling reservoir with the length 31 kilometers. The SRTM of Upper Citarum River can be seen in Figure 1.

* Corresponding author: rizawaloed@yahoo.com

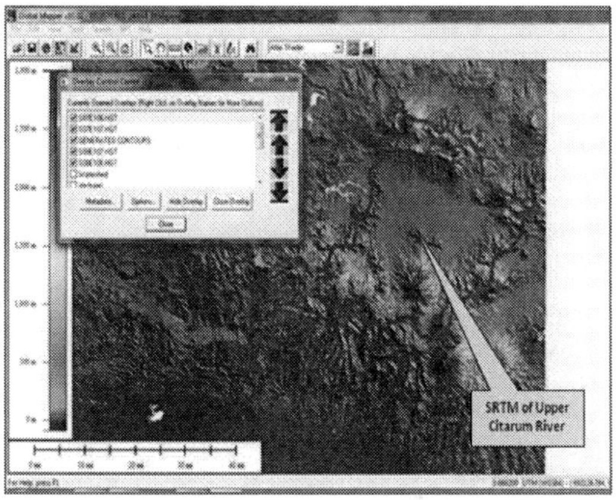

Fig.1. SRTM of Upper Citarum River and its catchment

Figure 1 shows that SRTM of Upper Citarum River and its catchment. The main purpose of the river data cross sections extracted from SRTM is to collect topographic data and makes data DEM (Digital Elevation Model). The SRTM with 90% confidence intervals and with global coverage area has absolute horizontal and vertical accuracy better than 20 and 16 meters [6].

SRTM is a satellite image depicting the height/elevation of each grid with each grid dimensions: 90 x 90 meters. The imaging project SRTM led by the National Geospatial-Intelligence Agency (NGA) and the National Aeronautics and Space Administration (NASA). Both are state-owned institutions United States [7]. However, the SRTM data can be used and developed the data and for public is free.

1.2. HEC-RAS and HEC-GEO RAS

HEC-RAS stands Hydrologic Engineering Centre River Analysis System software is an application program for one dimensional hydraulic modeling in open channels flow. One dimensional modeling is model that has flow direction, which is the direction of flow along main flow. A hydraulic modeling will analyze hydraulic basically looking for the depth and speed of the flow along the groove which is the result of the discharge at the input as a boundary condition [8].
Unsteady flow is a physical process in the flow in a channel by adopting the concept of conservation of mass and momentum conservation. This physical process can be described by mathematical equations, known as the equation St. Venant [8]. The equations consist of continuity equation (mass conservation principle) and the momentum equation (the principle of conservation of momentum), which is written in the form of partial differential equations as follows:

The continuity equation:

$$\frac{\partial A}{\partial t} + \frac{\partial Q}{\partial x} - q_l = 0 \quad (1)$$

The momentum equation:

$$\frac{\partial Q}{\partial t} + \frac{\partial QV}{\partial x} + gA\left(\frac{\partial z}{\partial x} + S_f\right) = 0 \quad (2)$$

Where: A = area (amount of flow in the main channel and overbank channel), Q = flow discharge, ql = lateral discharge per unit width of channel, V = velocity, g = acceleration of gravity, x = distance which measured in flow direction to downstream, z = mean depth, t = time of flow. HEC-Geo RAS is software to process geospatial data which is set of Arc GIS tools for used with the HEC-RAS. The extension allows the users with limited GIS experience to generate HEC-RAS import file containing geometric data from an existing DMT (Digital Terrain Model) and complementary data set [9].

2 Methodologies

The study uses HEC-Geo RAS and Arc GIS 10 and HEC-RAS 4.1.0 for modeling 1D hydrodynamic. Daily rainfall data are from 21 stations in 2011. Then the model is calibrated and validated with observed daily discharge in Nanjung which is a discharge measurement stations. The period of calibration is January 2011 to April 2011 and period of validation September 2011 to December 2011. There are two indicators namely observed values and simulated values each two models from surveyed data cross section and SRTM extracted data cross sections. The observed values consist of discharge and water level. [10].

The modeling of 2D hydrodynamic is supported by the topographic data with the high resolution, but now it is very difficult to obtain detailed supporting data for modeling. As a solution we used a Digital Elevation Model (DEM) which was generated from the ASTER data [11].

The RMSE index (Root Mean Squared Error) and the Nash Index are two criteria for calibration of the simulated models. RMSE and Nash index is used to evaluation of hydrological models with observed data [12]. Two performance indicators are NI (Nash Sutcliffe Index or Nash Index) and RMSE (Root Mean Square Error), which is written in the form of partial differential equations as follows:

NI (Nash-Sutcliffe Index):

$$NI = 1 - \frac{\sum_{i=1}^{n}(Q_o - Q_m)^2}{\sum_{i=1}^{n}(Q_o - \overline{Q_o})^2} \quad (3)$$

RMSE (Root Mean Square Error):

$$RMSE = \sqrt{\frac{\sum_{i=1}^{n}(Q_o - Q_m)^2}{n}} \quad (4)$$

Upper Citarum River length up to Nanjung (daily discharge station) which is surveyed cross sections length 31 kilometers. Survey was at intervals between points as far as 100 meters and cross sections data about 300 cross sections. Accuracy is expressed by the magnitude of the error. If getting close to the simulation results with cross sections extracted from SRTM with surveyed cross section, so the better method or process is used.

The important data required for HEC-Geo RAS is the TIN (terrain data) or DEM. land use information is needed to be additional data sets. Using data DEM for hydrology and hydraulic modeling usually involve some points: (1) preprocessing of data; (2) model execution; and (3) post-processing/ visualization of results [13].

The most important data to input in HEC-RAS are river data cross sections. The function of cross section cut lines extract the elevation data from TIN data to make ground profile across channel flow. The intersection of cut lines with other RAS layers such as centerline and flow path lines are used to compute HEC-RAS attributes such as bank stations (locations that separate main channel from the floodplain), downstream reach lengths (distance between cross sections) and Manning n. Hence, creating of data cross sections to create good representation of channel bed and floodplain is important. The best guidelines must be followed in creating cross sections cut lines: (1) the digitized process are perpendicular to the direction of river flow; (2) in flood extent must span to over the all section that can be modeled; and (3) the digitized process are from left to right (looking downstream). Even though it is not needed, but it is a good method to keep spacing between data cross-sections more consistent [14]. The cross sections extracted from SRTM can be seen in Figure 2.

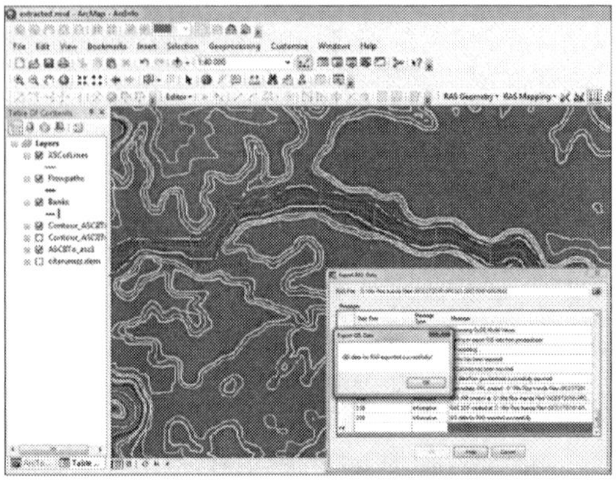

Fig.2. Upper Citarum River cross sections extracted from SRTM

Figure 2 shows that cross sections extracted from SRTM. The SRTM extracted river data cross sections could be used in hydraulic models to simulate stage and discharge hydrographs with considerable accuracy under the scarcity of measured cross sections data [15]. The geographic coordinate system of the DEM was converted to a projected coordinate system using the software package ArcGIS 10. The river branch was digitized along the center of the left and right bank. The cross-sections were automatically extracted along the river branch from the DEM data using the software HEC-Geo RAS [16].

3 Results and discussions

Two criteria are compared between observed discharge and simulated discharge with river surveyed data cross sections for period of calibration January 2011 to April 2011 in Nanjung. The validation period is from September 2011 to December 2011. Index of Performance indicator for calibration with NI (Nash–Sutcliffe Index) is 0.81 and RMSE index (Root Mean Square Error) is 0.96. Index of Performance indicator for validation with Nash–Sutcliffe Index (NI) is 0.7 and Root Mean Square Error (RMSE) is 1.72. Nash Index range from -∞ to 1, the model which perfect data cross section with the observed data has Nash index NI=1. The model predictions are as accurate as the mean of the observed data has Nash Index NI=0. If predictor better than model have Nash Index less than zero. RMSE index (Root Mean Square Error) is index used to gauge of the differences between values (sample and population values) predicted by a model or an estimator and the values observed. The calibration and validation can be seen in Figure 3.

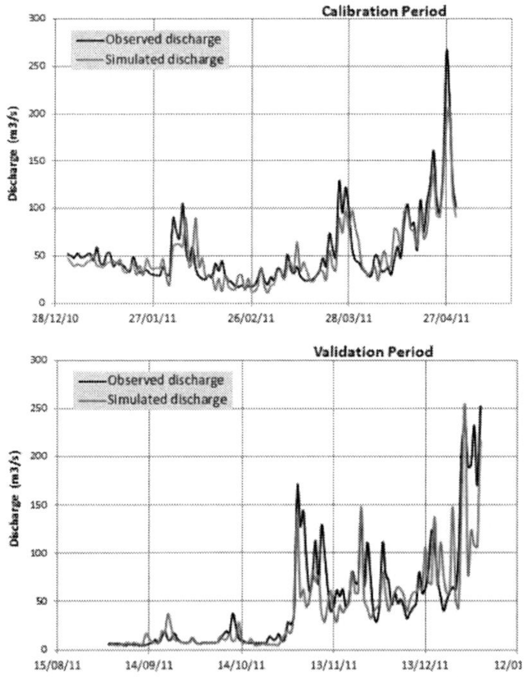

Fig.3. Calibration period and validation period

Figure 3 shows that discharge calibration discharges is simulated with period January to April and validation discharges is simulated with period September to December. It is simulation same with calibration period.

The comparison between simulated discharges and water depth models by surveyed data cross sections and SRTM cross sections can be seen in Figure 4.

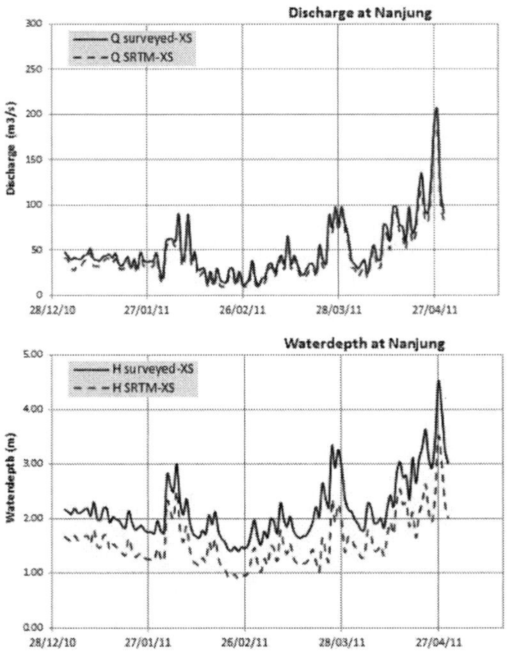

Fig.4. Simulated discharge between surveyed and SRTM; water depth between surveyed and SRTM

Fig. 4 shows that the result of discharges with simulation between surveyed and SRTM has small difference or almost the same and the result of simulated of water depth between surveyed and SRTM has a large difference. It is caused the difference of elevation cross sections surveyed-XS with the results generated SRTM-XS using HEC-Geo RAS and ArcGIS.

Index of Performance indicator for comparison of simulated discharges between models by surveyed cross sections and SRTM cross sections at Nanjung with NI (Nash–Sutcliffe Index) is 0.95 and RMSE index (Root Mean Square Error) is 0.41, it means discharge of simulated from model with extracted data cross sections SRTM is as good as from model with surveyed data cross sections. Index of Performance indicator for comparison of water depth between models by surveyed data cross sections and SRTM data cross sections at Nanjung with Nash–Sutcliffe Index (NI) is -0.21 and RMSE (Root Mean Square Error) is 0.04, it means simulated water depth from model with SRTM-extracted cross sections and from model built with surveyed cross sections are not good results. It occur because of low resolution and inadequate vertical accuracy of global data to cut lines data cross-section.

4 Conclusions

Based on the analysis that the output applied for Upper Citarum River show that SRTM data can be an alternative data to extract cross sections for Upper Citarum River modeling. Simulated daily discharge from model built with SRTM-extracted cross sections is as good as that of from model with surveyed cross sections. The comparison of simulated discharges between models by surveyed data cross sections and SRTM data cross sections have index of performance indicator for NI (Nash–Sutcliffe Index) is 0.95 and RMSE (Root Mean Square Error) is 0.41. Index of Performance indicator for comparison of simulated discharges between models by surveyed cross sections and SRTM cross sections at Nanjung with NI (Nash–Sutcliffe Index) is 0.95 and RMSE (Root Mean Square Error) is 0.41.

The authors gratefully acknowledge that the research is supported by University of North Sumatera, Indonesia.

References

1. V T Chow, *Open Channel Hydraulics*, McGraw-Hill Book Company Inc., Kogakusha Company, LTD., Tokyo, Japan, (1959)
2. P Ricardo, J Karsten, Tampa, Florida, (2007)
3. Sanders, F Brett., Elsevier, **30**, 1831-1843 (2007)
4. Gichamo, T.Z., Popescu, Jonoski, Solomatine, Elsevier, **31**, 37-46 (2011)
5. ADB and Bappenas, *Integrated Citarum Water Resources Management (ICWRM)*, Bandung, 2012.
6. N Kolecka, J Kozak, Pure and Applied Geophysics, Krakow, Poland, **171**, (2014)
7. ESRI, Environmental Systems Research Institute, USA.
8. US ARMY Corps of Engineers, US Army Corps of Engineer, Washington, (2010)
9. US ARMY Corps of Engineers, US Army Corps of Engineer, Davis, (2009)
10. Chidungdoan, Anhtuandao, S Liong, R Sanders, J Liu, and T Fewtrell, 10th International Conference on Hydroinformatics, Hamburg, Germany, (2012)
11. Tarekegn, Tesfaye, Alemseged, Tom, Reggiani, Dinand, Elsevier, **12**, 457-465 (2010)
12. V G Hoshin, K Harald, K. Y Koray, and F M Guillermo, Elsevier, **377**, 80-91, (2009)
13. M Venkatesh, *Tutorial on Using HEC-Geo RAS with Arc GIS 10 and HEC-RAS Modeling*, School of Civil Engineering, Purdue University, (2012)
14. L Arturo, School of Civil and Construction Engineering, Oregon State University, (2013)
15. Niranjan, Pramanik., Rabindra, Dhrubajyoti, *One Dimensional Hydrodynamic Modeling of River Flow Using DEM Extracted River Cross-Sections*, (2010)
16. Amir, Khan, Rasul, Sharma, and Akram, 19th Australasian Fluid Mechanics Conference, Melbourne, Australia, (2014)

Multi performance option in direct displacement based design

Ima Muljati[1,*], *Yonatan* [2], and *Adrian* Hartono[2]

[1]Civil Engineering Department, Petra Christian University, 60236 Surabaya, Indonesia
[2]Alumni of Civil Engineering Department, Petra Christian University, 60236 Surabaya, Indonesia

Abstract. Compare to traditional method, direct displacement based design (DDBD) offers the more rational design choice due to its compatibility with performance based design which is controlled by the targeted displacement in design. The objectives of this study are: 1) to explore the performance of DDBD for design Level-1, -2 and -3; 2) to determine the most appropriate design level based on material efficiency and damage risk; and 3) to verify the chosen design in order to check its performance under small-, moderate- and severe earthquake. As case study, it uses regular concrete frame structures consists of four- and eight-story with typical plan, located in low- and high-risk seismicity area. The study shows that design Level-2 (repairable damage) is the most appropriate choice. Nonlinear time history analysis is run for each case study in order to verify their performance based on parameter: story drift, damage indices, and plastic mechanism. It can be concluded that DDBD performed very well in predicting seismic demand of the observed structures. Design Level-2 can be chosen as the most appropriate design level. Structures are in safe plastic mechanism under all level of seismicity although some plastic hinges formed at some unexpected locations.

1 Introduction

In traditional seismic structural design, i.e. force based design (FBD), force becomes the most important parameter. FBD using the estimated initial stiffness to determine structural period and the distribution of forces among members. Force is directly related to structural stiffness and displacement during elastic stage. However, for structures in plastic condition, the relationship is complex. Engineers usually reduce the calculated elastic force level due to ductility possessed by the structures. Ductility enabling structures to deform inelastically to the required deformation without loss of strength. If the resistance of the structure is less than the applied force, or if their deformations exceed the limitation, then redesign should be taken.

The interdependency between strength and stiffness as well as between strength and ductility cause FBD hardly to be compatible with performance based design where structural performance is measured by specific displacement for a certain earthquake level. On the other hand, direct displacement based design (DDBD) has been developed over the past ten years to overcome the problems of FBD [1-2].

Unlike FBD, DDBD uses displacement as the designed target. Structures should be designed so that it could reach the targeted design displacement rather than to be bounded by a certain displacement limitation. Previous studies have indicated that DDBD performed well in many kind of structures [3-5]. However, they only observed the performance of structures under design level-2 only, repairable damage condition.

DDBD actually offers three level of design options, i.e. level-1 no damage, level-2 repairable damage, and level-3 no collapse for three design intensity levels as shown in Table 1 [6].

Table 1. Design intensity and performance criteria [6].

Importance Class	Earthquake Design Intensity		
	Level-1 No Damage	Level-2 Repairable Damage	Level-3 No Collapse
I	Not required	50% in 50yrs	10% in 50yrs
II	50% in 50yrs	10% in 50yrs	2% in 50yrs
III	20% in 50yrs	4% in 50yrs	1% in 50yrs
IV	10% in 50yrs	2% in 50yrs	1% in 50yrs
Drift Limit	0.01	0.025	No Limit

The design intensity level is defined as probability of exceedance for a given intensity level depends on the structural occupancy usage and damage consequences. Table 1 also shows the story drift limits for each design intensity under condition where buildings with non-structural elements detailed to sustain building displacement. For structures in moderate to high

* Corresponding author: imuljati@petra.ac.id

seismicity, performance criteria for levels 1, 2 and 3 intensities shall be met. While for structures in low seismicity, performance criteria for level 3 intensity shall be met [6].

Based on those conditions, this study is aimed to: 1. explore the design procedure of DDBD on regular concrete frame structures for level 1, 2 and 3; 2. determine the most appropriate design level based on material efficiency and damage risk; and 3. verify the controlled design in order to check the structural performance under small-, moderate- and severe earthquake.

2 Case study

The study uses regular concrete frame structure consist of 4 and 8 story with equal span 6 and 8 m (denote as variant A and B), equal story height of 4 m (Fig.1). Thus, there are eight case studies will be involved in this study. Assumption made for the design including: concrete strength of 30 MPa; reinforcement yield stress of 400 MPa; structures are located in Surabaya and Jayapura city representing low- and high-seismicity area; importance class is II for office building.

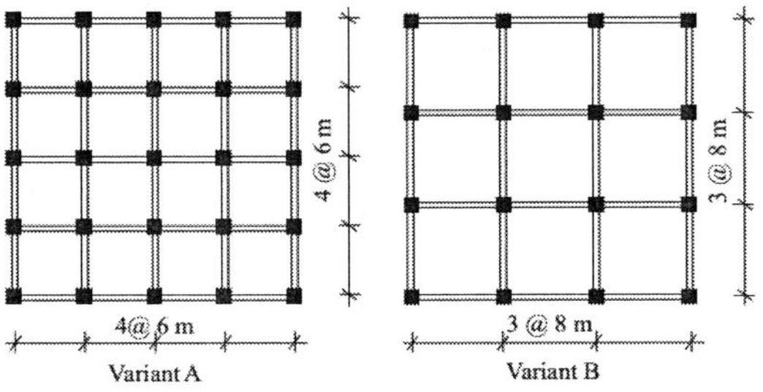

Fig. 1. Structure plans and variants.

3 Design and verification

3.1 DDBD Methodology

The eight observed structures are designed based on DDBD procedure [2] for three level of design, i.e. levels 1, 2 and 3 with probability of exceedance as much as 50%, 10% and 2% in 50 years (or equal to earthquake with 100, 500 and 2500 years return period respectively). The design drifts for Levels-1 and -2 are 0.010 and 0.025 [6]. Although there is no drift limitation for Level-3, but the study determines 0.040 as the design drift.

The elastic design response spectrum is constructed based on [7] for site class SE (soft soil) in Surabaya and Jayapura city. The study uses peak ground acceleration (PGA) calculation served by [8] to calculate the PGA of various return period, proportionally to the PGA of the maximum considered earthquake (MCE), i.e. 2500 years earthquake return period. The complete response spectrum for design Levels-1, -2 and -3 for Surabaya and Jayapura are shown in Fig.2.

The DDBD methodology [2] is illustrated in Fig.3. Starting by representing the building with a single degree of freedom (SDOF) model in Fig.3(a) and making a bi-linear relationship between lateral force – displacement in Fig.3(b). DDBD uses secant stiffness K_e to represent structural stiffness during inelastic at the design displacement Δ_d. In Fig.3(c) for a given level of ductility demand, the equivalent viscous damping ξ will be found.

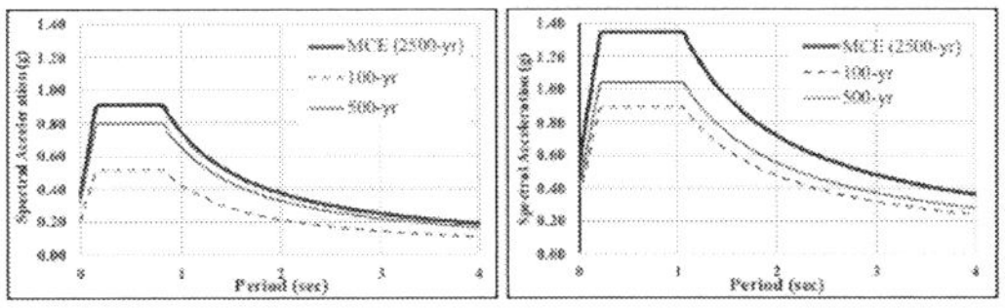

Fig. 2. Design response spectrum.

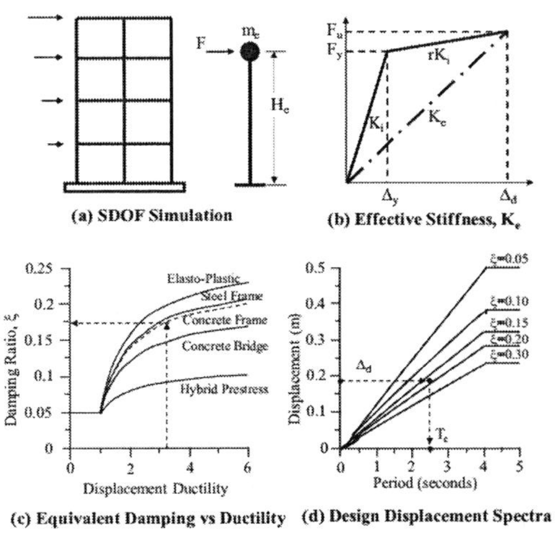

Fig. 3. Fundamental of DDBD [2].

The calculation of design displacement Δ_d is found from a normalized inelastic mode shape δ_i, where $i = 1$ to n are the stories, to the targeted displacement Δ_c (e.g. 0.010, 0.025 and 0.04) at the critical story following:

$$\Delta_i = \delta_i \, (\Delta_c / \delta_c) \tag{1}$$

where the mode shape depends on the floor height, H_i, and roof height H_n according to:

$$\text{for } n \leq 4: \; \delta_i = H_i / H_n \tag{2a}$$

$$\text{for } n > 4: \; \delta_i = (4/3)\,(H_i / H_n)\,\{1 - H_i / (4H_n)\} \tag{2b}$$

The equivalent design displacement is related to the story displacements by the relationship:

$$\Delta_d = \Sigma(m_i \Delta_i^2) / \Sigma(m_i \Delta_i) \tag{3}$$

where m_i is the mass at height H_i related with displacement Δ_i. The equivalent mass m_e and the effective height H_e of the SDOF are given by:

$$H_e = \Sigma(m_i \Delta_i H_i) / \Sigma(m_i \Delta_i) \tag{4}$$

The design displacement ductility factor can be related to the equivalent yield displacement Δ_y through:

$$\mu = \Delta_d / \Delta_y \qquad (5)$$

The yield drift of a story θ_y, in a frame structure depended on geometry, and is independent of strength [2]. And for reinforced concrete frame it is defined as:

$$\theta_y = 0.5\ \varepsilon_y L_b / h_b \qquad (6)$$

where L_b and h_b are the beam span between column centerline, and beam depth respectively, ε_y is the yield strain of the reinforcement steel.

The ductility demand can be estimated by assuming a linear yield displacement profile, and hence the yield displacement is given by:

$$\Delta_y = \theta_y / H_e \qquad (7)$$

The equivalent viscous damping of the SDOF for reinforced concrete frame can be related to the design displacement ductility demand as follows:

$$\xi_{eq} = 0.05 + 0.565\ (\mu - 1)/(\mu\pi) \qquad (8)$$

The effective period T_e of the SDOF from the peak displacement response can be found from the displacement spectrum, by relating the design displacement Δ_d to the corresponding equivalent viscous damping ξ_{eq} as illustrated in Fig.3(d).

The effective stiffness K_e of the SDOF at maximum displacement response Δ_d is given by:

$$K_e = F/\Delta_d = 4\pi^2 m_e/T_e^2 \qquad (9)$$

The design base shear for the multi degree of freedom (MDOF) structure is found from the SDOF structure as:

$$F = V_{base} = K_e \Delta_d \qquad (10)$$

The base shear force is then distributed to the floor level in proportion to the product of mass and displacement as:

$$F_i = V_{base}(m_i \Delta_i)/\Sigma(m_i \Delta_i) \qquad (11)$$

The structure is then analyzed under the simplified force vector referring to [2] to determine the required flexural strength at the beam plastic hinge locations. The method is based on simple equilibrium condition among members. It is not unnecessarily to add the full gravity-load moments to these seismic moments so that structures may develop its ductility capacity in achieving the design displacement. And finally, DDBD also apply capacity design procedure in order to assure the condition of "strong column weak beam" so that structures are within safe mechanism.

3.2 Design Verification

After choosing the most appropriate design level by considering material efficiency and damage level, each case study is modelled using [9]. Nonlinear time history analysis (NLTHA) is run for each case study using the acceleration record of El-Centro 1940 N-S, which has been modified in order to match with the targeted response spectrum for Surabaya dan Jayapura city. Since structures have symmetrically in plan, then the NLTHA is only run in one direction. The force – deformation relationship is modelled using [10]. The acceptance performance is measured based on parameter: story drift, plastic hinge rotation (represent level of damage) and structural plastic mechanism.

4 Results

4.1 Material consumption

From eight study cases, the average consumption of material relative to condition of design Level-2 are shown in Table 2. Material consumption are calculated by adding the volume of bending and shear reinforcement in beam and column, as well as the volume of beam and column for concrete volume. It can be said that design Level-1 is the most expensive design, 1.65- and 2.03-times of reinforcement and concrete consumption of design Level-2. Conversely, design Level-3 is the most efficient design, but not so much different from material consumption of design Level-2.

Table 2. Average consumption of material – relative to design Level-2.

Material	Surabaya			Jayapura		
	Lev-1	Lev-2	Lev-3	Lev-1	Lev-2	Lev-3
Steel reinf.	1.65	1.00	0.95	2.05	1.00	0.95
Concrete	2.03	1.00	0.90	2.65	1.00	1.00

From efficiency point of view, absolutely design Level-1 is too expensive to be applied. Considering damage risk, although design Level-2 and -3 are almost the same, but design Level-2 can be expected to perform better. Therefore, the study choses design Level-2 as the controlled design, and performs nonlinear time history analysis in order to check its structural performance due to small-, moderate- and severe-earthquake. The structural performance measured by parameter: story drift, damage indices and plastic mechanism are discussed in subsequent sections as well as the detailed results.

4.2 Story drift

The story drift from the eight case studies resulting from nonlinear time history analysis are shown in Fig.4. Each structure is codified according to its number of story, variant of geometry plan and its location, e.g. 4A SBY

DOI: 10.1051/matecconf/201710105020

means four story with variant A in plan, located in Surabaya.

The values shown in Fig.4 are taken from the maximum story drift during the considered excitation. Each of them are compared to the drift limit of each design level, i.e. 0.010, 0.025 and 0.040 due to small-, moderate- and severe-earthquake. It can be shown that all variants show excellent performance under story drift consideration. They are lower than the drift limit determined for each level of design. Thus, it can be said that the design Level-2 of DDBD successfully maintain the expected structural performance.

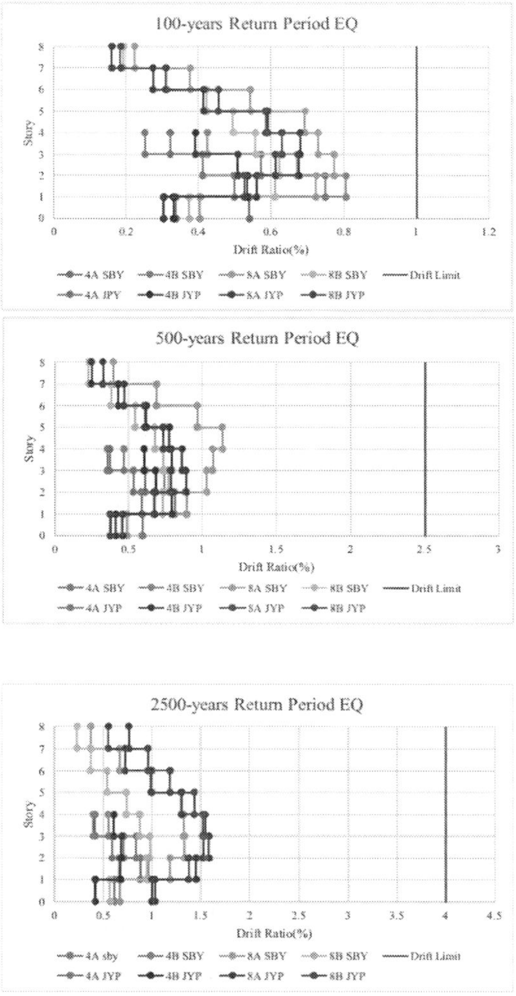

Fig. 4. Story drift.

4.3 Damage indices

Damage index is a parameter to measure the level of damage of the structure, defined as the plastic rotation of the member (beams and columns). Reference [6] determines the range of plastic rotation for each performance level. For beams, it is set as 0.004-0.008 rad, 0.008-0.01 rad and 0.01 rad – no limits for design Levels-1, -2 and -3 respectively. While for columns, it is set to 0.003-0.005 rad, 0.005-0.006 rad and 0.006 rad – no limits.

Table 3 and 4 show the structural performance based on the maximum plastic rotation during time history analysis on beams and columns. It can be shown that all beams have performed well under any earthquake level. Black blocked area in the table indicates unacceptable condition where the member rotation exceeding the limit.

Unfortunately, some columns experience excessive rotation due to small- to moderate-earthquake, e.g. structure 8A in Surabaya, 4A-, 8A- and 8B-in Jayapura, because columns experience load combination of axial tension and moment in a very short time. Despite these condition, all structures performed very well under severe-earthquake.

Table 3. Performance based on beam plastic rotation

Location	Return Period (Years)	First Yield	Lev-1 No Damage	Lev-2 Repairable Damage	Lev-3 No Collapse
Sura-baya	100	8A, 8B	4A, 4B		
	500		All		
	2500		4A, 4B, 8B	8A	
Jaya-pura	100	4A, 4B, 8B	8A		
	500	4A	4B, 8A, 8B		
	2500	4A	4B		8A, 8B
	Unacceptable performance				

Table 4. Performance based on column plastic rotation

Location	Return Period (Years)	First Yield	Lev-1 No Damage	Lev-2 Repairable Damage	Lev-3 No Collapse
Sura-baya	100	4A, 4B,8B			8A
	500	4A, 4B,8B			8A
	2500	4A, 4B,8B			8A
Jaya-pura	100	4B, 8B			4A, 8A
	500	4B, 8A			4A, 8B
	2500				4A, 4B, 8A, 8B
	Unacceptable performance				

DOI: 10.1051/matecconf/201710105020

4.4 Plastic mechanism

For the sake of safety, structures resulting from DDBD procedure are expected to experience "beam side sway mechanism" where the occurrence of plastic hinges are kept only at the beam, top floor columns and base columns as shown in Fig.5. To achieve this condition, designers apply capacity design in order to ensure condition of "strong column weak beam".

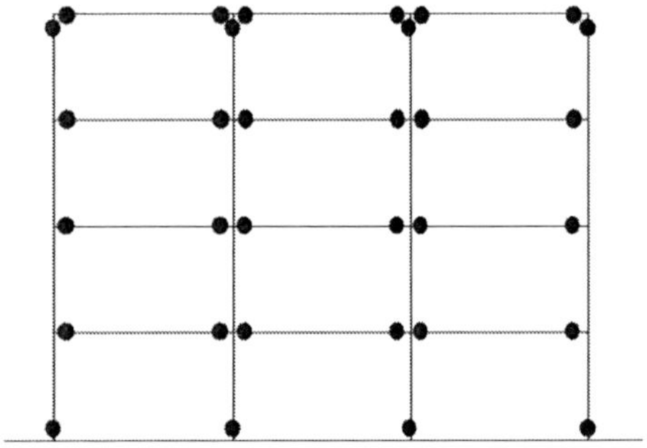

Fig. 5. Beam side sway mechanism.

Due to space limitation, this paper only presents the plastic mechanism of structure 8A SBY (Fig.6-8) [11]. The rotation performance of plastic hinge is indicated with color for each design level, i.e. yellow for first yield, green for Level-1 (no damage), light blue for Level-2 (repairable damage), and red for Level-3 (no collapse). The figures also show the condition of first yield and unacceptable condition (beyond Level-3 limitation), indicated with dark blue-color. The maximum rotation is indicated with square mark, while the plastic hinges in unexpected location are indicated with circle mark.

For low-earthquake excitation (Fig.6), the maximum plastic rotation in beam and column are in condition of first yield, some columns have already in no collapse criteria. For moderate-earthquake (Fig.7), the plastic rotation is worse than condition due to low-earthquake. Beam is in no damage condition while column is in no collapse condition. Finally, for severe-earthquake (Fig.8), structure experiences the worst plastic rotation. Beam and column are in repairable damage and no collapse respectively. These conditions are match with the plastic rotation performance as presented in Table 3 and 4. Capacity design successfully maintain the condition of 'strong column weak beam'. Yielding always initiate at the beams before at the columns.

Unfortunately, some unexpected yielding occurred at columns in story 1, 5 and 6 for low-earthquake, and in almost all stories for moderate- to severe-earthquake. It indicates that the beam side sway mechanism cannot be achieved perfectly. Nevertheless, the yielding of the column is still in first yield condition, and does not worsen the performance of the structures. The study does not identify any tendency of soft story mechanism. Thus, it can be concluded that structures are in safe mechanism. The complete plastic hinge formation of this study can be found in [11].

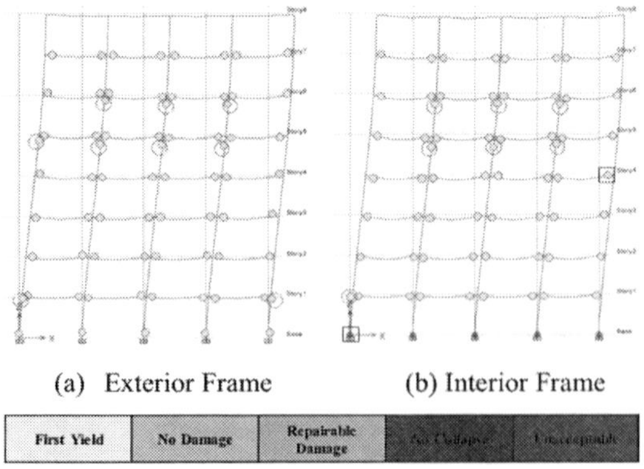

Fig.6. Plastic mechanism of 8A SBY, 100 yrs [11].

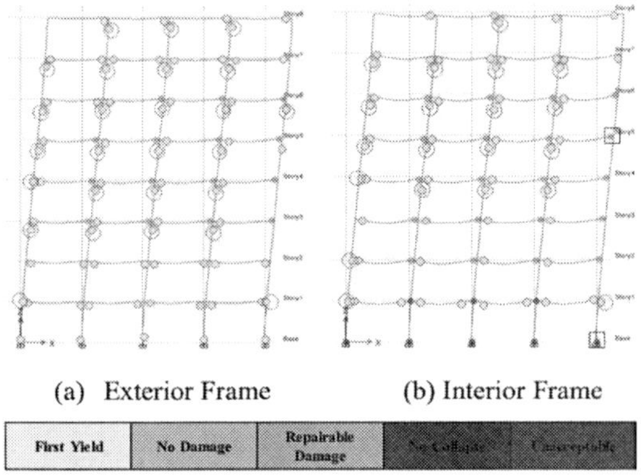

Fig.7. Plastic mechanism of 8A SBY, 500 yrs [11].

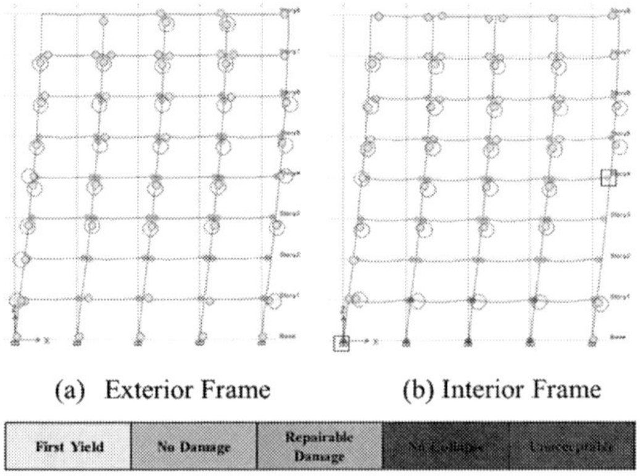

Fig.8. Plastic mechanism of 8A SBY, 2500 yrs [11].

5 Conclusion

It can be concluded that direct displacement based design (DDBD) performed very well in predicting seismic demand of regular concrete frame structures both in low- and highly risk seismicity area. Design

Level-2 (repairable damage), can be chosen as the most appropriate design level based on consideration of material efficiency and damage risk. Structures are in safe plastic mechanism under all level of seismicity although some plastic hinges formed at some unexpected locations.

References

1. M.J.N. Priestley. Bul. NZSEE, **26(3)**, 329-341 (1993).
2. M.J.N. Priestley, G.M. Calvi, M.J. Kowalsky. *Displacement-based seismic design of structures*. (IUSS Press, Pavia, Italy, 2007).
3. I. Muljati, F. Asisi, K. Willyanto. *Procedia Engineering*, **125**, 1050-1056 (2015).
4. I. Muljati, A. Kusuma, F. Hindarto. *Procedia Engineering*, 125 (2015) 1057-1064
5. I. Muljati, B. Lumantarna, R.P. Intan, A. Valentino. *Procedia Engineering* (to be published)
6. T.J. Sullivan, M.J.N. Priestley, G.M. Calvi. *A model code for the displacement-based seismic design of structures*. (IUSS Press, Pavia, Italy, 2012).
7. BSN, SNI-03-1726-2012. *Tata cara perencanaan ketahanan gempa untuk struktur bangunan gedung dan non gedung*. (Departemen Pekerjaan Umum, Jakarta 2012).
8. Indonesian Public Work Department, url: http://www.pusair-pu.go.id/pgacal (2011).
9. Computer and Structures, Inc. *CSI analysis reference manual for Sap2000, ETABS, SAFE, and CSI Bridge*. (Berkeley, California, USA, 2015).
10. L.A. Montejo. CUMBIA computer software. (North Carolina State University, (2007).
11. Yonatan, A. Hartono. Thesis No. **11012090** (Petra Christian University, Surabaya, Indonesia, 2016).

* Corresponding author: imuljati@petra.ac.id

Recycled aggregates in concrete production: engineering properties and environmental impact

Mohammed Seddik Meddah*

Department of Civil & Architectural Engineering, Sultan Qaboos University, 123 Al-Khod, Oman

Abstract: Recycled concrete aggregate is considered as the most abundant and used secondary aggregate in concrete production, other types of solid waste are also being used in concrete for specific purposes and to achieve some desired properties. Recycled aggregates and particularly, recycled concrete aggregate substantially affect the properties and mix design of concrete both at fresh and hardened states since it is known by high porosity due to the adhered layer of old mortar on the aggregate which results in a high water absorption of the recycled secondary aggregate. This leads to lower density and strength, and other durability related properties. The use of most recycled aggregate in concrete structures is still limited to low strength and non-structural applications due to important drop in strength and durability performances generated. Embedding recycled aggregates in concrete is now a current practice in many countries to enhance sustainability of concrete industry and reduce its environmental impacts. The present paper discusses the various possible recycled aggregates used in concrete production, their effect on both fresh and hardened properties as well as durability performances. The economic and environmental impacts of partially or fully substituting natural aggregates by secondary recycled aggregates are also discussed.

1 Introduction

Both natural and processed resources and energy are one of the most fundamental elements for the daily life of humankind. The tremendous and continuous development of modern societies has been accompanied by a fast and extensive use of natural resources that are considered as the locomotive of such social and economic growing. All humankind activities including industrial ones need the utilization of energy and large variety of natural resources.

Construction industry is considered as the engine of the economy and one of the main industrial activities. It is known to be the largest consumed of natural resources and one of the principal consumed of energy. The rapid and impressive human development during the last and current century, especially after the end of the second World War has led to a huge need of new buildings and infrastructures. The massive destruction induced by the WWII, especially in Europe has led to an extensive and abusive use of natural and non-renewable resources to not only rebuild and repair the destroyed infrastructures but also the huge need for new infrastructures. Moreover, the global population is in continuous increase, which increases demand for new housing and other related infrastructures.

In recent years, it has been recognized that increasing energy consumption from all sources could even contribute to the global climate change both directly and indirectly. It is anticipated that global warming phenomena will cause extremely serious problems in the future and should be addressed as early as possible. Not only the global climate change but also another very serious problem facing the modern society growing is the depletion of non-renewable resources due to their extensive use. The Statistical Review of World Energy released by British Petroleum indicated that the confirmed reserves-to-production ratio of oil, natural gas and coal are 53, 55 and 113 years respectively [1].

Although the direct greenhouse gases CO2, CH4, and N2O are naturally occurring in the atmosphere, intensive human activities have increased their atmospheric concentrations. It has been estimated that from the pre-industrial era (i.e., ending about 1750) to 2013, concentrations of these major greenhouse gases have increased globally by 43, 152, and 20 %, respectively [2,3]. It is generally agreed that current and most often used construction materials as well as building methods are unsustainable, having large environmental impacts such as CO_2 emissions, energy consumption, pollution, dusts, depletion of natural resources, generating considerable amount of solid waste. Among all existing construction materials, concrete is most used material because of many technical and economic factors. Concrete is also the second largest consumed material on earth just after water.

All countries are concerned by the global climate changes and greenhouse gas emissions including the developing countries and emerging economy such as Indonesia which contributes by around 1.9% of the

* Corresponding author: seddikm@squ.edu.om

global CO_2 emissions as per data published in 2011 (Fig.1) [4].

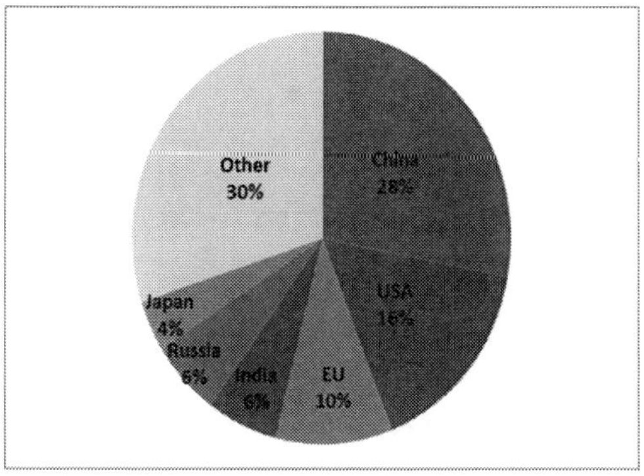

Fig.1. Breakdown of Total World Greenhouse Gas Emissions: 2015 [4]

Sustainable development was introduced to overcome those issues related to extensive use of energy and natural resources. It has been defined as "development that meets the need of actual generation without compromising the need of future generation. In other words, sustainable development is a development that will satisfy the need of people without neither depleting natural resources and energy, nor cause serious environmental impacts.

To address the issue of sustainability and assess its level, it is essential to define certain indexes that could be used to evaluate human activity in regards to sustainability concept. In construction industry, a sector that consumes a large part of natural resources, assessing sustainability is still a subjective matter.

Construction industry sector is intended to build various types of infrastructures that form the basis for social and economic activities of mankind. Over the years, infrastructures are built using mainly three principal materials: concrete, metal/steel and timber in addition to glass which becomes one of the main materials used in contemporary construction. As the world population is in continuous growing, demand for new housing and infrastructures is increasing annually. The amount of construction materials produced annually is also in incessant raise. According to the United States Geological Survey Report [5], the world total cement production has reached 4.193 billion tons in 2015 while CEMBUREAU [6]has estimated the total world cement production by 4.6 billion tons. While cement is not used only for making concrete but it is used also for other purposes, the concrete production was estimated to be around 32 billion tons based on an average of 300 kg of cement for $1m^3$ of concrete.

The production of steel was estimated to approximately 1.5 billion tons with about half of this is likely to be used in construction. As for wood, its production is estimated for approximately 3.5 billion m^2 with less than 50% of which is believed to be used in construction related purposes [7]. Except that the use of wood is very common as an interior material for housing, office buildings and others. These data show that concrete is the most used construction materials far exceeding in terms of the amount produced and consumed compared to the two other major construction materials steel and wood. Thought what made this possible for concrete to be the most used construction materials is mainly attributable to the fact that concrete as a whole, is made based on rocks and minerals that are mostly abundant and available in sufficient quantity everywhere.

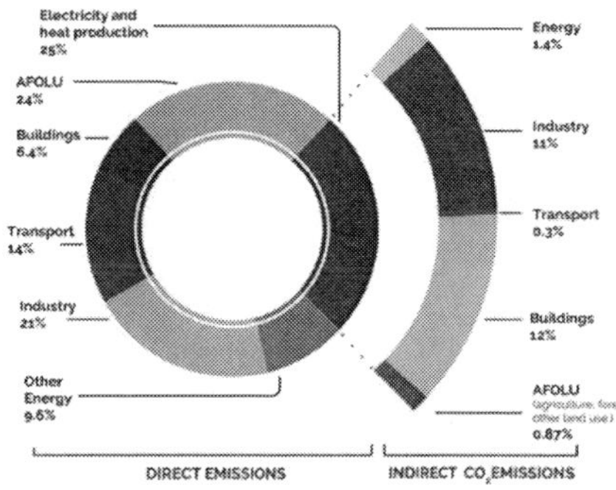

Fig.2. Direct and indirect CO_2 emissions per sector in 2010 [2]

Table 1. World steel/cement production [7]

Country	Crude steel Production (Million metric tons) (2015)	% Share of total	Cement Production (Billion ton) (2015) (from CEMBUREAU)	% Share of total
China	0.8038	49.6	2.5	51.3
India	0.896	5.5	0.280	5.9
Japan	0.1052	6.5	0.058	1.3
USA	0.0788	4.9	0.083	1.8
EU	0.1661	10.2	0.170	5.3
Indonesia	-	-	0.060	1.35
World	1.6211	100	4.6	100

The annual production of crude steel has been increasingly growing since 2006. A 30% increase was recorded for worldwide production and almost doubled for China from 2006 to 2014. Around 3% drop in the world annual steel production was noticed as compared to 2014 according to world steel association.

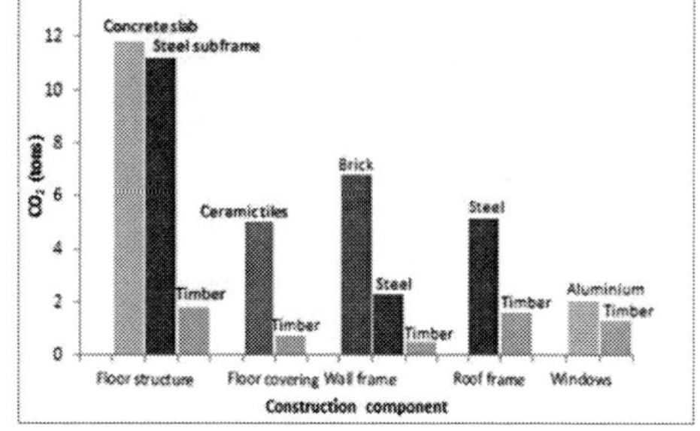

Fig.3. Greenhouse gas emission from the manufacturing of different building components in a family home [8]

Presently, energy efficiency concept is adopted to reduce operational energy requirements in the construction industry sector by selecting the suitable design and also using insulating materials. Reducing the global CO_2 emissions could be achieved also by selecting more sustainable materials that have lower environmental impact. Fig.4 showed that plastic, metal and even glass are generally requiring high embodied energy as compared to other materials such as concrete. However, the large amount of concrete used in construction sector makes it the largest consumed of energy among all construction materials. Extensive efforts are now made not only to design and operate buildings with less energy consumption but also considering other sustainability parameters. An estimated amount of energy of around 40% is used in Europe for activities related to the construction industry sector.

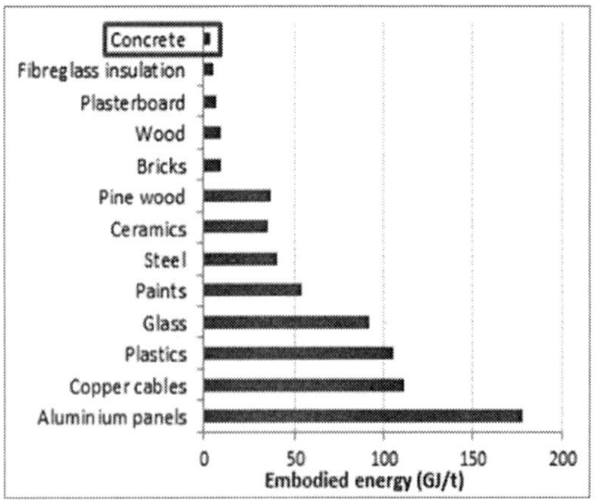

Fig.4. Comparison of embodied energy of construction materials per ton of product [9]

One of the construction industry sectors that consume a large amount of materials (aggregates) and energy is road construction.

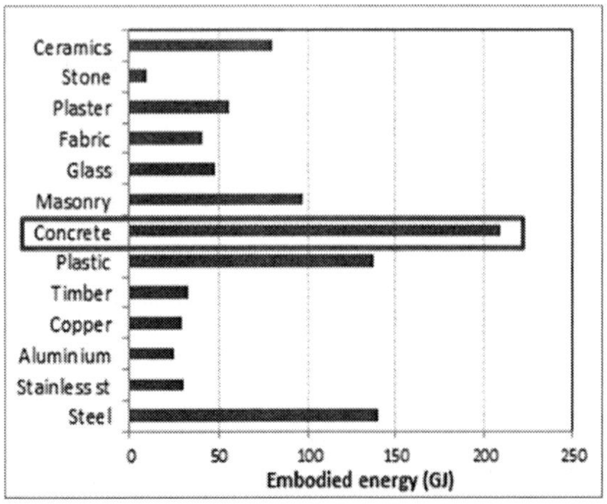

Fig.5. Embodied energy of construction materials based on their use in building [10]

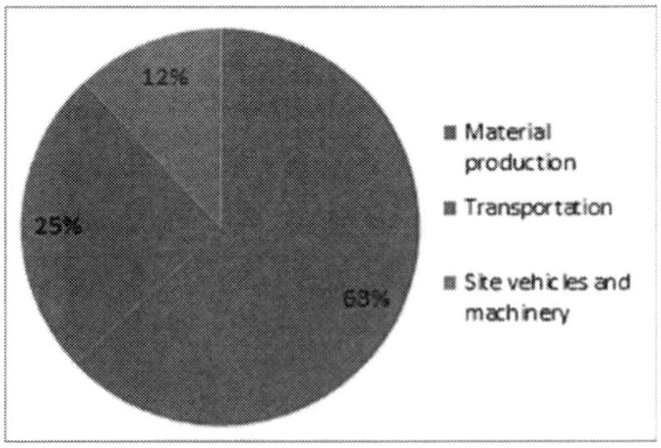

Fig.6. Relative contributions of CO_2 emissions in building roadways [11]

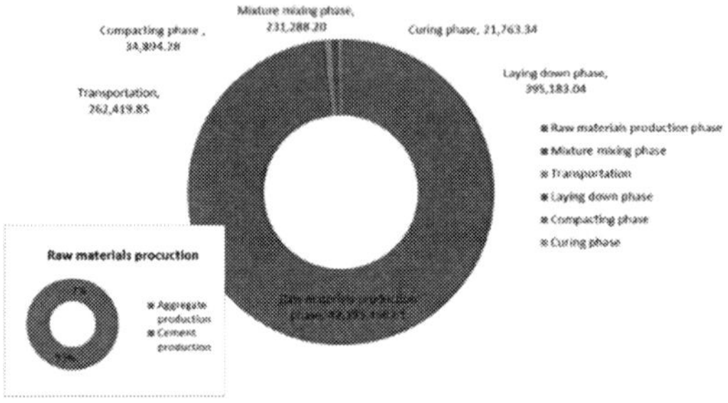

Fig.7. Greenhouse gas emission from Asphalt pavement construction in China [12]

2 Towards a sustainable concrete construction

For any socioeconomic development, buildings and infrastructures play a very important role. While developing new buildings and infrastructures has a positive effect on socioeconomic activities and grow, but also it has a significant negative impact on environment and natural resources depletion. The question, what progress and efforts have been undertaken to rationalize the use of natural resources and energy in construction industry? Still this sector is relying and consuming a very large amount of concrete, steel and wood as the three main building materials. As abovementioned, cement and concrete are the main construction materials worldwide used. Cement production has now reached more than 4.6 billion tons which is a double of its production in 2004 (2.11 billion tons). China alone accounts for 51.3%, India 5.9%, Japan 1.3%, and the other Asian countries account for 21.7%. This means that Asia actually produces more than 80% of the world's cement and also releases the 80% of the CO_2 generated by Portland cement production. In addition, construction as whole has been recognized to consume around 50% of existing raw materials and 40% of total energy but also creating 50% of total solid waste [13].

Given the rate of growth and need for new infrastructures in Asian developing countries and around the world, a direct reduction in world cement production would be difficult to achieved in the near future.

Therefore, it is necessary to reduce environmental impact by using current technology and practices as efficiently as possible and implementing some good habits in construction. Several recommendations have been made [Sakai, India] to reduce the environmental load of construction without reducing the cement production such as:

1. Lower CO_2 emissions levels in cement production to the level of developed countries.
2. Optimize the use of fly ash, blast furnace slag and other pozzolanic materials as replacements for cement.
3. Use a high-range water-reducing agent to reduce the amount of cement used.
4. Reduce the energy consumption required through improvement of construction efficiency.
5. Introduce high-performance concrete with high strength and durability to boost the strength of structural members, thereby reducing their size
6. Introduce technology to minimize energy consumption during the life cycle of a building.
7. Introduce ultra-high strength concrete to promote more advanced structural styles.
8. Use CO_2-absorbing concrete.
9. Implement optimal maintenance of buildings.

To produce concrete, with all its different types and performances, and even with considering the above measures, the use of aggregates (coarse and/or fine) in concrete production is indispensable.

For a long time and till now, virgin aggregates commonly called gravel is the main aggregate used to make concrete. Crushed aggregates are also using virgin rocks broken into fragments to obtain aggregates with different sizes. The extraction of these virgin natural resources and crushing them in quarries to produce aggregates is hugely damaging the environment by depletion of natural resources, changing the eco-system and releasing large amount of dusts and fine particles to the atmosphere.

It is recognized that huge quantity of solid wastes is generated annually as a results of demolishing old roads, buildings and infrastructures. Mostly, the concrete debris removed is often considered worthless and would be disposed in landfill. Since many decades ago, there has been attempts to collect the debris of old concrete resulted from demolition of buildings and subject it to a crushing and screening process, quite similarly to the normal aggregate, recycled concrete aggregate (RCA) is produced. The use of RCA to produce concrete for construction industry is not really a new technique but it date back to the after World War II, when there was, on the one hand, accumulation of large amount of waste generated by the excessive demolition of buildings and, on the other hand, an immense need to build new infrastructures and roads [12]. Since that time, several research work have been undertaken to assess the efficiency of using different types of recycled aggregates, including RCA as a partial or even a full replacement of natural aggregate in both structural and non-structural concrete applications. If the main reasons to use RCA in the era of post-World War II was shortage of raw materials and disposing the construction waste accumulated; the motivation now is mainly to make concrete more green and environmentally friendly material.

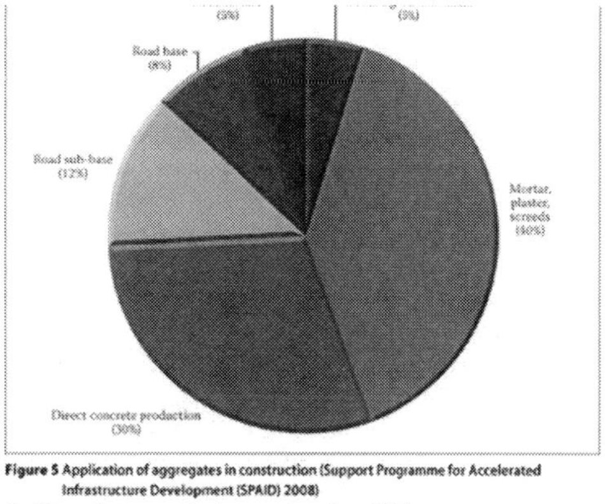

Fig.8. Use of aggregate in construction [15]

Whilst aggregate constitutes the largest component of concrete in terms of volume and lowest in terms of greenhouse gas emission as compared to cement, still its environmental impact is considerable. The environmental impact of a given material is not considered from the point of view of its greenhouse gas emission only but other types of related damages are also counted.

The main objective of this paper is to clarifying the current situation and issues of concrete production on sustainability, and proposes various uses of recycled aggregate for enhancing sustainability in construction. Furthermore, sustainable concrete construction technologies and systems are outlined, with respect to how the concrete sector, which uses a particularly large amount of material, especially aggregates in the construction industry, can contribute, and future design framework for concrete structures, by which we can promote their application, are given.

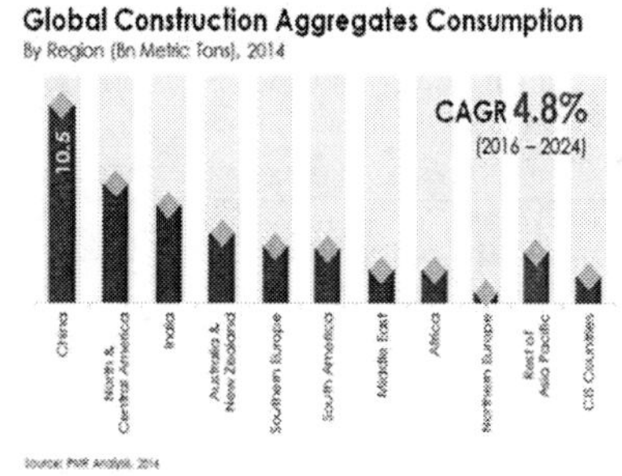

Fig.9. Global construction-aggregates-consumption-market [16]

Aggregates are the world`s most mined materials. The global market aggregate market is expected to expand by around 6% and may reach US$468.2 billion by 2021. The world construction aggregate demand for 2015 was estimated at 48.3 billion metric tons[16]. This

DOI: 10.1051/matecconf/201710105021

represents 10 times the production of cement and 30 times the production of crude steel.

3 Aggregates (coarse and fine)

Aggregates, both coarse and fine, are known to account for an average of around 70% of the total volume of concrete mixture. Large variety of aggregates sources are available including quarries, alluvial such as gravel and river sands, or recycled aggregates from different industrial waste materials mainly construction and demolition wastes. Currently and in most countries, aggregates from non-processed natural virgin resources such as gravel and sand pits or crushed rocks from quarries, constitute the main sources for raw materials and aggregates used in concrete production.

Although recycled aggregates from different sources are being used but in limited applications such as sub-base and base for pavement construction and non-structural applications still the proper use of recycled aggregate has a great potential to enhance the environmentally friendly aspect of concrete and promote concrete sustainability. The major difficulties that limit the use of recycled aggregate is the recycling facilities and equipment which requires a large investment, especially for high to medium quality of recycled aggregates. Recycled aggregates produced in recycling plants are also subjected to large variety in terms of their quality, especially if they are collected from different sources and sites. Also, standards and codes are still very conservative when it comes to use recycled aggregate in structural applications that requires certain level of durability and engineering properties. Several national standards have already included requirements in terms of using recycled aggregate for design of concrete structures (BS 8500-2, etc.).

3.1 Embodied CO_2 emissions from aggregates

Data on the embodied CO_2 emissions generated by the production of both natural fine and coarse aggregates are not readily available. The InEnergy report published in 2010 estimates that the production of one of ton aggregate including extraction and processing, generates in average 8.1 kg of CO_2 (Table 2).

Table 2. Co_2 emissions per ton of aggregate produced [17]

Activity	Energy source	Energy MJ/ton	Kg CO_2e/MJ	Kg CO_2e/ton
Quarrying	ANFO[c]	0.045[a]	0.044	0.002
Onsite transportation	Diesel	26.41[a]	0.073	1.928
Crushing, sieving & sorting	Electricity	28.80	0.119	3.43
Transportation to construction site (50 km)	Diesel	38	0.073	2.774
Total		93.3	0.309	8.1

a. Based on the assumption that diesel oil constitutes 99.9% of the energy and explosives are 0.1% during quarrying.
b. Typical transportation distances of materials to site; the capacity of the truck is estimated to be 25 t for aggregate.
c. ANFO – Ammonium Nitrate Fuel Oil
Note: The CO_2-e data on the cradle-to-gate phase of concrete constituents and concrete products.

3.2 Recycled aggregate

Properties of aggregate used to produce concrete are known to exercise a significant effect on the concrete performance. Recycled aggregates are available in large variety depending on local industries and materials used.

Fig.10. Recycled aggregates issued from various types of construction wastes

Recycled concrete aggregate, however, are the most abundant due to availability of their source and continuous demolishing of old buildings and pavements. Fig. 10 illustrates some types of recycled aggregates, and their sources, that have been employed in this study. The physical and mechanical properties of recycled aggregate are the main factor limiting the use of recycled aggregate in concrete. Lower properties are mostly resulting from using recycled aggregate as compared to their corresponding natural aggregates. Table III summarizes the main properties of three different types of recycled aggregates, namely recycled concrete aggregate (RCA), Porous ceramic aggregate (PCA) and Recycled crushed waste glass (RCG).

Table 3. Materials properties

Materials	Type	Properties
Cement	Normal Portland Cement	Specific gravity: 3.16 Surface area: 3260 cm^2/g
Chemical admixtures	Water reducing agent	Lignosulphonate Polymer
	High range water reducing agent	Polynaphtalene sulphonate
Recycled coarse aggregates	Recycled concrete aggregate	Specific gravity: 2.43 Water absorption: 5.3% Crush rate: 23.4% Aggregate size: 5-20 mm
Recycled fine aggregates	Recycled waste glass aggregate	Specific gravity: 2.49 Water absorption: 0.06% Aggregate size: 0-5 mm

3.3 Concrete design and testing procedures

As given in Table 1, three types of recycled aggregates (two coarse and one fine) were included in this study. Several concrete mixtures with various water-to-cement-ratio (w/c) were designed to elucidate the effect of types and properties of recycled aggregate used on concrete performances. For this study, compressive strength was measured on both cubic samples of 100 × 100 × 100 mm and cylindrical ones of 100 × 200 mm. Splitting tensile strength was measured on cylindrical specimens of 150 × 300 mm. Both compressive and splitting tensile strength tests were carried out at 28 days of curing in water at 20 ± 2 °C.

4 Recycled aggregate concrete performances

4.1 Strength performances

Tensile, and especially compressive strength of concrete are the two main mechanical performances required for structural design of concrete structures. The 28-day compressive and splitting tensile strengths are given in Fig. 11.

As could be seen in Fig. 11 (a) and (b), the partial or full replacement of natural coarse or fine aggregates by recycled aggregate lead to a strength reduction regardless of the type of recycled aggregate and their replacement level. The higher the replacement level of natural aggregate by recycled one, the largest the reduction in strength. This strength reduction is strongly affected by the recycled aggregate properties, their production process and the primary material from which they were recycled.

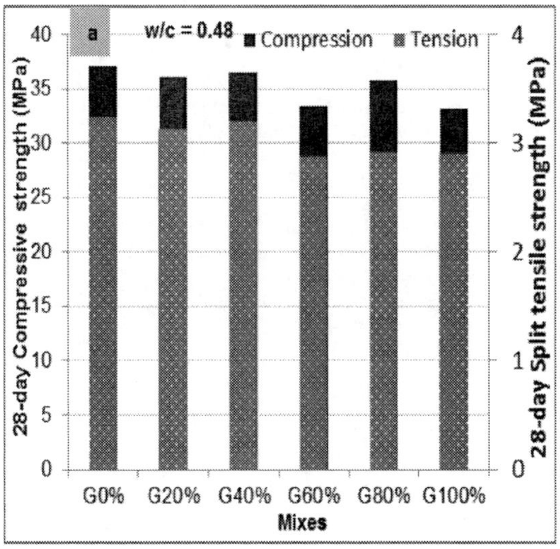

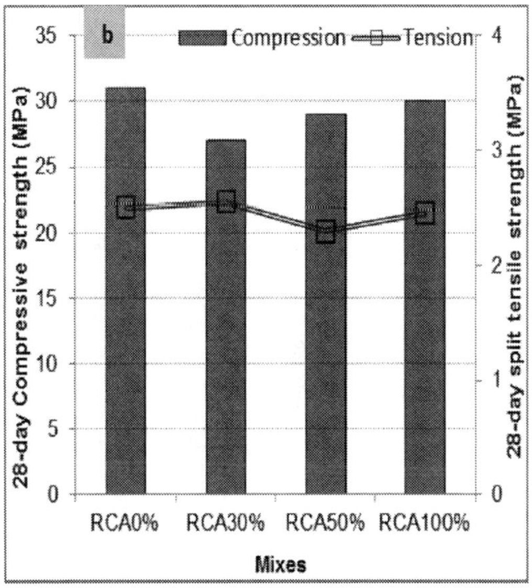

Fig.11. Effect of recycled glass fine aggregate (a) and recycled concrete coarse aggregate (b) on the 28-day compressive and tensile strengths of concrete

4.2 Durability performances

Durability performances of concrete can be evaluated by various tests and experiments. The most common tests used to assess the durability are resistance to aggressive attack such as chloride, sulfate or acids. Also, shrinkage (autogenous and drying) is one of the serious issues related to durability of concrete structures because an excessive shrinkage may lead to cracking of concrete. Cracks in concrete may lower the load carrying capacity of the structure and are also the major factor allowing penetration of harmful materials such as sulfate, chloride ions, carbon dioxide and acids.

Generally, it has been found that the partial replacement of natural aggregate by RCA could increase the drying shrinkage of concrete. The full replacement, however, of NA by the RCA may have substantially increase the total magnitude of drying shrinkage strains of concrete up to double even more as shown in Fig.12. This significant increase of shrinkage strains in concrete designed with RCA is completely attributed to the adhered old mortar on the RCA. One of the remedies proposed to overcome this excessive shrinkage strains is to reduce the volume of mortar and increase the portion of solid aggregate in concrete made with RCA.

Furthermore, total porosity and pore size distribution is one of the major factor affecting durability performances of concrete. Porosity is highly affected by numerous parameters related to both the cement matrix and the aggregates skeleton. With the same cement matrix characteristics, using recycled aggregates in partial/full replacement tends to increase the porosity of the resulting concrete. This effect is obviously function of the aggregates properties. Most RCA increase the total volume of pores due to the high porosity of the adhered mortar on RCA.

DOI: 10.1051/matecconf/201710105021

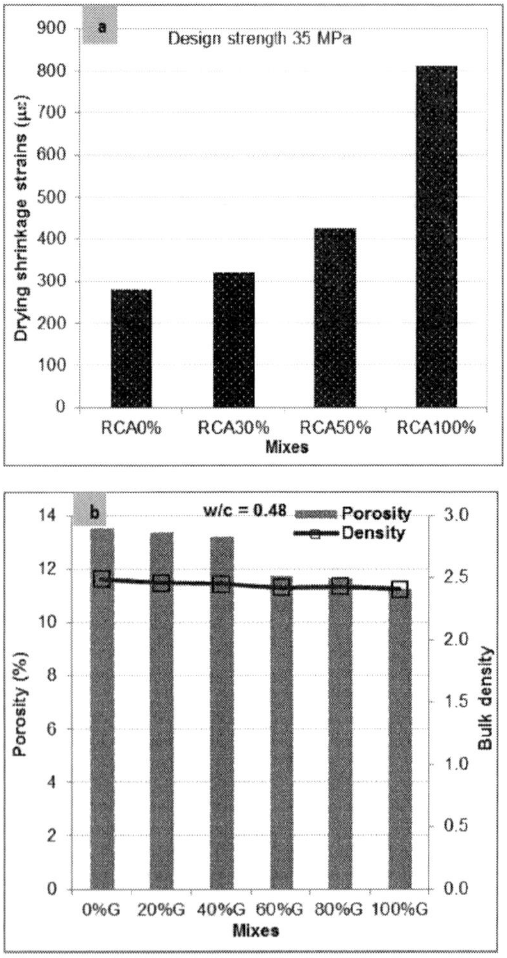

Fig.12. Effect of recycled concrete coarse aggregate on drying shrinkage (a) and waste glass fine aggregate on porosity (b) of concrete

5 Concluding remarks

Over the past few years, sustainable development and sustainability have emerged as the latest human innovation and become the motto of today's environmental protection. The concept of sustainability is now found everywhere and has been introduced as a pre-requisite for many socioeconomic activities. Construction industry has been recognized not only as one of the major contributors to energy and natural resources consumption but also as responsible of releasing a significant portion of the global carbon dioxide. The process of extracting, processing and manufacturing construction materials is causing damages to the environment. The process of manufacturing Portland cement alone is said to contribute to approximately 7% of the total CO_2 emissions. In addition to other related construction activities. Reducing clinker production and Portland cement consumption has been under the spot for a long time. Several genuine composite cements and pozzolanic/cementitious materials are now being developed and largely used in various concrete types and construction applications to reduce Portland cement consumption and hence, CO2 emissions.

Apart from cement, aggregate accounts for 65-80% of the total volume of concrete and could significantly affect the environmental load of concrete and its level of sustainability. Aggregates for concrete are whether collected as gravel or produced as a result of crushing rocks. In both cases, the excessive use of natural aggregate to respond to the high demand of construction leads to depletion of non-renewable resources. Introducing secondary aggregate from different sources such as aggregates collected from demolished buildings and subjected crushing and screening and then used in new concrete production has many environmental advantages. The experimental data available showed that properties of concrete made with RCA is depending on the properties of the RCA and the parent concrete from where they have been crushed. This concrete needs a special care and proper design parameters to obtain a nearly comparable performances to those made with natural aggregate. Overall, a moderate replacement content of around 30% should not be exceeded to avoid major negative effect of RCA on the new concrete properties. Other types of recycled aggregate have also proven to have less negative impact on strength and durability related performances such as recycled glass, and especially porous ceramic aggregate. Promoting the use of various types and contents of recycled aggregates in modern concrete may significantly reduce the amount of solid waste being disposed, safeguarding natural non-renewable resources and overall enhancing sustainability in construction industry sector.

References

1. British Petroleum (BP), Statistical review of world energy,
2. IPCC, Climate change, Syntheses report (2014).
3. National Oceanic and Atmospheric/Earth System Research Laboratory (NOAA/ESRL 2015), Trends in atmospheric carbon dioxide.
4. T. A. Boden, G. Marland, R. J. Andres. Carbon Dioxide Information Analysis Center, Oak Ridge National Laboratory, U.S. Department of Energy (2015). .
5. United States Geological Survey. USGS Mineral Program Cement Report, 44–45 (2015)
6. CEMBUREAU 2011. Available at: http://www.cembu-reau.be/
7. K. Sakai. Proceedings of the International UKIERI Concrete Congress Concrete Research Driving Profit and Sustainabilityp held at Dr Ambedkar National Institute of Technology Jalandhar, Punjab, India, on 2 - 5 November 2015pp. 3-17.
8. Cooperative Research Centre for Greenhouse Accounting, Report (2006).
9. H. Kara. Sustainable Construction Materials, Envirocities eMagazine, **10** 29-32 (2015)
10. CSIRO Commonwealth Scientific and Industrial Research Organisation, Report (2015).

11. RoadReady Newsletter, Pavement interactive, Newsletter (2012).
12. F. Ma , A. Sha , R. Lin , Y. Huang, C. Wang. International Journal of Environmental Research and Public Health, **13** 351(2016)
13. N. Oikonomou. Recycled concrete aggregates, Cement & Concrete Composites, **27** 315–318 (2005)
14. D. Buck. Recycled concrete as a source of aggregate, ACI Journal, **74** (1977) 212–219 (2005)
15. SPAID (Support Programme for Accelerated Infrastructure Development) 2008. *Research Report for the Infrastructure Inputs Monitoring Strategy.* Available at http://www.spaid.co.za/downloads/
16. Aggregates market accessed at https://www.pinterest .com/lehighhanson/aggregates/
17. World Energy Report 2010, International Energy Agency, https://www.iea.org/publications/freepublication s/publication/weo-2010.html

An analysis of activity timing and mode choice behavior for fixed time workers

Melawaty Agustien[1,*], *Ade* Sjafruddin[1], *Harun* Al Rasyid S. Lubis[1], *and Sony* S.Wibowo[1]

[1]Civil Engineering Department, Faculty of Civil and Environmental Engineering, Bandung Institute of Technology, 40132 Bandung, Indonesia

Abstract. Generally, the activity based travel demand modeling was resulted from various professions of individuals such as students, workers and non-workers. The model cannot properly represent travel behavior because their characteristics of activities timing and time allocation in a day significantly different. The purpose of this paper is to analyze travel behavior of out of home non-work activities of working groups who have fixed time working hours within a day in location study Palembang City, Indonesia. The effort to explains the travel behavior is conducted through developing activity timing and mode choice model for out of home non work activities. The activity timing and mode choice model are developed as multinomial logit model by adding the utility function of time allocation for non-work activities. There are 9 alternatives in the model in which the alternatives are the combination of 3 activity time schedule and 3 alternative modes. The result of the model significantly reveals that the characteristic of working individuals in deciding certain modes are not only caused by the mode attributes, such as travel time and cost, but also by the type and time allocation for non-working activities related to that travel.

1 Introduction

One of the advantages of the activity based travel demand model compared to the trip based and tour based approach is the ability of these models explicitly considering the time dimension in its analysis Pas and Harvey (1997), Shiftan Joram et.al (2003). Two dimensions of time are generally considered in the model is the activity timing that shows when the activity performed and time allocation shows how long the activity performed. Bhat (2009), Yagi S. and Mohammadian, A. (2006) argue that individual travel patterns is a result of the decision of the use of time to travel and activities. Research conducted by Fujiwara et.al (2010) indicates that the value of the utility of the activity based on the time allocation to do these activities can affect the pattern of travel in one day. Identification of changes in travel patterns made by individuals as a result of changes in the value of the utility of the activity can be used for prediction of the number of travel especially during rush hour morning, noon and evening Yamamoto and Kittamura (2009). Based on the research by Meloni, et al (2010), some one has to split time for activities and trips due to the time limitations in a day. In addition, modeling and analysis of the allocation of time is an important element in the development of transportation planning methodology.

This study contributes to the literature on activity and travel analysis by developing a comprehensive out-of-home non-work activity based travel demand model for fixed time workers. The periods of the day are defined based on the representation framework used by Bhat and Singh (2000) to describe the daily activity-travel patterns of workers. According to this framework and based on the temporal fixities of the work schedule, a worker's day is divided into the following three broad periods: (1) Home-to-work commute period, (2) Work-based period (3) Post home-arrival period and three modes of transport ; private car, private motorcycle and public transport. A multinomial logit model represents the alternative combination in conducting travel with respect to the aforementioned periods and modes of transport. The alternative attributes are divided into four attributes associated with the characteristics of the activities, characteristics of the travel, as well as individual and household social economic. Different from ordinary transport mode choice model, modification of multinomial logit model is conducted by substituting utility of time allocation component besides utility of travel. The utility of time allocation model is based on the utilitarian resource allocation theory, which states that an individual allocates available time to each activity such that the total utility derived from all activities will be maximized.

The next section will explain methodology and estimation procedure of the model. Data description is presented in the third section. The collected data is the result of interviews and questionnaires of the activities and travel performed in one day. The results of the

* Corresponding author: melawaty15@gmail.com

empirical analysis will be described in the fourth section, followed by a conclusion in the last section.

2 Modelling methodology and estimation procedure

The modification of utility function was used in the choice model of activity timing and transportation mode for out of home non work activities. The total utility of choosing alternatives is modeled to be the sum of the utility associated with the characteristic of travel and out of home non work activities. Individual chooses the activity timing and transport mode to maximize the utility associated with the travel and activities before and after the travel.

A unique feature of the model system of this study is that time allocation for non-work activity is analyzed in conjunction with the travel time and the time schedule of work activities before and after the non work activity. It is expected, that the separation of the travel time from the total time allotted for out of home non work activities can be used as a tool capable of capturing the direct consequences of modifications to an individual's time avaibility. For example, an improvement in the level of service of transport system will reduce travel time and the time saved can be spent in non work activities that generate trips.

The model system of non work activity and travel used to describe activity timing and mode choice is developed as a multinomial logit model based on utilitarian resource time allocation theory representing the combination alternatives between conducting non-work activities before work, after work and after returning home and three modes of transport, namely private car, private motorcycle and public transport.

The assumptions used in the model structure were developed based on the equations developed by Yamamoto and Kittamura (2009). In this study, the activity timing and mode choice are based on the utility value of the travel and out of home non work activity choices associated with the travel . The activity timing and mode utility (i) is:

$$U_i = UT_i + UA_i \quad (1)$$

$$UT_i = \gamma \; X_i + \varepsilon_i \quad (2)$$

$$UA_i = \sum_j UA_{ji} \quad (3)$$

$$UA_{ji} = \alpha_{Aji} \ln t_{ji} = \exp(\beta X_{Aji} + \varepsilon_{Aji}) \ln t_{ji} \quad (4)$$

where:

i : combination of mode used and activity timing
U_i : total utility
UT_i : the utility of the trip
γ : vector of Coefficient
X_i : vector of explanatory variable associated with the travel characteristic
ε_i : random error term of the utility of the travel
UAi : sum of the utilities of all activities associated with the alternative i
$UAji$: utility of engaging in the jth non work activity of the day associated with the alternative i
β : vector of coefficients
XAji : vector of explanatory variables
εAji : random error
tji : time spent on the jth activity associated with the alternative i

The value of parameter β in equation (4) is estimated by least squares regression assuming $\varepsilon_{Aji} - \varepsilon_{Aj'i}$ is a normal random variate with a mean of 0 and a variance of σ^2 :

$$\ln\left[\frac{t_{Aji}}{t_k - t_{k-1} - t_{trip}}\right] = \beta(X_{Aji} - X_{Aj'i}) + (\varepsilon_{Aji} - \varepsilon_{Aj'i}) \quad (5)$$

where:

t^*_{Aji} : the optimum time allocation of all non work activities for each alternatives
t_k : the time when the last activity starts after non work activivity
t_{k-1} : the time when the last activity ends before non work activivity
t_{trip} : travel time

The discrete choice of the alternative combination in conducting out of home non-work activity during the three commuting periods and the three modes of transport is formulated as a multinomial logit model assuming the random variables, ε_i, s, are independently & identically standard extreme value distributed (IID) with a Gumbel distribution. The parameter vector, γ, is estimated by the method of maximum likelihood. The probability of choosing an alternative i, Pr(i), is given as,

$$P_r(i) = \frac{\exp\left[\gamma X_i + \mu \sum_{j=1}^{J} \exp(\hat{\beta} X_{Aji}) \ln(t^*_{ji})\right]}{\sum_{i=1}^{12} \exp\left[\gamma X_i + \mu \sum_{j=1}^{J} \exp(\hat{\beta} X_{Aji}) \ln(t^*_{ji})\right]} \quad (6)$$

where:

γ : coefficient vector
X_{Aji} : vector of independent variables for out of home non work activity choice to j when the alternative i is chosen
$\hat{\beta}$: estimated coefficient vector
t^*_{ji} : optimum duration of out of home non work activity to j

There are 9 alternatives being considered in the multinomial logit model in which the alternatives are the combination of 3 activity time schedule and 3 alternative modes. The alternatives of activity time schedule are home-work period, work-based period and post home period. The alternative modes are consisted of private cars, private motorcycles, and public transport. Alternative combination of activiti timing and mode choice are explained on table I below.

Table 1. Alternative combination of Activiti timing and mode choice

No	Alternative Combination of Activiti Timing and Mode Choice	Variable
1	Before Work - private car	T_1m_1
2	Work Based - private car	T_2m_1
3	Post Home - private car	T_3m_1
4	Before Work - private motorcycle	T_1m_2
5	Work Based - private motorcycle	T_2m_2
6	Post Home - private motorcycle	T_3m_2
7	Before Work – public transport	T_1m_3
8	Work Based - public transport	T_2m_3
9	Post Home - public transport	T_3m_3

3 Data description

The data set is derived from a comprehensive survey of household demographical information, out of home non work activity and travel caracteristic that was collected in 2014 in Palembang. Palembang is capital city of South Sumatera and one of the cities in Indonesia. The survey was conducted using activity and travel diary quesioner survey method. The sampling location incorporated in this observation is spread across in 16 sub-districts in the city of Palembang. The number of samples collected are as many as 300 peoples.

The demographic data such as gender, age, occupation, household income, the number children in school age, the type and number of vehicle ownership. From the total number of respondents, 300 respondents, there were 129 male (43%) and 171 female (57%). The respondents were mostly of a working-age of 26 to 55 years old and not only have fixed time work schedules but also have non-work activities.

There are some variations in out of home non work activities, including social activities such as visiting family or friends, meetings, shopping, doing sports and entertainment. Some respondents have chosen more than one out of home non work activity. These activities were conducted in the morning before working, during their lunch break and after returning home.

Based on the survey results there were 505 out of home non work activities which are observed from 300 respondents. Figure 1 to 3 show that time allocated to perform the out of home non work activities is greater than the allocation of time allotted for travel time either the home-work, work based or post home periodes. The average of trip time during work based periode is 20 minutes. There was a relatively a short time, this was due to the shorter duration at lunch break which is ranging from 1 to 1.5 hours. The average of trip time in the post home periode for entertainment and socialization activities are 30 minutes while the average of trip time for meeting activity is 50 minutes. This phenomenon shows that each observation periods has their own travel and activity time use characteristic. This information is important to analyze the travel characteristic based on activity time allocation before and after the trip and to determine how people try to maximize utility of their time allocation to do some non-work activities and travel in one day.

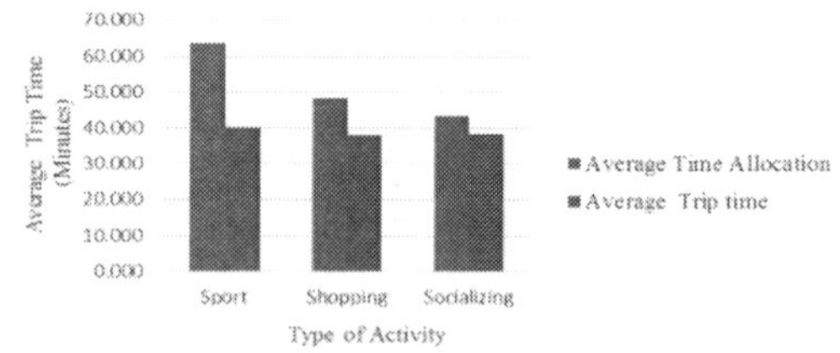

Fig. 1. Average time allocated for non work activity and travel time of before work (home-work) period (T_1)

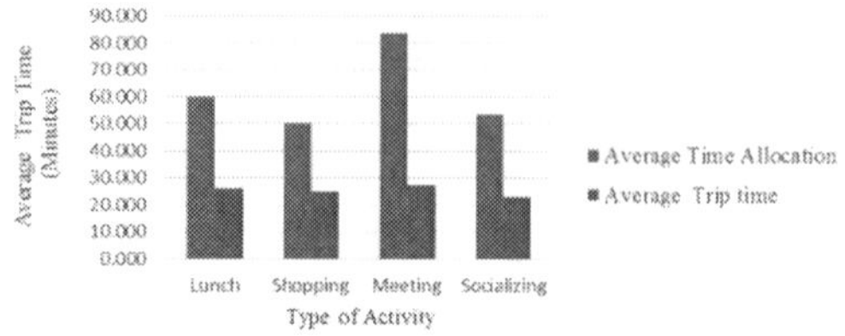

Fig. 2. Average time allocated for non work activity and travel time of work-based periode (T_2)

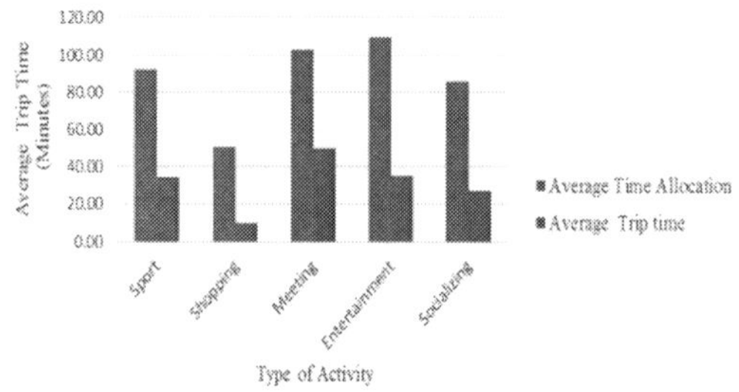

Fig. 3. Average time allocated for non work activity and travel time of post home periode (T_3)

Table 2. Number of activities based on mode choice and non work time allocation

Mode Choice	Out of home non work time allocation			Total
	< 1 Hour	1-3 Hour	> 3 Hour	
Private car	46	50	124	220
Private motorcycle	32	90	68	190
Public transport	28	41	26	95
Total	106	181	218	505

The relationship between alternative modes and out of home non work time allocation can be seen on table II. Respondents who allocate 1-3 hours for non work activities tend to use a private motorcycle, whereas respondents who allocate more than 3 hours tend to use a private car. This is caused by the type of activity and the number of family members who participated in the trip to perform these activities. The time allocation of more than 3 hours is generally used for entertainment activities that involve more family members so that car is generally used for these activity.

4 Estimation parameters of the activity timing and mode choice model

Model parameters are estimated using the household demographical information and daily out of home non work activity diary and travel caracteristic. In first phase, the data on the time allocation to out of home non workactivities are used to estimate the vector parameters β by using ordinary least-squares regression.

The time allocation of 505 non work activities are recapitulated to estimate the values of the vector parameters in equation 4. The explanatory variables in the model are summarized in Table 3.

Table 3. Explanatory variable used in time allocation model and multinomial logit

Alternative Attribute	Explanatory Variable	Variabel Name
Trip Attribute	Travel time from office	X_1
	The travel time from home	X_2
	Travel cost from office	X_3
	Travel cost from office	X_4
Individual and Household Attribute	Public Servants	X_5
	Private Employment	X_6
	Entrepreneurs	X_7
	Number of school age family member	X_8
	Household income	X_9
	Car ownership	X_{10}
	Motorcycle ownership	X_{11}
Activity Attribute	Hobby	X_{12}
	Social	X_{13}
	Entertainment	X_{14}

Not all of the explanatory variables in the model are considered significant. It is due to the low correlation between explanatory and dependent variable in time allocation model . Based on the t statistic value of explanatory variable obtained from the model, some individuals, households and activity attributes are found to significantly influence time allocation behavior. The explanation of each attributes which influences time allocation from the observed activities is outlined in table 4.

Among the explanatory variables, the number of family members attending school has a negative coefficient estimate and the largest t-statistic. It indicates that the respondents who have more family members attending school tend to have less time availablitity to do non work activities to maximize the utility of the activity time allocation.

Respondents who have a job in the private employment sector also tend to have less time availablitity to do non work activities than respondents who work as government employmees and entrepreneurs. This is due to the fixed work activity schedule for private employment so that the respective respondents have to allocate their time to do work and non work activities effectively.

Table 4. Parameter estimation of regression model of time allocation

Explanatory Variable	Variabel Name	Coefficient	t-Stat
Private Employment	X_6	-0.191	-3.979
Number of family member attending school	X_8	-0.159	-6.266
Household income	X_9	0.199	6.033
Car ownership	X_{10}	0.156	3.507
Motorcycle ownership	X_{11}	-0.058	-1.224
Hobby	X_{12}	0.003	0.068
Social	X_{13}	1.812	2.698
Entertainment	X_{14}	0.325	0.789

Household income has a statistically significant positive coefficient estimate, indicating that respondents who have a higher income tend to have more time available to do non work activities. It shows that based on the survey result number of trips is significantly influenced by household income. Families with higher income tend to do more out of home non work activities than respondents who have a lower income.

Car ownership variable also influences time allocation to do out of home non work activities. To maximize the utility of time allocation of the activities, respondents with a greater number of car ownership tend to allocate a greater expenditure of time to do the activities.

Besides individual and household attributes, activity attribut influences the utility of time allocation of non-work activities. The positive sign of variable social of doing the activity means that there is a longer time to do non work social activities.

Using the estimated parameters as presented in table IV, the optimum durations of the activities are calculated to be used in the estimation of a multinomial logit model. The estimation results are presented in table V(a) and V(b). The explanation about symbol of variabel can be seen on table I and table III. The likelihood ratio statistics indicate the model is highly significant. It means that the model can be used to predict the combination alternatives of acticity timing and transport modes based on social economic condition of individual and household, activity and travel characteristics.

The base alternative is doing out of home non work activity in the post home period using public transport. T-statistic value of each variables indicates that sosio-demographic, activity and trip characteristic variables are found to influence activity timing and mode choice. The alternative specific constant in the model shows that, relative to the using public transport at post home period, there is a greater likelihood of pursuing the out of home non wok activity in the home-work, work based and after returning home periods using car, motorcycle or public transport mode. All of the alternatives spesific constants are found to be positive and significant. The greatest likelihood occurs in the combination alternative cars–after returning home period.

Table 5 (a). Parameter estimation of multinomial logit model

Alter natif	Estimated parameter						
	C	*UAi*	X_1	X_2	X_3	X_4	X_5
T_1m_1	242.3	387.24		-206.7	-0.07	-0.05	27
T_2m_1	533.6	26.35	-114.3	-72.5	-0.01	-0.05	
T_3m_1	629.1	266.91		-5.7	-0.01	-0.03	41
T_1m_2	542.2	26.06	-53.81		-0.07		
T_2m_2	208.6	362.74	-207.6	-152.8		-0.02	18
T_3m_2	520.1	149.36	-86.4		-0.01		33
T_1m_3	281.6	281.63			-0.01		
T_2m_3	273.6	273.63			-0.02		32

Table 5 (b). Parameter estimation of multinomial logit model

Alter natif	X_6	X_7	X_8	X_9	X_{10}	X_{11}	X_{12}	X_{13}
T_1m_1	-26		-5		33	13	-5	-7
T_2m_1		41		5	47	17		4
T_3m_1	42			34	43	32	8	5
T_1m_2		79		6	43			
T_2m_2			-7	6		18	6	7
T_3m_2	19	47	57	4	62	18		4
T_1m_3	-87	27		4			4	8
T_2m_3			-7	11	53			
T_3m_3	17						9	7

Respondents in the private employment sector appear to show a greater propensity to pursue out of home non work activities in the after returning home periode with private car as a mode to conduct the activity.This may be due to the tight work schedule and the need to allocate more time to do out of home non work activity. Another significant variable is the value of utility of time allocation which has positive coefficient value for all alternatives. This means the alternative activity timing and mode of transport with the greater activity utility value tends to be preferred. Travel time from home to activity locations has negative coeficient variables, implying that the alternative activity timing and mode choices which facilitate for shorter time to get home tend to be preferred. Similarly for alternatives with a higher the travel costs which have negative coefficients, will change an individual's activity timing and mode choice.

The empirical result shows that the activity attributes which are considered in the model generate significant value. It shows for out of home non work activity destination traveller will consider the utility value of the activity when making decisssion about activity timing and transport modes for conducting the activity.

5 Conclusions

The results of the activities and trip data collection which were conducted in Palembang indicate a variation in the out of home non work activity schedule and selecting mode used to perform the activities. Activities at work-based period are dominated by lunch outside the office with public transportation as mode of transport. The activities conducted at post home period were dominated by entertainment activities using private cars.

The importance of time allocation to conduct activities is greater than trip time either in the morning before working, during the day at lunch break or post home periods. The results of utility time allocation model show that there are several variables involving individual and household social economic condition, as well as those of the activity characteristics, that influence the out of home non work activity time allocation. Some variables such as the number of family attending school and private employment as a type of job have a negative coefficient estimates and largest t-statistic values. It indicates that the respondents who have more family members attending school and private employment jobs tend to have less time available to do non work activities to maximize their utility time allocation for non work activities. The condition is may be due to the responsibility of respondents as a parents who have to take care their children and fix the time of their activity schedule for private employment. These respondents would then have to allocate their time more effectively to do work and non work activities.Similiarky for respondents who have a higher income and have more car ownership tend to allocate a greater expenditure of time for the activity. This indicate that the value of utility derived from out of home non-work activities increase in directly proportion to the level of individual and family income

Significance of multinomial logit model is showed by value of Pseudo R^2 is 0.36 which is equivalent with 0.65 in regression model. This indicates that the model is good enough to use for prediction. The significance of utility of time allocation variable and a positive value of the variable's coefficient show that the choice of activity timing and transport mode for non work activity are influenced by the utility of the activity. Respondents will choose the alternative with longer activity time allocation even though the cost of the transport mode alternative may be more expensive.

The results of the model show that the decision of choosing the activity timing and transport mode for travel-related non-work activities is not only influenced by attributes such as cost and travel time, but also by the type of activites that will be performed and the socio-economic condition of the travelers. Based on the results, it can be seen that the understanding of individuals' activity based travel behavior is more comprehensive compared to that of trip and tour based travel behavior. This understanding is thus expected to be relied upon in drafting policies that are associated with travel demand management at a particular time and location.

References

1. Bhat, C.R., and S.K. Singh., Transportation Research Part A **34**, p : 1-12, 1 (2000)
2. Bhat, C.R, et.al, Transportation Research Record, **2134**, p : 51-62, (2009)
3. Fujiwara, et. al., *Journal EASTS*, **8**, (2010)

4. Meloni, I., E. Spissu, and M. Bez., *Transportation Science* **41**, pp. 15-28, 1 (2007)
5. Pas, E.I. and Harvey, A.S, *Transport Policy*, 225-235(1997)
6. Yagi, S. and Mohammadian, A., *Proceedings of the 11th International Conference for Travel Behaviour Research Conference (IATBR),* Kyoto, Japan, (2006)
7. Yamamoto, T., dan R. Kitamura, *Transportation*, **26**, p : 211-2302, (2009)
8. Shiftan Yoram, Ben-Akiva, M.E., *10th Intrnational Confrnc of Travl Bhaviour Rsarch,* Lucrn (2003)

* Corresponding author: melawaty15@gmail.com

Floodplain simulation for Musi River using integrated 1D/2D hydrodynamic model

Muhammad B. Al Amin[1,*], *Sarino*[1], and *Helmi* Haki[1]

[1]Civil Engineering Department, Faculty of Engineering, Sriwijaya University, 30662 Indralaya, Indonesia

Abstract. This paper presents the simulation of floodplain at Musi River using integrated 1D and 2D hydrodynamic model. The 1D flow simulation was applied for the river channel with flow hydrograph as upstream boundary condition. The result of 1D flow simulation was integrated into 2D flow simulation in order to know the area and characteristics of flood inundation. The input data of digital terrain model which was used in this research had grid resolution of 10mx10m, but for 2D simulation the resolution was with grid resolution 50 m x 50 m so as to limit simulation time since the model size was big enough. The result of the simulation showed that the inundated area surrounding Musi River is about 107.44 km^2 with maximum flood depth is 3.24 m, water surface velocity ranges from 0.00 to 0.83 m/s. Most of floodplain areas varied from middle to high flood hazard level, and only few areas had very high level of flood hazard especially on river side. The structural flood control measurement to be recommended to Palembang is to construct flood dike and flood gate. The non structural measurement one is to improve watershed management and socialization of flood awareness.

1 Introduction

Musi River is one of big rivers in Indonesia flowing from Bengkulu to South Sumatra Province with a total length of about 760 km. The tides of Musi River normally cause inundation in several locations particularly around the river within Palembang which is one of cities crossed by Musi River with a low topography and relatively flat areas. Besides that, the inundation may be higher during rainfall season due to high rainfall intensities and also increased runoff which is caused by land use changes of most of undeveloped areas to be developed ones [1]. Besides that, Palembang does not have yet adequate flood control measures, such as flood dike and gates. These conditions make Palembang as the capital city of South Sumatra is prone to flood [2].

To design the flood control and management, the information like flooded area map is needed so that the area exposed to flood impact can be identified [3-5]. Unfortunately, such data and map are not available. This paper presents the simulation of floodplain of inundated areas around Musi River within Palembang City by using integrated 1D/2D hydrodynamic model. The simulation results are the estimation of the characteristics of flooded area (area, depth, and velocity) and flood hazard classification of floodplain around the river that can be used by Palembang Municipality regarding urban flood control and management program.

Flood is natural process and can be defined as water overflow over stream bank. Floodplain is an area at river side which is flooded during high water flow, but dry during normal condition [6-8]. The boundary of flooded area can be determined by 100-yr flood discharge [6, 7].

The 1D flow model in most practice is generally used in the case where flow passes through prismatic channel. But, for non prismatic channel (i.e., channel with varying cross section and alignment) such as river, flow of dam break, or lateral flow through dike failure, and river overflow, the use of 2D flow becomes better model which results in accurate estimation [9, 10]. Some of the examples of 1D and 2D in flood flow modelling were explained in detail in [11-19].

The advantages of 1D model is the ability to calculate quickly with small computer memory, whereas 2D model needs bigger memory especially for big one [18-20]. The 1D model is widely used for flood water level and flow rate simulation within channel, meanwhile 2D model is used for overland flow on floodplain [4, 33]. The channel and hydraulic structures requires the energy equation (also known as Bernoulli's equation) to calculate the head loss, while the floodplain requires the depth-averaged Saint-Venant equations to calculate multi-directional flow patterns [33]. Studies have showed the integration of 1D/2D model (1D model for river channel and 2D model for floodplain) can help calculate quickly and give accurate result in flood flow simulation specifically flood inundation [21, 22]. Besides that, 1D/2D flow simulation can provide information about spatial flood risk like the depth, inundation duration and flow velocity during the flood [23]. The illustration of general integrated 1D/2D flow model is shown in Figure 1. Some of the researches which were done previously concerning flood flow

* Corresponding author: baitullah@unsri.ac.id

simulation with integrated 1D/2D model are presented in [10, 20, 21, 23-25].

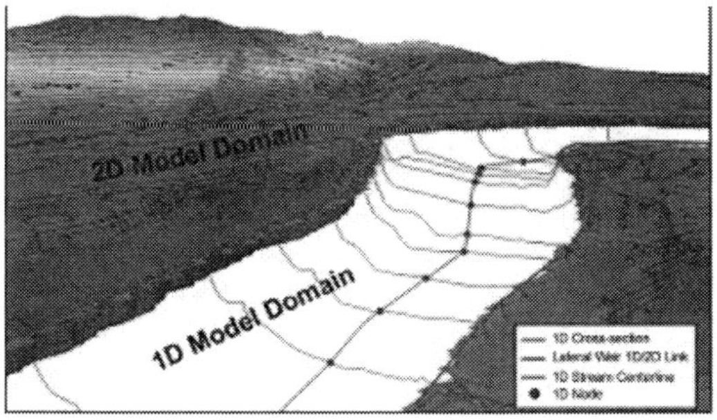

Fig. 1. Illustration of general 1D model of the river channel integrated with a 2D model of the floodplain [33]

2 Method

This research was done in Musi River within Palembang City (400,61 km^2), South Sumatra Province, Indonesia which is located between 2°52' - 3°5' South latitude and 104°37' - 104°52' East longitude as shown at Figure 2. Besides the river, there are other three big rivers flowing across Palembang namely: Keramasan, Ogan and Komering River whose flow rate and flood level were also calculated in this research. The width of Musi River crossing Palembang City ranges from 247.18 to 1,089.56 m, whereas Keramasan, Ogan, and Komering River consecutively ranges from 76.49 to 183.28 m, 144.44 to 295.65 m, and 85.03 to 189.60 m. In addition to that, there are also 18 subsystems or sub catchment areas in which the stream flow goes laterally to Musi River. Figure 3 shows the Musi River subsystems in the research location.

Most areas of Palembang City consist of housing and swamp. In 2013, about 52.56% area of Palembang was dominated by vegetated land, 25.34% built-up area, 8,4% barren land, and 13.70% water body [1]. Figure 4 shows the land use of Palembang City in detail. Topographic condition of Palembang is mostly flat area and the elevation is between 0.00 to 36.00 m above mean sea level (MSL). The average slope is 2.02 percent. Figure 5 shows the digital terrain model (DTM) map of Palembang.

Flood Modeller, a hydrodynamic model, had been released by CH2M was used in this research. The model can simulate flood inundation by integrating 1D and 2D flow model. Indeed there are other hydrodynamic models which can be used such as XP-SWMM [26, 27], TUFLOW [26], Flo-2D [28, 29], SOBEK-OVERLAND FLOW [4], Delft1D2D [21], LISFLOOD-FP [30], and the familiar HEC-RAS [10, 31]. But, Flood Modeller was applied as in the computation phase of 2D flow using solver alternating direction implicit (ADI) which was quick and stable [25].

Fig. 2. Location of research

Fig. 3. Musi subsystems within Palembang City

Fig. 4. Land use map

DOI: 10.1051/matecconf/201710105023

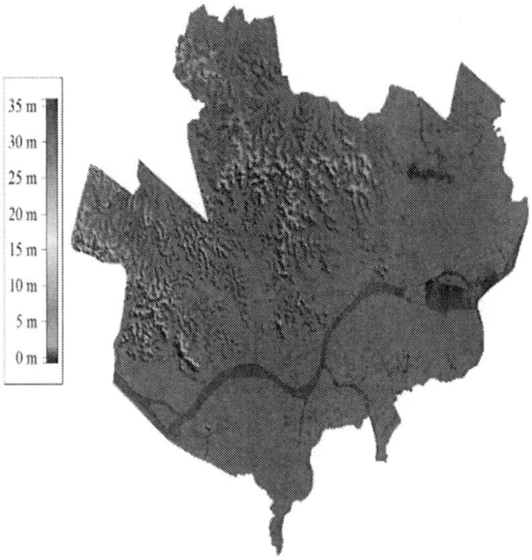

Fig. 5. Digital terrain model (DTM)

In general, phase of modelling included: 1) input of river geometry data such as river cross section, distance between cross sections, and river roughness coefficient; 2) entry of boundary conditions such as hydrograph for upstream boundary, stage hydrograph for downstream boundary and lateral inflow for boundary along the river; 3) 1D flow simulation; 4) input of 2D active area; 5) integration of 1D and 2D geometry; 6) 1D/2D model simulation; 7) visualization of simulation results.

The river cross sections were obtained by bathymetric survey with distance between cross sections of 200 – 600 m along 27 km Musi River, 2.4 km Keramasan River, 6.3 km Ogan River, and 2.5 km Komering River then the results were imposed into the model.

The flow velocities and water levels were measured in order to calculate flow rate of the rivers as upstream boundary condition, while the highest tide level was used as downstream boundary condition.

The 100-yr return period of direct runoff hydrographs from 18 Musi subsystems within Palembang were used as lateral inflows along the river. These hydrographs were resulted from hydrologic analysis.

The 1D simulation used 1 second time step within 24 hours. For 2D flow simulation, the grid size used in the model was 50 m x 50 m in order to limit time needed to do simulation due to big simulation model.

3 Result and discussion

The schematic of modelled system in Flood Modeller is presented in Figure 6, where the geometric model consisted of river sections, junctions, 2D active area, whereas flow component comprises of flow-time boundary, tidal boundary, and lateral inflow.

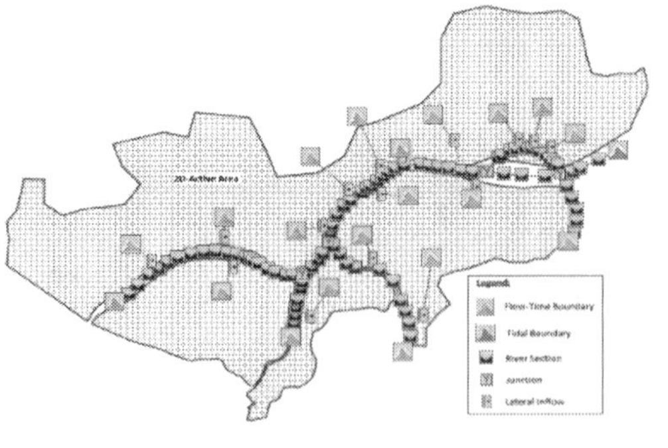

Fig. 6. Schematic of modelled system

3.1 Boundary conditions

The flow rate as flow hydrograph was used as boundary condition for upstream of the river, while the tide level as stage hydrograph was used for downstream one. The flow rates in the upstream of Musi River, Ogan River, and Komering River respectively are 1,557.74 m^3/s, 77.77 m^3/s, 302.89 m^3/s, and 77.09 m^3/s which were assumed to be constant, whereas the highest tide level of downstream was +2.95 m from mean sea level (MSL) which is also assumed constant.

Direct runoff (DRO) hydrographs from 18 subsystems along Musi River were used as lateral inflow hydrographs. The hydrographs were resulted from hydrologic analysis using SCS synthetic hydrograph with 100-yr design rainfall. The direct runoff of each subsystem is presented in Figure 7.

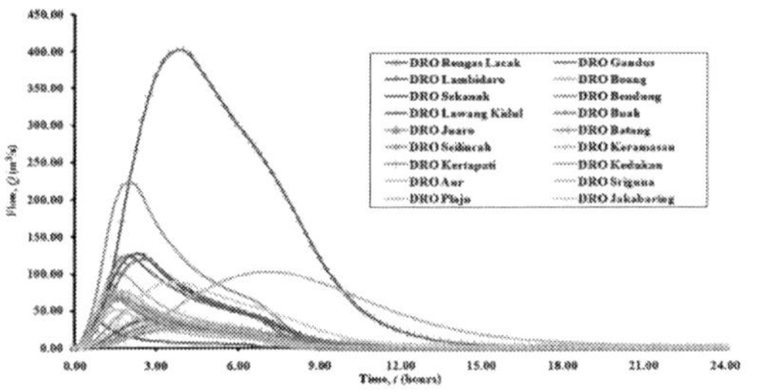

Fig. 7. Direct runoff hydrographs from subsystems

3.2 Area and depth of inundation

The result of simulation which shows distribution of inundation depth around Musi River is presented in Figure 8 to 11. Figure 8 shows the area and depth of inundation at 6th hour of the simulation, whereas Figures 9 to 11 show the area and depth of inundation at hour 12, 18, and 24 respectively.

Figure 12 shows area and depth of inundation at maximum condition. In detail, in Figures 8 to 11, it can be seen that the flooded area increasing. The maximum flooded area (Figure 12) resulted from analysis is about 107.44 km^2 with the flood depth ranges from 0.00 to 3.24 m.

Fig. 8. Flood depth at 6th hour (A = 64.89 km^2; d = 0.00 m – 3.01 m)

Fig. 9. Flood depth at 12th hour (A = 82.44 km^2; d = 0.00 m – 3.01 m)

Fig. 10. Flood depth at 18th hour (A = 94.19 km^2; d = 0.00 m – 3.01 m)

Fig. 11. Flood depth at 24th hour (A = 103.88 km^2; d = 0.00 m – 3.01 m)

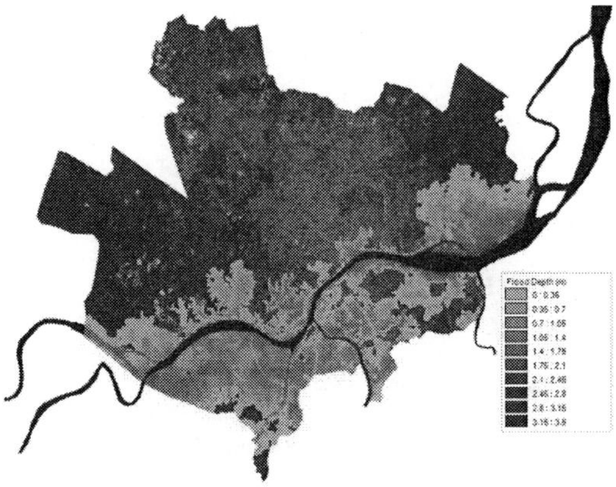

Fig. 12. Maximum flood depth (A = 107.44 km^2; d = 0.00 m – 3.24 m)

3.3 Velocity of flood flow

Fig. 13. Maximum velocity on floodplain (v = 0.00 m/s - 0.83 m/s)

The maximum velocity of the flood flow is showed in velocity vector as shown in Figure 13. The velocity ranges between 0.00 to 0.83 m/s which depends on the flood flow and water depth. The flow velocity is also influenced by the distance of certain point or location from the river. The farther the distance from the river, the lesser the velocity due to energy losses caused by friction with the bed surface.

3.4 Flood hazard

The classification of flood hazard which was used in this research is derived from [32], where flood hazard rate is as function of flood depth, flow velocity, and debris factor which can be expressed as:

$$HR = d \times (v + n) + DF \quad (1)$$

where:
HR : flood hazard rating
d : depth of flooding (m)
v : velocity of floodwaters (m/s)
DF : debris factor
n : a constant of 0.5

The debris factor varies and depends on flood depth, flow velocity, and land use. For urban land use, the debris factor was determined as:

$$DF = 0, \text{ if } d \leq 0.25 \text{ m} \quad (2)$$

$$DF = 1, \text{ if } d > 0.25 \text{ m and or if } v > 2 \text{ m/s} \quad (3)$$

The flood hazard classification was classified into four stages i.e., low, moderate, significant, and extreme. It was low if $HR < 0.75$, moderate if $0.75 < HR < 1.25$, significant if $1.25 < HR < 2.5$, and extreme if $HR > 2.5$.

The result of classification of flood hazard is shown in Figure 14. The figure shows that most of floodplain areas have moderate to significant flood hazard, while few have low flood hazard and the rest have extreme one particularly at areas exactly at riverside.

Fig. 14. Flood hazard classification

4 Conclusions and recommendation

From result and above discussion, it can be concluded that based on simulation of integrated 1D/2D hydrodynamic model, the floodplain area of the Musi River in Palembang City is about 107.44 km^2, with maximum flood depth between 0.00 to 3.24 m, and flow velocity between 0.00 to 0.83 m/s. Most of areas adjacent to Musi River have moderate to significant flood hazard level, and few areas have low one, while the rest have extreme one.

It is recommended for Palembang Municipality for urban flood control and management program to construct flood dike and tide gate to prevent overflow and backwater from Musi River as structural measurement, to control the land use and promote socialization of flood hazard awareness at flooded area as nonstructural measurements.

The authors highly appreciate Engineering Faculty of Sriwijaya University who had supported and funded this research. Then the authors also thanked Palembang City Board of Planning for having provided data.

References

1. M. B. Al Amin, Sarino, and H. Haki., *Proceeding of The 8th National Seminar of AVoER* (Engineering Faculty, Sriwijaya University, Palembang, Indonesia, 2016)
2. M. B. Al Amin, Sarino, and R. S. Ilmiaty. *Proceeding of 5th International Seminar of HATHI* (Denpasar, Indonesia, 2016)
3. S. Ghimire, *Climate*, **1**, 148-162 (2013)
4. M. G. F. Werner, Hydrology and Earth System Sciences, **8**(6), 1141-1152 (2004)
5. D. Zhang, et. al., Water Science and Engineering, **8**(3), 195-204 (2015)
6. S. L. Dingman. *Fluvial Hydraulics* (Oxford University Press, New York, 2009)
7. FEMA-Region 10. *Floodplain Management – NFIP Guidebook* (2009)
8. L. W. Mays. *Water Resources Engineering* (John Wiley & Sons, Inc., New York, 2001)
9. M. H. Chaudry. Second Edition Springer Science+Business Media, LLC, New York, (2008)
10. V. M. Quiroga, S. Kure, K. Udo, and A. Mano, RIBAGUA – Revista Iberoamericana del Agua, **3**, 25–33 (2016)
11. S. Ahmad, and S. P. Simonovic, Final Report Natural Resources Institute Facility for Intelligent Decision Support University of Manitoba, Canada, (1999)
12. R. J. Connell, D. J. Painter, and C. Beffa, Journal of Hydrologic Engineering, **6**(5), 406-415 (2001)
13. E. M. Ruji, Master Thesis International Institute for Geo-Information Science and Earth Observation, Enschede, The Netherlands (2007)
14. A. C. Cook, Master Thesis Purdue University, West Lafayette, Indiana (2008)
15. P. Vanderkimpen, E. Melger, and P. Peeters, Taylor & Francis Group, London, , 77-84, (2009)

16. D. Gilles, and M. Moore, University of Iowa, College of Engineering, (2010)
17. N. J. Lim, *Performance and uncertainty estimation of 1- and 2-dimensional flood models,* Bachelor Thesis (Akademin För Teknik Och Miljö, Högskolan I Gävle, 2011)
18. L. T. Chargel, and M. A. G. Massera da Hora, Global Journal of Researches in Engineering: E, Civil and Structural Engineering, **14**(2), 67-75 (2014)
19. M. Gharbi, A. Soualmia, D. Dartus, and L. Masbernat, J. Mater. Environ. Sci., **7**(8), 3017-3026 (2016)
20. J. Leandro, A. S. Chen, S. Djordjević, and D. A. Savić, Journal of Hydraulic Engineering, June, 495-504 (2009)
21. E. Frank, A. Ostan, M. Coccato, and G. S. Stelling, Ecology and Environment, **50** (2001)
22. D. Dutta, J. Alam, K. Umeda, M. Hayashi, and S. Hironaka, Hydrol. Process., **21**, 1223-1237 (2007)
23. X. Fang, and D. Su, Journal of American Water Resources Association, June, 713-724 (2006)
24. A. R. Faz, A. Meller, and G. B. L. Silva, *12nd International Conference on Urban Drainage* (Porto Alegre/Brazil, 2011)
25. R. Ahmadian, R. A. Falconer, J. Wicks, Journal of Flood Risk Management, 1-15 (2015)
26. B. C. Phillips, S. Yu, G. R. Thompson, N. de Silva, *10th International Conference on Urban Drainage* (Copenhagen/Denmark, 2005)
27. M. Morita, Water, **6**, 253-270 (2014)
28. J. S. O'Brien, P. Y. Julien, and W. T. Fullerton, Journal of Hydraulic Engineering, **119**(2), 244-261 (1993)
29. J. S. O'Brien, and W. T. Fullerton, *USDA Forest Service Proceedings RMRS-P-7*, 52-60 (1999)
30. M. Wilson, P. Bates, D. Alsdorf, B. Forsberg, M. Horrit, J. Melack, F. Frappart, and J. Famiglietti, American Geophysical Union, **34**, 1-10 (2007)
31. G. Di Baldassarre. *Floods in a Changing Climate* (Cambridge University Press, New York, 2012)
32. Defra, R&D Technical Report FD2320/TR2 (2005)
33. D. Gilles, et. al., Water, **4**(1), 85-106 (2012)

Impacts of long-span bamboo structure towards architectural-spatial experience

Revian Nathanael Wirabuana[1,*], *Anastasia* Maurina[1]

[1] Architecture Department, Faculty of Engineering, Universitas Katolik Parahyangan, 40141 Bandung, Indonesia

Abstract. Bamboo had recently been considered a significantly reliable material in constructing long-span structures due to its flexibility, durability, reduced dead weights, as well as its availability in most tropical regions. Contemporary architects favor its usage in constructing long-spanning structures as it allows freedom in designing spaces enclosed beneath it and creates a profound impact to the spatial experience of the building user. This research aims to provide an understanding regarding the impacts of bamboo structure to the spatial experience that the structure creates, through a description and comparison of several case studies. The research concludes that bamboos are mostly used in a planar truss system due to the feasibility in constructing its joints. Truss system made it so that bamboo members are placed within the structure in accordance to the forces that it needs to distribute and withstand, both vertically and laterally. The organization of multiple linear elements of a bamboo truss gave rise to a sense of orderly, organized complexity on the building which profoundly contributes to the spatial experience of its user.

1 The trend of bamboo architecture

Bamboo was a primary component of architecture in vast regions of South East Asia, Africa, and South America. The abundance of bamboo in that region on account of its tropical climate had given rise to a culture of architecture that revolves around the evergreen plant. Its usage had then significantly declined as technology developed. Bamboo buildings slowly changed to wooden constructions prior to colonization era. Industrial revolution introduced concrete and steel structures, which largely influenced the design of buildings in latter days.

However, bamboo had recently enjoyed a new era of global attention. The increasing level of concern towards climate change had forced architects and engineers to employ more sustainable building materials. Subsequently, a handful of them turned their attention to bamboo. Bamboo are known to have decent tensile and compressive strength but with significantly less economic and environmental drawbacks (carbon footprint). Further, bamboo grows to full size at a much faster rate than that of wood, making it much more renewable [1]. Such features are now driving architect to produce innovations revolving the integration of the traditional material to the contemporary style.

The bamboo trend grew exponentially faster in the Asian region, the place where it used to thrive in traditional era. Abundance of bamboo obviously contributed to the matter, just like it did in the past. However, another possible reason to the resilience of bamboo trend is the society's strong grasp on cultural heritage. That is to say that in spite of the rising popularity of modern construction materials, the knowledge of indigenous bamboo engineering are still passed down through generations, albeit in rural areas and villages. Perhaps it was this local wisdom that made it easy for Asian-based architect to adopt bamboo engineering into their contemporary designs.

As bamboo architecture grew and expand, its design are no longer used solely for its economic-and-environmental-sustainability. Bamboo had been an attraction on its own merit. Bamboo may have an inherent, unconventional beauty unique to it. That trait drove architects to compete in the creation of grand designs using the almost-forgotten material.

This research aims to analyse the trait of bamboo and the impacts of a bamboo structure to the architectural quality of a building. The method by which this research is conducted is an analysis on three case studies. Each of these case studies will be assessed based on five primary aspects that is considered to signify architectural impacts, namely *building exterior, interior structure, building function, light, and symbolism.*

2 Understanding the role of structure in architecture

There was a time in which a building's structure was considered a purely a load-bearing element to be judged separately from the building's façade and interior; supposedly a better representation to the aesthetics element of a building. However, the last few decade of architectural growth have shown significant improvement in the way people perceive aesthetics. In

* Corresponding author: revianwirabuana@gmail.com

light of that change, Bjorn Sandaker argued that a structure, being an object existing in space, will inevitably influence the space it exists in, consequently possessing a *spatial* role [2]. These spatial roles were then divided into several aspects which are, among others, *contextuality*, and *utility*.

1. *contextuality*; that a structural element interacts to the architecture of the building it supports
2. *utility*; that a structural element contributes the functionality of the space it exists in

That proposition can be compared with an earlier argument made by Andrew Charleson. His argument revolved around the role of structure as a part of architecture. He proposed, among others, five possible roles; namely *exterior, interior, light, and function* [3].

1. *Building exterior*; that a structural element can be an integral part of a building's façade by modulating, shading, or texturizing it.
2. *Interior Structure*; that a structural element can form compositions within a building's enclosure.
3. *Structure and Light*; that a structural element may contribute to lighting either by allowing light to enter the building, or by obstructing its entry to the building.
4. *Building Function*; that a structural element directly contributes to the building's intended function in accommodating activities by compartmentalizing spaces or articulating pathways.

On balance of both aforementioned arguments regarding the roles of structure, Charleson's proposition on architectural role can be considered as an elaboration to the specifics of Sandaker's proposition regarding spatial roles. The combination of the two arguments can be a basis of assessing the multiple variables of possible architectural-spatial impacts of a structural element, as explained in Table 1.

Table 1. Variables for Assessing Spatial-Architectural Impacts.

Sandaker	Charleson	Parameter
Contextuality	Building Exterior	The structure interacts with the the architecture of the building by being an integral part of the facade
	Interior Structure	The structure communicates with the users of the building by displaying compositions within the enclosure
Utility	Building Function	The structure contributes to the functionality of the building by accommodating its activities
	Structure and Light	The structure contributes to the functionality of the building by providing its users with natural lights

3 Characteristic of bamboo as long-span structure element

Bamboo's stem are hollow in the middle, making it fairly light in weight. Consequently, its stability is derived from the ring-shaped culm instead of the entire section. This fact proves to make bamboo a much more stable structure in comparison to wood as it has weaker outer-parts and firmer inner-parts instead of the other way around [4]

Fig. 1. Culm Section
(source: bamboobotanical.ca)

Fig. 2. Cross Section
(source: flaviodeslandes..com)

Hundreds of bamboo species are available as option for construction. There are 1012 species of Bamboo available in Asia-Pacific region alone [5]. The height of each species ranges between 2 meters to an astonishing 35 meters. Diameters of the trunks also ranges between 5 centimeters to 30 centimeters. It needs to be noted that the trunk are larger at the base and much smaller at the peak.

Given the limitation of its form in nature, multiple members of bamboo should be joined to allow freedom of form and span. The results of conjoining bamboo members would be bamboo monobeams or trusses. Standard joints employ bolts and knuckles in order to lock two, crossing bamboo members forming a fixed joint. However, other methods were also invented such as tying the members together to avoid breaking the bamboo due to the stress carried by bolts [6].

4 Long-span bamboo structures

4.1 The great hall, OBI eco campus, Jatiluhur

The Great Hall of OBI Eco Campus is an assembly hall of Outward Bound Indonesia facility. The structure spanned 20 meters across in the form of portal frame with straight elements [6].

Fig. 3. Interior
(source: outwardboundsindo.org)

Fig. 4. Exterior
(source: outwardboundsindo.org)

The portals are formed of joined bamboo members, forming a planar truss in which each members transfers compressive and tensile loads axially. The members were joined together using bolt-knuckle joints, strengthened by palm ropes knots. In addition to the trusses, bracings were positioned between each portals to increase lateral stability [6].

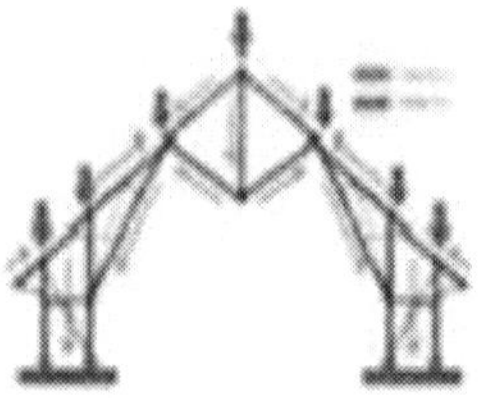

Fig. 5. Trussed Portal System
(source: Maurina et al, 2014)

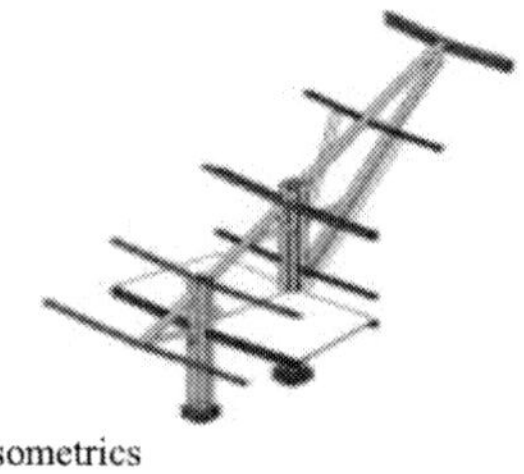

Fig. 6. Truss Isometrics
(source: Maurinaet al, 2014)

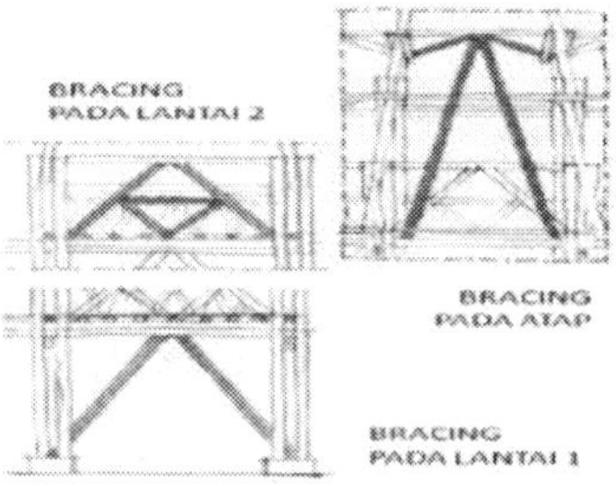

Fig. 7. Bracings
(source: Maurina et al , 2014)

Fig. 8. Joineries
(source: Maurina et al, 2014)

4.2 Mandala Agung, Puri Ahimsa, Bali

Mandala Agung[1] is a meditational facility located in the Puri Ahimsa Resort compound. Its span is derived from a ribbed-shell structure system, formed by seven vertical arches bound by horizontal hoops [6].

Fig. 9. Interior
(source: rooang.com)

Fig. 10. Exterior
(source: rooang.com)

Each of the arches were made of a truss. The trusses were configured in radial order, with a crown at the apex of the structure which unites all of the trusses. Singular bamboo hoops provided the shell with its lateral stability. Knuckle and bolts were used to join bamboo members together [6].

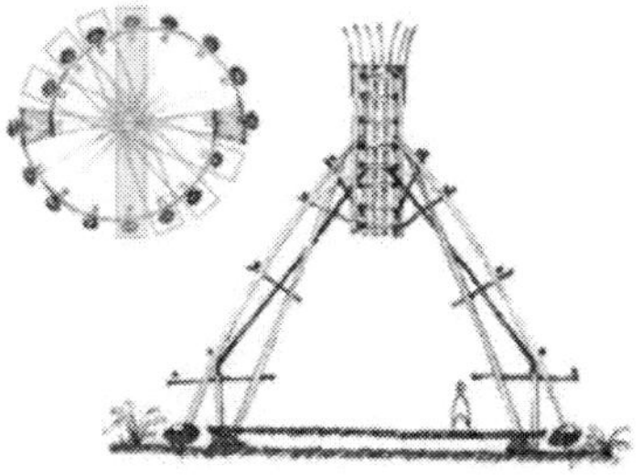

Fig. 11. Trussed Arch
(source: Maurina, Enggarsari, Krisanti, & Adhisaksana, 2014)

Fig. 12. Crown
(source: Maurina, Enggarsari, Krisanti, & Adhisaksana, 2014)

Fig. 13. Hoops
(source: Maurina, Enggarsari, Krisanti, & Adhisaksana, 2014)

Fig. 14. Joineries
(source: Maurina, Enggarsari, Krisanti, & Adhisaksana, 2014)

4.3 Wind & water bar, Binh Duong

Wind & Water Bar is a thatched dome structure built in the middle of a lake in Binh Duong Province of Vietnam. Similar to Mandala Agung, this building employs a ribbed-shell structure system, formed by vertical arches bound by horizontal hoops which forms its 15 meters-spanning dome.

Fig. 15. Interior
(source: inhabitat.com)

Fig. 16. Exterior
(source: votronghia.com)

Each of the arches were also made of a truss. The trusses were configured in radial order, with a compression ring uniting all of the trusses while serving as a skylight. Singular bamboo hoops provided the shell with its lateral stability with the help of bracings. Palm ropes were used to join bamboo members together.

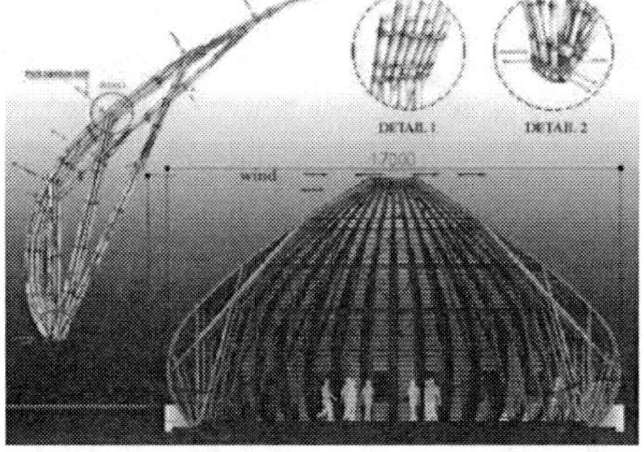

Fig. 17. Trussed Rib
(source: dezeen.com)

Fig. 18. Trusses
(sumber: floornature.com)

Fig. 19. Bracings
(sumber: 10plus1.jp)

Fig. 20. Joineries
(source: inhabitat.com)

4.5 Comparing the structures

The comparison provided above shows that most of the building employs trusses as their local structural form, most likely in account of making the most of bamboo's limited length and form. In forming these structural elements, joineries were generally employed by the buildings in order to allow a large, stable structural span.

Table 2. Structural Comparison

Building	Local Structure	
	Elements	**Joineries**
The Great Hall	Planar Trusses, supported by v-shaped bracings	Bolt & knuckle, palm rope joints
Mandala Agung	Trussed Arches, Trussed Ribs, supported by hoops	Bolt & knuckles
Wind & Water	Ribbed shell, Trussed Ribs, supported by hoops and bracings	Palm ropes joint

The decision to employ trusses also coincides with bamboo's natural ability to carry axial-loads. That is to say that a trussed structure would allow bamboo to transfer loads axially even when forces were globally perpendicular to the structure. Hence, the bamboo structure is capable of spanning impressive distances while carrying huge amounts of load.

5 Spatial experience in a bamboo structure

The research concludes that bamboo structures, as seen in the aforementioned buildings, have adverse impact to the architectural experience of those perceiving it. These impacts can be seen in the following aspects.

5.1 Building exterior

Repetition of vertical structural element can be seen in the discussed bamboo buildings, with some having significantly vast numbers of columns standing very close to each other.

This trait can be attributed to bamboo's lesser capability in handling compressive load in comparison to other material. The buildings should compensate for this deficit by increasing the number of elements supporting the building. Hence, the increased number of structural elements standing close to each other in bamboo buildings.

The presence of repeated elements in rhythm, according to Francis D.K. Ching, is one of the method to establish order in a composition [8]. Had this rhytmic array of structural element been exposed in the building's facade, it may serve as an impactful element to the building's exterior.

However in practice, two of the three buildings of discussion was designed not to expose its structural elements. This decision may be attributed the need to protect bamboo from the effects of weather (i.e. rainwater).

DOI: 10.1051/matecconf/201710105024

Fig. 21. Facade of The Great Hall
(source: outwardboundindo.org)

Fig. 22. Facade of Mandala Agung
(source: rooang.com)

Fig. 23. Facade of Wind & Water Bar
(source: arquitecturaviva.com)

In conclusion, buildings with bamboo structure does not have as much freedom in exposing its structure to the exterior as buildings with other, more weather–resistant materials for structure such as concrete.

5.2 Interior structure

While repetition of structures were rarely made visible from the outside, the array of structures were usually a dominant aspect in the interior of bamboo buildings, as seen in the interior of the three buildings in discussion.

The presence of repeated elements in rhythm, according to Francis D.K. Ching, is one of the method to establish order in a composition [8]. Had this rhytmic array of structural element been exposed in the building's facade, it may serve as an impactful element to the building's exterior.

The most noticeable pattern, however, is the vaulted shapes of the three buildings. A vaulted or arched shape carries with it the capability to convert tensile stresses into purely-compressive loads [9]. This unique trait was the reason why arched structrues, also called 'form active' by Angus Macdonald [10], were often employed to form long-spanning buildings. Bamboo, being efficient in carrying loads axially, benefits from having form-active or semi-form-active shapes.

The point of apex is often the orientation to which other structural elements (e.g. rafters, girders, hoops, bracings) refer to in thir positioning. Thus, the apex sets order to the structure, the way a datum that establishes order in a composition as argued by Francis D.K. Ching [8].

In conclusion, buildings with bamboo structure has the capability to form rhythmic composition in its interior, often oriented by the apex of the long-spanning building.

5.3 Structure and light

The apex of the spanning structures played another significant role in the spatial experience of the perceiver of the building ; the integration of natural lighting. Both Mandala Agung and the Wind & Water Bar sports a crown ring at the apex of their dome-like forms, above the center of the concentric configuration of the building.

Fig. 24. Skylighting in Mandala Agung
(source: fiveelement.org)

Fig. 25. Skylighting in Wind and Water
(source: votronghia.com)

The integration of skylighting comes naturally for both of those buildings. As both building has a dome-shaped form, the building will most likely experience compressive hoop stresses at its crown [11]. In account of that stress, the buildings would require the presence of a compression ring to carry the hoop stresses. Therefore the skylight serves multiple purposes in both buildings.

The shape of skylighting is different in The Great Hall ; it is linear in form and primarily made of the positioning of the thatch roof as opposed to the bamboo structures. The skylighting crevice was positioned on the point in the bamboo truss which carries the lowest amount of moment force [7].

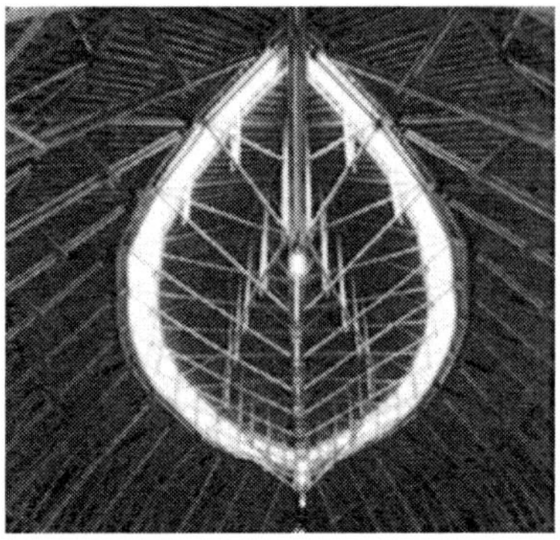

Fig. 26. Skylighting in The Great Hall
(source: outwardboundsindo.org)

The skylights accentuate the apex of the building, highlighting it as more than just a datum in the composition of structures, but also as hierarchically higher point in the composition of the structure. The establishment of hierarchy is also argued by Francis D.K. Ching as a key principle in establishing order amidst a composition [8].

As importantly, skylighting obviously contributes to the atmosphere felt within the space that the building encloses. Natural light offers higher degree of visual comfort to building users in comparison to artificial lighting.

In conclusion, bamboo structure offers vast possibilities of integration with skylighting. SUch intergation would improve the composition of structural elements as much as it increases the visual comfort in a particular building.

5.4 Building function

The spanning structure allows for a higher degree of freedom for activities. Both The Great Hall and Mandala Agung serves a an all-purpose meeting room, which requires the room to be free of obstructions oftenly caused by vertical structures. The Wind and Water Bar serves as a restaurant, which also reuires as little obstruction as possible in order to accomodate the circulation of customers and waiters alike.

Integration of bamboo structure with the building's function is especially significant in The Great Hall. Each of the portal frames were more than two-stories in height. The portals were used to house a second floor that serves as balconies to the meeting hall. The portals further accentuates the linear balcony by framing them.

Fig. 27. The Linear Balcony Framed by the Portals
(source: votronghia.com)

6 Conclusion

In summary, the role of structure is to contribute to the architecture of the building it serves. At some specific aspects, bamboo might not be able to do that role. For instance, with the currently available technology, bamboomight not be an appropriate material to contribute to the building's exterior due to the material's need to be protected from weather effects.

That being said, a structure made of bamboo still has the potential to bring unique impact to its perceivers. They have, among others, the capability to establish order in the composition of structure as seen from within the enclosure, allow skylighting to be integrated to the building's design, as well as allowing the building to better accomodate various functions and activities.

References

1. Bowyer, J. et al., *Bamboo Products and Their Environmental Impacts: Revisited*, 12 (2014)
2. Sandaker, B. N. *On Span and Space*, 5 (2008)
3. Charleson, A., *Sructure as Architecture,* (2005)
4. Rottke, E., 2002. *Mechanical Properties of Bamboo*, (2002)
5. Bystriakova, N., Kapos, V. & Lysenko, I., *Cambridge*, 5 (2001)
6. Frick, H., *Ilmu Konstruksi Bangunan Bambu*, 26-24 (2004).
7. Maurina, A., et al., *Journal Unpar,* (2014)
8. Ching, F. D. K., 2007. Form, Space, and Order, Third Edition, New Jersey: John Wiley & Sons, Inc..
9. Vaidyanathan, R. *Structural Analysis,* **2**, 127 (2015)
10. Macdonald, A., *Structure and Architecture*, (1994)
11. Schodek, D.L., *Structure,* 389, (1979)

MATEC Web of Conferences **101**, 05025 (2017) DOI: 10.1051/matecconf/201710105025
SICEST 2016

A topology of residents' based on preferences for sustainable riparian settlement in Palembang, Indonesia

Maya Fitri[1,2], *Sugeng* Triyadi[2] and *Ismet* B Harun[2]

[1]Architecture Department, Universitas Sriwijaya, 30662 Ogan Ilir, Indonesia
[2]School of Architecture, Planning, and Policy Development, Institut Teknologi Bandung (ITB), 40132 Bandung, Indonesia

Abstract. The riparian function as an ecosystem service is very important. However, town planning with the sustainable of ecological riparian approach is often in conflict with public interest. Planner needs to harmonize planning with community preference. Several studies show that conjoint analysis is capable of capturing public opinion as part of town planning. In this paper, in addition to residential preference research, a topology of resident profiles was also identified. This study demonstrates that sample size can break the overall average results into particular group characteristics. The analyses are composed through two-step approach. First, cluster analysis to categorize residents as their preference settlements and conjoint analysis to know the ideal settlement to each group. There are 150 respondents in a slum settlements of Musi River in Palembang, Indonesia. The cluster analysis identifies four respondent groups and all of them prefer house building as the very important attributes rather than residential environment attributes.

1 The principle of sustainable development in urban riparian

Riparian has an important function as a transition area to maintain a balance between two ecosystems: water and land. It is a wetland located in floodplains that directly affected by the river waters tidal and the surface flow occurs in this area [1]. Riparian has an ecosystem service to improve water quality [2], control the natural water cycle [3], and maintain diverse aquatic habitat [4]. In addition, the riparian landscape presents unique ecological character of the city [5] and it reflects the culture and characteristics of waterfront city society [6].

The quality of ecology and community's culture coexists. The loss of riparian water culture threatens the riparian environment and, this ecological damage also degrades the quality of community life [7]. The concept of sustainable development was developed to promote sustainable living in response to an economic and social pressures created by demands on the nature of human society. The balance of it is important to protect, enhance, or restore ecosystems. It should be assessed across the wide range of riparian ecosystem services to deliver tangible and quantifiable value of societal interests [8]. Therefore, the consideration of human preferences and behaviors is an important component of the science behind sustainable planning [9].

Palembang is a city that has low, flat topography and hundreds of rivers. The surface water flow spreads along floodplain forming wetlands area along the riverbanks. The city culture is formed in a unique ecosystem composed by harmonizing water and daily community activities with the riparian landscape. However, currently this riparian ecosystem has begun to disappear. The main problem is the rapid increase of new housing area built in the riparian. The reduction of natural riparian degrades the ability of ecosystem service for city's urban live [10].

The studied settlement attributes can be divided into two components, i.e. building and environment. The only building component is residential type. The sustainable development needs to decrease building mass in the riparian built-up area to enlarge natural green space area. The building foundation in riparian should use the open for daylighting such as piling, piers, floating. The building should raise public awareness on city landscape with riparian architecture character. The next attribute is green buffer area. It is residential distance from the river edge. Buffer area provides natural space along streamside and preserves wetland natural spaces. The third attribute is riverside construction. Ideally, riparian edges should remain without flow and tidal blockage for floodplain area. This option provides the conservation of riparian ecosystem service but it conflicts with the needs of communities to be free from flooding. The fourth attribute is accessibility. Public access will promote the significance of a riparian ecosystem as an important character element of urban landscape [11]. The last attribute is riverside open space. Riparian zones perform a range of functions with economic and social value to people. Riparian buffer area has potential space for urban activities or ecological conservation.

Table 1 List of attributes and their levels

Attributes	Level attributes
Residential type	1. Floating house 2. Stilt row house 3. Flats
Distance to river edge.	1. 0-10 m 2. 10-20 m 3. 20-30 m 4. > 30 m
Riverside construction	1. Natural floodplain 2. Polder 3. Waterfront platform
Accessibility	1. Boat dock 2. Promenade
Open space topology	1. Open green area 2. Riparian park with trees 3. Public playground field

This study addresses a better understanding of people preference of sustainable development residential profiles in riparian. The paper combines the use of cluster analysis in segmenting resident and a conjoint analysis to reveal their housing preference.

2 Methodology

2.1 The conjoint analysis

The conjoint analysis provides a robust method to understand the problem of multi attribute nature. The trade-offs involved several interactive variables in value changes of natural resources and environment [12]. In the conjoint survey, respondent chose residence like selecting housing brochures in property market. The conjoint model forces respondents to consider trade-offs among desirable and undesirable measures for the attributes rather than consider the importance of each attribute separately. The use of trade-off provides the important insight of type attributes residential that might motivate individuals to accept a planning of development [13].

The conjoint analysis method is one of the techniques to understand people behavioral intentions in housing preference. The residential and its environment have characteristics that can be seen as a choice of attributes that stimulates the cognition and behavior of a person [14)]. The attributes is measured by category and function usefulness that contains allegations part-worth or also called utility function. The conjoint analysis observes the part-worth utility of each level of the attributes that have the significance value in influencing decisions of a person's preference. The attributes value is different for each person. The more important the value the greater attention to it will be. Analyses such as regression and logit are usually used to determine the effect of utility. The procedure used in this study is the method of regression with dummy variables.

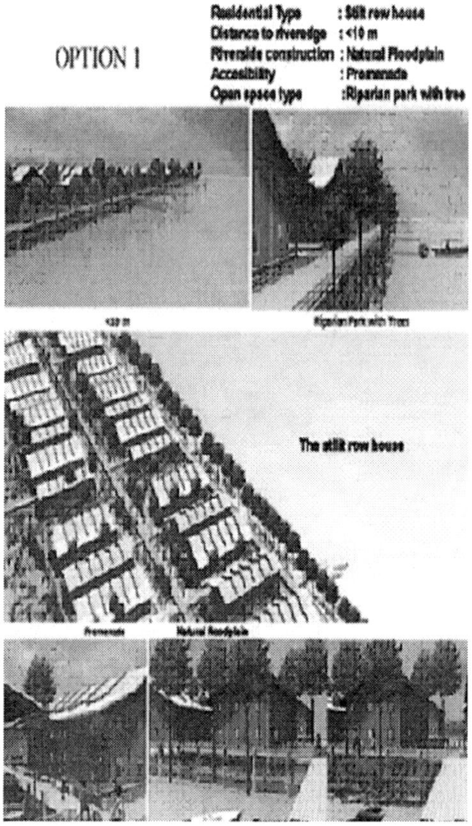

Fig. 1. An example of a residential profile set in conjoint rating question

This study uses part-worth utilities or coefficients of each respondent's utility function for segmenting respondents' preferences by means of clustering techniques. This approach focus on better representing heterogeneity in people preferences. Cluster analysis separates objects into several groups based on different characteristics among those groups. This analysis formed a group with homogeneous character among group members or similar to possessing character with as small variation as possible.

2.2 Design of residential profiles

The survey consists of five important attributes about water environment in residential area. These attributes were reselected by in-depth field observation and pre-tested survey among 35 arbitrarily selected respondents. Those five attributes were used in the study and each attribute was assigned two to four levels (see Table 1). Full-factorial experiments generate data with five attributes and fifteen levels is 3x3x3x2x4=216 based on all possible combinations of attribute levels.

That large number is impractical for surveys because subjects' cognitive and time had limitations. The conjoint survey can use orthogonal fraction factorial experimental design to reduce the alternatives. It minimizes orthogonality, the occurrences of any two levels of different attributes that are uncorrelated, and minimize overlap cases where attribute levels do not vary within a choice set.

An iterative computer algorithm produces an efficient design. The orthogonal fraction factorial experiment created sixteen alternative profiles. It was presented on simple rating choice cards. It is also ilustrated with visual graphic with virtual reality experiment for better measurement.

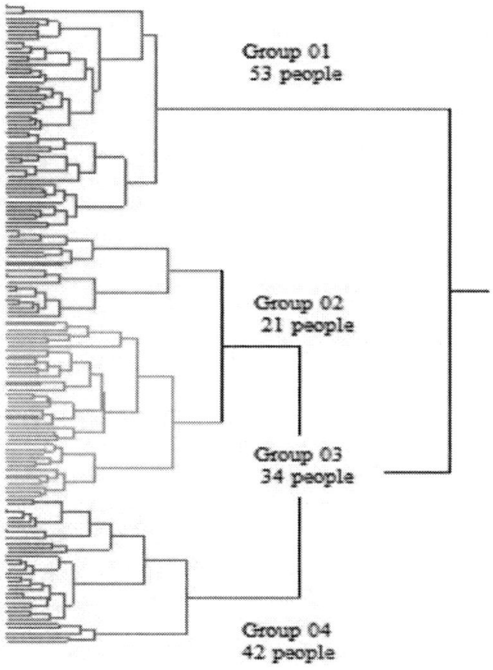

Fig. 2. Cluster analysis

2.3 Survey administration

The respondents chosen were those whose houses are within 250 meters from river edge. There were 150 samples. They were couples and individuals who were adult and eligible to make decision about the house. Each respondent answers a questionnaire within thirty minutes, including instructive information.

A trained surveyor conducted a door-to-door interview. They explained the content of sixteen profiles card one at a time and asked respondents to rate it. The rating scale is one to ten. It is numerical preference indicator in the continuous data.

3 Results and discussion

3.1 Respondent clusters based on preference

The analysis result divides respondents into four groups based on similarity on their preferences. This analysis uses clustering with ward method. In this method the distance between two clusters is the ANOVA sum of squares between two clusters for all variables. This method is used to combine small amounts of clusters (Milligan 1980 in [15].

3.2 Conjoint analysis to each group

3.2.1 Residential mass

The relative importance of residential type indicates this attribute receives a large portion of preference consideration. The value is more than half of all. The first cluster is about 55 %, while the second and the third clusters are 64.4% and 64.1%, respectively. The importance of attribute for the fourth group exceeds others groups. Their focus in preference is almost entirely for residential type. It shows that they less care about environmental of riparian residential (79.3%).

All groups have the same preference for the most attractive option. They all like stilts row house. Their preferences are different for the most undesirable residential type. The first cluster prefers floating house to flats.

3.2.2 Distance to river edge

The first group consider this attribute quite important. They put it as the second important attribute with 15,8 % attention to it. The first, second, and third groups are similar in this attribute preference. They consider twenty-one to thirty meters as the most ideal distance. These groups clearly dislike distance of more than thirty meters, except the the third. The third don't like most below ten meters. The fourth group preference analysis obtains different result. This cluster chose eleven to twenty as the best option, but same preference for the least option. This group also dislike residential distance that is too far from river edge (> 30m).

3.2.3 Riverside Construction

Riverside construction is an attribute associated with the tidal conditions of riparian, flood conditions, and convenience river access. The ecosystem approach requires constructions that do not block natural tidal. For people living in riparian residential area, this attribute usually get special attention. The first dan second groups agree that polder is a proper option for their needs. But, the third and fourth groups have different opinion. They like waterfront flatform. It improves the quality of residential environment by

Table 2 Conjoint Analysis to Each Cluster

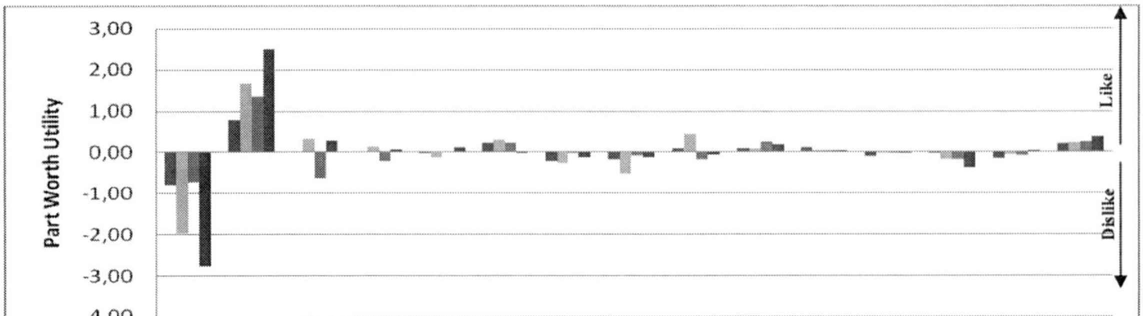

	Residential type			Distance to river edge.				Riverside Construction			Accessibility		Open space topology		
	Floating House	Stilt row house	Flats	<10 m	11-20 m	21-30 m	> 30 m	Natural	Polder System	Waterfront Platform	Boat dock	Promenade	Open green natural	Riparian park with trees	Public playground field
Group 01	-0,80	0,78	0,02	0,01	-0,02	0,23	-0,22	-0,18	0,10	0,09	0,10	-0,10	-0,04	-0,15	0,19
Group 02	-1,97	1,65	0,32	0,12	-0,14	0,30	-0,28	-0,53	0,45	0,07	0,03	-0,03	-0,17	-0,05	0,22
Group 03	-0,73	1,35	-0,62	-0,23	0,01	0,23	-0,01	-0,07	-0,17	0,24	0,03	-0,03	-0,18	-0,08	0,26
Group 04	-2,78	2,49	0,28	0,06	0,10	-0,03	-0,12	-0,13	-0,05	0,18	0,04	-0,04	-0,40	0,03	0,37

allocating a flood free open space. Three groups most dislike natural floodplain. This option does not protect.residence from flooding. Unlike the others, the third group chose a polder system (-0.17) as the least preferred

3.2.4 Open space topology

The third and the fourth groups put this attribute as second priority after residential type, while the first group put it in the third position. It is not much different with the second group opinion. They give it as the fourth important attribute position and just give 6.9% attention.

All groups most like publik playground. The part-worth utility of this option is quite significant (0.19, 0.22, 0.26 and 0.37). Open green area is the most unattractive option for the three groups except the third group, while the only the second group that most don't like riparian park wit tree.

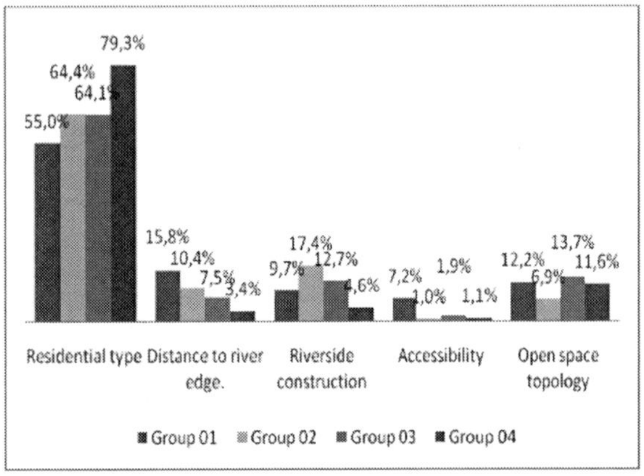

Fig. 3. Attribute importance for each group

3.2.5 Accessibility

Almost all groups do not concern with this attribute, except the first group. This group give this attribute with 7.2% value. Nevertheless, all groups tend to choose boat dock than promenade.

4 Conclusions

Town planners need to know the preferences of the residents to get a success planning. This study explains the importance of better understanding the people's preferences. In contrast to summarizing preferences into one set of attribute importance values and part-worth of level attribute utilities, clustering explores the differences among respondent that represents a resident segment. This approach is provided by multivariate analysis to get a topology of respondent. It does not focus on an easily preferences interpretable like the aggregation of preferences of all individuals within a region. It does not reveal all the information necessary for town planner. The opposing viewpoints can polarize all interpretation because it cancels out each other. Because of that, the preferences should be assessed separately for all relevant user groups.

References

1. Mitsch WJ, Gosselink JG. *Wetlands*. Fifth edition ed., John Wiley and Sons, Inc, Hoboken, NJ, (2015).
2. Fletcher TD, Vietz G, Walsh CJ. *Prog. Phys. Geogr.*, **38**, 543–555 (2014).
3. Spits J, Needham B, Smits T, Brinkhof T.. *Nat. Cult.*, **5**, (2010).
4. Meador MR, Goldstein RM. Water Physicochemistry, Riparian Condition, and Fish Community Structure. *Environ. Manage.*, **31**, 504–517 (2003).
5. Liao Q. Fuse. *Mod. Appl. Sci.*, **5**, (2011).

6. Echols SP, Nassar HF. *URBAN Des. Int.*, **11**, 203–212 (2006).
7. Davenport MA, Bridges CA, Mangun JC, Carver AD, Williard KWJ, Jones EO. *Environ. Manage.*, **45**, 711–722 (2010).
8. Everard M, Moggridge HL. *Urban Ecosyst.*, **15**, 293–314 (2012).
9. Scott M, Bullock C, Foley K. *Town Plan. Rev.*, **84**, 337–370 (2013).
10. Burton ML, Samuelson LJ. *Plant Ecol.*, **195**, 99–115 (2008).
11. Qiu Z, Prato T, Boehm G. *J. Am. Water Resour. Assoc.*, **42**, 1583–1596 (2006).
12. Adamowicz W, Louviere J, Swait J. *NOAA-Natl. Ocean. Athmospheric Adm. Wash. USA*, (1998).
13. Alriksson S, Öberg T. *Environ. Sci. Pollut. Res.*, **15**, 244–257 (2008).
14. Molin EJ., Oppewal H, Harry J.P Timmermans. *J. Hous. Built Environ.*, **16**, 267–284 (2001).
15. JMP, A Business Unit of SAS. SAS Institute Inc., North Carolina 27513., (2007)

Typology of accessibility between planned and unplanned settlement

Tin Budi Utami*, *Budi* Susetyo

[1]Architecture Department-Mercu Buana University, 11650 South Meruya-West Jakarta, Indonesia

Abstract. The development of physical boundaries in real estate created by developers has experienced changes that are dependent on time and environmental conditions of the vicinity. This also applies to the access of communities settled within the surrounding area into real estate. This study aims to describe the typology of accessibility of surrounding settlements (unplanned settlement) transitioning into a real estate environment (planned settlement) in South Tangerang, by using case studies of Bintaro Jaya real estate. The research method is done through surveys and interviews by individuals who understand the development process of access and the existing boundaries. Typology of accessibility will be described by using two variables, namely, whether or not there is access and its conditions. The results show that there are four types of access namely: access is planned and integrated within the existing; access is planned, but less integrated within the existing; access is made by dismantling most of the boundary; and no access. Although several typologies of accessibility can result in conflicts, yet job opportunities inside a real estate environment tend to be good with all access typologies. The physical boundary of real estate can potentially be good interface to be developed between two distinct settlements.

1 Introduction

Housing and settlements are considered elements of basic human needs. If the government focuses on the provision of housing for the lower class, private developers play an active role in providing for the middle class. The result of the national housing provision undertaken by private developers is still relatively small, which is less than 15%. This implies that, the other 85% remains the responsibility of the community itself without a business entity or other forms of organizations. (Koeswartojo et al., 2005: 44-45). Housings surrounding real estate which are built by the community without going through a private developer or government are called unplanned settlements. Whereas, a real estate is a term of settlements built by private developers, which is called a planned settlement. This can be in the form of closed or open systems.

Closed Real estates are commonly known as gated communities. Real estate can be present in the midst of existing settlements or instead it is able to stimulate the presence of surrounding settlements. Real estate has a wide range of physical boundaries that are designed with both a specific purpose and motive. The objectives in designing these boundaries are for security, lifestyle and prestige for people within the area.

The presence of real estate in South Tangerang (especially in Bintaro Jaya) cannot be separated from the existing surrounding settlements or new settlements that will later develop. Current regulation care aiming more to regulate the planning for real estates, so that real estate tends to establish an enclave for security, view, and social image. Physical boundaries come into form affecting the typology of accessibility from the surrounding settlements to real estate compounds.

The economic and social relations between the population of the surrounding settlements and the population of real estate led them towards a response upon the physical boundaries created by developers. This study aims to demonstrate the accessibility of the communities around the real estates to be (unplanned settlement) able to enter real estate compounds (planned settlement). The current phenomenon suggests that real estate boundaries are responded differently by the community around the real estates. Physical boundaries of real estates, in reality, are not necessarily beneficial for the people within the compound as their intended purpose. On one hand, real estate often have negative impacts for people living outside the real estates. From the 14,000 hectares of land in South Tangerang, 70 % is controlled by developers, the rest is controlled by the government and individuals. How to harmonize the region under the authority of developers and beyond in order for all residents to be comfortable (Andy Simarmata, 2010) is the reason for this research.

2 Literature review

2.1 Planned and unplanned settlement

Act. No. 1 of 2011 on Housing and Settlement stipulates that a housing environment unit is a residential area of

* Corresponding author: tinbudiutami@mercubuana.ac.id

various shapes and sizes with a structured arrangement of land and space, infrastructure and facilities. "Permukiman Swadaya" is a housing environment built by the community without a business entity or other organization forms. Real estate is a residential area that is formally constructed by private developers (Kwanda 2001: 53), labelled as planned settlement. The settlement (*permukiman swadaya*) around a real estate is labelled as unplanned settlement. Thus, the difference between planned and unplanned settlement is based on who builds the property.

Housing is a basic human need as the embodiment of personal self and/or of a group of residing and settling, as well as a place to foster and establish human personality and culture (Blaang, 1986: 9). Act. No. 1/2011 states that (1) every citizen has the right to occupy and/or enjoy and/or have a decent house in a healthy, safe, harmonious and an orderly mannered environment; (2) every citizen has the obligation and responsibility to participate in the development of housing and settlement.

The problem which often arises is that real estate presents itself as an exclusive zone within the existing unorganized settlement, or, real estates stimulate the spontaneous development of unplanned settlement, consequently creating mosaic patterns formed by the occurrence of planned and unplanned development (Pre BSD Studies, 1984: 26; Kuswartojo, 2005: 20). Soetomo (2009: 240) explains that the dualistic morphological structure can create a formal and informal symbiosis, thus realizing the power of symbiosis of two lives or activities that are interconnected as well as socially and economically dependent, however Maharika (2005), states that development that focuses only planning its territory will cause social problems due to lack of attention towards social networking and its space that exists outside of the real estates. Thus the relationship between real estates and informal settlements within the vicinity can become a social and economic mutual symbiosis benefit, but also increases the likelihood of physical and social problems. One of the factors that affect these conditions depends on how the physical boundaries of the real estates were planned.

2.2 The formation process of physical boundary of real estate

Act. of the Republic of Indonesia Number 4 of 1992 on Housing and Settlement states that in accordance to increase the value & dignity, quality of life and well-being for every Indonesian family, housing and settlement development as part of the national development needs to be improved and developed in an integrated, purposeful, planned and sustainable manner.

Due to the limited ability of the government to realize the fulfillment of housing needs and housing for its citizens, it encourages private developers to formally participate in fulfilling these needs. Citizens who cannot afford housings provided by private developers, take initiative to build houses and its surroundings independently (often referred to informally/unplanned). The spatial patterns of formal and informal settlements are formed because of the government's city planning growth rate is slower than the rate of independent societies as well as private developers in housing procurement. Understanding the different initiators of construction, the physical quality of their products is also different. In which, Haryo (2002) and Word (2004) states as the causes for both spatial and social segregation.

The motive of private developers besides aiming to help the government in the provision of homes for its citizens is certainly no other than for financial gains. The number of private developer increases rapidly, along with the sound profits it reaps and the view of property as a growing media of investment. It is no surprise, that since 1990 the growth of real estates as a product of developers increases rapidly, as that has happened in Jakarta (Leisch, 2002), in America (Blakely and Snyder, 1997). Even in Yogyakarta within the period of 2000-2005, the growth of settlements that were mostly fenced around has reached over 383 (Maharika, 2009).

However, when housing units shifted from a social welfare function (social welfare) to commodity, real estate planning principles used by developers in order for their products to be demanded by consumers, are the following:

1. Visual (Beer, 1990: 139), there are five best categories to assess the "view":
   - View: very good, should be kept open
   - View: good, should be kept open
   - View: good enough it can be used for things that benefit
   - View: unfavourable, may be closed
   - View: sorely lacking, should be closed
2. Security (Blakely and Snyder, 1998).
3. Exclusive image, social identity, lifestyle (Blakely and Snyder, 1998).

Those principles encourage the pattern of real estate, in the form of gated communities (cluster, gated community), which sometimes enforce have special rules or consensus (code of conduct) for its citizens to maintain their exclusivity (Blandy, 2003).

The conditions above create changes to their social life around real estates, from the weakening strong social ties, social de-cohesivity occur. This symptom is not a spatial transformation process, but as the impact itself of the spatial transformation process (Yunus, 2008: 292-293). Gold (2002: 35) said that development of real estate created interface between urban and rural systems that affected the community life. Yunus (2008: 293) adds that one of the symptoms of social de-cohesivity can occur because of the entry of new compounds into the environment of rural communities, raises socio-cultural values that are different than the values of socio-cultural that has long settled there.

Several impacts that arise varies from mild to severe, depending on the type of settlements built and its inhabitants. This is what Yunus states that should be considered that needs to be acted upon wisely by policy makers on the shape and pattern of settlement to be built, so that the negative impact on existing social cohesiveness does not pose a significant shock.

Maharika et al (2006 in Maharika, 2009) confirms that the most dominant motive for forging the physical boundaries in real estate are for the security of its residents. But the study's finding by Kim (2006, in Maharika 2009) in the US shows that the shape of the fenced settlement perceptually does cause a sense of security but does not reduce actual crime. While Maharika (2009) with research sites in Yogyakarta added that in the context of the spatial relationship between the typology of settlements with criminality seem the importance of comprehensive understanding that architectural intervention in the form of a fence does not unduly influence the perception of security. Thus, the narrow understanding that the fence can create a sense of security needs to be evaluated.

The phenomenon mentioned above indicates that the process of establishing physical boundaries of real estate that are initially designed for the sake of public security in real estate is beginning to be doubted. The many forms of physical boundaries will eventually be responded variously by the local community outside of the real estates.

2.3 The impact of real estate physical boundaries

The impact of physical boundary by real estate may cause the following:

2.3.1 The land price

Davis (1992, in Furlong Wu and Klaire Webber, 2004) stated that in Beijing, the price of land around the new residential development (real estate) increased more rapidly than others due to the increased facilities that exist in it directly affects the quality of the land. Jonah (2008: 245-246) stated the similar to a case in Indonesia.

2.3.2 The cohesion of social and economic

Albers (2008) stated that the meeting of two different settlement patterns in London allows social cohesion that mutually benefit each other, which is agreed by Soetomo (2009: 237-240) for a case in Indonesia where social and economic interdependence occur.

2.3.3 Social and spatial segregation

Leisch (2002), Word (2004) and Winarso (2005) on their research in the Greater Jakarta Area (Jakarta, Bogor, Depok, Tangerang, Bekasi) stated that the development of new areas or the presence of real estate strengthens the spatial and social segregation, which is also reinforced by Koeswartojo (2005: 190-192) and Yunus (2008: 248).

From the above explanations, it can be concluded that the boundaries of the real estate can cause a positive impact/synergy or negative/conflict to the environment. A conflict could arise if there are differences in interests that cannot be compromised (Yunus, 2008: 355-359). Conflicts of interest can be ended in two ways:

- There is a compromise
- There was the defeat of one party by the other party.

Disputes of land are very complex; it can be a matter of social, economic, political, cultural, spatial and environmental. Examples of land usage that have caused conflicts between community groups is the opposing interests among groups of people living in the new housing of real estate with a group of residents who live around the real estate. Conflicts because of land use, could lead to act of territory invasion, which among others are (Lyman and Scott, 1965 in Leboyer): violation, invasions, vandalism, obstruction and contamination.

2.4 Typology of accessibility to real estate

To answer the first problem of the various forms of typologies of physical boundaries of real estates in South Tangerang, it is necessary to approach the determination of typology. Steps to determine the type starts with observing the characteristic elements at the physical boundaries, then sort and determine the elements that always appear in similar boundaries. Naming the type of physical forms can be based on functional similarity and the similarity in appearance of the boundaries, whether there is access to the real estate or not.

Yunus (2008: 248) states that the construction of elite residential complex surrounded by a fence with one entry only, in some ways considered to be less suited for the conditions of Indonesian society. The presence of real estate creates symptoms of social exclusion and this will definitely generate a disruption of established socials sytems because of social interactions that cannot be implemented. Discontinuation of social interaction is able to trigger social de-cohesivity and it is actually a time bomb that is worrisome to the emergence of social unrest (social Upheaval). The condition is extremely vulnerable to social frictions that can trigger social Upheaval.

The residential of real estates, which have plenty of access from all directions into the environment of existing settlements will facilitate the process of social adaptation between old and new settlements. Gold (2002: 40) states that the planned road network in the real estates should be connected to existing roads and existing housings. Some of the job opportunities for the local population will be opened and the result will therefore impact on income improvements and an improved well-being.

Based on the review above, the context of boundaries can be distinguished by the following: a) there is access and b) there is no access. The seven indicators used as an instrument to determine the perception of the quality of construction (Thirion, 2008) related to access to important resources can be described as follows:

- access to formal and informal employment as a source of income for a decent living;
- access to health services; access to education and training of both formal and informal;

- access to cultural activities and entertainment as a means of representation and cultural diversity expression of citizens; and
- access to public transportation

Whereas the surrounding communities access to the real estate can be elaborated as the following: 1) The condition of access to the real estate, 2) The modes of transportation accommodated by the access, 3) access management, 4) the consensus of the access, 5) public facilities and social facilities within the real estate 6) public facilities and social facilities without the real estate.

3 Method

This explanatory study aims to find, understand and explain the phenomenon of the typology of accessibility from surrounding settlements to real estates. This study uses 18 samples of limited real estates, 11 non-clustered (non-gated community=NGC) and 7 clustered (gated community=GC). The method used is field surveys to observe the kind of access from surrounding neighborhoods into the real estate, and interviews with informants who know the development process of the boundaries.

Before determining the location of the study, researchers conducted a preliminary survey of some large-scale developers in South Tangerang such as Lippo Karawaci, Alam Sutera, Gading Serpong, Bumi Serpong Damai (BSD) and Bintaro Jaya. Lippo Karawaci, Gading Serpong, Alam Sutra and have relatively homogeneous residential patterns, the cluster (GC). While BSD and Bintaro Jaya has a diverse residential patterns (GC and NGC), the difference is the neighborhood around Bintaro Jaya denser than BSD.

In consideration of the above, therefore the location of the selected real estate is Bintaro Jaya for the following reasons:

1. It has a land area of 1700 hectares
2. It has been developed over 10 years,
3. Have lots of diverse spatial patterns in the form of an open grid and several variations of cluster
4. Residents of real estate is relatively dense and varied social strata
5. It is occupied by more than 60%

The Bintaro Jaya Real Estate is a residential area that was built in 1979 by PT. Jaya Real Property with a city garden concept. Development started in phases and was divided in clusters and non clusters. Up until present, there are still undergoing development. The market segmentation that is targeted is the middle upper class.

4 Results and discussion

The surrounding communities' access to the real estate can be elaborated as the following:1) The condition of access to the real estate, 2) The modes of transportation accommodated by the access, 3) access management, 4) the consensus of the access, 5) public facilities and social facilities within the real estate 6) public facilities and social facilities without the real estate.

4.1 The condition of access to the real estate

The existence and condition of the surrounding communities' access to the real estates can be divided into four categories, which are the following: access is planned and integrated within the existing, access is planned but poorly integrated within existing, and access with partially dismantled boundaries, and no access.

a. Access is planned and integrated within the existing

Access that are planned and integrated can be discovered in residential blocks built in the early years of the real estates development (1979-1980). Access is planned but less integrated is also found in cluster zones that are built in the early stage of the development (1980). Planning of integrated access and less integrated access was part of the initial strategy of the developer, so that the presence of the real estate can be accepted by the local community, as well as real estate residents, who can utilize the existing facilities in the neighborhood, because the facilities in real estate were not yet complete at the time.

b. Access is planned but less integrated within the existing

Access that is planned but less integrated within surrounding occurred as a result of the demolition of the boundaries by the villagers around, commonly found in residential blocks built after 1984, where the issue of security for the citizens of the real estate began to emerge, as good will of developers to protect consumers, as well as a part of the marketing strategy.

Access caused by the demolition of the boundary is the result of the compromise between the real estate with the villagers around. At first, developer closed all borders with massive walls as high as 3-6 meters, so that access of the villagers around to various sources were closed off. This condition triggered protests from villagers demanding that their access be restored. Despite having reached a compromise to open up access by dismantling the boundary, the conflict of interests between developers and occupants of residential real estate by residents around still occurred, due to security reasons and the right of citizens to access existing before the real estate was built.

c. Access is made by dismantling most of the boundary

Formerly, developers have appreciated the surrounding settlements, but with the passing of time, developers grow bigger and stronger which have led to their lack of attention towards the surrounding settlements' considerations for access. This condition triggers communities surrounding the real estate to dismantle boundaries of real estates to create their own access. This causes several citizens residing within the peripheral of the real estate compound having to move elsewhere, because their environment feels disrupted.

Providing accessibility solution by creating access for the surrounding villagers by the dismantling of the physical boundary has been implemented since 1984.

Conflicts became more frequent after 2000. Therefore, for reasons of security and business competition, developers opted for a more enclosed cluster, instead of residential non cluster.

d. No access

Access conditions of those built last by the developers do not provide direct access from neighborhoods surrounding communities. Learning from previous cases, since 2003 the developers chose to develop a cluster of residential blocks that are enclosed with high boundary walls, but without closing the access of the surrounding villages. As a logical consequence towards the implemented alternative solution, corridors exist between solid tall walls of real estates, which are visually and psychologically unpleasant.

The visual of typology of accessibility can be showed on table 1.

Table 1. Typology of accessibility

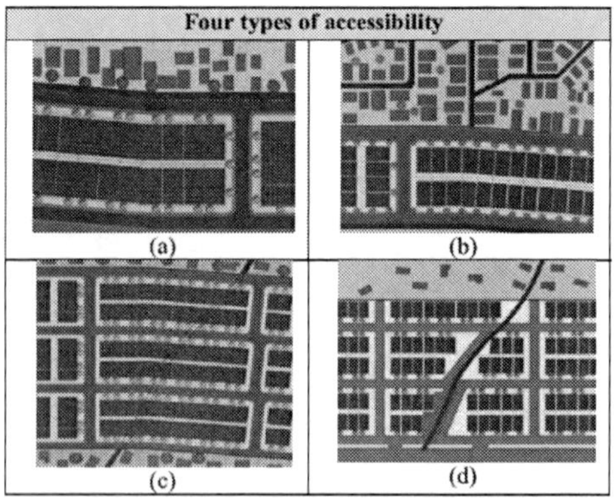

The visual of condition accessibility based on the fact, can be showed in table 2, (a) access is planned and integrated within existing; (b) access is planned and less integrated within existing; (c) access is made by dismantling most of the boundary; and (d) no access.

Table 2. The visual of accessibilty

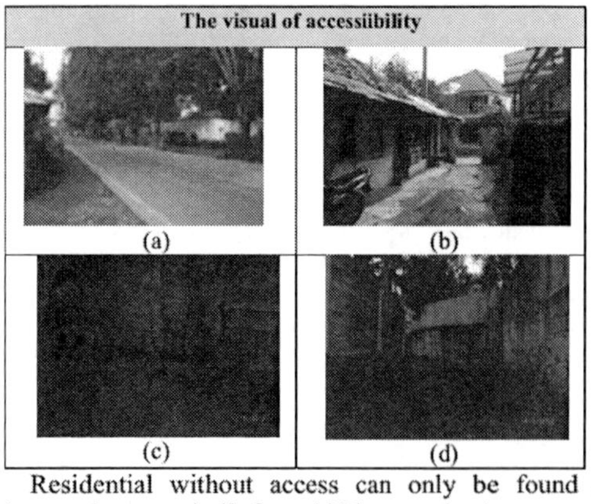

Residential without access can only be found in clusters that were built from 2000 onwards. A case study where there are no access exists only in cluster residence. This condition shows that current development, accessibility of the surrounding neighbourhoods in accessing real estate compounds experience development through integrated access, less integrated, dismantling existing boundaries, until having no access at all. Those actions was undertaken as developers were increasingly expanding the area of development, so that the residents of real estates continues to increase, even many local residents were displaced and marginalized by the expansion of construction by developers. While facilities that are made by developers also continue to increase in numbers and dominate the region, including transport rerouting from the main road to the village main street created by the developer.

The above situation is caused by the increase in competition of property industry, therefore they put more effort in improving the relationship with the surrounding communities, so that developers begin to improve on planning to avoid conflicts related to access. Conflicts due to obstruction of access of residents of the surrounding neighbourhood into RE are disrupting the convenience of various parties, from local RE surrounding residents, RE occupants as well as developers. In 2003, security threats have increased, which was the year where real estate development began to be designed without accessibility for the surrounding area into entering real estate compound. But learning from previous cases of conflicts often related to access, here the developers have learned not to completely close public access to the real estate. However, roads were enclosed by tall physical boundary, the roads became visually impairing and threaten the safety of passing vehicles. Hence, it is observed that there are changes in how the developers perceive the nature and function of access from 1979 to 2007.

Based on the description above showing the planning of physical boundaries of real estates, developers are not likely to provide access to the surrounding population into real estates. However, due to the fact that such access has existed far before the presence of the realestate developers, the closing of the access has only caused conflict and caused the surrounding communities to demolish the walls. The end of the conflict resulted in the demolition of most of the access that are closed off, so accessibility is no longer cut off or closed. It sometimes causes problems for residents who live near the peripheral area of the real estate compound, asthe demolished boundary walls increases traffic disturbance.

Up until the time of this research, there are cases where conflicts still do occur following the demolition of boundary walls, as RE residents requestthe access to real estate to be closed for security reason. Simultaneously, residents outside the compound want access to be opened because the road was initially an existing path before the real estate was even built. Resolution is usually achieved through discussions between the two parties without the role of the developer. However, the debate within real estate residents in several locations, whether access should be opened or not continues until present day. With the passing of time, the design of the physical boundaries in real estate follows the trend to form enclaves or exclude the real estate from surrounding settlements. From the lay out pattern of the

DOI: 10.1051/matecconf/201710105026

existing plan, recent developments indicate residential blocks have changed from an open, semi-open and closed often called a fenced or gated residential community.

4.2 The type of transportation modes available to go through the access

Access from surrounding settlements into real estate can be traversed by several types of transport modes. Types of transportation modes that can be accommodated access can be divided into three categories is accumulative, namely (a) cars, motorcycles and pedestrians, as well as (b) motorcycles and pedestrians, and (c) pedestrian.

Table 3. The Type of transportation modes

Transportation Modes	Non Cluster	Cluster	Amount
Car	3	0	3
Motorcycle	11	3	14
Pedestrian	11	3	14

Access modes of transportation that accommodates cars, motorcycles and pedestrians located in the residential non cluster. Access modes of transportation that accommodates the motorcycle and pedestrian are in the non cluster and cluster. Access modes of transportation that accommodates cars, motorcycles and pedestrians are in residential blocks built in the early development of the real estate. While access to modes of transportation that accommodates the motor and pedestrian contained entirely on the type of access that is not integrated (no access due partly to break the boundary).The visual of type of transportation modes that can be showed in table 4: (a) car; (b) motorcycle; (c) pedestrian.

Table 4. The type of transportation modes on boundary

4.3 Access Management

Each access of surrounding settlements into real estates have a variety of access management. Access management of surrounding settlements in real estate generally consists of the presence or absence of access, when to open or close the access, as well as the presence or absence of security personnel. Access management for residents around the real estate can be broadly grouped into five categories, namely (a) without portals, without security; (b) no portals, open continuously, without security; (c) no portals, open continuously, with a security guard; (d) No portal, opening and closing hours, without security; and (e) there is a portals, opening and closing hours, with the guard.

In the non-clustered residential blocks, there are five types of access management, ranging from the management of a very loose (without portals without guard) to the most restrictive (no portals, access schedules, with security guards). Inside clustered residential blocks there are only two types of access management, ranging from the management of the most loose (no portals, open/close schedules, without security personnel), up to more rigorous management (no doors, access schedules, with security guards).

The most strict access management are implemented at one of the non-clustered residencies, but after several security breache in the real estate, real estate residents adds doors. The presence of these doors initially caused conflict between the citizens and residents of the surrounding community, because residents suspect that the open access without a portal facing directly towards the settlementsis used a means for criminals to pass by. While people around the real estate were being suspected of a crime, in the end amicably reached an agreement that the doors areclosed from 22.00 until 05.00.

In one ofthe clustered residencies, access managed by scheduling when access it permitted and not. For clustered residences it is considered relatively loose. However, because there a revery narrow access, where even motorcycles dins find it difficult to pass, the road deemed as not a public road, so it is not necessary to provide security personnel.

The existing road for residents are in the middle of the cluster, thus not making it impossible to close the road completely. Initially the management, due to objections from the local community, it was decided to keep the access open.

4.4 The agreement of access

The agreement on the existence of access to the real estate, between the surrounding settlement it the developer or the real estates' residence can be grouped into two categories, namely (a) the agreement without being asked by citizens, and (b) an agreement reached after a conflict.

The scope of issues discussed in the agreement limits the development of real estate is generally associated with existing access to surrounding communities that were closed by the limits of real estate built by the developer. Conflicts related to the closing of public access by the limits of real estate built by the developer in some cases are still unresolved until the time of this study.

4.5 Public and social facilities in real estate

The presence of real estates in general has brought a benefit for the environment with social facilities and public facilities provided in it. Social facilities and public facilities are provided in the real estate varies from,(a)parks, (b) mosques, (c) sports center, (d) churches, (e) stores, (f) TK, (g) markets, and (h) terminals.

Parks area facilities that are always, both in the non-clustered residential and clustered residential. At the beginning of the real estate's development, social facilities and public facilities are fairly complete in one location (parks, mosques, shopping malls, sports centers) and the other location (parks, mosques, kindergartens, markets, terminals and shopping). In the clustered residential, public facility provided only park and sports center. There is only one cluster that provides a mosque, but is the result of non- residents of real estate, instead of the developer.

4.6 Public and social facilities outside real estate

Public and Social facilities in real estate built by developers that generally have not been able to meet all the needs of the residents of real estate. In some cases observed in the study, residents also take advantage of the real estate and public facilities beyond real estate. Social facilities and public facilities located outside real estate has varied types, among other things (a) street vendors, (b) the mosque, (c) stores, (d) TK, and (e) SD.

The street vendor is a facility that is most often found outside the real estates, followed by the mosques. Street vendor and mosques is a facility that can also be accessed and used by the citizens of real estate. Street vendor, usually becomes a medium for interacting between citizens real estate and the surrounding community, because in real estateno stalls that sell vegetables or daily necessities, though the shop most of whom are the servants. While the mosque is a place of interaction that makes local residents and citizens real estate become familiar with each other. Amenities shops, kindergarten, elementary school, only in the beginning development.

In some case studies, procurement of the mosque carried out jointly between the citizens and residents outside real estate, on land purchased from the villagers around the borders with real estate. The mosque served as a citizen interacts most good and safe, because all sorts of conflicts can be resolved through discussions at the mosque.

The mosque is also a meeting point neutral space and eliminate the impression of a clear boundary between real estate and non-real estate. Assistance such as maintaining mosques and distribution of sacrificial animals from residents of real estate is often given to the mosque built together outside the real estate. This situation illustrates that a neutral shared facilities, such as, places of worship such as mosques can be means to unite the real estate environment and the surrounding population.

In addition to the use of social facilities and public facilities in real estate, real estate presence could also create employment opportunities for the surrounding settlement. Job opportunities present in the real estate for the local population can be divided into several different categories, namely: (1) Domestic (PRT), (2) security personnel, (3) merchants, (4) employees.

4.7 Employment opportunities in real estate

Employment opportunities like a housemaid (PRT), security guards and merchants most often found in the real estate. PRT workers and security guards generally do not live with whom they work for stay overnight, so that the access of villagers surrounding environment into real estate is a primary need that is indispensable.

Employment opportunities for a people who live surrounding real estate varies in type, such as selling vegetables, fruit or cakes are allowed to enter into real estate in residential. Opportunities to trade in such situations usually occur because there is new construction or renovation of houses or other buildings that are in cluster.

Construction or renovation is generally carried out by the contractor by the number of 5-10 people builder. The carpenters were generally require a lunch provided by the villagers around by way of transfer to the project site. Types of available employees in general are shop keepers or administrative staff in an office. The distribution of employment opportunities can be showed on table 5.

Table 5. Employment opportunities in real estate

Employment Opportunities	Non Cluster	Cluster	Amount
Housemaid (PRT)	10	7	17
Security guard	10	7	17
Merchant	10	7	17
Staff in office	3	0	3

Job opportunities in the real estate for the villagers around are very open, both at non-cluster and cluster. Of the 18 residential blocks, 17 residential blocks provide employment opportunities, only one location which do not provide employment opportunities. That location is the residential blocks where villagers around the border of only a few houses (10 houses) with large and regular home type and partlya commercial area. Thevisual of employment opportunities can be showed on table 6.

Table 6. The type of employment opportunities in real estate

Employment opportunities in real estate	
 Housemaid (PRT)	 Merchant
 Scavenger	 Security guard

5 Conclusions

From the discussion it can be concluded that there are four types of typologies access of settlements around into real estate:

1. Access is planned and integrated within the existing
2. Access is planned, but less integrated within the existing
3. Access is made by dismantling most of the boundaries
4. No access

The types show that initially the developers appreciate the surrounding settlements, but with the along due time the developer grows and has more bargaining power thus paying less attention to the interests of the surrounding settlements in terms of access. The condition resulted in dismantling the boundaries of real estate to create their own access. This causes a loss of some residences who are uncomfortable with the fact. Learning from such experiences, developers are increasingly close the design of settlement in the form of clusters or gated community, without closing the access of settlements around.

Of the four settlements around typology access into real estate, there are six factors that affect community relations and public about real estate in the real estate, namely:

1. Type of transport modes that can be through access
2. Access Management
3. Agreement on access
4. Public and social facilities in real estate
5. Public and social facilities outside real estate
6. Employment opportunities in real estate

Of the six factors, five of which (a mode of transportation that can be through access, access management, an agreement on access, public and social facilities in real estate and outside the real estate) are strongly influenced by the typology of access, the better the quality of access, the better to the five factors for the surrounding community. But the employment opportunities in the real estate tend to be good for the condition of access to the typology of the real estate that is both good and bad.

The type of access typology are explored to explain how the relationship between real estate and surrounding. In general, the initial planning of real estate less attention to the surrounding existing condition that is usually directly be designed with boundary like high walls. The condition causes big problems mainly related to the accessibility. in fact, these societies need each other economically and socially, but there is no good spacial interaction.

Physical boundaries of real estate can be a common ground, potentially to be developed for two distinct settlements, otherwise if neglected, it can be a source of problems for all parties. To create the physical limits of real estate and public access around into good real estate and responsive to the interests of the surrounding community, it takes commitment and participation of all stakeholders, including governments, developers, civil society organizations, residents of real estate, and the surrounding community.

Thank you, we say to my lecturers Dr. Ir. Iwan Sudradjat, MSA.; Dr. Eng. Ir. Boedi Darma Sidi, MSA.; Dr. Ir. Woerjantari, MT; Dr. Ir. Hanson Endra Kusuma, M.Eng., A team of surveyors Nasrudin and Christy, people around Bintaro Jaya which has become informants in this study, faculty members at Department of Architecture, Mercu Buana University.

References

1. Appleyard, D., *Journal of American Planning Association*, **45** No. 2. page: 143-153 (1979)
2. Deasy, C.M., *Designing Places for People: A Handbook on Human Behavior for Architects, Designers and Facility Managers*, Watson-Guptill, New York (1990)
3. Hasan, A., Patel, S., Satterthwaite, D., *Environment & Urbanization*, **17** No. 1 (2005).
4. Khudori, D., *Menuju Kampung Pemerdekaan*. Yayasan Pondok Rakyat, Yogyakarta (2002)
5. Aalbers, B., *Urban Studies*, **45**, No.13, 243-261 (2008)
6. Ashihara, Y.,*Exterior Design in Architecture*. Van Nostrand Reinhold Company, New York(1970)
7. Beer, A. R., *Environmental Planning for Site Development*. E.& F. N. Spon, London, New York, Melbourn, Madras (1990)
8. Blakely, E. J. dan M.G. Snyder, Urban Land Institute, Washington, D.C.(1998)
9. Blakely, E. J. and M.G. Snyder, *Fortress America: Gated Communities in the United States*. Brookings Institution Press, Washington, D.C.(1997)
10. Firman, T., *Habitat International*, **28**, 349-368(2004).
11. GLC Studi,*Tata Letak Perumahan*, Terjemahan *An Introduction to Housing Lay Out*. Nichols Publishing Company, Intermatra, Bandung (1978).
12. Gold, H. *Urban Life and Society*. Prentice Hall, Upper Saddle River, New Jersey 07458 (2002)

13. Habraken, N.J., *The Sructure of the Ordinary.* Form and Control in the Built Environment, MIT Press, Cambridge, Massachutes, London, England (1998)
14. Hakim, R, *Unsur Perancangan dalam Arsitektur Lansekap*, cetakan ke 5. Bumi Aksara, Jakarta (1991)
15. Karnaya, *Merancang Kota yang Harmonis*. Makalahpada Seminar Nasional: Peran Swasta dalam Penataan Ruang dan Arsitektur Kota, 8-9 Februari 1990, UNPAR dan PU, Bandung (1990)
16. Koeswartojo, T., dkk. *Perumahan & Permukiman di Indonesia*. Penerbit ITB, Bandung (2005).
17. Kwanda, T., *Dimensi*, **29**, No.1, 52-63 (2001)
18. Leisch, H., *Cities*, **19**, No.5, 341-350. (2002)
19. Madanipour, A., *Design of Urban Space, An Inquiry into Socio-spatial Process*. John Wiley & Sons, New York (1996)
20. Madanipour, A.,*Public and Privat Spaces of the City*.Routledge, Taylor and Francvis Group, New York (2003)
21. Madanipour, A. dan Rostami, L., *Habitat International*, **30,** 361-378 (2006)
22. Maharika, I. F., *4th International Symposium of Nusantara Urban Research Institut (NURI), Change + Heritage in Architecture + Urban Development* Universitas Diponegoro, Semarang, Indonesia (2009)
23. Natalisa, A., *Ragam Desian Pagar Tembok.* Griya Kreasi, Jakarta (2007)
24. Soetomo, S., *Urbanisasi & Morfologi, Proses Perkembangan Peradaban & Wadah Ruang Fisiknya: Menuju Ruang Kehidupan yang Manusiawi.* Graha Ilmu, Yogyakarta (2009)
25. Thirion, *Samuel, No 20*, Council of Europe Publishing, 2008: 35-76 (2008)
26. Utami, Tin *Budi,Tipologi Batas Fisik Real Estat dengan Permukiman Swadaya Masyarakat, Pusat Penelitian*, Universitas Mercu Buana Jakarta (2010)
27. Utami, Tin Budi, *Kerangka Evalusi Batas Fisik Real Estate dengan Permukiman Swadaya Masyarakat*, Pusat Penelitian, Universitas Mercu Buana Jakarta(2011)
28. Winarso, H. dan Tommy F., **26**, 487-506 (2000)
29. Wu, F. dan K. Webber, Cities, **31**, 231-247 (2004)
30. Yunus, H. S., *Dinamika Wilayah Peri Urban, Determinan Masa Depan Kota.* Pustaka Pelajar, Jakarta (2008)

Architecture, space and power in historical multi-ethnic city Gresik

Dian Ariestadi[1,2,*], *Antariksa*[3], *Lisa* Dwi Wulandari[3], and *Surjono*[4]

[1]Civil Engineering, Faculty of Engineering Universitas Brawijaya, 65145 Malang, Indonesia
[2]Civil Engineering, Universitas Negeri Malang, 65145 Malang, Indonesia
[3]Architecture Department, Faculty of Engineering Universitas Brawijaya, 65145 Malang, Indonesia
[4]Urban and Regional Planning, Faculty of Engineering Universitas Brawijaya, 65145 Malang, Indonesia

Abstract. The study of historical-morphology cities is conducted to discover the socio-cultural characteristics which influence the formation and development of spatial patterns and architecture. Gresik as the historical multi-ethnic city on the north coast of East Java, is known as a major trading port, the center spread of the Islamic religion, and the government city in the colonial era. This research has been made to know the morphological phenomenon of the historical city, by using a qualitative method. The result has shown: 1) the urban structure with the segregation of settlements based on ethnicity indicates the authority power to control its territory, 2) the building form which is dominated by colonial architecture shows the authority power to control the physical changes, 3) the courtyard-house concept which is widely applied in Kampung Arab indicates about the ethnic power to arrange the environment, and 4) the use of landhuis type and luxury mansions in Kampong Kemasan represents the power of successful people as government officials and entrepreneur. The spatial patterns and architecture of the historical multi-ethnic city - Gresik were influenced by the power aspects in the form of efforts to dominate each other and self-defense in the personal and communal levels.

1 Introduction

Until the 1960s era, the learning of architectural theories in many topics related to originality, harmonious composition, and associated with the happened phenomena. In the 1970s era, Alexander Tzonis and Manfredo Tafuri introduced a theoretical framework which linked between architecture and the socio-economic-political contexts. Furthermore, Focoult, which is known as poststructuralist paradigm, explained how the expert knowledge affected the power of social working, and thus how the certain architectural configurations can play a role of disciplining people's minds and bodies. Based on Foucault's studies, many scholars conducted various studies about the diversity of architectural and social interaction patterns [1].

Currently, the development of a new theory of architecture studies is conducted to find more holistic relation using a variety of new paradigms including postcolonial. Postcolonial was originated from the emergence of movements against the colonialism paradigm, and the goal was toward the colonialism impacts in knowledge included the culture which influences individuals to dominate others. The concern could encourage increasingly chaotic world order. Postcolonialism is also oriented toward the realization of a new world relationship in the future. Postcolonial perspectives espouse a transformative agenda of architecture and urbanism that enables and empowers multiplicity in the processes of material production, the practices of their inhabitation, and the structures of representation. The attempt to break free of colonial hierarchies includes investigating the colonial dimensions of concepts like modernism, postmodernism, tradition, heritage, and sustainability. The interrogation of different demands for an investigation into the origins of the conceptual frameworks and spatial categories defines the discipline. Therein lies our future – and our hope – for a more just and equitable world which demands the globalizing knowledge production in architecture and planning [2].

The Development of Architecture and cities in Indonesia have been considering the historical development of coastal cities, especially in the north coast of Java. The cities in the north coast of Java are the trajectory of world trading and the spread of Islam. Gresik as one of the cities in the northern coastal region of East Java has a port and a major trading area, so that is where the influx of migrant ethnics of Chinese, Arabic and Dutch [3]. In general, the coastal cities evolved through the phenomenon of ethnic settlements as an important part of the formation of neighbourhood, city, and architecture. The city formed the heterogeneous multi-ethnic settlement with the close layout each other, and the power of space and architecture will affect the patterns of formation and development. In the northern

* Corresponding author: dian.ariestadi.ft@um.ac.id

coastal town of Java, there have formed the cultural characteristics of the coast, including space and architecture through a process of acculturation between Javanese culture and immigrant ethnic cultures or among the migrant cultures. The development of cities in the colonial era indicates the presence of political domination and power on the physical and cultural development of cities.

Political system and power can be concluded to have a good relationship directly or indirectly with the expression of architecture or the built environment. Moreover, the power is seen as the dominant factor determining the processes of social and cultural transactions between the roles of state on the one hand and the participation of people on the other side [4]. The control and maintain about the pattern of domination mastery become important to look at the characteristics of space and environment, particularly the establishment and development of city. The space and architecture pattern power and the patterns which maintain the dominance of power are essential to obtain the characteristics of space and environment in the processes of formation and development of city.

2 Methodology

The study has used the typology analysis through the qualitative description of spatial and architectural in cases of the building and the area [5]. The study has also used a qualitative approach with the consideration that the study is related to the historical region generally linked with the socio-cultural values that have the heterogeneous meaning and value and the symbol meanings which are the metaphorical tradition. Qualitative methods can be used to uncover the background of the socio-cultural-historical that is hidden behind the physical appearance of space, building and environment. Through in-depth interviews, this study repeated cross and would be disclosed what, why and how the background of the socio-cultural-historical observed objects [6].

The first step is the study of typology which conducted the exploratory of spatial patterns of residential environment and buildings in the area. The used data have been taken through the image map of area with transformed into a line mapping which shows the boundaries of building and environmental components. Typological analysis can be defined as a way to describe architecture in relation to its physical conditions by means of descriptive categories not dependent on "idealist" concepts of style and character [7].

The second step is to determine the study object in the form of case studies of spaces and buildings in Kampung Arab, Kampung Belanda, and Kampong Kemasan Gresik. Case studies will illustrate the historical development, the physical condition of the building and the environment, and other aspects of the socio-cultural-political accompaniments.

The final step is a qualitative in the depth descriptions of the study cases to find local concepts. Qualitative descriptions have been based on the data observations, interviews, and a comparison reference and theory. Researchers will make a complete illustration, researching archives textual, detailed reports from the perspective of the respondents, and conducted a study on the natural situation [8].

3 Results and discussion

3.1 The old town Gresik typologies as historical multi-ethnic city

Spatial typology at the old town of Gresik is dominated by the main functions of settlements, and some additional functions for trading facility, small office, and home industry. The existence of these functions was spread evenly across the Old Town area of Gresik. The old residential area in Gresik city consisted of: (1) Kampung Pecinan, the region which has the main function as a Chinese ethnic dwelling, (2) Kampung Arab, which is the region with the main function as a dwelling of Arab ethnic, with a pattern of enclosed settlement, and the activities of weaving craftsmen, (3) Kampong Kemasan, the region with the main function as a dwelling of Javanese ethnic with the strong ties of kinship patterns, and generally, it has a high socio-economic level, (4) Kampung Kepatihan, as the region which has the main function as a dwelling for the government employees, and (5) Kampung Pakelingan, an area with residential function, and characteristics of merchants community [9]. Multi Ethnic Settlement in The Old City of Gresik is shown in Figure 1.

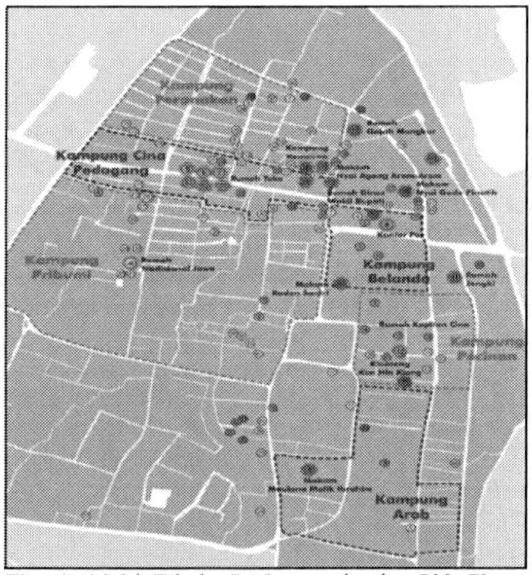

Fig. 1. Multi Ethnic Settlement in the Old City of Gresik (Source: The Green Map of Cultural Heritage Gresik, 2010).

The historical development of Java's cities relates to the historical development of trading cities in the northern coast of Java which was the entrance of cultures from outside. Several cities in the north coast of Java Island which have evolved since 1400 and earlier include Bintara, Jepara, Kudus, Lasem Tuban, Sedayu, Gresik,

Surabaya, Baremi, Gending, Panjarakan, Binor, Ketah, Patukangan, and Blambangan [10].

Based on the development of towns in the north coast region, which was recorded as a trading city and major port city is Gresik, Tuban, Lasem, Semarang, Cirebon, and Batavia. The trading port city is an initial entry of foreign migrants to trade and settle. The Cities in the north coast of East Java, such as Lasem, Tuban and Gresik are the cities which had a port which served as a trading city, so that the Chinese, Arabic and Dutch came through VOC in 1602 [11].

Gresik actually has developed long before the colonialism era in Indonesia. The historical development of Gresik town was divided into phases: (1) Phase I, the period of 1480-1487, (2) Phase II, the period of 1487-1605, (3) Phase III, the period of 1605-1748, (4) Phase IV, the period of 1748-1916, and (5) Phase V, the period of 1916-2008. The establishment and development of the city started from the establishment of the kingdom of Giri Kedaton stages which evolved as a religious center, was dominated by the colonial Dutch and grew rapidly as the industrial and trading port city, attracted many immigrants from various ethnics and shaped the city with a multi-ethnic neighbourhoods in the old town area of Gresik [12]. The old city of Gresik is shown in Figure 2.

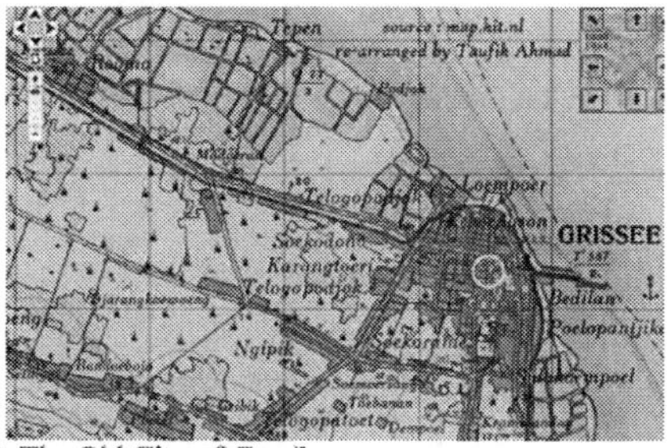

Fig. 2. The Old City of Gresik (Source:http://www.semboyan35.com/printthread.php?tid=53&page=6).

In the Dutch colonialism era, Gresik was a Afdeling region, which consisted of Gresik (the capital of residency), Lamongan, and the Sedayu district. The status was as an Afdeling only lasted until 1934, when it changed its status to be onderafdeling Gresik equivalent with Kawedanan. The port construction in Surabaya city in 1911 affected the trading track gradually moving to Tanjung Perak, and finally the merchants left the Gresik port. As a result, the Old Town of Gresik started to be abandoned. Thenceforth, trading activity in the area was more live during the day.

Gresik has a multi-ethnic settlement in the adjacent region to shape the character of specific environment. Gresik city structure formed from multi-ethnic settlements adjacent specific and can still be observed today. Heterogeneous structure composed by multi-ethnic enclave settlements, with a colonial style building that stands out, and the formation of open spaces in the middle of building homes and neighborhoods in Arab and China settlement. The multi-ethnic settlements of Gresik was formed and developed in the geographical environment and culture of the north coast of Java, in the colonialism era in Indonesia.

3.1.1 Landhuis typology in the old town of Gresik

The architectural style of existing buildings in all multi-ethnic settlement in the old city of Gresik is a colonial style. The condition was caused due to the development of Gresik town as a port city with many ethnic immigrants which began in the 1605-1748 period and rapidly grew in the 1748-1916 period. The architectural building style of region is dominated by the colonial-style which is shown in Figure 3.

Fig. 3. The typology of Colonial-style Building in Gresik (Source: The Green Map of Cultural Heritage Gresik, 2010).

The formation and development of the physical condition of Gresik town, as seen in the old city area nowadays mainly occurred in the 1748-1916 period and were marked by the growth of industrial and trading activities in Kampung Kemasan. This era included the 1800-1902 period which represented the period of Indonesia architecture development. The period was marked by the dissolution of VOC and the takeover of the Dutch East Indies by the Kingdom of the Netherlands.

The architectural style which was widely used is referred to as Neo-Classical, in the form of a classic style reapplied using a new material. The buildings which were built in this period are often referred to as the Empire Style. Building typologies in this architectural style have considered about the climate context in the design process. Building typologies that are typical such a symmetrical floor plan and a solid, thick and massive wall. The specific characteristics of colonial building in Indonesia include: (1) having arcades surround the outside spaces serve as a connecting line, and for thermal insulation and sunlight; (2) large roof with the steep slopes, sometimes comprising two layers with gaps for transferring heat or air, (3) high ceiling to avoid the heat on the room, (4) wall with large windows, (5) the size of windows and doors are large and high, and (6) the thick wall-room to reduce solar heat [13-14].

Furthermore, the colonial cultures interacted with local culture or acculturation, and established a new culture called Indies culture. The characteristics of Indies culture in Indonesia can be traced in: languages (pizjin), clothing (native king), home furnishings (furniture), and residential buildings (landhuis). Indies architecture is a style of architecture which blended between European architecture and local architecture to adapt the climate in Indonesia (tropical climates). The building characteristics of architectural Indies, including landhuis are: the high slope of roof to enlarge the volume of space under the roof, the high ceilings, vents in the top of doors and windows (fanlight or bovenlicht) perforated (ventilation block or krawangan); the front and back porches, located slightly to inside (the influence of Javanese house), and the attachment of natural stone plates in outside the bottom of buildings [15].

In the era of colonialism, the development of urban space and building structure governed by the laws of strict colonial rule. This shows the power of the colonial government in controlling the physical development of the region is very strong. There are two ages have great influence on morphological development of cities in Java, namely: 1) after 'Java Oorlog' / Java War in 1825-1830, followed by the system Cultivation (*Cultuurstelsel*), and 2) after the implementation of the Law of sugar 'in 1870 and the Law of Decentralization in 1905 [16].

As a political form of multi-ethnic and multi-culture, the colonial government developed the 'area of governance' (political domain) and 'trading area' (economical domain) [3]. Many wealthy people as a result of trade in coastal cities appointed rulers in the region. Furthermore, the environmental authorities have adopted a building officials and environmental conditions. The use of this building typology benefit from the government as a form of hegemony, and for society as a form of socio-economic success [17]

The characteristics of space and architecture in the area are dominated by the physical form of colonial architecture building which shows the dominance by the Dutch colonial government as an extremely strong power determinant. The building characteristics of official houses of landhuis types which are located in the Dutch settlement indicate an attempt to give the domination of spatial and architectural environment in the Old City of Gresik.

3.1.2 Courtyard in Kampung Arab Gresik

The existence of Kampung Arab Gresik that has been from the 1600s era, the patterns of residential and region were formed and developed in the colonial trading era, and survived until nowadays, influencing and forming the specific cultural community in Gresik today. Kampung Arab Gresik area is closed by the ethnic settlements and forms a Ethnic enclave pattern. The existence of ethnic enclave in this region is a picture of multiculturalism and diversity which can be used to maintain an ethnic culture, such as cultural activity and language [18].

Residential unit area in the center of Kampung Arab region, in general, is residential houses with a large area. The shape pattern of residential areas is a Cul de sac, with one access for entry and exit as shown in Figure 4. Generally, the dwelling has a front yard that serves as a receiving area after the main gate. The typology of courtyard building forms is located behind the main house. Furthermore, there are differences that separate the guest access to male and female guests. The access of male guests (public) is towards the main building, and the access of female guests (private) moves toward the courtyard space. The front yard can serve the private activity when the main gate is closed as shown in Figure 4.

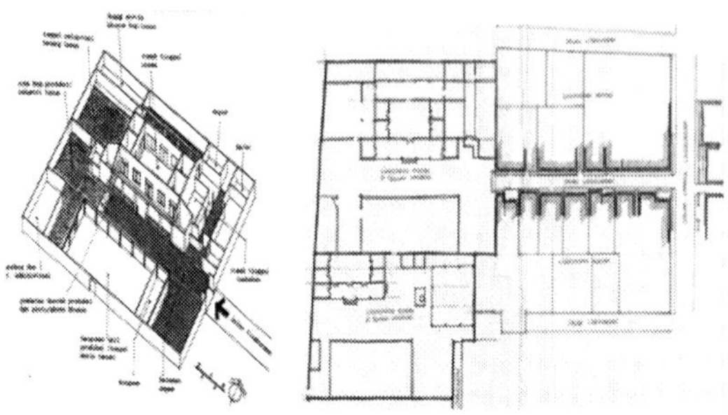

Fig. 4. Courtyard housing in the center area of Kampung Arab (Source: D. Ariestadi, 1999).

While the dwelling units in the road edge, both of the environment and the main street are a smaller residential area. In general, residential building has a boundary with the road environment, the main street of the area, or other dwellings. This building only serves as a dwelling house without the additional function of business activity. Courtyard space is located in the back or side, with separate access (entrance side). Thus, there are always two accesses namely the main access and the additional access which is marked by a door covered by a curtain or 'kerai'. Side entrance is located in the building edge, or, due to the limited of wide building, it can be located in the different street as shown in Figure 5.

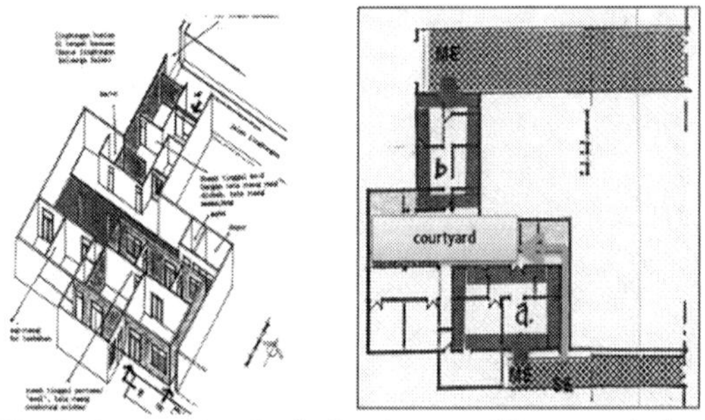

Fig. 5. Courtyard housing in the edge area of Kampung Arab (Source: D. Ariestadi, 1999).

Characteristics of courtyard houses in the Arab-Islamic architectural typology aim to meet the needs of protection level of the occupant privacy, particularly women in interaction with outsiders and male visitors, how to serve guests, attitude and responsibility towards the neighbours, and simplicity in life [19-21]. The typology of courtyard houses is related with the religious culture aspects, especially regarding to the privacy and separation activities, and visual access between men and women [22]. In traditional Islamic or Arab culture, it is important to separate men and women to protect the house against outsiders. The houses were designed to create a separate physical space between men and women, as well as to prevent the visual contact.

Courtyard house typologies in 'Kampung Arab' Gresik formed through a process of cultural adaptation by occupants in different settings. Courtyard house concept used in accordance with the demands of the inhabitants of the territory of gender space, privacy, and comfort. Environmental spatial patterns are formed in urban enclave settlements with the concept courtyard can arrange access system that supports the principle of the separation area of gender and privacy, as well as very efficient to control environmental thermal comfort. Nevertheless, the presence of the use of courtyard house concept which stands out in the residential neighborhood in Arab ethnic indicated the efforts to realize the domination power of strong ethnic in their own environment.

3.1.3 Luxury mansions in Kampong Kemasan Gresik

Kampung Kemasan is one of the alleys located in the old town area of Gresik. Historically, this village is called 'Kemasan' for its inhabitants work mostly gold handicraft makers. Currently, the craft of making gold was no longer found in this alley. Homeowner in this village are successful entrepreneurs, thus they built the unique and interesting house.

In general, residential buildings in Kampong Kemasan Gresik are composed of two building masses, which function the main building as the main activity for homeowners and the additional building as birds' nest. Two masses of the building are separated by a courtyard. In addition, there is also a building wing which is functioned as a service area [12]. The main building is a dwelling house typology mansion consists of 24 buildings. Identification of the typology of the building style is Dutch Colonial building with neo-classical style with influences of Chinese architecture and Arabic. Characterize the influence of Chinese vernacular house in the form of two storey building. The upper floor is made without partitions and used for storage and livelihood activities. Characterize the influence of Arab culture is spatial indicate an attempt to distinguish the activities of women and men, through the use of bamboo curtain covering the activities inside.

The most interesting element of buildings is the number of doors and windows which are more than a residential home in general. In front of the house, there is a form of stealth window, the window shaped ornaments which include the ventilation holes on it, but in reality the window deadlocked. The window was only made to decorate and beautify the building. In addition, the unique characteristics of residential buildings are also seen in the many ornaments in the buildings. The typology of residential buildings in Kampong Kemasan Gresik is shown in Figure 6 [23].

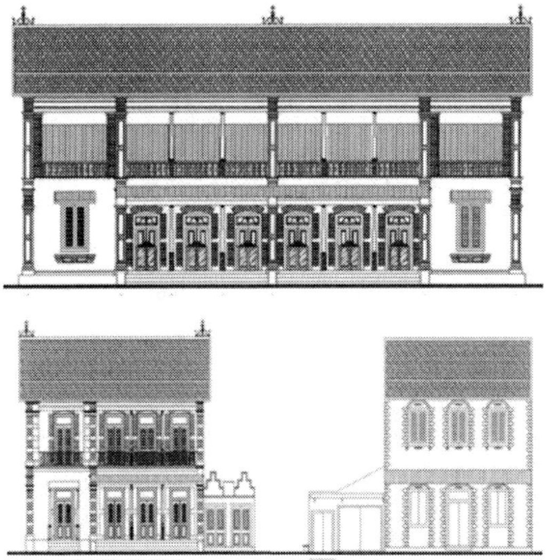

Fig. 6. Typology of residential buildings in Kampong Kemasan (Source: PU Ciptakarya Kabupaten Gresik, 2013).

The buildings and ornaments in residential houses in Kampong Kemasan were influenced by Chinese, Arab and European cultures. The Chinese influence is seen in the use of ornament to the wrought iron beams, which serves as the locking of doors and windows. The ornaments are generally found in Chinese style house. The Chinese influence is also shown in the dominance of use the red colour in the building. The red colour represents fire and blood which symbolize prosperity and luck, but red also is a symbol of virtue, truth and sincerity. The influence of Arabic culture is seen in: (a) the existence of patterned brick ornamentation in the pillars or columns and walls both exterior and interior building, with a triangular arrangement or pyramid, in which the composition of multiples-three, identic with the mosque roof; (2) the use of ornaments with the crenellation or Merlons motif is often found in the religious building ornaments; and (3) The use of the natural shape is the stylized-plant elements (flowers, leaves and branches). Floral motifs have been a medium of decorative expression due to the religious understanding of Islam that prohibits painting or creating the soulless creature sculptures, especially those can move or walk. The European cultural influences are seen in the peak of the roof ornaments, gevel and geveltoppen [13].

Building on the corridor Kampung Kemasan tend to have a colonial style and another in part Chinese-style architecture [12,23]. Dutch architectural style of the building is established earlier, because at the time there was no other building models that describe greatness and luxury. In the postcolonial perspective, the embodiment of the symbols indicate desire to showcase identity and the ability to live / to strugle of a group of sub-altern. The dynamics of the understanding and use of the building typology of luxury in Kampong Kemasan Old City Gresik is influenced by: 1) political power, in this case all the artifacts colonialism as a symbol of power,

2) attempts to show identity as minority groups were successful, and 3) efforts resilience to retains its identity [2].

The characteristics of the luxury building houses of a successful entrepreneur in Kampong Kemasan indicate an attempt to give the domination of spatial and architectural environment in the Old City of Gresik. The used ornaments were manifested with clear, very attractive, dominating, and even those are extremely interesting in its environment. The ornaments have also the symbols and meanings of success, prosperity, and positive values including virtue, truth and sincerity. Thereby, the existence of homeowner as a successful entrepreneur will gain prominence in the community.

3.2 Space and power in historical multi-ethnic city Gresik

Establishment of space and architecture in the multi-ethnic city of Gresik historically strongly influenced the dynamics of politics and power. Spatial and environmental building is based on a strong Islamic culture show the role of Islamic government as the initial formation of the town of Gresik. Spatial heterogeneous multi-ethnic describe government policy is the kingdom until the colonial government set Gresik city as a port city and growing as a center of commerce and government in the eastern region of the island of Java. While the physical dominance of colonial building with architectural influences show the strength of colonial government that controlled the region in the long term.

An overview of the formation of space and architecture shows that studies about architecture and multi-ethnic city is mostly conducted to find the concept of formation and use of spaces, because of the presence of complex interaction of diverse aspects including social, culture, and politic. This is in accordance with the theory of 'Production of Space' that emphasize social interaction as an important aspect in the formation of space [24], and the theory of 'Power of Place' which states that the use of space in the area of multi-ethnic influenced by aspects of domination and control of user groups space [25].

Postcolonial paradigm will be disclosed through the perspective of the opposing party (sub-altern). The dominance of power political forces that shape the space and architecture is balanced by the power-ethnic society as a user who wants to be able to control the social and cultural needs including space and architecture. The dynamics of space adjustment and colonial architecture with ethnic Chinese and Arabic influences shaping acculturation specific space and architecture, as seen through landhuis colonial building typologies in the entire area of the old town, courtyard houses pattern in Kampung Arab and luxury mansion in Kampung Kemasan. In the government that controls the political pressure, the adjustment of space and architecture in an effort to preserve the identity, privacy, and existence itself.

Space and built environment are instruments that reinforce hierarchies and the negotiation of relations of differences and adjustments. Postcolonial can show more unique ways and particles of forms and histories on the scope of global interconnection; henceforth considered to be universal, as well as the European modernism, which actually is a fact on the scope of provinces and particles [26]. Postcolonial thinking can contribute to the interpretation of architecture and city as a cultural artefact and as a symbolic landscape formed from the layers of meaning and identity. Postcolonial perspective can describe the political influence in architecture and urban governance [27].

4 Conclusion

The cities in Indonesia which developed in the colonial era greatly were influenced by the dominance of colonial government about the formation and development of the city physical character. The specific condition of environment and residential patterns in the old town of Gresik is the existence of multi-ethnic enclave settlement. The characteristics of space and architecture in the area were dominated by the physical form of colonial architecture building which showed the dominance of Dutch colonial government as an extremely strong power determinant. Nevertheless, the presence of the use of the courtyard-house concept which was applied in the ethnic Arab settlement indicates the efforts to realize the power of strong ethnic domination in their own environment. The characteristics of the house-officials of landhuis types which is located in the Dutch settlement, or a luxury mansion building of successful entrepreneur houses in Kampong Kemasan also indicates an attempt to give the domination of spatial and architectural environment in the Old City of Gresik. There is an attempt to establish the mutual power of space and architecture that will always occur in the establishment and development of space and the environment, while on the other hand, attempt to maintain the domination of power will always happen.

References

1. H. Heynen, G. *The book of Architectural Theory*. SAGE Publications Ltd. (2012)
2. J. Hosagrahar. *The book of Architectural Theory*. SAGE Publications Ltd. (2012)
3. R. Damayanti, Handinoto. *Dimensi Teknik Arsitektur*. **33**, No. 1 July (2005)
4. M. S. Barliana, http://file.upi.edu/Direktori/FPTK/JUR._PEND._TEKNIK_ARSITEKTUR/196302041031 MOKHAMAD_SYAOM_BARLIANA/Articles Makalah_Ilmiah/ArsitekturKekuasaan.pdf. Accessed, 10 October 2015
5. L. Groat, D. Wang. *Architectural Research Methods*. John Wiley & Sons, Inc. (2002)
6. G. E. Guba, S. Y. Lincoln. *Naturalistic Inquiry*. Beverly Hills: Sage Publications Inc. (1985)
7. A. Salama, *METU JFA*. **1,** 23: 1 (2006)

8. J. W. Cresswell. *Qualitative Inquiry & Research Design, Choosing Among Five Approaches*. SAGE Publications Ltd. (2007)
9. A.A.S.I. Widyastuty. *Jurnal Teknik WAKTU*. Vol. **09**, No. 02 (2011)
10. Sudaryono. *Laporan Jurnalistik Kompas*. 2008
11. B. Fauzy, Antariksa, P. Salura, *DIMENSI*. Vol. **38**, No. 2 (2011)
12. C. Riski, Antariksa, Surjono. *Arsitektur e-Journal*. **2** Nomor 2 (2009)
13. Y. Sumalyo. *Gadjah Mada University Press*. (1993)
14. Handinoto. *Perkembangan Kota dan Arsitektur Kolonial Belanda di Surabaya 1870-1940*. Penerbit Andi. (1996)
15. J. Soekiman. *Komunitas Bambu*. (2014)
16. Handinoto, *Dimensi Teknik Arsitektur*. Vol. **32**, No. 1. (2004)
17. D. Ariestadi. *Jurnal Teknologi dan Kejuruan*. FT UM. (1999)
18. S. Mazumdar, *EDRA* **36**-2005. (2010)
19. A. Shorbagy. *IJCEE-IJENS*. **10** No: 04. (2010)
20. F. A. Mustafa, A. S. Hassan. *American J. of Engineering and Applied Sciences*, **3** (2010)
21. A.Z. Karimi, B. Hosseini. *Proceedings: Archi-Cultural Translations through the Silk Road 2nd International Conference*, Mukogawa Women's Univ., Nishinomiya, Japan, July 14-16. (2012)
22. N. Qomariyyah, Antariksa, E. Y. Titisari. *Arsitektur e-Journal*. **2**, Nomor 2, Juli (2009) 79-89
23. PU Ciptakarya Kabupaten Gresik. Laporan Akhir Detail Pelaksanaan Tindak Konservasi Kawasan Jl. H.O.S. Cokroaminoto- Jl. Nyi Ageng Arem-Arem. (2013)
24. H. Lefebvre. *Book of Rethinking Architecture: A Reader in Cultural Theory*. Routledge (1997)
25. D. Hayden. *The MIT Pres*, (1995)
26. E. Akcan. *SAGE Publications Ltd*. (2014)
27. A. Kusno. *Behind the Postcolonial Architecture, urban space and political cultures in Indonesia*. Routledge. (2000)

* Corresponding author: dian.ariestadi.ft@um.ac.id

Gubeng bridge in Surabaya, a dutch architect c. citroen's work in the 1925 Paris exhibition

Joko Triwinarto Santoso*

Department of Architecture, Brawijaya University, 65145 Malang, Indonesia

Abstract. A Dutch architect Cosman Citroen (1881-1935) is recognized as one of the well-known architects in the Netherlands Indies or Indonesia nowadays. After leaving his job as an architect assistant in the architectural firm of B.J. Ouëndag in Amsterdam, the Netherlands, Citroen moved to Surabaya, Indonesia where he became a municipal architect. Here, he turned out to be a versatile architect whose projects range from furniture to town planning and from architectural designs to civil constructions. Discussion on Dutch colonial architecture in this city cannot be separated from his works. One of them was Gubeng Bridge that won an award at the 1925 Paris Exhibition. By using a combination of historical, descriptive, and archival methods, this paper reveals how initially Citroen involved in this civil project, how it was designed and built, which award is obtained, and why it could win the award. The result shows that design of the bridge is in accordance with the exhibition theme, namely Art Deco, and reflects the high aesthetic value of a civil construction.

1 Introduction

A Dutch architect Cosman Citroen (1881-1935) is recognized as one of the well-known architects in the Netherlands Indies or Indonesia nowadays (Fig. 1). After leaving his job as an architect assistant in the architectural firm of B.J. Ouëndag in Amsterdam, the Netherlands, Citroen moved to Surabaya, Indonesia where he became a municipal architect. He left a lot of works in this city when he is in his position and after his leaving it until his death. His works stretch from furniture design to town planning and from architectural design to construction of civil that are intended for the lower class to the upper-class society, both native and non-native (Dutch, English and Chinese). At the present, his works become historic architectural heritage that could form the townscape of Surabaya. The discourse about architecture in the city will not be separated from Citroen and his works. One of his works is Gubeng Bridge, which is the second civil construction work, after Kebondalem Wood Bridge, out of four projects.

Fig. 1. Cosman Citroen (1881-1935) [1]

Research on the Gubeng Bridge mentioning Citroen won an award in the 1925 Paris Exhibition was only an article written by De Zeeuw in 2001. Most likely, its reference source is an article at *Indisch Bouwkundig Tijdschrift - Locale Techniek* (IBT-LT) or Indies Architectural Magazine – Local Engineering, written by Heida (1933). But, until now there is no a study that is able to reveal which award is obtained and why it could win. This article will reveal the existence of the Gubeng Bridge from the beginning, starting from the idea to build a new Gubeng Bridge, Citroen involvement, design process, construction, award obtained, until the possible reasons why it could win the award.

2 Method

This study was conducted by using a combination of archival, historical, and descriptive methods.

Archives cover *Notulen van de Openbare Vergadering van den Gemeenteraad van Soerabaja* (Minutes of Public Meetings of the Surabaya City Council), *Gemeenteblad van Soerabaja* (Municipal Sheet of Surabaya), and old photographs of the Gubeng Bridge. The use of two first archives is based on the premise that involvement of Citroen, as the advisory architect of the Surabaya Municipality, had to be discussed and decided in the meetings of the city council as a legislative institution. All of this discussion is recorded in detail in the form of the minutes of the city council. Furthermore, conclusions and decisions of the city council are stated in a decree of the Surabaya mayor and written on a municipal sheet. Old photographs are used to compare form and details of the bridge in the past and present condition.

* Corresponding author: jokotris@yahoo.com, jokotris@ub.ac.id

 DOI: 10.1051/matecconf/201710105028

The above archives are strengthened by old publications (periodical, report, and book). They comprise *Indisch Bouwkundig Tijdschrift* (IBT) or Indies Architectural Magazine, which later became IBT-LT, report of *Oost Java Stoomtram Maatschappij* (East Java Tramway Company), and a book on Surabaya written by G.H. von Faber.

All of textual data is analyzed descriptively based on its content. It consists of diachronic analysis (chronological order or time sequence) and synchronic one (occurring at the same time).

3 Results and discussion

3.1 Idea

Gubeng Bridge crossed *Kali Mas* or Mas River (Fig. 2) and connected the Gubeng area and Ketabang, location of a New Surabaya Town Hall, via Simpang Street (now Pemuda Street). The idea was to replace the old bridge (Fig. 3) which was considered to be too narrow [2]. Gubeng Bridge was constructed out of steel which meant that it could not be made wider without having to remove it completely. Gubeng is one of the expansion areas of Surabaya. The Municipality bought the area for the amount of *f* 145,000 in 1908 [2]. Most of the area was used for housing for European. However, the existence of the train station in Gubeng, built by *Staatsspoorwegen* (SS) or State Railways, made it an important area. Not far to the west of this bridge, there was the old *Centrale Burgerlijke Ziekeninrichting* (CBZ) or Central Civil Patient Institution as well as Gubeng Sluice in the south (Fig. 4).

Fig. 2. View from Mas River

Fig. 3. Old Gubeng Bridge [5]

Fig. 4. Bird's-eye view [1]

The decision to build the new bridge was taken at a meeting of the Surabaya City Council on 5 October 1921, led by substitute Mayor Th.B.A. Faubel. All members of the Council accepted the proposal, after which the Municipality declared the proposal accepted in the *Gemeenteblad* (Municipal Sheet) 1921 number 248. The Municipal Sheet mentioned that the bridge was not only intended for cars and other vehicles, but also for electric trams operated by the *Oost-Java Stoomtram Maatschappij* (OJS) or East Java Tramway Company [6].

OJS's plan to connect Gubeng to Simpang by an electric tram line had been in preparation since at least 1911. This double track is part of the Gubeng-*Palmenlaan* route [7]. OJS built a tram stop on a platform of Gubeng Station.

3.2. Design

Based on an offer letter of the Director of *Dienst van Publieke Werken* (Public Work Service) to the Municipality, in 1922 Citroen was commissioned to design the bridge. In this project, he was assisted by engineers of the Public Work Service [1]. Unfortunately, the drawings cannot be found. *Het Hoofd van Constructie en Bruggenbouw* (the Head of Construction and Bridges) of the SS sent a letter dated 7 February 1923 number 1380/455 to the representative of OJS in Semarang concerning the design made by Citroen. OJS replied to it by sending a letter dated 22 February 1923 number 3568/13 which contained a proposal for design modifications and supplements to be approved by the authorities [8]. The Mayor of Surabaya received a letter concerning the design modifications from the Head of Construction and Bridges of the SS dated 17 September 1923 number 9800/455. The letter informed the Mayor that the modifications were still under consideration, but that the Director of Public Work Service did not have any objections. The technical and financial committees also approved it [8].

The new design was a bridge with a 47.4 m span, which consisted of one 22 m free span in the middle and two 13 m side spans [2], although Von Faber mentions a 22.5 m middle section and two 13.5 m side spans [1]. The total width was 16 m [9], 12 m for traffic and the rest for two sidewalks on either side. Traffic was divided

DOI: 10.1051/matecconf/201710105028

into three lines, one for double electric tram tracks in the centre and two lines for other vehicles [2] (Fig. 5).

Fig. 5. View from Gubeng Area [2]

The middle part of the railing is more 'transparent' [9]. It is made of iron bars, in contrast to the solid concrete in other parts [2]. The use of iron has the advantage of avoiding cracks in the middle section. There is a difference in the vibration between the middle and side sections because of the dynamic load from the electric trams and other vehicles; the farther the position from the bridge supports, the more substantial the vibration (Fig. 6).

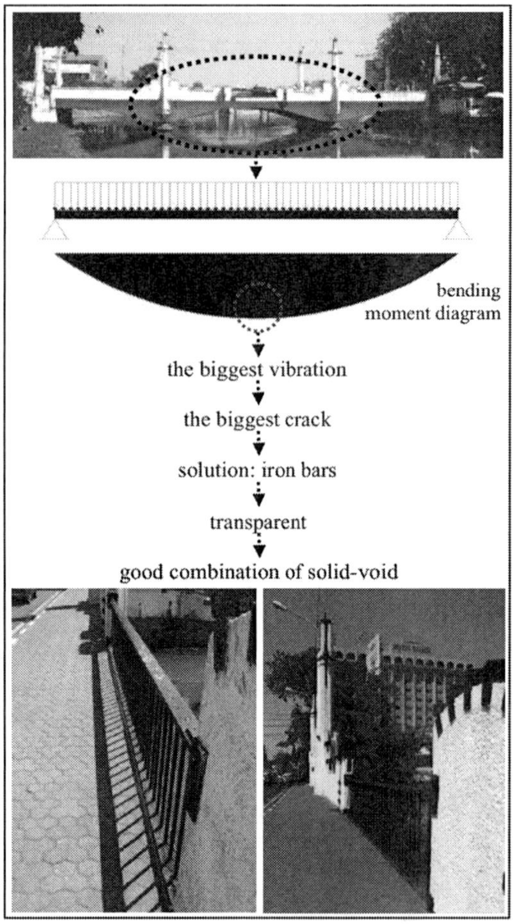

Fig. 6. Railing in the middle part

Several parts made of solid concrete are finished by iron 'fish-fin' or 'fish-scale' ornaments (Fig. 7, Fig. 8, and Fig. 9). Probably, selection of fish as a source of ideas is derived from the name of the city (Surabaya) which comes from Javanese words "*sura*" (= kind of fish) and "*baya*" (crocodile). Fish and crocodile are two animals found in the symbol of the city from time to time, both in the Dutch colonial period and in the present (Fig. 10).

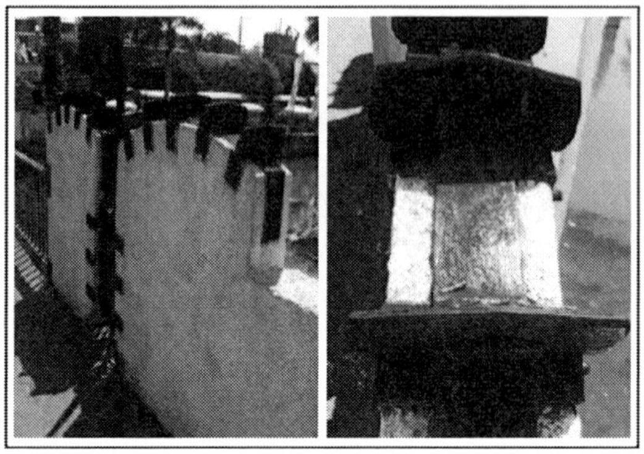

Fig. 7. Iron 'fish-fin' or 'fish-scale' ornaments on the railing

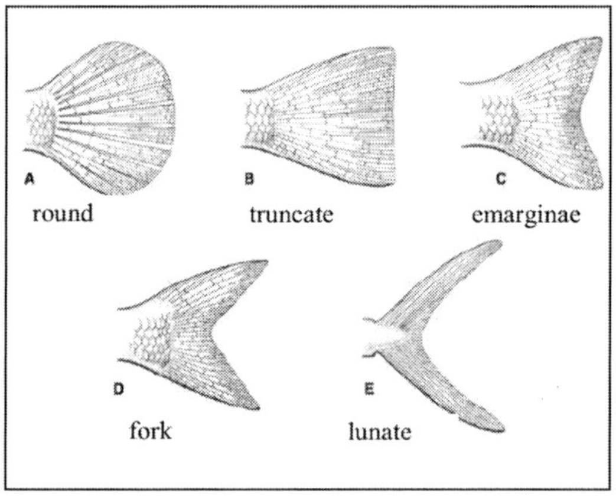

Fig. 8. Fish fin shape [10]

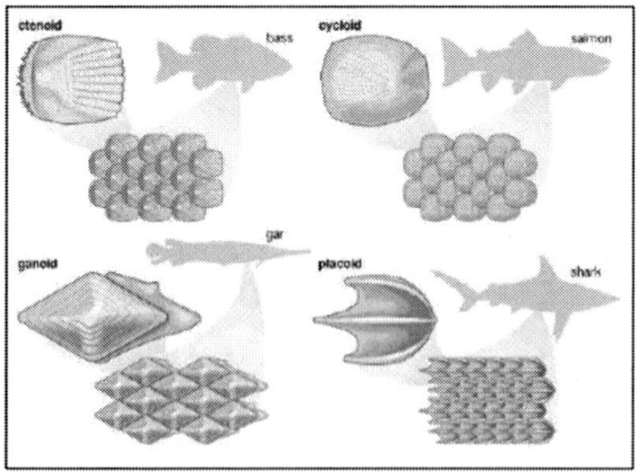

Fig. 9. Fish scale shape [11]

DOI: 10.1051/matecconf/201710105028

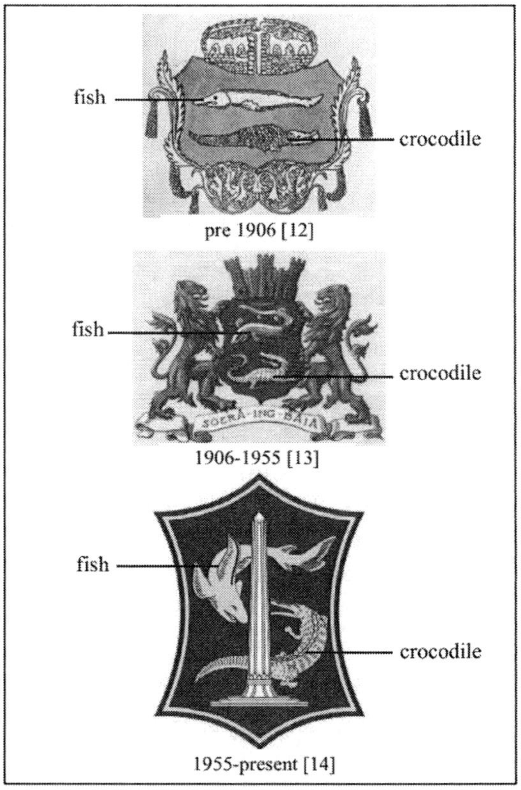

Fig. 10. Symbol of Surabaya City

The railing also features cylinders with pointed iron tips (Fig. 11), similar to the same elements in the Surabaya Town Hall, and a short iron 'spear' ornament (Fig. 12). The electric wire is attached to eight tall lamp posts [2] (Fig. 13). The top of each post has iron ornaments (Fig. 14), the form of which is also similar to ornaments in the town hall. Besides being aesthetic elements, these ornaments function as hooks for electric wires and street lamps. The positioning of the posts echoes the position of the supports of the bridge beams (Fig. 15).

Fig. 11. Iron ornaments on the railing and bottom part of a lamp post

Fig. 12. Iron 'spear' ornament on the railing

Fig. 13. The lamp post

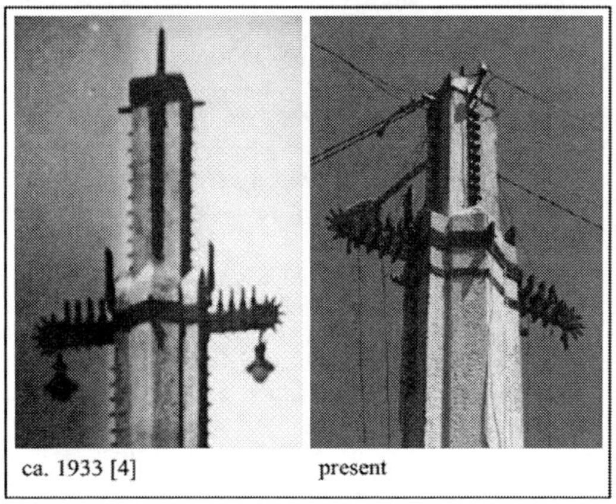

Fig. 14. Upper part of the lamp post

Fig. 15. The lamp post and middle pillar

The bridge's upper structure was made of reinforced concrete using the Gerber beam principle. The height of the beam in the middle section is only 90 cm, in order to allow boats to pass through. The maximum thickness or height of the beams is 2.90 m and the width is 35 cm. Each beam holds 13 steel bars, each 1.25" in diameter size. Each part of the foundation of the bridge has two 14 m wood piles, each of which is able to carry a maximum load of 12 tons. The thickness of the plate is 12 cm [2].

3.3 Cost and construction

After the municipality and the city council agreed to the proposal to build the bridge, and the design was finished by Citroen, the Director of Public Work Service sent a letter offering the total budget for building the new Gubeng bridge for the amount of *f* 254,000. OJS contributed *f* 115,000 to this budget. Furthermore, the municipality collected *f* 15,000 from the sale of iron or steel from the old bridge. At least another *f* 5,000 was raised from the sale of the wood used for the formwork (*bekisting*) of the reinforced concrete and a temporary dam. Thus, the municipality only raised *f* 119,000. This figure was included in the budget of 1923 point XIX B 202. The financial committee had approved the design and its budget, while recommendations from the technical committee would be delivered soon [15].

Based on the Municipal Sheet number 280 dated 19 December 1922, the bid for the construction work was ready to be held and the invitation would be announced publicly. No information can be found on the bidders. However, it is clear that the *Nederlandsche Aanneming Maatschappij* (Nedam) built the bridge in 1923-1924 [1].

3.4 Gubeng Bridge at the 1925 Paris exhibition and its award

In the design of the Gubeng Bridge Citroen shows his competence as an architect who is able to successfully combine aesthetic aspects and constructional considerations. Although he is helped by engineers from the Public Work Service, Citroen still plays an important role in the creation of a pleasing appearance for the bridge, so that it is not only strong enough to carry the traffic load (trams, cars, and other vehicles) but also represents a beautiful element of the urban infrastructure. He creates a harmonious blend between the solidity of concrete and the transparency of the iron balustrades, reflecting mechanical principles. An Amsterdam School touch is found in the iron details of the bridge, while the design of the concrete lamp posts tends to be in Art Deco style. Citroen was able to amalgamate different styles and materials in a harmonious composition. This is probably the reason why the design won an award at the 1925 Paris Exhibition. As we know, the 1920s were the period of Art Deco, expressed profusely in the 1925 International Paris Exhibition.

The International Paris Exhibition, opening on 28 April 1925 and closing in October 1925 with the theme "*Exposition Internationale des Arts Décoratifs et Industriels Modernes*" (International Exhibition of Modern Decorative and Industrial Arts) was the (international) exhibition where Citroen's works were showcased for the first time. His design was dedicated to the display of modern decorative arts. The exhibition brought together thousands of designs from all over Europe and beyond with over 16 million visitors attending. The exhibition was shaped by France's ambitions in the years immediately following the First World War. Its aim was to establish the preeminence of French taste and luxury goods. French displays dominated the exhibition, and Paris was presented as the most fashionable of cities [16].

J. de Bie Leuveling Tjeenk, Commissioner-General of the Dutch department at the exhibition, made it known that the Dutch Pavilion presented the works of 237 artists and designers, who were classified into nine groups: architecture, interior and furniture; sculpture and small plastic construction; wall decoration; stained glass and mosaic; textile; ceramics and glassware; processed metal and jewellery; book and print art; and applied graphics [17]. The Dutch pavilion received the special attention which was demonstrated by a letter of the president of the jury, Pontrémolly, to the Commissioner-General [18].

The Netherlands Indies, as a part of the Dutch pavilion, were represented by architect J.H. Antonisse, J.F.L. Blankenberg, C. Citroen, B.J.K. Cramer, J. van Hoytema, the Hulswit-Fermont-Ed. Cuypers architectural firm, the Karsten-Lutjens-Toussaint architectural firm, H. Maclaine Pont, and P.A.J. Moojen [17]. Among them, only Blankenberg, Citroen, and the Karsten-Lutjens-Toussaint firm were able to achieve an award in the exhibition. The two first earned a "Mention", the lowest category of awards, while the firm of Karsten-Lutjens-Toussaint got a bronze medal.

Citroen obtained the award for his design of the Gubeng Bridge [18].

4 Conclusion

Gubeng Bridge is a Dutch architect C. Citroen's civil construction work that has a touch of aesthetics, in addition to a structural aspect. The aesthetic touch here is a blend of Art Deco and Amsterdam School style. Art Deco itself was the architectural style which was popular in the 1920s and became the theme of the 1925 Paris Exhibition. Thus, the bridge reflects the high aesthetic value of a civil construction. That is why the bridge won the Mention award in the event.

References

1. G.H. von Faber. *Nieuw Soerabaia*, (H. van Ingen, Soerabaia, 1934).
2. R. Heida. Bruggenbouw in Indische Steden. IBT-LT, **2(2)**, 5-12 (1933).
3. Handinoto. *Perkembangan Kota dan Arsitektur Kolonial Belanda di Surabaya 1870-1940*, (Andi, Yogyakarta, 1996).
4. Nederlands Architectuurinstituut, archive, (Rotterdam, 1935).
5. G.H. Bartman. *Fotografien van Soerabaja*, (Thomas & Eras, 's-Gravenhage, 1975).
6. Gemeente Soerabaja. Notulen der Openbare Vergadering van den Gemeenteraad van Soerabaja, 349-352 (5 October 1921).
7. W.H. Trom. *Rapport Betreffende de Electrificatie der Stadslijnen der Oost Java Stoomtram Maatschappij te Soerabaia*, (Nederlandsch Indische Tram Mij., Batavia, 1911).
8. Gemeente Soerabaja. Gemeenteblad van Soerabaja, **265**, 911-913 (14 November 1923).
9. Wouter de Zeeuw. Cosman Citroen 1881-1935, typescript, (Nederlands Architectuurinstituut, Rotterdam, 2001).
10. Thomson Higher Education, (2007). Figure on: http://rodnbarrel.com/post/149743956200/fish-architecture-caudal-fins
11. Encyclopædia Britannica, (2009). Figure on: https://www.britannica.com/ science/scale-zoology
12. Nanang Purwono. *Mana Soerabaia Koe*, (Pustaka Eureka, Surabaya, 2006).
13. Hein Buitenweg. *Krokodillenstad*, (Servire B.V, Katwijk aan Zee, 1980).
14. Lambang Kota Surabaya, (2014). Figure on: http://blog.yoshikanji.web.id/2014/08/lambang-kota-surabaya.html
15. Gemeente Soerabaja. Gemeenteblad van Soerabaja, **280**, 1095-1097 (19 December 1922).
16. Victoria and Albert Museum. Art Deco: The 1925 Paris Exhibition. Information on: http://www.vam.ac.uk/content/articles/a/the-1925-paris-exhibition/
17. Anonymous. De Internationale Architectuur Tentoonstelling te Parijs in 1925, IBT, **28(12)**, 146-147 (June 1925).
18. Anonymous. Onderscheidingen op de Internationale Architectuurtentoonstelling te Parijs, IBT, **28(23)**, 262 (December 1925).

AUTHOR INDEX

AUTHOR INDEX

AUTHOR INDEX

AUTHOR INDEX